Regulating with RNA
in Bacteria and Archaea

SECTION VIII: RESOURCES

Regulating with RNA in Bacteria and Archaea

EDITED BY

Gisela Storz
Division of Molecular and Cellular Biology
Eunice Kennedy Shriver National Institute of Child Health
  and Human Development
Bethesda, Maryland

Kai Papenfort
Faculty of Biology
Department of Microbiology
Ludwig-Maximilians-University of Munich
Martinsried, Germany

ASM
PRESS
Washington, DC

Library of Congress Cataloging-in-Publication Data

Names: Storz, Gisela, editor. | Papenfort, Kai, editor.
Title: Regulating with RNA in bacteria and archaea / edited by Gisela Storz,
    Division of Molecular and Cellular Biology, Eunice Kennedy Shriver
    National Institute of Child Health and Human Development, Bethesda,
    Maryland, and Kai Papenfort, Faculty of Biology, Department of
    Microbiology, Ludwig-Maximilians-University of Munich, Martinsried,
    Germany.
Description: Washington, DC : ASM Press, [2019] | Includes index.
Identifiers: LCCN 2018037983 (print) | LCCN 2018038488 (ebook) | ISBN
    9781683670247 (ebook) | ISBN 9781683670230(hardcover)
Subjects: LCSH: RNA. | Bacteria. | Archaebacteria. | Molecular micrbiology.
    | Bacterial genetics.
Classification: LCC QP623 (ebook) | LCC QP623 .R425 2019 (print) | DDC
    572.8/8--dc23
LC record available at https://lccn.loc.gov/2018037983

10 9 8 7 6 5 4 3 2 1

Address editorial correspondence to
ASM Press, 1752 N St., N.W.,
Washington, DC 20036-2904, USA

Send orders to ASM Press, P.O. Box 605, Herndon, VA 20172, USA
Phone: 800-546-2416; 703-661-1593
Fax: 703-661-1501
E-mail: books@asmusa.org
Online: http://www.asmscience.org

Cover artwork courtesy of Sandy R. Pernitzsch, SCIGRAPHIX.

Contents

SECTION III: CIS-ENCODED BASE PAIRING RNAS

SECTION IV: TRANS-ENCODED BASE PAIRING RNAS

Contributors

SHOSHY ALTUVIA
Department of Microbiology and Molecular Genetics, IMRIC
The Hebrew University-Hadassah Medical School
Jerusalem, Israel

PAUL BABITZKE
Department of Biochemistry and Molecular Biology
Center for RNA Molecular Biology
The Pennsylvania State University
University Park, Pennsylvania

ROLF BACKOFEN
Bioinformatics Group
Center for Biological Signaling Studies (BIOSS)
University of Freiburg
Freiburg, Germany

KATARZYNA J. BANDYRA
Department of Biochemistry
University of Cambridge
Cambridge, United Kingdom

HEATHER J. BECK
Max F. Perutz Laboratories, Center for Molecular Biology
Department of Microbiology, Immunology and Genetics
University of Vienna
Vienna Biocenter
Vienna, Austria

COLLEEN BIANCO
Department of Microbiology
University of Illinois
Urbana, Illinois

LIONELLO BOSSI
Institute for Integrative Biology of the Cell (I2BC)
CEA, CNRS, University of Paris-Sud
University of Paris-Saclay
Gif-sur-Yvette, France

RONALD R. BREAKER
Howard Hughes Medical Institute
Department of Molecular, Cellular and Developmental Biology
Department of Molecular Biophysics and Biochemistry
Yale University
New Haven, Connecticut

ANAÏS BROSSE
CNRS UMR8261
Associated with University Paris Diderot, Sorbonne Paris Cité
Institut de Biologie Physico-Chimique
Paris, France

SYLVIA CHAREYRE
Aix Marseille Université-CNRS
Institut de Microbiologie de la Méditéranée
Laboratoire de Chimie Bactérienne
Marseille, France

CIARÁN CONDON
UMR8261 CNRS
Université Paris Diderot (Sorbonne Paris Cité)
Institut de Biologie Physico-Chimique
Paris, France

LYDIA M. CONTRERAS
Institute of Cellular and Molecular Biology
Department of Chemical Engineering
The University of Texas at Austin
Austin, Texas

FABIEN DARFEUILLE
ARNA Laboratory
INSERM U1212, CNRS UMR 5320
University of Bordeaux
Bordeaux, France

SYLVAIN DURAND
UMR8261 CNRS
Université Paris Diderot (Sorbonne Paris Cité)
Institut de Biologie Physico-Chimique
Paris, France

SVETLANA DURICA-MITIC
Department of Microbiology, Immunobiology and Genetics
Max F. Perutz Laboratories
University of Vienna
Vienna Biocenter
Vienna, Austria

H. AUGUSTE DUTCHER
Department of Biology and Center for Life in Extreme Environments
Portland State University
Portland, Oregon

HANNES EICHNER
Microbiology, Tumor and Cell Biology
Karolinska Institutet
Stockholm, Sweden

JINGYI FEI
Department of Biochemistry and Molecular Biology
Institute for Biophysical Dynamics
The University of Chicago
Chicago, Illinois

NARA FIGUEROA-BOSSI
Institute for Integrative Biology of the Cell (I2BC)
CEA, CNRS, University of Paris-Sud
University of Paris-Saclay
Gif-sur-Yvette, France

KATHRIN S. FRÖHLICH
Department of Biology I, Microbiology
LMU Munich
Martinsried, Germany

JENS GEORG
University of Freiburg, Faculty of Biology
Institute of Biology III
Genetics and Experimental Bioinformatics
Freiburg, Germany

JONATHAN R. GOODSON
The University of Maryland
Department of Cell Biology and Molecular Genetics
College Park, Maryland

YVONNE GÖPEL
Department of Microbiology, Immunobiology and Genetics
Max F. Perutz Laboratories
University of Vienna
Vienna Biocenter
Vienna, Austria

BORIS GÖRKE
Department of Microbiology, Immunobiology and Genetics
Max F. Perutz Laboratories
University of Vienna
Vienna Biocenter
Vienna, Austria

SUSAN GOTTESMAN
Laboratory of Molecular Biology
Center for Cancer Research
National Cancer Institute
Bethesda, Maryland

MAUDE GUILLIER
CNRS UMR8261
Associated with University Paris Diderot, Sorbonne Paris Cité
Institut de Biologie Physico-Chimique
Paris, France

MING C. HAMMOND
Department of Chemistry
Department of Molecular & Cell Biology
University of California, Berkeley
Berkeley, California

KIMBERLY A. HARRIS
Howard Hughes Medical Institute
Department of Molecular, Cellular and Developmental Biology
Yale University
New Haven, Connecticut

TINA M. HENKIN
Department of Microbiology and Center for RNA Biology
The Ohio State University
Columbus, Ohio

WOLFGANG R. HESS
University of Freiburg, Faculty of Biology
Institute of Biology III
Genetics and Experimental Bioinformatics
Freiburg, Germany

JAY C.D. HINTON
Institute of Integrative Biology
University of Liverpool
Liverpool, United Kingdom

KATHARINA HÖFER
Institute of Pharmacy and Molecular Biotechnology
Im Neuenheimer Feld 364
Heidelberg University
Heidelberg, Germany

ANDRES JÄSCHKE
Institute of Pharmacy and Molecular Biotechnology
Im Neuenheimer Feld 364
Heidelberg University
Heidelberg, Germany

ALISA KING
Department of Microbiology
University of Illinois
Urbana, Illinois

KIEL D. KREUZER
Department of Microbiology and Center for RNA Biology
Molecular, Cellular and Developmental Biology Graduate Program
The Ohio State University
Columbus, Ohio

SIDNEY R. KUSHNER
Department of Genetics
Department of Microbiology
University of Georgia
Athens, Georgia

EDMUND LOH
Microbiology, Tumor and Cell Biology
Karolinska Institutet
Stockholm, Sweden
SCELSE
Nanyang Technological University
Singapore

THEA S. LOTZ
Synthetic Genetic Circuits
Department of Biology
TU Darmstadt
Darmstadt, Germany

BEN F. LUISI
Department of Biochemistry
University of Cambridge
Cambridge, United Kingdom

PIERRE MANDIN
Aix Marseille Université-CNRS
Institut de Microbiologie de la Méditéranée
Laboratoire de Chimie Bactérienne
Marseille, France

ALEXANDER S. MANKIN
Center for Biomolecular Sciences
University of Illinois at Chicago
Chicago, Illinois

MARTIN MANN
Bioinformatics Group
University of Freiburg
Freiburg, Germany

SARA MASACHIS
ARNA Laboratory
INSERM U1212, CNRS UMR 5320
University of Bordeaux
Bordeaux, France

SEZEN MEYDAN
Center for Biomolecular Sciences
University of Illinois at Chicago
Chicago, Illinois

MICHELLE M. MEYER
Department of Biology
Boston College
Chestnut Hill, Massachusetts

BIJOY K. MOHANTY
Department of Genetics
University of Georgia
Athens, Georgia

ISABELLA MOLL
Max F. Perutz Laboratories, Center for Molecular Biology
Department of Microbiology, Immunology and Genetics
University of Vienna
Vienna Biocenter
Vienna, Austria

FRANZ NARBERHAUS
Microbial Biology
Ruhr University
Bochum, Germany

SUBRATA PANJA
T.C. Jenkins Department of Biophysics
Johns Hopkins University
Baltimore, Maryland

KAI PAPENFORT
Faculty of Biology
Department of Microbiology
Ludwig-Maximilians-University of Munich
Martinsried, Germany

BLANCA M. PEREZ-SEPULVEDA
Institute of Integrative Biology
University of Liverpool
Liverpool, United Kingdom

DANIELA PRASSE
Christian-Albrechts-University Kiel
Institute of General Microbiology
Kiel, Germany

RAHUL RAGHAVAN
Department of Biology and Center for Life in Extreme Environments
Portland State University
Portland, Oregon

MEDHA RAINA
Division of Molecular and Cellular Biology
Eunice Kennedy Shriver National Institutes of Child Health and Human
Development
Bethesda, Maryland

FRANCESCO RIGHETTI
Microbiology, Tumor and Cell Biology
Karolinska Institutet
Stockholm, Sweden

TONY ROMEO
Department of Microbiology and Cell Science
Institute of Food and Agricultural Sciences
University of Florida
Gainesville, Florida

ANDREW SANTIAGO-FRANGOS
Program in Cell, Molecular and Developmental Biology and Biophysics
Johns Hopkins University
Baltimore, Maryland
Department of Microbiology and Immunology
Montana State University
Bozeman, Montana

RUTH A. SCHMITZ
Christian-Albrechts-University Kiel
Institute of General Microbiology
Kiel, Germany

CYNTHIA M. SHARMA
Chair of Molecular Infection Biology II
Institute of Molecular Infection Biology (IMIB)
University of Würzburg
Würzburg, Germany

SOYEONG SIM
RNA Biology Laboratory
Center for Cancer Research
National Cancer Institute
National Institutes of Health
Frederick, Maryland

GISELA STORZ
Division of Molecular and Cellular Biology
Eunice Kennedy Shriver National Institute of Child Health and Human Development
Bethesda, Maryland

YICHI SU
Department of Chemistry
University of California, Berkeley
Berkeley, California

BEATRIX SUESS
Synthetic Genetic Circuits
Department of Biology
TU Darmstadt
Darmstadt, Germany

SINE LO SVENNINGSEN
Department of Biology
University of Copenhagen
Copenhagen, Denmark

CHRISTIAN TWITTENHOFF
Microbial Biology
Ruhr University
Bochum, Germany

CARIN K. VANDERPOOL
Department of Microbiology
University of Illinois
Urbana, Illinois

NORA VÁZQUEZ-LASLOP
Center for Biomolecular Sciences
University of Illinois at Chicago
Chicago, Illinois

JORDAN K. VILLA
Institute of Cellular and Molecular Biology
The University of Texas at Austin
Austin, Texas

KAREN M. WASSARMAN
Department of Bacteriology
University of Wisconsin-Madison
Madison, Wisconsin

ALEXANDER J. WESTERMANN
Institute of Molecular Infection Biology
University of Würzburg
Helmholtz Institute for RNA-Based Infection Research
Würzburg, Germany

WADE C. WINKLER
The University of Maryland
Department of Cell Biology and Molecular Genetics
College Park, Maryland

SANDRA L. WOLIN
RNA Biology Laboratory
Center for Cancer Research
National Cancer Institute
National Institutes of Health
Frederick, Maryland

SARAH A. WOODSON
T.C. Jenkins Department of Biophysics
Johns Hopkins University
Baltimore, Maryland

PATRICK R. WRIGHT
Bioinformatics Group
University of Freiburg
Freiburg, Germany

Foreword

It took time to realize, but now it's clear: the field of prokaryotic RNA biology is here. The science of how RNA is made, processed, regulated, modified, translated, and turned over has established itself as a core discipline in molecular microbiology. A central aspect is the understanding that RNA molecules function as regulators and sensors across both the archaea and bacteria, often with striking similarity to the complex world of noncoding RNA of eukaryotes.

It has developed in waves, starting with the discoveries of basic mechanisms and factors of gene expression in the 1960s and 1970s, when bacteria were the workhorses of the emerging field of molecular biology. As molecular tools advanced, the 1980s and 1990s not only provided insights into the underlying molecular mechanisms but also saw bacteriologists stumble upon the first RNA molecules with regulatory functions, creating a sense that there might be more in the transcriptomes of these seemingly simple organisms than the established triumvirate of mRNA, tRNA, and rRNA.

As the old century gave way to the new, the next wave broke: systematic genome-wide searches unearthed small RNAs, riboswitches, and RNA thermometers in stunning numbers and diversity. Concomitantly, investigation of the mechanisms and cellular targets of these new *cis*- and *trans*-acting RNAs showed that they played by defined molecular rules and used protein cofactors such as the protein Hfq or, more recently, ProQ. It became increasingly obvious that there is a whole layer of gene expression control above that of transcriptional regulation. Again, bacteria were the trailblazers, but it soon emerged that their sister group in the noneukaryotic world, the archaea, were also full of interesting regulatory RNA molecules that share features with both bacterial and eukaryotic noncoding RNA. And of course, there was CRISPR (clustered regularly interspaced short palindromic repeat)-Cas, whose rise as a revolutionary genome-editing tool in biomedicine and biotechnology originated in part in deciphering the function of particularly enigmatic noncoding RNAs in bacteria.

Many of these recent discoveries in prokaryotic RNA biology were made possible by new technology, be it genome sequencing, which fed global biocomputational searches, or technology for global transcript profiling—microarrays at first, now largely replaced by RNA deep sequencing. The latter now drives the next wave, in which RNA biology has gone global in new ways: through methodology that can track individual transcripts from birth to death, with high temporal resolution and in concert with the behavior of all other transcripts in the same cell; by drafting comprehensive RNA maps that can immediately highlight important RNA players under previously ignored physiological conditions or in an organism never looked at before; and global in the sense that we have so far investigated only a tiny sliver of the microbial world, and our attention is increasingly drawn to the astoundingly diverse bacteria of the human microbiota and environmental communities, promising new surprises for RNA biology.

Against the backdrop of this ongoing RNA revolution, the editors of *Regulating with RNA in Bacteria and Archaea* should be congratulated on having put together an excellent collection of chapters that in their sum easily convey the excitement of this field. Catering to principal investigators, postdocs, and advanced students alike, this book gives a comprehensive account of the state of the art of the prokaryotic RNA inventory and underlying molecular mechanisms. It also provides a sense of what the next decade may bring, with regard to global discovery on the genome scale, enhanced structural and molecular resolution, and a deeper mechanistic understanding of cellular RNA molecules; the reader will find all of these aspects covered. It is a pleasure to see that the author list is a healthy mix of established researchers known for their seminal contributions to prokaryotic RNA biology and young scientists who have only recently started independent work on regulatory RNA. On top of that, several chapters are focused on general aspects of bacterial gene expression that are crucially relevant to our understanding of the activities and consequences of RNA-based regulation. I hope the reader will find *Regulating with RNA in Bacteria and Archaea* as exciting as I, and the authors of the following chapters, do.

Jörg Vogel
Helmholtz Institute for RNA-based Infection Research (HIRI);
Institute for Molecular Infection Biology,
University of Würzburg,
Würzburg, Germany

Preface

In 1961, Jacob and Monod hypothesized that the regulator of the *lac* operon might be an RNA (1). When it was discovered that the Lac repressor was a protein, the possibility that an RNA could be a regulator was largely forgotten, with a few notable exceptions. The first "unusual" RNA regulators found included antisense RNAs that controlled plasmid replication as well as transposition (reviewed in references 2 and 3). Subsequently, a handful of chromosomally encoded small regulatory RNAs (sRNAs) that act by base-pairing with *trans*-encoded mRNAs were discovered in bacteria. Typically, identification was serendipitous, for example, due to overexpression phenotypes or the detection of bands by phosphate labeling or Northern blot analysis (reviewed in references 3 and 4). The realization that sRNAs are a large class of regulators came from bioinformatic searches for conserved intergenic regions as well as systematic sequencing of cDNAs corresponding to small transcripts (reviewed in references 5 and 6). Another major step in the appreciation that RNAs are widespread and important regulators was the discovery that the 5′ untranslated regions of mRNAs can function as sensory elements responding to the binding of tRNAs or small molecules with connections to the functions of the downstream genes (reviewed in references 7 and 8).

The initial characterization of individual RNAs on a "gene by gene" basis followed by the recent expansion to genome-wide analysis exploiting deep sequencing have made it clear that RNA regulators rival transcription factors with respect to their regulatory scope (reviewed in reference 9). These studies have been revealing that regulatory RNAs are part of an incredibly intricate regulatory network. For example, sRNAs frequently regulate multiple and sometimes dozens of transcripts. At the same time, sRNA concentrations can be affected by their binding to chaperone proteins as well as RNA "sponges," which can be independent sRNAs, degradation products of mRNAs, or tRNA fragments. While many of the initial studies of regulatory RNAs were carried out in model bacterial organisms, the advances in sequencing technologies have facilitated their discovery in a wide range of microbial species, indicating that regulatory RNAs are present in all organisms.

Given the amazing progress that has occurred in the study of regulatory RNAs in bacteria and archaea in the past 20 years, we thought a summary of current knowledge would be a useful resource. Thus, the chapters in this book cover well-characterized *cis*-encoded RNA thermometers, T-box regulators, riboswitches, and regulatory RNA elements within mRNA transcripts, antisense RNAs, as well as *trans*-encoded base-pairing and protein-binding sRNAs. These chapters illustrate how regulatory RNAs are an integral part of most microbial responses, including adaptation to stressful environments and changes in nutrient availability, and contribute to pathogenesis.

We hope the book also will focus attention on open questions in the field and stimulate further research in these areas. Despite the significant progress, there are aspects of regulatory RNAs in bacteria and archaea that are still poorly studied. The improved sequencing technology has revealed that much of bacterial and archaeal genomes is transcribed and that there are regulatory RNAs in "blind spots" that were previously ignored. For example, it is becoming increasing clear that 3' untranslated region-derived transcripts are a major class of base-pairing sRNAs. There additionally are many RNA regulators about which less is known in general, including protein-binding RNAs, RNAs with dual functions, and larger RNAs whose structural complexity rivals that of ribosomes. Future studies should illuminate the molecular underpinnings of what distinguishes different classes of regulatory RNAs and whether clear distinctions are appropriate. New areas of research necessarily will involve not only the RNA components themselves and their modifications but also their associated protein partners and the spatiotemporal parameters underlying their interactions in the cell.

As for many fields of research, our understanding of microbial RNA-based gene regulation comes largely from a few model bacterial organisms such as *Escherichia coli* and *Bacillus subtilis*. Although it is tempting to apply lessons learned from these organisms to other species, it is likely that there are important differences. The availability of an exponentially increasing number of data sets for total RNA or RNAs that coimmunoprecipitate with particular proteins or associate with other RNAs in a wide range of organisms undoubtedly will uncover unique features as well as further common principles of RNA-mediated gene control in bacteria and archaea. Understanding and generalizing these principles will be key for the design of synthetic RNA regulators for applications in biotechnology and medicine. Cross-species comparisons also should facilitate the development of hypotheses about the evolution of regulatory RNA elements and whether the evolution differs from that of protein counterparts.

The final chapters of the book discuss how the remarkable expansion of data necessitates new ways of analyzing and visualizing the information. This includes new strategies to extract and present relevant information from genome-wide gene expression analysis, which should help to uncover common principles of RNA-mediated gene control in bacteria and archaea. Given that standard bacterial genome annotations typically fail to include regulatory RNAs, annotation is still incomplete and the number of microbial regulatory RNAs is unknown.

In summary, these are exciting times for microbiologists, particularly for those studying the regulatory RNA complement encoded by microbes. We think this book summarizes the most significant information gained from studies on RNA-based gene regulation in prokaryotes over the past few decades and will serve as a useful resource for researchers new to the field. Furthermore, the book summarizes open questions that hopefully will inspire new research directions and approaches. We would like to conclude by thanking our many colleagues who so willingly contributed chapters and provided comments that significantly improved the content of this book.

Gisela Storz
Kai Papenfort

References

1. **Jacob F, Monod J.** 1961. Genetic regulatory mechanisms in the synthesis of proteins. *J Mol Biol* 3:318–356.
2. **Simons RW, Kleckner N.** 1988. Biological regulation by antisense RNA in prokaryotes. *Annu Rev Genet* 22:567–600.
3. **Wagner EG, Simons RW.** 1994. Antisense RNA control in bacteria, phages, and plasmids. *Annu Rev Microbiol* 48:713–742.
4. **Wassarman KM, Zhang A, Storz G.** 1999. Small RNAs in *Escherichia coli*. *Trends Microbiol* 7:37–45.
5. **Waters LS, Storz G.** 2009. Regulatory RNAs in bacteria. *Cell* 136:615–628.
6. **Sharma CM, Vogel J.** 2009. Experimental approaches for the discovery and characterization of regulatory small RNA. *Curr Opin Microbiol* 12:536–546.
7. **Winkler WC, Breaker RR.** 2005. Regulation of bacterial gene expression by riboswitches. *Annu Rev Microbiol* 59:487–517.
8. **Grundy FJ, Henkin TM.** 2004. Regulation of gene expression by effectors that bind to RNA. *Curr Opin Microbiol* 7:126–131.
9. **Storz G, Vogel J, Wassarman KM.** 2011. Regulation by small RNAs in bacteria: expanding frontiers. *Mol Cell* 43:880–891.

Acknowledgments

We thank all of the authors for their contributions as well as Shoshy Altuvia, Katarzyna Bandyra, Chase Beisel, Sabine Brantl, Ron Breaker, Allen Buskirk, Claude Chiaruttini, Nicholas De Lay, Sylvain Durand, Sven Findeiß, Konrad Förstner, Elizabeth Fozo, Boris Görke, Susan Gottesman, Ming Hammond, Roland Hartmann, Andres Jäschke, Christine Jacobs-Wagner, Eugene Koonin, Iñigo Lasa, Stephen Lory, Pierre Mandin, Hanah Margalit, Eric Massé, Bryce Nickels, Wai-Leung Ng, Evgeny Nudler, Mikołaj Olejniczak, Rahul Raghavan, Lennart Randau, Marina Rodnina, Tony Romeo, Cynthia Sharma, Alejandro Toledo-Arana, Jai Tree, Julia van Kessel, Joseph Wade, Gerhart Wagner, Karen Wassarman, Wade Winkler, and Jinwei Zhang for their comments which improved all of the chapters. We also thank Sandy Pernitzsch for the lovely cover design and are extremely grateful for all of the help from Megan Angelini of ASM Press.

About the Editors

Gisela Storz is an NIH Distinguished Investigator in the Eunice Kennedy Shriver National Institute of Child Health and Human Development in Bethesda, Maryland. She carried out graduate work with Dr. Bruce Ames at the University of California, Berkeley and postdoctoral work with Dr. Sankar Adhya at the National Cancer Institute and Dr. Fred Ausubel at Harvard Medical School. As a result of the serendipitous discovery of the peroxide-induced OxyS RNA in *E. coli*, one of the first small, regulatory RNAs to be found, much of the work in her lab has focused on the genome-wide identification of small RNAs and their characterization.

Kai Papenfort is a Professor of Microbiology at the Ludwig Maximilians University of Munich, Germany. He received a diploma in biology from the University of Marburg and carried out graduate work with Dr. Jörg Vogel at the Max Planck Institute for Infection Biology and the Humboldt University of Berlin. In his postdoctoral work at the University of Würzburg and Princeton University, Dr. Papenfort studied the regulatory functions of small RNA in bacterial pathogens and their involvement in bacterial communication processes such as quorum sensing. His laboratory focuses on the molecular mechanisms underlying the regulation by small RNAs in the major human pathogen, *Vibrio cholerae*.

RNases
and Helicases

I

Regulating with RNA in Bacteria and Archaea
Edited by Gisela Storz and Kai Papenfort
© 2018 American Society for Microbiology, Washington, DC
doi:10.1128/microbiolspec.RWR-0008-2017

RNase E and the High-Fidelity Orchestration of RNA Metabolism

1

Katarzyna J. Bandyra[1] and Ben F. Luisi[1]

INTRODUCTION

It may seem surprising that in almost all known life-forms, information-encoding transcripts are actively annihilated. Although at first glance this seems to be a potential waste of resources and loss of information, the anticipated advantages of restricting transcript lifetimes include fast response rates and a capacity to rapidly redirect gene expression pathways. In this way, destroying individual transcripts in a modulated manner might effectively enhance the collective information capacity of the living system. *Escherichia coli* has proven to be a useful model system to study such processes, and nearly 45 years ago, a hypothetical endoribonuclease was proposed by Apirion as the key missing factor that might account for the observed degradation patterns of mRNA in that bacterium. At the time this hypothesis was formulated, transcript decay in *E. coli* was best described as a series of endonucleolytic cleavages and subsequent fragment scavenging by 3′ exonucleases (1). A few years later, Apirion and colleagues reported the discovery of the endoribonuclease RNase E and showed it to be involved in processing of rRNA precursors (2–4), and the enzyme was subsequently discovered to also cleave an mRNA from T4 phage into a stable intermediate (5). Thus, RNase E seemed to be an ideal candidate for the proposed endonuclease factor to initiate RNA decay in bacteria. What made these findings surprising was that it had previously been thought that RNases might be specialized, with one set presumed responsible for mRNA decay and another set dedicated to stable RNA processing, whereas RNase E could perform both of these distinct tasks (6). This broad functionality has been found to be a recurrent feature of other RNases in *E. coli* and evolutionarily distant bacteria (7).

In the ensuing decades following the discovery of RNase E, more evidence and deeper insights have been gained into the function and importance of the enzyme in RNA metabolism. The data corroborate the numerous roles played by the RNase, including the initiation of turnover for many mRNA species (8–12) and the maturation of precursors of tRNA (13, 14) and rRNA (15, 16). The roles for RNase E have been expanded to include processing and degradation of regulatory RNAs (17, 18) and rRNA quality control (19).

It is important to note that RNase E is not the sole RNase that can initiate turnover in *E. coli*, as others can catalyze the initial cleavage of mRNAs, including RNase G, RNase P, the double-strand-specific RNase III, and RNases from the toxin/antitoxin families (20) (see also chapter 2). Whether any particular RNA will be engaged by RNase E or another RNase is determined by enzyme specificity, substrate accessibility, and which arrives first at the scene. RNase E appears to have privileged access to substrates, and despite the apparent functional overlap with other RNases for all RNA metabolic processes, the enzyme is essential under most growth conditions (21, 22), implicating a unique and dominating role.

The access of RNase E and other RNases to substrates can be modulated by RNA-binding proteins (23). For instance, the ribosome protein S1 can shield RNase E recognition sites and protect mRNAs against cleavage (24). Ribosomes can protect mRNAs from enzyme attack during translation, but speculatively these might become accessible in a process of cotranslational decay. RNase E recognition sites can become either buried or exposed in locally formed RNA structures (25). These local structures can be induced or remodeled by base-pairing interactions formed in *cis* or

[1]Department of Biochemistry, University of Cambridge, Cambridge CB2 1GA, United Kingdom.

trans, or by other binding proteins and the unwinding/remodeling activity of helicases. The actions of all these factors modulate substrate access.

In the degradation pathway of mRNA for *E. coli*, the initial cleavage of a transcript by RNase E is followed closely by exonucleolytic degradation of the products by PNPase (polynucleotide phosphorylase), RNase II, or RNase R (26) (Fig. 1, left). Depending on the organism, RNase E forms a complex with some of these exoribonucleases as part of a cooperative system referred to as the RNA degradosome (27). These assemblies have some mechanistic parallels and, for one component in particular—namely, PNPase—evolutionary

relationship to the exosome of archaea and eukaryotes (28). The exosome complex, like RNase E, recruits RNA helicases and accessory RNases that help to achieve efficient and complete substrate degradation (29, 30) (Table 1).

In *E. coli* and numerous other bacteria that are phylogenetically diverse, RNase E is membrane associated (31–33), and this compartmentalization is expected to give an intrinsic temporal delay between transcription and the onset of decay (34). mRNAs encoding membrane-bound proteins were found to have shorter half-lives due to cotranslational migration of the mRNA to the membrane [33], where they might potentially

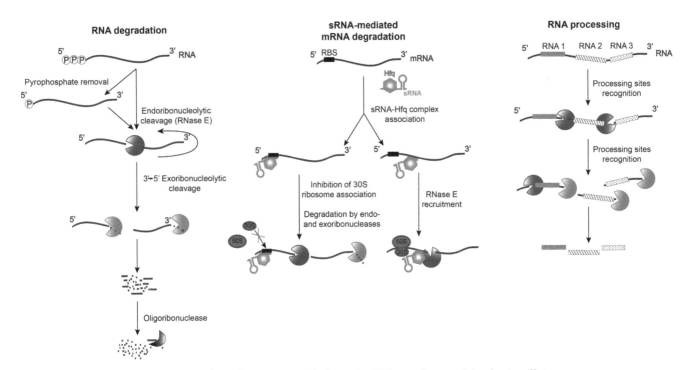

Figure 1 RNase-dependent processes in bacteria. RNases play crucial roles in efficient removal of defective or unnecessary RNAs, regulation of gene expression by sRNAs, and processing of various types of RNAs. (Left) RNA degradation is initiated by endoribonucleolytic cleavage, which can be preceded by pyrophosphate removal from the primary transcript. The majority of degradation initiation events are RNase E dependent. The initial cleavage generates monophosphorylated RNA fragments that can either boost subsequent RNase E cleavage or become substrates for cellular exoribonucleases. Fragments resulting from exoribonucleolytic degradation are further converted to nucleotides by oligoribonuclease. (Middle) When RNA degradation is mediated by sRNA, sRNA-chaperone complexes (such as sRNA-Hfq) can recognize a complementary sequence near the translation initiation region and prevent ribosome association on the transcript (left branch). Naked mRNA is rapidly scavenged by endo- and exoribonucleases. The sRNA-Hfq complex can also bind within the coding region of mRNA, recruiting RNase E and promoting transcript decay (right branch). (Right) In the case of substrates for processing, the order of RNA processing can be defined by the structure of precursors and the specificity of the RNases. The processing can form a cascade of interdependent events where some target sites are being revealed only upon specific initial cleavage. RNA, dark blue; endoribonucleases, purple; exoribonucleases, light blue; sRNA, red; ribosomes, gray ovals; Hfq, orange.

Table 1 Components of the bacterial RNA degradosome and analogous or homologous assemblies from archaea and eukaryotes

Organism/organelle	RNase(s)	RNA helicase(s)	Exosome-like protein(s)	Metabolic enzyme	Chaperone	Other proteins
Escherichia coli	RNase E	RhlB	PNPase	Enolase	Hfq[a]	RapZ, CsrD, RraA, RraB
Pseudomonas aeruginosa	RNase E	DeaD	PNPase		Hfq	Dip
Caulobacter crescentus	RNase E, RNase D	RhlB	PNPase	Aconitase		
Saccharomyces cerevisiae	Rrp44[b]	Mtr4, Ski2	Rrp41, Rrp45, Rrp46, Rrp43, Mtr3, Rrp42, Rrp40, Rrp4, Csl4			Ski complex
Mitochondrial exosome	Dss1	Suv3				

[a]Interaction with Hfq might be indirect and mediated by RNA.
[b]Hydrolytic RNase component of the exosome, a member of the RNase R family and unrelated to RNase E.

become more accessible substrates for the membrane-associated degradosome.

RNase E can work in conjunction with small regulatory RNAs (sRNAs) and RNA chaperones like the Hfq protein to identify specific transcripts (35) (Fig. 1, middle). The recognition specificity is achieved through base pairing of an element in the sRNA, referred to as the seed, to the complementary region of the target. This mode of recognition resembles the use of seed pairing for target recognition by the microRNAs and small interfering RNAs of eukaryotes, and by the bacterial CRISPR (clustered regularly interspaced short palindromic repeat)-Cas9 antiphage system (36). The silencing process also bears some mechanistic analogy to the eukaryotic regulatory RNA-induced silencing complex (RISC) formed by proteins of the Argonaute family, which uses ~22-nucleotide guide RNAs. However, bacterial sRNAs are much larger than the guide RNAs used by eukaryotes, and include structural and sequence elements used for recognition by Hfq and other facilitator proteins, such as ProQ and cold shock proteins (37, 38). The question arises how the seed-target pairings are guiding RNase E and the degradosome machinery to recognize transcripts tagged for degradation.

RNase E is also implicated in RNA processing, releasing mRNA from polycistronic transcripts and participating in sRNA, tRNA, and rRNA maturation (Fig. 1, right). In these processes RNase E is often assisted by other endo- and exoribonucleases, in tandem with which it leads to release of ready-made RNA (Fig. 1, right).

An overview is provided here of the salient structural, functional, and mechanistic features of RNase E from the perspective of accounting for the many *in vivo* functions of the enzyme. The accumulating data provide insights into how RNase E might operate in a cellular context as an intricate machine and interconnected hub of regulatory networks that finds and acts upon general substrates as well as specific targets with optimal speed and accuracy.

A BRIEF EVOLUTIONARY HISTORY OF RNASE E

At 1,061 amino acids, RNase E is one of the largest proteins encoded by the *E. coli* genome (Fig. 2). The enzyme has a conserved N-terminal domain (NTD) of roughly 510 residues that encompasses the endonucleolytic active site. The remaining C-terminal portion is predicted to be predominantly natively unstructured, and consequently this region is anticipated to lack a compact and globular character (39). Again, this gives RNase E the distinction of containing one of the largest encoded segments with this predicted disordered property within *E. coli* (40) (Fig. 2). *E. coli* and many other species also express the nonessential RNase G, a paralog of RNase E. In *E. coli*, RNase G is 489 amino acids in length and has roughly 30% identity and close to 50% similarity in amino acid sequence over their common 430 residues, with maintenance of the key residues involved in substrate recognition and catalysis. RNase G and E are likely to have diverged after a duplication within a chromosome early in the evolution of the gammaproteobacteria, but the enzymes still share some similar activities, including rRNA processing and mRNA turnover (34). Orthologs of RNase E and RNase G are found in roughly 80% of all bacterial genomes sequenced to date (40). Although some Gram-positive bacterial species such as *Bacillus subtilis* encode neither RNase E nor RNase G, they nonetheless express functionally equivalent enzymes such as RNase J1/J2 or RNase Y (41–43) (see also chapter 3).

In the chloroplasts of plants, an RNase E homolog is implicated in polycistronic RNA cleavage (44), and counterintuitively, deficiency of this enzyme appears to have a destabilizing effect on some transcripts (45).

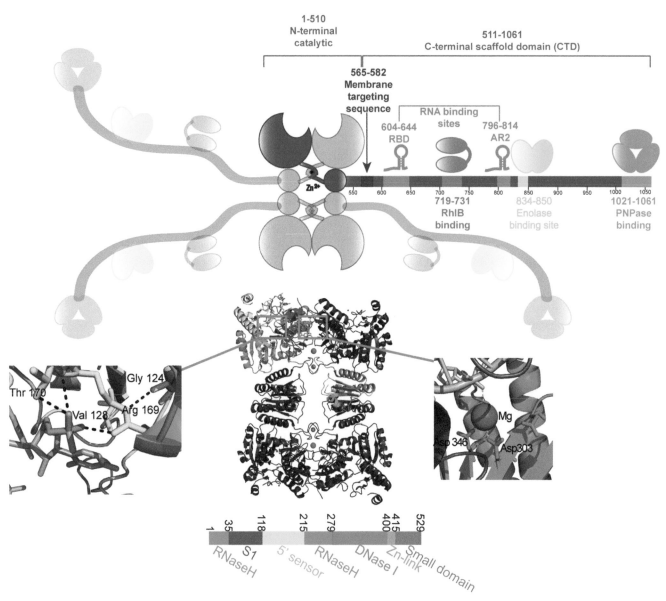

Figure 2 RNase E catalytic domain and a model of the organization of the *E. coli* RNA degradosome. (Top) RNase E is a tetramer (purple, with a single protomer highlighted in dark purple), and the quaternary organization is secured through zinc coordination (black spot) linking NTDs. The CTD is predicted to be predominantly unstructured and provides binding sites for the other degradosome components: RhlB (green), enolase (yellow), and PNPase (blue). The C terminus also harbors two RNA-binding sites (red) and a membrane anchor (dark gray). (Bottom) Structure of the RNase E catalytic domain, with the subdomains of one protomer color-coded. Close view of the phosphate binding pocket (left) and the active site (right), with the main amino acids of functional importance labeled (47).

Perhaps RNase E and its plant homolog could have similar roles with respect to noncoding RNAs. Some *E. coli* noncoding RNAs are also destabilized in the absence of RNase E, which may be due to direct interactions of RNase E and sRNA or result from changes in other mRNAs or unidentified interactors (46).

DOMAINS, MICRODOMAINS, AND PSEUDODOMAINS OF RNASE E

X-ray crystallography analysis reveals a patchwork of domains for the N-terminal catalytic region of RNase E. There is an RNase H-like subdomain at the N terminus that probably fulfills a structural function (residues

1 to 35 and 215 to 278), an RNA-binding S1 domain (36 to 118), a 5′ sensor responsible for the enzyme preference of monophosphorylated substrates (119 to 214), a DNase I domain (279 to 400), a Zn-binding domain (401 to 414) that stabilizes dimer formation, and a small domain (415 to 510) that mediates tetramer formation at the dimer-dimer interface (47) (Fig. 2).

The crystal structure of the catalytic domain of RNase E identifies two aspartic acid residues involved in metal ion coordination at the active site: D303 and D346 (Fig. 2). The metal bound at this site was proposed to be magnesium and/or manganese, which also supports RNA cleavage (48). These metals are likely to help activate water for nucleophilic attack of the phosphate backbone. From the crystallographic data and simulations, it is inferred that the scissile phosphate must approach the nucleophile in a defined geometry that is incompatible with the A-form conformation found in duplex regions (12, 47). This requirement accounts for the observation that RNase E prefers single-stranded substrates. The sites recognized by RNase E can be accompanied by stem-loop structures, situated either upstream or downstream of the cleavage site.

The C-terminal domain of RNase E (CTD; amino acids 511 to 1061) forms a scaffold region that serves as a platform for the degradosome complex (Fig. 2). Because it is predicted to be predominantly unstructured, it is not a conventional domain with compacted, folded character bearing defined secondary structural elements like strands and helices. However, the CTD is punctuated by small microdomains that have structural propensity and that recruit protein partners (39). Biophysical analyses indicate that recruitment of the partner proteins may help to partially compact the natively unstructured portion of the CTD into a conformation that may facilitate RNA interactions (49).

THE INTERACTION PARTNERS OF RNASE E

As the scaffolding core of the RNA degradosome, RNase E interacts with several types of proteins; its repertoire of partners can vary depending on growth conditions (50–53). RNase E homologs in divergent bacterial lineages have different interaction modules and partners, making the degradosome a variable machinery with capacity to perform specialized tasks depending on biological context (27, 54, 55) (Table 1).

The *E. coli* RNase E forms a degradosome assembly in which the canonical components associated with the CTD are a DEAD-box RNA helicase (RhlB), the glycolytic enzyme enolase, and the exoribonuclease PNPase (Fig. 2 and 3; Table 1). The *Caulobacter crescentus*

RNase E forms an RNA degradosome complex together with RhlB, the metabolic enzyme aconitase, PNPase, and the exoribonuclease RNase D (56, 57) (Table 1). As with the *E. coli* degradosome, the proteins of the *C. crescentus* degradosome bind to RNase E mainly through its unstructured CTD, with the exception of RhlB, which binds to a partially helical insert in the S1 domain within the globular N-terminal part of RNase E. We describe the degradosomal interactions and their functional consequences in the following subsections.

The CTD of *E. coli* RNase E has two RNA-binding domains flanking the helicase binding site (Fig. 2). Accumulating evidence indicates that this C-terminal portion of RNase E plays an important role in sRNA-mediated regulation. *In vivo*, truncation of RNase E to disrupt the degradosome assembly impacts on the kinetics of substrate cleavage, as shown from results of single-molecule studies of the action of sRNA SgrS on the *ptsG* transcript (58). Removing the RNase E CTD diminishes the efficiency of codegradation of the sRNA-mRNA pair RyhB-*sodB*, implying that the degradosome might be important for presenting the RNA duplex to the catalytic domain of RNase E (17, 59). The CTD also has a membrane-association motif that compartmentalizes the assembly to the cytoplasmic membrane (27, 31, 32, 34) (Fig. 2 and 4B).

The Exosome-like PNPase

The very C terminus of RNase E encodes a small segment that recruits PNPase, a 3′-to-5′ exoribonuclease (60) (Fig. 2). PNPase uses phosphate to attack the backbone of the RNA substrate, generating nucleoside diphosphates as products. This mechanism requires magnesium as a cofactor, and it is likely that the metal plays a role in stabilizing the charge in the transition state (60). PNPase contributes to RNA quality control and can degrade improperly folded tRNA and rRNA (61). As PNPase requires a 3′ single-stranded region to bind to its substrate, polyadenylation of RNAs may increase their degradation rate by this enzyme (62). In mammalian mitochondria mRNA is polyadenylated, and the degradation rate by the Suv3-PNPase complex depends on the length of the poly(A) tail (63).

The first experimental structure of PNPase, from *Streptomyces antibioticus*, was used to propose a model for the archaeal and eukaryotic exosome (28, 64). The eukaryotic exosome is not a phosphorolytic enzyme like PNPase, but acts as a scaffold to recruit exo- and endonucleases on the periphery of the central, PNPase-like channel, through which some substrates are threaded as single strands for delivery to the active sites (65). Analogous cooperation of the RhlB helicase with

RNase E and PNPase is described in the following subsection. Crystal structures have been obtained for PNPase from numerous other species, and the structures of the *C. crescentus* homolog revealed the full domain architecture of this enzyme, and illuminated the path for the RNA to thread into the central channel to the active sites and how the KH RNA-binding domains engage single-stranded RNA substrate (66).

PNPase may play a chaperone role for some sRNAs (67–70). The chaperone property opens the possibility that regulatory RNA recruitment by PNPase could modulate the activity of the degradosome.

RNA Helicase Partners of RNase E and Their Role in Substrate Channeling

The degradosome's RhlB bears the conserved "DExD/H box" sequence motif found in RNA helicases from bacteria, archaea, and eukaryotes. Like other RNA helicases, RhlB can harness the energy of nucleoside triphosphate binding and hydrolysis to dynamically remodel RNA structures and protein-RNA complexes. The interaction site for RhlB has been mapped on RNase E by limited proteolysis and includes a highly conserved motif (40, 49) (Fig. 2). The physical interaction of RhlB with its binding site on RNase E boosts the helicase ATPase and unwinding activities. In turn, association of the helicase with RNase E results in efficient unfolding of structured RNAs and can facilitate the action of PNPase and RNase E to degrade such substrates (71, 72) (Fig. 3B).

RNA helicases are associated with phylogenetically diverse RNase E homologs, with some notable variations. In stationary phase or in response to cold stress, other RNA helicases are recruited to the *E. coli* degradosome to replace RhlB (73–76). A helicase expressed in response to cold shock is also associated with the degradosome in *C. crescentus* (77).

The association of the RhlB helicase with the degradosome has some interesting mechanistic and structural parallels with the complex formed by RNA helicase Mtr4 and the nuclear exosome of the yeast *Saccharomyces cerevisiae* (29) (Table 1). There are also similarities to the eukaryotic Ski complex, which is a conserved multiprotein assembly required for the cytoplasmic functions of the exosome, including RNA turnover, surveillance, and interference (30) (Table 1). The crystal

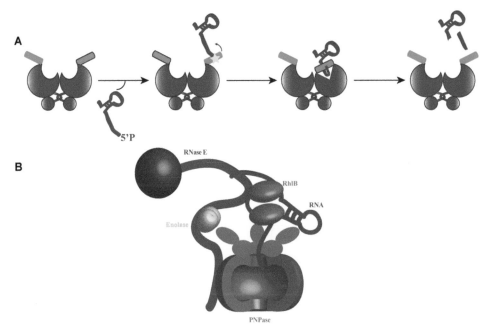

Figure 3 RNase E and interactions with RNA substrates. (A) RNase E activation by 5′-monophosphorylated substrate binding (5′P depicted as a yellow star). Only the principle dimer of the RNase E tetramer (purple) is shown for clarity. The S1 domain together with 5′ sensor (red bar) is capturing the substrate (dark blue) and aligning it in the active site by structural changes induced by RNA binding. (B) Substrate (dark blue) channeling by the ATP helicase (green) to the active site of PNPase (blue). Its action may thread substrate down the channel into the active site, as occurs for the exosome and the mitochondrial exoribonuclease-helicase complex of yeast. The helicase is also likely to provide the same threading function for RNase E (purple).

structure of *S. cerevisiae* core complex shows that the Ski3 N-terminal arm and a Ski2 insertion domain allosterically modulate the ATPase and helicase activities of the complex. Interactions with the C-terminal RecA domain of the ATPase may have some functional analogy to the interaction of RNase E with RhlB, and the helicase-mediated operation of threading substrate into the exosome is analogous to substrate channeling by RhlB to PNPase or the catalytic domain of RNase E.

Another analogy can be drawn with the yeast mitochondrial exoribonuclease complex (mtEXO), in which the helicase Suv3 acts as a motor that feeds the 3′ end of the RNA into the catalytic channel of the 3′-to-5′ exoribonuclease Dss1 for efficient processive degradation (78) (Table 1). This is particularly important for structured RNAs that cannot be degraded by the nuclease on its own and for which helicase unwinding activity is required. In higher eukaryotes, including humans, the functional equivalent of the mitochondrial exosome is a complex of human SUV3 and PNPase (79, 80). All of the described exosome and helicase-exoribonuclease complexes suggest that there might be similar cooperation between RhlB and PNPase, which is not capable of degradation of structured RNAs on its own but can efficiently hydrolyze RNA unwound by RhlB (Fig. 3B).

Metabolic Enzymes

A long-standing puzzle is why metabolic enzymes are associated with the RNA degradosome. In numerous gammaproteobacteria, the glycolytic enzyme enolase is a conserved, canonical component of the RNA degradosome assembly (27), whereas in *C. crescentus*, RNase E is associated with the Krebs cycle enzyme aconitase (56) (Table 1). Intriguingly, the Gram-positive bacterium *Bacillus subtilis* lacks an RNase E homolog, but does have an analogous RNase that interacts with glycolytic enzymes under certain growth conditions (81). The evolutionary convergence of stable interactions between RNases and metabolic enzymes indicates an important biological function. Moreover, enolase has also been identified as a component of the RNA interactome in human cells (82), suggesting that it might have a secondary role in supporting RNA metabolism.

An earlier study probing enolase function in the degradosome showed that under conditions of phosphosugar stress in *E. coli*, the depletion of enolase suppresses the sRNA-induced degradation of the *ptsG* transcript by RNase E (83), indicating that enolase may play a role in riboregulation. Another study observed that the deletion of enolase is associated with increased levels of mRNAs encoding proteins that participate in the uptake and utilization of multiple carbon sources

(84). However, as the proportion of enolase associated with the degradosome is relatively small (~5 to 10%) (53), removing this key enzyme entirely from the cell could have strong secondary effects unrelated to its function in degradosome assembly. Such perturbations can be circumvented by deleting the enolase binding site in the RNase E gene on the bacterial chromosome. In such a mutated strain of *E. coli*, the growth phenotype is surprisingly little affected under aerobic conditions (85). However, under anaerobic conditions the mutant shows abnormal cell morphology that is likely to arise from destabilization of the sRNA DicF and higher levels of its target mRNA *ftsZ* encoding a critical cell-division protein (85). Biophysical studies indicate that enolase helps to compact an adjacent RNA-binding motif in the RNase E CTD, known as AR2 (49) (Fig. 2). Degradosome-associated enolase could therefore be indirectly involved in efficient RNA binding by inducing a conformational change in the AR2 RNA-binding domain of the RNase E C terminus. Such structural change could favor RNase E interaction with DicF and confer a certain degree of protection to the sRNA.

INTERACTION OF RNASE E WITH REGULATORY RNAs AND CHAPERONES

Ever since the discovery of the sRNA MicF, which represses the expression of the major outer membrane porin of *E. coli*, *ompF* (86), hundreds of small regulatory RNAs have been discovered that help to regulate gene expression in bacteria (87). Prokaryotic sRNAs vary tremendously in size and are involved in various aspects of gene regulation. These sRNAs are generally synthesized in response to stress or metabolic conditions and act by pairing to target mRNAs and regulating their translation and stability with the assistance of Hfq, a conserved RNA-binding protein from the extensive Lsm/Sm protein family. Many sRNAs studied so far lead to translational repression by binding to the translation initiation region of target mRNAs and occluding the ribosome-binding site (RBS) (88, 89) (Fig. 1, middle). sRNA-mediated degradation of target mRNAs often involves RNase E (17, 90, 91), and sRNAs paired with target mRNAs can be degraded together in an RNase E-dependent manner (17). Hfq, along with Hfq-binding sRNAs, associates with RNase E, resulting in the formation of an RNP effector complex that allows for the tethering of RNase E near the base-pairing region of target mRNAs (92).

Bacterial sRNAs that bind to their target within the coding region can alter transcript stability by creating a RNase cleavage site (93). Additionally, sRNAs can differentially alter transcript accumulation within an operon. An example is RyhB, which downregulates the *iscSUA* genes within the *iscRSUA* operon to allow independent accumulation of *iscR* (94). sRNAs can be degraded by RNase E and PNPase, or by RNase III if the sRNA has long stretches of complementarity (95). sRNA can act in *trans* to allosterically activate RNase E, which we will describe further in the next section (96).

The question arises where and how the RNase E-dependent degradation of the mRNA-sRNA hybrid is initiated (97). The key challenge faced in answering this question is that the complexes are formed transiently and are therefore difficult to capture. One experimental approach to overcome the challenge is to cross-link the RNase E-RNA complexes *in vivo*, in their cellular context, and then identify the components. Such an approach was employed with a variant of RNA sequencing, the CLASH methodology (UV-cross-linking, ligation, and sequencing of hybrids) (35). The data obtained using this method support an interaction/displacement model whereby RNase E binds closely to Hfq interaction sites on sRNAs, thereby displacing Hfq from the sRNA-target RNA pair. This is followed by cleavage that occurs maximally 13 nucleotides downstream of the pairing site. The model is consistent with the RNase E cleavage 6 nucleotides downstream of the MicC-*ompD* sRNA-mRNA duplex (93, 96).

In addition to Hfq, RNase E can cooperate with several other known RNA-binding proteins to specifically regulate sRNA stability. One example is RapZ, which was identified in *E. coli* as an adaptor guiding RNase E for processing of the sRNA GlmZ, as part of the mechanism controlling amino-sugar metabolism. As this metabolic pathway provides essential components required for cell envelope biogenesis, it must be synchronized with other cell division processes and part of a highly interconnected regulatory network. It has been hypothesized that RapZ guides RNase E through modification of the structure of the sRNA so it can be recognized by RNase E, or by functioning as an interaction platform by delivering the sRNA to RNase E (98, 99). Another protein cooperating with RNase E is CsrD, an RNA-binding protein that destabilizes the sRNAs CsrB and CsrC in an RNase E-dependent way, and it was suggested that CsrD might induce structural changes in the RNAs that make them more susceptible to attack by the RNase (100). RNase E cleavage sites have been mapped in RsmZ, an analog of CsrB (101). Most likely, there are more proteins like RapZ and CsrD that bind sRNAs and target them for degradation by RNase E and other enzymes.

The CTD of RNase E can also interact with regulators such as the inhibitory proteins RraA and RraB, which bind the CTD at distinct sites (102). RraA was shown to modulate degradosome activity by altering its composition through interaction with RhlB and the RNA-binding regions of the CTD (98). RraB binds within amino acids 694 to 727 of the C terminus of RNase E and also influences the degradosome composition *in vivo* (103). In *Pseudomonas aeruginosa*, the RNase E CTD is targeted by a phage protein, Dip, that inhibits its activity during phage infection (104, 105). Crystallographic and biophysical data suggest that the interaction site resides within the RNA-binding region of the CTD.

SUBSTRATE PREFERENCES OF RNASE E

Given the pivotal role of RNase E in RNA processing, turnover, and RNA-mediated regulation, the question naturally arises whether the enzyme has cleavage preferences and if these might encode information on the lifetime of the substrate. One salient signature of RNase E cleavage sites is that they are AU rich and single stranded (106–108), with strong preference for U at position +2 with respect to the scissile phosphate. U+2 is predicted to make favorable and conserved interactions with the enzyme (12). The site of cleavage by RNase E was originally mapped to a consensus sequence, A/GAUUA/U (5), in single-stranded region (5, 109, 110). This consensus sequence has been confirmed and extended by the recent mapping of >22,000 cleavage sites in *Salmonella* (12). This preference for single-stranded substrates is in accord with the crystallographic data, as the catalytic site cannot fit double-stranded substrates. Recently it was noted that RNase E cleavage sites are flanked by stem-loop structures (111).

A notable feature of RNase E is its 5′-end dependence for certain substrates, for which it strongly prefers 5′-mono- over the -triphosphorylated group (34, 112). This feature explains why stable 5′ stem-loops that mask the 5′ end from the enzyme protect mRNAs from decay. Crystallographic studies show that the terminal phosphate in a single-stranded region is recognized by hydrogen bonding interactions with R169 and T170 in a 5′ sensor domain, and this interaction is proposed to favor a closed conformational state that boosts enzyme activity (Fig. 2 and 3A). Comparison of crystal structures of enzyme with and without an RNA substrate analog bound shows that RNase E seems to be in an open form in the latter that closes upon substrate binding (113). This movement enables optimal orientation of substrate for catalysis and is favored by the interaction of the 5′ end with the sensing pocket (Fig. 3A). While the 5′ sensing pocket can accommodate 5′ triphosphate, due to steric hindrance the enzyme cannot close upon the substrate, and in consequence the catalytic activity is impeded. Studies with short oligonucleotides indicate that the apparent boost in catalytic efficiency for substrates with 5′ monophosphate compared to those with a 5′ hydroxyl group arises principally from the reduced Michaelis-Menten parameter, K_m (114), suggesting that the effect of 5′-end sensing is mostly to contribute to substrate binding.

A second, potentially large class of RNase E substrates exists for which the enzyme activity is not affected at all by the chemical status of the 5′ end (106, 114, 115). The existence of this class has led to the hypothesis of two potential pathways for substrate recognition by RNase E: 5′-end-dependent and internal entry (116). The first pathway relies on 5′-monophosphate recognition, described above. Data from experiments with substrates with complex secondary structure suggest that the NTD can engage such substrates as a potential mechanism that contributes to the 5′ bypass mode of operation (114). Moreover, for some substrates, the internal entry pathway could be favored by interactions of RNA substrates with the NTD and arginine-rich segments present in the CTD (39, 117). While two different ways of recognizing the substrate exist, neither of those pathways is strictly essential and the pathways may work in a cooperative manner (21, 116). It is interesting in this regard to note that the truncation of the CTD of RNase E is lethal when combined with mutations in the 5′ sensing pocket in the catalytic domain, suggesting that the pathways of degradation involving 5′-end activation and RNA fold recognition are not redundant and there is a mechanistically important interplay between the CTD and NTD during substrate recognition and cleavage (21).

Early studies of RNA processing of complex substrates by RNase E indicate a role for secondary structure recognition by the enzyme. The 9S gene product is cleaved by RNase E twice to yield the 5S precursor, which is further trimmed by nucleases to give the mature 5S rRNA (15, 110, 118). Evidence indicates that a secondary structure in the 5′ region of 9S is essential for its recognition by RNase E (110). More recently, high-throughput sequencing analysis on transcriptome-wide scale in *E. coli* revealed that, in many mRNAs that are RNase E substrates, a stem-loop is present upstream of the cleavage site (111). It was also found that RNase E cleaves *ompD* mRNA in the presence of a small noncoding RNA, MicC, which guides mRNA degradation in a specific manner (96). Those two RNAs

form a duplex by imperfect base pairing upstream of the cleavage site, and changes to this duplex influence RNase E activity (K. Bandyra, K. Froehlich, J. Vogel, B. Luisi, in preparation). These findings indicate that structural motifs in RNA substrates might be crucial for recognition by RNase E to help align single-stranded regions for cleavage (119, 120).

ENCOUNTERING AND ACTING UPON SUBSTRATES

RNases act on any RNA to which they have access and a match to their specificity. As they have redundant activities, the RNase that performs the first cleavage will depend on which first encounters the substrate in a permissive conformation (7, 121). As RNase E seems to be the main enzyme for initiating the onset of degradation, it must have some type of privileged access. Since degradosomes are tethered on the cytoplasmic membrane during aerobic growth, RNAs destined for processing or decay must be delivered to them, or encounter them by random diffusion.

It has been noted that the overexpression of certain transcripts results in rapid degradation by RNase E, most likely because they are produced faster than could be accommodated by the ribosomes and therefore are exposed (122). Regulatory RNA-binding proteins can modulate the efficiency of translation initiation by directly competing with ribosomes for binding to the ribosome interaction region or by initiating a change in the secondary structure of the mRNA sequence near this region (123–125). The resulting reduction in translation initiation efficiency often decreases mRNA stability as well. Moreover, emerging data suggest that codon optimality may also regulate mRNA degradation pace by influencing the rate of ribosome elongation (126). Data support a model in which mRNA degradation is prevented by efficient translation and operates through close coordination of transcription with translation. However, it might be expected that at some stage in committed translation, the transcript is no longer needed as a template, and consequently the mRNA would move off a polysome into the degradation pathway. One model for substrate access is that RNase E might encounter the emerging 5′ end of transcripts as it spools off the end of a polysome. This could be activated by an sRNA and would lead to a process of cotranslational decay (127) (Fig. 4A). Another mode of degradation might occur when the rates of translation initiation efficiency or elongation are reduced, so that the spacing of the translating ribosomes on the mRNA is less compact. In this case, it is more likely that RNase recognition sites in the mRNA would become exposed, causing transcript decay (128) (Fig. 4A). Such a degradation mode would demand the cooperation of rescue mechanisms such as the nonstop-decay process.

CELLULAR LOCALIZATION OF RNASE E AND THE DEGRADOSOME

Adjacent to the N-terminal catalytic domain is an amphipathic α-helix that tethers the RNA degradosome to the bacterial cell membrane in *E. coli* and is expected to affect the way that the four natively unstructured C-terminal regions would extend outwards the tetrameric catalytic center (Fig. 4B). Strikingly, the functionally analogous (but not homologous) enzyme of *Staphylococcus aureus*, RNase Y, is also membrane associated (129, 130). Under anaerobic conditions, RNase E becomes destabilized and may come off the membrane (85).

Although the association of RNase E with the cytoplasmic membrane is required for optimal cell growth

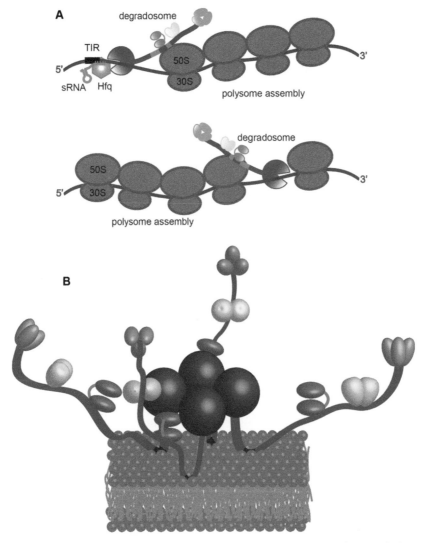

Figure 4 Model of the degradosome interaction with polysomes and the cellular membrane. (A) Speculative model of degradosome interaction with the polysome. RNase E can gain access to translated transcripts (dark blue) upon sRNA action (top), when an sRNA (red) in complex with Hfq (orange) targets the translation initiation region (TIR, black) and by inhibiting assembly of ribosomes provides access for RNase E. The enzyme can also gain access to translated mRNA on its own (bottom). Adapted from reference 140. (B) Association with the inner membrane (gray) is mediated by an amphipathic helix (dark gray) localized in the CTD of RNase E. RNase E, purple; RhlB, green; enolase, yellow; PNPase, blue.

in *E. coli*, for other bacteria the RNA degradosome is not membrane localized. For example, in the alphaproteobacterium *C. crescentus*, RNase E forms patchy foci that associate with the nucleoid instead of the membrane (131).

The membrane-associated *E. coli* RNase E generates transient foci that form on transcripts *in vivo* (32). These foci may be cooperative degradation centers formed by several degradosome particles, and they share remarkable similarities and functional analogy with the eukaryotic RNP granules formed by RNA-binding and -processing enzymes (132). These RNP granules are microscopic structures resembling phase-separated droplets and are proposed to act as "nano-organelles" that are partitioned from the cytoplasm without the requirement for a lipid membrane. The liquid-liquid phase separation is postulated to be mediated by disordered regions of RNA-binding proteins that can form new interactions within such droplets. This phase separation not only brings about compartmentalization of the enzymes and RNA-binding proteins but also influences their specificities for nucleic acids. In the context of the degradosome, extensive disordered regions in the C-terminal tail of RNase E could promote phase separation through self-interaction or distributed contacts with RNA and association with unstructured regions of other degradosome components (97). Thus, clustering of the RNase on the cytoplasmic membrane may be a two-dimensional analog of the phase-transition behavior proposed for RNP granules and could yield highly cooperative activities on a bound substrate. The CTD of RNase E has capacity to accommodate numerous RNA species of different sequence and structure, and could contribute to the formation of such proposed granule-like foci. The CTD might help to intercept polysomes, free RNA, or structured RNA precursors for interaction with the degradosome, and orchestrate the constantly alternating RNA universe.

SUMMARY AND PERSPECTIVES

Analyses of bacterial RNA metabolism continue to reveal its numerous links to key regulatory processes. Many of the mechanisms that enable efficient RNA processing and degradation in bacteria have analogous processes in organisms of other domains of life, including metazoans. RNase E has been a key paradigm in understanding the complexity of RNA-mediated regulation and metabolism in bacteria. RNA-binding proteins, which often act in concert with RNase E, can act as global regulators to help orchestrate complex behavior.

Detailed profiling of cellular targets of the sRNA chaperone Hfq, the translational repressor CsrA, the ProQ protein, and cold shock proteins has shown that there are extensive posttranscriptional networks in bacteria (38, 133–138).

There is growing appreciation of the contribution of RNA metabolism in mediating complex behavior of individual cells and cooperative communities. This behavior is also manifested during bacterial infection—for example, in the rapid response provided by sRNAs in pathogenic bacteria to coordinate invasion steps and adjust quickly to demanding and hostile environments inside the host (139). Although understanding of the processes of RNA metabolism and riboregulation is growing, there is still scope to discover some extraordinary solutions that nature has developed to improve cellular efficiency and capacity to adapt, develop, and evolve.

Acknowledgments. We thank our colleagues for many stimulating discussions over the years: AJ Carpousis, Kenny McDowall, Steven Hardwick, Gigi Storz, Heather Bruce, Tom Dendooven, Jörg Vogel, Yanjie Chao, Kathie Fröhlich, and Kai Papenfort. We are supported by the Wellcome Trust.

Citation. Bandyra KJ, Luisi BF. 2018. RNase E and the high-fidelity orchestration of RNA metabolism. Microbiol Spectrum 6(2):RWR-0008-2017.

References

1. **Apirion D.** 1973. Degradation of RNA in *Escherichia coli*. A hypothesis. *Mol Gen Genet* **122**:313–322.

2. **Apirion D.** 1978. Isolation, genetic mapping and some characterization of a mutation in *Escherichia coli* that affects the processing of ribonuleic acid. *Genetics* **90**:659–671.

3. **Ghora BK, Apirion D.** 1979. Identification of a novel RNA molecule in a new RNA processing mutant of *Escherichia coli* which contains 5 S rRNA sequences. *J Biol Chem* **254**:1951–1956.

4. **Misra TK, Apirion D.** 1979. RNase E, an RNA processing enzyme from *Escherichia coli*. *J Biol Chem* **254**:11154–11159.

5. **Ehretsmann CP, Carpousis AJ, Krisch HM.** 1992. Specificity of *Escherichia coli* endoribonuclease RNase E: *in vivo* and *in vitro* analysis of mutants in a bacteriophage T4 mRNA processing site. *Genes Dev* **6**:149–159.

6. **Carpousis AJ, Vanzo NF, Raynal LC.** 1999. mRNA degradation. A tale of poly(A) and multiprotein machines. *Trends Genet* **15**:24–28.

7. **Deutscher MP.** 2006. Degradation of RNA in bacteria: comparison of mRNA and stable RNA. *Nucleic Acids Res* **34**:659–666.

8. **Arraiano CM, Yancey SD, Kushner SR.** 1988. Stabilization of discrete mRNA breakdown products in *ams pnp rnb* multiple mutants of *Escherichia coli* K-12. *J Bacteriol* **170**:4625–4633.

9. Wellington CL, Greenberg ME, Belasco JG. 1993. The destabilizing elements in the coding region of c-*fos* mRNA are recognized as RNA. *Mol Cell Biol* **13**: 5034–5042.

10. Ow MC, Liu Q, Kushner SR. 2000. Analysis of mRNA decay and rRNA processing in *Escherichia coli* in the absence of RNase E-based degradosome assembly. *Mol Microbiol* **38**:854–866.

11. Coburn GA, Mackie GA. 1999. Degradation of mRNA in *Escherichia coli*: an old problem with some new twists. *Prog Nucleic Acid Res Mol Biol* **62**:55–108.

12. Chao Y, Li L, Girodat D, Förstner KU, Said N, Corcoran C, Śmiga M, Papenfort K, Reinhardt R, Wieden HJ, Luisi BF, Vogel J. 2017. *In vivo* cleavage map illuminates the central role of RNase E in coding and non-coding RNA pathways. *Mol Cell* **65**:39–51.

13. Li Z, Deutscher MP. 2002. RNase E plays an essential role in the maturation of *Escherichia coli* tRNA precursors. *RNA* **8**:97–109.

14. Ow MC, Kushner SR. 2002. Initiation of tRNA maturation by RNase E is essential for cell viability in *E. coli*. *Genes Dev* **16**:1102–1115.

15. Ghora BK, Apirion D. 1978. Structural analysis and *in vitro* processing to p5 rRNA of a 9S RNA molecule isolated from an *rne* mutant of *E. coli*. *Cell* **15**:1055–1066.

16. Li Z, Pandit S, Deutscher MP. 1999. RNase G (CafA protein) and RNase E are both required for the 5′ maturation of 16S ribosomal RNA. *EMBO J* **18**:2878–2885.

17. Massé E, Escorcia FE, Gottesman S. 2003. Coupled degradation of a small regulatory RNA and its mRNA targets in *Escherichia coli*. *Genes Dev* **17**:2374–2383.

18. Kim KS, Lee Y. 2004. Regulation of 6S RNA biogenesis by switching utilization of both sigma factors and endoribonucleases. *Nucleic Acids Res* **32**:6057–6068.

19. Sulthana S, Basturea GN, Deutscher MP. 2016. Elucidation of pathways of ribosomal RNA degradation: an essential role for RNase E. *RNA* **22**:1163–1171.

20. Mohanty BK, Kushner SR. 2016. Regulation of mRNA decay in bacteria. *Annu Rev Microbiol* **70**:25–44.

21. Garrey SM, Mackie GA. 2011. Roles of the 5′-phosphate sensor domain in RNase E. *Mol Microbiol* **80**:1613–1624.

22. Tamura M, Moore CJ, Cohen SN. 2013. Nutrient dependence of RNase E essentiality in *Escherichia coli*. *J Bacteriol* **195**:1133–1141.

23. Moll I, Afonyushkin T, Vytvytska O, Kaberdin VR, Bläsi U. 2003. Coincident Hfq binding and RNase E cleavage sites on mRNA and small regulatory RNAs. *RNA* **9**:1308–1314.

24. Komarova AV, Tchufistova LS, Dreyfus M, Boni IV. 2005. AU-rich sequences within 5′ untranslated leaders enhance translation and stabilize mRNA in *Escherichia coli*. *J Bacteriol* **187**:1344–1349.

25. Barria C, Malecki M, Arraiano CM. 2013. Bacterial adaptation to cold. *Microbiology* **159**:2437–2443.

26. Kaberdin VR, Singh D, Lin-Chao S. 2011. Composition and conservation of the mRNA-degrading machinery in bacteria. *J Biomed Sci* **18**:23.

27. Aït-Bara S, Carpousis AJ. 2015. RNA degradosomes in bacteria and chloroplasts: classification, distribution and evolution of RNase E homologs. *Mol Microbiol* **97**: 1021–1135.

28. Symmons MF, Jones GH, Luisi BF. 2000. A duplicated fold is the structural basis for polynucleotide phosphorylase catalytic activity, processivity, and regulation. *Structure* **8**:1215–1226.

29. Wasmuth EV, Zinder JC, Zattas D, Das M, Lima CD. 2017. Structure and reconstitution of yeast Mpp6-nuclear exosome complexes reveals that Mpp6 stimulates RNA decay and recruits the Mtr4 helicase. *eLife* **6**: e29062.

30. Halbach F, Reichelt P, Rode M, Conti E. 2013. The yeast Ski complex: crystal structure and RNA channeling to the exosome complex. *Cell* **154**:814–826.

31. Khemici V, Poljak L, Luisi BF, Carpousis AJ. 2008. The RNase E of *Escherichia coli* is a membrane-binding protein. *Mol Microbiol* **70**:799–813.

32. Strahl H, Turlan C, Khalid S, Bond PJ, Kebalo JM, Peyron P, Poljak L, Bouvier M, Hamoen L, Luisi BF, Carpousis AJ. 2015. Membrane recognition and dynamics of the RNA degradosome. *PLoS Genet* **11**: e1004961.

33. Moffitt JR, Pandey S, Boettiger AN, Wang S, Zhuang X. 2016. Spatial organization shapes the turnover of a bacterial transcriptome. *eLife* **5**:e13065.

34. Mackie GA. 2013. RNase E: at the interface of bacterial RNA processing and decay. *Nat Rev Microbiol* **11**: 45–57.

35. Waters SA, McAteer SP, Kudla G, Pang I, Deshpande NP, Amos TG, Leong KW, Wilkins MR, Strugnell R, Gally DL, Tollervey D, Tree JJ. 2017. Small RNA interactome of pathogenic *E. coli* revealed through crosslinking of RNase E. *EMBO J* **36**:374–387.

36. Gorski SA, Vogel J, Doudna JA. 2017. RNA-based recognition and targeting: sowing the seeds of specificity. *Nat Rev Mol Cell Biol* **18**:215–228.

37. Smirnov A, Wang C, Drewry LL, Vogel J. 2017. Molecular mechanism of mRNA repression in *trans* by a ProQ-dependent small RNA. *EMBO J* **36**:1029–1045.

38. Michaux C, Holmqvist E, Vasicek E, Sharan M, Barquist L, Westermann AJ, Gunn JS, Vogel J. 2017. RNA target profiles direct the discovery of virulence functions for the cold-shock proteins CspC and CspE. *Proc Natl Acad Sci U S A* **114**:6824–6829.

39. Callaghan AJ, Aurikko JP, Ilag LL, Günter Grossmann J, Chandran V, Kühnel K, Poljak L, Carpousis AJ, Robinson CV, Symmons MF, Luisi BF. 2004. Studies of the RNA degradosome-organizing domain of the *Escherichia coli* ribonuclease RNase E. *J Mol Biol* **340**: 965–979.

40. Aït-Bara S, Carpousis AJ, Quentin Y. 2015. RNase E in the γ-Proteobacteria: conservation of intrinsically disordered noncatalytic region and molecular evolution of microdomains. *Mol Genet Genomics* **290**:847–862.

41. Even S, Pellegrini O, Zig L, Labas V, Vinh J, Bréchemmier-Baey D, Putzer H. 2005. Ribonucleases J1 and J2: two novel endoribonucleases in *B.subtilis* with

functional homology to *E.coli* RNase E. *Nucleic Acids Res* 33:2141–2152.

42. Shahbabian K, Jamalli A, Zig L, Putzer H. 2009. RNase Y, a novel endoribonuclease, initiates riboswitch turnover in *Bacillus subtilis*. *EMBO J* 28:3523–3533.

43. Durand S, Tomasini A, Braun F, Condon C, Romby P. 2015. sRNA and mRNA turnover in Gram-positive bacteria. *FEMS Microbiol Rev* 39:316–330.

44. Walter M, Piepenburg K, Schöttler MA, Petersen K, Kahlau S, Tiller N, Drechsel O, Weingartner M, Kudla J, Bock R. 2010. Knockout of the plastid RNase E leads to defective RNA processing and chloroplast ribosome deficiency. *Plant J* 64:851–863.

45. Hotto AM, Schmitz RJ, Fei Z, Ecker JR, Stern DB. 2011. Unexpected diversity of chloroplast noncoding RNAs as revealed by deep sequencing of the *Arabidopsis* transcriptome. *G3 (Bethesda)* 1:559–570.

46. Stead MB, Marshburn S, Mohanty BK, Mitra J, Pena Castillo L, Ray D, van Bakel H, Hughes TR, Kushner SR. 2011. Analysis of *Escherichia coli* RNase E and RNase III activity *in vivo* using tiling microarrays. *Nucleic Acids Res* 39:3188–3203.

47. Callaghan AJ, Marcaida MJ, Stead JA, McDowall KJ, Scott WG, Luisi BF. 2005. Structure of *Escherichia coli* RNase E catalytic domain and implications for RNA turnover. *Nature* 437:1187–1191.

48. Thompson KJ, Zong J, Mackie GA. 2015. Altering the divalent metal ion preference of RNase E. *J Bacteriol* 197:477–482.

49. Bruce HA, Du D, Matak-Vinkovic D, Bandyra KJ, Broadhurst RW, Martin E, Sobott F, Shkumatov AV, Luisi BF. 2018. Analysis of the natively unstructured RNA/protein-recognition core in the *Escherichia coli* RNA degradosome and its interactions with regulatory RNA/Hfq complexes. *Nucleic Acids Res* 46:387–402.

50. Carabetta VJ, Silhavy TJ, Cristea IM. 2010. The response regulator SprE (RssB) is required for maintaining poly(A) polymerase I-degradosome association during stationary phase. *J Bacteriol* 192:3713–3721.

51. Carpousis AJ, Van Houwe G, Ehretsmann C, Krisch HM. 1994. Copurification of *E. coli* RNAase E and PNPase: evidence for a specific association between two enzymes important in RNA processing and degradation. *Cell* 76:889–900.

52. Miczak A, Kaberdin VR, Wei CL, Lin-Chao S. 1996. Proteins associated with RNase E in a multicomponent ribonucleolytic complex. *Proc Natl Acad Sci U S A* 93:3865–3869.

53. Py B, Higgins CF, Krisch HM, Carpousis AJ. 1996. A DEAD-box RNA helicase in the *Escherichia coli* RNA degradosome. *Nature* 381:169–172.

54. Marcaida MJ, DePristo MA, Chandran V, Carpousis AJ, Luisi BF. 2006. The RNA degradosome: life in the fast lane of adaptive molecular evolution. *Trends Biochem Sci* 31:359–365.

55. Górna MW, Carpousis AJ, Luisi BF. 2012. From conformational chaos to robust regulation: the structure and function of the multi-enzyme RNA degradosome. *Q Rev Biophys* 45:105–145.

56. Hardwick SW, Chan VSY, Broadhurst RW, Luisi BF. 2011. An RNA degradosome assembly in *Caulobacter crescentus*. *Nucleic Acids Res* 39:1449–1459.

57. Voss JE, Luisi BF, Hardwick SW. 2014. Molecular recognition of RhlB and RNase D in the *Caulobacter crescentus* RNA degradosome. *Nucleic Acids Res* 42:13294–13305.

58. Fei J, Singh D, Zhang Q, Park S, Balasubramanian D, Golding I, Vanderpool CK, Ha T. 2015. RNA biochemistry. Determination of *in vivo* target search kinetics of regulatory noncoding RNA. *Science* 347:1371–1374.

59. Massé E, Gottesman S. 2002. A small RNA regulates the expression of genes involved in iron metabolism in *Escherichia coli*. *Proc Natl Acad Sci U S A* 99:4620–4625.

60. Nurmohamed S, Vaidialingam B, Callaghan AJ, Luisi BF. 2009. Crystal structure of *Escherichia coli* polynucleotide phosphorylase core bound to RNase E, RNA and manganese: implications for catalytic mechanism and RNA degradosome assembly. *J Mol Biol* 389:17–33.

61. Cheng ZF, Deutscher MP. 2003. Quality control of ribosomal RNA mediated by polynucleotide phosphorylase and RNase R. *Proc Natl Acad Sci U S A* 100:6388–6393.

62. O'Hara EB, Chekanova JA, Ingle CA, Kushner ZR, Peters E, Kushner SR. 1995. Polyadenylylation helps regulate mRNA decay in *Escherichia coli*. *Proc Natl Acad Sci U S A* 92:1807–1811.

63. Chujo T, Ohira T, Sakaguchi Y, Goshima N, Nomura N, Nagao A, Suzuki T. 2012. LRPPRC/SLIRP suppresses PNPase-mediated mRNA decay and promotes polyadenylation in human mitochondria. *Nucleic Acids Res* 40:8033–8047.

64. Symmons MF, Williams MG, Luisi BF, Jones GH, Carpousis AJ. 2002. Running rings around RNA: a superfamily of phosphate-dependent RNases. *Trends Biochem Sci* 27:11–18.

65. Bonneau F, Basquin J, Ebert J, Lorentzen E, Conti E. 2009. The yeast exosome functions as a macromolecular cage to channel RNA substrates for degradation. *Cell* 139:547–559.

66. Hardwick SW, Gubbey T, Hug I, Jenal U, Luisi BF. 2012. Crystal structure of *Caulobacter crescentus* polynucleotide phosphorylase reveals a mechanism of RNA substrate channelling and RNA degradosome assembly. *Open Biol* 2:120028.

67. Cameron TA, De Lay NR. 2016. The phosphorolytic exoribonucleases polynucleotide phosphorylase and RNase PH stabilize sRNAs and facilitate regulation of their mRNA targets. *J Bacteriol* 198:3309–3317.

68. De Lay N, Gottesman S. 2011. Role of polynucleotide phosphorylase in sRNA function in *Escherichia coli*. *RNA* 17:1172–1189.

69. Bandyra KJ, Sinha D, Syrjanen J, Luisi BF, De Lay NR. 2016. The ribonuclease polynucleotide phosphorylase can interact with small regulatory RNAs in both protective and degradative modes. *RNA* 22:360–372.

70. Condon C. 2015. Airpnp: auto- and integrated regulation of polynucleotide phosphorylase. *J Bacteriol* 197:3748–3750.

71. Chandran V, Poljak L, Vanzo NF, Leroy A, Miguel RN, Fernandez-Recio J, Parkinson J, Burns C, Carpousis AJ, Luisi BF. 2007. Recognition and cooperation between the ATP-dependent RNA helicase RhlB and ribonuclease RNase E. *J Mol Biol* **367**:113–132.

72. Khemici V, Poljak L, Toesca I, Carpousis AJ. 2005. Evidence *in vivo* that the DEAD-box RNA helicase RhlB facilitates the degradation of ribosome-free mRNA by RNase E. *Proc Natl Acad Sci U S A* **102**:6913–6918.

73. Khemici V, Toesca I, Poljak L, Vanzo NF, Carpousis AJ. 2004. The RNase E of *Escherichia coli* has at least two binding sites for DEAD-box RNA helicases: functional replacement of RhlB by RhlE. *Mol Microbiol* **54**:1422–1430.

74. Prud'homme-Généreux A, Beran RK, Iost I, Ramey CS, Mackie GA, Simons RW. 2004. Physical and functional interactions among RNase E, polynucleotide phosphorylase and the cold-shock protein, CsdA: evidence for a 'cold shock degradosome'. *Mol Microbiol* **54**:1409–1421.

75. Purusharth RI, Klein F, Sulthana S, Jäger S, Jagannadham MV, Evguenieva-Hackenberg E, Ray MK, Klug G. 2005. Exoribonuclease R interacts with endoribonuclease E and an RNA helicase in the psychrotrophic bacterium *Pseudomonas syringae* Lz4W. *J Biol Chem* **280**:14572–14578.

76. Aït-Bara S, Carpousis AJ. 2010. Characterization of the RNA degradosome of *Pseudoalteromonas haloplanktis*: conservation of the RNase E-RhlB interaction in the gammaproteobacteria. *J Bacteriol* **192**:5413–5423.

77. Aguirre AA, Vicente AM, Hardwick SW, Alvelos DM, Mazzon RR, Luisi BF, Marques MV. 2017. Association of the cold shock DEAD-box RNA helicase RhlE to the RNA degradosome in *Caulobacter crescentus*. *J Bacteriol* **199**:e00135-17.

78. Razew M, Warkocki Z, Taube M, Kolondra A, Czarnocki-Cieciura M, Nowak E, Labedzka-Dmoch K, Kawinska A, Piatkowski J, Golik P, Kozak M, Dziembowski A, Nowotny M. 2018. Structural analysis of mtEXO mitochondrial RNA degradosome reveals tight coupling of nuclease and helicase components. *Nat Commun* **9**:97.

79. Wang DD, Shu Z, Lieser SA, Chen PL, Lee WH. 2009. Human mitochondrial SUV3 and polynucleotide phosphorylase form a 330-kDa heteropentamer to cooperatively degrade double-stranded RNA with a 3′-to-5′ directionality. *J Biol Chem* **284**:20812–20821.

80. Borowski LS, Dziembowski A, Hejnowicz MS, Stepien PP, Szczesny RJ. 2013. Human mitochondrial RNA decay mediated by PNPase-hSuv3 complex takes place in distinct foci. *Nucleic Acids Res* **41**:1223–1240.

81. Commichau FM, Rothe FM, Herzberg C, Wagner E, Hellwig D, Lehnik-Habrink M, Hammer E, Völker U, Stülke J. 2009. Novel activities of glycolytic enzymes in *Bacillus subtilis*: interactions with essential proteins involved in mRNA processing. *Mol Cell Proteomics* **8**:1350–1360.

82. Castello A, Fischer B, Eichelbaum K, Horos R, Beckmann BM, Strein C, Davey NE, Humphreys DT, Preiss T, Steinmetz LM, Krijgsveld J, Hentze MW. 2012. Insights into RNA biology from an atlas of mammalian mRNA-binding proteins. *Cell* **149**:1393–1406.

83. Morita T, Kawamoto H, Mizota T, Inada T, Aiba H. 2004. Enolase in the RNA degradosome plays a crucial role in the rapid decay of glucose transporter mRNA in the response to phosphosugar stress in *Escherichia coli*. *Mol Microbiol* **54**:1063–1075.

84. Bernstein JA, Lin PH, Cohen SN, Lin-Chao S. 2004. Global analysis of *Escherichia coli* RNA degradosome function using DNA microarrays. *Proc Natl Acad Sci U S A* **101**:2758–2763.

85. Murashko ON, Lin-Chao S. 2017. *Escherichia coli* responds to environmental changes using enolasic degradosomes and stabilized DicF sRNA to alter cellular morphology. *Proc Natl Acad Sci U S A* **114**:E8025–E8034.

86. Mizuno T, Chou MY, Inouye M. 1984. A unique mechanism regulating gene expression: translational inhibition by a complementary RNA transcript (micRNA). *Proc Natl Acad Sci U S A* **81**:1966–1970.

87. Wagner EGH, Romby P. 2015. Small RNAs in bacteria and archaea: who they are, what they do, and how they do it. *Adv Genet* **90**:133–208.

88. Waters LS, Storz G. 2009. Regulatory RNAs in bacteria. *Cell* **136**:615–628.

89. Gottesman S, Storz G. 2011. Bacterial small RNA regulators: versatile roles and rapidly evolving variations. *Cold Spring Harb Perspect Biol* **3**:a003798.

90. Vanderpool CK, Gottesman S. 2004. Involvement of a novel transcriptional activator and small RNA in post-transcriptional regulation of the glucose phosphoenolpyruvate phosphotransferase system. *Mol Microbiol* **54**:1076–1089.

91. Kawamoto H, Morita T, Shimizu A, Inada T, Aiba H. 2005. Implication of membrane localization of target mRNA in the action of a small RNA: mechanism of post-transcriptional regulation of glucose transporter in *Escherichia coli*. *Genes Dev* **19**:328–338.

92. Morita T, Maki K, Aiba H. 2005. RNase E-based ribonucleoprotein complexes: mechanical basis of mRNA destabilization mediated by bacterial noncoding RNAs. *Genes Dev* **19**:2176–2186.

93. Pfeiffer V, Papenfort K, Lucchini S, Hinton JCD, Vogel J. 2009. Coding sequence targeting by MicC RNA reveals bacterial mRNA silencing downstream of translational initiation. *Nat Struct Mol Biol* **16**:840–846.

94. Desnoyers G, Morissette A, Prévost K, Massé E. 2009. Small RNA-induced differential degradation of the polycistronic mRNA *iscRSUA*. *EMBO J* **28**:1551–1561.

95. Saramago M, Bárria C, Dos Santos RF, Silva IJ, Pobre V, Domingues S, Andrade JM, Viegas SC, Arraiano CM. 2014. The role of RNases in the regulation of small RNAs. *Curr Opin Microbiol* **18**:105–115.

96. Bandyra KJ, Said N, Pfeiffer V, Górna MW, Vogel J, Luisi BF. 2012. The seed region of a small RNA drives the controlled destruction of the target mRNA by the endoribonuclease RNase E. *Mol Cell* **47**:943–953.

97. Dendooven T, Luisi BF. 2017. RNA search engines empower the bacterial intranet. *Biochem Soc Trans* 45: 987–997.

98. Göpel Y, Papenfort K, Reichenbach B, Vogel J, Görke B. 2013. Targeted decay of a regulatory small RNA by an adaptor protein for RNase E and counteraction by an anti-adaptor RNA. *Genes Dev* 27:552–564.

99. Gonzalez GM, Durica-Mitic S, Hardwick SW, Moncrieffe MC, Resch M, Neumann P, Ficner R, Görke B, Luisi BF. 2017. Structural insights into RapZ-mediated regulation of bacterial amino-sugar metabolism. *Nucleic Acids Res* 45:10845–10860.

100. Suzuki K, Babitzke P, Kushner SR, Romeo T. 2006. Identification of a novel regulatory protein (CsrD) that targets the global regulatory RNAs CsrB and CsrC for degradation by RNase E. *Genes Dev* 20:2605–2617.

101. Duss O, Michel E, Yulikov M, Schubert M, Jeschke G, Allain FH. 2014. Structural basis of the non-coding RNA RsmZ acting as a protein sponge. *Nature* 509:588–592.

102. Górna MW, Pietras Z, Tsai YC, Callaghan AJ, Hernández H, Robinson CV, Luisi BF. 2010. The regulatory protein RraA modulates RNA-binding and helicase activities of the *E. coli* RNA degradosome. *RNA* 16:553–562.

103. Gao J, Lee K, Zhao M, Qiu J, Zhan X, Saxena A, Moore CJ, Cohen SN, Georgiou G. 2006. Differential modulation of *E. coli* mRNA abundance by inhibitory proteins that alter the composition of the degradosome. *Mol Microbiol* 61:394–406.

104. Van den Bossche A, Hardwick SW, Ceyssens PJ, Hendrix H, Voet M, Dendooven T, Bandyra KJ, De Maeyer M, Aertsen A, Noben JP, Luisi BF, Lavigne R. 2016. Structural elucidation of a novel mechanism for the bacteriophage-based inhibition of the RNA degradosome. *eLife* 5:e16413.

105. Dendooven T, Van den Bossche A, Hendrix H, Ceyssens PJ, Voet M, Bandyra KJ, De Maeyer M, Aertsen A, Noben JP, Hardwick SW, Luisi BF, Lavigne R. 2017. Viral interference of the bacterial RNA metabolism machinery. *RNA Biol* 14:6–10.

106. Clarke JE, Kime L, Romero AD, McDowall KJ. 2014. Direct entry by RNase E is a major pathway for the degradation and processing of RNA in *Escherichia coli*. *Nucleic Acids Res* 42:11733–11751.

107. McDowall KJ, Lin-Chao S, Cohen SN. 1994. A+U content rather than a particular nucleotide order determines the specificity of RNase E cleavage. *J Biol Chem* 269:10790–10796.

108. McDowall KJ, Kaberdin VR, Wu SW, Cohen SN, Lin-Chao S. 1995. Site-specific RNase E cleavage of oligonucleotides and inhibition by stem-loops. *Nature* 374: 287–290.

109. Mackie GA. 1992. Secondary structure of the mRNA for ribosomal protein S20. Implications for cleavage by ribonuclease E. *J Biol Chem* 267:1054–1061.

110. Cormack RS, Mackie GA. 1992. Structural requirements for the processing of *Escherichia coli* 5 S ribosomal RNA by RNase E *in vitro*. *J Mol Biol* 228: 1078–1090.

111. Del Campo C, Bartholomäus A, Fedyunin I, Ignatova Z. 2015. Secondary structure across the bacterial transcriptome reveals versatile roles in mRNA regulation and function. *PLoS Genet* 11:e1005613.

112. Mackie GA. 1998. Ribonuclease E is a 5′-end-dependent endonuclease. *Nature* 395:720–723.

113. Koslover DJ, Callaghan AJ, Marcaida MJ, Garman EF, Martick M, Scott WG, Luisi BF. 2008. The crystal structure of the *Escherichia coli* RNase E apoprotein and a mechanism for RNA degradation. *Structure* 16: 1238–1244.

114. Kime L, Jourdan SS, Stead JA, Hidalgo-Sastre A, McDowall KJ. 2010. Rapid cleavage of RNA by RNase E in the absence of 5′ monophosphate stimulation. *Mol Microbiol* 76:590–604.

115. Kime L, Clarke JE, Romero AD, Grasby JA, McDowall KJ. 2014. Adjacent single-stranded regions mediate processing of tRNA precursors by RNase E direct entry. *Nucleic Acids Res* 42:4577–4589.

116. Bouvier M, Carpousis AJ. 2011. A tale of two mRNA degradation pathways mediated by RNase E. *Mol Microbiol* 82:1305–1310.

117. Kaberdin VR, Walsh AP, Jakobsen T, McDowall KJ, von Gabain A. 2000. Enhanced cleavage of RNA mediated by an interaction between substrates and the arginine-rich domain of *E. coli* ribonuclease E. *J Mol Biol* 301:257–264.

118. Christiansen J. 1988. The 9S RNA precursor of *Escherichia scoli* 5S RNA has three structural domains: implications for processing. *Nucleic Acids Res* 16: 7457–7476.

119. Kim D, Song S, Lee M, Go H, Shin E, Yeom JH, Ha NC, Lee K, Kim YH. 2014. Modulation of RNase E activity by alternative RNA binding sites. *PLoS One* 9: e90610.

120. Go H, Moore CJ, Lee M, Shin E, Jeon CO, Cha CJ, Han SH, Kim SJ, Lee SW, Lee Y, Ha NC, Kim YH, Cohen SN, Lee K. 2011. Upregulation of RNase E activity by mutation of a site that uncompetitively interferes with RNA binding. *RNA Biol* 8:1022–1034.

121. Deutscher MP. 2015. Twenty years of bacterial RNases and RNA processing: how we've matured. *RNA* 21: 597–600.

122. Iost I, Dreyfus M. 1995. The stability of *Escherichia coli* lacZ mRNA depends upon the simultaneity of its synthesis and translation. *EMBO J* 14:3252–3261.

123. Dubey AK, Baker CS, Romeo T, Babitzke P. 2005. RNA sequence and secondary structure participate in high-affinity CsrA-RNA interaction. *RNA* 11:1579–1587.

124. Baker CS, Eöry LA, Yakhnin H, Mercante J, Romeo T, Babitzke P. 2007. CsrA inhibits translation initiation of *Escherichia coli* hfq by binding to a single site overlapping the Shine-Dalgarno sequence. *J Bacteriol* 189: 5472–5481.

125. Irie Y, Starkey M, Edwards AN, Wozniak DJ, Romeo T, Parsek MR. 2010. *Pseudomonas aeruginosa* biofilm matrix polysaccharide Psl is regulated transcriptionally by RpoS and post-transcriptionally by RsmA. *Mol Microbiol* 78:158–172.

126. Boël G, Letso R, Neely H, Price WN, Wong KH, Su M, Luff J, Valecha M, Everett JK, Acton TB, Xiao R, Montelione GT, Aalberts DP, Hunt JF. 2016. Codon influence on protein expression in *E. coli* correlates with mRNA levels. *Nature* **529**:358–363.

127. Tsai YC, Du D, Domínguez-Malfavón L, Dimastrogiovanni D, Cross J, Callaghan AJ, García-Mena J, Luisi BF. 2012. Recognition of the 70S ribosome and polysome by the RNA degradosome in *Escherichia coli*. *Nucleic Acids Res* **40**:10417–10431.

128. Deana A, Belasco JG. 2005. Lost in translation: the influence of ribosomes on bacterial mRNA decay. *Genes Dev* **19**:2526–2533.

129. Khemici V, Prados J, Linder P, Redder P. 2015. Decay-initiating endoribonucleolytic cleavage by RNase Y is kept under tight control via sequence preference and sub-cellular localisation. *PLoS Genet* **11**:e1005577.

130. Koch G, Wermser C, Acosta IC, Kricks L, Stengel ST, Yepes A, Lopez D. 2017. Attenuating *Staphylococcus aureus* virulence by targeting flotillin protein scaffold activity. *Cell Chem Biol* **24**:845–857.e6.

131. Montero Llopis P, Jackson AF, Sliusarenko O, Surovtsev I, Heinritz J, Emonet T, Jacobs-Wagner C. 2010. Spatial organization of the flow of genetic information in bacteria. *Nature* **466**:77–81.

132. Lin Y, Protter DS, Rosen MK, Parker R. 2015. Formation and maturation of phase-separated liquid droplets by RNA-binding proteins. *Mol Cell* **60**:208–219.

133. Chao Y, Papenfort K, Reinhardt R, Sharma CM, Vogel J. 2012. An atlas of Hfq-bound transcripts reveals 3′ UTRs as a genomic reservoir of regulatory small RNAs. *EMBO J* **31**:4005–4019.

134. Holmqvist E, Wright PR, Li L, Bischler T, Barquist L, Reinhardt R, Backofen R, Vogel J. 2016. Global RNA recognition patterns of post-transcriptional regulators Hfq and CsrA revealed by UV crosslinking *in vivo*. *EMBO J* **35**:991–1011.

135. Sittka A, Lucchini S, Papenfort K, Sharma CM, Rolle K, Binnewies TT, Hinton JC, Vogel J. 2008. Deep sequencing analysis of small noncoding RNA and mRNA targets of the global post-transcriptional regulator, Hfq. *PLoS Genet* **4**:e1000163.

136. Tree JJ, Granneman S, McAteer SP, Tollervey D, Gally DL. 2014. Identification of bacteriophage-encoded anti-sRNAs in pathogenic *Escherichia coli*. *Mol Cell* **55**:199–213.

137. Zhang A, Wassarman KM, Rosenow C, Tjaden BC, Storz G, Gottesman S. 2003. Global analysis of small RNA and mRNA targets of Hfq. *Mol Microbiol* **50**:1111–1124.

138. Smirnov A, Förstner KU, Holmqvist E, Otto A, Günster R, Becher D, Reinhardt R, Vogel J. 2016. Grad-seq guides the discovery of ProQ as a major small RNA-binding protein. *Proc Natl Acad Sci U S A* **113**:11591–11596.

139. Westermann AJ, Förstner KU, Amman F, Barquist L, Chao Y, Schulte LN, Müller L, Reinhardt R, Stadler PF, Vogel J. 2016. Dual RNA-seq unveils noncoding RNA functions in host-pathogen interactions. *Nature* **529**:496–501.

140. Bandyra KJ, Bouvier M, Carpousis AJ, Luisi BF. 2013. The social fabric of the RNA degradosome. *Biochim Biophys Acta* **1829**:514–522.

Regulating with RNA in Bacteria and Archaea
Edited by Gisela Storz and Kai Papenfort
© 2018 American Society for Microbiology, Washington, DC
doi:10.1128/microbiolspec.RWR-0011-2017

Enzymes Involved in Posttranscriptional RNA Metabolism in Gram-Negative Bacteria

2

Bijoy K. Mohanty[1] and Sidney R. Kushner[1,2]

INTRODUCTION

All living organisms, including the Gram-negative bacteria, have two major classes of RNA molecules. mRNAs contain the information for the synthesis of the various proteins that are required for a living cell. The so-called nontranslated RNAs, which include tRNAs, rRNAs, and small regulatory RNAs (sRNAs), provide the RNA components for ribosome assembly, protein synthesis, and the regulation of mRNA functionality based on RNA/RNA interactions. The highly diverse functions that these RNAs perform within the cell are possible due to numerous enzymes that are involved in posttranscriptional RNA metabolism. However, many of these enzymes have overlapping activities. Besides the normal cellular complement of enzymes that carry out the above functions, there are ribonucleases that are specifically associated with particular stress conditions as part of toxin/antitoxin (TA) systems.

Based on our current knowledge of bacterial RNases, there are no dedicated RNases for either RNA degradation or processing. Many of the RNases and RNA helicases participate in more than one pathway, which is visually described in Fig. 1. For example, endoribonuclease E (RNase E) is involved in almost all aspects of RNA metabolism. RNase III is very important for initiation of rRNA processing, but it also participates in mRNA decay and sRNA degradation. Similarly, although RNase P is essential for tRNA 5′ end maturation, it also participates in tRNA processing and mRNA decay. The RNase activities of most of the exonucleases are highly redundant, since they can complement each other. Furthermore, as single-stranded RNAs can rapidly fold into more-complex forms containing secondary and tertiary structures, RNA helicases play an important role in various aspects of posttranscriptional processing and decay. In this review, RNases and RNA helicases found in various Gram-negative bacteria are discussed in the context of their *in vivo* biological functions. The basic properties of all the RNases and RNA helicases are outlined in Table 1 (see also chapter 3).

GENERAL mRNA DECAY

Initiation of mRNA Decay by Endonucleases

The decay of mRNAs plays a major role in the posttranscriptional regulation of gene expression, since it helps to control the steady-state level of each individual mRNA and in turn the protein level. As a result, considerable effort has been devoted to document the mechanisms of mRNA decay in *Escherichia coli*, which still serves as the model organism for Gram-negative bacteria. Apirion was the first to propose a model for mRNA decay in 1973 (1), which involved a combination of endoribonucleases and exoribonucleases. However, at that time the evidence supporting his hypothesis was not particularly strong. Subsequently, RNase E was initially identified based on its role in rRNA processing (2), as discussed later in this review. At about the same time as RNase E was discovered, the *ams-1* allele (altered mRNA stability) was isolated and shown to affect cell viability. Strains carrying the *ams-1* mutation demonstrated an increase in the half-life of bulk mRNA at the nonpermissive temperature (3, 4). Subsequently, the *ams* and *rne* loci were shown to encode the same protein, which is now called RNase E (5).

---

[1]Department of Genetics; [2]Department of Microbiology, University of Georgia, Athens, GA 30602.

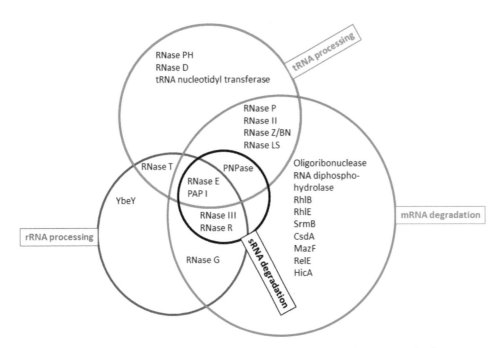

Figure 1 Venn diagram of RNases in *E. coli* showing their involvement in the four major RNA metabolic pathways in Gram-negative bacteria. The participation of the various proteins is only included in pathways where it has been established that they play a significant role. In addition, it is possible that some proteins, such as YbeY, are involved in additional pathways.

RNase E, an essential enzyme in *E. coli*, is one of its largest proteins. It is divided into an N-terminal catalytic region and a C-terminal scaffold region. Experiments have shown that the catalytic region of RNase E is located in the first 500 amino acids of the protein (6) and that it is associated with the inner membrane of the cell through a short amino acid sequence that is immediately downstream of the catalytic region (7). A very important advance in the analysis of mRNA decay pathways was the finding that RNase E was associated with several other enzymes involved in RNA degradation. First, it was shown that RNase E forms a complex with polynucleotide phosphorylase (PNPase), a 3′→5′ exonuclease (8). Subsequently, it was determined that the RhlB RNA helicase and the glycolytic enzyme enolase were also included in this multiprotein complex, which was called the "degradosome" (8, 9). However, mRNA decay is not significantly affected in the absence of either degradosome assembly or membrane attachment, as shown by deletions of the carboxy-terminal region that lack either the degradosome scaffold region (10) or both the degradosome scaffold region and the membrane attachment site (11). The degradosome is discussed in more detail in chapter 1.

It appears that RNase E can recognize its substrates in more than one way. Mackie demonstrated that the enzyme was inhibited by the triphosphate moiety found at the 5′ terminus of RNA transcripts (12). Thus, RNase E processing is significantly enhanced by the removal of the pyrophosphate from the 5′-terminal triphosphate of many primary transcripts. While conversion of the 5′ triphosphate to a 5′ monophosphate was believed to be catalyzed by the RNA pyrophosphohydrolase (RppH) encoded by *rppH* (13), a recent report suggests that an unidentified enzyme converts the triphosphate to a diphosphate, which is the substrate of choice for the RppH protein (14). Although RNase E prefers to bind to substrates containing a 5′ monophosphate, a number of experiments have demonstrated that it can also cleave both mRNAs and tRNAs using a direct entry mechanism (15–19).

Several laboratories attempted to determine if there was sequence specificity associated with RNase E cleavages by comparing the sequences of known cleavage sites (20–22). These studies revealed that the enzyme prefers to cleave RNA in single-stranded A/U-rich regions. Further work has shown that the cleavage sites are usually found either upstream or downstream of secondary structures. A recent analysis of the transcriptome employing RNA-seq suggests a minimal 5-nucleotide (nt) RNase E consensus cleavage site as "RN↓WUU" (with R as G/A, W as A/U, and N as any nucleotide) with a strong preference for uridine at the +2 position (23).

Table 1 Proteins involved in posttranscriptional RNA metabolism in *E. coli*

Name	Gene(s)	Mode of action	Substrate preference	Functions[a]
RNase E	*rne*	Endonuclease	Single-stranded RNA	mRNA decay, tRNA processing, rRNA processing, sRNA degradation
RNase III	*rnc*	Endonuclease	Double-stranded RNA	rRNA processing, mRNA decay, sRNA degradation
RNase P	*rnpA*, *rnpB*	Endonuclease	Single-stranded RNA	tRNA processing, mRNA decay, rRNA processing
RNase G	*rng*	Endonuclease	Single-stranded RNA	rRNA processing, mRNA decay, tRNA processing
YbeY	*ybeY*	Endonuclease	Single-stranded RNA	rRNA processing
RNase Z/BN	*rnz*	Endonuclease/ $3' \rightarrow 5'$ exonuclease	Single-stranded RNA	mRNA decay, tRNA processing
RNase LS	*rnlA*	Endonuclease	Single-stranded RNA	mRNA decay
RNase T	*rnt*	$3' \rightarrow 5'$ exonuclease	Single-stranded RNA	tRNA processing, rRNA processing
RNase PH	*rph*	$3' \rightarrow 5'$ exonuclease	Single-stranded RNA	tRNA processing
RNase D	*rnd*	$3' \rightarrow 5'$ exonuclease	Single-stranded RNA	tRNA processing
PNPase	*pnp*	$3' \rightarrow 5'$ exonuclease	Single-stranded RNA	mRNA decay, sRNA degradation, tRNA processing
RNase II	*rnb*	$3' \rightarrow 5'$ exonuclease	Single-stranded RNA	mRNA decay, tRNA processing
RNase R	*rnr*	$3' \rightarrow 5'$ exonuclease	Single-stranded RNA	rRNA processing, sRNA degradation
Oligoribonuclease	*orn*	$3' \rightarrow 5'$ exonuclease	Single-stranded RNA	mRNA decay
Poly(A) polymerase	*pcnB*	Polymerase	Single-stranded RNA	mRNA decay, tRNA maturation
Hfq	*hfq*	RNA-binding protein	Single-stranded RNA	mRNA decay, polyadenylation, sRNA metabolism
tRNA nucleotidyltransferase	*cca*	Polymerase	tRNAs	tRNA maturation
RNA diphosphohydrolase	*rppH*	Phosphatase	5′ triphosphate or 5′ diphosphate	mRNA decay
RhlB	*rhlB*	Helicase	Double-stranded RNA	mRNA decay
RhlE	*rhlE*	Helicase	Double-stranded RNA	mRNA decay
SrmB	*srmB*	Helicase	Double-stranded RNA	mRNA decay
CsdA	*csdA*	Helicase	Double-stranded RNA	mRNA decay under low temperature
DbpA	*dbpA*	Helicase	Double-stranded RNA	Ribosome biogenesis
Enolase	*eno*	Enolase	?[b]	RNA metabolism?[b]
MazF	*mazF*	mRNA interferase	Single-stranded RNA	mRNA decay, stress management
RelE	*relE*	mRNA interferase	Single-stranded RNA	mRNA decay, stress management
HicA	*hicA*	mRNA interferase	Single-stranded RNA	mRNA decay, stress management
Colicin E5		tRNA-targeting enzyme	Single-stranded RNA	Stress management
Colicin D		tRNA-targeting enzyme	Single-stranded RNA	Stress management
Colicin E3		tRNA-targeting enzyme	Single-stranded RNA	Stress management
Cas6A	*cas6*	Endonuclease	Single-stranded RNA	Immune defense

[a]Listed in order of the importance of the enzyme.
[b]The role of enolase in RNA metabolism is yet to be defined.

High-density tiling arrays as well as RNA-seq experiments have demonstrated that RNase E is responsible for the initiation of mRNA decay for >50% of all the transcripts generated during exponential growth in *E. coli* (18, 24). However, besides RNase E, there are a significant number of other endonucleases that initiate the decay of mRNAs as minor players. For example, RNase III, which will be discussed in more detail later in the context of rRNA processing, has been shown to affect the steady-state levels of up to 10% of the mRNAs in exponentially growing cells of *E. coli*,

resulting in either destabilization or stabilization of the transcripts (24, 25). RNase P, which is a ribozyme that contains an RNA catalytic subunit (26), specifically cleaves a small number of polycistronic mRNAs in intercistronic regions (27).

In addition, many Gram-negative bacteria have an RNase E ortholog called RNase G. A major distinction between the two enzymes is that RNase G lacks the degradosome scaffold region (28, 29). RNase G has been most studied in *E. coli*. It has been shown to target specific mRNAs and work in tandem with RNase E

(30). Similar to RNase E, the enzyme prefers single-stranded AU-rich sequences and is 5′ end dependent (31), but is present in much smaller amounts than RNase E (32). *In vitro*, both RNase E and RNase G have similar substrate specificity. However, RNase G cleaves the 5′ terminus of pre-16S rRNA precursors at a different site than RNase E (28, 29). Although it was originally thought that RNase G could complement the conditional lethality associated with RNase E mutants (32, 33), it has now been shown that complementation only takes place in the presence of a mutationally altered RNase G protein (34).

RNase Z, another endonuclease found in *E. coli* and other Gram-negative bacteria, was initially identified in eukaryotes based on its ability to cleave tRNA precursors endonucleolytically to generate a 3′ terminus that was a substrate for tRNA nucleotidyltransferase, which adds the CCA determinant (35). In *E. coli*, the enzyme seems to function primarily in mRNA decay (36) but can also serve in a backup role in some aspects to tRNA processing (see "Enzymes Involved in tRNA Processing" below). Interestingly, RNase Z, which was originally identified as RNase BN (37), also has 3′→5′ exonuclease activity and will be discussed below.

RNase LS, another endonuclease encoded by *rnlA*, has been shown to play a very limited role in the decay in both bacterial mRNAs and phage-encoded mRNAs in *E. coli* (38) and is also part of the *rnlAB* TA module (39).

The Role of 3′→5′ Exonucleases in mRNA Decay

RNase II, RNase R, and PNPase are the three major 3′→5′ exonucleases affecting mRNAs and rRNAs in most Gram-negative bacteria. RNase II and RNase R degrade single-stranded RNA employing a hydrolytic mechanism, releasing mononucleotides (40, 41). It has been shown that RNase II accounts for ~95% of the hydrolytic RNase activity in *E. coli* (42). RNase II is strongly inhibited by secondary structures (43), while RNase R can easily degrade RNA molecules containing secondary structures (44) but requires a single-stranded region of at least 7 nt in order to bind (45). Interestingly, RNase R is more important in stationary-phase cells than RNase II (46).

In contrast, PNPase degrades RNA using a phosphorolytic mechanism that requires inorganic phosphate, releasing nucleoside diphosphates (47). Since the equilibrium constant for this reaction is 1, the enzyme can also synthesize RNA in an untemplated reaction employing nucleoside diphosphates to generate single-stranded RNA containing all four nucleotides (47). In fact, the enzyme can work both biosynthetically and degradatively in *E. coli* (48) and other Gram-negative bacteria. PNPase exists in at least two multiprotein complexes: (i) the degradosome, which contains RNase E, PNPase, the RhlB RNA helicase, and enolase; and (ii) the polyadenylation complex, which contains poly (A) polymerase I (PAP I), PNPase, and the RNA-binding protein Hfq (49). RhlB is also reported to be associated with PNPase independently of the degradosome assembly, helping its exonucleolytic activity *in vitro* (50), although such an effect has yet to be observed *in vivo* (51).

While none of these three RNases are essential for cell viability by themselves, inactivation of both RNase II and PNPase results in synthetic lethality (52). Of most significance is that at the nonpermissive temperature (44°C) there is a large accumulation of partially degraded mRNAs in a *pnp-7 rnb-500* double mutant, demonstrating a significant role for these two enzymes in the mRNA turnover (52). It has also been shown that a *pnp-7 rnr* double mutant is a synthetic lethal but an *rnb rnr* strain is viable (41).

While the decay of the majority of mRNAs is initiated via endonucleolytic cleavages employing a combination of RNase E, RNase III, RNase G, RNase P, RNase Z, and RNase LS, the degradation of a significant number of transcripts is also initiated by exonucleases PNPase, RNase II, and RNase R. A series of experiments employing either macroarrays (53) or RNA-seq (54) have demonstrated that inactivation of any of the three exonucleases leads to significant changes in the steady-state levels of between 5 and 10% of the mRNAs. It is also thought that RNase R is more important in ribosome quality control by degrading nonfunctional rRNAs and tRNAs rather than mRNAs. In contrast, PNPase and RNase II are more involved in mRNA decay (53) and some aspects of tRNA processing (51) and rRNA degradation (55). It should also be noted that unlike Gram-positive bacteria (56), there do not appear to be any 5′→3′ exonucleases in Gram-negative species.

A common feature of all three of these exonucleases is that they cannot degrade an RNA substrate completely, leaving short oligonucleotides of 2 to 4 nt in length (57–59). Many bacteria contain another 3′→5′ exoribonuclease called oligoribonuclease, which specifically degrades these short oligonucleotides using a hydrolytic mechanism (60, 61). The complete degradation of the short oligonucleotides that remain after the action of RNase II, RNase R, and PNPase seems to be essential, since there is evidence that oligoribonuclease is essential for cell viability (62). It is not clear at this time if an enzyme such as RNase T can also function on short oligoribonucleotides. Figure 2 presents a model for general mRNA decay in *E. coli* initiated by RNase E and RNase G.

ENZYMES INVOLVED IN tRNA PROCESSING

In *E. coli* and many other Gram-negative bacteria, all of the tRNAs are encoded with extra nucleotides at both their 5′ and 3′ ends. They occur as either monocistronic or polycistronic transcripts that contain only tRNAs, tRNAs and mRNAs, or tRNAs and rRNAs. Figure 3 presents some of the pathways involved in tRNA processing. Many of these transcripts are terminated in a Rho-independent manner, which generates a

stem-loop at the 3′ terminus. The processing of the majority of these tRNA transcripts is initiated by RNase E, which either removes the Rho-independent transcription terminators (63–65) or cleaves in the intercistronic regions (11, 66). In some cases, RNase G and RNase Z can inefficiently substitute for RNase E (65) (Fig. 3A). In at least one well-documented case, the Rho-independent transcription terminator on the *leuX* primary transcript is removed exonucleolytically by PNPase (51) (Fig. 3B).

Interestingly, many polycistronic tRNA transcripts do not utilize RNase E, but rather are dependent on RNase P for their initial processing (67) (Fig. 3C). Some require the initial removal of the Rho-independent transcription terminator by RNase E before RNase P can

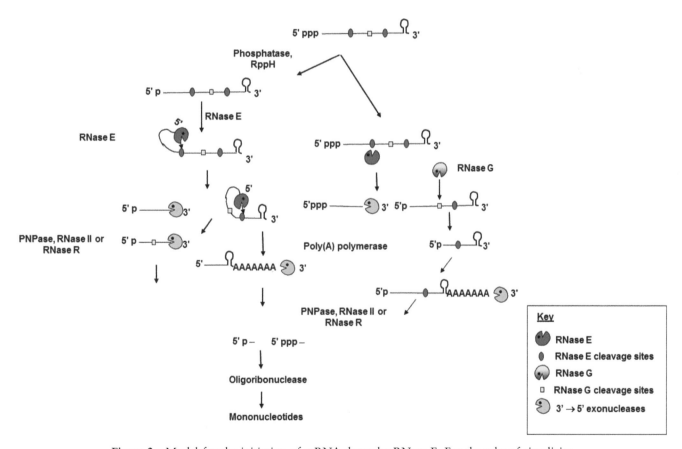

Figure 2 Model for the initiation of mRNA decay by RNase E. For the sake of simplicity, the other proteins associated with the RNase E-based degradosome are not shown. In addition, this model is independent of whether RNase E is associated with the inner membrane of *E. coli*. 5′ monophosphate RNA, a preferred substrate for RNase E, is degraded via a 5′-end-dependent pathway. In contrast, 5′ triphosphate RNA is degraded via an RNase E internal entry mechanism. Any endonucleolytically cleaved fragments with strong secondary structures, such as one containing a Rho-independent transcription terminator shown here, undergo polyadenylation by PAP I. Subsequently, all decay intermediates are degraded by 3′→5′ exonucleases (PNPase, RNase II, and RNase R) followed by oligoribonuclease to mononucleotides. Figure is not drawn to scale. p, phosphomonoester; ppp, triphosphate.

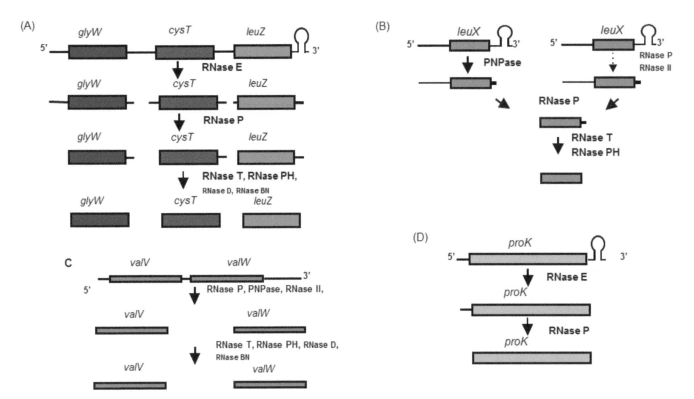

Figure 3 Diagrammatic representation of four independent pathways of tRNA processing. (A) Processing of the *glyW cysT leuZ* polycistronic operon. RNase E initiates processing by cleaving the polycistronic transcript to release pre-tRNAs (11). Processing at the 5′ termini is carried out by RNase P. Maturation of the 3′ termini is usually carried out by RNase T and/or RNase PH. If these two enzymes are not present, RNase D and/or RNase BN can complete the process. (B) Processing of the monocistronic *leuX* transcript (51). The Rho-independent transcription terminator is removed exonucleolytically by PNPase. In the absence of PNPase, a combination of RNase P and RNase II can digest the terminator. Subsequently, RNase P matures the 5′ terminus, while RNase T and RNase PH complete the process at the 3′ terminus. (C) Processing of the *valV valW* polycistronic operon (67). RNase P separates *valV* and *valW* pre-tRNAs by cleaving at their respective mature 5′ ends, while PNPase and RNase II shorten the 3′ Rho-dependent terminator. Subsequently, 3′→5′ exonucleases (RNase T/RNase PH/RNase D/RNase BN) mature the 3′ ends. (D). Processing of the monocistronic *proK* transcript (63). RNase E removes the Rho-independent transcription terminator to generate the mature 3′ terminus without the need of any of the 3′→5′ exonucleases. RNase P cleaves at the mature 5′ end. Figure is not drawn to scale.

process the rest of the transcript (64, 65, 67). The RNase P cleavages occur at the mature 5′ termini (65), while RNase E cleavages, with the exception of the three proline tRNAs (63), leave extra nucleotides downstream of the CCA determinant (11, 66). A surprising observation from the RNase P-processed transcripts was that the enzyme cleaves the polycistronic transcripts starting from the 3′ terminus and not the 5′ terminus (64).

An interesting feature of tRNA maturation in Gram-negative bacteria is that there is only one ribonuclease, RNase P, that generates the mature 5′ terminus of all the tRNA species (26). In contrast, maturation of the 3′ terminus can be carried out by a variety of 3′→5′ exonucleases including RNase T, RNase PH, RNase D, RNase BN, and RNase II (42). PNPase and RNase II can also remove extra nucleotides from the 3′ terminus of a pre-tRNA, but they cannot complete the final maturation process to expose the CCA determinant (51, 65).

The bulk (79 of 86) of the pre-tRNAs in *E. coli* employ a combination of RNase T and RNase PH for their final 3′ end maturation (68). However, RNase T appears to be the most important enzyme of the two based on its unique substrate specificity. Specifically, unlike other ribonucleases, it specifically stops at the terminal CCA because of its inhibition by the presence of C ribonucleotides within its catalytic site (69). Thus, tRNAs containing C residues downstream of CCA are more dependent on RNase PH, which is not inhibited by C residues (64). Inactivation of both RNase T and RNase PH leads to rapid accumulation of 3′ immature tRNAs, which become substrates for PAP I (68). Contrary to mRNAs, PAP I adds short poly(A) tails (generally <5 nt) to the immature tRNAs (68). However, instead of undergoing degradation as defective tRNAs (70), a majority of the polyadenylated tRNAs are matured slowly by inefficient exonucleases, such as RNase D and RNase BN/Z (68). These results suggest that PAP I helps regulate functional tRNA levels in *E. coli* (68). However, excess PAP I polyadenylates mature tRNAs, inhibiting aminoacylation and protein synthesis, resulting in rapid cell death (71).

Recently, it has been shown that the three proline tRNAs do not require exonucleolytic processing at their 3′ termini (63) (Fig. 3D). Rather, RNase E removes the Rho-independent transcription terminator by cleaving immediately downstream of the CCA determinant (63). It is not known at this time what accounts for the altered substrate specificity of RNase E in the case of the proline tRNA transcripts.

It has now been shown that the 3′→5′ exonuclease RNase BN and the endoribonuclease RNase Z are encoded by the same gene (72), and the enzyme does not remove the CCA determinant from pre-tRNAs (73). Although it has been shown that RNase D and RNase BN can participate in 3′ end maturation (37, 74), it is not clear at this time how significant a role they play in cells that contain RNase T and RNase PH. As described above, RNase Z has been shown to play a role in mRNA decay (36).

With the possible exception of RNase T and RNase BN, the other 3′→5′ exonucleases can conceivably partially degrade into the CCA determinant that is required for aminoacylation. The repair of such termini can be carried out by the enzyme tRNA nucleotidyltransferase, which can add C, CC, or CCA in an untemplated reaction (75). Interestingly, this enzyme is not essential for cell viability in *E. coli* (76), most likely due to its overlapping biosynthetic activity with PNPase and PAP I (77). Recently, it has been suggested that the enzyme plays a role in tRNA quality control by adding the CCACCA tag to defective tRNA molecules, which are preferentially degraded by RNase R (78).

ENZYMES INVOLVED IN rRNA PROCESSING

All functional rRNAs in Gram-negative bacteria are matured from large (>3 kb) 30S rRNA polycistronic transcripts that contain the coding sequences for the 16S, 23S, and 5S rRNAs as well as at least one tRNA and a significant amount of spacer sequences in between (Fig. 4). Inverted repeats lead to two large stem-loop structures containing the 16S and 23S species as part of large loops (Fig. 4). Initial processing by RNase III, an enzyme that is specific for double-stranded RNA (79, 80), within the double-stranded stems releases 17S (pre-16S), 25S (pre-23S), and 9S (pre-5S) rRNA precursors (81–83), as well as the embedded tRNAs.

The pre-16S species containing 115 extra nucleotides at its 5′ terminus is initially cleaved endonucleolytically by RNase E, removing 60 nucleotides. Subsequently, RNase G endonucleolytically removes the remaining 55 nucleotides to generate the mature 5′ terminus of the 16S rRNA (28, 29). The extra 33 nucleotides found at the 3′ terminus of the pre-16S rRNA are removed by either the YbeY endonuclease or a combination of RNase II, RNase PH, PNPase, and RNase R (84–87).

The pre-5S species is cleaved by RNase E to within 3 nucleotides on each side of the mature sequence (81, 88–91). The three extra nucleotides at the 3′ end of the pre-5S rRNA are removed by the 3′→5′ exonuclease RNase T. Nothing is currently known about how the three extra nucleotides at the 5′ end of 5S rRNA species are processed.

In the case of the pre-23S species, the RNase III cleavages leave 7 to 9 extra nucleotides at the 3′ terminus and either 3 or 7 extra nucleotides at the 5′ terminus (89, 92). The final maturation of the 3′ terminus can be carried out either by RNase T alone or by a combination of PAP I, RNase II, RNase PH, and RNase T (91). In fact, the pre-23S rRNA species is an excellent target for PAP I in the wild-type cells (49, 93). Although it is not clear at this time how the 5′ end is matured, since Gram-negative bacteria do not contain 5′→3′ exonucleases, this step most likely is carried out by an endonuclease. Recent experiments suggest that RNase III might in fact carry out this reaction (G. Chardhuri, M. Stead, V. Maples, B. Mohanty, and S.R. Kushner, submitted for publication.)

The tRNAs embedded in the rRNA operons are released as pre-tRNAs following RNase III cleavage of the primary transcript and RNase E action on the 9S precursor (Fig. 4). Presumably, these pre-tRNAs employ the various enzymes described earlier (see "Enzymes Involved in tRNA Processing").

The ribosome, an essential component of protein synthetic machinery, is assembled using mature 23S, 16S, and 5S rRNAs. While it has been shown that pre-23S rRNAs can be incorporated to form functional

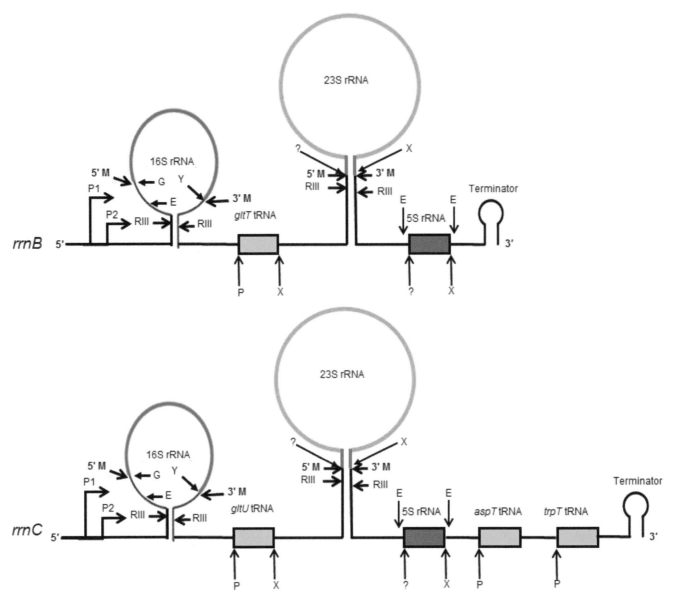

Figure 4 Processing of rRNA operons in *E. coli*. The *rrnB* and *rrnC* operons are shown as model operons. RNase III (RIII) cleaves the 30S rRNA transcript first within the double-stranded stems formed by the spacer sequences adjacent to the mature 16S and 23S rRNAs, generating 17S, 25S, and 9S pre-rRNAs. The functional mature 16S rRNA is generated from 17S pre-rRNA after initial RNase E (E) cleavage, followed by RNase G (G) at the mature (M) 5′ end and removal of an extra 33 nucleotides at the 3′ ends by YbeY (Y) along with multiple exoribonucleases (not shown). A p5S precursor is generated from the 9S precursor by initial RNase E cleavage at 3 nucleotides upstream (E) and downstream (E) of the mature termini of the mature 5S rRNA. The mature 5′ ends of the tRNAs are generated by RNase P (P) cleavage. Exoribonucleases (X) (primarily RNase T) are responsible for the 3′ end maturation of the tRNAs, 23S rRNA, and 5S rRNA, but the RNase(s) (?) responsible for the maturation of the 5′ ends of 23S and 5S rRNAs remain unidentified. The model is not drawn to scale.

ribosomes (81), incorporation of pre-16S species does not result in functional ribosomes (94). Thus, it was predicted that RNase III would be an essential enzyme because of the need to separate the 16S, 23S, and 5S species from the larger polycistronic transcript. However, it has been shown that *rnc* deletion mutants are viable and show only small defects in their growth rates (90, 95). Clearly, there is an alternative mechanism for generating rRNAs that can be incorporated into functional ribosomes in the absence of RNase III.

PROCESSING AND DECAY OF sRNAS

Recent experiments have shown that in a wide range of organisms sRNAs play more widespread regulatory roles than previously envisioned. In prokaryotes, sRNAs generally range in size from ~50 to 400 nt in length. Gram-negative bacteria contain >100 sRNAs, which are encoded on plasmids or the genome (96–99). In fact, high-density tiling array analysis of RNase E and RNase III mutants has suggested that there may in fact be as many as 300 sRNAs in *E. coli* (24, 100). While sRNAs that do not undergo any processing to be functional cannot be ruled out, all sRNAs identified to date require some initial processing by RNase E for them to be functional (101, 102). In the case of MicC, conversion of 5′ triphosphate to 5′ monophosphate enhanced RNase E-mediated decay of *ompD* (103).

Based on their interactions with their target mRNAs, sRNAs can inhibit translation by masking the ribosome binding site (104) or in some cases binding outside of the ribosome binding site (105), followed by rapid turnover of the transcripts. Other sRNAs control Rho-dependent termination at the 5′ untranslated region of the transcripts by blocking the action of Rho and resulting in increased translation of the transcripts (106). The association of many sRNAs with a functionally important cognate RNA-binding protein (Hfq) is critical for its functionality. For example, RyhB directly promotes mRNA instability by forming an RNA-protein complex containing Hfq that attracts RNase E (107). A recent study suggests that a large number of sRNAs in *Salmonella enterica* are associated with a conserved RNA-binding protein, ProQ (108). In some cases, RNase III is responsible for the decay of an mRNA targeted by an sRNA (109).

An interesting question relates to the degradation of sRNAs. Since these are relatively small molecules and highly structured, it was not expected that their degradation would be initiated by endonucleolytic attack. However, both RNase E and RNase III play important roles in degrading sRNAs when they are bound to their target

mRNAs (110, 111). The association of Hfq with sRNA-mRNA complexes facilitates this degradation process (112), but the experiments of Andrade et al. (113) have shown that sRNAs not associated with Hfq were not substrates for either RNase E or RNase III. Rather, PNPase seemed to be required for their degradation (113), a surprising result since these sRNAs are highly structured. However, it has previously been shown that PNPase can degrade the Rho-independent transcription terminator associated with the *leuX* primary transcript (51). The recent demonstration that PNPase can both protect and degrade sRNAs is consistent with these observations (114). However, since PNPase has been shown to be inhibited by strong secondary structures (43), it is likely that either polyadenylation by PAP I or unwinding of the secondary structures by one of the DEAD-box RNA helicases (see below) facilitates degradation of sRNAs by PNPase (113, 115). In contrast, sRNAs, such as RyhB, SgrS, and CyaR, are destabilized in the absence of PNPase, presumably by RNase E (116).

RNA HELICASES

Although RNAs are transcribed as single-stranded molecules, both nontranslated and translated transcripts readily form secondary and tertiary structures, which are critical for their proper functioning and stability. For example, many mRNAs are terminated with Rho-independent transcription terminators, a double-stranded stem-loop structure, which also serve as stability elements for mRNAs. All tRNAs assume their tertiary cloverleaf structures as soon as they are transcribed, a step that is crucial for their maturation by various RNases, posttranscriptional modification, and their functionality as amino acid carriers. The formation of large stem-loops in rRNA precursors generates cleavage sites for RNase III. The mature 16S and 23S rRNAs form complex arrays of secondary and tertiary structures in the generation of functional ribosomes. In addition, changes in growth temperatures, such as cold and heat shock, result in altered mRNA structure, which has a significant effect on translation efficiency. It is thus not surprising that bacteria contain helicases that use the energy derived from the hydrolysis of ATP to alter the structure of various RNA molecules. This class of enzyme is characterized by the DEXD/H (usually referred to as the DEAD-box) amino acid motif.

The best-characterized of the DEAD-box RNA helicases are the five paralogs (*rhlB*, *rhlE*, *srmB*, *dbpA*, and *csdA*, formerly called *deaD*) found in *E. coli*. A strain lacking all the helicases is still viable (117). Only loss of DeaD or SrmB causes significant growth defects and alterations in rRNA processing and maturation at both 30 and 37°C (117–119).

RhlB is involved in mRNA decay through its association with the RNase E-based degradosome (9, 120), as well as separately with PNPase (50) and is discussed in more detail in chapter 1. It has also been shown that RhlE can also associate with the RNase E-based degradosome under certain conditions (121). It aids in the degradation of mRNAs and seems to regulate the roles of other RNA helicases associated with ribosome maturation (121, 122). The SrmB and CsdA helicases function in the process of ribosome biogenesis (118, 119, 123). CsdA has also been associated with a "cold shock degradosome" (124). The DbpA helicase activity is dependent on 23S rRNA (125–127).

SPECIALIZED RNASES

TA Systems

The phenomenon of bacterial persistence was originally discovered by Bigger in 1944 (128). Simply put, in any bacterial population, some cells grow considerably more slowly than others. These slow-growing cells are inherently more resistant to antibiotics, which are more effective in killing rapidly growing bacteria. In 1983, Moyed et al. (129) showed that mutations in the *hipA* locus led to increased levels of persister cells. These mutations were part of the *hipBA* TA locus. Since then, multiple TA systems have been identified, which are broadly classified into five types. Generally, most toxins target an mRNA to inhibit translation in either a ribosome-dependent or -independent manner. These are sequence-specific endonucleases and have been called "mRNA interferases." See chapter 10 for more detail.

E. coli contains a large number of TA loci including *mazEF* (two copies), *relBE* (six copies), *hipAB*, *rnlAB*, and *hicAB* (130). In the case of RelE, the protein cleaves an mRNA positioned at the ribosomal A-site, between the second and third base of the A-site codon (131–133), leading to translation inhibition. The MazF endonuclease probably has the most sequence specificity of any *E. coli* endonuclease, with the possible exception of the RNases (bacteriocins) that target tRNAs (see next section). It cleaves mRNAs site-specifically at ACA base motifs independent of ribosomes (134). In contrast, the HicA ribonuclease, a protein of only 58 amino acids, cleaves mRNAs without any ribosomal involvement in an apparent random fashion (135).

RNase LS, which was discussed earlier (see "General mRNA Decay"), is part of the *rnlAB* TA module (39). It is not clear whether there is any cleavage specificity

associated with this enzyme, whether it has specific targets, or whether it plays any role in persistence.

Bacteriocins Function as RNases

Bacteria can produce bacteriocins that can kill other bacteria that are living in close proximity. Colicin, the earliest-discovered bacteriocin, is encoded by the Col plasmids. *E. coli* cells not carrying the same or cognate plasmid are killed by the released colicin (136). Colicin production is induced by nutrient starvation, the stringent response, the SOS response, and various other stress responses (136). Many of the colicins use the BtuB receptor for cell entry, which is involved in vitamin B_{12} uptake. There are several types of colicins, but here we only focus on those encoding an RNase activity.

Of the RNase type, colicin E3 is the most studied. It interacts with intact ribosomes, cleaving the 3′ region of the 16S rRNA between A1493 and G1494 in the decoding A-site, leading to functional inactivation (137–139). In contrast, colicin E5 and colicin D use different cell surface receptors and specifically cleave selected tRNAs. For example, colicin E5 cleaves between the Q and U residues (positions 34 and 35) in the anticodon arm of $tRNA^{His}$, $tRNA^{Asn}$, $tRNA^{Asp}$, and $tRNA^{Tyr}$ (140). Colicin D cleaves four of the six $tRNA^{Arg}$ species between positions 38 and 39 at the 3′ end of the anticodon loop (140).

Additional tRNA-cleaving ribonucleases have been identified in *Shigella* and *Salmonella* as part of the *vapBC* TA system (141). In this case, translation is inhibited by cleavage of the initiator $tRNA^{Met}$ at the 3′ end of the anticodon loop (142). While it had been suggested that the tRNA cleavages lead to cell death, recent work has shown that cells are not immediately killed (140). Rather, the tRNA cleavages serve to help regulate cell growth in order to survive a variety of stress conditions (140).

CRISPR-Cas Systems

Unlike the well-characterized restriction/modification systems that provide bacteria broad-range protection from phage attack (143), CRISPR (clustered regularly interspaced short palindromic repeat)-Cas systems provide adaptive immunity to the bacterial strains that carry them. The term CRISPR was first used by Jansen et al. (144). These systems are found in almost all archaea and about half of the bacteria that have been examined (145). The pathway found in *E. coli* K-12 falls into the type IE system. Simply put, bacteria carrying a CRISPR-Cas system have one or more CRISPR arrays, which contain short spacer sequences derived from previous bacteriophage infections or exposure to foreign bacterial plasmids, that are separated by identical repeat sequences (146). For any of the spacers to function properly in terms of targeting an invading bacteriophage, the CRISPR array must be transcribed and processed. In *E. coli*, five Cas proteins form a multiprotein complex called Cascade (147). The *cas6a* gene encodes a metal-independent endoribonuclease that cleaves in the repeat sequences to yield 61- to 62-nt CRISPR RNAs that have extra nucleotides at both their 5′ and 3′ ends. These species remain bound to the Cascade complex to facilitate its interaction with the target DNA (146). Interestingly, the Cas6a ribonuclease is normally not expressed in *E. coli* (146). Readers are encouraged to read the excellent recent review by Mohanraju et al. (148) on diverse CRISPR-Cas systems for more details.

CONTROL OF RNASES

There are many different ways that bacteria respond to the ever-changing environments in which they live. The regulation of various RNase activities is one of them. The levels of RNase E, RNase III, and PNPase are autoregulated based on specific circumstances by modulating the stability of their respective mRNAs (149–151). Under certain stress conditions, the activity of RNase E is also controlled by global protein regulators, such as RraA and RraB (152, 153). RraA has also been shown to interact with RhlB helicase, a component of the degradosome, to inhibit its RNA-binding and helicase activities (154). The protein YmdB, which is expressed upon cold shock or entry into stationary phase, has been shown to inhibit RNase III activity by preventing its dimerization (155). The level of the 3′→5′ exoribonuclease RNase R increases significantly upon entry into stationary phase or cold shock compared to the exponential-phase cells due to reduced proteolysis (156).

CONCLUSIONS

Although there has been tremendous progress in our understanding of the mechanisms of posttranscriptional RNA processing and decay over the past 25 years, many questions remain to be answered. With the discovery of new enzymes, some aspects of RNA processing have gotten closer attention and some conventional models have had to be significantly revised. However, even after the complete sequencing of multiple *E. coli* genomes, the functions for more than one-third of the genes are still unknown. Are there additional ribonucleases present in the cell that have not yet been identified?

Perhaps all of the enzymes involved in mRNA decay have been identified, but the regulation of the process is still not well understood (157) and all the targets of the various enzymes have yet to be identified. For example, RNase III initiates the processing of the 30S rRNA, but the enzyme is not essential for cell viability. Clearly, there must be an alternative processing pathway to generate the 16S, 23S, and 5S species that are required to produce functional ribosomes. In addition, the RNase (s) required for the 5′ end maturation of both 23S and 5S rRNAs are still a mystery.

Our understanding of the role of PAP I and PNPase as polyadenylating enzymes in Gram-negative bacteria is also still limited, especially the function of the polynucleotide tails synthesized by PNPase. First of all, it is not yet clear how these two enzymes select their substrates. The available data (49, 158, 159) suggest that PAP I can add poly(A) tails to both the full-length and decay intermediates. However, in the case of full-length transcripts, those terminated with a Rho-independent transcription terminator are the preferred substrates and the poly(A) tails are mostly added downstream of the terminator. In contrast, PNPase adds polynucleotide tails to both full-length transcripts terminating in a Rho-dependent fashion and decay intermediates. While poly(A) tail addition clearly targets transcripts for decay (160), no such data are available for polynucleotide tails. However, the extremely long, unstructured polynucleotide tails added by PNPase (49, 159) suggest that they also facilitate degradation of its substrates by exonucleases by providing them unstructured substrates.

The notion of cellular compartmentalization in bacteria (161), similar to what occurs in eukaryotes, has further complicated our understanding of the process of posttranscriptional regulation by various enzymes. Since the degradosome is associated with the inner membrane, does that mean that all mRNA decay has to occur at this location? What about the transcripts that are decayed by enzymes other than RNase E? In fact, limited dispersion of some transcripts from the site of transcription was observed in a recent study (162). Clearly, identification of new players and their regulatory processes will help us develop a better understanding of RNA processing and degradation.

Acknowledgments. This work was supported by National Institutes of Health grant GM081554 to S.R.K.

Citation. Mohanty BK, Kushner SR. 2018. Enzymes involved in posttranscriptional RNA metabolism in Gram-negative bacteria. Microbiol Spectrum 6(2):RWR-0011-2017.

References

1. **Apirion D.** 1973. Degradation of RNA in *Escherichia coli*. A hypothesis. *Mol Gen Genet* **122**:313–322.

2. **Misra TK, Apirion D.** 1979. RNase E, an RNA processing enzyme from *Escherichia coli*. *J Biol Chem* **254**: 11154–11159.

3. **Kuwano M, Ono M, Endo H, Hori K, Nakamura K, Hirota Y, Ohnishi Y.** 1977. Gene affecting longevity of messenger RNA: a mutant of *Escherichia coli* with altered mRNA stability. *Mol Gen Genet* **154**:279–285.

4. **Ono M, Kuwano M.** 1979. A conditional lethal mutation in an *Escherichia coli* strain with a longer chemical lifetime of messenger RNA. *J Mol Biol* **129**:343–357.

5. **Babitzke P, Kushner SR.** 1991. The Ams (altered mRNA stability) protein and ribonuclease E are encoded by the same structural gene of *Escherichia coli*. *Proc Natl Acad Sci U S A* **88**:1–5.

6. **McDowall KJ, Cohen SN.** 1996. The N-terminal domain of the *rne* gene product has RNase E activity and is non-overlapping with the arginine-rich RNA-binding site. *J Mol Biol* **255**:349–355.

7. **Khemici V, Poljak L, Luisi BF, Carpousis AJ.** 2008. The RNase E of *Escherichia coli* is a membrane-binding protein. *Mol Microbiol* **70**:799–813.

8. **Carpousis AJ, Van Houwe G, Ehretsmann C, Krisch HM.** 1994. Copurification of *E. coli* RNAase E and PNPase: evidence for a specific association between two enzymes important in RNA processing and degradation. *Cell* **76**:889–900.

9. **Py B, Higgins CF, Krisch HM, Carpousis AJ.** 1996. A DEAD-box RNA helicase in the *Escherichia coli* RNA degradosome. *Nature* **381**:169–172.

10. **Ow MC, Liu Q, Kushner SR.** 2000. Analysis of mRNA decay and rRNA processing in *Escherichia coli* in the absence of RNase E-based degradosome assembly. *Mol Microbiol* **38**:854–866.

11. **Ow MC, Kushner SR.** 2002. Initiation of tRNA maturation by RNase E is essential for cell viability in *E. coli*. *Genes Dev* **16**:1102–1115.

12. **Mackie GA.** 1998. Ribonuclease E is a 5′-end-dependent endonuclease. *Nature* **395**:720–723.

13. **Deana A, Celesnik H, Belasco JG.** 2008. The bacterial enzyme RppH triggers messenger RNA degradation by 5′ pyrophosphate removal. *Nature* **451**:355–358.

14. **Luciano DJ, Vasilyev N, Richards J, Serganov A, Belasco JG.** 2017. A novel RNA phosphorylation state enables 5′ end-dependent degradation in *Escherichia coli*. *Mol Cell* **67**:44–54.e46.

15. **Garrey SM, Mackie GA.** 2011. Roles of the 5′-phosphate sensor domain in RNase E. *Mol Microbiol* **80**:1613–1624.

16. **Baker KE, Mackie GA.** 2003. Ectopic RNase E sites promote bypass of 5′-end-dependent mRNA decay in *Escherichia coli*. *Mol Microbiol* **47**:75–88.

17. **Kime L, Clarke JE, Romero AD, Grasby JA, McDowall KJ.** 2014. Adjacent single-stranded regions mediate processing of tRNA precursors by RNase E direct entry. *Nucleic Acids Res* **42**:4577–4589.

18. **Clarke JE, Kime L, Romero AD, McDowall KJ.** 2014. Direct entry by RNase E is a major pathway for the degradation and processing of RNA in *Escherichia coli*. *Nucleic Acids Res* **42**:11733–11751.

19. **Kime L, Jourdan SS, Stead JA, Hidalgo-Sastre A, McDowall KJ.** 2010. Rapid cleavage of RNA by RNase E in the absence of 5′ monophosphate stimulation. *Mol Microbiol* **76**:590–604.

20. **Ehretsmann CP, Carpousis AJ, Krisch HM.** 1992. Specificity of *Escherichia coli* endoribonuclease RNase E: *in vivo* and *in vitro* analysis of mutants in a bacteriophage T4 mRNA processing site. *Genes Dev* **6**:149–159.

21. **McDowall KJ, Kaberdin VR, Wu SW, Cohen SN, Lin-Chao S.** 1995. Site-specific RNase E cleavage of oligonucleotides and inhibition by stem-loops. *Nature* **374**:287–290.

22. **McDowall KJ, Lin-Chao S, Cohen SN.** 1994. A+U content rather than a particular nucleotide order determines the specificity of RNase E cleavage. *J Biol Chem* **269**: 10790–10796.

23. **Chao Y, Li L, Girodat D, Förstner KU, Said N, Corcoran C, Śmiga M, Papenfort K, Reinhardt R, Wieden HJ, Luisi BF, Vogel J.** 2017. *In vivo* cleavage map illuminates the central role of RNase E in coding and non-coding RNA pathways. *Mol Cell* **65**:39–51.

24. **Stead MB, Marshburn S, Mohanty BK, Mitra J, Pena Castillo L, Ray D, van Bakel H, Hughes TR, Kushner SR.** 2011. Analysis of *Escherichia coli* RNase E and RNase III activity *in vivo* using tiling microarrays. *Nucleic Acids Res* **39**:3188–3203.

25. **Gordon GC, Cameron JC, Pfleger BF.** 2017. RNA sequencing identifies new RNase III cleavage sites in *Escherichia coli* and reveals increased regulation of mRNA. *MBio* **8**:e00128-17.

26. **Altman S.** 1989. Ribonuclease P: an enzyme with a catalytic RNA subunit. *Adv Enzymol Relat Areas Mol Biol* **62**:1–36.

27. **Li Y, Altman S.** 2003. A specific endoribonuclease, RNase P, affects gene expression of polycistronic operon mRNAs. *Proc Natl Acad Sci U S A* **100**:13213–13218.

28. **Wachi M, Umitsuki G, Shimizu M, Takada A, Nagai K.** 1999. *Escherichia coli cafA* gene encodes a novel RNase, designated as RNase G, involved in processing of the 5′ end of 16S rRNA. *Biochem Biophys Res Commun* **259**:483–488.

29. **Li Z, Pandit S, Deutscher MP.** 1999. RNase G (CafA protein) and RNase E are both required for the 5′ maturation of 16S ribosomal RNA. *EMBO J* **18**:2878–2885.

30. **Ow MC, Perwez T, Kushner SR.** 2003. RNase G of *Escherichia coli* exhibits only limited functional overlap with its essential homologue, RNase E. *Mol Microbiol* **49**:607–622.

31. **Tock MR, Walsh AP, Carroll G, McDowall KJ.** 2000. The CafA protein required for the 5′-maturation of 16 S rRNA is a 5′-end-dependent ribonuclease that has context-dependent broad sequence specificity. *J Biol Chem* **275**:8726–8732.

32. **Lee K, Bernstein JA, Cohen SN.** 2002. RNase G complementation of *rne* null mutation identifies functional

interrelationships with RNase E in *Escherichia coli*. *Mol Microbiol* **43**:1445–1456.

33. **Deana A, Belasco JG.** 2004. The function of RNase G in *Escherichia coli* is constrained by its amino and carboxyl termini. *Mol Microbiol* **51**:1205–1217.

34. **Chung DH, Min Z, Wang BC, Kushner SR.** 2010. Single amino acid changes in the predicted RNase H domain of *E. coli* RNase G lead to the complementation of RNase E mutants. *RNA* **16**:1371–1385.

35. **Schiffer S, Rösch S, Marchfelder A.** 2002. Assigning a function to a conserved group of proteins: the tRNA 3′-processing enzymes. *EMBO J* **21**:2769–2777.

36. **Perwez T, Kushner SR.** 2006. RNase Z in *Escherichia coli* plays a significant role in mRNA decay. *Mol Microbiol* **60**:723–737.

37. **Asha PK, Blouin RT, Zaniewski R, Deutscher MP.** 1983. Ribonuclease BN: identification and partial characterization of a new tRNA processing enzyme. *Proc Natl Acad Sci U S A* **80**:3301–3304.

38. **Otsuka Y, Yonesaki T.** 2005. A novel endoribonuclease, RNase LS, in *Escherichia coli*. *Genetics* **169**:13–20.

39. **Koga M, Otsuka Y, Lemire S, Yonesaki T.** 2011. *Escherichia coli rnlA* and *rnlB* compose a novel toxin-antitoxin system. *Genetics* **187**:123–130.

40. **Nossal NG, Singer MF.** 1968. The processive degradation of individual polynucleotide chains. *J Biol Chem* **243**:913–922.

41. **Cheng ZF, Zuo Y, Li Z, Rudd KE, Deutscher MP.** 1998. The *vacB* gene required for virulence in *Shigella flexneri* and *Escherichia coli* encodes the exoribonuclease RNase R. *J Biol Chem* **273**:14077–14080.

42. **Kelly KO, Deutscher MP.** 1992. The presence of only one of five exoribonucleases is sufficient to support the growth of *Escherichia coli*. *J Bacteriol* **174**:6682–6684.

43. **Spickler C, Mackie GA.** 2000. Action of RNase II and polynucleotide phosphorylase against RNAs containing stem-loops of defined structure. *J Bacteriol* **182**:2422–2427.

44. **Hossain ST, Malhotra A, Deutscher MP.** 2016. How RNase R degrades structured RNA: role of the helicase activity and the S1 domain. *J Biol Chem* **291**:7877–7887.

45. **Vincent HA, Deutscher MP.** 2006. Substrate recognition and catalysis by the exoribonuclease RNase R. *J Biol Chem* **281**:29769–29775.

46. **Andrade JM, Cairrão F, Arraiano CM.** 2006. RNase R affects gene expression in stationary phase: regulation of *ompA*. *Mol Microbiol* **60**:219–228.

47. **Grunberg-Manago M.** 1963. Polynucleotide phosphorylase. *Prog Nucl Acids Res* **1**:93–133.

48. **Mohanty BK, Kushner SR.** 2000. Polynucleotide phosphorylase functions both as a 3′ → 5′ exonuclease and a poly(A) polymerase in *Escherichia coli*. *Proc Natl Acad Sci U S A* **97**:11966–11971.

49. **Mohanty BK, Maples VF, Kushner SR.** 2004. The Sm-like protein Hfq regulates polyadenylation dependent mRNA decay in *Escherichia coli*. *Mol Microbiol* **54**:905–920.

50. **Lin PH, Lin-Chao S.** 2005. RhlB helicase rather than enolase is the β-subunit of the *Escherichia coli* polynucleotide phosphorylase (PNPase)-exoribonucleolytic complex. *Proc Natl Acad Sci U S A* **102**:16590–16595.

51. **Mohanty BK, Kushner SR.** 2010. Processing of the *Escherichia coli leuX* tRNA transcript, encoding tRNA^{Leu5}, requires either the 3′→5′ exoribonuclease polynucleotide phosphorylase or RNase P to remove the Rho-independent transcription terminator. *Nucleic Acids Res* **38**:597–607.

52. **Donovan WP, Kushner SR.** 1986. Polynucleotide phosphorylase and ribonuclease II are required for cell viability and mRNA turnover in *Escherichia coli* K-12. *Proc Natl Acad Sci U S A* **83**:120–124.

53. **Mohanty BK, Kushner SR.** 2003. Genomic analysis in *Escherichia coli* demonstrates differential roles for polynucleotide phosphorylase and RNase II in mRNA abundance and decay. *Mol Microbiol* **50**:645–658.

54. **Pobre V, Arraiano CM.** 2015. Next generation sequencing analysis reveals that the ribonucleases RNase II, RNase R and PNPase affect bacterial motility and biofilm formation in *E. coli*. *BMC Genomics* **16**:72.

55. **Basturea GN, Zundel MA, Deutscher MP.** 2011. Degradation of ribosomal RNA during starvation: comparison to quality control during steady-state growth and a role for RNase PH. *RNA* **17**:338–345.

56. **Durand S, Tomasini A, Braun F, Condon C, Romby P.** 2015. sRNA and mRNA turnover in Gram-positive bacteria. *FEMS Microbiol Rev* **39**:316–330.

57. **Amblar M, Barbas A, Gomez-Puertas P, Arraiano CM.** 2007. The role of the S1 domain in exoribonucleolytic activity: substrate specificity and multimerization. *RNA* **13**:317–327.

58. **Singer MF.** 1958. Phosphorolysis of oligoribonucleotides by polynucleotide phosphorylase. *J Biol Chem* **232**:211–228.

59. **Cannistraro VJ, Kennell D.** 1994. The processive reaction mechanism of ribonuclease II. *J Mol Biol* **243**:930–943.

60. **Datta AK, Niyogi K.** 1975. A novel oligoribonuclease of *Escherichia coli*. II. Mechanism of action. *J Biol Chem* **250**:7313–7319.

61. **Niyogi SK, Datta AK.** 1975. A novel oligoribonuclease of *Escherichia coli*. I. Isolation and properties. *J Biol Chem* **250**:7307–7312.

62. **Ghosh S, Deutscher MP.** 1999. Oligoribonuclease is an essential component of the mRNA decay pathway. *Proc Natl Acad Sci U S A* **96**:4372–4377.

63. **Mohanty BK, Petree JR, Kushner SR.** 2016. Endonucleolytic cleavages by RNase E generate the mature 3′ termini of the three proline tRNAs in *Escherichia coli*. *Nucleic Acids Res* **44**:6350–6362.

64. **Agrawal A, Mohanty BK, Kushner SR.** 2014. Processing of the seven valine tRNAs in *Escherichia coli* involves novel features of RNase P. *Nucleic Acids Res* **42**:11166–11179.

65. **Mohanty BK, Kushner SR.** 2008. Rho-independent transcription terminators inhibit RNase P processing of the *secG leuU* and *metT* tRNA polycistronic transcripts in *Escherichia coli*. *Nucleic Acids Res* **36**:364–375.

66. **Li Z, Deutscher MP.** 2002. RNase E plays an essential role in the maturation of *Escherichia coli* tRNA precursors. *RNA* **8:**97–109.

67. **Mohanty BK, Kushner SR.** 2007. Ribonuclease P processes polycistronic tRNA transcripts in *Escherichia coli* independent of ribonuclease E. *Nucleic Acids Res* **35:** 7614–7625.

68. **Mohanty BK, Maples VF, Kushner SR.** 2012. Polyadenylation helps regulate functional tRNA levels in *Escherichia coli*. *Nucleic Acids Res* **40:**4589–4603.

69. **Zuo Y, Deutscher MP.** 2002. Mechanism of action of RNase T. I. Identification of residues required for catalysis, substrate binding, and dimerization. *J Biol Chem* **277:**50155–50159.

70. **Li Z, Reimers S, Pandit S, Deutscher MP.** 2002. RNA quality control: degradation of defective transfer RNA. *EMBO J* **21:**1132–1138.

71. **Mohanty BK, Kushner SR.** 2013. Deregulation of poly (A) polymerase I in *Escherichia coli* inhibits protein synthesis and leads to cell death. *Nucleic Acids Res* **41:** 1757–1766.

72. **Ezraty B, Dahlgren B, Deutscher MP.** 2005. The RNase Z homologue encoded by *Escherichia coli elaC* gene is RNase BN. *J Biol Chem* **280:**16542–16545.

73. **Dutta T, Deutscher MP.** 2010. Mode of action of RNase BN/RNase Z on tRNA precursors: RNase BN does not remove the CCA sequence from tRNA. *J Biol Chem* **285:**22874–22881.

74. **Zhang JR, Deutscher MP.** 1988. Transfer RNA is a substrate for RNase D *in vivo*. *J Biol Chem* **263:**17909–17912.

75. **Deutscher MP, Evans JA.** 1977. Transfer RNA nucleotidyltransferase repairs all transfer RNAs randomly. *J Mol Biol* **109:**593–597.

76. **Zhu L, Deutscher MP.** 1987. tRNA nucleotidyltransferase is not essential for *Escherichia coli* viability. *EMBO J* **6:**2473–2477.

77. **Reuven NB, Zhou Z, Deutscher MP.** 1997. Functional overlap of tRNA nucleotidyltransferase, poly(A) polymerase I, and polynucleotide phosphorylase. *J Biol Chem* **272:**33255–33259.

78. **Wellner K, Czech A, Ignatova Z, Betat H, Mörl M.** 2017. Examining tRNA 3′-ends in *Escherichia coli*: teamwork between CCA-adding enzyme, RNase T, and RNase R. *RNA* **24:**361–370.

79. **Kindler P, Keil TU, Hofschneider PH.** 1973. Isolation and characterization of a ribonuclease 3 deficient mutant of *Escherichia coli*. *Mol Gen Genet* **126:**53–59.

80. **Robertson HD, Webster RE, Zinder ND.** 1967. A nuclease specific for double-stranded RNA. *Virology* **32:** 718–719.

81. **King TC, Sirdeshmukh R, Schlessinger D.** 1984. RNase III cleavage is obligate for maturation but not for function of *Escherichia coli* pre-23S rRNA. *Proc Natl Acad Sci U S A* **81:**185–188.

82. **Srivastava AK, Schlessinger D.** 1988. Coregulation of processing and translation: mature 5′ termini of *Escherichia coli* 23S ribosomal RNA form in polysomes. *Proc Natl Acad Sci U S A* **85:**7144–7148.

83. **Gutgsell NS, Jain C.** 2010. Coordinated regulation of 23S rRNA maturation in *Escherichia coli*. *J Bacteriol* **192:**1405–1409.

84. **Sulthana S, Deutscher MP.** 2013. Multiple exoribonucleases catalyze maturation of the 3′ terminus of 16S ribosomal RNA (rRNA). *J Biol Chem* **288:**12574–12579.

85. **Rasouly A, Schonbrun M, Shenhar Y, Ron EZ.** 2009. YbeY, a heat shock protein involved in translation in *Escherichia coli*. *J Bacteriol* **191:**2649–2655.

86. **Jacob AI, Köhrer C, Davies BW, RajBhandary UL, Walker GC.** 2013. Conserved bacterial RNase YbeY plays key roles in 70S ribosome quality control and 16S rRNA maturation. *Mol Cell* **49:**427–438.

87. **Davies BW, Köhrer C, Jacob AI, Simmons LA, Zhu J, Aleman LM, Rajbhandary UL, Walker GC.** 2010. Role of *Escherichia coli* YbeY, a highly conserved protein, in rRNA processing. *Mol Microbiol* **78:**506–518.

88. **Roy MK, Singh B, Ray BK, Apirion D.** 1983. Maturation of 5-S rRNA: ribonuclease E cleavages and their dependence on precursor sequences. *Eur J Biochem* **131:** 119–127.

89. **Sirdeshmukh R, Krych M, Schlessinger D.** 1985. *Escherichia coli* 23S ribosomal RNA truncated at its 5′ terminus. *Nucleic Acids Res* **13:**1185–1192.

90. **Babitzke P, Granger L, Olszewski J, Kushner SR.** 1993. Analysis of mRNA decay and rRNA processing in *Escherichia coli* multiple mutants carrying a deletion in RNase III. *J Bacteriol* **175:**229–239.

91. **Li Z, Pandit S, Deutscher MP.** 1999. Maturation of 23S ribosomal RNA requires the exoribonuclease RNase T. *RNA* **5:**139–146.

92. **Bram RJ, Young RA, Steitz JA.** 1980. The ribonuclease III site flanking 23S sequences in the 30S ribosomal precursor RNA of *E. coli*. *Cell* **19:**393–401.

93. **Mohanty BK, Kushner SR.** 1999. Analysis of the function of *Escherichia coli* poly(A) polymerase I in RNA metabolism. *Mol Microbiol* **34:**1094–1108.

94. **Wireman JW, Sypherd PS.** 1974. *In vitro* assembly of 30S ribosomal particles from precursor 16S RNA of *Escherichia coli*. *Nature* **247:**552–554.

95. **Takiff HE, Baker T, Copeland T, Chen SM, Court DL.** 1992. Locating essential *Escherichia coli* genes by using mini-Tn10 transposons: the *pdxJ* operon. *J Bacteriol* **174:**1544–1553.

96. **Storz G, Vogel J, Wassarman KM.** 2011. Regulation by small RNAs in bacteria: expanding frontiers. *Mol Cell* **43:**880–891.

97. **Raghavan R, Groisman EA, Ochman H.** 2011. Genomewide detection of novel regulatory RNAs in *E. coli*. *Genome Res* **21:**1487–1497.

98. **Thomason MK, Bischler T, Eisenbart SK, Förstner KU, Zhang A, Herbig A, Nieselt K, Sharma CM, Storz G.** 2015. Global transcriptional start site mapping using differential RNA sequencing reveals novel antisense RNAs in *Escherichia coli*. *J Bacteriol* **197:**18–28.

99. **Bilusic I, Popitsch N, Rescheneder P, Schroeder R, Lybecker M.** 2014. Revisiting the coding potential of the *E. coli* genome through Hfq co-immunoprecipitation. *RNA Biol* **11:**641–654.

100. Lybecker M, Zimmermann B, Bilusic I, Tukhtubaeva N, Schroeder R. 2014. The double-stranded transcriptome of *Escherichia coli*. *Proc Natl Acad Sci U S A* **111**: 3134–3139.

101. Soper T, Mandin P, Majdalani N, Gottesman S, Woodson SA. 2010. Positive regulation by small RNAs and the role of Hfq. *Proc Natl Acad Sci U S A* **107**: 9602–9607.

102. Guo MS, Updegrove TB, Gogol EB, Shabalina SA, Gross CA, Storz G. 2014. MicL, a new σ^E-dependent sRNA, combats envelope stress by repressing synthesis of Lpp, the major outer membrane lipoprotein. *Genes Dev* **28**:1620–1634.

103. Bandyra KJ, Said N, Pfeiffer V, Górna MW, Vogel J, Luisi BF. 2012. The seed region of a small RNA drives the controlled destruction of the target mRNA by the endoribonuclease RNase E. *Mol Cell* **47**:943–953.

104. Opdyke JA, Kang JG, Storz G. 2004. GadY, a small-RNA regulator of acid response genes in *Escherichia coli*. *J Bacteriol* **186**:6698–6705.

105. Desnoyers G, Bouchard MP, Massé E. 2013. New insights into small RNA-dependent translational regulation in prokaryotes. *Trends Genet* **29**:92–98.

106. Sedlyarova N, Shamovsky I, Bharati BK, Epshtein V, Chen J, Gottesman S, Schroeder R, Nudler E. 2016. sRNA-mediated control of transcription termination in *E. coli*. *Cell* **167**:111–121.e113.

107. Massé E, Gottesman S. 2002. A small RNA regulates the expression of genes involved in iron metabolism in *Escherichia coli*. *Proc Natl Acad Sci U S A* **99**:4620–4625.

108. Smirnov A, Förstner KU, Holmqvist E, Otto A, Günster R, Becher D, Reinhardt R, Vogel J. 2016. Grad-seq guides the discovery of ProQ as a major small RNA-binding protein. *Proc Natl Acad Sci U S A* **113**:11591–11596.

109. Viegas SC, Silva IJ, Saramago M, Domingues S, Arraiano CM. 2011. Regulation of the small regulatory RNA MicA by ribonuclease III: a target-dependent pathway. *Nucleic Acids Res* **39**:2918–2930.

110. Afonyushkin T, Vecerek B, Moll I, Bläsi U, Kaberdin VR. 2005. Both RNase E and RNase III control the stability of *sodB* mRNA upon translational inhibition by the small regulatory RNA RyhB. *Nucleic Acids Res* **33**: 1678–1689.

111. Morita T, Maki K, Aiba H. 2005. RNase E-based ribonucleoprotein complexes: mechanical basis of mRNA destabilization mediated by bacterial noncoding RNAs. *Genes Dev* **19**:2176–2186.

112. Aiba H. 2007. Mechanism of RNA silencing by Hfq-binding small RNAs. *Curr Opin Microbiol* **10**:134–139.

113. Andrade JM, Pobre V, Matos AM, Arraiano CM. 2012. The crucial role of PNPase in the degradation of small RNAs that are not associated with Hfq. *RNA* **18**:844–855.

114. Bandyra KJ, Sinha D, Syrjanen J, Luisi BF, De Lay NR. 2016. The ribonuclease polynucleotide phosphorylase can interact with small regulatory RNAs in both protective and degradative modes. *RNA* **22**:360–372.

115. Viegas SC, Pfeiffer V, Sittka A, Silva IJ, Vogel J, Arraiano CM. 2007. Characterization of the role of ribonucleases in *Salmonella* small RNA decay. *Nucleic Acids Res* **35**:7651–7664.

116. De Lay N, Gottesman S. 2011. Role of polynucleotide phosphorylase in sRNA function in *Escherichia coli*. *RNA* **17**:1172–1189.

117. Jagessar KL, Jain C. 2010. Functional and molecular analysis of *Escherichia coli* strains lacking multiple DEAD-box helicases. *RNA* **16**:1386–1392.

118. Charollais J, Dreyfus M, Iost I. 2004. CsdA, a cold-shock RNA helicase from *Escherichia coli*, is involved in the biogenesis of 50S ribosomal subunit. *Nucleic Acids Res* **32**:2751–2759.

119. Charollais J, Pflieger D, Vinh J, Dreyfus M, Iost I. 2003. The DEAD-box RNA helicase SrmB is involved in the assembly of 50S ribosomal subunits in *Escherichia coli*. *Mol Microbiol* **48**:1253–1265.

120. Miczak A, Kaberdin VR, Wei CL, Lin-Chao S. 1996. Proteins associated with RNase E in a multicomponent ribonucleolytic complex. *Proc Natl Acad Sci U S A* **93**: 3865–3869.

121. Khemici V, Toesca I, Poljak L, Vanzo NF, Carpousis AJ. 2004. The RNase E of *Escherichia coli* has at least two binding sites for DEAD-box RNA helicases: functional replacement of RhlB by RhlE. *Mol Microbiol* **54**: 1422–1430.

122. Jain C. 2008. The *E. coli* RhlE RNA helicase regulates the function of related RNA helicases during ribosome assembly. *RNA* **14**:381–389.

123. Peil L, Virumäe K, Remme J. 2008. Ribosome assembly in *Escherichia coli* strains lacking the RNA helicase DeaD/CsdA or DbpA. *FEBS J* **275**:3772–3782.

124. Prud'homme-Généreux A, Beran RK, Iost I, Ramey CS, Mackie GA, Simons RW. 2004. Physical and functional interactions among RNase E, polynucleotide phosphorylase and the cold-shock protein, CsdA: evidence for a 'cold shock degradosome.' *Mol Microbiol* **54**:1409–1421.

125. Fuller-Pace FV, Nicol SM, Reid AD, Lane DP. 1993. DbpA: a DEAD box protein specifically activated by 23S rRNA. *EMBO J* **12**:3619–3626.

126. Nicol SM, Fuller-Pace FV. 1995. The "DEAD box" protein DbpA interacts specifically with the peptidyl-transferase center in 23S rRNA. *Proc Natl Acad Sci U S A* **92**:11681–11685.

127. Tsu CA, Uhlenbeck OC. 1998. Kinetic analysis of the RNA-dependent adenosinetriphosphatase activity of DbpA, an *Escherichia coli* DEAD protein specific for 23S ribosomal RNA. *Biochemistry* **37**:16989–16996.

128. Bigger JW. 1944. Treatment of staphylococcal infections with penicillin by intermittent sterilisation. *Lancet* **294**:497–500.

129. Moyed HS, Bertrand KP. 1983. *hipA*, a newly recognized gene of *Escherichia coli* K-12 that affects frequency of persistence after inhibition of murein synthesis. *J Bacteriol* **155**:768–775.

130. Gerdes K, Maisonneuve E. 2012. Bacterial persistence and toxin-antitoxin loci. *Annu Rev Microbiol* **66**:103–123.

131. Christensen SK, Gerdes K. 2003. RelE toxins from Bacteria and Archaea cleave mRNAs on translating ribosomes, which are rescued by tmRNA. *Mol Microbiol* **48:**1389–1400.

132. Neubauer C, Gao YG, Andersen KR, Dunham CM, Kelley AC, Hentschel J, Gerdes K, Ramakrishnan V, Brodersen DE. 2009. The structural basis for mRNA recognition and cleavage by the ribosome-dependent endonuclease RelE. *Cell* **139:**1084–1095.

133. Pedersen K, Zavialov AV, Pavlov MY, Elf J, Gerdes K, Ehrenberg M. 2003. The bacterial toxin RelE displays codon-specific cleavage of mRNAs in the ribosomal A site. *Cell* **112:**131–140.

134. Zhang Y, Zhang J, Hoeflich KP, Ikura M, Qing G, Inouye M. 2003. MazF cleaves cellular mRNAs specifically at ACA to block protein synthesis in *Escherichia coli*. *Mol Cell* **12:**913–923.

135. Jørgensen MG, Pandey DP, Jaskolska M, Gerdes K. 2009. HicA of *Escherichia coli* defines a novel family of translation-independent mRNA interferases in bacteria and archaea. *J Bacteriol* **191:**1191–1199.

136. Cascales E, Buchanan SK, Duché D, Kleanthous C, Lloubès R, Postle K, Riley M, Slatin S, Cavard D. 2007. Colicin biology. *Microbiol Mol Biol Rev* **71:**158–229.

137. Senior BW, Holland IB. 1971. Effect of colicin E3 upon the 30S ribosomal subunit of *Escherichia coli*. *Proc Natl Acad Sci U S A* **68:**959–963.

138. Bowman CM, Dahlberg JE, Ikemura T, Konisky J, Nomura M. 1971. Specific inactivation of 16S ribosomal RNA induced by colicin E3 *in vivo*. *Proc Natl Acad Sci U S A* **68:**964–968.

139. Boon T. 1972. Inactivation of ribosomes *in vitro* by colicin E 3 and its mechanism of action. *Proc Natl Acad Sci U S A* **69:**549–552.

140. Ogawa T. 2016. tRNA-targeting ribonucleases: molecular mechanisms and insights into their physiological roles. *Biosci Biotechnol Biochem* **80:**1037–1045.

141. Pandey DP, Gerdes K. 2005. Toxin-antitoxin loci are highly abundant in free-living but lost from host-associated prokaryotes. *Nucleic Acids Res* **33:**966–976.

142. Winther KS, Gerdes K. 2011. Enteric virulence associated protein VapC inhibits translation by cleavage of initiator tRNA. *Proc Natl Acad Sci U S A* **108:**7403–7407.

143. Roberts RJ, Belfort M, Bestor T, Bhagwat AS, Bickle TA, Bitinaite J, Blumenthal RM, Degtyarev SK, Dryden DT, Dybvig K, Firman K, Gromova ES, Gumport RI, Halford SE, Hattman S, Heitman J, Hornby DP, Janulaitis A, Jeltsch A, Josephsen J, Kiss A, Klaenhammer TR, Kobayashi I, Kong H, Krüger DH, Lacks S, Marinus MG, Miyahara M, Morgan RD, Murray NE, Nagaraja V, Piekarowicz A, Pingoud A, Raleigh E, Rao DN, Reich N, Repin VE, Selker EU, Shaw PC, Stein DC, Stoddard BL, Szybalski W, Trautner TA, Van Etten JL, Vitor JM, Wilson GG, Xu SY. 2003. A nomenclature for restriction enzymes, DNA methyltransferases, homing endonucleases and their genes. *Nucleic Acids Res* **31:**1805–1812.

144. Jansen R, Embden JD, Gaastra W, Schouls LM. 2002. Identification of genes that are associated with DNA repeats in prokaryotes. *Mol Microbiol* **43:**1565–1575.

145. Grissa I, Vergnaud G, Pourcel C. 2007. The CRISPRdb database and tools to display CRISPRs and to generate dictionaries of spacers and repeats. *BMC Bioinformatics* **8:**172.

146. Westra ER, Swarts DC, Staals RH, Jore MM, Brouns SJ, van der Oost J. 2012. The CRISPRs, they are a-changin': how prokaryotes generate adaptive immunity. *Annu Rev Genet* **46:**311–339.

147. Brouns SJ, Jore MM, Lundgren M, Westra ER, Slijkhuis RJ, Snijders AP, Dickman MJ, Makarova KS, Koonin EV, van der Oost J. 2008. Small CRISPR RNAs guide antiviral defense in prokaryotes. *Science* **321:**960–964.

148. Mohanraju P, Makarova KS, Zetsche B, Zhang F, Koonin EV, van der Oost J. 2016. Diverse evolutionary roots and mechanistic variations of the CRISPR-Cas systems. *Science* **353:**aad5147.

149. Jain C, Belasco JG. 1995. RNase E autoregulates its synthesis by controlling the degradation rate of its own mRNA in *Escherichia coli*: unusual sensitivity of the *rne* transcript to RNase E activity. *Genes Dev* **9:**84–96.

150. Jarrige AC, Mathy N, Portier C. 2001. PNPase autocontrols its expression by degrading a double-stranded structure in the *pnp* mRNA leader. *EMBO J* **20:**6845–6855.

151. Matsunaga J, Simons EL, Simons RW. 1996. RNase III autoregulation: structure and function of *rncO*, the posttranscriptional "operator." *RNA* **2:**1228–1240.

152. Lee K, Zhan X, Gao J, Qiu J, Feng Y, Meganathan R, Cohen SN, Georgiou G. 2003. RraA. a protein inhibitor of RNase E activity that globally modulates RNA abundance in *E. coli*. *Cell* **114:**623–634.

153. Gao J, Lee K, Zhao M, Qiu J, Zhan X, Saxena A, Moore CJ, Cohen SN, Georgiou G. 2006. Differential modulation of *E. coli* mRNA abundance by inhibitory proteins that alter the composition of the degradosome. *Mol Microbiol* **61:**394–406.

154. Górna MW, Pietras Z, Tsai YC, Callaghan AJ, Hernández H, Robinson CV, Luisi BF. 2010. The regulatory protein RraA modulates RNA-binding and helicase activities of the *E. coli* RNA degradosome. *RNA* **16:**553–562.

155. Kim KS, Manasherob R, Cohen SN. 2008. YmdB: a stress-responsive ribonuclease-binding regulator of *E. coli* RNase III activity. *Genes Dev* **22:**3497–3508.

156. Liang W, Malhotra A, Deutscher MP. 2011. Acetylation regulates the stability of a bacterial protein: growth stage-dependent modification of RNase R. *Mol Cell* **44:**160–166.

157. Mohanty BK, Kushner SR. 2016. Regulation of mRNA decay in bacteria. *Annu Rev Microbiol* **70:**25–44.

158. Mohanty BK, Kushner SR. 2006. The majority of *Escherichia coli* mRNAs undergo post-transcriptional modification in exponentially growing cells. *Nucleic Acids Res* **34:**5695–5704.

159. **Mohanty BK, Kushner SR.** 2011. Bacterial/archaeal/organellar polyadenylation. *Wiley Interdiscip Rev RNA* 2:256–276.

160. **Mohanty BK, Kushner SR.** 1999. Residual polyadenylation in poly(A) polymerase I (*pcnB*) mutants of *Escherichia coli* does not result from the activity encoded by the *f310* gene. *Mol Microbiol* 34:1109–1119.

161. **Keiler KC.** 2011. RNA localization in bacteria. *Curr Opin Microbiol* 14:155–159.

162. **Montero Llopis P, Jackson AF, Sliusarenko O, Surovtsev I, Heinritz J, Emonet T, Jacobs-Wagner C.** 2010. Spatial organization of the flow of genetic information in bacteria. *Nature* 466:77–81.

Regulating with RNA in Bacteria and Archaea
Edited by Gisela Storz and Kai Papenfort
© 2018 American Society for Microbiology, Washington, DC
doi:10.1128/microbiolspec.RWR-0003-2017

RNases and Helicases in Gram-Positive Bacteria

3

Sylvain Durand[1] and Ciarán Condon[1]

INTRODUCTION

Posttranscriptional regulation is a key modulator of gene expression in bacteria and allows their rapid adaptation to the environment. This regulation can be performed by proteins and/or RNA, by modifying either mRNA stability and/or translation. RNases are key enzymes in these processes. There are two main classes of RNases: endoribonucleases, which cleave directly in the "body" of the RNA; and exoribonucleases, which attack RNA from either its 5′ or 3′ end. Although RNases play a central role in RNA metabolism, these enzymes are not identical in Gram-negative and Gram-positive bacteria. For example, endoribonuclease E (RNase E) initiates bulk mRNA degradation and is essential in *Escherichia coli*, but this RNase is absent in many *Firmicutes*, such as *Bacillus subtilis*.

As in *E. coli*, mRNA degradation in Gram-positive bacteria is often initiated by an endoribonucleolytic cleavage followed by the 3′-to-5′ exoribonuclease digestion of the upstream cleavage product (Fig. 1A). However, in contrast to *E. coli* and other *Gammaproteobacteria*, most Gram-positive bacteria also encode at least one exoribonuclease with 5′-to-3′ activity (RNase J), which contributes to the degradation of the downstream product of endoribonucleolytic cleavage (Fig. 1A). Alternatively, the 5′-to-3′ activity of RNase J also permits mRNA degradation directly from its 5′ end, without a prior cleavage by an endoribonuclease. However, RNase J activity is generally inhibited by the 5′ triphosphate of the primary transcripts (see below), and consequently this degradation pathway requires dephosphorylation of the RNA by a pyrophosphohydrolase (such as RppH) prior to RNase J degradation (Fig. 1B). Although the set of RNases in Gram-positive bacteria is well conserved (Table 1), global analyses

performed in the last decade by tiling array and RNA deep sequencing in different *Firmicutes* mutated for the main RNases have established that the relative importance of these RNases in each organism is variable.

In *E. coli*, RNase E binds to a 3′-to-5′ exoribonuclease (polynucleotide phosphorylase [PNPase]), an RNA helicase (RhlB), and a glycolytic enzyme (enolase) to form a complex called the "degradosome." A degradosome-like network (DLN) of RNases with different factors such as helicases and glycolytic enzymes has also been detected in several Gram-positive bacteria and could potentially modulate RNase activities in these organisms. This review will summarize the most recent discoveries about these RNases, their cofactors, and how they participate in the regulation of gene expression in selected organisms.

THE MAIN RIBONUCLEASES OF GRAM-POSITIVE BACTERIA

Gram-positive bacteria encode >20 RNases that are relatively well conserved (Table 1). Half of them are essentially devoted to the maturation of stable RNA (rRNA, tRNA) and will not be detailed here. We will also exclude from this review RNases that are not well conserved in Gram-positive bacteria, such as endoribonucleases belonging to type II toxin/antitoxin (TA) systems (Table 1). Indeed, the distribution of this type of RNase is very variable in Gram-positive bacteria (79 in *Mycobacterium tuberculosis* versus 1 in *B. subtilis*), and they play specific roles in stress conditions. For a review on these toxins in *M. tuberculosis*, see reference 1. We will focus here on the main RNases identified in Gram-positive bacteria that play a role in regulation with RNA.

[1]UMR8261 CNRS, Université Paris Diderot (Sorbonne Paris Cité), Institut de Biologie Physico-Chimique, Paris, France.

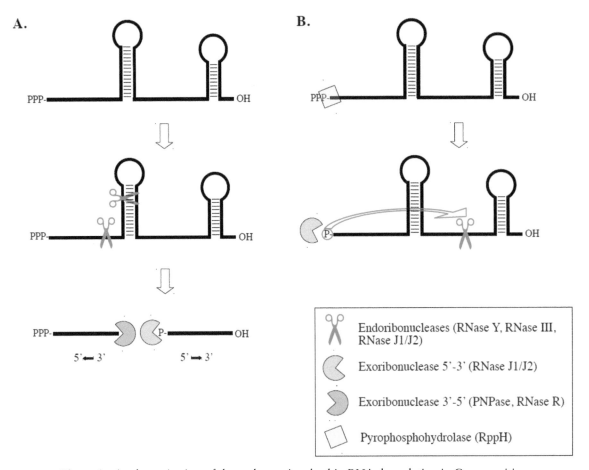

Figure 1   A schematic view of the pathways involved in RNA degradation in Gram-positive bacteria. (A) Primary mRNA transcripts in bacteria are protected at their 5′ end by a triphosphate group. Initiation of mRNA degradation can involve an endoribonuclease cut (RNase Y or RNase III), which is the limiting step. This step generates a downstream product with a 5′ monophosphate extremity, which can be attacked by the 5′-to-3′ exoribonuclease RNase J (in blue). The 3′ end of the upstream cleavage product is degraded by 3′-to-5′ exoribonucleases (in green), principally PNPase in *B. subtilis*. (B) In the alternative degradation pathway, the 5′ triphosphate of the mRNA can be converted to a 5′ monophosphate by an RNA pyrophosphohydrolase (e.g., RppH [yellow square]). After removal of the triphosphate, the mRNA can be degraded by the 5′-to-3′ exoribonuclease RNase J or by RNase Y in cases where initial cleavage by RNase Y is sensitive to the 5′ status of the mRNA.

The Endoribonucleases

Most bacterial mRNAs have 5′ triphosphate extremities and Rho-independent terminators (stem-loop structures) at their 3′ end that protect them from degradation. One way of bypassing this problem is to cleave endonucleolytically to start mRNA degradation. The single-strand-specific RNase Y and the double-strand-specific RNase III are the main endoribonucleases involved in this process and are among the most conserved in Gram-positive bacteria. These RNases and some of their key roles in the regulation of gene expression will be presented in this section.

RNase Y

RNase Y is a dimeric multidomain endoribonuclease with (i) an N-terminal hydrophobic α-helical domain, which anchors RNase Y to the membrane; (ii) an intrinsically disordered domain that participates in RNase Y interactions with other proteins through formation of a coiled-coil structure (see below); (iii) a KH domain responsible for RNA binding; and (iv) an HD domain containing the catalytic site (Fig. 2) (2). The first two domains are the most important in RNase Y dimerization (2). Like RNase E in Gram-negative bacteria, RNase Y localizes to the membrane.

Table 1 RNases and their functions[a]

RNases	Activity	Present in				Substrates[b]
		B. subtilis	S. aureus	S. pyogenes	M. tuberculosis	
PNPase	Exo 3′ → 5′	x	x	x	x	tRNA, rRNA, mRNA
YhaM	Exo 3′ → 5′	x	x	x		mRNA, tRNA, DNA
RNase PH	Exo 3′ → 5′	x			x	tRNA, mRNA
RNase R	Exo 3′ → 5′	x	x	x	x	mRNA, rRNA
RNase D	Exo 3′ → 5′				x	tRNA
RNase Z	Endo	x	x	x	x	tRNA (3′), tmRNA
RNase P	Endo	x	x	x	x	tRNA (5′), tmRNA
RNase M5	Endo (ds)	x	x	x		5S rRNA
RNase III	Endo (ds)	x	x	x	x	16S and 23S rRNA, mRNA, scRNA
Mini-III	Endo (ds)	x	x	x		23S rRNA, intron recycling
RNase E	Endo				x	16S and 5S rRNA, mRNA, tRNA
RNase Y	Endo	x	x	x		mRNA, RNase P RNA, scRNA
RNase J1	Exo 5′ → 3′ / Endo?	x	x	x	x	16S rRNA, mRNA
RNase J2	Exo 5′ → 3′ / Endo?	x	x	x		mRNA
MazF/EndoA	Endo	x	x		x	mRNA, rRNA
RelE	Endo		x		x	mRNA
ParD	Endo				x	mRNA
HigB	Endo				x	mRNA
YoeB	Endo		x		x	mRNA
VapC	Endo				x	mRNA
Orn	Exo 3′ → 5′				x	mRNA
NrnA	Exo 3′ → 5′	x	x	x	x	mRNA
NrnB	Exo 3′ → 5′	x				mRNA
RNase HII	Endo	x	x	x	x	RNA primer of replication, R-loops?
RNase HIII	Endo	x	x	x		RNA primer of replication, R-loops?
YacP/RaeI	Endo	x	x	x		mRNA
YhcR	Endo	x				Nonspecific RNA and DNA (extracellular)
RNase Bsn	Endo	x				RNA (extracellular)
RNase AS	Exo 3′ → 5′				x	?

[a]Activity (endo- vs. exoribonucleolytic), substrates of RNases, and conservation in several Gram-positive bacteria (B. subtilis, S. aureus, S. pyogenes, and M. tuberculosis) are indicated (80, 81). In this review, only RNases involved in mRNA degradation are discussed in detail, such as RNase J1/J2, RNase Y, PNPase, and RNase III. Type II toxins are indicated in red. mRNA, messenger RNA; rRNA, ribosomal RNA; tmRNA, transfer-messenger RNA.
[b]Most substrates have been identified in B. subtilis.

RNase Y is involved in maturation of RNase P RNA in *Streptococcus pyogenes* and *B. subtilis*, and small cytoplasmic RNA (scRNA) in *B. subtilis*, a component of the signal recognition particle complex that targets specific proteins to the plasma membrane. RNase Y also plays a major role in mRNA degradation in Gram-positive bacteria (3–6). The deletion of RNase Y severely impacts growth of *B. subtilis*, with a 2-fold increase in its doubling time and pleiotropic effects on competence, sporulation, and biofilm formation (7, 8). In *Clostridium perfringens*, a depletion of RNase Y also drastically affects its growth (6). Several transcriptome analyses performed in *B. subtilis* and *C. perfringens* found that the depletion of RNase Y has a large impact on gene expression, with 13 to 23% of transcripts affected (2, 4–6). Although there is almost no impact of *rny* deletion on *S. pyogenes* growth, this mutation has

significant impact on gene expression (14% of mRNAs affected) in stationary phase (9). In *Staphylococcus aureus*, the role of RNase Y in mRNA degradation seems to be subtler than in *B. subtilis*, *S. pyogenes*, or *C. perfringens*. Indeed, in *S. aureus*, only ∼100 to 200 mRNAs are affected by the deletion of the *rny* gene (4 to 8%) and the doubling time of this mutant is similar to the wild-type strain (10, 11). The reason for the differential impact of RNase Y in these bacteria is unclear. Comparison of the RNase Y sequence from *B. subtilis* and *S. aureus* shows that the KH and HD domains are very well conserved. Most of the differences between both proteins appear in the N-terminal part of the protein (transmembrane and coiled-coil domain). These variations in sequence could affect the composition of the DLN and modify the specificity of RNase Y for RNA (see below). RNase Y has been

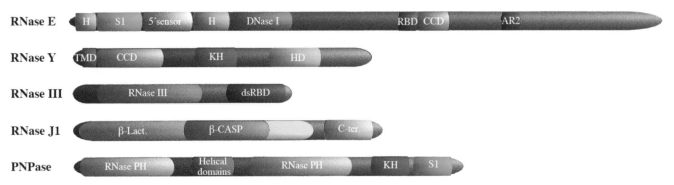

Figure 2   Comparison of the domain structure of RNases described in this review. All structures are based on RNases found in *B. subtilis* except RNase E (structure from *E. coli*). Abbreviations: H, RNase H domain; CCD, coiled-coil domain; TMD, transmembrane domain; β-Lact., β-lactamase domain. RNA binding domains: S1, S1 domain; KH, KH domain; AR2, AR2 domain; RBD, RNA binding domain; dsRBD, double-stranded RNA binding domain.

shown to cleave single-stranded A/U-rich sequences, and, in contrast to RNase J (see below), the 5′ monophosphate extremity is not a requirement, but can stimulate its activity in some specific cases (12). Two recent studies in *S. aureus* suggest that RNase Y may require G in the middle of the A/U-rich sequence and/or could recognize a secondary structure at the 3′ side of the cleavage site (11, 13).

Key role of RNase Y in the regulation of gene expression with RNA

RNase Y was shown to be important for the virulence of *S. aureus*, *C. perfringens*, and *S. pyogenes* (6, 9–11). For example, RNase Y in *S. aureus* allows the differential expression of the *saePQRS* operon. Cleavage downstream of the *saeP* coding sequence by RNase Y leads to stabilization of the downstream fragment encoding the two-component system SaeRS, controlling the major virulence genes of *S. aureus* (10, 13). RNase Y also plays an important role in regulating virulence factors in *C. perfringens*, such as the κ-toxin collagenase encoded by the *colA* gene (6). The *colA* mRNA has a long 5′ untranslated region (5′ UTR) with a secondary structure hindering ribosome binding, which impairs translation and renders the *colA* mRNA unstable (Fig. 3). The binding of a regulatory RNA called VR-RNA modifies the secondary structure of the 5′ UTR of the *colA* mRNA, which both releases the ribosome binding site to stimulate translation and triggers an RNase Y cleavage. This cleavage creates an mRNA with a stem-loop structure at its 5′ end protecting it from degradation by 5′-to-3′ exoribonucleases (Fig. 3). In a contrasting mechanism in *B. subtilis*, the binding of the regulatory RNA RoxS to the target mRNAs *ppnkB* (NAD/NADH kinase) and *sucD* (succinate dehydrogenase) interferes

with their translation and leads to their degradation by RNase Y (Fig. 3) (14).

The turnover of at least 30% of 5′ *cis*-acting regulatory RNAs (e.g., riboswitches) is handled by RNase Y in *B. subtilis* (5). For example, this endoribonuclease is involved in the regulation of the *trp* operon in *B. subtilis*, encoding genes necessary for the biosynthesis of tryptophan. When tryptophan is in excess, the TRAP protein (*trp* RNA binding attenuation protein) binds to the 5′ UTR of the operon mRNA to prevent formation of an antiterminator structure and trigger termination of transcription (Fig. 4). Both RNase Y and RNase J1 (see below) have been implicated in initiation of the degradation of this terminated transcript to allow the recycling of the RNA binding protein TRAP (15–17). RNase Y also largely participates in the turnover of the *S*-adenosylmethionine, T-box, and riboflavin riboswitches in *B. subtilis* and *S. aureus* (4, 5, 11, 12).

Lastly, RNase Y is involved in setting the levels of the *trans*-acting regulatory RNAs themselves. Indeed, the VR-RNA in *C. perfringens* and FasX in *S. pyogenes* are downregulated after depletion or deletion of this RNase. In these cases, the effect is probably indirect (6, 9). In contrast, the RsaA small RNA (sRNA) in *S. aureus* and RoxS in *B. subtilis* are stabilized after deletion of RNase Y (4, 10, 11). Thus, similar to the role played by RNase E in *E. coli*, RNase Y has an important function in mRNA degradation and in control by regulatory RNAs in Gram-positive bacteria.

RNase III

RNase III (encoded by *rnc* gene) is an endoribonuclease with a double-stranded RNA binding domain (Fig. 2) that cleaves double-stranded RNA and exists *in vivo* as a dimer. This enzyme was shown to be responsible for

A. **B.**

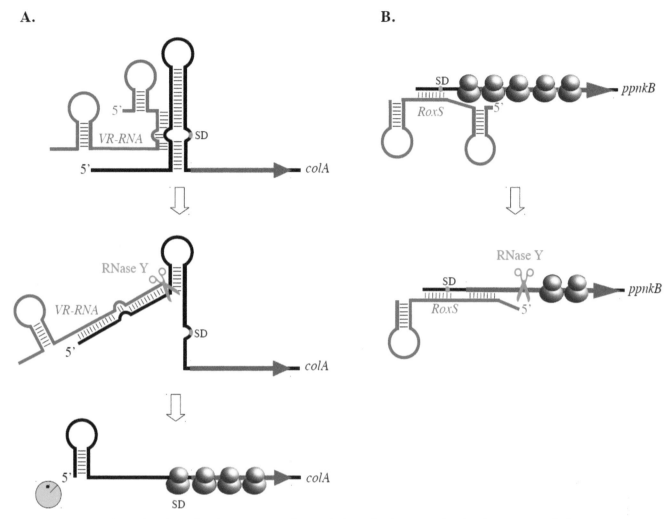

Figure 3 (A) Role of RNase Y in the regulation of *colA* expression by the *trans*-encoded sRNA VR-RNA. *C. perfringens* VR-RNA binds the 5′ UTR of the *colA* mRNA, encoding a collagenase, and triggers cleavage of the *colA* mRNA by RNase Y. This cleavage in turn stabilizes the mRNA by creating a stem-loop structure at the 5′ end of the mRNA. The binding of the sRNA also stimulates *colA* translation by releasing the SD sequence (83). (B) The RoxS sRNA binds to the SD sequence of *ppnkB* mRNA, inhibiting its translation. The reduction of the ribosome trafficking on the *ppnkB* mRNA uncovers RNase Y cleavage sites. The sRNA is in blue, the mRNA target is colored in black, the SD sequence is in gray, ribosomes are in blue, and RNase Y is represented by red scissors.

16S and 23S maturation in *B. subtilis* and *S. aureus*, and scRNA maturation in *B. subtilis* (18–20). It is not essential in most bacteria, with some exceptions such as *B. subtilis* (see below).

RNase III plays a specific role in mRNA degradation by virtue of its targeting of double-stranded RNA. Several mRNA targets of RNase III have been captured by coimmunoprecipitation in *S. aureus* (20). Besides rRNA and tRNA, 58 different noncoding RNAs and numerous antisense RNAs were found to bind RNase III. A depletion of RNase III in *B. subtilis* was shown to

affect only 12% of transcripts *in vivo* with numerous indirect transcriptional effects, similar to what was observed in *E. coli* (4, 21). However, it is a key enzyme in the control of gene expression by specific regulatory RNAs.

Key role of RNase III in the control of gene expression by regulatory RNAs

As mentioned above, RNase III is essential in *B. subtilis*, but this is more an exception than the rule. This singularity has been recently understood and is due

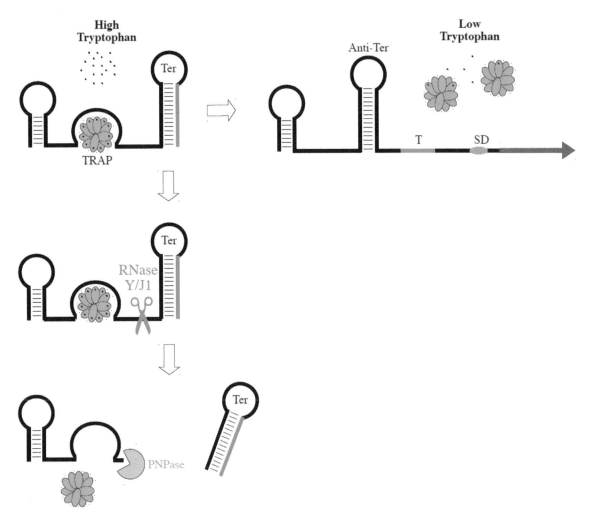

Figure 4 Regulation of the expression of the *trp* operon. When tryptophan is not limiting, the TRAP protein binds to the 5′ UTR of the *trp* operon and facilitates transcriptional termination at the terminator. The aborted transcript is probably cleaved by either RNase Y or J1 followed by the attack of the new 3′ end by PNPase. This degradation allows release of the TRAP protein for further regulation (left). When tryptophan is limiting, TRAP complex does not bind to the *trp* operon and an antiterminator structure can be formed to allow transcription of the *trp* mRNA (right). The mRNA is colored in black, the SD sequence is in gray, Ter is for terminator and anti-Ter for antiterminator, TRAP proteins are colored in green, PNPase is represented by a light green Pacman symbol, and RNase Y/J1 by red scissors.

to the presence of specific type I TA systems on the *B. subtilis* chromosome (see also chapter 11). Briefly, type I TA systems encode a toxin and at the same locus a regulatory antitoxin RNA encoded on the opposite strand (Fig. 5A). Both RNAs interact due to their complementarity, which leads to downregulation of toxin expression at either the translational or mRNA stability level. Most of the duplexes between the mRNA and antisense RNA of these modules are subjected to RNase III cleavages in *B. subtilis* (4). Downregulation can no longer occur in cells lacking RNase III, leading to toxin overproduction and in some cases cell death. The deletion of two of these modules (TxpA/RatA and YonT/as-YonT) encoded by the Skin and SPβ prophages, respectively, restores the viability of the RNase III mutant strain to the point that there is no perceptible effect on growth in rich medium at 37°C (4). This result shows that cleavage of the duplex RNA by RNase III is crucial to prevent expression of the TxpA and YonT toxins and that the base-pairing with their respective antisense RNAs (RatA and as-YonT) is not by itself sufficient to block the translation of these toxins.

In *Listeria monocytogenes*, recent data also suggest that RNase III is involved in the regulation of gene

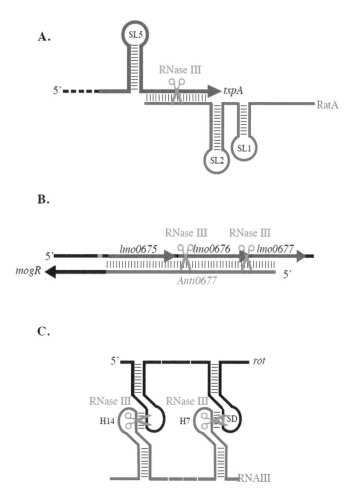

Figure 5   Role of RNase III in the regulation of gene expression by antisense and *trans*-encoded sRNA. (A) *B. subtilis* type I TA system TxpA/RatA. The 3′ end of the RatA sRNA forms a large duplex with the *txpA* mRNA, which is then cleaved by RNase III. (B) The *mogR-lmo074/lmo0675-0676-0677* excludon in *L. monocytogenes*. One of the transcripts of the *mogR* operon starts with the Lmo0677 open reading frame on the opposite strand. The long 5′ UTR of the *mogR* operon (Anti0677) is antisense to the *lmo0675-0676-0677* operon. The duplex RNA is probably cleaved by RNase III, although this has not been shown directly. (C) *S. aureus* RNAIII represses translation of the *rot* mRNA by sequestering the SD sequence (84). The sRNA is in blue, the mRNA target is colored in black, the SD sequence is in gray, and RNase III represented by red scissors.

expression by antisense RNAs. Indeed, in *L. monocytogenes*, particular loci called excludons encode two cistrons in opposite orientations with mutually exclusive functions (Fig. 5B). The transcript of one cistron overlaps that of the divergently oriented cistron, creating a long antisense RNA. The *mogR-lmo074/lmo0675-0676-0677* locus represents one of such excludon in *L. monocytogenes*. The *mogR* gene, encoding a transcriptional repressor of motility genes, can be transcribed from two promoters. One of the resulting transcripts has a long 5′ UTR (Anti0677) complementary to the *lmo0675-0676-0677* operon encoding genes of the flagellar ex-

port apparatus. Thus, transcription of Anti0677 represses expression of the *lmo0675-0676-0677* operon (22). Accordingly, RNase III cleavages have been detected in the overlapping region (Fig. 5B) (23).

RNase III has also been proposed to remove antisense RNA generated by pervasive transcription and also to play a key role in regulation by *trans*-encoded regulatory RNAs such as RNAIII in *S. aureus* (23). RNAIII is a dual-function RNA (see chapter 27) both encoding a small protein of 26 amino acids (δ-hemolysin) and acting as a regulatory RNA through its long 3′ UTR. The 3′ UTR of RNAIII contains several stem-loop structures with C-rich regions used to base-pair with G-rich targets typically found in ribosome binding sites. It regulates the expression of numerous virulence genes in response to the quorum-sensing system *agr* (24), including the virulence factor protein A (*spa*), coagulase (*coA*), and SA1000, and one mRNA encoding a master transcriptional regulator (Rot). In all these cases, RNAIII blocks translation of the target mRNA by complementarity with the Shine-Dalgarno (SD) sequence, and the resulting duplex is recognized and cleaved by RNase III, triggering its complete degradation (Fig. 5C) (for a recent review, see reference 82). The same kind of regulation was observed with the *psm-mec* sRNA in *S. aureus* (25). Like RNAIII, this is a dual-function RNA that binds to the *agrA* transcript and induces an RNase III cleavage, destabilizing the *agrA* mRNA. In *B. subtilis*, RNase III is also used to help to degrade mRNA targets of the *trans*-encoded RNA RoxS (14).

Another important role of RNase III in the Grampositive pathogen *S. pyogenes* is the activation of the type II CRISPR (clustered regularly interspaced short palindromic repeat)-Cas immune system (26). Indeed, the CRISPR RNA (crRNA) encoded by the CRISPR-Cas system is essential to guide effector nucleases to the invading nucleic acid and to destroy it. The crRNA is synthesized as a precursor (pre-crRNA) that is processed by endoribonucleases to become functional. In some bacteria, the endoribonucleases responsible for this maturation are specifically encoded at the CRISPR-Cas locus (e.g., Cas6). In *S. pyogenes*, these enzymes are absent but a noncoding RNA (tracrRNA) with a complementary sequence to the pre-crRNA is encoded upstream of the CRISPR-Cas system. The pre-crRNA/tracrRNA duplex is cleaved by RNase III, resulting in maturation of the crRNA.

B. subtilis and other Gram-positive bacteria have a second RNase related to RNase III called Mini-III. This enzyme possesses only the catalytic domain of the RNase III and was shown to be involved in 23S rRNA maturation (27, 28). Recently, this enzyme was shown to be involved in both 23S rRNA processing and the turnover of group I introns in *Arabidopsis thaliana* chloroplasts, prompting a demonstration that it could perform a similar role in two prophage-encoded group I introns of the *nrdEB* and *yosP* mRNAs in *B. subtilis* (29). This result leaves open the possibility that Mini-III could have yet other substrates in Grampositive bacteria, but this has not yet been addressed.

RNase E/G

RNase E/G is a family of endoribonucleases discovered in Gram-negative bacteria such as *E. coli* and present in >95% of *Gammaproteobacteria*. RNase E of *E. coli* has two distinct regions (Fig. 2): (i) the N-terminal catalytic region, composed of a small domain and a large domain encompassing a 5′ sensor domain and RNase H, S1, and DNase I structural domains; and (ii) the C-terminal noncatalytic region, which interacts with components of the degradosome and which also has two RNA binding domains (namely, RBD and AR2). *E. coli* also has a paralog of RNase E called RNase G that contains only the catalytic domain. Other organisms have just one ortholog (RNase E/G) equally similar to RNase G and the N-terminal domain of RNase E.

Although RNase E is essential and has a key role in mRNA turnover in *E. coli*, this enzyme is absent in *B. subtilis* and its function is thought to be ensured by RNase Y in this and related bacteria. However, RNase E/G and RNase Y are not exclusively restricted to Gram-negative and Gram-positive bacteria, respectively. Indeed, RNase E/G is present in some *Firmicutes* (*Clostridia* and *Bacillales*) and *Actinobacteria* (such as mycobacteria). Similarly, RNase Y can be found in Gram-negative bacteria like *Helicobacter pylori* (30). Remarkably, some organisms such as the *Clostridia* have both endoribonucleases.

The RNase E function in Gram-positive bacteria is not well defined. The primary structure of RNase E/G in *Firmicutes* is closer to RNase G than RNase E of *E. coli* since there is no conservation of the C-terminal part that interacts with the degradosome partners. This suggests that *Firmicutes* orthologs of RNase E may have fewer interaction partners and a more limited set of targets. In mycobacteria, RNase E (MyRNase E) is involved in maturation of rRNA and is responsible for the processing of the *furA-katG* operon (31, 32). MyRNase E is also sensitive to the 5′-end phosphorylation status and exists in a dimeric and tetrameric state, as in *E. coli*, but the specificity of the enzyme seems to differ between these two orthologs (31). However, the actinobacterial MyRNase E has an N-terminal extension involved in the interaction with an NAD kinase and an acetyltransferase (33). The relevance of these interactions for RNA degradation is unknown.

The key role of RNase E in mRNA degradation and its involvement in the regulation of gene expression by noncoding RNA has been largely demonstrated in Gram-negative bacteria. In *Salmonella enterica* serovar Typhimurium, the ternary complex between Hfq, the MicC sRNA, and the *ompD* mRNA recruit RNase E to initiate the degradation of both RNAs. Moreover, the 5′ monophosphate extremity of the sRNA can activate RNase E cleavage (34, 35). A similar role for RNase

E/G has not yet been demonstrated in Gram-positive bacteria, but there is nothing that would exclude this type of regulation. Alternatively, RNase E could be replaced by RNase Y for this role in some organisms.

Although endoribonucleases like RNase Y and RNase III clearly play key roles in regulation with RNA in Gram-positive bacteria and often perform the rate-limiting step initiating mRNA degradation, exoribonucleases are nonetheless essential for some control mechanisms. The main exoribonucleases and their key roles in RNA regulation in Gram-positive bacteria will be presented below.

The Exoribonucleases

Three well-conserved 3′-to-5′ exoribonucleases have been identified in Gram-positive bacteria with roles in mRNA turnover: PNPase, RNase R, and YhaM. *B. subtilis* and *M. tuberculosis* have a fourth enzyme of this family named RNase PH, which plays a role in tRNA maturation primarily (36), and a minor role in mRNA degradation is not excluded (37). Moreover, in contrast to the paradigm of the Gram-negative bacterium *E. coli*, which only has 3′-to-5′ exoribonuclease activity, most Gram-positive organisms encode enzyme(s) with 5′-to-3′ exoribonucleolytic activity, named RNase J. Because our knowledge of the roles of RNase R, RNase PH, and YhaM in mRNA degradation is relatively limited, only RNases J and PNPase will be detailed in this section.

RNase J1 and J2

RNase J comprises a metallo-β-lactamase domain typically containing two active-site zinc ions with an intervening β-CASP domain (Fig. 2). In most low-GC Gram-positive bacteria there are at least two orthologs of RNase J (RNase J1 and J2) (38). In contrast, in Gram-positive bacteria with high genome GC content like the mycobacteria, only one RNase J is typically found. Of note, RNase J is also present in some Gram-negative bacteria such as *H. pylori*.

RNase J exoribonuclease activity is processive on substrates with 5 nucleotides or more (39). It has also been shown that RNase J has endoribonucleolytic activity *in vitro*. However, the endoribonucleolytic activity of RNase J in *B. subtilis in vivo* is still a matter of debate. There are no data unambiguously showing its endoribonucleolytic activity *in vivo*, and structural investigations of RNase J in *B. subtilis* and *Thermus thermophilus* suggest that the enzyme is set up to act mainly as an exoribonuclease (39, 40). Indeed, the dimer of RNase J1 would require large reorganization to allow an endoribonucleolytic substrate to reach the catalytic site. The main complex *in vivo*, at least in *B. subtilis*, is probably an RNase J1/J2 heterodimer (or heterotetramer), since both enzymes were initially copurified as a complex (38). Moreover, it is also possible to copurify stoichiometric amounts of RNase J2 using a FLAG-tagged RNase J1 protein. The interaction between RNase J1 and J2 was also seen with yeast and bacterial two-hybrid assays, and interestingly, both RNases are present in similar numbers in *B. subtilis* (41, 42). The role played by RNase J2 in this complex is still not fully understood, and until the structure of that complex is solved, it cannot be fully excluded that it is slightly different from that of the homodimer of RNase J1 and allows these RNases to act as endoribonucleases on some substrates. Relevant to this possibility, the specificity of the endoribonucleolytic activity can be modified *in vitro* when RNase J1 and J2 are in complex (41).

The structure of the protein shows that RNase J1 in *B. subtilis* has a 5′ monophosphate-binding pocket, limiting the enzyme to substrates with 5′ monophosphate or 5′ hydroxyl for exonucleolytic degradation (39, 43, 44). In this mode, RNase J1 has been implicated in 16S rRNA maturation in *B. subtilis* and *S. aureus* and scRNA maturation in *B. subtilis*, but also plays an important role in mRNA degradation. In *B. subtilis*, a depletion of this enzyme affects the levels of ~21% of mRNAs (4). A deletion of *rnjA* (RNase J1) drastically impairs growth, competence, and its ability to sporulate (8). The cells are also cold sensitive. In contrast, an *rnjB* (BsRNase J2) deletion strain grows like a wild-type strain in rich medium. This is likely explained by the fact that the 5′ exoribonucleolytic activity of RNase J2 is 2 orders of magnitude weaker than that of RNase J1. The reason for this difference is not well understood.

In *S. aureus*, it has also been shown that RNase J1 and RNase J2 are not essential, but deletion of either enzyme has a significant effect on growth (45). In contrast, a mutation in the catalytic site of RNase J2 has no impact on growth. This observation suggests that, like in *B. subtilis*, RNase J2 plays more of a structural role in stabilizing the complex with RNase J1 or modulating its activity on some targets. Intriguingly, a recent study shows that the activity of SaRNase J1 is stimulated by manganese *in vitro*. One possibility is that SaRNase J1 uses manganese rather than zinc ions like in *B. subtilis* or *T. thermophilus* for catalysis (46). However, manganese could also bind elsewhere on the protein or the RNA substrate to facilitate cleavage. Moreover, in contrast to the BsRNase J1, the enzyme from *S. aureus* seems to be able to digest 5′ triphosphorylated RNA as efficiently as 5′ hydroxyl or 5′ monophosphorylated transcripts (46). The explanation for the lack of susceptibility of SaRNase J1 to the 5′ end of the RNA is unknown.

In *S. pyogenes*, both RNase J1 and J2 are essential and a depletion of RNase J2 impacts mRNA stability. Moreover, RNase J2 is expressed at higher levels than RNase J1 (47). In *Streptococcus mutans*, neither RNase J1 nor J2 is essential. However, the mutation of each enzyme individually has pleiotropic effects, and the deletion of RNase J2 in *S. mutans* has a more pronounced phenotype than a mutant lacking RNase J1 (48). Interestingly, the catalytic center of RNase J2 in *S. pyogenes* and *S. mutans* is closer to RNase J1 than that of BsRNase J2 and may contribute to the more important role of RNase J2 in these organisms. Nevertheless, a comparison of the RNase J2 sequences from *B. subtilis* and *S. mutans* shows numerous variations in addition to the catalytic site of the enzyme that could also contribute to the differential role of RNase J2 in these bacteria.

A recent study in *Enterococcus faecalis* showed that RNase J2 also has a weaker activity than RNase J1 in this organism. Nonetheless, the E*frnjB* (RNase J2) deletion strain has a decreased growth rate, and the expression of 62 genes was affected, among them, the *ebpABC* operon, which is important for pilus production (49). Even though this regulation is probably indirect, an inactive mutant of RNase J2 that is not affected for its binding to RNase J1 does not restore the normal expression of the *ebp* operon, suggesting that *rnjB* not only has a structural role but may also have a dedicated function in *E. faecalis*.

Key role of RNase J1/J2 in the regulation of gene expression with RNA

As mentioned above, RNase Y was shown to be involved in the degradation of the *trp* leader mRNA (Fig. 4). However, before the identification of this endoribonuclease, it was shown that RNase J1 also plays a role in the degradation of this transcript (16) (Fig. 4). Interestingly, deletion of either RNase Y or RNase J1 has a similar impact on the half-life of the *trp* leader (17; also see above). Three hypotheses can explain these data. (i) RNase J1 and RNase Y may have redundant activities on this RNA. (ii) Deletion of one enzyme may indirectly affect the activity of the other. (iii) The pleiotropic defects of these mutants may affect the synthesis/activity of a third enzyme responsible for the cleavage, although the fact that both RNase Y and RNase J1 are able to cleave at the correct internal site in the *trp* leader *in vitro* weakens the case for an involvement of a

third endoribonuclease. It is still not fully clear whether the effect of RNase J1 observed *in vivo* is through its endo- or 5′-exoribonucleolytic activity. Interestingly, the stability of other full-length mRNAs seems to depend on both RNase Y and RNase J1 *in vivo*, e.g., the *sigW-rsiW* operon and the *yflS* gene (19, 50), suggesting that two independent pathways of degradation (endo- and 5′-exoribonucleolytic) can coexist for some mRNAs.

The activity of the RNase J1 enzyme can be affected by the binding of regulatory RNA to its substrates. Indeed, the RoxS regulatory RNA controls the expression of numerous genes of central metabolism to balance the NAD/NADH ratio in *B. subtilis*. This *trans*-encoded RNA has also been shown to stabilize the *yflS* mRNA, encoding a malate transporter. The RoxS RNA binds to the extreme 5′ end of the *yflS* mRNA, and this base-pairing blocks the attack by RNase J1 (Fig. 6A) (51). In parallel, *yflS* translation is stimulated, which protects the mRNA further from cleavages by endoribonucleases such as RNase Y within the coding sequence (Fig. 6A). Interestingly, in the human group A pathogen *Streptococcus*, another regulatory RNA named FasX base-pairs at the extreme 5′ end of *ska* mRNA, encoding the streptokinase, and stabilizes it. The RNase involved in this control mechanism has not yet been identified, but RNase J1 is the best candidate for this regulation.

To date, only one substrate directly targeted by RNase J2 has been characterized in Gram-positive bacteria. The 5′ UTR of the *S. mutans irvA* mRNA (encoding a putative transcriptional repressor) base-pairs within the open reading frame of the *gbpC* mRNA (encoding a surface-exposed lectin) under several stress conditions. This binding directly interferes with an internal cleavage site for RNase J2 in this region, allowing stabilization of the *gpbC* mRNA (52) (Fig. 6B). It is interesting to note that the sole target of RNase J2 identified so far in *S. mutans* is cleaved by the endoribonucleolytic activity of this enzyme. This result suggests that RNase J2 from *S. mutans* could have a stronger endoribonucleolytic activity than the enzyme from *B. subtilis*.

PNPase

PNPase is considered to be the main phosphorolytic 3′-to-5′ exoribonuclease in *B. subtilis* (53). The protein is composed of two RNase PH domains, an S1 RNA binding domain, and a KH domain (Fig. 2). A *B. subtilis* strain deleted for the *pnp* gene has several phenotypes. Indeed, cells are impaired in competence, are cold sensitive, are filamentous, present a swarming de-

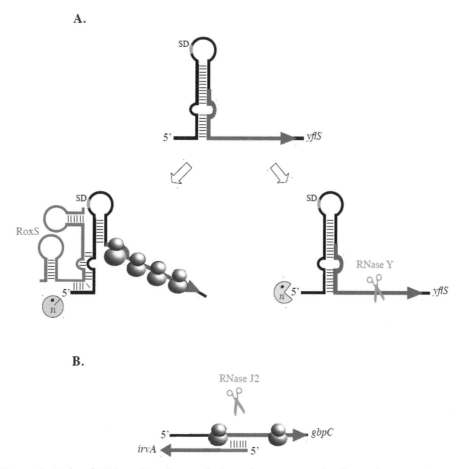

Figure 6 Role of RNase J in the regulation of gene expression by antisense and *trans*-encoded sRNA. (A) RoxS sRNA forms base-pairing interactions with the first 7 nucleotides of the *yflS* mRNA to create a stable RNA helix at the 5′ end of the mRNA and protect it from degradation by RNase J1 (left) (50). RoxS binding also stimulates translation by rendering the SD sequence more accessible. This increase of translation protects the *yflS* mRNA from degradation by RNase Y (left). In contrast, when RoxS does not bind to *yflS*, the 5′ end of this mRNA is free and can be attacked by RNase J1. The SD sequence of *yflS* mRNA also stays embedded in a stem-loop structure that reduces its translation efficiency and promotes degradation by RNase Y (right). Exoribonucleolytic activity of RNase J1 is represented by a light blue Pacman symbol. The sRNA is in blue, the mRNA target is colored in black, the SD sequence is in gray, and ribosomes are in blue. (B) The 5′ UTR of the *S. mutans irvA* mRNA base-pairs with the coding sequence of the *gbpC* mRNA to block endoribonucleolytic cleavage by RNase J2 (52). The inhibition of the endoribonucleolytic activity of RNase J2 is represented by red scissors.

ficiency, and are more sensitive to tetracycline (8, 50, 54, 55). Liu et al. observed that sigma factor D (SigD) expression is lower in a Δ*pnp* strain and that this defect can explain the filamentous and swarming deficiency phenotypes (55).

PNPase is involved both in tRNA maturation and in bulk mRNA degradation (37, 56). The *pnp* gene is not essential in *B. subtilis* or *S. aureus*. A transcriptome analysis performed in a *pnp* mutant strain shows that at least 10% of *B. subtilis* mRNAs are direct substrates

of PNPase and this degradation cannot be done by any of the other three 3′-to-5′ exoribonucleases (55). However, it cannot be excluded that PNPase has more direct targets in *B. subtilis* but the redundancy with the three other exoribonucleases hides these targets. This study also confirms that most PNPase substrates are the upstream fragments resulting from cleavage of an mRNA by an endoribonuclease. This can be due in part to the fact that, as in *E. coli*, PNPase is sensitive to the RNA secondary structure of Rho-independent transcription terminators, preventing direct access to mRNA 3′ ends (57).

Key role of PNPase in the regulation of gene expression with RNA

In *B. subtilis*, the degradation of the *trp* mRNA leader depends on a first cleavage by RNase Y and/or J1 (Fig. 4; also see above). However, the digestion of the endoribonucleolytic product by PNPase is indispensable for the recycling of the TRAP protein and to prevent massive derepression of several TRAP-regulated genes (Fig. 4) (15). PNPase has also been shown to be involved in sRNA regulation in *E. coli* (58, 59). Indeed, this exoribonuclease is able to sequester and protect some sRNAs from degradation. To date, there is no evidence of a similar role in Gram-positive bacteria, and the half-life of at least one sRNA, the RoxS sRNA in *B. subtilis*, is similar in a wild-type and a *pnp* mutant strain (unpublished results).

HELICASES

In bacteria, RNA helicases have been shown to be involved in ribosome biogenesis, translation initiation, and facilitation of RNA processing and decay (for a recent review, see reference 60).

Helicases are classified in six superfamilies, and most of the RNA helicases found in bacteria belong to DEAD-box RNA helicase class in superfamily 2 (61). There are also some RNA helicases belonging to the DEAH-box and Ski2-like DExH families, which are less well studied. For the few DEAH-box helicases studied, the work has been done in Gram-negative bacteria (62–64). Moreover, the role of the recently characterized Ski2-like DexH helicase in *Mycobacterium smegmatis* is unknown (65).

The DEAD-box RNA helicase core is composed of two domains with a RecA fold that allows ATP-dependent RNA binding and RNA-dependent ATP hydrolysis. Four such RNA helicases have been identified in *B. subtilis* (CshA, CshB, YmfL, and DeaD). These helicases have different C-terminal domains, which likely confer their substrate specificity and partner interactions (66). Indeed, the deletion of each helicase gene presents different phenotypes, suggesting that they fulfill different functions in the cell. A single deletion of the *cshA*, *cshB*, or *ymfL* gene impairs growth at 16°C in *B. subtilis*, *L. monocytogenes*, and *Bacillus cereus*, with the greatest defect observed in the *cshA* mutant (67–69). The main role of the CshA, CshB, and YmfL enzymes is probably to participate in the biogenesis of the ribosome (YmfL and CshA participate in the assem-

bly of the 50S ribosomal subunit and CshB is important for the proper assembly of the 70S particle) (68), and this could explain the cold-sensitive phenotype of these mutants. These helicases have also been involved in two other major cellular processes: translation initiation (60, 70) and RNA degradation (see below). In *B. subtilis*, the CshA protein probably has the most predominant role among RNA helicases, since it is constantly expressed from exponential to stationary phase and the mutant strain has morphological defects. In contrast, the levels of the three other DEAD-box RNA helicases decrease upon the transition to stationary phase (68).

Several interactions between bacterial RNA helicases and RNases have been identified. Bacterial two-hybrid screening in *B. subtilis* allowed detection of a strong nonreciprocal interaction between PNPase and the DeaD helicase (68). The role of this interaction is unknown. In contrast, no interaction was detected between YmfL and any RNases tested, but the fact that YmfL has no C-terminal extension probably limits its ability to interact with other partners. Interestingly, like the DEAD-box RNA helicase RhlB in *E. coli*, which is a part of the RNA degradosome, CshA was found in *B. subtilis* and in *S. aureus* to bind several components of the RNA degradation machinery such as RNase J1, RNase Y, and PNPase (66, 71, 72). RNase J1 was co-purified with CshA in *B. subtilis* and in *S. aureus*, but it has been suggested that this interaction may be indirect through ribosomal proteins, since RNase J1 is known to bind to ribosomes (38) and CshA to be involved in ribosome biogenesis. Furthermore, RNase J1 and CshA purified proteins from *S. aureus* do not directly interact *in vitro* (66). The interaction of CshA with PNPase, RNase Y, and two glycolytic enzymes suggests that this helicase could be a component of a Gram-positive DLN (see below).

All these interactions strongly suggest that helicases play a central role in RNA maturation and degradation. In agreement with this, a deletion of the *cshA* gene affects the levels of >200 mRNAs in *B. subtilis* (68) and 100 mRNAs in *S. aureus* (66), including the quorum-sensing system *agr*, which regulates the switch from adhesive to dispersal behavior (73).

A GRAM-POSITIVE DEGRADOSOME-LIKE NETWORK

In *E. coli* and its relatives, a large, stable complex called the degradosome has been identified and is localized to the membrane. In this complex, RNases can interact with several partners, affecting their activities. In *E. coli*, the degradosome is organized around the

C-terminal half of RNase E and involves stable interactions with PNPase, RhlB helicase, and enolase, a glycolytic enzyme (*see* chapter 1). Although the set of RNases in Gram-positive bacteria is different, an equivalent of the degradosome has been proposed in *B. subtilis* and *S. aureus* (Fig. 7) (2, 42, 71, 72, 74).

The results obtained in *B. subtilis* came from bacterial two-hybrid system experiments, copurification after cross-linking, and surface plasmon resonance experiments and suggest that the degradosome-like complex is formed by RNase Y, which binds to a DEAD-box helicase (CshA), PNPase, and two glycolytic enzymes that play an undefined role in this complex, enolase and phosphofructokinase (42, 71, 74). The direct interaction between CshA and RNase Y was also confirmed by the transfer of the C-terminal domain of CshA to CshB, which allows CshB to interact with RNase Y (71).

Several of these interactions have also been observed in *S. aureus* in similar bacterial two-hybrid assays (72). CshA and enolase were identified in this assay to directly interact with RNase Y. The interaction between PNPase and RNase Y in *S. aureus* has not been confirmed (72), suggesting variability in this complex in different bacteria.

Moreover, a recent structural analysis identified the region of PNPase interaction with RNase Y in *B. subtilis*. This region is very similar to the one used by PNPase of *E. coli* to interact with RNase E (75). By mutating residues in this critical region, the authors constructed a PNPase mutant unable to interact with RNase Y but that remains catalytically active. This mutant does not have a phenotype comparable to the

Δ*pnpA* strain, suggesting that this interaction with RNase Y is not necessary for most PNPase activity. Moreover, the degradation pathway of two mRNAs (from the *rpsO* and *cggR* operons) known to be cleaved by RNase Y followed by the degradation of the upstream fragment by PNPase is also not affected in this mutant strain. These results suggest that the interaction between the two partners may only be necessary to degrade a very specific set of targets, which remains to be identified.

B. subtilis RNase Y was found to interact with RNase J1 by bacterial two-hybrid study (42), but this interaction is weak and no interaction was detected by Mathy et al. (41) or Newman et al. (74). This interaction was also not observed in two independent studies in *S. aureus* (72, 74).

A recent examination of the localization of the RNA degradosome components in *B. subtilis* yielded unexpected results. Although RNase Y was localized at the membrane as expected, none of the other proposed components colocalized. RNase J1/J2 and CshA were found in the cytoplasm and excluded from the nucleoid, and PNPase along with the two glycolytic enzymes were uniformly distributed throughout the cytoplasm (76). These results do not exclude the interaction of these proteins with RNase Y, but the lack of colocalization suggests that the interactions between the RNA degradosome components in Gram-positive bacteria, if they exist, are very transitory. All of these protein partners are likely to be involved in other degradosome-independent processes and by consequence need to localize in different parts of the cell. In agreement with

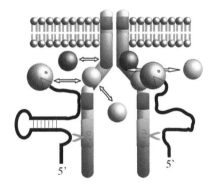
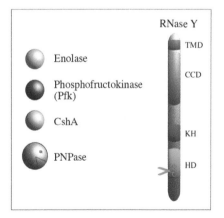

Figure 7 Schematic representation of the DLN. RNase Y is anchored at the membrane and can transiently interact with metabolic enzymes (enolase and phosphofructokinase) and the 3′-to-5′ exoribonuclease PNPase. Domains of interaction between RNase Y and each partner are not characterized. Transient interactions are represented by two-headed arrows. RNase Y domains are indicated (TMD, transmembrane domain; CCD, coiled-coil domain; KH, KH domain; HD, HD domain with the catalytic site represented by red scissors).

this, most of the two-partner interactions found by two-hybrid assays are weak, and to date the complex with all partners of the degradosome has not been purified. These data strongly suggest that the degradosome of Gram-positive bacteria is not stable and is better described as a degradosome-like network (DLN) that may be dedicated to the degradation of only a subset of mRNAs. One of the privileged targets of this DLN could be RNAs with specific secondary structures recognized by the CshA helicase. In agreement with this hypothesis, a transcriptome analysis of a strain containing a deletion of the C-terminal extension of CshA involved in the interaction with RNase Y was similar to that of a complete deletion of the helicase in *S. aureus* (66). Although this result is consistent with the possibility that the effect of CshA on RNA stability is through its interaction with RNase Y, other explanations are also possible, since this deletion mutant only retains 30% of its ATP hydrolysis activity *in vitro*.

CROSS-REGULATION AND AUTOREGULATION OF RNASE EXPRESSION

The effect of the depletion or the deletion of individual RNases on RNA regulation may be blurred by auto- or cross-regulation between RNases. Indeed, it cannot be excluded that some mutations indirectly affect the level or the specificity of other RNases. An example of the effect of cross-regulation observed in *B. subtilis* is that between RNase J1 and J2, probably to maintain an equimolar level of each subunit to form heterodimers. Indeed, the level of the RNase J1 protein expression is stimulated in a Δ*rnjB* strain (77). Reciprocally, the levels of the *rnjB* mRNA increase upon depletion of *rnjA* (4). The expression of RNase J1 is also slightly auto-controlled in *B. subtilis* (77). In contrast, autoregulation of the expression of RNase III in both *B. subtilis* and *S. aureus* is much stronger (20, 78). Furthermore, depletion of RNase Y and J1 affects the levels of the 5′ end of the *rnc* transcript encoding RNase III. The 5′ UTR of the *pnp* mRNA is also affected by a depletion of RNase J1 (4). In *S. aureus*, RNase Y was found to cleave in mRNAs encoding proteins of the DLN (*pfk*: phosphofructokinase; *rnjB*: RNase J2; *rnjA*: RNase J1; and *rny*: RNase Y). However, the effect of these cleavages on RNase expression has not been addressed (11).

PERSPECTIVES

Over the past decade, the identification of the key RNases in Gram-positive bacteria combined with RNA

sequencing analyses have allowed significant progress in the characterization of the RNA maturation and degradation pathways in Gram-positive bacteria. Although the set of RNases differs from that found in *E. coli* and its relatives, most notably the presence of a 5′-to-3′ exoribonuclease in Gram-positive bacteria, some similarities exist between *E. coli* and *B. subtilis*. It is intriguing that RNase Y is able to bind to the same region of PNPase as RNase E, but also to interact with a DEAD-box helicase and enolase as observed in *E. coli*. However, interactions between the different partners of this DLN seem to be far less stable than the complex identified in *E. coli*, and its role, if any, in RNA maturation and degradation remains unresolved. Regulatory RNAs could be a target for the cooperative degradation, as has been shown in *E. coli*. Indeed, an RNase Y cleavage site in a double-stranded region of the RoxS regulatory RNA has been identified (14), and even transiently interacting partners such as CshA may allow access to RNase Y, which has a preference for single-stranded RNA. Intriguingly, CshA was also found by pulldown to bind to the regulatory RNAIII in *S. aureus* (79).

Another interesting observation from the most recent RNA sequencing studies in Gram-positive bacteria is the difference in the importance of the RNase orthologs in different organisms. Indeed, a mutation of RNase Y has far fewer consequences in *S. aureus* than in *B. subtilis*. One explanation for this observation could be that the 5′ exonuclease activity of RNase J1 has a greater role in RNA turnover in *S. aureus* than in *B. subtilis* since SaRNase J1 is less sensitive to the 5′-end status of RNA than is BsRNase J1 (see above). Along the same lines, the mutation of RNase J1 and J2 does not have the same impact on cell growth in *B. subtilis*, *S. aureus*, *S. pyogenes*, and *S. mutans* and the numbers of substrates of RNase J1 and J2 are different in these organisms.

The role of the complex between the two orthologs RNase J1 and J2 is also not understood. This can affect the specificity of these RNases in endoribonucleolytic mode *in vitro*, but this has not yet been demonstrated *in vivo*. RNase J2 seems to have more of a structural role in *B. subtilis* and in *S. aureus*, but seems to play a more important and specific role in mRNA degradation in *E. faecalis* and in the streptococci. These observations suggest an adaptability of these enzymes to ensure a proper regulation of gene expression in functions related to the ecological niches of these bacteria.

Acknowledgments. *The authors wish to acknowledge the following funding sources: CNRS-UMR8261, Université Paris Diderot, Equipex (Cacsice) and LABEX program (Dynamo).*

Citation. Durand S, Condon C. 2018. RNases and helicases in Gram-positive bacteria. Microbiol Spectrum 6(2):RWR-0003-2017.

References

1. Sala A, Bordes P, Genevaux P. 2014. Multiple toxin-antitoxin systems in *Mycobacterium tuberculosis*. *Toxins (Basel)* **6**:1002–1020.

2. Lehnik-Habrink M, Newman J, Rothe FM, Solovyova AS, Rodrigues C, Herzberg C, Commichau FM, Lewis RJ, Stülke J. 2011. RNase Y in *Bacillus subtilis*: a natively disordered protein that is the functional equivalent of RNase E from *Escherichia coli*. *J Bacteriol* **193**:5431–5441.

3. Gilet L, DiChiara JM, Figaro S, Bechhofer DH, Condon C. 2015. Small stable RNA maturation and turnover in *Bacillus subtilis*. *Mol Microbiol* **95**:270–282.

4. Durand S, Gilet L, Bessières P, Nicolas P, Condon C. 2012. Three essential ribonucleases—RNase Y, J1, and III—control the abundance of a majority of *Bacillus subtilis* mRNAs. *PLoS Genet* **8**:e1002520.

5. Laalami S, Bessières P, Rocca A, Zig L, Nicolas P, Putzer H. 2013. *Bacillus subtilis* RNase Y activity *in vivo* analysed by tiling microarrays. *PLoS One* **8**:e54062.

6. Obana N, Nakamura K, Nomura N. 2016. Role of RNase Y in *Clostridium perfringens* mRNA decay and processing. *J Bacteriol* **199**:e00703-16.

7. Delougery A, Dengler V, Chai Y, Losick R. 2011. A multiprotein complex required for biofilm formation by *Bacillus subtilis*. *Mol Microbiol* **99**:425–437.

8. Figaro S, Durand S, Gilet L, Cayet N, Sachse M, Condon C. 2013. *Bacillus subtilis* mutants with knockouts of the genes encoding ribonucleases RNase Y and RNase J1 are viable, with major defects in cell morphology, sporulation, and competence. *J Bacteriol* **195**:2340–2348.

9. Chen Z, Itzek A, Malke H, Ferretti JJ, Kreth J. 2013. Multiple roles of RNase Y in *Streptococcus pyogenes* mRNA processing and degradation. *J Bacteriol* **195**:2585–2594.

10. Marincola G, Schäfer T, Behler J, Bernhardt J, Ohlsen K, Goerke C, Wolz C. 2012. RNase Y of *Staphylococcus aureus* and its role in the activation of virulence genes. *Mol Microbiol* **85**:817–832.

11. Khemici V, Prados J, Linder P, Redder P. 2015. Decay-initiating endoribonucleolytic cleavage by RNase Y is kept under tight control via sequence preference and sub-cellular localisation. *PLoS Genet* **11**:e1005577.

12. Shahbabian K, Jamalli A, Zig L, Putzer H. 2009. RNase Y, a novel endoribonuclease, initiates riboswitch turnover in *Bacillus subtilis*. *EMBO J* **28**:3523–3533.

13. Marincola G, Wolz C. 2017. Downstream element determines RNase Y cleavage of the *saePQRS* operon in *Staphylococcus aureus*. *Nucleic Acids Res* **45**:5980–5994.

14. Durand S, Braun F, Lioliou E, Romilly C, Helfer AC, Kuhn L, Quittot N, Nicolas P, Romby P, Condon C. 2015. A nitric oxide regulated small RNA controls expression of genes involved in redox homeostasis in *Bacillus subtilis*. *PLoS Genet* **11**:e1004957.

15. Deikus G, Babitzke P, Bechhofer DH. 2004. Recycling of a regulatory protein by degradation of the RNA to which it binds. *Proc Natl Acad Sci U S A* **101**:2747–2751.

16. Deikus G, Bechhofer DH. 2009. *Bacillus subtilis trp* Leader RNA: RNase J1 endonuclease cleavage specificity and PNPase processing. *J Biol Chem* **284**:26394–26401.

17. Deikus G, Bechhofer DH. 2011. 5′ end-independent RNase J1 endonuclease cleavage of *Bacillus subtilis* model RNA. *J Biol Chem* **286**:34932–34940.

18. Panganiban AT, Whiteley HR. 1983. Purification and properties of a new *Bacillus subtilis* RNA processing enzyme. Cleavage of phage SP82 mRNA and *Bacillus subtilis* precursor rRNA. *J Biol Chem* **258**:12487–12493.

19. Oguro A, Kakeshita H, Nakamura K, Yamane K, Wang W, Bechhofer DH. 1998. *Bacillus subtilis* RNase III cleaves both 5′- and 3′-sites of the small cytoplasmic RNA precursor. *J Biol Chem* **273**:19542–19547.

20. Lioliou E, Sharma CM, Caldelari I, Helfer AC, Fechter P, Vandenesch F, Vogel J, Romby P. 2012. Global regulatory functions of the *Staphylococcus aureus* endoribonuclease III in gene expression. *PLoS Genet* **8**:e1002782.

21. Stead MB, Marshburn S, Mohanty BK, Mitra J, Pena Castillo L, Ray D, van Bakel H, Hughes TR, Kushner SR. 2011. Analysis of *Escherichia coli* RNase E and RNase III activity *in vivo* using tiling microarrays. *Nucleic Acids Res* **39**:3188–3203.

22. Toledo-Arana A, Dussurget O, Nikitas G, Sesto N, Guet-Revillet H, Balestrino D, Loh E, Gripenland J, Tiensuu T, Vaitkevicius K, Barthelemy M, Vergassola M, Nahori MA, Soubigou G, Régnault B, Coppée JY, Lecuit M, Johansson J, Cossart P. 2009. The *Listeria* transcriptional landscape from saprophytism to virulence. *Nature* **459**:950–956.

23. Lasa I, Toledo-Arana A, Dobin A, Villanueva M, de los Mozos IR, Vergara-Irigaray M, Segura V, Fagegaltier D, Penadés JR, Valle J, Solano C, Gingeras TR. 2011. Genome-wide antisense transcription drives mRNA processing in bacteria. *Proc Natl Acad Sci U S A* **108**:20172–20177.

24. Novick RP, Ross HF, Projan SJ, Kornblum J, Kreiswirth B, Moghazeh S. 1993. Synthesis of staphylococcal virulence factors is controlled by a regulatory RNA molecule. *EMBO J* **12**:3967–3975.

25. Kaito C, Saito Y, Ikuo M, Omae Y, Mao H, Nagano G, Fujiyuki T, Numata S, Han X, Obata K, Hasegawa S, Yamaguchi H, Inokuchi K, Ito T, Hiramatsu K, Sekimizu K. 2013. Mobile genetic element SCC*mec*-encoded *psm-mec* RNA suppresses translation of *agrA* and attenuates MRSA virulence. *PLoS Pathog* **9**:e1003269.

26. Deltcheva E, Chylinski K, Sharma CM, Gonzales K, Chao Y, Pirzada ZA, Eckert MR, Vogel J, Charpentier E. 2011. CRISPR RNA maturation by *trans*-encoded small RNA and host factor RNase III. *Nature* **471**:602–607.

27. Redko Y, Bechhofer DH, Condon C. 2008. Mini-III, an unusual member of the RNase III family of enzymes, catalyses 23S ribosomal RNA maturation in *B. subtilis*. *Mol Microbiol* **68**:1096–1106.

28. Olmedo G, Guzmán P. 2008. Mini-III, a fourth class of RNase III catalyses maturation of the *Bacillus subtilis* 23S ribosomal RNA. *Mol Microbiol* **68**:1073–1076.

29. Hotto AM, Castandet B, Gilet L, Higdon A, Condon C, Stern DB. 2015. Arabidopsis chloroplast mini-ribonuclease III participates in rRNA maturation and intron recycling. *Plant Cell* **27**:724–740.

30. Aït-Bara S, Carpousis AJ. 2015. RNA degradosomes in bacteria and chloroplasts: classification, distribution and evolution of RNase E homologs. *Mol Microbiol* **97**:1021–1135.

31. Zeller ME, Csanadi A, Miczak A, Rose T, Bizebard T, Kaberdin VR. 2007. Quaternary structure and biochemical properties of mycobacterial RNase E/G. *Biochem J* **403**:207–215.

32. Taverniti V, Forti F, Ghisotti D, Putzer H. 2011. *Mycobacterium smegmatis* RNase J is a 5′-3′ exo-/endoribonuclease and both RNase J and RNase E are involved in ribosomal RNA maturation. *Mol Microbiol* **82**:1260–1276.

33. Kovacs L, Csanadi A, Megyeri K, Kaberdin VR, Miczak A. 2005. Mycobacterial RNase E-associated proteins. *Microbiol Immunol* **49**:1003–1007.

34. Pfeiffer V, Papenfort K, Lucchini S, Hinton JC, Vogel J. 2009. Coding sequence targeting by MicC RNA reveals bacterial mRNA silencing downstream of translational initiation. *Nat Struct Mol Biol* **16**:840–846.

35. Bandyra KJ, Said N, Pfeiffer V, Górna MW, Vogel J, Luisi BF. 2012. The seed region of a small RNA drives the controlled destruction of the target mRNA by the endoribonuclease RNase E. *Mol Cell* **47**:943–953.

36. Wen T, Oussenko IA, Pellegrini O, Bechhofer DH, Condon C. 2005. Ribonuclease PH plays a major role in the exonucleolytic maturation of CCA-containing tRNA precursors in *Bacillus subtilis*. *Nucleic Acids Res* **33**:3636–3643.

37. Oussenko IA, Abe T, Ujiie H, Muto A, Bechhofer DH. 2005. Participation of 3′-to-5′ exoribonucleases in the turnover of *Bacillus subtilis* mRNA. *J Bacteriol* **187**: 2758–2767.

38. Even S, Pellegrini O, Zig L, Labas V, Vinh J, Bréchemmier-Baey D, Putzer H. 2005. Ribonucleases J1 and J2: two novel endoribonucleases in *B.subtilis* with functional homology to *E.coli* RNase E. *Nucleic Acids Res* **33**:2141–2152.

39. Dorléans A, Li de la Sierra-Gallay I, Piton J, Zig L, Gilet L, Putzer H, Condon C. 2011. Molecular basis for the recognition and cleavage of RNA by the bifunctional 5′-3′ exo/endoribonuclease RNase J. *Structure* **19**:1252–1261.

40. Newman JA, Hewitt L, Rodrigues C, Solovyova A, Harwood CR, Lewis RJ. 2011. Unusual, dual endo- and exonuclease activity in the degradosome explained by crystal structure analysis of RNase J1. *Structure* **19**: 1241–1251.

41. Mathy N, Hébert A, Mervelet P, Bénard L, Dorléans A, Li de la Sierra-Gallay I, Noirot P, Putzer H, Condon C. 2010. *Bacillus subtilis* ribonucleases J1 and J2 form a complex with altered enzyme behaviour. *Mol Microbiol* **75**:489–498.

42. Commichau FM, Rothe FM, Herzberg C, Wagner E, Hellwig D, Lehnik-Habrink M, Hammer E, Völker U, Stülke J, Volker U, Stulke J. 2009. Novel activities of gly-

colytic enzymes in *Bacillus subtilis*: interactions with essential proteins involved in mRNA processing. *Mol Cell Proteomics* **8**:1350–1360.

43. Mathy N, Bénard L, Pellegrini O, Daou R, Wen T, Condon C. 2007. 5′-to-3′ exoribonuclease activity in bacteria: role of RNase J1 in rRNA maturation and 5′ stability of mRNA. *Cell* **129**:681–692.

44. Li de la Sierra-Gallay I, Zig L, Jamalli A, Putzer H, de la Sierra-Gallay IL. 2008. Structural insights into the dual activity of RNase J. *Nat Struct Mol Biol* **15**:206–212.

45. Linder P, Lemeille S, Redder P. 2014. Transcriptome-wide analyses of 5′-ends in RNase J mutants of a Gram-positive pathogen reveal a role in RNA maturation, regulation and degradation. *PLoS Genet* **10**:e1004207.

46. Hausmann S, Guimarães VA, Garcin D, Baumann N, Linder P, Redder P, Redder P. 2017. Both exo- and endo-nucleolytic activities of RNase J1 from *Staphylococcus aureus* are manganese dependent and active on triphosphorylated 5′-ends. *RNA Biol* **14**:1431–1443.

47. Bugrysheva JV, Scott JR. 2010. The ribonucleases J1 and J2 are essential for growth and have independent roles in mRNA decay in *Streptococcus pyogenes*. *Mol Microbiol* **75**:731–743.

48. Chen X, Liu N, Khajotia S, Qi F, Merritt J, Merritt J. 2015. RNases J1 and J2 are critical pleiotropic regulators in *Streptococcus mutans*. *Microbiology* **161**:797–806.

49. Gao P, Pinkston KL, Bourgogne A, Murray BE, van Hoof A, Harvey BR. 2017. Functional studies of *E. faecalis* RNase J2 and its role in virulence and fitness. *PLoS One* **12**:e0175212.

50. Luttinger A, Hahn J, Dubnau D. 1996. Polynucleotide phosphorylase is necessary for competence development in *Bacillus subtilis*. *Mol Microbiol* **19**:343–356.

51. Durand S, Braun F, Helfer AC, Romby P, Condon C. 2017. sRNA-mediated activation of gene expression by inhibition of 5′-3′ exonucleolytic mRNA degradation. *eLife* **6**:e23602.

52. Liu N, Niu G, Xie Z, Chen Z, Itzek A, Kreth J, Gillaspy A, Zeng L, Burne R, Qi F, Merritt J. 2015. The *Streptococcus mutans irvA* gene encodes a *trans*-acting riboregulatory mRNA. *Mol Cell* **57**:179–190.

53. Deutscher MP, Reuven NB. 1991. Enzymatic basis for hydrolytic versus phosphorolytic mRNA degradation in *Escherichia coli* and *Bacillus subtilis*. *Proc Natl Acad Sci U S A* **88**:3277–3280.

54. Wang ZF, Whitfield ML, Ingledue TC III, Dominski Z, Marzluff WF. 1996. The protein that binds the 3′ end of histone mRNA: a novel RNA-binding protein required for histone pre-mRNA processing. *Genes Dev* **10**:3028–3040.

55. Liu B, Deikus G, Bree A, Durand S, Kearns DB, Bechhofer DH. 2014. Global analysis of mRNA decay intermediates in *Bacillus subtilis* wild-type and polynucleotide phosphorylase-deletion strains. *Mol Microbiol* **94**:41–55.

56. Wang W, Bechhofer DH. 1996. Properties of a *Bacillus subtilis* polynucleotide phosphorylase deletion strain. *J Bacteriol* **178**:2375–2382.

57. Deikus G, Bechhofer DH. 2007. Initiation of decay of *Bacillus subtilis trp* leader RNA. *J Biol Chem* **282**: 20238–20244.

58. De Lay N, Gottesman S. 2011. Role of polynucleotide phosphorylase in sRNA function in *Escherichia coli*. *RNA* 17:1172–1189.

59. Bandyra KJ, Sinha D, Syrjanen J, Luisi BF, De Lay NR. 2016. The ribonuclease polynucleotide phosphorylase can interact with small regulatory RNAs in both protective and degradative modes. *RNA* 22:360–372.

60. Khemici V, Linder P. 2016. RNA helicases in bacteria. *Curr Opin Microbiol* 30:58–66.

61. Fairman-Williams ME, Guenther UP, Jankowsky E. 2010. SF1 and SF2 helicases: family matters. *Curr Opin Struct Biol* 20:313–324.

62. Koo JT, Choe J, Moseley SL. 2004. HrpA, a DEAH-box RNA helicase, is involved in mRNA processing of a fimbrial operon in *Escherichia coli*. *Mol Microbiol* 52:1813–1826.

63. Salman-Dilgimen A, Hardy PO, Radolf JD, Caimano MJ, Chaconas G. 2013. HrpA, an RNA helicase involved in RNA processing, is required for mouse infectivity and tick transmission of the Lyme disease spirochete. *PLoS Pathog* 9:e1003841.

64. Granato LM, Picchi SC, Andrade MO, Takita MA, de Souza AA, Wang N, Machado MA. 2016. The ATP-dependent RNA helicase HrpB plays an important role in motility and biofilm formation in *Xanthomonas citri* subsp. *citri*. *BMC Microbiol* 16:55.

65. Uson ML, Ordonez H, Shuman S. 2015. *Mycobacterium smegmatis* HelY is an RNA-activated ATPase/dATPase and 3′-to-5′ helicase that unwinds 3′-tailed RNA duplexes and RNA:DNA hybrids. *J Bacteriol* 197:3057–3065.

66. Giraud C, Hausmann S, Lemeille S, Prados J, Redder P, Linder P. 2015. The C-terminal region of the RNA helicase CshA is required for the interaction with the degradosome and turnover of bulk RNA in the opportunistic pathogen *Staphylococcus aureus*. *RNA Biol* 12:658–674.

67. Pandiani F, Brillard J, Bornard I, Michaud C, Chamot S, Nguyen-the C, Broussolle V. 2010. Differential involvement of the five RNA helicases in adaptation of *Bacillus cereus* ATCC 14579 to low growth temperatures. *Appl Environ Microbiol* 76:6692–6697.

68. Lehnik-Habrink M, Rempeters L, Kovács ÁT, Wrede C, Baierlein C, Krebber H, Kuipers OP, Stülke J. 2013. DEAD-box RNA helicases in *Bacillus subtilis* have multiple functions and act independently from each other. *J Bacteriol* 195:534–544.

69. Bäreclev C, Vaitkevicius K, Netterling S, Johansson J. 2014. DExD-box RNA-helicases in *Listeria monocytogenes* are important for growth, ribosomal maturation, rRNA processing and virulence factor expression. *RNA Biol* 11:1457–1466.

70. Redder P, Hausmann S, Khemici V, Yasrebi H, Linder P. 2015. Bacterial versatility requires DEAD-box RNA helicases. *FEMS Microbiol Rev* 39:392–412.

71. Lehnik-Habrink M, Pförtner H, Rempeters L, Pietack N, Herzberg C, Stülke J. 2010. The RNA degradosome in *Bacillus subtilis*: identification of CshA as the major RNA helicase in the multiprotein complex. *Mol Microbiol* 77:958–971.

72. Roux CM, DeMuth JP, Dunman PM. 2011. Characterization of components of the *Staphylococcus aureus* mRNA degradosome holoenzyme-like complex. *J Bacteriol* 193:5520–5526.

73. Oun S, Redder P, Didier JP, François P, Corvaglia AR, Buttazzoni E, Giraud C, Girard M, Schrenzel J, Linder P. 2013. The CshA DEAD-box RNA helicase is important for quorum sensing control in *Staphylococcus aureus*. *RNA Biol* 10:157–165.

74. Newman JA, Hewitt L, Rodrigues C, Solovyova AS, Harwood CR, Lewis RJ. 2012. Dissection of the network of interactions that links RNA processing with glycolysis in the *Bacillus subtilis* degradosome. *J Mol Biol* 416:121–136.

75. Salvo E, Alabi S, Liu B, Schlessinger A, Bechhofer DH. 2016. Interaction of *Bacillus subtilis* polynucleotide phosphorylase and RNase Y: structural mapping and effect on mRNA turnover. *J Biol Chem* 291:6655–6663.

76. Cascante-Estepa N, Gunka K, Stülke J. 2016. Localization of components of the RNA-degrading machine in *Bacillus subtilis*. *Front Microbiol* 7:1492.

77. Jamalli A, Hébert A, Zig L, Putzer H. 2014. Control of expression of the RNases J1 and J2 in *Bacillus subtilis*. *J Bacteriol* 196:318–324.

78. DiChiara JM, Liu B, Figaro S, Condon C, Bechhofer DH. 2016. Mapping of internal monophosphate 5′ ends of *Bacillus subtilis* messenger RNAs and ribosomal RNAs in wild-type and ribonuclease-mutant strains. *Nucleic Acids Res* 44:3373–3389.

79. Zhang X, Zhu Q, Tian T, Zhao C, Zang J, Xue T, Sun B. 2015. Identification of RNAIII-binding proteins in *Staphylococcus aureus* using tethered RNAs and streptavidin aptamers based pull-down assay. *BMC Microbiol* 15:102.

80. Condon C, Putzer H. 2002. The phylogenetic distribution of bacterial ribonucleases. *Nucleic Acids Res* 30:5339–5346.

81. Kaberdin VR, Singh D, Lin-Chao S. 2011. Composition and conservation of the mRNA-degrading machinery in bacteria. *J Biomed Sci* 18:23.

82. Bronesky D, Wu Z, Marzi S, Walter P, Geissmann T, Moreau K, Vandenesch F, Caldelari I, Romby P. 2016. *Staphylococcus aureus* RNAIII and its regulon link quorum sensing, stress responses, metabolic adaptation, and regulation of virulence gene expression. *Annu Rev Microbiol* 70:299–316.

83. Obana N, Shirahama Y, Abe K, Nakamura K. 2010. Stabilization of *Clostridium perfringens* collagenase mRNA by VR-RNA-dependent cleavage in 5′ leader sequence. *Mol Microbiol* 77:1416–1428.

84. Boisset S, Geissmann T, Huntzinger E, Fechter P, Bendridi N, Possedko M, Chevalier C, Helfer AC, Benito Y, Jacquier A, Gaspin C, Vandenesch F, Romby P. 2007. Staphylococcus aureus RNAIII coordinately represses the synthesis of virulence factors and the transcription regulator Rot by an antisense mechanism. *Genes Dev* 21:1353–1366.

Cis-Acting RNAs

II

Regulating with RNA in Bacteria and Archaea
Edited by Gisela Storz and Kai Papenfort
© 2018 American Society for Microbiology, Washington, DC
doi:10.1128/microbiolspec.RWR-0012-2017

RNA Thermometers in Bacterial Pathogens

4

Edmund Loh,[1,2] Francesco Righetti,[1] Hannes Eichner,[1] Christian Twittenhoff,[3] and Franz Narberhaus[3]

INTRODUCTION

Temperature is an environmental cue that affects essentially every cellular process. To cope with sudden temperature changes, all living cells closely survey their ambient temperature through numerous sensory mechanisms, which involve regulatory proteins, changes in membrane fluidity, and impacts on DNA topology and RNA structures (1, 2). Most of these mechanisms were initially discovered in studies of the heat shock response, which protects the cell from serious damage after a drastic shift to high temperatures. However, it is now established that subtle temperature changes already induce cellular responses. One process that involves reversible temperature changes is the entry and exit of mammalian pathogens into and from the host. A temperature of ~37°C serves as a very good indicator to the bacterium that it is in a mammalian host. Various mechanisms regulating gene expression in response to host body temperature have been discovered, with some involving regulatory proteins and others utilizing sensory and regulatory RNAs. In this review, the main focus will be on RNA-mediated mechanisms; however, when the regulation involves a multicomponent regulatory network, protein-dependent regulatory events will be discussed.

A common regulatory principle acting in response to temperature in bacterial pathogens is translational control by RNA thermometers/thermosensors (RNATs). RNATs are elements usually located in the 5′ untranslated region (UTR) of mRNAs. They operate by changing their secondary structures in response to temperature fluctuation. Due to the close proximity to the protein-coding region, a change in the RNA secondary structure exerts a major effect on translation efficiency of the downstream gene. Generally, at low temperature (<30°C), an RNAT forms a stable structure masking the ribosome binding site (RBS) and blocking translation of the downstream gene. Upon encountering higher temperature, such as the host body temperature of 37°C, the higher thermodynamic energy weakens the RNAT structure, liberating the previously inaccessible RBS and allowing translation initiation (Fig. 1A).

Like many temperature-responsive mechanisms, RNATs were discovered as regulatory elements in the heat shock response. In *Escherichia coli*, an RNAT with a very complex secondary structure involving >200 nucleotides (nt) of the coding region regulates the synthesis of its heat shock master regulator Sigma-32 (σ^{32}, RpoH) (3). As a result, RpoH protects the bacterium through upregulation of a large group of heat shock proteins such as molecular chaperones and proteases. In many bacteria, the translation of a particular class of heat shock proteins termed small heat shock proteins provides the first line of defense against heat-induced protein aggregation. The synthesis of these proteins is under the direct control of an RNAT in their mRNA (4). Later, it emerged that RNATs are equally well suited for mounting a virulence response when a pathogen experiences a rise in temperature upon ingestion by a mammalian host (5). Acting through a posttranscriptional mechanism, RNATs allow a rapid response as the mRNA—although in an inactive conformation—is already present when the bacterium encounters its mammalian host. An instantaneous melting of the RNA structure followed by ribosome binding and translation initiation enables the immediate production of virulence factors.

Since the discovery of a virulence-related RNA thermosensor, similar elements have been identified in the

[1]Microbiology, Tumor and Cell Biology, Karolinska Institutet, 17177 Stockholm, Sweden; [2]SCELSE, Nanyang Technological University, 639798, Singapore; [3]Microbial Biology, Ruhr University, Bochum, Germany.

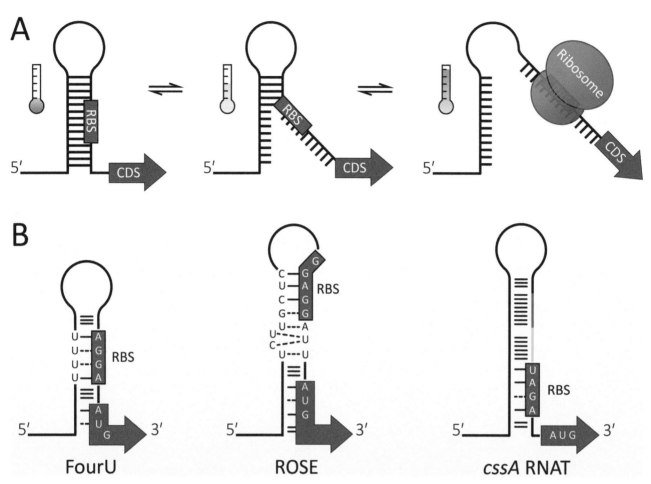

Figure 1 (A) RNATs are structural elements located within the 5′ UTR of protein-coding mRNAs and control its translation by operating as reversible molecular zippers that mask or unmask the RBS in response to temperature changes. (B) Three examples of RNAT secondary structures: FourU element of *S. enterica agsA*, ROSE element of *Bradyrhizobium japonicum hspA*, and the 8-bp tandem repeats of *N. meningitidis cssA* (blue and yellow lines indicate repeats). CDS, coding sequence.

translation initiation region of many other bacterial genes directly or indirectly related to virulence (6, 7). All currently reported virulence-associated RNATs are summarized in Fig. 2 and Table 1. Interestingly, the nucleotide sequences and secondary structures of these thermosensors are very diverse and poorly conserved. Only a few recurring nucleotide motifs have been observed (Fig. 1B) (for reviews, see references 8 and 9). One example is the FourU motif, composed of four uridines that base-pair with the Shine-Dalgarno (SD) sequence. It is present in various virulence and heat shock thermometers (10–12). The diversity of RNAT sequences and structures suggests that the conceptually simple principle of sequestering parts of the RBS by complementary base pairing has evolved multiple times independently during evolution. In the following sec-

tions, the currently known virulence-associated RNATs will be described and grouped according to their function in virulence.

CONTROL OF MASTER REGULATORS OF VIRULENCE

Listeria monocytogenes—Two Regulatory RNAs Team Up

Controlling the virulence master regulator with an RNAT is a strategy adopted by at least three different human pathogens. The first virulence-related RNAT was discovered upstream of the *prfA* gene in *Listeria monocytogenes* (5). This Gram-positive, rod-shaped, and nonsporulating bacterium is commonly found in

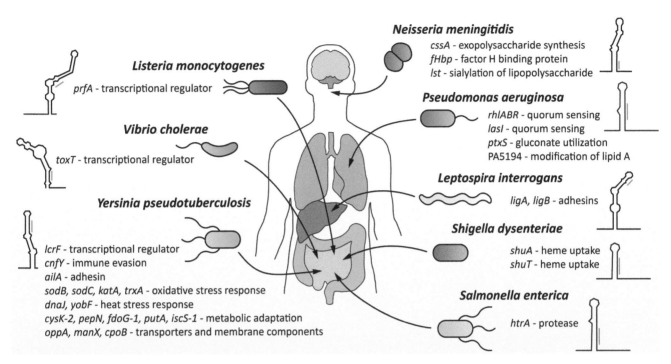

Figure 2 Virulence-associated RNATs in bacterial pathogens. For details, see text and Table 1.

decaying soil and in food. In humans, *L. monocytogenes* is responsible for causing meningitis, meningoencephalitis, fetal infection, and neonatal abortion. The disease termed listeriosis is common among individuals with weakened immune systems, pregnant women, and the elderly. *L. monocytogenes* possesses the ability to cross the host's intestinal, fetal placental, and blood-brain barrier. In addition, *L. monocytogenes* is a serious threat to food production industries as it can tolerate high concentrations of salt and acidic conditions and is able to multiply at refrigeration temperatures.

To survive and thrive, *L. monocytogenes* alters its gene expression rapidly in adaptation to new environments. The expression of major listerial virulence factors is regulated by its transcriptional regulator PrfA. Due to its pivotal role in the bacterium's survival, *prfA* expression itself is tightly regulated through multiple levels: transcriptional, translational, and posttranslational through the availability of glutathione and other factors (13–15).

Temperature plays a decisive role in the expression of *prfA*. Translation is controlled by an unusually long RNAT located within the 5′ UTR of the mRNA. Outside of the host and at temperatures such as 26°C, the RNA forms a stem-loop masking the RBS and blocking translation. During transition to the human host temperature of 37°C, the stem-loop structure unwinds, thus allowing translation of the *prfA* mRNA, as demonstrated through structure probing (5). Reporter gene and toeprint experiments showed that the first 20 codons within the *prfA* mRNA form a flexible hairpin-loop required for ribosome binding and efficient translation (16).

In addition to sensing temperature, the *prfA* RNAT is able to directly base-pair with a prematurely terminated S-adenosylmethionine (SAM) riboswitch. Two *trans*-acting SAM riboswitch-derived small RNAs have been demonstrated to be able to interact with the *prfA* 5′ UTR to inhibit translation (17). It has been speculated that upon base pairing, the noninteracting nucleotides of the *prfA* mRNA form an alternative structure occluding the RBS. Interestingly, this interaction between SAM riboswitch and *prfA* 5′ UTR is only observed at 37°C and not at low temperature, possibly due to the inaccessible *prfA* thermosensor stem-loop (Fig. 3). The intracellular concentration of SAM could play a role in controlling *prfA* expression during infection, especially during growth within the intestine, where PrfA-controlled virulence factors are not required (17). Overall, the *prfA* RNAT poses as a unique example as it is able to sense changes in temperature and SAM concentration for its efficient translation and activation of virulence factors.

Table 1 Summary of currently known virulence-associated RNATs in bacterial pathogens

Regulated gene	Gene function	Role in pathogenicity	Organism	Characteristics of the RNAT	Experimental evidence	Reference(s)
prfA	Transcriptional regulator	Activates transcription of major virulence factors	L. monocytogenes	Single stem-loop structure occluding the RBS	In vitro characterization via cross-linking, structure probing, and ribosome toeprinting assays	5, 16, 17
				Two prematurely terminated SAM riboswitches base-pair to the RNAT region, inhibiting translation	Reporter gene studies	
					Virulence assays in a mouse infection model	
toxT	Transcriptional regulator	Activates transcription of the toxin-coregulated pilus and the cholera toxin	V. cholerae	Single stem-loop structure containing a FourU element	In vitro characterization via enzymatic structure probing and ribosome toeprinting assays	12
					Reporter gene studies	
					Virulence assays in a mouse infection model	
lcrF	Transcriptional regulator	Activates transcription of plasmid-encoded virulence factors, including the T3SS apparatus, Yersinia outer protein genes, and regulatory factors	Y. pseudotuberculosis	The intergenic region of the cotranscribed genes yscW and lcrF folds into two hairpin structures; the second contains a FourU element	In vitro characterization via enzymatic structure probing and ribosome toeprinting assays	11, 29
					Reporter gene studies	
					Virulence assays in a mouse infection model	
cnfY	Cytotoxic necrotizing factor	Toxin that modulates the host cell cytoskeleton, increases inflammatory responses, protects the bacteria from attacks of innate immune effectors, and enhances the severity of a Yersinia infection	Y. pseudotuberculosis	Single stem-loop structure partially occluding the RBS	Ectopic reporter gene studies	32
ailA	Adhesin	Mediates the attachment and killing of neutrophils, increases the delivery of Yops, and reduces the complement-mediated immune response	Y. pseudotuberculosis	Single stem-loop structure occluding the RBS	In vitro characterization via enzymatic structure probing and ribosome toeprinting assays	32
					Ectopic reporter gene studies	
sodB	Cytoplasmic superoxide dismutase	Protects from superoxide by converting it into hydrogen peroxide and water	Y. pseudotuberculosis	Two stem-loop structures; the second occludes the RBS	Ectopic reporter gene studies	32
sodC	Periplasmic superoxide dismutase	Protects from superoxide by converting it into hydrogen peroxide and water	Y. pseudotuberculosis	Two stem-loop structures; the second partially occludes the RBS	Ectopic reporter gene studies	32
katA	Catalase	Protects from hydrogen peroxide by converting it into water and oxygen	Y. pseudotuberculosis	Two stem-loop structures; the second partially occludes the RBS	Ectopic reporter gene studies	32

trxA	Thioredoxin	Reduces oxidized proteins and other molecules	Y. pseudotuberculosis	Two stem-loop structures; the second partially occludes the RBS	Ectopic reporter gene studies	32
dnaJ	DnaK cochaperone	Prevents protein aggregation and assists protein folding, in particular during stress conditions	Y. pseudotuberculosis	Complex architecture with base-paired AUG	Ectopic reporter gene studies	32
yobF	Putative heat shock protein	Unknown function	Y. pseudotuberculosis	Single stem-loop structure occluding the RBS	Ectopic reporter gene studies	32
cysK-2	Cysteine synthase A component	Enzyme involved in the synthesis of cysteine	Y. pseudotuberculosis	Single stem-loop structure occluding the RBS	Ectopic reporter gene studies	32
pepN	Aminopeptidase N	Involved in the degradation of peptides generated by protein breakdown	Y. pseudotuberculosis	Two stem-loop structures; the second partially occludes the RBS	Ectopic reporter gene studies	32
fdoG-1	Formate dehydrogenase	Catalyzes the oxidation of formate to carbon dioxide, donating the electrons	Y. pseudotuberculosis	Complex architecture with partially base-paired RBS	Ectopic reporter gene studies	32
putA	Δ^1-Pyrroline-5-carboxylate dehydrogenase	Transcriptional repressor and membrane-associated enzyme involved in proline degradation	Y. pseudotuberculosis	Two stem-loop structures; the second partially occludes the RBS	Ectopic reporter gene studies	32
iscS-1	Component of cysteine desulfurase	Enzyme involved in the synthesis of Fe-S cluster	Y. pseudotuberculosis	Complex architecture with partially base-paired RBS	Ectopic reporter gene studies	32
oppA	Oligopeptide ABC transporter periplasmic binding protein	Uptake of oligopeptides	Y. pseudotuberculosis	Complex architecture with partially base-paired RBS	Ectopic reporter gene studies	32
manX	Component of mannose PTS^a permease	Subunit of the mannose PTS permease	Y. pseudotuberculosis	Three stem-loop structures; the third partially occludes the RBS	Ectopic reporter gene studies	32
cpoB	Component of Tol-Pal cell envelope complex	Periplasmic protein involved in the coordination of peptidoglycan synthesis and cell division	Y. pseudotuberculosis	Four stem-loop structures; the fourth partially occludes the RBS	Ectopic reporter gene studies	32
cssA	UDP-N-acetylglucosamine 2-epimerase	Polysialic acid capsule biosynthesis, which confers resistance to immune system detection	N. meningitidis	Single stem-loop structure occluding the RBS	In vitro characterization via enzymatic structure probing and ribosome toeprinting assays	41, 44
fHbp	Factor H binding protein	Confers protection against complement defense	N. meningitidis	Single stem-loop structure occluding the RBS	Ectopic reporter gene studies	41

(Continued)

Table 1   (*Continued*)

Regulated gene	Gene function	Role in pathogenicity	Organism	Characteristics of the RNAT	Experimental evidence	Reference(s)
lst	Sialyltransferase	Sialylates surface lipooligosaccharide	*N. meningitidis*	Single stem-loop structure occluding the RBS	Ectopic reporter gene studies	41
ligA/*ligB*	Surface-exposed lipoprotein	Binds human complement factor H and C4-binding proteins	*L. interrogans*	Single stem-loop structure occluding the RBS	Ribosome toeprinting assays Ectopic reporter gene studies	57
rhlR	Transcriptional regulator	Mediates the QS response by controlling the expression of multiple genes, including virulence factors	*P. aeruginosa*	ROSE element located within the 5′ UTR of *rhlABR* mRNA and controlling RhlR synthesis through a polar effect	Reporter gene studies	63
lasI	Acylhomoserine lactone synthase	Enzyme involved in the synthesis of an effector molecule that triggers the QS response	*P. aeruginosa*	ROSE-like element	Reporter gene studies	63
ptxS	Transcriptional regulator	Controls gluconate transport and usage	*P. aeruginosa*	Two stem-loop structures; the second partially occludes the RBS	*In vitro* characterization via enzymatic and chemical structure probing Reporter gene studies	66
PA5194 (*lpxT*)	Putative lipid A 1-diphosphate synthase	Enzyme putatively involved in the phosphorylation of the lipid A	*P. aeruginosa*	Two stem-loop structures; the second contains a FourU-like element that partially occludes the RBS	*In vitro* characterization via enzymatic and chemical structure probing Reporter gene studies	66
shuA	Heme uptake protein A	Outer membrane receptor required for heme uptake and iron scavenging	*S. dysenteriae*	Short single stem-loop containing a FourU element	Reporter gene studies	70
shuT	Heme uptake protein T	Periplasmic heme binding protein	*S. dysenteriae*	Single stem-loop structure occluding the RBS and part of the AUG start codon	*In vitro* characterization via enzymatic structure probing Reporter gene studies	71
htrA	Protease	Periplasmic protease involved in protein quality control and stress response	*S. enterica*	FourU element	Reporter gene studies and extensive *in vitro* characterization via chemical structure probing (SHAPE)	83, 84

[a]PTS, phosphotransferase system.

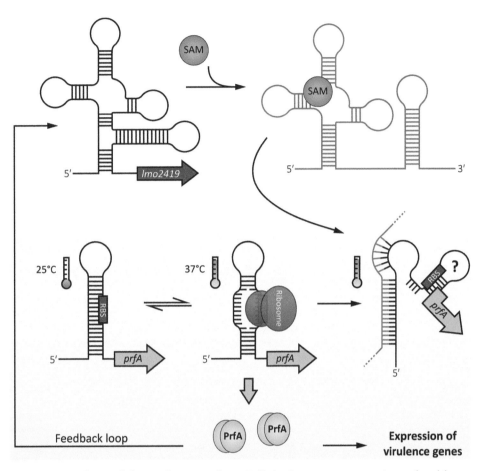

Figure 3 Synthesis of the virulence regulator PrfA in *L. monocytogenes* is regulated by an RNAT in the 5′ UTR of *prfA* and a *trans*-acting SAM riboswitch element. *L. monocytogenes* detects temperature increase to 37°C as the signal of host entry, and as a consequence, PrfA is synthesized. A prematurely terminated riboswitch produced in the presence of SAM inhibits the translation of *prfA* via base pairing with its 5′ UTR.

Vibrio cholerae—From the Water and Back into the Water

Vibrio cholerae is a facultative human pathogen and the causative agent of the diarrheal disease cholera. The bacterium resides in aquatic environments and is well adapted to the many challenges in this habitat (18). It experiences temperature fluctuations due to seasonal changes, after transmission from the environment into a human host, and when shed from the host back into the environment. Within the small intestine, *V. cholerae* initiates a virulence program that is dependent on its transcriptional regulators ToxR and ToxT (19). This program ultimately results in the production of two ToxT-induced factors: the toxin-coregulated pilus, required for intestinal colonization, and the cholera toxin, responsible for the profuse watery diarrhea.

Translation of the *toxT* mRNA is inhibited at environmental temperatures through a FourU-type ther-

mometer that pairs with the AGAG of the SD sequence (12). Liberation of the SD sequence and binding of the ribosome at 37°C were demonstrated by structure probing and toeprint experiments. Most importantly, *V. cholerae* strains carrying point mutations that strengthen the *toxT* thermometer structure and prevent its opening at 37°C were unable to colonize the mouse intestine and to produce CT. These findings provided compelling evidence for the relevance of the temperature-responsive RNA element in disease development.

Upon returning to the water reservoir through fecal contamination, the reversible nature of this zipper-like RNA structure is expected to diminish ToxT production. *V. cholerae* closes the transmission cycle by inducing a genetic program that depends on the signaling molecule cyclic-di-GMP and favors a sessile lifestyle and biofilm formation (20).

Yersinia Species—Temperature Sensing at Multiple Levels

The third example of an RNAT-controlled virulence transcription factor derives from *Yersinia* species and is best studied in *Yersinia pseudotuberculosis*. This foodborne pathogen infects humans and animals, causing enteritis and lymphadenitis. It is closely related to *Yersinia pestis*, the causative agent of bubonic and pneumonic plague. *Y. enterocolitica* is another human pathogen belonging to the *Yersinia* genus and infects the gastrointestinal tract, leading to enteritis and diarrhea.

Despite adopting diverse lifestyles and infection strategies, all three *Yersinia* species possess a similar virulence plasmid encoding the type 3 secretion system (T3SS) apparatus, *Yersinia* outer protein genes (*yop*), and regulatory factors (21, 22). Yops are either components of the translocation pore or effector proteins, which are injected into target eukaryotic cells via the T3SS. Yop effectors inhibit phagocytosis and the innate and adaptive immune responses by disrupting several signaling systems and interfering with the host cytoskeleton assembly (23).

Y. pseudotuberculosis is an excellent model system to study the correlation between temperature and virulence, as many of its genes are under temperature control. A transcriptomic study revealed that >300 genes are differentially regulated in response to temperature fluctuations typical of warm-blooded host infection (24). For most of these genes, the detailed regulatory mechanisms remain unknown.

The best-studied temperature-controlled gene is *lcrF* (*virF* in *Y. enterocolitica*), coding for a transcription factor responsible for the expression of the majority of the plasmid-encoded virulence factors (25). It has long been observed that the expression of *Yersinia* virulence factors is thermoregulated in response to a transition from a colder external environment to the host body temperature of 37°C (22). This regulation is mediated by LcrF, which is produced at high temperature due to a multilayered control cascade (Fig. 4) (26, 27).

The first level of temperature control acts on transcription of the *lcrF* gene, which is repressed by YmoA, a histone-like protein that binds to the *yscW-lcrF* promoter at low environmental temperature (11, 22). A temperature increase causes alteration of the DNA architecture, resulting in the liberation of the *yscW-lcrF* promoter and subsequent increase in transcription efficiency. Moreover, YmoA is stable at low temperature,

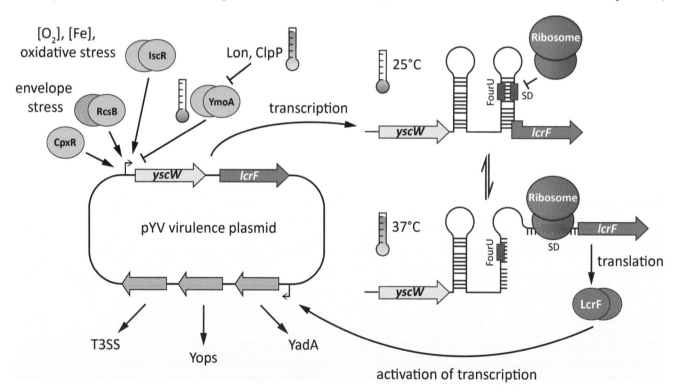

Figure 4 Environmental regulation of *lcrF* expression in *Yersinia* species. Multiple stimuli are integrated and influence LcrF synthesis on the transcriptional and translational level. Temperature affects both transcription and translation of *lcrF*, via the histone-like protein YmoA and the *cis*-encoded FourU RNAT, respectively.

while it is rapidly degraded by Clp and Lon proteases at 37°C (11, 28). In addition to temperature, other environmental signals are involved in the regulation of *lcrF* transcription (Fig. 4). These include oxygen stress via the IscR regulator, extracytoplasmic stress via the Rcs phosphorelay system, and the CpxA-CpxR two-component system (26, 27).

It was hypothesized in 1993 that the thermoregulation of *Y. pestis lcrF* occurs at the translational level due to an RNA structure sequestering the ribosome binding region (29). Only recently, experimental evidence was reported for a functional FourU-type RNAT upstream of *lcrF* in *Y. pseudotuberculosis* (11). In contrast to the RNATs discussed above, this thermosensor is not positioned in a 5′ UTR but located within the 124-nt-long intergenic region of the cotranscribed *yscW* and *lcrF* genes (Fig. 4). Its functionality is context independent, as it confers temperature regulation regardless of whether it is present in the full-length bicistronic transcript or as a short, isolated fragment. The *lcrF* thermometer consists of two hairpin structures. The second hairpin contains the FourU motif that occludes the SD region, thus impairing translation at 25°C. Strains of *Y. pseudotuberculosis* carrying stabilizing or destabilizing mutations in the RNAT that alter its thermosensing ability are either noninfectious or attenuated in a murine infection model, showing that a functional thermosensor is critically important for virulence.

Recently developed deep sequencing-based RNA structure-probing approaches enabled the identification of RNATs on a transcriptome-wide scale. By parallel analysis of RNA structures (PARS) the conformations of all detectable RNA species extracted from a bacterial culture can be mapped (30, 31). The analysis of conformational changes in response to temperature changes provided insights into the dynamic RNA structurome of *Y. pseudotuberculosis*. At least 16 novel RNATs have been discovered by PARS. They regulate genes for two major virulence factors, oxidative stress protection, and metabolic functions (32, 33). Several of these candidates are currently under investigation.

CONTROL OF IMMUNE EVASION AND EXTRACELLULAR FACTORS

Neisseria meningitidis—Three RNATs To Undermine the Host Immune Response

The capsular polysaccharide of the obligate human pathogen *Neisseria meningitidis* is the major virulence factor of the bacterium. A variety of at least 13 different capsular structures has so far been identified and is used for classification of meningococcal serogroups. Six serogroups are associated with outbreaks, namely, A, B, C, W-135, X, and Y (34, 35). Isolates from the nasopharynx of carriers often are unencapsulated, while *N. meningitidis* isolates found in the circulatory or central nervous system possess capsule (36). The carbohydrate layer protects the bacterium from complement-mediated killing; and antimicrobial peptides are less effective, thus enabling the bacteria to survive intracellularly (37, 38). On the contrary, overproduction of capsule in the nasopharynx could mask adhesins important for attaching to epithelial cells and consequently impede its colonization ability (39).

Production of this capsular polysaccharide involves multiple steps and is energy demanding. The *css* locus, coding for capsule production, is highly conserved among the same serogroup (40). An RNAT has been discovered that modulates the translation of proteins involved in capsular biosynthesis (Fig. 1B). When *Neisseria* resides in the nasopharynx, where the temperatures range from 32 to 34°C, the *cssA* RNAT adopts a closed structure that covers the RBS (41–43). As a consequence, less capsular polysaccharide is produced, and thus adhesins are more exposed, allowing better attachment to epithelial cells.

Upon dissemination into the circulatory system, the temperature rises to 37°C and the *cssA* RNAT stem-loop becomes more flexible, allowing ribosomes to bind, thus increasing the production of capsule (41). Through nuclear magnetic resonance (NMR) and SHAPE (selective 2′-hydroxyl acylation analyzed by primer extension) analyses, the *cssA*-RNAT structure was shown to start melting from top down at lower temperatures. The RBS, located at the bottom of the stem-loop, becomes partially accessible only at 37°C (44).

It is noteworthy that two tandem repeats of 8 nt in the 5′ UTR of the *cssA* RNAT are important for its functionality. Absence of one of these repeats leads to increased capsule biosynthesis even at the thermal conditions of the nasopharynx due to a weaker RNA secondary structure (41). Through gradual accessibility of the RBS, *N. meningitidis* fine-tunes capsule production in order to colonize and/or not detach from the epithelial cells but also to evade immune responses. Hence, the delicate and reversible RNAT control mechanism can be involved in virulence but also is important for colonization as a commensal.

Another RNAT-controlled immune evasive factor of *N. meningitidis* is the surface-located factor H binding protein (fHbp). Factor H is a negative regulator of the innate immune system's alternative complement pathway. Neisserial fHbp is able to bind to the host factor

H, thus supporting C3b inactivation and C3-convertase degradation, ultimately blocking membrane attack complex formation and lysis (45, 46). During an invasive infection, *N. meningitidis* is exposed to this host protection mechanism; thus, binding factor H enables the bacterium to survive in human blood (47, 48). The *fHbp* RNAT possesses two anti-RBSs within the coding region. In the slightly colder nasopharynx, the bacterium is not in direct contact with such defenses and therefore downregulates fHbp (49). In addition, fHbp is surface exposed and highly immunogenic; therefore, reducing its presentation during colonization is a necessary consequence to not provoke an immune cell response.

The production of the sialyltransferase Lst, which modifies the lipopolysaccharide molecules on the outer membrane of *N. meningitidis*, is also directly controlled by an RNAT (41). Sialylation of lipopolysaccharide molecules protects *N. meningitidis* from complement-mediated serum killing as well as phagocytic killing by neutrophils.

Infection with viral pathogens, such as influenza virus, could lead to local inflammation (temperature increases in the nasopharynx) and recruitment of immune effectors onto the surface. Temperature therefore acts as a warning signal to *N. meningitidis*, enhancing its defense against human immune killing. Thus, *N. meningitidis* poses as a unique example where the bacterium has three independently evolved RNATs to counteract immune killing. In addition, *N. meningitidis* is currently the only known obligate human pathogen, with no host other than humans or external habitat, to possess RNAT-mediated virulence gene regulation.

Leptospira interrogans—Two Identical RNATs To Colonize the Host

Leptospira interrogans is a Gram-negative, obligate aerobic spirochete bacterium that possesses periplasmic flagella. These spirochetes, as well as other pathogenic members of the genus *Leptospira*, are the causative agents of the globally endemic zoonotic infection termed leptospirosis. The most common mammalian reservoir of *L. interrogans* is the rodent, where it starts its life cycle within the renal tubules. Transmission usually occurs through the rodents' urine coming in contact with abraded human skin, eyes, or mouth. Human patients usually experience a sudden onset of fever as the bacteria disseminate within their circulatory system and replicate within the internal organs. This multifaceted pathogen encounters various environmental signals and has evolved sensing mechanisms to modulate its gene expression to facilitate its survival within hosts.

Such regulation is exemplified in the expression of surface-exposed lipoproteins allowing the bacterium to bind to various host proteins such as fibronectin. The *L. interrogans* Lig proteins encoded by *ligA* and *ligB* have previously been shown to facilitate colonization of the bacterium among components of the host extracellular matrix (50–54). In addition, both Lig proteins are able to directly bind human complement factor H and C4-binding protein (55), avoiding complement-mediated killing. Lig proteins were also shown to bind to other complement proteins such as C3b and C4b (56).

The expression of both *ligA* and *ligB* is temperature dependent and is controlled by two identical RNATs (57). These RNATs are predicted to form an extended secondary structure of two stem-loops, with the second loop occluding the RBS. Due to difficulties in the genetic manipulation of *L. interrogans*, all molecular studies on *lig* RNATs have so far been conducted in *E. coli*. The RNATs were shown to control expression in a temperature-dependent manner both *in vivo* and *in vitro*. Various point and deletion mutations within the 5′ UTR confirmed that the nucleotides predicted to be positioned opposite to the RBS are crucial for translation efficiency (57). In addition to responding to temperature, Lig protein production is upregulated during osmotic stress (58). However, the precise mechanisms of how the osmolarity and temperature signals are integrated and coordinated remain unknown.

QUORUM SENSING AND IRON ACQUISITION

Pseudomonas aeruginosa—RNATs Coordinate Lifestyle Decisions

Pseudomonas aeruginosa is an opportunistic pathogen that causes nosocomial infections, especially in immunocompromised patients. It is the primary cause of mortality in cystic fibrosis patients, where the bacterium establishes chronic lung infections and forms biofilms that are difficult to eradicate due to intrinsic antibiotic resistance (59). *P. aeruginosa* possesses a remarkably versatile lifestyle and is capable of proliferating in diverse environments such as soil, water, and animal and plant tissues.

Posttranscriptional temperature responses in *P. aeruginosa* rely on various ROSE (repression of heat shock gene expression)-like RNATs. In addition to the FourU motif, ROSE elements constitute a second class of moderately conserved RNATs (60). The NMR structure of the founding member of this family revealed a temperature-labile hairpin with several unusual base pairs that form a network of weak hydrogen bonds and thereby facilitate opening of the structure (61) (Fig. 1B). As in many other proteobacteria, the small heat shock gene *ibpA* is under the control of a ROSE element in *P. aeruginosa* (62).

Two additional ROSE-like RNATs in *P. aeruginosa* are responsible for regulating key components of the quorum sensing (QS) cascade (63). QS is a communication mechanism used by bacteria to sense the population density and adjust their lifestyle accordingly. *Pseudomonas* uses this strategy to modulate metabolism, biofilm formation, motility, competition with other commensal microorganisms, and, importantly, expression of several virulence factors, including pyocyanin, rhamnolipid biosurfactants, and the cytotoxic PA-IL lectin (64).

Three interconnected regulatory cascades govern the *P. aeruginosa* QS response, with RhlR acting as one of the master transcriptional regulators. The gene *rhlR* can be cotranscribed in a tricistronic operon, *rhlABR*, together with two genes involved in the biosynthesis of rhamnolipids. A ROSE element in the 5′ UTR of the first open reading frame, *rhlA*, controls the expression of *rhlR* through a polar effect on transcription elongation of the operon. At 30°C, the ROSE element inhibits translation of *rhlA*. As a consequence, transcription is prematurely terminated due to the formation of possible RNA structures that fold within the *rhlA* coding region. Consequently, the downstream *rhlB* and *rhlR* genes are not expressed. At 37°C, the ROSE structure melts, *rhlA* is translated, and transcription of the full-length tricistronic operon is enabled. Concomitantly, the *rhlA* 5′ UTR was identified as the target of the small RNA NrsZ (65). NrsZ is expressed under nitrogen limitation and binds to the anti-SD region of the *rhlA* 5′ UTR, liberating the RBS and thus promoting translation. The interplay between the RNAT-mediated thermoregulation of *rhlABR* and the small RNA action in response to nitrogen deprivation has not yet been investigated.

Another thermoregulated gene is *lasI*, coding for an enzyme involved in the synthesis of an important effector molecule that triggers the QS response (63). The RBS is partially occluded in the predicted structure of the *lasI* 5′ UTR in a ROSE-like conformation. The regulatory potential of the *lasI* RNAT was verified in reporter gene studies; however, the temperature-dependent accumulation of LasI only leads to a minor increase in the production of the effector molecules and expression of the associated virulence traits.

To identify additional RNATs in *P. aeruginosa*, a genetic screening approach was developed and led to the identification of four candidates that responded to a temperature shift from 28 to 37°C (66). One of the regulated genes codes for PtxS, a transcription factor that controls gluconate transport and usage; and another codes for PA5194, an ortholog of *E. coli* LpxT, which phosphorylates lipid A. The structure of the *ptxS* RNAT does not resemble any conserved class of RNATs, while PA5194 contains a partial FourU motif comprising three uridines that pair with the SD sequence. Interestingly, the predicted structure of the 5′ UTR of *E. coli lpxT* might also form a stem-loop that involves the ribosome binding region. The physiological role of the newly identified *P. aeruginosa* RNATs remains to be elucidated. Interestingly, PtxS has been shown to bind to PtxR, which regulates the expression of ToxA, an important virulence factor (67). Moreover, modifications of the lipopolysaccharide, such as the phosphorylation of lipid A, are known to impact its stability and endotoxicity in several pathogenic bacteria (68).

Shigella—Host Temperature Facilitates Iron Acquisition

Shigella species are Gram-negative, facultative anaerobic, pathogenic enterobacteria that cause dysentery in humans. Throughout the infection process, pathogens experience numerous environmental changes, among them iron limitation. Freely available iron is scarce in the human body because most of it is complexed by heme and other iron-binding proteins. To cope with this situation, *Shigella* species employ several iron acquisition systems to take up ferric iron (Fe^{3+}), heme, or ferrous iron (Fe^{2+}) (69). These systems are regulated at the transcriptional level by Fur, an iron-binding repressor protein. Since heme is an iron source that is typically encountered in the host but not in the environment, the expression of heme acquisition systems should ideally be coordinated with the presence in a mammalian host. This is in fact the case in *Shigella dysenteriae* and in enteropathogenic *E. coli* strains, where translation of *shuA* and *chuA*, respectively, is under dual control, namely, by the Fur regulator and an RNAT (70). ShuA and its ortholog ChuA are outer membrane receptors for heme. Toward the 3′ end of their ~300-nt-long 5′ UTR, the *shuA* and *chuA* transcripts are able to fold into a short and fairly simple hairpin structure that can be classified as FourU thermometer. It confers temperature-dependent translation control. The isolated RNAT region is sufficient for this regulation, and dysregulated expression by stabilization and destabilizing point mutations supported that it acts in a zipper-like manner. Whether other environmental signals are integrated by the rest of the long 5′ UTR remains an interesting open question.

The periplasmic binding protein ShuT is another component of the *Shigella* heme uptake system, and expression of the *shuT* gene is also under transcriptional control by the Fur regulator and under translational control by an RNAT (71). Interestingly, this RNAT has nothing in common with the *shuA* thermosensor. The full-length *shuT* 5′ UTR is only 42 nt long and folds into a single hairpin that differs from the *shuA* RNAT in both sequence and structure. The existence of two different RNATs in the same physiological pathway suggests that temperature-regulated acquisition of the host-specific iron source is fundamentally important and that it can easily be achieved by the evolution of RNA-based thermosensors.

OTHER VIRULENCE-RELATED FUNCTIONS

Salmonella—Heat Shock Thermometers Might Contribute to Virulence

Members of the genus *Salmonella* are motile, Gram-negative pathogenic bacteria and closely related to *E. coli*. The genus consists of two members, namely, *S. bongori* and *S. enterica*, which can be subdivided into >2,400 serovars, based on their antigenic presentation (72). Infection via ingestion of contaminated food or liquids causes enteric diseases. *S. enterica* serovars Typhi and Paratyphi are restricted to the human host and responsible for a systemic disease termed enteric (typhoid) fever. Interestingly, nontyphoidal *Salmonella* serovars, such as *S. enterica* serovar Typhimurium or *S. enterica* serovar Enteritidis, have a broad host range and are causative agents of enteritis in immunocompetent as well as bacteremia in immunocompromised individuals (73, 74). Consequently, infection strategies of nontyphoidal *Salmonella* and typhoidal *S. enterica* species are different for the most part, although *S.* Typhi and *S.* Typhimurium species share 89% of their genes (75, 76).

S. enterica is a facultative intracellular pathogen that is able to survive and replicate in phagocytic and nonphagocytic cells (77). For this purpose, *S. enterica* has evolved a huge arsenal of virulence factors. Among these factors are two T3SSs (T3SS1 and T3SS2) that encode transport machineries for injection of effector proteins (78, 79). Among the numerous virulence-related factors is the high temperature requirement A (HtrA or DegP) protease, which plays a key role in protein quality control and stress response through refolding or degrading misfolded proteins in the periplasm (80). Mutants that carry a null mutation in *htrA* (and *htrA* homologs) have been created in several bacterial spe-

cies and showed overlapping phenotypes, such as increased thermosensitivity or sensitivity to oxidative as well as osmotic stress (81). A deletion of *htrA* impairs growth of *S.* Typhimurium at higher temperatures (42°C), and additionally, the deletion strain is less viable in a mouse model compared to the wild type (82).

Translation of the *S.* Typhimurium *htrA* gene is controlled by a FourU RNAT that forms a single hairpin occluding the RBS at low temperatures (83). A recent SHAPE analysis systematically characterized the dynamics of the *htrA* thermometer at single-nucleotide resolution and confirmed the proposed hairpin structure, which unfolds in a cooperative fashion, with nucleotides from the upper and lower part of the stem gaining flexibility at a common transition temperature (84). Mutational analysis revealed three U-G base pairings, including two uridines of the FourU motif, to be essential for the thermometer functionality.

Several non-virulence-associated RNATs have also been identified and characterized in *Salmonella*. The small heat shock gene *agsA* (aggregation suppressing A) is regulated by the founding member of the FourU family (10) (Fig. 1B). The cooperative melting behavior of this RNAT has been studied at nucleotide resolution by NMR, and magnesium binding to the FourU motif was found to stabilize the structure (85, 86). A recent in-cell SHAPE analysis revealed that only a small fraction of the RNA population undergoes structural changes after temperature upshift, and that subtle changes in the RNA helix are sufficient to increase translation efficiency. An active contribution of the ribosome to the melting process was proposed (87).

An RNAT upstream of the *groES* gene confers differential temperature control to the *groESL* heat shock operon coding for a major molecular chaperone system (88). In addition to these structurally and functionally characterized thermometers, a biocomputational approach predicted various new RNATs in the genomes of 25 *S. enterica* isolates (89). The identified potential hairpin structures were further analyzed by secondary structure prediction in order to retrieve the melting temperature at the physiological ionic strength (89). If and how these diverse RNAT candidates, often upstream of genes with unknown function, are involved in virulence processes in *Salmonella* remains to be determined.

MUCH MORE TO COME?

The concept of RNAT-mediated translational control established almost 20 years ago (3, 90) has paved the way to the discovery of numerous heat shock and virulence genes under the control of such structured RNA elements (91). Challenges in the identification of new thermosensors are the poor conservation of RNAT sequences and structures as well as the existence of noncanonical base pairs in the temperature-responsive region, as was shown for the ROSE element (61). This makes genome-wide searches for temperature-regulated RNA structures difficult and often leads to false-positive candidates showing no temperature regulation *in vivo* (92). The implementation of algorithms able to predict temperature-modulated RNA structures such as RNA tips (temperature-induced perturbation of structure) or RNAthermsw (93, 94) has improved the reliability of computational approaches but nonetheless requires labor-intensive experimental validation of the candidates.

Fundamentally new strategies have recently been established to experimentally map temperature-sensitive RNA structures on a transcriptome-wide scale with nucleotide resolution. "RNA structuromics" approaches combine structure-probing protocols with the power of next-generation sequencing (95, 96). Typical PARS experiments have been performed at one given temperature, for example, in yeast (30) and *E. coli* (97). In this *in vitro* method, total RNA samples are partially digested *in vitro* with nucleases that have a preference for nucleotides in either paired or unpaired conformation. After adapter ligation, cDNA synthesis, and next-generation sequencing, the cleavage sites are mapped onto the transcriptome and PARS-guided structure calculations reveal the structure of thousands of transcripts. Extending the protocol to samples refolded at different temperatures allows the transcriptome-wide mapping of temperature-responsive RNA structures. Such PARTE (PARS with temperature elevation) experiments have been conducted with yeast (31) and *Y. pseudotuberculosis* (32). Thousands of temperature-sensitive RNA structures in yeast and hundreds in *Yersinia* promise the discovery of novel RNA thermosensors involved in the regulation of temperature-modulated processes. Although not all PARTE-derived RNAT candidates turned out to be true thermosensors *in vivo*, the success rate of this approach was ~75% (32). Given that only subtle changes in the dynamic RNAT structure can be sufficient to foster translation initiation, it is well possible that some temperature-responsive RNA structures will escape detection by global structure-probing approaches (87). Nonetheless, in *Y. pseudotuberculosis*, the PARTE approach revealed numerous functional RNATs upstream of bona fide virulence genes, as well as upstream of genes responsible for oxidative stress adaptation and various other metabolic pathways (32). These findings suggest that numerous cellular pathways in bacterial pathogens are under RNAT control to allow induction at host body temperature.

An interesting open question is whether RNAT-modulated virulence gene expression also occurs in eukaryotes such as fungal pathogens. Given the abundance of RNATs in bacterial pathogens (Fig. 2) and the presence of myriads of yeast mRNA structures that change their conformation with increasing temperature (31, 98), it might be rewarding to investigate the dynamic RNA structurome of microbial eukaryotic pathogens that transition between environmental temperatures and warm-blooded animals or humans.

In summary, the rapid temperature sensing through RNATs utilized by bacterial pathogens is an exquisite but simple concept. Due to the lack of sequence conservation, these RNA elements have independently evolved based on their functional secondary structures. The combination of structure probing and next-generation sequencing promises the discovery of more of these RNA elements in all kingdoms of life in the future.

Acknowledgments. E.L., F.R., and H.E. were supported by the Knut and Alice Wallenberg Foundation, Karolinska Institutet, Swedish Research Council grant 2014–2050, and the Swedish Foundation for Strategic Research. F.N. is supported by the German Research Foundation (NA 240/10-1).

Citation. Loh E, Righetti F, Eichner H, Twittenhoff C, Narberhaus F. 2018. RNA thermometers in bacterial pathogens. Microbiol Spectrum 6(2):RWR-0012-2017.

References

1. **Klinkert B, Narberhaus F.** 2009. Microbial thermosensors. *Cell Mol Life Sci* **66:**2661–2676.

2. **Roncarati D, Scarlato V.** 2017. Regulation of heat-shock genes in bacteria: from signal sensing to gene expression output. *FEMS Microbiol Rev* **41:**549–574.

3. **Morita MT, Tanaka Y, Kodama TS, Kyogoku Y, Yanagi H, Yura T.** 1999. Translational induction of heat shock transcription factor σ³²: evidence for a built-in RNA thermosensor. *Genes Dev* **13:**655–665.

4. **Nocker A, Hausherr T, Balsiger S, Krstulovic NP, Hennecke H, Narberhaus F.** 2001. A mRNA-based thermosensor controls expression of rhizobial heat shock genes. *Nucleic Acids Res* **29:**4800–4807.

5. **Johansson J, Mandin P, Renzoni A, Chiaruttini C, Springer M, Cossart P.** 2002. An RNA thermosensor controls expression of virulence genes in *Listeria monocytogenes*. *Cell* **110:**551–561.

6. **Grosso-Becera MV, Servín-González L, Soberón-Chávez G.** 2015. RNA structures are involved in the thermoregulation of bacterial virulence-associated traits. *Trends Microbiol* **23:**509–518.

7. **Ignatov D, Johansson J.** 2017. RNA-mediated signal perception in pathogenic bacteria. *Wiley Interdiscip Rev RNA* **8:**e1429.

8. Kortmann J, Narberhaus F. 2012. Bacterial RNA thermometers: molecular zippers and switches. *Nat Rev Microbiol* **10**:255–265.

9. Krajewski SS, Narberhaus F. 2014. Temperature-driven differential gene expression by RNA thermosensors. *Biochim Biophys Acta* **1839**:978–988.

10. Waldminghaus T, Heidrich N, Brantl S, Narberhaus F. 2007. FourU: a novel type of RNA thermometer in *Salmonella*. *Mol Microbiol* **65**:413–424.

11. Böhme K, Steinmann R, Kortmann J, Seekircher S, Heroven AK, Berger E, Pisano F, Thiermann T, Wolf-Watz H, Narberhaus F, Dersch P. 2012. Concerted actions of a thermo-labile regulator and a unique intergenic RNA thermosensor control *Yersinia* virulence. *PLoS Pathog* **8**:e1002518.

12. Weber GG, Kortmann J, Narberhaus F, Klose KE. 2014. RNA thermometer controls temperature-dependent virulence factor expression in *Vibrio cholerae*. *Proc Natl Acad Sci U S A* **111**:14241–14246.

13. Reniere ML, Whiteley AT, Hamilton KL, John SM, Lauer P, Brennan RG, Portnoy DA. 2015. Glutathione activates virulence gene expression of an intracellular pathogen. *Nature* **517**:170–173.s

14. Hall M, Grundström C, Begum A, Lindberg MJ, Sauer UH, Almqvist F, Johansson J, Sauer-Eriksson AE. 2016. Structural basis for glutathione-mediated activation of the virulence regulatory protein PrfA in *Listeria*. *Proc Natl Acad Sci U S A* **113**:14733–14738.

15. de las Heras A, Cain RJ, Bielecka MK, Vázquez-Boland JA. 2011. Regulation of *Listeria* virulence: PrfA master and commander. *Curr Opin Microbiol* **14**:118–127.

16. Loh E, Memarpour F, Vaitkevicius K, Kallipolitis BH, Johansson J, Sondén B. 2012. An unstructured 5′-coding region of the *prfA* mRNA is required for efficient translation. *Nucleic Acids Res* **40**:1818–1827.

17. Loh E, Dussurget O, Gripenland J, Vaitkevicius K, Tiensuu T, Mandin P, Repoila F, Buchrieser C, Cossart P, Johansson J. 2009. A *trans*-acting riboswitch controls expression of the virulence regulator PrfA in *Listeria monocytogenes*. *Cell* **139**:770–779.

18. Conner JG, Teschler JK, Jones CJ, Yildiz FH. 2016. Staying alive: *Vibrio cholerae*'s cycle of environmental survival, transmission, and dissemination. *Microbiol Spectr* **4**:VMBF-0015-2015.

19. DiRita VJ, Parsot C, Jander G, Mekalanos JJ. 1991. Regulatory cascade controls virulence in *Vibrio cholerae*. *Proc Natl Acad Sci U S A* **88**:5403–5407.

20. Townsley L, Yildiz FH. 2015. Temperature affects c-di-GMP signalling and biofilm formation in *Vibrio cholerae*. *Environ Microbiol* **17**:4290–4305.

21. Wren BW. 2003. The yersiniae—a model genus to study the rapid evolution of bacterial pathogens. *Nat Rev Microbiol* **1**:55–64.

22. Cornelis G, Sluiters C, de Rouvroit CL, Michiels T. 1989. Homology between *virF*, the transcriptional activator of the *Yersinia* virulence regulon, and AraC, the *Escherichia coli* arabinose operon regulator. *J Bacteriol* **171**:254–262.

23. Chung LK, Bliska JB. 2016. *Yersinia* versus host immunity: how a pathogen evades or triggers a protective response. *Curr Opin Microbiol* **29**:56–62.

24. Nuss AM, Heroven AK, Waldmann B, Reinkensmeier J, Jarek M, Beckstette M, Dersch P. 2015. Transcriptomic profiling of *Yersinia pseudotuberculosis* reveals reprogramming of the Crp regulon by temperature and uncovers Crp as a master regulator of small RNAs. *PLoS Genet* **11**:e1005087.

25. Wattiau P, Cornelis GR. 1994. Identification of DNA sequences recognized by VirF, the transcriptional activator of the *Yersinia yop* regulon. *J Bacteriol* **176**:3878–3884.

26. Schwiesow L, Lam H, Dersch P, Auerbuch V. 2015. *Yersinia* type III secretion system master regulator LcrF. *J Bacteriol* **198**:604–614.

27. Chen S, Thompson KM, Francis MS. 2016. Environmental regulation of *Yersinia* pathophysiology. *Front Cell Infect Microbiol* **6**:25.

28. Jackson MW, Silva-Herzog E, Plano GV. 2004. The ATP-dependent ClpXP and Lon proteases regulate expression of the *Yersinia pestis* type III secretion system via regulated proteolysis of YmoA, a small histone-like protein. *Mol Microbiol* **54**:1364–1378.

29. Hoe NP, Goguen JD. 1993. Temperature sensing in *Yersinia pestis*: translation of the LcrF activator protein is thermally regulated. *J Bacteriol* **175**:7901–7909.

30. Kertesz M, Wan Y, Mazor E, Rinn JL, Nutter RC, Chang HY, Segal E. 2010. Genome-wide measurement of RNA secondary structure in yeast. *Nature* **467**:103–107.

31. Wan Y, Qu K, Ouyang Z, Kertesz M, Li J, Tibshirani R, Makino DL, Nutter RC, Segal E, Chang HY. 2012. Genome-wide measurement of RNA folding energies. *Mol Cell* **48**:169–181.

32. Righetti F, Nuss AM, Twittenhoff C, Beele S, Urban K, Will S, Bernhart SH, Stadler PF, Dersch P, Narberhaus F. 2016. Temperature-responsive *in vitro* RNA structurome of *Yersinia pseudotuberculosis*. *Proc Natl Acad Sci U S A* **113**:7237–7242.

33. Nuss AM, Heroven AK, Dersch P. 2017. RNA regulators: formidable modulators of *Yersinia* virulence. *Trends Microbiol* **25**:19–34.

34. Jafri RZ, Ali A, Messonnier NE, Tevi-Benissan C, Durrheim D, Eskola J, Fermon F, Klugman KP, Ramsay M, Sow S, Zhujun S, Bhutta ZA, Abramson J. 2013. Global epidemiology of invasive meningococcal disease. *Popul Health Metr* **11**:17.

35. Tzeng YL, Thomas J, Stephens DS. 2016. Regulation of capsule in *Neisseria meningitidis*. *Crit Rev Microbiol* **42**:759–772.

36. Harrison LH, Trotter CL, Ramsay ME. 2009. Global epidemiology of meningococcal disease. *Vaccine* **27**(Suppl 2):B51–B63.

37. Jarvis GA, Vedros NA. 1987. Sialic acid of group B *Neisseria meningitidis* regulates alternative complement pathway activation. *Infect Immun* **55**:174–180.

38. Spinosa MR, Progida C, Talà A, Cogli L, Alifano P, Bucci C. 2007. The *Neisseria meningitidis* capsule is

important for intracellular survival in human cells. *Infect Immun* 75:3594–3603.

39. Jones CH, Mohamed N, Rojas E, Andrew L, Hoyos J, Hawkins JC, McNeil LK, Jiang Q, Mayer LW, Wang X, Gilca R, De Wals P, Pedneault L, Eiden J, Jansen KU, Anderson AS. 2016. Comparison of phenotypic and genotypic approaches to capsule typing of *Neisseria meningitidis* by use of invasive and carriage isolate collections. *J Clin Microbiol* 54:25–34.

40. Harrison OB, Claus H, Jiang Y, Bennett JS, Bratcher HB, Jolley KA, Corton C, Care R, Poolman JT, Zollinger WD, Frasch CE, Stephens DS, Feavers I, Frosch M, Parkhill J, Vogel U, Quail MA, Bentley SD, Maiden MC. 2013. Description and nomenclature of *Neisseria meningitidis* capsule locus. *Emerg Infect Dis* 19:566–573.

41. Loh E, Kugelberg E, Tracy A, Zhang Q, Gollan B, Ewles H, Chalmers R, Pelicic V, Tang CM. 2013. Temperature triggers immune evasion by *Neisseria meningitidis*. *Nature* 502:237–240.

42. Abbas AK, Heimann K, Jergus K, Orlikowsky T, Leonhardt S. 2011. Neonatal non-contact respiratory monitoring based on real-time infrared thermography. *Biomed Eng Online* 10:93.

43. Keck T, Leiacker R, Schick M, Rettinger G, Kühnemann S. 2000. Temperature and humidity profile of the paranasal sinuses before and after mucosal decongestion by xylometazolin. *Laryngorhinootologie* 79:749–752. (In German.)

44. Barnwal RP, Loh E, Godin KS, Yip J, Lavender H, Tang CM, Varani G. 2016. Structure and mechanism of a molecular rheostat, an RNA thermometer that modulates immune evasion by *Neisseria meningitidis*. *Nucleic Acids Res* 44:9426–9437.

45. Ferreira VP, Herbert AP, Hocking HG, Barlow PN, Pangburn MK. 2006. Critical role of the C-terminal domains of factor H in regulating complement activation at cell surfaces. *J Immunol* 177:6308–6316.

46. Józsi M, Oppermann M, Lambris JD, Zipfel PF. 2007. The C-terminus of complement factor H is essential for host cell protection. *Mol Immunol* 44:2697–2706.

47. Schneider MC, Exley RM, Chan H, Feavers I, Kang YH, Sim RB, Tang CM. 2006. Functional significance of factor H binding to *Neisseria meningitidis*. *J Immunol* 176:7566–7575.

48. Seib KL, Serruto D, Oriente F, Delany I, Adu-Bobie J, Veggi D, Aricò B, Rappuoli R, Pizza M. 2009. Factor H-binding protein is important for meningococcal survival in human whole blood and serum and in the presence of the antimicrobial peptide LL-37. *Infect Immun* 77:292–299.

49. Loh E, Lavender H, Tan F, Tracy A, Tang CM. 2016. Thermoregulation of meningococcal fHbp, an important virulence factor and vaccine antigen, is mediated by anti-ribosomal binding site sequences in the open reading frame. *PLoS Pathog* 12:e1005794.

50. Choy HA, Kelley MM, Chen TL, Møller AK, Matsunaga J, Haake DA. 2007. Physiological osmotic induction of *Leptospira interrogans* adhesion: LigA and LigB bind extracellular matrix proteins and fibrinogen. *Infect Immun* 75:2441–2450.

51. Lin YP, McDonough SP, Sharma Y, Chang YF. 2010. The terminal immunoglobulin-like repeats of LigA and LigB of *Leptospira* enhance their binding to gelatin binding domain of fibronectin and host cells. *PLoS One* 5:e11301.

52. Lin YP, Greenwood A, Nicholson LK, Sharma Y, McDonough SP, Chang YF. 2009. Fibronectin binds to and induces conformational change in a disordered region of leptospiral immunoglobulin-like protein B. *J Biol Chem* 284:23547–23557.

53. Lin YP, Lee DW, McDonough SP, Nicholson LK, Sharma Y, Chang YF. 2009. Repeated domains of *Leptospira* immunoglobulin-like proteins interact with elastin and tropoelastin. *J Biol Chem* 284:19380–19391.

54. Choy HA, Kelley MM, Croda J, Matsunaga J, Babbitt JT, Ko AI, Picardeau M, Haake DA. 2011. The multifunctional LigB adhesin binds homeostatic proteins with potential roles in cutaneous infection by pathogenic *Leptospira interrogans*. *PLoS One* 6:e16879.

55. Castiblanco-Valencia MM, Fraga TR, Silva LB, Monaris D, Abreu PA, Strobel S, Józsi M, Isaac L, Barbosa AS. 2012. Leptospiral immunoglobulin-like proteins interact with human complement regulators factor H, FHL-1, FHR-1, and C4BP. *J Infect Dis* 205:995–1004.

56. Choy HA. 2012. Multiple activities of LigB potentiate virulence of *Leptospira interrogans*: inhibition of alternative and classical pathways of complement. *PLoS One* 7:e41566.

57. Matsunaga J, Schlax PJ, Haake DA. 2013. Role for *cis*-acting RNA sequences in the temperature-dependent expression of the multiadhesive Lig proteins in *Leptospira interrogans*. *J Bacteriol* 195:5092–5101.

58. Matsunaga J, Medeiros MA, Sanchez Y, Werneid KF, Ko AI. 2007. Osmotic regulation of expression of two extracellular matrix-binding proteins and a haemolysin of *Leptospira interrogans*: differential effects on LigA and Sph2 extracellular release. *Microbiology* 153:3390–3398.

59. Davis PB. 2001. Cystic fibrosis. *Pediatr Rev* 22:257–264.

60. Waldminghaus T, Fippinger A, Alfsmann J, Narberhaus F. 2005. RNA thermometers are common in α- and γ-proteobacteria. *Biol Chem* 386:1279–1286.

61. Chowdhury S, Maris C, Allain FH, Narberhaus F. 2006. Molecular basis for temperature sensing by an RNA thermometer. *EMBO J* 25:2487–2497.

62. Krajewski SS, Nagel M, Narberhaus F. 2013. Short ROSE-like RNA thermometers control IbpA synthesis in *Pseudomonas* species. *PLoS One* 8:e65168.

63. Grosso-Becerra MV, Croda-García G, Merino E, Servín-González L, Mojica-Espinosa R, Soberón-Chávez G. 2014. Regulation of *Pseudomonas aeruginosa* virulence factors by two novel RNA thermometers. *Proc Natl Acad Sci U S A* 111:15562–15567.

64. Winstanley C, Fothergill JL. 2009. The role of quorum sensing in chronic cystic fibrosis *Pseudomonas aeruginosa* infections. *FEMS Microbiol Lett* 290:1–9.

65. Wenner N, Maes A, Cotado-Sampayo M, Lapouge K. 2014. NrsZ: a novel, processed, nitrogen-dependent, small

non-coding RNA that regulates *Pseudomonas aeruginosa* PAO1 virulence. *Environ Microbiol* **16**:1053–1068.

66. Delvillani F, Sciandrone B, Peano C, Petiti L, Berens C, Georgi C, Ferrara S, Bertoni G, Pasini ME, Dehò G, Briani F. 2014. Tet-Trap, a genetic approach to the identification of bacterial RNA thermometers: application to *Pseudomonas aeruginosa*. *RNA* **20**:1963–1976.

67. Daddaoua A, Fillet S, Fernández M, Udaondo Z, Krell T, Ramos JL. 2012. Genes for carbon metabolism and the ToxA virulence factor in *Pseudomonas aeruginosa* are regulated through molecular interactions of PtxR and PtxS. *PLoS One* **7**:e39390.

68. Needham BD, Trent MS. 2013. Fortifying the barrier: the impact of lipid A remodelling on bacterial pathogenesis. *Nat Rev Microbiol* **11**:467–481.

69. Wei Y, Murphy ER. 2016. *Shigella* iron acquisition systems and their regulation. *Front Cell Infect Microbiol* **6**:18.

70. Kouse AB, Righetti F, Kortmann J, Narberhaus F, Murphy ER. 2013. RNA-mediated thermoregulation of iron-acquisition genes in *Shigella dysenteriae* and pathogenic *Escherichia coli*. *PLoS One* **8**:e63781.

71. Wei Y, Kouse AB, Murphy ER. 2017. Transcriptional and posttranscriptional regulation of *Shigella shuT* in response to host-associated iron availability and temperature. *MicrobiologyOpen* **6**:e00442.

72. Brenner FW, Villar RG, Angulo FJ, Tauxe R, Swaminathan B. 2000. *Salmonella* nomenclature. *J Clin Microbiol* **38**:2465–2467.

73. Keestra-Gounder AM, Tsolis RM, Bäumler AJ. 2015. Now you see me, now you don't: the interaction of *Salmonella* with innate immune receptors. *Nat Rev Microbiol* **13**:206–216.

74. LaRock DL, Chaudhary A, Miller SI. 2015. Salmonellae interactions with host processes. *Nat Rev Microbiol* **13**:191–205.

75. McClelland M, Sanderson KE, Spieth J, Clifton SW, Latreille P, Courtney L, Porwollik S, Ali J, Dante M, Du F, Hou S, Layman D, Leonard S, Nguyen C, Scott K, Holmes A, Grewal N, Mulvaney E, Ryan E, Sun H, Florea L, Miller W, Stoneking T, Nhan M, Waterston R, Wilson RK. 2001. Complete genome sequence of *Salmonella enterica* serovar Typhimurium LT2. *Nature* **413**:852–856.

76. Parkhill J, Dougan G, James KD, Thomson NR, Pickard D, Wain J, Churcher C, Mungall KL, Bentley SD, Holden MT, Sebaihia M, Baker S, Basham D, Brooks K, Chillingworth T, Connerton P, Cronin A, Davis P, Davies RM, Dowd L, White N, Farrar J, Feltwell T, Hamlin N, Haque A, Hien TT, Holroyd S, Jagels K, Krogh A, Larsen TS, Leather S, Moule S, O'Gaora P, Parry C, Quail M, Rutherford K, Simmonds M, Skelton J, Stevens K, Whitehead S, Barrell BG. 2001. Complete genome sequence of a multiple drug resistant *Salmonella enterica* serovar Typhi CT18. *Nature* **413**:848–852.

77. de Jong HK, Parry CM, van der Poll T, Wiersinga WJ. 2012. Host-pathogen interaction in invasive Salmonellosis. *PLoS Pathog* **8**:e1002933.

78. Galán JE. 1999. Interaction of *Salmonella* with host cells through the centisome 63 type III secretion system. *Curr Opin Microbiol* **2**:46–50.

79. Galán JE, Curtiss R III. 1989. Cloning and molecular characterization of genes whose products allow *Salmonella typhimurium* to penetrate tissue culture cells. *Proc Natl Acad Sci U S A* **86**:6383–6387.

80. Clausen T, Kaiser M, Huber R, Ehrmann M. 2011. HTRA proteases: regulated proteolysis in protein quality control. *Nat Rev Mol Cell Biol* **12**:152–162.

81. Pallen MJ, Wren BW. 1997. The HtrA family of serine proteases. *Mol Microbiol* **26**:209–221.

82. Mo E, Peters SE, Willers C, Maskell DJ, Charles IG. 2006. Single, double and triple mutants of *Salmonella enterica* serovar Typhimurium *degP* (*htrA*), *degQ* (*hhoA*) and *degS* (*hhoB*) have diverse phenotypes on exposure to elevated temperature and their growth in vivo is attenuated to different extents. *Microb Pathog* **41**:174–182.

83. Klinkert B, Cimdins A, Gaubig LC, Roßmanith J, Aschke-Sonnenborn U, Narberhaus F. 2012. Thermogenetic tools to monitor temperature-dependent gene expression in bacteria. *J Biotechnol* **160**:55–63.

84. Choi EK, Ulanowicz KA, Nguyen YA, Frandsen JK, Mitton-Fry RM. 2017. SHAPE analysis of the *htrA* RNA thermometer from *Salmonella enterica*. *RNA* **23**:1569–1581.

85. Rinnenthal J, Klinkert B, Narberhaus F, Schwalbe H. 2010. Direct observation of the temperature-induced melting process of the *Salmonella* fourU RNA thermometer at base-pair resolution. *Nucleic Acids Res* **38**:3834–3847.

86. Rinnenthal J, Klinkert B, Narberhaus F, Schwalbe H. 2011. Modulation of the stability of the *Salmonella* fourU-type RNA thermometer. *Nucleic Acids Res* **39**:8258–8270.

87. Meyer S, Carlson PD, Lucks JB. 2017. Characterizing the structure-function relationship of a naturally occurring RNA thermometer. *Biochemistry* **56**:6629–6638.

88. Cimdins A, Roßmanith J, Langklotz S, Bandow JE, Narberhaus F. 2013. Differential control of *Salmonella* heat shock operons by structured mRNAs. *Mol Microbiol* **89**:715–731.

89. Limanskaia OI, Murtazaeva LA, Limanskiĭ AP. 2013. Search for new potential RNA thermometers in the *Salmonella enterica* genome. *Mikrobiologiia* **82**:69–78. (In Russian.)

90. Storz G. 1999. An RNA thermometer. *Genes Dev* **13**:633–636.

91. Righetti F, Narberhaus F. 2014. How to find RNA thermometers. *Front Cell Infect Microbiol* **4**:132.

92. Waldminghaus T, Gaubig LC, Narberhaus F. 2007. Genome-wide bioinformatic prediction and experimental evaluation of potential RNA thermometers. *Mol Genet Genomics* **278**:555–564.

93. Chursov A, Kopetzky SJ, Bocharov G, Frishman D, Shneider A. 2013. RNAtips: analysis of temperature-induced changes of RNA secondary structure. *Nucleic Acids Res* **41**(Web Server Issue):W486–W491.

94. Churkin A, Avihoo A, Shapira M, Barash D. 2014. RNA thermsw: direct temperature simulations for predicting the location of RNA thermometers. *PLoS One* **9**:e94340.

95. Wan Y, Kertesz M, Spitale RC, Segal E, Chang HY. 2011. Understanding the transcriptome through RNA structure. *Nat Rev Genet* **12**:641–655.

96. Kwok CK, Tang Y, Assmann SM, Bevilacqua PC. 2015. The RNA structurome: transcriptome-wide structure probing with next-generation sequencing. *Trends Biochem Sci* **40**:221–232.

97. Del Campo C, Bartholomäus A, Fedyunin I, Ignatova Z. 2015. Secondary structure across the bacterial transcriptome reveals versatile roles in mRNA regulation and function. *PLoS Genet* **11**:e1005613.

98. Qi F, Frishman D. 2017. Melting temperature highlights functionally important RNA structure and sequence elements in yeast mRNA coding regions. *Nucleic Acids Res* **45**:6109–6118.

Regulating with RNA in Bacteria and Archaea
Edited by Gisela Storz and Kai Papenfort
© 2018 American Society for Microbiology, Washington, DC
doi:10.1128/microbiolspec.RWR-0025-2018

Small-Molecule-Binding Riboswitches

5

Thea S. Lotz[1] and Beatrix Suess[1]

INTRODUCTION

Traditionally, the functional role of RNA was thought to be restricted to transferring genetic information from DNA to protein. However, the discovery of RNA elements mediating gene control, chemical reaction catalysis, and signal transduction has changed this perception fundamentally. Its ability to form complex three-dimensional structures that precisely present chemical moieties is imperative in enabling RNA to function as a biological catalyst, regulator, or structural scaffold.

In the last 15 years, many different types of regulatory RNA molecules have been discovered in nature. An abundance of riboswitches, ribozymes, RNA thermometers, and short and long noncoding RNAs have been found in all three domains of life. However, plenty of questions regarding how RNA influences cell physiology, differentiation, and development still remain unanswered and the subject of extensive research (1). There are several advantages when gene expression is regulated via RNA alone as opposed to regulation in combination with proteins. It allows (i) faster regulatory responses, (ii) easier transfer of a single-step genetic control element to other organisms, and (iii) flexible combination with different downstream readout platforms for a maximum of regulatory outputs (2).

Riboswitches are protein-independent, RNA-based gene regulatory elements—short, structured RNA elements able to regulate gene expression in response to binding a small-molecule ligand. Their use allows temporal and dosage control over gene expression. Following the discovery and validation of the first riboswitch in 2002 (3), they have been the subject of intense research. Their mechanism of action and importance for their host organism have been explored (4–6), along with the generation of new, synthetic riboswitches by genetic engineers (7, 8).

In this review, we have compiled information on all the natural riboswitches discovered so far and discuss their mechanisms, occurrence, and potential to develop genetic control elements for genetic analyses and synthetic biology in the future.

RIBOSWITCHES—LOCATION, MECHANISM, AND DISTRIBUTION

Riboswitches are highly structured RNA sequence elements typically located in the 5′ untranslated region (5′ UTR) of many bacterial mRNAs that control a plethora of metabolic processes. They act as molecular switches regulating gene expression via conformational changes in their three-dimensional structure upon direct interaction ("binding") with a specific small molecule, typically a metabolite. They consist of two domains, the so-called aptamer domain and the expression platform. The aptamer domain selectively recognizes its small-molecule ligand while at the same time discriminating against closely related variants of the ligand. Binding of the ligand to the aptamer domain leads to structural changes in the following expression platform, inducing altered expression of the downstream mRNA. The changes in gene expression typically include regulating transcription termination or translation initiation (Fig. 1), more rarely, ribozyme-mediated mRNA degradation or the control of splicing.

Riboswitches are widely distributed throughout the bacterial world and are increasingly found in eukaryotes and archaea. They are divided into so-called classes, groups of riboswitches that respond to the same ligand and that show a similar conserved core structure. There can be more than one class of riboswitches binding to the same ligand. However, the classes then have distinctly different core structures. For example, among the prequeuosine-1 (preQ$_1$) binding ribo-

[1]Synthetic Genetic Circuits, Department of Biology, TU Darmstadt, 64287 Darmstadt, Germany.

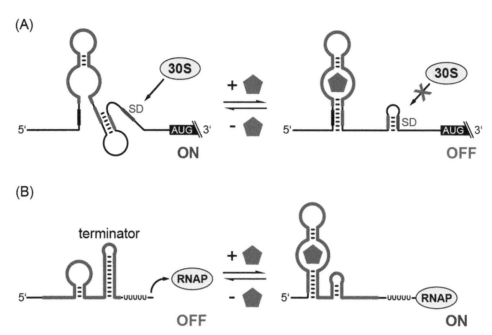

Figure 1 Common mechanism of riboswitches in bacteria. (A) Regulation of translation initiation: In the absence of the ligand, a stem-loop structure is formed between the aptamer domain and a sequence complementary to the Shine-Dalgarno (SD) sequence. Thus, the SD sequence is accessible for 30S binding, and translation initiation occurs. As a consequence of ligand binding (pentagon) and the folding of the aptamer domain, an alternative stem-loop is formed, which sequesters the SD sequence, and the binding of the 30S ribosomal subunit is blocked. (B) Regulation of transcription termination: The aptamer domain is followed by a sequence complementary to the 3′ part of the aptamer and a U stretch. In the absence of the ligand, the complementary 3′ part is base-paired with the aptamer, forming a terminator structure. Thus, RNA polymerase (RNAP) dissociates and transcription is blocked. Upon ligand binding, terminator structure formation is inhibited and transcription can proceed, resulting in expression of the reporter gene.

switches, there are major differences between classes, as, e.g., preQ$_1$-I riboswitches are small (25 to 45 nucleotides long) whereas preQ$_1$-II have a larger, more complex core consensus sequence (average, 58 nucleotides). Between the two classes, there are no detectable structure or sequence similarities (9, 10). In the meantime, close to 40 riboswitch classes have been identified and many more await discovery. The variety of ligands ranges from vitamins, amino acids, and nucleotides to second messengers and a number of ions.

Several excellent reviews about riboswitches have been published in recent years (4, 11–13). For a comprehensive overview, we have compiled a table (Table 1) summarizing all riboswitch classes described to date. In addition to the recognized ligands, we list the commonly regulated genes and the main organisms in which the riboswitches occur. In addition, we include the year of discovery of the ligand-riboswitch pair and the first description of a three-dimensional structure if available. Albeit many RNA motifs have been justly

suspected to act as riboswitches since their discovery (14, 15), some "orphan riboswitches" have not yet been assigned a ligand. For this reason, the initial reference column of riboswitches in Table 1 lists the first publication in which a riboswitch could be successfully linked to its specific ligand.

Despite the fact that riboswitches are widely distributed, some riboswitch classes are much more prevalent than others. The thiamine pyrophosphate (TPP) riboswitch is the most widely distributed class and the only one known to date to occur in all three domains of life: *Bacteria* (16, 17), *Archaea* (18), and *Eukarya* (fungi [19, 20], plants [19, 21–24], and algae [25]). Cobalamin riboswitches (AdoCbl) belong to another widely distributed riboswitch class and are located in the 5′ UTRs of vitamin B$_{12}$-related genes across several strains of bacteria. In contrast, there are riboswitch classes that are extremely rare, corresponding to either recently evolved pathways or such that are on the way to extinction (12). For instance, only four examples of the

Table 1   Naturally occurring riboswitches

Riboswitch	Ligand	Ligand group	Regulated gene(s) or pathways[a]	Distribution; organism in which described first[b]	Year first described	Reference(s)	Reference(s) (3D)[c]	PDB ID
AdoCbl	Adenosylcobalamin (vitamin B$_{12}$)	Coenzyme	Transport and biosynthesis of cobalamin or similar metabolites; btuB and cob operon	Extremely widely distributed; Escherichia coli	2002	28	86	4GMA
TPP	Thiamine pyrophosphate (vitamin B$_1$)	Coenzyme	Transport and biosynthesis of thiamine; thi operon	Extremely widely distributed; E. coli, Bacillus subtilis, Neurospora crassa, Arabidopsis thaliana, and Volvox carteri	2002	16, 17, 25	30–32	2HOJ, 2GDI, 2CKY
FMN	Flavin mononucleotide (vitamin B$_2$)	Coenzyme	Biosynthesis and transport of riboflavin; rfn operon	Extremely widely distributed; B. subtilis	2002	3	87, 88	3F2Q, 2YIF
Purine	Adenine, Guanine	Nucleotide derivative	Purine metabolism; xpt-pbuX and ydhL	Bacilli, clostridia, etc.; B. subtilis	2003	52	89	2EES
Lysine	L-Lysine	Amino acid	Lysin synthesis, catabolism, and transport; lysC	Widely distributed; B. subtilis	2003	22	90, 91	3DOU, 3DIL
SAM-I	S-Adenosylmethionine (SAM)	Coenzyme	Sulfur metabolism (biosynthesis of cysteine, methionine, and SAM)	Extremely widely distributed; B. subtilis	2003	92	54, 93	2GIS, 3IQN
AqCbl	Aquacobalamin	Coenzyme	Cobalamin biosynthesis; cob operon	Widely distributed; Salmonella enterica serovar Typhimurium	2004	94	86	4FRN
GlmS	Glucosamine-6-phosphate	Others	glmS	Widely distributed; B. subtilis	2004	36	95, 96	2NZ4, 2Z75
Glycine	Glycine	Amino acid	Glycine catabolism; gcvT operon	Widely distributed; B. subtilis	2004	97	98	3OWW
SAM-II	SAM	Coenzyme	Methionine biosynthesis; metA and metC	Widely distributed; Agrobacterium tumefaciens	2005	99	100	2QWY
SAM-III	SAM	Coenzyme	SAM synthetase; metK	Enterococcus, Streptococcus, and Lactococcus species	2006	101	102	3E5C
Mg^{2+}-II	Magnesium ions, proline	Ion	mgtA and mgtL	Gammaproteobacteria; S. enterica	2006	103, 104		
Mg^{2+}-I	Magnesium ions	Ion	Magnesium homeostasis	Widely distributed; B. subtilis	2007	105	106	3PDR
2′-dG-I (mfl motif)	2′-Deoxyguanosine	Nucleotide derivative	Subunit β of ribonucleotide reductase	Mesoplasma florum	2007	26	107	3SKI
PreQ$_1$-I type 1–3	Prequeuosine-1	Nucleotide derivative	Queuosine biosynthesis and transport; queCDEF operon and queT	Widely distributed; B. subtilis, E. coli, and Shigella dysenteriae	2007	9, 108–110	111	3Q50

(Continued)

Table 1 (*Continued*)

Riboswitch	Ligand	Ligand group	Regulated gene(s) or pathways[a]	Distribution; organism[b] in which described first	Year first described	Reference(s)	Reference(s) (3D)[c]	PDB ID
PreQ₁-II	Prequeuosine-1	Nucleotide derivative	Queuosine biosynthesis; COG4708 (proposed transporter)	Bacilli and clostridia; *Streptococcus pneumoniae*	2007	9, 10		
Moco	Molybdenum cofactor	Coenzyme	Molybdate transporters, Moco biosynthesis, and Moco-containing proteins	Widely distributed; *E. coli*	2008	112		
SAH	S-Adenosylhomocysteine (SAH)	Coenzyme	SAM recycling (converting SAH to L-methionine); *metH*	Widely distributed; *Pseudomonas syringae*	2008	113	114	3NPN, 3NPQ
SAM-IV	SAM	Coenzyme	Sulfur metabolism	Actinomycetes; *Mycobacterium tuberculosis*	2008	115		
Wco	Tungsten cofactor	Coenzyme	Riboswitch suspected	Widely distributed	2008	112		
c-di-GMP-I	3′-5′-Cyclic-di-GMP	Signaling molecule	Various cellular processes	Widely distributed; *Clostridium difficile* and *Vibrio cholerae*	2008	116	53, 117	3IRN, 3IWN
SAM-V	SAM	Coenzyme	Sulfur metabolism	Marine alphaproteobacteria; *Pelagibacter ubique*	2009	118		
c-di-GMP-II	3′-5′-Cyclic-di-GMP	Signaling molecule	c-di-GMP biosynthesis, degradation, and signaling	*C. difficile*	2010	119	120	3Q3Z
SAM-I/IV	SAM	Coenzyme	SAM synthetase; *metK*	Marine phyla	2010	15		
SAM-SAH	SAM, SAH	Coenzyme		Alphaproteobacteria; *Roseobacter* sp.	2010	15		
THF	Tetrahydrofolate	Coenzyme	Folate transport and biosynthesis; *folT*, *folE*, *folC*, and *folQPBK*	*Firmicutes*, clostridia, and lactobacilli	2010	121	122	3SD3, 4LVV
Glutamine	L-Glutamine	Amino acid	Nitrogen metabolism	Rare cyanobacteria and marine metagenomic sequences; *Synechococcus elongatus*	2011	123	124	5DDP
Fluoride (*crcB* motif)	Fluoride ions	Ion	Fluoride-sensitive enzymes and fluoride exporter; *CrcB*	Extremely widely distributed; *Pseudomonas syringae*	2012	125	126	4ENC
c-di-AMP (*ydaO* motif)	3′-5′-Cyclic-di-AMP	Signaling molecule	Cell wall metabolism, osmotic stress, and sporulation	*B. subtilis*	2013	14, 127	128	4QLM
PreQ₁-III	Prequeuosine-1	Nucleotide derivative	Transport of queuosine derivatives; *queT*	Rare; *Faecalibacterium prausnitzii* and *Shigella dysenteriae*	2014	129	130	4RZD

Mn²⁺ (yybP motif)	Manganese ions	Ion	Metal ion household, tellurium resistance, and cation-transport ATPase; yybP-ykoY	Widely distributed; *Lactococcus lactis*, *E. coli*, and *B. subtilis*	2015	131	132	4Y1I
cAMP-GMP	3′-5′-Cyclic AMP-GMP	Signaling molecule	Extracellular electron transfer; pgcA	Rare; *Geobacter sulfurreducens*	2015	133, 134	135	4YAZ
NiCo	Nickel or cobalt ions	Ion	Heavy metal ion sensing	Rare; *Clostridium scindens* and *Clostridium botulinum*	2015	136	136	4RUM
ZTP (pfl motif)	5-Aminoimidazole-4-carboxamide riboside 5′-triphosphate	Signaling molecule	De novo purine biosynthesis and one-carbon metabolism	Widely distributed; *Streptococcus thermophilus*	2015	137	138	4ZNP
Aza-aromatic (yjdF motif)	Aza-aromatics	Others	yjdF	*Firmicutes; B. subtilis*	2016	139		
Guanidine-I (ykkC motif)	Guanidine	Others	Urea carboxylases and multidrug efflux pumps (e.g., EmrE and SugE)	Widely distributed; *B. subtilis* and *E. coli*	2016	140	141	5T83
2′-dG-II	2′-Deoxyguanosine	Signaling molecule	Phosphate transporter, phospholipase D, endonuclease I, and signal receiver domain	Metagenomic sequence data	2017	47		
Guanidine-II (mini-ykkC)	Guanidine	Others	SMR efflux pumps; EmrE and SugE	Rather rare, mostly proteobacteria	2017	142	143	5NDI
Guanidine-III (ykkC-III)	Guanidine	Others	SMR efflux pumps; EmrE and SugE	Rare, mostly actinobacteria	2017	144	145	5NWQ

[a]Most riboswitches regulate a multitude of genes; here we try to list the ones described first.
[b]Distribution is based on reference 12; in addition, the table lists the organisms in which the riboswitches were first described.
[c]Publication of the first three-dimensional structure of the riboswitch.

2′-dG-I riboswitch class (sensing 2′-deoxyguanosine) have been found, all of which occur in *Mesoplasma florum* (26).

Remarkably, a look across all known classes of riboswitches reveals that most of them bind selectively to second-messenger signal molecules like cyclic di-GMP or to coenzymes like TPP or *S*-adenosylmethionine (SAM). Both of these molecules can be derived from RNA nucleotides or their precursors, implying a heavy reliance of RNA on RNA-like substances or, in other words, an independence from proteins as part of a regulatory system within an organism. This is concurrent with the now widely accepted RNA world theory, which proposes a phase of evolution where there were no proteins and DNA and all their current roles were carried out by RNA on its own. Modern riboswitches could be descendants of these ancient regulatory systems (6, 27, 28).

THE TPP RIBOSWITCH—AN EXAMPLE ACROSS ALL DOMAINS OF LIFE

TPP-binding riboswitches belong to the most widely distributed riboswitch class. They are found in all domains of life; the structure of their aptamer domain is highly conserved. Important differences between the TPP riboswitches can be found in their regulation mechanisms. In bacteria, ligand binding to the aptamer domain leads to changed transcription termination and translation initiation, respectively. Occasionally, an organism also uses both mechanisms. *Escherichia coli*,

for example, controls the *thiM* operon via a TPP riboswitch regulated by translation initiation, whereas the *thiC* operon is regulated both at the translational and transcriptional level (17, 29). In eukarya, TPP riboswitches are associated with splicing or alternative splicing events. Notably, the TPP riboswitch from land plants is the only example of gene control by means of a 3′ UTR-positioned riboswitch in any organism to date (23).

Several atomic-resolution structures of the TPP-binding domain have been solved (30–32). The structure reveals a complex folded RNA with two subdomains. One subdomain forms an intercalation pocket for the pyrimidine moiety of TPP (Fig. 2, J5-4), whereas another subdomain forms a wider pocket that employs bivalent metal ions and water molecules to make bridging contacts to the pyrophosphate moiety of the ligand (J3-2). The two pockets are positioned to function as a molecular measuring device that recognizes TPP in an extended conformation (Fig. 2). The central thiazole moiety is not recognized by the RNA, which explains why the antimicrobial compound pyrithiamine pyrophosphate targets this riboswitch and downregulates the expression of thiamine metabolic genes. The stabilization of the RNA structure by the natural ligand (but also its drug-like analog) is then transferred to the expression platform, which then controls the synthesis of the proteins coded by the mRNA.

Although there are different genes regulated by TPP riboswitches, they are commonly associated with thiamine metabolism or transport (33). TPP is an active

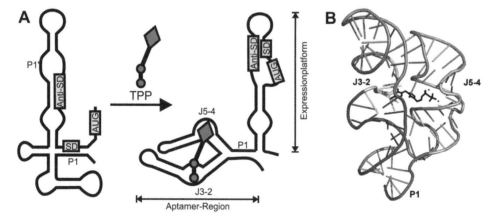

Figure 2 Structure of the TPP riboswitch. (A) Schematic depiction of the secondary structure of the *thiM* riboswitch with and without TPP (marked in pink). TPP stabilizes the P1-P1′ helix, which leads to secondary structure changes. The formation of the expression platform follows as a consequence, so that the Shine-Dalgarno (SD) sequence is sequestered in another stem, inhibiting any further gene expression. (B) The X-ray crystal structure of the *E. coli thiM* aptamer bound to TPP (black sticks, center) and Mg²⁺ ions (black spheres). PDB ID 2GDI (31); annotations on the structures refer to helices (P) and junctions (J). Adapted from reference 85 with permission.

form of vitamin B_1, an essential participant in many protein-catalyzed reactions. TPP itself is indirectly involved in several metabolic steps of the electron transport chain, contributes to energy extraction from carbohydrate sources (34, 35), and acts as a coenzyme in the conversion of glucose to ATP. Therefore, it is crucial for cells that the metabolism is correctly controlled and regulated. The fact that riboswitches carry out such a vital function throughout the kingdoms of life stresses their versatility and reliability across the generations of evolution already undergone.

VERSATILITY OF RIBOSWITCHES

In nature, there are several combinatorial approaches to solve complex regulatory problems. An excellent example is the *glmS* riboswitch discovered in *Bacillus subtilis*, as it can respond to more than one ligand. It acts as a ribozyme that self-cleaves once it binds to glucosamine-6-phosphate. The glucosamine-6-phosphate-binding riboswitch thus represents a classic negative feedback loop, as it regulates the gene responsible for its own production (*glmS*) (36). However, it has been shown that the *glms* riboswitch can also bind to glucose-6-phosphate; yet this binding leads to an inhibition of ribozyme function and stops subsequent self-cleavage. The riboswitch is thus able to react to the overall metabolic state of the cell and to regulate gene expression accordingly (37). Another example of a single riboswitch processing two input signals is the adenine riboswitch encoded by the *add* gene from *Vibrio vulnificus*, a bacterium that causes severe gastrointestinal infections in humans when ingested (38). As these bacteria live in two very different environments, i.e., the human body and marine conditions, they need to regulate their metabolism according to environmental conditions. Its adenine riboswitch acts in a translational manner and can sense temperature and ligand concentration to mediate efficient gene expression regulation over the range of physiologically relevant temperatures, allowing the organism to adapt to the differing temperatures experienced in the free-living state and within the human intestine (39). Other dual-input systems in nature favor a two-pronged approach to gene regulation, such as the prokaryote *Bacillus clausii*. Two different independent riboswitches that respond to SAM and vitamin B_{12}, respectively, can be found in the 5′ UTR of the *metE* mRNA. Analysis of this system showed that it fits the theory behind a Boolean NOR gate, as transcription termination can be achieved in the presence of only one or both ligands (40).

Although riboswitches act as *cis*-regulatory elements of their downstream mRNAs for the most part, some combine gene expression in *cis* and *trans*. They are able to act as processed, terminated riboswitches that operate like a small RNA (41), as full-length transcripts containing a binding site for sequestering an activated RNA-binding protein (42, 43) or a regulating part of antisense RNA, designating the antisense RNAs function (44). These *trans*-acting abilities of riboswitches further expand their regulatory abilities. The ability of a riboswitch to not only serve as a regulator of its associated downstream gene but also to control gene expression in *trans* makes the functional RNA-protein comparison even more appropriate. It will be interesting to see whether future research uncovers more of these dual-function riboswitches, which represent a complex, efficiently organized RNA regulation system.

HOW TO FIND NEW RIBOSWITCHES

All along, identification of new riboswitches has been supported by and is nowadays almost exclusively carried out by bioinformatics tools. Software that used sequence analysis for secondary structure and folding energy calculations and subsequently compared these values to *B. subtilis* RNA motif variants was published as early as 2004 (45). Since then, complex bioinformatics search pipelines have been developed, which not only can predict novel riboswitch candidates (46) but can search entire genome databases for the presence of known riboswitches or variants whose ligand-binding specificities might have been altered (47). However, it is important to note that bioinformatics predictions cannot replace experimental validation of those switches. An early example of an RNA structure correctly identified as a riboswitch that did not fit in its predicted class is the adenine riboswitch. Attributed to the class of guanine riboswitches (19), it contains a single nucleotide substitution in the conserved aptamer core that leads to a change in ligand recognition. The riboswitch recognizes adenine as its ligand and discriminates against guanine and other purine analogs with a high selectivity of several orders of magnitude (48). This example shows that riboswitches, which might have evolved to have a different ligand specificity but still retain much of their structural heritage, might be incorrectly assigned in bioinformatics searches.

GETTING A GLIMPSE OF RIBOSWITCH STRUCTURE

As structure is closely tied to function, and the binding of a specific ligand to the riboswitch induces a conformational change, the study of riboswitch structures is pivotal. A challenge in structure determination of ribo-switches is their intrinsic flexibility. In most cases, riboswitches without their ligand present can adopt several different conformations, and only the binding of the ligand leads to the adoption of a rigid and relatively inflexible structure (49). There are a number of methods to determine the secondary structure of nucleic acids in general, e.g., in-line probing (50, 51). This type of probing assay makes use of the natural instability of RNA with single-stranded regions, which renders these more flexible and unstable so that they will degrade over time. The method involves incubation of structural or functional RNAs over a long period of time (up to several days) and the subsequent visualization of the bands by gel electrophoresis. In-line probing is also used to determine structural changes due to ligand binding. Application of different ligand concentrations allows rapid estimation of the dissociation constant (K_d), making it a method supplying information on riboswitch secondary structure (single- or double-stranded regions), nucleotides involved in ligand binding, and binding affinity. Structural probing has the additional benefit that it can be carried out in aqueous solutions, which renders the resulting structures more likely to resemble their natural state of the RNA inside their organisms. Due to its low-tech approach and fast appraisal of results, many riboswitch publications first apply in-line probing to obtain information on the secondary structure of putative riboswitches (52).

To get an overview of the general riboswitch structure, more complex methods than biochemical probing of the secondary structure must be employed. Small-angle X-ray scattering (SAXS) analysis (53, 54) as well as X-ray crystallography (55–57) and nuclear magnetic resonance (NMR) spectroscopy (58, 59) have been applied successfully, but these approaches are laborious and depend on specialized knowledge and experimental equipment. Each method is associated with specific challenges. While SAXS analysis can deliver information on different conformations of a riboswitch in solution, it does not yield high-resolution insights. In contrast, the resolution of X-ray crystallographic analysis is very high, but as it only shows the details of one "frozen" conformation, this method can deliver an incomplete picture of all the different conformations possible in solution, particularly when analyzing ligand-free riboswitches. However, it must be emphasized that in recent years, the structure of the aptamer domains of many riboswitches has been successfully determined by X-ray crystallography, providing an interesting insight into the ligand-binding state of these regulators (see Table 1, last column). NMR spectroscopy has the scope

to gather atomic-resolution information on different conformational states in solution, but there are limits to the molecular size suitable for analysis, with riboswitches already close to or above the upper size threshold.

As many structures to date only provide a "time-frozen" conformation of aptamer and ligand together, the process of ligand-induced folding of the riboswitch itself generally remains unexplored. To tackle this challenge, the method of single-molecule fluorescence resonance energy transfer has recently become a tool to study the dynamics of the ligand-free riboswitch structure in particular (49, 60, 61). Its benefits include the analysis of one single riboswitch molecule only, which is advantageous when studying dynamic biomolecular systems that can exist in multiple distinct conformations. In summary, riboswitch structure determination remains a complex issue. Although a plethora of methods are available, a full understanding of individual riboswitch structure and its dynamics still requires application of a combination of several of these methods.

SYNTHETIC RIBOSWITCHES—NOVEL MEANS FOR THE CONDITIONAL CONTROL OF GENE EXPRESSION

An outstanding characteristic of riboswitches is that RNA acts as both sensor and effector, demonstrating that a protein cofactor for sensing the ligand is not an obligate requirement for regulation. Motivated by this, researchers have engineered a versatile set of synthetic riboswitches by combining *in vitro* generated aptamer domains (using the SELEX method) with expression platforms (62, 63). These now represent promising and powerful tools in synthetic biology. Several approaches based on the assembly of individual riboswitch building blocks *de novo*-designed into functional regulators have been realized. The best working synthetic riboswitches in bacteria exploit the theophylline aptamer as their sensing domain. Through the use of screening systems, synthetic riboswitches that can sequester the ribosome binding site depending on the presence of the ligand have been developed. A series of different variants allow the efficient application in a variety of bacteria (64, 65). Using an *in silico* approach combined with additional optimization steps, the same aptamer and the tetracycline aptamer were used to develop synthetic riboswitches that control transcription termination (66). Another elegant strategy using a modular "mix-and-match" approach was applied to construct transcriptional switches. In this work, expression platforms from natural riboswitches were combined with aptamer domains both derived from riboswitches and selected *in vitro* (67, 68). Other approaches tried to change the ligand specificity of natural riboswitches (69, 70).

In contrast to the rather rare examples of eukaryotic riboswitches in nature, various synthetic riboswitches have been developed for application in eukaryotes. The easiest strategy is the simple insertion of small-molecule-binding aptamers into the 5′ UTR of a gene of interest. Efficient conditional gene expression can be obtained using *in vitro*-selected aptamers binding to tetracycline, neomycin, or ciprofloxacin (71–73). However, aptamers can also be applied to control pre-mRNA splicing (74–76), microRNA processing (77, 78), or internal ribosome entry site-mediated translation initiation (79). Several independent approaches have sought to regulate gene expression by ribozymes by coupling aptamer domains (theophylline, tetracycline, and neomycin aptamer) to various ribozymes (hammerhead, HDV, and twister) (80, 81). These methods allow efficient induction and inhibition, respectively, of gene expression in bacteria (82), yeast (83), and even human cell lines (84), to name only a few.

Currently, gaps in the riboswitch toolbox are being filled with new aptamer domains, new expression modules, and combinations thereof. In addition, advanced *in vitro* and *in vivo* screening systems and improved *in silico* and structure-based design approaches have added to our riboswitch engineering principles and methods, so that the riboswitches we generate should more closely resemble their natural counterparts. Consolidation of the knowledge gained in all of these and future experiments will create positive feedback and ultimately permit the design of even better tools that should finally allow standardized "off-the-shelf" components to be utilized in "plug-and-play" approaches for the generation of synthetic genetic circuits of adjustable complexity, sophisticated biosensors, "intelligent" responsive metabolic pathways, and optimized diagnostic and therapeutic tools.

CONCLUDING REMARKS

Although there has been considerable research exploring new natural or synthetic riboswitches, it is still a new field of study. Benefiting from the steady decrease in cost and complexity of high-throughput analysis of entire genomes, rapid advancements in bioinformatics analysis, and growing understanding of the relation of RNA structure and function, future discovery of new riboswitches will likely be less expensive and less time-consuming. Despite the fact that plenty of highly abundant riboswitch classes have already been identified,

Breaker and colleagues predict that there could potentially be thousands of additional undiscovered riboswitch classes, albeit many of these are rare (12). In this review, we have tried to provide a comprehensive overview of the timeline of natural riboswitch discovery, as well as a summary of the validated classes of riboswitches and their distinguishing features to date, and conclude with an overview of the progress in the field of synthetic riboswitches.

Acknowledgments. We thank Dr. Florian Groher for the preparation of Figure 1. This work was supported by the Deutsche Forschungsgemeinschaft (SFB902/A2) and LOEWE CompuGene.

Citation. Lotz TS, Suess B. 2018. Small-molecule-binding riboswitches. Microbiol Spectrum 6(4):RWR-0025-2018.

References

1. **Morris KV, Mattick JS.** 2014. The rise of regulatory RNA. *Nat Rev Genet* **15:**423–437.

2. **Etzel M, Mörl M.** 2017. Synthetic riboswitches: from plug and pray toward plug and play. *Biochemistry* **56:**1181–1198.

3. **Winkler WC, Cohen-Chalamish S, Breaker RR.** 2002. An mRNA structure that controls gene expression by binding FMN. *Proc Natl Acad Sci U S A* **99:**15908–15913.

4. **Serganov A, Nudler E.** 2013. A decade of riboswitches. *Cell* **152:**17–24.

5. **Wachter A.** 2014. Gene regulation by structured mRNA elements. *Trends Genet* **30:**172–181.

6. **Breaker RR.** 2012. Riboswitches and the RNA world. *Cold Spring Harb Perspect Biol* **4:**a003566.

7. **Wittmann A, Suess B.** 2012. Engineered riboswitches: expanding researchers' toolbox with synthetic RNA regulators. *FEBS Lett* **586:**2076–2083.

8. **Wieland M, Ausländer D, Fussenegger M.** 2012. Engineering of ribozyme-based riboswitches for mammalian cells. *Methods* **56:**351–357.

9. **Roth A, Winkler WC, Regulski EE, Lee BW, Lim J, Jona I, Barrick JE, Ritwik A, Kim JN, Welz R, Iwata-Reuyl D, Breaker RR.** 2007. A riboswitch selective for the queuosine precursor preQ1 contains an unusually small aptamer domain. *Nat Struct Mol Biol* **14:**308–317.

10. **Meyer MM, Roth A, Chervin SM, Garcia GA, Breaker RR.** 2008. Confirmation of a second natural preQ$_1$ aptamer class in Streptococcaceae bacteria. *RNA* **14:**685–695.

11. **Batey RT.** 2015. Riboswitches: still a lot of undiscovered country. *RNA* **21:**560–563.

12. **McCown PJ, Corbino KA, Stav S, Sherlock ME, Breaker RR.** 2017. Riboswitch diversity and distribution. *RNA* **23:**995–1011.

13. **Jones CP, Ferré-D'Amaré AR.** 2017. Long-range interactions in riboswitch control of gene expression. *Annu Rev Biophys* **46:**455–481.

14. Barrick JE, Corbino KA, Winkler WC, Nahvi A, Mandal M, Collins J, Lee M, Roth A, Sudarsan N, Jona I, Wickiser JK, Breaker RR. 2004. New RNA motifs suggest an expanded scope for riboswitches in bacterial genetic control. *Proc Natl Acad Sci U S A* **101**: 6421–6426.

15. Weinberg Z, Wang JX, Bogue J, Yang J, Corbino K, Moy RH, Breaker RR. 2010. Comparative genomics reveals 104 candidate structured RNAs from bacteria, archaea, and their metagenomes. *Genome Biol* **11**:R31.

16. Mironov AS, Gusarov I, Rafikov R, Lopez LE, Shatalin K, Kreneva RA, Perumov DA, Nudler E. 2002. Sensing small molecules by nascent RNA: a mechanism to control transcription in bacteria. *Cell* **111**:747–756.

17. Winkler W, Nahvi A, Breaker RR. 2002. Thiamine derivatives bind messenger RNAs directly to regulate bacterial gene expression. *Nature* **419**:952–956.

18. Rodionov DA, Vitreschak AG, Mironov AA, Gelfand MS. 2002. Comparative genomics of thiamin biosynthesis in procaryotes. New genes and regulatory mechanisms. *J Biol Chem* **277**:48949–48959.

19. Sudarsan N, Barrick JE, Breaker RR. 2003. Metabolite-binding RNA domains are present in the genes of eukaryotes. *RNA* **9**:644–647.

20. Cheah MT, Wachter A, Sudarsan N, Breaker RR. 2007. Control of alternative RNA splicing and gene expression by eukaryotic riboswitches. *Nature* **447**:497–500.

21. Kubodera T, Watanabe M, Yoshiuchi K, Yamashita N, Nishimura A, Nakai S, Gomi K, Hanamoto H. 2003. Thiamine-regulated gene expression of *Aspergillus oryzae thiA* requires splicing of the intron containing a riboswitch-like domain in the 5′-UTR. *FEBS Lett* **555**: 516–520.

22. Sudarsan N, Wickiser JK, Nakamura S, Ebert MS, Breaker RR. 2003. An mRNA structure in bacteria that controls gene expression by binding lysine. *Genes Dev* **17**:2688–2697.

23. Wachter A, Tunc-Ozdemir M, Grove BC, Green PJ, Shintani DK, Breaker RR. 2007. Riboswitch control of gene expression in plants by splicing and alternative 3′ end processing of mRNAs. *Plant Cell* **19**:3437–3450.

24. Bocobza SE, Aharoni A. 2014. Small molecules that interact with RNA: riboswitch-based gene control and its involvement in metabolic regulation in plants and algae. *Plant J* **79**:693–703.

25. Croft MT, Moulin M, Webb ME, Smith AG. 2007. Thiamine biosynthesis in algae is regulated by riboswitches. *Proc Natl Acad Sci U S A* **104**:20770–20775.

26. Kim JN, Roth A, Breaker RR. 2007. Guanine riboswitch variants from *Mesoplasma florum* selectively recognize 2′-deoxyguanosine. *Proc Natl Acad Sci U S A* **104**:16092–16097.

27. Gilbert W. 1986. Origin of life: the RNA world. *Nature* **319**:618.

28. Nahvi A, Sudarsan N, Ebert MS, Zou X, Brown KL, Breaker RR. 2002. Genetic control by a metabolite binding mRNA. *Chem Biol* **9**:1043–1049.

29. Ontiveros-Palacios N, Smith AM, Grundy FJ, Soberon M, Henkin TM, Miranda-Ríos J. 2008. Molecular basis

30. of gene regulation by the THI-box riboswitch. *Mol Microbiol* **67**:793–803.

30. Edwards TE, Ferré-D'Amaré AR. 2006. Crystal structures of the *thi*-box riboswitch bound to thiamine pyrophosphate analogs reveal adaptive RNA-small molecule recognition. *Structure* **14**:1459–1468.

31. Serganov A, Polonskaia A, Phan AT, Breaker RR, Patel DJ. 2006. Structural basis for gene regulation by a thiamine pyrophosphate-sensing riboswitch. *Nature* **441**: 1167–1171.

32. Thore S, Leibundgut M, Ban N. 2006. Structure of the eukaryotic thiamine pyrophosphate riboswitch with its regulatory ligand. *Science* **312**:1208–1211.

33. Manzetti S, Zhang J, van der Spoel D. 2014. Thiamin function, metabolism, uptake, and transport. *Biochemistry* **53**:821–835.

34. Franken JF, Stapert FP. 1954. Restoration of pyruvate breakdown in pigeon muscle homogenates impaired by thiamine deficiency. *Biochim Biophys Acta* **14**:293–294.

35. Asakawa T, Wada H, Yamano T. 1968. Enzymatic conversion of phenylpyruvate to phenylacetate. *Biochim Biophys Acta* **170**:375–391.

36. Winkler WC, Nahvi A, Roth A, Collins JA, Breaker RR. 2004. Control of gene expression by a natural metabolite-responsive ribozyme. *Nature* **428**:281–286.

37. Watson PY, Fedor MJ. 2011. The *glmS* riboswitch integrates signals from activating and inhibitory metabolites in vivo. *Nat Struct Mol Biol* **18**:359–363.

38. Farmer JJ III. 1979. *Vibrio* ("*Beneckea*") *vulnificus*, the bacterium associated with sepsis, septicaemia, and the sea. *Lancet* **2**:903.

39. Loh E, Righetti F, Eichner H, Twittenhoff C, Narberhaus F. 2018. RNA thermometers in bacterial pathogens. *Microbiol Spectr* **6**:RWR-0012-2017.

40. Sudarsan N, Hammond MC, Block KF, Welz R, Barrick JE, Roth A, Breaker RR. 2006. Tandem riboswitch architectures exhibit complex gene control functions. *Science* **314**:300–304.

41. Loh E, Dussurget O, Gripenland J, Vaitkevicius K, Tiensuu T, Mandin P, Repoila F, Buchrieser C, Cossart P, Johansson J. 2009. A *trans*-acting riboswitch controls expression of the virulence regulator PrfA in *Listeria monocytogenes*. *Cell* **139**:770–779.

42. DebRoy S, Gebbie M, Ramesh A, Goodson JR, Cruz MR, van Hoof A, Winkler WC, Garsin DA. 2014. Riboswitches. A riboswitch-containing sRNA controls gene expression by sequestration of a response regulator. *Science* **345**:937–940.

43. Mellin JR, Koutero M, Dar D, Nahori MA, Sorek R, Cossart P. 2014. Riboswitches. Sequestration of a two-component response regulator by a riboswitch-regulated noncoding RNA. *Science* **345**:940–943.

44. Mellin JR, Tiensuu T, Bécavin C, Gouin E, Johansson J, Cossart P. 2013. A riboswitch-regulated antisense RNA in *Listeria monocytogenes*. *Proc Natl Acad Sci U S A* **110**:13132–13137.

45. Bengert P, Dandekar T. 2004. Riboswitch finder—a tool for identification of riboswitch RNAs. *Nucleic Acids Res* **32**(Web Server issue):W154–W159.

46. Chang TH, Huang HD, Wu LC, Yeh CT, Liu BJ, Horng JT. 2009. Computational identification of riboswitches based on RNA conserved functional sequences and conformations. *RNA* **15**:1426–1430.

47. Weinberg Z, Nelson JW, Lünse CE, Sherlock ME, Breaker RR. 2017. Bioinformatic analysis of riboswitch structures uncovers variant classes with altered ligand specificity. *Proc Natl Acad Sci U S A* **114**:E2077–E2085.

48. Mandal M, Breaker RR. 2004. Adenine riboswitches and gene activation by disruption of a transcription terminator. *Nat Struct Mol Biol* **11**:29–35.

49. Liberman JA, Wedekind JE. 2012. Riboswitch structure in the ligand-free state. *Wiley Interdiscip Rev RNA* **3**:369–384.

50. Soukup GA, Breaker RR. 1999. Relationship between internucleotide linkage geometry and the stability of RNA. *RNA* **5**:1308–1325.

51. Regulski EE, Breaker RR. 2008. In-line probing analysis of riboswitches. *Methods Mol Biol* **419**:53–67.

52. Mandal M, Boese B, Barrick JE, Winkler WC, Breaker RR. 2003. Riboswitches control fundamental biochemical pathways in *Bacillus subtilis* and other bacteria. *Cell* **113**:577–586.

53. Kulshina N, Baird NJ, Ferré-D'Amaré AR. 2009. Recognition of the bacterial second messenger cyclic diguanylate by its cognate riboswitch. *Nat Struct Mol Biol* **16**:1212–1217.

54. Stoddard CD, Montange RK, Hennelly SP, Rambo RP, Sanbonmatsu KY, Batey RT. 2010. Free state conformational sampling of the SAM-I riboswitch aptamer domain. *Structure* **18**:787–797.

55. Scott WG, Finch JT, Grenfell R, Fogg J, Smith T, Gait MJ, Klug A. 1995. Rapid crystallization of chemically synthesized hammerhead RNAs using a double screening procedure. *J Mol Biol* **250**:327–332.

56. Gilbert SD, Montange RK, Stoddard CD, Batey RT. 2006. Structural studies of the purine and SAM binding riboswitches. *Cold Spring Harb Symp Quant Biol* **71**:259–268.

57. Edwards AL, Garst AD, Batey RT. 2009. Determining structures of RNA aptamers and riboswitches by X-ray crystallography. *Methods Mol Biol* **535**:135–163.

58. Noeske J, Buck J, Fürtig B, Nasiri HR, Schwalbe H, Wöhnert J. 2007. Interplay of 'induced fit' and preorganization in the ligand induced folding of the aptamer domain of the guanine binding riboswitch. *Nucleic Acids Res* **35**:572–583.

59. Ottink OM, Rampersad SM, Tessari M, Zaman GJ, Heus HA, Wijmenga SS. 2007. Ligand-induced folding of the guanine-sensing riboswitch is controlled by a combined predetermined induced fit mechanism. *RNA* **13**:2202–2212.

60. Haller A, Soulière MF, Micura R. 2011. The dynamic nature of RNA as key to understanding riboswitch mechanisms. *Acc Chem Res* **44**:1339–1348.

61. Suddala KC, Walter NG. 2014. Riboswitch structure and dynamics by smFRET microscopy. *Methods Enzymol* **549**:343–373.

62. Ellington AD, Szostak JW. 1990. In vitro selection of RNA molecules that bind specific ligands. *Nature* **346**:818–822.

63. Tuerk C, Gold L. 1990. Systematic evolution of ligands by exponential enrichment: RNA ligands to bacteriophage T4 DNA polymerase. *Science* **249**:505–510.

64. Topp S, Reynoso CM, Seeliger JC, Goldlust IS, Desai SK, Murat D, Shen A, Puri AW, Komeili A, Bertozzi CR, Scott JR, Gallivan JP. 2010. Synthetic riboswitches that induce gene expression in diverse bacterial species. *Appl Environ Microbiol* **76**:7881–7884.

65. Berens C, Suess B. 2015. Riboswitch engineering—making the all-important second and third steps. *Curr Opin Biotechnol* **31**:10–15.

66. Wachsmuth M, Findeiß S, Weissheimer N, Stadler PF, Mörl M. 2013. *De novo* design of a synthetic riboswitch that regulates transcription termination. *Nucleic Acids Res* **41**:2541–2551.

67. Ceres P, Garst AD, Marcano-Velázquez JG, Batey RT. 2013. Modularity of select riboswitch expression platforms enables facile engineering of novel genetic regulatory devices. *ACS Synth Biol* **2**:463–472.

68. Ceres P, Trausch JJ, Batey RT. 2013. Engineering modular 'ON' RNA switches using biological components. *Nucleic Acids Res* **41**:10449–10461.

69. Robinson CJ, Vincent HA, Wu MC, Lowe PT, Dunstan MS, Leys D, Micklefield J. 2014. Modular riboswitch toolsets for synthetic genetic control in diverse bacterial species. *J Am Chem Soc* **136**:10615–10624.

70. Porter EB, Polaski JT, Morck MM, Batey RT. 2017. Recurrent RNA motifs as scaffolds for genetically encodable small-molecule biosensors. *Nat Chem Biol* **13**:295–301.

71. Suess B, Hanson S, Berens C, Fink B, Schroeder R, Hillen W. 2003. Conditional gene expression by controlling translation with tetracycline-binding aptamers. *Nucleic Acids Res* **31**:1853–1858.

72. Weigand JE, Sanchez M, Gunnesch EB, Zeiher S, Schroeder R, Suess B. 2008. Screening for engineered neomycin riboswitches that control translation initiation. *RNA* **14**:89–97.

73. Groher F, Bofill-Bosch C, Schneider C, Braun J, Jager S, Geißler K, Hamacher K, Suess B. 2018. Riboswitching with ciprofloxacin-development and characterization of a novel RNA regulator. *Nucleic Acids Res* **46**:2121–2132.

74. Kim DS, Gusti V, Pillai SG, Gaur RK. 2005. An artificial riboswitch for controlling pre-mRNA splicing. *RNA* **11**:1667–1677.

75. Vogel M, Weigand JE, Kluge B, Grez M, Suess B. 2018. A small, portable RNA device for the control of exon skipping in mammalian cells. *Nucleic Acids Res* **46**:e48.

76. Kötter P, Weigand JE, Meyer B, Entian KD, Suess B. 2009. A fast and efficient translational control system for conditional expression of yeast genes. *Nucleic Acids Res* **37**:e120.

77. Beisel CL, Chen YY, Culler SJ, Hoff KG, Smolke CD. 2011. Design of small molecule-responsive microRNAs based on structural requirements for Drosha processing. *Nucleic Acids Res* **39**:2981–2994.

78. An CI, Trinh VB, Yokobayashi Y. 2006. Artificial control of gene expression in mammalian cells by modulating RNA interference through aptamer-small molecule interaction. *RNA* **12**:710–716.

79. Ogawa A. 2011. Rational design of artificial riboswitches based on ligand-dependent modulation of internal ribosome entry in wheat germ extract and their applications as label-free biosensors. *RNA* **17**:478–488.

80. Nomura Y, Zhou L, Miu A, Yokobayashi Y. 2013. Controlling mammalian gene expression by allosteric hepatitis delta virus ribozymes. *ACS Synth Biol* **2**: 684–689.

81. Felletti M, Stifel J, Wurmthaler LA, Geiger S, Hartig JS. 2016. Twister ribozymes as highly versatile expression platforms for artificial riboswitches. *Nat Commun* **7**:12834.

82. Wieland M, Benz A, Klauser B, Hartig JS. 2009. Artificial ribozyme switches containing natural riboswitch aptamer domains. *Angew Chem Int Ed Engl* **48**:2715–2718.

83. Win MN, Smolke CD. 2008. Higher-order cellular information processing with synthetic RNA devices. *Science* **322**:456–460.

84. Beilstein K, Wittmann A, Grez M, Suess B. 2015. Conditional control of mammalian gene expression by tetracycline-dependent hammerhead ribozymes. *ACS Synth Biol* **4**:526–534.

85. Cressina E, Chen L, Moulin M, Leeper FJ, Abell C, Smith AG. 2011. Identification of novel ligands for thiamine pyrophosphate (TPP) riboswitches. *Biochem Soc Trans* **39**:652–657.

86. Johnson JE Jr, Reyes FE, Polaski JT, Batey RT. 2012. B$_{12}$ cofactors directly stabilize an mRNA regulatory switch. *Nature* **492**:133–137.

87. Serganov A, Huang L, Patel DJ. 2009. Coenzyme recognition and gene regulation by a flavin mononucleotide riboswitch. *Nature* **458**:233–237.

88. Vicens Q, Mondragón E, Batey RT. 2011. Molecular sensing by the aptamer domain of the FMN riboswitch: a general model for ligand binding by conformational selection. *Nucleic Acids Res* **39**:8586–8598.

89. Gilbert SD, Love CE, Edwards AL, Batey RT. 2007. Mutational analysis of the purine riboswitch aptamer domain. *Biochemistry* **46**:13297–13309.

90. Garst AD, Héroux A, Rambo RP, Batey RT. 2008. Crystal structure of the lysine riboswitch regulatory mRNA element. *J Biol Chem* **283**:22347–22351.

91. Serganov A, Huang L, Patel DJ. 2008. Structural insights into amino acid binding and gene control by a lysine riboswitch. *Nature* **455**:1263–1267.

92. Winkler WC, Nahvi A, Sudarsan N, Barrick JE, Breaker RR. 2003. An mRNA structure that controls gene expression by binding S-adenosylmethionine. *Nat Struct Biol* **10**:701–707.

93. Montange RK, Batey RT. 2006. Structure of the S-adenosylmethionine riboswitch regulatory mRNA element. *Nature* **441**:1172–1175.

94. Nahvi A, Barrick JE, Breaker RR. 2004. Coenzyme B$_{12}$ riboswitches are widespread genetic control elements in prokaryotes. *Nucleic Acids Res* **32**:143–150.

95. Cochrane JC, Lipchock SV, Strobel SA. 2007. Structural investigation of the GlmS ribozyme bound to its catalytic cofactor. *Chem Biol* **14**:97–105.

96. Klein DJ, Wilkinson SR, Been MD, Ferré-D'Amaré AR. 2007. Requirement of helix P2.2 and nucleotide G1 for positioning the cleavage site and cofactor of the *glmS* ribozyme. *J Mol Biol* **373**:178–189.

97. Mandal M, Lee M, Barrick JE, Weinberg Z, Emilsson GM, Ruzzo WL, Breaker RR. 2004. A glycine-dependent riboswitch that uses cooperative binding to control gene expression. *Science* **306**:275–279.

98. Huang L, Serganov A, Patel DJ. 2010. Structural insights into ligand recognition by a sensing domain of the cooperative glycine riboswitch. *Mol Cell* **40**:774–786.

99. Corbino KA, Barrick JE, Lim J, Welz R, Tucker BJ, Puskarz I, Mandal M, Rudnick ND, Breaker RR. 2005. Evidence for a second class of S-adenosylmethionine riboswitches and other regulatory RNA motifs in alpha-proteobacteria. *Genome Biol* **6**:R70.

100. Gilbert SD, Rambo RP, Van Tyne D, Batey RT. 2008. Structure of the SAM-II riboswitch bound to S-adenosylmethionine. *Nat Struct Mol Biol* **15**:177–182.

101. Fuchs RT, Grundy FJ, Henkin TM. 2006. The S(MK) box is a new SAM-binding RNA for translational regulation of SAM synthetase. *Nat Struct Mol Biol* **13**:226–233.

102. Lu C, Smith AM, Fuchs RT, Ding F, Rajashankar K, Henkin TM, Ke A. 2008. Crystal structures of the SAM-III/S(MK) riboswitch reveal the SAM-dependent translation inhibition mechanism. *Nat Struct Mol Biol* **15**:1076–1083.

103. Park SY, Cromie MJ, Lee EJ, Groisman EA. 2010. A bacterial mRNA leader that employs different mechanisms to sense disparate intracellular signals. *Cell* **142**: 737–748.

104. Cromie MJ, Shi Y, Latifi T, Groisman EA. 2006. An RNA sensor for intracellular Mg^{2+}. *Cell* **125**:71–84.

105. Dann CE III, Wakeman CA, Sieling CL, Baker SC, Irnov I, Winkler WC. 2007. Structure and mechanism of a metal-sensing regulatory RNA. *Cell* **130**:878–892.

106. Ramesh A, Wakeman CA, Winkler WC. 2011. Insights into metalloregulation by M-box riboswitch RNAs via structural analysis of manganese-bound complexes. *J Mol Biol* **407**:556–570.

107. Pikovskaya O, Polonskaia A, Patel DJ, Serganov A. 2011. Structural principles of nucleoside selectivity in a 2′-deoxyguanosine riboswitch. *Nat Chem Biol* **7**:748–755.

108. Kang M, Peterson R, Feigon J. 2009. Structural insights into riboswitch control of the biosynthesis of queuosine, a modified nucleotide found in the anticodon of tRNA. *Mol Cell* **33**:784–790.

109. Klein DJ, Edwards TE, Ferré-D'Amaré AR. 2009. Cocrystal structure of a class I preQ$_1$ riboswitch reveals a pseudoknot recognizing an essential hypermodified nucleobase. *Nat Struct Mol Biol* **16**:343–344.

110. Spitale RC, Torelli AT, Krucinska J, Bandarian V, Wedekind JE. 2009. The structural basis for recognition

of the preQ$_0$ metabolite by an unusually small riboswitch aptamer domain. *J Biol Chem* **284**:11012–11016.

111. Jenkins JL, Krucinska J, McCarty RM, Bandarian V, Wedekind JE. 2011. Comparison of a preQ$_1$ riboswitch aptamer in metabolite-bound and free states with implications for gene regulation. *J Biol Chem* **286**:24626–24637.

112. Regulski EE, Moy RH, Weinberg Z, Barrick JE, Yao Z, Ruzzo WL, Breaker RR. 2008. A widespread riboswitch candidate that controls bacterial genes involved in molybdenum cofactor and tungsten cofactor metabolism. *Mol Microbiol* **68**:918–932.

113. Wang JX, Lee ER, Morales DR, Lim J, Breaker RR. 2008. Riboswitches that sense *S*-adenosylhomocysteine and activate genes involved in coenzyme recycling. *Mol Cell* **29**:691–702.

114. Edwards AL, Reyes FE, Héroux A, Batey RT. 2010. Structural basis for recognition of *S*-adenosylhomocysteine by riboswitches. *RNA* **16**:2144–2155.

115. Weinberg Z, Regulski EE, Hammond MC, Barrick JE, Yao Z, Ruzzo WL, Breaker RR. 2008. The aptamer core of SAM-IV riboswitches mimics the ligand-binding site of SAM-I riboswitches. *RNA* **14**:822–828.

116. Sudarsan N, Lee ER, Weinberg Z, Moy RH, Kim JN, Link KH, Breaker RR. 2008. Riboswitches in eubacteria sense the second messenger cyclic di-GMP. *Science* **321**:411–413.

117. Smith KD, Lipchock SV, Ames TD, Wang J, Breaker RR, Strobel SA. 2009. Structural basis of ligand binding by a c-di-GMP riboswitch. *Nat Struct Mol Biol* **16**:1218–1223.

118. Poiata E, Meyer MM, Ames TD, Breaker RR. 2009. A variant riboswitch aptamer class for *S*-adenosylmethionine common in marine bacteria. *RNA* **15**:2046–2056.

119. Lee ER, Baker JL, Weinberg Z, Sudarsan N, Breaker RR. 2010. An allosteric self-splicing ribozyme triggered by a bacterial second messenger. *Science* **329**:845–848.

120. Smith KD, Shanahan CA, Moore EL, Simon AC, Strobel SA. 2011. Structural basis of differential ligand recognition by two classes of bis-(3′-5′)-cyclic dimeric guanosine monophosphate-binding riboswitches. *Proc Natl Acad Sci U S A* **108**:7757–7762.

121. Ames TD, Rodionov DA, Weinberg Z, Breaker RR. 2010. A eubacterial riboswitch class that senses the coenzyme tetrahydrofolate. *Chem Biol* **17**:681–685.

122. Trausch JJ, Ceres P, Reyes FE, Batey RT. 2011. The structure of a tetrahydrofolate-sensing riboswitch reveals two ligand binding sites in a single aptamer. *Structure* **19**:1413–1423.

123. Ames TD, Breaker RR. 2011. Bacterial aptamers that selectively bind glutamine. *RNA Biol* **8**:82–89.

124. Ren A, Xue Y, Peselis A, Serganov A, Al-Hashimi HM, Patel DJ. 2015. Structural and dynamic basis for low-affinity, high-selectivity binding of L-glutamine by the glutamine riboswitch. *Cell Rep* **13**:1800–1813.

125. Baker JL, Sudarsan N, Weinberg Z, Roth A, Stockbridge RB, Breaker RR. 2012. Widespread genetic switches and toxicity resistance proteins for fluoride. *Science* **335**:233–235.

126. Ren A, Rajashankar KR, Patel DJ. 2012. Fluoride ion encapsulation by Mg^{2+} ions and phosphates in a fluoride riboswitch. *Nature* **486**:85–89.

127. Nelson JW, Sudarsan N, Furukawa K, Weinberg Z, Wang JX, Breaker RR. 2013. Riboswitches in eubacteria sense the second messenger c-di-AMP. *Nat Chem Biol* **9**:834–839.

128. Ren A, Patel DJ. 2014. c-di-AMP binds the *ydaO* riboswitch in two pseudo-symmetry-related pockets. *Nat Chem Biol* **10**:780–786.

129. McCown PJ, Liang JJ, Weinberg Z, Breaker RR. 2014. Structural, functional, and taxonomic diversity of three preQ$_1$ riboswitch classes. *Chem Biol* **21**:880–889.

130. Liberman JA, Suddala KC, Aytenfisu A, Chan D, Belashov IA, Salim M, Mathews DH, Spitale RC, Walter NG, Wedekind JE. 2015. Structural analysis of a class III preQ$_1$ riboswitch reveals an aptamer distant from a ribosome-binding site regulated by fast dynamics. *Proc Natl Acad Sci U S A* **112**:E3485–E3494.

131. Dambach M, Sandoval M, Updegrove TB, Anantharaman V, Aravind L, Waters LS, Storz G. 2015. The ubiquitous *yybP-ykoY* riboswitch is a manganese-responsive regulatory element. *Mol Cell* **57**:1099–1109.

132. Price IR, Gaballa A, Ding F, Helmann JD, Ke A. 2015. Mn^{2+}-sensing mechanisms of *yybP-ykoY* orphan riboswitches. *Mol Cell* **57**:1110–1123.

133. Kellenberger CA, Wilson SC, Hickey SF, Gonzalez TL, Su Y, Hallberg ZF, Brewer TF, Iavarone AT, Carlson HK, Hsieh YF, Hammond MC. 2015. GEMM-I riboswitches from *Geobacter* sense the bacterial second messenger cyclic AMP-GMP. *Proc Natl Acad Sci U S A* **112**:5383–5388.

134. Nelson JW, Sudarsan N, Phillips GE, Stav S, Lünse CE, McCown PJ, Breaker RR. 2015. Control of bacterial exoelectrogenesis by c-AMP-GMP. *Proc Natl Acad Sci U S A* **112**:5389–5394.

135. Ren A, Wang XC, Kellenberger CA, Rajashankar KR, Jones RA, Hammond MC, Patel DJ. 2015. Structural basis for molecular discrimination by a 3′,3′-cGAMP sensing riboswitch. *Cell Rep* **11**:1–12.

136. Furukawa K, Ramesh A, Zhou Z, Weinberg Z, Vallery T, Winkler WC, Breaker RR. 2015. Bacterial riboswitches cooperatively bind Ni^{2+} or Co^{2+} ions and control expression of heavy metal transporters. *Mol Cell* **57**:1088–1098.

137. Kim PB, Nelson JW, Breaker RR. 2015. An ancient riboswitch class in bacteria regulates purine biosynthesis and one-carbon metabolism. *Mol Cell* **57**:317–328.

138. Ren A, Rajashankar KR, Patel DJ. 2015. Global RNA fold and molecular recognition for a *pfl* riboswitch bound to ZMP, a master regulator of one-carbon metabolism. *Structure* **23**:1375–1381.

139. Li S, Hwang XY, Stav S, Breaker RR. 2016. The *yjdF* riboswitch candidate regulates gene expression by binding diverse azaaromatic compounds. *RNA* **22**:530–541.

140. Nelson JW, Atilho RM, Sherlock ME, Stockbridge RB, Breaker RR. 2017. Metabolism of free guanidine in

bacteria is regulated by a widespread riboswitch class. *Mol Cell* **65**:220–230.

141. **Reiss CW, Xiong Y, Strobel SA.** 2017. Structural basis for ligand binding to the guanidine-I riboswitch. *Structure* **25**:195–202.

142. **Sherlock ME, Malkowski SN, Breaker RR.** 2017. Biochemical validation of a second guanidine riboswitch class in bacteria. *Biochemistry* **56**:352–358.

143. **Huang L, Wang J, Lilley DM.** 2017. The structure of the guanidine-II riboswitch. *Cell Chem Biol* **24**:695–702.e2.

144. **Sherlock ME, Breaker RR.** 2017. Biochemical validation of a third guanidine riboswitch class in bacteria. *Biochemistry* **56**:359–363.

145. **Huang L, Wang J, Wilson TJ, Lilley DM.** 2017. Structure of the guanidine III riboswitch. *Cell Chem Biol* **24**:1407–1415.e2.

Regulating with RNA in Bacteria and Archaea
Edited by Gisela Storz and Kai Papenfort
© 2018 American Society for Microbiology, Washington, DC
doi:10.1128/microbiolspec.RWR-0028-2018

The T-Box Riboswitch: tRNA as an Effector to Modulate Gene Regulation

6

Kiel D. Kreuzer[1,2] and Tina M. Henkin[1]

INTRODUCTION

Bacteria have evolved a wide array of mechanisms to control gene expression in response to environmental changes. These regulatory mechanisms ensure that specific genes are expressed under the appropriate physiological conditions, and they regulate every step of expression from transcription initiation to posttranslational modification and protein stability. The discovery of the T-box mechanism revealed that an uncharged tRNA can interact with an mRNA to regulate expression of the downstream coding region (1) (Fig. 1). This mechanism was the first of many regulatory systems to be discovered in which *cis*-encoded RNA responds directly to a physiological signal to control gene expression through structural rearrangements. Regulatory RNAs of this type, termed riboswitches, have become an intense focus of research, and to date dozens of riboswitch classes that respond to various signals have been identified and characterized, including those that respond to temperature, pH, and metabolites such as enzyme cofactors, amino acids, and nucleotides (2, see also chapter 5).

The T-box riboswitch is a unique class of regulatory RNA in part because its tRNA ligand is itself a complex structured RNA. All nonmitochondrial tRNAs share a similar L-shaped tertiary structure, but they vary considerably in sequence, stability, flexibility, and posttranscriptional modifications both across species and within the same organism (3). Unlike metabolite-binding riboswitches, which generally fold to form a compact binding pocket that recognizes specific chemical functional groups, T-box RNAs contact their cognate tRNA ligand at several distant sites through different RNA-RNA interactions. Binding specificity and affinity for

cognate tRNAs and discrimination against noncognate tRNAs are major challenges for T-box riboswitches that mirror those for aminoacyl-tRNA synthetases (aaRSs). Significant progress has been made toward understanding the biochemical and structural features of T-box regulation and tRNA binding, yet many aspects are not fully understood. Here, the current knowledge about the T-box riboswitch is reviewed, with an emphasis on recent developments and discoveries.

DISCOVERY AND MECHANISM

The T-box regulatory system was first identified by recognition of a conserved 14-nucleotide (nt) sequence in the upstream region of several aaRS genes in *Bacillus subtilis* (4). This sequence precedes an intrinsic transcription terminator helix in the 5′ untranslated region, or leader region, of these genes, which suggested a common regulatory mechanism involving transcription attenuation (see chapter 8). Mutational analysis of this conserved sequence, designated the T box, revealed that it was required for expression of the downstream genes. Intriguingly, expression of each gene could be induced only by limitation for its cognate amino acid and not by general amino acid starvation. The key discovery in understanding the specificity of this regulatory mechanism was identification of a 3-nt codon-like sequence that corresponds to the amino acid identity of the downstream gene (1) (Fig. 2A). This element was termed the Specifier Sequence, and mutation of the UAC sequence in a *tyrS-lacZ* transcriptional fusion to match the UUC phenylalanine sequence was sufficient to promote expression in response to limitation for phenylalanine instead of tyro-

[1]Department of Microbiology and Center for RNA Biology; [2]Molecular, Cellular and Developmental Biology Graduate Program, The Ohio State University, Columbus, OH 43210.

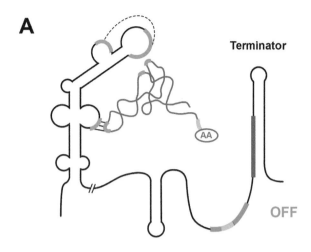

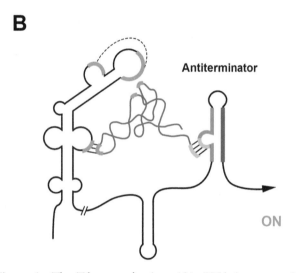

Figure 1 The T-box mechanism. (A) tRNA interacts with Stem I of the T-box leader RNA at two locations. The tRNA anticodon base-pairs with the Specifier Sequence (green), and the tRNA elbow stacks with the Stem I platform (orange), which is formed by interactions between conserved sequence motifs. The presence of an amino acid (AA) at the 3′ end of a charged tRNA blocks the base-pairing interaction with a bulge in the antiterminator helix. The terminator helix forms and transcription terminates, which turns gene expression off. (B) Uncharged tRNA also interacts with the Specifier Sequence and Stem I platform, and the acceptor arm base-pairs with a bulge in the antiterminator helix (cyan). The stabilization of the antiterminator prevents formation of the competing terminator helix, and RNA polymerase continues to transcribe the downstream coding sequence.

sine. This result provided the first evidence that tRNA was likely to be the effector in the T-box mechanism, and the tRNA anticodon was proposed to base-pair with the Specifier Sequence (Fig. 2B). Mutation of the UAC sequence to amber (UAG) and ochre (UAA) nonsense

codons resulted in expression dependent on a suppressor tRNA, which further supported the crucial role for tRNA as the effector (5). To test if tRNA interacts with the leader RNA in the absence of the ribosome, a single nucleotide was inserted immediately upstream from the Specifier Sequence, which would cause a frameshift in any potential open reading frame. This mutation did not significantly affect *tyrS-lacZ* expression or induction, indicating that tRNA promotes antitermination independent of Specifier Sequence translation (1).

The T-box sequence was predicted to form the 5′ side of an antiterminator helix, comprising two short helices flanking a 7-nt bulge, that competes with formation of the more thermodynamically stable terminator helix. The antiterminator element was predicted to be stabilized when the bulge nucleotides interact with the acceptor arm of uncharged tRNA, thereby preventing formation of the terminator helix (5) (Fig. 1). Interaction of charged tRNA promotes termination indirectly, as it is unable to stabilize the antiterminator due to the amino acid at the 3′ acceptor end but can compete with uncharged tRNA for binding to the Specifier Sequence (6). The T-box system therefore can monitor the charging ratio of a tRNA, such that only uncharged tRNA can stabilize the antiterminator and promote expression. Subsequent studies demonstrated tRNA-dependent antitermination in purified *in vitro* systems for *B. subtilis glyQS* and *Clostridium acetobutylicum alaS*, which revealed that regulation of these genes can occur without additional cellular factors (7, 8).

Whereas most T-box riboswitches regulate expression at the level of premature transcription termination, a subclass regulates at the level of translation initiation (9). Sequence analysis indicated that the T-box leader regions in some *Actinobacteria* do not contain a terminator helix; instead, the leader RNA was predicted to form a sequestrator helix that sequesters the Shine-Dalgarno sequence to prevent translation initiation (10). In this regulatory model, the acceptor arm of uncharged tRNA binds and stabilizes an antisequestrator element, which prevents formation of the competing sequestrator helix and therefore promotes translation. Biochemical studies of the *Nocardia farcinica ileS* leader RNA provide evidence for specific tRNA binding and tRNA-dependent structural changes that are consistent with the proposed regulatory model (9).

CONSERVED ELEMENTS AND STRUCTURAL ORGANIZATION

T-box leader RNAs are characterized by three helical regions (Stems I, II, and III) and a pseudoknot (Stem

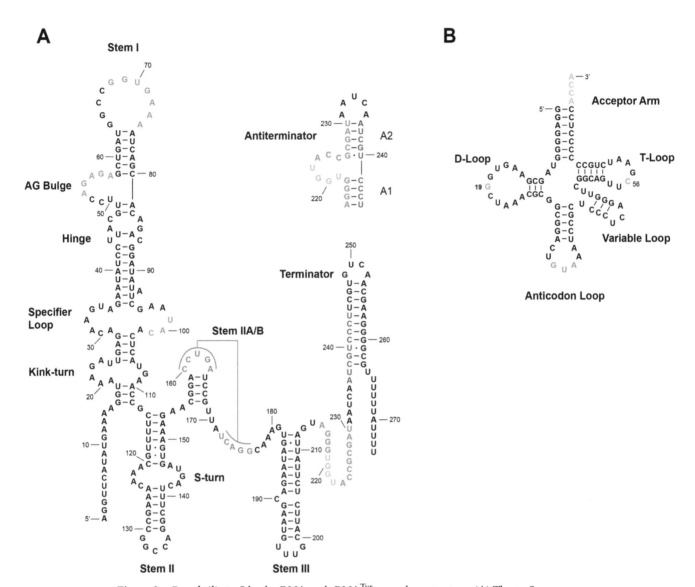

Figure 2 *B. subtilis tyrS* leader RNA and tRNA^Tyr secondary structure. (A) The *tyrS* sequence is numbered from the transcriptional start site to the end of the transcriptional terminator element. Conserved structural domains are labeled, including Stems I, II, IIA/B, and III and mutually exclusive terminator and antiterminator helices. The Stem IIB pseudoknot interaction is shown in magenta. The Specifier Sequence in the Specifier Loop and residues in the antiterminator bulge that base-pair with tRNA are shown in green and cyan, respectively. The orange sequences in the AG bulge and Stem I terminal loop interact to form the Stem I platform, which contacts the tRNA elbow. The red- and blue-labeled sequences interact to form the antiterminator element shown above the terminator conformation. The antiterminator is composed of helices A1 and A2. (B) Cloverleaf structure of tRNA^Tyr. The anticodon sequence is shown in green, and the nucleotides in the acceptor arm that base-pair with the *tyrS* antiterminator bulge are shown in cyan. The orange residues in the D-loop and T-loop (G19 and C56) interact to form the outermost tertiary interaction of the elbow and stack with the Stem I platform.

IIA/B) preceding the mutually exclusive terminator and antiterminator helices (or sequestrator and antisequestrator helices) (1) (Fig. 2A). Within Stem I, the Specifier Sequence is found in an internal loop called the Specifier Loop, which in most T-box RNAs contains an S-turn (loop E) element. The distal region of Stem I includes conserved sequence motifs in the AG bulge and terminal loop. In *glyQS* leader RNAs, these motifs interact to form a platform that contacts the tRNA D-/T-loops through a stacking interaction (11, 12). A structurally conserved internal loop or small bulge between the Specifier Loop and terminal region acts as a hinge to allow Stem I to bend and contact the tRNA at the anticodon and the D-/T-loop region. Below the Specifier Loop is a kink-turn (K-turn or GA motif) (13), which is a common RNA structural element that introduces an ~120° angle in the backbone (14). The Stem II domain, which usually stacks onto the base of the K-turn, varies in size and typically contains an S-turn within an internal loop (10). A pseudoknot element is usually found immediately downstream of Stem II and is formed when nucleotides in the loop of Stem IIA pair with the downstream single-stranded region to form Stem IIB (15). In *tyrS*, both the S-turn in Stem II and the pseudoknot are required for efficient antitermination, but their roles in the T-box mechanism are not fully understood. Mutational analysis indicated that both the secondary structure and primary sequence of the pseudoknot element are required for antitermination (15). Stem III is found just upstream of the antiterminator element and varies considerably in sequence and length but is conserved in nearly all T-box leader RNAs.

Although T-box RNAs contain many conserved structural elements, not all elements are found in every leader RNA. All glycyl T-box genes lack Stem II and the pseudoknot, which are replaced by a single-stranded linker region between Stem I and Stem III (7). Alanyl T-box genes are split into two classes that either lack (e.g., *Clostridium acetobutylicum alaS*) or include (e.g., *Bacillus subtilis alaS*) Stem II and the pseudoknot (8). Threonyl genes lack the S-turn motif in the Specifier Loop, which suggests an alternate presentation of the Specifier Sequence. The *ileS* genes in *Actinobacteria* contain Stem I variants that lack the motifs in the Stem I terminal region that comprise the tRNA-binding platform (9). In the ultrashort class of *ileS* RNAs, the Specifier Sequence is in the terminal loop, while another class contains sequences above the Specifier Loop that do not resemble those in the canonical Stem I. Studies of these structural variants demonstrated that all are functional *in vitro*, *in vivo*, or both. A structural element that is absent from one T-box gene is often essential in another, which suggests that the variant classes achieve proper tRNA binding and regulation through other compensatory elements.

PHYLOGENETIC DISTRIBUTION

Transcriptional units regulated by the T-box system have been identified based on the conservation of the T-box sequence as well as other structural elements (10, 16). The T-box mechanism is proposed to have arisen in a common ancestor to the *Firmicutes*, *Actinobacteria*, *Chloroflexi*, and *Deinococcus-Thermus* groups, with scattered examples of horizontal gene transfer to members of *Deltaproteobacteria*. This regulatory mechanism is most prevalent in *Firmicutes*, where many species contain multiple genes and operons under T-box regulation, in some cases representing >1% of transcriptional units. Of the gene classes that are regulated by the T-box system, aaRS genes comprise the vast majority, followed by amino acid biosynthesis, amino acid transport, and regulatory genes. All 20 canonical amino acids are represented in the T-box family, and this mechanism is the most common way of regulating aaRS genes in *Firmicutes*.

T-BOX LEADER RNA-tRNA INTERACTIONS

Specifier Sequence-tRNA Anticodon Interaction

The main specificity determinant of the T-box system is the base-pairing of the Specifier Sequence with the tRNA anticodon. This leader RNA-tRNA interaction was first demonstrated genetically in the initial study of the *tyrS* T-box mechanism (1, 5). Further studies examined tRNA specificity in more detail by testing additional mutations in the Specifier Sequence (and antiterminator bulge) in a *tyrS-lacZ* fusion, which enabled response to many different uncharged tRNA species (17). Several noncognate tRNAs can promote *tyrS* antitermination (with reduced efficiency relative to tRNA^{Tyr}) with the appropriate changes to the Specifier Sequence and antiterminator bulge, but some tRNAs did not promote readthrough at a detectable level. Of the tRNAs that can promote readthrough, none are able to reach the induction levels of wild-type tRNA^{Tyr}, which indicates that there are additional specificity determinants in the tRNA and/or leader RNA that contribute to maximal antitermination. Studies of the *B. subtilis ilv-leu*, *valS*, and *proBA* systems yielded similar results, where mutations to the Specifier Sequence resulted in a response to limitation for a different amino acid (18–20). Additionally, the wild-type tRNA-leader RNA interaction for all systems promoted higher antitermination compared to the noncognate tRNAs, consistent with the results from *tyrS*.

In *C. acetobutylicum*, the *aspS2o-gatCABo* operon is regulated by a T-box riboswitch with overlapping Specifier Sequences that bind both tRNA^{Asn} and tRNA^{Glu} (21). The indirect sensing of asparagine and glutamate levels permits appropriate expression of a nondiscriminating AspRS and the GatCAB complex of the tRNA-dependent transamidation pathway, which connects the metabolism of these amino acids. *In silico* analysis suggests that T-box genes or operons involved in the metabolism of multiple amino acids are more likely to contain a larger Specifier Loop capable of binding multiple physiologically relevant tRNA species. The ability of such T-box RNAs to respond to the charging levels of multiple tRNAs may be a mechanism to fine-tune the expression of genes involved in complex metabolic networks.

The base-pairing of the tRNA anticodon with the Specifier Sequence resembles a codon-anticodon interaction, but it occurs independently of the ribosome. The tRNA-mRNA interaction is strictly monitored during translation, which requires base-pairing at the first two positions with tolerance for mismatches and non-Watson-Crick interactions at the third (wobble) position. Phylogenetic analysis of T-box family genes revealed a bias for C at the third position of the Specifier Sequence, which does not correlate with codon bias or abundance of tRNA isoacceptors (1, 10). This observation suggested that the constraints that govern the Specifier Sequence-anticodon interaction differ from those for the anticodon-codon interaction in the context of the ribosome. The rules governing the Specifier Sequence-anticodon interaction were tested *in vitro* using the *B. subtilis glyQS* system and tRNA^{Gly}, which natively does not contain modifications in the anticodon loop (22). Mismatches of all possible combinations were introduced at every position of the base-pairing in both the tRNA and Specifier Sequence, and the effects on tRNA binding and antitermination were determined. In general, mismatches are least well tolerated at the second position, and mismatches are better tolerated at the first and third positions. Nearly all mismatches are better tolerated for tRNA binding than for antitermination, indicating that antitermination has more stringent requirements for the Specifier Sequence-anticodon pairing. Overall, the binding affinity is related to the predicted stability of the interaction between the Specifier Sequence and anticodon, which

differs from the mechanisms used by the ribosome to ensure accurate translation.

Antiterminator-tRNA Acceptor Arm Interaction

In addition to the base-pairing between the Specifier Sequence and anticodon, another base-pairing interaction occurs between the acceptor arm of uncharged tRNA and the antiterminator bulge. This base-pairing interaction, which stabilizes the antiterminator helix and prevents formation of the terminator helix, is essential for gene expression. Pairing between these sequences was proposed initially due to the complementarity and covariation between residues at the 5′ end of the antiterminator bulge (5′-UGGN-3′) and the acceptor end of tRNA (5′-NCCA-3′) (5). The specificity of this interaction was first shown in *tyrS*, where mutational analysis of the tRNA discriminator base and antiterminator bulge revealed that base-pairing is required for maximal expression and induction. Additional studies provided biochemical evidence for the specificity of this interaction *in vitro* and indicated that tRNA tertiary structure is also important for this interaction (23).

The specificity of the antiterminator-acceptor arm interaction is achieved by base-pairing, but the interaction is stabilized primarily by coaxial stacking (24). The tRNA acceptor end-antiterminator bulge pairing stacks with the antiterminator helix below the bulge, which extends the coaxial stacking of the tRNA T stem and acceptor stem. In this way, uncharged tRNA becomes an integral part of the antiterminator domain to prevent formation of the terminator helix. Most charged tRNAs in the cell are bound to EF-Tu, but the antiterminator discriminates against charged tRNA in the absence of EF-Tu by directly sensing the presence of the amino acid at the 3′ end (24). Even the smallest amino acid, glycine (57 Da), destabilizes the antiterminator-acceptor arm interaction through steric rejection and causes maximal termination. The competition between charged and uncharged tRNAs has been studied *in vitro* using a charged tRNA mimic that contains an additional C residue at the 3′ end (EX1C), which prevents stabilization of the antiterminator (6). Both charged and uncharged tRNAs interact with the Specifier Sequence in Stem I, so charged tRNA acts as a competitive inhibitor to prevent binding of uncharged tRNA. These studies revealed that tRNA-EX1C can interact with the leader RNA until the antiterminator helix is fully transcribed, at which point the transcriptional fate of the gene is determined. This ability to monitor both charged and uncharged tRNA indicates that the T-box riboswitch senses the tRNA charging ratio rather than the absolute levels of uncharged tRNA, which is appropriate for regulation of genes the products of which are involved in conversion of uncharged tRNA to charged tRNA, allowing both induction by the substrate and feedback repression by the product of the pathway.

The T-box antiterminator domain was the first RNA shown to perform several functions previously observed only in proteins, particularly aaRSs. With as few as 30 nucleotides, this RNA recognizes the tRNA discriminator base as a specificity determinant and directly evaluates tRNA charging status. Using these functions as inputs, the antiterminator then acts as a transcriptional switch to control gene expression, where it exploits the structure of the tRNA ligand to promote its own stabilization. The variety of functions accomplished through this RNA-RNA interaction in the absence of proteins highlights the versatility of RNA and adds to the growing list of biochemical functions of this macromolecule.

Stem I Platform-tRNA Elbow Interaction

For many years, the two sites of base-pairing between T-box leader RNA and tRNA were the only known intermolecular contacts. The conservation of secondary structure and sequence motifs in T-box RNAs suggested that there were likely to be additional leader RNA-tRNA interactions. In particular, the highly conserved residues within the Stem I terminal region were known to be essential for antitermination, but their mechanistic role was not understood (15). Crystalliza-

tion of the *glyQS* Stem I terminal region revealed a structure that was predicted to interact with tRNA (25), and this interaction was confirmed by cocrystallization of the Stem I domain with tRNA (11, 12) (Fig. 3). The conserved sequence motifs in the AG bulge and Stem I terminal loop interact to form a docking platform that contacts the tRNA D-/T-loop elbow region. The tRNA elbow is a structural feature of the canonical tRNA L-shape and is formed by tertiary interactions between residues in the D-loop and T-loop.

The Stem I domain monitors both the anticodon sequence and the tertiary structure of the cognate tRNA, and these two interactions comprise the majority of the tRNA binding affinity for *glyQS* leader RNA (12). The helical region between the Specifier Loop and the terminal region tracks along the tRNA body from the anticodon stem to the elbow, and the Stem I domain has been proposed to act as a molecular ruler for tRNA (26). Extending the length of Stem I by a full helical turn (+11 bp) eliminates tRNA binding, which can be restored by extending the tRNA anticodon stem by the same length, suggesting that both Stem I contacts are important for binding (11). The order of interaction for the two Stem I binding sites remains unknown, but the Stem I platform is transcribed first and may interact with the pool of cellular tRNAs to facilitate Specifier Sequence-anticodon pairing. Alternatively, tRNA may bind the Specifier Sequence first, followed by anchoring and stabilization by the Stem I platform. Currently,

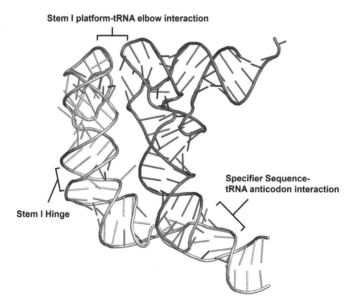

Figure 3 Cocrystal structure of the *glyQS* Stem I-tRNA complex. The *G. kaustophilus glyQS* Stem I (gray) is bound to tRNA^{Gly} (purple). The Specifier Sequence-anticodon interaction is shown in green, and the Stem I platform that stacks with the tRNA elbow is shown in orange. Data from reference 11 (PDB ID: 4MGN).

all structural data for the T-box Stem I platform are from glycyl T-box genes (e.g., *glyQS*), which contain a unique set of residues that comprise that platform that are not present in the majority of T-box family genes. These glycyl-specific residues may contribute to tRNA^{Gly} specificity, either by selecting for cognate tRNA^{Gly} or discriminating against noncognate tRNAs. The role of the platform in tRNA specificity is currently under investigation, and the structural differences between the consensus and glycyl Stem I platforms remain unknown.

The interaction between the tRNA elbow and T-box Stem I platform is structurally analogous to similar interactions that occur in the ribosome and RNase P (27). Both the ribosome and RNase P are essential in all three domains of life, and their RNA-tRNA elbow interactions are crucial to their cellular functions. In 23S rRNA, an interlocking T-loop module in the L1 stalk stacks with the tRNA elbow during the P/E-site transition (28). This interaction is essential for directing the translocation and ejection of deacylated tRNA during the elongation cycle. RNase P, an RNP required for tRNA 5′-end maturation (see chapter 2), contains an interlocking T-loop module in the J11/12-J12/11 domain (29). The interlocking T-loop motif is defined primarily by structure and not sequence, so although these three distinct RNAs all utilize this motif to bind the tRNA elbow, they do not share a common evolutionary origin or a conserved sequence. Furthermore, the docking platform in the ribosome and RNase P approaches the elbow from the tRNA D-loop side, while the T-box Stem I approaches from the tRNA T-loop side. This indicates that each double T-loop–tRNA elbow interaction arose independently through a striking example of convergent evolution.

Stem II and Pseudoknot Interactions with tRNA

Most *in vitro* studies of T-box regulation have relied on the *glyQS* system, which lacks the conserved Stem II and Stem IIA/B pseudoknot elements found in the majority of T-box genes. In *tyrS*, the S-turn in Stem II and the pseudoknot are required for maximal antitermination *in vivo* (15), but the exact roles of these elements are not understood. A role for the region between Stem I and Stem III in positioning of the tRNA acceptor end for interaction with the antiterminator was suggested by studies with *B. subtilis glyQS* in which *in vitro* antitermination was disrupted by extension of the tRNA acceptor arm by a half turn of the helix but restored by extension by a full turn (+11 bp) (30). Introduction of a second full turn (+22 bp) was not tolerated. As this region is a single-stranded linker in *glyQS* genes, it appears that it allows some flexibility in its measurement of tRNA length, but positioning of the tRNA in the correct orientation for interaction with the antiterminator is required.

The translational *ileS* systems from *Actinobacteria* permit the biochemical analysis of these domains and their role in tRNA-dependent regulation (9). The Stem II S-turn and the pseudoknot of *Mycobacterium smegmatis ileS* contribute substantially to tRNA binding affinity (31). Furthermore, SHAPE (selective 2′-hydroxyl acylation analyzed by primer extension) and tRNA cross-linking experiments provide evidence that the Stem II S-turn contacts the tRNA T-arm and the pseudoknot contacts the D-loop. This is the first biochemical evidence that shows that these elements contribute to the molecular recognition of the tRNA ligand. While this *ileS* RNA contains the structural elements lacking in *glyQS*, the Stem I domain belongs to the ultrashort class, which lacks the Stem I platform and contains the Specifier Sequence within the terminal loop. It remains to be discovered if these same interactions occur in T-box RNAs that contain the canonical Stem I, Stem II S-turn, and pseudoknot elements together, although their conservation and sensitivity to mutation is consistent with the hypothesis that they are utilized in RNAs in which they are present.

STRUCTURAL ANALYSES OF T-BOX LEADER RNAs

Antiterminator Domain

The structures of riboswitch RNAs are essential to their biological function, but high-resolution structural analysis can be particularly challenging if such RNAs are inherently flexible. To address this, riboswitches are often divided into discrete structural domains and studied in isolation. The first structural study of the T-box riboswitch was the nuclear magnetic resonance (NMR) analysis of a model antiterminator RNA based on the *tyrS* sequence (32). This RNA is composed of two short helices (A1 and A2) separated by a 7-nt bulge that contains the 5′-UGGA-3′ motif that pairs with the acceptor arm of uncharged tRNA. The remaining three nucleotides of the bulge (5′-ACC-3′) are highly conserved throughout the T-box family, and mutations in this sequence disrupt *tyrS* antitermination *in vivo* (15). The NMR analysis revealed a structural basis for the observed conservation; the residues in the 3′ portion of the bulge participate in a stacking interaction that introduces an ~80° kink between the helices. In contrast to the relatively rigid 3′ bulge residues, the bases that pair with the tRNA acceptor arm are more flexible and are likely to sample different conformations to achieve an induced fit with tRNA. Studies of fluorescently labeled antiterminator RNAs indicate that the bulge structure changes in the presence of Mg^{2+} (33). This structural change is necessary for proper base-pairing with the tRNA acceptor arm, and additional fluorescence changes upon tRNA binding support the induced-fit model.

Specifier Loop and K-Turn Motif

The canonical Stem I domain contains several RNA structural motifs, including an S-turn motif within the Specifier Loop and a K-turn motif at the base of Stem I. The first structural information for the S-turn was from NMR studies of the lower portion of *tyrS* Stem I (34). Within the Specifier Loop, the S-turn is directly above the Specifier Sequence, and the conserved 5′-AGUA-3′ motif on the 5′ side of Stem I forms noncanonical interactions with the 5′-GAA-3′ motif on the opposite side of the loop. These interactions generate a distortion in the phosphate backbone that causes the Specifier Sequence nucleotides to rotate into the minor groove. A rotation toward the minor groove exposes more of the Watson-Crick face, such that the S-turn motif assists in the presentation of the Specifier Sequence for tRNA binding. Structural analyses of this region in complex with tRNA revealed that the conserved purine 3′ of the Specifier Sequence stabilizes the interaction by stacking with the leader RNA-anticodon duplex (11, 12). Additionally, the structural perturbation caused by the S-turn allows the Specifier Loop to accommodate large posttranscriptional modifications that are common at position 37 in the anticodon loop (35).

The K-turn element is found just below the Specifier Loop and comprises flanking 5′-GA-3′ residues on both sides of an asymmetric internal loop that introduces a kink in the helical axis (13). This structural motif is found in many RNAs, including other riboswitches, 23S rRNA, and RNase P, where it is typically a binding site for an L7Ae family protein (36, see also chapter 7). The T-box mechanism does not appear to require this RNA-protein interaction, as *tyrS* expression is not affected in strains in which L7Ae homologs are inactivated by mutation (F. J. Grundy and T. M. Henkin, unpublished data), and other leader RNAs function *in vitro* without additional protein factors (7–9). The K-turn in T-box leader RNAs is predicted to determine the relative orientation of the Stem I domain with the downstream RNA, which would facilitate simultaneous tRNA binding with Stem I and the antiterminator bulge. NMR

analysis of the K-turn in *tyrS* revealed the sheared G-A pairs that are characteristic of this motif, but the RNA is in an extended conformation rather than the canonical kinked conformation (37). The crystal structure of *glyQS* Stem I in complex with tRNA contains the K-turn, but the kinked conformation is stabilized by the L7Ae homolog used to promote cocrystallization (12). The structure of this motif is likely to be influenced by the context of the upstream and downstream RNA (38), so further studies of this structural element in the context of the full-length leader RNA are necessary to validate its role in the T-box mechanism.

Stem I Platform and Hinge

The crystal structures of *glyQS* Stem I from *Geobacillus kaustophilus* (11, 25) and *Oceanobacillus iheyensis* (12) revealed key details of Stem I architecture. The most striking observation was the formation of a platform at the terminal region of Stem I that interacts with the tRNA elbow (Fig. 3). This platform is formed by the interaction of two highly conserved sequence motifs; one of these (5′-AGAGA-3′; 5′-AGCGA-3′ in *glyQS*) is within the AG bulge and the other (5′-GSUGNRA-3′; S = G or C; R = A or G) is within the terminal loop. The structure formed by this interaction alters the direction of base-stacking by 90° relative to the Stem I helix to allow coaxial stacking with the tRNA. The third residue in the AG bulge motif (C in most glycyl T-box genes) forms a *cis*-Watson-Crick interaction with a glycyl-specific G at the 3′ end of the terminal loop, and an additional Hoogsteen interaction from a residue (G or A) in the terminal loop forms a base triple. This triplet is the outermost platform layer of a total of five base layers that stack with tRNA. In the majority of T-box family genes, the residues that comprise the base triple that contacts the tRNA elbow either are conserved as other residues or are absent. Further studies are needed to investigate the structural differences between glycyl and nonglycyl T-box Stem I platforms.

The helical region between the AG bulge and Specifier Loop is always disrupted by either a bulge or internal loop that is not conserved in sequence. The disruption to the helix introduces structural flexibility that acts as a hinge to bend the flanking helices and form a cradle-shaped Stem I domain. In the *G. kaustophilus glyQS* Stem I, the hinge is an asymmetric internal loop with three nucleotides opposite a single nucleotide (11), which is similar to the predicted *B. subtilis glyQS* hinge. In contrast, the hinge in *O. iheyensis* comprises only two bulged nucleotides on the 5′ side (12). The additional unpaired nucleotides on the 5′ side of both Stem I structures cause a length discrepancy that kinks the distal portion of Stem I toward the tRNA elbow. This hinge element is likely to be crucial for Stem I to properly orient the Specifier Loop and terminal platform for optimal interaction with tRNA, and mutations to this region decrease tRNA binding (12).

Leader RNA-tRNA Complexes

The high-resolution crystal structures have provided extensive insight into the Stem I-tRNA interactions. Comparisons of these complexes with the structures of both RNAs in the unbound state reveal the changes that occur upon binding. The anticodon loop of the free tRNA^{Gly} adopts an extended conformation (39), while the bound tRNA^{Gly} anticodon loop is more ordered to interact properly with the Specifier Sequence (11, 12). Stem I also induces tRNA bending at the 26-44 pair at the junction of the anticodon stem and the D-stem. This junction is inherently flexible, and its flexibility is important for tRNA bending during translation, particularly in the P/E hybrid state where it is bent up to 37° (40). Flexibility in tRNA structure was shown to be important for both glyQS and alaS antitermination in vitro (8; L. C. Liu, F. J. Grundy, and T. M. Henkin, personal communication), consistent with the structural studies. These comparisons reveal that there are both localized and global structural changes in tRNA that occur upon interaction with Stem I, and there are likely to be additional changes when more of the leader RNA is present.

The Specifier Loop also undergoes changes upon tRNA binding, such that the Specifier Sequence residues rotate farther out from the minor groove to expose the Watson-Crick face (12). This rotation also affects the position of the conserved 3′ A residue, allowing it to stack with the Specifier Sequence-anticodon pairing. In the G. kaustophilus Specifier Loop, the S-turn element is separated from the Specifier Sequence by two nonconserved nucleotides, each of which forms a hydrogen bond with nucleotides in the anticodon loop (11). These additional nucleotides may contribute to tRNA specificity in the T-box genes that contain them, or they may generate overlapping Specifier Sequences that allow response to multiple tRNA species (21, 41). The structural changes that occur in both tRNA and Stem I in addition to the flexible hinges in these RNAs support a mutually induced-fit model of binding.

The crystal structure of the Stem I-tRNA complex revealed new aspects of the structural basis for tRNA recognition, but it lacks important downstream RNA elements such as Stem III and the antiterminator. Recent studies report low-resolution structural models of the full complex in solution using small-angle X-ray scattering (SAXS). SAXS-derived molecular envelopes of the B. subtilis glyQS leader RNA (up to the antiterminator) in the absence of tRNA suggest an elongated structure (42). Molecular envelopes were also determined for mutants containing extensions or deletions in helical regions to determine the positions of Stem I, Stem III, and the antiterminator. The leader RNA-tRNA complex resembles a flattened disc shape, which suggests that all RNA elements lie in the same plane (43). Modeling the RNAs to fit this envelope suggests that Stem III stacks with the A1 helix of the antiterminator. The additional stabilization of the antiterminator conformation by Stem III would explain why Stem III is highly conserved but variable in sequence and length.

The O. iheyensis glyQS leader RNA-tRNA complex (42) forms a different structure compared to the B. subtilis complex (43). The O. iheyensis complex is more linear and elongated, and molecular modeling indicates that the Stem I platform-elbow interaction does not occur in the antiterminator conformation. The interaction is restored when the tRNA is unable to pair with the antiterminator (by removal of the 5′-UCCA-3′ end), which led to a proposed capture-and-release mechanism by the Stem I platform. This mechanism is not supported by the structural model of the B. subtilis complex, where the Stem I platform and antiterminator bulge interact with tRNA simultaneously. Additional studies are needed to investigate the equilibrium binding state of the leader RNA-tRNA complex to resolve the apparent structural discrepancies.

THE T-BOX REGULATORY SYSTEM AS AN ANTIBIOTIC TARGET

Bacterial riboswitch RNAs, including the T-box riboswitch, have drawn interest as a potential target for the development of novel antibiotic compounds (44). Many T-box family genes, such as aaRS genes, are essential, and disruption of the regulation of these genes can reduce competitive fitness or be lethal. Many human pathogens in Firmicutes contain multiple T-box-regulated genes; for example, B. anthracis is predicted to contain a T-box regulon of 39 transcriptional units, 21 of which include aaRS genes (10). The presence of multiple essential targets in combination with the default "off" state of the T-box riboswitch would help to prevent mutations that confer antibiotic resistance. Mutations in the T-box leader RNA that prevent interaction with the antibiotic compound may also prevent interaction with tRNA, resulting in loss of expression. The identification and characterization of candidate compounds that target the T-box regulatory system is ongoing (45). The primary target in these studies is a model antiterminator RNA based on the *B. subtilis tyrS* sequence (23). The antiterminator is highly conserved and is essential for gene expression. The goal of these studies is to identify compounds that interact with the antiterminator in a way that prevents it from binding the tRNA acceptor end but does not excessively stabilize the antiterminator independent of tRNA binding. Several classes of chemical compounds have been identified that bind the antiterminator RNA with structural specificity and low micromolar affinity (46, 47) and disrupt binding of the tRNA acceptor end (48).

Several commonly used antibiotics that are known to interact with structured RNA have been shown to interact with T-box leader RNA and affect regulation (49). Many antibiotics that inhibit protein synthesis by binding to rRNA can also bind to other noncoding RNAs with potentially antagonistic or synergistic antibiotic effects (50). Some antibiotics, such as tigecycline, increase tRNA-dependent antitermination, while others, such as linezolid, decrease tRNA-dependent antitermination (49). Molecular modeling of the binding of the antibiotics to the RNAs suggests that they interact with crucial RNA sequences such as the tRNA elbow, anticodon, Specifier Loop, and Stem I terminal region. Additional studies of T-box riboswitch structure and function will aid in the development of antibiotic compounds that specifically target this widespread regulatory mechanism.

CONCLUSIONS

For many bacterial species, riboswitches are one of the primary mechanisms to control gene expression in response to changing environmental conditions. The T-box riboswitch is a widespread mechanism that regulates the expression of amino acid-related genes, most commonly aaRS genes, by directly monitoring tRNA aminoacylation. The phylogenetic distribution of this system highlights its evolutionary adaptability to monitor tRNAs of all canonical amino acid classes (10). It remains unknown why T-box regulons vary so much among different species, but it is likely related to the different ecological niches that require optimal responses to changes in the environment and nutrient availability.

With a typical length of <300 nt, the T-box leader RNA recognizes the sequence and structure of tRNA and determines its aminoacylation status to control premature transcription termination or translation initiation. A combination of genetic, biochemical, and

structural techniques has revealed many fundamental aspects of the leader RNA-tRNA interactions. These interactions include anticodon recognition and stacking with the tRNA elbow, which are similar to interactions with tRNA in other cellular RNAs, including the ribosome and RNase P. Furthermore, the T-box RNAs utilize tRNA recognition determinants analogous to those often utilized by aaRS enzymes. Despite the significant progress in understanding this regulatory system since its discovery, many aspects remain to be investigated. In particular, it is evident that variations in RNA structure yield alternate modes of tRNA recognition for which structural information is currently lacking. Further studies of both canonical and variant T-box RNAs and their interactions with tRNA will contribute to our understanding of modes of RNA-RNA recognition.

Acknowledgments. This work was supported by National Institutes of Health National Institute of General Medical Sciences grant R01 GM047823. We thank members of the Henkin lab past and present for helpful discussions.

Citation. Kreuzer KD, Henkin TM. 2018. The T-box riboswitch: tRNA as an effector to modulate gene regulation. Microbiol Spectrum 6(4):RWR-0028-2018.

References

1. **Grundy FJ, Henkin TM.** 1993. tRNA as a positive regulator of transcription antitermination in *B. subtilis. Cell* **74:**475–482.

2. **Sherwood AV, Henkin TM.** 2016. Riboswitch-mediated gene regulation: novel RNA architechtures dictate gene expression responses. *Annu Rev Microbiol* **70:**361–374.

3. **Giegé R, Jühling F, Pütz J, Stadler P, Sauter C, Florentz C.** 2012. Structure of transfer RNAs: similarity and variability. *Wiley Interdiscip Rev RNA* **3:**37–61.

4. **Henkin TM, Glass BL, Grundy FJ.** 1992. Analysis of the *Bacillus subtilis tyrS* gene: conservation of a regulatory sequence in multiple tRNA synthetase genes. *J Bacteriol* **174:**1299–1306.

5. **Grundy FJ, Rollins SM, Henkin TM.** 1994. Interaction between the acceptor end of tRNA and the T box stimulates antitermination in the *Bacillus subtilis tyrS* gene: a new role for the discriminator base. *J Bacteriol* **176:** 4518–4526.

6. **Grundy FJ, Yousef MR, Henkin TM.** 2005. Monitoring uncharged tRNA during transcription of the *Bacillus subtilis glyQS* gene. *J Mol Biol* **346:**73–81.

7. **Grundy FJ, Winkler WC, Henkin TM.** 2002. tRNA-mediated transcription antitermination *in vitro*: codon-anticodon pairing independent of the ribosome. *Proc Natl Acad Sci U S A* **99:**11121–11126.

8. **Liu LC, Grundy FJ, Henkin TM.** 2015. Non-conserved residues in *Clostridium acetobutylicum* tRNA^Ala contribute to tRNA tuning for efficient antitermination of the *alaS* T box riboswitch. *Life (Basel)* **5:**1567–1582.

9. Sherwood AV, Grundy FJ, Henkin TM. 2015. T box riboswitches in *Actinobacteria*: translational regulation via novel tRNA interactions. *Proc Natl Acad Sci U S A* 112:1113–1118.

10. Gutiérrez-Preciado A, Henkin TM, Grundy FJ, Yanofsky C, Merino E. 2009. Biochemical features and functional implications of the RNA-based T-box regulatory mechanism. *Microbiol Mol Biol Rev* 73:36–61.

11. Grigg JC, Ke A. 2013. Structural determinants for geometry and information decoding of tRNA by T box leader RNA. *Structure* 21:2025–2032.

12. Zhang J, Ferré-D'Amaré AR. 2013. Co-crystal structure of a T-box riboswitch Stem I domain in complex with its cognate tRNA. *Nature* 500:363–366.

13. Winkler WC, Grundy FJ, Murphy BA, Henkin TM. 2001. The GA motif: an RNA element common to bacterial antitermination systems, rRNA, and eukaryotic RNAs. *RNA* 7:1165–1172.

14. Klein DJ, Schmeing TM, Moore PB, Steitz TA. 2001. The kink-turn: a new RNA secondary structure motif. *EMBO J* 20:4214–4221.

15. Rollins SM, Grundy FJ, Henkin TM. 1997. Analysis of *cis*-acting sequence and structural elements required for antitermination of the *Bacillus subtilis tyrS* gene. *Mol Microbiol* 25:411–421.

16. Vitreschak AG, Mironov AA, Lyubetsky VA, Gelfand MS. 2008. Comparative genomic analysis of T-box regulatory systems in bacteria. *RNA* 14:717–735.

17. Grundy FJ, Hodil SE, Rollins SM, Henkin TM. 1997. Specificity of tRNA-mRNA interactions in *Bacillus subtilis tyrS* antitermination. *J Bacteriol* 179:2587–2594.

18. Marta PT, Ladner RD, Grandoni JA. 1996. A CUC triplet confers leucine-dependent regulation of the *Bacillus subtilis ilv-leu* operon. *J Bacteriol* 178:2150–2153.

19. Luo D, Leautey J, Grunberg-Manago M, Putzer H. 1997. Structure and regulation of expression of the *Bacillus subtilis* valyl-tRNA synthetase gene. *J Bacteriol* 179: 2472–2478.

20. Brill J, Hoffmann T, Putzer H, Bremer E. 2011. T-box-mediated control of the anabolic proline biosynthetic genes of *Bacillus subtilis*. *Microbiology* 157:977–987.

21. Saad NY, Stamatopoulou V, Brayé M, Drainas D, Stathopoulos C, Becker HD. 2013. Two-codon T-box riboswitch binding two tRNAs. *Proc Natl Acad Sci U S A* 110:12756–12761.

22. Caserta E, Liu LC, Grundy FJ, Henkin TM. 2015. Codon-anticodon recognition in the *Bacillus subtilis glyQS* T box riboswitch: RNA-dependent codon selection outside the ribosome. *J Biol Chem* 290:23336–23347.

23. Gerdeman MS, Henkin TM, Hines JV. 2002. *In vitro* structure-function studies of the *Bacillus subtilis tyrS* mRNA antiterminator: evidence for factor-independent tRNA acceptor stem binding specificity. *Nucleic Acids Res* 30:1065–1072.

24. Zhang J, Ferré-D'Amaré AR. 2014. Direct evaluation of tRNA aminoacylation status by the T-box riboswitch using tRNA-mRNA stacking and steric readout. *Mol Cell* 55:148–155.

25. Grigg JC, Chen Y, Grundy FJ, Henkin TM, Pollack L, Ke A. 2013. T box RNA decodes both the information content and geometry of tRNA to affect gene expression. *Proc Natl Acad Sci U S A* 110:7240–7245.

26. Zhang J, Ferré-D'Amaré AR. 2016. The tRNA elbow in structure, recognition and evolution. *Life (Basel)* 6:3.

27. Lehmann J, Jossinet F, Gautheret D. 2013. A universal RNA structural motif docking the elbow of tRNA in the ribosome, RNAse P and T-box leaders. *Nucleic Acids Res* 41:5494–5502.

28. Trabuco LG, Schreiner E, Eargle J, Cornish P, Ha T, Luthey-Schulten Z, Schulten K. 2010. The role of L1 stalk-tRNA interaction in the ribosome elongation cycle. *J Mol Biol* 402:741–760.

29. Reiter NJ, Osterman A, Torres-Larios A, Swinger KK, Pan T, Mondragón A. 2010. Structure of a bacterial ribonuclease P holoenzyme in complex with tRNA. *Nature* 468:784–789.

30. Yousef MR, Grundy FJ, Henkin TM. 2003. tRNA requirements for *glyQS* antitermination: a new twist on tRNA. *RNA* 9:1148–1156.

31. Sherwood AV, Frandsen JK, Grundy FJ, Henkin TM. 2018. New tRNA contacts facilitate ligand binding in a *Mycobacterium smegmatis* T box riboswitch. *Proc Natl Acad Sci U S A* 115:3894–3899.

32. Gerdeman MS, Henkin TM, Hines JV. 2003. Solution structure of the *Bacillus subtilis* T-box antiterminator RNA: seven nucleotide bulge characterized by stacking and flexibility. *J Mol Biol* 326:189–201.

33. Means JA, Simson CM, Zhou S, Rachford AA, Rack JJ, Hines JV. 2009. Fluorescence probing of T box antiterminator RNA: insights into riboswitch discernment of the tRNA discriminator base. *Biochem Biophys Res Commun* 389:616–621.

34. Wang J, Henkin TM, Nikonowicz EP. 2010. NMR structure and dynamics of the Specifier Loop domain from the *Bacillus subtilis tyrS* T box leader RNA. *Nucleic Acids Res* 38:3388–3398.

35. Chang AT, Nikonowicz EP. 2013. Solution NMR determination of hydrogen bonding and base pairing between the *glyQS* T box riboswitch Specifier domain and the anticodon loop of tRNA^{Gly}. *FEBS Lett* 587:3495–3499.

36. Schroeder KT, McPhee SA, Ouellet J, Lilley DMJ. 2010. A structural database for k-turn motifs in RNA. *RNA* 16:1463–1468.

37. Wang J, Nikonowicz EP. 2011. Solution structure of the K-turn and Specifier Loop domains from the *Bacillus subtilis tyrS* T-box leader RNA. *J Mol Biol* 408:99–117.

38. Lilley DM. 2014. The K-turn motif in riboswitches and other RNA species. *Biochim Biophys Acta* 1839:995–1004.

39. Chang AT, Nikonowicz EP. 2012. Solution nuclear magnetic resonance analyses of the anticodon arms of proteinogenic and nonproteinogenic tRNA^{Gly}. *Biochemistry* 51:3662–3674.

40. Dunkle JA, Wang L, Feldman MB, Pulk A, Chen VB, Kapral GJ, Noeske J, Richardson JS, Blanchard SC, Cate JH. 2011. Structures of the bacterial ribosome in classical and hybrid states of tRNA binding. *Science* 332:981–984.

41. **Zhang J, Ferré-D'Amaré AR.** 2015. Structure and mechanism of the T-box riboswitches. *Wiley Interdiscip Rev RNA* 6:419–433.

42. **Fang X, Michnicka M, Zhang Y, Wang YX, Nikonowicz EP.** 2017. Capture and release of tRNA by the T-loop receptor in the function of the T-box riboswitch. *Biochemistry* 56:3549–3558.

43. **Chetnani B, Mondragón A.** 2017. Molecular envelope and atomic model of an anti-terminated *glyQS* T-box regulator in complex with tRNA^Gly. *Nucleic Acids Res* 45:8079–8090.

44. **Thomas JR, Hergenrother PJ.** 2008. Targeting RNA twith small molecules. *Chem Rev* 108:1171–1224.

45. **Zhou S, Means JA, Acquaah-Harrison G, Bergmeier SC, Hines JV.** 2012. Characterization of a 1,4-disubstituted 1,2,3-triazole binding to T box antiterminator RNA. *Bioorg Med Chem* 20:1298–1302.

46. **Means JA, Hines JV.** 2005. Fluorescence resonance energy transfer studies of aminoglycoside binding to a T box antiterminator RNA. *Bioorg Med Chem Lett* 15:2169–2172.

47. **Orac CM, Zhou S, Means JA, Boehm D, Bergmeier SC, Hines JV.** 2011. Synthesis and stereospecificity of 4,5-disubstituted oxazolidinone ligands binding to T-box riboswitch RNA. *J Med Chem* 54:6786–6795.

48. **Zhou S, Acquaah-Harrison G, Bergmeier SC, Hines JV.** 2011. Anisotropy studies of tRNA-T box antiterminator RNA complex in the presence of 1,4-disubstituted 1,2,3-triazoles. *Bioorg Med Chem Lett* 21:7059–7063.

49. **Stamatopoulou V, Apostolidi M, Li S, Lamprinou K, Papakyriakou A, Zhang J, Stathopoulos C.** 2017. Direct modulation of T-box riboswitch-controlled transcription by protein synthesis inhibitors. *Nucleic Acids Res* 45:10242–10258.

50. **Dar D, Shamir M, Mellin JR, Koutero M, Stern-Ginossar N, Cossart P, Sorek R.** 2016. Term-seq reveals abundant ribo-regulation of antibiotics resistance in bacteria. *Science* 352:aad9822.

Regulating with RNA in Bacteria and Archaea
Edited by Gisela Storz and Kai Papenfort
© 2018 American Society for Microbiology, Washington, DC
doi:10.1128/microbiolspec.RWR-0006-2017

rRNA Mimicry in RNA Regulation of Gene Expression

7

Michelle M. Meyer[1]

Despite the many roles for RNA as a regulator in eukaryotes, archaea, and bacteria, the rRNA is the most abundant cellular RNA and the size of the rRNA outstrips nearly all other functional RNAs. Furthermore, the ribosome is also composed of >50 ribosomal proteins (r-proteins), the majority of which directly contact the rRNA, forming specific interactions with RNA (1). Since most regulatory RNAs in bacteria appear to be relatively recent inventions (2–5), they most certainly have evolved in the context of abundant rRNA and r-proteins, and thus have been shaped by them. Many regulatory RNA structures contain portions that bear strong resemblance to motifs within the rRNA. Some of this similarity is due to the role that rRNA plays in our understanding of RNA structure, and in other cases it is due to interaction with an r-protein. This review will first illustrate the role of the ribosome in our understanding of RNA structures generally and subsequently examine how r-proteins may interact with RNA outside the ribosome to act in a regulatory capacity.

THE rRNA AS A SOURCE OF RNA STRUCTURAL MOTIFS

The rRNA plays an outsized role in our general understanding of RNA structure. Despite more than a decade since publication of the initial high-resolution ribosome structures and significant growth in the number and diversity of RNA structures in the Protein Data Bank, the rRNA still represents a significant proportion of the three-dimensional structure information available for RNA and RNA-protein complexes. Of the 3,692 structures containing RNA, 1,082 contain segments derived from the rRNA or otherwise associated with the ribosome. The ribosome has also significantly influenced the

development of RNA structure descriptions (6). Many recurring RNA structure motifs, such as kink-turn (k-turn), loop-E, and loop-C motifs (7–11), were first recognized in the context of the ribosome, and our knowledge of the sequences that may fold into many such features is heavily influenced by rRNA alignments (12–14). These structural motifs form the basis of not only the rRNA but many other structured RNAs including riboswitches (15), T boxes (16), as well as other catalytic RNAs such as the group I and II introns (17, 18). Several reviews specifically addressing the roles such motifs play in RNA structure are available (19–21).

r-PROTEINS AS AUTOGENOUS REGULATORS

Many r-proteins have secondary functions (22, 23) as negative regulators of their own synthesis. r-Proteins and other protein components necessary for translation can account for up to 40% of cellular proteins (24) and 41% of active translation in actively growing cells in rich medium (25). Thus, maintaining stoichiometry among the >60 ribosome components is essential for efficient resource utilization, and the mRNA structures responsible for implementing regulation are only one of several regulatory layers. In *Escherichia coli*, over half of the r-protein operons are regulated by autogenous regulatory mechanisms where an individual r-protein will bind to a portion of its own transcript to inhibit transcription or translation. Often the mRNA will take a structure that bears significant similarity to the rRNA; however, there are several different paradigms for RNA-protein recognition that are embodied by the mRNA structures that mediate r-protein autogenous regulation.

---

[1]Department of Biology, Boston College, Chestnut Hill, MA 02467.

Discovery of r-Protein Autogenous Regulatory mRNA Structures

The mRNA structures enabling regulation of r-protein synthesis in *E. coli* were among the first mRNA regulatory sites discovered. Many distinct *E. coli* examples were described based on similar observations and using the same experimental approaches. Initial studies demonstrated that overexpression of specific r-proteins resulted in inhibited synthesis of entire r-protein operons (26, 27), and that these effects were operon specific (28). Using *in vitro* transcription/translation systems as well as reporter gene assays, the inhibitory properties for several r-proteins including L1, L4, S4, S7, S8, and the L10(L12)$_4$ complex were uncovered (29–33). Most mechanisms involve inhibition of translation (30, 34–37); however, alterations to the mRNA decay rate (27, 38–40) and attenuation (premature transcription termination) mechanisms also occur in conjunction with translational inhibition (41, 42).

In many cases, mimicry between the mRNA regulatory sites responding to an r-protein and its rRNA binding site was proposed as soon as a DNA sequence became available (e.g., S4, S7, S8, L1, L4, and L10 [35, 43–46]). However, demonstration of direct RNA-protein contacts that such similarity would imply lagged behind the speculation considerably (47–49). In several cases, proposed similarities were merely the result of sequence gazing, and it has become apparent that the rRNA and mRNA binding sites do not have structural similarity (e.g., S4 and L4) (50, 51). In other cases, the initially observed similarity between the mRNA and rRNA was verified when three-dimensional structural data became available (e.g., S8 and L1) (52, 53).

Since the initial discoveries of r-protein autogenous mRNA structures in *E. coli*, an additional 9 mRNA structures responding to r-proteins (S1, S2, S15, S20, L19, L20 [2 sites], and L25) (54–60) or r-protein complexes (S6:S18) (61–63) have been described in *E. coli*, and today there are a total of 15 r-protein-interacting mRNA structures described in *E. coli* (Fig. 1A; Table 1). Many have been extensively characterized, but for others the mechanisms of action, or even whether a direct RNA-protein interaction occurs, remain undetermined. With some exceptions, the complement of r-proteins and organization of r-protein operons are largely conserved across bacterial species (64, 65). However, many of the structures allowing regulation in *E. coli* are not widely distributed to organisms outside of a few orders of gammaproteobacteria (66–71). Furthermore, most enterobacterial endosymbionts appear to have lost these structures during the course of genome reduction

(71, 72). The only organism with significant study of r-protein regulation other than *E. coli* is the Gram-positive model bacterium *Bacillus subtilis*. This organism shares the mRNA binding sites that interact with r-proteins L1, L10, S2, and S6 with *E. coli* (Fig. 1; Table 1), but the other 11 structures known in *E. coli* are not apparent in *B. subtilis* or its relatives. Alternative regulatory structures that respond to S4, L20, and S15 have been described (73–75) (Fig. 1B; Table 1).

With the growing number and diversity of sequenced bacterial genomes, comparative genomics has also proved to be a powerful approach for discovery. The combination of RNA-specific homology search tools (76) and the availability of RNA structural families corresponding to most known r-protein-responsive structures (71, 75, 77) enables accurate annotation of these structures in bacterial genomes. In addition to characterized mRNA structures, hundreds of novel putative *cis*-regulatory mRNA motifs have been identified in bacterial genomes, many of which are associated with r-proteins or bear resemblance to the rRNA (78–82). The low cost of sequencing has also enabled the direct discovery of regulatory RNAs through comparative transcriptomics (4), 5′-end sequencing (83), and RNA-protein immunoprecipitations (84). However, relatively few such motifs have been experimentally validated.

The S8-Interacting mRNA Structure: A Prototype r-Protein *cis*-Regulatory RNA

The mRNA segment bound by r-protein S8 to regulate the *spc* operon is the prototype mRNA binding motif that embodies all the properties initially hypothesized for all mRNA structures bound by r-proteins. S8 is a primary rRNA binding protein that interacts with the rRNA early during ribosome assembly. The interaction site for S8 on the mRNA is within the intergenic region between *rplX* and *rplE* (encoding L24 and L5) and the coding region of *rplE* (Fig. 1A). S8 inhibits translation of several proteins following the protein binding site (L5, S14, S8, L6, L18, S5, L20, and L15), and there is evidence that the two genes upstream of the S8 binding region (*rplN* and *rplX*, encoding L14 and L24) are also downregulated in response to S8 due to increased mRNA degradation (40, 85).

The initial observed sequence similarity between the rRNA and mRNA binding sites for S8 extends to shared secondary structure (35, 86–88) and three-dimensional structure (52, 86) (Fig. 2). The S8 binding site consists of an internal loop. The motif centers on two internal Watson-Crick base pairs that are separated from the rest of the pairing element by bulged bases on either side,

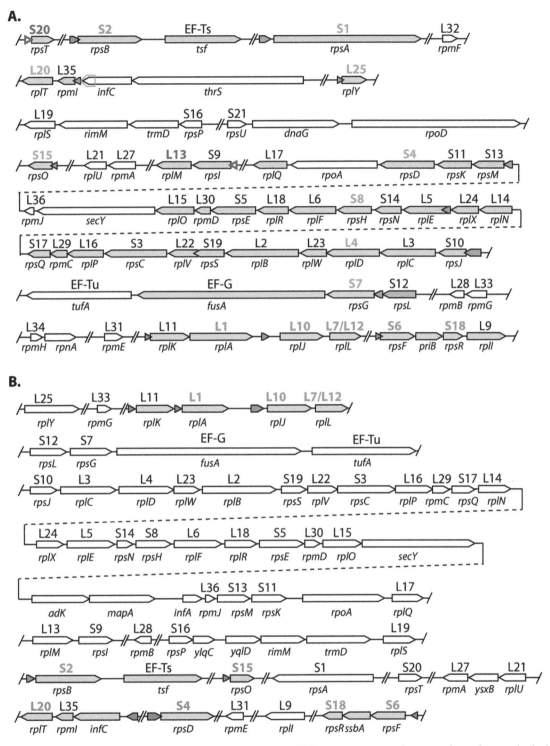

Figure 1 Diagrams of r-protein operons from *E. coli* (A) and *B. subtilis* (B). Genes are shown in the order in which they appear in the genome and to scale. Gray genes are subject to r-protein autogenous regulation; white genes have no described autogenous regulation. Colored arrows represent r-protein RNA binding structures. Red arrows indicate structures that are widely distributed to many bacterial phyla, blue arrows indicate RNA structures that are confined to *Gammaproteobacteria*, green arrows indicate RNA structures confined to *Firmicutes*, and purple arrows indicate presumed r-protein binding sites where no explicit RNA secondary structure has been described. For each operon the effector protein is colored to match the RNA site with which it interacts.

Table 1 Summary of r-protein-interacting mRNAs that allow regulation of r-protein genes

Binding partner	Regulated genes	Position	Species distribution[a]
L1[b,c,d,e]	rplA, rplK,[f] rplP1[f]	Varied	Archaea/Bacteria
L4[b,c,d,e]	rpsJ, rplC, rplD, rplW, rplB, rpsS, rplV, rpsC, rplP, rpmC, rpsQ	rpsJ 5′ UTR	Gammaproteobacteria
L10/L10(L12)₄[b,c,d,e]	rplJ, rplL	rplJ 5′ UTR	Bacteria
L13[b]	rplM, rpsI	rplM 5′ UTR	Escherichia coli
L20[b,c,d,e]	rpmI, rplT	infC 5′ UTR	Firmicutes
L20[b,c,d,e]	rpmI, rplT	infC-rpmI intergenic	Gammaproteobacteria
L20[b,c,d,e]	rpmI, rplT	infC coding/infC-rpmI intergenic	Escherichia coli
L25[b,d]	rplY	rplY 5′ UTR	Gammaproteobacteria
S1[b,c,e]	rpsA	rpsA 5′ UTR	Gammaproteobacteria
S2[b,d]	rpsB	rpsB 5′ UTR	Bacteria
S4[b,c,d,e]	rpsM, rpsK, rpsD, rplQ	rpsM 5′ UTR	Gammaproteobacteria
S4[b,d]	rpsD	rpsD 5′ UTR	Firmicutes
S6:S18[b,c,d]	rpsF, rpsR, rplI[f]	rpsF 5′ UTR	Bacteria
S7[b,c,d]	rpsL, rpsG, fusA	rpsL-rpsG intergenic	Gammaproteobacteria
S8[b,c,d,e]	rplN, rplX, rplE, rpsN, rpsH, rplF, rplR, rpsE, rpmD, rplO, secY, rpmJ	rplX-rplE intergenic	Gammaproteobacteria
S15[b,c,d,e]	rpsO	rpsO 5′ UTR	Gammaproteobacteria
S15[c,d]	rpsO	rpsO 5′ UTR	Firmicutes
S15[c,d]	rpsO	rpsO 5′ UTR	Thermus thermophilus
S15[b,c,d]	rpsO	rpsO 5′ UTR	Alphaproteobacteria
S20[b,c]	rpsT	rpsT 5′ UTR	Escherichia coli

[a]Where a single species is listed for distribution, either no structure is available or no comparative genomic work has been conducted for the RNA and only the species of characterization is given.
[b]Regulation demonstrated using in vitro transcription/translation system or reporter gene assays.
[c]Direct RNA-protein interaction demonstrated in vitro.
[d]Structure of mRNA binding site characterized.
[e]Mechanism of action known.
[f]May only be regulated in some species.

although many of the base identities are not strongly conserved in the case of the rRNA (Fig. 2B). S8 itself directly contacts the minor groove of the internal loop. Structures of the mRNA and rRNA are directly superimposable (Fig. 2C) (52). The major difference between the rRNA and mRNA binding sites is an additional bulged base in the mRNA structure a few nucleotides away from the S8 recognition sites (orange). While this base decreases binding affinity by about 10-fold, it does not directly interact with S8. Despite a highly conserved rRNA-S8 interface across all bacteria (88–92) and archaea (93), the S8-responsive regulatory RNA structure observed in E. coli is narrowly distributed to a few orders of Gammaproteobacteria (71). What, if any, regulation occurs in other organisms has not yet been characterized, and the causes of the narrow distribution are unclear. The phylogenetic distribution of the S8-interacting mRNA structure is similar to those of many r-protein mRNA regulators identified in E. coli, suggesting that similar selective constraints influenced the evolution of all the regulatory structures. The preponderance of known structures in E. coli is likely due to a significant discovery bias. Ribosome assembly and stoichiometry is by far the best studied in E. coli. Similar regulators may be present but as of yet unidentified in other bacteria. The narrow distribution displayed by most of the E. coli structures makes it more difficult to utilize comparative genomic approaches for discovery, and it is likely that several of the characterized motifs in E. coli would not be easily rediscovered using state-of-the-art comparative genomic tools.

The L10(L12)₄-Interacting Regulatory Structure: Homologous Binding Sites, Different Mechanisms of Action

The L10(L12)₄-interacting mRNA structure also represents a mimic of the rRNA (47, 48, 94–96) and participates in the regulation of translation initiation in E. coli (34, 97), directly impacting only rplJ and rplL (Fig. 1A). Sequence similarity between the mRNA and rRNA binding sites has been described (46, 68), but the L10(L12)₄ complex is typically not resolved in ribo-

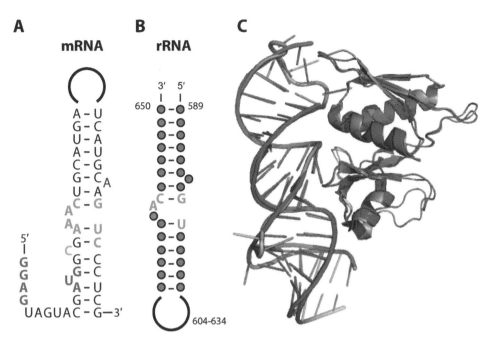

Figure 2 S8 mRNA binding site in *E. coli* mRNA (A) and rRNA (B) consensus structure. Green nucleotides indicate Shine-Dalgarno sequence and translational start; red nucleotides directly contact S8 in the three-dimensional structure (52). rRNA nucleotides conserved <90% are shown as filled circles; nucleotides conserved ≥90% are indicated by letters. Numbering corresponds to bacterial consensus sequence (129). (C) Aligning structural data for each site based on the S8 protein backbone shows that the two binding sites are superimposable. The structure of the S8 with its mRNA binding site is shown in green (1s03.cif [52]), and the structure of S8 interacting with the rRNA is shown in blue (4v9d.cif [130]). Bases of the mRNA directly contacting S8 are colored red; a bulged base in the mRNA that differentiates the rRNA and mRNA binding sites is colored orange.

some crystal structures and three-dimensional data for an mRNA-L10(L12)₄ complex are not available. The L10(L12)₄ binding site consists of a k-turn motif that is four base pairs away from an internal loop containing a pair of adenosines. In the rRNA the internal loop is a multi-stem junction (Fig. 3), while in the mRNA the structure is often a bulge, but may be a multi-stem junction (71). In both the rRNA and the mRNA the adenosines are highly conserved, and mutating them reduces binding affinity substantially (68).

In contrast to the S8-interacting mRNA structure, the RNA structure responsible for interacting with L10 in *E. coli* is widely conserved throughout many bacterial species (68, 71, 80). However, the mechanism of action is not the same across all species. In many Gram-positive species, the L10-interacting structure is followed by an intrinsic transcription terminator (80), and the mechanism of regulation in *B. subtilis* is regulation of transcription termination (98). Thus, r-protein binding structures are similar to riboswitches where homologous sensor domains may utilize different mechanisms of action in diverse species (99).

The L1-Interacting mRNA Structure: Convergence on the Same Binding Determinants

Like the L10- and S8-responsive mRNA structures, the L1-interacting mRNA structure shows obvious similarity to the rRNA (45, 53, 100), and examples of the L1 recognition site are found across nearly all bacterial phyla (71) as well as archaea (101–103). The binding determinants for L1 are often accommodated in a short hairpin of <30 nucleotides and consist of a base-paired region containing an asymmetric internal loop closed by a noncanonical A•G pairing (Fig. 4A). In three dimensions this corresponds to two canonical helixes, one of which is capped by the noncanonical A · G pair, that are separated by a sharp turn (53). Diverse L1 homologs are able to interact with an example of the mRNA binding site from *Methanococcus vannielii* (103), and structural data show that the rRNA and mRNA sites are nearly superimposable (53).

In *E. coli*, the binding site is within the 5′ untranslated region (5′ UTR) of the transcript encoding both *rplK* and *rplA* and L1 regulates translation initiation

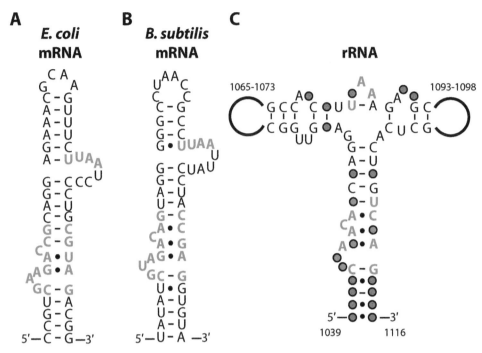

Figure 3 L10(L12)₄ mRNA binding sites from (A) *E. coli*, (B) *B. subtilis,* and the rRNA consensus structure (C). Red nucleotides are implicated in binding; rRNA nucleotides conserved <90% are shown as filled circles; nucleotides conserved ≥90% are indicated by letters. Numbering corresponds to bacterial consensus sequence (129).

of L11 and L1 (45) (Fig. 4). Surprisingly, in the archaea *M. vannielii, Methanocaldococcus jannaschii,* and *Methanococcus thermolithotrophicus,* the L1 binding site appears ~30 nucleotides inside the coding region for L1 and regulates translation of L1, L10, and P1 (homolog of L12 [104]), and the gene encoding L11 occurs elsewhere in the genome (101, 103). In *Sulfolobus solfataricus,* the L1 binding site is found within the L11 coding region, which directly precedes genes encoding L1, L10, and P1 (Fig. 4B) (103).

In addition to examples that have been explicitly examined, a systematic homology search for L1 binding sites in bacterial genomes identified the site within transcripts encoding L1 and L11 in many bacterial species (71). However, like the examples identified in *Archaea,* the location of the binding site relative to the coding regions is not consistent. In *Cyanobacteria, Actinobacteria,* and *Chloroflexi,* the L1 binding site precedes *rplA,* typically between *rplK* and *rplA.* In *Proteobacteria, Spirochaetes, Thermotogae,* and *Tenericutes,* the binding site precedes *rplK,* presumably to regulate both *rplA* and *rplK.* Furthermore, in many species of *Firmicutes,* L1 binding sites appear preceding both *rplA* and *rplK.* In *Geobacillus kaustophilus,* both sites are capable of binding L1 *in vitro* (71). Interestingly, there is evidence of loss for each individual bind-

ing site within species scattered throughout *Firmicutes* (Fig. 4B). The combination of the wide distribution and changing position of the L1 binding site relative to the regulated genes suggests that the site may have evolved convergently in many species.

L20-Interacting mRNA Regulatory Structures: Diverse Scaffolds Support the Same Binding Determinants

In addition to cases where there is a single mRNA binding site that mimics the rRNA, there are also cases where homologous r-proteins interact with distinct mRNA secondary structures in different bacterial species. Three L20-interacting mRNA structures are known, two in *E. coli* and one in *B. subtilis.* Each structure mimics the rRNA, but uses a different arrangement of secondary structure to support the necessary bases in the correct geometry required for recognition (Fig. 5). In *E. coli,* two L20-responsive mRNA structures control the IF3 operon (*infC, rpmI,* and *rplT,* encoding IF3, L35, and L20). One structure is found within the intergenic region between *infC* and *rpmI* (70) (Fig. 5A), and consists of a relatively straightforward bulged stem-loop where the binding site includes a pair of adenosines within the bulge and a set of consecutive G-C base pairs just after the closing base pair of the loop (Fig. 5C, mRNA-

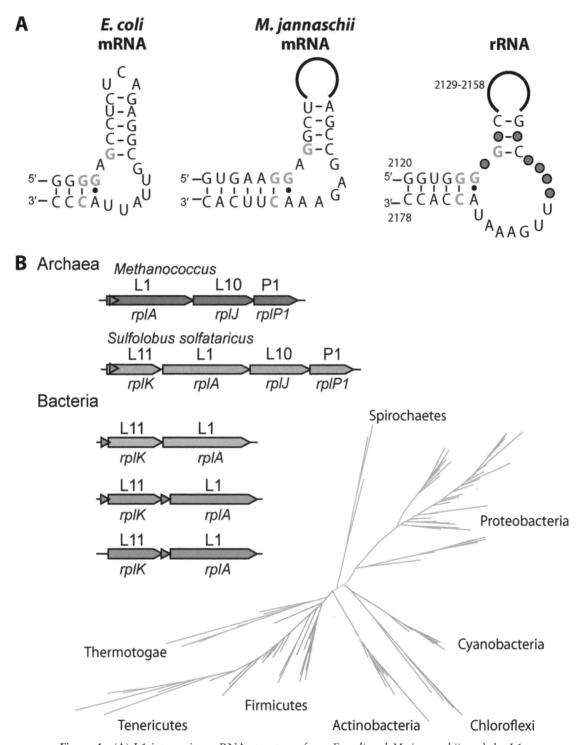

Figure 4 (A) L1-interacting mRNA structures from *E. coli* and *M. jannaschii*, and the L1 rRNA binding site (bacterial consensus). Red nucleotides directly contact L1 in the three-dimensional structure (53). rRNA nucleotides conserved <90% are shown as filled circles; nucleotides conserved ≥90% are indicated by letters. Numbering corresponds to bacterial consensus sequence (129). (B) Diagrams indicating the genomic position of L1 mRNA binding sites in two *Archaea* clades (several *Methanococcus* species and *Sulfolobus solfataricus*) and in bacteria. Proteins encoded by rplJ and rplP0 are homologous. Bacterial genomic positions of L1 binding site are mapped to a 16S rRNA tree.

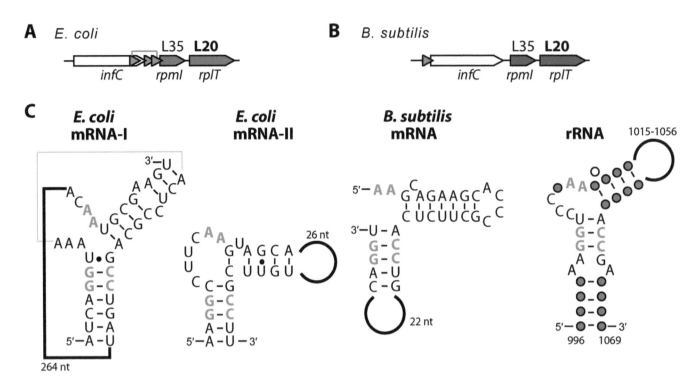

Figure 5 Diagram of *infC* operons showing genomic positions of L20-interacting mRNA structures (red arrows) in *E. coli* (A) and *B. subtilis* (B). Genes regulated by the RNA structure are colored. (C) L20-interacting mRNA structures from *E. coli* (mRNA-I and mRNA-II) and *B. subtilis* and the consensus rRNA L20 binding site. Red nucleotides are important for L20 interaction. rRNA nucleotides conserved <90% are shown as filled circles; nucleotides conserved ≥90% are indicated by letters. Numbering corresponds to bacterial consensus sequence (129).

II). This arrangement is the closest mimic of the rRNA. The second structure comprises a pseudoknot formed by long-range interactions between a sequence within *infC* and a sequence adjacent to the start of the *rpmI* coding region (54, 70) (Fig. 5C, mRNA-I). In this structure, the pair of adenosines is found in the single-stranded region just prior to the 3′-most portion of the pseudoknot. Both of these structures are required for full translational repression of the operon *in vivo*, and L20 binds independently to each (70, 105). A high-quality alignment and phylogenetic distribution is only available for the mRNA structure preceding *rpmI*. The pseudoknotted binding site is challenging to identify using RNA-specific homology search tools (106) due to its significant overlap with coding sequence, long-range interactions, and pseudoknotted structure. However, the structure preceding *rpmI* is narrowly distributed to *Gammaproteobacteria*.

In addition to the two L20-responsive structures in *E. coli*, L20 binds to a regulatory structure in *B. subtilis* that precedes *infC* (Fig. 5B and C). While this structure shares many features with the L20-inter-acting structures identified in *E. coli*, the binding features present near the multi-stem junction are supported by a different arrangement of secondary structure, and the ordering of the elements with respect to one another in a linear sequence is distinct. A potential intrinsic transcription terminator follows this mRNA structure, and the mechanism of action is L20-induced structural change resulting in early transcription termination (74). In this structure, the conserved adenosines are in a single-stranded region just 5′ of the first hairpin and the G-C pairs in the second hairpin. This structure is found in most *Firmicutes* (75, 80), although more frequently in the class *Bacilli* than in *Clostridia*, and the transcription attenuation mechanism appears conserved in these organisms. While *infC* is part of this operon, in *B. subtilis* IF3 levels are decoupled from those of L20 and L35 through the presence of a second upstream promoter. The transcript produced from this promoter is cleaved by RNase Y and only allows trans-lation of L20 and L35 translation (107).

The three L20 sites all present the same effective binding geometry for L20 recognition (Fig. 5C). Both *E. coli* structures are capable of interacting with an L20 homolog from *Aquifex aeolicus* to repress gene expression from *rpmI'-'lacZ* reporter constructs (108), and the *E. coli* homolog is able to stimulate premature transcription termination during *in vitro* assays with the *B. subtilis* mRNA structure (74). Thus, the L20 regulators serve as an example of how the same three-dimensional geometry may be supported by different arrangements of secondary structure elements. This example also illuminates how challenging identification of common binding sites may be. Despite similar binding determinants, the distinct arrangements of the necessary recognition elements make automatic detection difficult or impossible.

S15-Interacting Regulatory Structures: Diverse Binding Determinants Produce Diverse Structures

r-Protein S15 also regulates gene expression using multiple distinct mRNA binding sites in diverse bacterial species. To date, four different S15-interacting mRNA structures spanning several bacterial phyla have been experimentally characterized (109–112) and several additional putative structures identified (112) (Fig. 6). Each of these structures directly precedes and controls expression of *rpsO*, the gene encoding S15. In *E. coli* the mechanism of action is through entrapment of the translation initiation complex (110), but in other species the mechanism has not explicitly been characterized.

The structures share very little in the way of a single recognizable sequence or structural motif. This is partially due to the bidentate nature of the S15 binding site on the rRNA. S15 recognizes two portions of the 16S rRNA: a multi-stem junction and a stem containing a slight defect characterized by a G•U/G-C set of base pairs directly adjacent to the junction (Fig. 6). The mRNA regulatory structures that interact with S15 often only partially mimic this binding site. For example, the mRNA from *Thermus thermophilus* includes a three-stem junction formed by the bases of three adjacent pairing elements. However, the pairing elements themselves show no evidence for the G•U/G-C defect recognized by S15 in the rRNA. In contrast, mRNA structures described from *E. coli*, *Rhizobium radiobacter* (formerly *Agrobacterium radiobacter*), and *Geobacillus stearothermophilus* (previously *Bacillus stearothermophilus*) include more obvious mimics of the G•U/G-C elements, and require this element for interaction. In several cases the mRNAs have additional recognition elements that are necessary but do not directly mimic the rRNA (Fig. 6) (109, 113, 114).

The differences between the mRNAs are sufficiently large such that specificity of interaction has been reported (113, 115). For example, the S15 homolog from *G. kaustophilus* does not interact with the mRNA structure originating from *E. coli* and the S15 homolog from *T. thermophilus* does not interact with several of the mRNA structures containing only the G•U/G-C motif and no mimic of the three-stem junction (113, 115). Mutagenesis studies indicate that the same face of S15 appears to be used for interaction (115, 116). However, in *E. coli* different S15 amino acids are implicated in rRNA and mRNA binding (114, 117). Furthermore, selective recognition of the *Geobacillus* and *E. coli* mRNA structures may be traced to specific amino acids that are differentially conserved in S15 homologs originating from organisms containing RNAs of each type (113). Thus, the diversity of S15-interacting structures is due not only to the bidentate recognition site that may allow a larger set of potential interaction partners but also to differences in the protein homologs that change recognition. These findings suggest that despite very similar rRNA recognition sites across all bacteria, the r-proteins and their mRNA binding sites are influencing each other's evolution.

BEYOND AUTOGENOUS REGULATION: L7Ae

Archaeal r-protein L7Ae participates in processes well beyond its role in the ribosome. L7Ae interacts with k-turn and k-loop motifs as a component of the ribosome (1), RNase P (118), the C/D box and H/ACA box snoRNPs responsible for site-selective 2'-O-methylation (119, 120), and in mammals an L7Ae homolog binds to the U4 snRNP of the spliceosome (121). There is no r-protein that directly corresponds to L7Ae in prokaryotes. Two L7Ae homologs in *B. subtilis* exist, and both bind to k-turns (122), but their biological function is unknown. The role of the k-turn as a fundamental RNA structural building block has already been discussed. L7Ae specifically recognizes this motif, and therefore has a role in many RNA complexes, primarily to stabilize RNA structure.

A recent RNA immunoprecipitation sequencing (RIP-seq) study of L7Ae in *Sulfolobus acidocaldarius* identified several mRNA fragments in addition to the expected interaction partners (84). Many of these mRNA fragments contained sequences corresponding to the consensus sequences for a k-turn, suggesting a biologically relevant interaction. Among the mRNAs identified are those encoding L7Ae, Nop5, and fibrillarin (other components of the snoRNP), a hypothetical DNA binding protein, and a hypothetical glycosyltransferase. Subsequent reporter gene assays and phylogenetic analysis showed that L7Ae negatively regulates the transcript encoding L7Ae in *S. acidocaldarius* and

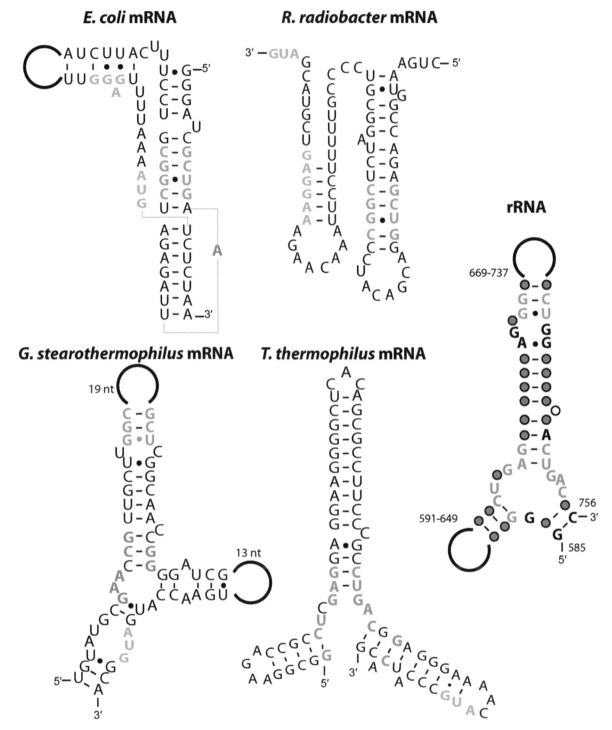

Figure 6 S15-interacting mRNA structures in different bacterial phyla and the consensus S15 rRNA binding site. Red nucleotides correspond to the rRNA three-stem junction and its direct mimics. Blue nucleotides correspond to G•U/G-C helix imperfection in the rRNA binding site and its mimics in mRNA structures. Purple nucleotides are important for S15 recognition but do not directly correspond to any rRNA motif. Green nucleotides correspond to Shine-Dalgarno or translational start sequences. rRNA nucleotides conserved <90% are shown as filled circles; nucleotides conserved ≥90% are indicated by letters. Numbering corresponds to bacterial consensus sequence (129).

several other diverse *Archaea* species. The presence of k-turn motifs preceding several genes and L7Ae interaction with these motifs suggests that L7Ae may regulate not only its own synthesis but also synthesis of its interaction partners in snoRNPs, Nop5 and fibrillarin (84).

ENGINEERED r-PROTEIN-RESPONSIVE REGULATORY RNA SYSTEMS

r-Protein binding motifs have also been used to create synthetic regulatory systems. Repressing systems designed for eukaryotic cells were created by placing the L7Ae binding site near the translational start site (123), allowing L7Ae to prevent translation initiation. Systems with the L7Ae binding site within the coding region proved more effective than those where the binding site was placed in the 5′ UTR, both in an *in vitro* translation system and within HeLa cells. Activating systems in which L7Ae binding removes a *trans*-acting RNA to prevent translation also proved effective *in vitro*. In addition, L7Ae-mediated activation was achieved in HeLa cells by adding an L7Ae binding site to a synthetic shRNA (short hairpin RNA); thus L7Ae binding prevented shRNA-mediated mRNA degradation (124). These examples demonstrate how the L7Ae protein binding site may be easily transferred to an alternative context and harnessed for gene expression in a modular manner.

Indeed, creation of synthetic regulatory systems responding to r-proteins within cells appears to be facile in comparison to the creation of many types of synthetic regulators where the transition from *in vitro* to *in vivo* can be challenging (125). Several synthetic regulatory systems responding to r-protein S15 have also been created (126). Unlike the L7Ae examples, these regulators were created through *in vitro* selection of RNA aptamers interacting with r-protein S15 from *G. kaustophilus* rather than transplantation of a known binding site. One striking observation from this work is that even without explicit selection for regulation, a high proportion of aptamers enable regulation when positioned correctly relative to the start codon. A second finding is that r-protein S15 can interact with a wide variety of different binding sites (127). This observation is echoed in previous work where *in vitro* selection to r-protein S8 yielded both aptamers similar to the natural RNA binding partners as well as those showing substantial differences (128).

CONCLUDING REMARKS

Regulatory RNA structures displaying motifs found in the rRNA are commonly identified. While in some cases similarity may be due to shared RNA tertiary structure motifs, in other cases structural similarity can imply a shared r-protein binding partner. Many r-proteins have a secondary role as negative regulators of their own synthesis, and while it was postulated that all such regulatory structures would resemble the rRNA, this has proved true only in some cases. This review illustrates a range of different regulatory mRNA structures that display similarity to the rRNA, but it is by no means exhaustive. While the mRNA structures controlling r-protein synthesis in *E. coli* remain the best characterized, r-protein-responsive mRNA structures hail from nearly all species of bacteria and several archaea. From these examples it is apparent that r-protein-responsive mRNA structures can be direct and obvious mimics of the rRNA, but they do not have to be. Many r-protein-interacting mRNA structures display no similarity to their cognate rRNA sites (e.g., *E. coli* S4 regulator), while others share only partial similarity. Second, very similar binding sites can appear in diverse organisms, but they may use alternative mechanisms to regulate gene expression or display different positioning relative to regulated genes. Third, due to the structural plasticity of RNA, a geometrically similar binding site may be displayed in several very different manners. Finally, from the diversity of natural regulatory mRNA structures, *in vitro* selection of aptamers, and design of r-protein-responsive regulatory mechanisms, it is clear that the sequence space that allows for r-protein binding and subsequent gene regulation may be quite large. This conclusion combined with the lack of knowledge of r-protein regulation outside of *E. coli* suggests that many r-protein-responsive mRNA structures, including those not directly associated with r-protein operons, remain undiscovered or unverified.

Acknowledgments. This material is based on work partially supported by National Science Foundation grants MCB-1411970 and -1715440 to M.M. Special thanks to Arianne Babina and Betty Slinger for their hard work and insightful discussions of this topic. The author declares no conflicts.

Citation. Meyer MM. 2018. rRNA mimicry in RNA regulation of gene expression. Microbiol Spectrum 6(2):RWR-0006-2017.

References

1. **Ban N, Nissen P, Hansen J, Moore PB, Steitz TA.** 2000. The complete atomic structure of the large ribosomal subunit at 2.4 A resolution. *Science* **289:**905–920.

2. **Peer A, Margalit H.** 2014. Evolutionary patterns of *Escherichia coli* small RNAs and their regulatory interactions. *RNA* **20:**994–1003.

3. **Hoeppner MP, Gardner PP, Poole AM.** 2012. Comparative analysis of RNA families reveals distinct repertoires for each domain of life. *PLoS Comput Biol* **8:**e1002752.

4. Lindgreen S, Umu SU, Lai AS, Eldai H, Liu W, McGimpsey S, Wheeler NE, Biggs PJ, Thomson NR, Barquist L, Poole AM, Gardner PP. 2014. Robust identification of noncoding RNA from transcriptomes requires phylogenetically-informed sampling. *PLoS Comput Biol* 10:e1003907.

5. Kacharia FR, Millar JA, Raghavan R. 2017. Emergence of new sRNAs in enteric bacteria is associated with low expression and rapid evolution. *J Mol Evol* 84:204–213.

6. Leontis NB, Westhof E. 2002. The annotation of RNA motifs. *Comp Funct Genomics* 3:518–524.

7. Klein DJ, Schmeing TM, Moore PB, Steitz TA. 2001. The kink-turn: a new RNA secondary structure motif. *EMBO J* 20:4214–4221.

8. Lescoute A, Leontis NB, Massire C, Westhof E. 2005. Recurrent structural RNA motifs, isostericity matrices and sequence alignments. *Nucleic Acids Res* 33:2395–2409.

9. Fox GE, Woese CR. 1975. 5S RNA secondary structure. *Nature* 256:505–507.

10. Leontis NB, Westhof E. 1998. A common motif organizes the structure of multi-helix loops in 16 S and 23 S ribosomal RNAs. *J Mol Biol* 283:571–583.

11. Correll CC, Freeborn B, Moore PB, Steitz TA. 1997. Metals, motifs, and recognition in the crystal structure of a 5S rRNA domain. *Cell* 91:705–712.

12. Woese CR, Winker S, Gutell RR. 1990. Architecture of ribosomal RNA: constraints on the sequence of "tetraloops". *Proc Natl Acad Sci U S A* 87:8467–8471.

13. Gutell RR, Larsen N, Woese CR. 1994. Lessons from an evolving rRNA: 16S and 23S rRNA structures from a comparative perspective. *Microbiol Rev* 58:10–26.

14. Wu JC, Gardner DP, Ozer S, Gutell RR, Ren P. 2009. Correlation of RNA secondary structure statistics with thermodynamic stability and applications to folding. *J Mol Biol* 391:769–783.

15. Serganov A, Huang L, Patel DJ. 2008. Structural insights into amino acid binding and gene control by a lysine riboswitch. *Nature* 455:1263–1267.

16. Wang J, Nikonowicz EP. 2011. Solution structure of the K-turn and specifier loop domains from the *Bacillus subtilis tyrS* T-box leader RNA. *J Mol Biol* 408:99–117.

17. Strobel SA, Adams PL, Stahley MR, Wang J. 2004. RNA kink turns to the left and to the right. *RNA* 10:1852–1854.

18. Keating KS, Toor N, Perlman PS, Pyle AM. 2010. A structural analysis of the group II intron active site and implications for the spliceosome. *RNA* 16:1–9.

19. Garst AD, Edwards AL, Batey RT. 2011. Riboswitches: structures and mechanisms. *Cold Spring Harb Perspect Biol* 3:a003533.

20. Huang L, Lilley DM. 2016. The kink turn, a key architectural element in RNA structure. *J Mol Biol* 428(5 Pt A):790–801.

21. Chan CW, Chetnani B, Mondragón A. 2013. Structure and function of the T-loop structural motif in noncoding RNAs. *Wiley Interdiscip Rev RNA* 4:507–522.

22. Aseev LV, Boni IV. 2011. Extraribosomal functions of bacterial ribosomal proteins. *Mol Biol* 45:739.

23. Warner JR, McIntosh KB. 2009. How common are extraribosomal functions of ribosomal proteins? *Mol Cell* 34:3–11.

24. Schmidt A, Kochanowski K, Vedelaar S, Ahrné E, Volkmer B, Callipo L, Knoops K, Bauer M, Aebersold R, Heinemann M. 2016. The quantitative and condition-dependent *Escherichia coli* proteome. *Nat Biotechnol* 34:104–110.

25. Li GW, Burkhardt D, Gross C, Weissman JS. 2014. Quantifying absolute protein synthesis rates reveals principles underlying allocation of cellular resources. *Cell* 157:624–635.

26. Fallon AM, Jinks CS, Yamamoto M, Nomura M. 1979. Expression of ribosomal protein genes cloned in a hybrid plasmid in *Escherichia coli*: gene dosage effects on synthesis of ribosomal proteins and ribosomal protein messenger ribonucleic acid. *J Bacteriol* 138:383–396.

27. Fallon AM, Jinks CS, Strycharz GD, Nomura M. 1979. Regulation of ribosomal protein synthesis in *Escherichia coli* by selective mRNA inactivation. *Proc Natl Acad Sci U S A* 76:3411–3415.

28. Lindahl L, Zengel JM. 1979. Operon-specific regulation of ribosomal protein synthesis in *Escherichia coli*. *Proc Natl Acad Sci U S A* 76:6542–6546.

29. Dean D, Yates JL, Nomura M. 1981. *Escherichia coli* ribosomal protein S8 feedback regulates part of *spc* operon. *Nature* 289:89–91.

30. Yates JL, Arfsten AE, Nomura M. 1980. *In vitro* expression of *Escherichia coli* ribosomal protein genes: autogenous inhibition of translation. *Proc Natl Acad Sci U S A* 77:1837–1841.

31. Dean D, Nomura M. 1980. Feedback regulation of ribosomal protein gene expression in *Escherichia coli*. *Proc Natl Acad Sci U S A* 77:3590–3594.

32. Brot N, Caldwell P, Weissbach H. 1980. Autogenous control of *Escherichia coli* ribosomal protein L10 synthesis *in vitro*. *Proc Natl Acad Sci U S A* 77:2592–2595.

33. Holowachuk EW, Friesen JD, Fiil NP. 1980. Bacteriophage λ vehicle for the direct cloning of *Escherichia coli* promoter DNA sequences: feedback regulation of the *rplJL-rpoBC* operon. *Proc Natl Acad Sci U S A* 77:2124–2128.

34. Robakis N, Meza-Basso L, Brot N, Weissbach H. 1981. Translational control of ribosomal protein L10 synthesis occurs prior to formation of first peptide bond. *Proc Natl Acad Sci U S A* 78:4261–4264.

35. Olins PO, Nomura M. 1981. Translational regulation by ribosomal protein S8 in *Escherichia coli*: structural homology between rRNA binding site and feedback target on mRNA. *Nucleic Acids Res* 9:1757–1764.

36. Dean D, Yates JL, Nomura M. 1981. Identification of ribosomal protein S7 as a repressor of translation within the *str* operon of *E. coli*. *Cell* 24:413–419.

37. Jinks-Robertson S, Nomura M. 1982. Ribosomal protein S4 acts in *trans* as a translational repressor to regu-

late expression of the α operon in *Escherichia coli*. *J Bacteriol* **151**:193–202.

38. Singer P, Nomura M. 1985. Stability of ribosomal protein mRNA and translational feedback regulation in *Escherichia coli*. *Mol Gen Genet* **199**:543–546.

39. Cole JR, Nomura M. 1986. Changes in the half-life of ribosomal protein messenger RNA caused by translational repression. *J Mol Biol* **188**:383–392.

40. Mattheakis LC, Nomura M. 1988. Feedback regulation of the *spc* operon in *Escherichia coli*: translational coupling and mRNA processing. *J Bacteriol* **170**:4484–4492.

41. Friedman DI, Schauer AT, Baumann MR, Baron LS, Adhya SL. 1981. Evidence that ribosomal protein S10 participates in control of transcription termination. *Proc Natl Acad Sci U S A* **78**:1115–1118.

42. Zengel JM, Lindahl L. 1990. *Escherichia coli* ribosomal protein L4 stimulates transcription termination at a specific site in the leader of the S10 operon independent of L4-mediated inhibition of translation. *J Mol Biol* **213**:67–78.

43. Nomura M, Yates JL, Dean D, Post LE. 1980. Feedback regulation of ribosomal protein gene expression in *Escherichia coli*: structural homology of ribosomal RNA and ribosomal protein mRNA. *Proc Natl Acad Sci U S A* **77**:7084–7088.

44. Olins PO, Nomura M. 1981. Regulation of the S10 ribosomal protein operon in *E. coli*: nucleotide sequence at the start of the operon. *Cell* **26**:205–211.

45. Baughman G, Nomura M. 1983. Localization of the target site for translational regulation of the L11 operon and direct evidence for translational coupling in *Escherichia coli*. *Cell* **34**:979–988.

46. Lindahl L, Zengel JM. 1986. Ribosomal genes in *Escherichia coli*. *Annu Rev Genet* **20**:297–326.

47. Johnsen M, Christensen T, Dennis PP, Fiil NP. 1982. Autogenous control: ribosomal protein L10-L12 complex binds to the leader sequence of its mRNA. *EMBO J* **1**:999–1004.

48. Christensen T, Johnsen M, Fiil NP, Friesen JD. 1984. RNA secondary structure and translation inhibition: analysis of mutants in the *rplJ* leader. *EMBO J* **3**:1609–1612.

49. Deckman IC, Draper DE. 1985. Specific interaction between ribosomal protein S4 and the alpha operon messenger RNA. *Biochemistry* **24**:7860–7865.

50. Tang CK, Draper DE. 1989. Unusual mRNA pseudoknot structure is recognized by a protein translational repressor. *Cell* **57**:531–536.

51. Shen P, Zengel JM, Lindahl L. 1988. Secondary structure of the leader transcript from the *Escherichia coli* S10 ribosomal protein operon. *Nucleic Acids Res* **16**:8905–8924.

52. Merianos HJ, Wang J, Moore PB. 2004. The structure of a ribosomal protein S8/*spc* operon mRNA complex. *RNA* **10**:954–964.

53. Nevskaya N, Tishchenko S, Gabdoulkhakov A, Nikonova E, Nikonov O, Nikulin A, Platonova O, Garber M, Nikonov S, Piendl W. 2005. Ribosomal protein L1 recognizes the same specific structural motif in its target sites on the autoregulatory mRNA and 23S rRNA. *Nucleic Acids Res* **33**:478–485.

54. Lesage P, Truong HN, Graffe M, Dondon J, Springer M. 1990. Translated translational operator in *Escherichia coli*. Auto-regulation in the *infC-rpmI-rplT* operon. *J Mol Biol* **213**:465–475.

55. Portier C, Dondon L, Grunberg-Manago M. 1990. Translational autocontrol of the *Escherichia coli* ribosomal protein S15. *J Mol Biol* **211**:407–414.

56. Parsons GD, Donly BC, Mackie GA. 1988. Mutations in the leader sequence and initiation codon of the gene for ribosomal protein S20 (*rpsT*) affect both translational efficiency and autoregulation. *J Bacteriol* **170**:2485–2492.

57. Aseev LV, Levandovskaya AA, Tchufistova LS, Scaptsova NV, Boni IV. 2008. A new regulatory circuit in ribosomal protein operons: S2-mediated control of the *rpsB-tsf* expression *in vivo*. *RNA* **14**:1882–1894.

58. Aseev LV, Bylinkina NS, Boni IV. 2015. Regulation of the *rplY* gene encoding 5S rRNA binding protein L25 in *Escherichia coli* and related bacteria. *RNA* **21**:851–861.

59. Skouv J, Schnier J, Rasmussen MD, Subramanian AR, Pedersen S. 1990. Ribosomal protein S1 of *Escherichia coli* is the effector for the regulation of its own synthesis. *J Biol Chem* **265**:17044–17049.

60. Aseev LV, Koledinskaya LS, Boni IV. 2016. Regulation of ribosomal protein operons *rplM-rpsI*, *rpmB-rpmG*, and *rplU-rpmA* at the transcriptional and translational levels. *J Bacteriol* **198**:2494–2502.

61. Matelska D, Purta E, Panek S, Boniecki MJ, Bujnicki JM, Dunin-Horkawicz S. 2013. S6:S18 ribosomal protein complex interacts with a structural motif present in its own mRNA. *RNA* **19**:1341–1348.

62. Fu Y, Deiorio-Haggar K, Soo MW, Meyer MM. 2014. Bacterial RNA motif in the 5′ UTR of *rpsF* interacts with an S6:S18 complex. *RNA* **20**:168–176.

63. Babina AM, Soo MW, Fu Y, Meyer MM. 2015. An S6:S18 complex inhibits translation of *E. coli rpsF*. *RNA* **21**:2039–2046.

64. Fujita K, Baba T, Isono K. 1998. Genomic analysis of the genes encoding ribosomal proteins in eight eubacterial species and *Saccharomyces cerevisiae*. *Genome Inform Ser Workshop Genome Inform* **9**:3–12.

65. Coenye T, Vandamme P. 2005. Organisation of the *S10*, *spc* and *alpha* ribosomal protein gene clusters in prokaryotic genomes. *FEMS Microbiol Lett* **242**:117–126.

66. Allen T, Shen P, Samsel L, Liu R, Lindahl L, Zengel JM. 1999. Phylogenetic analysis of L4-mediated autogenous control of the S10 ribosomal protein operon. *J Bacteriol* **181**:6124–6132.

67. Allen TD, Watkins T, Lindahl L, Zengel JM. 2004. Regulation of ribosomal protein synthesis in *Vibrio cholerae*. *J Bacteriol* **186**:5933–5937.

68. Iben JR, Draper DE. 2008. Specific interactions of the L10(L12)$_4$ ribosomal protein complex with mRNA, rRNA, and L11. *Biochemistry* **47**:2721–2731.

69. Aseev LV, Levandovskaya AA, Skaptsova NV, Boni IV. 2009. Conservation of regulatory elements controlling the expression of the *rpsB-tsf* operon in γ-proteobacteria. *Mol Biol* **43**:101–107.

70. Guillier M, Allemand F, Raibaud S, Dardel F, Springer M, Chiaruttini C. 2002. Translational feedback regulation of the gene for L35 in *Escherichia coli* requires binding of ribosomal protein L20 to two sites in its leader mRNA: a possible case of ribosomal RNA-messenger RNA molecular mimicry. *RNA* **8**:878–889.

71. Fu Y, Deiorio-Haggar K, Anthony J, Meyer MM. 2013. Most RNAs regulating ribosomal protein biosynthesis in *Escherichia coli* are narrowly distributed to Gammaproteobacteria. *Nucleic Acids Res* **41**:3491–3503.

72. Matelska D, Kurkowska M, Purta E, Bujnicki JM, Dunin-Horkawicz S. 2016. Loss of conserved noncoding RNAs in genomes of bacterial endosymbionts. *Genome Biol Evol* **8**:426–438.

73. Grundy FJ, Henkin TM. 1991. The *rpsD* gene, encoding ribosomal protein S4, is autogenously regulated in *Bacillus subtilis*. *J Bacteriol* **173**:4595–4602.

74. Choonee N, Even S, Zig L, Putzer H. 2007. Ribosomal protein L20 controls expression of the *Bacillus subtilis* *infC* operon via a transcription attenuation mechanism. *Nucleic Acids Res* **35**:1578–1588.

75. Deiorio-Haggar K, Anthony J, Meyer MM. 2013. RNA structures regulating ribosomal protein biosynthesis in bacilli. *RNA Biol* **10**:1180–1184.

76. Nawrocki EP, Kolbe DL, Eddy SR. 2009. Infernal 1.0: inference of RNA alignments. *Bioinformatics* **25**:1335–1337.

77. Griffiths-Jones S, Bateman A, Marshall M, Khanna A, Eddy SR. 2003. Rfam: an RNA family database. *Nucleic Acids Res* **31**:439–441.

78. Weinberg Z, Wang JX, Bogue J, Yang J, Corbino K, Moy RH, Breaker RR. 2010. Comparative genomics reveals 104 candidate structured RNAs from bacteria, archaea, and their metagenomes. *Genome Biol* **11**:R31.

79. Weinberg Z, Barrick JE, Yao Z, Roth A, Kim JN, Gore J, Wang JX, Lee ER, Block KF, Sudarsan N, Neph S, Tompa M, Ruzzo WL, Breaker RR. 2007. Identification of 22 candidate structured RNAs in bacteria using the CMfinder comparative genomics pipeline. *Nucleic Acids Res* **35**:4809–4819.

80. Yao Z, Barrick J, Weinberg Z, Neph S, Breaker R, Tompa M, Ruzzo WL. 2007. A computational pipeline for high-throughput discovery of *cis*-regulatory noncoding RNA in prokaryotes. *PLoS Comput Biol* **3**:e126.

81. Tseng HH, Weinberg Z, Gore J, Breaker RR, Ruzzo WL. 2009. Finding non-coding RNAs through genome-scale clustering. *J Bioinform Comput Biol* **7**:373–388.

82. Weinberg Z, Lünse CE, Corbino KA, Ames TD, Nelson JW, Roth A, Perkins KR, Sherlock ME, Breaker RR. 2017. Detection of 224 candidate structured RNAs by comparative analysis of specific subsets of intergenic regions. *Nucleic Acids Res* **45**:10811–10823.

83. Dar D, Shamir M, Mellin JR, Koutero M, Stern-Ginossar N, Cossart P, Sorek R. 2016. Term-seq reveals abundant ribo-regulation of antibiotics resistance in bacteria. *Science* **352**:aad9822.

84. Daume M, Uhl M, Backofen R, Randau L. 2017. RIP-Seq suggests translational regulation by L7Ae in *Archaea*. *mBio* **8**:e00730–17.

85. Mattheakis L, Vu L, Sor F, Nomura M. 1989. Retro-regulation of the synthesis of ribosomal proteins L14 and L24 by feedback repressor S8 in *Escherichia coli*. *Proc Natl Acad Sci U S A* **86**:448–452.

86. Gregory RJ, Cahill PB, Thurlow DL, Zimmermann RA. 1988. Interaction of *Escherichia coli* ribosomal protein S8 with its binding sites in ribosomal RNA and messenger RNA. *J Mol Biol* **204**:295–307.

87. Cerretti DP, Mattheakis LC, Kearney KR, Vu L, Nomura M. 1988. Translational regulation of the *spc* operon in *Escherichia coli*. Identification and structural analysis of the target site for S8 repressor protein. *J Mol Biol* **204**:309–329.

88. Wu H, Jiang L, Zimmermann RA. 1994. The binding site for ribosomal protein S8 in 16S rRNA and *spc* mRNA from *Escherichia coli*: minimum structural requirements and the effects of single bulged bases on S8-RNA interaction. *Nucleic Acids Res* **22**:1687–1695.

89. Davies C, Ramakrishnan V, White SW. 1996. Structural evidence for specific S8-RNA and S8-protein interactions within the 30S ribosomal subunit: ribosomal protein S8 from *Bacillus stearothermophilus* at 1.9 Å resolution. *Structure* **4**:1093–1104.

90. Vysotskaya V, Tischenko S, Garber M, Kern D, Mougel M, Ehresmann C, Ehresmann B. 1994. The ribosomal protein S8 from *Thermus thermophilus* VK1. Sequencing of the gene, overexpression of the protein in *Escherichia coli* and interaction with rRNA. *Eur J Biochem* **223**:437–445.

91. Kalurachchi K, Uma K, Zimmermann RA, Nikonowicz EP. 1997. Structural features of the binding site for ribosomal protein S8 in *Escherichia coli* 16S rRNA defined using NMR spectroscopy. *Proc Natl Acad Sci U S A* **94**:2139–2144.

92. Nevskaya N, Tischenko S, Nikulin A, al-Karadaghi S, Liljas A, Ehresmann B, Ehresmann C, Garber M, Nikonov S. 1998. Crystal structure of ribosomal protein S8 from *Thermus thermophilus* reveals a high degree of structural conservation of a specific RNA binding site. *J Mol Biol* **279**:233–244.

93. Tishchenko S, Nikulin A, Fomenkova N, Nevskaya N, Nikonov O, Dumas P, Moine H, Ehresmann B, Ehresmann C, Piendl W, Lamzin V, Garber M, Nikonov S. 2001. Detailed analysis of RNA-protein interactions within the ribosomal protein S8-rRNA complex from the archaeon *Methanococcus jannaschii*. *J Mol Biol* **311**:311–324.

94. Friesen JD, Tropak M, An G. 1983. Mutations in the rplJ leader of *Escherichia coli* that abolish feedback regulation. *Cell* **32**:361–369.

95. Climie SC, Friesen JD. 1987. Feedback regulation of the *rplJL-rpoBC* ribosomal protein operon of *Escherichia coli* requires a region of mRNA secondary structure. *J Mol Biol* **198**:371–381.

96. Climie SC, Friesen JD. 1988. *In vivo* and *in vitro* structural analysis of the *rplJ* mRNA leader of *Escherichia coli*. Protection by bound L10-L7/L12. *J Biol Chem* **263**:15166–15175.

97. Yates JL, Dean D, Strycharz WA, Nomura M. 1981. *E. coli* ribosomal protein L10 inhibits translation of L10 and L7/L12 mRNAs by acting at a single site. *Nature* **294**:190–192.

98. Yakhnin H, Yakhnin AV, Babitzke P. 2015. Ribosomal protein L10(L12)$_4$ autoregulates expression of the *Bacillus subtilis rplJL* operon by a transcription attenuation mechanism. *Nucleic Acids Res* **43**:7032–7043.

99. Barrick JE, Breaker RR. 2007. The distributions, mechanisms, and structures of metabolite-binding riboswitches. *Genome Biol* **8**:R239.

100. Baughman G, Nomura M. 1984. Translational regulation of the L11 ribosomal protein operon of *Escherichia coli*: analysis of the mRNA target site using oligonucleotide-directed mutagenesis. *Proc Natl Acad Sci U S A* **81**:5389–5393.

101. Hanner M, Mayer C, Köhrer C, Golderer G, Gröbner P, Piendl W. 1994. Autogenous translational regulation of the ribosomal *MvaL1* operon in the archaebacterium *Methanococcus vannielii*. *J Bacteriol* **176**:409–418.

102. Kraft A, Lutz C, Lingenhel A, Gröbner P, Piendl W. 1999. Control of ribosomal protein L1 synthesis in mesophilic and thermophilic Archaea. *Genetics* **152**:1363–1372.

103. Köhrer C, Mayer C, Neumair O, Gröbner P, Piendl W. 1998. Interaction of ribosomal L1 proteins from mesophilic and thermophilic Archaea and Bacteria with specific L1-binding sites on 23S rRNA and mRNA. *Eur J Biochem* **256**:97–105.

104. Shimmin LC, Ramirez C, Matheson AT, Dennis PP. 1989. Sequence alignment and evolutionary comparison of the L10 equivalent and L12 equivalent ribosomal proteins from archaebacteria, eubacteria, and eucaryotes. *J Mol Evol* **29**:448–462.

105. Haentjens-Sitri J, Allemand F, Springer M, Chiaruttini C. 2008. A competition mechanism regulates the translation of the *Escherichia coli* operon encoding ribosomal proteins L35 and L20. *J Mol Biol* **375**:612–625.

106. Nawrocki EP, Eddy SR. 2013. Infernal 1.1: 100-fold faster RNA homology searches. *Bioinformatics* **29**:2933–2935.

107. Bruscella P, Shahbabian K, Laalami S, Putzer H. 2011. RNase Y is responsible for uncoupling the expression of translation factor IF3 from that of the ribosomal proteins L35 and L20 in *Bacillus subtilis*. *Mol Microbiol* **81**:1526–1541.

108. Guillier M, Allemand F, Dardel F, Royer CA, Springer M, Chiaruttini C. 2005. Double molecular mimicry in *Escherichia coli*: binding of ribosomal protein L20 to its two sites in mRNA is similar to its binding to 23S rRNA. *Mol Microbiol* **56**:1441–1456.

109. Scott LG, Williamson JR. 2001. Interaction of the *Bacillus stearothermophilus* ribosomal protein S15 with its 5′-translational operator mRNA. *J Mol Biol* **314**:413–422.

110. Philippe C, Eyermann F, Bénard L, Portier C, Ehresmann B, Ehresmann C. 1993. Ribosomal protein S15 from *Escherichia coli* modulates its own translation by trapping the ribosome on the mRNA initiation loading site. *Proc Natl Acad Sci U S A* **90**:4394–4398.

111. Serganov A, Polonskaia A, Ehresmann B, Ehresmann C, Patel DJ. 2003. Ribosomal protein S15 represses its own translation via adaptation of an rRNA-like fold within its mRNA. *EMBO J* **22**:1898–1908.

112. Slinger BL, Deiorio-Haggar K, Anthony JS, Gilligan MM, Meyer MM. 2014. Discovery and validation of novel and distinct RNA regulators for ribosomal protein S15 in diverse bacterial phyla. *BMC Genomics* **15**:657.

113. Slinger BL, Newman H, Lee Y, Pei S, Meyer MM. 2015. Co-evolution of bacterial ribosomal protein S15 with diverse mRNA regulatory structures. *PLoS Genet* **11**:e1005720.

114. Mathy N, Pellegrini O, Serganov A, Patel DJ, Ehresmann C, Portier C. 2004. Specific recognition of *rpsO* mRNA and 16S rRNA by *Escherichia coli* ribosomal protein S15 relies on both mimicry and site differentiation. *Mol Microbiol* **52**:661–675.

115. Scott LG, Williamson JR. 2005. The binding interface between *Bacillus stearothermophilus* ribosomal protein S15 and its 5′-translational operator mRNA. *J Mol Biol* **351**:280–290.

116. Serganov A, Ennifar E, Portier C, Ehresmann B, Ehresmann C. 2002. Do mRNA and rRNA binding sites of *E. coli* ribosomal protein S15 share common structural determinants? *J Mol Biol* **320**:963–978.

117. Ehresmann C, Ehresmann B, Ennifar E, Dumas P, Garber M, Mathy N, Nikulin A, Portier C, Patel D, Serganov A. 2004. Molecular mimicry in translational regulation: the case of ribosomal protein S15. *RNA Biol* **1**:66–73.

118. Cho IM, Lai LB, Susanti D, Mukhopadhyay B, Gopalan V. 2010. Ribosomal protein L7Ae is a subunit of archaeal RNase P. *Proc Natl Acad Sci U S A* **107**:14573–14578.

119. Kuhn JF, Tran EJ, Maxwell ES. 2002. Archaeal ribosomal protein L7 is a functional homolog of the eukaryotic 15.5kD/Snu13p snoRNP core protein. *Nucleic Acids Res* **30**:931–941.

120. Rozhdestvensky TS, Tang TH, Tchirkova IV, Brosius J, Bachellerie JP, Hüttenhofer A. 2003. Binding of L7Ae protein to the K-turn of archaeal snoRNAs: a shared RNA binding motif for C/D and H/ACA box snoRNAs in Archaea. *Nucleic Acids Res* **31**:869–877.

121. Nottrott S, Hartmuth K, Fabrizio P, Urlaub H, Vidovic I, Ficner R, Lührmann R. 1999. Functional interaction of a novel 15.5kD [U4/U6.U5] tri-snRNP protein with the 5′ stem-loop of U4 snRNA. *EMBO J* **18**:6119–6133.

122. Baird NJ, Zhang J, Hamma T, Ferré-D'Amaré AR. 2012. YbxF and YlxQ are bacterial homologs of L7Ae and bind K-turns but not K-loops. *RNA* **18**:759–770.

123. Saito H, Kobayashi T, Hara T, Fujita Y, Hayashi K, Furushima R, Inoue T. 2010. Synthetic translational regulation by an L7Ae-kink-turn RNP switch. *Nat Chem Biol* **6**:71–78.

124. Saito H, Fujita Y, Kashida S, Hayashi K, Inoue T. 2011. Synthetic human cell fate regulation by protein-driven RNA switches. *Nat Commun* 2:160–168.

125. Berens C, Suess B. 2015. Riboswitch engineering—making the all-important second and third steps. *Curr Opin Biotechnol* 31:10–15.

126. Slinger BL, Meyer MM. 2016. RNA regulators responding to ribosomal protein S15 are frequent in sequence space. *Nucleic Acids Res* 44:9331–9341.

127. Pei S, Slinger BL, Meyer MM. 2017. Recognizing RNA structural motifs in HT-SELEX data for ribosomal protein S15. *BMC Bioinformatics* 18:298.

128. Moine H, Cachia C, Westhof E, Ehresmann B, Ehresmann C. 1997. The RNA binding site of S8 ribosomal protein of *Escherichia coli*: Selex and hydroxyl radical probing studies. *RNA* 3:255–268.

129. Cannone JJ, Subramanian S, Schnare MN, Collett JR, D'Souza LM, Du Y, Feng B, Lin N, Madabusi LV, Müller KM, Pande N, Shang Z, Yu N, Gutell RR. 2002. The Comparative RNA Web (CRW) site: an online database of comparative sequence and structure information for ribosomal, intron, and other RNAs. *BMC Bioinformatics* 3:2.

130. Dunkle JA, Wang L, Feldman MB, Pulk A, Chen VB, Kapral GJ, Noeske J, Richardson JS, Blanchard SC, Cate JH. 2011. Structures of the bacterial ribosome in classical and hybrid states of tRNA binding. *Science* 332:981–984.

Regulating with RNA in Bacteria and Archaea
Edited by Gisela Storz and Kai Papenfort
© 2018 American Society for Microbiology, Washington, DC
doi:10.1128/microbiolspec.RWR-0031-2018

Processive Antitermination

8

Jonathan R. Goodson[1] and Wade C. Winkler[1]

INTRODUCTION

An extraordinarily diverse range of genetic regulatory mechanisms has been discovered in the half century since Francois Jacob and Jacques Monod first proposed the operon model of gene regulation (1). Studies based on this model identified a soluble regulator, located distally from the targeted operon, that acts to repress transcription initiation of the *lac* operon. This discovery led to the identification and characterization of many more repressor proteins, each acting in modestly different ways to reduce the efficiency of transcription initiation. Soon followed discoveries of other types of transcriptional regulators, including those that activate gene expression by enhancing transcription initiation. And now, in an era in which bacterial genome sequences can be acquired and draft-annotated in mere days and at low cost, it is clear that all bacteria encode dozens or hundreds of proteins that regulate transcription initiation and that this "layer" of genetic regulation is both ubiquitous and profoundly important. However, perhaps because transcription initiation is so universally recognized as a key point of regulatory influence (2), later stages of transcription elongation have not yet been sufficiently analyzed for genetic regulation. While the molecular mechanisms of transcription have been, and continue to be, intensively investigated, the biological extent of postinitiation regulatory mechanisms has been incompletely analyzed. Transcription initiation is only the first stage of gene expression. The stages that follow include transcription elongation, transcription termination, translation, and mRNA degradation; each of these stages can be subjected to genetic regulatory control (3).

While riboswitches, which control transcription attenuation in a signal-dependent manner, are widely used by bacteria, their initial discoveries have been significantly aided by the extensive conservation of their sequences and secondary structures (4, 5). This level of sequence conservation is not observed for many other types of transcription elongation regulatory strategies, a limitation that may have slowed discovery of the latter. How, then, may other transcription elongation-based regulatory strategies be systematically discovered if experimentalists cannot rely primarily on bioinformatics searches of highly conserved regulatory RNAs? And what kinds of transcription elongation regulatory mechanisms have not yet been found? One type of regulatory mechanism that might still be understudied, but yet has been identified through a variety of experimental approaches, is called processive antitermination (PA). These systems offer a convenient and powerful mechanism for altering the efficiency of transcription elongation (6–8).

In PA mechanisms, antitermination factors associate with a bacterial RNA polymerase (RNAP) elongation complex, leading to read-through of termination sites (6). Termination signals normally induce rapid dissociation of the transcription elongation complex (TEC) and are most often located at the ends of operons (9). However, when placed within operons, they can serve as key points of regulatory control (10). In bacteria, there are two known classes of termination signals: intrinsic and Rho-dependent terminators (9). In many bacteria, intrinsic terminators consist of a GC-rich RNA hairpin followed by a polyuridine tract. Alone (11), or enhanced by a factor such as NusA (12, 13), these RNA elements promote pausing of the TEC, followed by release of the nascent transcript and dissociation of polymerase (14). In contrast, Rho-dependent termination depends upon the ATP-dependent translocase Rho associating with Rho utilization (*rut*) sites on a nascent mRNA and translocating the RNA to eventually promote TEC dissociation (15, 16). Both classes of termination sites may be specifically regulated by signal-responsive riboswitches (5, 17) or *trans*-encoded small RNAs (18, 19). However, whereas riboswitches exert control over a single intrinsic terminator site or a particular entry point for Rho, PA

[1]Department of Cell Biology and Molecular Genetics, The University of Maryland, College Park, MD 20742.

systems differ in that they modify TECs to render them generally resistant to downstream termination sites (8). PA systems, therefore, are capable of causing read-through of multiple termination sites, even over long genomic distances. While only a few classes of PA mechanisms have been discovered in the past 4 decades, they vary widely in the molecular mechanisms they utilize and in their biological applications. Several new examples of PA mechanisms have been discovered more recently, which appear to be broadly used by bacteria for regulation of diverse sets of genes. We extrapolate from these discoveries that many new PA mechanisms still await discovery.

PROCESSIVE ANTITERMINATION

Termination of transcription at any given location is rarely 100% complete, with some proportion of elongation complexes proceeding past the point of termination. In general, two types of mechanisms can control transcription elongation to affect the efficiency of termination: transcription attenuation and PA. For the former, regulatory mechanisms determine the formation of either Rho-dependent or Rho-independent termination sites (10). Importantly, transcription attenuation-based regulatory mechanisms exert their influence on only a single, defined terminator region. In other words, a regulatory RNA that promotes transcription attenuation by definition evolved in concert with the terminator region that it targets—it does not affect other terminator regions. Riboswitches, which are signal-responsive, *cis*-acting regulatory RNAs, oftentimes affect gene expression via transcription attenuation-based mechanisms (20). As discussed elsewhere, riboswitches are widespread in bacteria and offer localized control of transcription termination sites throughout bacterial genomes. In many instances, these transcription attenuation-based regulatory elements can be considered modular, with a signal-responsive portion followed by a portion responsible for premature transcription termination (21).

In contrast, PA mechanisms do not necessarily target a specific terminator region, but instead manipulate elongating RNAP complexes to avoid termination signals throughout an individual transcript (6). These PA strategies do not take a single form and may reduce transcript termination through a variety of direct and indirect effects. For example, some PA strategies rely on direct interference with factor-mediated termination (22). Alternatively, they can modify recruitment of transcription elongation factors, such as NusA, to affect nascent RNA behavior (23, 24). Additionally, they may alter recruitment of ribosomes in a manner that

affects termination within coding regions (25). Furthermore, some PA systems have evolved to utilize multiple strategies simultaneously (23, 26).

Phage Lambda Antitermination

During lytic growth, phage λ transcription temporally progresses from one large set of genes to another (27). To switch from intermediate-early gene expression to delayed-early gene expression, the phage utilizes a unique protein, λN, to promote antitermination, which enables expression of downstream genes (Fig. 1A) (28). λN is a small protein that is intrinsically disordered alone (29) but is stabilized by protein and RNA contacts in the final, λN antitermination complex (Fig. 2A) (23). Formation of the λN antitermination complex is triggered by synthesis of a *nut* sequence, composed of two RNA elements. The first, *boxB*, is a 15-nucleotide motif that resembles a GNRA tetraloop structure (Fig. 3) (30, 31) and serves as the substrate for λN binding (23, 24). In addition to binding λN, *boxB* also interacts with NusA. Formation of the antitermination complex occurs in steps, with initial association of λN to *boxB* followed by binding of NusA to the λN-*boxB* complex (32). This minimal λN-*boxB*-NusA complex is sufficient for antitermination of *nut*-proximal terminator sequences (6), although it is generally believed that the full antitermination complex *in vivo* relies on additional elongation proteins loaded at the second RNA element. This second RNA element, *boxA*, acts as a loading site for the NusB-NusE (S10) complex (33). Binding of the NusB-NusE (S10) complex to *boxA* promotes additional contacts between λN and NusA. This results in a unique complex of factors that are associated with RNAP near the RNA exit channel and remain together as an RNP complex (Fig. 2B) (23).

Binding of λN alone to RNAP modifies transcription elongation both *in vitro* and *in vivo*, promoting antitermination by modulating RNA exit channel elements and by suppressing melting of the RNA-DNA hybrid after terminator hairpin formation or in response to Rho activity (23, 24, 34, 35). However, formation of the complex with the full complement of transcription elongation factors is thought to further stabilize the interaction of λN with RNAP and increase its duration of occupancy—and, therefore, the overall processivity of λN antitermination (36). In "standard" TECs, NusA binds RNAP near the RNA exit channel, where it can enhance intrinsic termination (37). Indeed, NusA affects transcription termination at many locations across the genome and is even required for formation of some NusA-dependent termination sites (13). However, λN is thought to counteract the direct effects of NusA

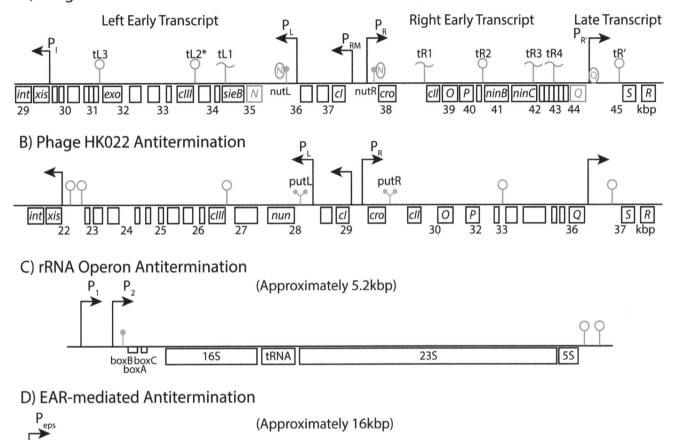

Figure 1 Genomic context of PA systems. This figure schematically illustrates the transcripts regulated by the λN, *put*, rRNA, and EAR RNA-based antitermination systems. (A) Phage λ early transcripts are initiated from two divergently facing promoters with *boxA/B nut* elements found early in the transcripts. The λN protein is encoded by the first gene in the left early transcript. RNAP complexes associated with λN bypass multiple terminators in both transcripts. Using a different mechanism, the λQ protein promotes antitermination of the late transcript by binding to DNA near the late promoter and promoting a terminator-resistant configuration of RNAP. (B) Phage HK022 early transcripts are similar to phage λ, although they include *put* elements early in each transcript, which trigger λN-independent antitermination. Additional Rho-dependent terminators are likely present in these transcripts, although they have not been specifically characterized and are therefore not indicated here. (C) A representative *E. coli* rRNA operon is shown, containing *boxA/B/C* elements immediately downstream of the P₂ promoter. These elements promote read-through of Rho-dependent termination in the noncoding rRNA genes. (D) Several intrinsic terminators have been demonstrated in the *B. subtilis eps* operon, which codes for biosynthesis of biofilm exopolysaccharides. The *eps*-associated RNA (EAR) is found within the *epsBC* intergenic region and promotes read-through of the terminators within the operon. Intrinsic terminators are shown as sticks with empty circles, and Rho termination regions are shown as sticks with wavy lines, both in red. RNA elements involved in antitermination are shown in blue, and proteins and protein-coding genes involved in antitermination are shown in green. Elements are not shown to scale.

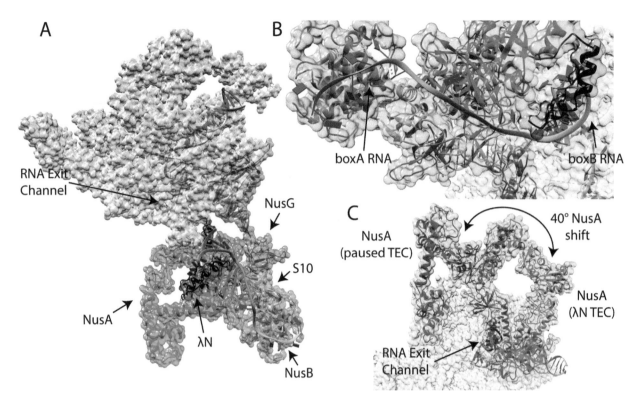

Figure 2 Cryo-EM reveals details of the antitermination mechanism. This figure contains structural models generated from cryo-EM data on TECs (PDB IDs: 5MS0 and 6FLQ). (A) The λN antitermination complex (PDB ID: 5MS0) comprising λN (black), NusA (magenta), NusB (red), S10 (orange), NusG (green), and *boxA/B* RNA (blue), in addition to RNAP (gray). (B) A zoom-in on the *boxA/B* and λN complex shows extended binding of the *nut* RNA sequence with multiple protein components, with *boxA* bound to the NusB-S10 dimer and the *boxB* hairpin bound to λN and NusA. (C) Formation of the λN antitermination complex shifts the position of NusA (magenta) by 40° away from the RNA exit channel, as compared to NusA (purple) in a TEC constructed with the *E. coli his* hairpin-mediated pause sequence (PDB ID: 6FLQ). Nascent RNA is shown in green.

on terminator hairpin folding (24). A recent high-resolution structural model of the λN antitermination complex revealed that the C-terminal RNA-binding domains of NusA are repositioned such that they redirect nascent RNA away from the RNA exit channel (Fig. 2C). This is predicted to reduce formation of terminator hairpins, thereby essentially reprogramming NusA into a transcription antitermination factor (23). Formation of the λN complex also inhibits Rho-dependent termination. In "standard" elongation complexes, NusG helps recruit Rho to nascent RNA and thereby aids in Rho termination (38, 39). In contrast, the λN antitermination complex is likely to restrict NusG-mediated recruitment of Rho by instead promoting association of factors that compete for binding to NusG (e.g., S10-NusB), and also because of restricted access to the nascent RNA as it is looped out of the antitermination complex (23). Therefore, the λN com-

plex acts as a physical roadblock to prevent Rho translocation and helps occlude access to *rut* sites.

Phage λ also contains a second antitermination system, which relies upon another unique protein (λQ) to promote antitermination of late-expressed genes (6, 40). However, unlike the N-antitermination system, λQ protein is a DNA-binding protein that associates with RNAP within the promoter region during transcription initiation and triggers formation of an antitermination complex that is different from the N complex (41).

rRNA Operon Antitermination

Dissociation of TECs by Rho helicase underlies the polarity that occurs when nonsense mutations reduce transcript abundance of downstream genes (42). Rho is capable of loading onto RNA molecules via C-rich binding sequences (*rut* sites), but the presence of ribosomes during coupled transcription-translation gener-

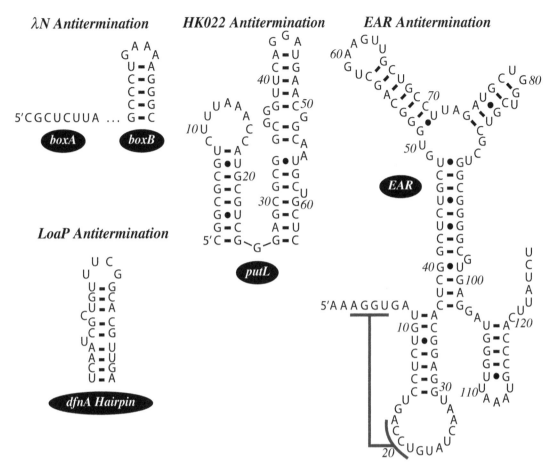

Figure 3 RNA elements involved in PA. This figure shows the sequence and secondary structure of RNA elements known or predicted to be utilized in PA mechanisms. Shown are the *boxA* and *boxB* elements forming the λN *nut* sequence as well as rRNA antitermination signal, the *put* RNA element from phage HK022, EAR from the *B. subtilis* exopolysaccharide pathway, and a UNCG-type hairpin implicated in LoaP antitermination.

ally reduces Rho loading and translocation (43). Given that rRNA operons are not translated and are thereby not protected by ribosomes, their transcripts must be protected from Rho termination by other means. This protection may be partially explained by the extensive secondary structure of rRNAs, which acts to reduce loading of Rho at potential *rut* sites (44, 45). However, in *Escherichia coli* and many other bacteria, these operons are also subjected to an antitermination system that closely resembles the λN antitermination mechanism (44–46). For example, the 5′ leader regions of *E. coli* rRNA operons contain *boxA* as well as a *boxB*-like hairpin, although only *boxA* appears to be essential for antitermination activity (Fig. 1C) (33, 47). Binding of the NusB-NusE (S10) complex to *boxA* RNA occurs in a manner similar to N-mediated antitermination, ultimately promoting a conformational state that strongly disfavors association of Rho (33).

In contrast to λN antitermination, which requires N protein in addition to host Nus proteins, rRNA antitermination requires an additional host factor, SuhB (48). The complete elongation complex containing NusB-NusE, NusA, NusG, and SuhB is required not only for full rRNA antitermination activity *in vitro* but for correct rRNA maturation *in vivo* (48). In addition to regulation of rRNA transcription, *boxA* and Nus factors directly repress *suhB* translation in enterobacteria in a manner reminiscent of λN autoregulation and have been implicated in regulation of additional genes (49). Therefore, the rRNA antitermination system relies exclusively on general transcription elongation factors and their recruitment to the *boxA* RNA element. This system serves a dual purpose in rRNA operons, promoting both antitermination and RNA folding, and may regulate yet additional transcripts. Together, these observations suggest that N-antitermination may have arisen as a modification of the host Nus protein antitermination system, where λN protein evolved to reconfigure and further manipulate host transcription elongation factors.

RNA Elements that Promote PA

In addition to the role that RNA elements (boxA and boxB) play in antitermination of phage λ and rRNA operons, a few PA systems have been discovered that involve larger and more complicated RNA elements. Many if not most lambdoid phages utilize PA systems related to both N- or Q-antitermination (6). However, phage HK022 differs in that it encodes λQ yet lacks λN, despite the fact that it still requires antitermination of early-expressed genes (50). Moreover, HK022 does not utilize nut sites for antitermination. Instead, early gene antitermination is mediated directly by a larger RNA motif called put, found in regions analogous to λ nut sites (Fig. 1B) (51). HK022 put forms a two-hairpin RNA element ~65 nucleotides in length that is critical for antitermination activity (Fig. 3) (51, 52). This element appears to directly affect RNAP elongation activity through pause suppression, potentially requiring no additional elements to promote antitermination (50). Evolution of this mechanism is likely interrelated with the evolution of a λN-like protein, Nun, which is also produced by HK022 (53, 54). Nun, found in the same relative genomic position as λN in phage λ, instead promotes Nun termination at nut elements by binding to boxB and inhibiting RNAP translocation (55, 56). HK022 put promotes antitermination of both Rho-dependent termination and Nun-dependent transcription arrest in the HK022 early transcripts (55) as well as intrinsic terminators (57). While some mechanistic details of put-mediated antitermination are still lacking, its discovery was significant, as it demonstrated proof in principle that PA could be driven primarily by RNA elements.

More recently, an even larger and more structurally complicated RNA element was discovered to trigger PA in bacteria. This RNA element, which is at least ~125 nucleotides in length and is constructed from an array of at least five helical elements and a characteristic pseudoknot, was discovered to be broadly conserved in Bacillales (Fig. 3) (58). Coined the EAR element, for eps-associated RNA, it is almost always associated with operons that code for biosynthesis of biofilm or capsule exopolysaccharides (Fig. 1D). Either mutagenesis of conserved residues or deletion of EAR resulted in incomplete transcription of the Bacillus subtilis eps operon. Instead, transcripts were found to be prematurely truncated at the site of intrinsic terminators, located in the middle region of the eps operon. Indeed, placement of EAR directly upstream of this terminator site resulted in nearly complete read-through of the terminators in vivo, whereas, conversely, mutagenesis of conserved EAR residues resulted in termination. Moreover, placement of EAR upstream of unrelated intrinsic terminators, originating from sources other than the eps operon, still resulted in their read-through, demonstrating that EAR promotes general PA of intrinsic terminators. That EAR promoted read-through of intrinsic terminators is strikingly different than the biological utilization of the λN and rRNA PA systems, which are believed to function primarily for read-through of Rho termination. However, EAR PA has not yet been recapitulated in vitro or in a heterologous host, indicating that at least one additional factor may be required for its antitermination activity, in contrast to HK022 put. Regardless, discovery of EAR demonstrated that structurally complicated RNAs, with the size and apparent complexity resembling that of riboswitches, are sometimes used to promote PA. Moreover, the distribution of EAR PA determinants further showcases how PA mechanisms can be broadly important for biologically important functions such as biofilm formation.

SPECIALIZED NusG PARALOGS

RfaH

Although most known PA systems are found in phage genomes or are reliant on general transcription elongation factors, some Gammaproteobacteria code for the specialized PA and translation factor RfaH (26). RfaH is a paralog of NusG. NusG is an elongation factor generally associated with TECs and is an integral component of the λ and rRNA PA systems (59). RfaH, encoded by an essential gene in E. coli, is required for the expression of a regulon of virulence-related pathways—including synthesis of hemolysin, lipopolysaccharide, and the F-factor sex pilus (59, 60)—as well as additional targets involved in the production of membrane or extracellular components (61).

As a paralog of NusG, RfaH is a small protein containing two conserved domains. In general, the core domains of NusG homolog proteins exhibit strongly conserved structure (62, 63) and interface with RNAP in a similar fashion (63–65). The first domain is an N-terminal domain (NTD) unique to the NusG/Spt5 family of proteins (66). This domain is responsible for binding of RfaH to RNAP at the same site normally occupied by NusG. The C-terminal domain (CTD) contains a KOW motif found in several ribosomal proteins in addition to NusG (67). This characteristic CTD is shared among nearly all NusG homologs as well as several ribosomal proteins (67), and is believed to function as a tether that can interact with additional proteins (68).

While RfaH and NusG have distinct regulatory consequences, they rely on similar mechanisms to improve transcriptional processivity (65). The NTDs of both proteins share highly similar sequences and structures (61, 63) and suppress pausing at many sites when added to purified transcription complexes in vitro (22, 69, 70). Both proteins are believed to suppress pausing by binding to the β′ clamp and β pincer and stabilizing the active closed conformation of RNAP (63, 71). Recently, single-molecule cryo-electron microscopy (cryo-EM) studies have clarified how stabilization of RNAP structure can promote processive elongation. Certain types of transcriptional pauses are affected by a swiveling of the RNAP β′ pincer elements, resulting in an increase in pause lifetimes (72). However, binding of NusG or RfaH to RNAP disfavors this "swiveled" conformation, thereby suppressing pausing (65). Additional mechanisms for antipausing activity of NusG proteins have been proposed, including stabilization of the elongation complex by direct binding to nontemplate DNA (22, 70, 73) as well as upstream DNA (74–77). Indeed, both NusG and RfaH interact with the upstream DNA fork and promote reannealing of the upstream DNA, although the specific effects of this activity on RNAP activity are unclear (65, 74). These mechanisms are conserved between NusG and RfaH and are likely shared to varying degrees with other NusG paralogs.

RfaH is specifically recruited to the operons that comprise its regulon by a DNA element called the operon polarity suppressor, or ops (Fig. 4A). Deletion of this 8-bp conserved element reduces downstream gene expression (60); correspondingly, introduction of ops to other transcripts increases their expression (59). Depletion of RfaH mirrors these results, indicating that RfaH and ops are both required for expression of target operons (78). RfaH is specifically recruited to TECs by binding to the nontemplate DNA strand of the ops element; this occurs during the lifetime of a programmed transcriptional pause (22). The ops element forms both a consensus pause sequence as well as a DNA hairpin loop that makes specific, direct contacts to the RfaH NTD (79). RfaH and NusG are mutually exclusive, as both homologs share the same binding site on RNAP (80, 81). Moreover, once recruited, RfaH exhibits increased affinity for RNAP relative to NusG, allowing

for extended association of RfaH with TECs (82). This increased affinity may also be responsible for the more pronounced effects of RfaH NTD on RNAP as compared to NusG (65). In this way, RfaH exerts its regulatory effects specifically on those operons that include the ops element.

RfaH in solution differs from RNAP-associated RfaH. Instead of the common β-barrel fold found in most high-resolution structures of KOW domains, the CTD of free RfaH forms a dramatically different α-helical structure (80). This α-helical CTD interacts with the NTD, partially masking the RNAP-binding portion and thereby resulting in an autoinhibited form of the protein (Fig. 5) (82). After a conformational change is triggered, the NTD can associate fully with the transcription complex, which in turn promotes refolding of the CTD to the β-barrel structure found in NusG (Fig. 5) (26). Because of this structural mechanism, RfaH adopts the classical NusG KOW domain structure only after the NTD has fully associated with RNAP.

Though NusG and RfaH display nearly identical antipausing effects on transcription complexes in vitro, their overall regulatory outcomes are different. In some instances, NusG may promote pausing in vivo (83), perhaps as a result of increased affinity for certain nontemplate DNA strand sequences (70). More importantly, NusG is known to directly bind Rho (68). This interaction is likely to broadly promote Rho-dependent termination activity, possibly by increasing the rate at which Rho successfully binds RNA and forms a closed, translocation-capable conformation (39). Ultimately, association of NusG results in Rho-dependent termination and suppression of transcription, particularly in genomic regions that feature foreign DNA (25). This activity is essential in most E. coli strains primarily due to suppression of toxic genes in prophage DNA (25). However, in addition to its interaction with Rho, the NusG CTD can associate with NusE (S10), as well as NusA (84, 85). Similar to NusG, RfaH can associate with NusE (S10); however, in contrast to NusG, RfaH is incapable of binding Rho (26, 84). Because of this, RfaH strongly discourages Rho termination within its targeted operons (86).

Finally, the remaining mechanism by which RfaH may promote antitermination is through recruitment of ribosomes to nascent transcripts. NusG proteins are thought to couple transcription and translation by facilitating macromolecular interactions between both of these machines (84). RfaH in particular has been shown to exhibit much stronger polarity effects in vivo than its effects on transcription in vitro (82). Also, genes that are known to be regulated by RfaH display

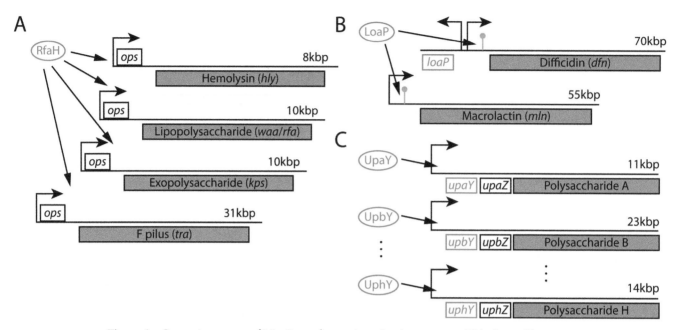

Figure 4 Genomic context of NusG paralog antitermination systems. This figure illustrates the transcripts regulated by the RfaH, LoaP, and UpxY antitermination systems. (A) RfaH regulates multiple pathways in *E. coli*, including the hemolysin, F pilus, and lipo- and exopolysaccharide operons. Each regulated transcript includes the DNA *ops* element for RfaH recruitment. RfaH promotes antitermination of Rho-dependent promoters. (B) LoaP regulates two polyketide antibiotic operons in *B. velezensis*: the *dfn* difficidin operon and the *mln* macrolactin operon. LoaP is found divergently oriented upstream of the *dfn* operon. Each transcript includes a required sequence region in the 5′ leader region, which might include a functionally important hairpin followed by an intrinsic terminator. Additional intrinsic terminator sites have been implicated within the *dfn* and *mln* operons, although they are not shown in this figure. (C) UpxY proteins regulate multiple capsular polysaccharide pathways in *B. fragilis*. Each polysaccharide operon includes both a UpxY and UpxZ protein involved in targeted regulation, with the 5′ leader sequence required for antitermination. *B. fragilis* has eight distinct polysaccharide operons containing UpxY proteins. Gray rectangles represent multigene operons. RNA elements potentially involved in antitermination are shown in blue, and proteins and protein-coding genes involved in antitermination are shown in green.

particularly poor ribosome binding sites, suggesting that translational enhancement is likely to be a key feature of RfaH regulation (86). It is possible that binding of NusG or RfaH to ribosomal S10 (NusE) may assist ribosome recruitment, thereby increasing local concentration of ribosomes and promoting translation initiation on nascent RNA (26, 87). This functional interaction might also affect transcription processivity. Indeed, recent data suggest that the leading ribosome—which conducts translation immediately upstream of RNAP, and that may participate in the RNAP-ribosome "expressome" (88, 89)—improves transcription processivity by directly blocking RNAP backtracking (90) and by obstructing Rho access (84, 91).

Through these aggregate mechanisms, RfaH acts as a specialized elongation factor that exhibits antipausing activity, prevents NusG-mediated Rho termination, and

encourages ribosome recruitment, for each of the operons that display *ops* elements.

Other NusG Paralogs (ActX, TaA, UpxY, and LoaP)

Although RfaH is the most prominent and best-studied NusG paralog, other examples have been identified, several of which have been predicted to function in transcription antitermination (92–95). All of these homologs share significant sequence similarity to NusG and RfaH and undoubtedly share conserved structural features. Moreover, for those NusG paralogs in which a functional role has been demonstrated, they have inevitably been found to affect transcription of certain targeted transcripts, suggesting that NusG paralogs are broadly used by bacteria as specialized transcription regulators (93–95).

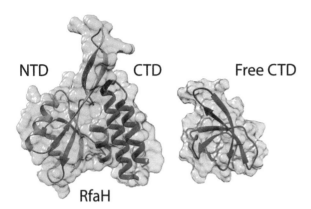

Figure 5 The RfaH CTD undergoes a large conformational shift from an α-helix to a β-barrel. Full-length RfaH (left) exists as an autoinhibited structure with the CTD (blue) in an α-helix conformation bound to the NTD (red) (PDB ID: 2OUG). Upon binding to RNAP and the *ops* DNA, the CTD (right) is released and forms the β-barrel conformation characteristic of NusG KOW domains (PDB ID: 2LCL).

ActX and TraB proteins are most phylogenetically similar to RfaH (92, 96) and are found in a variety of conjugative plasmids conferring antibiotic resistance in *Gammaproteobacteria* (92, 97). Though a function has not been demonstrated for these proteins, they are often transcribed as the first open reading frame in long pilus biosynthesis operons and are suspected to be involved in the transcription of conjugation genes (98).

Myxococcus xanthus, a Gram-negative soil bacterium, produces the well-studied polyketide antibiotic TA (also called myxovirescin) (99, 100). The first open reading frame of the TA-producing gene cluster is *taA*, which encodes a NusG paralog (93). Disruption of the *taA* gene eliminated antibiotic production, suggesting a regulatory relationship. However, the specific role of TaA in expression of the TA gene cluster is unknown, although as a NusG paralog and relative of other known NusG specialized paralogs, it has been proposed to regulate transcription elongation, perhaps through PA.

More recently, a NusG paralog called UpxY has been proposed to function as a family of regulators for complex polysaccharide pathways in *Bacteroidetes* (94, 101). They are widely used by these microorganisms. Indeed, many *Bacteroides* spp. encode between six and nine copies of the UpxY proteins. The genes encoding these proteins, initially described in *Bacteroides fragilis*, are each associated with a different capsular polysaccharide gene cluster (Fig. 4C) (101). These proteins have been shown to affect transcription of their associated gene cluster, and it has been proposed that they participate in antitermination-based regulatory mechanisms that involve unique sequence features located within the 5′ leader regions of their respective operons. Additionally, while these regulators might be cotranscribed with the operons they affect, they can also affect gene expression when moved to a distal location, supporting the claim that they are recruited to their targeted operons, perhaps via sequence elements within the transcript leader regions. Yet, despite these observations, little is known regarding the molecular mechanisms of UpxY proteins. Adding a new wrinkle to the overall family of NusG paralogous proteins, *B. fragilis* also encodes a set of unique proteins (UpxZ) alongside genes encoding UpxY proteins. The UpxZ proteins can act as *trans*-inhibitors of UpxY proteins, and have been hypothesized to hierarchically regulate the expression of different sets of capsular polysaccharides, although the underlying mechanism of this inhibition is also unknown (94).

The most recently described NusG paralog is LoaP. Genes encoding LoaP are consistently positioned adjacent to polyketide biosynthesis pathways in *Firmicutes*, or near polysaccharide biosynthesis gene clusters in certain *Firmicutes*, *Actinobacteria*, and *Spirochaetes* (95). While proximity alone is not evidence of a functional relationship, *Bacillus velezensis loaP* was shown to affect expression of an adjacent polyketide synthesis gene cluster. More specifically, *B. velezensis loaP* is situated adjacent to a gene cluster (*dfn*) that codes for production of the polyketide antibiotic difficidin (102). Deletion of *loaP* resulted in low abundance across the *dfn* transcript, whereas complementation of *loaP* from an ectopic locus restored *dfn* expression. Moreover, these global expression experiments revealed that the difficidin operon is not the only region affected by LoaP. LoaP is also required for transcription elongation of a second polyketide gene cluster, which codes for production of another antibiotic, macrolactin, indicating that LoaP controls a regulon of antibiotic biosynthesis operons in *B. velezensis* (Fig. 4B).

Upon depletion of LoaP, transcript abundance dramatically decreases at intrinsic terminator sites located within the targeted operons, whereas induction of LoaP restored read-through of these terminators (95). These initial data suggest that LoaP antitermination may function primarily on intrinsic termination, in contrast to the suppression of Rho-dependent termination known for RfaH (22, 95). However, while this might be due to the preference of Gram-positive bacteria for Rho-independent antitermination, it is also true that Rho termination has been insufficiently characterized in *B. velezensis* and other *Firmicutes* (103). Therefore, the relationship between LoaP proteins and Rho termination is still unknown. Nor have the full determinants for LoaP PA yet been described. While LoaP affects transcript abundance across the length of the targeted operon, it appears to require sequence elements located somewhere within the 5′ leader region. Indeed, when placed upstream of terminator signals and upstream of a reporter gene, the *dfn* leader region alone is sufficient for promoting LoaP-dependent PA activity (95). Therefore, the recruitment signals for LoaP are located fully within this leader region. Interestingly, a small UNCG-type hairpin was identified in the leader regions of both the difficidin and macrolactin operons in a sequence region required for antitermination, although its exact relationship to LoaP regulation has not yet been investigated (Fig. 3) (95).

The discovery of LoaP, along with the initial description of TaA, suggests that transcription elongation may be a broad point of regulatory control for secondary metabolite gene clusters in bacteria. Therefore, it is important to study PA mechanisms in order to improve discovery and production of new natural products from bacteria.

Phylogenetic Overview of the NusG Family of Proteins

NusG paralogs putatively involved in antitermination have been identified in a variety of bacteria, including but not limited to *Alphaproteobacteria*, *Gammaproteobacteria*, *Deltaproteobacteria*, *Bacteroidetes*, and most recently *Firmicutes*, *Actinobacteria*, and *Spirochaetes* (86, 95). Of the general transcription elongation factors, only NusG is found in all three domains of life, suggesting that its function is important in all organisms. Therefore, essentially all bacteria code for a core NusG protein, while archaea and eukaryotes code for a similar protein, Spt5 (80). As a result, all NusG family proteins share core conserved sequence and structure features (80).

Although analysis of the paralogs supports grouping them within the overall NusG family, each subgrouping displays significant sequence diversity, with some subgroups displaying very limited overall sequence identity despite sharing remarkably conserved structural elements (95). In recent work, the phylogenetic analysis of the NusG family was extended to include as many distant homologs as are detectable by Hidden Markov Model-based homology modeling (95). This large-scale phylogenetic analysis utilized structural modeling to efficiently align specialized NusG paralogs with limited sequence similarity, and focused on comprehensively covering the diversity of paralog sequences without restriction to the known subgroups. The resulting phylo-

genetic tree (Fig. 6) confirmed that each set of NusG paralogs forms its own distinct group, separate from core bacterial NusG and archaeal Spt5, while also revealing a few new candidate subgroups (82, 95). It is likely that each subgroup will be defined by specific sequence differences. Indeed, a number of characteristic differences between sequences—such as between RfaH or UpxY and core NusG—have been identified as being important for the distinct activities of those specialized paralogs (82, 101, 104).

As NusG paralogs were found in a variety of distinct genetic contexts (82), it was important to systematically identify associations between these genes and potential target pathways. Overall, they were found in diverse genomic contexts, with some positioned alone, at the beginning of complex polysaccharide or secondary metabolite gene clusters, at the end of operons, or in unique contexts (82, 95). For example, NusG paralogous sequences from *Betaproteobacteria* and *Bacteroides* are located in or near large polysaccharide pathways. TaA and LoaP sequences are generally present in or near large polyketide biosynthesis pathways, which suggests that they share a broad relationship to secondary metabolites (82, 95). Indeed, there appears to be a general association of NusG specialized paralogs with polysaccharide biosynthesis gene clusters, and to a lesser extent polyketide synthase gene clusters. In fact, of all the paralog groups, only the gammaproteobacterial RfaH and its related ActX gene sequences were not frequently identified near or in these classes of gene clusters (95).

There also appear to be several subgroups of NusG paralogs with interesting genomic association and evolutionary distribution but that have not been characterized or named. For example, a group of sequences closely related to RfaH and found in *Alpha-*, *Beta-*, and *Gammaproteobacteria* is oftentimes associated with polysaccharide gene clusters. Similarly, at least two more uncharacterized and unnamed putative groups of sequences are consistently associated with polysaccharide and polyketide biosynthesis gene clusters. From this, it can be tentatively speculated that NusG specialized paralogs evolved as regulators of these long operons (polysaccharides and secondary metabolite biosynthesis genes) and became further specialized into RfaH in *Gammaproteobacteria*. Finally, an additional set of paralog sequences in *Alphaproteobacteria* was not found in a consistent genomic context, and remains unnamed. Ultimately, the evolutionary relationship between all these different NusG paralogs remains unclear, as bootstrap support for early branches after divergence from core NusG is low, likely due to the

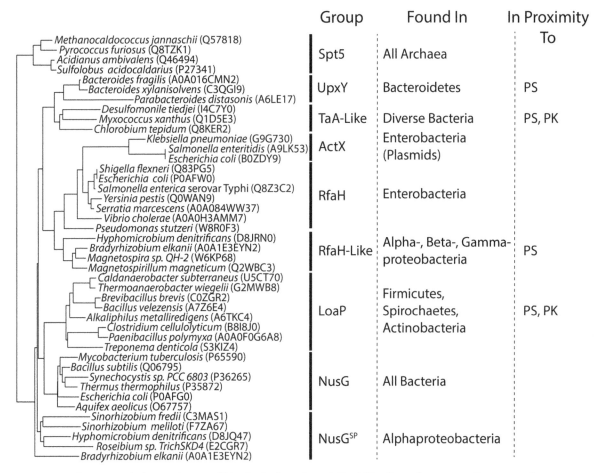

Figure 6 Phylogenetic tree of NusG, Spt5, and specialized NusG paralog groups. Represented NusG sequences selected from the subgroups discussed in this review form the NusG family. Bacterial sequences from core NusG proteins are found in all bacteria, while a variety of paralogs are found in diverse bacteria phyla. Some groups of NusG paralogs are commonly found in or adjacent to large gene clusters coding for production of polysaccharides (PS) or polyketides (PK).

extensive sequence divergence in this family. Elucidating the true history of this family may require different approaches, integrating more information about the structural changes and sequence insertions and deletions during evolution of the NusG paralogs. However, it is already clear that the NusG family of proteins is widely used in bacteria as specialized transcription elongation regulatory factors, and that they regulate expression of fundamentally important pathways, albeit through largely unexplored molecular mechanisms.

OUTLOOK

The past few years have uncovered a few new examples of PA mechanisms, as well as remarkable new insight into the structural basis of antitermination activity.

However, it is possible that these findings still only represent a small proportion of what remains to be discovered. Therefore, what is now needed is a systematic exploration of the molecular mechanisms used by NusG paralogs, combined with new bioinformatics searches for RNA elements that promote PA. From this, an accurate portrayal of the extent of PA usage in bacteria will emerge, which will resolve whether transcription elongation is a much broader point of regulatory control than has historically been perceived. Furthermore, studying new PA systems will help uncover the diversity of their molecular mechanisms and shed important light on when and why PA mechanisms are employed by bacteria.

Citation. Goodson JR, Winkler WC. 2018. Processive antitermination. Microbiol Spectrum 6(5):RWR-0031-2018.

References

1. Jacob F, Monod J. 1961. Genetic regulatory mechanisms in the synthesis of proteins. *J Mol Biol* **3**:318–356.

2. Walsh C. 2003. *Antibiotics: Actions, Origins, Resistance*, p 159–174. ASM Press, Washington, DC.

3. Waters LS, Storz G. 2009. Regulatory RNAs in bacteria. *Cell* **136**:615–628.

4. Barrick JE, Breaker RR. 2007. The distributions, mechanisms, and structures of metabolite-binding riboswitches. *Genome Biol* **8**:R239.

5. Breaker RR. 2011. Prospects for riboswitch discovery and analysis. *Mol Cell* **43**:867–879.

6. Weisberg RA, Gottesman ME. 1999. Processive antitermination. *J Bacteriol* **181**:359–367.

7. Roberts JW, Shankar S, Filter JJ. 2008. RNA polymerase elongation factors. *Annu Rev Microbiol* **62**:211–233.

8. Santangelo TJ, Artsimovitch I. 2011. Termination and antitermination: RNA polymerase runs a stop sign. *Nat Rev Microbiol* **9**:319–329.

9. Ray-Soni A, Bellecourt MJ, Landick R. 2016. Mechanisms of bacterial transcription termination: all good things must end. *Annu Rev Biochem* **85**:319–347.

10. Merino E, Yanofsky C. 2005. Transcription attenuation: a highly conserved regulatory strategy used by bacteria. *Trends Genet* **21**:260–264.

11. Gusarov I, Nudler E. 1999. The mechanism of intrinsic transcription termination. *Mol Cell* **3**:495–504.

12. Greenblatt J, McLimont M, Hanly S. 1981. Termination of transcription by *nusA* gene protein of *Escherichia coli*. *Nature* **292**:215–220.

13. Mondal S, Yakhnin AV, Sebastian A, Albert I, Babitzke P. 2016. NusA-dependent transcription termination prevents misregulation of global gene expression. *Nat Microbiol* **1**:15007.

14. Epshtein V, Cardinale CJ, Ruckenstein AE, Borukhov S, Nudler E. 2007. An allosteric path to transcription termination. *Mol Cell* **28**:991–1001.

15. Holmes WM, Platt T, Rosenberg M. 1983. Termination of transcription in *E. coli*. *Cell* **32**:1029–1032.

16. Epshtein V, Dutta D, Wade J, Nudler E. 2010. An allosteric mechanism of Rho-dependent transcription termination. *Nature* **463**:245–249.

17. Proshkin S, Mironov A, Nudler E. 2014. Riboswitches in regulation of Rho-dependent transcription termination. *Biochim Biophys Acta* **1839**:974–977.

18. DebRoy S, Gebbie M, Ramesh A, Goodson JR, Cruz MR, van Hoof A, Winkler WC, Garsin DA. 2014. Riboswitches. A riboswitch-containing sRNA controls gene expression by sequestration of a response regulator. *Science* **345**:937–940.

19. Sedlyarova N, Shamovsky I, Bharati BK, Epshtein V, Chen J, Gottesman S, Schroeder R, Nudler E. 2016. sRNA-mediated control of transcription termination in *E. coli*. *Cell* **167**:111–121.e13.

20. Montange RK, Batey RT. 2008. Riboswitches: emerging themes in RNA structure and function. *Annu Rev Biophys* **37**:117–133.

21. Ceres P, Garst AD, Marcano-Velázquez JG, Batey RT. 2013. Modularity of select riboswitch expression platforms enables facile engineering of novel genetic regulatory devices. *ACS Synth Biol* **2**:463–472.

22. Artsimovitch I, Landick R. 2002. The transcriptional regulator RfaH stimulates RNA chain synthesis after recruitment to elongation complexes by the exposed nontemplate DNA strand. *Cell* **109**:193–203.

23. Said N, Krupp F, Anedchenko E, Santos KF, Dybkov O, Huang YH, Lee CT, Loll B, Behrmann E, Bürger J, Mielke T, Loerke J, Urlaub H, Spahn CM, Weber G, Wahl MC. 2017. Structural basis for λN-dependent processive transcription antitermination. *Nat Microbiol* **2**:17062.

24. Gusarov I, Nudler E. 2001. Control of intrinsic transcription termination by N and NusA: the basic mechanisms. *Cell* **107**:437–449.

25. Cardinale CJ, Washburn RS, Tadigotla VR, Brown LM, Gottesman ME, Nudler E. 2008. Termination factor Rho and its cofactors NusA and NusG silence foreign DNA in *E. coli*. *Science* **320**:935–938.

26. Burmann BM, Knauer SH, Sevostyanova A, Schweimer K, Mooney RA, Landick R, Artsimovitch I, Rösch P. 2012. An α helix to β barrel domain switch transforms the transcription factor RfaH into a translation factor. *Cell* **150**:291–303.

27. Friedman DI. 1988. Regulation of phage gene expression by termination and antitermination of transcription, p 262–319. *In* Calendar R (ed), *The Bacteriophages*, Vol 2. Springer US, Plenum Press, New York, NY.

28. Patterson TA, Zhang Z, Baker T, Johnson LL, Friedman DI, Court DL. 1994. Bacteriophage lambda N-dependent transcription antitermination. Competition for an RNA site may regulate antitermination. *J Mol Biol* **236**:217–228.

29. Mogridge J, Legault P, Li J, Van Oene MD, Kay LE, Greenblatt J. 1998. Independent ligand-induced folding of the RNA-binding domain and two functionally distinct antitermination regions in the phage lambda N protein. *Mol Cell* **1**:265–275.

30. Legault P, Li J, Mogridge J, Kay LE, Greenblatt J. 1998. NMR structure of the bacteriophage λ N peptide/*boxB* RNA complex: recognition of a GNRA fold by an arginine-rich motif. *Cell* **93**:289–299.

31. Thapar R, Denmon AP, Nikonowicz EP. 2014. Recognition modes of RNA tetraloops and tetraloop-like motifs by RNA-binding proteins. *Wiley Interdiscip Rev RNA* **5**:49–67.

32. Mogridge J, Mah TF, Greenblatt J. 1998. Involvement of *boxA* nucleotides in the formation of a stable ribonucleoprotein complex containing the bacteriophage λ N protein. *J Biol Chem* **273**:4143–4148.

33. Nodwell JR, Greenblatt J. 1993. Recognition of *boxA* antiterminator RNA by the *E. coli* antitermination factors NusB and ribosomal protein S10. *Cell* **72**:261–268.

34. Mason SW, Li J, Greenblatt J. 1992. Host factor requirements for processive antitermination of transcription and suppression of pausing by the N protein of bacteriophage λ. *J Biol Chem* **267**:19418–19426.

35. Rees WA, Weitzel SE, Das A, von Hippel PH. 1997. Regulation of the elongation-termination decision at intrinsic terminators by antitermination protein N of phage λ. *J Mol Biol* **273**:797–813.

36. Nudler E, Gottesman ME. 2002. Transcription termination and anti-termination in *E. coli*. *Genes Cells* **7**:755–768.

37. Liu K, Zhang Y, Severinov K, Das A, Hanna MM. 1996. Role of *Escherichia coli* RNA polymerase alpha subunit in modulation of pausing, termination and antitermination by the transcription elongation factor NusA. *EMBO J* **15**:150–161.

38. Peters JM, Mooney RA, Grass JA, Jessen ED, Tran F, Landick R. 2012. Rho and NusG suppress pervasive antisense transcription in *Escherichia coli*. *Genes Dev* **26**:2621–2633.

39. Valabhoju V, Agrawal S, Sen R. 2016. Molecular basis of NusG-mediated regulation of Rho-dependent transcription termination in bacteria. *J Biol Chem* **291**:22386–22403.

40. Herskowitz I, Signer ER. 1970. A site essential for expression of all late genes in bacteriophage λ. *J Mol Biol* **47**:545–556.

41. Yarnell WS, Roberts JW. 1992. The phage λ gene *Q* transcription antiterminator binds DNA in the late gene promoter as it modifies RNA polymerase. *Cell* **69**:1181–1189.

42. Lowery C, Richardson JP. 1977. Characterization of the nucleoside triphosphate phosphohydrolase (ATPase) activity of RNA synthesis termination factor ρ. II. Influence of synthetic RNA homopolymers and random copolymers on the reaction. *J Biol Chem* **252**:1381–1385.

43. Guérin M, Robichon N, Geiselmann J, Rahmouni AR. 1998. A simple polypyrimidine repeat acts as an artificial Rho-dependent terminator *in vivo* and *in vitro*. *Nucleic Acids Res* **26**:4895–4900.

44. Li SC, Squires CL, Squires C. 1984. Antitermination of *E. coli* rRNA transcription is caused by a control region segment containing lambda *nut*-like sequences. *Cell* **38**:851–860.

45. Squires CL, Greenblatt J, Li J, Condon C, Squires CL. 1993. Ribosomal RNA antitermination in vitro: requirement for Nus factors and one or more unidentified cellular components. *Proc Natl Acad Sci U S A* **90**:970–974.

46. Arnvig KB, Zeng S, Quan S, Papageorge A, Zhang N, Villapakkam AC, Squires CL. 2008. Evolutionary comparison of ribosomal operon antitermination function. *J Bacteriol* **190**:7251–7257.

47. Berg KL, Squires C, Squires CL. 1989. Ribosomal RNA operon anti-termination. Function of leader and spacer region box B-box A sequences and their conservation in diverse micro-organisms. *J Mol Biol* **209**:345–358.

48. Singh N, Bubunenko M, Smith C, Abbott DM, Stringer AM, Shi R, Court DL, Wade JT. 2016. SuhB associates with Nus factors to facilitate 30S ribosome biogenesis in *Escherichia coli*. *mBio* **7**:e00114.

49. Baniulyte G, Singh N, Benoit C, Johnson R, Ferguson R, Paramo M, Stringer AM, Scott A, Lapierre P, Wade JT. 2017. Identification of regulatory targets for the bacterial Nus factor complex. *Nat Commun* **8**:2027.

50. Clerget M, Jin DJ, Weisberg RA. 1995. A zinc-binding region in the β′ subunit of RNA polymerase is involved in antitermination of early transcription of phage HK022. *J Mol Biol* **248**:768–780.

51. King RA, Banik-Maiti S, Jin DJ, Weisberg RA. 1996. Transcripts that increase the processivity and elongation rate of RNA polymerase. *Cell* **87**:893–903.

52. Banik-Maiti S, King RA, Weisberg RA. 1997. The antiterminator RNA of phage HK022. *J Mol Biol* **272**:677–687.

53. Robert J, Sloan SB, Weisberg RA, Gottesman ME, Robledo R, Harbrecht D. 1987. The remarkable specificity of a new transcription termination factor suggests that the mechanisms of termination and antitermination are similar. *Cell* **51**:483–492.

54. Hung SC, Gottesman ME. 1997. The Nun protein of bacteriophage HK022 inhibits translocation of *Escherichia coli* RNA polymerase without abolishing its catalytic activities. *Genes Dev* **11**:2670–2678.

55. King RA, Weisberg RA. 2003. Suppression of factor-dependent transcription termination by antiterminator RNA. *J Bacteriol* **185**:7085–7091.

56. Vitiello CL, Kireeva ML, Lubkowska L, Kashlev M, Gottesman M. 2014. Coliphage HK022 Nun protein inhibits RNA polymerase translocation. *Proc Natl Acad Sci U S A* **111**:E2368–E2375.

57. Oberto J, Clerget M, Ditto M, Cam K, Weisberg RA. 1993. Antitermination of early transcription in phage HK022. Absence of a phage-encoded antitermination factor. *J Mol Biol* **229**:368–381.

58. Irnov I, Winkler WC. 2010. A regulatory RNA required for antitermination of biofilm and capsular polysaccharide operons in Bacillales. *Mol Microbiol* **76**:559–575.

59. Bailey MJ, Hughes C, Koronakis V. 1996. Increased distal gene transcription by the elongation factor RfaH, a specialized homologue of NusG. *Mol Microbiol* **22**:729–737.

60. Bailey MJ, Koronakis V, Schmoll T, Hughes C. 1992. *Escherichia coli* HlyT protein, a transcriptional activator of haemolysin synthesis and secretion, is encoded by the *rfaH (sfrB)* locus required for expression of sex factor and lipopolysaccharide genes. *Mol Microbiol* **6**:1003–1012.

61. Bailey MJ, Hughes C, Koronakis V. 1997. RfaH and the *ops* element, components of a novel system controlling bacterial transcription elongation. *Mol Microbiol* **26**:845–851.

62. Reay P, Yamasaki K, Terada T, Kuramitsu S, Shirouzu M, Yokoyama S. 2004. Structural and sequence comparisons arising from the solution structure of the transcription elongation factor NusG from *Thermus thermophilus*. *Proteins* **56**:40–51.

63. Martinez-Rucobo FW, Sainsbury S, Cheung AC, Cramer P. 2011. Architecture of the RNA polymerase-Spt4/5 complex and basis of universal transcription processivity. *EMBO J* **30**:1302–1310.

64. Liu B, Steitz TA. 2017. Structural insights into NusG regulating transcription elongation. *Nucleic Acids Res* **45**:968–974.

65. Kang JY, Mooney RA, Nedialkov Y, Saba J, Mishanina TV, Artsimovitch I, Landick R, Darst SA. 2018. Structural basis for transcript elongation control by NusG family universal regulators. *Cell* **173**:1650–1662.e14.

66. Ponting CP. 2002. Novel domains and orthologues of eukaryotic transcription elongation factors. *Nucleic Acids Res* **30**:3643–3652.

67. Kyrpides NC, Woese CR, Ouzounis CA. 1996. KOW: a novel motif linking a bacterial transcription factor with ribosomal proteins. *Trends Biochem Sci* **21**:425–426.

68. Mooney RA, Schweimer K, Rösch P, Gottesman M, Landick R. 2009. Two structurally independent domains of *E. coli* NusG create regulatory plasticity via distinct interactions with RNA polymerase and regulators. *J Mol Biol* **391**:341–358.

69. Burova E, Hung SC, Sagitov V, Stitt BL, Gottesman ME. 1995. *Escherichia coli* NusG protein stimulates transcription elongation rates in vivo and in vitro. *J Bacteriol* **177**:1388–1392.

70. Yakhnin AV, Murakami KS, Babitzke P. 2016. NusG is a sequence-specific RNA polymerase pause factor that binds to the non-template DNA within the paused transcription bubble. *J Biol Chem* **291**:5299–5308.

71. Weixlbaumer A, Leon K, Landick R, Darst SA. 2013. Structural basis of transcriptional pausing in bacteria. *Cell* **152**:431–441.

72. Kang JY, Mishanina TV, Bellecourt MJ, Mooney RA, Darst SA, Landick R. 2018. RNA polymerase accommodates a pause RNA hairpin by global conformational rearrangements that prolong pausing. *Mol Cell* **69**:802–815.e1.

73. Crickard JB, Fu J, Reese JC. 2016. Biochemical analysis of yeast suppressor of Ty 4/5 (Spt4/5) reveals the importance of nucleic acid interactions in the prevention of RNA polymerase II arrest. *J Biol Chem* **291**:9853–9870.

74. Nedialkov Y, Svetlov D, Belogurov GA, Artsimovitch I. 2018. Locking the non-template DNA to control transcription. *Mol Microbiol*.

75. Guo G, Gao Y, Zhu Z, Zhao D, Liu Z, Zhou H, Niu L, Teng M. 2015. Structural and biochemical insights into the DNA-binding mode of *Mj*Spt4p:Spt5 complex at the exit tunnel of RNAPII. *J Struct Biol* **192**:418–425.

76. Ehara H, Yokoyama T, Shigematsu H, Yokoyama S, Shirouzu M, Sekine SI. 2017. Structure of the complete elongation complex of RNA polymerase II with basal factors. *Science* **357**:921–924.

77. Turtola M, Belogurov GA. 2016. NusG inhibits RNA polymerase backtracking by stabilizing the minimal transcription bubble. *eLife* **5**:e18096.

78. Beutin L, Manning PA, Achtman M, Willetts N. 1981. *sfrA* and *sfrB* products of *Escherichia coli* K-12 are transcriptional control factors. *J Bacteriol* **145**:840–844.

79. Zuber PK, Artsimovitch I, NandyMazumdar M, Liu Z, Nedialkov Y, Schweimer K, Rösch P, Knauer SH. 2018. The universally-conserved transcription factor RfaH is recruited to a hairpin structure of the non-template DNA strand. *eLife* **7**:e36349.

80. Belogurov GA, Vassylyeva MN, Svetlov V, Klyuyev S, Grishin NV, Vassylyev DG, Artsimovitch I. 2007. Structural basis for converting a general transcription factor into an operon-specific virulence regulator. *Mol Cell* **26**:117–129.

81. Yakhnin AV, Yakhnin H, Babitzke P. 2008. Function of the *Bacillus subtilis* transcription elongation factor NusG in hairpin-dependent RNA polymerase pausing in the *trp* leader. *Proc Natl Acad Sci U S A* **105**:16131–16136.

82. Belogurov GA, Mooney RA, Svetlov V, Landick R, Artsimovitch I. 2009. Functional specialization of transcription elongation factors. *EMBO J* **28**:112–122.

83. Yakhnin AV, Babitzke P. 2014. NusG/Spt5: are there common functions of this ubiquitous transcription elongation factor? *Curr Opin Microbiol* **18**:68–71.

84. Burmann BM, Schweimer K, Luo X, Wahl MC, Stitt BL, Gottesman ME, Rösch P. 2010. A NusE:NusG complex links transcription and translation. *Science* **328**:501–504.

85. Strauß M, Vitiello C, Schweimer K, Gottesman M, Rösch P, Knauer SH. 2016. Transcription is regulated by NusA:NusG interaction. *Nucleic Acids Res* **44**:5971–5982.

86. Tomar SK, Artsimovitch I. 2013. NusG-Spt5 proteins—universal tools for transcription modification and communication. *Chem Rev* **113**:8604–8619.

87. Saxena S, Myka KK, Washburn R, Costantino N, Court DL, Gottesman ME. 2018. *Escherichia coli* transcription factor NusG binds to 70S ribosomes. *Mol Microbiol* **108**:495–504.

88. Kohler R, Mooney RA, Mills DJ, Landick R, Cramer P. 2017. Architecture of a transcribing-translating expressome. *Science* **356**:194–197.

89. Demo G, Rasouly A, Vasilyev N, Svetlov V, Loveland AB, Diaz-Avalos R, Grigorieff N, Nudler E, Korostelev AA. 2017. Structure of RNA polymerase bound to ribosomal 30S subunit. *eLife* **6**:e28560.

90. Proshkin S, Rahmouni AR, Mironov A, Nudler E. 2010. Cooperation between translating ribosomes and RNA polymerase in transcription elongation. *Science* **328**:504–508.

91. Banerjee S, Chalissery J, Bandey I, Sen R. 2006. Rho-dependent transcription termination: more questions than answers. *J Microbiol* **44**:11–22.

92. Núñez B, Avila P, de la Cruz F. 1997. Genes involved in conjugative DNA processing of plasmid R6K. *Mol Microbiol* **24**:1157–1168.

93. Paitan Y, Orr E, Ron EZ, Rosenberg E. 1999. A NusG-like transcription anti-terminator is involved in the biosynthesis of the polyketide antibiotic TA of *Myxococcus xanthus*. *FEMS Microbiol Lett* **170**:221–227.

94. Chatzidaki-Livanis M, Weinacht KG, Comstock LE. 2010. *Trans* locus inhibitors limit concomitant polysaccharide synthesis in the human gut symbiont *Bacteroides fragilis*. *Proc Natl Acad Sci U S A* **107**:11976–11980.

95. Goodson JR, Klupt S, Zhang C, Straight P, Winkler WC. 2017. LoaP is a broadly conserved antiterminator protein that regulates antibiotic gene clusters in *Bacillus amyloliquefaciens*. *Nat Microbiol* **2**:17003.

96. Arutyunov D, Arenson B, Manchak J, Frost LS. 2010. F plasmid TraF and TraH are components of an outer membrane complex involved in conjugation. *J Bacteriol* **192**:1730–1734.

97. Jones CS, Osborne DJ, Stanley J. 1993. Molecular comparison of the IncX plasmids allows division into IncX1 and IncX2 subgroups. *J Gen Microbiol* **139**:735–741.

98. NandyMazumdar M, Artsimovitch I. 2015. Ubiquitous transcription factors display structural plasticity and diverse functions. *BioEssays* **37**:324–334.

99. Varon M, Fuchs N, Monosov M, Tolchinsky S, Rosenberg E. 1992. Mutation and mapping of genes involved in production of the antibiotic TA in *Myxococcus xanthus*. *Antimicrob Agents Chemother* **36**:2316–2321.

100. Simunovic V, Zapp J, Rachid S, Krug D, Meiser P, Müller R. 2006. Myxovirescin A biosynthesis is directed by hybrid polyketide synthases/nonribosomal peptide synthetase, 3-hydroxy-3-methylglutaryl-CoA synthases, and trans-acting acyltransferases. *Chembiochem* **7**:1206–1220.

101. Chatzidaki-Livanis M, Coyne MJ, Comstock LE. 2009. A family of transcriptional antitermination factors necessary for synthesis of the capsular polysaccharides of *Bacteroides fragilis*. *J Bacteriol* **191**:7288–7295.

102. Chen XH, Vater J, Piel J, Franke P, Scholz R, Schneider K, Koumoutsi A, Hitzeroth G, Grammel N, Strittmatter AW, Gottschalk G, Süssmuth RD, Borriss R. 2006. Structural and functional characterization of three polyketide synthase gene clusters in *Bacillus amyloliquefaciens* FZB 42. *J Bacteriol* **188**:4024–4036.

103. Mitra P, Ghosh G, Hafeezunnisa M, Sen R. 2017. Rho protein: roles and mechanisms. *Annu Rev Microbiol* **71**:687–709.

104. Shi D, Svetlov D, Abagyan R, Artsimovitch I. 2017. Flipping states: a few key residues decide the winning conformation of the only universally conserved transcription factor. *Nucleic Acids Res* **45**:8835–8843.

Regulating with RNA in Bacteria and Archaea
Edited by Gisela Storz and Kai Papenfort
© 2018 American Society for Microbiology, Washington, DC
doi:10.1128/microbiolspec.RWR-0020-2018

Genes within Genes in Bacterial Genomes

9

Sezen Meydan[1], Nora Vázquez-Laslop[1], and Alexander S. Mankin[1]

INTRODUCTION

The most common result of the translation of a gene is the production of a single protein product (Fig. 1a). However, the redundancy of the genetic code and the plasticity of the mRNA structure allow for expansions of the proteome by unorthodox interpreting of genetic information. A familiar strategy leading to unusual interpretation of genetic information is collectively known as recoding and has been discussed in several excellent reviews (1–4). The most conventional recoding involves programmed ribosomal frameshifting and can generate two gene products that are identical in their N-terminal segments but differ in the sequences of their C termini (Fig. 1b).

A much less studied and discussed type of unconventional translation strategy leads to the production of two (or in rare cases more than two) distinct proteins from one gene that differ in their N termini but have identical C-terminal structures. This scenario occurs when translation of a gene is initiated not only at the primary translation initiation site (pTIS) but also at an additional, in-frame, internal translation initiation site (iTIS) (Fig. 1c). Furthermore, if the iTIS directs the ribosome to begin protein synthesis at an out-of-frame (OOF) start codon, the amino acid sequence of the resulting protein would be principally different from that of the main-frame protein (Fig. 1c).

Internal initiation is controlled by the mRNA structure in the immediate neighborhood of the iTIS. Additionally, it may also be influenced by the ribosome traffic through the iTIS, which depends on the structure of a remote mRNA segment in the vicinity of the pTIS.

Internal initiation can lead to production of functional proteins that could play beneficial or even essential roles in cell physiology. Therefore, discovery and analysis of unconventional translation scenarios and identification of the functions of the alternative protein products are critical for understanding both the principles of translation regulation and the complexity of physiological networks. In the following pages we will discuss the known examples of bacterial genes that use internal initiation to encode more than one polypeptide. Most of these examples have been discovered serendipitously and have been analyzed with a varying degree of scrutiny. For consistency, we retained the designations of the protein isoforms used by the authors of the original papers. We will also discuss the emerging strategies that may pave the way for a more systematic identification of genes with more than one start codon. Although extensive literature exists about overlapping genes in bacteriophages (5–7), we are unaware of any published compilations of bacterial genes with alternative sites of translation initiation. We apologize to our colleagues whose findings relevant to the topic of this review have evaded our quest.

INITIATION OF TRANSLATION IN BACTERIA

Initiation of translation in bacteria is driven by the small ribosomal subunit (30S). The 30S subunit locates a ribosome binding site in mRNA and, with the help of translation initiation factors, binds the initiator fMet-tRNA and positions the start codon of the open reading frame (ORF) in the small subunit P-site (reviewed in reference 8). Upon the large subunit (50S) joining and dissociation of the initiation factors, an elongator tRNA binds in the ribosomal A-site, the first peptide bond is formed, and the ribosome transitions into the elongation phase of protein synthesis (8, 9).

Identifying the start codon of the protein-coding sequence in mRNA is one of the most critical tasks of the

---

[1]Center for Biomolecular Sciences, University of Illinois at Chicago, Chicago, IL 60607.

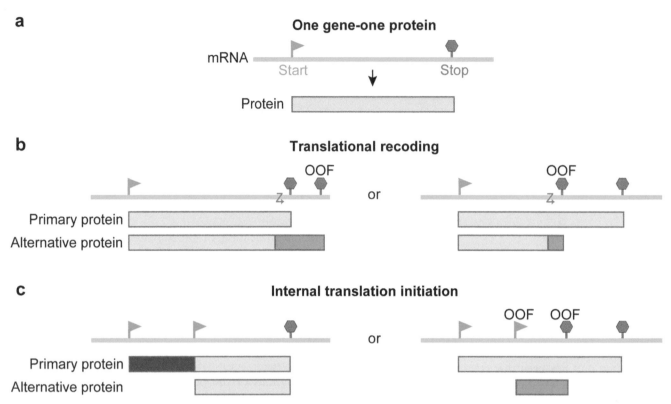

Figure 1 Different strategies for encoding more than one protein in one gene. (a) The conventional one gene-one protein scenario. (b) Programmed ribosomal frameshifting results in translation of two proteins whose N-terminal sequences are the same but differ in their C-terminal segments. The alternative protein could be shorter or longer than the primary one depending on the position of the OOF stop codon. (c) Presence of an internal in-frame start codon within the ORF results in production of a truncated protein devoid of the N-terminal segment of the full-size translation product. If the internal start codon is OOF relative to the main ORF, the sequence of the alternative polypeptide is completely different from that of the protein encoded in the main ORF.

ribosome (8, 10). Not only does the start codon define the boundary of the mRNA segment to be translated, but it also sets the frame in which the ribosome reads the genetic message. AUG is the most commonly used start codon, but other triplets (GUG, UUG, CUG, AUU, AUC, and AUA) can also be employed for translation initiation (8, 11). Utilization of a codon as a translation initiation site (TIS) is often assisted by the Shine-Dalgarno (SD) sequence (with the consensus AGGAGG), which is complementary to the 3′ end of the 16S rRNA of the 30S subunit (12). The SD sequence is positioned 4 to 9 nucleotides (nt) upstream of the start codon and plays an important role in modulating the efficiency of initiation of translation. However, its presence is not sufficient or even required for defining the start codon of an ORF (8). In fact, some bacterial phyla make little use of the SD sequence (13). But even in bacteria that widely exploit the SD sequence,

some SD-lacking mRNAs, as well as leaderless mRNAs devoid of the 5′ untranslated region (5′ UTR), can also be efficiently translated (14, see also chapter 10). The efficiency of translation initiation is also significantly affected by the accessibility of the TIS (9, 15). Sequestering the TIS in mRNA secondary or tertiary structure could dramatically reduce the initiation rate, whereas highly accessible TISs are more conducive to active translation. Binding of small, *trans*-acting RNAs or proteins to the TIS provides an additional layer of regulation of translation initiation (16).

The pTIS and the first in-frame stop codon delineate the boundaries of an ORF. The constellation of the structural elements favoring translation initiation, the known sequence of the encoded protein, the evolutionary conservation, and the codon-related nucleotide bias are usually employed to properly annotate the placement of the start codons of the bacterial genes. However, any of the in-frame or OOF internal codons that can be decoded by fMet-tRNA$_i^{fMet}$ may potentially define an additional site (iTIS) where translation could start. Utilization of unwanted iTISs can be precluded by occluding them in unfavorable higher-order structure of mRNA (17). Some iTISs, however, can be exploited for expanding the proteome by making two proteins from a single gene.

Throughout this review, we will refer to the proteins whose translation is initiated at the pTIS as the *primary* or *full-size protein*. We will call the polypeptide product translated from an iTIS the *alternative protein*, which in most cases means a second protein expressed in addition to, not instead of, the primary product (Fig. 1c). Some ORFs even carry multiple iTISs and therefore encode several alternative proteins.

Most of the known examples of internal initiation have been discovered inadvertently, when the production of two polypeptides from a single gene was noticed *in vivo* or *in vitro*. A special effort was required to demonstrate that the production of the smaller protein was not the result of proteolytic cleavage of the full-size polypeptide. In several cases the function of the internal initiation product has been established, but in many instances the purpose of production of the alternative protein remains obscure. We will present these examples according to the functionality of the alternative polypeptide in regard to that of the primary protein, with the full understanding that the assignment to a specific category is inevitably arbitrary.

INTERNAL INITIATION FROM IN-FRAME START CODONS

The Function of the Alternative Polypeptide Is Closely Related to That of the Primary Protein

Alternative IF2 isoforms function as initiation factors

Production of more than one protein from the *Escherichia coli infB* gene, which encodes the translation initiation factor 2 (IF2), presents one of the best-characterized examples of exploiting internal initiation for generating functional isoforms of the main translation product. The existence of two IF2 isoforms, IF2-1 (formerly IF2α), which is the primary protein, and IF2-2 (IF2β), the alternative product, was noticed during the chromatographic analysis of the purified protein (18). Sequencing of the *infB* gene helped establish the GUG$_{158}$ codon as the iTIS that directs translation of IF2-2 (19) (Table 1). The use of GUG$_{158}$ as a true start codon was additionally verified when the N-terminal

amino acid sequences of the purified IF2 isoforms were directly determined by Edman degradation (20). Subsequent experiments revealed the presence of the third, IF2-3 isoform (IF2γ), whose translation is initiated at AUG$_{165}$, 7 codons downstream from the iTIS of IF2-2 (21, 22). The conservation of iTISs in *infB* and the expression of two or three IF2 isoforms have been demonstrated in a wide range of bacteria (23, 24) (Table 1).

In vitro assays suggested that not only IF2-1 but also the IF2-2/3 variants are true initiation factors: they all associate with the 30S subunit and stimulate binding of fMet-tRNA (25). However, the activities of IF2-1 and IF2-2/3 are likely not redundant. Both major isoforms (IF2-1 and IF2-2) are necessary for optimal cell fitness, since deletion of *infB* is fully complemented only when both IF2 isoforms are expressed. Consistently, synthesis of either only IF2-1 or IF2-2 is insufficient to support maximal cell growth and results in cold sensitivity (22).

The *infB* gene belongs to the cold stress regulon, and its expression is additionally stimulated when cells are exposed to low temperatures. While IF2-1 and IF2-2 proteins are produced in approximately equimolar amounts in *E. coli* grown at 37°C, shifting temperature to 20°C results in higher expression of IF2-2 compared to IF2-1 (26). This suggests that usage of the *infB* TISs could be regulated in order to adjust the relative abundance of the IF2 isoforms for specific needs of the cell.

It is not known, however, whether there are differences in the functional properties of the IF2 variants. Conceivably, the IF2 isoforms could play particular roles in expressing mRNAs with different translation initiation signals, e.g., SD-containing cistrons versus leaderless ones or those exploiting certain start codons. It is also possible that the reasons for the existence of IF2 variants are not even related to the function of the factor in translation: IF2 could be also involved in facilitating the restart of DNA replication after exposing cells to DNA-damaging agents (27, 28). The exact role of IF2-1 in this process is unclear, but it is remarkable that only IF2-1, but not IF2-2/3, is capable of promoting replication restart (29).

Several attempts have been made to understand the regulation of internal initiation in *infB*. It has been proposed that the presence of rare codons upstream of the iTIS could result in reduced ribosome traffic through the internal start site, thereby increasing the probability of translation initiation at the iTIS (19). However, ribosome profiling (Ribo-seq) experiments yielded no evidence of such an effect (30). The propensity of mRNA to adopt an unfolded state (23) or its possible folding

Table 1 Bacterial genes with multiple TISs and the mRNA elements promoting translation initiation

Gene	Organism	Length of the primary protein (amino acids)	Length of the alternative protein (amino acids)	Primary SD sequence	Spacer length (nt)	Primary start codon	Internal SD sequence	Spacer length (nt)	Internal start codon
infB	*E. coli*	890	733	AAGGA	5	AUG	GGAAAAAG	5	GUG
infB	*E. coli*	890	726	AAGGA	5	AUG	AAG	5	AUG
infB	*B. subtilis*	716	623	GGGGUG	5	AUG	AAGAAG	6	GUG
rpoS	*E. coli* K-12 ZK126	330	277	AGGAG	7	AUG	AGGAG	11	GUG
mrcB	*E. coli*	844	799	GGAG	7	AUG	GAGG	11	AUG
mip	*C. burnetii*	230	140	AGGAG	8	AUG	GAAAAA	15	AUG
mip	*C. burnetii*	230	136	AGGAG	8	AUG	AAAAAA	11	AUG
pikAIV	*S. venezuelae*	1,346	1,146	GGAAG	11	AUG	GGAGGG	7	GUG
trfA	*E. coli*	382	285	AGGAGG	4	AUG	GAGG	4	AUG
clpB	*E. coli*	857	709	GGAGG	4	AUG	GGAGG	7	GUG
clpB	Synechococcus 7942	874	724	-	-	AUG	GAGG	10	GUG
clpA	*E. coli*	758	590	GGAGG	5	AUG	GGAGG	5	AUG
cvaA	Avian pathogenic *E. coli*	413	253	GGAG	7	UUG	AGGAAG	11	AUG
lcnD	*L. lactis*	474	429	AGGAG	6	AUG	AGGAG	0	UUG
mcrB	*E. coli*	459	298	AGGAAG	9	AUG	AAGAG	13	AUG
ccmM	Synechococcus 7942	539	324	GGAGG	7	AUG	GAGG	8	GUG
ssaQ	S. Typhimurium	322	106	GGAG	7	AUG	AGAGG	8	AUG
cheA	*E. coli*	654	557	GAGG	7	GUG	AAGGA	4	AUG
lysC	*B. subtilis*	408	163	AAAGG	10	AUG	AGGAGG	8	AUG
tilS/hprT	*L. monocytogenes*	648	179	AGGAAG	10	AUG	AGGAGG	10	AUG
pgaM	Streptomyces sp. PGA64	784	253	GGAAG	12	AUG	GGAG	7	AUG
incC	*E. coli*	364	259	GGAG	6	AUG	AGGA	9	AUG
petH	Synechocystis 6803	413	301	GGAG	10	AUG	-	-	AUG
fliO	S. Typhimurium	125	104	GAG	6	AUG	-	-	GUG
mvp	*B. brevis* 47	1,084	1,053	AGGAGG	10	UUG	AGGAG	11	AUG
safA	*B. subtilis*	387	224	AGGAGG	8	UUG	AGGAGG	11	AUG
hlyB	*E. coli*	707	422	GGAG	5	AUG	GCGG	1	AUG
plcR	*P. aeruginosa*	207	151	AGGAG	8	AUG	CCGAG	8	AUG
srfAB[a]	*B. subtilis*	3,583	46	AGAGG	6	AUG	AGGAGG	9	UUG
rmpA[a]	*T. thermophilus*	163	77	GGAGG	3	AUG	GGAGG	7	AUG
gnd[a]	*E. coli*	468	36–53	AGGAG	6	AUG	?	?	?

[a]Indicates the OOF internal start sites.

into a pseudoknot (31) have also been considered as factors facilitating initiation at the *infB* iTISs, but no experimental evidence supporting these proposals has been reported. Thus, at the moment, the principles of regulation of the internal initiation event in *infB*, in spite of it being one of the best-studied examples, remain unclear.

A truncated RNA polymerase σ^S factor allows survival of mutant cells

The *rpoS* gene codes for the RNA polymerase σ^S factor, a key transcription master regulator of the genes whose products are needed in the stationary phase and under stress. Puzzlingly, the *rpoS* gene of several *E. coli* K-12 strains carries a nonsense mutation replacing the CAG_{33} (or, in some strains, GAG) codon with the UAG stop codon (32, 33). Nevertheless, growth of the mutant cells was unaffected despite the presence of the premature stop codon in *rpoS*. This is possible because the ribosomes can initiate translation from an iTIS located in *rpoS* a short distance downstream from the premature stop codon (at GUG_{54}). Internal initiation leads to production of an N-terminally truncated, but nevertheless functional, variant of the σ^S factor (34) (Table 1). No expression of truncated σ^S was observed in the strains expressing *rpoS* lacking the premature stop codon (35). This may indicate that the iTIS becomes available for initiation only in the absence of translating ribosome traffic. However, no systematic analysis of potential coexpression of two *rpoS* isoforms under different physiological conditions has been undertaken, and it remains possible that the presence of an iTIS even in the wild-type *rpoS* gene is physiologically meaningful.

The redundant PBP-1b isoforms

The *E. coli mrcB* gene encodes one of the major penicillin binding proteins, known as PBP-1b or MrcB, a peptidoglycan synthase with transglycosylation and transpeptidation activities. Analysis of the purified MrcB protein by SDS-PAGE revealed three distinct bands designated as MrcB-α, -β, and -γ (36, 37). While the presence of the β variant was likely an artifact of isolation resulting from protease cleavage of MrcB-α (38), MrcB-γ was shown to be the product of internal initiation at the *mrcB* AUG_{46} codon (39) (Table 1). The α and γ isoforms assemble into α_2 or γ_2 homodimers (but not an αγ heterodimer) that are found in the membrane fraction (37, 40). They also show similar enzymatic activities (37). However, the biological reasons for coproduction of the full-length MrcB-α and its truncated MrcB-γ variant remain enigmatic, moreover that ex-

pression of the γ variant is sufficient to fully compensate for the *mrcB* deletion (36).

Virulence factor Mip variants with similar activities but distinct cellular localization

Macrophage infectivity potentiator (Mip) proteins are virulence factors of many pathogenic bacteria. These proteins possess peptidyl-prolyl *cis/trans* isomerase activity and are involved in protein folding, maturation, and targeting. Bacteria often carry several different *mip* genes, some of which encode proteins that are retained in the cytoplasm whereas the products of others are targeted to the outer membrane or the periplasm (41). In contrast, the compact genome of the obligate intracellular parasite *Coxiella burnetii* contains only one *mip* gene (*cbmip*), encoding a Mip-like protein, CbMip (42). Expression of *cbmip* in *E. coli* led to the production of three polypeptides (42). Sequencing of the N termini of these products and mutational analysis of the gene demonstrated that the two shorter forms of CbMip were generated due to initiation of translation at two iTISs present within *cbmip* (42) (Table 1). Similar CbMip isoforms were also detected in the native host. Even though all three CbMip variants have comparable enzymatic activities, the full-size CbMip carries a signal sequence whereas the products of internal initiation lack it. Therefore, the primary CbMip is exported from the cell, but the shorter isoforms are retained in the cytoplasm (42) (Fig. 2). Being an obligate intracellular parasite, *C. burnetii* has a rather streamlined genome, encoding merely 2,000 genes (43). Therefore, it could be that *C. burnetii* employs internal initiation to diversify the repertoire of the Mip proteins without expanding the size of its genome. It is yet to be investigated whether the relative expression of the intracellular and secreted variants of CbMip is regulated.

Polyketide synthase isoforms produce two different antibiotics

Streptomyces venezuelae produces two major macrolide antibiotics, the 14-atom macrolactone-ring pikromycin and a smaller, 12-atom macrolactone, methymycin (Fig. 3) (44). Both antibiotics inhibit protein synthesis by targeting the bacterial ribosome (45). The *pikAI* to -*IV* genes of the biosynthetic operon encode modular polyketide synthases that are required for generating the macrolactone core. The product of the *pikAIV* gene is responsible for the last condensation step during biosynthesis of the 14-atom macrolactone of pikromycin. How the smaller methymycin core is generated by the proteins encoded in the same operon remained an enigma for many years. The puzzle was solved when

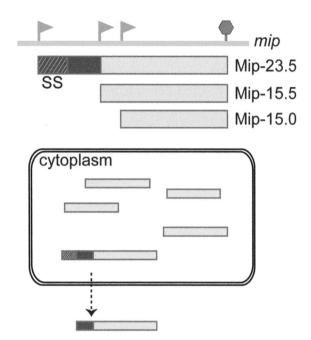

Figure 2 Internal initiation alters protein localization. Full-size Mip-23.5 protein carries a signal sequence (SS) at its N terminus, which predisposes it to be secreted. The products of internal initiation, Mip-15.5 and Mip-15.0, lack the signal sequence and, as a result, remain in the cytoplasm.

genetic and biochemical experiments revealed that initiation of translation from an iTIS within the *pikAIV* gene generates an N-terminally truncated version of the PikAIV protein (46) (Table 1). The truncated PikAIV contains intact acyl carrier protein, acyltransferase, and thioesterase domains of the full-size PikAIV, but only a portion of the first, ketosynthase domain. As a result, the last condensation step of the macrolactone synthesis does not take place and, instead, the 12-atom macrolactone of methymycin is produced (46) (Fig. 3). Remarkably, the composition of the growth media defines which of the two antibiotics is preferentially synthesized in *S. venezuelae* cells (47, 48), suggesting that the use of the pTIS or the iTIS of *pikAIV* is regulated in response to environmental cues. However, the mechanism controlling such regulation remains unknown.

RK2 plasmid copy number in different hosts is controlled by TrfA isoforms

Replication of the broad-range Gram-negative bacterial plasmid RK2 relies on the coordinated activity of plasmid-encoded replication initiation protein TrfA and host proteins (reviewed in reference 49). TrfA is

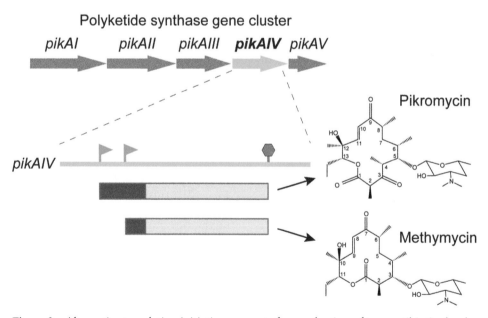

Figure 3 Alternative translation initiation accounts for production of two antibiotics by the same biochemical pathway. Production of the full-size PikAIV polyketide synthase module supports synthesis of the 14-atom macrolactone ring of pikromycin. The N-terminally truncated PikAIV variant leads to the synthesis of the smaller (12-atom) macrolactone ring of methymycin.

expressed in two versions, the full-size TrfA-44 and its N-terminally truncated TrfA-33 isoform, whose translation is initiated at the internal *trfA* codon AUG$_{98}$ (50, 51) (Table 1). Each of the individual isoforms is sufficient to support plasmid replication in several bacterial species (52). Nevertheless, TrfA-44 and TrfA-33 probably play distinct roles in controlling plasmid copy number in specific bacterial hosts. For example, the lack of TrfA-44 negatively affects RK2 replication in *Pseudomonas aeruginosa*, likely because the DnaB protein of this host that is required for RK2 replication is unable to properly interact with TrfA-33 (53–55). Conversely, the lack of TrfA-33 reduces the RK2 copy number in *E. coli* and *Azotobacter vinelandii* (56), suggesting that in these bacteria the internal initiation product fulfills an important supportive function. It is unknown if the relative efficiency of utilization of the pTIS and iTIS is regulated and whether it is host specific.

Alternative Protein Facilitates the Function of the Primary Protein

Isoforms of ClpB and ClpA aid in handling the misfolded and aggregated proteins

Chaperones and proteases help cells deal with misfolded and aggregated proteins that accumulate under stress conditions (57). One such protein is ClpB, a chaperone that utilizes the energy of ATP hydrolysis to disaggregate polypeptides. Gel-electrophoretic analysis of the *E. coli* proteins induced during heat shock identified a shorter isoform of ClpB, ClpB79, in addition to the full-length ClpB93. Mutagenesis studies demonstrated that an iTIS at codon GUG$_{149}$ of *clpB* was required for production of ClpB79 (58, 59) (Table 1). While both isoforms retain the two ATP-binding sites, ClpB79 lacks the N-terminal domain of its full-size counterpart, which is involved in substrate recognition (58). Nevertheless, the ATPase, chaperone, and oligomerization activities of ClpB79 seem to be largely unaffected (60). ClpB functions as a homotetrameric complex in which both isoforms could be present simultaneously (58). Both ClpB versions are also produced in several other bacteria (60, 61) (Table 1). The mechanistic advantage of their coproduction is unclear, but it has been shown that *E. coli* or *Synechococcus* sp. handles thermal stress more efficiently when both variants are coordinately expressed (62–64), while in other species, the internal initiation product alone can afford the same level of thermoprotection as the simultaneous expression of both ClpBs (63, 65). Interestingly,

the ratio of the ClpB isoforms varies depending on the severity of the heat shock, suggesting that the relative activity of the pTIS and iTIS could be regulated (62).

A protein-dependent ATPase, ClpA, shows significant sequence similarity to ClpB. ClpA associates with the proteolytic subunit ClpP to form the Clp protease. Similar to the existence of the ClpB isoforms, two ClpA variants are expressed from the *clpA* gene, the main protein ClpA84 and the alternative polypeptide ClpA65, whose translation is initiated at the AUG$_{169}$ codon of *clpA* (66) (Table 1). In the absence of substrates, the Clp complex is prone to autodegradation because idling ClpP can cleave the ClpA84 subunit. However, ClpA65 represses Clp self-proteolysis by acting as a decoy, suggesting that the product of internal initiation plays a regulatory role in maintaining protein homeostasis (67).

While the precise physiological roles of the internal initiation products of *clpA* and *clpB* genes are yet to be fully elucidated, the conservation of the iTISs in these genes in several bacterial species argues that the ability to produce N-terminally truncated versions of these heat shock proteins is important for coping with thermal shock. An attractive hypothesis is that temperature-dependent alterations in the mRNA structure could affect the relative efficiency of translation initiation at the primary and internal start codons (68, see also chapter 4). However, this possibility has not been explored.

Alternative transport proteins facilitate secretion of bacteriocins

Several *E. coli* strains secrete the antibacterial toxin colicin V (ColV). The secretion of ColV requires the activity of the membrane-associated transport proteins CvaA and CvaB (69, 70). Besides full-size CvaA, the shorter CvaA* variant is translated from the same *cvaA* gene, due to translation initiation at the AUG$_{161}$ codon (69, 70) (Table 1). Coexpression of CvaA and CvaA* is required for optimal secretion of ColV. Similar to the differential targeting of the Mip isoforms discussed above (Fig. 2), the primary protein CvaA is bound to the membrane, but CvaA*, which lacks the hydrophobic N-terminal segment of CvaA, remains in the cytoplasm (70). How the CvaA variants cooperate in secretion of the ColV toxin is unknown, but reminiscent of the stabilization of ClpA by its shorter isoform, CvaA* protects the full-length CvaA from degradation (70). The regulation of relative expression of CvaA and CvaA* has not been yet investigated.

A comparable mechanism is likely involved in production of another bacteriocin, lactococcin A (LcnA), secreted by strains of *Lactococcus lactis*. Export of LcnA relies on the activity of two membrane proteins, LcnC and LcnD. As for CvaA*, the existence of an iTIS in *lcnD* leads to expression of a full-size membrane-associated LcnD and a shorter version, LcnD*, which is retained in the cytoplasm (71) (Table 1). Functions of LcnD variants in LcnA secretion have not been characterized.

The alternative McrB protein tunes the activity of a restriction enzyme

The *E. coli* McrBC restriction enzyme cleaves foreign DNA containing 5-methylcytosine (72). The active enzyme is formed by association of the nucleolytic McrC polypeptide with seven copies of the GTPase subunit McrB (73) (Fig. 4).

The *mcrB* gene directs production of two polypeptides, the full-size protein McrB$_L$ and its N-terminally truncated variant McrB$_S$, whose translation is initiated

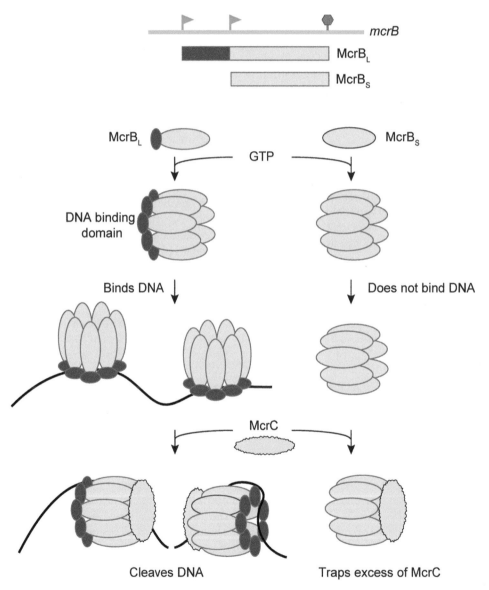

Figure 4 Alternative protein restores stoichiometry of the functional restriction enzyme. Full-size McrB$_L$ binds DNA and associates with the nucleolytic McrC to cleave DNA. McrB$_S$ lacks the N-terminal DNA-binding domain (dark blue) but is able to titrate out the excess McrC to maintain optimal activity of the restriction enzyme complex. The general scheme of the figure was adapted from reference 72.

at the AUG_{162} codon (74) (Table 1). Because $McrB_S$ lacks the DNA-binding domain of $McrB_L$, its complex with McrC is enzymatically inactive, which raised the question of the biological role of $McrB_S$ (73). The possible solution to the puzzle came from the observation that an excess of McrC leads to reduced activity of the enzyme, whereas the simultaneous production of $McrB_S$ remedies the problem (75, 76). Overproduction of McrC distorts the ratio of the subunits in the McrB complex, thus decreasing its overall enzymatic activity. $McrB_S$ may trap the excessive amounts of McrC, thereby restoring the optimal stoichiometry of the functional McrBC enzyme (Fig. 4).

A delicate balance of the amounts of $McrB_L$, $McrB_S$, and McrC is likely required for optimal activity of the restriction nuclease in the cell (75), but whether such balance is achieved by regulating the relative expression of the McrB isoforms is unknown.

The internal initiation product of CcmM is required for organizing the structure of β-carboxysomes

RubisCO plays the central role in CO_2 fixation during photosynthesis. To stimulate this reaction, several cyanobacterial strains form specialized microcompartments called β-carboxysomes that help increase the local concentration of CO_2 (77). Immunoblot analysis of the β-carboxysome polypeptides in *Synechococcus* sp. strain PCC 7942 revealed the presence of two isoforms of the protein CcmM, full-size CcmM-58 and the N-terminally truncated CcmM-35. They are translated from the same gene due to the presence of an iTIS at the *ccmM* GUG_{216} codon (78, 79) (Table 1). The CcmM isoforms carry out specific and distinct structural functions (78). The N-terminal domain of CcmM-58 facilitates the formation of protein trimers and localizes them to the inner shell of the carboxysome; the C-terminal segments of CcmM-58 interlink RubisCO holoenzymes (Fig. 5). The shorter CcmM-35, which lacks the N-terminal domain of CcmM-58, is located in the lumen of the β-carboxysome, where it organizes RubisCO in a paracrystalline array (77, 78, 80) (Fig. 5). The *ccmM* genes of several other cyanobacterial species also contain a similarly positioned iTIS (77), suggesting that the use of internal initiation is a successful strategy for expressing polypeptides with related functions but unique patterns of interprotein interactions.

Assembly of type III secretion systems relies upon expression of the C-ring protein variant

Some pathogenic bacteria, including *Salmonella enterica* serovar Typhimurium, use a type III secretion system

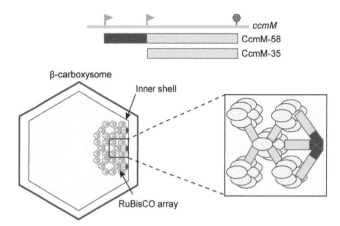

Figure 5 The isoforms of CcmM help to differentially organize RubisCO in β-carboxysome. The N-terminal segment of the full-size CcmM-58 (dark blue) anchors it to the inner shell of the β-carboxysome, while its C-terminal segment (light blue) arranges the first layer of RubisCO. The N-terminally truncated Ccm-35 organizes RubisCO into a lattice in the lumen of the β-carboxysome.

(T3SS), also known as an injectisome, to export effector proteins into the invaded cells. The core of the T3SS is a highly complex, multiprotein, needle-like structure with rings of specific proteins in the inner and outer membranes (reviewed in reference 81). The protein product of the *ssaQ* gene oligomerizes to form the cytoplasmic ring (C-ring) of the injectisome. However, two polypeptides are expressed from *ssaQ*, the full-size $SsaQ_L$ and the shorter $SsaQ_S$, the latter due to internal initiation at the AUG_{217} codon (82) (Table 1). *Salmonella* strains unable to express $SsaQ_S$ are attenuated in their virulence (82). A similar organization of the genes analogous to *ssaQ* of *S.* Typhimurium has been found in *Yersinia*, *Shigella*, and *Thermotoga* strains (83, 84) (Table 1). In all these bacteria, expression of the N-terminally truncated protein is necessary for the formation of a functional T3SS. Although the exact function of $SsaQ_S$ remains obscure, it has been shown that the internal initiation product stabilizes $SsaQ_L$ and augments the activity of the T3SS (82). An unexpected validation of the importance of coexpression of the two isoforms of the C-ring protein came from the attempts of synthetic biologists to assemble a functional injectisome using a minimal set of synthetic genes resembling the T3SS genes of the *Salmonella* pathogenicity island 1 (85). While the initial gene assembly was inactive, subsequent debugging pinned down the problem to the inadvertent elimination of an iTIS in the *spaO* gene (the functional homolog of *ssaQ*). Subsequent restoration of the iTIS allowed the expression of the functional T3SS (85).

The T3SS injectisome is evolutionarily and structurally related to the flagellar machinery of Gram-negative bacteria (81). Interestingly, while SsaQ$_L$ and SsaQ$_S$ (and the equivalent injectisome components in other bacteria) are expressed from a single gene due to the presence of iTISs, their structural equivalents FliM/FliN of the flagellar C-ring are expressed from two individual genes. It remains to be investigated why coexpression of the C-ring components from a single gene is beneficial for the formation and function of the injectisome but not the flagella.

Involvement of CheA$_L$ and CheA$_S$ in chemotaxis

Many Gram-negative bacteria sense and respond to gradients of chemical attractants or repellents by functionally coupling membrane-bound receptors with the rotation of their flagella. A central regulatory hub of this mechanism is the signal transduction system, which operates through phosphorylation of the response regulator CheY. CheA is the sensory histidine kinase responsible for CheY phosphorylation, and CheZ is a phosphatase that dephosphorylates CheY (reviewed in reference 86). The CheY-regulated complex biochemical cascade eventually helps bacteria swim toward the attractant.

The *cheA* gene encodes two polypeptides, the full-size CheA$_L$ and the smaller CheA$_S$, whose translation is driven by the internal AUG$_{98}$ codon (87, 88) (Table 1). Although CheA$_S$ lacks His$_{48}$, the site of CheA$_L$ autophosphorylation, various functions have been attributed to this CheA isoform: from facilitating phosphorylation of CheA$_L$ (89) to mediating protein-protein interactions important for chemotaxis (90). Nevertheless, despite decades of research, the precise role of CheA$_S$ in chemotaxis remains obscure. Interestingly, the internal start codon of *E. coli cheA* is located within the loop of a hairpin formed by the mRNA segments flanking the iTIS, suggesting that the access to it and the iTIS usage could be modulated by mRNA structure (87). To the best of our knowledge, no evidence for such regulation has been reported so far.

The alternative product of the *lysC* gene expands the amino acid-mediated regulation of aspartate kinases

Aspartate kinases (AKs) catalyze the first step of the pathways leading to the synthesis of several essential amino acids, including lysine and threonine. The catalytic domain of AKs is linked to the regulatory domain responsible for the negative feedback regulation by specific amino acids, the end product(s) of the pathway. In many organisms, AKs assemble into multimers, where the binding

site(s) for the regulatory amino acid(s) is formed at the interface of the regulatory domains (reviewed in reference 91). The amino acid-binding pockets of homomultimeric class I AKs afford feedback inhibition by lysine. In contrast, the heterotetrameric class II AKs display an expanded allosteric control because nonequivalent binding sites for two amino acids, lysine and threonine, are formed at the interface of the α and β subunits (Fig. 6). Remarkably, in some bacteria, the α and β subunits of LysC are translated from the same gene: the α subunit represents the full-size LysC, whereas the β subunit corresponds to the N-terminally truncated form of the protein, initiated at the *lysC* iTIS (92–95) (Table 1; Fig. 6). While the α and β subunits are expressed in equimolar amounts in exponentially growing *Bacillus subtilis*, the α subunit is produced in 2-fold excess in germinating *B. subtilis* spores (96), implying the involvement of a yet unknown regulatory mechanism that modulates the relative expression of LysC isoforms.

The function of the TilS-HprT protein complex in *Listeria* relies on the iTIS of the fused genes

The genes of the tRNA^{Ile} lysidine synthetase, TilS, involved in biogenesis of tRNA^{Ile}, and of hypoxanthine-

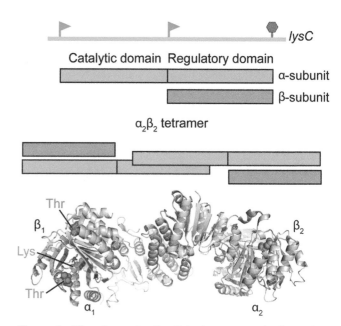

Figure 6 The alternative LysC isoform expands the amino acid-mediated regulation of the enzyme. The full-size (α) and N-terminally truncated (β) isoforms of LysC associate into a functional α$_2$β$_2$ tetramer, in which the binding sites for the regulatory amino acids lysine (orange) and threonine (red) are formed at the interface of the α-β subunits. Shown is the structure of *Corynebacterium glutamicum* LysC (PDB ID: 3AAW) (95).

guanine phosphoribosyltransferase HprT, which participates in the purine salvage pathway, are often organized into an operon (Fig. 7). In *B. subtilis*, these two proteins form a complex, which binds to the promoter and regulates the expression of the *ftsH* gene, whose product is involved in membrane quality control (97). In many bacterial genomes, the *hprT* start codon overlaps with the stop codon of *tilS*. However, in *Listeria monocytogenes*, the stop codon of *tilS* is missing and translation of the in-frame *tilS-hprT* ORFs results in the expression of the single hybrid protein TilS-HprT (Fig. 7). While the hybrid protein can perform the functions of both individual enzymes (97), proteomics analysis showed that HprT is additionally expressed from the internal AUG start codon of the fused *tilS-hprT* gene (61). Interestingly, it was impossible to isolate mutants with inactivated iTIS (61), suggesting that the presence of individual (not TilS-fused) HprT may be required for stabilizing the TilS-HprT hybrid in *L. monocytogenes* (97). Alternatively, translation from the iTIS may increase the stability of the hybrid mRNA.

Cooperation of the alternative products of the *pgaM* gene is required for optimal synthesis of the antibiotic gaudimycin C

The biosynthetic gene cluster *pgaEFLM* of *Streptomyces* sp. PGA64 is responsible for production of the angucyclinone polyketide antibiotic gaudimycin C (98). Similar to the *tilS-hprT* fusion (Fig. 7), the gene *pgaM* of the cluster is the result of the fusion of two independent genes, where the upstream ORF lost its stop codon and the downstream ORF retained its own

TIS. The full-size protein, PgaM, possesses both oxygenase (in its N-terminal portion) and reductase (in its C-terminal segment) activities. In contrast, the shorter variant, PgaMred, whose translation starts at the iTIS, contains only the reductase portion (99) (Table 1). Expressed together, PgaM and PgaMred form a stable heterotetramer, conducive to the production of gaudimycin C (99). In contrast, if iTIS is mutated and expression of PgaMred is abolished, PgaM undergoes self-oligomerization, forming a less active complex. Unexpectedly, the iTIS mutation decreased the yield of PgaM by 90%. Conversely, inactivation of the *pgaM* pTIS abolished production not only of the full-size protein but also of PgaMred (99). This is a clear demonstration of interdependent expression of nested genes that could be potentially regulated either at the level of translation or due to altered mRNA stability.

Putative iTISs were also identified between the domains of fused angucycline oxygenase-reductase genes in other *Streptomyces* species (99). This observation, together with the example of the *pikAIV* gene in *S. venezuelae* discussed earlier, suggests that optimization of secondary metabolite biosynthesis by producing two proteins from one gene could be a widespread strategy.

The alternative IncC variant facilitates plasmid partitioning

We discussed earlier the isoforms of the TrfA protein encoded in the RK2 plasmid that help control plasmid copy number in different hosts. Another RK2 gene, *incC* (also known as *parA*), also directs synthesis of

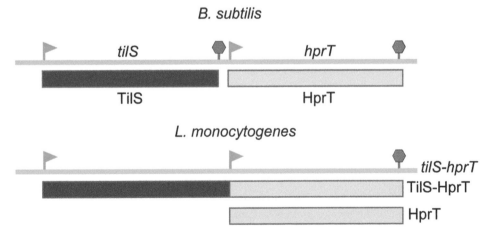

Figure 7 Elimination of the stop codon creates a fused gene with two translation starts. Two genes, *tilS* and *hprT*, which encode functionally distinct proteins and associate to form a functional complex, are organized in an operon in *B. subtilis*. Due to the elimination of the *tilS* stop codon in *L. monocytogenes*, the two ORFs are fused into a single gene and can be expressed from the primary and the internal TIS.

two polypeptides involved in plasmid partitioning, IncC1 and IncC2, via the pTIS and an iTIS, respectively (100) (Table 1). Both proteins are detected in bacterial cells carrying the plasmid, but their relative production is host specific (101). The N-terminal domain of IncC1, absent in IncC2, plays a role in the ADP-stimulated dimerization of the protein required for its activity. Interestingly, however, mixing IncC1 and IncC2 leads to protein polymerization as well as improved DNA binding. Thus, cooperation of both variants may be required for optimal plasmid partitioning (101).

The Function of the Alternative Protein Differs Significantly from That of the Primary Protein

FNR enzymes with opposite functions in cyanobacteria

Ferredoxin:NADP oxidoreductases (FNRs) facilitate CO_2 fixation during photosynthesis by reducing $NADP^+$ to NADPH. However, nonphotosynthetic plastids contain a distinct version of this protein that is able to perform the opposite reaction and oxidize NADPH to provide electrons for nitrogen assimilation (102,

103). During nitrogen starvation, the cyanobacterium *Synechocystis* sp. strain PCC 6803 can switch from phototrophic to chemoheterotrophic metabolism. Remarkably, instead of carrying two distinct genes encoding the photosynthetic and the heterotrophic versions of FNR, *Synechocystis* sp. has a single gene, *petH*, which codes for two FNR variants capable of carrying the opposite reactions (Fig. 8). During photosynthetic growth, the major product of the *petH* gene is the full-length FNR_L, while under conditions of heterotrophic metabolism, the alternative AUG_{113} codon is used to produce the shorter variant FNR_S (104, 105) (Table 1). The relative expression of the FNR isoforms is controlled using alternative promoters. A transcript initiated at the upstream promoter has a long 5′ UTR, which folds into a stable structure that hinders access to the pTIS, downregulating synthesis of FNR_L and favoring production of FNR_S (Fig. 8). In the transcript with a shorter 5′ UTR, initiated at the promoter more proximal to the primary start codon of the gene, the pTIS becomes accessible and FNR_L is preferentially synthesized. Similar mechanisms are used for controlled utilization of alternative initiation sites in the *petH*

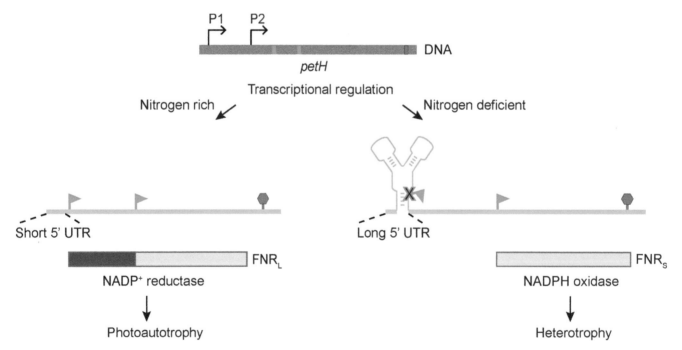

Figure 8 The use of different promoters can regulate the utilization of alternative start sites within a gene. Two promoters (P1 and P2) precede the *petH* gene in *Synechocystis* 6803. The transcript initiated at P2 has a short 5′ UTR; the accessible pTIS favors translation of the full-size FNR_L. Folding of the long 5′ UTR of the transcripts initiated at the P1 promoter occludes the pTIS, thereby shifting the relative expression of the FNR isoforms in favor of the shorter FNR_S. While photosynthetic growth favors synthesis of FNR_L, heterotrophic conditions stimulate production of the FNR_S isoform.

genes in a range of cyanobacteria (106). This elegant regulatory scheme illustrates how mRNA conformations enable ribosomes to choose between different TISs.

Alternative Translation Products with an Unknown Raison d'Être

Either of the two FliO variants can help assemble functional flagella

The transmembrane protein FliO is involved in assembly of the flagella of Gram-negative bacteria. Two FliO isoforms are produced from the alternative TISs proposed to operate in the *fliO* gene: the primary one that overlaps with the stop codon of the upstream *fliN* gene or an iTIS located 20 (in *E. coli*) or 21 (in *S.* Typhimurium) codons downstream from the pTIS (107, 108) (Table 1). The *fliO* null mutants could be complemented by either version of FliO, although the shorter version was less efficient (108). While the physiological role of the short FliO is unclear, it is interesting that the iTIS is located within the loop of an mRNA hairpin, whose presence, but not the sequence, is conserved between the *fliO* genes of *E. coli* and *S.* Typhimurium (108).

Two translation start sites could be operating in the middle wall protein gene of *Bacillus brevis*

The cell wall of some *B. brevis* strains is composed of three layers: outer wall protein, middle wall protein (MWP), and peptidoglycan (109). Two TISs seem to operate within the *mwp* gene of the *B. brevis* 47 strain, leading to production of the putative longer and shorter variants of the MWP polypeptide (109) (Table 1). A similar arrangement of two TISs is found in the cell wall protein genes of several other *B. brevis* strains (110). The N terminus of the short MWP contains a conventional signal sequence, while the N-terminal extension present in the longer MWP variant is rich in charged residues and does not resemble any signal sequence. Deletion of the pTIS does not preclude expression of a functional MWP (109), arguing that the shorter MWP variant produced from iTIS is actively used in the cell. Mutating the iTIS from AUG to AUC reduced, but did not preclude, the expression of the protein (111). Although this result was interpreted as an indication that the functional protein could be also translated from the upstream pTIS, this conclusion must be taken with caution because AUC can function as initiator codon in bacteria (8). To the best of our knowledge, the expression of two MWP variants has not been directly demonstrated.

Two SafA protein isoforms in the *B. subtilis* spore coat

Upon nutrition stress, *B. subtilis* differentiates into two cell types. One of these, the endospore, matures within the mother cell to sequester and protect the genome until the environmental conditions improve. SafA protein is incorporated in the endospore coat at the early stages of development. However, a shorter SafA-C_{30} isoform, translated from the AUG_{164} codon of the *safA* gene, can additionally be found in the coat of the mature spore (112) (Table 1). While mutating the iTIS does not affect spore coat formation, overexpression of SafA-C_{30} blocks sporulation (112), hinting that the internal initiation product may play a yet to be determined regulatory role in spore formation.

Protein isoforms might be involved in the export of hemolytic enzymes

Several pathogenic Gram-negative bacteria secrete hemolytic enzymes (113). For example, hemolysin A is produced by uropathogenic *E. coli* strains. The export machinery of hemolysin A is composed of several proteins, including the ABC-type transporter HlyB. However, the *hlyB* gene from the uropathogenic *E. coli* plasmid pGL570 could direct expression of two HlyB variants, with the synthesis of the shorter polypeptide initiating at the internal AUG_{286} codon (114, 115) (Table 1). The function of the shorter HlyB isoform or its importance for secretion of hemolysin A is unknown. It is noteworthy that the mRNA secondary structure in the vicinity of the pTIS was proposed to attenuate translation of the full-length HlyB (116) and thus could be a part of the mechanism controlling the relative expression of the HlyB isoforms.

Another hemolytic enzyme, phospholipase C (PlcH), is secreted by *P. aeruginosa* and is important for the virulence of this organism (reviewed in reference 117). The operation of the PlcH secretion apparatus is not fully understood, but the product of the *plcR* gene is known to be important for this process because its deletion interferes with PlcH secretion (118). Two polypeptides are translated from the *plcR* gene, the full-length PlcR1 and the product of internal initiation at the AUG_{57} codon, PlcR2 (119) (Table 1). Like the differential localization of the CbMip (Fig. 2) and CvaA isoforms mentioned earlier, PlcR1 and PlcR2, when expressed in *E. coli*, are localized in the periplasm and the cytoplasm, respectively, because the signal sequence present in PlcR1 is lacking in PlcR2. On the basis of differential compartmentalization, it was proposed that PlcR2 might act as a PlcH chaperone, helping its folding or translocation through the inner membrane, while the periplasm-located PlcR1 could help export PlcH across the outer membrane (118). Noteworthy, the PlcR1 segment preceding the iTIS contains surprisingly many proline residues (10 out of 30 codons encode proline). Translation of clusters of prolines could present a serious problem for the ribosome (reviewed in reference 120). It is possible that translation of the proline-rich segment modulates the expression of PlcR2 by slowing down ribosomes immediately upstream of the iTIS.

INTERNAL INITIATION OF TRANSLATION THAT DIRECTS SYNTHESIS OF A PROTEIN FROM AN ALTERNATIVE READING FRAME

When an iTIS directs translation from an OOF start codon, the sequence of the translated protein will be entirely different from the sequence of the polypeptide encoded in the main frame (Fig. 1). The stop codon of the frame corresponding to the OOF iTIS is often located at a relatively short distance from it, and as a result, the product of internal initiation would be a fairly short protein. Because identification and characterization of small proteins is experimentally difficult and bioinformatically challenging, to the best of our knowledge, only three examples of this type of genetic coding in bacteria are known.

The Competence Protein ComS Is Encoded within the *srfA* Gene of *B. subtilis*

The ability to take up exogenous DNA allows *B. subtilis* to survive unfavorable conditions. The timing of establishing the competent state is controlled by quorum sensing and is regulated by several genes (reviewed in reference 121). Genetic mapping suggested that one of the determinants of competence, *comS*, is located within the *srfA* operon, which contains genes responsible for production of the antibiotic surfactin (122). Although secretion of surfactin is coordinated with the establishment of competence, none of the activities associated with surfactin production are important for competence per se. Instead, an OOF ORF encoding the 46-amino-acid-long ComS protein was identified within the second gene of the *srfA* operon (122, 123). Translation of the nested *comS* gene is initiated at an OOF UUG codon preceded by a strong SD sequence (Table 1) (note that UUG is a common start codon in *B. subtilis* genes [11]). Embedding the *comS* gene within one of the surfactin operon genes is an elegant strategy to coregulate two distinct but related activities contributing to DNA uptake: starvation induces expression and secretion of surfactin, which causes

lysis of the cells of other competing bacterial species, whereas ComS facilitates uptake and utilization of the released genetic material.

The Gene of Ribosomal Protein L34 Is Nested within the RNase P Protein Gene of *Thermus thermophilus*

RNase P directs maturation of the tRNA 5′ end. In bacteria, this enzyme is composed of a catalytic RNA and a small accessory protein encoded by the *rnpA* gene (124). In many bacterial genomes, the *rnpA* gene is positioned immediately downstream of the *rpmH* gene, encoding ribosomal protein L34, and both genes are usually cotranscribed within one operon (125). In *T. thermophilus*, however, no *rpmH* ORF was found upstream of the unusually long *rnpA* gene. Instead, the entire 49-codon-long *rpmH* is completely embedded within the *rnpA* ORF, but in an alternative reading frame (126) (Fig. 9). The start codons of *rpmH* and *rnpA* are separated by only one nucleotide, and expression of both proteins is likely supported by the same SD sequence (126) (Table 1).

Because bacterial cells contain a large (~100-fold) excess of ribosomes over RNase P, *rpmH* is likely translated much more efficiently than *rnpA*. One of the factors contributing to the difference in expression of the two overlapping ORFs is the distance of their respective start codons from SD: the AUG start codon of

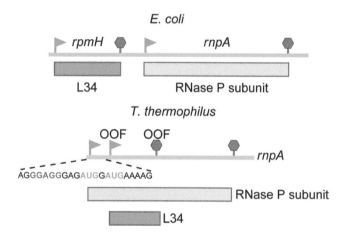

Figure 9 Elimination of a start codon leads to gene occlusion. In *E. coli*, the neighboring genes of ribosomal protein L34 (*rpmH*) and RNase P protein (*rnpA*) are independently translated. In *T. thermophilus*, the mutation of the *rnpA* translation initiation codon and appearance of an additional start site upstream of *rpmH* lead to occlusion of *rpmH* within the *rnpA* gene, but in an alternative reading frame. Start sites of both genes utilize the same SD sequence (blue), whose optimal spacing with the *rpmH* start codon shifts the balance of translation in favor of the ribosomal protein.

rpmH is located at the optimal distance of 7 nt from the GGAGG SD sequence, whereas the *rnpA* AUG start codon is only 3 nt away from it (126) (Fig. 9). Short spacing between the SD sequence and the start codon likely decreases the efficiency of initiation of *rnpA* translation (127). Furthermore, the highly efficient initiation of *rpmH* translation might compete with initiation at the *rnpA* start codon. Similar to other ribosomal protein genes, the codon usage in *rpmH* has been evolutionarily optimized for highly active translation. In contrast, the codon usage of *rnpA* is less favorable. The presence of rare codons in the vicinity of the *rnpA* TIS could lead to abortive translation, which would additionally decrease the expression level of RNase P protein relative to that of L34.

Occlusion of *rpmH* within the extended version of the *rnpA* gene likely occurred due to a mutation leading to appearance of a start codon upstream of *rpmH* (in frame with the downstream *rnpA*, but OOF relative to *rpmH*) and subsequent elimination of the *rnpA* authentic start. A similar overlapping arrangement of *rnpA* and *rpmH* genes has been found in other members of the genus *Thermus* (126). It is worth noting that occlusion of an OOF gene within another gene should be more frequent in the high-GC-content organisms, including the *Thermus* species, due to reduced frequency of occurrence of OOF stop codons.

A Heat Shock Protein Is Encoded in an Alternative Reading Frame within the *gnd* Gene of *E. coli*

To survive abrupt increase in temperature, bacteria express heat shock proteins. However, not all heat shock proteins have been fully characterized, especially low-abundance, small, membrane-associated polypeptides. A recently developed approach for quantitative membrane proteomics identified a tryptic peptide apparently belonging to a previously unknown small heat shock protein named GndA (128). The identified tryptic peptide mapped to a short ORF located entirely within the *gnd* gene encoding 6-phosphogluconate dehydrogenase, but in the -1 frame (128). At what codon the translation of the small *gndA* ORF is initiated is unknown because two in-frame ATG codons are found upstream of the sequence encoding the mapped peptide (Table 1). It was previously suggested that sequestration of the pTIS of the main *gnd* ORF in the mRNA secondary structure could modulate the expression of 6-phosphogluconate dehydrogenase in response to changes in the cell growth rate (129). The proposed mRNA folding may also involve the *gndA* ORF start site, and it is possible that regulation of two genes is interconnected.

NEW TOOLS FOR IDENTIFYING ALTERNATIVE TRANSLATION START SITES

Most of the examples of internal initiation discussed in the previous sections have been discovered serendipitously, mainly while analyzing the purified protein or characterizing its functions. Although many bacterial genes could potentially contain functional iTISs, their experimental identification is a challenging problem.

Powerful proteomics techniques potentially could be employed for detecting the N termini of the proteins and thus mapping TISs (130). Unfortunately, neither conventional top-down proteomics nor the traditional bottom-up techniques can reliably distinguish the products of internal initiation from the protein fragments generated by proteolysis. Newly emerging proteomics methodologies are better suited for revealing iTISs. The most promising of them relies on the use of peptide deformylase inhibitors, such as actinonin. Formylmethionine at the N termini of peptides helps to distinguish translation initiation products from those of proteolytic degradation and to determine the true pTISs or iTISs (131). Nineteen iTISs have been recently mapped in the genome of *L. monocytogenes* using this technique (61). This approach, however, has its intrinsic limitations because neither very short nor very long peptides can be easily detected by mass spectrometry and, in addition, low-abundance proteins (a common scenario for many of the internally initiated polypeptides) evade detection. A recently developed methodology combines bottom-up mass spectrometry with analysis of hypothetical polypeptides encoded in all six frames in the genome of interest (128). Applying this approach to membrane proteins helped to identify the novel OOF internal ORF *gndA* within the *gnd* gene of *E. coli* (128) that we described earlier. The downside of the approach is that it does not immediately determine the codon used for initiation of translation.

A more systematic and sensitive strategy is to use the gene-oriented ribosome-profiling approach (30, 132). This technique involves isolation and deep sequencing of ribosome-protected mRNA fragments ("ribosome footprints"). Mapped to the genome, ribosome footprints reflect the distribution of ribosomes along the translated mRNAs (Fig. 10a). If the iTIS is sufficiently strong, it should increase the ribosome traffic through the downstream portion of the ORF and lead to higher occupancy of the distal mRNA segment (133). The changes in the ribosome density could be, however, too subtle when the iTIS promotes translation with moderate or, moreover, low efficiencies. For example, ribosome profiling does not immediately reveal internal

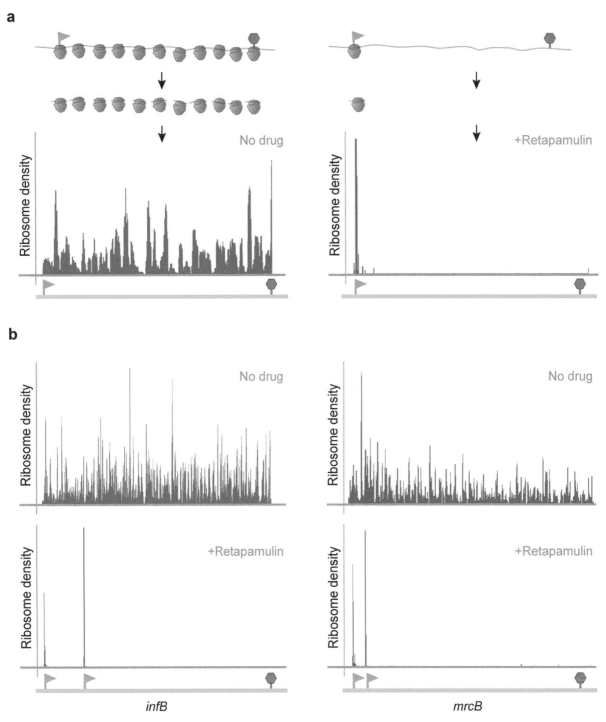

Figure 10 Retapamulin-assisted ribosome profiling illuminates sites of internal translation initiation. (a) (Left) Ribosome profiling in untreated cells shows the distribution of translating ribosomes along the mRNAs (30). (Right) Brief pretreatment of cells with the translation initiation inhibitor retapamulin arrests the ribosomes at the translation start site of the ORFs. (b) Examples of primary and internal TISs revealed by retapamulin-assisted ribosome profiling in *E. coli* genes known to contain functional internal start codons. Note that the presence of the iTISs is difficult to detect when ribosome profiling is performed in untreated cells.

initiation within the *E. coli* genes *infB* or *mrcB* that we discussed earlier (30) (Fig. 10b).

Fortunately, using ribosome-targeting antibiotics that specifically inhibit translation initiation can increase the sensitivity of ribosome profiling for detecting TISs. Such an approach has been successfully used for mapping TISs in eukaryotic genomes and identified thousands of previously unknown in-frame and OOF iTISs as well as upstream TISs that would result in N-terminal extensions of known ORFs (134–136). The paucity of antibiotics specifically targeting the initiation step of bacterial translation delayed the application of this approach for bacteria. Nevertheless, it has been shown that tetracycline could be potentially suitable for mapping TISs in bacterial genes (137). This finding was unexpected because tetracycline has been known as an inhibitor of translation elongation and yet, in profiling experiments, the majority of the ribosome density peaks in the tetracycline-treated cells were observed at the start codons of the genes (137). Tetracycline-assisted ribosome profiling helped correct the assignment of some of the previously misannotated TISs of *E. coli* genes. Furthermore, several of the observed peaks of ribosome density also suggested that some *E. coli* genes have more than one start codon, with the additional sites located commonly in the vicinity of the pTIS (137). Nevertheless, the uncertainty of why tetracycline preferentially inhibits translation initiation leaves open the possibility that some of the peaks of ribosome density observed at the internal codons of the gene could result from inhibition of elongation rather than initiation of translation.

A more promising approach relies on the use of an authentic translation initiation inhibitor, retapamulin, in ribosome-profiling experiments. Retapamulin belongs to the pleuromutilin class of protein synthesis inhibitors. It targets the peptidyltransferase center, and its binding is incompatible with the presence of a nascent peptide within the ribosome (138). Therefore, retapamulin is capable of acting exclusively during translation initiation and efficiently arrests the ribosome at the start codons by blocking formation of the first peptide bond, probably by altering the placement of the CCA end of initiator tRNA bound at the ribosomal P-site (139). Indeed, we have found that retapamulin can be used as an excellent tool for revealing both pTISs and iTISs in bacteria (unpublished data). In retapamulin-treated cells, well-defined ribosome density peaks could be readily seen at previously known iTISs in the *E. coli* genome (Fig. 10). In addition, a number of other iTIS candidates have been identified and are being currently analyzed. Additional tools, for example, the use of spe-

cific inhibitors of translation termination (140) in conjunction with initiation inhibitors, might facilitate the detection of OOF internal ORFs. In addition, advanced RNA-sequencing methods, such as differential RNA sequencing (141), can be used to identify alternative transcript isoforms that can also result in expression of internal ORFs. In the future, combining antibiotic-enhanced ribosome profiling with transcriptome sequencing and actinonin-based proteomics could form an even more robust framework for revealing the alternative proteome originated from unusual translation events.

CONCLUDING REMARKS

In our review, we attempted to present a snapshot of the currently known examples of internal translation initiation in bacteria. We have intentionally left out bacteriophages, which present many more examples of genes-within-genes encoding as well as unconfirmed suggestions that internal initiation occurs within some archaeal genes, and instead focused our attention primarily on intrinsic bacterial genes.

In this review, we decided to keep the names given to the internal initiation products by the authors who originally discovered them. For some of the in-frame cases, the molecular weights (in kDa) of the proteins were indicated next to their names (for example, ClpB93 and ClpB79). We find this to be the most straightforward designation for identification of the different protein isoforms and suggest using this unifying nomenclature to annotate the products of in-frame alternative initiation.

Most of the few currently known examples of translation initiation at internal start codons of genes have been identified serendipitously. Many questions about the pervasiveness of this mode of gene expression, functions of the alternative protein products, or the mechanisms of regulation remain unanswered. We do not know how the ribosome traffic influences the utilization of the iTIS or what the influence is of the local and global mRNA structure on the relative use of the primary and secondary start codons. We also poorly understand the advantage of expressing a second protein from an internal codon of a gene instead of maintaining its independent gene in a bacterial genome.

Emerging tools will likely rapidly expand our knowledge about the scenarios of unconventional protein coding in a more systematic and time- and cost-effective way. Studies of the alternative proteome cryptically encoded in the genome will most likely illuminate new layers of gene regulation and inform us of the possible functions of previously unknown proteins. This new knowledge may contribute to developing new medicines that could specifically target the alternative proteome or the mechanisms of unconventional decoding of genetic information.

Acknowledgments. We thank Dorota Klepacki and James Marks for help in performing and interpreting the ribosome-profiling experiments mentioned in this review and Elizabeth Woods for proofreading the manuscript. The work on translation regulation in the Mankin/Vázquez-Laslop laboratory is supported by grant MCB 161585 from the National Science Foundation.

Citation. Meydan S, Vázquez-Laslop N, Mankin AS. 2018. Genes within genes in bacterial genomes. Microbiol Spectrum 6(4):RWR-0020-2018.

References

1. **Atkins JF, Loughran G, Bhatt PR, Firth AE, Baranov PV.** 2016. Ribosomal frameshifting and transcriptional slippage: from genetic steganography and cryptography to adventitious use. *Nucleic Acids Res* **44:**7007–7078.

2. **Baranov PV, Atkins JF, Yordanova MM.** 2015. Augmented genetic decoding: global, local and temporal alterations of decoding processes and codon meaning. *Nat Rev Genet* **16:**517–529.

3. **Caliskan N, Peske F, Rodnina MV.** 2015. Changed in translation: mRNA recoding by -1 programmed ribosomal frameshifting. *Trends Biochem Sci* **40:**265–274.

4. **Dinman JD.** 2012. Mechanisms and implications of programmed translational frameshifting. *Wiley Interdiscip Rev RNA* **3:**661–673.

5. **Miller ES, Kutter E, Mosig G, Arisaka F, Kunisawa T, Rüger W.** 2003. Bacteriophage T4 genome. *Microbiol Mol Biol Rev* **67:**86–156.

6. **Doore SM, Baird CD, 2012 University of Arizona Virology Undergraduate Lab, Roznowski AP, Fane BA.** 2014. The evolution of genes within genes and the control of DNA replication in microviruses. *Mol Biol Evol* **31:**1421–1431.

7. **Shcherbakov DV, Garber MB.** 2000. Overlapping genes in bacterial and bacteriophage genomes. *Mol Biol (Mosk)* **34:**572–583. (In Russian.)

8. **Gualerzi CO, Pon CL.** 2015. Initiation of mRNA translation in bacteria: structural and dynamic aspects. *Cell Mol Life Sci* **72:**4341–4367.

9. **Kozak M.** 2005. Regulation of translation via mRNA structure in prokaryotes and eukaryotes. *Gene* **361:**13–37.

10. **Laursen BS, Sørensen HP, Mortensen KK, Sperling-Petersen HU.** 2005. Initiation of protein synthesis in bacteria. *Microbiol Mol Biol Rev* **69:**101–123.

11. **Vellanoweth RL, Rabinowitz JC.** 1992. The influence of ribosome-binding-site elements on translational efficiency in *Bacillus subtilis* and *Escherichia coli in vivo*. *Mol Microbiol* **6:**1105–1114.

12. **Shine J, Dalgarno L.** 1975. Determinant of cistron specificity in bacterial ribosomes. *Nature* **254:**34–38.

13. **Accetto T, Avguštin G.** 2011. Inability of *Prevotella bryantii* to form a functional Shine-Dalgarno interaction reflects unique evolution of ribosome binding sites in *Bacteroidetes*. *PLoS One* **6:**e22914.

14. **Moll I, Grill S, Gualerzi CO, Bläsi U.** 2002. Leaderless mRNAs in bacteria: surprises in ribosomal recruitment and translational control. *Mol Microbiol* **43:**239–246.

15. **Salis HM.** 2011. The ribosome binding site calculator. *Methods Enzymol* **498:**19–42.

16. **Valverde C, Haas D.** 2008. Small RNAs controlled by two-component systems. *Adv Exp Med Biol* **631:**54–79.

17. **Whitaker WR, Lee H, Arkin AP, Dueber JE.** 2015. Avoidance of truncated proteins from unintended ribosome binding sites within heterologous protein coding sequences. *ACS Synth Biol* **4:**249–257.

18. **Miller MJ, Wahba AJ.** 1973. Chain initiation factor 2. Purification and properties of two species from *Escherichia coli* MRE 600. *J Biol Chem* **248:**1084–1090.

19. **Sacerdot C, Dessen P, Hershey JW, Plumbridge JA, Grunberg-Manago M.** 1984. Sequence of the initiation factor IF2 gene: unusual protein features and homologies with elongation factors. *Proc Natl Acad Sci U S A* **81:**7787–7791.

20. **Plumbridge JA, Deville F, Sacerdot C, Petersen HU, Cenatiempo Y, Cozzone A, Grunberg-Manago M, Hershey JW.** 1985. Two translational initiation sites in the *infB* gene are used to express initiation factor IF2α and IF2β in *Escherichia coli*. *EMBO J* **4:**223–229.

21. **Nyengaard NR, Mortensen KK, Lassen SF, Hershey JW, Sperling-Petersen HU.** 1991. Tandem translation of *E. coli* initiation factor IF2β: purification and characterization *in vitro* of two active forms. *Biochem Biophys Res Commun* **181:**1572–1579.

22. **Sacerdot C, Vachon G, Laalami S, Morel-Deville F, Cenatiempo Y, Grunberg-Manago M.** 1992. Both forms of translational initiation factor IF2 (α and β) are required for maximal growth of *Escherichia coli*: evidence for two translational initiation codons for IF2β. *J Mol Biol* **225:**67–80.

23. **Laursen BS, de A Steffensen SA, Hedegaard J, Moreno JM, Mortensen KK, Sperling-Petersen HU.** 2002. Structural requirements of the mRNA for intracistronic translation initiation of the enterobacterial *infB* gene. *Genes Cells* **7:**901–910.

24. **Shazand K, Tucker J, Chiang R, Stansmore K, Sperling-Petersen HU, Grunberg-Manago M, Rabinowitz JC, Leighton T.** 1990. Isolation and molecular genetic characterization of the *Bacillus subtilis* gene (*infB*) encoding protein synthesis initiation factor 2. *J Bacteriol* **172:** 2675–2687.

25. **Caserta E, Tomsic J, Spurio R, La Teana A, Pon CL, Gualerzi CO.** 2006. Translation initiation factor IF2 interacts with the 30 S ribosomal subunit via two separate binding sites. *J Mol Biol* **362:**787–799.

26. **Giuliodori AM, Brandi A, Gualerzi CO, Pon CL.** 2004. Preferential translation of cold-shock mRNAs during cold adaptation. *RNA* **10:**265–276.

27. **Madison KE, Abdelmeguid MR, Jones-Foster EN, Nakai H.** 2012. A new role for translation initiation

factor 2 in maintaining genome integrity. *PLoS Genet* **8:** e1002648.

28. **Madison KE, Jones-Foster EN, Vogt A, Kirtland Turner S, North SH, Nakai H.** 2014. Stringent response processes suppress DNA damage sensitivity caused by deficiency in full-length translation initiation factor 2 or PriA helicase. *Mol Microbiol* **92:**28–46.

29. **North SH, Kirtland SE, Nakai H.** 2007. Translation factor IF2 at the interface of transposition and replication by the PriA-PriC pathway. *Mol Microbiol* **66:**1566–1578.

30. **Li GW, Burkhardt D, Gross C, Weissman JS.** 2014. Quantifying absolute protein synthesis rates reveals principles underlying allocation of cellular resources. *Cell* **157:**624–635.

31. **Hénaut A, Lisacek F, Nitschké P, Moszer I, Danchin A.** 1998. Global analysis of genomic texts: the distribution of AGCT tetranucleotides in the *Escherichia coli* and *Bacillus subtilis* genomes predicts translational frameshifting and ribosomal hopping in several genes. *Electrophoresis* **19:**515–527.

32. **Atlung T, Nielsen HV, Hansen FG.** 2002. Characterisation of the allelic variation in the *rpoS* gene in thirteen K12 and six other non-pathogenic *Escherichia coli* strains. *Mol Genet Genomics* **266:**873–881.

33. **Subbarayan PR, Sarkar M.** 2004. A comparative study of variation in codon 33 of the *rpoS* gene in *Escherichia coli* K12 stocks: implications for the synthesis of σ^S. *Mol Genet Genomics* **270:**533–538.

34. **Subbarayan PR, Sarkar M.** 2004. *Escherichia coli rpoS* gene has an internal secondary translation initiation region. *Biochem Biophys Res Commun* **313:**294–299.

35. **Subbarayan PR, Sarkar M.** 2004. A stop codon-dependent internal secondary translation initiation region in *Escherichia coli rpoS*. *RNA* **10:**1359–1365.

36. **Kato J, Suzuki H, Hirota Y.** 1984. Overlapping of the coding regions for alpha and gamma components of penicillin-binding protein 1 b in *Escherichia coli*. *Mol Gen Genet* **196:**449–457.

37. **Nakagawa J, Matsuhashi M.** 1982. Molecular divergence of a major peptidoglycan synthetase with transglycosylase-transpeptidase activities in *Escherichia coli*—penicillin-binding protein 1Bs. *Biochem Biophys Res Commun* **105:**1546–1553.

38. **Henderson TA, Dombrosky PM, Young KD.** 1994. Artifactual processing of penicillin-binding proteins 7 and 1b by the OmpT protease of *Escherichia coli*. *J Bacteriol* **176:**256–259.

39. **Broome-Smith JK, Edelman A, Yousif S, Spratt BG.** 1985. The nucleotide sequences of the *ponA* and *ponB* genes encoding penicillin-binding protein 1A and 1B of *Escherichia coli* K12. *Eur J Biochem* **147:**437–446.

40. **Zijderveld CA, Aarsman ME, den Blaauwen T, Nanninga N.** 1991. Penicillin-binding protein 1B of *Escherichia coli* exists in dimeric forms. *J Bacteriol* **173:**5740–5746.

41. **Ünal CM, Steinert M.** 2015. FKBPs in bacterial infections. *Biochim Biophys Acta* **1850:**2096–2102.

42. **Mo YY, Seshu J, Wang D, Mallavia LP.** 1998. Synthesis in *Escherichia coli* of two smaller enzymatically active analogues of *Coxiella burnetii* macrophage infectivity potentiator (CbMip) protein utilizing a single open reading frame from the *cbmip* gene. *Biochem J* **335:**67–77.

43. **Seshadri R, Paulsen IT, Eisen JA, Read TD, Nelson KE, Nelson WC, Ward NL, Tettelin H, Davidsen TM, Beanan MJ, Deboy RT, Daugherty SC, Brinkac LM, Madupu R, Dodson RJ, Khouri HM, Lee KH, Carty HA, Scanlan D, Heinzen RA, Thompson HA, Samuel JE, Fraser CM, Heidelberg JF.** 2003. Complete genome sequence of the Q-fever pathogen *Coxiella burnetii*. *Proc Natl Acad Sci U S A* **100:**5455–5460.

44. **Kittendorf JD, Sherman DH.** 2009. The methymycin/pikromycin pathway: a model for metabolic diversity in natural product biosynthesis. *Bioorg Med Chem* **17:**2137–2146.

45. **Almutairi MM, Svetlov MS, Hansen DA, Khabibullina NF, Klepacki D, Kang HY, Sherman DH, Vázquez-Laslop N, Polikanov YS, Mankin AS.** 2017. Co-produced natural ketolides methymycin and pikromycin inhibit bacterial growth by preventing synthesis of a limited number of proteins. *Nucleic Acids Res* **45:**9573–9582.

46. **Xue Y, Sherman DH.** 2000. Alternative modular polyketide synthase expression controls macrolactone structure. *Nature* **403:**571–575.

47. **Hori T, Maezawa I, Nagahama N, Suzuki M.** 1971. Isolation and structure of narbonolide, narbomycin, aglycone, from *Streptomyces venezuelae* and its biological transformation into picromycin via narbomycin. *J Chem Soc D Chem Commun* **0:**304–305.

48. **Lambalot RH, Cane DE.** 1992. Isolation and characterization of 10-deoxymethynolide produced by *Streptomyces venezuelae*. *J Antibiot (Tokyo)* **45:**1981–1982.

49. **Zabrocka E, Wegrzyn K, Konieczny I.** 2014. Two replication initiators—one mechanism for replication origin opening? *Plasmid* **76:**72–78.

50. **Shingler V, Thomas CM.** 1984. Analysis of the *trfA* region of broad host-range plasmid RK2 by transposon mutagenesis and identification of polypeptide products. *J Mol Biol* **175:**229–249.

51. **Smith CA, Thomas CM.** 1984. Nucleotide sequence of the *trfA* gene of broad host-range plasmid RK2. *J Mol Biol* **175:**251–262.

52. **Kittell BL, Helinski DR.** 1991. Iteron inhibition of plasmid RK2 replication *in vitro*: evidence for intermolecular coupling of replication origins as a mechanism for RK2 replication control. *Proc Natl Acad Sci U S A* **88:**1389–1393.

53. **Durland RH, Helinski DR.** 1987. The sequence encoding the 43-kilodalton *trfA* protein is required for efficient replication or maintenance of minimal RK2 replicons in *Pseudomonas aeruginosa*. *Plasmid* **18:**164–169.

54. **Konieczny I.** 2003. Strategies for helicase recruitment and loading in bacteria. *EMBO Rep* **4:**37–41.

55. **Zhong Z, Helinski D, Toukdarian A.** 2003. A specific region in the N terminus of a replication initiation protein of plasmid RK2 is required for recruitment of *Pseudomonas aeruginosa* DnaB helicase to the plasmid origin. *J Biol Chem* **278:**45305–45310.

56. Fang FC, Helinski DR. 1991. Broad-host-range properties of plasmid RK2: importance of overlapping genes encoding the plasmid replication initiation protein TrfA. *J Bacteriol* **173**:5861–5868.

57. Doyle SM, Genest O, Wickner S. 2013. Protein rescue from aggregates by powerful molecular chaperone machines. *Nat Rev Mol Cell Biol* **14**:617–629.

58. Park SK, Kim KI, Woo KM, Seol JH, Tanaka K, Ichihara A, Ha DB, Chung CH. 1993. Site-directed mutagenesis of the dual translational initiation sites of the *clpB* gene of *Escherichia coli* and characterization of its gene products. *J Biol Chem* **268**:20170–20174.

59. Squires CL, Pedersen S, Ross BM, Squires C. 1991. ClpB is the *Escherichia coli* heat shock protein F84.1. *J Bacteriol* **173**:4254–4262.

60. Beinker P, Schlee S, Groemping Y, Seidel R, Reinstein J. 2002. The N terminus of ClpB from *Thermus thermophilus* is not essential for the chaperone activity. *J Biol Chem* **277**:47160–47166.

61. Impens F, Rolhion N, Radoshevich L, Bécavin C, Duval M, Mellin J, García Del Portillo F, Pucciarelli MG, Williams AH, Cossart P. 2017. N-terminomics identifies Prli42 as a membrane miniprotein conserved in Firmicutes and critical for stressosome activation in *Listeria monocytogenes*. *Nat Microbiol* **2**:17005.

62. Chow IT, Baneyx F. 2005. Coordinated synthesis of the two ClpB isoforms improves the ability of *Escherichia coli* to survive thermal stress. *FEBS Lett* **579**:4235–4241.

63. Eriksson MJ, Clarke AK. 1996. The heat shock protein ClpB mediates the development of thermotolerance in the cyanobacterium *Synechococcus* sp. strain PCC 7942. *J Bacteriol* **178**:4839–4846.

64. Nagy M, Guenther I, Akoyev V, Barnett ME, Zavodszky MI, Kedzierska-Mieszkowska S, Zolkiewski M. 2010. Synergistic cooperation between two ClpB isoforms in aggregate reactivation. *J Mol Biol* **396**:697–707.

65. Clarke AK, Eriksson MJ. 2000. The truncated form of the bacterial heat shock protein ClpB/HSP100 contributes to development of thermotolerance in the cyanobacterium *Synechococcus* sp. strain PCC 7942. *J Bacteriol* **182**:7092–7096.

66. Seol JH, Yoo SJ, Kim KI, Kang MS, Ha DB, Chung CH. 1994. The 65-kDa protein derived from the internal translational initiation site of the *clpA* gene inhibits the ATP-dependent protease Ti in *Escherichia coli*. *J Biol Chem* **269**:29468–29473.

67. Seol JH, Yoo SJ, Kang MS, Ha DB, Chung CH. 1995. The 65-kDa protein derived from the internal translational start site of the *clpA* gene blocks autodegradation of ClpA by the ATP-dependent protease Ti in *Escherichia coli*. *FEBS Lett* **377**:41–43.

68. Loh E, Righetti F, Eichner H, Twittenhoff C, Narberhaus F. 2018. RNA thermometers in bacterial pathogens. *Microbiol Spectr* **6**:RWR-0012-2017.

69. Gilson L, Mahanty HK, Kolter R. 1987. Four plasmid genes are required for colicin V synthesis, export, and immunity. *J Bacteriol* **169**:2466–2470.

70. Hwang J, Manuvakhova M, Tai PC. 1997. Characterization of in-frame proteins encoded by *cvaA*, an essential gene in the colicin V secretion system: CvaA* stabilizes CvaA to enhance secretion. *J Bacteriol* **179**:689–696.

71. Varcamonti M, Nicastro G, Venema G, Kok J. 2001. Proteins of the lactococcin A secretion system: *lcnD* encodes two in-frame proteins. *FEMS Microbiol Lett* **204**:259–263.

72. Loenen WA, Raleigh EA. 2014. The other face of restriction: modification-dependent enzymes. *Nucleic Acids Res* **42**:56–69.

73. Pieper U, Groll DH, Wünsch S, Gast FU, Speck C, Mücke N, Pingoud A. 2002. The GTP-dependent restriction enzyme McrBC from *Escherichia coli* forms high-molecular mass complexes with DNA and produces a cleavage pattern with a characteristic 10-base pair repeat. *Biochemistry* **41**:5245–5254.

74. Ross TK, Achberger EC, Braymer HD. 1989. Nucleotide sequence of the McrB region of *Escherichia coli* K-12 and evidence for two independent translational initiation sites at the *mcrB* locus. *J Bacteriol* **171**:1974–1981.

75. Beary TP, Braymer HD, Achberger EC. 1997. Evidence of participation of McrB$_S$ in McrBC restriction in *Escherichia coli* K-12. *J Bacteriol* **179**:7768–7775.

76. Panne D, Raleigh EA, Bickle TA. 1998. McrBs, a modulator peptide for McrBC activity. *EMBO J* **17**:5477–5483.

77. Rae BD, Long BM, Badger MR, Price GD. 2013. Functions, compositions, and evolution of the two types of carboxysomes: polyhedral microcompartments that facilitate CO_2 fixation in cyanobacteria and some proteobacteria. *Microbiol Mol Biol Rev* **77**:357–379.

78. Long BM, Badger MR, Whitney SM, Price GD. 2007. Analysis of carboxysomes from *Synechococcus* PCC7942 reveals multiple Rubisco complexes with carboxysomal proteins CcmM and CcaA. *J Biol Chem* **282**:29323–29335.

79. Price GD, Sültemeyer D, Klughammer B, Ludwig M, Badger MR. 1998. The functioning of the CO_2 concentrating mechanism in several cyanobacterial strains: a review of general physiological characteristics, genes, proteins, and recent advances. *Can J Bot* **76**:973–1002.

80. Long BM, Tucker L, Badger MR, Price GD. 2010. Functional cyanobacterial β-carboxysomes have an absolute requirement for both long and short forms of the CcmM protein. *Plant Physiol* **153**:285–293.

81. Abrusci P, McDowell MA, Lea SM, Johnson S. 2014. Building a secreting nanomachine: a structural overview of the T3SS. *Curr Opin Struct Biol* **25**:111–117.

82. Yu XJ, Liu M, Matthews S, Holden DW. 2011. Tandem translation generates a chaperone for the *Salmonella* type III secretion system protein SsaQ. *J Biol Chem* **286**:36098–36107.

83. Bzymek KP, Hamaoka BY, Ghosh P. 2012. Two translation products of *Yersinia yscQ* assemble to form a complex essential to type III secretion. *Biochemistry* **51**:1669–1677.

84. McDowell MA, Marcoux J, McVicker G, Johnson S, Fong YH, Stevens R, Bowman LA, Degiacomi MT, Yan J, Wise A, Friede ME, Benesch JL, Deane JE, Tang

CM, Robinson CV, Lea SM. 2016. Characterisation of *Shigella* Spa33 and *Thermotoga* FliM/N reveals a new model for C-ring assembly in T3SS. *Mol Microbiol* **99**: 749–766.

85. Song M, Sukovich DJ, Ciccarelli L, Mayr J, Fernandez-Rodriguez J, Mirsky EA, Tucker AC, Gordon DB, Marlovits TC, Voigt CA. 2017. Control of type III protein secretion using a minimal genetic system. *Nat Commun* **8**:14737.

86. Typas A, Sourjik V. 2015. Bacterial protein networks: properties and functions. *Nat Rev Microbiol* **13**:559–572.

87. Kofoid EC, Parkinson JS. 1991. Tandem translation starts in the *cheA* locus of *Escherichia coli*. *J Bacteriol* **173**:2116–2119.

88. Smith RA, Parkinson JS. 1980. Overlapping genes at the *cheA* locus of *Escherichia coli*. *Proc Natl Acad Sci U S A* **77**:5370–5374.

89. Wolfe AJ, Stewart RC. 1993. The short form of the CheA protein restores kinase activity and chemotactic ability to kinase-deficient mutants. *Proc Natl Acad Sci U S A* **90**:1518–1522.

90. Wang H, Matsumura P. 1997. Phosphorylating and dephosphorylating protein complexes in bacterial chemotaxis. *J Bacteriol* **179**:287–289.

91. Dumas R, Cobessi D, Robin AY, Ferrer JL, Curien G. 2012. The many faces of aspartate kinases. *Arch Biochem Biophys* **519**:186–193.

92. Chen NY, Paulus H. 1988. Mechanism of expression of the overlapping genes of *Bacillus subtilis* aspartokinase II. *J Biol Chem* **263**:9526–9532.

93. Kalinowski J, Cremer J, Bachmann B, Eggeling L, Sahm H, Pühler A. 1991. Genetic and biochemical analysis of the aspartokinase from *Corynebacterium glutamicum*. *Mol Microbiol* **5**:1197–1204.

94. Lo CC, Bonner CA, Xie G, D'Souza M, Jensen RA. 2009. Cohesion group approach for evolutionary analysis of aspartokinase, an enzyme that feeds a branched network of many biochemical pathways. *Microbiol Mol Biol Rev* **73**:594–651.

95. Yoshida A, Tomita T, Kuzuyama T, Nishiyama M. 2010. Mechanism of concerted inhibition of $\alpha_2\beta_2$-type hetero-oligomeric aspartate kinase from *Corynebacterium glutamicum*. *J Biol Chem* **285**:27477–27486.

96. Bondaryk RP, Paulus H. 1985. Expression of the gene for *Bacillus subtilis* aspartokinase II in *Escherichia coli*. *J Biol Chem* **260**:592–597.

97. Lin TH, Hu YN, Shaw GC. 2014. Two enzymes, TilS and HprT, can form a complex to function as a transcriptional activator for the cell division protease gene *ftsH* in *Bacillus subtilis*. *J Biochem* **155**:5–16.

98. Kharel MK, Pahari P, Shepherd MD, Tibrewal N, Nybo SE, Shaaban KA, Rohr J. 2012. Angucyclines: biosynthesis, mode-of-action, new natural products, and synthesis. *Nat Prod Rep* **29**:264–325.

99. Kallio P, Liu Z, Mäntsälä P, Niemi J, Metsä-Ketelä M. 2008. A nested gene in *Streptomyces* bacteria encodes a protein involved in quaternary complex formation. *J Mol Biol* **375**:1212–1221.

100. Thomas CM, Smith CA. 1986. The *trfB* region of broad host range plasmid RK2: the nucleotide sequence reveals *incC* and key regulatory gene *trfB/korA/korD* as overlapping genes. *Nucleic Acids Res* **14**:4453–4469.

101. Batt SM, Bingle LE, Dafforn TR, Thomas CM. 2009. Bacterial genome partitioning: N-terminal domain of IncC protein encoded by broad-host-range plasmid RK2 modulates oligomerisation and DNA binding. *J Mol Biol* **385**:1361–1374.

102. Bowsher CG, Dunbar B, Emes MJ. 1993. The purification and properties of ferredoxin-NADP$^+$-oxidoreductase from roots of *Pisum sativum* L. *Protein Expr Purif* **4**: 512–518.

103. Neuhaus HE, Emes MJ. 2000. Nonphotosynthetic metabolism in plastids. *Annu Rev Plant Physiol Plant Mol Biol* **51**:111–140.

104. Omairi-Nasser A, de Gracia AG, Ajlani G. 2011. A larger transcript is required for the synthesis of the smaller isoform of ferredoxin:NADP oxidoreductase. *Mol Microbiol* **81**:1178–1189.

105. Thomas JC, Ughy B, Lagoutte B, Ajlani G. 2006. A second isoform of the ferredoxin:NADP oxidoreductase generated by an in-frame initiation of translation. *Proc Natl Acad Sci U S A* **103**:18368–18373.

106. Omairi-Nasser A, Galmozzi CV, Latifi A, Muro-Pastor MI, Ajlani G. 2014. NtcA is responsible for accumulation of the small isoform of ferredoxin:NADP oxidoreductase. *Microbiology* **160**:789–794.

107. Malakooti J, Ely B, Matsumura P. 1994. Molecular characterization, nucleotide sequence, and expression of the *fliO*, *fliP*, *fliQ*, and *fliR* genes of *Escherichia coli*. *J Bacteriol* **176**:189–197.

108. Schoenhals GJ, Kihara M, Macnab RM. 1998. Translation of the flagellar gene *fliO* of *Salmonella typhimurium* from putative tandem starts. *J Bacteriol* **180**: 2936–2942.

109. Yamagata H, Adachi T, Tsuboi A, Takao M, Sasaki T, Tsukagoshi N, Udaka S. 1987. Cloning and characterization of the 5′ region of the cell wall protein gene operon in *Bacillus brevis* 47. *J Bacteriol* **169**:1239–1245.

110. Ebisu S, Tsuboi A, Takagi H, Naruse Y, Yamagata H, Tsukagoshi N, Udaka S. 1990. Conserved structures of cell wall protein genes among protein-producing *Bacillus brevis* strains. *J Bacteriol* **172**:1312–1320.

111. Adachi T, Yamagata H, Tsukagoshi N, Udaka S. 1990. Use of both translation initiation sites of the middle wall protein gene in *Bacillus brevis* 47. *J Bacteriol* **172**: 511–513.

112. Ozin AJ, Costa T, Henriques AO, Moran CP Jr. 2001. Alternative translation initiation produces a short form of a spore coat protein in *Bacillus subtilis*. *J Bacteriol* **183**:2032–2040.

113. Thomas S, Holland IB, Schmitt L. 2014. The type 1 secretion pathway—the hemolysin system and beyond. *Biochim Biophys Acta* **1843**:1629–1641.

114. Felmlee T, Pellett S, Welch RA. 1985. Nucleotide sequence of an *Escherichia coli* chromosomal hemolysin. *J Bacteriol* **163**:94–105.

115. Mackman N, Nicaud JM, Gray L, Holland IB. 1985. Identification of polypeptides required for the export of haemolysin 2001 from *E. coli*. *Mol Gen Genet* **201**: 529–536.

116. Blight MA, Menichi B, Holland IB. 1995. Evidence for post-transcriptional regulation of the synthesis of the *Escherichia coli* HlyB haemolysin translocator and production of polyclonal anti-HlyB antibody. *Mol Gen Genet* **247**:73–85.

117. Flores-Díaz M, Monturiol-Gross L, Naylor C, Alape-Girón A, Flieger A. 2016. Bacterial sphingomyelinases and phospholipases as virulence factors. *Microbiol Mol Biol Rev* **80**:597–628.

118. Cota-Gomez A, Vasil AI, Kadurugamuwa J, Beveridge TJ, Schweizer HP, Vasil ML. 1997. PlcR1 and PlcR2 are putative calcium-binding proteins required for secretion of the hemolytic phospholipase C of *Pseudomonas aeruginosa*. *Infect Immun* **65**:2904–2913.

119. Shen BF, Tai PC, Pritchard AE, Vasil ML. 1987. Nucleotide sequences and expression in *Escherichia coli* of the in-phase overlapping *Pseudomonas aeruginosa plcR* genes. *J Bacteriol* **169**:4602–4607.

120. Buskirk AR, Green R. 2017. Ribosome pausing, arrest and rescue in bacteria and eukaryotes. *Philos Trans R Soc Lond B Biol Sci* **372**:20160183.

121. Hamoen LW, Venema G, Kuipers OP. 2003. Controlling competence in *Bacillus subtilis*: shared use of regulators. *Microbiology* **149**:9–17.

122. D'Souza C, Nakano MM, Zuber P. 1994. Identification of *comS*, a gene of the *srfA* operon that regulates the establishment of genetic competence in *Bacillus subtilis*. *Proc Natl Acad Sci U S A* **91**:9397–9401.

123. Hamoen LW, Eshuis H, Jongbloed J, Venema G, van Sinderen D. 1995. A small gene, designated *comS*, located within the coding region of the fourth amino acid-activation domain of *srfA*, is required for competence development in *Bacillus subtilis*. *Mol Microbiol* **15**:55–63.

124. Evans D, Marquez SM, Pace NR. 2006. RNase P: interface of the RNA and protein worlds. *Trends Biochem Sci* **31**:333–341.

125. Ogasawara N, Yoshikawa H. 1992. Genes and their organization in the replication origin region of the bacterial chromosome. *Mol Microbiol* **6**:629–634.

126. Feltens R, Gossringer M, Willkomm DK, Urlaub H, Hartmann RK. 2003. An unusual mechanism of bacterial gene expression revealed for the RNase P protein of *Thermus* strains. *Proc Natl Acad Sci U S A* **100**:5724–5729.

127. Osterman IA, Evfratov SA, Sergiev PV, Dontsova OA. 2013. Comparison of mRNA features affecting translation initiation and reinitiation. *Nucleic Acids Res* **41**: 474–486.

128. Yuan P, D'Lima NG, Slavoff SA. 2018. Comparative membrane proteomics reveals a nonannotated *E. coli* heat shock protein. *Biochemistry* **57**:56–60.

129. Chang JT, Green CB, Wolf RE Jr. 1995. Inhibition of translation initiation on *Escherichia coli gnd* mRNA by formation of a long-range secondary structure involving the ribosome binding site and the internal complementary sequence. *J Bacteriol* **177**:6560–6567.

130. Berry IJ, Steele JR, Padula MP, Djordjevic SP. 2016. The application of terminomics for the identification of protein start sites and proteoforms in bacteria. *Proteomics* **16**:257–272.

131. Bienvenut WV, Giglione C, Meinnel T. 2015. Proteome-wide analysis of the amino terminal status of *Escherichia coli* proteins at the steady-state and upon deformylation inhibition. *Proteomics* **15**:2503–2518.

132. Ingolia NT, Ghaemmaghami S, Newman JR, Weissman JS. 2009. Genome-wide analysis in vivo of translation with nucleotide resolution using ribosome profiling. *Science* **324**:218–223.

133. Schrader JM, Zhou B, Li GW, Lasker K, Childers WS, Williams B, Long T, Crosson S, McAdams HH, Weissman JS, Shapiro L. 2014. The coding and noncoding architecture of the *Caulobacter crescentus* genome. *PLoS Genet* **10**:e1004463.

134. Fritsch C, Herrmann A, Nothnagel M, Szafranski K, Huse K, Schumann F, Schreiber S, Platzer M, Krawczak M, Hampe J, Brosch M. 2012. Genome-wide search for novel human uORFs and N-terminal protein extensions using ribosomal footprinting. *Genome Res* **22**:2208–2218.

135. Ingolia NT, Lareau LF, Weissman JS. 2011. Ribosome profiling of mouse embryonic stem cells reveals the complexity and dynamics of mammalian proteomes. *Cell* **147**:789–802.

136. Lee S, Liu B, Lee S, Huang SX, Shen B, Qian SB. 2012. Global mapping of translation initiation sites in mammalian cells at single-nucleotide resolution. *Proc Natl Acad Sci U S A* **109**:E2424–E2432.

137. Nakahigashi K, Takai Y, Kimura M, Abe N, Nakayashiki T, Shiwa Y, Yoshikawa H, Wanner BL, Ishihama Y, Mori H. 2016. Comprehensive identification of translation start sites by tetracycline-inhibited ribosome profiling. *DNA Res* **23**:193–201.

138. Davidovich C, Bashan A, Auerbach-Nevo T, Yaggie RD, Gontarek RR, Yonath A. 2007. Induced-fit tightens pleuromutilins binding to ribosomes and remote interactions enable their selectivity. *Proc Natl Acad Sci U S A* **104**:4291–4296.

139. Yan K, Madden L, Choudhry AE, Voigt CS, Copeland RA, Gontarek RR. 2006. Biochemical characterization of the interactions of the novel pleuromutilin derivative retapamulin with bacterial ribosomes. *Antimicrob Agents Chemother* **50**:3875–3881.

140. Florin T, Maracci C, Graf M, Karki P, Klepacki D, Berninghausen O, Beckmann R, Vázquez-Laslop N, Wilson DN, Rodnina MV, Mankin AS. 2017. An antimicrobial peptide that inhibits translation by trapping release factors on the ribosome. *Nat Struct Mol Biol* **24**: 752–757.

141. Thomason MK, Bischler T, Eisenbart SK, Förstner KU, Zhang A, Herbig A, Nieselt K, Sharma CM, Storz G. 2015. Global transcriptional start site mapping using differential RNA sequencing reveals novel antisense RNAs in *Escherichia coli*. *J Bacteriol* **197**:18–28.

Regulating with RNA in Bacteria and Archaea
Edited by Gisela Storz and Kai Papenfort
© 2018 American Society for Microbiology, Washington, DC
doi:10.1128/microbiolspec.RWR-0016-2017

Leaderless mRNAs in the Spotlight: Ancient but Not Outdated!

10

Heather J. Beck[1] and Isabella Moll[1]

In bacteria and archaea, translation initiates with a 30S ribosomal subunit interacting with an initiator tRNA at the ribosome binding site on a canonical mRNA to form a stable translation initiation complex that is primed for elongation. Canonical mRNAs contain both 5′ and 3′ untranslated regions (UTRs) containing information that will influence the stability and translation efficiency of the mRNA. Within the 5′ UTR, these signals can include ribosome recognition regions such as purine-rich Shine-Dalgarno (SD) sequences that are complementary to the anti-SD (aSD) sequence near the 16S rRNA 3′ terminus (1), AU-rich sequences that interact with ribosomal protein (r-protein) bS1 (2, 3) and prevent the formation of secondary structures, and enhancer regions. Additionally, 5′ UTRs may contain sequences that can be bound by *trans*-acting elements (i.e., proteins, antisense and small regulatory RNAs, or low-molecular-weight effectors) to change secondary structures or block translation initiation regions. Therefore, the regulatory and translation initiation signals are primarily contained within the 5′ UTR. Despite this functional importance of the 5′ UTR, there exists a class of mRNAs that are completely devoid of 5′ UTRs or possess very short 5′ UTRs. These mRNAs lack the SD sequence and any other translational signals and are so named leaderless mRNAs (lmRNAs). Thus, the mechanism underlying their recognition and binding by the translational apparatus is still not entirely elucidated.

lmRNAs can be naturally encoded on the chromosome and occur as a result of transcription and translation initiating at the same position or as a result of posttranscriptional or cotranslational processing events (Fig. 1). Intriguingly, lmRNAs are found in every domain of life and are universally translatable by heterologous hosts (4), signifying a fundamental lmRNA recognition capability of all translation systems. These results suggest that they represent ancestral RNA remnants. Hence, understanding their translatability could provide insight into the evolution of the ribosome.

WIDESPREAD OCCURRENCE OF lmRNAs

As the number of annotated genomes has drastically increased due to advancements in sequencing technology, we are beginning to uncover the vast number of lmRNAs. Recent *in silico* analyses examining translation initiation regions have made predictions as to the prevalence of lmRNAs across numerous bacterial and archaeal genomes (5–7). However, these studies can only estimate quantities based on promoter sequence logos and analysis of intergenic regions. Therefore, while they provide useful information regarding the high probability of lmRNAs, they cannot determine exact reproducible numbers of leaderless transcripts. Rather, transcriptome sequencing (RNA-seq)-based techniques, while limited in the number of transcriptomes analyzed in one study, can provide precise measurements of lmRNAs, which have been directly sequenced with high confidence. Such studies have uncovered a wide variety of lmRNAs in bacteria and archaea ranging from <1% up to ~70% of primary and secondary transcripts (Table 1). lmRNAs are not confined to bacteria and archaea either. They are also found in eukaryotes, with all mammalian mitochondrial mRNAs being leaderless (8, 9).

lmRNAs may represent a functional relic of an earlier evolutionary age, as bacteria and archaea closer to the root of the phylogenetic tree commonly have a higher proportion of lmRNAs. As determined by an *in silico* analysis in *Gammaproteobacteria*, lmRNA genes are most found in *Xanthomonadales* and *Legionellales*, which are

[1]Max F. Perutz Laboratories, Center for Molecular Biology, Department of Microbiology, Immunology and Genetics, University of Vienna, Vienna Biocenter, A-1030 Vienna, Austria.

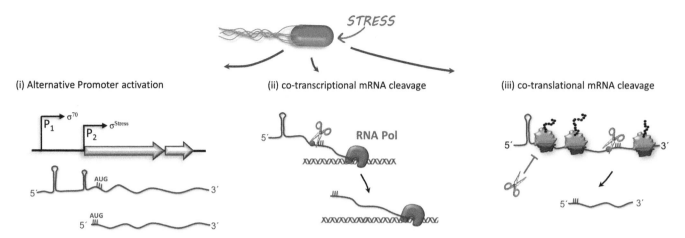

Figure 1 Mechanisms leading to the generation of lmRNAs in bacteria. Besides genes that are generally transcribed as lmRNAs, bacteria can generate lmRNAs in response to adverse environmental conditions (i) by activation of alternative promoters, where the transcriptional start point coincides with the A of the AUG start codon; (ii) by cotranscriptional cleavage, when the 5′ UTR is removed by RNases during the process of transcription; or (iii) cotranslationally. Here, the cleavage can be regulated by translating ribosomes that might either protect mRNAs from cleavage or expose specific sites for the processing event by RNases. Cleavage sites and potential RNases are indicated by red spheres and scissors, respectively.

located near the root of *Gammaproteobacteria* (6). Furthermore, phylogenetically close species tend to have a relatively equal proportion of lmRNA genes. Analysis of 16S rRNA revealed that archaeal genomes with the highest level of leaderless transcripts clustered together (10).

Additionally, in archaea, the majority of single genes or the proximal cistrons in operons encode lmRNAs (11–13). Operon structure appears to have coevolved with translation initiation signals, requiring additional SD sequences for an increasing number of cistrons. The proportion of operon distal genes shows positive correlation with the proportion of SD-led genes in both bacteria and archaea, supporting this notion (6). Furthermore, a transcriptome analysis in *Escherichia coli* uncovered previously uncharacterized 5′-AUGs at the end of canonical SD-led mRNAs (5′-uAUGs) (14). This genetic structure in which the first open reading frame (ORF) in a polycistronic mRNA is leaderless, parallels the operon structure in archaea (11). In *E. coli*, these 5′-uAUGs were shown to bind 70S ribosomes and were translated at biologically relevant levels, leading to the conclusion that these transcripts should also be classified as lmRNAs and are likely to exist in many bacterial species (14). This notion is supported by the characterization of the *acuR-acuI-dddL* operon in the Gram-negative alphaproteobacterium *Rhodobacter sphaeroides* (15) and the *sptA* mRNA in the haloarchaeal *Natrinema* sp. strain J7-2 (16), in which these mRNAs were found to be regulated by their ability to

be alternatively transcribed as an lmRNA. The regulation ascribed to each of these leaderless transcripts is described below.

lmRNAs are not only naturally encoded on the chromosome but can also be a result of posttranscriptional processing (Fig. 1). In one such instance, leadered mRNAs are cleaved to generate a subset of lmRNAs that are specialized for translation during suboptimal growth conditions (17, 18). This phenomenon involves a conditionally active toxin whose endonucleolytic capability transforms the RNA landscape to allow for certain mRNAs to maintain translatability. This illustrates the presence of previously overlooked lmRNAs that are not evident from genome analysis but play a role in regulation. The exact mechanism of this processing is detailed below.

Therefore, with the advances in global RNA sequencing and classification of the translatome, the prevalence of lmRNAs is becoming apparent, demonstrating the need to understand their mechanism of translation and the role they play in cellular physiology and regulation.

MAIN PATHWAYS OF TRANSLATION INITIATION

Translation Initiation on Canonical mRNAs

Despite the fact that lmRNAs are rather infrequent in *E. coli*, the majority of studies addressing the path-

Table 1 Compilation of published transcriptome analyses outlining the number of leaderless mRNAs in a variety of bacterial and archaeal genomes[a]

	Nucleotides upstream of start codon	Total no. of transcription start sites or mRNAs identified	No. of lmRNAs	% of lmRNAs	Reference
Bacteria					
Alphaproteobacteria					
Caulobacter crescentus NA1000	0	2,201	375	17	84
Sinorhizobium meliloti 1021	0	4,430	171	4	119
Deltaproteobacteria					
Geobacter sulfurreducens ATCC 51573	≤5	3,487	52	1.5	120
Epsilonproteobacteria					
Campylobacter jejuni NCTC 11168	0	992	12	1.2	121
Campylobacter jejuni	0	3,241	48	1.48	122
Gammaproteobacteria					
Escherichia coli BW25113	0	728	5	0.7	123
Escherichia coli MG1655	0	4,261	14		124
Klebsiella pneumoniae MGH 78578	≤5	5,194	83	1.6	125
Salmonella enterica serovar Typhimurium SL1344	0	1,873	23	1.2	126
Xanthomonas campestris pv. campestris B100	≤2	3,067	130	8.4	127
Legionella pneumophila strain Paris	0	1,905	12	0.63	128
Deinococcus-Thermus					
Deinococcus deserti RD19	0	1,958	916	47	91
Firmicutes					
Clostridium acetobutylicum ATCC 824	0	616	212	34.4	129
Bacillus amyloliquefaciens XH7	≤5	1,064	57	5.4	130
Actinobacteria					
Mycobacterium tuberculosis H37Rv	≤5	2,524	505	22	106
Mycobacterium tuberculosis H37Rv	≤5	4,979	1,098	22	79
Mycobacterium smegmatis mc^{2}155	0	1,098	206	19	79
Mycobacterium avium TMC724	≤3	844	278	33	131
Actinoplanes sp. SE50/110	0	661	135	20	132
Corynebacterium glutamicum ATCC 13032	0	2,454	546	22	133
Corynebacterium glutamicum ATCC 13032	0	2,147	707	33	134
Streptomyces tsukubaensis NRRL 18488	0	3,678	581	15.8	135
Cyanobacteria					
Prochlorococcus MED4	0	4,126	30	0.7	136
Prochlorococcus MIT913	0	8,587	41	0.5	136
Archaea					
Haloferax volcanii H26	≤5	4,749	1,329	72	137
Thermococcus onnurineus NA1	≤5	1,082	117	10.8	13
Pyrococcus abyssi GE5	≤5	1,893	27	1.4	138
Sulfolobus solfataricus P2	≤3	1,040	718	69	139

[a]The total number of lmRNAs identified in the different studies as well as their percentage (%) are given. mRNAs harboring up to 5 nucleotides upstream of the AUG start codon are included for completeness.

way of translation initiation on these peculiar mRNAs have been performed in this model organism. In *E. coli*, the canonical pathway of translation initiation (Fig. 2A) entails first the initiation factor-assisted recognition and binding of the mRNA and the recruitment of the fMet-tRNA^{fMet} to the 30S subunit. Next, the codon-anticodon interaction in the ribosomal P-site is established to form the 30S initiation complex. The efficiency of this step is affected by the strength of the SD-aSD interaction, secondary structures within the ribosome binding site, and the nature of the start codon. The three initiation factors (IF1, IF2, and IF3) contribute synergistically to the kinetics and fidelity of translation initiation (19–22). IF1 binds at the ribosomal A-site. Thus, it prevents elongator tRNA binding but also controls the conformational dynamics of the 30S subunit

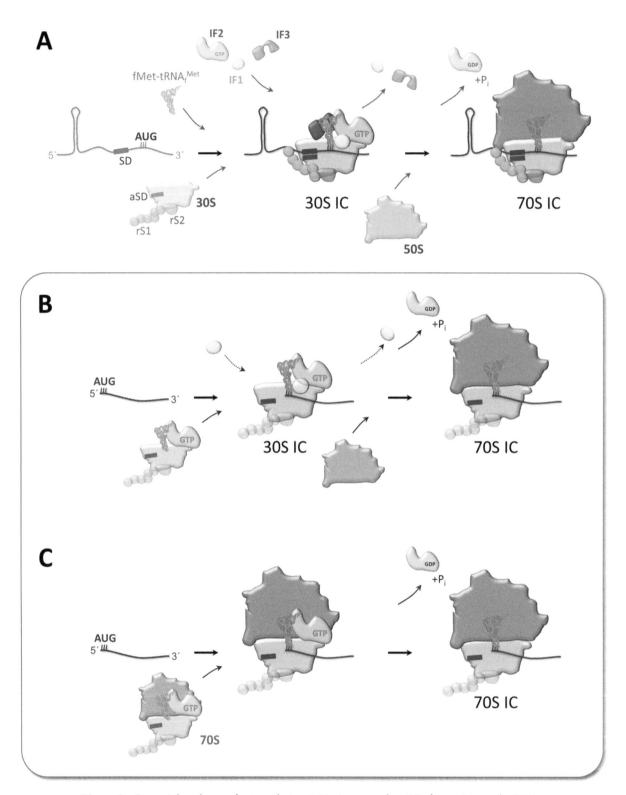

Figure 2 Potential pathways for translation initiation complex (IC) formation on lmRNAs. (A) Schematic showing the main steps during canonical initiation. (B and C) Potential steps during translation initiation on lmRNAs via 30S subunits and 70S monosomes, respectively. r-Proteins bS1 and uS2 are transparent, indicating their dispensability during this process. See text for details.

(21). Further, by providing the anchoring point for IF2 and IF3 on the 30S subunit, IF1 enhances their activity (23). IF2 selects for the correct initiator tRNA by directly binding to the N-formyl-methionine and aminoacyl acceptor stem of fMet-tRNA^{fMet} (24). Hence, IF2 increases the affinity of fMet-tRNA^{fMet} to the 30S subunit, whereas IF3 reduces the affinity of the fMet-tRNA^{fMet} (25) and discriminates against noncognate initiation complexes by introducing conformational dynamics during complex formation (19). With respect to subunit joining, IF2 and IF3 likewise show opposite activities. Due to the localization of the C-domain of IF3 at the interface side of the platform of the 30S subunit, the factor sterically blocks subunit joining at the B2 bridge between helix 69 of the 23S rRNA and the 16S rRNA (26). After release of IF3, a dynamic conformational switch introduced by binding of GTP to IF2 stimulates the rate of subunit joining (20). Finally, binding of the 50S subunit triggers GTP hydrolysis, which causes a conformational change in IF2 and drives an intersubunit rotation. Thereby, IF2 is released and the 70S complex is stabilized and transitions into the elongation-competent state (27).

In addition to the initiation factors, some r-proteins were shown to play key roles during translation initiation complex formation. The multidomain r-protein bS1 is essential for canonical initiation (28). It interacts via the N-terminal helix with r-protein uS2 on the 30S subunit (Fig. 1) (29). The flexible and dynamic protein is suggested to scan the vicinity of the ribosome for mRNAs with its C-terminal RNA-binding domains (30). Further, protein bS1 promotes RNA unwinding by binding to single-stranded RNA during thermal breathing (31). Together, these activities suggest that the protein promotes mRNA recruitment and facilitates positioning of structured mRNA in the RNA track of the ribosome by unwinding secondary structures. r-Protein bS21, which is likewise essential for ribosome binding during initiation of MS2 RNA and canonical *E. coli* mRNA translation, is localized on the platform of the 30S subunit close to the 16S rRNA 3′ terminus. It is suggested to stimulate the base-pairing potential of the SD-aSD interaction by exposing the 16S rRNA 3′ terminus (32–34).

Factors and r-Proteins That Affect Translation Initiation on lmRNAs

In contrast to the established pathway for translation initiation on canonical mRNAs in bacteria, initiation on lmRNAs can occur with a reduced set of factors and/or r-proteins (Fig. 2B). Several lines of evidence indicate that for translation initiation on lmRNA r-proteins

bS1 and uS2 are both dispensable, as the *cI* lmRNA is more efficiently translated in their absence, suggesting they negatively impact lmRNA translation (35–37). Upon bS1 depletion, lmRNAs become more abundant and more stable and their translation is stimulated, whereas translation of canonical mRNAs is greatly reduced, an effect that can be reverted by the addition of exogenous bS1 (38, 39). This positive impact of the lack of bS1 on lmRNA translation is likely due to the loss of competition with canonical mRNAs. As structured, leadered mRNAs can no longer compete for bS1-depleted ribosomes, this allows the ribosomes to selectively translate only unstructured or leaderless transcripts. Additionally, loss of bS1 will eliminate the negative impact of IF3 because bS1 is required for the discriminatory activity of IF3 against lmRNA translation (35). Interestingly, bS1 has been shown to be dispensable for the formation of the initiation complex on leadered mRNAs if they contain a strong SD sequence and a weakly structured ribosome binding site (40). Moreover, some Gram-positive bacteria with low G+C content, the majority of which contain relatively high levels of lmRNA (6) (Table 1), either lack a bS1 homolog or contain a homolog missing the first domain, rendering it unable to bind to the ribosome (41). The lack of a functional bS1 homolog in archaea also correlates with their high number of leaderless transcripts. Together, these data support the notion that bS1 is in general not essential for translation.

The role of r-protein uS2 in lmRNA translation might be rather indirect, as it represents the primary anchoring point for bS1 on the ribosome (29). Therefore, the effects seen as a result of uS2 depletion are the same as deleting bS1 alone (38, 42), as only translation of canonical mRNAs is affected while the translational efficiency of lmRNAs increases 8- to 10-fold (43). Additionally, bS2 mutants contain increased levels of 70S ribosomes, in line with the stimulation of lmRNA expression (44).

In addition to bS1 and uS2, r-protein bS21 is also expendable for lmRNA translation. Here, promoting the base-pairing potential of the SD-aSD interaction (33) is irrelevant for ribosome binding. However, the absence of bS21 did not affect fMet-tRNA^{fMet} binding necessary for lmRNA translation, or poly(U) translation (32), which resembles lmRNAs in that it is translated under conditions in which 70S ribosomes are prevalent (44).

As the 5′-AUG initiation codon appears to act as the unique ribosome binding signal on lmRNAs, their translation is more dependent on the presence of fMet-tRNA^{fMet} than of canonical mRNA. This notion is supported by the observation that IF2 promotes lmRNA translation, which can effectively compete with canonical mRNAs for 30S occupation when IF2 levels are increased (4, 45). In contrast, IF3, which decreases the affinity of the fMet-tRNA^{fMet} (25), discriminates against lmRNA translation by recognizing it as "noncanonical" and destabilizing the initiation complex (45–49). This discrimination could be attributed to conformational dynamics introduced in the ribosome by IF3 during initiation complex formation rather than direct scanning of the initiation codon (19, 50) and requires the presence of r-protein bS1 (35). Therefore, the ratio between IF2 and IF3 can modulate the efficiency of lmRNA translation (45). These ratios are equal at a steady state (51) but can vary in response to physiological changes, as IF3 concentration increases during increased growth rate (52), while IF2 levels increase under cold shock (53). This suggests that the efficiency of lmRNA translation might also fluctuate in response to environmental changes.

Together, these observations propose the first pathway for initiation on lmRNAs in bacteria, where a 30S subunit with a prebound IF2-fMet-tRNA^{fMet} complex binds to the 5′ terminus of the lmRNA in the absence of IF3. Here, the 5′-terminal AUG initiation codon and the 5′ triphosphate appear to act as ribosome binding signals (Fig. 2B). Intriguingly, this process is analogous to the eukaryotic initiation pathway in which the 40S subunit must first be loaded with tRNA$_i$ to scan for the start codon (54).

The second model suggests that the lmRNA is directly bound by an intact 70S ribosome through codon-anticodon base pairing with a prebound IF2-fMet-tRNA^{fMet} complex (Fig. 2C). 70S ribosomes show a high preference for a 5′-terminal AUG (55) and bind to lmRNAs more strongly than 30S subunits (46). Furthermore, 70S initiation complexes are intrinsically more stable on lmRNAs, with a 5- to 10-fold-higher retention with the addition of fMet-tRNA^{fMet} compared to 30S subunits (56), and chemically cross-linked 70S complexes, unable to dissociate into 30S and 50S subunits, retained the ability to translate leaderless but not canonical mRNAs (44). Also, inactivation of ribosome recycling factor to prevent subunit dissociation resulted in increased lmRNA translation, while bulk mRNA translation greatly diminished (44). Considering that IF3 prevents subunit association (26), the 70S initiation pathway on lmRNA can also be reconciled with the reduced translation initiation when IF3 levels are increased (35, 45). Finally, conditions that create a prevalence of 70S monosomes *in vivo* correlate with preferential translation of lmRNA and loss of canonical

translation (44). While evidence suggests that the use of the 70S pathway is more likely, the two models are not mutually exclusive, making it possible that both the 30S and 70S initiation mechanisms are operating *in vivo* depending on the physiological conditions.

Although lmRNAs are generally more prominent in archaea, less is known about their initiation mechanism. It is considered to closely parallel the proposed bacterial pathway, with 70S-bound a/eIF2 recruiting Met-tRNA$_i^{Met}$ and directing it to bind specifically to the initiation codon (Fig. 2C) (57, 58). However, there are likely additional auxiliary factors that contribute to archaeal lmRNA initiation, as they contain >10 genes encoding proteins homologous to eukaryotic initiation factors, including aIF1, aIF1A, aIF2, a/eIF2, and aIF6 (59–62). Moreover, archaea utilize a hybrid system where most factors are more similar to the eukaryotic mechanism; however, they follow the basic bacterial mechanism for initiation codon selection (61). Nonetheless, in both bacteria and archaea, the initiator tRNA is absolutely essential for stable initiation complex formation on lmRNA (4, 44, 56, 57).

Together, the working model for lmRNA translation in archaea suggests that a 70S-aIF1-aIF1A-a/eIF2-GTP-Met-tRNA$_i^{Met}$ complex binds the mRNA (58). Although the exact order of events to form this complex is unknown, it is anticipated that a/eIF2 initially binds to the ribosome. The binding is stimulated by the concerted activities of aIF1 and aIF1A, which are hypothesized to induce structural changes in the 30S to facilitate a/eIF2-GTP binding to, in turn, recruit the Met-tRNA$_i^{Met}$ (58, 63, 64). a/eIF2 performs distinct functions depending on whether or not it is bound to the ribosome. Off the ribosome it has a high affinity for 5′-triphosphorylated ends to protect mRNA from degradation (62, 65). However, this binding negatively impacts lmRNA, as it will block translation when a/eIF2 is bound. Once bound by the ribosome, a/eIF2 has a higher affinity for Met-tRNA$_i^{Met}$ and instead positively contributes to translation initiation (66). This is especially important for lmRNAs, which are dependent on Met-tRNA$_i^{Met}$ to be prebound to the 70S ribosome for efficient translation initiation complex formation.

Additional initiation factors aIF2 and aIF6, common only between archaea and eukaryotes, play indirect roles in archaeal lmRNA translation. aIF2 acts as a ribosome-dependent thermophilic GTPase that interacts preferentially with 50S subunits and 70S monosomes (67). It stimulates the binding of Met-tRNA$_i^{Met}$ to the ribosome and thereby increases the translation of both leadered and leaderless mRNAs (67, 68). aIF6 localizes at the nucleation core of the 50S subunit to prevent formation of the 70S ribosome. It is upregulated during cold and heat shock to act as translational repressor under unfavorable conditions (62, 69). Therefore, it is possible that the ratios between archaeal initiation factors vary in response to external conditions, thereby similarly modulating lmRNA expression.

As demonstrated, the various initiation factors can modulate lmRNA translation in distinct ways when compared to canonical translation. Nevertheless, the reduced number of translational factors needed for lmRNA translation further illustrates their ancestral nature by following a simpler mechanism of initiation established prior to the addition of certain r-proteins and initiation factors to increase complexity.

Signals Intrinsic to the lmRNA That Affect Translation Initiation

Due to the absence of 5′-UTR signals in lmRNA to direct ribosome binding, it was hypothesized that sequences downstream of the start codon may instead fulfill this role. However, extensive examination has failed to produce a consensus sequence that may be responsible and has disproven the existence of a potential "downstream box" that was thought to interact with the 16S rRNA (70, 71). The presence of CA repeats or A-richness has been shown to increase the translational efficiency on both lmRNAs and leadered mRNAs (72, 73), and these sequences are thought to contribute to the reduction of secondary structures, resulting in a more open complex and therefore fewer constraints on ribosome binding.

The remaining factors present on every lmRNA are then reduced to only the initiation codon and the 5′ triphosphate, both of which play a role in lmRNA initiation. The 5′ triphosphate was found to influence ternary complex formation and stability on lmRNAs, with a 5′ hydroxyl abolishing translation of the naturally leaderless *cI* mRNA in *E. coli* but not the SD leadered version (74). Brock et al. demonstrated that in *E. coli*, the 70S ribosome can distinguish mRNAs by the presence or absence of a 5′-AUG and that this signal is both necessary and sufficient for ribosome binding and subsequent translation (75). Most convincing was the demonstration that the addition of an AUG triplet to the 5′ end of an internal fragment of the *lacZ* mRNA restored its ability to be bound by ribosomes and successfully translated *in vivo* (75).

It is the 5′-AUG itself, and not the codon-anticodon base pairing with the fMet-tRNA^{fMet}, that contributes to the initial ribosome interaction with lmRNAs. The codon-anticodon interaction further stabilizes the complex (75, 76). To support the notion that the 5′-AUG

itself is important, in many organisms, mutation of the AUG to another initiation codon (GUG, UUG, CUG, AUU) completely abolishes lmRNA translation (77, 78). In mycobacteria, a 5′-AUG or 5′-GUG is necessary and sufficient for lmRNA translation, with no other initiation codon enriched in the leaderless transcripts (79). Therefore, an AUG may be highly preferred but not absolutely required. The proximity of the AUG to the 5′ terminus also affects translation efficiency, as a further recessed AUG results in reduced translation (80). It has been proposed that the ribosome is able to examine 5′ termini for the presence of a start codon even on mRNAs that are not naturally leaderless (81, 82). This may occur as a result of coupling transcription and translation through interaction of the RNA polymerase (Pol) and the ribosome (83). Thereby the ribosome may directly scan each mRNA as it exits the RNA Pol. However, this would only account for the first round of ribosome binding on each lmRNA, after which free ribosomes would be responsible for lmRNA recognition. Without the benefit of direct binding facilitated through RNA Pol-ribosome interactions, the ribosome binding efficiency to lmRNA is likely decreased, accounting for the less efficient translation of lmRNA in general.

REGULATION UNDER VARIOUS PHYSIOLOGICAL CONDITIONS

Due to the lack of a 5′ UTR and canonical initiation signals, lmRNAs are typically regarded to be less efficiently translated during exponential growth. This observation may be an artifact of using *E. coli* as the model organism for the majority of lmRNA translation studies despite its relatively low number of lmRNAs. Conversely, the use of ribosome profiling data and reporter assays has shown that lmRNAs are translated with equal efficiency to non-SD or SD-led mRNAs in *Caulobacter crescentus* (84) and *Mycobacterium* spp. (79). Also, in both bacterial and archaeal systems, in some cases the translational efficiency is higher when deleting the entire 5′ UTR compared to mutations that disrupt the SD sequence (76, 85, 86). However, one has to carefully consider the impact of secondary structures introduced by addition or removal of 5′ UTRs.

Continued Translation of lmRNAs during Stress Conditions

Regardless of whether lmRNAs are less efficiently translated during exponential growth, lmRNA translation appears to be adapted to adverse physiological conditions. They can be translated during stress (17, 18, 87), without any initiation factors (55, 88), at low temperatures (89), and their translation is immune to various antibiotics (90). A higher occurrence of lmRNA genes also appears in many extremophiles such as radiation-resistant *Deinococcus radiodurans*, thermophilic *Thermus thermophilus*, and many archaea (6). In *Deinococcus deserti*, 60% of the transcriptome is leaderless and proteomics have confirmed that the lmRNAs are efficiently translated and encode the most abundant proteins present (91) (Table 1). Some of these lmRNAs code for proteins that are essential for viability, such as the HU proteins used for nucleoid compaction (92). The HU proteins in *D. deserti* were shown to be more efficiently synthesized from lmRNAs in contrast to the leadered variants, which resulted in protein levels that were insufficient to provide their essential function (92). Many of the leaderless transcripts code for small peptides that have homologs in other *Deinococcus* species (91). These small peptides are hypothesized to act as antioxidants, resulting in the radiation and desiccation tolerance displayed by *Deinococcus* species (91).

Numerous genes encoding resistance mechanisms against antibiotics targeting the protein synthesis machinery are transcribed as lmRNAs (82, 93–95). It is tempting to speculate that this mechanism is required for continued translation even in the presence of the antibiotic. The aminoglycoside antibiotic kasugamycin (Ksg) targets the ribosome in a position overlapping the kink between the P- and E-site codons and therefore interferes with the path of the mRNA immediately upstream of the start codon (96, 97). However, Ksg only affects 30S subunits without impairing fMet-tRNA^{fMet} binding to the P-site in 70S ribosomes due to 50S stabilization (96). Therefore, lmRNA translation is resistant to Ksg treatment due to the lack of competition between the mRNA and Ksg for binding sites (90, 98). This mechanism further supports the 70S initiation pathway for lmRNA translation containing no 30S intermediate steps (98). Intriguingly, the mRNAs encoding the ABC transporter that exports Ksg in the producing actinobacterium *Streptomyces kasugaensis* correspondingly are leaderless or translationally coupled (99). By this means the synthesis of the export mechanism continues even when the intracellular concentration of Ksg increases, ensuring the survival of the producer cell.

Generation of lmRNAs during Stress

During adverse physiological conditions, lmRNAs are even produced via endoribonucleolytic cleavage, allowing for their continued translation. Stress conditions such as treatment with antibiotics affecting transcription or translation (100–102) or amino acid starvation

(103) cause a reduction in the expression of the toxin-antitoxin system *mazEF* in *E. coli*. The labile antitoxin is quickly degraded and therefore no longer represses the stable MazF toxin. The active sequence-specific endoribonuclease MazF subsequently cleaves at single-stranded ACA sites to cause bulk mRNA degradation (Fig. 1). However, in some cases, the only available ACA site resides within the 5′ UTR just upstream of the initiation codon (17). This cleavage therefore generates a subset of mRNAs including both lmRNA and mRNA with shortened leaders that specifically contain a 5′-terminal hydroxyl group (104).

Concurrently, MazF also cleaves the 16S rRNA within mature 70S ribosomes upstream of position A1500, removing 43 nucleotides from the 3′ terminus, including the aSD sequence (17). Therefore, MazF cleavage results not only in a subset of short leadered/leaderless mRNAs but also in a subpopulation of so-called 70S$^{\Delta 43}$ ribosomes that is specialized for lmRNA translation. Subsequent ribosome profiling has revealed that the transcripts that are processed by MazF upstream of the start codon are bound and selectively translated by the modified 70S$^{\Delta 43}$ ribosomes (18). Whether the ribosome is able to select for the mRNAs in the MazF regulon via the absence of a complex leader sequence or through recognition of the 5′ hydroxyl remains to be elucidated. However, recent data suggest that the 70S$^{\Delta 43}$ ribosome, in contrast to the 70S, is more promiscuous, as it binds both 5′-triphosphate and 5′-hydroxyl termini (H. J. Beck, M. Sauert, and I. Moll, unpublished data), presenting a new potential pathway for lmRNA initiation.

The ~330 mRNAs that make up the MazF regulon do not appear to be functionally clustered, but there is an overrepresentation of essential genes, suggesting that the mRNAs that are not degraded by MazF are important for cell survival during stress and likely post-stress recovery (18). During recovery, the RNA ligase RtcB catalyzes the religation of the 43-nucleotide cleaved fragment to the 3′ end of the 16S rRNA, successfully restoring the ribosomes' ability to translate canonical leadered mRNAs (105). Therefore, the cell has processes in place to not only reduce its translatome to move into dormancy but also to transition back to steady state by utilizing the lmRNA initiation mechanism.

Differential Expression in Response to Stress

Other organisms also exploit the use of lmRNAs to adapt to stress. In *Mycobacterium tuberculosis* the majority of toxin-antitoxin genes is transcribed as lmRNA (106). A transcriptome analysis revealed a significant increase in lmRNAs during starvation, and notably, lmRNAs were the most strongly upregulated. Although lmRNAs are typically thought to be less efficiently expressed, in *M. tuberculosis* they were shown to have longer half-lives than the SD-led mRNAs, resulting in nearly equivalent protein production. These results suggest that lmRNAs, while playing a secondary role during exponential growth, are key in the physiology of nonreplicating cells (106). *Mycobacterium* spp. also produce a MazF toxin shown to target both the 23S rRNA helix loop 70, to completely inactivate translation, and the 3′ end of the 16S rRNA, to alter ribosomal specificity by removing the aSD sequence (107, 108). Mycobacterial MazF recognizes a different specific sequence than *E. coli* MazF but targets similar functional regions of the translational machinery, suggesting that the orthologous toxins have evolved their relative specificities in an effort to exploit essential and accessible ribosomal regions, which could lead to an adaptive response to various conditions (108).

Shigella spp. utilize alternative promoters within one gene to transcribe a leadered and a leaderless variant of the *virF* mRNA, translation of which results in the major full-length and an N-terminally truncated VirF protein, respectively (109). By directly binding to consensus *virF* promoter elements, the truncated protein negatively regulates the levels of full-length VirF that triggers a regulatory cascade, leading to virulence and subsequent invasion of the intestinal epithelial barrier in humans (109). The conditions triggering the transcription of *virF* lmRNA from a secondary promoter within the canonical *virF* coding sequence remain elusive, but it has been speculated that environmental cues signaling unsuitable conditions for an invasive response may play a role. Alternative transcription in response to butanol and butyrate stress also leads to the generation of lmRNA in *Clostridium acetobutylicum* (110). It is thus conceivable that numerous organisms employ leaderless transcripts to ensure continuous translation, albeit with lower efficiency, during adverse physiological conditions.

Further Benefits of 5′-Terminal AUGs

The ability of ribosomes to interact with the 5′-AUG can additionally allow for indirect regulation of downstream ORFs in polycistronic mRNAs. As previously mentioned, a recent transcriptome analysis brought to light the prevalence of 5′-uAUGs, which were able to bind ribosomes and in some cases allow for efficient translation (14). Similar 5′-uAUGs were also discovered in *D. deserti* and in some cases were found to be involved in transcriptional attenuation (91). In the *acuR-acuI-dddL* operon in *R. sphaeroides*, the first cis-

tron encodes a regulatory protein that represses the entire operon but is translated at a much lower rate than the downstream cistrons because it lacks a ribosome binding site (15). This genetic makeup generates equal transcriptional rates but disproportionate translational rates, resulting in the desired regulation. The pairwise arrangement of *acuR-acuI*, lacking a 5′ UTR, was observed in several individual strains of alpha-, beta-, and gammaproteobacteria (15), suggesting that utilizing an lmRNA to control protein expression is a widespread phenomenon.

In some cases, the additional upstream start codon is in frame with the canonical start codon, which could result in two independently translated protein isoforms from one transcript. Such cases of alternative translation initiation have been well documented in eukaryotes (reviewed in reference 110) and haloarchaea (16). In the haloarchaeon *Natrinema* sp. strain J7-2, the alternative initiation site on the *sptA* mRNA is 5 nucleotides from the 5′ terminus, thereby characterizing the transcript as both leadered and leaderless depending on the start codon utilized. Here, greater translation initiation efficiency occurs at the 5′-terminal AUG, with the internal AUG serving as a remedial initiation site in case the 5′-terminal AUG is disrupted (16). This observation is not surprising, as lmRNAs form the majority of transcripts in haloarchaea and are known to be efficiently translated (111). However, the SptA protein is processed posttranslationally to remove the N terminus when translated from the leaderless mRNA; thus, regardless of the initiation codon used, the final protein product remains the same (111). The presence of both AUGs does provide benefits, positively affecting mRNA stability to enhance expression. Redundancy in initiation as either a leadered or leaderless mRNA ensures its translation in response to different cellular signals or stress. Alternative translation initiation also occurs in *M. tuberculosis*, where the extracytoplasmatic sigma factor σ^E (*sigE*) is differentially translated to produce different isoforms depending on environmental cues. In response to surface stress or an alkaline pH, transcription of the leaderless variant of the *sigE* transcript is upregulated while the leadered variant, usually selected for during normal growth, is repressed (112). This genome structure is conserved in other mycobacteria, suggesting that this regulatory mechanism evolved early in mycobacteria history.

The translation of the upstream ORF can similarly affect downstream expression through translational coupling (113). A high percentage of leaderless small proteins are coupled to downstream genes in mycobacteria (79). Since translation of the first ORF can be initiated with a 70S ribosome, this may allow for increased expression of the entire operon under conditions of increased 70S concentration. 70S ribosomes can reinitiate downstream without dissociating by scanning using unidimensional diffusion to search for an SD sequence (114). This scanning can take place in the absence of initiation factors if fMet-tRNA^{fMet} is present. Therefore, under conditions selecting for lmRNAs, translational coupling through reinitiation could allow the entire operon to maintain efficient expression.

One example in which the 70S scanning mechanism appears to be essential for downstream expression occurs in *E. coli ptrB* mRNA encoding a serine protease. The *ptrB* transcript contains a 5′-uAUG upstream of its canonical start site, and mutation of this 5′-uAUG drastically reduces downstream *ptrB* expression by >90% (115). The 5′-uAUG does not act as an initiation codon for translation, but rather as a 70S loading site, necessary during optimal growth conditions due to *ptrB*'s weak SD sequence and also to ensure continued translation during stress. In this case, regulation of the mRNA is dependent upon the inherent ability of a 70S ribosome to recognize a 5′-AUG via the 70S lmRNA initiation pathway (115).

Apart from initiating translation, an upstream leaderless AUG can elicit indirect effects from a bound ribosome. 70S ribosome binding at the 5′ terminus can stabilize the mRNA (2); block the action of exonucleases (116); prevent premature transcriptional termination (117); or block the binding of other ribosomes or low-molecular-weight effectors such as amino acids, coenzymes, vitamins, or cyclic di-GMP whose binding could affect the mRNA secondary structure. Increasing the mRNA's half-life or protecting it from degradation will have a positive impact on surrounding ORFs. However, prohibiting ribosomal loading through steric hindrance will ultimately reduce downstream expression. Furthermore, occupying an area on the mRNA can also affect secondary structure, which in turn can lead to opening or occluding downstream ribosomal binding regions. Therefore, upstream ribosomal binding through recognition of the 5′-AUG can indirectly modulate the expression of the other cistrons in the operon in a variety of ways.

These regulation mechanisms are able to utilize an intrinsic property that may be an ancestral remnant used to recognize lmRNAs for their own specialized purposes. This "recycling" of inherent interaction mechanisms has already been shown: aside from assisting in translation initiation, the aSD sequence of the ribosome can interact with the SD sequence within coding sequences in archaea to induce pausing (113) and in bacteria for programmed frameshifting (114). In this way, intrinsic properties of the ribosome are used not only for initiation but during elongation as well.

PERSPECTIVE

Although it is impossible to definitively conclude the evolutionary origins of any biological process, there is substantial evidence to suggest that lmRNAs are ancestral in nature. From their ability to be translated by ribosomes of any domain of life, the utilization of a scaled-down mechanism of initiation, and their prevalence in organisms close to the root of the phylogenetic tree and in extremophiles, which are resistant to conditions reflecting the traits of prehistoric times, lmRNAs appear to predate SD-led mRNAs. lmRNAs can provide insight into the minimal requirements for translation. More-evolved prokaryotes and eukaryotes appear to have built upon the primitive method by increasing complexity of mRNA structure and number of initiation factors and other ancillary components. However, ribosomes have retained the intrinsic ability to recognize and bind the 5′-AUG signature of lmRNAs, allowing them to remain active in today's cells. This suggests that ribosomes may inspect the 5′ termini of all mRNAs as a remnant of an ancestral mechanism that has evolved into the eukaryotic mechanism of 5′-cap binding. The existence of capped lmRNAs in the protozoan parasite *Giardia lamblia* that use the lmRNA recognition mechanism without ribosome scanning might provide a bridge between prokaryotic and eukaryotic mechanisms with parallels to archaea (118). Furthermore, 5′-end recognition may continue to be utilized in bacteria and archaea for ribosomal loading for downstream translation or to protect the mRNA from degradation. These observations illustrate the importance of studying lmRNAs to provide insight into the evolution and function of the ribosome.

lmRNAs are not simply a functional relic of ancient times. They are still being used to control translational efficiency and to modulate the stress response. The ability to withstand adverse conditions makes lmRNAs beneficial in combating stress and transitioning into dormancy. The use of lmRNAs provides a way to control protein production without the need for auxiliary factors, such as small RNAs or RNA-binding proteins. Therefore, no further energy expenditure is required to modulate translational efficiency, making lmRNAs a valuable tool during cellular shutdown. Considering that the threat of antibiotic resistance and persistence formation is increasing, elucidating the role of lmRNA and the mechanisms underlying its translation will be crucial in controlling bacterial infections. As the vast numbers of lmRNAs throughout bacteria and archaea will be identified, further examples of their various roles in multiple cellular processes are expected to be elucidated, which may provide insights into the intrinsic abilities of the ribosome.

Citation. Beck HJ, Moll I. 2018. Leaderless mRNAs in the spotlight: ancient but not outdated! Microbiol Spectrum 6(4): RWR-0016-2017.

References

1. Shine J, Dalgarno L. 1974. The 3′-terminal sequence of *Escherichia coli* 16S ribosomal RNA: complementarity to nonsense triplets and ribosome binding sites. *Proc Natl Acad Sci U S A* **71:**1342–1346.

2. Komarova AV, Tchufistova LS, Dreyfus M, Boni IV. 2005. AU-rich sequences within 5′ untranslated leaders enhance translation and stabilize mRNA in *Escherichia coli*. *J Bacteriol* **187:**1344–1349.

3. Boni IV, Isaeva DM, Musychenko ML, Tzareva NV. 1991. Ribosome-messenger recognition: mRNA target sites for ribosomal protein S1. *Nucleic Acids Res* **19:**155–162.

4. Grill S, Gualerzi CO, Londei P, Bläsi U. 2000. Selective stimulation of translation of leaderless mRNA by initiation factor 2: evolutionary implications for translation. *EMBO J* **19:**4101–4110.

5. Chang B, Halgamuge S, Tang SL. 2006. Analysis of SD sequences in completed microbial genomes: non-SD-led genes are as common as SD-led genes. *Gene* **373:**90–99.

6. Zheng X, Hu GQ, She ZS, Zhu H. 2011. Leaderless genes in bacteria: clue to the evolution of translation initiation mechanisms in prokaryotes. *BMC Genomics* **12:**361.

7. Srivastava A, Gogoi P, Deka B, Goswami S, Kanaujia SP. 2016. *In silico* analysis of 5′-UTRs highlights the prevalence of Shine-Dalgarno and leaderless-dependent mechanisms of translation initiation in bacteria and archaea, respectively. *J Theor Biol* **402:**54–61.

8. Montoya J, Ojala D, Attardi G. 1981. Distinctive features of the 5′-terminal sequences of the human mitochondrial mRNAs. *Nature* **290:**465–470.

9. Jones CN, Wilkinson KA, Hung KT, Weeks KM, Spremulli LL. 2008. Lack of secondary structure characterizes the 5′ ends of mammalian mitochondrial mRNAs. *RNA* **14:**862–871.

10. Torarinsson E, Klenk HP, Garrett RA. 2005. Divergent transcriptional and translational signals in Archaea. *Environ Microbiol* **7:**47–54.

11. Tolstrup N, Sensen CW, Garrett RA, Clausen IG. 2000. Two different and highly organized mechanisms of translation initiation in the archaeon *Sulfolobus solfataricus*. *Extremophiles* **4:**175–179.

12. Slupska MM, King AG, Fitz-Gibbon S, Besemer J, Borodovsky M, Miller JH. 2001. Leaderless transcripts of the crenarchaeal hyperthermophile *Pyrobaculum aerophilum*. *J Mol Biol* **309:**347–360.

13. Cho S, Kim MS, Jeong Y, Lee BR, Lee JH, Kang SG, Cho BK. 2017. Genome-wide primary transcriptome analysis of H_2-producing archaeon *Thermococcus onnurineus* NA1. *Sci Rep* **7:**43044.

14. Beck HJ, Fleming IMC, Janssen GR. 2016. 5′-Terminal AUGs in *Escherichia coli* mRNAs with Shine-Dalgarno sequences: identification and analysis of their roles in non-canonical translation initiation. *PLoS One* **11:**e0160144.

15. Sullivan MJ, Curson AR, Shearer N, Todd JD, Green RT, Johnston AW. 2011. Unusual regulation of a leaderless operon involved in the catabolism of dimethylsulfoniopropionate in *Rhodobacter sphaeroides*. *PLoS One* **6:**e15972.

16. Tang W, Wu Y, Li M, Wang J, Mei S, Tang B, Tang XF. 2016. Alternative translation initiation of a haloarchaeal serine protease transcript containing two in-frame start codons. *J Bacteriol* **198:**1892–1901.

17. Vesper O, Amitai S, Belitsky M, Byrgazov K, Kaberdina AC, Engelberg-Kulka H, Moll I. 2011. Selective translation of leaderless mRNAs by specialized ribosomes generated by MazF in *Escherichia coli*. *Cell* **147:**147–157.

18. Sauert M, Wolfinger MT, Vesper O, Müller C, Byrgazov K, Moll I. 2016. The MazF-regulon: a toolbox for the post-transcriptional stress response in *Escherichia coli*. *Nucleic Acids Res* **44:**6660–6675.

19. Hussain T, Llácer JL, Wimberly BT, Kieft JS, Ramakrishnan V. 2016. Large-scale movements of IF3 and tRNA during bacterial translation initiation. *Cell* **167:**133–144.e13.

20. Caban K, Pavlov M, Ehrenberg M, Gonzalez RL Jr. 2017. A conformational switch in initiation factor 2 controls the fidelity of translation initiation in bacteria. *Nat Commun* **8:**1475.

21. Milón P, Rodnina MV. 2012. Kinetic control of translation initiation in bacteria. *Crit Rev Biochem Mol Biol* **47:**334–348.

22. Milón P, Maracci C, Filonava L, Gualerzi CO, Rodnina MV. 2012. Real-time assembly landscape of bacterial 30S translation initiation complex. *Nat Struct Mol Biol* **19:**609–615.

23. Antoun A, Pavlov MY, Lovmar M, Ehrenberg M. 2006. How initiation factors maximize the accuracy of tRNA selection in initiation of bacterial protein synthesis. *Mol Cell* **23:**183–193.

24. Simonetti A, Marzi S, Jenner L, Myasnikov A, Romby P, Yusupova G, Klaholz BP, Yusupov M. 2009. A structural view of translation initiation in bacteria. *Cell Mol Life Sci* **66:**423–436.

25. Antoun A, Pavlov MY, Lovmar M, Ehrenberg M. 2006. How initiation factors tune the rate of initiation of protein synthesis in bacteria. *EMBO J* **25:**2539–2550.

26. Dallas A, Noller HF. 2001. Interaction of translation initiation factor 3 with the 30S ribosomal subunit. *Mol Cell* **8:**855–864.

27. Marshall RA, Aitken CE, Puglisi JD. 2009. GTP hydrolysis by IF2 guides progression of the ribosome into elongation. *Mol Cell* **35:**37–47.

28. Sørensen MA, Fricke J, Pedersen S. 1998. Ribosomal protein S1 is required for translation of most, if not all, natural mRNAs in *Escherichia coli in vivo*. *J Mol Biol* **280**:561–569.

29. Byrgazov K, Grishkovskaya I, Arenz S, Coudevylle N, Temmel H, Wilson DN, Djinovic-Carugo K, Moll I. 2015. Structural basis for the interaction of protein S1 with the *Escherichia coli* ribosome. *Nucleic Acids Res* **43**:661–673.

30. Lauber MA, Rappsilber J, Reilly JP. 2012. Dynamics of ribosomal protein S1 on a bacterial ribosome with cross-linking and mass spectrometry. *Mol Cell Proteomics* **11**: 1965–1976.

31. Qu X, Lancaster L, Noller HF, Bustamante C, Tinoco I Jr. 2012. Ribosomal protein S1 unwinds double-stranded RNA in multiple steps. *Proc Natl Acad Sci U S A* **109**: 14458–14463.

32. Van Duin J, Wijnands R. 1981. The function of ribosomal protein S21 in protein synthesis. *Eur J Biochem* **118**:615–619.

33. Backendorf C, Ravensbergen CJ, Van der Plas J, van Boom JH, Veeneman G, Van Duin J. 1981. Basepairing potential of the 3′ terminus of 16S RNA: dependence on the functional state of the 30S subunit and the presence of protein S21. *Nucleic Acids Res* **9**:1425–1444.

34. Odom OW, Stöffler G, Hardesty B. 1984. Movement of the 3′-end of 16 S RNA towards S21 during activation of 30 S ribosomal subunits. *FEBS Lett* **173**:155–158.

35. Moll I, Resch A, Bläsi U. 1998. Discrimination of 5′-terminal start codons by translation initiation factor 3 is mediated by ribosomal protein S1. *FEBS Lett* **436**: 213–217.

36. Tedin K, Resch A, Bläsi U. 1997. Requirements for ribosomal protein S1 for translation initiation of mRNAs with and without a 5′ leader sequence. *Mol Microbiol* **25**:189–199.

37. Shean CS, Gottesman ME. 1992. Translation of the prophage λ *cl* transcript. *Cell* **70**:513–522.

38. Moll I, Grill S, Gründling A, Bläsi U. 2002. Effects of ribosomal proteins S1, S2 and the DeaD/CsdA DEAD-box helicase on translation of leaderless and canonical mRNAs in *Escherichia coli*. *Mol Microbiol* **44**:1387–1396.

39. Delvillani F, Papiani G, Dehò G, Briani F. 2011. S1 ribosomal protein and the interplay between translation and mRNA decay. *Nucleic Acids Res* **39**:7702–7715.

40. Duval M, Korepanov A, Fuchsbauer O, Fechter P, Haller A, Fabbretti A, Choulier L, Micura R, Klaholz BP, Romby P, Springer M, Marzi S. 2013. *Escherichia coli* ribosomal protein S1 unfolds structured mRNAs onto the ribosome for active translation initiation. *PLoS Biol* **11**:e1001731.

41. Salah P, Bisaglia M, Aliprandi P, Uzan M, Sizun C, Bontems F. 2009. Probing the relationship between Gram-negative and Gram-positive S1 proteins by sequence analysis. *Nucleic Acids Res* **37**:5578–5588.

42. Byrgazov K, Manoharadas S, Kaberdina AC, Vesper O, Moll I. 2012. Direct interaction of the N-terminal domain of ribosomal protein S1 with protein S2 in *Escherichia coli*. *PLoS One* **7**:e32702.

43. Aseev LV, Chugunov AO, Efremov RG, Boni IV. 2013. A single missense mutation in a coiled-coil domain of *Escherichia coli* ribosomal protein S2 confers a thermo-sensitive phenotype that can be suppressed by ribosomal protein S1. *J Bacteriol* **195**:95–104.

44. Moll I, Hirokawa G, Kiel MC, Kaji A, Bläsi U. 2004. Translation initiation with 70S ribosomes: an alternative pathway for leaderless mRNAs. *Nucleic Acids Res* **32**:3354–3363.

45. Grill S, Moll I, Hasenöhrl D, Gualerzi CO, Bläsi U. 2001. Modulation of ribosomal recruitment to 5′-terminal start codons by translation initiation factors IF2 and IF3. *FEBS Lett* **495**:167–171.

46. O'Donnell SM, Janssen GR. 2002. Leaderless mRNAs bind 70S ribosomes more strongly than 30S ribosomal subunits in *Escherichia coli*. *J Bacteriol* **184**: 6730–6733.

47. Tedin K, Moll I, Grill S, Resch A, Graschopf A, Gualerzi CO, Bläsi U. 1999. Translation initiation factor 3 antagonizes authentic start codon selection on leaderless mRNAs. *Mol Microbiol* **31**:67–77.

48. Maar D, Liveris D, Sussman JK, Ringquist S, Moll I, Heredia N, Kil A, Bläsi U, Schwartz I, Simons RW. 2008. A single mutation in the IF3 N-terminal domain perturbs the fidelity of translation initiation at three levels. *J Mol Biol* **383**:937–944.

49. O'Connor M, Gregory ST, Rajbhandary UL, Dahlberg AE. 2001. Altered discrimination of start codons and initiator tRNAs by mutant initiation factor 3. *RNA* **7**: 969–978.

50. Moazed D, Samaha RR, Gualerzi C, Noller HF. 1995. Specific protection of 16 S rRNA by translational initiation factors. *J Mol Biol* **248**:207–210.

51. Howe JG, Hershey JW. 1983. Initiation factor and ribosome levels are coordinately controlled in *Escherichia coli* growing at different rates. *J Biol Chem* **258**:1954–1959.

52. Liveris D, Klotsky RA, Schwartz I. 1991. Growth rate regulation of translation initiation factor IF3 biosynthesis in *Escherichia coli*. *J Bacteriol* **173**:3888–3893.

53. Jones PG, VanBogelen RA, Neidhardt FC. 1987. Induction of proteins in response to low temperature in *Escherichia coli*. *J Bacteriol* **169**:2092–2095.

54. Hinnebusch AG. 2017. Structural insights into the mechanism of scanning and start codon recognition in eukaryotic translation initiation. *Trends Biochem Sci* **42**: 589–611.

55. Balakin AG, Skripkin EA, Shatsky IN, Bogdanov AA, Belozersky AN. 1992. Unusual ribosome binding properties of mRNA encoding bacteriophage λ repressor. *Nucleic Acids Res* **20**:563–571.

56. Udagawa T, Shimizu Y, Ueda T. 2004. Evidence for the translation initiation of leaderless mRNAs by the intact 70 S ribosome without its dissociation into subunits in eubacteria. *J Biol Chem* **279**:8539–8546.

57. Benelli D, Maone E, Londei P. 2003. Two different mechanisms for ribosome/mRNA interaction in archaeal translation initiation. *Mol Microbiol* 50:635–643.

58. Hasenöhrl D, Fabbretti A, Londei P, Gualerzi CO, Bläsi U. 2009. Translation initiation complex formation in the crenarchaeon *Sulfolobus solfataricus*. *RNA* 15:2288–2298.

59. Bell SD, Jackson SP. 1998. Transcription and translation in Archaea: a mosaic of eukaryal and bacterial features. *Trends Microbiol* 6:222–228.

60. Dennis PP. 1997. Ancient ciphers: translation in Archaea. *Cell* 89:1007–1010.

61. La Teana A, Benelli D, Londei P, Bläsi U. 2013. Translation initiation in the crenarchaeon *Sulfolobus solfataricus*: eukaryotic features but bacterial route. *Biochem Soc Trans* 41:350–355.

62. Benelli D, Londei P. 2011. Translation initiation in Archaea: conserved and domain-specific features. *Biochem Soc Trans* 39:89–93.

63. Hasenöhrl D, Benelli D, Barbazza A, Londei P, Bläsi U. 2006. *Sulfolobus solfataricus* translation initiation factor 1 stimulates translation initiation complex formation. *RNA* 12:674–682.

64. Pedullà N, Palermo R, Hasenöhrl D, Bläsi U, Cammarano P, Londei P. 2005. The archaeal eIF2 homologue: functional properties of an ancient translation initiation factor. *Nucleic Acids Res* 33:1804–1812.

65. Arkhipova V, Stolboushkina E, Kravchenko O, Kljashtorny V, Gabdulkhakov A, Garber M, Nikonov S, Märtens B, Bläsi U, Nikonov O. 2015. Binding of the 5′-triphosphate end of mRNA to the γ-subunit of translation initiation factor 2 of the crenarchaeon *Sulfolobus solfataricus*. *J Mol Biol* 427:3086–3095.

66. Hasenöhrl D, Lombo T, Kaberdin V, Londei P, Bläsi U. 2008. Translation initiation factor a/eIF2(-γ) counteracts 5′ to 3′ mRNA decay in the archaeon *Sulfolobus solfataricus*. *Proc Natl Acad Sci U S A* 105:2146–2150.

67. Londei P. 2005. Evolution of translational initiation: new insights from the archaea. *FEMS Microbiol Rev* 29:185–200.

68. Yatime L, Schmitt E, Blanquet S, Mechulam Y. 2004. Functional molecular mapping of archaeal translation initiation factor 2. *J Biol Chem* 279:15984–15993.

69. Benelli D, Marzi S, Mancone C, Alonzi T, la Teana A, Londei P. 2009. Function and ribosomal localization of aIF6, a translational regulator shared by archaea and eukarya. *Nucleic Acids Res* 37:256–267.

70. Resch A, Tedin K, Gründling A, Mündlein A, Bläsi U. 1996. Downstream box-anti-downstream box interactions are dispensable for translation initiation of leaderless mRNAs. *EMBO J* 15:4740–4748.

71. Moll I, Huber M, Grill S, Sairafi P, Mueller F, Brimacombe R, Londei P, Bläsi U. 2001. Evidence against an interaction between the mRNA downstream box and 16S rRNA in translation initiation. *J Bacteriol* 183:3499–3505.

72. Martin-Farmer J, Janssen GR. 1999. A downstream CA repeat sequence increases translation from leadered and unleadered mRNA in *Escherichia coli*. *Mol Microbiol* 31:1025–1038.

73. Brock JE, Paz RL, Cottle P, Janssen GR. 2007. Naturally occurring adenines within mRNA coding sequences affect ribosome binding and expression in *Escherichia coli*. *J Bacteriol* 189:501–510.

74. Giliberti J, O'Donnell S, Etten WJ, Janssen GR, Van Etten WJ, Janssen GR. 2012. A 5′-terminal phosphate is required for stable ternary complex formation and translation of leaderless mRNA in *Escherichia coli*. *RNA* 18:508–518.

75. Brock JE, Pourshahian S, Giliberti J, Limbach PA, Janssen GR. 2008. Ribosomes bind leaderless mRNA in *Escherichia coli* through recognition of their 5′-terminal AUG. *RNA* 14:2159–2169.

76. Van Etten WJ, Janssen GR. 1998. An AUG initiation codon, not codon-anticodon complementarity, is required for the translation of unleadered mRNA in *Escherichia coli*. *Mol Microbiol* 27:987–1001.

77. O'Donnell SM, Janssen GR. 2001. The initiation codon affects ribosome binding and translational efficiency in *Escherichia coli* of *cI* mRNA with or without the 5′ untranslated leader. *J Bacteriol* 183:1277–1283.

78. Hering O, Brenneis M, Beer J, Suess B, Soppa J. 2009. A novel mechanism for translation initiation operates in haloarchaea. *Mol Microbiol* 71:1451–1463.

79. Shell SS, Wang J, Lapierre P, Mir M, Chase MR, Pyle MM, Gawande R, Ahmad R, Sarracino DA, Ioerger TR, Fortune SM, Derbyshire KM, Wade JT, Gray TA. 2015. Leaderless transcripts and small proteins are common features of the mycobacterial translational landscape. *PLoS Genet* 11:e1005641.

80. Krishnan KM, Van Etten WJ III, Janssen GR. 2010. Proximity of the start codon to a leaderless mRNA's 5′ terminus is a strong positive determinant of ribosome binding and expression in *Escherichia coli*. *J Bacteriol* 192:6482–6485.

81. Wu CJ, Janssen GR. 1996. Translation of *vph* mRNA in *Streptomyces lividans* and *Escherichia coli* after removal of the 5′ untranslated leader. *Mol Microbiol* 22:339–355.

82. Wu CJ, Janssen GR. 1997. Expression of a streptomycete leaderless mRNA encoding chloramphenicol acetyltransferase in *Escherichia coli*. *J Bacteriol* 179:6824–6830.

83. Kohler R, Mooney RA, Mills DJ, Landick R, Cramer P. 2017. Architecture of a transcribing-translating expressome. *Science* 356:194–197.

84. Schrader JM, Zhou B, Li GW, Lasker K, Childers WS, Williams B, Long T, Crosson S, McAdams HH, Weissman JS, Shapiro L. 2014. The coding and noncoding architecture of the *Caulobacter crescentus* genome. *PLoS Genet* 10:e1004463.

85. Sartorius-Neef S, Pfeifer F. 2004. *In vivo* studies on putative Shine-Dalgarno sequences of the halophilic archaeon *Halobacterium salinarum*. *Mol Microbiol* 51:579–588.

86. Condò I, Ciammaruconi A, Benelli D, Ruggero D, Londei P. 1999. *cis*-Acting signals controlling translational initiation in the thermophilic archaeon *Sulfolobus solfataricus*. *Mol Microbiol* 34:377–384.

87. Amitai S, Kolodkin-Gal I, Hananya-Meltabashi M, Sacher A, Engelberg-Kulka H. 2009. *Escherichia coli* MazF leads to the simultaneous selective synthesis of both "death proteins" and "survival proteins." *PLoS Genet* **5**:e1000390.

88. Andreev DE, Terenin IM, Dunaevsky YE, Dmitriev SE, Shatsky IN. 2006. A leaderless mRNA can bind to mammalian 80S ribosomes and direct polypeptide synthesis in the absence of translation initiation factors. *Mol Cell Biol* **26**:3164–3169.

89. Grill S, Moll I, Giuliodori AM, Gualerzi CO, Bläsi U. 2002. Temperature-dependent translation of leaderless and canonical mRNAs in *Escherichia coli*. *FEMS Microbiol Lett* **211**:161–167.

90. Chin K, Shean CS, Gottesman ME. 1993. Resistance of λ *cI* translation to antibiotics that inhibit translation initiation. *J Bacteriol* **175**:7471–7473.

91. de Groot A, Roche D, Fernandez B, Ludanyi M, Cruveiller S, Pignol D, Vallenet D, Armengaud J, Blanchard L. 2014. RNA sequencing and proteogenomics reveal the importance of leaderless mRNAs in the radiation-tolerant bacterium *Deinococcus deserti*. *Genome Biol Evol* **6**: 932–948.

92. Bouthier de la Tour C, Blanchard L, Dulermo R, Ludanyi M, Devigne A, Armengaud J, Sommer S, de Groot A. 2015. The abundant and essential HU proteins in *Deinococcus deserti* and *Deinococcus radiodurans* are translated from leaderless mRNA. *Microbiology* **161**: 2410–2422.

93. Baumeister R, Flache P, Melefors O, von Gabain A, Hillen W. 1991. Lack of a 5′ non-coding region in Tn*1721* encoded *tetR* mRNA is associated with a low efficiency of translation and a short half-life in *Escherichia coli*. *Nucleic Acids Res* **19**:4595–4600.

94. Jones RL III, Jaskula JC, Janssen GR. 1992. *In vivo* translational start site selection on leaderless mRNA transcribed from the *Streptomyces fradiae aph* gene. *J Bacteriol* **174**:4753–4760.

95. August PR, Flickinger MC, Sherman DH. 1994. Cloning and analysis of a locus (*mcr*) involved in mitomycin C resistance in *Streptomyces lavendulae*. *J Bacteriol* **176**:4448–4454.

96. Schluenzen F, Takemoto C, Wilson DN, Kaminishi T, Harms JM, Hanawa-Suetsugu K, Szaflarski W, Kawazoe M, Shirouzu M, Nierhaus KH, Yokoyama S, Fucini P. 2006. The antibiotic kasugamycin mimics mRNA nucleotides to destabilize tRNA binding and inhibit canonical translation initiation. *Nat Struct Mol Biol* **13**: 871–878.

97. Schuwirth BS, Day JM, Hau CW, Janssen GR, Dahlberg AE, Cate JH, Vila-Sanjurjo A. 2006. Structural analysis of kasugamycin inhibition of translation. *Nat Struct Mol Biol* **13**:879–886.

98. Moll I, Bläsi U. 2002. Differential inhibition of 30S and 70S translation initiation complexes on leaderless mRNA by kasugamycin. *Biochem Biophys Res Commun* **297**: 1021–1026.

99. Ikeno S, Yamane Y, Ohishi Y, Kinoshita N, Hamada M, Tsuchiya KS, Hori M. 2000. ABC transporter genes, *kasKLM*, responsible for self-resistance of a kasugamycin producer strain. *J Antibiot (Tokyo)* **53**:373–384.

100. Müller C, Sokol L, Vesper O, Sauert M, Moll I. 2016. Insights into the stress response triggered by kasugamycin in *Escherichia coli*. *Antibiotics (Basel)* **5**:E19.

101. Hazan R, Sat B, Engelberg-Kulka H. 2004. *Escherichia coli mazEF*-mediated cell death is triggered by various stressful conditions. *J Bacteriol* **186**:3663–3669.

102. Sat B, Hazan R, Fisher T, Khaner H, Glaser G, Engelberg-Kulka H. 2001. Programmed cell death in *Escherichia coli*: some antibiotics can trigger *mazEF* lethality. *J Bacteriol* **183**:2041–2045.

103. Aizenman E, Engelberg-Kulka H, Glaser G. 1996. An *Escherichia coli* chromosomal "addiction module" regulated by guanosine [corrected] 3′,5′-bispyrophosphate: a model for programmed bacterial cell death. *Proc Natl Acad Sci U S A* **93**:6059–6063.

104. Zhang Y, Zhang J, Hara H, Kato I, Inouye M. 2005. Insights into the mRNA cleavage mechanism by MazF, an mRNA interferase. *J Biol Chem* **280**:3143–3150.

105. Temmel H, Müller C, Sauert M, Vesper O, Reiss A, Popow J, Martinez J, Moll I. 2017. The RNA ligase RtcB reverses MazF-induced ribosome heterogeneity in *Escherichia coli*. *Nucleic Acids Res* **45**:4708–4721.

106. Cortes T, Schubert OT, Rose G, Arnvig KB, Comas I, Aebersold R, Young DB. 2013. Genome-wide mapping of transcriptional start sites defines an extensive leaderless transcriptome in *Mycobacterium tuberculosis*. *Cell Rep* **5**:1121–1131.

107. Schifano JM, Edifor R, Sharp JD, Ouyang M, Konkimalla A, Husson RN, Woychik NA. 2013. Mycobacterial toxin MazF-mt6 inhibits translation through cleavage of 23S rRNA at the ribosomal A site. *Proc Natl Acad Sci U S A* **110**:8501–8506.

108. Schifano JM, Vvedenskaya IO, Knoblauch JG, Ouyang M, Nickels BE, Woychik NA. 2014. An RNA-seq method for defining endoribonuclease cleavage specificity identifies dual rRNA substrates for toxin MazF-mt3. *Nat Commun* **5**:3538.

109. Di Martino ML, Romilly C, Wagner EG, Colonna B, Prosseda G. 2016. One gene and two proteins: a leaderless mRNA supports the translation of a shorter form of the *Shigella* VirF regulator. *mBio* **7**:e01860-16.

110. Kochetov AV. 2008. Alternative translation start sites and hidden coding potential of eukaryotic mRNAs. *BioEssays* **30**:683–691.

111. Brenneis M, Hering O, Lange C, Soppa J. 2007. Experimental characterization of *cis*-acting elements important for translation and transcription in halophilic archaea. *PLoS Genet* **3**:e229.

112. Donà V, Rodrigue S, Dainese E, Palù G, Gaudreau L, Manganelli R, Provvedi R. 2008. Evidence of complex transcriptional, translational, and posttranslational regulation of the extracytoplasmic function sigma factor σ^E in *Mycobacterium tuberculosis*. *J Bacteriol* **190**:5963–5971.

113. Rex G, Surin B, Besse G, Schneppe B, McCarthy JE. 1994. The mechanism of translational coupling in *Escherichia coli*. Higher order structure in the *atpHA* mRNA

acts as a conformational switch regulating the access of *de novo* initiating ribosomes. *J Biol Chem* **269**:18118–18127.

114. Yamamoto H, Wittek D, Gupta R, Qin B, Ueda T, Krause R, Yamamoto K, Albrecht R, Pech M, Nierhaus KH. 2016. 70S-scanning initiation is a novel and frequent initiation mode of ribosomal translation in bacteria. *Proc Natl Acad Sci U S A* **113**:E1180–E1189.

115. Beck HJ, Janssen GR. 2017. Novel translation initiation regulation mechanism in *Escherichia coli ptrB* mediated by a 5′-terminal AUG. *J Bacteriol* **199**:e00091-17.

116. Deana A, Belasco JG. 2005. Lost in translation: the influence of ribosomes on bacterial mRNA decay. *Genes Dev* **19**:2526–2533.

117. Eriksen M, Sneppen K, Pedersen S, Mitarai N. 2017. Occlusion of the ribosome binding site connects the translational initiation frequency, mRNA stability and premature transcription termination. *Front Microbiol* **8**:362.

118. Li L, Wang CC. 2004. Capped mRNA with a single nucleotide leader is optimally translated in a primitive eukaryote, *Giardia lamblia*. *J Biol Chem* **279**:14656–14664.

119. Schlüter JP, Reinkensmeier J, Barnett MJ, Lang C, Krol E, Giegerich R, Long SR, Becker A. 2013. Global mapping of transcription start sites and promoter motifs in the symbiotic α-proteobacterium *Sinorhizobium meliloti* 1021. *BMC Genomics* **14**:156.

120. Qiu Y, Cho BK, Park YS, Lovley D, Palsson BØ, Zengler K. 2010. Structural and operational complexity of the *Geobacter sulfurreducens* genome. *Genome Res* **20**:1304–1311.

121. Porcelli I, Reuter M, Pearson BM, Wilhelm T, van Vliet AH. 2013. Parallel evolution of genome structure and transcriptional landscape in the Epsilonproteobacteria. *BMC Genomics* **14**:616.

122. Dugar G, Herbig A, Förstner KU, Heidrich N, Reinhardt R, Nieselt K, Sharma CM. 2013. High-resolution transcriptome maps reveal strain-specific regulatory features of multiple *Campylobacter jejuni* isolates. *PLoS Genet* **9**:e1003495.

123. Romero DA, Hasan AH, Lin YF, Kime L, Ruiz-Larrabeiti O, Urem M, Bucca G, Mamanova L, Laing EE, van Wezel GP, Smith CP, Kaberdin VR, McDowall KJ. 2014. A comparison of key aspects of gene regulation in *Streptomyces coelicolor* and *Escherichia coli* using nucleotide-resolution transcription maps produced in parallel by global and differential RNA sequencing. *Mol Microbiol* **94**:963–987.

124. Thomason MK, Bischler T, Eisenbart SK, Förstner KU, Zhang A, Herbig A, Nieselt K, Sharma CM, Storz G. 2015. Global transcriptional start site mapping using differential RNA sequencing reveals novel antisense RNAs in *Escherichia coli*. *J Bacteriol* **197**:18–28.

125. Seo JH, Hong JS, Kim D, Cho BK, Huang TW, Tsai SF, Palsson BO, Charusanti P. 2012. Multiple-omic data analysis of *Klebsiella pneumoniae* MGH 78578 reveals its transcriptional architecture and regulatory features. *BMC Genomics* **13**:679.

126. Kröger C, Dillon SC, Cameron ADS, Papenfort K, Sivasankaran SK, Hokamp K, Chao Y, Sittka A, Hébrard M, Händler K, Colgan A, Leekitcharoenphon P, Langridge GC, Lohan AJ, Loftus B, Lucchini S, Ussery DW, Dorman CJ, Thomson NR, Vogel J, Hinton JC. 2012. The transcriptional landscape and small RNAs of *Salmonella enterica* serovar Typhimurium. *Proc Natl Acad Sci U S A* **109**:E1277–E1286.

127. Alkhateeb RS, Vorhölter FJ, Rückert C, Mentz A, Wibberg D, Hublik G, Niehaus K, Pühler A. 2016. Genome wide transcription start sites analysis of *Xanthomonas campestris* pv. campestris B100 with insights into the *gum* gene cluster directing the biosynthesis of the exopolysaccharide xanthan. *J Biotechnol* **225**:18–28.

128. Sahr T, Rusniok C, Dervins-Ravault D, Sismeiro O, Coppee JY, Buchrieser C. 2012. Deep sequencing defines the transcriptional map of *L. pneumophila* and identifies growth phase-dependent regulated ncRNAs implicated in virulence. *RNA Biol* **9**:503–519.

129. Venkataramanan KP, Min L, Hou S, Jones SW, Ralston MT, Lee KH, Papoutsakis ET. 2015. Complex and extensive post-transcriptional regulation revealed by integrative proteomic and transcriptomic analysis of metabolite stress response in *Clostridium acetobutylicum*. *Biotechnol Biofuels* **8**:81.

130. Liao Y, Huang L, Wang B, Zhou F, Pan L. 2015. The global transcriptional landscape of *Bacillus amyloliquefaciens* XH7 and high-throughput screening of strong promoters based on RNA-seq data. *Gene* **571**:252–262.

131. Ignatov D, Malakho S, Majorov K, Skvortsov T, Apt A, Azhikina T. 2013. RNA-Seq analysis of *Mycobacterium avium* non-coding transcriptome. *PLoS One* **8**:e74209.

132. Schwientek P, Neshat A, Kalinowski J, Klein A, Rückert C, Schneiker-Bekel S, Wendler S, Stoye J, Pühler A. 2014. Improving the genome annotation of the acarbose producer *Actinoplanes* sp. SE50/110 by sequencing enriched 5′-ends of primary transcripts. *J Biotechnol* **190**:85–95.

133. Albersmeier A, Pfeifer-Sancar K, Rückert C, Kalinowski J. 2017. Genome-wide determination of transcription start sites reveals new insights into promoter structures in the actinomycete *Corynebacterium glutamicum*. *J Biotechnol* **257**:99–109.

134. Pfeifer-Sancar K, Mentz A, Rückert C, Kalinowski J. 2013. Comprehensive analysis of the *Corynebacterium glutamicum* transcriptome using an improved RNAseq technique. *BMC Genomics* **14**:888.

135. Bauer JS, Fillinger S, Förstner K, Herbig A, Jones AC, Flinspach K, Sharma C, Gross H, Nieselt K, Apel AK. 2017. dRNA-seq transcriptional profiling of the FK506 biosynthetic gene cluster in *Streptomyces tsukubaensis* NRRL18488 and general analysis of the transcriptome. *RNA Biol* **14**:1617–1626.

136. Voigt K, Sharma CM, Mitschke J, Lambrecht SJ, Voß B, Hess WR, Steglich C. 2014. Comparative transcriptomics of two environmentally relevant cyanobacteria reveals unexpected transcriptome diversity. *ISME J* **8**:2056–2068.

137. Babski J, Haas KA, Näther-Schindler D, Pfeiffer F, Förstner KU, Hammelmann M, Hilker R, Becker A,

Sharma CM, Marchfelder A, Soppa J. 2016. Genome-wide identification of transcriptional start sites in the haloarchaeon *Haloferax volcanii* based on differential RNA-Seq (dRNA-Seq). *BMC Genomics* **17**:629.

138. Toffano-Nioche C, Ott A, Crozat E, Nguyen AN, Zytnicki M, Leclerc F, Forterre P, Bouloc P, Gautheret D. 2013. RNA at 92°C: the non-coding transcriptome of the hyperthermophilic archaeon *Pyrococcus abyssi*. *RNA Biol* **10**:1211–1220.

139. Wurtzel O, Sapra R, Chen F, Zhu Y, Simmons BA, Sorek R. 2010. A single-base resolution map of an archaeal transcriptome. *Genome Res* **20**:133–141.

Cis-Encoded Base Pairing RNAs

III

Regulating with RNA in Bacteria and Archaea
Edited by Gisela Storz and Kai Papenfort
© 2018 American Society for Microbiology, Washington, DC
doi:10.1128/microbiolspec.RWR-0030-2018

Type I Toxin-Antitoxin Systems: Regulating Toxin Expression via Shine-Dalgarno Sequence Sequestration and Small RNA Binding

11

Sara Masachis[1] and Fabien Darfeuille[1]

INTRODUCTION

Toxin-antitoxin (TA) systems are, by definition, simple genetic loci composed of two genes: a toxin and an antitoxin that counteracts either the toxin's action or its expression. Usually, toxin synthesis leads to growth arrest or death of the bacterium that produces it by inhibiting essential cellular processes such as replication, translation, or cell division (1). Analogies can be made with other gene pairs with a similar bicistronic operon organization, such as bacteriocins, restriction-modification systems, and type VI secretion system effector/immunity proteins (2). For instance, they are often found within mobile genetic elements and pathogenicity islands. However, canonical TA systems generally act on the cell that produces them, and in contrast to bacteriocins, they are not involved in interbacterial competition. Thus, the presence of TA-encoded toxins in bacterial genomes strongly depends on the expression of their cognate antitoxins (3). TA systems can be classified into six different types depending on the nature and mode of action of the antitoxin (for a review, see reference 4). While the toxins are always proteins, antitoxins can be either proteins (types II, IV, V, and VI) or small RNAs (sRNAs, types I and III TA systems). Antitoxins can act by (i) sequestering the toxin (types II and III), (ii) inhibiting its expression (types I and V), or (iii) counterbalancing its activity (type IV). In this review, we will focus on type I TA systems, in which an antisense RNA plays the role of antitoxin to prevent the synthesis of its cognate toxin by directly base-pairing to the mRNA.

The *hok*/Sok and the *par* loci were the first type I TA systems discovered on the R1 and pAD1 plasmids, respectively, where they confer stability through post-segregational killing (PSK) (5–11). This complex mechanism relies on the differential stability between the stable, toxin-encoding mRNA and the labile antitoxin RNA. Consequently, upon loss of the TA locus (e.g., plasmid loss), antitoxin levels quickly drop while the stable toxin mRNA remains present, eventually leading to toxin synthesis and cell death. Thus, the toxin kills plasmid-free cells, whereas plasmid-containing cells are unaffected due to continuous *de novo* synthesis of the antitoxin. Although homologs of these TA systems were predicted on bacterial chromosomes (12–15), their role in this context remains enigmatic. It became even more intriguing with the later discovery of additional TA systems found exclusively on bacterial chromosomes (16–20), and with the identification of new type I TA systems during a systematic search for sRNA in bacteria (21). For instance, the *Escherichia coli* chromosome can contain up to 19 type I TA loci (22), but the function of only 3 of them has been reported (20, 23–25). Indeed, while toxicity could be easily assessed by toxin overexpression, single or even multiple TA deletions (in earlier studies) did not show any obvious phenotype (26, 27). The question of the function of TA systems, if any, when present on bacterial chromosomes is still elusive (Table 1).

Many excellent reviews covering the field of type I TA systems have already been published (10, 28–30). However, recent work on the characterization of a new

---

[1]ARNA Laboratory, INSERM U1212, CNRS UMR 5320, University of Bordeaux, F-33000 Bordeaux, France.

Table 1 Type I TA systems identified in bacteria

System	Bacteria	Present on:	Expression	Role	Copy number	Reference(s)
Gram Positive						
fst/RNAII	*Enterococcus faecalis*	pAD1, VRSAp; SaPlbov2 pathogenicity island	Constitutive	PSK	1	5, 7, 81, 82
fst-like/RNAII	*E. faecalis, Staphylococcus saprophyticus, Streptococcus mutans,* and *Lactobacillus casei*	Chromosome	Unknown	Carbohydrate metabolism (?)	1	14, 15, 83
sprA1/SprA1AS[a]	*Staphylococcus aureus* N315	νΣαβ pathogenicity island	Constitutive	Pathogenicity (?)	1	62, 99
sprG1/SprF1[b]	*S. aureus* N315	MGE[g] ΦSa3 pathogenicity island	Constitutive	Unknown	1	67
txpA/RatA	*Bacillus subtilis*	Skin element	Constitutive	PSK-like of Skin element, sporulation (?)	Up to 6	63, 97
bsrG/SR4	*B. subtilis*	SPβ prophage region	ResD O_2 limitation response	Unknown	1	64
bsrE/SR5	*B. subtilis*	Prophage-like region P6	Multistress responsive	Unknown	1	65, 100
bsrH/as-bsrH	*B. subtilis*	Skin element	ResD O_2 limitation response	Unknown	1	19
yonT/SR6	*B. subtilis*	SPβ prophage region	Multistress responsive	Unknown	1	68
lpt/RNAII	*Lactobacillus rhamnosus*	Plasmid	Constitutive	PSK (?)	1	84
CDS2517.1/RCd8	*Clostridium difficile*	Chromosome CRISPR (clustered regularly interspaced short palindromic repeat) 12 region	Coregulated with CRISPR	Unknown	Up to 6	66
Gram Negative						
hok/Sok	*Escherichia coli*	Plasmid R1	Constitutive	PSK	1	8, 10
hok/Sok	Enterobacteria[c]	Chromosome	Unknown	Persistence	Up to 15	12
hokB/SokB	*E. coli* K-12	Chromosome	ppGpp induced, Obg (?)	Persistence	1	23
srnB/SrnC-RNA	*E. coli*	Plasmid F	Constitutive	PSK	1	11
pndA/PndB-RNA	*E. coli*	Plasmid R483	Constitutive	PSK	1	11
Ldr/Rdl	*E. coli* K-12, *Salmonella,* and *Shigella*	Chromosome	Unknown	Unknown	Up to 10	17
Ibs/Sib	Enterobacteria[d] and *Helicobacter pylori*	Chromosome	Unknown	Unknown	Up to 7	16, 21
shoB/OhsC	*E. coli* and *Shigella*	Chromosome	Unknown	Unknown	1	16, 79
zorO/OrzO	*E. coli*	Chromosome	Unknown	Unknown	2	38, 59
symE/ISymR	*E. coli*	Chromosome	SOS response (LexA)	RNA recycling upon SOS (?)	1	36
tisB/IstR	Enterobacteria[e]	Chromosome	SOS response (LexA)	Persistence	1	18, 58
dinQ/AgrAB	*E. coli*	Chromosome	SOS response (LexA)	Nucleoid compaction	1	20, 60
dinQ-like/AgrAB	Enterobacteria[f]	Chromosome	Unknown	Unknown	1	60
ralR/RalA	*E. coli*	Cryptic prophage Rac	Unknown	Biofilm maintenance (?)	1	35
aapA1/IsoA1	*H. pylori*	Chromosome	Constitutive	Unknown	Up to 2	31
aapA/IsoA	*Helicobacter* and *Campylobacter*	Chromosome and plasmid	Unknown	Unknown	Up to 9	31

[a]This system has been predicted to be a homolog of *fst*/RNAII, initially discovered in *E. faecalis* (13–15).
[b]This system has been predicted to be a homolog of *txpA*/RatA, initially discovered in *B. subtilis* (22, 97).
[c]*Escherichia* (4–15), *Shigella* (7–12), *Enterobacter* (6), *Klebsiella* (2), *Serratia* (5), *Vibrio* (1), *Yersinia* (1), *Shewanella, Photobacterium,* and *Photorhabdus* (3). Numbers in parentheses indicate copy number.
[d]*Escherichia* (3–7), *Shigella* (3–6), *Citrobacter* (1), *Salmonella* (3), and *H. pylori* (2, 3). Numbers in parentheses indicate copy number.
[e]*E. coli, Salmonella* (large insertion), *Citrobacter, Shigella, Enterobacter,* and *Klebsiella.*
[f]*E. coli, Salmonella enterica* serovar Typhimurium, *Citrobacter koseri, Shigella sonnei, Shigella flexneri, Shigella dysenteriae,* and *Shigella boydii.*
[g]MGE, mobile genetic element.

family of type I TA systems in *Helicobacter pylori* (31) pointed to the importance of *cis*-encoded mRNA functional elements that act together with the sRNA antitoxin to prevent toxin expression. For this reason, this review will focus on the comparison of the various modes of regulation adopted by TA loci, and, more specifically, on the role played by mRNA structure to regulate toxin expression at the translation initiation step.

EXTENSIVE REGULATION OF TYPE I TA SYSTEMS

A major consequence of toxins being lethal to the cell that produces them is that unwanted expression can induce bacterial cell death. Therefore, this lethality applies a strong selection pressure to control the expression of toxin-encoding genes in bacterial genomes. Except for type I, TA systems have an operon organization in which both the toxin and the antitoxin are usually produced from the same promoter, the antitoxin being always encoded upstream from the toxic gene (1). This ensures the availability of an antitoxin molecule ready to counteract toxin activity as soon as it is produced, via either sequestration or an indirect mechanism (antagonism) (Fig. 1A and B). Although this operon organization appears to be sufficient to control toxicity, several additional layers control toxin expression at both the transcriptional and posttranscriptional levels. For instance, in type II TA loci, transcriptional autoregulation by the TA complex is combined with translational coupling of the upstream antitoxin, ensuring that for each toxin molecule produced, an antitoxin molecule has been synthesized as well (Fig. 1A). A reverse gene order has been described for some type II TA systems (32), but it correlates with the use of a noncanonical start codon (GUG) or a short leader region of the toxin-encoding mRNA, probably leading to poor translation efficiency (33, 34).

In type I TA systems, on the contrary, the lack of direct interaction between the toxic peptide and the RNA antitoxin implies that toxicity can only be controlled at the level of toxin expression and not toxin activity (Fig. 1C). This regulation is achieved via the base-pairing of an untranslated RNA with the toxin-encoding mRNA, which inhibits toxin translation and/or promotes its degradation (Fig. 1C) (10, 28). Both RNAs are transcribed from their own promoters. Another shared feature of many type I TA systems is that toxins are small proteins (peptide size, <60 amino acids). With the exceptions of RalR and SymE toxins (Table 1) (35, 36), they target the bacterial membrane (28), leading to membrane depolarization and ATP loss (20, 27, 37, 38)

Regulation modes

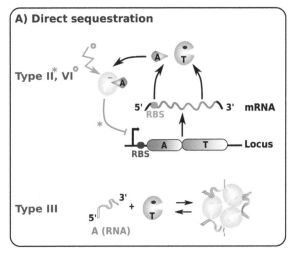

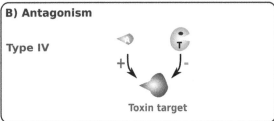

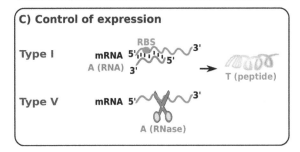

Figure 1 Various modes of antitoxin-mediated regulation. The three main types of toxicity regulation by antitoxins. (A) Direct sequestration. In types II and VI TA systems, toxin inactivation involves a direct interaction between the toxin (T) and the antitoxin (A). The formation of an inactive TA heterocomplex (T-A) can, in its turn (for type II), lead to the transcriptional repression of the operon (red star). In type VI, the formation of the inactive TA complex favors toxin degradation by cellular proteases (yellow circle). In type III, the antitoxin is an RNA that directly binds to the toxin to prevent its toxic activity. (B) Antagonism. Both toxin and antitoxin compete for binding to the same target. The interaction can additionally have opposite functional (antagonistic) effects. (C) Control of expression. In types I and V, regulation occurs at the posttranscriptional level. In type I, antitoxins are antisense RNA molecules that base-pair to the toxin-encoded mRNA to alter its expression by either inhibiting translation initiation or promoting its degradation. In type V, the antitoxin is an RNase that cleaves the toxin-encoded mRNA.

or membrane invagination (39). However, some families of toxins, such as Fst/Ldr, have been shown to lead to nucleoid condensation as the most immediate effect upon production (17, 40). Remarkably, in the case of the type V *ghoT*/GhoS TA locus, the toxin is also predicted to be a small membrane protein (57 amino acids), but in this case, the antitoxin GhoS is an RNase that specifically cleaves the toxin mRNA (Fig. 1C) (41). Hence, similarly to type I TA systems, GhoS regulates toxicity indirectly at the level expression.

Shine-Dalgarno Sequestration Is Required To Uncouple Transcription and Translation

The absence of nuclear compartmentalization is one of the key features distinguishing prokaryotic from eukaryotic cells. One consequence is that it allows the coupling, in time and space, of two major cellular processes: transcription and translation (42) (Fig. 2A). Although it does not mean a complete absence of physical compartmentalization in bacteria, as some specific RNases (RNase E and RNase Y) have been observed to be at the membrane (43, 44), it is clear that a physical link does exist between transcription and translation, at least during the first round of protein synthesis (45). This physical interaction between both machineries is thought to occur not only at early stages of translation to allow cellular colocalization for efficient translation initiation (46) but also during translation elongation to prevent RNA polymerase from backtracking (42). This coupling between transcription and translation of nascent mRNAs is important for some regulatory mechanisms. It can be very useful to prevent the accumulation of nonfunctional transcripts via nonsense polarity (47). However, when it comes to toxin-encoding mRNAs, transcription/translation coupling can be lethal and make difficult the regulation by an antisense RNA.

In bacteria, initiation is usually the rate-limiting step of translation. The canonical mechanism of translation initiation involves the binding of the 30S ribosomal subunit to the translation initiation region (TIR) of an mRNA. The TIR comprises all the elements required for translation initiation to occur (48). The translation initiation of many, though not all, mRNAs depends on the interaction between a sequence element upstream from the start codon (the Shine-Dalgarno sequence [SD]) and a complementary sequence at the 3′ end of the 16S rRNA (anti-SD sequence [aSD]) (49, 50). Additionally, a downstream AUG (or GUG/UUG) start codon, optimally spaced (at around 4 to 9 nucleotides) from the SD sequence (51), binds fMet-tRNA^{fMet} and GTP-bound initiation factor IF2 to set the reading frame. Besides these two regulatory elements acting at the primary sequence level, mRNAs having a highly structured TIR can also possess so-called translational enhancer sequences (often AC-, AU-, or U-rich), proposed to facilitate translation initiation by protein factors such as the ribosomal protein S1 (52–54). In the absence of an SD sequence, alternative modes of translation initiation have been described. An extensive description of translation of leaderless mRNAs (lacking SD sequence) can be found in chapter 10. For instance, translation could initiate via direct 70S ribosome binding or, alternatively, for polycistrons, through a 70S-scanning initiation mechanism (55). This latter mode could be used by type II TA systems to translate their toxins that lack a proper SD sequence (Fig. 1A).

Usually, the efficiency of translation initiation has been linked to ribosome accessibility and mRNA folding around the TIR (54). However, in the case of toxic genes, ribosome accessibility must be tightly controlled. Several studies have highlighted the presence of one common feature of the toxin-encoding mRNAs: the primary transcript is usually translationally inert due to the intramolecular sequestration of the SD sequence by a partially or totally complementary aSD sequence. This has been observed for many TA systems in both Gram-negative and Gram-positive bacteria (see some examples in Fig. 3). Experimental evidence of this sequestration has been shown for Hok (56, 57), TisB (58), ZorO (59), DinQ (60), and AapA1 (31) mRNAs in Gram-negative bacteria, and for Fst (61), SprA1 (62), TxpA (63), BsrE, and BrsG (64, 65) mRNAs in Gram-positive bacteria. More importantly, if we look carefully at the mRNA sequence of some TA systems for which no SD sequestration has been reported yet, we can easily find an aSD sequence located a few nucleotides upstream from the SD sequence for many of them. This is the case for the newly identified type I TA systems in *Clostridium difficile*, for which a 5′-CCUCCC-3′ sequence can be found 11 nucleotides upstream from the SD sequence of the toxin mRNA (66). We also identified a complementary sequence just upstream from the SD in the mRNA of the SprG1 toxin in *Staphylococcus aureus* (67). Remarkably, the aSD sequence is always encoded upstream from the SD sequence to immediately mask the SD after its synthesis, uncoupling transcription from translation (Fig. 2B and 3). It would be interesting to reverse this order to see if this is lethal or not.

An exception to this paradigm does exist. In the case of YonT toxin-encoding mRNA, the SD sequence is surprisingly not sequestered (63). In this particular case, the start codon is GUG instead of the canonical AUG (63, 68). This could reduce the translation initiation rate by altering the binding affinity of the fMet-

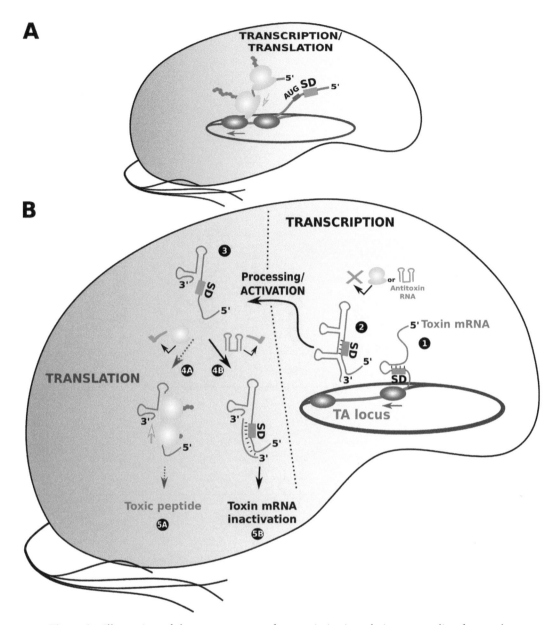

Figure 2 Illustration of the consequences of transcription/translation uncoupling for regulation of type I TA expression. (A) The lack of nuclear compartmentalization in bacteria leads to the coupling in time and space of transcription and translation processes. (B) Transcription/translation coupling of type I toxin-encoding mRNAs would be lethal. TA systems can be conserved in bacterial genomes thanks to the decoupling of such processes through the sequestration of the SD sequence (red) during transcription (**1**). This SD sequestration is conserved in primary transcripts, making them unable to interact with both ribosomes and antitoxin sRNAs (**2**). In the cases where SD sequence sequestration involves a 5′-3′ LDI, the formation of successive metastable structures ensuring SD inaccessibility to both ribosomes and antitoxin during transcription is essential to prevent premature toxin expression and mRNA degradation, respectively. Translational activation is achieved by the enzymatic processing (of the 5′ or 3′ mRNA end, depending on the TA system) of the primary transcript followed by a structural rearrangement (**3**) that renders the mRNA able to interact with both ribosomes (**4A**) and the antitoxin (**4B**). Ribosome binding to the accessible SD sequence (green) leads to toxin production, inducing either growth arrest or cell death (**5A**). On the opposite, antitoxin binding efficiently inhibits toxin translation and promoter mRNA degradation, allowing cell survival (**5B**).

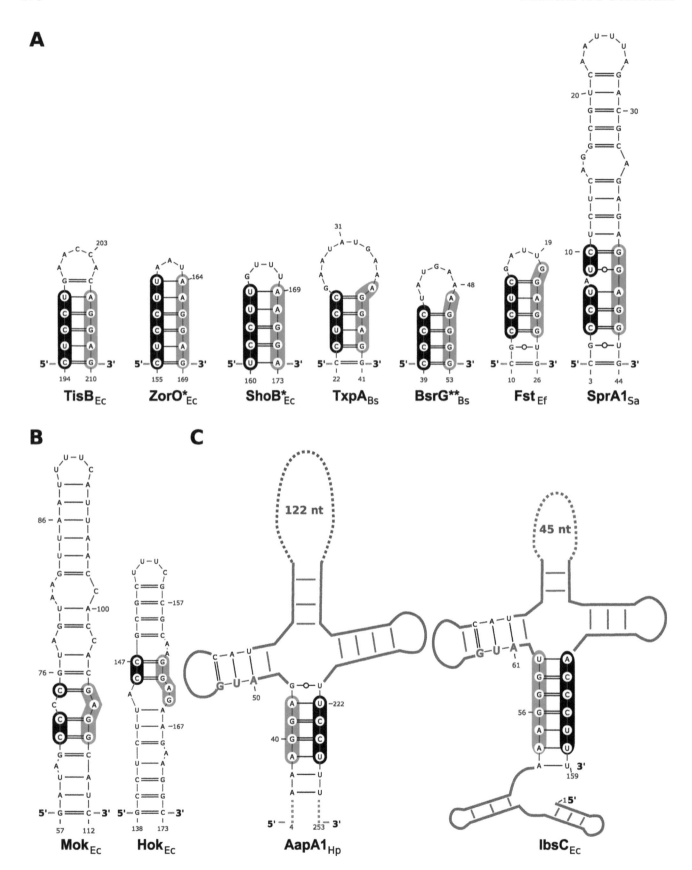

tRNA^{fMet} to the start codon. Another strategy to decrease translation is to display a strong complementarity between the toxin SD sequence and the 16S rRNA aSD sequence. This has been described for TxpA and YonT toxin mRNAs, which have 11- and 12-nucleotide-long SD sequences, respectively. A strong SD sequence increases ribosome pausing, leading to a reduction of the translation initiation rate and an increase of the time window the sRNA antitoxin has to bind to the 3′ end of the toxin mRNA, preventing its translation (63, 68).

In some cases, the intramolecular sequestration of the SD sequence in primary transcripts involves a long-distance interaction (LDI) between the 5′ and 3′ ends of the toxin mRNA (as described for AapA1 and Hok, and Ibs mRNAs in *H. pylori* and *E. coli*, respectively) (Fig. 3) (31, 69, 70). In this case, considering the context of transcription/translation coupling, a new question arises. What prevents premature toxin translation of nascent transcripts in which the most energetically stable mRNA conformation sequestering the SD sequence has not been synthesized yet?

Previous studies on the MS2 phage maturation protein suggested a kinetic model of translational control mediated by delayed RNA folding (71). Indeed, this mRNA is untranslatable because the leader adopts a well-defined cloverleaf structure in which the SD sequence of the maturation gene is taken up in a long-distance base-pairing with an upstream complementary sequence. To allow the synthesis of the maturation protein, a transient RNA hairpin is formed before the sequence that will create the inhibitory structure is transcribed (71). In fact, such transient RNA structures have been shown to form during Hok mRNA transcription to mask the SD sequence temporally, ensuring the inaccessibility of ribosomes to the nascent mRNA (72).

More recently, by using the Kinefold stochastic simulations that predict cotranscriptional folding paths of mRNAs (73), we predicted the formation of two successive metastable hairpins in the AapA1 toxin mRNA. These transient structures sequester the SD sequence during transcription and before the RNA polymerase reaches the end of the transcript (31). We recently confirmed the existence of such structures *in vivo* for the AapA3 toxin mRNA (from the same family as AapA1) by the use of genetic approaches (S. Masachis, N.J. Tourasse, C. Lays, M. Faucher, S. Chabas, I. Iost, and F. Darfeuille, unpublished data).

Global mRNA folding may itself play key roles in the regulation of type I TA systems. A compact mRNA secondary structure has often been described as a key feature of toxin-encoding mRNAs (e.g., Hok, TisB, AapA1, Ibs, Fst, SprA, and BsrG). For instance, the AapA1 and Ibs mRNAs adopt, as in the case of MS2, a well-defined cloverleaf structure in which only a few nucleotides are in single-stranded regions (Fig. 3C) (31, 70). This comes together with the co- and posttranscriptional SD sequestration, but it can go way beyond it. Indeed, in the absence of transcription/translation coupling, the nascent mRNA can base-pair to the DNA template and form so-called R-loops. It is well known that highly structured RNAs are less prone to R-loop formation. Hence, for TA systems, avoiding R-loop formation may prevent premature transcription termination and enable the completion of transcription until the translationally inactive full-length mRNA is made. Further, considering that R-loops can often lead to an increased mutation rate during RNA synthesis (74), their prevention through the conservation of highly structured mRNAs may represent an evolutionary advantage for the conservation of type I TA systems.

Figure 3 Examples of secondary structures sequestering the SD sequence of toxin-encoding mRNAs. (A) Stem-loop sequences in which the SD sequence is totally or partially sequestered by an upstream aSD sequence. These secondary structures have been experimentally validated *in vitro* (58, 61–64) or predicted (indicated by *) (105). For BsrG, the stem-loop sequestering the SD is shown in absence of SR4 antitoxin (**). Indeed, when the antitoxin binds to the mRNA, the stem-loop sequestering the SD is extended by 4 additional base pairs (64). (B) Stem-loop structures sequestering the SD sequence of the Mok leader peptide and the Hok toxin. In this case, the formation of the Mok SD-sequestering stem-loop is dependent on a 5′-3′ LDI (69, 106). (C) Examples of SD sequestration achieved by a stable LDI between both mRNA ends, creating a cloverleaf structure (31, 70). The start codon (AUG, shown in green) can additionally be partially or totally sequestered in one of the cloverleaf structures. Dotted gray lines indicate the presence of unrepresented structures/lengths. Full gray lines schematically represent structures and base pairs. SD sequences are shown in red; aSD sequences are shown in black. Positions are relative to the transcription start site of the mRNAs (+1). 5′ and 3′ indicate orientation of the mRNA. Toxin mRNA names are indicated under each structure. The small index next to each name indicates the host organisms: Ec, *E. coli*; Bs, *B. subtilis*; Ef, *E. faecalis*; Sa, *S. aureus*; and Hp, *H. pylori*.

In addition, mRNA folding is a key determinant in controlling mRNA stability. As described above, most primary transcripts of type I TAs are translationally inactive. It is well known that the absence of translating ribosomes causes mRNA destabilization in *E. coli* (75). Surprisingly, these untranslatable primary transcripts are particularly highly stable compared with other cellular mRNAs, displaying an extremely long RNA half-life (31, 76). A compact folding of most toxin-encoding mRNAs is probably very efficient in protecting these mRNAs from degradation by cellular RNases. Whether some specific RNA-binding proteins play a role has not been reported yet (77).

In summary, all these regulatory threats (premature toxin synthesis, R-loop formation, and RNA degradation) likely represent a significant selective pressure to preserve a compact mRNA structure in toxin-encoding mRNAs and may lead to a possible domestication process ensuring the conservation of TA loci in bacterial genomes.

Toxin-Encoding mRNAs Require an Activation Step To Be Translated

During the study of the *aapA1*/IsoA1 TA locus in *H. pylori* (31), we realized that the most abundant mRNA transcript was untranslatable. In contrast to classical mRNAs, the "active" toxin-encoding mRNA species could not be detected under normal growth conditions. This translatable transcript was revealed during rifampin assays. This antibiotic is classically used to block transcription in bacteria in order to determine RNA half-life. One hour after rifampin addition, a shorter transcript was accumulating while the pool of RNA antitoxin had been completely degraded. We further showed that this transcript underwent a two-step 3′-end nucleolytic activation process involving the 3′-5′ exonucleolytic activity of the polynucleotide phosphorylase (PNPase) (S. Masachis, H. Arnion, S. Chabas, F. Boissier, and F. Darfeuille, unpublished). This result is reminiscent of the 3′-end mRNA activation described more than 20 years ago for the *hok*/Sok system in *E. coli* (76). Indeed, in both cases, the primary transcript cannot be translated due to the stable sequestration of the SD sequence by an LDI between the 5′ and 3′ ends, and requires an activation step via 3′ processing (Fig. 4). Although the *hok*/Sok and *aapA1*/IsoA1 TA systems show no sequence homology, it was really striking to see that an activation process involving the PNPase was conserved in two evolutionarily distant Gram-negative bacteria. However, although this 3′-end trimming induces a strong structural rearrangement of the 5′ untranslated region (5′ UTR) of both AapA1

and Hok mRNAs, translational activation of the Hok mRNA requires the translational coupling of an upstream open reading frame (ORF) encoding the Mok peptide (Fig. 3B and 4) (78).

More surprisingly, for TA systems whose antitoxin is encoded in a nonoverlapping fashion (Fig. 4, right panel), translation activation happens at the 5′ end of the toxin mRNA. Multiple mRNA 5′ ends have been mapped for the *tisB*/IstR (58), *dinQ*/AgrAB (20), *shoB*/OhsC (79), and *zorO*/OrzO (59) TA systems. However, the enzyme(s) responsible for this processing is still unknown. Irrespective of how the activation step occurs, it always triggers a structural rearrangement that opens the possibility to interact with both ribosomes and antitoxin, entering into an essential decision step: toxin translation or mRNA inactivation (Fig. 2B). For the *dinQ*/AgrAB system, this cleavage leads to a strong rearrangement between various complementary sequences that renders the SD site directly accessible to ribosomes (60). In this case, after 5′-end processing, the mRNA becomes translatable.

In the case of Gram-positive bacteria (Fig. 5), it is so far unclear how the activation step occurs, or, eventually, if it is required at all. Indeed, similarly to what has been observed in Gram negative-bacteria, SD sequestration has also been reported for several toxin-encoding mRNAs, such as BsrG (80), BsrE (65), TxpA (63), Fst (61), and SprA1 (62). Although extensive studies have been carried out on the *par* locus in *Enterococcus faecalis*, no processing could be detected for the Fst mRNA (5, 7, 61, 81, 82). Nevertheless, the accumulation of a truncated transcript after rifampin addition was observed during the characterization of Fst-like TA systems in *Streptococcus mutans* and *Lactobacillus rhamnosus* (83, 84). However, although this observation is really intriguing, it will require further characterization. Indeed, while the *sprA1*/asSprA1 TA locus has been shown to be a member of the Fst family, the stem-loop masking the SD sequence in the SprA1 mRNA seems very different than the one in the Fst mRNA (Fig. 3). So, while some TA systems share strong homologies, they may undergo different pathways of activation.

In summary, Fig. 2B illustrates the main regulatory mechanisms occurring in Gram-negative bacteria, which could probably be extended to Gram-positive bacteria: during the course of transcription, translation of the nascent transcript, as well as antitoxin binding, is impaired by the formation of successive RNA secondary structures sequestering the SD sequence. Upon transcription termination, the full-length mRNA is translationally inactive due to the stable sequestration of

Gram-negative

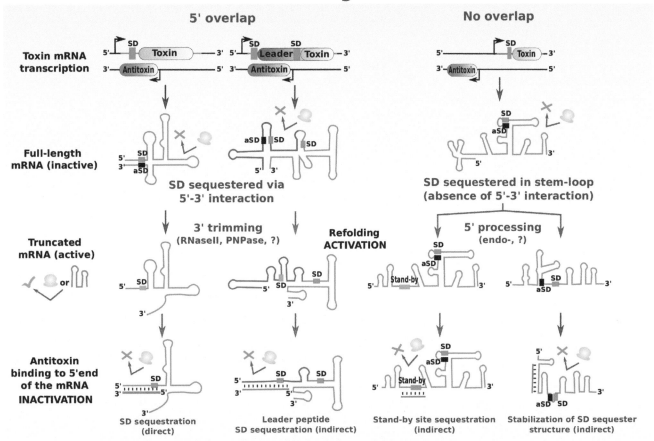

Figure 4 Type I operon organization in Gram-negative bacteria and mechanistic consequences of its regulation. Type I antitoxin sRNAs in Gram-negative bacteria can be encoded in two main fashions: (i) overlapping the 5′ end of the toxin mRNA, the ORF, or a leader ORF (left panel) or (ii) not overlapping (right panel). In both cases, transcription/translation coupling forces the sequestration of the SD sequence by partially or totally complementary sequences called anti-SD (aSD). This sequestration starts during transcription but is maintained upon transcription termination, leading to the generation of a translationally inert and sRNA-inaccessible primary transcript (full-length mRNA). Location of the aSD sequence will determine whether the sequestration occurs via 5′-3′ LDI (5′-overlapping TA loci) or in a stem-loop (nonoverlapping TA loci). In both cases, an enzymatic activation step is required for the generation of the truncated (active) mRNA. When the SD sequestration involves 5′-3′ interaction, this activation step often occurs via 3′ trimming by 3′-5′ exonucleases (RNaseII, PNPase). In contrast, when SD is sequestered in a stem-loop, activation occurs via 5′-end processing by endonucleases. In either case, a light or strong structural rearrangement (refolding) is needed upon processing to render the SD accessible to both ribosomes and sRNA binding. Next, in noninduced conditions, antitoxin sRNAs outcompete the ribosomes for binding to the 5′ end of the toxin mRNA and render it translationally inert. This inactivation step can occur via direct sequestration of the SD sequence or the leader ORF SD sequence (5′-overlapping TA loci), or indirectly via the sequestration of the ribosome standby site (stand-by) or the stabilization of an SD-trapped structure (nonoverlapping TA loci). In most cases, sRNA binding to the toxin mRNA leads to RNase III-mediated toxin mRNA decay and cell survival.

Gram-positive

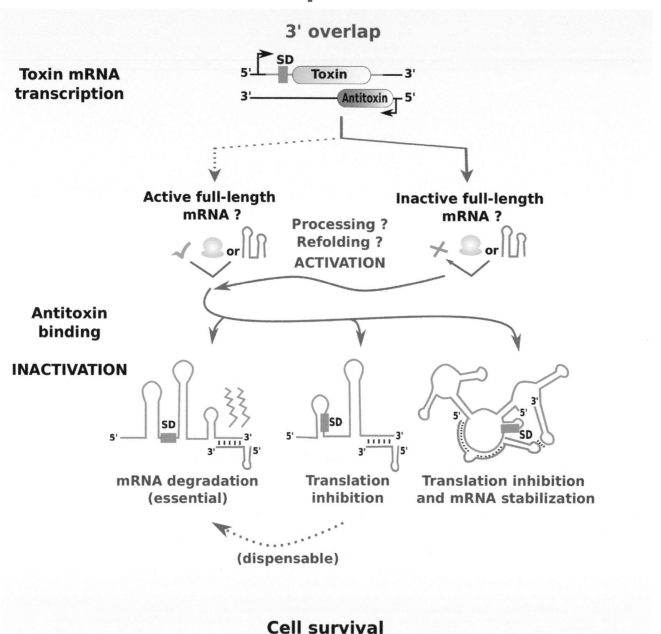

Figure 5 Mechanistic regulatory consequences of the 3′-overlapping antitoxin sRNAs in Gram-positive bacteria. In all type I TAs described so far in Gram-positive bacteria, antitoxin sRNAs are encoded in 3′-overlapping fashion to the toxin mRNAs. As in Gram-negative bacteria, transcription/translation coupling forces the sequestration of the SD sequence during transcription, the nascent transcript being accessible neither for the ribosome nor the antitoxin. Upon transcription termination, the full-length mRNA becomes targeted by the antitoxin RNA that binds to the 3′ end of the toxin mRNA. The question marks represent how mRNA activation could occur in Gram-positive bacteria (little is known about the mRNA activation, processing, or refolding steps compared to Gram-negative bacteria). In some cases (TxpA and YonT), sRNA binding is not sufficient to impede toxin translation, and thus, mRNA degradation by RNase III is essential to avoid toxicity. In some cases, sRNA binding leads to a structural rearrangement that enhances the sequestration of the SD sequence. This interaction can in its turn lead to mRNA degradation by RNase III or on the contrary to the stabilization of the translationally inert complex. In all cases, antitoxin binding to the toxin mRNA hampers its expression and allows cell survival.

the ribosome binding site (RBS). Finally, the mRNA undergoes a nucleolytic activation process followed by a structural refolding that renders it able to interact with the ribosomes or the antitoxin RNA. As a result, either ribosomes bind and the toxic peptide is produced, leading to growth arrest or cell death, or the sRNA antitoxin binds, primarily inhibiting translation initiation and leading to the irreversible degradation of the toxin mRNA (Fig. 2B).

Standby RBSs Can Be Required for Translation

For some type I TA systems, the structural rearrangement induced upon the nucleolytic activation step does not change the accessibility of the SD sequence in the toxin mRNA. Those mRNAs require an additional element to be optimally translated: a ribosome standby site. While studying the translation of the highly structured MS2 coat protein mRNA, de Smit and van Duin (85, 86) encountered a paradox. They showed that ribosomes compete with mRNA folding and need to capture the SD in an unfolded state to initiate translation. However, the calculated energetic ΔG value of the SD-sequestering stem-loop was a problem: the fraction of time the stem-loop would spend in an unfolded state is in the microsecond range. Considering the time in an unfolded state and the ribosomal 30S subunit association rate, the calculations predicted a coat protein translation rate >10,000-fold lower than the observed one. This paradox was solved by the "standby" model (86). In this model, 30S subunits preloaded to a yet undefined single-stranded region can access the SD as soon as the structure is transiently unfolded ("breathing"), thus following a kinetically driven (first-order reaction) instead of a thermodynamically driven model (second-order reaction) (87).

This mechanism has been identified for the TisB toxin mRNA (58). The TisAB primary transcript (+1) is processed into a 5′-end shorter transcript (+42). However, the SD in both mRNA species is sequestered within a stable stem-loop (Fig. 3). The structural rearrangement upon processing of the TisAB mRNA exposes a region in the 5′ UTR, the standby site, thus allowing 30S accessibility to the sequestered RBS as soon as the stem "breathes" (Fig. 4). A similar translation mechanism has been recently suggested for the ZorO toxin mRNA (59); however, further studies are needed to confirm this hypothesis.

As mentioned above, "translational enhancer" sequences (often AC-, AU-, or U-rich) have been proposed to facilitate translation initiation mediated by protein factors such as the ribosomal protein S1. Strik-ingly, AC-rich motifs can be found in several toxin-encoding mRNAs, including the TisB standby site and the ZorO putative standby site, suggesting a potential relationship between the presence of such sequences and the translation efficiency mediated by standby sites (59, 87).

Antisense RNAs Bind to and Repress Activated mRNAs

Until now, we have highlighted the importance of RNA structural elements embedded in toxin mRNAs to prevent toxin expression. However, by definition, the inhibition of toxin synthesis in type I TA systems is achieved by direct base-pairing of an antisense RNA to the toxin-encoding mRNA. Depending on whether the antisense is encoded in *cis* or *trans*, the region and length of base-pairing can vary. Several organizations have been described in Gram-negative bacteria (Fig. 4). For instance, in *E. coli*, four TA loci have the antisense RNA encoded in a divergent orientation to the toxin-encoding gene: *shoB*/OhsC (16), *zorO*/OrzO (38), *tisB*/IstR (18), and *dinQ*/AgrAB (20). Remarkably, the antisense RNA of these TA systems targets a similar region in the 5′ UTR of the toxin-encoding mRNA displaying a short region of perfect complementarity (18 to 23 nucleotides). Indeed, a minimum of 15 continuous base pairs has been shown to be necessary for efficient repression of ZorO toxin expression (38). In the case of the DinQ toxin, the region of complementarity can be extended to 31 nucleotides containing 1 mismatch (60). However, antisense RNA antitoxins are usually encoded in *cis*, overlapping either the 5′ (in Gram-negative bacteria) or the 3′ end (in Gram-positive bacteria) of their toxin mRNA counterparts (Fig. 4 and 5). Although this organization can lead to the extensive base-pairing between the antisense RNA and its target mRNA, the initial pairing is often mediated by single-stranded regions such as kissing-loop interactions (31, 63). These single-stranded regions are often formed after the processing of the primary toxin-encoding mRNA (31, 76) and facilitate the pairing between both toxin and antitoxin RNAs. In particular, for some TA loci, the presence of a U-turn (YUNR) motif in one of the loops greatly enhances the pairing rate between the toxin and antitoxin RNAs (5, 7, 64, 82, 88). This initial short pairing is usually propagated into an extensive base-pairing forming a long RNA intermolecular duplex with increased stability. In most of the cases, this duplex is sufficient to mask the RBS, leading to the inhibition of toxin translation. Remarkably, in the case of the BsrG toxin-encoding mRNA, the SR4 antitoxin binding does not directly interfere with ribosome binding but instead generates an extended intramolecular SD sequestration (64). More surprisingly, RNAI, the Fst toxin mRNA in *E. faecalis*, displays two complementary regions, named DR (direct repeat) a and b, with the antisense RNAII (5, 7, 82). These two regions are in close proximity to the SD sequence of the RNAI. After the initial pairing of RNAII with RNAI in the mRNA 3′ end, this interaction is followed by an extended pairing within this region of complementarity in the 5′ UTR. As a consequence, the binding of RNAII leads to the formation of a stable duplex that blocks toxin translation. Surprisingly, the homolog of the *par* locus in *S. aureus* (Table 1), *sprA1*/asSprA1, seems to retain only the interaction with the 5′ UTR but not with the terminator stem-loop, to inhibit translation initiation (62). Another important point is that antitoxins seem to specifically target the activated toxin mRNAs (processed at the 5′ or 3′ end), the primary transcript being inert for interaction, as shown for the processed Hok (89), AapA1 (31), and TisB (58) mRNAs. Thus, in both Gram-negative and Gram-positive bacteria, RNA antitoxins act primarily at the translational level by efficiently impeding ribosome binding either directly or indirectly.

Besides translation inhibition, antisense RNA binding also leads to mRNA degradation (Fig. 4 and 5). Indeed, these duplexes are recognized and cleaved by the double-stranded specific RNase III, as reported, for instance, for the AapA1 (31), Hok (89), TisB (58), and BsrG (64) toxin-encoding mRNAs. However, this degradation is, in most cases, not essential to suppress toxin expression. This feature can be easily confirmed, for instance, by the viability of strains deleted for RNase III, as shown for *H. pylori* and *E. coli* (18, 31). Nevertheless, there are two exceptions to this paradigm. In *Bacillus subtilis*, the antitoxin binding is not always sufficient to avoid toxin expression. The binding of RatA and SR6 antitoxins to TxpA and YonT mRNAs, respectively, does not block toxin translation (63, 68). In contrast, the antisense pairing to the toxin mRNA 3′ end leads to the formation of a long, extended duplex cleaved by RNase III, whose activity is essential to avoid toxin expression (Fig. 5).

Type I TA Systems May Also Be Modulated by RNA Editing

RNA editing is a widespread posttranscriptional regulation layer known to recode RNAs and proteins in plants, animals, fungi, protists, bacteria, and viruses (90). The most common modification is adenosine (A)-to-inosine (I) conversion, mediated by enzymes of the ADAR (adenosine deaminase, RNA-specific) family. In bacteria, RNA editing was so far thought to occur exclusively in a single nucleotide site within a tRNA for arginine (tRNA^{Arg}). It was known to be mediated by a tRNA-specific adenosine deaminase (*tadA*) that recognizes a 4-base-long motif, TACG, with the edited A in second position (91). Interestingly, TadA activity requires a specific mRNA secondary structure on which the edited base is embedded within a loop (92).

Remarkably, a recent study in *E. coli* revealed 15 novel A-to-I RNA-editing events (93). Twelve of these occur within ORFs, contrasting with the mostly noncoding RNA editing reported in eukaryotes. Four such events occur within ORFs belonging to the Hok family of host-killing toxins, which may not be a surprise, as the original *tadA* mutant strain was obtained during the search for suppressors upon overexpression of HokC toxin (*gef* protein) (91, 94). The HokB mRNA showed the highest levels of RNA editing. Sequence alignment reflected that HokB mRNA editing could occur even outside *E. coli*, in different bacterial species such as *Klebsiella pneumoniae* and *Yersinia enterocolitica*, thanks to the conservation of the TadA enzyme. Interestingly, HokB orthologs either have a tyrosine (Tyr) encoded by the editable TAC codon at position 29 and a cysteine (Cys) at position 46, or on the opposite, a Cys at position 29 and an editable Tyr codon at position 46. Notably, both positions reside in one of the two β-strands predicted in the HokB peptide, suggesting that they may play a functional role. Interestingly, all the editable positions lie within the TACG TadA-recognition motif in a loop that is the only mRNA region identical to the tRNA^{Arg}.

Overexpression experiments to study the effect of the different variants at position 29 showed that all three studied variants (Tyr encoded by the editable TAC codon; Tyr encoded by the noneditable TAT codon; and Cys TGC, mimicking the postediting situation) are toxic. However, the strongest effect was observed for the peptide mimicking the postediting situation (Cys at position 29), indicating that RNA editing in HokB mRNA enhances peptide toxicity.

These results established RNA editing as an alternative mode to regulate toxin activity, one more posttranscriptional regulation layer in bacteria, which importantly may contribute to proteome diversity and phenotypic heterogeneity between genetically identical cells. Remarkably, it also represents one more example in which an RNA secondary structure rules a posttranscriptional regulatory event. In the case of HokB, it additionally opens the question of whether RNA editing could represent a key adaptation mechanism, which could be implied in situations where phenotypic diversity is advantageous, such as antibiotic persistence, thus reflecting a more complex bet-hedging mechanism than the previously suggested one (23). However, while RNA editing has been shown for HokB, it still remains to be seen how widespread such a modification may be in other type I toxins.

PERSPECTIVES

Identifying Type I TA Systems

As already mentioned, type I TA systems were initially discovered on plasmids based on their biological function (i.e., PSK). However, only a few families of type I TA systems have been identified on plasmids, while many have been found on chromosomes (*hok*/Sok, *fst*/RNAI, and, more recently, *aapA*/IsoA) (15, 31, 95) (Table 1). As highlighted by Coray and colleagues (96), there is a narrower distribution of type I and III TA systems compared to type II TA systems. Type I systems are more likely found within single clades of bacteria rather than dispersed across phyla (96).

Much of the known distribution of TA systems relies on *in silico* searches based on sequence conservation (12, 15). Historically, type I systems have been difficult to predict *in silico*, often due to the small size of the toxin (<60 amino acids) and to the lack of a prediction tool for their RNA antitoxin counterpart. Pioneering work done by Fozo and colleagues greatly expanded the number of known TA loci across 774 bacterial genomes and predicted new families of TA systems (22). They based their search on toxin hydrophobicity and highly structured sRNA based on RNA-folding energy profiles. They also looked for loci duplication (based on *ibs*/Sib and *ldr*/Rdl examples) and polar C-terminal residues of the potential toxins.

However, even if some strategies have succeeded in the *de novo* prediction of type I TA systems (22), it remains a big challenge. Our group has recently implemented a new database that includes all the known type I TA loci (Table 1) (N. J. Tourasse and F. Darfeuille, unpublished data). The search for new type I TA systems focused on the mRNA folding features of the toxin-encoding transcripts rather than on the hydrophobicity and sequence of the predicted toxin. The mRNA features common to each family of type I TA systems (AapA, BsrE/G/H, DinQ, Fst, Hok, Ibs, Ldr, ShoB, TisB, TxpA, YonT, and ZorO) included the mRNA length, SD sequence sequestration by an upstream complementary sequence (within a stem-loop or via 5′-to-3′ LDI), as well as the antitoxin sRNA secondary structure. This algorithm independently recovered the mRNAs and sRNAs of most TA loci that were previously identified based on pep-

tide features and greatly increased the number of newly identified ones, demonstrating that RNA features are key determinants for the identification of type I TA systems.

Understanding the Functions of Chromosomally Encoded TA Systems

As we have seen in the first part of this review, toxin expression needs to be tightly regulated and this regulation strongly depends on antitoxin RNA levels. For PSK function, as in the *hok*/Sok and *fst*/RNAII systems, constitutive expression from both toxin and antitoxin promoters is required (10). The same is true for the two chromosomally encoded *txpA*/RatA and *yonT*/SR6 loci in *B. subtilis* that are involved in the stabilization of the Skin and SPβ prophages, respectively (63, 97). However, the link between chromosomal TA systems and PSK has not yet been shown for other TA systems. For instance, we observed constitutive expression of both toxin-encoding mRNA and antitoxin sRNA in the *aapA1*/IsoA1 TA system in *H. pylori* (31). However, we have not been able to demonstrate its role in PSK despite its location near integrative and conjugative elements.

While the function of TA systems encoded on plasmids or near mobile genetic elements can be easily associated with PSK, elucidating their function when they are encoded on the chromosome is still challenging. However, from a biological point of view, TA systems could potentially be beneficial at the population level (e.g., increase phenotypic heterogeneity, which enhances fitness and adaptation to adverse environmental conditions). Therefore, bacteria may have domesticated TA systems, allowing them to fully take over the control of their expression and make use of it for their own advantage. One way to address this question is to search for possible regulation of the antitoxin RNA expression. There are a number of examples where TA expression may be related to stress responses. In *E. coli tisB*, *symE*, and *dinQ*, toxin promoters are preceded by LexA boxes, making them inducible during SOS response via the LexA transcriptional repressor, whereas their sRNA antitoxins are constitutively expressed (18, 20, 36) (Table 1). While both toxin-encoding mRNA and antitoxin RNA of the *par* locus (pAD1 plasmid) have been found to be constitutively expressed, the antitoxin RNA of its chromosomal copy in *E. faecalis* (98), as well as the *S. mutans* and *S. aureus* homologs, are regulated in response to various stress conditions (13, 62, 83). SprA1 antitoxin expression decreases under acidic pH stress and in response to oxidative stress (99). In *B. subtilis*, the expression of several TA loci is modulated in response to multiple stresses including ethanol, heat shock, and anaerobic stresses (65, 80, 100) (Table 1). Finally, it has recently been shown that upstream of the promoter of IsoA3 antitoxin in *H. pylori* there is a binding site for the essential orphan response regulator HP1043, which has been related to oxidative stress and growth phase-dependent responses (101). This is an interesting observation, and further studies need to be performed to confirm a potential transcriptional regulation of IsoA3 expression.

The levels of RNA antitoxins are also adjusted at the posttranscriptional level by RNA degradation. Depending on the bacterial species, several RNases are responsible for antitoxin turnover. RNase E controls Sok RNA degradation in *E. coli* (102), RNase J controls IsoA3 RNA stability in *H. pylori* (103), and RNase Y controls many antitoxin sRNAs (e.g., SR4 and RatA) in *B. subtilis*. Therefore, it could be very exciting to evaluate to what extent bacteria can modulate the level of antitoxin RNA by adjusting the accessibility of these degrading enzymes to their substrate. For instance, several antitoxin sRNAs (IstR, Sib, and Rdl) were found to be associated with the RNA-binding protein ProQ in *Salmonella* (77). Whether this interaction affects their stability or their ability to regulate their toxin-encoding mRNA counterpart is not known yet.

CONCLUSIONS

Overall, in this review we wish to highlight that for type I TA systems, antisense RNA binding is not sufficient to control toxin expression. Due to the transcription/translation coupling that occurs in bacteria, tight control of toxin expression requires the presence of *cis*-encoded elements, which co- and/or posttranscriptionally occlude the RBS, as already discussed in earlier reviews (10, 13, 28). This sequestration can occur within the stem-loop or by an LDI between both mRNA ends. In the first case, the aSD sequence is encoded just upstream from the SD sequence in order to form a stable stem-loop, leading to the sequestration of the RBS at early stages of transcription. In the second case, the SD sequestration requires sequences that are encoded at the 3′ end of the mRNA. Thus, when SD sequestration occurs by mRNA end-pairing (LDI strategy), additional intermediate structures must be formed to impede premature toxin translation (72). These unstable and successive interactions lead to the formation of so-called metastable structures, avoiding cotranscriptional translation of the toxin mRNA. Why and how these two mechanisms of SD sequestration evolved and were kept

is still unknown. From an energetic point of view, the stem-loop strategy may be advantageous, as the first SD-sequestering structure formed will be the most stable one. On the opposite, the LDI strategy may be more difficult to make, as one or several metastable structures need to be successively formed before reaching the final and stable folding state of the mRNA. Additionally, the stem-loop strategy may impede premature RNA antitoxin binding in a more efficient way than the metastable structures formed in the LDI strategy. From an evolutionary point of view, the stem-loop strategy seems to offer greater sequence flexibility, as only one aSD sequence needs to be conserved, contrary to the LDI strategy. Nevertheless, a potential advantage of the LDI strategy may be that it allows the formation of a highly stable primary toxin-encoding mRNA (1). This consequence might be really advantageous for the PSK mechanism, as the primary transcript needs to be transmitted to the daughter cell. Irrespective of the employed mechanism, both SD-sequestering strategies require an activation step. While it is clear that the activation needs 3′ exonucleolytic activity in the case of the LDI strategy, no general mechanism has been shown for the stem-loop strategy. Yet an activation step via 5′ endonucleolytic cleavage has been shown or suggested for only a few cases (25, 58–60). In many cases, no processing event has been described (13, 63–65, 68). The combination of new sequencing technologies to precisely map the mRNA 3′ end (Term-seq) (104) with classical techniques, such as rifampin assays to analyze RNA processing (31), should help in deciphering these mechanisms. Alternatively, no processing event may be required to activate these primary transcripts. The involvement of RNA helicases or RNA chaperones (such as ProQ), to facilitate the removal of the antisense RNA paired to the toxin-encoding mRNA, could be a good alternative to RNA processing. Further studies will be required to address this question in more detail.

Acknowledgments. Work in our lab was supported by funds from INSERM (U1212), CNRS (UMR 5320), Université de Bordeaux, and Agence Nationale de la Recherche (ARNA; http://www.agence-nationale-recherche.fr/) grants Bactox1 and asSUPYCO. S.M.G. has received funding from the European Union's Horizon 2020 research and innovation program under the Marie Sklodowska-Curie grant agreement No. 642738. We thank Nicolas J. Tourasse for critical reading of the manuscript and all present and past members of the ARNA laboratory for helpful discussions.

Citation. Masachis S, Darfeuille F. 2018. Type I toxin-antitoxin systems: regulating toxin expression via Shine-Dalgarno sequence sequestration and small RNA binding. Microbiol Spectrum 6(4):RWR-0030-2018.

References

1. Harms A, Brodersen DE, Mitarai N, Gerdes K. 2018. Toxins, targets, and triggers: an overview of toxin-antitoxin biology. *Mol Cell* **70:**768–784.

2. Benz J, Meinhart A. 2014. Antibacterial effector/immunity systems: it's just the tip of the iceberg. *Curr Opin Microbiol* **17:**1–10.

3. Van Melderen L. 2010. Toxin-antitoxin systems: why so many, what for? *Curr Opin Microbiol* **13:**781–785.

4. Goeders N, Van Melderen L. 2014. Toxin-antitoxin systems as multilevel interaction systems. *Toxins (Basel)* **6:** 304–324.

5. Greenfield TJ, Ehli E, Kirshenmann T, Franch T, Gerdes K, Weaver KE. 2000. The antisense RNA of the *par* locus of pAD1 regulates the expression of a 33-amino-acid toxic peptide by an unusual mechanism. *Mol Microbiol* **37:**652–660.

6. Weaver KE. 2012. The *par* toxin-antitoxin system from *Enterococcus faecalis* plasmid pAD1 and its chromosomal homologs. *RNA Biol* **9:**1498–1503.

7. Greenfield TJ, Franch T, Gerdes K, Weaver KE. 2001. Antisense RNA regulation of the *par* post-segregational killing system: structural analysis and mechanism of binding of the antisense RNA, RNAII and its target, RNAI. *Mol Microbiol* **42:**527–537.

8. Gerdes K, Bech FW, Jørgensen ST, Løbner-Olesen A, Rasmussen PB, Atlung T, Boe L, Karlstrom O, Molin S, von Meyenberg K. 1986. Mechanism of postsegregational killing by the *hok* gene product of the *parB* system of plasmid Rl and its homology with the *relF* gene product of the *E. coli relB* operon. *EMBO J* **5:**2023–2029.

9. Gerdes K, Gultyaev AP, Franch T, Pedersen K, Mikkelsen ND. 1997. Antisense RNA-regulated programmed cell death. *Annu Rev Genet* **31:**1–31.

10. Gerdes K, Wagner EG. 2007. RNA antitoxins. *Curr Opin Microbiol* **10:**117–124.

11. Nielsen AK, Thorsted P, Thisted T, Wagner EG, Gerdes K. 1991. The rifampicin-inducible genes *srnB* from F and *pnd* from R483 are regulated by antisense RNAs and mediate plasmid maintenance by killing of plasmid-free segregants. *Mol Microbiol* **5:**1961–1973.

12. Pedersen K, Gerdes K. 1999. Multiple *hok* genes on the chromosome of *Escherichia coli*. *Mol Microbiol* **32:** 1090–1102.

13. Weaver KE. 2015. The type I toxin-antitoxin *par* locus from *Enterococcus faecalis* plasmid pAD1: RNA regulation by both *cis-* and *trans-*acting elements. *Plasmid* **78:** 65–70.

14. Kwong SM, Jensen SO, Firth N. 2010. Prevalence of Fst-like toxin-antitoxin systems. *Microbiology* **156:**975–977, discussion 977.

15. Weaver KE, Reddy SG, Brinkman CL, Patel S, Bayles KW, Endres JL. 2009. Identification and characterization of a family of toxin-antitoxin systems related to the *Enterococcus faecalis* plasmid pAD1 *par* addiction module. *Microbiology* **155:**2930–2940.

16. Fozo EM, Kawano M, Fontaine F, Kaya Y, Mendieta KS, Jones KL, Ocampo A, Rudd KE, Storz G. 2008. Repression of small toxic protein synthesis by the Sib and OhsC small RNAs. *Mol Microbiol* **70:**1076–1093.

17. Kawano M, Oshima T, Kasai H, Mori H. 2002. Molecular characterization of long direct repeat (LDR) sequences expressing a stable mRNA encoding for a 35-amino-acid cell-killing peptide and a *cis*-encoded small antisense RNA in *Escherichia coli*. *Mol Microbiol* **45:**333–349.

18. Vogel J, Argaman L, Wagner EG, Altuvia S. 2004. The small RNA IstR inhibits synthesis of an SOS-induced toxic peptide. *Curr Biol* **14:**2271–2276.

19. Durand S, Jahn N, Condon C, Brantl S. 2012. Type I toxin-antitoxin systems in *Bacillus subtilis*. *RNA Biol* **9:** 1491–1497.

20. Weel-Sneve R, Kristiansen KI, Odsbu I, Dalhus B, Booth J, Rognes T, Skarstad K, Bjørås M. 2013. Single transmembrane peptide DinQ modulates membrane-dependent activities. *PLoS Genet* **9:**e1003260.

21. Sharma CM, Hoffmann S, Darfeuille F, Reignier J, Findeiss S, Sittka A, Chabas S, Reiche K, Hackermüller J, Reinhardt R, Stadler PF, Vogel J. 2010. The primary transcriptome of the major human pathogen *Helicobacter pylori*. *Nature* **464:**250–255.

22. Fozo EM, Makarova KS, Shabalina SA, Yutin N, Koonin EV, Storz G. 2010. Abundance of type I toxin-antitoxin systems in bacteria: searches for new candidates and discovery of novel families. *Nucleic Acids Res* **38:**3743–3759.

23. Verstraeten N, Knapen WJ, Kint CI, Liebens V, Van den Bergh B, Dewachter L, Michiels JE, Fu Q, David CC, Fierro AC, Marchal K, Beirlant J, Versées W, Hofkens J, Jansen M, Fauvart M, Michiels J. 2015. Obg and membrane depolarization are part of a microbial bet-hedging strategy that leads to antibiotic tolerance. *Mol Cell* **59:** 9–21.

24. Dörr T, Vulić M, Lewis K. 2010. Ciprofloxacin causes persister formation by inducing the TisB toxin in *Escherichia coli*. *PLoS Biol* **8:**e1000317.

25. Berghoff BA, Hoekzema M, Aulbach L, Wagner EG. 2017. Two regulatory RNA elements affect TisB-dependent depolarization and persister formation. *Mol Microbiol* **103:**1020–1033.

26. Tsilibaris V, Maenhaut-Michel G, Mine N, Van Melderen L. 2007. What is the benefit to *Escherichia coli* of having multiple toxin-antitoxin systems in its genome? *J Bacteriol* **189:**6101–6108.

27. Unoson C, Wagner EGH. 2008. A small SOS-induced toxin is targeted against the inner membrane in *Escherichia coli*. *Mol Microbiol* **70:**258–270.

28. Fozo EM, Hemm MR, Storz G. 2008. Small toxic proteins and the antisense RNAs that repress them. *Microbiol Mol Biol Rev* **72:**579–589.

29. Wen J, Fozo EM. 2014. sRNA antitoxins: more than one way to repress a toxin. *Toxins (Basel)* **6:**2310–2335.

30. Brantl S, Jahn N. 2015. sRNAs in bacterial type I and type III toxin-antitoxin systems. *FEMS Microbiol Rev* **39:**413–427.

31. Arnion H, Korkut DN, Masachis Gelo S, Chabas S, Reignier J, Iost I, Darfeuille F. 2017. Mechanistic insights

into type I toxin antitoxin systems in *Helicobacter pylori*: the importance of mRNA folding in controlling toxin expression. *Nucleic Acids Res* **45**:4782–4795.

32. Turnbull KJ, Gerdes K. 2017. HicA toxin of *Escherichia coli* derepresses *hicAB* transcription to selectively produce HicB antitoxin. *Mol Microbiol* **104**:781–792.

33. Christensen-Dalsgaard M, Gerdes K. 2006. Two *higBA* loci in the *Vibrio cholerae* superintegron encode mRNA cleaving enzymes and can stabilize plasmids. *Mol Microbiol* **62**:397–411.

34. Wang X, Lord DM, Hong SH, Peti W, Benedik MJ, Page R, Wood TK. 2013. Type II toxin/antitoxin MqsR/MqsA controls type V toxin/antitoxin GhoT/GhoS. *Environ Microbiol* **15**:1734–1744.

35. Guo Y, Quiroga C, Chen Q, McAnulty MJ, Benedik MJ, Wood TK, Wang X. 2014. RalR (a DNase) and RalA (a small RNA) form a type I toxin-antitoxin system in *Escherichia coli*. *Nucleic Acids Res* **42**:6448–6462.

36. Kawano M, Aravind L, Storz G. 2007. An antisense RNA controls synthesis of an SOS-induced toxin evolved from an antitoxin. *Mol Microbiol* **64**:738–754.

37. Mok WW, Patel NH, Li Y. 2010. Decoding toxicity: deducing the sequence requirements of IbsC, a type I toxin in *Escherichia coli*. *J Biol Chem* **285**:41627–41636.

38. Wen J, Won D, Fozo EM. 2014. The ZorO-OrzO type I toxin-antitoxin locus: repression by the OrzO antitoxin. *Nucleic Acids Res* **42**:1930–1946.

39. Jahn N, Brantl S, Strahl H. 2015. Against the mainstream: the membrane-associated type I toxin BsrG from *Bacillus subtilis* interferes with cell envelope biosynthesis without increasing membrane permeability. *Mol Microbiol* **98**:651–666.

40. Patel S, Weaver KE. 2006. Addiction toxin Fst has unique effects on chromosome segregation and cell division in *Enterococcus faecalis* and *Bacillus subtilis*. *J Bacteriol* **188**:5374–5384.

41. Wang X, Lord DM, Cheng HY, Osbourne DO, Hong SH, Sanchez-Torres V, Quiroga C, Zheng K, Herrmann T, Peti W, Benedik MJ, Page R, Wood TK. 2012. A new type V toxin-antitoxin system where mRNA for toxin GhoT is cleaved by antitoxin GhoS. *Nat Chem Biol* **8**:855–861.

42. Proshkin S, Rahmouni AR, Mironov A, Nudler E. 2010. Cooperation between translating ribosomes and RNA polymerase in transcription elongation. *Science* **328**:504–508.

43. Strahl H, Turlan C, Khalid S, Bond PJ, Kebalo JM, Peyron P, Poljak L, Bouvier M, Hamoen L, Luisi BF, Carpousis AJ. 2015. Membrane recognition and dynamics of the RNA degradosome. *PLoS Genet* **11**:e1004961.

44. Khemici V, Prados J, Linder P, Redder P. 2015. Decay-initiating endoribonucleolytic cleavage by RNase Y is kept under tight control via sequence preference and sub-cellular localisation. *PLoS Genet* **11**:e1005577.

45. McGary K, Nudler E. 2013. RNA polymerase and the ribosome: the close relationship. *Curr Opin Microbiol* **16**:112–117.

46. Sanamrad A, Persson F, Lundius EG, Fange D, Gynnå AH, Elf J. 2014. Single-particle tracking reveals that free ribosomal subunits are not excluded from the *Escherichia coli* nucleoid. *Proc Natl Acad Sci U S A* **111**:11413–11418.

47. Richardson JP. 1991. Preventing the synthesis of unused transcripts by Rho factor. *Cell* **64**:1047–1049.

48. Simonetti A, Marzi S, Jenner L, Myasnikov A, Romby P, Yusupova G, Klaholz BP, Yusupov M. 2009. A structural view of translation initiation in bacteria. *Cell Mol Life Sci* **66**:423–436.

49. Shine J, Dalgarno L. 1974. The 3′-terminal sequence of *Escherichia coli* 16S ribosomal RNA: complementarity to nonsense triplets and ribosome binding sites. *Proc Natl Acad Sci U S A* **71**:1342–1346.

50. Steitz JA, Jakes K. 1975. How ribosomes select initiator regions in mRNA: base pair formation between the 3′ terminus of 16S rRNA and the mRNA during initiation of protein synthesis in *Escherichia coli*. *Proc Natl Acad Sci U S A* **72**:4734–4738.

51. Chen H, Bjerknes M, Kumar R, Jay E. 1994. Determination of the optimal aligned spacing between the Shine-Dalgarno sequence and the translation initiation codon of *Escherichia coli* mRNAs. *Nucleic Acids Res* **22**:4953–4957.

52. Komarova AV, Tchufistova LS, Dreyfus M, Boni IV. 2005. AU-rich sequences within 5′ untranslated leaders enhance translation and stabilize mRNA in *Escherichia coli*. *J Bacteriol* **187**:1344–1349.

53. Duval M, Korepanov A, Fuchsbauer O, Fechter P, Haller A, Fabbretti A, Choulier L, Micura R, Klaholz BP, Romby P, Springer M, Marzi S. 2013. *Escherichia coli* ribosomal protein S1 unfolds structured mRNAs onto the ribosome for active translation initiation. *PLoS Biol* **11**:e1001731.

54. Duval M, Simonetti A, Caldelari I, Marzi S. 2015. Multiple ways to regulate translation initiation in bacteria: mechanisms, regulatory circuits, dynamics. *Biochimie* **114**:18–29.

55. Yamamoto H, Wittek D, Gupta R, Qin B, Ueda T, Krause R, Yamamoto K, Albrecht R, Pech M, Nierhaus KH. 2016. 70S-scanning initiation is a novel and frequent initiation mode of ribosomal translation in bacteria. *Proc Natl Acad Sci U S A* **113**:E1180–E1189.

56. Franch T, Gerdes K. 1996. Programmed cell death in bacteria: translational repression by mRNA end-pairing. *Mol Microbiol* **21**:1049–1060.

57. Gultyaev AP, Franch T, Gerdes K. 1997. Programmed cell death by *hok/sok* of plasmid R1: coupled nucleotide covariations reveal a phylogenetically conserved folding pathway in the *hok* family of mRNAs. *J Mol Biol* **273**:26–37.

58. Darfeuille F, Unoson C, Vogel J, Wagner EG. 2007. An antisense RNA inhibits translation by competing with standby ribosomes. *Mol Cell* **26**:381–392.

59. Wen J, Harp JR, Fozo EM. 2017. The 5′ UTR of the type I toxin ZorO can both inhibit and enhance translation. *Nucleic Acids Res* **45**:4006–4020.

60. Kristiansen KI, Weel-Sneve R, Booth JA, Bjørås M. 2016. Mutually exclusive RNA secondary structures regulate translation initiation of DinQ in *Escherichia coli*. *RNA* **22**:1739–1749.

61. Shokeen S, Patel S, Greenfield TJ, Brinkman C, Weaver KE. 2008. Translational regulation by an intramolecular stem-loop is required for intermolecular RNA regulation of the *par* addiction module. *J Bacteriol* 190:6076–6083.

62. Sayed N, Jousselin A, Felden B. 2011. A *cis*-antisense RNA acts in *trans* in *Staphylococcus aureus* to control translation of a human cytolytic peptide. *Nat Struct Mol Biol* 19:105–112.

63. Durand S, Gilet L, Condon C. 2012. The essential function of *B. subtilis* RNase III is to silence foreign toxin genes. *PLoS Genet* 8:e1003181.

64. Jahn N, Brantl S. 2013. One antitoxin—two functions: SR4 controls toxin mRNA decay and translation. *Nucleic Acids Res* 41:9870–9880.

65. Müller P, Jahn N, Ring C, Maiwald C, Neubert R, Meißner C, Brantl S. 2016. A multistress responsive type I toxin-antitoxin system: *bsrE/SR5* from the *B. subtilis* chromosome. *RNA Biol* 13:511–523.

66. Maikova A, Peltier J, Boudry P, Hajnsdorf E, Kint N, Monot M, Poquet I, Martin-Verstraete I, Dupuy B, Soutourina O. 2018. Discovery of new type I toxin-antitoxin systems adjacent to CRISPR arrays in *Clostridium difficile*. *Nucleic Acids Res* 46:4733–4751.

67. Pinel-Marie ML, Brielle R, Felden B. 2014. Dual toxic-peptide-coding *Staphylococcus aureus* RNA under antisense regulation targets host cells and bacterial rivals unequally. *Cell Rep* 7:424–435.

68. Reif C, Löser C, Brantl S. 2018. *Bacillus subtilis* type I antitoxin SR6 promotes degradation of toxin *yonT* mRNA and is required to prevent toxic *yoyJ* overexpression. *Toxins (Basel)* 10:74.

69. Thisted T, Sørensen NS, Gerdes K. 1995. Mechanism of post-segregational killing: secondary structure analysis of the entire Hok mRNA from plasmid R1 suggests a fold-back structure that prevents translation and antisense RNA binding. *J Mol Biol* 247:859–873.

70. Han K, Kim KS, Bak G, Park H, Lee Y. 2010. Recognition and discrimination of target mRNAs by Sib RNAs, a *cis*-encoded sRNA family. *Nucleic Acids Res* 38:5851–5866.

71. van Meerten D, Girard G, van Duin J. 2001. Translational control by delayed RNA folding: identification of the kinetic trap. *RNA* 7:483–494.

72. Møller-Jensen J, Franch T, Gerdes K. 2001. Temporal translational control by a metastable RNA structure. *J Biol Chem* 276:35707–35713.

73. Xayaphoummine A, Bucher T, Isambert H. 2005. Kinefold web server for RNA/DNA folding path and structure prediction including pseudoknots and knots. *Nucleic Acids Res* 33(Web Server issue):W605–W610.

74. Gan W, Guan Z, Liu J, Gui T, Shen K, Manley JL, Li X. 2011. R-loop-mediated genomic instability is caused by impairment of replication fork progression. *Genes Dev* 25:2041–2056.

75. Iost I, Dreyfus M. 1995. The stability of *Escherichia coli lacZ* mRNA depends upon the simultaneity of its synthesis and translation. *EMBO J* 14:3252–3261.

76. Franch T, Gultyaev AP, Gerdes K. 1997. Programmed cell death by *hok/sok* of plasmid R1: processing at the *hok* mRNA 3′-end triggers structural rearrangements that allow translation and antisense RNA binding. *J Mol Biol* 273:38–51.

77. Smirnov A, Förstner KU, Holmqvist E, Otto A, Günster R, Becher D, Reinhardt R, Vogel J. 2016. Grad-seq guides the discovery of ProQ as a major small RNA-binding protein. *Proc Natl Acad Sci U S A* 113:11591–11596.

78. Thisted T, Gerdes K. 1992. Mechanism of post-segregational killing by the *hok/sok* system of plasmid R1. Sok antisense RNA regulates *hok* gene expression indirectly through the overlapping *mok* gene. *J Mol Biol* 223:41–54.

79. Kawano M, Reynolds AA, Miranda-Rios J, Storz G. 2005. Detection of 5′- and 3′-UTR-derived small RNAs and *cis*-encoded antisense RNAs in *Escherichia coli*. *Nucleic Acids Res* 33:1040–1050.

80. Jahn N, Preis H, Wiedemann C, Brantl S. 2012. *BsrG/SR4* from *Bacillus subtilis*—the first temperature-dependent type I toxin-antitoxin system. *Mol Microbiol* 83:579–598.

81. Shokeen S, Greenfield TJ, Ehli EA, Rasmussen J, Perrault BE, Weaver KE. 2009. An intramolecular upstream helix ensures the stability of a toxin-encoding RNA in *Enterococcus faecalis*. *J Bacteriol* 191:1528–1536.

82. Greenfield TJ, Weaver KE. 2000. Antisense RNA regulation of the pAD1 *par* post-segregational killing system requires interaction at the 5′ and 3′ ends of the RNAs. *Mol Microbiol* 37:661–670.

83. Koyanagi S, Lévesque CM. 2013. Characterization of a *Streptococcus mutans* intergenic region containing a small toxic peptide and its *cis*-encoded antisense small RNA antitoxin. *PLoS One* 8:e54291.

84. Folli C, Levante A, Percudani R, Amidani D, Bottazzi S, Ferrari A, Rivetti C, Neviani E, Lazzi C. 2017. Toward the identification of a type I toxin-antitoxin system in the plasmid DNA of dairy *Lactobacillus rhamnosus*. *Sci Rep* 7:12051.

85. de Smit MH, van Duin J. 1990. Secondary structure of the ribosome binding site determines translational efficiency: a quantitative analysis. *Proc Natl Acad Sci U S A* 87:7668–7672.

86. de Smit MH, van Duin J. 2003. Translational standby sites: how ribosomes may deal with the rapid folding kinetics of mRNA. *J Mol Biol* 331:737–743.

87. Sterk M, Romilly C, Wagner EG. 2018. Unstructured 5′-tails act through ribosome standby to override inhibitory structure at ribosome binding sites. *Nucleic Acids Res* 46:4188–4199.

88. Franch T, Petersen M, Wagner EG, Jacobsen JP, Gerdes K. 1999. Antisense RNA regulation in prokaryotes: rapid RNA/RNA interaction facilitated by a general U-turn loop structure. *J Mol Biol* 294:1115–1125.

89. Thisted T, Sørensen NS, Wagner EG, Gerdes K. 1994. Mechanism of post-segregational killing: Sok antisense RNA interacts with Hok mRNA via its 5′-end single-stranded leader and competes with the 3′-end of Hok mRNA for binding to the *mok* translational initiation region. *EMBO J* 13:1960–1968.

90. Knoop V. 2011. When you can't trust the DNA: RNA editing changes transcript sequences. *Cell Mol Life Sci* 68:567–586.

91. Wolf J, Gerber AP, Keller W. 2002. *tadA*, an essential tRNA-specific adenosine deaminase from *Escherichia coli*. *EMBO J* **21**:3841–3851.

92. Keller W, Wolf J, Gerber A. 1999. Editing of messenger RNA precursors and of tRNAs by adenosine to inosine conversion. *FEBS Lett* **452**:71–76.

93. Bar-Yaacov D, Mordret E, Towers R, Biniashvili T, Soyris C, Schwartz S, Dahan O, Pilpel Y. 2017. RNA editing in bacteria recodes multiple proteins and regulates an evolutionarily conserved toxin-antitoxin system. *Genome Res* **27**:1696–1703.

94. Poulsen LK, Larsen NW, Molin S, Andersson P. 1992. Analysis of an *Escherichia coli* mutant strain resistant to the cell-killing function encoded by the *gef* gene family. *Mol Microbiol* **6**:895–905.

95. Gerdes K, Rasmussen PB, Molin S. 1986. Unique type of plasmid maintenance function: postsegregational killing of plasmid-free cells. *Proc Natl Acad Sci U S A* **83**: 3116–3120.

96. Coray DS, Wheeler NE, Heinemann JA, Gardner PP. 2017. Why so narrow: distribution of anti-sense regulated, type I toxin-antitoxin systems compared with type II and type III systems. *RNA Biol* **14**:275–280.

97. Silvaggi JM, Perkins JB, Losick R. 2005. Small untranslated RNA antitoxin in *Bacillus subtilis*. *J Bacteriol* **187**: 6641–6650.

98. Michaux C, Hartke A, Martini C, Reiss S, Albrecht D, Budin-Verneuil A, Sanguinetti M, Engelmann S, Hain T, Verneuil N, Giard J-C. 2014. Involvement of *Enterococcus faecalis* small RNAs in stress response and virulence. *Infect Immun* **82**:3599–3611.

99. Sayed N, Nonin-Lecomte S, Réty S, Felden B. 2012. Functional and structural insights of a *Staphylococcus aureus* apoptotic-like membrane peptide from a toxin-antitoxin module. *J Biol Chem* **287**:43454–43463.

100. Meißner C, Jahn N, Brantl S. 2016. *In vitro* characterization of the type I toxin-antitoxin system *bsrE/SR5* from *Bacillus subtilis*. *J Biol Chem* **291**:560–571.

101. Pelliciari S, Pinatel E, Vannini A, Peano C, Puccio S, De Bellis G, Danielli A, Scarlato V, Roncarati D. 2017. Insight into the essential role of the *Helicobacter pylori* HP1043 orphan response regulator: genome-wide identification and characterization of the DNA-binding sites. *Sci Rep* **7**:41063.

102. Gerdes K. 2016. Hypothesis: type I toxin-antitoxin genes enter the persistence field—a feedback mechanism explaining membrane homoeostasis. *Philos Trans R Soc Lond B Biol Sci* **371**:20160189.

103. Redko Y, Galtier E, Arnion H, Darfeuille F, Sismeiro O, Coppée JY, Médigue C, Weiman M, Cruveiller S, De Reuse H. 2016. RNase J depletion leads to massive changes in mRNA abundance in *Helicobacter pylori*. *RNA Biol* **13**:243–253.

104. Dar D, Shamir M, Mellin JR, Koutero M, Stern-Ginossar N, Cossart P, Sorek R. 2016. Term-seq reveals abundant ribo-regulation of antibiotics resistance in bacteria. *Science* **352**:aad9822.

105. Fozo EM. 2012. New type I toxin-antitoxin families from "wild" and laboratory strains of *E. coli*: Ibs-Sib, ShoB-OhsC and Zor-Orz. *RNA Biol* **9**:1504–1512.

106. Nagel JH, Gultyaev AP, Gerdes K, Pleij CW. 1999. Metastable structures and refolding kinetics in *hok* mRNA of plasmid R1. *RNA* **5**:1408–1418.

Regulating with RNA in Bacteria and Archaea
Edited by Gisela Storz and Kai Papenfort
© 2018 American Society for Microbiology, Washington, DC
doi:10.1128/microbiolspec.RWR-0029-2018

Widespread Antisense Transcription in Prokaryotes

12

Jens Georg[1] and Wolfgang R. Hess[1]

INTRODUCTION: ANTISENSE RNAs—PERVASIVE TRANSCRIPTION OR FUNCTIONAL IMPORTANCE?

The first documented *cis*-encoded antisense RNAs (asRNAs) in bacteria were the RNA I, controlling ColE1 replication (1), and the OOP asRNA of bacteriophage λ (2–4). However, until the year 2007 a mere ~10 bacterial asRNAs had been characterized (5). It was only with the advent of global approaches for the analysis of bacterial transcriptomes that it was recognized that actually a substantial fraction of transcripts, in fact, constitute asRNAs. The first hints obtained with high-density microarrays pointed at antisense transcription linked to possibly as many as 3,000 to 4,000 open reading frames in *Escherichia coli* (6), more recently reinforced by the finding that asRNAs originate from 37% of all transcription start sites (TSSs) (7), which might still be an underestimation of the initial level of antisense transcription (8). By the hybridization of directly labeled RNA instead of cDNA to high-density microarrays, a high number of strongly expressed asRNAs were experimentally identified in the model cyanobacterium *Synechocystis* sp. PCC 6803 (9). The direct labeling of RNA avoided the artificial second-strand synthesis in the production of cDNA, a step at which experimental artifacts might be introduced (10). In agreement with the initial evidence, numerous transcriptome studies have demonstrated more recently that a substantial fraction of the discovered TSS in vastly different bacterial taxa is not associated with a protein-coding gene. Internal parts of coding regions in sense and antisense orientation are massively transcribed, a phenomenon often referred to as pervasive transcription (11, 12).

The detection of pervasive transcription has been followed by questions about its possible physiological meaning. Intergenic transcripts are relatively well investigated and could be, e.g., *trans*-acting regulatory small RNAs (sRNAs) (recently reviewed by Wagner and Romby [13]) or the mRNAs of unannotated small proteins (14). However, for asRNAs, as well as for sense transcripts from within a coding region, the data are less distinct. Genes within genes are reviewed in chapter 9 (15), while we here are focusing on asRNAs.

Over time, specific functions have been assigned to several individual asRNAs, but experiments have failed to do so in many other instances. Hence, it became an object of debate whether asRNAs are functional or a mere by-product of insufficiently controlled noisy transcription (11, 12, 16). Insights obtained in a number of recent studies indicate surprisingly that antisense transcription can be important in itself, e.g., in transcription-coupled DNA-repair processes (17), transcriptional interference (18–22), or potential functional RNA-DNA hybrid formation (23, 24). This review summarizes the recent progress in the field of asRNAs in bacteria and presents a scenario for the evolution of these kinds of transcripts.

Some Characteristics of Bacterial asRNAs

Aside from the fact that they originate from the antisense strand of a known transcriptional unit, bacterial asRNAs differ greatly in length, genomic location, regulation, and especially in abundance. An overview on the main categories is given in Fig. 1. There are short asRNAs with a defined 3′ end that show up as distinct single bands in a Northern blot, and long asRNAs, potentially lacking a strong terminator, that usually produce multiple and more diffuse bands. The shorter asRNAs, such as SymR (25), GadY (26), and IsrR (27), are between 100 and 300 nucleotides (nt), but many asRNAs are also substantially longer. One example in *Listeria monocytogenes* overlaps as many as three genes and is >2,000 nt long (28). The asRNA

---

[1]University of Freiburg, Faculty of Biology, Institute of Biology III, Genetics and Experimental Bioinformatics, D-79104 Freiburg, Germany.

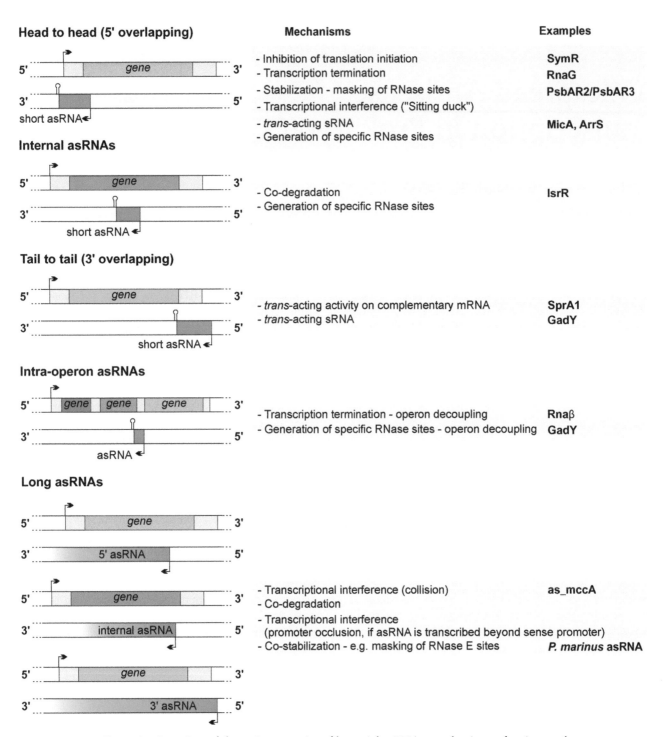

Figure 1   Overview of the main categories of bacterial asRNAs, mechanisms of action, and selected examples.

in *Prochlorococcus* sp. MED4, which protects a set of host mRNAs and viral transcripts from RNase E degradation during phage infection, is up to 7 kb long (29, 30). Regarding their location, asRNAs can be roughly classified into long forms spanning several genes, shorter 5′-overlapping (divergent, head-to-head), 3′-overlapping (convergent, tail-to-tail), and internally located asRNAs. Another specific case are asRNAs to internal genes within an operon, which may cause decoupled expression of genes 5′ and 3′ of the asRNA binding site. Based on the type of the antisense overlap, some of the regulatory scenarios described below are more or less pronounced (summarized in Fig. 1). A short internal asRNA, for example, is likely to act via the codegradation mechanism, while repression by all kinds of transcriptional interference mechanisms is either not possible or very unlikely.

There are also overlapping mRNAs from protein-coding genes that can mutually influence their expression by the same mechanisms as bona fide noncoding asRNAs. This class includes transcripts from classical, well-annotated genes for which transcriptomics approaches revealed long 5′ or 3′ untranslated regions (UTRs) that overlap with neighboring genes or UTRs. Based on transcriptomic studies in *Listeria* spp. (28, 31), the excludon concept has emerged for examples of these overlapping mRNAs. Excludons are made up by overlapping divergently oriented genes that code for proteins of potentially antagonistic function (32). This genomic organization establishes a regulatory relationship that results in the exclusive expression of only one of both coding regions. In *L. monocytogenes*, transcription from a σ^B-dependent promoter generates a long 5′ UTR of the *mogR-lmo0673* operon (Fig. 2A) (32). The 5′ UTR overlaps with an operon coding for flagellum export apparatus proteins (*lmo0675-fliPQR* to -*lmo0689*). The genomic organization results in an antisense transcript-dependent repression of the flagellum operon if the long 5′ UTR of *mogR-lmo0673* is expressed. Additionally, MogR is a transcription factor that represses flagellum and motility genes on the transcriptional level. In return, a strong expression of the flagellum operon posttranscriptionally represses the σ^B-dependent *mogR-lmo0673* transcripts. Thus, the excludon supports a strict separation of motility and a sessile lifestyle that is encoded within the genome archi-

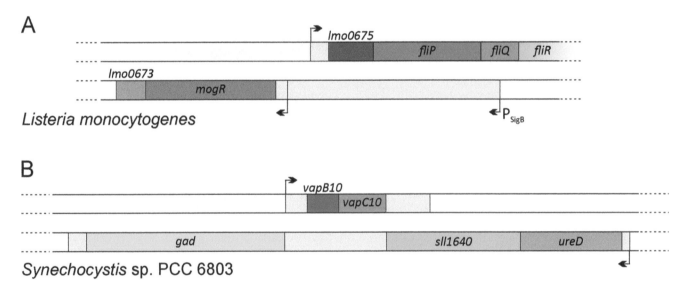

Figure 2 Excludons, instances of long overlapping mRNAs that inhibit the expression of one set of genes by the expression of a second overlapping set of genes. (A) Excludon in *L. monocytogenes* formed by the overlap between the *lmo0675-fliPQR* motility operon transcript and the *mogR-lmo0673* transcript, with its long 5′ UTR originating from the distal σ^B-dependent promoter (32). MogR is a repressor of flagellum and motility gene transcription. Therefore, the arrangement of these two transcriptional units in an excludon ensures the exclusive expression of only one of both coding regions, which is of direct relevance for the motile or nonmotile lifestyle. Note that there is also a proximal σ^B-independent *mogR-lmo0673* promoter. (B) Arrangement of the VapBC10-type toxin-antitoxin system genes *slr1767* and *ssr2962* in *Synechocystis* 6803 in an excludon with the *sll1639* to -*41* genes encoding urease accessory protein UreD, nitrilase (*sll1640*), and glutamate decarboxylase (34). The genes *sll1639* to -*41* are transcribed in the form of a long mRNA that is transcribed from TSSs upstream from *ureD*. The resulting transcriptional unit overlaps *slr1767* and *ssr2962* just between the final and the penultimate genes. This arrangement contributes to silence expression of this toxin-antitoxin system under most conditions in addition to the autoregulatory transcriptional and the proteolytic regulation (33). The scheme has been redrawn according to primary transcriptome information from *Synechocystis* 6803 (106).

tecture. Other examples in *L. monocytogenes* include two likely permease-efflux pump excludons and a putative carbon source utilization excludon (32).

Some excludons were found to be conserved beyond *Listeria* spp., e.g., also in the more distantly related bacterium *Clostridium difficile* (31). An example of another excludon architecture, in which one operon completely overlaps the other operon, is shown in Fig. 2B. In the cyanobacterium *Synechocystis* 6803, the genes *slr1767* and *ssr2962* encode a VapBC10 type II toxin-antitoxin system (33). Over their entire length, these two genes are covered by a long asRNA, which is the mRNA of the three genes *sll1639* to *-41*, encoding the urease accessory protein UreD, nitrilase, and glutamate decarboxylase (34). This arrangement contributes to silence the VapBC10 system, as can be seen from the fact that the only detectable transcript accumulation occurred during nitrogen starvation, when *sll1639* to *-41* were at minimum expression. Silencing of toxin-antitoxin systems by asRNA is a more prominent feature of type I toxin-antitoxin systems and is covered in more detail in chapter 11 (35).

Additionally, an underestimated category of asRNAs might be those containing a small open reading frame, hence constituting actual instances of overlapping mRNAs. This is illustrated by the *Pseudomonas fluorescens* PF0-1 gene *iiv14* (*cosA*), which lies nearly completely antisense to *Pfl_0939* (~1-kb overlap). Translation of both mRNAs into proteins was shown, and the shorter, Iiv14, was still ~37 kDa (36). Another instance was suggested for the *aatS* gene, which is encoded opposite to the *aatC* gene in an enterotoxigenic *E. coli* strain (37). Hence, similar to intergenically encoded alternative proteins (15), short transcripts that are annotated as asRNAs might turn out as overlapping mRNAs and an underestimated source of novel small proteins. In summary, asRNAs come in various types and different lengths and forms; some even act as both *cis*- and *trans*-encoded regulators. Therefore, highly divergent mechanisms can be found associated with asRNA functions in bacteria.

Mapping the Extent of Antisense Transcription in Bacteria

Numerous studies have reported asRNAs to encompass significant fractions of bacterial transcriptomes, ranging up to 75% of all genes to be associated with asRNAs (Table 1).

The differences in the percentages of antisense transcripts in the various bacteria are striking. These differences may be real, i.e., due to biological reasons and strain-specific characteristics, but variations in the applied analysis techniques and also likely diverse definitions of

asRNAs in different labs certainly have contributed as well. For example, it matters if the percentage is calculated based on the numbers of TSSs or the number of genes that are overlapped by asRNAs. This question was addressed by Voigt et al. (38), comparing three high-throughput approaches (differential transcriptome sequencing [RNA-seq] utilizing pyrosequencing and Illumina sequencing platforms as well as hybridization of directly labeled total RNA to Affymetrix high-density microarrays). Indeed, for the identification of asRNAs, the overlap between the three methods was smaller than expected. Schlüter et al. observed a similar phenomenon for asRNAs and sRNAs by comparing transcriptome data from microarrays and different RNA-seq methods in *Sinorhizobium meliloti* (39). The above two examples compare different techniques that were applied in one particular lab. Here we can assume that factors such as growth conditions, sample preparation, library preparation, sequencing depth, asRNA definition, and expression thresholds have been comparable, pointing at distinct biases of the applied techniques and the likely underestimation of antisense transcription in many studies. However, all these factors surely differ from lab to lab, which makes it hard to compare numbers in different studies. To answer the question of organism-specific variations, the experimental and bioinformatics parts of transcriptome studies in different organisms need to be done in a standardized way. Such a standardized approach, applied to 18 different prokaryotic organisms, yielded very diverse asRNA TSS numbers, ranging from 1 to 642 (40). Another study used a standardized bioinformatic analysis for transcriptome data from 21 organisms generated by various groups using various sequencing protocols (16). They found clearly distinct asRNA numbers that correlated with the AT content in the genomes of the different bacteria (16). Based on the above findings, we can conclude that the observed differences in asRNA numbers exemplarily shown in Table 1 have both technical and biological reasons. Nevertheless, it is evident that antisense transcription is as common in bacteria as it is in the other two domains of life and represent a substantial fraction of the transcriptome.

Are asRNAs and Their Sense Partners Expressed at the Same Time in the Same Cell at Sufficient Levels?

The most prominent asRNA-based regulatory mechanisms, such as codegradation or costabilization, require the duplex formation of a sense RNA and its antisense partner. However, basically all transcriptome studies detecting asRNAs integrate over a large number of cells

Table 1 Overview of selected transcriptome analyses performed in different bacteria and the reported share in asRNAs

Strain	Taxon	% asRNAs and comments	Reference
Sinorhizobium meliloti 1021	*Alphaproteobacteria*	21.8% of all TSSs, 35% of protein-coding genes had asRNAs	108
Neisseria meningitidis 8013	*Betaproteobacteria*	39% of all TSSs, 573 genes possess at least one asRNA	51
Alteromonas macleodii Te101	*Gammaproteobacteria*	6% of chromosomal and 16% of all plasmid-located TSSs	109
Escherichia coli	*Gammaproteobacteria*	37% of all TSSs	7
Salmonella enterica serovar Typhimurium SL1344	*Gammaproteobacteria*	13% of all TSSs	110
Vibrio cholerae	*Gammaproteobacteria*	47% of all TSSs	111
Xanthomonas campestris pv. campestris B100	*Gammaproteobacteria*	24.6% of all TSSs, 16.8% of all genes had asRNAs	112
Geobacter sulfurreducens ATCC 5157	*Deltaproteobacteria*	5.6% of all genes had asRNAs	113
Campylobacter jejuni	*Epsilonproteobacteria*	~44 to 47% of the TSSs were classified as antisense TSSs, testing four different strains	114
Helicobacter pylori	*Epsilonproteobacteria*	41% of all TSSs, 46% of all protein-coding genes	115
Synechocystis sp. PCC 6803	*Cyanobacteria*	26.8% of all TSSs gave rise to asRNAs, one growth condition	116
Synechocystis sp. PCC 6803	*Cyanobacteria*	41.6% of all transcriptional units, 10 different growth conditions	106
Anabaena (Nostoc) sp. PCC 7120	*Cyanobacteria*	30.4% of all TSSs; asRNAs detected for 39% of all protein-coding genes	103
Trichodesmium erythraeum IMS101	*Cyanobacteria*	14% of all TSSs, 15% of protein-coding genes had asRNAs	117
Prochlorococcus sp. MED4 and MIT9313	*Cyanobacteria*	75% of all genes were associated with asRNAs; asRNAs were only counted when verified in least two of three independent approaches	38
Mycoplasma pneumoniae	*Terrabacteria*	13% of coding genes were covered by asRNAs	118
Staphylococcus aureus NCTC 8325	*Firmicutes*	≥75% of genes were associated with asRNAs	78
Staphylococcus aureus RN6390	*Firmicutes*	44% of the annotated mRNAs were associated with asRNAs in a coimmunoprecipitation strategy using two catalytically impaired RNase III proteins	44

that were pooled to extract sufficient amounts of RNA. Assuming some degree of intrapopulation diversity of gene expression and considering the partly low transcription levels of antisense and/or sense RNAs, it was a matter of debate if a substantial fraction of complementary RNAs actually coexist in individual cells. Following this rationale, asRNAs could be rather noise and not functional if both partners were mutually exclusive. Single-cell transcriptomics would be a perfect tool to analyze this aspect, but this is not an established technique for bacteria yet. Nevertheless, the coexistence of sense and antisense RNAs in one cell was proven in various studies using RNase III mutants and an antibody against double-stranded RNA (41–43), as described in more detail below. Another aspect is the ratios in the expression levels of asRNAs and the respective sense partners. Assuming a mechanism that requires stoichiometric RNA-RNA interaction, it appears obvious that the interaction partners need to exist simultaneously in the same cell and that the asRNA is present in a concentration that is high enough to affect its respective mRNA. At this point, it is worth mentioning that not all known asRNA-based mechanisms (e.g.,

transcriptional interference) require RNA-RNA interaction. Furthermore, some mechanisms might not be known to date. Also, different mechanisms, e.g., interaction-based regulation and transcript-independent transcriptional interference, can be additive (18). Regarding the expression levels, Lloréns-Rico et al. used a deterministic model to calculate that the threshold expression level for a functional posttranscriptional regulation by asRNAs is higher than the actual measured levels in *Mycoplasma pneumoniae* (16). Nevertheless, asRNAs clearly have a global impact on the transcriptomes of various bacteria, e.g., by RNase III-dependent codegradation (44). In another example, the estimated cellular transcript numbers of the asRNA PsbA2R from *Synechocystis* 6803 are 300 times lower than those of the *psbA2* mRNA; nonetheless, the knockdown of the asRNA resulted in a measurable survival phenotype (16). Furthermore, one should keep in mind that all conclusions drawn on the basis of a transcriptome experiment result from both RNA synthesis and degradation, including the possible effects of codegradation of sense and antisense RNAs by an RNase. Thus, we only see what is left after potential regulatory events. Therefore, studies in RNase III deletion strains yield a more realistic picture of the actual asRNA levels independent of codegradation (41, 44).

SPECIFIC MECHANISMS OF asRNA ACTION IN BACTERIA

Many specific asRNAs have been characterized as important players in the regulation of gene expression. Among the best-studied class of such interactions are the antitoxin asRNAs that inhibit the formation of proteins that are toxic for the cell (25, 35, 45–48). Similar to *trans*-encoded regulatory sRNAs, asRNAs may affect the initiation of translation or modify target stability by the formation of hydrogen bonds with complementary bases in their RNA targets. However, asRNAs differ from sRNAs due to the transcription from the same genomic loci as their targets. The much longer and more pronounced complementarity leads to divergent and additional properties. In the following, the main mechanisms of specific asRNA actions will be reviewed by means of specific examples. An overview of the examples mentioned is given in Table 2, and a more detailed explanation for some mechanisms exists in a previous review (49).

Hfq-Associated and *trans*-Acting, *cis*-Encoded asRNAs

Hfq is an RNA chaperone widely conserved throughout the bacterial domain. It is of crucial importance for many sRNA-mediated regulatory processes by restructuring both the sRNAs and the mRNAs into more interaction-favorable conformations (reviewed by Vogel and Luisi [50]). Hfq's RNA-binding capacity has been addressed in many different bacteria, revealing hundreds of bound mRNAs and *trans*-encoded sRNAs in most of the studied species (51–54). Interaction with Hfq is considered an indicator of functionality for a given sRNA. Therefore, it is intriguing that *cis*-encoded asRNAs have also been identified to interact with Hfq. By coimmunoprecipitation and deep sequencing of RNAs bound to Hfq in *E. coli in vivo*, Bilusic et al. also identified 67 asRNAs (55). Hence, the interaction of these asRNAs with their targets might be facilitated by Hfq. However, only nine corresponding mRNAs were found to be coenriched with the cognate asRNAs via Hfq immunoprecipitation (55). Therefore, these findings might point alternatively to a previously only rarely considered possibility: that *cis*-encoded asRNAs can act also as *trans* regulators. Indeed, there are reports in the literature that certain asRNAs can also function on *trans*-encoded targets. MicA, one of the

best-characterized sRNAs in enterobacteria, is a *trans*-encoded translational inhibitor of OmpA protein synthesis (56, 57). Transcription of MicA overlaps with the 5′ UTR of the *luxS* gene. Higher expression of MicA leads to the occurrence of a cleaved *luxS* mRNA species exhibiting enhanced stability (58).

Further supporting the possibility of *trans* targets for bona fide asRNAs, Melamed et al. identified two *E. coli* asRNAs, GadY and ArrS, simultaneously bound to Hfq and their *trans* interaction partners using the RNA interaction by ligation and sequencing method (RIL-seq) (54). The likely *trans* targets of GadY and ArrS were functionally enriched for genes involved in acid stress response (54). This matches the fact that overexpression of ArrS increases the acid resistance of *E. coli* (59). On the one hand, ArrS stabilizes its cognate mRNA encoding the transcription factor GadE in *cis* (59), whereas it appears to upregulate the cyclopropane fatty acyl phospholipid synthase (*cfa*) mRNA in *trans*. GadY, the other asRNA identified by Melamed et al., was already previously found as binding Hfq (26). GadY originates from a gene antisense to the 3′ end of *gadX*, which is an activator of the glutamate-dependent acid resistance system (26). Upon binding of GadY, cleavage of the *gadXW* dicistronic mRNA is promoted by RNase E (60), generating the more stable monocistronic *gadX* mRNA (61). Hence, with ArrS and GadY, two asRNAs have been identified that act both on a *cis*-encoded and on at least one *trans*-encoded target, supporting survival of the cell under acid stress conditions. Another interesting case was observed in *Staphylococcus aureus*. The 3′ end of the mRNA *sprA1*, coding for a catalytic virulence factor, overlaps by 35 nt with the 3′ end of its asRNA, SprA1$_{AS}$. Not unexpectedly, SprA1$_{AS}$ represses its sense mRNA, *sprA1*. However, it appeared quite unusual that the perfectly complementary 3′ part of the asRNA is not necessary for the regulation. Instead, the asRNA binds like a *trans*-acting sRNA via an imperfect complementarity of its "non-antisense" 5′ part to the 5′ end of *sprA1* and prevents translation (62). In general, a "*trans*" function is more likely for 5′- or 3′-overlapping sRNAs, which can evolve partly independently without the constraint of keeping a functional protein-coding gene in the complementary strand.

Mechanisms That Require RNA-RNA Interaction

Tampering with the stability of target RNAs

The interaction of an asRNA with its target by base-pairing alters the secondary structure of both interact-

Table 2 Names and characteristic features of functionally characterized asRNAs discussed in the text

Name of asRNA	Host	Phylum	Length	Functional mechanism	Outcome	Reference(s)
As1_flv4	Synechocystis sp. PCC 6803	Cyanobacteria	500 and 280 nt	Codegradation with flv4 mRNA	Delayed induction of flv4 under stress and filtering of environmental fluctuation	63
GadY	Escherichia coli	Gammaproteobacteria	105, 90, and 59 nt	Cleavage of the dicistronic gadXW mRNA, likely by RNase E	Enhanced stability of gadX mRNA	26
IsrR	Synechocystis sp. PCC 6803	Cyanobacteria	177 nt	Codegradation with isiA mRNA	Threshold linear response: delayed induction of isiA under stress and faster recovery	27, 76
MtlS	Vibrio cholerae	Gammaproteobacteria	120 nt	Occlusion of ribosome binding site	Repression of mtlA translation	69
OOP asRNA	Bacteriophage λ	Siphoviridae	77 nt	Codegradation: RNase III-dependent cleavage of the mRNA cII	Repression of cII; involved in lysis-lysogeny decision; favors lysis	2–4
PsbA2R and PsbA3R	Synechocystis sp. PCC 6803	Cyanobacteria	130 and 220 nt (PsbA2R), ca. 160 and 180 nt (PsbA3R)	Protection from RNase E cleavage	Stabilized psbA2 and psbA3 mRNA levels under conditions requiring maximum gene expression	66
RNAα	Vibrio anguillarum	Gammaproteobacteria	650 nt	Complementary to fatB gene of the iron transport-biosynthesis operon	Repression of fatA and fatB under iron-rich conditions	70, 71
RNAβ	Vibrio anguillarum	Gammaproteobacteria	427 nt	Complementary to the 3′ region of fatA and the 5′ end of angR	Termination of transcription	70, 71
RnaG	Shigella flexneri	Gammaproteobacteria	450 nt	Complementary to 5′ UTR of icsA (virG) virulence gene; transcriptional interference with convergent icsA promoter and transcription attenuation of icsA mRNA	Premature termination of transcription and direct reduction of icsA transcription	73
SymR	Escherichia coli	Gammaproteobacteria	77 nt	Overlaps the symE mRNA 5′ end and ribosome binding site, preventing the initiation of translation	Control of SymE translation	25, 47
SprA1AS	Staphylococcus aureus	Firmicutes	60 nt	Interacts with its 5′ sequence as a trans-acting sRNA to the 5′ part of its complementary gene and prevents translation	Repression of the virulence factor SprA1	62

ing molecules and results in duplex formation. Those structural changes influence the stability and half-life of the RNAs, with various outcomes. In some cases, duplex formation leads to codegradation, i.e., the rapid and complete degradation of both transcripts.

A prominent and early-described example is the 77-nt OOP asRNA of phage λ. The RNA is complementary to the 3′ end of the λ *c*II repressor mRNA. Overexpression of OOP led to the RNase III-dependent cleavage of the *c*II mRNA, initially at two sites, one in the 3′ end of the coding region and one in the *c*II and O gene intergenic region (2–4). Hence, codegradation between an asRNA and its target transcript became the archetype of asRNA function. Another example of codegradation is the *isiA*/IsrR sense-antisense pair in the cyanobacterium *Synechocystis* 6803 (27). The accumulation of these transcripts is strictly inverse to each other; therefore, both RNAs exist as almost mutually exclusive species. The transcription of *isiA* is induced upon iron, redox, or light stress. When *isiA* becomes induced, the two transcripts can interact and become codegraded. Therefore, *isiA* is not detected until the number of *isiA* mRNA molecules has titrated the number of IsrR molecules. As a consequence, IsrR causes a delay of *isiA* expression in the early stress phase as well as a faster depletion of *isiA* mRNA during the recovery from stress. This mode of asRNA function leads to the "threshold linear response" (49).

The interaction of mRNA and asRNA does not necessarily result in rapid degradation. This is illustrated by the As1_flv4 asRNA in *Synechocystis* 6803. Similar to IsrR and in the same organism, it prevents the premature expression of the *flv4-2* operon by codegradation (63). This operon is induced upon shifts in inorganic carbon supply, effectively opening alternative electron valves for the photosynthetic machinery (64). However, the As1_flv4-mRNA interaction appears not as tight as for IsrR-*isiA* mRNA, indicated by the incomplete degradation of the asRNA (63).

An asRNA-mRNA duplex can also generate a specific processing site, which may lead to the formation of a stabilized form of the mRNA. Examples include the *E. coli* GadY asRNA, which is involved in the response to acid stress by selective stabilization of the *gadX* mRNA (61), as outlined above. Other examples may include the stabilization of the *luxS* mRNA via MicA (58) and of the *gadE* mRNA via the asRNA ArrS (59).

The majority of distinct functions have been described for strongly expressed bacterial asRNAs. However, many occur at relatively low abundance, raising the question of whether these low-abundance asRNAs

can perform specific functions in addition to the possibly global functions. Evidence for an elaborate interplay between asRNAs and RNase E was found with asRNAs overlapping the 5′ region of an mRNA inhibiting RNase E-dependent decay. In photosynthetic organisms, the *psbA* genes encode the D1 reaction center protein of water-splitting photosystem II. The D1 protein shows faster turnover than any other proteins of the photosynthetic machinery and needs to be continuously synthesized, especially under conditions of high irradiance or photooxidative stress. However, the highly abundant *psbA* mRNA becomes degraded rapidly in the dark. Besides transcriptional repression, Horie et al. showed several closely spaced recognition sites for RNase E in an AU-rich element surrounding the ribosome binding site, which trigger mRNA turnover in the dark (65). In *Synechocystis* 6803, this site is protected by relatively low-abundance asRNAs in the light, which originate within the 5′ UTR and effectively block RNase E access together with recruitment by the ribosome. Hence these asRNAs provide an example of mRNA stabilization and positive regulation of *psbA* gene expression by *cis*-encoded asRNAs (66). Although the mechanism has remained unknown, another asRNA, RblR, was identified that exerts a positive effect on gene expression of *rbcL*, encoding the large chain of RubisCO, the enzyme that catalyzes carbon fixation in cyanobacteria (67).

Modulation of translation

In some cases, the degradation of the RNA is a secondary effect of the suppressed gene expression. The SOS response-induced protein SymE in enterobacteria is a type I toxin RNA endonuclease. One of at least three repression mechanisms under standard conditions involves the asRNA SymR (25). SymR overlaps the 5′ end of the *symE* mRNA, including the ribosome binding site and the start codon. Therefore, formation of a *symE* mRNA-SymR duplex leads to its occlusion, preventing the initiation of translation and triggering the enhanced degradation of the *symR* mRNA (25, 47).

The *Vibrio cholerae* MtlS asRNA represses the synthesis of MtlA, a mannitol transporter protein produced exclusively in the presence of mannitol (68). The TSS of MtlS is only 5 nt upstream from the *mtlA* start codon. Hence, MtlS fully covers the ribosome binding site, and its occlusion as a mechanism for translational repression was suggested and subsequently experimentally confirmed (69). One should note that in the case of such 5′ UTR-covering asRNAs a few nucleotides' difference in the distance between the TSS of the asRNA and the start codon can make a big difference. MtlS starts 5 nt upstream from the *mtlA* start codon and is repressive; in contrast, the asRNAs PsbAR2 and PsbAR3 start 19 nt upstream from the respective start codons (66) and stabilize the mRNA, together with the bound ribosomes (Fig. 3 and see below). In the case of MtlS, other mechanisms in addition to the occlusion of the ribosome binding site may be involved. Proximity to the target was found as an important element of the regulation and interpreted to be due to the high local concentrations of the interacting molecules. However, transcriptional interference could also explain this effect.

Direct influence of asRNAs on the transcription of target genes

There are also mechanisms that specifically act on the target gene's transcription. The *Vibrio anguillarum* iron transport-biosynthesis operon includes four ferric siderophore transport genes (*fatDCBA*), two siderophore

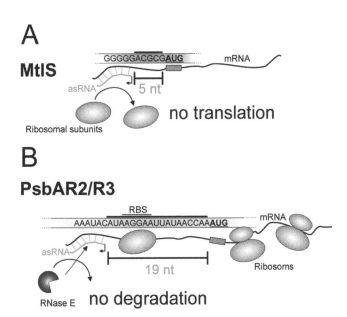

Figure 3 Distance does matter. The divergent effects of two asRNAs (colored red) initiated within the 5′ UTR of a gene (black) are compared. (A) MtlS, an asRNA in *V. cholerae*, starts 5 nt upstream from the *mtlA* start codon in inverse orientation and is repressive (68). (B) PsbAR2 and PsbAR3, two asRNAs in *Synechocystis* 6803, start 19 nt upstream from the respective start codons (66). The target genes, *psbA2* and *psbA3*, are in the shown region identical to each other. The 5′ UTR of the *psbA2* and *psbA3* mRNAs is a substrate for the RNase E endoribonuclease. The cleavage occurs in an AU-rich element, preferably at the sites indicated by the dashed arrows (65), which was recently confirmed in an independent study (107). The ribosome binding site (RBS) was defined previously (65). As a consequence, PsbAR2 and PsbAR3 stabilize the mRNA, together with the bound ribosomes (66).

biosynthesis genes (*angR* and *angT*), and two asRNAs, called RNAα and RNAβ (70, 71). RNAβ is complementary to the 3′ region of *fatA* and the 5′ end of *angR*. Binding of RNAβ to the growing polycistronic *fatDCBA* message was found to lead to termination of transcription close to the stop codon of *fatA*. Alternative mechanisms like codegradation or transcriptional interference could be excluded because the regulation could be reproduced in an *in vitro* transcription assay (72).

Another example is the *Shigella flexneri* virulence gene *icsA* (*virG*), which is controlled by the asRNA RnaG. In addition to transcriptional interference, this asRNA represses *icsA* transcription by transcription attenuation (73). It was suggested that this regulation proceeds through two long hairpin motifs, which resemble an antiterminator structure within the *icsA* mRNA 5′ segment. The binding of RnaG to the nascent mRNA interferes with the formation of the antiterminator. Instead, a terminator stem-loop can form (73). VirF, the master activator of virulence genes in *Shigella*, activates expression of *icsA* at the transcriptional level. Interestingly, the transcription factor VirF not only binds to DNA but can also bind RNA. By binding to interaction sites within the RNA molecules of *icsA* and its asRNA RnaG, VirF hampers the sense-antisense interaction and hence alleviates the RnaG-mediated termination of *icsA* transcription (74).

GLOBAL FUNCTIONS THAT HAVE BEEN ASSOCIATED WITH ANTISENSE TRANSCRIPTION IN BACTERIA

In view of the substantial pervasive antisense transcription detected for the majority of studied bacteria, it is tempting to find general global mechanisms for asRNAs or to use experimental and computational approaches to address the importance of asRNAs in a more general way. Following the rationale that functionally important genes should be evolutionarily conserved, the conservation of asRNAs between closely related species was analyzed in different studies. The compared organisms were *E. coli* and *Salmonella enterica*, different *Shewanella* strains, the two closely related marine cyanobacteria *Prochlorococcus* sp. MED4 and *Prochlorococcus* sp. MIT9313, and the freshwater cyanobacteria *Synechocystis* 6803 and *Synechocystis* 6714 (12, 38, 75, 76). All four studies showed that the degree of conservation of asRNAs or of the TSSs driving asRNA transcription is lower than the conservation of protein-coding genes, intergenic sRNAs, or intragenic sense RNAs. This could be interpreted as evidence that the majority of asRNAs might be rather transcriptional noise. However, while conservation is clearly a strong indicator for importance, the lack of conservation does not rule out functionality, as shown in the following example. The high-light-induced PsbA2R asRNA in *Synechocystis* 6803 protects the *psbA2* mRNA, coding for a major component of the photosystem II, against RNase E-dependent degradation. This contributes to the light/dark acclimation of the photosynthetic organism (66). Nevertheless, due to a single nucleotide polymorphism in the -10 element, the promoter of the asRNA is not functional in the closely related *Synechocystis* 6714 and the asRNA is not conserved (76). This observation fits with the fact that bacterial promoters are rather simple. The main requirements are a -10 element or Pribnow box with the consensus 5′-ANAAT-3′ and a low melting temperature (16). Consequently, Lloréns-Rico et al. argued that random promoters should appear more frequently in organisms with a higher genomic AT content. Indeed, they showed that the numbers of asRNAs per megabase were positively correlated with the AT percentage, while the numbers of intergenic-encoded sRNAs were not (16). The above findings support the transcriptional noise hypothesis, but they also indicate that bacterial genomes have a high flexibility to gain and lose asRNAs. These asRNAs could be a valuable resource for the fast (strain-specific) generation of regulatory elements to cope with individual challenges. In an alternative view, maybe only the existence of asRNAs is important, but not the exact positioning, e.g., if asRNAs are required to globally control the half-life of mRNAs (41, 44, 77–80) or the transcription of asRNAs is important to enable transcription-coupled DNA repair (17). Experimental evidence for a very high share of asRNAs in an AT-rich genome was obtained for *Prochlorococcus* MED4 (38), in which, even when counted very conservatively, 75% of all genes were associated with asRNAs (Table 1). Interestingly, *Prochlorococcus* MED4 also exhibits the fastest ever measured global RNA turnover, with a global median half-life of only 2.4 min and a median decay rate of 2.6 min (81).

To boil everything down: excessive pervasive transcription is likely detrimental to cell viability and thus controlled by various mechanisms. This includes Rho and NusG (11, 82, 83), which prevent asRNA formation at the transcriptional level, while RNase H degrades (as)RNA in DNA-RNA hybrids (8). In *E. coli*, the function of Rho seems to be supported by signals within the RNA transcript, so-called inhibitory RNA aptamers (iRAPs) (84). The sequences of iRAPs are enriched opposite to the 3′ ends of coding sequences

and likely prevent transcriptional interference by transcription from antisense promoters (84). Nevertheless, substantial antisense transcription exists and many functional asRNAs have been described. The difference between transcriptional noise and functional antisense transcription appears to be quite fluid, and they are hardly distinguishable from one another at a global scale.

In the following, we address some "global" functions of asRNAs in more detail. In a strict definition, that should be only functions that require a certain amount of antisense RNA or antisense transcription independent of the sense gene. However, we also include examples that act on individual targets but by a globally shared mechanism and with a global cellular impact.

Transcription-Associated Mechanisms

For some mechanisms, the process of transcription itself is important and the resulting asRNAs might be nonfunctional by-products. An example is the transcription-coupled DNA repair described below or regulation by transcriptional interference. Transcription together with the *de novo* transcript can also lead to the formation of DNA-RNA hybrids such as R-loops (23) or G-loops (85). A more speculative function of asRNA transcription is that some of these R-loops might have beneficial functions.

Mechanisms involving RNA-DNA hybrids

A newly transcribed RNA molecule can form a hybrid with the codogenic DNA strand while the other DNA strand is displaced, a structure called an R-loop (23). R-loop formation is more pronounced from weakly structured noncoding transcripts, because hybrid formation is not counteracted by translating ribosomes or intramolecular secondary structures (23). R-loops are detrimental to cell viability as they can interfere with replication, increase DNA damage, and reduce genome stability (23). Consequently, massive R-loop formation is prevented by, e.g., Rho- and NusG-dependent transcription termination (86) as well as by the degradation of the RNA in RNA-DNA hybrids by RNase H (8). Combined Rho and RNase H deficiency resulted in >500 antisense transcription-dependent R-loops in *E. coli* (8).

However, individual R-loops can have beneficial functions. In eukaryotes, R-loops are involved in, for example, hypermutation (23), activation of transcription (87), or the class-switching recombination in B cells (88). All mechanisms are also thinkable in bacteria. Local R-loop formation might, e.g., facilitate transcription factor binding or transcription initiation in GC-rich promoter regions with an otherwise high melting temperature. At other positions R-loops might represent roadblocks that inhibit transcription. An interesting aspect is that R-loop-dependent mechanisms would require only a very low amount of asRNAs. Basically, only one asRNA that forms a stable hybrid with the codogenic DNA strand is required.

In bacteria, an RNA-DNA hybrid was shown to be required to prime a homologous recombination event. Studying the antigenic variation of PilE, a major pilin, an sRNA originating immediately upstream from the *pilE* gene was identified in *Neisseria gonorrhoeae*. This sRNA forms an RNA-DNA hybrid that is required for the formation of a guanine quartet structure in the displaced DNA strand and subsequent recombination (24). Although not strictly an asRNA, we have included it here because it functions in *cis*, and analogous mechanisms involving asRNAs are possible. Interestingly, the amount of antigenic variation seems to be adjustable by an asRNA that overlaps the whole *pilE* transcript and the promoter of the above-mentioned guanine quartet structure-associated sRNA (89).

Transcriptional interference

Simultaneous transcription from convergent promoters of overlapping transcripts can interfere due to the collision of both RNA polymerases (RNAPs), a phenomenon called transcriptional interference (TI). Very likely TI is not an isolated phenome, but frequently works in addition to other mechanisms that require RNA-RNA interaction, such as codegradation. In an elegant approach, Brophy and Voigt tested the performance of different artificial asRNA promoter and terminator combinations in *E. coli* (18). In the test setup, the asRNA originated from the 3′ side of a transcriptional repressor that in turn represses the synthesis of Yfp. The strength of the asRNA-based regulation could then be measured by the level of Yfp fluorescence. The authors could furthermore disentangle TI-based and interaction-based posttranscriptional regulation by comparing the effect of the asRNAs transcribed in *cis* or in *trans*. They conclude that both effects act synergistically and contribute roughly equally to the regulation (18).

Basically, three variations of the theme have been described. (i) The collision of two divergently elongating RNAP complexes results in the premature termination of one or both transcription events. (ii) Promoter occlusion occurs when an elongating RNAP passes over another promoter element and thus prevents formation of an initiation complex at this promoter. (iii) In the case of "sitting duck" interference, a preinitiation RNAP at an open complex is removed by the collision with an elongating RNAP complex. Depending on the charac-

teristics of the opposing promoters, e.g., spacing and strength, one or the other type of TI might dominate (22). In brief, the collision-type TI is most pronounced when the overlapping region is long so that the probability of an encounter of the two RNAPs is increased, a situation that often appears in, e.g., excludon-type arrangements. Promoter occlusion requires either a strong promoter that ensures frequent passage of an RNAP over the controlled promoter or a pausing of the RNAP in the controlled promoter region. Sitting duck TI is strongest at closely spaced promoters of moderate strength (22). The consequences of TI-based regulation can be tuned by various parameters including relative promoter strength, the distance between the promoters, the sequence content, the kinetics of involved transcriptional regulators, and the combination with posttranscriptional mechanisms (18–21).

TI-based regulation was also shown to be involved in the generation of bistable switches in bacteria. The excludon-like convergent overlapping expression of the antagonistic genes *prgX* and *prgQ* in *Enterococcus faecalis* establishes a switch that controls conjugative transfer of a resistance plasmid (90). For *Streptomyces coelicolor*, Chatterjee et al. showed that TI-based regulation of the *scbA-scbR* system is required for the switch between either an antibiotic-producing lifestyle or a lifestyle in which no antibiotics are made (91). In a modeling approach, Bordoy and Chatterjee generalized that combined asRNA-dependent TI and posttranscriptional regulation can give rise to hypersensitive switches, which can show bistable behavior if protein-based feedback loops are involved (21). Other examples for TI are the regulation of the *ubiG-mccBA* operon in *Clostridium acetobutylicum* (92), the interference of the divergent λ phage P_R and P_{RE} promoters (93), and the interference of the lytic-phase promoter (pR) with the lysogenic promoter (pL) of the bacteriophage 186 (94).

Transcriptome Reshaping, Repression of General Gene Expression, Cell Fate Decision, and Phenotypic Variation

Another global function of asRNA might be to accelerate the replacement of a "standard growth condition" transcriptome by another, largely different, "stress-specific" transcriptome. In *Bacillus subtilis*, asRNAs are less often expressed from SigA (vegetative sigma factor)-dependent promoters than mRNAs and, furthermore, the expression of sense-antisense pairs is often anticorrelated (95). While this can be interpreted in a way that asRNAs are more often the result of transcriptional noise or randomly generated promoters (95), an interesting hypothesis is that these asRNAs are activated by alternative sigma factors at stress conditions to repress or enhance the turnover of mRNAs from SigA-dependent genes. Under severe stress conditions like, e.g., ethanol stress, in *B. subtilis* the gene expression machinery for transcription and translation is repressed (95). In *Bacillus* the stress-dependent

SigB regulon consists of ~200 protein-coding genes and 136 putative regulatory RNAs (96). This includes an asRNA that is responsible for the ethanol stress-dependent repression of the essential *rpsD* gene, thereby contributing to the reduction in the number of ribosomes (96) (Fig. 4A). Other bacteria respond to severe stress conditions in a similar way. After pro-

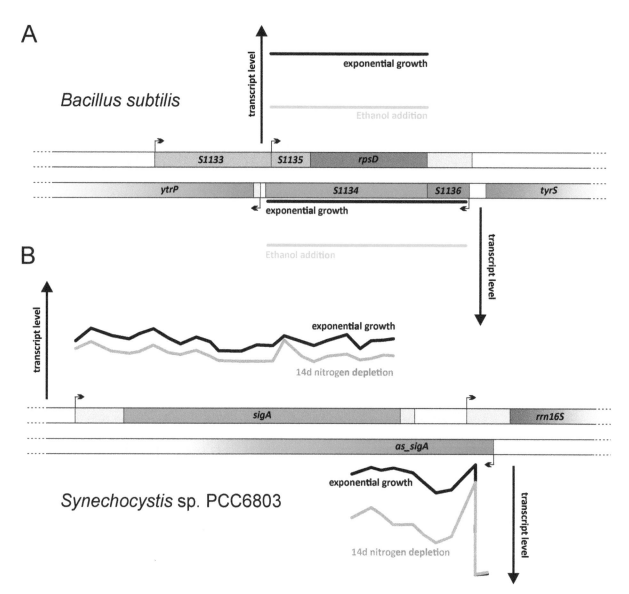

Figure 4 Stress-induced asRNAs functioning in global transcriptome remodeling. (A) Ethanol addition triggers the SigB-dependent transcription of the S1136-S1134 asRNA in *B. subtilis* (96). This asRNA contributes to the reduction in the number of ribosomes during ethanol stress by repressing *rpsD*, encoding the ribosomal protein S4 (96). (B) Expression of asRNAs overlapping the *sigA* gene in *Synechocystis* 6803, which become strongly induced upon long-term nitrogen depletion. The figure has been redrawn according to information about the *Synechocystis* 6803 primary transcriptome (106) and the transcriptome analysis during prolonged nitrogen starvation (97). Note the location of this asRNA linking one of the ribosomal RNA operons with *sigA*, encoding the vegetative sigma factor.

longed nitrogen depletion, *Synechocystis* 6803 falls into a dormant state where, e.g., the photosynthetic apparatus is completely degraded and the transcription and translation machinery becomes strongly repressed (97). This is associated with a strong decrease in the total cellular levels of the housekeeping sigma factor SigA in the soluble and RNAP-bound forms, whereas the levels of the alternative sigma factor SigC increase (98). Interestingly, 252 asRNAs have significantly higher levels under this condition, while only 84 asRNAs show lower levels. This includes the >30-fold induction of an asRNA that overlaps the *sigA* gene and the gene coding for the 16S rRNA (Fig. 4B) and strongly exceeds the expression of the mRNA at nitrogen depletion (97).

In many bacteria, termination factor Rho is involved in the repression of pervasive, mostly antisense, transcription (82, 83). However, there is good evidence that Rho is involved in the differential accumulation of certain asRNAs. Conditions that lead to lower cellular Rho concentrations should induce increased asRNA transcription, with global consequences. Based on insight obtained in a Rho inactivation strain, Bidnenko et al. (82) showed that the control of specific transcripts by Rho contributes to the balanced regulation of three mutually exclusive differentiation programs in *B. subtilis*. When Rho was lacking, several sense and antisense transcripts became upregulated. This, in turn, affected key elements of the genetic programs leading to sporulation, cell motility, or biofilm formation (82). The motility phenotype could, for example, be partially linked to an asRNA that is normally Rho-dependently terminated but covers the whole *flhO-flhP* operon encoding important flagellum components in the mutant strain (82). It is interesting to note that this observation is reminiscent of a long asRNA in the *L. monocytogenes* flagellum biosynthesis excludon (32). Furthermore, an asRNA called RrrR(+) is involved in biofilm formation in the archaeon *Sulfolobus acidocaldarius* (119).

"Noisy" antisense transcription might contribute to phenotypic heterogeneity within a population (49). The study of Bidnenko et al. indicates that stochastic, Rho-dependent transcription termination is an additional source for cell-to-cell differences in RNA content, beyond transcriptional bursts. Indeed, Rho overexpression and Rho-null *B. subtilis* cells are locked in specific cellular states and have lost the ability to show bistable behavior, which could be partly linked to asRNAs (82, 99).

Hence, antisense transcription contributes not only to phenotypic heterogeneity but even to morphologically and physiologically deeply divergent states.

asRNAs and Overlapping Transcripts Together with RNase III as a Means for RNA Maturation

Analyzing the transcriptome of *S. aureus*, Lasa et al. (78) fractionated total RNA into one sample containing longer molecules (>50 nt) and another one with shorter RNA molecules (<50 nt). Both fractions were analyzed separately by cDNA sequencing, revealing a collection of short, ~20-nt-long RNA fragments that accumulated in every genome region where overlapping transcription was detected and which derived from sense-antisense RNA processing by the double-stranded endoribonuclease, RNase III (78). Using a totally different approach, the genome-wide role of RNase III in RNA processing in *S. aureus* was confirmed (44). Combining coimmunoprecipitation of a catalytically dead version of RNase III with deep RNA-seq, Lioliou et al. found that RNase III was bound to many different asRNAs, covering as much as 44% of the annotated coding and noncoding genes. This mechanism is not restricted to *S. aureus* or to Gram-positive bacteria. Many mRNAs enriched together with RNase III were shown to also have cognate asRNAs in *Streptomyces* (42, 43). Using a monoclonal antibody that recognizes double-stranded RNA molecules, Lybecker et al. performed immunoprecipitation experiments with total RNA from *E. coli* wild type and a strain with a catalytically inactive RNase III (41), demonstrating, among other findings, that RNase III mediates the digestion of sense-antisense hybrids also in *E. coli*. While the observed processing by RNase III is fully consistent with earlier results on the codegradation of individual RNA-asRNA pairs and suggests also a mechanism for the excludon-mediated type of gene silencing (32), it raises new questions due to its ubiquity. Under which scenario would the almost global degradation of transcripts make sense? Lasa and Villanueva (77) suggested that this mechanism could turn over mainly mRNA molecules that had already been translated, hence constituting a mechanism of RNA decay.

Antisense Transcription and DNA-Repair Processes

Transcription-coupled repair is the enhanced rate of DNA repair of the actually transcribed strand of a transcription unit relative to the nontranscribed strand and other, nontranscribed regions of the genome. An important factor in this mechanism is the transcription repair-coupling factor encoded by the *mfd* gene (100). This factor binds to RNAP-RNA complexes stalled at DNA lesions and recruits the UvrAB system (101). Using

A   Global Mechanisms

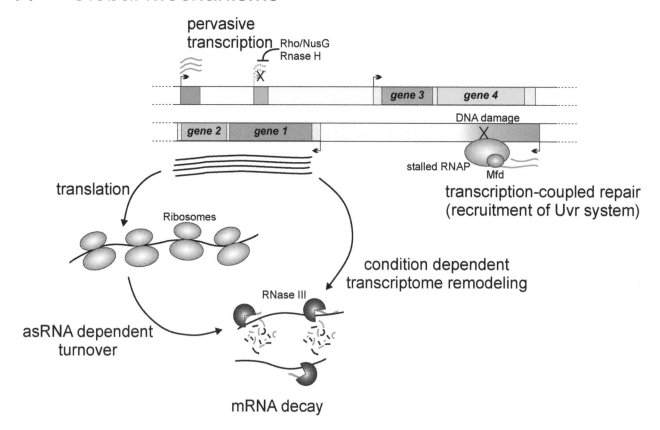

pervasive transcription

Rho/NusG
Rnase H

gene 3   gene 4

gene 2   gene 1

DNA damage

stalled RNAP   Mfd

translation

Ribosomes

transcription-coupled repair
(recruitment of Uvr system)

condition dependent
transcriptome remodeling

RNase III

asRNA dependent
turnover

mRNA decay

B   Specific Mechanisms

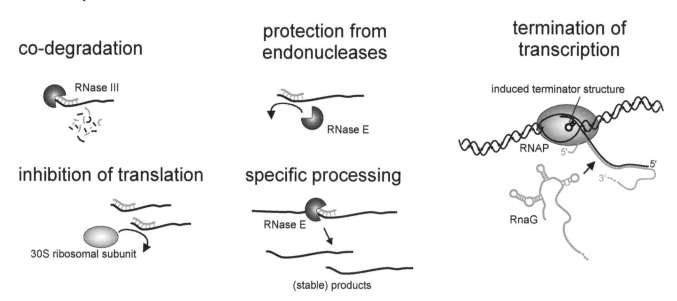

co-degradation

RNase III

protection from
endonucleases

RNase E

termination of
transcription

induced terminator structure

RNAP

5'
3'

5'

RnaG

inhibition of translation

30S ribosomal subunit

specific processing

RNase E

(stable) products

the eXcision Repair-sequencing (XR-seq) approach to map excision repair events along the entire *E. coli* genome, Adebali et al. showed that Mfd-dependent transcription-coupled repair occurs widely throughout the *E. coli* genome (17). Moreover, by comparing the repair data with the published transcriptome data for *E. coli* (7), the authors found a considerable effect of antisense transcription on transcription-coupled repair, exemplified for the *insL1* gene (17). The transcription-repair coupling factor, a large protein of >1,000 amino acids, is widely conserved throughout the bacterial domain. It is therefore tempting to conclude that pervasive transcription serves as a genome-wide scanning mechanism for DNA lesions and repair. However, the effects of antisense transcription would in such a scenario be particularly helpful for genes that are under most conditions not transcribed, e.g., because their products are only needed for responses to specific environmental cues.

Pervasive (Antisense) Transcription as a Driver of Evolution and Genome Flexibility?

This is surely a more speculative role, but pervasive transcription from randomly generated promoters could also generate a pool of transcripts that can be tested for functionality at the population level by natural evolution. As described above, the evolution of asRNA genes seems to be particularly fast, as indicated by the low conservation between closely related species (12, 38, 75, 76) and the positive correlation to the genomic AT percentage (16). Even if the majority of newly generated asRNAs are nonfunctional, the sheer number increases the odds for single beneficial transcripts. The fast evolution might allow a rather rapid fine-tuning of gene expression to fluctuating conditions or different lifestyles. Two asRNAs from cyanobacteria might support this hypothesis. As mentioned above, *Synechocystis* 6803 asRNA PsbA2R enhances resilience to light stress, while the closely related species *Synechocystis* 6714 lacks the asRNA due to a single nucleotide polymorphism (66, 76). In contrast, both strains contain the asRNA IsrR (76), which enables an expression threshold for the iron stress-induced protein A (IsiA) (27, 102). In both organisms, the expression of IsiA is mostly restricted to iron-deplete stress conditions (76). However, in other cyanobacteria such as *Anabaena* or *Nodularia*, the expression of IsiA homologs seems to be differently regulated. Although strains of these genera possess *isiA*, homologs to IsrR could not be detected (103, 104). In these and other closely related strains, *isiA* homologs are embedded in a gene cluster with three to four additional paralogous genes (55), which is also expressed at standard conditions. Because these strains share physiological and developmental peculiarities such as the propensity for the fixation of atmospheric N_2, the formation of heterocysts, and true multicellularity, there is a substantial difference in lifestyle compared to *Synechocystis* cyanobacteria, which is reflected here in the lack of an asRNA.

Brophy and Voigt showed that asRNA transcription coming from the 3′ end of a sense transcript can tune repression of the respective gene in a >30-fold range depending on the asRNA promoter strength. Based on these findings, the authors discuss that it might be easier for an organism to establish an asRNA-based regulation to control gene expression than to mutate the sense promoter, which furthermore could possibly interfere with signal integration through the different transcription factors (18).

Figure 5 Scenario of the evolutionary processes in bacteria leading from pervasive transcription to global functions of asRNAs and to highly specific roles and mechanisms. (A) Global mechanisms. TSSs (arrows) originate relatively easily due to the simplicity of bacterial promoters. They give rise to various transcript types, including mRNAs (black) and asRNAs (red). These transcripts are not automatically functional. The TSSs with detrimental effects will rapidly be selected out by evolution or the pervasive transcription is counteracted by diverse safety mechanisms involving e.g., Rho, NusG and RNAse H (cross). However, in many instances transcription is beneficial. Thus, global functions of antisense transcription can be exerted at the DNA level as well as the RNA level. Examples at the DNA level include transcription-coupled repair; at the RNA level, asRNAs contribute to transcriptome remodeling and possibly mRNA decay after translation. (B) Specific roles. The rich pool of existing asRNAs is a resource from which some become associated with a specific role (only selected examples are shown). These specific roles may interfere with the transcription of specific genes, here exemplified by the RnaG asRNA, which upon base-pairing to the *icsA* mRNA inhibits the formation of an antiterminator, leading to termination of transcription. Multiple examples exist for the involvement of asRNAs in hampering the translation of specific mRNAs, in codegradation by recruiting RNase III, or in providing protection from cleavage by masking RNase E cleavage sites.

Newly generated asRNA genes might not only act as regulators of the corresponding sense genes; they are also sources for novel *trans*-acting sRNAs (54, 55, 58, 59) or (small) proteins (36, 37). The generation and evolution of *trans*-acting sRNAs is discussed in chapter 28 (105). In terms of genome flexibility, even more-exotic hypothetical functions are thinkable, e.g., the generation of predetermined breaking points for targeted recombination by R-loop formation, as described above (24, 88).

OUTLOOK AND FURTHER DIRECTIONS

The understanding of antisense transcription and the involved evolutionary processes has substantially changed within recent years (Fig. 5). Transcriptome analyses performed in many different bacteria demonstrated massive and pervasive transcription including substantial antisense transcription. The very process of pervasive transcription can be functional, as demonstrated e.g., by the recent findings on transcription-coupled repair. Antisense transcription appears to be of particular global relevance in setting thresholds for gene expression together with RNase III and possibly as an element of a mechanism for rapid turnover of RNA molecules that are not, or no longer, translated. The vast numbers of asRNAs provide ample material from which further, specific antisense regulators can evolve. These may become regulated in specific ways but keep the mechanism of codegradation, hence acting on a single target under particular conditions. They may also evolve additional functions and alternative mechanisms. In this way, asRNAs interfering with the initiation of translation, playing a role in transcriptional interference, serving as beacons for RNA processing or the termination of transcription may have evolved. Hence, asRNA appears much more versatile and to be involved in a larger variety of processes than thought 5 years ago. Certainly, antisense transcription and asRNAs have to be taken seriously.

Acknowledgments. W.R.H. was supported during the writing phase of this review by the Freiburg Institute of Advanced Studies (FRIAS).

Citation. Georg J, Hess WR. 2018. Widespread antisense transcription in prokaryotes. Microbiol Spectrum 6(4):RWR-0029-2018.

References

1. Itoh T, Tomizawa J. 1980. Formation of an RNA primer for initiation of replication of ColE1 DNA by ribonuclease H. *Proc Natl Acad Sci U S A* 77:2450–2454.

2. Krinke L, Mahoney M, Wulff DL. 1991. The role of the OOP antisense RNA in coliphage λ development. *Mol Microbiol* 5:1265–1272.

3. Krinke L, Wulff DL. 1990. RNase III-dependent hydrolysis of λcII-O gene mRNA mediated by λ OOP antisense RNA. *Genes Dev* 4:2223–2233.

4. Krinke L, Wulff DL. 1987. OOP RNA, produced from multicopy plasmids, inhibits λ *cII* gene expression through an RNase III-dependent mechanism. *Genes Dev* 1:1005–1013.

5. Brantl S. 2007. Regulatory mechanisms employed by *cis*-encoded antisense RNAs. *Curr Opin Microbiol* 10:102–109.

6. Selinger DW, Cheung KJ, Mei R, Johansson EM, Richmond CS, Blattner FR, Lockhart DJ, Church GM. 2000. RNA expression analysis using a 30 base pair resolution *Escherichia coli* genome array. *Nat Biotechnol* 18:1262–1268.

7. Thomason MK, Bischler T, Eisenbart SK, Förstner KU, Zhang A, Herbig A, Nieselt K, Sharma CM, Storz G. 2015. Global transcriptional start site mapping using differential RNA sequencing reveals novel antisense RNAs in *Escherichia coli*. *J Bacteriol* 197:18–28.

8. Raghunathan N, Kapshikar RM, Leela JK, Mallikarjun J, Bouloc P, Gowrishankar J. 2018. Genome-wide relationship between R-loop formation and antisense transcription in *Escherichia coli*. *Nucleic Acids Res* 46:3400–3411.

9. Georg J, Voss B, Scholz I, Mitschke J, Wilde A, Hess WR. 2009. Evidence for a major role of antisense RNAs in cyanobacterial gene regulation. *Mol Syst Biol* 5:305.

10. Perocchi F, Xu Z, Clauder-Münster S, Steinmetz LM. 2007. Antisense artifacts in transcriptome microarray experiments are resolved by actinomycin D. *Nucleic Acids Res* 35:e128.

11. Wade JT, Grainger DC. 2014. Pervasive transcription: illuminating the dark matter of bacterial transcriptomes. *Nat Rev Microbiol* 12:647–653.

12. Raghavan R, Sloan DB, Ochman H. 2012. Antisense transcription is pervasive but rarely conserved in enteric bacteria. *mBio* 3:e00156–e12.

13. Wagner EG, Romby P. 2015. Small RNAs in bacteria and archaea: who they are, what they do, and how they do it. *Adv Genet* 90:133–208.

14. Storz G, Wolf YI, Ramamurthi KS. 2014. Small proteins can no longer be ignored. *Annu Rev Biochem* 83:753–777.

15. Meydan S, Vázquez-Laslop N, Mankin AS. 2018. Genes within genes in bacterial genomes. *Microbiol Spectr* 6:RWR-0020-2018.

16. Lloréns-Rico V, Cano J, Kamminga T, Gil R, Latorre A, Chen WH, Bork P, Glass JI, Serrano L, Lluch-Senar M. 2016. Bacterial antisense RNAs are mainly the product of transcriptional noise. *Sci Adv* 2:e1501363.

17. Adebali O, Chiou YY, Hu J, Sancar A, Selby CP. 2017. Genome-wide transcription-coupled repair in *Escherichia coli* is mediated by the Mfd translocase. *Proc Natl Acad Sci U S A* 114:E2116–E2125.

18. Brophy JA, Voigt CA. 2016. Antisense transcription as a tool to tune gene expression. *Mol Syst Biol* 12:854.

19. Bordoy AE, Varanasi US, Courtney CM, Chatterjee A. 2016. Transcriptional interference in convergent pro-

moters as a means for tunable gene expression. *ACS Synth Biol* **5**:1331–1341.

20. **Hao N, Palmer AC, Ahlgren-Berg A, Shearwin KE, Dodd IB.** 2016. The role of repressor kinetics in relief of transcriptional interference between convergent promoters. *Nucleic Acids Res* **44**:6625–6638.

21. **Bordoy AE, Chatterjee A.** 2015. *cis*-antisense transcription gives rise to tunable genetic switch behavior: a mathematical modeling approach. *PLoS One* **10**:e0133873.

22. **Sneppen K, Dodd IB, Shearwin KE, Palmer AC, Schubert RA, Callen BP, Egan JB.** 2005. A mathematical model for transcriptional interference by RNA polymerase traffic in *Escherichia coli*. *J Mol Biol* **346**:399–409.

23. **Gowrishankar J, Leela JK, Anupama K.** 2013. R-loops in bacterial transcription: their causes and consequences. *Transcription* **4**:153–157.

24. **Cahoon LA, Seifert HS.** 2013. Transcription of a *cis*-acting, noncoding, small RNA is required for pilin antigenic variation in *Neisseria gonorrhoeae*. *PLoS Pathog* **9**:e1003074.

25. **Kawano M, Aravind L, Storz G.** 2007. An antisense RNA controls synthesis of an SOS-induced toxin evolved from an antitoxin. *Mol Microbiol* **64**:738–754.

26. **Opdyke JA, Kang JG, Storz G.** 2004. GadY, a small-RNA regulator of acid response genes in *Escherichia coli*. *J Bacteriol* **186**:6698–6705.

27. **Dühring U, Axmann IM, Hess WR, Wilde A.** 2006. An internal antisense RNA regulates expression of the photosynthesis gene *isiA*. *Proc Natl Acad Sci U S A* **103**:7054–7058.

28. **Toledo-Arana A, Dussurget O, Nikitas G, Sesto N, Guet-Revillet H, Balestrino D, Loh E, Gripenland J, Tiensuu T, Vaitkevicius K, Barthelemy M, Vergassola M, Nahori MA, Soubigou G, Régnault B, Coppée JY, Lecuit M, Johansson J, Cossart P.** 2009. The *Listeria* transcriptional landscape from saprophytism to virulence. *Nature* **459**:950–956.

29. **Stazic D, Lindell D, Steglich C.** 2011. Antisense RNA protects mRNA from RNase E degradation by RNA-RNA duplex formation during phage infection. *Nucleic Acids Res* **39**:4890–4899.

30. **Stazic D, Pekarski I, Kopf M, Lindell D, Steglich C.** 2016. A novel strategy for exploitation of host RNase E activity by a marine cyanophage. *Genetics* **203**:1149–1159.

31. **Wurtzel O, Sesto N, Mellin JR, Karunker I, Edelheit S, Bécavin C, Archambaud C, Cossart P, Sorek R.** 2012. Comparative transcriptomics of pathogenic and non-pathogenic *Listeria* species. *Mol Syst Biol* **8**:583.

32. **Sesto N, Wurtzel O, Archambaud C, Sorek R, Cossart P.** 2013. The excludon: a new concept in bacterial antisense RNA-mediated gene regulation. *Nat Rev Microbiol* **11**:75–82.

33. **Ning D, Liu S, Xu W, Zhuang Q, Wen C, Tang X.** 2013. Transcriptional and proteolytic regulation of the toxin-antitoxin locus *vapBC10* (*ssr2962/slr1767*) on the chromosome of *Synechocystis* sp. PCC 6803. *PLoS One* **8**:e80716.

34. **Kopfmann S, Roesch SK, Hess WR.** 2016. Type II toxin-antitoxin systems in the unicellular cyanobacterium *Synechocystis* sp. PCC 6803. *Toxins (Basel)* **8**:E228.

35. **Masachis Gelo S, Darfeuille F.** 2018. Type I toxin-antitoxin systems: regulating toxin expression via Shine-Dalgarno sequestration and small RNA binding. *Microbiol Spectr* **6**:RWR-0030-2018.

36. **Silby MW, Levy SB.** 2008. Overlapping protein-encoding genes in *Pseudomonas fluorescens* Pf0-1. *PLoS Genet* **4**:e1000094.

37. **Haycocks JRJ, Grainger DC.** 2016. Unusually situated binding sites for bacterial transcription factors can have hidden functionality. *PLoS One* **11**:e0157016.

38. **Voigt K, Sharma CM, Mitschke J, Lambrecht SJ, Voß B, Hess WR, Steglich C.** 2014. Comparative transcriptomics of two environmentally relevant cyanobacteria reveals unexpected transcriptome diversity. *ISME J* **8**:2056–2068.

39. **Schlüter JP, Reinkensmeier J, Daschkey S, Evguenieva-Hackenberg E, Janssen S, Jänicke S, Becker JD, Giegerich R, Becker A.** 2010. A genome-wide survey of sRNAs in the symbiotic nitrogen-fixing alpha-proteobacterium *Sinorhizobium meliloti*. *BMC Genomics* **11**:245.

40. **Cohen O, Doron S, Wurtzel O, Dar D, Edelheit S, Karunker I, Mick E, Sorek R.** 2016. Comparative transcriptomics across the prokaryotic tree of life. *Nucleic Acids Res* **44**:W46–W53.

41. **Lybecker M, Zimmermann B, Bilusic I, Tukhtubaeva N, Schroeder R.** 2014. The double-stranded transcriptome of *Escherichia coli*. *Proc Natl Acad Sci U S A* **111**:3134–3139.

42. **Gatewood ML, Bralley P, Weil MR, Jones GH.** 2012. RNA-Seq and RNA immunoprecipitation analyses of the transcriptome of *Streptomyces coelicolor* identify substrates for RNase III. *J Bacteriol* **194**:2228–2237.

43. **Šetinová D, Šmídová K, Pohl P, Musić I, Bobek J.** 2018. RNase III-binding-mRNAs revealed novel complementary transcripts in *Streptomyces*. *Front Microbiol* **8**:2693.

44. **Lioliou E, Sharma CM, Caldelari I, Helfer A-C, Fechter P, Vandenesch F, Vogel J, Romby P.** 2012. Global regulatory functions of the *Staphylococcus aureus* endoribonuclease III in gene expression. *PLoS Genet* **8**:e1002782.

45. **Durand S, Gilet L, Condon C.** 2012. The essential function of *B. subtilis* RNase III is to silence foreign toxin genes. *PLoS Genet* **8**:e1003181.

46. **Fozo EM, Hemm MR, Storz G.** 2008. Small toxic proteins and the antisense RNAs that repress them. *Microbiol Mol Biol Rev* **72**:579–589.

47. **Kawano M.** 2012. Divergently overlapping *cis*-encoded antisense RNA regulating toxin-antitoxin systems from *E. coli*: *hok/sok*, *ldr/rdl*, *symE/symR*. *RNA Biol* **9**:1520–1527.

48. **Coray DS, Wheeler NE, Heinemann JA, Gardner PP.** 2017. Why so narrow: distribution of anti-sense regulated, type I toxin-antitoxin systems compared with type II and type III systems. *RNA Biol* **14**:275–280.

49. **Georg J, Hess WR.** 2011. *cis*-antisense RNA, another level of gene regulation in bacteria. *Microbiol Mol Biol Rev* **75**:286–300.

50. **Vogel J, Luisi BF.** 2011. Hfq and its constellation of RNA. *Nat Rev Microbiol* **9**:578–589.

51. **Heidrich N, Bauriedl S, Barquist L, Li L, Schoen C, Vogel J.** 2017. The primary transcriptome of *Neisseria*

meningitidis and its interaction with the RNA chaperone Hfq. *Nucleic Acids Res* **45:**6147–6167.

52. Chao Y, Papenfort K, Reinhardt R, Sharma CM, Vogel J. 2012. An atlas of Hfq-bound transcripts reveals 3′ UTRs as a genomic reservoir of regulatory small RNAs. *EMBO J* **31:**4005–4019.

53. Sittka A, Lucchini S, Papenfort K, Sharma CM, Rolle K, Binnewies TT, Hinton JC, Vogel J. 2008. Deep sequencing analysis of small noncoding RNA and mRNA targets of the global post-transcriptional regulator, Hfq. *PLoS Genet* **4:**e1000163.

54. Melamed S, Peer A, Faigenbaum-Romm R, Gatt YE, Reiss N, Bar A, Altuvia Y, Argaman L, Margalit H. 2016. Global mapping of small RNA-target interactions in bacteria. *Mol Cell* **63:**884–897.

55. Bilusic I, Popitsch N, Rescheneder P, Schroeder R, Lybecker M. 2014. Revisiting the coding potential of the *E. coli* genome through Hfq co-immunoprecipitation. *RNA Biol* **11:**641–654.

56. Rasmussen AA, Eriksen M, Gilany K, Udesen C, Franch T, Petersen C, Valentin-Hansen P. 2005. Regulation of *ompA* mRNA stability: the role of a small regulatory RNA in growth phase-dependent control. *Mol Microbiol* **58:**1421–1429.

57. Udekwu KI, Darfeuille F, Vogel J, Reimegård J, Holmqvist E, Wagner EG. 2005. Hfq-dependent regulation of OmpA synthesis is mediated by an antisense RNA. *Genes Dev* **19:**2355–2366.

58. Udekwu KI. 2010. Transcriptional and post-transcriptional regulation of the *Escherichia coli luxS* mRNA; involvement of the sRNA MicA. *PLoS One* **5:**e13449.

59. Aiso T, Kamiya S, Yonezawa H, Gamou S. 2014. Overexpression of an antisense RNA, ArrS, increases the acid resistance of *Escherichia coli*. *Microbiology* **160:**954–961.

60. Takada A, Umitsuki G, Nagai K, Wachi M. 2007. RNase E is required for induction of the glutamate-dependent acid resistance system in *Escherichia coli*. *Biosci Biotechnol Biochem* **71:**158–164.

61. Tramonti A, De Canio M, De Biase D. 2008. GadX/GadW-dependent regulation of the *Escherichia coli* acid fitness island: transcriptional control at the *gadY*-*gadW* divergent promoters and identification of four novel 42 bp GadX/GadW-specific binding sites. *Mol Microbiol* **70:**965–982.

62. Sayed N, Jousselin A, Felden B. 2011. A *cis*-antisense RNA acts in *trans* in *Staphylococcus aureus* to control translation of a human cytolytic peptide. *Nat Struct Mol Biol* **19:**105–112.

63. Eisenhut M, Georg J, Klähn S, Sakurai I, Mustila H, Zhang P, Hess WR, Aro EM. 2012. The antisense RNA As1_flv4 in the cyanobacterium *Synechocystis* sp. PCC 6803 prevents premature expression of the *flv4-2* operon upon shift in inorganic carbon supply. *J Biol Chem* **287:**33153–33162.

64. Shimakawa G, Shaku K, Nishi A, Hayashi R, Yamamoto H, Sakamoto K, Makino A, Miyake C. 2015. FLAVODIIRON2 and FLAVODIIRON4 proteins mediate an oxygen-dependent alternative electron flow in *Synechocystis* sp. PCC 6803 under CO_2-limited conditions. *Plant Physiol* **167:**472–480.

65. Horie Y, Ito Y, Ono M, Moriwaki N, Kato H, Hamakubo Y, Amano T, Wachi M, Shirai M, Asayama M. 2007. Dark-induced mRNA instability involves RNase E/G-type endoribonuclease cleavage at the AU-box and SD sequences in cyanobacteria. *Mol Genet Genomics* **278:**331–346.

66. Sakurai I, Stazic D, Eisenhut M, Vuorio E, Steglich C, Hess WR, Aro EM. 2012. Positive regulation of *psbA* gene expression by *cis*-encoded antisense RNAs in *Synechocystis* sp. PCC 6803. *Plant Physiol* **160:**1000–1010.

67. Hu J, Li T, Xu W, Zhan J, Chen H, He C, Wang Q. 2017. Small antisense RNA RblR positively regulates RuBisCo in *Synechocystis* sp. PCC 6803. *Front Microbiol* **8:**231.

68. Mustachio LM, Aksit S, Mistry RH, Scheffler R, Yamada A, Liu JM. 2012. The *Vibrio cholerae* mannitol transporter is regulated posttranscriptionally by the MtlS small regulatory RNA. *J Bacteriol* **194:**598–606.

69. Chang H, Replogle JM, Vather N, Tsao-Wu M, Mistry R, Liu JM. 2015. A *cis*-regulatory antisense RNA represses translation in *Vibrio cholerae* through extensive complementarity and proximity to the target locus. *RNA Biol* **12:**136–148.

70. Chen Q, Crosa JH. 1996. Antisense RNA, Fur, iron, and the regulation of iron transport genes in *Vibrio anguillarum*. *J Biol Chem* **271:**18885–18891.

71. Waldbeser LS, Chen Q, Crosa JH. 1995. Antisense RNA regulation of the *fatB* iron transport protein gene in *Vibrio anguillarum*. *Mol Microbiol* **17:**747–756.

72. Stork M, Di Lorenzo M, Welch TJ, Crosa JH. 2007. Transcription termination within the iron transport-biosynthesis operon of *Vibrio anguillarum* requires an antisense RNA. *J Bacteriol* **189:**3479–3488.

73. Giangrossi M, Prosseda G, Tran CN, Brandi A, Colonna B, Falconi M. 2010. A novel antisense RNA regulates at transcriptional level the virulence gene *icsA* of *Shigella flexneri*. *Nucleic Acids Res* **38:**3362–3375.

74. Giangrossi M, Giuliodori AM, Tran CN, Amici A, Marchini C, Falconi M. 2017. VirF relieves the transcriptional attenuation of the virulence gene *icsA* of *Shigella flexneri* affecting the *icsA* mRNA-RnaG complex formation. *Front Microbiol* **8:**650.

75. Shao W, Price MN, Deutschbauer AM, Romine MF, Arkin AP. 2014. Conservation of transcription start sites within genes across a bacterial genus. *mBio* **5:**e01398–e14.

76. Kopf M, Klähn S, Scholz I, Hess WR, Voß B. 2015. Variations in the non-coding transcriptome as a driver of inter-strain divergence and physiological adaptation in bacteria. *Sci Rep* **5:**9560.

77. Lasa I, Villanueva M. 2014. Overlapping transcription and bacterial RNA removal. *Proc Natl Acad Sci U S A* **111:**2868–2869.

78. Lasa I, Toledo-Arana A, Dobin A, Villanueva M, de los Mozos IR, Vergara-Irigaray M, Segura V, Fagegaltier D, Penadés JR, Valle J, Solano C, Gingeras TR. 2011. Genome-wide antisense transcription drives mRNA

processing in bacteria. *Proc Natl Acad Sci U S A* **108:** 20172–20177.

79. Lasa I, Toledo-Arana A, Gingeras TR. 2012. An effort to make sense of antisense transcription in bacteria. *RNA Biol* **9:**1039–1044.

80. Lybecker M, Bilusic I, Raghavan R. 2014. Pervasive transcription: detecting functional RNAs in bacteria. *Transcription* **5:**e944039.

81. Steglich C, Lindell D, Futschik M, Rector T, Steen R, Chisholm SW. 2010. Short RNA half-lives in the slow-growing marine cyanobacterium *Prochlorococcus*. *Genome Biol* **11:**R54.

82. Bidnenko V, Nicolas P, Grylak-Mielnicka A, Delumeau O, Auger S, Aucouturier A, Guerin C, Repoila F, Bardowski J, Aymerich S, Bidnenko E. 2017. Termination factor Rho: from the control of pervasive transcription to cell fate determination in *Bacillus subtilis*. *PLoS Genet* **13:** e1006909.

83. Peters JM, Mooney RA, Grass JA, Jessen ED, Tran F, Landick R. 2012. Rho and NusG suppress pervasive antisense transcription in *Escherichia coli*. *Genes Dev* **26:**2621–2633.

84. Sedlyarova N, Rescheneder P, Magán A, Popitsch N, Rziha N, Bilusic I, Epshtein V, Zimmermann B, Lybecker M, Sedlyarov V, Schroeder R, Nudler E. 2017. Natural RNA polymerase aptamers regulate transcription in *E. coli*. *Mol Cell* **67:**30–43.e6.

85. Duquette ML, Handa P, Vincent JA, Taylor AF, Maizels N. 2004. Intracellular transcription of G-rich DNAs induces formation of G-loops, novel structures containing G4 DNA. *Genes Dev* **18:**1618–1629.

86. Leela JK, Syeda AH, Anupama K, Gowrishankar J. 2013. Rho-dependent transcription termination is essential to prevent excessive genome-wide R-loops in *Escherichia coli*. *Proc Natl Acad Sci U S A* **110:**258–263.

87. Boque-Sastre R, Soler M, Oliveira-Mateos C, Portela A, Moutinho C, Sayols S, Villanueva A, Esteller M, Guil S. 2015. Head-to-head antisense transcription and R-loop formation promotes transcriptional activation. *Proc Natl Acad Sci U S A* **112:**5785–5790.

88. Aguilera A, Gaillard H. 2014. Transcription and recombination: when RNA meets DNA. *Cold Spring Harb Perspect Biol* **6:**a016543.

89. Tan FY, Wörmann ME, Loh E, Tang CM, Exley RM. 2015. Characterization of a novel antisense RNA in the major pilin locus of *Neisseria meningitidis* influencing antigenic variation. *J Bacteriol* **197:**1757–1768.

90. Chatterjee A, Johnson CM, Shu CC, Kaznessis YN, Ramkrishna D, Dunny GM, Hu WS. 2011. Convergent transcription confers a bistable switch in *Enterococcus faecalis* conjugation. *Proc Natl Acad Sci U S A* **108:** 9721–9726.

91. Chatterjee A, Drews L, Mehra S, Takano E, Kaznessis YN, Hu WS. 2011. Convergent transcription in the butyrolactone regulon in *Streptomyces coelicolor* confers a bistable genetic switch for antibiotic biosynthesis. *PLoS One* **6:**e21974.

92. André G, Even S, Putzer H, Burguière P, Croux C, Danchin A, Martin-Verstraete I, Soutourina O. 2008.

S-box and T-box riboswitches and antisense RNA control a sulfur metabolic operon of *Clostridium acetobutylicum*. *Nucleic Acids Res* **36:**5955–5969.

93. Palmer AC, Ahlgren-Berg A, Egan JB, Dodd IB, Shearwin KE. 2009. Potent transcriptional interference by pausing of RNA polymerases over a downstream promoter. *Mol Cell* **34:**545–555.

94. Callen BP, Shearwin KE, Egan JB. 2004. Transcriptional interference between convergent promoters caused by elongation over the promoter. *Mol Cell* **14:**647–656.

95. Nicolas P, Mäder U, Dervyn E, Rochat T, Leduc A, Pigeonneau N, Bidnenko E, Marchadier E, Hoebeke M, Aymerich S, Becher D, Bisicchia P, Botella E, Delumeau O, Doherty G, Denham EL, Fogg MJ, Fromion V, Goelzer A, Hansen A, Härtig E, Harwood CR, Homuth G, Jarmer H, Jules M, Klipp E, Le Chat L, Lecointe F, Lewis P, Liebermeister W, March A, Mars RA, Nannapaneni P, Noone D, Pohl S, Rinn B, Rügheimer F, Sappa PK, Samson F, Schaffer M, Schwikowski B, Steil L, Stülke J, Wiegert T, Devine KM, Wilkinson AJ, van Dijl JM, Hecker M, Völker U, Bessières P, Noirot P. 2012. Condition-dependent transcriptome reveals high-level regulatory architecture in *Bacillus subtilis*. *Science* **335:** 1103–1106.

96. Mars RA, Mendonça K, Denham EL, van Dijl JM. 2015. The reduction in small ribosomal subunit abundance in ethanol-stressed cells of *Bacillus subtilis* is mediated by a SigB-dependent antisense RNA. *Biochim Biophys Acta* **1853:**2553–2559.

97. Klotz A, Georg J, Bučinská L, Watanabe S, Reimann V, Januszewski W, Sobotka R, Jendrossek D, Hess WR, Forchhammer K. 2016. Awakening of a dormant cyanobacterium from nitrogen chlorosis reveals a genetically determined program. *Curr Biol* **26:**2862–2872.

98. Heilmann B, Hakkila K, Georg J, Tyystjärvi T, Hess WR, Axmann IM, Dienst D. 2017. 6S RNA plays a role in recovery from nitrogen depletion in *Synechocystis* sp. PCC 6803. *BMC Microbiol* **17:**229.

99. Bidnenko E, Bidnenko V. 2017. Transcription termination factor Rho and microbial phenotypic heterogeneity. *Curr Genet* **64:**541–546.

100. Selby CP, Sancar A. 1993. Molecular mechanism of transcription-repair coupling. *Science* **260:**53–58.

101. Fan J, Leroux-Coyau M, Savery NJ, Strick TR. 2016. Reconstruction of bacterial transcription-coupled repair at single-molecule resolution. *Nature* **536:**234–237.

102. Legewie S, Dienst D, Wilde A, Herzel H, Axmann IM. 2008. Small RNAs establish delays and temporal thresholds in gene expression. *Biophys J* **95:**3232–3238.

103. Mitschke J, Vioque A, Haas F, Hess WR, Muro-Pastor AM. 2011. Dynamics of transcriptional start site selection during nitrogen stress-induced cell differentiation in *Anabaena* sp. PCC7120. *Proc Natl Acad Sci U S A* **108:**20130–20135.

104. Voss B, Bolhuis H, Fewer DP, Kopf M, Möke F, Haas F, El-Shehawy R, Hayes P, Bergman B, Sivonen K, Dittmann E, Scanlan DJ, Hagemann M, Stal LJ, Hess WR. 2013. Insights into the physiology and ecology of the brackish-water-adapted cyanobacterium *Nodularia*

spumigena CCY9414 based on a genome-transcriptome analysis. *PLoS One* **8**:e60224.

105. Dutcher HA, Raghavan R. 2018. Origin, evolution, and loss of bacterial small RNAs. *Microbiol Spectr* **6**:RWR-0004-2017.

106. Kopf M, Klähn S, Scholz I, Matthiessen JK, Hess WR, Voß B. 2014. Comparative analysis of the primary transcriptome of *Synechocystis* sp. PCC 6803. *DNA Res* **21**: 527–539.

107. Behler J, Sharma K, Reimann V, Wilde A, Urlaub H, Hess WR. 2018. The host-encoded RNase E endonuclease as the crRNA maturation enzyme in a CRISPR-Cas subtype III-Bv system. *Nat Microbiol* **3**:367–377.

108. Schlüter JP, Reinkensmeier J, Barnett MJ, Lang C, Krol E, Giegerich R, Long SR, Becker A. 2013. Global mapping of transcription start sites and promoter motifs in the symbiotic α-proteobacterium *Sinorhizobium meliloti* 1021. *BMC Genomics* **14**:156.

109. Hou S, López-Pérez M, Pfreundt U, Belkin N, Stüber K, Huettel B, Reinhardt R, Berman-Frank I, Rodriguez-Valera F, Hess WR. 2018. Benefit from decline: the primary transcriptome of *Alteromonas macleodii* str. Te101 during *Trichodesmium* demise. *ISME J* **12**:981–996.

110. Kröger C, Dillon SC, Cameron AD, Papenfort K, Sivasankaran SK, Hokamp K, Chao Y, Sittka A, Hébrard M, Händler K, Colgan A, Leekitcharoenphon P, Langridge GC, Lohan AJ, Loftus B, Lucchini S, Ussery DW, Dorman CJ, Thomson NR, Vogel J, Hinton JC. 2012. The transcriptional landscape and small RNAs of *Salmonella enterica* serovar Typhimurium. *Proc Natl Acad Sci U S A* **109**:E1277–E1286.

111. Papenfort K, Förstner KU, Cong JP, Sharma CM, Bassler BL. 2015. Differential RNA-seq of *Vibrio cholerae* identifies the VqmR small RNA as a regulator of biofilm formation. *Proc Natl Acad Sci U S A* **112**:E766–E775.

112. Alkhateeb RS, Vorhölter FJ, Rückert C, Mentz A, Wibberg D, Hublik G, Niehaus K, Pühler A. 2016. Genome wide transcription start sites analysis of *Xantho-monas campestris* pv. campestris B100 with insights into the gum gene cluster directing the biosynthesis of the exopolysaccharide xanthan. *J Biotechnol* **225**: 18–28.

113. Qiu Y, Cho BK, Park YS, Lovley D, Palsson BØ, Zengler K. 2010. Structural and operational complexity of the *Geobacter sulfurreducens* genome. *Genome Res* **20**:1304–1311.

114. Dugar G, Herbig A, Förstner KU, Heidrich N, Reinhardt R, Nieselt K, Sharma CM. 2013. High-resolution transcriptome maps reveal strain-specific regulatory features of multiple *Campylobacter jejuni* isolates. *PLoS Genet* **9**:e1003495.

115. Sharma CM, Hoffmann S, Darfeuille F, Reignier J, Findeiss S, Sittka A, Chabas S, Reiche K, Hackermüller J, Reinhardt R, Stadler PF, Vogel J. 2010. The primary transcriptome of the major human pathogen *Helicobacter pylori*. *Nature* **464**:250–255.

116. Mitschke J, Georg J, Scholz I, Sharma CM, Dienst D, Bantscheff J, Voss B, Steglich C, Wilde A, Vogel J, Hess WR. 2011. An experimentally anchored map of transcriptional start sites in the model cyanobacterium *Synechocystis* sp. PCC6803. *Proc Natl Acad Sci U S A* **108**: 2124–2129.

117. Pfreundt U, Kopf M, Belkin N, Berman-Frank I, Hess WR. 2014. The primary transcriptome of the marine diazotroph *Trichodesmium erythraeum* IMS101. *Sci Rep* **4**:6187.

118. Güell M, van Noort V, Yus E, Chen WH, Leigh-Bell J, Michalodimitrakis K, Yamada T, Arumugam M, Doerks T, Kühner S, Rode M, Suyama M, Schmidt S, Gavin AC, Bork P, Serrano L. 2009. Transcriptome complexity in a genome-reduced bacterium. *Science* **326**:1268–1271.

119. Orell A, Tripp V, Aliaga-Tobar V, Albers SV, Maracaja-Coutinho V, Randau L. 2018. A regulatory RNA is involved in RNA duplex formation and biofilm regulation in Sulfolobus acidocaldarius. *Nucleic Acids Res* **46**: 4794–4806.

Trans-Encoded Base Pairing RNAs

IV

Regulating with RNA in Bacteria and Archaea
Edited by Gisela Storz and Kai Papenfort
© 2018 American Society for Microbiology, Washington, DC
doi:10.1128/microbiolspec.RWR-0022-2018

Small Regulatory RNAs in the Enterobacterial Response to Envelope Damage and Oxidative Stress

13

Kathrin S. Fröhlich[1] and Susan Gottesman[2]

INTRODUCTION

One major paradigm for RNA-based regulation in both eukaryotes and prokaryotes is small regulatory RNAs (sRNAs) that pair with mRNAs, leading to changes in translation and mRNA stability. In bacteria, rather than the highly processed very short microRNAs found in eukaryotes, these sRNAs are generally on the order of 50 to 200 nucleotides (nt) long, and in the Gram-negative organisms that are the major focus of this review, annealing of sRNAs to their target mRNAs is usually dependent on the RNA chaperone, Hfq. Annealing can lead to positive regulation of translation, by remodeling inhibitory RNA structures or blocking access of negative regulators (for instance, RNases or the Rho transcription termination factor), or negative regulation, by inhibiting translation, recruiting RNases, or both. A given sRNA can have multiple targets, and can carry out both negative and positive regulation (1–3).

A member of the conserved family of Sm and Sm-like (LSm) proteins, Hfq assembles as a stable, homo-hexameric ring that offers three principal binding sites for RNA: the proximal and distal surfaces of the ring, as well as the lateral rim. Hfq binds sRNAs on the proximal face, recognizing the uridine stretch at the 3′ end of the sRNA's Rho-independent terminator sequence. In the absence of the chaperone, almost all of these Hfq-binding sRNAs become quite unstable, and that may be sufficient to explain the loss of sRNA function in *hfq* mutants (see, for instance, reference 4). In addition, Hfq binds to mRNAs, frequently but not always via its distal surface. *In vitro*, Hfq promotes pairing of sRNAs and mRNAs, suggesting that it is likely to do that as well *in vivo* (5–7). Overall, for the discussion here, the phenotypes of *hfq* mutants serve as a starting point for understanding the role of sRNA-based regulation in *Escherichia coli* and *Salmonella*. Many but not all of these phenotypes are now understood and point to major roles for sRNAs in the use of alternative sigma factors and the response to stress—including envelope and oxidative damage—in bacteria.

Loss of Hfq-Dependent Regulation Leads to Low Levels of RpoS

The first descriptions of phenotypes of a mutation in *hfq* in *E. coli* (8), including osmosensitivity and elongated cell shape, were noted as consistent with the phenotypes of mutants of *rpoS* (encoding the alternative sigma subunit of RNA polymerase, RpoS [also called σ^S]). We now know that at least three sRNAs, DsrA, RprA, and ArcZ, each produced under different conditions, interact with Hfq to positively regulate RpoS translation (9) (Fig. 1A). Being a major stress sigma factor, RpoS controls >10% of all protein-coding genes in *E. coli* (10–12); during stationary phase, the RpoS-mediated general stress response provides resistance to a variety of cell-damaging conditions, including, for example, oxidative stress, low pH, and high osmolarity (9). The demonstration that the *hfq* mutant was defective in the production of RpoS (13, 14) thus explained many of its phenotypes, including its sensitivity to high osmolarity and low pH.

In addition to the direct effect of sRNAs and Hfq on RpoS synthesis, a set of sRNAs are RpoS-dependent (see Fig. 1A), and thus any phenotypes associated with

[1]Department of Biology I, Microbiology, LMU Munich, 82152 Martinsried, Germany; [2]Laboratory of Molecular Biology, Center for Cancer Research, National Cancer Institute, Bethesda, MD 20892.

A   **B**

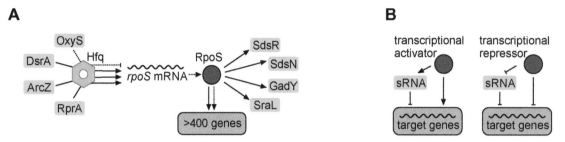

Figure 1 Activity of sRNAs in the general stress response. (A) Together with Hfq, the sRNAs DsrA, ArcZ, and RprA activate translation of the *rpoS* transcript by alleviating a self-inhibitory structure within the 5′ UTR of the mRNA. The sRNA OxyS functions as an indirect, negative regulator of *rpoS* expression. The major alternative sigma factor RpoS governs the general stress response and controls >400 genes in *E. coli* and related enterobacteria, including at least four sRNAs (SdsR, SdsN, GadY, and SraL). (B) Transcription factors which are restricted to function as either activators or repressors utilize sRNAs to facilitate opposite regulation.

those RNAs will also be Hfq-dependent, both for their expression and for their own function and stability. For instance, GadY, an Hfq-dependent sRNA whose synthesis is controlled by RpoS in *E. coli* and *Shigella*, positively regulates GadX and GadW, transcriptional activators of genes for glutamate-dependent acid resistance (15, 16), and the *E. coli*-specific SdsN represses genes involved in the metabolism of oxidized nitrogen compounds (17). More widely conserved are the RpoS-controlled sRNAs SraL and SdsR. A single target has been identified for SraL (*tig* mRNA, encoding the chaperone trigger factor [18]), while SdsR interferes with translation of >20 transcripts in *E. coli* and *Salmonella* and has been implicated in the response to antibiotics and in mismatch repair upon DNA damage (19–22).

Loss of Hfq-Dependent Regulation Leads to High Levels of RpoE

A second major phenotype of *hfq* mutants was first highlighted in studies examining the transcriptional changes when Hfq was absent (23). These experiments were carried out specifically under conditions under which RpoS is not abundant, i.e., early exponential phase. Transcripts encoding outer membrane (OM) β-barrel proteins were significantly overrepresented and generally were upregulated in the *hfq* mutant, while RpoE (or σ^E, encoded by the *rpoE* gene), the alternative sigma factor that regulates OM stress responses, was induced. Combined with studies by others in *Salmonella* (24), and studies of specific sRNAs and their targets (25), these observations have led to evidence that loss of sRNA downregulation of OM protein synthesis leads to an RpoE-inducing stress. Thus, *hfq* mutants express RpoE-dependent genes at a high level, discussed further below.

Specialized Sigma Factors and Hfq: Changing the Sign of Regulation

Worth noting here is that the two major phenotypes of *hfq* mutants discussed above are due to effects on the levels of two specialized sigma factors. Sigma factors act with core RNA polymerase to direct it to particular promoters, and thus, other than competing for core, are not themselves capable of carrying out negative regulation. Therefore, any negative regulation dependent upon a specialized sigma factor is likely indirect, by positive regulation of a negative regulator, including, in many cases, sRNAs (Fig. 1B). Thus, in the initial studies of *hfq* mutants in the RpoE response (23), many upregulated genes have now been shown to be negatively regulated by RpoE-dependent sRNAs (26). This switch in the sign of regulation is also seen for repressors (for instance, positive regulation by the Fur repressor [27]; see Fig. 1B and chapter 16), and thus an unexpected direction of regulation should lead to examining the possible involvement of Hfq and sRNAs.

CONSTRUCTING sRNA REGULATORY NETWORKS: GENERAL PRINCIPLES AND EXPECTATIONS

For any stress, one can consider the roles of sRNAs and what they tell us about how the stress is sensed and responded to. As shown in Fig. 1A for the RpoS general stress response, sRNAs can act upstream, to regulate expression of a transcriptional regulator, or downstream, as part of the regulon. In some cases, sRNAs do both, providing feedback regulatory loops (see chapter 26). Possibly because sRNAs oftentimes work in a stoichiometric fashion (28), i.e., the regulatory RNA is degraded together with the mRNA it is pairing with, the promoters of sRNAs are frequently among the best regulated and most robust in a given stress regulon (see, for instance, reference 29). Thus, they can also serve as excellent reporters for the stress response. The general expectations discussed here are primarily relevant to sRNAs expressed as part of a stress response (i.e., those whose expression is dependent on the transcriptional signals for the response) and are outlined here to provide some guidelines for considering the role of sRNAs within regulons.

sRNAs as Guides to a Stress Response

What sRNAs are expressed in response to the stress (and/or are regulated by the known transcriptional regulators for that stress)? Presumably these sRNAs have effects that help in repairing or avoiding the damage from the stress. Can that contribution to avoiding or overcoming stress be demonstrated? If not, do the sRNA targets suggest novel components of the stress response, not previously appreciated?

Are there sRNAs expressed as part of other regulons that contribute, positively or negatively, to the stress under consideration? sRNAs can provide interactions between different regulons, modulating or setting hierarchies for regulon expression.

A few clear examples of the types of sRNA functions in stress responses are noted here; some of these are discussed in more detail elsewhere.

Reinforcing or Helping Implement the Stress Response (Positive Feedback Loops)

Spot 42 sRNA synthesis is negatively regulated by cyclic AMP (cAMP)/cAMP receptor protein (CRP) and negatively regulates many operons involved in alternative carbon source use that are dependent upon cAMP (30). Thus, Spot 42 contributes to reducing the basal levels of these operons when cAMP is low (favoring efficient use of glucose/favored carbon sources).

Minimizing Stress Signals (Negative Feedback Loops)

The negative feedback loop in which RpoE-dependent sRNAs (as well as others) downregulate translation of many OM proteins is discussed further below. In another, more indirect example of negative feedback, RyhB, made when iron is limiting, inhibits synthesis of nonessential iron-binding proteins, thus helping to overcome the stress by increasing the availability of iron (31).

Connecting Regulons/Stress Responses/ Setting Hierarchies

sRNAs connect different regulons, for purposes that are not always yet well understood. In two cases noted here, sRNAs modulate the interaction between specific regulatory responses and a specialized sigma factor. For instance, the Hfq-dependent OxyS sRNA, synthesized under the control of OxyR, negatively regulates RpoS (32). Given that OxyR regulates genes involved in the response to oxidative stress, which RpoS does as well, it would seem that OxyS helps the cell to use the specific (OxyR-dependent) response rather than the general stress response under some conditions. PhoQ-PhoP, a two-component system (TCS) that is activated at low Mg^{2+} concentrations and leads to the synthesis of genes regulating Mg^{2+} homeostasis as well as lipopolysaccharide (LPS) modifications, is negatively regulated by the RpoE-dependent MicA sRNA (33).

INVOLVEMENT OF sRNAs IN THE ENVELOPE STRESS RESPONSE

Cell Envelope Structure and Function in Gram-Negative Bacteria

The cell envelope represents a barrier shielding the bacterium from its environment and allowing the selective passage of both harmful and beneficial molecules (34). In Gram-negative bacteria, the envelope is composed of two concentric membrane layers that enclose the periplasmic space containing a thin peptidoglycan (PG) cell wall (35). The inner membrane (IM) separating the cytoplasm from the periplasm is a phospholipid bilayer, and the proteins associated with or integrated in the IM are frequently involved in key cellular processes including energy generation, signal transduction, metabolism, transport, and cell division (35). The periplasm is an aqueous cellular compartment densely packed with proteins and harbors the mesh-like PG layer, which is formed from linear amino-sugar polymers cross-connected via oligopeptide chains (36). Structural integrity of the cell envelope is ensured by the tight linkage of the PG layer to the cellular OM via the lipoprotein Lpp (also referred to as Braun's lipoprotein) (37). Lpp is the most abundant protein in *E. coli* (>500,000 copies/cell) (38), and lipids attached to the N-terminus of Lpp embed it into the OM. Concomitantly, Lpp can be covalently attached to the peptide cross-bridges of the PG layer via its carboxy-terminal end (39). In contrast to the IM, the OM is an asymmetric bilayer consisting of phospholipids in the inner and LPS in the surface-exposed leaflet (35). LPS is a complex glycolipid composed of lipid A (a glucosamine disaccharide decorated with fatty acids anchoring LPS to the membrane), an oligosaccharide core, and an extended polysaccharide chain commonly referred to as

O-antigen (40). Tightly packed LPS serves as an effective permeability barrier to hydrophobic substances (41), but certain other molecules are able to cross the OM through protein transporters. Small, hydrophilic compounds can diffuse through the lumen of porins, which are highly abundant β-barrel OM proteins (OMPs) that only discriminate their substrates by size (34). Gated, high-affinity uptake of ligands including siderophores, vitamins, and carbohydrates is mediated by an additional class of larger, integral β-barrel OMPs. Active transport via so-called TonB-dependent receptors into the periplasmic space involves coupling to a protein complex localized in the IM (42).

The integrity of the cellular envelope is essential for bacterial survival, and consequently, its architecture and composition are tightly regulated. Bacteria have evolved a suite of stress responses that function in monitoring impairment or deficiencies of the envelope and govern adaptation of the bacterial gene expression to alleviate stress (43). In *E. coli* and related enterobacteria, at least five envelope stress responses coordinately function to maintain membrane homeostasis (44). While no sRNAs have yet been associated with the minor membrane stress responses coordinated via the BaeS/R or the Psp system, the three major pathways (regulated through RpoE and the Cpx and Rcs systems, respectively) all rely on the activity of regulatory RNAs. Intriguingly, sRNAs function not only as effectors regulating individual target genes but also in mediating the extensive crosstalk and adaptation of individual stress responses.

sRNAs in the RpoE-Mediated Envelope Stress Response

Maintenance of OM homeostasis in *E. coli* and related bacteria relies on the dedicated activity of the alternative sigma factor RpoE, which is encoded by the *rpoE* gene and cotranscribed with *rseABC* from a RpoD-dependent promoter (45). Under nonstress conditions, RpoE is only expressed at a low basal level (38) and sequestered to the plasma membrane in an inactive state by its cognate anti-sigma factor, RseA (46) (Fig. 2A). The activity of RpoE is under complex control and tuned to perturbations of OMP folding, the status of LPS, as well as nutrient availability and growth phase (47). The accumulation of misfolded OMPs within the periplasmic space is the most thoroughly characterized signal triggering RpoE activation (47) (Fig. 2A). The C-termini of unfolded OMPs—which are inaccessible in properly assembled porins—are recognized by the periplasmic PDZ domain of a serine protease, DegS (48). The resulting conformational change initiates a

protease cascade resulting in degradation of RseA and consequent release of RpoE into the cytoplasm (47). Maximal induction of RpoE requires the integration of a second activating signal. In response to mislocalized LPS (49), the RseB protein, which protects the RseA anti-sigma factor from cleavage by DegS, is displaced.

The RpoE-orchestrated envelope stress response possesses two parallel branches, one acting through proteins and one mediated by sRNAs. In *E. coli*, RpoE drives transcription of ∼100 genes, including genes encoding all components required to assemble and transport OMPs and LPS to the OM (29, 50, 51). However, the rate at which new OM components are synthesized may easily exceed the capacity of chaperones and transporters. While RpoE, as for all sigma factors, is intrinsically restricted to function as a transcriptional activator, it employs regulatory RNAs to function as repressors of gene expression at the posttranscriptional level.

The strongest promoters within the RpoE regulon control RpoE itself and two Hfq-dependent sRNAs, RybB and MicA (29, 52, 53). Both sRNAs act as global regulators of the envelope stress response and together govern expression of >30 targets in *E. coli*. Most prominently, RybB and/or MicA repress all major OMPs, as well as several lipoproteins and transporters (26). Rapid decay of these target transcripts results in an immediate relief for the periplasmic folding machinery. In addition, the target suites of RybB and MicA also encompass genes involved in production of OM vesicles and the response regulator *phoP* (26, 33) (see below).

Mutants of *rpoE* in *E. coli* are not viable, but cells can be depleted for RpoE by overexpression of its antagonists, RseA and RseB (54, 55). The important contribution of the two regulatory RNAs, RybB and MicA, to cell homeostasis is reflected by their ability to counteract the growth and viability phenotypes associated with loss of RpoE (26).

Mechanistically, both RybB and MicA exert their regulatory activity by a variety of means. RybB employs a conserved seed region of 16 nt located at the very 5′ end of the molecule to interact with its target genes (56). Depending on the location of the base-pairing site on the target mRNA, RybB is able to interfere with ribosome association at the translational start site (57), to promote RNase cleavage in the coding sequence of target mRNAs, or to interrupt structural elements within the 5′ untranslated region (5′ UTR) (56). In the latter case, base-pairing can abrogate the protective effect of 5′-terminal structures, which have been shown to stabilize transcripts by restricting access of nucleases, including RNase E (58, 59). Similar to

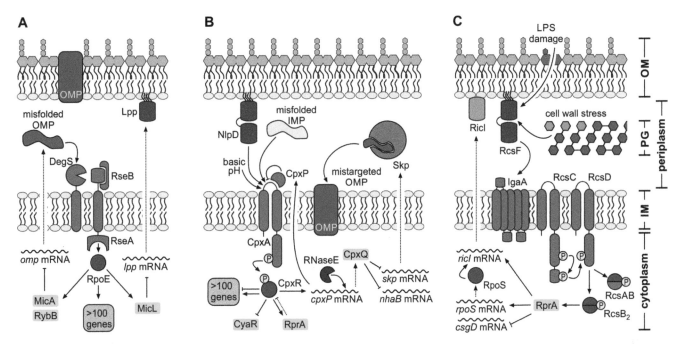

Figure 2 The role of sRNAs in the major envelope stress responses. Gram-negative bacteria are diderm, with the OM and IM being separated by the periplasmic space containing the PG cell wall. (A) OM homeostasis is regulated by the RpoE response. A series of proteolysis steps results in the degradation of the anti-sigma factor RseA and concomitant release of RpoE. The large regulon of the alternative sigma factor also comprises at least three sRNAs: MicA and RybB function to downregulate the transcripts of all major OMPs to reduce the accumulation of misfolded porins within the periplasm. MicL specifically represses translation of the *lpp* mRNA. (B) Maintenance of the IM relies on the CpxA-CpxR TCS, which amongst other targets controls expression of at least three sRNAs, CyaR, RprA, and CpxQ. CpxQ is a stable fragment released by RNase E processing from the 3′ end of the *cpxP* mRNA. In association with Hfq, CpxQ functions to repress translation of several transcripts including *skp* mRNA, which encodes a periplasmic chaperone promoting the mistargeting of OMPs into the IM. (C) The IM-associated histidine kinase RcsC, phosphotransfer protein RcsD, and response regulator RcsB constitute the core of the Rcs system. The sRNA RprA is one highly induced component of the Rcs response, which is activated by LPS damage and perturbations of the cell wall. While acting as a negative regulator of the *csgD* mRNA, RprA also promotes translation of both the *rpoS* and the *ricI* messages. As transcription of *ricI* (encoding an inhibitor of the conjugation machinery) is dependent on RpoS, RprA functions at the heart of a posttranscriptional feedforward loop for RicI activity.

RybB, the first 24 nt of MicA are hyperconserved and involved in regulation of its target genes. In contrast to RybB, the seed region of MicA is less stringently defined, and the first 7 nt of the molecule are dispensable for regulation of approximately half of its targets (26). In the majority of cases, predicted interaction sites of MicA overlap the translation initiation region of its target transcripts; base-pairing of MicA close to the start codon for one noteworthy target, *phoP* (also see below), has been confirmed (26, 33). The regulatory mechanisms underlying the observed repression of other targets are yet to be experimentally validated (26).

Both RybB and MicA are highly conserved in numerous enterobacteria and likely contribute to OMP homeostasis in these species (60). Albeit not conserved at the sequence level, the human pathogen *Vibrio cholerae* encodes a functional homolog of these *E. coli* sRNAs. The 140-nt VrrA is under strict control of RpoE and represses translation of the major OMPs OmpA and OmpT as well as the biofilm matrix component RbmC and the ribosome binding protein Vrp in response to perturbations of the OM (61). Of note, accumulation of the VrrA sRNA is not affected in an *hfq* deletion mutant of *V. cholerae*, and possibly for this reason, the RNA chaperone is not strictly required for regulation of all targets (61–64).

In *E. coli*, a third sRNA is transcribed from an RpoE-dependent promoter positioned within the *cutC* coding sequence. The 308-nt MicL (also known as RyeF [53] and SlrA [65]) is processed by a currently unknown cleavage mechanism to a smaller, ~80-nt transcript (MicL-S), and both isoforms associate with the Hfq chaperone (66). In stark contrast to the many targets of RybB and MicA, MicL appears to interact with only few transcripts (6), and the only experimentally confirmed target of this sRNA is currently the *lpp* gene (66). By binding to a target sequence located within the beginning of the coding sequence, MicL inhibits translation of *lpp* mRNA and triggers accelerated degradation of the transcript. Why is repression of Lpp synthesis beneficial to the RpoE-dependent envelope stress response? As an OM lipoprotein, folding and transport of Lpp are dependent on the LolAB system. The same machinery is, however, also required for the installation of other lipoproteins, including BamD and LptE, which are essential to chaperone OM insertion of OMPs and LPS, respectively (35). Consequently, MicL-mediated downregulation of Lpp may relieve envelope stress by reducing the demand on the Lol machinery, indirectly promoting the correct assembly and localization of other OM components.

sRNAs in the Cpx-Mediated Envelope Stress Response

Maintenance of OM integrity via the RpoE response is complemented by another major regulon primarily controlling homeostasis of the periplasm and the IM. The central hub of the Cpx envelope stress response is a TCS consisting of the IM-localized histidine kinase CpxA and its cognate DNA-binding response regulator, CpxR (67). While the molecular characteristics of the stimulus perceived by CpxA remain to be determined, numerous environmental cues triggering the signaling cascade have been identified, including alterations in IM composition (68), alkaline pH (69), and overexpression of the OM lipoprotein NlpE (70). Recent work also reports that the Cpx system may respond to perturbations of the PG cell wall (71). Activation of CpxA results in autophosphorylation and phosphotransfer to CpxR, which consequently enables transcriptional control of the Cpx regulon, comprising >100 genes (69) (Fig. 2B). In the absence of an inducing signal, CpxA acts as a phosphatase on CpxR~P to rapidly inactivate the response regulator. Different from sigma factors, CpxR functions as both an activator and a repressor (69). It downregulates the expression of envelope-associated, macromolecular complexes, including cellular appendages (72) and respiratory complexes (73), and at the same time fosters transcription of periplasmic proteases and chaperones to alleviate the burden of protein

folding (67). Another highly upregulated gene controlled by CpxR is *cpxP*, encoding a periplasmic inhibitor of the Cpx pathway that likely exerts its negative feedback control by masking the CpxA sensory domain (74, 75).

Processing of the 3′ UTR of the *cpxP* transcript by the major endonuclease RNase E liberates a stable, ~60-nt long mRNA fragment, termed CpxQ, which associates with the Hfq chaperone (52, 76). A transcriptomic approach in *Salmonella* revealed that CpxQ acts as a *trans*-encoded sRNA negatively regulating multiple targets (76). Strikingly, the CpxQ regulon is enriched for proteins localizing to the IM, including those controlling the proton motive force (PMF). Employing one of two seed regions, CpxQ also represses translation of *skp* (encoding a periplasmic chaperone) and *nhaB* mRNAs (encoding a proton/sodium antiporter). Different from other chaperones, Skp is able to mistarget OMPs into the IM if the OM insertion machinery is compromised. By downregulating *skp*, CpxQ prevents the unrestricted flux over the IM via ion-permeable pores that would result from OMPs within the IM, and thus protects cells from collapse of the PMF. CpxQ addresses a similar problem by repression of *nhaB*, as overexpression of the protein, and concomitant increase in proton uptake, results in a loss of membrane polarization. In protecting the integrity of the IM and the PMF, CpxQ appears to play a major function as a repressive arm of the Cpx response (76, 77).

Additional sRNAs have also been shown to be integrated into this complex network. For example, CyaR and RprA sRNAs appear to be both directly and indirectly controlled by the Cpx response, with some differences depending on whether enteropathogenic *E. coli* or *E. coli* K-12 was examined (78).

Expression of CyaR (formerly RyeE [79]), a conserved, Hfq-associated sRNA, is under complex regulation by both the Crp and Cpx regulons. Crp induces CyaR under conditions when cAMP levels are high (80–82). CpxR functions as a transcriptional repressor of the *cyaR* promoter (72); CyaR in turn functions as a posttranscriptional repressor of several target genes, including the *yqaE* transcript (80). As expression of *yqaE*, encoding an IM protein of currently unknown function, is at the same time induced by CpxR~P at the transcriptional level, repression of *cyaR* by CpxR integrates the sRNA into a coherent feedforward loop within the Cpx regulon (72, 80). The physiological importance of this regulation has not been examined; presumably, the CpxR effect on the *cyaR* promoter may not be significant under conditions when cAMP levels are high (i.e., growth in poorer carbon sources), and repression of *yqaE* may not be important under those conditions.

The promoter of *rprA*, a conserved enterobacterial sRNA, is activated under conditions of high CpxR~P and is directly bound by CpxR~P. Overexpression of RprA in turn feeds back to repress the Cpx response, indirectly via a currently undefined target (78). However, *rprA* expression is primarily controlled by an additional envelope stress response, the Rcs pathway, discussed below.

Integration of the Cpx and Rcs Envelope Stress Responses via RprA sRNA

Damage of the surface-exposed LPS, mutations in genes required for disulfide bond formation in the periplasm, and perturbations of PG cell wall biosynthesis are all cues triggering the Rcs phosphorelay (83, 84) (Fig. 2C). Signal transduction in this stress response system is more complex than in conventional TCSs: under inducing conditions, the surface-exposed lipoprotein, RcsF, likely inactivates the periplasmic IgaA repressor to trigger a phosphorelay. Upon activation, RcsC is autophosphorylated, and phosphotransfer via RcsD relays the signal to the cognate response regulator, RcsB. RcsB in turn controls the transcription of the Rcs stress response by binding to target promoters either as a heterodimer in cooperation with an additional DNA-binding protein (RcsA in the case of activation of colanic acid production), or as a homodimer (as is the case for the *rprA* gene). Originally identified in *E. coli* as one of the sRNAs activating *rpoS* translation (85), RprA has recently been demonstrated to posttranscriptionally modulate expression of >60 additional genes in *Salmonella* (86), although how many of these are direct targets remains to be explored. RprA is a substrate of RNase E, and both the full-length (107 nt) and the cleaved versions of the sRNA (~50 nt) lacking the 5′ end are present in the cell (86, 87). Interestingly, both variants associate with Hfq (52) and control different sets of mRNAs (86).

One of the mRNA targets activated by RprA is the *Salmonella*-specific *ricI* transcript (86). Similar to the positive regulation of *rpoS*, RprA also promotes translation of *ricI* by interference with a self-inhibitory structure within the 5′ UTR of the mRNA. Of note, while only full-length RprA harbors the site required for base-pairing with *rpoS* mRNA, the *ricI* transcript is recognized via a conserved sequence stretch located downstream of the cleavage site of RprA. As expression of *ricI* is controlled by RpoS at the transcriptional level, RprA functions as the centerpiece of a posttranscrip-

tional feedforward loop. RicI acts as an inhibitor for conjugative transfer of the *Salmonella* virulence plasmid pSLT by binding to the conjugation apparatus at the cytoplasmic membrane. With regard to envelope homeostasis, the Rcs regulon might employ RprA to prevent assembly of the complex conjugation machinery when membrane integrity is compromised.

In *E. coli*, RprA is one of six currently known sRNAs (together with OmrA/B, McaS, GcvB, and RydC [88–91, 142]) to repress translation of *csgD*, encoding a transcriptional regulator of curli fimbriae and cellulose production (92). In addition, RprA also downregulates expression of *ydaM*, which encodes a diguanylate cyclase involved in activating *csgD* transcription. While curli are required for efficient adhesion of bacterial cells in growing biofilms, massive synthesis of surface-exposed curli fimbriae may be detrimental to cells experiencing envelope stress (90). In addition, it has been speculated that RprA may function to balance expression of curli/cellulose and colanic acid, an additional biofilm matrix component directly controlled by the Rcs pathway (93).

Although none of the signaling components of the Cpx system has been shown to be directly controlled by RprA, overexpression of the sRNA exerts negative feedback onto the stress response in a CpxR-dependent manner (78). Further investigation regarding the integration of the sRNA into the Cpx regulon is required, but two tempting hypotheses may explain the observed phenotype. First, a yet-to-be-identified auxiliary factor modulating CpxR activity could be under control of RprA (43). Alternatively, RprA could indirectly reduce induction of the Cpx response by contributing to stress relief. Intriguingly, one of the most upregulated genes following pulse overexpression of RprA in *Salmonella* is *dsbG*. Together with the CpxR-controlled *dsbB*, *dsbG* functions in disulfide bond formation within the periplasm (86, 94), and upregulation of the gene would be consistent with the induction of the Rcs phosphorelay upon loss of DsbA (83).

Additional Pathways Mediating Envelope Homeostasis

The activity of the major envelope stress response is complemented by several additional regulatory pathways modulating membrane homeostasis and modifications. In many cases, sRNAs constitute central nodes of these systems.

The EnvZ/OmpR system is one of the most thoroughly studied TCSs and contributes to the maintenance of the OM by controlling expression of multiple OMPs. The environmental cues triggering activation of the sensor kinase, EnvZ, include increased growth temperature, acidic pH, and, most importantly, increased osmolarity (95). Phosphotransfer from activated EnvZ to its cognate response regulator, OmpR, in conditions of high osmolarity controls (among other genes) the ratio of the major OMPs OmpC and OmpF (96). The opposite regulation of the two OMP genes by OmpR~P, i.e., induction of *ompC* and repression of *ompF* transcription, respectively, is reinforced at the posttranscriptional level. While the promoter of the MicC sRNA (which represses *ompC* mRNA) is repressed by OmpR~P, the transcription factor activates production of MicF sRNA (which represses *ompF* mRNA) (97, 98).

In addition, the two homologous sRNAs OmrA and OmrB, encoded in tandem orientation on the *E. coli* chromosome, are under positive control of OmpR~P (99). Together, OmrA/OmrB repress the synthesis of the TonB-dependent receptors CirA, FecA, and FepA; the OM protease OmpT; as well as the transcription factor CsgD (88, 99). In addition, OmrA/B autoregulates its own transcription by negatively regulating the *ompR-envZ* mRNA (100).

Involvement of sRNAs in Modification of LPS

In pathogenic bacteria, the cell surface provides numerous exposed epitopes recognized by the host's immune response after infection (101). Moreover, given its essential functions, the bacterial cell envelope is an effective target for antimicrobial peptides (AMPs). To evade the response of the host immune system, bacteria are able to tune the composition of OMPs within the OM. In addition, modifications of the LPS also contribute to the bacterial survival strategy in the presence of the host immune defense and AMPs (41).

Several sRNAs are integrated into the regulatory loops governing LPS modifications, either directly by controlling expression of modifying enzymes or indirectly by influencing the activity of transcriptional regulators (Fig. 3).

The LPS component lipid A is a common site for modifications and can, for example, be subject to dephosphorylation, deacylation, or hydroxylation (102). One of the enzymes responsible for lipid A deacylation, encoded by *lpxR* in *Salmonella* and some other bacteria, including the pathogenic *E. coli* O157:H7, is posttranscriptionally repressed by MicF (103). Of note, MicF, which itself is controlled by the EnvZ-OmpR TCS, uses two RNA stretches to form base-pairing interactions both at the translation initiation site and within the coding sequence of *lpxR* mRNA.

A major regulon controlling the physiology of LPS is the PhoQ-PhoP TCS, which is activated by low levels

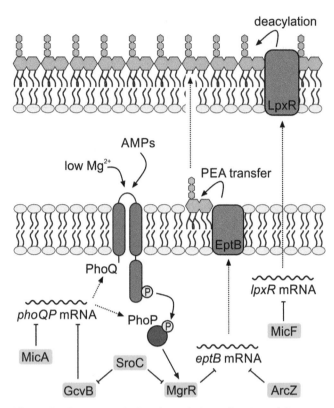

Figure 3 Posttranscriptional regulation of LPS modification. The PhoQ-PhoP TCS, a major determinant of LPS modifications, is activated in response to Mg^{2+} starvation as well as by AMPs. Translation of the *phoPQ* bicistronic transcript is repressed by two sRNAs, MicA and GcvB. PhoQ-PhoP controls expression of MgrR, which, together with ArcZ, inhibits phosphoethanolamine (PEA) addition to the LPS oligosaccharide core by EptB. Both GcvB and MgrR are regulated at the posttranscriptional level by the sRNA SroC, which acts as a sponge and induces decay of its target sRNAs. Downregulation of *lpxR* mRNA by MicF decreases lipid A deacylation.

of Mg^{2+} ions, as well as by AMPs (104). The PhoQ-PhoP system has been extensively studied in the enteric pathogen *Salmonella*, where its integrity is essential for the infection process (105). The TCS is, however, widely conserved in several enterobacterial species, where it is involved in the adaptation to low-Mg^{2+} environments and/or the regulation of virulence factors (104). One of the several dozen genes directly controlled via PhoQ-PhoP encodes an sRNA, MgrR (termed Stnc560 in *Salmonella*), which was originally identified in Hfq coimmunoprecipitation experiments (53, 106). At the posttranscriptional level, the activity of MgrR is counteracted by yet another sRNA, SroC, which acts as a sponge RNA to sequester and trigger the decay of MgrR (107). SroC, processed from the *gltIJKL* mRNA, encoding a glutamate/aspartate ABC transporter, was first described to directly base-pair and repress GcvB sRNA (108), where it provides a feed-forward loop regulating amino acid transport.

MgrR represses at least two mRNA targets, *ygdQ* (encoding an IM protein of unknown function) and *eptB* (*pmrC* in *Salmonella*) (109). The *eptB* transcript is one of the most highly deregulated genes in *hfq* mutant strains of *Salmonella* (24) and encodes a phospho-ethanolamine transferase modifying the outer Kdo (3-deoxy-D-manno-octulosonic acid) unit of the LPS core (110). Thus, while directly activating other LPS-modifying enzymes, the PhoQ-PhoP TCS employs the sRNA MgrR to negatively act on *eptB* expression. EptB is barely expressed under standard laboratory growth conditions but is transcriptionally activated by the RpoE response, hinting at the benefit of the EptB-mediated LPS modification during OM stress. The deletion of *mgrR* from the *E. coli* chromosome, and the consequent expression of *eptB*, result in increased resistance to the AMP polymyxin B due to modification of the LPS structure (109). In addition, *eptB* mRNA is also repressed by the sRNA ArcZ, further specifying timing and degree of EptB-mediated LPS modification (111). Since ArcZ is preferentially expressed under aerobic conditions (112), the cooperative activity of both ArcZ and MgrR allows the expression of EptB only when cells encounter an Mg^{2+}/Ca^{2+}-rich, anaerobic (or microaerobic) environment (111).

Additional sRNAs mediate the crosstalk between the PhoQ-PhoP TCS and other regulons. The *phoQ* and *phoP* genes are encoded in a bicistronic operon, and the *phoQP* mRNA has been shown to be posttranscriptionally controlled by both MicA and GcvB sRNAs. However, given that *phoQP* mRNA levels are elevated in the absence of *hfq* even when *micA* and *gcvB* have been deleted from the genome, additional, yet-to-be-identified sRNAs might contribute to the complex regulation of the TCS (113), or possibly Hfq alone can repress this mRNA (114).

MicA, induced by the RpoE response, represses translation of the TCS by base-pairing within the translation initiation site of *phoP* (33). Repression of PhoP activity by MicA is consistent with and reinforces the activation of *eptB* (see above) under RpoE-inducing conditions. Similarly, GcvB sRNA inhibits translation initiation of *phoP* by binding a region in close proximity to the MicA pairing site (113). The posttranscriptional activity of GcvB links the PhoQ-PhoP regulon to cell metabolism, as the sRNA is mainly involved in limiting amino acid and peptide uptake under nutrient-rich conditions (115). It is intriguing, given the negative regulation of *phoP* by GcvB, that SroC provides another

link between GcvB and the PhoP-dependent MgrR sRNA. With the SroC sponge acting as a competitor for mRNA targets of the different sRNAs, gene expression will depend not only on the binding affinities and relative concentrations of cognate sRNA/mRNA pairs but also critically on the expression level of the sponge RNA. Consequently, SroC might serve to fine-tune the coordination of posttranscriptional control via the MgrR and GcvB networks (see chapter 25).

sRNAs AND THE RESPONSE TO OXIDATIVE STRESS AND DNA DAMAGE

Under aerobic growth conditions, reactive oxygen species (ROS), including, for example, superoxide and hydrogen peroxide, are generated as natural by-products of bacterial metabolic activity (116). ROS are able to harm the cell by damaging DNA, iron-sulfur (Fe-S) clusters, and other enzymes (117). Exploiting these toxic effects, one of the host defense mechanisms to counteract infection with *Salmonella* and other intracellular bacteria is the production of ROS (118). In response, bacteria have evolved mechanisms to detoxify ROS and to respond to and help the cell repair oxidative damage. As is the case for many major stress responses of Gram-negative bacteria, sRNAs are embedded in these networks (Fig. 4). However, exactly what these sRNAs do is not entirely clear.

One major contributor to resistance to oxidative stress in *E. coli* and *Salmonella*, as well as other bacteria, is the general stress sigma factor RpoS and the genes under its control, including catalase (encoded by *katE*), required for the detoxification of hydrogen peroxide. As noted above, RpoS levels are low in *hfq* mutants because sRNAs are required to activate its translation (Fig. 1A). To what extent increased sensitivity to oxidative stress of *hfq* mutants is due to the RpoS defect has not been investigated. However, ArcZ, RprA, and DsrA, the sRNAs that promote RpoS translation, certainly should contribute to resistance to oxidative stress by inducing RpoS.

Aside from RpoS, microbes use additional, distinct mechanisms to detect different forms of oxidative stress. For example, the SoxR/S and OxyR regulons mediate the response to superoxide and hydrogen peroxide stresses, respectively, in *E. coli* and related enterobacteria, and each of the systems also involves the activity of sRNAs.

sRNAs Induced by Oxidative Stress

MicF, one of the first-described chromosomally encoded antisense sRNAs (98), is not only part of the OmpR-mediated envelope stress response (see above)

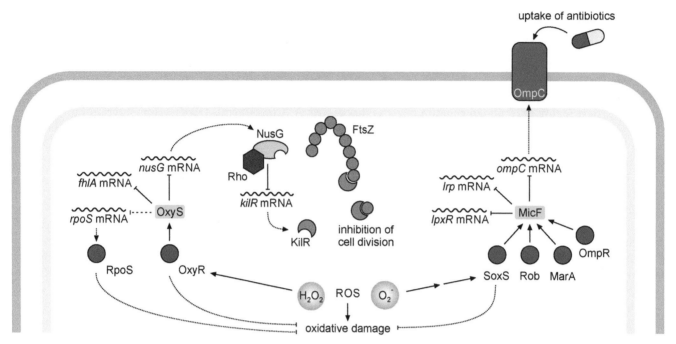

Figure 4 The OxyS and MicF sRNAs are integrated into the enterobacterial response to oxidative stress. OxyS, induced by the hydrogen peroxide-responsive OxyR, downregulates *fhlA* mRNA (encoding a transcription factor regulating formate metabolism) and indirectly represses *rpoS* expression. In addition, OxyS-mediated repression of *nusG* results in increased expression of *kilR*, encoded in the cryptic Rac prophage. KilR sequesters FtsZ, thereby leading to inhibition of cell division and growth arrest, which allows the cell to facilitate DNA damage repair. MicF contributes to increased bacterial resistance against antibiotics of different classes by repressing the major porin OmpF. Additional targets of MicF include *lpxR* mRNA (encoding an LPS modification enzyme), as well as *lrp* mRNA (encoding a transcriptional regulator of amino acid metabolism and transport). Expression of MicF is positively controlled by the transcription factors OmpR, MarA, Rob, and SoxS, with the last being induced in the presence of superoxide.

but is also integrated into the cellular program to defeat oxidative damage. Expression of MicF is induced by three homologs of the AraC family of transcription factors with overlapping activity, SoxS, Rob, and MarA, all of which help the bacteria respond to a range of toxic molecules, including antibiotics, as well as free radicals such as superoxide and nitric oxide (119). SoxS is activated upon oxidation of an Fe-S cluster in its regulator, SoxR, in response to treating cells with the superoxide-generating drug paraquat; repression of MarA by MarR is relieved in the presence of certain phenolic compounds (120); and the activity of Rob is posttranscriptionally increased when bacteria encounter the iron chelators 2,2'- or 4,4'-dipyridyl (121). Under these conditions, downregulation of OmpF by MicF contributes to increased bacterial resistance against antibiotics entering the cell through this porin. In addition, the SoxR/S regulon, and consequently MicF, is activated by nitric oxide, produced by activated macrophages during the infection process. Deletion of *micF*,

and thus loss of MicF-mediated repression of OmpF, results in similar hypersensitivity to killing of *E. coli* by murine macrophages, as observed for *soxR/S* mutants (122). Combined with its activity in repressing the LPS modification enzyme LpxR (see above) and the transcriptional regulator Lrp (103, 123), MicF can be considered a bacterial virulence factor, acting by restricting the entry of a variety of harmful molecules, some of which cause oxidative stress.

One of the first Hfq-dependent sRNAs described in *E. coli* was OxyS, which is induced as part of the OxyR regulon. Initial studies found no evidence that OxyS directly affects the resistance of bacteria to hydrogen peroxide; however, expression of the sRNA confers a protective effect against spontaneous and hydrogen peroxide-induced mutagenesis (32).

A major effect of OxyS overproduction is reduced expression of RpoS (32, 124). As noted above, RpoS contributes significantly to resistance to oxidative stress in stationary phase, e.g., via synthesis of the *katE*-

encoded catalase or the *gor*-encoded glutathione oxido-reductase (125, 126). However, RpoS can also compete with RpoD for RNA polymerase, and one can imagine that the OxyS-mediated reduction of RpoS might help favor the RpoD-dependent OxyR response to oxidative stress. Repression of RpoS once OxyR is activated might avoid redundant activation of stress genes. No direct pairing of OxyS with the *rpoS* mRNA has been detected. The current interpretation is that OxyS might compete with other sRNAs and/or the *rpoS* leader itself for binding to Hfq, thereby blocking the sRNA-dependent activation of RpoS translation (127, 128). Deletion of *oxyS* from the *E. coli* chromosome results in significantly higher intracellular levels of both hydrogen peroxide and superoxide compared to the wild type (129). This phenotype was suggested to result from the ability of OxyS to restrict cellular respiration, and thus to reduce the burden of ROS produced during metabolic activities (129). A model in which OxyS helps reduce the endogenous sources of the inducing stress may be characteristic of many of the sRNA arms of stress regulons. Consistent with this idea, the best-characterized OxyS target is *fhlA*, which is repressed by the sRNA through the formation of a kissing-loop interaction targeting sites in both the 5′ UTR and the coding sequence of the transcript (130). FhlA is a transcriptional activator of complexes involved in formate metabolism (131), whose metal cofactors likely increase cellular damage under oxidative stress conditions (132). In addition, confirmed targets of OxyS also include *wrbA*, encoding an NAD(P)H:quinone oxidoreductase (133).

The ability of OxyS to function as an antimutator is the subject of a very recent study (134, see also chapter 29). One of the transcripts downregulated by OxyS is *nusG* mRNA, coding for a highly conserved regulator of RNA polymerase (Fig. 4). Together with Rho, NusG aids termination at a subset of sites and plays an important role in silencing horizontally acquired genetic elements by inhibiting their transcription (135). One of the loci silenced by NusG and Rho is the *E. coli* cryptic prophage *rac*, encoding amongst others the *kilR* gene (136). When NusG levels are low, i.e., when OxyS is active, *rac* genes are transcribed and KilR is produced. Interfering with assembly of the cell division machinery, KilR inhibits cell cycle progression, which allows the cell to facilitate damage repair. In line with this model, cells expressing OxyS and KilR appear elongated, and the decreased rate of mutations observed in the presence of OxyS is likewise dependent on KilR (134). Repression of *nusG* by OxyS, and thus loss of silencing of cryptic prophages and other horizontally acquired elements, is reminiscent of prophage induction during the SOS response. In both cases, DNA damage may act as a signal for repressed prophages to jump ship and seek new, undamaged hosts.

RNA-Based Regulation of DNA Mutagenesis

OxyS is not the only known posttranscriptional regulator of DNA damage in bacteria. Cells strictly control the rates at which mutations can occur, presumably to balance beneficial and detrimental changes of the genome. One system to limit the integration of mutations upon DNA damage is the mismatch repair (MMR) system, which is dedicated to the recognition and repair of mismatches in the genome. One central component of the *E. coli* MMR system is MutS, which detects and binds to DNA mismatches, thus initiating the repair process (137). While it had long been known that the cellular levels of MutS decrease in an Hfq-dependent manner when *E. coli* enters stationary growth (138, 139), the molecular principles underlying this regulation were only recently discovered. Overexpression of two sRNAs, the RpoS-dependent SdsR and the RpoS-activating ArcZ (see Fig. 1A), results in significant repression of a posttranscriptional *mutS* reporter (114). More importantly, however, Hfq itself binds to the 5′ UTR of *mutS* mRNA independently of base-pairing sRNAs, and thereby inhibits translation of the transcript. In turn, the cell is able to react to changing environments by limiting *mutS* expression and thus raising the mutagenesis rate under specific stress conditions (through induction of ArcZ or SdsR), as well as by titration of Hfq (through high levels of sRNAs or transcripts competing for binding).

RpoS contributes to stationary-phase mutagenesis and break repair by different mechanisms, including the induction of error-prone polymerases and the SdsR sRNA (9, 21, 22, 140). Cells lacking the sRNA GcvB were found to limit mutagenic break repair and mutagenesis in stationary phase; surprisingly, this phenotype of GcvB was entirely suppressed by loss of RpoE (141). The authors find a modest induction of RpoE in cells devoid of GcvB and suggest that the higher level of RpoE may compete with RpoS access to core polymerase. While the mode of action of GcvB and the basis for loss of RpoS activity will need further investigation, the results do point out the degree to which stress responses and the roles of sRNAs in regulating them are entangled in organisms like *E. coli*.

CONCLUSIONS AND FUTURE DIRECTIONS

sRNAs exist in essentially all bacterial stress responses, and their major roles within the regulons of specialized sigma factors have been studied extensively. In many other cases, the physiological functions of the sRNAs within a given regulon are only partially understood but can provide new insight into what cells perceive as stress and the pathways they can use to overcome stress or minimize intrinsic sources of stress. The recent discovery of many sRNAs encoded within the 3′ UTRs of genes has expanded the identification of sRNA-dependent arms for stress responses. The sRNAs also play critical roles in crosstalk between regulons, in setting regulatory hierarchies, and in providing feedback loops that allow rapid response to stress and rapid return to equilibrium when stresses are dealt with. Although we highlight here many examples of sRNA-mediated control that we understand, the sRNAs frequently have multiple targets for which the physiological role of the regulation remains to be understood, necessary for a full understanding of the role of the regulon.

Acknowledgments. We thank members of our laboratories for comments on the manuscript and thank S. Altuvia for sharing results prior to publication and for comments on the manuscript. Preparation of this review was supported in part by the Intramural Research Program of the NIH, National Cancer Institute, Center for Cancer Research (S.G.), and the BioMentoring Program of the LMU Faculty of Biology (K.S.F.).

Citation. Fröhlich KS, Gottesman S. 2018. Small regulatory RNAs in the enterobacterial response to envelope damage and oxidative stress. Microbiol Spectrum 6(4):RWR-0022-2018.

References

1. Papenfort K, Vanderpool CK. 2015. Target activation by regulatory RNAs in bacteria. *FEMS Microbiol Rev* **39:**362–378.

2. Waters LS, Storz G. 2009. Regulatory RNAs in bacteria. *Cell* **136:**615–628.

3. Storz G, Vogel J, Wassarman KM. 2011. Regulation by small RNAs in bacteria: expanding frontiers. *Mol Cell* **43:**880–891.

4. Zhang A, Schu DJ, Tjaden BC, Storz G, Gottesman S. 2013. Mutations in interaction surfaces differentially impact E. coli Hfq association with small RNAs and their mRNA targets. *J Mol Biol* **425:**3678–3697.

5. Kawamoto H, Koide Y, Morita T, Aiba H. 2006. Base-pairing requirement for RNA silencing by a bacterial small RNA and acceleration of duplex formation by Hfq. *Mol Microbiol* **61:**1013–1022.

6. Melamed S, Peer A, Faigenbaum-Romm R, Gatt YE, Reiss N, Bar A, Altuvia Y, Argaman L, Margalit H. 2016. Global mapping of small RNA-target interactions in bacteria. *Mol Cell* **63:**884–897.

7. Soper TJ, Woodson SA. 2008. The rpoS mRNA leader recruits Hfq to facilitate annealing with DsrA sRNA. *RNA* **14:**1907–1917.

8. Tsui HC, Leung HC, Winkler ME. 1994. Characterization of broadly pleiotropic phenotypes caused by an hfq insertion mutation in *Escherichia coli* K-12. *Mol Microbiol* **13:**35–49.

9. Battesti A, Majdalani N, Gottesman S. 2011. The RpoS-mediated general stress response in *Escherichia coli*. *Annu Rev Microbiol* **65:**189–213.

10. Weber H, Polen T, Heuveling J, Wendisch VF, Hengge R. 2005. Genome-wide analysis of the general stress response network in *Escherichia coli*: σS-dependent genes, promoters, and sigma factor selectivity. *J Bacteriol* **187:** 1591–1603.

11. Dong T, Schellhorn HE. 2009. Control of RpoS in global gene expression of *Escherichia coli* in minimal media. *Mol Genet Genomics* **281:**19–33.

12. Patten CL, Kirchhof MG, Schertzberg MR, Morton RA, Schellhorn HE. 2004. Microarray analysis of RpoS-mediated gene expression in *Escherichia coli* K-12. *Mol Genet Genomics* **272:**580–591.

13. Muffler A, Fischer D, Hengge-Aronis R. 1996. The RNA-binding protein HF-I, known as a host factor for phage Qβ RNA replication, is essential for rpoS translation in *Escherichia coli*. *Genes Dev* **10:**1143–1151.

14. Muffler A, Traulsen DD, Fischer D, Lange R, Hengge-Aronis R. 1997. The RNA-binding protein HF-I plays a global regulatory role which is largely, but not exclusively, due to its role in expression of the σS subunit of RNA polymerase in *Escherichia coli*. *J Bacteriol* **179:**297–300.

15. Opdyke JA, Fozo EM, Hemm MR, Storz G. 2011. RNase III participates in GadY-dependent cleavage of the gadX-gadW mRNA. *J Mol Biol* **406:**29–43.

16. Opdyke JA, Kang JG, Storz G. 2004. GadY, a small-RNA regulator of acid response genes in *Escherichia coli*. *J Bacteriol* **186:**6698–6705.

17. Hao Y, Updegrove TB, Livingston NN, Storz G. 2016. Protection against deleterious nitrogen compounds: role of σS-dependent small RNAs encoded adjacent to sdiA. *Nucleic Acids Res* **44:**6935–6948.

18. Silva IJ, Ortega AD, Viegas SC, García-Del Portillo F, Arraiano CM. 2013. An RpoS-dependent sRNA regulates the expression of a chaperone involved in protein folding. *RNA* **19:**1253–1265.

19. Parker A, Gottesman S. 2016. Small RNA regulation of TolC, the outer membrane component of bacterial multidrug transporters. *J Bacteriol* **198:**1101–1113.

20. Fröhlich KS, Haneke K, Papenfort K, Vogel J. 2016. The target spectrum of SdsR small RNA in *Salmonella*. *Nucleic Acids Res* **44:**10406–10422.

21. Fröhlich KS, Papenfort K, Berger AA, Vogel J. 2012. A conserved RpoS-dependent small RNA controls the synthesis of major porin OmpD. *Nucleic Acids Res* **40:** 3623–3640.

22. Gutierrez A, Laureti L, Crussard S, Abida H, Rodríguez-Rojas A, Blázquez J, Baharoglu Z, Mazel D, Darfeuille F, Vogel J, Matic I. 2013. β-Lactam antibiotics promote bacterial mutagenesis via an RpoS-mediated reduction in replication fidelity. *Nat Commun* **4:**1610.

23. Guisbert E, Rhodius VA, Ahuja N, Witkin E, Gross CA. 2007. Hfq modulates the σE-mediated envelope

stress response and the σ^{32}-mediated cytoplasmic stress response in *Escherichia coli*. *J Bacteriol* **189**:1963–1973.

24. **Figueroa-Bossi N, Lemire S, Maloriol D, Balbontín R, Casadesús J, Bossi L.** 2006. Loss of Hfq activates the σ^E-dependent envelope stress response in *Salmonella enterica*. *Mol Microbiol* **62**:838–852.

25. **Vogel J, Papenfort K.** 2006. Small non-coding RNAs and the bacterial outer membrane. *Curr Opin Microbiol* **9**:605–611.

26. **Gogol EB, Rhodius VA, Papenfort K, Vogel J, Gross CA.** 2011. Small RNAs endow a transcriptional activator with essential repressor functions for single-tier control of a global stress regulon. *Proc Natl Acad Sci U S A* **108**:12875–12880.

27. **Massé E, Gottesman S.** 2002. A small RNA regulates the expression of genes involved in iron metabolism in *Escherichia coli*. *Proc Natl Acad Sci U S A* **99**:4620–4625.

28. **Schu DJ, Zhang A, Gottesman S, Storz G.** 2015. Alternative Hfq-sRNA interaction modes dictate alternative mRNA recognition. *EMBO J* **34**:2557–2573.

29. **Mutalik VK, Nonaka G, Ades SE, Rhodius VA, Gross CA.** 2009. Promoter strength properties of the complete σ^E regulon of *Escherichia coli* and *Salmonella enterica*. *J Bacteriol* **191**:7279–7287.

30. **Beisel CL, Storz G.** 2011. The base-pairing RNA Spot 42 participates in a multioutput feedforward loop to help enact catabolite repression in *Escherichia coli*. *Mol Cell* **41**:286–297.

31. **Massé E, Salvail H, Desnoyers G, Arguin M.** 2007. Small RNAs controlling iron metabolism. *Curr Opin Microbiol* **10**:140–145.

32. **Altuvia S, Weinstein-Fischer D, Zhang A, Postow L, Storz G.** 1997. A small, stable RNA induced by oxidative stress: role as a pleiotropic regulator and antimutator. *Cell* **90**:43–53.

33. **Coornaert A, Lu A, Mandin P, Springer M, Gottesman S, Guillier M.** 2010. MicA sRNA links the PhoP regulon to cell envelope stress. *Mol Microbiol* **76**:467–479.

34. **Nikaido H.** 2003. Molecular basis of bacterial outer membrane permeability revisited. *Microbiol Mol Biol Rev* **67**:593–656.

35. **Silhavy TJ, Kahne D, Walker S.** 2010. The bacterial cell envelope. *Cold Spring Harb Perspect Biol* **2**:a000414.

36. **Vollmer W, Blanot D, de Pedro MA.** 2008. Peptidoglycan structure and architecture. *FEMS Microbiol Rev* **32**:149–167.

37. **Braun V.** 1975. Covalent lipoprotein from the outer membrane of *Escherichia coli*. *Biochim Biophys Acta* **415**:335–377.

38. **Li GW, Burkhardt D, Gross C, Weissman JS.** 2014. Quantifying absolute protein synthesis rates reveals principles underlying allocation of cellular resources. *Cell* **157**:624–635.

39. **Inouye M, Shaw J, Shen C.** 1972. The assembly of a structural lipoprotein in the envelope of *Escherichia coli*. *J Biol Chem* **247**:8154–8159.

40. **Whitfield C, Trent MS.** 2014. Biosynthesis and export of bacterial lipopolysaccharides. *Annu Rev Biochem* **83**:99–128.

41. **Maldonado RF, Sá-Correia I, Valvano MA.** 2016. Lipopolysaccharide modification in Gram-negative bacteria during chronic infection. *FEMS Microbiol Rev* **40**:480–493.

42. **Braun V, Endriss F.** 2007. Energy-coupled outer membrane transport proteins and regulatory proteins. *Biometals* **20**:219–231.

43. **Grabowicz M, Silhavy TJ.** 2017. Envelope stress responses: an interconnected safety net. *Trends Biochem Sci* **42**:232–242.

44. **Bury-Moné S, Nomane Y, Reymond N, Barbet R, Jacquet E, Imbeaud S, Jacq A, Bouloc P.** 2009. Global analysis of extracytoplasmic stress signaling in *Escherichia coli*. *PLoS Genet* **5**:e1000651.

45. **Raina S, Missiakas D, Georgopoulos C.** 1995. The *rpoE* gene encoding the σ^E (σ^{24}) heat shock sigma factor of *Escherichia coli*. *EMBO J* **14**:1043–1055.

46. **Ades SE, Grigorova IL, Gross CA.** 2003. Regulation of the alternative sigma factor σ^E during initiation, adaptation, and shutoff of the extracytoplasmic heat shock response in *Escherichia coli*. *J Bacteriol* **185**:2512–2519.

47. **Ades SE.** 2004. Control of the alternative sigma factor σ^E in *Escherichia coli*. *Curr Opin Microbiol* **7**:157–162.

48. **Walsh NP, Alba BM, Bose B, Gross CA, Sauer RT.** 2003. OMP peptide signals initiate the envelope-stress response by activating DegS protease via relief of inhibition mediated by its PDZ domain. *Cell* **113**:61–71.

49. **Lima S, Guo MS, Chaba R, Gross CA, Sauer RT.** 2013. Dual molecular signals mediate the bacterial response to outer-membrane stress. *Science* **340**:837–841.

50. **Skovierova H, Rowley G, Rezuchova B, Homerova D, Lewis C, Roberts M, Kormanec J.** 2006. Identification of the σ^E regulon of *Salmonella enterica* serovar Typhimurium. *Microbiology* **152**:1347–1359.

51. **Rhodius VA, Suh WC, Nonaka G, West J, Gross CA.** 2006. Conserved and variable functions of the σ^E stress response in related genomes. *PLoS Biol* **4**:e2.

52. **Chao Y, Papenfort K, Reinhardt R, Sharma CM, Vogel J.** 2012. An atlas of Hfq-bound transcripts reveals 3′ UTRs as a genomic reservoir of regulatory small RNAs. *EMBO J* **31**:4005–4019.

53. **Zhang A, Wassarman KM, Rosenow C, Tjaden BC, Storz G, Gottesman S.** 2003. Global analysis of small RNA and mRNA targets of Hfq. *Mol Microbiol* **50**:1111–1124.

54. **Hayden JD, Ades SE.** 2008. The extracytoplasmic stress factor, σ^E, is required to maintain cell envelope integrity in *Escherichia coli*. *PLoS One* **3**:e1573.

55. **De Las Peñas A, Connolly L, Gross CA.** 1997. σ^E is an essential sigma factor in *Escherichia coli*. *J Bacteriol* **179**:6862–6864.

56. **Papenfort K, Bouvier M, Mika F, Sharma CM, Vogel J.** 2010. Evidence for an autonomous 5′ target recognition domain in an Hfq-associated small RNA. *Proc Natl Acad Sci U S A* **107**:20435–20440.

57. Bouvier M, Sharma CM, Mika F, Nierhaus KH, Vogel J. 2008. Small RNA binding to 5′ mRNA coding region inhibits translational initiation. *Mol Cell* **32**:827–837.

58. Celesnik H, Deana A, Belasco JG. 2007. Initiation of RNA decay in *Escherichia coli* by 5′ pyrophosphate removal. *Mol Cell* **27**:79–90.

59. Emory SA, Bouvet P, Belasco JG. 1992. A 5′-terminal stem-loop structure can stabilize mRNA in *Escherichia coli*. *Genes Dev* **6**:135–148.

60. Papenfort K, Pfeiffer V, Mika F, Lucchini S, Hinton JC, Vogel J. 2006. σ^E-Dependent small RNAs of *Salmonella* respond to membrane stress by accelerating global *omp* mRNA decay. *Mol Microbiol* **62**:1674–1688.

61. Song T, Mika F, Lindmark B, Liu Z, Schild S, Bishop A, Zhu J, Camilli A, Johansson J, Vogel J, Wai SN. 2008. A new *Vibrio cholerae* sRNA modulates colonization and affects release of outer membrane vesicles. *Mol Microbiol* **70**:100–111.

62. Song T, Sabharwal D, Wai SN. 2010. VrrA mediates Hfq-dependent regulation of OmpT synthesis in *Vibrio cholerae*. *J Mol Biol* **400**:682–688.

63. Song T, Sabharwal D, Gurung JM, Cheng AT, Sjöström AE, Yildiz FH, Uhlin BE, Wai SN. 2014. *Vibrio cholerae* utilizes direct sRNA regulation in expression of a biofilm matrix protein. *PLoS One* **9**:e101280.

64. Sabharwal D, Song T, Papenfort K, Wai SN. 2015. The VrrA sRNA controls a stationary phase survival factor Vrp of *Vibrio cholerae*. *RNA Biol* **12**:186–196.

65. Klein G, Kobylak N, Lindner B, Stupak A, Raina S. 2014. Assembly of lipopolysaccharide in *Escherichia coli* requires the essential LapB heat shock protein. *J Biol Chem* **289**:14829–14853.

66. Guo MS, Updegrove TB, Gogol EB, Shabalina SA, Gross CA, Storz G. 2014. MicL, a new σ^E-dependent sRNA, combats envelope stress by repressing synthesis of Lpp, the major outer membrane lipoprotein. *Genes Dev* **28**:1620–1634.

67. Raivio TL. 2014. Everything old is new again: an update on current research on the Cpx envelope stress response. *Biochim Biophys Acta* **1843**:1529–1541.

68. Mileykovskaya E, Dowhan W. 1997. The Cpx two-component signal transduction pathway is activated in *Escherichia coli* mutant strains lacking phosphatidylethanolamine. *J Bacteriol* **179**:1029–1034.

69. Raivio TL, Silhavy TJ. 1997. Transduction of envelope stress in *Escherichia coli* by the Cpx two-component system. *J Bacteriol* **179**:7724–7733.

70. Snyder WB, Davis LJ, Danese PN, Cosma CL, Silhavy TJ. 1995. Overproduction of NlpE, a new outer membrane lipoprotein, suppresses the toxicity of periplasmic LacZ by activation of the Cpx signal transduction pathway. *J Bacteriol* **177**:4216–4223.

71. Delhaye A, Collet JF, Laloux G. 2016. Fine-tuning of the Cpx envelope stress response is required for cell wall homeostasis in *Escherichia coli*. *mBio* **7**:e00047-e16.

72. Vogt SL, Nevesinjac AZ, Humphries RM, Donnenberg MS, Armstrong GD, Raivio TL. 2010. The Cpx envelope stress response both facilitates and inhibits elaboration of the enteropathogenic *Escherichia coli* bundle-forming pilus. *Mol Microbiol* **76**:1095–1110.

73. Guest RL, Wang J, Wong JL, Raivio TL. 2017. A bacterial stress response regulates respiratory protein complexes to control envelope stress adaptation. *J Bacteriol* **199**:e00153-17.

74. Danese PN, Silhavy TJ. 1998. CpxP, a stress-combative member of the Cpx regulon. *J Bacteriol* **180**:831–839.

75. Raivio TL, Popkin DL, Silhavy TJ. 1999. The Cpx envelope stress response is controlled by amplification and feedback inhibition. *J Bacteriol* **181**:5263–5272.

76. Chao Y, Vogel J. 2016. A 3′ UTR-derived small RNA provides the regulatory noncoding arm of the inner membrane stress response. *Mol Cell* **61**:352–363.

77. Grabowicz M, Koren D, Silhavy TJ. 2016. The CpxQ sRNA negatively regulates Skp to prevent mistargeting of β-barrel outer membrane proteins into the cytoplasmic membrane. *mBio* **7**:e00312-16.

78. Vogt SL, Evans AD, Guest RL, Raivio TL. 2014. The Cpx envelope stress response regulates and is regulated by small noncoding RNAs. *J Bacteriol* **196**:4229–4238.

79. Wassarman KM, Repoila F, Rosenow C, Storz G, Gottesman S. 2001. Identification of novel small RNAs using comparative genomics and microarrays. *Genes Dev* **15**:1637–1651.

80. De Lay N, Gottesman S. 2009. The Crp-activated small noncoding regulatory RNA CyaR (RyeE) links nutritional status to group behavior. *J Bacteriol* **191**:461–476.

81. Johansen J, Eriksen M, Kallipolitis B, Valentin-Hansen P. 2008. Down-regulation of outer membrane proteins by noncoding RNAs: unraveling the cAMP-CRP- and σ^E-dependent CyaR-*ompX* regulatory case. *J Mol Biol* **383**:1–9.

82. Papenfort K, Pfeiffer V, Lucchini S, Sonawane A, Hinton JC, Vogel J. 2008. Systematic deletion of *Salmonella* small RNA genes identifies CyaR, a conserved CRP-dependent riboregulator of OmpX synthesis. *Mol Microbiol* **68**:890–906.

83. Majdalani N, Gottesman S. 2005. The Rcs phosphorelay: a complex signal transduction system. *Annu Rev Microbiol* **59**:379–405.

84. Laubacher ME, Ades SE. 2008. The Rcs phosphorelay is a cell envelope stress response activated by peptidoglycan stress and contributes to intrinsic antibiotic resistance. *J Bacteriol* **190**:2065–2074.

85. Majdalani N, Chen S, Murrow J, St John K, Gottesman S. 2001. Regulation of RpoS by a novel small RNA: the characterization of RprA. *Mol Microbiol* **39**:1382–1394.

86. Papenfort K, Espinosa E, Casadesús J, Vogel J. 2015. Small RNA-based feedforward loop with AND-gate logic regulates extrachromosomal DNA transfer in *Salmonella*. *Proc Natl Acad Sci U S A* **112**:E4772–E4781.

87. Madhugiri R, Basineni SR, Klug G. 2010. Turn-over of the small non-coding RNA RprA in *E. coli* is influenced by osmolarity. *Mol Genet Genomics* **284**:307–318.

88. Holmqvist E, Reimegård J, Sterk M, Grantcharova N, Römling U, Wagner EG. 2010. Two antisense RNAs

target the transcriptional regulator CsgD to inhibit curli synthesis. *EMBO J* **29**:1840–1850.

89. **Thomason MK, Fontaine F, De Lay N, Storz G.** 2012. A small RNA that regulates motility and biofilm formation in response to changes in nutrient availability in *Escherichia coli*. *Mol Microbiol* **84**:17–35.

90. **Mika F, Busse S, Possling A, Berkholz J, Tschowri N, Sommerfeldt N, Pruteanu M, Hengge R.** 2012. Targeting of *csgD* by the small regulatory RNA RprA links stationary phase, biofilm formation and cell envelope stress in *Escherichia coli*. *Mol Microbiol* **84**:51–65.

91. **Jørgensen MG, Nielsen JS, Boysen A, Franch T, Møller-Jensen J, Valentin-Hansen P.** 2012. Small regulatory RNAs control the multi-cellular adhesive lifestyle of *Escherichia coli*. *Mol Microbiol* **84**:36–50.

92. **Ogasawara H, Yamamoto K, Ishihama A.** 2011. Role of the biofilm master regulator CsgD in cross-regulation between biofilm formation and flagellar synthesis. *J Bacteriol* **193**:2587–2597.

93. **Boehm A, Vogel J.** 2012. The *csgD* mRNA as a hub for signal integration via multiple small RNAs. *Mol Microbiol* **84**:1–5.

94. **Pogliano J, Lynch AS, Belin D, Lin EC, Beckwith J.** 1997. Regulation of *Escherichia coli* cell envelope proteins involved in protein folding and degradation by the Cpx two-component system. *Genes Dev* **11**:1169–1182.

95. **Pratt LA, Hsing W, Gibson KE, Silhavy TJ.** 1996. From acids to *osmZ*: multiple factors influence synthesis of the OmpF and OmpC porins in *Escherichia coli*. *Mol Microbiol* **20**:911–917.

96. **Alphen WV, Lugtenberg B.** 1977. Influence of osmolarity of the growth medium on the outer membrane protein pattern of *Escherichia coli*. *J Bacteriol* **131**:623–630.

97. **Chen S, Zhang A, Blyn LB, Storz G.** 2004. MicC, a second small-RNA regulator of Omp protein expression in *Escherichia coli*. *J Bacteriol* **186**:6689–6697.

98. **Mizuno T, Chou MY, Inouye M.** 1984. A unique mechanism regulating gene expression: translational inhibition by a complementary RNA transcript (micRNA). *Proc Natl Acad Sci U S A* **81**:1966–1970.

99. **Guillier M, Gottesman S.** 2006. Remodelling of the *Escherichia coli* outer membrane by two small regulatory RNAs. *Mol Microbiol* **59**:231–247.

100. **Brosse A, Korobeinikova A, Gottesman S, Guillier M.** 2016. Unexpected properties of sRNA promoters allow feedback control via regulation of a two-component system. *Nucleic Acids Res* **44**:9650–9666.

101. **Singh SP, Williams YU, Klebba PE, Macchia P, Miller S.** 2000. Immune recognition of porin and lipopolysaccharide epitopes of *Salmonella typhimurium* in mice. *Microb Pathog* **28**:157–167.

102. **Needham BD, Trent MS.** 2013. Fortifying the barrier: the impact of lipid A remodelling on bacterial pathogenesis. *Nat Rev Microbiol* **11**:467–481.

103. **Corcoran CP, Podkaminski D, Papenfort K, Urban JH, Hinton JC, Vogel J.** 2012. Superfolder GFP reporters validate diverse new mRNA targets of the classic porin regulator, MicF RNA. *Mol Microbiol* **84**:428–445.

104. **Groisman EA.** 2001. The pleiotropic two-component regulatory system PhoP-PhoQ. *J Bacteriol* **183**:1835–1842.

105. **Miller SI, Kukral AM, Mekalanos JJ.** 1989. A two-component regulatory system (*phoP phoQ*) controls *Salmonella typhimurium* virulence. *Proc Natl Acad Sci U S A* **86**:5054–5058.

106. **Sittka A, Lucchini S, Papenfort K, Sharma CM, Rolle K, Binnewies TT, Hinton JC, Vogel J.** 2008. Deep sequencing analysis of small noncoding RNA and mRNA targets of the global post-transcriptional regulator, Hfq. *PLoS Genet* **4**:e1000163.

107. **Acuña LG, Barros MJ, Peñaloza D, Rodas PI, Paredes-Sabja D, Fuentes JA, Gil F, Calderón IL.** 2016. A feed-forward loop between SroC and MgrR small RNAs modulates the expression of *eptB* and the susceptibility to polymyxin B in *Salmonella* Typhimurium. *Microbiology* **162**:1996–2004.

108. **Miyakoshi M, Chao Y, Vogel J.** 2015. Cross talk between ABC transporter mRNAs via a target mRNA-derived sponge of the GcvB small RNA. *EMBO J* **34**:1478–1492.

109. **Moon K, Gottesman S.** 2009. A PhoQ/P-regulated small RNA regulates sensitivity of *Escherichia coli* to antimicrobial peptides. *Mol Microbiol* **74**:1314–1330.

110. **Reynolds CM, Kalb SR, Cotter RJ, Raetz CR.** 2005. A phosphoethanolamine transferase specific for the outer 3-deoxy-D-manno-octulosonic acid residue of *Escherichia coli* lipopolysaccharide. Identification of the *eptB* gene and Ca^{2+} hypersensitivity of an *eptB* deletion mutant. *J Biol Chem* **280**:21202–21211.

111. **Moon K, Six DA, Lee HJ, Raetz CR, Gottesman S.** 2013. Complex transcriptional and post-transcriptional regulation of an enzyme for lipopolysaccharide modification. *Mol Microbiol* **89**:52–64.

112. **Mandin P, Gottesman S.** 2010. Integrating anaerobic/aerobic sensing and the general stress response through the ArcZ small RNA. *EMBO J* **29**:3094–3107.

113. **Coornaert A, Chiaruttini C, Springer M, Guillier M.** 2013. Post-transcriptional control of the *Escherichia coli* PhoQ-PhoP two-component system by multiple sRNAs involves a novel pairing region of GcvB. *PLoS Genet* **9**:e1003156.

114. **Chen J, Gottesman S.** 2017. Hfq links translation repression to stress-induced mutagenesis in *E. coli*. *Genes Dev* **31**:1382–1395.

115. **Sharma CM, Papenfort K, Pernitzsch SR, Mollenkopf HJ, Hinton JC, Vogel J.** 2011. Pervasive post-transcriptional control of genes involved in amino acid metabolism by the Hfq-dependent GcvB small RNA. *Mol Microbiol* **81**:1144–1165.

116. **Storz G, Imlay JA.** 1999. Oxidative stress. *Curr Opin Microbiol* **2**:188–194.

117. **Imlay JA.** 2013. The molecular mechanisms and physiological consequences of oxidative stress: lessons from a model bacterium. *Nat Rev Microbiol* **11**:443–454.

118. **Frawley ER, Fang FC.** 2014. The ins and outs of bacterial iron metabolism. *Mol Microbiol* **93**:609–616.

119. **Alekshun MN, Levy SB.** 1997. Regulation of chromosomally mediated multiple antibiotic resistance: the

mar regulon. *Antimicrob Agents Chemother* **41**:2067–2075.

120. Demple B. 1996. Redox signaling and gene control in the *Escherichia coli* *soxRS* oxidative stress regulon—a review. *Gene* **179**:53–57.

121. Rosner JL, Dangi B, Gronenborn AM, Martin RG. 2002. Posttranscriptional activation of the transcriptional activator Rob by dipyridyl in *Escherichia coli*. *J Bacteriol* **184**:1407–1416.

122. Nunoshiba T, deRojas-Walker T, Wishnok JS, Tannenbaum SR, Demple B. 1993. Activation by nitric oxide of an oxidative-stress response that defends *Escherichia coli* against activated macrophages. *Proc Natl Acad Sci U S A* **90**:9993–9997.

123. Lee HJ, Gottesman S. 2016. sRNA roles in regulating transcriptional regulators: Lrp and SoxS regulation by sRNAs. *Nucleic Acids Res* **44**:6907–6923.

124. Zhang A, Altuvia S, Tiwari A, Argaman L, Hengge-Aronis R, Storz G. 1998. The OxyS regulatory RNA represses *rpoS* translation and binds the Hfq (HF-I) protein. *EMBO J* **17**:6061–6068.

125. Schellhorn HE, Hassan HM. 1988. Transcriptional regulation of *katE* in *Escherichia coli* K-12. *J Bacteriol* **170**:4286–4292.

126. Becker-Hapak M, Eisenstark A. 1995. Role of *rpoS* in the regulation of glutathione oxidoreductase (*gor*) in *Escherichia coli*. *FEMS Microbiol Lett* **134**:39–44.

127. Moon K, Gottesman S. 2011. Competition among Hfq-binding small RNAs in *Escherichia coli*. *Mol Microbiol* **82**:1545–1562.

128. Hämmerle H, Večerek B, Resch A, Bläsi U. 2013. Duplex formation between the sRNA DsrA and *rpoS* mRNA is not sufficient for efficient RpoS synthesis at low temperature. *RNA Biol* **10**:1834–1841.

129. González-Flecha B, Demple B. 1999. Role for the *oxyS* gene in regulation of intracellular hydrogen peroxide in *Escherichia coli*. *J Bacteriol* **181**:3833–3836.

130. Argaman L, Altuvia S. 2000. *fhlA* repression by OxyS RNA: kissing complex formation at two sites results in a stable antisense-target RNA complex. *J Mol Biol* **300**:1101–1112.

131. Schlensog V, Böck A. 1990. Identification and sequence analysis of the gene encoding the transcriptional activator of the formate hydrogenlyase system of *Escherichia coli*. *Mol Microbiol* **4**:1319–1327.

132. Altuvia S, Zhang A, Argaman L, Tiwari A, Storz G. 1998. The *Escherichia coli* OxyS regulatory RNA represses *fhlA* translation by blocking ribosome binding. *EMBO J* **17**:6069–6075.

133. Tjaden B, Goodwin SS, Opdyke JA, Guillier M, Fu DX, Gottesman S, Storz G. 2006. Target prediction for small, noncoding RNAs in bacteria. *Nucleic Acids Res* **34**:2791–2802.

134. Barshishat S, Elgrably-Weiss M, Edelstein J, Georg J, Govindarajan S, Haviv M, Wright PR, Hess WR, Altuvia S. 2017. OxyS small RNA induces cell cycle arrest to allow DNA damage repair. *EMBO J* **37**:413–426.

135. Ray-Soni A, Bellecourt MJ, Landick R. 2016. Mechanisms of bacterial transcription termination: all good things must end. *Annu Rev Biochem* **85**:319–347.

136. Conter A, Bouché JP, Dassain M. 1996. Identification of a new inhibitor of essential division gene *ftsZ* as the *kil* gene of defective prophage Rac. *J Bacteriol* **178**:5100–5104.

137. Li GM. 2008. Mechanisms and functions of DNA mismatch repair. *Cell Res* **18**:85–98.

138. Feng G, Tsui HC, Winkler ME. 1996. Depletion of the cellular amounts of the MutS and MutH methyl-directed mismatch repair proteins in stationary-phase *Escherichia coli* K-12 cells. *J Bacteriol* **178**:2388–2396.

139. Tsui HC, Feng G, Winkler ME. 1997. Negative regulation of *mutS* and *mutH* repair gene expression by the Hfq and RpoS global regulators of *Escherichia coli* K-12. *J Bacteriol* **179**:7476–7487.

140. Layton JC, Foster PL. 2003. Error-prone DNA polymerase IV is controlled by the stress-response σ factor, RpoS, in *Escherichia coli*. *Mol Microbiol* **50**:549–561.

141. Barreto B, Rogers E, Xia J, Frisch RL, Richters M, Fitzgerald DM, Rosenberg SM. 2016. The small RNA GcvB promotes mutagenic break repair by opposing the membrane stress response. *J Bacteriol* **198**:3296–3308.

142. Bordeau V, Felden B. 2014. Curli synthesis and biofilm formation in enteric bacteria are controlled by a dynamic small RNA module made up of a pseudoknot assisted by an RNA chaperone. *Nucleic Acids Res* **42**:4682–4696.

Regulating with RNA in Bacteria and Archaea
Edited by Gisela Storz and Kai Papenfort
© 2018 American Society for Microbiology, Washington, DC
doi:10.1128/microbiolspec.RWR-0013-2017

Carbohydrate Utilization in Bacteria: Making the Most Out of Sugars with the Help of Small Regulatory RNAs

14

Svetlana Durica-Mitic[1,*], Yvonne Göpel[1,*], and Boris Görke[1]

INTRODUCTION

Carbohydrates are degraded in central metabolic pathways, namely, glycolysis, the pentose phosphate pathway, and the tricarboxylic acid (TCA) cycle, to fuel cells with energy and building blocks to synthesize all biomolecules. A functional carbohydrate metabolism requires sufficient supply with carbon sources but also coordination with the availability of other nutrients and cellular activities. Hence, bacterial carbohydrate metabolism is controlled at all levels by large and densely interconnected regulatory networks (1). In recent years, posttranscriptional mechanisms involving small regulatory RNAs (sRNAs) have emerged as an additional layer in these networks. Extensive cross talk of sRNAs with transcriptional regulators ensures a fine-tuned and coordinated metabolism.

Bacterial sRNAs come in two flavors. (i) *cis*-Encoded sRNAs are transcribed from the opposite strand of their target genes. Due to their perfect complementarity, they form extensive RNA duplexes with their target transcripts, influencing transcription, translation, or degradation of the target (2). (ii) *trans*-Encoded sRNAs regulate distantly encoded targets that can be either RNA or protein. They regulate translation or RNA stability, either negatively or positively, through imperfect base-pairing (3). In addition, modulation of transcription termination by sRNAs has also been observed (4). In Gram-negative bacteria, *trans*-encoded sRNAs often require protein Hfq for protection from degradation and RNA duplex formation (5, 6). The activities of sRNAs are tightly controlled, at the level of either biogenesis or their decay (7–9). A recently emerging

mechanism is decoy and sponge RNAs that are capable of sequestering sRNAs by base-pairing (10).

Here, we review the manifold roles of sRNAs in regulation of carbohydrate metabolism. The outline of the review is illustrated in Fig. 1. First, specialized sRNAs that regulate consumption of particular carbohydrates are described (Fig. 1A). Bacteria residing in mixed environments often select the energetically most favorable carbon source. The regulatory contribution of sRNAs to substrate prioritization will be discussed subsequently (Fig. 1B). Next, we will dissect how transcriptional and posttranscriptional mechanisms cooperate to coordinate metabolic activities with carbohydrate availability and other cues such as iron and oxygen availability (Fig. 1C). Finally, we will discuss the amino sugar pathway generating precursors for cell envelope synthesis as an exemplary anabolic pathway regulated by sRNAs (Fig. 1D). To date, sRNAs are most thoroughly investigated in the Gram-negative model bacteria *Escherichia coli* and *Salmonella*, but knowledge from unrelated species has increased and is incorporated.

sRNAs REGULATING UTILIZATION OF PARTICULAR CARBON SOURCES

Heterotrophic bacteria such as *E. coli* and *Salmonella* can grow on a plethora of compounds as sole sources of carbon and energy (11). To preserve resources, genes required for uptake and utilization of a particular carbon source are tightly regulated by substrate availability. Traditionally this task is thought to be achieved by dedicated transcription factors, for which the lactose

---

[1]Department of Microbiology, Immunobiology and Genetics, Max F. Perutz Laboratories, University of Vienna, Vienna Biocenter, Vienna, Austria.
*These authors contributed equally to the manuscript.

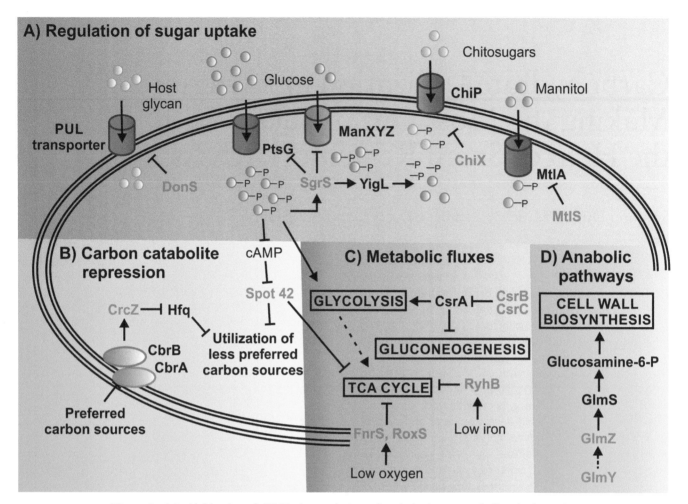

Figure 1 Manifold roles of sRNAs in regulation of carbohydrate metabolism in bacteria. The figure summarizes the major roles of sRNAs (depicted in red) in regulation of carbohydrate metabolism in bacteria. (A) Regulation of uptake and utilization of particular carbohydrates by sRNAs in various species. In *Enterobacteriaceae*, the *trans*-encoded sRNA SgrS counteracts phosphosugar stress through repression of glucose transporters and activation of the sugar phosphatase YigL. sRNA ChiX downregulates the chitosugar-specific porin ChiP, setting the threshold concentration for induction of degrading enzymes. Further examples include regulation of host glycan and mannitol uptake by *cis*-encoded sRNAs in *Bacteroides* and *Vibrio* species, respectively. (B) Role of sRNAs in CCR. In *Enterobacteriaceae* and *Vibrionaceae*, the sRNA Spot 42 represses genes for utilization of secondary carbon sources. Spot 42 is repressed by cAMP-CRP and therefore only active in the presence of preferred sugars generating low cAMP levels. In *Pseudomonas*, translation of mRNAs for utilization of secondary carbon sources is repressed by Hfq. In the absence of preferred substrates, the CbrA/CbrB TCS activates expression of the decoy sRNA CrcZ, titrating Hfq from target transcripts. (C) sRNAs coordinate carbohydrate metabolism with carbohydrate, oxygen, and iron availability. The RNA-binding protein CsrA activates glycolysis and represses gluconeogenesis by binding to corresponding RNAs. CsrA activity is counteracted through sequestration by sRNAs CsrB/CsrC, whose levels are regulated by signals from metabolism. In the absence of oxygen, sRNAs such as FnrS in *E. coli* and RoxS in *B. subtilis* redirect metabolism from oxidative phosphorylation to anaerobic respiration or fermentation. Upon iron starvation, sRNA RyhB represses TCA cycle enzymes to save iron for essential processes. (D) Example of an anabolic pathway regulated by sRNAs. In *Enterobacteriaceae*, two homologous sRNAs regulate the key enzyme GlmS to achieve homeostasis of glucosamine-6-phosphate (glucosamine-6-P), an essential precursor for cell envelope synthesis.

repressor provides the classical paradigm (12). In fact, Jacob and Monod speculated that the Lac repressor could be an RNA acting at the posttranscriptional level (13), but this idea was largely forgotten until the first report of gene regulation by a small antisense RNA in bacteria (14). Meanwhile, several bacterial carbohydrate utilization systems are known to be fine-tuned by sRNAs. The sRNAs in these circuits may limit the response to substrate availability, set the threshold concentration, or modulate the delay time required for activation and shutdown of the system.

Posttranscriptional Regulation of Glucose Uptake

Many bacteria, including *E. coli*, preferentially utilize glucose when growing in a mixture of carbon sources (15, 16), which also holds true for many enterobacterial pathogens when residing in mammalian host cells (17). In *E. coli*, glucose is internalized by the glucose transporter PtsG and to a minor degree by mannose transporter ManXYZ (18). Both transporters belong to the phosphotransferase system (PTS). PTS transporters generate phosphosugars during transport. The phosphoryl groups derived from phosphoenolpyruvate (PEP) are transferred via phosphotransferases enzyme I and HPr to the transporters, including the EIIA^{Glc} protein, which phosphorylates PtsG.

While phosphosugars are a primary energy source, high intracellular concentrations are toxic (19, 20). Such conditions cause rapid degradation of *ptsG* mRNA, limiting further glucose uptake, which relieves stress (21, 22). The dedicated transcriptional regulator SgrR senses phosphosugar stress and induces expression of sRNA SgrS (23, 24). Hfq-assisted base-pairing of SgrS with *ptsG* inhibits translation and recruits endoribonuclease RNase E to degrade *ptsG* mRNA at the cytoplasmic membrane (23, 25–27). SgrT, a short peptide encoded by SgrS, contributes to stress relief by blocking glucose transport via direct inhibition of PtsG independently of SgrS base-pairing (28–30). In *E. coli*, SgrS regulates at least eight mRNAs by direct base-pairing (31, 32). Downregulation of the *manXYZ* mRNA through a dual base-pairing mechanism prevents leaky glucose uptake (33, 34). Stabilization of the *yigL* mRNA, encoding a sugar phosphatase, by masking an RNase E cleavage site upon base-pairing allows export of sugars following their dephosphorylation (35–37).

While regulation of *manXYZ* and *yigL* clearly contributes to phosphosugar stress relief (36), the roles of the remaining targets *adiY*, *asd*, *folE*, *ptsI*, and *purR* are less obvious (31, 32). The *ptsI* gene encodes enzyme I, which delivers phosphoryl groups to all 21

PTS transporters in *E. coli* (38, 39). Some of these transporters internalize sugars that generate glucose-6-phosphate or other phosphosugars upon catabolism. Thus, global deceleration of PTS activity may contribute to the phosphosugar stress response. Interestingly, phosphosugar stress elicited by a glucose analog or a block in glycolysis, e.g., by *pgi* mutation, can be rescued by addition of glycolytic intermediates downstream of the block (21, 40). This suggests that toxicity results from the depletion of a downstream metabolite, most likely PEP, rather than from accumulation of glucose-6-phosphate itself. This could explain the physiological roles of SgrS targets such as *asd*, which encodes an enzyme that converts aspartate to other amino acids. Downregulation of *asd* may preserve aspartate to replenish PEP and relieve stress (31).

The SgrS-mediated phosphosugar stress response seems conserved in *Enterobacteriaceae* and *Aeromonas* species (41, 42). However, PTS-type glucose transporters are much more widespread (18, 43). Do these bacteria also encounter phosphosugar stress, and how do they cope? Downregulation of the *ptsG* transcript by glucose in a *pgi* mutant has also been observed for the Gram-positive *Corynebacterium glutamicum* (44). *C. glutamicum* lacks Hfq, and therefore the underlying mechanism must differ from *E. coli*. A phosphosugar stress response has not been reported for any of the Gram-positive *Firmicutes* species. However, as demonstrated for *Bacillus subtilis*, these bacteria may activate a glycolytic bypass, the methylglyoxal pathway, to prevent deleterious accumulation of phosphosugars (45).

Regulation of Chitin and Chitosugar Utilization by sRNAs

Chitin is one of the most abundant polysaccharides on earth and particularly ample in aquatic environments, representing an important carbon source for aquatic bacteria such as *Vibrionaceae*. In *Vibrio cholerae*, an important facultative human pathogen, chitin even serves as signal for natural competence. Chitin is sensed by the orphan sensor kinase ChiS, which activates expression of chitin utilization genes by a still-unknown mechanism (46, 47). ChiS further activates transcription factor TfoS, which is necessary for expression of the Hfq-dependent sRNA TfoR. This sRNA stimulates translation of TfoX, a regulator required for induction of competence (48, 49). In addition, TfoX induces expression of type VI secretion systems for killing of nonimmune cells and subsequent acquisition of the released DNA (50). This mechanism provides a mechanistic basis for the high degree of genomic diversity observed in *V. cholerae*.

For *E. coli* and *Salmonella*, chitin-derived carbohydrates represent a secondary carbon source as they become sporadically available as part of the hosts' diet. These species rely on excreted chitinases of other bacteria to convert chitin to chitosugars. Multiple transcriptional regulators and the sRNA ChiX are employed to restrict expression of chitosugar utilization genes to conditions of substrate sufficiency (51–53). Chitoporin ChiP, required for uptake of chitosugars across the outer membrane, is encoded in the *chiPQ* operon. ChiX inhibits *chiP* translation initiation by base-pairing with its 5′ untranslated region (5′ UTR) (54, 55) but also represses the distal cistron *chiQ* by facilitating Rho-dependent transcriptional termination (56). Interestingly, ChiX is not codegraded with its target *chiP* but with a decoy RNA derived from the *chb* operon (54, 57). The *chb* operon encodes a PTS transporter and enzymes for chitosugar uptake and degradation. Expression of the *chb* operon is activated by the operon-specific transcription regulator ChbR in response to chitosugar availability. When *chb* transcription rates are sufficiently high, base-pairing with the *chb* RNA trap sequesters ChiX and relieves *chiPQ* repression, boosting synthesis of chitoporin ChiP. Thereby, ChiX likely sets the delay time and threshold concentration for chitosugar utilization. ChiX only affects the *chb* transcript under noninducing conditions, leading to efficient silencing of the mRNA (53).

In the chitinolytic bacterium *Serratia marcescens*, ChiX coordinates synthesis of ChiP and chitin-degrading chitinases (58). Whereas *chiP*/ChiX base-pairing is conserved, the ChiX target site within the *chb* mRNA is lacking. In contrast, ChiX represses *chiR*, encoding a transcriptional activator of chitinase genes. Upon induction of *chiP* expression, ChiX is sequestered by base-pairing and repression of *chiR* is relieved (58). Thereby, ChiX couples induction of degrading enzymes to the expression of the specific transporters, coordinating extracellular breakdown of chitin with uptake of the products.

Regulation of Mannitol Uptake by a *cis*-Encoded sRNA in *V. cholerae*

In addition to chitin, mannitol represents an important carbon source for *V. cholerae*, as it is produced in large quantities by marine algae. In *V. cholerae*, synthesis of the mannitol PTS transporter MtlA is controlled by the *cis*-encoded sRNA MtlS through an Hfq-independent mechanism (59). MtlS is transcribed antisense to the mannitol *mtlADR* operon and shares 71 nucleotides of perfect complementarity with the 5′ UTR of *mtlA*. MtlS and *mtlA* form a stable duplex inhibiting *mtlA*

translation without impairing transcript stability. How this affects the cotranscribed *mtlDR* genes is unknown. Close proximity of the *mtlA* and *mtlS* loci is required to efficiently repress *mtlA*, presumably by enabling rapid formation of the RNA duplex (60). Mannitol represses *mtlS* transcription, but the responsible regulator has not been identified. The MtlR repressor protein, which is encoded in the *mtlADR* operon itself, and MtlS appear to operate independently from each other (59, 61). MtlA was shown to activate biofilm formation, suggesting that mannitol serves as extracellular signal for *V. cholerae* to colonize beneficial habitats (62). Mannitol may also act as compatible solute, helping *V. cholerae* to withstand the high osmolarity in the human intestine (59). How these additional roles are integrated into the mannitol operon remains to be addressed.

Regulation of Polysaccharide Utilization Genes by *cis*-Encoded sRNAs in *Bacteroides*

Gram-negative *Bacteroidetes* is a dominating phylum of the microbiota in the human colon (63) and specializes in utilizing a wide variety of dietary polysaccharides and glycans derived from the mucosa of the gut (64). To this end, *Bacteroides* spp. carry a large number of polysaccharide utilization loci (PULs), each one dedicated to the uptake and utilization of a specific glycan or polysaccharide. Each PUL encodes its own protein regulators for substrate-dependent induction of the locus. Transcriptome sequencing analysis of *Bacteroides fragilis* revealed that many of these PULs transcribe sRNAs from the opposite strand (65). The antisense RNAs seem to be conserved, as they are also observed in other *Bacteroides*. Overexpression of such an sRNA, DonS in *B. fragilis*, triggers loss of the corresponding *pul* transcript, causing disability to utilize corresponding host glycans. DonS may target the cognate *pul* mRNA towards degradation or act through transcriptional interference by RNA polymerase collision, as observed for other antisense RNAs (2). Regulation by DonS might become relevant when the concentration of the inducing substrate declines, leading to an excess of constitutively produced DonS over the *pul* transcript. Interestingly, the PULs shown to include antisense sRNAs are all involved in the utilization of host-derived glycans (65). Species like *Bacteroides thetaiotaomicron* preferentially utilize dietary polysaccharides if available and consequently repress the PULs for glycan utilization (66). Hence, it is possible that the DonS-like sRNAs mediate substrate prioritization in *Bacteroides*.

CARBON CATABOLITE REPRESSION AT THE POSTTRANSCRIPTIONAL LEVEL

In mixed environments, bacteria often selectively utilize the carbon source favoring fastest growth (15, 67). In *E. coli*, uptake of the preferred substrate glucose triggers dephosphorylation of the PTS, which activates mechanisms that prevent uptake and utilization of less-preferred carbon sources—collectively known as carbon catabolite repression (CCR) (43, 68). Accumulation of nonphosphorylated EIIA^{Glc} inhibits uptake of less-preferred carbon sources by inducer exclusion. It also impedes production of cyclic AMP (cAMP), thereby preventing activation of carbohydrate utilization genes by the global transcription regulator cAMP receptor protein (CRP). While these mechanisms are well studied, the involvement of posttranscriptional mechanisms in CCR emerged only recently. A global omics study in *E. coli* found >90 genes to be posttranscriptionally regulated by CCR (69). In the evolutionary distant pseudomonads, CCR even appears to operate solely at the posttranscriptional level (70).

Spot 42: the Third Pillar of CCR in *Enterobacteriaceae*

The sRNA Spot 42 cooperates with cAMP-CRP in coherent feedforward loops to regulate multiple carbohydrate metabolic genes (Fig. 2A) (71). It might therefore be considered the third pillar of CCR, working in addition to the well-established mechanisms involving inducer exclusion and cAMP (43). Spot 42, encoded by *spf*, is one of few genes that are repressed by cAMP-CRP in *E. coli* (72, 73). The first Spot 42 target discovered was *galK*, encoding galactokinase for galactose utilization (74). In the presence of glucose, Spot 42 accumulates and selectively downregulates GalK without affecting the other proteins encoded in the *galETKM* operon, which have additional functions for synthesis of UDP-sugars (Fig. 3) (74). Subsequent work revealed that Spot 42 has a global role in carbohydrate metabolism (75, 76). To date, its validated regulon contains 29 genes (Fig. 2B), but is expected to increase further as many additional potential Spot 42 targets were identified by RNA interaction by ligation and sequencing (RIL-seq) (77). This novel methodology identifies sRNA/mRNA pairs by Hfq pulldown and subsequent ligation of bound RNAs. Impressively, RIL-seq recovered 11 previously validated Spot 42 targets, emphasizing the reliability of the method (Fig. 2B). Again, many of the newly identified candidate targets have roles in carbohydrate metabolism (77).

Most targets are repressed by Spot 42, and where known, their transcription is activated by cAMP-CRP,

fostering the hypothesis that Spot 42 cooperates with cAMP-CRP in coherent feedforward loops to regulate carbohydrate utilization genes (Fig. 2) (75). In the presence of glucose, Spot 42 prevents leaky expression of these genes by targeting the few mRNAs produced despite inactivity of CRP. Furthermore, Spot 42 also shapes the dynamics of gene expression when cells shuttle between CCR and CCR-free conditions. Upon a shift to glucose-rich growth conditions, Spot 42 accelerates repression of the secondary carbohydrate utilization genes, which may facilitate adaptation to the more favorable growth condition. Vice versa, upon activation of CRP by cAMP, Spot 42 delays target activation, perhaps to prevent their premature activation in case glucose reappears (71, 75). Interestingly, many secondary carbon sources whose utilization is repressed by Spot 42 are available in the mucosa of mammalian guts (e.g., arabinose, *N*-acetylneuraminic acid, and L-fucose; Fig. 2A), *E. coli*'s natural habitat (78), where Spot 42 may be particularly important for carbon source selection (71).

Until recently, cAMP-CRP was the only known regulator of Spot 42. However, RIL-seq identified a sponge sRNA, PspH, whose overexpression reduces Spot 42 levels and thus derepresses its targets (77). The role of this interaction is unknown. Another study reported induction of *spf* expression by pyruvate independent of cAMP-CRP (79), suggesting that the *spf* promoter is controlled by additional transcription factor(s) that remain to be identified.

Spot 42 base-pairing sites overlap or are close to the ribosomal binding site (RBS) (75, 76), and work on *galK* demonstrated that Spot 42 inhibits translation in an Hfq-dependent manner (74, 80). However, Spot 42 also alters target mRNA levels (75). In-depth study of the *galETKM* operon revealed that Spot 42 stimulates Rho-dependent transcription termination at the *galT-galK* junction, recapitulating observations for ChiX (56, 81). A noncanonical mechanism of Spot 42 action was observed for the *sdhCDAB* mRNA, encoding succinate dehydrogenase (82). Spot 42 pairs far upstream of the *sdhC* RBS and merely recruits Hfq, which inhibits translation. A direct role for Hfq as a translation repressor has also been reported for other mRNAs (83, 84), even in species beyond *E. coli* (70). Spot 42 contains three unstructured regions, each of them involved in target regulation, explaining the high conservation of the entire Spot 42 sequence (75). In some cases, Spot 42 employs multiple base-pairing sites to regulate a single target, which might improve regulatory strength (76). Perhaps multisite pairing provides the flexibility required to regulate multiple targets, as observed for

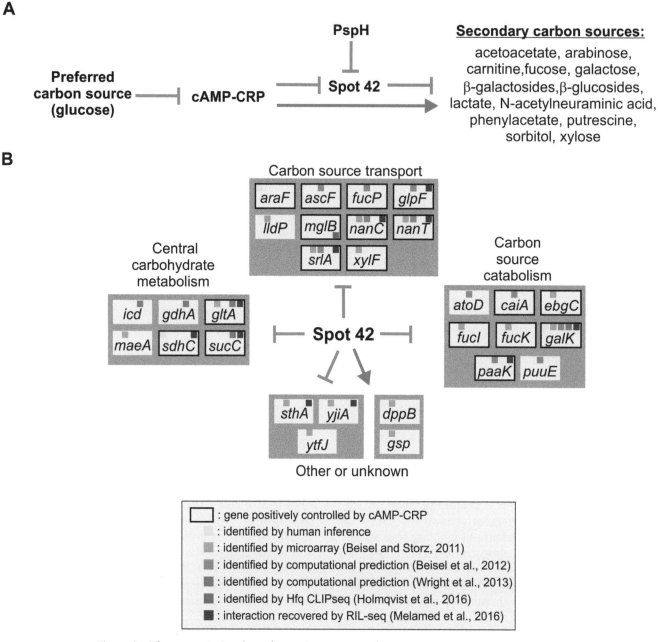

Figure 2 The transcriptional regulator cAMP-CRP and sRNA Spot 42 cooperate to trigger CCR in *Enterobacteriaceae*. (A) CRP and Spot 42 participate in coherent feedforward loops to prevent utilization of the indicated secondary carbon sources when the preferred carbon source glucose is present. In addition to cAMP-CRP, Spot 42 is regulated by base-pairing with the sponge RNA PspH. (B) The validated Spot 42 regulon to date. Target genes that are also positively controlled by cAMP-CRP at the level of transcription are boxed. Microarray analysis of Spot 42 pulse expression (75) and improved software prediction algorithms (32, 76) fostered the identification of most targets. Additional targets were identified by human inference or by a CLIP-seq approach mapping Hfq binding sites on a global scale (74, 82, 173, 181). Several of these targets were recovered by RIL-seq (77).

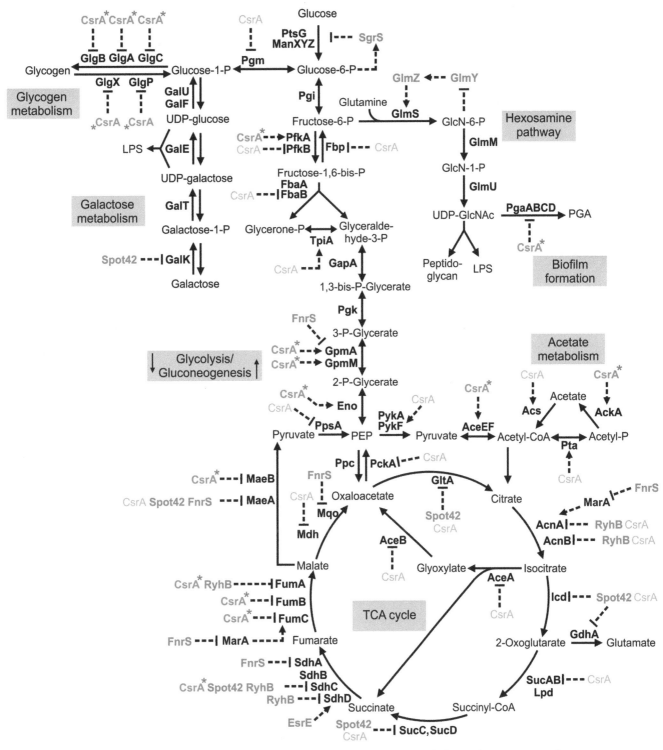

Figure 3 Posttranscriptional regulation of central carbon metabolic pathways in *E. coli*. Effects of sRNAs (depicted in red) and of the RNA-binding protein CsrA (blue) on synthesis of enzymes involved in glycolysis, gluconeogenesis, and the TCA cycle. A green asterisk and bold letters indicate direct regulation by CsrA (106). Anabolic pathways directing synthesis of glycogen, UDP-sugars, and the biofilm compound PGA are also shown.

other sRNAs controlling exceptionally large regulons (77, 85, 86).

Outside of the *Enterobacteriaceae*, the *spf* gene is found in four additional orders of *Gammaproteobacteria*, including *Vibrionaceae* (87). In the latter order, the role of Spot 42 was also studied in the fish pathogen *Aliivibrio salmonicida* and in *Vibrio parahaemolyticus*, which causes diarrhea and gastroenteritis in humans through consumption of contaminated seafood (88–90). In *A. salmonicida*, *spf* expression is negatively regulated by cAMP-CRP, like in *E. coli* (88). Microarray analysis of an *spf* deletion mutant revealed upregulation of genes involved in sugar catabolism, motility, and chemotaxis (88). Thus, *A. salmonicida* Spot 42 may also impact on carbohydrate metabolism, but by targeting different genes than in *E. coli*, and carrying out additional roles. In *V. parahaemolyticus*, Spot 42 is strongly upregulated during infection and impacts on the activities of two type III secretion systems (89, 90). It is tempting to speculate that Spot 42 coordinates the activities of the type III secretion systems with carbohydrate availability in the host.

CCR in *Pseudomonas* by Hfq-Mediated Translational Repression

In pseudomonads, CCR is regulated exclusively at the posttranscriptional level. Contrary to *E. coli*, pseudomonads do not prefer glucose. Rather, succinate elicits the highest degree of CCR in *Pseudomonas aeruginosa* but only weakly affects CCR in *Pseudomonas putida*, which prefers alkanes and branched-chain amino acids (91). CCR requires sRNA antagonists, the catabolite repression control protein Crc, and Hfq as master regulator. Hfq binds the 5′ UTRs of mRNAs encoding transporters and catabolic enzymes for less-preferred carbon sources and directly blocks translation initiation (70). Protein Crc is required for efficient CCR and forms stable ternary complexes with Hfq and RNAs containing A-rich motifs (92, 93). Crc contributes to the stability of these complexes through interaction with both Hfq and RNA (93).

Availability of the preferred carbon source coincides with low levels of sRNA antagonists for Hfq. *P. aeruginosa* possesses one such sRNA, CrcZ. *P. putida* encodes two, CrcY and CrcZ, and other species possess similar sRNAs (94–96). Expression of Crc sRNAs is induced by the two-component system (TCS) CbrA/CbrB in the absence of preferred carbon sources (92, 94–97). Kinase CbrA presumably senses internal stimuli reflecting the energetic state of the cell, such as the α-ketoglutarate/glutamine ratio (91, 97). CrcZ of *P. aeruginosa* and CrcY/CrcZ of *Pseudomonas fluorescens* bind Hfq with an ∼5-to-20-fold-higher affinity as compared to the mRNAs targeted by Hfq (70, 95, 98). Consequently, increased levels of the Crc sRNAs effectively sequester Hfq from its target transcripts, relieving repression (92, 94). Through competition for Hfq, CrcZ may also interfere with riboregulation exerted by other sRNAs and thus indirectly impact their regulatory potential (98).

POSTTRANSCRIPTIONAL CONTROL OF CENTRAL CARBOHYDRATE METABOLISM

Posttranscriptional Mechanisms Coordinating Central Metabolism with Carbohydrate Availability

The activity of central carbohydrate metabolism is tightly coordinated with carbon supply by adjusting the amounts of corresponding enzymes in response to key metabolites, namely, fructose-1,6-bisphosphate (FBP) and the PEP/pyruvate ratio (1, 99). In *E. coli*, FBP is sensed by the transcriptional regulator Cra, which represses glycolytic genes and activates genes involved in gluconeogenesis (100, 101). The PEP/pyruvate ratio determines the phosphorylation state of EIIA^{Glc}, which regulates adenylate cyclase and thus cAMP synthesis (102, 103). Importantly, FBP decreases the PEP/pyruvate ratio through feedforward activation of pyruvate kinase and PEP carboxylase (1). A high FBP level activates glycolysis through inactivation of Cra and decreases activity of cAMP-CRP by inhibiting phosphorylation of EIIA^{Glc}. Of note, cAMP-CRP activates expression of TCA cycle enzymes (104), whose transcripts are repressed by Spot 42 (Fig. 3). Therefore, cAMP-CRP and Spot 42 also cooperate to redirect metabolism from oxidative phosphorylation to fermentation when glycolytic carbon sources are available. Similar to CRP, Cra possesses a counterpart at the posttranscriptional level, which is the carbon storage regulatory (Csr) system.

Regulation of Glycolysis and Gluconeogenesis by the Csr System

Protein CsrA represents a global posttranscriptional regulator of diverse activities across bacterial species. In *E. coli* and other Gram-negative bacteria, CsrA controls carbohydrate metabolic pathways, carbon source and nutrient acquisition, biofilm formation, motility, stress responses, and virulence (105, 106). CsrA binds mRNA substrates at GGA motifs and mostly represses translation, but examples of positive regulation also exist (107). For instance, CsrA activates glycolysis by positively regulating mRNAs of several glycolytic enzymes while repressing synthesis of enzymes for gluconeogenesis and the TCA cycle (Fig. 3) (106, 108, 109). In fact, CsrA is essential for growth on glycolytic substrates, reflecting its crucial role in an undisturbed carbohydrate metabolism (110). Flux analysis showed that stabilization of the *pfkA* mRNA, encoding phosphofructokinase, is crucial for regulation of glycolytic activity by CsrA (106, 109). CsrA also inhibits accumulation of the carbon storage compound glycogen and synthesis of the exopolysaccharide poly-β-1,6-N-acetyl-D-glucosamine (PGA), a major component of biofilm matrices (Fig. 3). Of note, CsrA was recently shown to bind Spot 42 and to activate target genes that are repressed by this sRNA (106). It remains to be shown whether CsrA binding inhibits Spot 42 base-pairing, thereby also influencing CCR. Further, binding to *cra* mRNA was also demonstrated, but the physiological consequences are so far unclear. These sophisticated interconnections may expand the already complex regulatory network governing carbohydrate metabolism.

CsrA activity is antagonized by the decoy sRNAs CsrB and CsrC, which sequester CsrA by presenting multiple binding sites (111, 112). Transcription and decay of these sRNAs are controlled by signals derived from carbohydrate metabolism. The BarA/UvrY TCS activates transcription of both sRNAs in response to short-chained carboxylic acids, e.g., acetate and formate, which accumulate when cells have expended glycolytic carbon sources and transition into stationary phase (Fig. 4) (113, 114). Degradation of the sRNAs by RNase E requires the protein CsrD (115, 116). CsrD is activated by interaction with nonphosphorylated EIIA^{Glc} in the presence of glycolytic substrates (102, 117). Together, activation of *csrB/csrC* transcription and slowdown of CsrB/CsrC decay increases abundance of these sRNAs when preferred carbon sources have been consumed (Fig. 4). The resulting shutdown of CsrA activity promotes the shift to stationary-phase metabolism by repression of glycolytic genes and derepression of gluconeogenetic and glycogen biosynthetic mRNAs (117, 118).

EIIA^{Glc} has an additional role in the activity of the Csr system in *E. coli*, as it also controls transcription of CsrB/CsrC through cAMP-CRP (119). cAMP-CRP represses *csrB* indirectly and *csrC* directly by blocking access of response regulator UvrY to the *csrC* promoter. Thus, nonphosphorylated EIIA^{Glc} has opposing effects, as it activates the turnover but also transcription of these sRNAs (Fig. 4), creating an incoherent feed-forward loop with the potential to integrate further cues (119). For instance, the activity of adenylate cyclase is also inhibited by α-ketoglutarate, signaling nitrogen limitation (Fig. 4) (120), which may affect CsrB/CsrC expression, but not degradation.

The Csr system is conserved in *Proteobacteria*, albeit the number of CsrA paralogs and Csr sRNAs may vary (105). The regulatory links between Csr and EIIA^{Glc}/CRP may likewise differ, as CsrD is absent in most *Proteobacteria* beyond the families *Enterobacteriaceae*, *Shewanellaceae*, and *Vibrionaceae* (116). Similarly, control of *csrB/csrC* transcription may be different; e.g., in *Yersinia pseudotuberculosis*, a close relative of *E. coli*, expression of *csrC* is activated by PhoP/PhoQ rather than the BarA/UvrY TCS (121). In sum, the Csr system provides a further tier of controlling fluxes through central carbohydrate metabolic pathways in response to carbohydrate availability. In addition, CsrA may cross-talk to CCR and integrate information on the metabolic status into other intricately regulated processes, namely biofilm formation, motility, and pathogenicity. For more information on this topic, see chapter 19 (182).

Regulation of TCA Cycle Activity by sRNAs

ATP can be produced either by substrate-level phosphorylation or by oxidative phosphorylation. Respiration yields more ATP but is also more costly, as it requires more proteins. Bacteria sense the availability of carbon, oxygen, and energy to efficiently regulate the TCA cycle and respiration. As already discussed, in *E. coli* the information on carbohydrate availability is integrated into the TCA cycle by CsrA, cAMP-CRP, and sRNA Spot 42. sRNAs with comparable functions may also exist in unrelated bacteria. For instance, pathogenic *Neisseria* species employ two homologous sRNAs to repress transcripts of TCA cycle enzymes (122, 123). Overexpression of these sRNAs impairs growth of *Neisseria meningitidis* in cerebrospinal fluid but not in blood, suggesting that they integrate information about the metabolic status into the decision to colonize different niches in the host (122). In addition, activity of the TCA cycle is strongly shaped by availability of iron and oxygen. Again, sRNAs play prominent roles in these adaptations.

Downregulation of TCA Cycle Activity by sRNAs in Response to Iron Limitation

Iron is indispensable for activity of numerous enzymes operating within major metabolic pathways. Upon limitation, bacteria redirect iron from nonessential to essential processes with the aid of transcription factor Fur and sRNA RyhB (124). Fur represses transcription of *ryhB* under iron sufficiency (125). However, upon

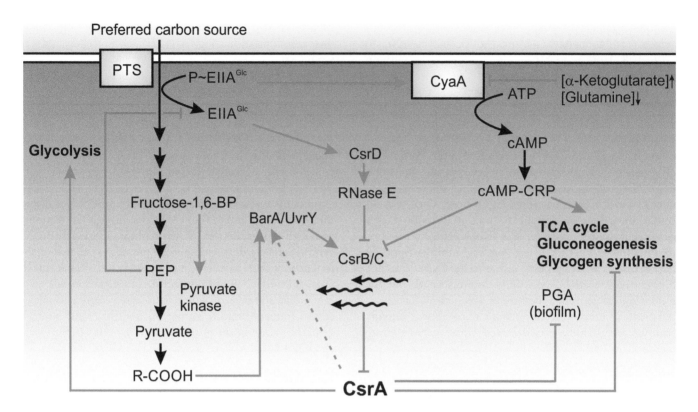

Figure 4 Model of the interconnection of the CsrA system with central carbon metabolism. Decoy sRNAs CsrB and CsrC regulate CsrA activity by sequestering the protein from its target mRNAs. CsrA indirectly activates *csrB/csrC* transcription, creating a negative feedback loop. In fast-growing cells, when CsrB/C levels are low, CsrA activates glycolytic genes and represses the TCA cycle, gluconeogenesis, and glycogen synthesis. Metabolism of glycolytic carbon sources causes accumulation of FBP, which activates pyruvate kinase, thereby reducing the PEP/pyruvate ratio. Intake of PTS substrates and a low PEP/pyruvate ratio trigger dephosphorylation of EIIA^{Glc}, leading to activation of CsrD, which triggers degradation of CsrB/CsrC by RNase E. Upon accumulation of short carboxylic acids (R-COOH) as metabolic end products, expression of *csrB/csrC* is induced by the BarA/UvrY TCS. Deceleration of glycolytic activity elevates the PEP/pyruvate level and increases EIIA^{Glc} phosphorylation, leading to stabilization of CsrB/CsrC and titration of CsrA. EIIA^{Glc}~P stimulates adenylate cyclase CyaA, which converts ATP to cAMP. The cAMP-CRP complex inhibits transcription of *csrB/csrC*. Involvement of EIIA^{Glc} in regulation of CsrB/CsrC synthesis as well as decay allows integration of further cues.

iron starvation, RyhB is relieved from repression and downregulates nonessential iron-containing proteins, including TCA cycle enzymes (Fig. 3) (126), prompting cells to resort to fermentation (127). This trade-off enables essential pathways involving iron-dependent enzymes to remain functional when iron is scarce. Iron limitation in particular is encountered by pathogenic bacteria within the host (128). *Staphylococcus aureus* was shown to switch to fermentation inside the host, thereby producing lactate, which lowers the surrounding pH. This increases iron availability through release from host iron storage proteins (129). Switch to fermentation is restricted to bacteria that can grow anaer-

obically. In the obligate aerobe *Azotobacter vinelandii*, the functional analog of RyhB named ArrF does not affect TCA cycle-related enzymes, but rather represses genes involved in nitrogen fixation, a nonessential process (130).

Coordination of Carbon Metabolism with Oxygen Availability by sRNA FnrS in *Enterobacteriaceae*

Enterobacteriaceae are facultative anaerobes. In the absence of oxygen, *E. coli* uses alternative electron acceptors to procure anaerobic respiration. If oxygen is not available, NAD$^+$ is regenerated by fermenting carbon sources to mixed acids and ethanol (131). Two global transcription factors, ArcA and Fnr, reprogram metabolism in response to anaerobiosis (132, 133). Fnr senses oxygen directly, whereas response regulator ArcA is activated by its cognate kinase ArcB when the redox state of the quinone pool changes. Upon anaerobiosis, Fnr and ArcA collectively activate genes of alternative electron transport chains and repress functions of aerobic metabolism, including the TCA cycle, the glyoxylate shunt, and respiratory NADH dehydrogenases (134–137).

Notably, Fnr and ArcA also employ sRNAs in their regulons. One of them, sRNA FnrS, is conserved among *Enterobacteriaceae*. FnrS is only detectable in the absence of oxygen, as its transcription strictly depends on Fnr and to a minor extent on ArcA (86, 138). Globally, FnrS appears to extend the regulons of Fnr and ArcA by acting as a noncoding regulator to repress functions that are not required in absence of oxygen, including enzymes of aerobic carbohydrate metabolism (Fig. 3) (86, 138). For other targets, e.g., *mqo* (Fig. 3), FnrS cooperates in coherent feedforward loops, as these genes are also directly repressed by Fnr or ArcA (86, 139). This also applies to *acnA* and *fumC*, but here FnrS acts indirectly through repression of MarA, which is a transcriptional activator of these TCA cycle genes—a regulatory scenario known as a multistep coherent feedforward loop (Fig. 3) (32, 138). RIL-seq revealed many additional metabolism-related transcripts putatively base-pairing with FnrS, including the *fnr* mRNA itself (77), hinting at a feedback loop balancing Fnr and FnrS levels. FnrS is Hfq dependent and appears to act primarily by inhibition of translation initiation (138). Interestingly, FnrS uses distinct sequences to base-pair with subsets of its targets. Transcripts linked to oxidative stress and folate metabolism appear to base-pair with the 5′ end of FnrS, whereas mRNAs of central metabolic enzymes are regulated by a single-stranded region in the sRNA body (86). This functional specialization may reflect evolution of FnrS by fusion of two originally distinct sRNAs (86).

The *E. coli* and *Salmonella* ArcA regulon contains an additional sRNA, ArcZ, which is encoded downstream of the *arcB* gene and is only expressed under aerobic conditions. ArcZ limits accumulation of active ArcA through destabilization of the *arcB* mRNA (140) and targets further diverse functions, but is apparently not involved in regulation of carbohydrate metabolism (140–143). Recently, the sRNA EsrE was shown to activate synthesis of subunit SdhD of succinate dehydrogenase in *E. coli* (Fig. 3) (144). EsrE appears somewhat

as an aerobic opponent of FnrS, as it is essential for aerobic growth on TCA cycle substrates, but the signal to which it responds remains unknown.

An sRNA activated in response to anaerobiosis was also identified in pathogenic *Neisseria* species (145, 146). These bacteria, which likely face oxygen limitation during host colonization, are capable of anaerobic respiration (147–149). The anaerobically induced sRNA was named AniS in *N. meningitidis* and FnrS in *Neisseria gonorrhoeae*, and both clearly belong to the Fnr regulon, albeit sequence homology to enterobacterial FnrS is lacking (145, 146). So far, only a few targets for these sRNAs are known, and they do not contribute to a common metabolic process (146, 150), leaving it open whether these sRNAs are indeed functional equivalents of enterobacterial FnrS.

RsaE: a Functional Equivalent of FnrS in Gram-Positive Bacteria?

sRNAs also play a role in regulation of central carbohydrate metabolism in Gram-positive *Firmicutes*. RsaE—later renamed RoxS in *B. subtilis* (151)—is besides the ubiquitous 6S RNA the sole *trans*-acting sRNA known to be conserved between staphylococci and *Bacillaceae* (152). Two independent studies linked RsaE of *S. aureus* to regulation of carbohydrate metabolism, amino acid transport, and the folate pathway for one-carbon metabolism (152, 153). In particular, RsaE represses pyruvate dehydrogenase and several TCA cycle enzymes (152, 153). Consistently, downregulation of TCA cycle enzymes was also observed for RoxS in *B. subtilis* (151, 154).

Expression of RsaE/RoxS is induced by the response regulator ResD of the ResD/ResE TCS (151). *S. aureus* and *B. subtilis* are facultative anaerobes and can switch to fermentation or nitrate respiration in absence of oxygen. The ResD/ResE TCS (named SrrA/SrrB in *S. aureus*) responds to oxygen limitation or increased nitric oxide (NO) levels and activates genes required for anaerobic metabolism and NO detoxification (155). Nitrate respiration produces NO as a by-product, which is likely sensed as an indicator of nitrate availability and leads to induction of RsaE/RoxS expression through ResD (SrrA) (151). Therefore, RoxS (RsaE) may extend the regulon of the ResD/ResE TCS, contributing to adaptation to anoxia. *B. subtilis* RoxS is additionally controlled by transcription factor Rex, which represses genes for fermentation under oxic conditions when the NADH/NAD$^+$ ratio is low (154). RoxS is transiently released from Rex repression when malate is utilized, which generates NADH in the early steps of catabolism (154). By stimulating synthesis of the malate transporter YflS, RoxS ensures continuous uptake of malate (154).

Detailed analysis of *yflS* regulation by RoxS revealed a novel mechanism for how RNA degradation may be counteracted by sRNAs in Gram-positive bacteria. RoxS base-pairs with the 5′ end of the *yflS* mRNA, thereby protecting it from RNase J1, which degrades RNA in 5′-to-3′ direction—an activity absent in *Enterobacteriaceae* (154). Among the negatively regulated RoxS targets, the *ppnKB* mRNA was studied in detail (151). Base-pairing inhibits translation but also creates an RNase III cleavage site destabilizing the mRNA. RoxS uses a C-rich motif for base-pairing—a feature shared by many Gram-positive sRNAs to prevent ribosome recruitment (156). RoxS is cleaved by endoribonuclease RNase Y. Intriguingly, processed RoxS and full-length RoxS exhibit distinct regulatory potentials, albeit the physiological meaning of this functional specialization remains unclear (151).

Posttranscriptional Regulation of Anabolic Carbohydrate Pathways

A number of anabolic pathways using carbohydrates as substrates are regulated at the posttranscriptional level. One example is provided by CsrA, which regulates gluconeogenesis. Another important example is provided by the posttranscriptional control of biosynthesis of cell wall precursors, which must be safeguarded in growing cells, regardless of the nature of the carbon source and the catabolic pathway. In *Enterobacteriaceae*, this task is achieved by two hierarchically acting sRNAs, GlmY and GlmZ.

Regulation of the Hexosamine Pathway by sRNAs GlmY and GlmZ

Glucosamine-6-phosphate (GlcN6P) synthase (GlmS) catalyzes the first and rate-limiting step in the hexosamine biosynthesis pathway by converting fructose-6-phosphate to GlcN6P (Fig. 3), an essential precursor for cell wall and outer membrane biogenesis (157). Intracellular GlcN6P levels dictate the need for GlmS, whose amount is fine-tuned by posttranscriptional regulatory mechanisms. In Gram-positive bacteria, GlcN6P serves as cofactor for a ribozyme present in the 5′ UTR of the *glmS* mRNA (158, 159). Following self-cleavage, the *glmS* mRNA is rapidly degraded by RNase J1 (160). In *Enterobacteriaceae*, GlmS levels are feedback-regulated by two homologous sRNAs: GlmY and GlmZ (Fig. 5) (161, 162). Only GlmZ is a direct activator of *glmS* translation (163–165). When GlcN6P is plentiful, GlmZ is inactivated by RNase E cleavage, which requires the dedicated adaptor protein RapZ (166). How-

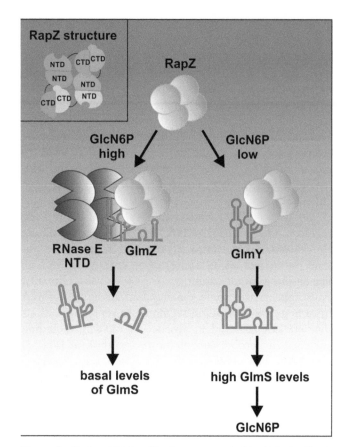

Figure 5 Role of RNase E adaptor protein RapZ in feedback regulation of GlmS synthesis in *E. coli*. When GlcN6P is plentiful in the cell, RapZ prevents *glmS* upregulation by targeting its activating sRNA GlmZ to cleavage by RNase E. Within the tripartite complex formed, the sRNA is envisioned to be sandwiched between the tetrameric RapZ protein and the N-terminal domain (NTD) of RNase E, which also forms a tetramer (168). Processing results in functional inactivation of GlmZ and subsequent decline in GlmS levels. Conversely, under GlcN6P depletion, RapZ is predominantly sequestered in complexes with the homologous sRNA GlmY, whose levels increase under this condition. Consequently, GlmZ remains in its active full-length form and stimulates *glmS* expression. Higher levels of GlmS replenish GlcN6P levels in the cell. Whether RapZ has an active role in sensing GlcN6P via direct binding of the metabolite within its C-terminal domain (CTD) is currently under investigation. The unusual tetrameric structure of RapZ is schematically depicted in the box in the upper left. Each monomer is represented by one color and consists of two globular domains, NTD and CTD, connected via flexible linkers. Three distinct surfaces involved in self-interaction can be discerned: CTD-CTD, NTD-NTD, as well as CTD-NTD (168).

ever, under GlcN6P depletion, cleavage of GlmZ is counteracted to elevate GlmS amounts and replenish the GlcN6P pool. This is achieved through sequestration of RapZ by the decoy sRNA GlmY, whose levels increase when amounts of the metabolite decline (Fig. 5) (164, 166). Consequently, full-length GlmZ base-pairs with the *glmS* leader and activates expression by disrupting an inhibitory stem-loop structure, thereby exposing the RBS (163, 165). GlmZ is an Hfq-dependent sRNA and a substrate of RNase E, whereas GlmY is not recognized by either of the two proteins (166, 167).

RapZ represents a highly specialized RNA-binding protein, as it exclusively binds GlmY and GlmZ (Fig. 5) (166). Upon binding, no major structural rearrangements are observable in the sRNAs, suggesting that RapZ stimulates cleavage of GlmZ by RNase E through protein-protein interaction (for a detailed discussion of RNase E, see chapter 1 [183]). Recently, the crystal structure of RapZ revealed an unusual quaternary structure comprising a domain-swapped dimer of dimers—an arrangement that is a prerequisite for RapZ activity *in vivo* (Fig. 5) (168). The RNA-binding function is located in the C terminus, which bears homology to a subdomain of 6-phosphofructokinase, implying that RapZ may have evolved through repurposing of enzyme components from central metabolism. Putative RNA-binding residues are surface exposed and form basic patches around an extended loop. Intriguingly, a binding pocket for a nonprotein ligand is observed in close vicinity to the presumptive RNA-binding domain (168). It remains to be seen whether this site binds GlcN6P, potentially interfering with sRNA binding. Identification of the GlcN6P binding site may foster the rational design of artificial ligands that can be used for antimicrobial chemotherapy (162).

CONCLUSION AND PERSPECTIVES

A decade ago, when a first review on the current topic was published, only a single target had been identified for SgrS and Spot 42, and together with GlmZ these were the only base-pairing sRNAs known to regulate carbohydrate metabolic genes (169). Meanwhile, such sRNAs have become common, and further examples are expected to follow. For instance, CCR in Gram-positive bacteria may also include posttranscriptional mechanisms, as two sRNAs are controlled by the CCR master regulator CcpA in *Streptococcus mutans* (170). Even though the contribution of posttranscriptional mechanisms to regulation of metabolism is evident, they are usually neglected in studies assessing metabolic flux control and CCR (171, 172). In addition, the regulatory mechanisms employed by sRNAs are much more diverse than previously envisioned. Novel principles include modulation of target accessibility to degrading RNases, regulation of Rho-dependent transcription termination, recruitment of Hfq as translational repressor, and employment of decoy RNAs sequestering sRNAs or their interacting proteins.

RNA-seq has pushed the development of sophisticated omics approaches facilitating assessment of posttranscriptional regulators on a global scale. RNA immunoprecipitation sequencing (RIP-seq) and UV cross-linking and immunoprecipitation sequencing (CLIP-seq) provide snapshots of substrates bound to RNA-binding proteins at a given time and also identify RNA-binding sites (see, e.g., reference 173). CLASH (cross-linking, ligation, and sequencing of hybrids) and RIL-seq enable the recovery of sRNA-mRNA duplexes, revealing whole RNA networks in a single experiment (77, 174). Most recently, CLIP-seq, ribosome profiling, transcriptomics, and proteomics are combined in "multi-omics" approaches exploring several layers of regulation in parallel and genome-wide. Application of multi-omics to *E. coli* CsrA revealed novel targets and physiological roles but also confirmed the global character of this posttranscriptional regulator in coordinating bacterial lifestyles with metabolic cues (106). The additional integration of metabolic flux analyses could reveal the specific contribution of sRNAs such as Spot 42 to reprogramming of metabolism.

There is an intimate connection between metabolism and virulence for which carbohydrate-related sRNAs and their protein interaction partners play an important role (175, 176). In fact, bacterial pathogenesis can be regarded as a developmental program granting access to nutrients in a hostile host environment. For instance, mutants lacking CsrA are severely compromised in establishing an infection, which not only is a consequence of dysregulated metabolism but may also result from discoordinated expression of virulence factors (177, 178). A recurrent theme observed in pathogenic bacteria is that sRNAs from the core genome are recruited to regulate horizontally acquired virulence functions, which also applies to SgrS, Spot 42, and GlmY/GlmZ (89, 90, 179, 180). In line with these observations, SgrS and GlmY are strongly upregulated in *Y. pseudotuberculosis* during infection (177). For more information on this interesting topic, see chapter 18 (184).

Acknowledgments. *Work in the laboratory of B.G. is supported by the Austrian Science Fund (FWF) (grant numbers P 26681-B22, F4317 to B.G.). The authors declare no conflicts.*

Citation. Durica-Mitic S, Göpel Y, Görke B. 2018. Carbohydrate utilization in bacteria: making the most out of sugars with the help of small regulatory RNAs. Microbiol Spectrum 6(2):RWR-0013-2017.

References

1. **Chubukov V, Gerosa L, Kochanowski K, Sauer U.** 2014. Coordination of microbial metabolism. *Nat Rev Microbiol* **12**:327–340.

2. **Georg J, Hess WR.** 2011. *cis*-Antisense RNA, another level of gene regulation in bacteria. *Microbiol Mol Biol Rev* **75**:286–300.

3. **Storz G, Vogel J, Wassarman KM.** 2011. Regulation by small RNAs in bacteria: expanding frontiers. *Mol Cell* **43**:880–891.

4. **Sedlyarova N, Shamovsky I, Bharati BK, Epshtein V, Chen J, Gottesman S, Schroeder R, Nudler E.** 2016. sRNA-mediated control of transcription termination in *E. coli. Cell* **167**:111–121.e113.

5. **Vogel J, Luisi BF.** 2011. Hfq and its constellation of RNA. *Nat Rev Microbiol* **9**:578–589.

6. **Kavita K, de Mets F, Gottesman S.** 2017. New aspects of RNA-based regulation by Hfq and its partner sRNAs. *Curr Opin Microbiol* **42**:53–61.

7. **Göpel Y, Görke B.** 2012. Rewiring two-component signal transduction with small RNAs. *Curr Opin Microbiol* **15**:132–139.

8. **Mandin P, Guillier M.** 2013. Expanding control in bacteria: interplay between small RNAs and transcriptional regulators to control gene expression. *Curr Opin Microbiol* **16**:125–132.

9. **Arraiano CM, Andrade JM, Domingues S, Guinote IB, Malecki M, Matos RG, Moreira RN, Pobre V, Reis FP, Saramago M, Silva IJ, Viegas SC.** 2010. The critical role of RNA processing and degradation in the control of gene expression. *FEMS Microbiol Rev* **34**:883–923.

10. **Ziebuhr W, Vogel J.** 2015. The end is not the end: remnants of tRNA precursors live on to sponge up small regulatory RNAs. *Mol Cell* **58**:389–390.

11. **Baumler DJ, Peplinski RG, Reed JL, Glasner JD, Perna NT.** 2011. The evolution of metabolic networks of *E. coli. BMC Syst Biol* **5**:182.

12. **Müller-Hill B.** 1996. *The* lac *Operon: a Short History of a Genetic Paradigm.* de Gruyter, Berlin, Germany.

13. **Jacob F, Monod J.** 1961. Genetic regulatory mechanisms in the synthesis of proteins. *J Mol Biol* **3**:318–356.

14. **Mizuno T, Chou MY, Inouye M.** 1984. A unique mechanism regulating gene expression: translational inhibition by a complementary RNA transcript (micRNA). *Proc Natl Acad Sci U S A* **81**:1966–1970.

15. **Monod J.** 1942. *Recherches sur la Croissance des Cultures Bacteriennes.* Hermann et Cie, Paris, France.

16. **Deutscher J, Francke C, Postma PW.** 2006. How phosphotransferase system-related protein phosphorylation regulates carbohydrate metabolism in bacteria. *Microbiol Mol Biol Rev* **70**:939–1031.

17. **Eisenreich W, Rudel T, Heesemann J, Goebel W.** 2017. To eat and to be eaten: mutual metabolic adaptations of immune cells and intracellular bacterial pathogens upon infection. *Front Cell Infect Microbiol* **7**:316.

18. **Jahreis K, Pimentel-Schmitt EF, Brückner R, Titgemeyer F.** 2008. Ins and outs of glucose transport systems in eubacteria. *FEMS Microbiol Rev* **32**:891–907.

19. **Irani MH, Maitra PK.** 1977. Properties of *Escherichia coli* mutants deficient in enzymes of glycolysis. *J Bacteriol* **132**:398–410.

20. **Kadner RJ, Murphy GP, Stephens CM.** 1992. Two mechanisms for growth inhibition by elevated transport of sugar phosphates in *Escherichia coli. J Gen Microbiol* **138**:2007–2014.

21. **Kimata K, Tanaka Y, Inada T, Aiba H.** 2001. Expression of the glucose transporter gene, *ptsG*, is regulated at the mRNA degradation step in response to glycolytic flux in *Escherichia coli. EMBO J* **20**:3587–3595.

22. **Morita T, El-Kazzaz W, Tanaka Y, Inada T, Aiba H.** 2003. Accumulation of glucose 6-phosphate or fructose 6-phosphate is responsible for destabilization of glucose transporter mRNA in *Escherichia coli. J Biol Chem* **278**:15608–15614.

23. **Vanderpool CK, Gottesman S.** 2004. Involvement of a novel transcriptional activator and small RNA in post-transcriptional regulation of the glucose phosphoenolpyruvate phosphotransferase system. *Mol Microbiol* **54**:1076–1089.

24. **Vanderpool CK, Gottesman S.** 2007. The novel transcription factor SgrR coordinates the response to glucose-phosphate stress. *J Bacteriol* **189**:2238–2248.

25. **Kawamoto H, Morita T, Shimizu A, Inada T, Aiba H.** 2005. Implication of membrane localization of target mRNA in the action of a small RNA: mechanism of post-transcriptional regulation of glucose transporter in *Escherichia coli. Genes Dev* **19**:328–338.

26. **Morita T, Maki K, Aiba H.** 2005. RNase E-based ribonucleoprotein complexes: mechanical basis of mRNA destabilization mediated by bacterial noncoding RNAs. *Genes Dev* **19**:2176–2186.

27. **Morita T, Mochizuki Y, Aiba H.** 2006. Translational repression is sufficient for gene silencing by bacterial small noncoding RNAs in the absence of mRNA destruction. *Proc Natl Acad Sci U S A* **103**:4858–4863.

28. **Wadler CS, Vanderpool CK.** 2007. A dual function for a bacterial small RNA: SgrS performs base pairing-dependent regulation and encodes a functional polypeptide. *Proc Natl Acad Sci U S A* **104**:20454–20459.

29. **Lloyd CR, Park S, Fei J, Vanderpool CK.** 2017. The small protein SgrT controls transport activity of the glucose-specific phosphotransferase system. *J Bacteriol* **199**:e00869-16.

30. **Kosfeld A, Jahreis K.** 2012. Characterization of the interaction between the small regulatory peptide SgrT and the EIICB^{Glc} of the glucose-phosphotransferase system of *E. coli* K-12. *Metabolites* **2**:756–774.

31. **Bobrovskyy M, Vanderpool CK.** 2016. Diverse mechanisms of post-transcriptional repression by the small RNA regulator of glucose-phosphate stress. *Mol Microbiol* **99**:254–273.

32. **Wright PR, Richter AS, Papenfort K, Mann M, Vogel J, Hess WR, Backofen R, Georg J.** 2013. Comparative genomics boosts target prediction for bacterial small RNAs. *Proc Natl Acad Sci U S A* **110**:E3487–E3496.

33. **Rice JB, Vanderpool CK.** 2011. The small RNA SgrS controls sugar-phosphate accumulation by regulat-

ing multiple PTS genes. *Nucleic Acids Res* 39:3806–3819.

34. Rice JB, Balasubramanian D, Vanderpool CK. 2012. Small RNA binding-site multiplicity involved in translational regulation of a polycistronic mRNA. *Proc Natl Acad Sci U S A* 109:E2691–E2698.

35. Papenfort K, Sun Y, Miyakoshi M, Vanderpool CK, Vogel J. 2013. Small RNA-mediated activation of sugar phosphatase mRNA regulates glucose homeostasis. *Cell* 153:426–437.

36. Sun Y, Vanderpool CK. 2013. Physiological consequences of multiple-target regulation by the small RNA SgrS in *Escherichia coli*. *J Bacteriol* 195:4804–4815.

37. Maleki S, Hrudikova R, Zotchev SB, Ertesvåg H. 2016. Identification of a new phosphatase enzyme potentially involved in the sugar phosphate stress response in *Pseudomonas fluorescens*. *Appl Environ Microbiol* 83: e02361-16.

38. Tchieu JH, Norris V, Edwards JS, Saier MH Jr. 2001. The complete phosphotransferase system in *Escherichia coli*. *J Mol Microbiol Biotechnol* 3:329–346.

39. Reichenbach B, Breustedt DA, Stülke J, Rak B, Görke B. 2007. Genetic dissection of specificity determinants in the interaction of HPr with enzymes II of the bacterial phosphoenolpyruvate:sugar phosphotransferase system in *Escherichia coli*. *J Bacteriol* 189:4603–4613.

40. Richards GR, Patel MV, Lloyd CR, Vanderpool CK. 2013. Depletion of glycolytic intermediates plays a key role in glucose-phosphate stress in *Escherichia coli*. *J Bacteriol* 195:4816–4825.

41. Wadler CS, Vanderpool CK. 2009. Characterization of homologs of the small RNA SgrS reveals diversity in function. *Nucleic Acids Res* 37:5477–5485.

42. Horler RS, Vanderpool CK. 2009. Homologs of the small RNA SgrS are broadly distributed in enteric bacteria but have diverged in size and sequence. *Nucleic Acids Res* 37:5465–5476.

43. Görke B, Stülke J. 2008. Carbon catabolite repression in bacteria: many ways to make the most out of nutrients. *Nat Rev Microbiol* 6:613–624.

44. Lindner SN, Petrov DP, Hagmann CT, Henrich A, Krämer R, Eikmanns BJ, Wendisch VF, Seibold GM. 2013. Phosphotransferase system-mediated glucose uptake is repressed in phosphoglucoisomerase-deficient *Corynebacterium glutamicum* strains. *Appl Environ Microbiol* 79:2588–2595.

45. Landmann JJ, Busse RA, Latz JH, Singh KD, Stülke J, Görke B. 2011. Crh, the paralogue of the phosphocarrier protein HPr, controls the methylglyoxal bypass of glycolysis in *Bacillus subtilis*. *Mol Microbiol* 82:770–787.

46. Klancher CA, Hayes CA, Dalia AB. 2017. The nucleoid occlusion protein SlmA is a direct transcriptional activator of chitobiose utilization in *Vibrio cholerae*. *PLoS Genet* 13:e1006877.

47. Meibom KL, Li XB, Nielsen AT, Wu CY, Roseman S, Schoolnik GK. 2004. The *Vibrio cholerae* chitin utilization program. *Proc Natl Acad Sci U S A* 101:2524–2529.

48. Dalia AB, Lazinski DW, Camilli A. 2014. Identification of a membrane-bound transcriptional regulator that links

chitin and natural competence in *Vibrio cholerae*. *mBio* 5:e01028-e13.

49. Yamamoto S, Izumiya H, Mitobe J, Morita M, Arakawa E, Ohnishi M, Watanabe H. 2011. Identification of a chitin-induced small RNA that regulates translation of the *tfoX* gene, encoding a positive regulator of natural competence in *Vibrio cholerae*. *J Bacteriol* 193:1953–1965.

50. Borgeaud S, Metzger LC, Scrignari T, Blokesch M. 2015. The type VI secretion system of *Vibrio cholerae* fosters horizontal gene transfer. *Science* 347:63–67.

51. Göpel Y, Görke B. 2014. Lies and deception in bacterial gene regulation: the roles of nucleic acid decoys. *Mol Microbiol* 92:641–647.

52. Bossi L, Figueroa-Bossi N. 2016. Competing endogenous RNAs: a target-centric view of small RNA regulation in bacteria. *Nat Rev Microbiol* 14:775–784.

53. Plumbridge J, Bossi L, Oberto J, Wade JT, Figueroa-Bossi N. 2014. Interplay of transcriptional and small RNA-dependent control mechanisms regulates chitosugar uptake in *Escherichia coli* and *Salmonella*. *Mol Microbiol* 92:648–658.

54. Figueroa-Bossi N, Valentini M, Malleret L, Fiorini F, Bossi L. 2009. Caught at its own game: regulatory small RNA inactivated by an inducible transcript mimicking its target. *Genes Dev* 23:2004–2015.

55. Rasmussen AA, Johansen J, Nielsen JS, Overgaard M, Kallipolitis B, Valentin-Hansen P. 2009. A conserved small RNA promotes silencing of the outer membrane protein YbfM. *Mol Microbiol* 72:566–577.

56. Bossi L, Schwartz A, Guillemardet B, Boudvillain M, Figueroa-Bossi N. 2012. A role for Rho-dependent polarity in gene regulation by a noncoding small RNA. *Genes Dev* 26:1864–1873.

57. Overgaard M, Johansen J, Møller-Jensen J, Valentin-Hansen P. 2009. Switching off small RNA regulation with trap-mRNA. *Mol Microbiol* 73:790–800.

58. Suzuki K, Shimizu M, Sasaki N, Ogawa C, Minami H, Sugimoto H, Watanabe T. 2016. Regulation of the chitin degradation and utilization system by the ChiX small RNA in *Serratia marcescens* 2170. *Biosci Biotechnol Biochem* 80:376–385.

59. Mustachio LM, Aksit S, Mistry RH, Scheffler R, Yamada A, Liu JM. 2012. The *Vibrio cholerae* mannitol transporter is regulated posttranscriptionally by the MtlS small regulatory RNA. *J Bacteriol* 194:598–606.

60. Chang H, Replogle JM, Vather N, Tsao-Wu M, Mistry R, Liu JM. 2015. A *cis*-regulatory antisense RNA represses translation in *Vibrio cholerae* through extensive complementarity and proximity to the target locus. *RNA Biol* 12:136–148.

61. Byer T, Wang J, Zhang MG, Vather N, Blachman A, Visser B, Liu JM. 2017. MtlR negatively regulates mannitol utilization by *Vibrio cholerae*. *Microbiology*.

62. Ymele-Leki P, Houot L, Watnick PI. 2013. Mannitol and the mannitol-specific enzyme IIB subunit activate *Vibrio cholerae* biofilm formation. *Appl Environ Microbiol* 79: 4675–4683.

63. White BA, Lamed R, Bayer EA, Flint HJ. 2014. Biomass utilization by gut microbiomes. *Annu Rev Microbiol* **68**: 279–296.

64. Grondin JM, Tamura K, Déjean G, Abbott DW, Brumer H. 2017. Polysaccharide utilization loci: fueling microbial communities. *J Bacteriol* **199**:e00860-16.

65. Cao Y, Förstner KU, Vogel J, Smith CJ. 2016. *cis*-Encoded small RNAs, a conserved mechanism for repression of polysaccharide utilization in *Bacteroides*. *J Bacteriol* **198**:2410–2418.

66. Pudlo NA, Urs K, Kumar SS, German JB, Mills DA, Martens EC. 2015. Symbiotic human gut bacteria with variable metabolic priorities for host mucosal glycans. *mBio* **6**:e01282-e15.

67. Aidelberg G, Towbin BD, Rothschild D, Dekel E, Bren A, Alon U. 2014. Hierarchy of non-glucose sugars in *Escherichia coli*. *BMC Syst Biol* **8**:133.

68. Rothe FM, Bahr T, Stülke J, Rak B, Görke B. 2012. Activation of *Escherichia coli* antiterminator BglG requires its phosphorylation. *Proc Natl Acad Sci U S A* **109**:15906–15911.

69. Borirak O, Rolfe MD, de Koning LJ, Hoefsloot HC, Bekker M, Dekker HL, Roseboom W, Green J, de Koster CG, Hellingwerf KJ. 2015. Time-series analysis of the transcriptome and proteome of *Escherichia coli* upon glucose repression. *Biochim Biophys Acta* **1854**(10 Pt A): 1269–1279.

70. Sonnleitner E, Bläsi U. 2014. Regulation of Hfq by the RNA CrcZ in *Pseudomonas aeruginosa* carbon catabolite repression. *PLoS Genet* **10**:e1004440.

71. Beisel CL, Storz G. 2011. Discriminating tastes: physiological contributions of the Hfq-binding small RNA Spot 42 to catabolite repression. *RNA Biol* **8**:766–770.

72. Rice PW, Dahlberg JE. 1982. A gene between *polA* and *glnA* retards growth of *Escherichia coli* when present in multiple copies: physiological effects of the gene for spot 42 RNA. *J Bacteriol* **152**:1196–1210.

73. Polayes DA, Rice PW, Garner MM, Dahlberg JE. 1988. Cyclic AMP-cyclic AMP receptor protein as a repressor of transcription of the *spf* gene of *Escherichia coli*. *J Bacteriol* **170**:3110–3114.

74. Møller T, Franch T, Udesen C, Gerdes K, Valentin-Hansen P. 2002. Spot 42 RNA mediates discoordinate expression of the *E. coli* galactose operon. *Genes Dev* **16**: 1696–1706.

75. Beisel CL, Storz G. 2011. The base-pairing RNA spot 42 participates in a multioutput feedforward loop to help enact catabolite repression in *Escherichia coli*. *Mol Cell* **41**:286–297.

76. Beisel CL, Updegrove TB, Janson BJ, Storz G. 2012. Multiple factors dictate target selection by Hfq-binding small RNAs. *EMBO J* **31**:1961–1974.

77. Melamed S, Peer A, Faigenbaum-Romm R, Gatt YE, Reiss N, Bar A, Altuvia Y, Argaman L, Margalit H. 2016. Global mapping of small RNA-target interactions in bacteria. *Mol Cell* **63**:884–897.

78. Conway T, Cohen PS. 2015. Commensal and pathogenic *Escherichia coli* metabolism in the gut. *Microbiol Spectr* **3**:MBP-0006-2014.

79. Wu J, Li Y, Cai Z, Jin Y. 2014. Pyruvate-associated acid resistance in bacteria. *Appl Environ Microbiol* **80**: 4108–4113.

80. Møller T, Franch T, Højrup P, Keene DR, Bächinger HP, Brennan RG, Valentin-Hansen P. 2002. Hfq: a bacterial Sm-like protein that mediates RNA-RNA interaction. *Mol Cell* **9**:23–30.

81. Wang X, Ji SC, Jeon HJ, Lee Y, Lim HM. 2015. Two-level inhibition of *galK* expression by Spot 42: degradation of mRNA mK2 and enhanced transcription termination before the *galK* gene. *Proc Natl Acad Sci U S A* **112**:7581–7586.

82. Desnoyers G, Massé E. 2012. Noncanonical repression of translation initiation through small RNA recruitment of the RNA chaperone Hfq. *Genes Dev* **26**:726–739.

83. Ellis MJ, Trussler RS, Haniford DB. 2015. Hfq binds directly to the ribosome-binding site of IS*10* transposase mRNA to inhibit translation. *Mol Microbiol* **96**:633–650.

84. Salvail H, Caron MP, Bélanger J, Massé E. 2013. Antagonistic functions between the RNA chaperone Hfq and an sRNA regulate sensitivity to the antibiotic colicin. *EMBO J* **32**:2764–2778.

85. Sharma CM, Papenfort K, Pernitzsch SR, Mollenkopf HJ, Hinton JC, Vogel J. 2011. Pervasive post-transcriptional control of genes involved in amino acid metabolism by the Hfq-dependent GcvB small RNA. *Mol Microbiol* **81**:1144–1165.

86. Durand S, Storz G. 2010. Reprogramming of anaerobic metabolism by the FnrS small RNA. *Mol Microbiol* **75**: 1215–1231.

87. Bækkedal C, Haugen P. 2015. The Spot 42 RNA: a regulatory small RNA with roles in the central metabolism. *RNA Biol* **12**:1071–1077.

88. Hansen GA, Ahmad R, Hjerde E, Fenton CG, Willassen NP, Haugen P. 2012. Expression profiling reveals Spot 42 small RNA as a key regulator in the central metabolism of *Aliivibrio salmonicida*. *BMC Genomics* **13**:37.

89. Livny J, Zhou X, Mandlik A, Hubbard T, Davis BM, Waldor MK. 2014. Comparative RNA-Seq based dissection of the regulatory networks and environmental stimuli underlying *Vibrio parahaemolyticus* gene expression during infection. *Nucleic Acids Res* **42**:12212–12223.

90. Tanabe T, Miyamoto K, Tsujibo H, Yamamoto S, Funahashi T. 2015. The small RNA Spot 42 regulates the expression of the type III secretion system 1 (T3SS1) chaperone protein VP1682 in *Vibrio parahaemolyticus*. *FEMS Microbiol Lett* **362**:fnv173.

91. Rojo F. 2010. Carbon catabolite repression in *Pseudomonas*: optimizing metabolic versatility and interactions with the environment. *FEMS Microbiol Rev* **34**:658–684.

92. Moreno R, Hernández-Arranz S, La Rosa R, Yuste L, Madhushani A, Shingler V, Rojo F. 2015. The Crc and Hfq proteins of *Pseudomonas putida* cooperate in catabolite repression and formation of ribonucleic acid complexes with specific target motifs. *Environ Microbiol* **17**: 105–118.

93. Sonnleitner E, Wulf A, Campagne S, Pei XY, Wolfinger MT, Forlani G, Prindl K, Abdou L, Resch A, Allain FH, Luisi BF, Urlaub H, Bläsi U. 2017. Interplay between the catabolite repression control protein Crc, Hfq and RNA in Hfq-dependent translational regulation in *Pseudomonas aeruginosa*. *Nucleic Acids Res*.

94. Sonnleitner E, Abdou L, Haas D. 2009. Small RNA as global regulator of carbon catabolite repression in *Pseudomonas aeruginosa*. *Proc Natl Acad Sci U S A* 106: 21866–21871.

95. Liu Y, Gokhale CS, Rainey PB, Zhang XX. 2017. Unravelling the complexity and redundancy of carbon catabolic repression in *Pseudomonas fluorescens* SBW25. *Mol Microbiol* 105:589–605.

96. Filiatrault MJ, Stodghill PV, Wilson J, Butcher BG, Chen H, Myers CR, Cartinhour SW. 2013. CrcZ and CrcX regulate carbon source utilization in *Pseudomonas syringae* pathovar *tomato* strain DC3000. *RNA Biol* 10:245–255.

97. Valentini M, García-Mauriño SM, Pérez-Martínez I, Santero E, Canosa I, Lapouge K. 2014. Hierarchical management of carbon sources is regulated similarly by the CbrA/B systems in *Pseudomonas aeruginosa* and *Pseudomonas putida*. *Microbiology* 160:2243–2252.

98. Sonnleitner E, Prindl K, Bläsi U. 2017. The *Pseudomonas aeruginosa* CrcZ RNA interferes with Hfq-mediated riboregulation. *PLoS One* 12:e0180887.

99. Kochanowski K, Gerosa L, Brunner SF, Christodoulou D, Nikolaev YV, Sauer U. 2017. Few regulatory metabolites coordinate expression of central metabolic genes in *Escherichia coli*. *Mol Syst Biol* 13:903.

100. Shimada T, Yamamoto K, Ishihama A. 2011. Novel members of the Cra regulon involved in carbon metabolism in *Escherichia coli*. *J Bacteriol* 193:649–659.

101. Saier MH Jr, Ramseier TM. 1996. The catabolite repressor/activator (Cra) protein of enteric bacteria. *J Bacteriol* 178:3411–3417.

102. Hogema BM, Arents JC, Bader R, Eijkemans K, Yoshida H, Takahashi H, Aiba H, Postma PW. 1998. Inducer exclusion in *Escherichia coli* by non-PTS substrates: the role of the PEP to pyruvate ratio in determining the phosphorylation state of enzyme IIA^{Glc}. *Mol Microbiol* 30:487–498.

103. Bettenbrock K, Sauter T, Jahreis K, Kremling A, Lengeler JW, Gilles ED. 2007. Correlation between growth rates, EIIA^{Crr} phosphorylation, and intracellular cyclic AMP levels in *Escherichia coli* K-12. *J Bacteriol* 189:6891–6900.

104. Matsuoka Y, Shimizu K. 2011. Metabolic regulation in *Escherichia coli* in response to culture environments via global regulators. *Biotechnol J* 6:1330–1341.

105. Vakulskas CA, Potts AH, Babitzke P, Ahmer BM, Romeo T. 2015. Regulation of bacterial virulence by Csr (Rsm) systems. *Microbiol Mol Biol Rev* 79:193–224.

106. Potts AH, Vakulskas CA, Pannuri A, Yakhnin H, Babitzke P, Romeo T. 2017. Global role of the bacterial post-transcriptional regulator CsrA revealed by integrated transcriptomics. *Nat Commun* 8:1596.

107. Van Assche E, Van Puyvelde S, Vanderleyden J, Steenackers HP. 2015. RNA-binding proteins involved in post-transcriptional regulation in bacteria. *Front Microbiol* 6:141.

108. Sabnis NA, Yang H, Romeo T. 1995. Pleiotropic regulation of central carbohydrate metabolism in *Escherichia coli* via the gene *csrA*. *J Biol Chem* 270:29096–29104.

109. Morin M, Ropers D, Letisse F, Laguerre S, Portais JC, Cocaign-Bousquet M, Enjalbert B. 2016. The post-transcriptional regulatory system CSR controls the balance of metabolic pools in upper glycolysis of *Escherichia coli*. *Mol Microbiol* 100:686–700.

110. Timmermans J, Van Melderen L. 2009. Conditional essentiality of the *csrA* gene in *Escherichia coli*. *J Bacteriol* 191:1722–1724.

111. Weilbacher T, Suzuki K, Dubey AK, Wang X, Gudapaty S, Morozov I, Baker CS, Georgellis D, Babitzke P, Romeo T. 2003. A novel sRNA component of the carbon storage regulatory system of *Escherichia coli*. *Mol Microbiol* 48:657–670.

112. Liu MY, Gui G, Wei B, Preston JF III, Oakford L, Yüksel U, Giedroc DP, Romeo T. 1997. The RNA molecule CsrB binds to the global regulatory protein CsrA and antagonizes its activity in *Escherichia coli*. *J Biol Chem* 272:17502–17510.

113. Chavez RG, Alvarez AF, Romeo T, Georgellis D. 2010. The physiological stimulus for the BarA sensor kinase. *J Bacteriol* 192:2009–2012.

114. Lawhon SD, Maurer R, Suyemoto M, Altier C. 2002. Intestinal short-chain fatty acids alter *Salmonella typhimurium* invasion gene expression and virulence through BarA/SirA. *Mol Microbiol* 46:1451–1464.

115. Vakulskas CA, Leng Y, Abe H, Amaki T, Okayama A, Babitzke P, Suzuki K, Romeo T. 2016. Antagonistic control of the turnover pathway for the global regulatory sRNA CsrB by the CsrA and CsrD proteins. *Nucleic Acids Res* 44:7896–7910.

116. Suzuki K, Babitzke P, Kushner SR, Romeo T. 2006. Identification of a novel regulatory protein (CsrD) that targets the global regulatory RNAs CsrB and CsrC for degradation by RNase E. *Genes Dev* 20:2605–2617.

117. Leng Y, Vakulskas CA, Zere TR, Pickering BS, Watnick PI, Babitzke P, Romeo T. 2016. Regulation of CsrB/C sRNA decay by EIIA^{Glc} of the phosphoenolpyruvate: carbohydrate phosphotransferase system. *Mol Microbiol* 99:627–639.

118. Morin M, Ropers D, Cinquemani E, Portais JC, Enjalbert B, Cocaign-Bousquet M. 2017. The Csr system regulates *Escherichia coli* fitness by controlling glycogen accumulation and energy levels. *mBio* 8:e01628-17.

119. Pannuri A, Vakulskas CA, Zere T, McGibbon LC, Edwards AN, Georgellis D, Babitzke P, Romeo T. 2016. Circuitry linking the catabolite repression and Csr global regulatory systems of *Escherichia coli*. *J Bacteriol* 198: 3000–3015.

120. You C, Okano H, Hui S, Zhang Z, Kim M, Gunderson CW, Wang YP, Lenz P, Yan D, Hwa T. 2013. Coordination of bacterial proteome with metabolism by cyclic AMP signalling. *Nature* 500:301–306.

121. Nuss AM, Schuster F, Kathrin Heroven A, Heine W, Pisano F, Dersch P. 2014. A direct link between the

global regulator PhoP and the Csr regulon in *Y. pseudo-tuberculosis* through the small regulatory RNA CsrC. *RNA Biol* **11**:580–593.

122. Pannekoek Y, Huis In 't Veld RA, Schipper K, Bovenkerk S, Kramer G, Brouwer MC, van de Beek D, Speijer D, van der Ende A. 2017. *Neisseria meningitidis* uses sibling small regulatory RNAs to switch from cataplerotic to anaplerotic metabolism. *mBio* **8**:e02293-16.

123. Bauer S, Helmreich J, Zachary M, Kaethner M, Heinrichs E, Rudel T, Beier D. 2017. The sibling sRNAs NgncR_162 and NgncR_163 of *Neisseria gonorrhoeae* participate in the expression control of metabolic, transport and regulatory proteins. *Microbiology* **163**:1720–1734.

124. Oglesby-Sherrouse AG, Murphy ER. 2013. Iron-responsive bacterial small RNAs: variations on a theme. *Metallomics* **5**:276–286.

125. Salvail H, Massé E. 2012. Regulating iron storage and metabolism with RNA: an overview of posttranscriptional controls of intracellular iron homeostasis. *Wiley Interdiscip Rev RNA* **3**:26–36.

126. Massé E, Gottesman S. 2002. A small RNA regulates the expression of genes involved in iron metabolism in *Escherichia coli*. *Proc Natl Acad Sci U S A* **99**:4620–4625.

127. Seo SW, Kim D, Latif H, O'Brien EJ, Szubin R, Palsson BO. 2014. Deciphering Fur transcriptional regulatory network highlights its complex role beyond iron metabolism in *Escherichia coli*. *Nat Commun* **5**:4910.

128. Porcheron G, Dozois CM. 2015. Interplay between iron homeostasis and virulence: Fur and RyhB as major regulators of bacterial pathogenicity. *Vet Microbiol* **179**:2–14.

129. Friedman DB, Stauff DL, Pishchany G, Whitwell CW, Torres VJ, Skaar EP. 2006. *Staphylococcus aureus* redirects central metabolism to increase iron availability. *PLoS Pathog* **2**:e87.

130. Jung YS, Kwon YM. 2008. Small RNA ArrF regulates the expression of *sodB* and *feSII* genes in *Azotobacter vinelandii*. *Curr Microbiol* **57**:593–597.

131. de Graef MR, Alexeeva S, Snoep JL, Teixeira de Mattos MJ. 1999. The steady-state internal redox state (NADH/NAD) reflects the external redox state and is correlated with catabolic adaptation in *Escherichia coli*. *J Bacteriol* **181**:2351–2357.

132. Green J, Paget MS. 2004. Bacterial redox sensors. *Nat Rev Microbiol* **2**:954–966.

133. Sawers G. 1999. The aerobic/anaerobic interface. *Curr Opin Microbiol* **2**:181–187.

134. Constantinidou C, Hobman JL, Griffiths L, Patel MD, Penn CW, Cole JA, Overton TW. 2006. A reassessment of the FNR regulon and transcriptomic analysis of the effects of nitrate, nitrite, NarXL, and NarQP as *Escherichia coli* K12 adapts from aerobic to anaerobic growth. *J Biol Chem* **281**:4802–4815.

135. Myers KS, Yan H, Ong IM, Chung D, Liang K, Tran F, Keleş S, Landick R, Kiley PJ. 2013. Genome-scale analysis of *Escherichia coli* FNR reveals complex features of transcription factor binding. *PLoS Genet* **9**:e1003565.

136. Liu X, De Wulf P. 2004. Probing the ArcA-P modulon of *Escherichia coli* by whole genome transcriptional analysis and sequence recognition profiling. *J Biol Chem* **279**:12588–12597.

137. Park DM, Akhtar MS, Ansari AZ, Landick R, Kiley PJ. 2013. The bacterial response regulator ArcA uses a diverse binding site architecture to regulate carbon oxidation globally. *PLoS Genet* **9**:e1003839.

138. Boysen A, Møller-Jensen J, Kallipolitis B, Valentin-Hansen P, Overgaard M. 2010. Translational regulation of gene expression by an anaerobically induced small non-coding RNA in *Escherichia coli*. *J Biol Chem* **285**:10690–10702.

139. van der Rest ME, Frank C, Molenaar D. 2000. Functions of the membrane-associated and cytoplasmic malate dehydrogenases in the citric acid cycle of *Escherichia coli*. *J Bacteriol* **182**:6892–6899.

140. Mandin P, Gottesman S. 2010. Integrating anaerobic/aerobic sensing and the general stress response through the ArcZ small RNA. *EMBO J* **29**:3094–3107.

141. Papenfort K, Said N, Welsink T, Lucchini S, Hinton JC, Vogel J. 2009. Specific and pleiotropic patterns of mRNA regulation by ArcZ, a conserved, Hfq-dependent small RNA. *Mol Microbiol* **74**:139–158.

142. Parker A, Cureoglu S, De Lay N, Majdalani N, Gottesman S. 2017. Alternative pathways for *Escherichia coli* biofilm formation revealed by sRNA overproduction. *Mol Microbiol* **105**:309–325.

143. Chen J, Gottesman S. 2017. Hfq links translation repression to stress-induced mutagenesis in *E. coli*. *Genes Dev* **31**:1382–1395.

144. Xia H, Yang X, Tang Q, Ye J, Wu H, Zhang H. 2017. EsrE—a *yigP* locus-encoded transcript—is a 3′ UTR sRNA involved in the respiratory chain of *E. coli*. *Front Microbiol* **8**:1658.

145. Isabella VM, Clark VL. 2011. Deep sequencing-based analysis of the anaerobic stimulon in *Neisseria gonorrhoeae*. *BMC Genomics* **12**:51.

146. Fantappiè L, Oriente F, Muzzi A, Serruto D, Scarlato V, Delany I. 2011. A novel Hfq-dependent sRNA that is under FNR control and is synthesized in oxygen limitation in *Neisseria meningitidis*. *Mol Microbiol* **80**:507–523.

147. Knapp JS, Clark VL. 1984. Anaerobic growth of *Neisseria gonorrhoeae* coupled to nitrite reduction. *Infect Immun* **46**:176–181.

148. Barth KR, Isabella VM, Clark VL. 2009. Biochemical and genomic analysis of the denitrification pathway within the genus *Neisseria*. *Microbiology* **155**:4093–4103.

149. Phillips NJ, Steichen CT, Schilling B, Post DM, Niles RK, Bair TB, Falsetta ML, Apicella MA, Gibson BW. 2012. Proteomic analysis of *Neisseria gonorrhoeae* biofilms shows shift to anaerobic respiration and changes in nutrient transport and outermembrane proteins. *PLoS One* **7**:e38303.

150. Tanwer P, Bauer S, Heinrichs E, Panda G, Saluja D, Rudel T, Beier D. 2017. Post-transcriptional regulation of target genes by the sRNA FnrS in *Neisseria gonorrhoeae*. *Microbiology* **163**:1081–1092.

151. Durand S, Braun F, Lioliou E, Romilly C, Helfer AC, Kuhn L, Quittot N, Nicolas P, Romby P, Condon C. 2015. A nitric oxide regulated small RNA controls expression of genes involved in redox homeostasis in *Bacillus subtilis*. *PLoS Genet* 11:e1004957.

152. Geissmann T, Chevalier C, Cros MJ, Boisset S, Fechter P, Noirot C, Schrenzel J, François P, Vandenesch F, Gaspin C, Romby P. 2009. A search for small noncoding RNAs in *Staphylococcus aureus* reveals a conserved sequence motif for regulation. *Nucleic Acids Res* 37:7239–7257.

153. Bohn C, Rigoulay C, Chabelskaya S, Sharma CM, Marchais A, Skorski P, Borezée-Durant E, Barbet R, Jacquet E, Jacq A, Gautheret D, Felden B, Vogel J, Bouloc P. 2010. Experimental discovery of small RNAs in *Staphylococcus aureus* reveals a riboregulator of central metabolism. *Nucleic Acids Res* 38:6620–6636.

154. Durand S, Braun F, Helfer AC, Romby P, Condon C. 2017. sRNA-mediated activation of gene expression by inhibition of 5′-3′ exonucleolytic mRNA degradation. *eLife* 6:e23602.

155. Kinkel TL, Roux CM, Dunman PM, Fang FC. 2013. The *Staphylococcus aureus* SrrAB two-component system promotes resistance to nitrosative stress and hypoxia. *mBio* 4:e00696-e13.

156. Fechter P, Caldelari I, Lioliou E, Romby P. 2014. Novel aspects of RNA regulation in *Staphylococcus aureus*. *FEBS Lett* 588:2523–2529.

157. Durand P, Golinelli-Pimpaneau B, Mouilleron S, Badet B, Badet-Denisot MA. 2008. Highlights of glucosamine-6P synthase catalysis. *Arch Biochem Biophys* 474:302–317.

158. Winkler WC, Nahvi A, Roth A, Collins JA, Breaker RR. 2004. Control of gene expression by a natural metabolite-responsive ribozyme. *Nature* 428:281–286.

159. Bingaman JL, Zhang S, Stevens DR, Yennawar NH, Hammes-Schiffer S, Bevilacqua PC. 2017. The GlcN6P cofactor plays multiple catalytic roles in the *glmS* ribozyme. *Nat Chem Biol* 13:439–445.

160. Collins JA, Irnov I, Baker S, Winkler WC. 2007. Mechanism of mRNA destabilization by the *glmS* ribozyme. *Genes Dev* 21:3356–3368.

161. Göpel Y, Lüttmann D, Heroven AK, Reichenbach B, Dersch P, Görke B. 2011. Common and divergent features in transcriptional control of the homologous small RNAs GlmY and GlmZ in *Enterobacteriaceae*. *Nucleic Acids Res* 39:1294–1309.

162. Khan MA, Göpel Y, Milewski S, Görke B. 2016. Two small RNAs conserved in *Enterobacteriaceae* provide intrinsic resistance to antibiotics targeting the cell wall biosynthesis enzyme glucosamine-6-phosphate synthase. *Front Microbiol* 7:908.

163. Kalamorz F, Reichenbach B, März W, Rak B, Görke B. 2007. Feedback control of glucosamine-6-phosphate synthase GlmS expression depends on the small RNA GlmZ and involves the novel protein YhbJ in *Escherichia coli*. *Mol Microbiol* 65:1518–1533.

164. Reichenbach B, Maes A, Kalamorz F, Hajnsdorf E, Görke B. 2008. The small RNA GlmY acts upstream of the sRNA GlmZ in the activation of *glmS* expression and is subject to regulation by polyadenylation in *Escherichia coli*. *Nucleic Acids Res* 36:2570–2580.

165. Urban JH, Vogel J. 2008. Two seemingly homologous noncoding RNAs act hierarchically to activate *glmS* mRNA translation. *PLoS Biol* 6:e64.

166. Göpel Y, Papenfort K, Reichenbach B, Vogel J, Görke B. 2013. Targeted decay of a regulatory small RNA by an adaptor protein for RNase E and counteraction by an anti-adaptor RNA. *Genes Dev* 27:552–564.

167. Göpel Y, Khan MA, Görke B. 2016. Domain swapping between homologous bacterial small RNAs dissects processing and Hfq binding determinants and uncovers an aptamer for conditional RNase E cleavage. *Nucleic Acids Res* 44:824–837.

168. Gonzalez GM, Durica-Mitic S, Hardwick SW, Moncrieffe MC, Resch M, Neumann P, Ficner R, Görke B, Luisi BF. 2017. Structural insights into RapZ-mediated regulation of bacterial amino-sugar metabolism. *Nucleic Acids Res* 45:10845–10860.

169. Görke B, Vogel J. 2008. Noncoding RNA control of the making and breaking of sugars. *Genes Dev* 22:2914–2925.

170. Zeng L, Choi SC, Danko CG, Siepel A, Stanhope MJ, Burne RA. 2013. Gene regulation by CcpA and catabolite repression explored by RNA-seq in *Streptococcus mutans*. *PLoS One* 8:e60465.

171. Erickson DW, Schink SJ, Patsalo V, Williamson JR, Gerland U, Hwa T. 2017. A global resource allocation strategy governs growth transition kinetics of *Escherichia coli*. *Nature* 551:119–123.

172. Westermayer SA, Fritz G, Gutiérrez J, Megerle JA, Weißl MP, Schnetz K, Gerland U, Rädler JO. 2016. Single-cell characterization of metabolic switching in the sugar phosphotransferase system of *Escherichia coli*. *Mol Microbiol* 100:472–485.

173. Holmqvist E, Wright PR, Li L, Bischler T, Barquist L, Reinhardt R, Backofen R, Vogel J. 2016. Global RNA recognition patterns of post-transcriptional regulators Hfq and CsrA revealed by UV crosslinking *in vivo*. *EMBO J* 35:991–1011.

174. Waters SA, McAteer SP, Kudla G, Pang I, Deshpande NP, Amos TG, Leong KW, Wilkins MR, Strugnell R, Gally DL, Tollervey D, Tree JJ. 2017. Small RNA interactome of pathogenic *E. coli* revealed through crosslinking of RNase E. *EMBO J* 36:374–387.

175. Heroven AK, Dersch P. 2014. Coregulation of host-adapted metabolism and virulence by pathogenic *yersiniae*. *Front Cell Infect Microbiol* 4:146.

176. Papenfort K, Vogel J. 2014. Small RNA functions in carbon metabolism and virulence of enteric pathogens. *Front Cell Infect Microbiol* 4:91.

177. Nuss AM, Beckstette M, Pimenova M, Schmühl C, Opitz W, Pisano F, Heroven AK, Dersch P. 2017. Tissue dual RNA-seq allows fast discovery of infection-specific functions and riboregulators shaping host-pathogen transcriptomes. *Proc Natl Acad Sci U S A* 114:E791–E800.

178. Heroven AK, Nuss AM, Dersch P. 2016. RNA-based mechanisms of virulence control in *Enterobacteriaceae*. *RNA Biol* 14:471–487.

179. **Papenfort K, Podkaminski D, Hinton JC, Vogel J.** 2012. The ancestral SgrS RNA discriminates horizontally acquired *Salmonella* mRNAs through a single G-U wobble pair. *Proc Natl Acad Sci U S A* **109**:E757–E764.

180. **Gruber CC, Sperandio V.** 2015. Global analysis of posttranscriptional regulation by GlmY and GlmZ in enterohemorrhagic *Escherichia coli* O157:H7. *Infect Immun* **83**:1286–1295.

181. **Chen J, Gottesman S.** 2016. Spot 42 sRNA regulates arabinose-inducible *araBAD* promoter activity by repressing synthesis of the high-affinity low-capacity arabinose transporter. *J Bacteriol* **199**:e00691-16.

182. **Romeo T, Babitzke P.** 2018. Global regulation by CsrA and its RNA antagonists. *Microbiol Spectrum* **6**(2):RWR-0009-2017.

183. **Bandyra KJ, Luisi BF.** 2018. RNase E and the high fidelity orchestration of RNA metabolism. *Microbiol Spectrum* **6**(2):RWR-0008-2017.

184. **Westermann A.** 2018. Regulatory RNAs in virulence and host-microbe interactions. *Microbiol Spectrum* **6**(2):RWR-0002-2017.

Regulating with RNA in Bacteria and Archaea
Edited by Gisela Storz and Kai Papenfort
© 2018 American Society for Microbiology, Washington, DC
doi:10.1128/microbiolspec.RWR-0018-2018

Small RNAs Involved in Regulation of Nitrogen Metabolism

15

Daniela Prasse[1] and Ruth A. Schmitz[1]

INTRODUCTION: REGULATION OF NITROGEN METABOLISM

Global metabolic regulatory networks allow microorganisms to survive periods of nutrient starvation or stress resulting from changes in the environment. Besides the metabolic regulatory network controlling uptake and metabolism of carbon and energy sources, the respective regulatory system responsible for acquisition of different nitrogen sources is highly important for surviving nutrient starvation. Bacteria and archaea developed several strategies for uptake and utilization of various nitrogen sources, like ammonium, amino acids, and inorganic nitrogen compounds. In the absence of any available nitrogen source, nitrogen-fixing microorganisms—so-called diazotrophs—are able to use molecular nitrogen (N_2), a process exclusively limited to prokaryotes but found in both domains, *Bacteria* and *Archaea* (for reviews, see references 1–5). Uptake and utilization of alternative nitrogen sources is strictly regulated due to higher energy demands, and respective enzymes are thus tightly regulated in response to the available nitrogen source to minimize energy costs for the nitrogen metabolism. The well-studied and -documented regulation occurs mainly at the transcriptional and posttranslational level, where the molecular mechanisms differ from microorganism to microorganism as well as between bacteria and archaea (e.g., see references 6–9). In bacteria, the transcription of nitrogen-relevant components is generally regulated by major global transcriptional regulators, most of which are constitutively expressed (9). The two-component regulatory system NtrB/NtrC—present in numerous proteobacteria—regulates the transcriptional activation of σ^{54} promoters recognized by the alternative RNA polymerase (RNAP) containing the sigma factor σ^{54} (RpoN) (8). In cyanobacteria, transcriptional regulation of central nitrogen metabolism is mediated by the master regulator NtcA, a transcription factor belonging to the CRP (cyclic AMP receptor protein) family. NtcA forms dimers capable of binding to its specific promoter motifs in a nitrogen-dependent way, and thus coordinates the cellular response to nitrogen availability at the transcriptional level (reviewed in references 10 and 11).

Ammonium is the preferred nitrogen source, as it can be directly incorporated into glutamate by glutamate dehydrogenase at low energy costs but with low binding affinity for ammonium (see Fig. 1). Under nitrogen limitation, often a combination of glutamine synthetase (GS) showing very high binding affinities for ammonium and a GOGAT system (glutamine oxoglutarate aminotransferase) are responsible for ammonium assimilation (Fig. 1). However, the cost of this assimilation is one ATP per ammonium; thus the expression of GS is mainly induced under nitrogen limitation and GS is deactivated by covalent modifications or direct interactions with proteins in response to an ammonium upshift (for details, see reference 12). The key nitrogen sensor proteins for this regulation are the so-called PII-like proteins, which are highly conserved in the three domains of life and sense the nitrogen status in response to the three main nitrogen metabolites glutamate, glutamine, and 2-oxoglutarate (Fig. 1) (7, 9, 13). PII-like proteins transmit the signal of the internal cellular nitrogen status to the target proteins (Fig. 1), often resulting in posttranslational modifications in response to the nitrogen availability (e.g., adenylylation of the GS by the ATase in *Escherichia coli*) (12), or transfer the signal toward two-component regulatory systems (e.g., NtrB/NtrC in enterobacteria) (14).

The ability of diazotrophs to use molecular nitrogen as a sole nitrogen source is an important process to the environment and agriculture since it is part of the nitro-

---

[1]Christian-Albrechts-University Kiel, Institute of General Microbiology, D-24118 Kiel, Germany.

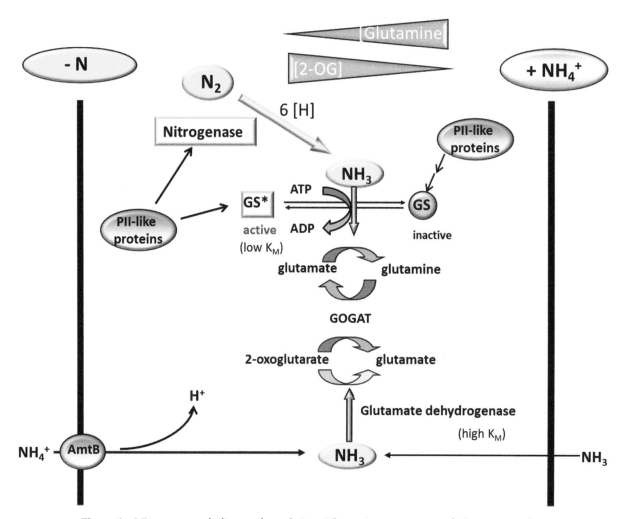

Figure 1 Nitrogen metabolism and regulation. The main components of nitrogen metabolism are depicted in a simplified way. Under nitrogen limitation (left), the GS with high binding affinity to ammonium (NH_4^+) is activated. The remaining NH_4^+ is actively transferred by ammonia transporters (AmtB) into the cell and dinitrogen (N_2) directly reduced into ammonium by the key enzyme nitrogenase. Subsequently, NH_4^+ is assimilated by the GS and the GOGAT system. Under nitrogen sufficiency (right), the GS is inhibited and ammonia diffuses over the cytoplasm membrane into the cell, which can be directly assimilated by the glutamate dehydrogenase (low affinity). PII-like proteins are involved in most regulatory ways, sensing the current nitrogen status within the cell and forwarding the respective signal to GS, inducing reversible deactivation of GS by covalent modifications or direct interactions with proteins in response to the nitrogen availability, but also to nitrogenase and ammonia transporters (for details, see reference 9).

gen cycle and helps prevent the loss of nitrogen from the biosphere by denitrification (8, 15). For example, in the open ocean, nitrogen-fixing marine cyanobacteria are important providers of fixed nitrogen from atmospheric dinitrogen (15). The key enzyme complex for nitrogen fixation, nitrogenase, is composed of two enzymes: dinitrogenase (encoded by *nifD* and *nifK*) and dinitrogenase reductase (encoded by *nifH*) (16, 17). The process of reducing molecular nitrogen is mainly

energy driven, which makes it costly for the cell. Thus, nitrogen fixation is tightly regulated at the transcriptional as well as the posttranslational level and is well studied for several diazotrophic proteobacteria, often involving PII-like proteins for sensing the internal nitrogen status (see above, and reviewed in references 8, 9, 18, and 19). However, in contrast to bacteria, little is known about the regulation of nitrogen metabolism and N_2 fixation in archaea. Moreover, as the archaeal transcription and translation machineries have features more similar to their eukaryotic than their bacterial counterparts, novel non-bacterial-like regulatory mechanisms appear likely (20, 21). The best-studied examples for nitrogen regulation in archaea are the methanogenic model organisms *Methanococcus maripaludis* and *Methanosarcina mazei*. In both species, it has been shown that regulation occurs mainly at the transcriptional level: the global master regulator NrpR regulates transcription of its target genes by binding to its corresponding operator under nitrogen sufficiency, leading to repression of transcription by blocking the recruitment of the RNAP at the promoter (22–26). Nutritional change to nitrogen-limiting conditions causes an increase of the intracellular 2-oxoglutarate concentration, which signals the depletion of nitrogen. 2-Oxoglutarate binds to NrpR, resulting in a conformational change of the repressor, causing the release of its target promoters (e.g., *nif* operon, *glnA₁*, *glnK₁-amtB₁* operon, and *nrpA*) (23–28). Homologs of NrpR have been found in various N_2-fixing methanoarchaeal genomes, indicating that this regulatory mechanism might be a common feature (29). Recently, a second *nif*-specific transcriptional regulator was identified in *M. mazei*, NrpA, which itself is under direct control of NrpR. Under nitrogen depletion, NrpA is expressed and activates transcription of the *nif* operon (30).

Only in recent years has it become more evident that besides transcriptional and posttranslational regulation, a third level of regulation is of high importance for fine-tuning in metabolic networks. Small noncoding RNAs (sRNAs) were identified, playing major roles in posttranscriptional regulation (reviewed in reference 31). Surprisingly, in contrast to their involvement in most metabolic and stress regulation, sRNAs involved in nitrogen metabolism are only rarely reported and investigated. In the following section, we summarize the current knowledge, focusing on those sRNAs directly and indirectly involved in regulation of nitrogen metabolism.

POSTTRANSCRIPTIONAL REGULATION OF NITROGEN METABOLISM BY sRNAs

sRNAs directly participating in the response to environmental nitrogen fluctuations, or particularly in regulation of N_2 fixation, have not been reported until very recently, though the expression of *nif* genes and proteins is tightly controlled at the transcriptional and posttranscriptional level. However, several examples of indirect involvement of sRNAs in nitrogen metabolism have been described. The current knowledge on respec-

tive sRNAs and their characterization is reviewed in the following sections, summarized in Table 1 and schematically depicted in Fig. 2.

sRNAs INDIRECTLY INVOLVED IN NITROGEN METABOLISM

sRNAs in Cyanobacteria

In 2011, Mitschke et al. (32) performed a genome-wide transcriptome sequencing (RNA-seq) analysis of the filamentous cyanobacterium *Anabaena* sp. PCC 7120 in response to nitrogen availability. About 600 transcriptional start sites potentially corresponding to *cis*- or *trans*-encoded sRNAs were identified, strongly suggesting a prominent role of sRNAs in regulating nitrogen assimilation in this species. One example of these *trans*-encoded sRNAs, which was investigated in more detail, is the heterocyst-specific sRNA NsiR1 (33). Under nitrogen starvation, oxygen-producing photosynthetic vegetative cells transform into nitrogen-fixing cells (heterocysts), providing fixed nitrogen to the rest of the nondifferentiated cells within the filament (34, 35). The HetR-dependent sRNA NsiR1 is highly conserved in many genomes of heterocyst-generating cyanobacteria (36). Hereby, HetR is a regulatory protein that is required for cell differentiation in *Anabaena* (37, 38). Reporter fusion studies of filaments growing under atmospheric nitrogen show that synthesis of NsiR1 occurs not only in morphologically distinct heterocysts but also in prospective heterocysts still showing characteristics of vegetative cells, strongly indicating that NsiR1 might be an early marker for cell differentiation in filamentous cyanobacteria under nitrogen-limiting conditions (36).

The noncoding 6S RNA, one of the first identified and characterized sRNAs (39–42), is known to be involved in regulating RNAP activity in stationary growth phase (43–48, see also chapter 20). For *Synechocystis* sp. PCC 6803, evidence was obtained that 6S RNA is involved in promoting the recovery processes from nitrogen-starving growth conditions (49). Comparing the chromosomal deletion strain of 6S RNA ($\Delta ssaA$) and the *Synechocystis* wild type under nitrogen acclimatization conditions after nitrogen depletion demonstrated that transcriptional as well as physiological response is generally decelerated in the deletion strain. For example, physiological studies showed that reassembly of photosynthetic pigments is delayed in the deletion strain, which shows a much stronger bleaching phenotype compared to the wild type resulting from delayed phycobilisome production after nitrogen recovery. More-

Table 1 sRNAs involved in nitrogen metabolism and their specific characteristics

sRNA	Organism	Target	Mechanism	Directly/indirectly affecting nitrogen metabolism	sRNA conservation	Transcription regulator	Reference(s)
NsiR1	*Anabaena* sp. PCC 7120	Heterocyst formation	Not known	More directly	Highly conserved in heterocyst-forming cyanobacteria	HetR	33, 36
6S RNA	*Synechocystis* sp. PCC 6803	RNAP	σ factor recruitment	More indirectly	Highly conserved in bacteria	Not known	49
CyaR	*Escherichia coli*	NadE (NAD synthetase)	Translation inhibition	Indirectly	Conserved in enterobacteria	Crp	68
GcvB	*E. coli, Salmonella enterica* serovar Typhimurium	Amino acid transporters (e.g., *dppABCDF*)	Translation inhibition	Indirectly	Highly conserved in Gram-negative bacteria	GcvA	72, 73
SdsN	*E. coli*	NarP, NfsA, HmpA	Translation inhibition	More directly	Not conserved	Crl, NarP	74
ArrF	*Azotobacter vinelandii*	PhbR / FeSII	Translation activation / Translation inhibition	Indirectly / Indirectly	RhyB analog	Fur-FeII complex	80, 82
MmgR	*Sinorhizobium meliloti*	PhaP1, PhaP2 (phasin 1, phasin 2)	Translation inhibition	Indirectly	Highly conserved in alphaproteobacteria	NtrB/C two-component regulatory system, AniA	83, 84, 86
NalA	*Pseudomonas aeruginosa* PAO1	Nitrate-assimilation operon (*nirBD-PA1779-cobA* operon)	RNA leader-mediated antitermination (with NasT)	Directly	Not conserved	NtrB/C two-component regulatory system	53, 54
NrsZ	*P. aeruginosa* PAO1	RhlA (rhamnolipid)	Translation activation	Indirectly	Highly conserved in pseudomonads	NtrB/C two-component regulatory system	63
sRNA$_{41}$	*Methanosarcina mazei* Gö1	ACDS complex	Translation inhibition	Indirectly	Highly conserved in *Methanosarcinales*	Constitutively expressed	87, 88
NsiR4	*Synechocystis* 6803	IF7 (GS-inactivating factor 7)	Translation inhibition	Directly	Highly conserved in all five cyanobacterial sections	NtcA	92
NfiS	*Pseudomonas stutzeri* A1501	NifK	mRNA stabilization	Directly	Highly conserved in *P. stutzeri* strains	NtrB/C two-component regulatory system	104
sRNA$_{154}$	*M. mazei* Gö1	NifH, NrpA, GlnA$_1$, GlnA$_2$	mRNA stabilization / Translation inhibition	Directly / Directly	Highly conserved in *Methanosarcinales*	NrpR	87, 106

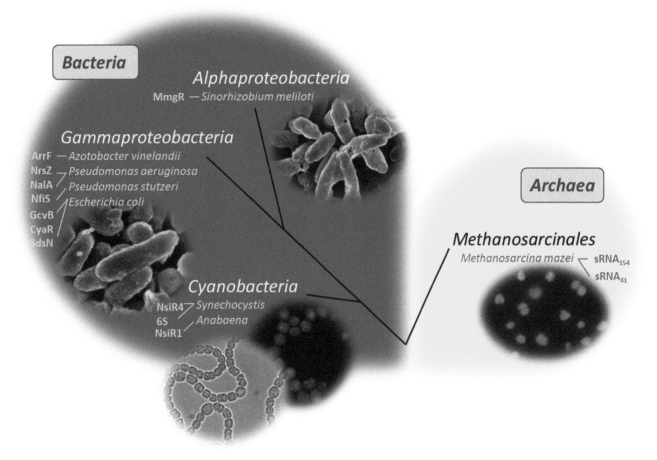

Figure 2 Visualization of microorganisms for which currently known sRNAs are directly or indirectly involved in nitrogen metabolism and their phylogenetic distribution. Organisms belonging to the two different phylogenetic domains of *Bacteria* and *Archaea* are colored and single organisms are exemplarily indicated. Each organism is depicted with its attendant sRNAs. *Synechocystis* sp. image obtained with confocal microscopy; *Anabaena*, light microscopy; *M. mazei*, fluorescence phase-contrast microscopy; and *P. aeruginosa* and *S. meliloti*, scanning electron microscopy.

over, photosynthetic activity was highly affected and up to 15% lower compared to the wild type. Since there are two major phases in the recovery process, it was shown that within the first phase basal cellular functions were regenerated, which includes expression of the ATP synthase. Within the second phase photosynthetic activity and glycogen degradation were upregulated (50). Consistent with that, it was observed by Heilmann and coworkers that storage of carbon polymers typical for *Synechocystis* under nitrogen depletion was rapidly degraded in the wild type in response to a shift to nitrogen sufficiency but with a significant delay in the 6S RNA deletion strain (49). Further transcriptome studies using genome-wide microarrays nicely showed that transcription of genes encoding subunits

of ATP synthase and ribosomal subunits were upregulated after nitrogen addition in the wild type, but still delayed in Δ*ssaA*, indicating an advantageous role of 6S RNA for the native acclimation system as mentioned above (50). In addition, some evidence was obtained that sigma factor recruitment was also influenced by 6S RNA during nitrogen recovery. Protein pulldown assays, using a His-tagged version of RNAP in the wild type and Δ*ssaA* background, showed that under nitrogen depletion, recruitment of the SigA sigma factor into the RNAP complex is reduced in Δ*ssaA*. In contrast, group 2 sigma factors like SigB and SigC were more rapidly recruited in the 6S deletion strain after nitrogen recovery compared to the wild type, demonstrating that the presence of 6S RNA restricts the recruitment of SigB and SigC into the RNAP core enzyme. It is suggested that SigC might act as regulator in nitrogen metabolism in stationary growth phase and somehow prevent the cells from quitting the stress-adapted state as well, as is the case for SigB, which prevents the cells from being more vulnerable to stress (51, 52). Consistent with that, the authors show that cells that harbor SigC (deletion of *sigBDE*) as the only group 2 sigma factor had an extended recovery phase compared to the wild type. In summary, 6S RNA appears not to be essential in *Synechocystis*, but it promotes the recovery process from nitrogen starvation by regulating the switch from group 2 sigma factor transcription to SigA-dependent transcription. Hence, 6S RNA is a newly identified candidate for sigma factor recruitment in *Synechocystis* 6803 (49).

sRNAs in Gammaproteobacteria

In gammaproteobacteria, several sRNAs have been identified that are regulated in response to nitrogen status but mostly involved in different biosynthetic pathways rather than directly in nitrogen metabolism.

A nitrogen-dependent, *cis*-acting regulatory RNA element was identified in 2012 in *Pseudomonas aeruginosa* PAO1 (53, 54). First described as a putative sRNA, NalA was identified directly in front of the *nirB* gene of the *nas* operon in a bioinformatics screen for σ[54]-dependent promoters and consequently designated as nitrate assimilation leader A (NalA) (53, 54). The gene products of the *nas* operon are known to be involved in nitrate assimilation to convert nitrate to nitrite and subsequently to ammonium, the biologically preferred form of nitrogen (55–58). Romeo and coworkers studied the nitrate assimilation operon in *P. aeruginosa* and the *cis*-acting RNA NalA in detail using biochemical and genetic approaches (54). The promoter sequence of NalA revealed putative binding sites for σ[54] and the nitrogen response regulator NtrC, strongly indicating nitrogen-dependent transcriptional upregulation after a shift to nitrogen depletion. Transcriptional fusions of the NalA promoter to the *lacZ* gene confirmed very low NalA promoter activity in a σ[54] deletion strain, whereas an increase of expression was detectable in the wild type. Consequently, the authors further tested the *nalA* promoter fusion under different nitrogen sources, demonstrating very low activity in the presence of ammonium or glutamine, whereas the activity was high under growth conditions with nitrate, nitrite, or glutamate. These results clearly showed that NalA transcription is controlled by a σ[54]- and NtrC-dependent promoter. *Azotobacter vinelandii* and *P. aeruginosa* seem to be phylogenetically closely related,

since it was already shown that several functions of housekeeping proteins are highly conserved within both strains (59). Accordingly, the nitrate/nitrite assimilation operons in *A. vinelandii*, *Klebsiella pneumoniae*, and *P. aeruginosa* are also highly conserved. To shed further light into the relationship between both strains, the authors tested whether antitermination within the NalA sequence is required for expression of the complete nitrate assimilation operon in *P. aeruginosa*, as it is already known for *Klebsiella* and *A. vinelandii* (57, 60–62). Transcription profiling from strains grown in minimal medium, with and without nitrate, showed that cotranscription of the *nalA-nirB* cistron only occurred in the presence of nitrate. The important sequence required for full expression seemed to be the full-length sequence of *nalA* including the promoter and the first 12 nucleotides (nt) of *nirB*, since no expression was detectable in a *nalA* deletion background. These results strongly support the hypothesis that transcription of the nitrate assimilation operon depends on antitermination within the NalA leader sequence. Furthermore, a homolog of NasT, which represents an antiterminator protein known from *A. vinelandii* (61), was identified in the *P. aeruginosa* genome, suggesting a similar function in *P. aeruginosa* PAO1. Overall, this study strongly supports the hypothesis that the nitrate assimilation process is regulated by a conserved metabolic mechanism in *P. aeruginosa* and *A. vinelandii*, further supporting the assumption of the phylogenetic relationship between both strains (54).

In 2014, using a bioinformatics approach searching in the intergenic regions for RpoN (σ[54]) consensus binding sequences to identify novel sRNA candidates, Wenner et al. identified a second highly conserved nitrogen-dependent sRNA in *P. aeruginosa* PAO1, designated as NrsZ (63). They found 12 well-conserved RpoN-related motifs in intergenic regions, one of which was located in the intergenic region between the *ntrC* and PA5126 genes. Evaluating *lacZ* fusions of this sRNA promoter with respect to different nitrogen availabilities showed that the promoter activity is triggered by nitrogen depletion, in particular, in response to nitrate or in the stationary growth phase with increasing cell density and decreasing nitrogen availability. Northern blot analysis as well as data analysis of RNA deep-sequencing experiments revealed that the NrsZ transcript is transcribed under nitrogen-limiting conditions. Analyzing total RNA derived from an *rpoN* or *ntrC* deletion strain showed no detectable transcript of the respective sRNA, confirming that this sRNA is transcribed in an RpoN- and NtrC-dependent manner, which was verified by complementing with *ntrC* or *rpoN*, fully restoring the wild-type transcription pattern of NrsZ (63). To analyze the biological relevance of NrsZ, phenotypic assays were performed. A deletion strain of the sRNA NrsZ showed a complete defect in swarming motility. Complementation with the full-length and only the first 60 nt of NrsZ fully restored the swarming defect. In *P. aeruginosa*, rhamnolipids synthesized by the *rhlAB* gene cluster are essential for the swarming ability (64–67). A conserved motif was identified in the first stem-loop of NrsZ, which is able to bind upstream of the ribosome binding site (RBS) of *rhlA* mRNA. Expression analysis of *rhlA* promoter activity under different growth conditions in the wild type and the *nrsZ* deletion background demonstrated that *rhlA* production is high in the wild type, with nitrate as the nitrogen source and under increasing cell densities, whereas in the absence of NrsZ, *rhlA* expression is completely repressed. Again, complementation with the first 60 nt of NrsZ expressed from a plasmid fully restored *rhlA* expression, clearly showing that regulation takes place at the posttranscriptional level (63). As it is known that nutrient limitation triggers swarming motility of *P. aeruginosa*, the physiological role of NrsZ might be to enhance the capability to adapt to environmental niches or colonize eukaryotic hosts under nitrogen limitation, ensuring an optimal physiological response to prevent nitrogen starvation.

A few sRNAs that appear to be regulated by nitrogen availability were identified in *E. coli*. One is the sRNA CyaR, under direct control of the global regulator Crp, well known to mediate catabolite repression in *E. coli* (68–71). It has been shown that CyaR directly inhibits translation of *nadE*, encoding an essential NAD synthetase, which uses ammonia to catalyze the last step in NAD synthesis. De Lay and Gottesmann further showed that the targeted sequence is located directly in front of the translational start site of *nadE*. Based on their findings, they propose that under carbon-limiting growth conditions, downregulation of NadE by CyaR helps to limit the use of ammonia (68).

A second sRNA that has to be mentioned, though it is not directly regulated by nitrogen, is GcvB, which indirectly acts as a regulatory factor in nitrogen metabolism through the repression of amino acid uptake transporters (72). GcvB is one of the most highly conserved sRNAs in gammaproteobacteria and was extensively investigated, for example, in *Salmonella*. Pulse-expression experiments followed by RNA-sequencing analysis revealed that GcvB controls genes involved in amino acid metabolism at the posttranscriptional level. GcvB is highly expressed in rich media, and regulation occurs via a highly conserved C/A-rich single-stranded region within the GcvB sequence. Interaction with the 5′ untranslated region (5′ UTR) of several ABC transporters represses translation initiation by blocking ribosome binding, thus altering the uptake of amino acids in response to the availability of nitrogen. Repression of ABC transporters is suggested to save energy for unnecessary amino acid uptake when sufficient nutrients are provided for amino acid synthesis (72, 73).

The Hfq-dependent sRNA SdsN in *E. coli* is transcribed in a σ^S-dependent manner and is highly induced during stationary growth phase (74). The nitrogen-dependent and σ^S-related assembly factor Crl is also involved in facilitating transcription of SdsN. As it is known that expression of Crl is reduced under nitrogen limitation (75), it was suggested that SdsN is also somehow influenced by nitrogen availability. Northern blots confirmed that SdsN levels are increased under rich nitrogen growth conditions. Seeking target mRNAs, microarray analysis was conducted to compare wild-type and a pulse-expressed overproduction strain of SdsN. Intriguingly, NarP, the response regulator of the NarP-NarQ two-component system, is negatively regulated by SdsN. The NarP-NarQ regulatory system mediates nitrate/nitrite-responsive transcriptional regulation during anaerobic respiration (76). Since SdsN levels are upregulated under nitrogen-limiting growth conditions in a *narP* deletion background, a negative feedback loop is indicated. Moreover, SdsN represses genes involved in the metabolism of oxidized nitrogen compounds. The above-mentioned microarray analysis unraveled NfsA as second nitrogen-relevant target for SdsN. The synthesis of NfsA is repressed by SdsN, which leads to resistance to nitrofuran compounds, a group of antibiotics. In an overproduction strain of SdsN, treatment with nitrofurazone leads to stronger resistance, whereas strains containing mutations in the regulatory binding site of SdsN are unable to regulate *nfsA* expression and are therefore more sensitive to nitrofurazone as well as the complete deletion of SdsN (74).

Interestingly, SdsN is present in three different isoforms, of which the longest isoform, $SdsN_{178}$, is bound with higher affinity to Hfq than the two shorter forms. One explanation for that is that the longer 5′ end of $SdsN_{178}$ shows a very AU-rich stretch that is preferably bound by Hfq and cleaved by RNase E. Furthermore, the long isoform is unable to regulate the identified target mRNAs compared to the short isoforms $SdsN_{137}$ and $SdsN_{124}$, due to an inhibitory effect of the AU-rich 5′ end. Cleavage of the 5′ end to the shorter isoform $SdsN_{124}$ is able to regulate the *nfsA* target mRNA. One physiological explanation for having different isoforms of an sRNA is that this allows an additional level of regulation by altering, for example, the secondary structure (74). This phenomenon was observed for many other sRNAs that were activated by posttranscriptional cleavage of the 5′ end (77–79).

In *A. vinelandii*, the nitrogen-associated and highly conserved sRNA ArrF is strongly induced under iron depletion and negatively regulated by Fur-Fe^{2+}, the master regulator of iron metabolism tightly controlling the uptake and storage of iron (80). The promoter region of ArrF contains an "iron box" that binds Fur-Fe^{2+}, leading to repression of transcription under iron-rich conditions. To identify potential target mRNAs of ArrF, mRNA transcript levels in a *fur* and a *arrF* mutant were compared to the wild type. This approach identified FeSII mRNA as a target for ArrF. In the *arrF* strain, FeSII was upregulated under iron limitation, suggesting a negative, posttranscriptional regulation by ArrF. In agreement with these findings, a possible interaction between the ArrF core sequence and the FeSII transcript was predicted. FeSII plays an essential role in nitrogen fixation by protecting the nitrogenase enzyme from oxygen inactivation by forming a protective complex with the enzyme (81). These results support the assumption that ArrF might have a regulatory role associated with nitrogen fixation (80). In a recent study, it was shown that ArrF additionally regulates expression of PhbR, the transcriptional activator of the *phbBAC* operon, responsible for polyhydroxybutyrate (PHB) synthesis, at the posttranscriptional level. PHB is catabolized as an energy and carbon source upon nutrient depletion in *A. vinelandii*. Under iron limitation it was shown that transcription of the *phbBAC* operon is activated by PhbR, which is posttranscriptionally upregulated by ArrF under Fe-limiting conditions. Furthermore, it was found that elevated ArrF positively regulates translation of PhbR by binding at the 5′ UTR of *phbR* mRNA and melting an inhibitory secondary structure, allowing free access to the RBS. This example of a bacterial sRNA nicely emphasizes that the two main cycles, the carbon and nitrogen cycle, are closely linked (82).

sRNAs in Alphaproteobacteria

A second sRNA that bridges central nitrogen and carbon metabolism is MmgR from *Sinorhizobium meliloti*. A recently published study showed that transcription of MmgR RNA in *S. meliloti* is dependent on the cellular nitrogen status and is only activated under nitrogen limitation. It has been proposed that transcription of MmgR is regulated upstream from the transcriptional start site by a nitrogen-dependent protein, potentially

NtrC, part of the well-known NrtB/C two-component regulatory system (83). Moreover, in 2017 it was discovered that MmgR, similarly to ArrF of *A. vinelandii*, is involved in regulating PHB accumulation (84). In response to nitrogen depletion and carbon-surplus growth conditions, MmgR posttranscriptionally regulates the expression of the phasin genes, encoding the PHB granule-associated proteins 1 and 2 (PhaP1 and PhaP2). A deletion strain of the *mmgR* core region shows an imbalance in PHB accumulation under C surplus, resulting in ~20% greater amounts of PHB granules and thus causing an uncontrolled accumulation of PHB and biomass increase. The respective complementation of the *mmgR* deletion mutant fully restored the wild-type behavior. Proteome studies with whole-cell extracts of *mmgR* deletion strain and wild type confirmed an overexpression of PhaP1 and PhaP1 in the absence of MmgR. Overall, the nitrogen-regulated MmgR appears to be a negative regulator of phasin gene expression under carbon excess (84). Accordingly, a recently published study revealed that the MmgR promoter possesses a conserved heptamer motif that is responsible for transcription initiation under nitrogen limitation. Interestingly, many nitrogen-relevant genes in *S. meliloti*, like *nifH*, and the promoters of several enterobacterial genes, like *glnA*, also harbor such a motif in their promoter sequence and are under transcriptional control of NtrC (85). Reporter fusions of *mmgR* promoter to *gfp* showed decreased promoter activity in an *ntrC* mutant background. These findings argue for a direct NtrC-dependent activation of MmgR transcription under nitrogen-limiting growth conditions. Moreover, the authors clearly showed that MmgR is also under negative transcriptional control of AniA, the global regulator of carbon flow, and thus is repressed under carbon-limiting conditions. Accordingly, MmgR represents an sRNA apparently involved in carbon and nitrogen metabolism, fine-tuning the expression of target genes at the posttranscriptional level with respect to varying carbon and nitrogen availabilities (86).

sRNAs in Methanoarchaea

One sRNA of *M. mazei* Gö1 denoted sRNA$_{41}$ is induced 100-fold under nitrogen sufficiency compared to nitrogen limitation (87) and is highly conserved amongst different *Methanosarcina* species. Bioinformatics target predictions revealed several potential target mRNAs, for which interactions were mostly predicted within the regions of translation initiation (88). Two of the most probable targets are two homologous operons encoding an acetyl-coenzyme A decarbonylase/synthase complex

(ACDS), for which multiple interactions within the polycistronic mRNA targeting several RBSs were predicted. Reverse transcription-quantitative PCR analysis comparing RNA levels of ACDS in the deletion and overexpression strain of sRNA$_{41}$ showed no significant changes in transcript levels, indicating posttranscriptional regulation by sRNA$_{41}$. The binding between sRNA$_{41}$ and RBS of the two polycistronic mRNAs was confirmed using microscale thermophoresis, showing strong binding affinities. Mutated binding sites in sRNA$_{41}$ or the target mRNA inhibited binding, whereas complementary mutants were able to restore the interaction. A global proteome approach comparing the protein levels of the chromosomal deletion strain and the wild type demonstrated that subunits of both ACDS operons are upregulated in the sRNA$_{41}$ deletion strain, experimentally confirmed by Western blot analysis, clearly arguing for posttranscriptional regulation of ACDS complex by sRNA$_{41}$. During growth on methanol, the ACDS complex is essential for generation of acetyl-coenzyme A, a major precursor of amino acids, via the Wood-Ljungdahl pathway. We propose that the significantly reduced amounts of sRNA$_{41}$ under nitrogen depletion and consequently the upregulation of the ACDS complex have a positive impact on providing large amounts of amino acids for nitrogenase synthesis. Thus, sRNA$_{41}$ represents the first sRNA in archaea to have an important regulatory impact linking the two major metabolic cycles (carbon and nitrogen) (88).

sRNAs DIRECTLY INVOLVED IN REGULATING NITROGEN METABOLISM

Several sRNAs with direct involvement in nitrogen metabolism have been identified and characterized (see Fig. 3).

sRNA NsiR4 in *Synechocystis* 6803

In cyanobacteria, several sRNAs are upregulated during nitrogen starvation, such as in *Anabaena* 7120 (32, 33) and *Synechocystis* 6803 and 6714 (89–91), making their regulatory function very likely. The first functionally characterized sRNA determined to be directly involved in nitrogen metabolism was the sRNA NsiR4 (nitrogen stress-induced RNA 4) of *Synechocystis* 6803. This sRNA uses an intriguing mechanism to affect nitrogen assimilation through the enzyme GS. NsiR4 was first identified by comparative transcriptomics using differential RNA sequencing, which showed strong upregulation under nitrogen depletion, a first hint of an important regulatory role in nitrogen metabolism (90).

NsiR4 is highly conserved in several cyanobacterial genomes, distributed over all five morphologically distinct sections, suggesting its positive selection. A 16-nt single-stranded region is almost identical among all NsiR4 homologs and therefore is likely involved in interactions with potential target mRNAs. A binding motif for NtcA, which is the global transcriptional regulator of nitrogen assimilation in cyanobacteria (11), was predicted and experimentally confirmed within the NsiR4 promoter (92). The location of this NtcA binding site in an activating position, 48 to 35 nt upstream from the NsiR4 transcriptional start site in *Synechocystis* 6803, is consistent with the fact that NsiR4 transcription is highly induced under nitrogen depletion. NtcA activity is regulated by the metabolite 2-oxoglutarate (10, 93). Under nitrogen sufficiency, the 2-oxoglutarate level is low and thus NtcA is inactive, with low affinity to its target promoters. To predict potential targets of NsiR4, the authors used a combination of bioinformatics using the tool CopraRNA (94, 95) and experimental transcriptome studies by performing microarrays with RNA isolated from different *nsiR4* mutation strains (overexpression, deletion, and compensatory strain of NsiR4 [Δ*nsiR4*::oex], including pulse expression). Both assays revealed the *gifA* mRNA as the most probable target. The *gifA* gene encodes the GS-inactivating factor (IF) 7 and is under direct negative control of NtcA. Hence, *gifA* transcription is downregulated under nitrogen depletion. The *gifA* mRNA level was found to be increased in the chromosomal deletion strain of NsiR4, whereas in the overexpression strain *gifA* mRNA levels were reduced. The potential interaction with NsiR4 is located in the *gifA* 5′ UTR. Reporter assays using the 5′ UTR, which was fused to the superfolder green fluorescence protein coding sequence, revealed reduced fluorescence in the presence of wild-type NsiR4, whereas the mutated interaction sites in the 5′ UTR did not affect the protein levels in the two-plasmid system in *E. coli* developed for identifying sRNA-target mRNA interactions by Urban and Vogel (96, 97). These experiments verified a direct interaction between *gifA* mRNA and NsiR4, which affects the protein levels of IF7 posttranscriptionally. Additionally, Western blot analyses of *Synechocystis* 6803 confirmed that protein levels of IF7 decreased during overexpression of NsiR4 compared to the wild type, whereas in the deletion strain protein levels of IF7 were about 60% higher. Examination of substrate and product kinetics after an upshift of ammonia supplementation also shed light on the global impact of this regulation. It was shown that higher IF7 levels in the deletion strain of NsiR4 significantly con-

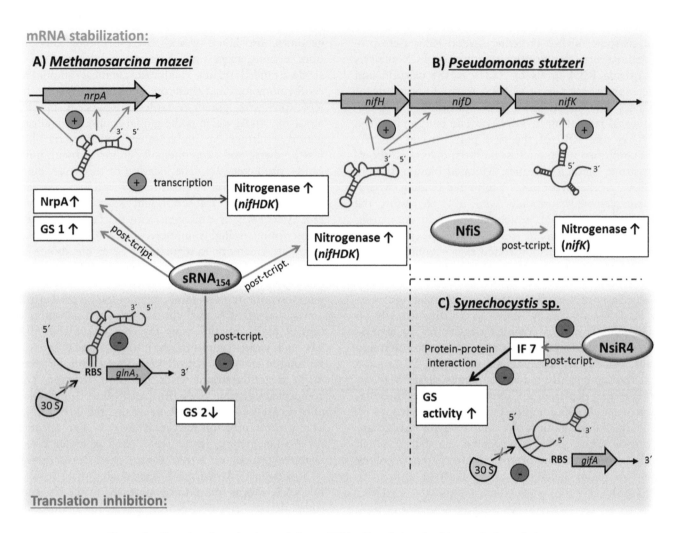

Figure 3 Functional mechanisms of three sRNAs directly involved in regulation of nitrogen fixation. (A) The sRNA$_{154}$ is a central regulatory component within the regulatory network of nitrogen fixation in *M. mazei* strain Gö1. It was shown that sRNA$_{154}$ acts on several target mRNAs by either stabilizing the mRNA by direct binding (in case of *glnA$_1$*, *nrpA*, and *nif* operon) or repressing translation initiation by blocking the RBS (in case of *glnA$_2$*). In addition, sRNA$_{154}$-mediated regulation allows feedforward regulation of *nif* expression via *nrpA* mRNA stabilization. (B) In *P. stutzeri* A1501, sRNA NfiS targets *nifK* mRNA (possibly multiple times) and enhances transcript stability, very similar to sRNA$_{154}$ in *M. mazei*, which also targets *nrpA* and *nif* operon, possibly at multiple regions. (C) NsiR4 from *Synechocystis* 6803 interacts with *gifA* mRNA, which encodes GS-inactivation factor 7, and blocks translation initiation by targeting the 5′ UTR of *gifA*, consequently leading to a positive effect on GS activity under nitrogen limitation. This mode of action is very similar to that of sRNA$_{154}$, which targets the 5′ UTR of the *glnA$_2$* transcript as well and represses translation initiation. post-tcript, posttranscriptional.

tributed to GS inactivation. Consistently, the authors observed decreased IF7 levels and diminished GS deactivation during NsiR4 overexpression, while the compensatory strain (Δ*nsiR4*::oex) behaved like the wild type. The latter indicated that the altered rate of glutamine/glutamate pool restoration in Δ*nsiR4* was compensated through NsiR4 expression. The mechanism

used by NsiR4 is intriguing: If an sRNA targeted GS synthesis directly, the effects would be delayed due to the very high abundance of GS. Instead, regulation is efficiently executed at the level of GS enzyme activity through the two labile inactivating factors, IF7 and the second inactivation factor IF17 (encoded by *gifB*). Consequently, it is more effective to target the biosynthesis of these factors rather than GS *de novo* synthesis.

In contrast to IF7, the regulation of IF17 is so far unclear. There are very likely many more sRNAs and even other types of noncoding RNAs, e.g., riboswitches, involved in the regulation of nitrogen assimilation in bacteria (98, 99). However, thus far, NsiR4 represents the first characterized sRNA involved in the regulation of nitrogen metabolism in bacteria, mediating the acclimation of cyanobacteria to rapid changes in available nitrogen sources (92).

sRNA NfiS in *Pseudomonas stutzeri* A1501

Very recently, the second sRNA found to directly participate in nitrogen metabolism was identified in *P. stutzeri* A1501, a root-associated bacterium, which is able to fix nitrogen due to the acquisition of the *nif* operon by horizontal gene transfer (100–103). This sRNA, NfiS, is σ^{54}-dependently transcribed and under direct control of the NtrC/NifA regulatory cascade, resulting in high induction under nitrogen depletion or sorbitol stress (104). Though NfiS is highly conserved in *P. stutzeri* strains, it is not found in other bacterial species. Furthermore, transcript stability appears to be affected by Hfq, since in experiments in an *hfq* deletion background the NfiS transcript is hardly detectable compared to the wild type. Knockout of NfiS results in a strong decrease of nitrogenase activity, whereas overexpression leads to an increase of up to 150% in nitrogenase activity, strongly indicating that NfiS plays a crucial role in the regulation of nitrogen fixation. Overall, expression levels of proteins encoded by *nif* and *nif*-related genes were positively affected in the overexpression mutant of NfiS and negatively affected in the corresponding deletion strain. Structure predictions revealed a stem-loop in the secondary structure of NfiS predicted to be able to base-pair at the 5′ end of *nifK* mRNA (nucleotides 207 to 217). Consistently, the *nifK* gene encodes a subunit of the key enzyme nitrogenase. Further, mRNA half-life assays showed that *nifK* mRNA was more rapidly degraded in the absence of NfiS compared to the wild type, strongly arguing that interaction between NfiS and *nifK* enhances mRNA half-life and thus translation efficiency. Microscale thermophoresis experiments further confirmed *in vitro* that *nifK* is directly targeted by NfiS. The secondary structure of *nifK* mRNA contains an inhibitory hairpin structure that is apparently destroyed when interacting with NfiS, consequently inducing translation and mRNA stability. The fact that NfiS is located in the core genome and regulates components of the *nif* operons that were obtained by horizontal gene transfer strongly

indicates that the sRNA was recruited by *nifK* as a new regulatory element to integrate the *nif* island into the basic regulons of the host (104, see also chapter 28).

sRNA$_{154}$ in *M. mazei* strain Gö1

The first reported nitrogen-regulated sRNA in archaea is sRNA$_{154}$ of *M. mazei* strain Gö1, which regulates expression of the *nif* promoter-specific activator NrpA, nitrogenase, and the two GSs by influencing the transcript stability. Identified by a differential RNA-seq approach comparing nitrogen sufficiency and nitrogen-limiting conditions, a regulatory function in nitrogen fixation was already predicted in 2009 (87). The sequence and structure of sRNA$_{154}$ are highly conserved amongst the *Methanosarcinales*, as well as the transcriptional control by global nitrogen repressor NrpR, thus pointing to an important direct regulatory role of sRNA$_{154}$ in nitrogen metabolism. Bioinformatics target predictions using the tool IntaRNA (94, 105) revealed potential interactions with several mRNAs encoding nitrogen-associated enzymes (106). Furthermore, secondary structure predictions revealed two conserved stem-loops of sRNA$_{154}$ with the ability to interact with those target mRNAs. To validate these predictions, a chromosomal deletion strain of sRNA$_{154}$ was constructed and a differential RNA-seq (107) approach was performed under nitrogen-fixing growth conditions, comparing the transcript levels of the potential target mRNAs in wild type and the sRNA$_{154}$ deletion strain. This demonstrated that transcript levels of genes encoding GS1, PII-like protein GlnK$_1$, nitrogenase subunit NifH, and the transcriptional activator of the *nif* operon, NrpA, were simultaneously downregulated in the sRNA$_{154}$ deletion strain. To further determine whether these interactions influence transcript stabilities of translation initiation, an mRNA stability assay was performed, revealing that transcript stability of *nrpA*, *nifH*, and *glnA$_1$* is highly decreased in the absence of sRNA$_{154}$, arguing for direct stabilization by sRNA$_{154}$. In contrast, levels of the *glnA$_2$* mRNA increased slightly over time in the absence of sRNA$_{154}$, which suggests a negative impact of sRNA$_{154}$ on *glnA$_2$* expression. Electrophoretic mobility shift assay analysis demonstrated direct interactions between sRNA$_{154}$ and *glnA$_1$*, *glnA$_2$*, and *nrpA* mRNAs, whereas no interaction was observed between sRNA$_{154}$ and *nifH*. Western blot analysis confirmed decreased protein levels of NifH in the chromosomal deletion and an increase in the corresponding overexpression of sRNA$_{154}$. In addition, a slightly negative effect on GS1 levels and a slightly positive effect on GS2 protein amounts were observed in the chromosomal deletion strain compared to the wild-type strain (106). To clarify the regulatory mechanism on *nifH*, the full-length *nif* operon (5 kbp) was reinspected using the IntaRNA tool (94, 105), revealing that sRNA$_{154}$ most likely interacts several times with the polycistronic mRNA of the *nif* operon, which explains the positive effect on transcript stability and finally NifH protein levels. Based on these findings, it was proposed that binding of both loops of several sRNA$_{154}$ molecules stabilizes the polycistronic *nif* mRNA, most likely inhibiting endonucleolytic cleavage by masking specific recognition sites for RNases. Furthermore, an additional small scaffold protein might be required to effectively stabilize those interactions *in vivo*.

Overall, these findings clearly demonstrate a global role of sRNA$_{154}$ in nitrogen metabolism in *M. mazei*. The current working model is the following: sRNA$_{154}$ stabilizes the transcripts of the polycistronic *nif* operon and *nrpA*, which consequently leads to strong expression of nitrogenase mediated by direct interaction between sRNA$_{154}$ and the *nif* transcript and additionally by higher NrpA amounts based on sRNA$_{154}$-induced stability of *nrpA* mRNA. Thus, it is proposed that besides direct stabilization of the *nif* transcript, sRNA$_{154}$ facilitates feedforward regulation of *nif* gene expression by stabilizing the *nrpA* transcript. Furthermore, it seems that exclusive transcription of sRNA$_{154}$ under nitrogen limitation leads to pleiotropic effects on various genes involved in nitrogen metabolism (e.g., *glnA$_1$*, *glnA$_2$*, and *glnK$_1$*) in *M. mazei*, demonstrating its central regulatory role in nitrogen metabolism (106).

ARE REGULATORY RNAs SENSING NITROGEN STATUS?

To respond to nitrogen availability, it is crucial to perceive the internal nitrogen status of the cell. The well-studied sensing mechanisms generally based on the internal metabolite pools of 2-oxoglutarate, glutamate, and glutamine often involve PII-like proteins transmitting the signal to target proteins (see also Fig. 1) (12, 14). Due to the fact that all nitrogen metabolites are charged, one can speculate that regulatory RNAs might be able to bind to those metabolites directly (particularly to the amino groups of glutamine and glutamate), which in turn might affect the regulatory function of the RNA. In this context, RNA binding of a nitrogen metabolite by a riboswitch is an attractive mechanism to directly sense the nitrogen status internally, and the first evidence has been obtained: naturally occurring glutamine aptamers have been identified in cyanobacteria and marine

metagenomes upstream from genes involved in nitrogen metabolism (98, 99) (for aptamers, see also chapter 5). Most of the reported sRNAs in bacteria and archaea involved in nitrogen-dependent posttranscriptional regulation have been shown to be expressed in a nitrogen-dependent manner, induced or repressed by a transcriptional (often global) regulator, which itself perceives the nitrogen status—e.g., NrpR in *M. mazei*, binding oxoglutarate (106); NtrC in *P. aeruginosa* (108); and NtcA in *Synechocystis* (10) (see above). However, it is highly attractive to speculate that in addition to those sRNAs, sRNAs might be able to sense the signal directly by binding the nitrogen metabolites (see above) or by interacting with a protein. In the latter case, one can speculate that in binding the metabolites directly (as is the case for PII-like proteins) the protein changes its binding affinity to the sRNA, which in turn affects the respective sRNA to interact with its target(s) or might lead to the degradation of the sRNA itself (as proposed for sRNA$_{41}$ in *M. mazei* [see above] [88]). Until today, however, to our knowledge, no nitrogen metabolite-binding sRNA has been reported.

Small proteins translated from small open reading frames are often encoded by small mRNAs, which have been frequently originally identified as sRNAs. Specific screening by reinspecting existing data from transcriptome analyses revealed first examples of small open reading frames, some of which might be involved in regulation of nitrogen metabolism. A functional analysis of *Synechocystis* 6803 revealed one previously annotated sRNA as a small protein-encoding mRNA (NsiR6) (90, 109). Transcription of NsiR6 is highly induced under nitrogen-limiting conditions, and a conserved NtcA binding site 41 nt upstream of its transcription start site was identified. Moreover, NsiR6 is highly conserved in cyanobacteria; sequence alignment revealed two conserved cysteine pair residues that might be involved in protein-protein interactions, structure formation, or metal binding (110). Likewise, in the methanoarchaeal model organism *M. mazei*, the nitrogen-dependent small proteome was analyzed, identifying among others small protein 36, encoded by spRNA36, highly conserved in other *Methanosarcina* strains (111, 112) Transcription as well as protein amount is upregulated under nitrogen-fixing growth conditions. In this context, it is attractive to speculate that small protein 36 of *M. mazei*, which contains a predicted SAM domain (known to be a protein-interaction domain), might interact with a target protein that is part of the nitrogen metabolism under nitrogen-fixing conditions, for example, the GS or the PII-like protein GlnK (111).

CONCLUSION

As outlined above, nitrogen metabolism in prokaryotes is diversely regulated on all molecular levels, including posttranscriptional regulation by sRNAs, which mostly act as fine-tuning regulators, enabling the cell to rapidly react to nitrogen fluctuations. Most interestingly, several sRNAs have been reported (sRNA$_{41}$ in *M. mazei*, MmgR in *S. meliloti*, and ArrF in *A. vinelandii* [see Table 1]) that appear to line the two main metabolic cycles, carbon and nitrogen, emphasizing the close interconnection of the two cycles in *Prokarya*. Surprisingly, sRNAs directly involved in the regulatory network of nitrogen assimilation or nitrogen fixation are still scarce, and knowledge regarding their roles in nitrogen metabolism is lacking compared to that for other metabolic and cellular prokaryotic pathways. However, taking into account that the nitrogen cycle is one of the important and central metabolic pathways, one can expect a plethora of sRNAs and other types of regulatory RNA elements, e.g., riboswitches, as well as small regulatory proteins translated from sRNAs, to be involved in the regulation of nitrogen metabolism, which represents a fascinating field and remains to be uncovered in the future.

Acknowledgments. We thank Dr. Scarlett Sett and Dr. Sarah Habig from our laboratory for helpful discussions on the manuscript. We further thank Prof. Dr. Wolfgang Hess (University of Freiburg) for helpful discussions and comments on the NsiR4 section. We particularly thank Elvira Olmedo-Verd (Spanish National Research Council) for providing the Anabaena *image, Prof. Dr. Conrad W. Mullineaux (University of London) for providing the image of* Synechocystis, *and Dr. Ines Krohn-Molt (University of Hamburg) for providing the* P. aeruginosa *and* S. meliloti *images; all images are shown in Fig. 2. This work was financially supported by the Deutsche Forschungsgesellschaft (SCHM1052/9-2).*

Citation. Prasse D, Schmitz RA. 2018. Small RNAs involved in regulation of nitrogen metabolism. Microbiol Spectrum 6(4):RWR-0018-2018.

References

1. **Fischer HM.** 1994. Genetic regulation of nitrogen fixation in rhizobia. *Microbiol Rev* **58:**352–386.

2. **Kessler PS, McLarnan J, Leigh JA.** 1997. Nitrogenase phylogeny and the molybdenum dependence of nitrogen fixation in *Methanococcus maripaludis. J Bacteriol* **179:**541–543.

3. **Smith DR, Doucette-Stamm LA, Deloughery C, Lee H, Dubois J, Aldredge T, Bashirzadeh R, Blakely D, Cook R, Gilbert K, Harrison D, Hoang L, Keagle P, Lumm W, Pothier B, Qiu D, Spadafora R, Vicaire R, Wang Y, Wierzbowski J, Gibson R, Jiwani N, Caruso A, Bush D, Reeve JN.** 1997. Complete genome sequence of *Methanobacterium thermoautotrophicum* ΔH: functional analysis and comparative genomics. *J Bacteriol* **179:**7135–7155.

4. **Leigh JA.** 2000. Nitrogen fixation in methanogens: the archaeal perspective. *Curr Issues Mol Biol* **2:**125–131.

5. **Dos Santos PC, Fang Z, Mason SW, Setubal JC, Dixon R.** 2012. Distribution of nitrogen fixation and nitrogenase-like sequences amongst microbial genomes. *BMC Genomics* **13:**162.

6. **Moure VR, Costa FF, Cruz LM, Pedrosa FO, Souza EM, Li XD, Winkler F, Huergo LF.** 2015. Regulation of nitrogenase by reversible mono-ADP-ribosylation. *Curr Top Microbiol Immunol* **384:**89–106.

7. **Merrick M.** 2015. Post-translational modification of P II signal transduction proteins. *Front Microbiol* **5:**763.

8. **Dixon R, Kahn D.** 2004. Genetic regulation of biological nitrogen fixation. *Nat Rev Microbiol* **2:**621–631.

9. **Leigh JA, Dodsworth JA.** 2007. Nitrogen regulation in Bacteria and Archaea. *Annu Rev Microbiol* **61:**349–377.

10. **Muro-Pastor MI, Reyes JC, Florencio FJ.** 2005. Ammonium assimilation in cyanobacteria. *Photosynth Res* **83:**135–150.

11. **Herrero A, Muro-Pastor AM, Flores E.** 2001. Nitrogen control in cyanobacteria. *J Bacteriol* **183:**411–425.

12. **van Heeswijk WC, Westerhoff HV, Boogerd FC.** 2013. Nitrogen assimilation in *Escherichia coli*: putting molecular data into a systems perspective. *Microbiol Mol Biol Rev* **77:**628–695.

13. **Forchhammer K, Lüddecke J.** 2016. Sensory properties of the PII signalling protein family. *FEBS J* **283:**425–437.

14. **Schumacher J, Behrends V, Pan Z, Brown DR, Heydenreich F, Lewis MR, Bennett MH, Razzaghi B, Komorowski M, Barahona M, Stumpf MP, Wigneshweraraj S, Bundy JG, Buck M.** 2013. Nitrogen and carbon status are integrated at the transcriptional level by the nitrogen regulator NtrC *in vivo. mBio* **4:**e00881-e13.

15. **Zehr JP.** 2011. Nitrogen fixation by marine cyanobacteria. *Trends Microbiol* **19:**162–173.

16. **Hoffman BM, Lukoyanov D, Yang ZY, Dean DR, Seefeldt LC.** 2014. Mechanism of nitrogen fixation by nitrogenase: the next stage. *Chem Rev* **114:**4041–4062.

17. **Seefeldt LC, Hoffman BM, Dean DR.** 2009. Mechanism of Mo-dependent nitrogenase. *Annu Rev Biochem* **78:**701–722.

18. **Huergo LF, Pedrosa FO, Muller-Santos M, Chubatsu LS, Monteiro RA, Merrick M, Souza EM.** 2012. PII signal transduction proteins: pivotal players in posttranslational control of nitrogenase activity. *Microbiology* **158:**176–190.

19. **Masepohl B, Hallenbeck PC.** 2010. Nitrogen and molybdenum control of nitrogen fixation in the phototrophic bacterium *Rhodobacter capsulatus. Adv Exp Med Biol* **675:**49–70.

20. **Dodsworth JA, Leigh JA.** 2006. Regulation of nitrogenase by 2-oxoglutarate-reversible, direct binding of a PII-like nitrogen sensor protein to dinitrogenase. *Proc Natl Acad Sci U S A* **103:**9779–9784.

21. **Leigh JA.** 1999. Transcriptional regulation in Archaea. *Curr Opin Microbiol* **2:**131–134.

22. Cohen-Kupiec R, Blank C, Leigh JA. 1997. Transcriptional regulation in Archaea: *in vivo* demonstration of a repressor binding site in a methanogen. *Proc Natl Acad Sci U S A* **94**:1316–1320.

23. Lie TJ, Leigh JA. 2003. A novel repressor of *nif* and *glnA* expression in the methanogenic archaeon *Methanococcus maripaludis*. *Mol Microbiol* **47**:235–246.

24. Weidenbach K, Glöer J, Ehlers C, Sandman K, Reeve JN, Schmitz RA. 2008. Deletion of the archaeal histone in *Methanosarcina mazei* Gö1 results in reduced growth and genomic transcription. *Mol Microbiol* **67**:662–671.

25. Lie TJ, Hendrickson EL, Niess UM, Moore BC, Haydock AK, Leigh JA. 2010. Overlapping repressor binding sites regulate expression of the *Methanococcus maripaludisglnK₁* operon. *Mol Microbiol* **75**:755–762.

26. Weidenbach K, Ehlers C, Kock J, Schmitz RA. 2010. NrpRII mediates contacts between NrpRI and general transcription factors in the archaeon *Methanosarcina mazei* Gö1. *FEBS J* **277**:4398–4411.

27. Lie TJ, Wood GE, Leigh JA. 2005. Regulation of *nif* expression in *Methanococcus maripaludis*: roles of the euryarchaeal repressor NrpR, 2-oxoglutarate, and two operators. *J Biol Chem* **280**:5236–5241.

28. Lie TJ, Leigh JA. 2007. Genetic screen for regulatory mutations in *Methanococcus maripaludis* and its use in identification of induction-deficient mutants of the euryarchaeal repressor NrpR. *Appl Environ Microbiol* **73**:6595–6600.

29. Lie TJ, Dodsworth JA, Nickle DC, Leigh JA. 2007. Diverse homologues of the archaeal repressor NrpR function similarly in nitrogen regulation. *FEMS Microbiol Lett* **271**:281–288.

30. Weidenbach K, Ehlers C, Schmitz RA. 2014. The transcriptional activator NrpA is crucial for inducing nitrogen fixation in *Methanosarcina mazei* Gö1 under nitrogen-limited conditions. *FEBS J* **281**:3507–3522.

31. Wagner EG, Romby P. 2015. Small RNAs in bacteria and archaea: who they are, what they do, and how they do it. *Adv Genet* **90**:133–208.

32. Mitschke J, Vioque A, Haas F, Hess WR, Muro-Pastor AM. 2011. Dynamics of transcriptional start site selection during nitrogen stress-induced cell differentiation in *Anabaena* sp. PCC7120. *Proc Natl Acad Sci U S A* **108**:20130–20135.

33. Ionescu D, Voss B, Oren A, Hess WR, Muro-Pastor AM. 2010. Heterocyst-specific transcription of NsiR1, a non-coding RNA encoded in a tandem array of direct repeats in cyanobacteria. *J Mol Biol* **398**:177–188.

34. Zhao J, Wolk CP. 2008. Developmental biology of heterocysts, 2006, p 397–418. *In* Whitworth DE (ed), *Myxobacteria: Multicellularity and Differentiation*. ASM Press, Washington, DC.

35. Muro-Pastor AM, Hess WR. 2012. Heterocyst differentiation: from single mutants to global approaches. *Trends Microbiol* **20**:548–557.

36. Muro-Pastor AM. 2014. The heterocyst-specific NsiR1 small RNA is an early marker of cell differentiation in cyanobacterial filaments. *mBio* **5**:e01079-e14.

37. Olmedo-Verd E, Muro-Pastor AM, Flores E, Herrero A. 2006. Localized induction of the *ntcA* regulatory gene in developing heterocysts of *Anabaena* sp. strain PCC 7120. *J Bacteriol* **188**:6694–6699.

38. Rajagopalan R, Callahan SM. 2010. Temporal and spatial regulation of the four transcription start sites of *hetR* from *Anabaena* sp. strain PCC 7120. *J Bacteriol* **192**:1088–1096.

39. Barrick JE, Sudarsan N, Weinberg Z, Ruzzo WL, Breaker RR. 2005. 6S RNA is a widespread regulator of eubacterial RNA polymerase that resembles an open promoter. *RNA* **11**:774–784.

40. Steuten B, Hoch PG, Damm K, Schneider S, Köhler K, Wagner R, Hartmann RK. 2014. Regulation of transcription by 6S RNAs: insights from the *Escherichia coli* and *Bacillus subtilis* model systems. *RNA Biol* **11**:508–521.

41. Cavanagh AT, Wassarman KM. 2014. 6S RNA, a global regulator of transcription in *Escherichia coli*, *Bacillus subtilis*, and beyond. *Annu Rev Microbiol* **68**:45–60.

42. Burenina OY, Elkina DA, Hartmann RK, Oretskaya TS, Kubareva EA. 2015. Small noncoding 6S RNAs of bacteria. *Biochemistry (Mosc)* **80**:1429–1446.

43. Wassarman KM, Storz G. 2000. 6S RNA regulates *E. coli* RNA polymerase activity. *Cell* **101**:613–623.

44. Trotochaud AE, Wassarman KM. 2004. 6S RNA function enhances long-term cell survival. *J Bacteriol* **186**:4978–4985.

45. Cavanagh AT, Klocko AD, Liu X, Wassarman KM. 2008. Promoter specificity for 6S RNA regulation of transcription is determined by core promoter sequences and competition for region 4.2 of σ^{70}. *Mol Microbiol* **67**:1242–1256.

46. Neusser T, Polen T, Geissen R, Wagner R. 2010. Depletion of the non-coding regulatory 6S RNA in *E. coli* causes a surprising reduction in the expression of the translation machinery. *BMC Genomics* **11**:165.

47. Cavanagh AT, Sperger JM, Wassarman KM. 2012. Regulation of 6S RNA by pRNA synthesis is required for efficient recovery from stationary phase in *E. coli* and *B. subtilis*. *Nucleic Acids Res* **40**:2234–2246.

48. Cabrera-Ostertag IJ, Cavanagh AT, Wassarman KM. 2013. Initiating nucleotide identity determines efficiency of RNA synthesis from 6S RNA templates in *Bacillus subtilis* but not *Escherichia coli*. *Nucleic Acids Res* **41**:7501–7511.

49. Heilmann B, Hakkila K, Georg J, Tyystjärvi T, Hess WR, Axmann IM, Dienst D. 2017. 6S RNA plays a role in recovery from nitrogen depletion in *Synechocystis* sp. PCC 6803. *BMC Microbiol* **17**:229.

50. Klotz A, Georg J, Bučinská L, Watanabe S, Reimann V, Januszewski W, Sobotka R, Jendrossek D, Hess WR, Forchhammer K. 2016. Awakening of a dormant cyanobacterium from nitrogen chlorosis reveals a genetically determined program. *Curr Biol* **26**:2862–2872.

51. Asayama M, Imamura S, Yoshihara S, Miyazaki A, Yoshida N, Sazuka T, Kaneko T, Ohara O, Tabata S, Osanai T, Tanaka K, Takahashi H, Shirai M. 2004. SigC, the group 2 sigma factor of RNA polymerase,

contributes to the late-stage gene expression and nitrogen promoter recognition in the cyanobacterium *Synechocystis* sp. strain PCC 6803. *Biosci Biotechnol Biochem* **68**:477–487.

52. **Tuominen I, Tyystjärvi E, Tyystjärvi T.** 2003. Expression of primary sigma factor (PSF) and PSF-like sigma factors in the cyanobacterium *Synechocystis* sp. strain PCC 6803. *J Bacteriol* **185**:1116–1119.

53. **Livny J, Brencic A, Lory S, Waldor MK.** 2006. Identification of 17 *Pseudomonas aeruginosa* sRNAs and prediction of sRNA-encoding genes in 10 diverse pathogens using the bioinformatic tool sRNAPredict2. *Nucleic Acids Res* **34**:3484–3493.

54. **Romeo A, Sonnleitner E, Sorger-Domenigg T, Nakano M, Eisenhaber B, Bläsi U.** 2012. Transcriptional regulation of nitrate assimilation in *Pseudomonas aeruginosa* occurs via transcriptional antitermination within the *nirBD-PA1779-cobA* operon. *Microbiology* **158**:1543–1552.

55. **Moreno-Vivián C, Cabello P, Martínez-Luque M, Blasco R, Castillo F.** 1999. Prokaryotic nitrate reduction: molecular properties and functional distinction among bacterial nitrate reductases. *J Bacteriol* **181**:6573–6584.

56. **Richardson DJ.** 2001. Introduction: nitrate reduction and the nitrogen cycle. *Cell Mol Life Sci* **58**:163–164.

57. **Lin JT, Stewart V.** 1996. Nitrate and nitrite-mediated transcription antitermination control of *nasF* (nitrate assimilation) operon expression in *Klebsiella pheumoniae* M5al. *J Mol Biol* **256**:423–435.

58. **Setubal JC, dos Santos P, Goldman BS, Ertesvåg H, Espin G, Rubio LM, Valla S, Almeida NF, Balasubramanian D, Cromes L, Curatti L, Du Z, Godsy E, Goodner B, Hellner-Burris K, Hernandez JA, Houmiel K, Imperial J, Kennedy C, Larson TJ, Latreille P, Ligon LS, Lu J, Maerk M, Miller NM, Norton S, O'Carroll IP, Paulsen I, Raulfs EC, Roemer R, Rosser J, Segura D, Slater S, Stricklin SL, Studholme DJ, Sun J, Viana CJ, Wallin E, Wang B, Wheeler C, Zhu H, Dean DR, Dixon R, Wood D.** 2009. Genome sequence of *Azotobacter vinelandii*, an obligate aerobe specialized to support diverse anaerobic metabolic processes. *J Bacteriol* **191**:4534–4545.

59. **Rediers H, Vanderleyden J, De Mot R.** 2004. *Azotobacter vinelandii*: a *Pseudomonas* in disguise? *Microbiology* **150**:1117–1119.

60. **Chai W, Stewart V.** 1998. NasR, a novel RNA-binding protein, mediates nitrate-responsive transcription antitermination of the *Klebsiella oxytoca* M5al *nasF* operon leader *in vitro*. *J Mol Biol* **283**:339–351.

61. **Gutierrez JC, Ramos F, Ortner L, Tortolero M.** 1995. *nasST*, two genes involved in the induction of the assimilatory nitrite-nitrate reductase operon (*nasAB*) of *Azotobacter vinelandii*. *Mol Microbiol* **18**:579–591.

62. **Stülke J.** 2002. Control of transcription termination in bacteria by RNA-binding proteins that modulate RNA structures. *Arch Microbiol* **177**:433–440.

63. **Wenner N, Maes A, Cotado-Sampayo M, Lapouge K.** 2014. NrsZ: a novel, processed, nitrogen-dependent, small non-coding RNA that regulates *Pseudomonas aeruginosa* PAO1 virulence. *Environ Microbiol* **16**:1053–1068.

64. **Maier RM, Soberón-Chávez G.** 2000. *Pseudomonas aeruginosa* rhamnolipids: biosynthesis and potential applications. *Appl Microbiol Biotechnol* **54**:625–633.

65. **Soberón-Chávez G, Lépine F, Déziel E.** 2005. Production of rhamnolipids by *Pseudomonas aeruginosa*. *Appl Microbiol Biotechnol* **68**:718–725.

66. **Köhler T, Curty LK, Barja F, van Delden C, Pechère JC.** 2000. Swarming of *Pseudomonas aeruginosa* is dependent on cell-to-cell signaling and requires flagella and pili. *J Bacteriol* **182**:5990–5996.

67. **Déziel E, Lépine F, Milot S, Villemur R.** 2003. *rhlA* is required for the production of a novel biosurfactant promoting swarming motility in *Pseudomonas aeruginosa*: 3-(3-hydroxyalkanoyloxy)alkanoic acids (HAAs), the precursors of rhamnolipids. *Microbiology* **149**:2005–2013.

68. **De Lay N, Gottesman S.** 2009. The Crp-activated small noncoding regulatory RNA CyaR (RyeE) links nutritional status to group behavior. *J Bacteriol* **191**:461–476.

69. **Saier MH Jr.** 1998. Multiple mechanisms controlling carbon metabolism in bacteria. *Biotechnol Bioeng* **58**:170–174.

70. **Görke B, Stülke J.** 2008. Carbon catabolite repression in bacteria: many ways to make the most out of nutrients. *Nat Rev Microbiol* **6**:613–624.

71. **Deutscher J.** 2008. The mechanisms of carbon catabolite repression in bacteria. *Curr Opin Microbiol* **11**:87–93.

72. **Sharma CM, Darfeuille F, Plantinga TH, Vogel J.** 2007. A small RNA regulates multiple ABC transporter mRNAs by targeting C/A-rich elements inside and upstream of ribosome-binding sites. *Genes Dev* **21**:2804–2817.

73. **Sharma CM, Papenfort K, Pernitzsch SR, Mollenkopf HJ, Hinton JC, Vogel J.** 2011. Pervasive post-transcriptional control of genes involved in amino acid metabolism by the Hfq-dependent GcvB small RNA. *Mol Microbiol* **81**:1144–1165.

74. **Hao Y, Updegrove TB, Livingston NN, Storz G.** 2016. Protection against deleterious nitrogen compounds: role of σ^S-dependent small RNAs encoded adjacent to *sdiA*. *Nucleic Acids Res* **44**:6935–6948.

75. **Zafar MA, Carabetta VJ, Mandel MJ, Silhavy TJ.** 2014. Transcriptional occlusion caused by overlapping promoters. *Proc Natl Acad Sci U S A* **111**:1557–1561.

76. **Stewart V.** 1994. Dual interacting two-component regulatory systems mediate nitrate- and nitrite-regulated gene expression in *Escherichia coli*. *Res Microbiol* **145**:450–454.

77. **Durand S, Braun F, Lioliou E, Romilly C, Helfer AC, Kuhn L, Quittot N, Nicolas P, Romby P, Condon C.** 2015. A nitric oxide regulated small RNA controls expression of genes involved in redox homeostasis in *Bacillus subtilis*. *PLoS Genet* **11**:e1004957.

78. **Bandyra KJ, Said N, Pfeiffer V, Górna MW, Vogel J, Luisi BF.** 2012. The seed region of a small RNA drives

the controlled destruction of the target mRNA by the endoribonuclease RNase E. *Mol Cell* **47**:943–953.

79. Papenfort K, Espinosa E, Casadesús J, Vogel J. 2015. Small RNA-based feedforward loop with AND-gate logic regulates extrachromosomal DNA transfer in *Salmonella*. *Proc Natl Acad Sci U S A* **112**:E4772–E4781.

80. Jung YS, Kwon YM. 2008. Small RNA ArrF regulates the expression of *sodB* and *feSII* genes in *Azotobacter vinelandii*. *Curr Microbiol* **57**:593–597.

81. Moshiri F, Kim JW, Fu C, Maier RJ. 1994. The FeSII protein of *Azotobacter vinelandii* is not essential for aerobic nitrogen fixation, but confers significant protection to oxygen-mediated inactivation of nitrogenase *in vitro* and *in vivo*. *Mol Microbiol* **14**:101–114.

82. Muriel-Millán LF, Castellanos M, Hernandez-Eligio JA, Moreno S, Espín G. 2014. Posttranscriptional regulation of PhbR, the transcriptional activator of polyhydroxybutyrate synthesis, by iron and the sRNA ArrF in *Azotobacter vinelandii*. *Appl Microbiol Biotechnol* **98**:2173–2182.

83. Ceizel Borella G, Lagares A Jr, Valverde C. 2016. Expression of the *Sinorhizobium meliloti* small RNA gene *mmgR* is controlled by the nitrogen source. *FEMS Microbiol Lett* **363**:fnw069.

84. Lagares A Jr, Ceizel Borella G, Linne U, Becker A, Valverde C. 2017. Regulation of polyhydroxybutyrate accumulation in *Sinorhizobium meliloti* by the *trans*-encoded small RNA MmgR. *J Bacteriol* **199**:e00776-16.

85. Ow DW, Sundaresan V, Rothstein DM, Brown SE, Ausubel FM. 1983. Promoters regulated by the *glnG* (*ntrC*) and *nifA* gene products share a heptameric consensus sequence in the −15 region. *Proc Natl Acad Sci U S A* **80**:2524–2528.

86. Ceizel Borella G, Lagares A Jr, Valverde C. 2018. Expression of the small regulatory RNA gene *mmgR* is regulated negatively by AniA and positively by NtrC in *Sinorhizobium meliloti* 2011. *Microbiology* **164**:88–98.

87. Jäger D, Sharma CM, Thomsen J, Ehlers C, Vogel J, Schmitz RA. 2009. Deep sequencing analysis of the *Methanosarcina mazei* Gö1 transcriptome in response to nitrogen availability. *Proc Natl Acad Sci U S A* **106**: 21878–21882.

88. Buddeweg A, Sharma K, Urlaub H, Schmitz RA. 2018. sRNA$_{41}$ affects ribosome binding sites within polycistronic mRNAs in *Methanosarcina mazei* Gö1. *Mol Microbiol* **107**:595–609.

89. Kopf M, Klähn S, Scholz I, Hess WR, Voß B. 2015. Variations in the non-coding transcriptome as a driver of inter-strain divergence and physiological adaptation in bacteria. *Sci Rep* **5**:9560.

90. Kopf M, Klähn S, Pade N, Weingärtner C, Hagemann M, Voß B, Hess WR. 2014. Comparative genome analysis of the closely related *Synechocystis* strains PCC 6714 and PCC 6803. *DNA Res* **21**:255–266.

91. Giner-Lamia J, Robles-Rengel R, Hernández-Prieto MA, Muro-Pastor MI, Florencio FJ, Futschik ME. 2017. Identification of the direct regulon of NtcA during early acclimation to nitrogen starvation in the cyanobacterium *Synechocystis* sp. PCC 6803. *Nucleic Acids Res* **45**: 11800–11820.

92. Klähn S, Schaal C, Georg J, Baumgartner D, Knippen G, Hagemann M, Muro-Pastor AM, Hess WR. 2015. The sRNA NsiR4 is involved in nitrogen assimilation control in cyanobacteria by targeting glutamine synthetase inactivating factor IF7. *Proc Natl Acad Sci U S A* **112**: E6243–E6252.

93. Golden JW, Yoon HS. 2003. Heterocyst development in *Anabaena*. *Curr Opin Microbiol* **6**:557–563.

94. Wright PR, Georg J, Mann M, Sorescu DA, Richter AS, Lott S, Kleinkauf R, Hess WR, Backofen R. 2014. CopraRNA and IntaRNA: predicting small RNA targets, networks and interaction domains. *Nucleic Acids Res* **42**(Web Server issue):W119–W123.

95. Wright PR, Richter AS, Papenfort K, Mann M, Vogel J, Hess WR, Backofen R, Georg J. 2013. Comparative genomics boosts target prediction for bacterial small RNAs. *Proc Natl Acad Sci U S A* **110**:E3487–E3496.

96. Urban JH, Vogel J. 2009. A green fluorescent protein (GFP)-based plasmid system to study post-transcriptional control of gene expression in vivo. *Methods Mol Biol* **540**:301–319.

97. Urban JH, Vogel J. 2007. Translational control and target recognition by *Escherichia coli* small RNAs *in vivo*. *Nucleic Acids Res* **35**:1018–1037.

98. Ames TD, Breaker RR. 2011. Bacterial aptamers that selectively bind glutamine. *RNA Biol* **8**:82–89.

99. Ren A, Xue Y, Peselis A, Serganov A, Al-Hashimi HM, Patel DJ. 2015. Structural and dynamic basis for low-affinity, high-selectivity binding of l-glutamine by the glutamine riboswitch. *Cell Rep* **13**:1800–1813.

100. Yan Y, Yang J, Dou Y, Chen M, Ping S, Peng J, Lu W, Zhang W, Yao Z, Li H, Liu W, He S, Geng L, Zhang X, Yang F, Yu H, Zhan Y, Li D, Lin Z, Wang Y, Elmerich C, Lin M, Jin Q. 2008. Nitrogen fixation island and rhizosphere competence traits in the genome of root-associated *Pseudomonas stutzeri* A1501. *Proc Natl Acad Sci U S A* **105**:7564–7569.

101. Yu H, Yuan M, Lu W, Yang J, Dai S, Li Q, Yang Z, Dong J, Sun L, Deng Z, Zhang W, Chen M, Ping S, Han Y, Zhan Y, Yan Y, Jin Q, Lin M. 2011. Complete genome sequence of the nitrogen-fixing and rhizosphere-associated bacterium *Pseudomonas stutzeri* strain DSM4166. *J Bacteriol* **193**:3422–3423.

102. Bentzon-Tilia M, Severin I, Hansen LH, Riemann L. 2015. Genomics and ecophysiology of heterotrophic nitrogen-fixing bacteria isolated from estuarine surface water. *mBio* **6**:e00929.

103. Yan Y, Lu W, Chen M, Wang J, Zhang W, Zhang Y, Ping S, Elmerich C, Lin M. 2013. Genome transcriptome analysis and functional characterization of a nitrogen-fixation island in root-associated *Pseudomonas stutzeri*, p 851–863. *In* de Bruijn FJ (ed), *Molecular Microbial Ecology of the Rhizosphere*. John Wiley & Sons, Inc, New York, NY.

104. Zhan Y, Yan Y, Deng Z, Chen M, Lu W, Lu C, Shang L, Yang Z, Zhang W, Wang W, Li Y, Ke Q, Lu J, Xu Y, Zhang L, Xie Z, Cheng Q, Elmerich C, Lin M. 2016. The novel regulatory ncRNA, NfiS, optimizes nitrogen fixation via base pairing with the nitrogenase gene *nifK*

mRNA in *Pseudomonas stutzeri* A1501. *Proc Natl Acad Sci U S A* **113**:E4348–E4356.

105. **Busch A, Richter AS, Backofen R.** 2008. IntaRNA: efficient prediction of bacterial sRNA targets incorporating target site accessibility and seed regions. *Bioinformatics* **24**:2849–2856.

106. **Prasse D, Förstner KU, Jäger D, Backofen R, Schmitz RA.** 2017. sRNA$_{154}$ a newly identified regulator of nitrogen fixation in *Methanosarcina mazei* strain Gö1. *RNA Biol* **14**:1544–1558.

107. **Sharma CM, Vogel J.** 2014. Differential RNA-seq: the approach behind and the biological insight gained. *Curr Opin Microbiol* **19**:97–105.

108. **Hervás AB, Canosa I, Little R, Dixon R, Santero E.** 2009. NtrC-dependent regulatory network for nitrogen assimilation in *Pseudomonas putida*. *J Bacteriol* **191**: 6123–6135.

109. **Baumgartner D, Kopf M, Klähn S, Steglich C, Hess WR.** 2016. Small proteins in cyanobacteria provide a paradigm for the functional analysis of the bacterial microproteome. *BMC Microbiol* **16**:285.

110. **Miseta A, Csutora P.** 2000. Relationship between the occurrence of cysteine in proteins and the complexity of organisms. *Mol Biol Evol* **17**:1232–1239.

111. **Prasse D, Thomsen J, De Santis R, Muntel J, Becher D, Schmitz RA.** 2015. First description of small proteins encoded by spRNAs in *Methanosarcina mazei* strain Gö1. *Biochimie* **117**:138–148.

112. **Cassidy L, Prasse D, Linke D, Schmitz RA, Tholey A.** 2016. Combination of bottom-up 2D-LC-MS and semi-top-down GelFree-LC-MS enhances coverage of proteome and low molecular weight short open reading frame encoded peptides of the archaeon *Methanosarcina mazei*. *J Proteome Res* **15**:3773–3783.

Regulating with RNA in Bacteria and Archaea
Edited by Gisela Storz and Kai Papenfort
© 2018 American Society for Microbiology, Washington, DC
doi:10.1128/microbiolspec.RWR-0010-2017

Bacterial Iron Homeostasis Regulation by sRNAs

16

Sylvia Chareyre[1] and Pierre Mandin[1]

INTRODUCTION

Iron: Life and Death Metal

Iron is one of the most abundant elements on earth. Due to its chemical properties, in particular its redox potential, it was used as a cofactor in a large number of proteins since the emergence of life. Before the appearance of an oxidative atmosphere, iron was found mainly in its reduced, ferrous form (Fe^{2+}). Fe^{2+} is typically the bioreactive form of iron that is found in proteins, as an isolated ion, in the center of porphyrin to form heme, or in coordination with sulfur atoms to constitute so-called Fe-S cluster cofactors (1–3). Bacteria contain many iron-using proteins involved in a plethora of reactions, mainly, but not limited to, aerobic and anaerobic respiration, the tricarboxylic acid (TCA) cycle, photosynthesis, N_2 fixation, and DNA biosynthesis.

Despite their ubiquity and usefulness, the utilization of iron cofactors poses two main problems to living organisms, especially in an oxidative atmosphere: (i) The oxidated form of iron, ferric iron (Fe^{3+}), is mostly insoluble at neutral pH and has to be reduced before being incorporated into proteins and used in biochemical reactions. To circumvent this problem, living organisms produce siderophores that scavenge Fe^{3+} from the environment, the metal being reduced upon its entry (4). However, this provokes a struggle for iron in between species that share the same ecological niches. For instance, eukaryotic hosts starve bacterial cells for iron as a defense mechanism by producing iron-binding proteins such as lactoferrin or transferrin (5). (ii) H_2O_2, which is adventitiously produced by aerobic respiration, can react with ferric iron via the Fenton or Haber-Weiss reaction and produce reactive oxygen species that can cause damage to biomolecules and eventually provoke cell death (6–8). Iron homeostasis is thus key in maintaining a sufficient iron pool to sustain the production of iron-utilizing proteins while avoiding excess iron to generate reactive oxygen species. Bacteria have thus developed exquisite ways to control iron homeostasis, particularly through the control of gene expression.

RyhB, the Fur-Sparing Partner

In many bacteria, iron homeostasis is controlled mainly by the master transcriptional regulator named Fur (9, 10). Fur is an Fe^{2+}-binding metalloprotein that acts as classical homeostatic regulator: when bound to iron, Fur binds DNA at so-called Fur boxes and represses expression of close to 100 genes involved mainly in iron uptake (e.g., siderophores) (11, 12). This prevents unnecessary iron accumulation and toxicity. When iron concentrations lower, Fur dissociates from DNA and repression is abolished, allowing for import of Fe in the cell.

While Fur is chiefly a transcriptional repressor of gene expression, early studies in *Escherichia coli* found that expression of certain genes was repressed in *fur* mutants, raising the possibility that Fur could act as an activator in certain cases (13, 14). This phenomenon was clarified in the early 2000s when the Gottesman lab found that a small RNA (sRNA) named RyhB, itself repressed by Fur, was actually responsible for the downregulation of these genes, and many others, during iron starvation (15). Since then RyhB has been extensively studied, in large part thanks to the pioneering work of Eric Massé's laboratory, and RyhB has been shown to play a key role in iron homeostasis by controlling >50 genes involved in iron utilization and import, thereby establishing an "iron-sparing response." This concerted action, together with Fur, remodels iron

[1]Aix Marseille Université-CNRS, Institut de Microbiologie de la Méditéranée, Laboratoire de Chimie Bactérienne, Marseille 13009, France.

content and utilization in the cell, helping the cell cope with iron scarcity.

RyhB uses multiple posttranscriptional mechanisms to exert its function, either by repressing gene expression through translation inhibition and mRNA destabilization or, inversely, by promoting the expression of certain of its targets. In fact, the amount of information gathered over the years on RyhB's modes of action and physiological roles has defined it as a "paradigmatic" sRNA that exemplifies general principles of sRNA regulation. While RyhB has been mainly studied in *E. coli*, homologs of RyhB that are also Fur regulated are found in other enterobacteria. Furthermore, other sRNAs that function in a parallel manner by regulating iron homeostasis have now also been found in more-distant species. In these bacteria, RyhB homologs and functional analogs play major roles not only in the iron starvation response but also in other processes such as virulence and photosynthesis.

ONE sRNA, MANY MODES OF ACTION

The first studies on RyhB brought to light what is now considered as the canonical mode of action for repression by sRNAs. Indeed, soon after its discovery, overexpression of this 90-nucleotide (nt) sRNA was found to induce the disappearance of the mRNAs of various target genes, among them the mRNAs encoding the iron-containing superoxide dismutase (*sodB*) and succinate dehydrogenase (*sdh*) (15, 16). The first step of this mechanism involves the base-pairing of the sRNA to the mRNA by imperfect complementarity (Fig. 1A). Like many other sRNAs, RyhB base-pairing to its mRNA targets, as well as its stability, requires the RNA chaperone Hfq (17, 18). Base-pairing usually happens in the translation initiation region of the target mRNA, in a region of ~50 nt encompassing the ribosome binding site (RBS) and the translation start codon. As with many other Hfq-dependent sRNAs, base-pairing of RyhB to this region leads to inhibition of translation initiation through ribosome occlusion (19, 20). Consequently to this translation inhibition, RyhB recruits the RNase E and the RNA degradosome, inducing the degradation of the target mRNA and of the sRNA in a stoichiometric manner, without affecting transcription of the gene (16, 21). It is thought that in this way, if iron becomes available again, the fast clearance of RyhB through its degradation will ensure rapid *de novo* synthesis of its targets.

Remarkably, for at least three mRNA targets of RyhB, initial cleavage by RNase E happens distantly from the base-pairing region (e.g., >350 nt in the case of *sodB* mRNA) (Fig. 1A) (20). Distal cleavage is most likely dictated by the absence of ribosomes, consequent to translation inhibition, and structural accessibility of the mRNA to the degradosome, although precise requirements are not completely elucidated. Such a mechanism would likely ensure that translating ribosomes finish translation before degrading the mRNA, avoiding stalled ribosomes and incomplete proteins.

In the vast majority of the cases studied thus far, RyhB repression occurs on the first gene of an operon, affecting the ensemble of the genes encoded downstream of the base-pairing site on the mRNA. However, in the case of *iscRSUA*, RyhB binds inside this polycistronic mRNA, provoking the degradation of its 3′ part while the uncleaved 5′ part remains stable (Fig. 1B) (22). The presence of a rich stem-loop structure that resembles iron-responsive elements found in eukaryotic RNAs is thought to be responsible for the protection of the 5′ part of the mRNA against degradation. In this way, RyhB uncouples the regulation of this particular set of genes, blocking the synthesis of an Fe-S cluster biogenesis machinery, while allowing for the transcriptional regulator encoded by the first gene of the operon to exert its function (see below).

While most RyhB targets are negatively regulated (see Table 1), RyhB is also able to promote expression of other genes. Only two such targets have been thoroughly studied as yet: *shiA*, which encodes the transporter of shikimate, a precursor in siderophore production; and *cirA*, encoding a TonB-dependent transporter of certain Fe^{3+}-bound siderophores (23, 24). In absence of RyhB, the *shiA* mRNA forms a stem-loop structure in its 5′ untranslated region that masks the RBS and hinders ribosome binding (Fig. 1C) (23). RyhB base-pairs to the 5′ part of this stem-loop structure, thereby opening it and allowing for ribosomes to access and initiate translation of *shiA*. Remarkably, this case of positive regulation is very similar to what was previously described for the sRNAs activating the expression of RpoS (25, 26).

Activation of *cirA* translation involves yet another novel mechanism implicating Hfq. Indeed, in the absence of RyhB, Hfq binds the *cirA* mRNA and prevents 30S ribosomal subunit binding (Fig. 1D) (24). This results in inhibition of *cirA* translation initiation and ultimately degradation of the unprotected mRNA. Binding of RyhB to the *cirA* mRNA promotes a change in the structure that displaces Hfq from the mRNA. This improves binding of the ribosome to the mRNA and ultimately activates translation of *cirA*.

Finally, a recent study has shown that a "decoy" target of RyhB serves to modulate its activity (27). Using

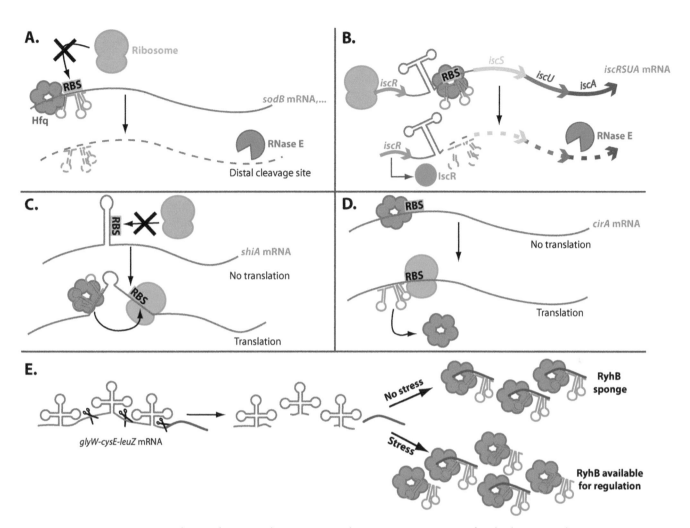

Figure 1 RyhB regulatory mechanisms. (A) RyhB represses expression of multiple mRNAs by inhibiting translation initiation and inducing mRNA degradation. RyhB base-pairing blocks ribosome attachment to the RBS. Consequently, the mRNA is degraded by RNase E recruitment at sites that can be distant from the base-pairing region. (B) RyhB promotes the degradation of the *iscRSUA* transcript by base-pairing to the translation initiation region of *iscS* while the 5′ part of the mRNA, encoding *iscR*, remains stable and is translated. (C) RyhB positively regulates *shiA* expression by opening a stem-loop structure that otherwise inhibits ribosome attachment to the RBS. (D) RyhB activates translation of *cirA* by displacing Hfq, which otherwise blocks ribosome attachment. (E) RyhB's activity can be modulated thanks to the 3′ external transcribed spacer of the *glyW-cysE-leuZ* tRNA (in purple).

an MS2 affinity purification to find RNAs interacting with sRNAs, it was shown that RyhB base-pairs to the 3′ part of the external transcribed spacer of the *glyW-cysT-leuZ* tRNA precursor (Fig. 1E). The reason for such an *a priori* "odd" target is that it would serve as a "sponge" to absorb excess of the sRNA in undesired conditions, for instance, transcriptional noise of the *PryhB* promoter. Whether this interaction may have a function in other conditions remains to be elucidated. Nevertheless, what is true for RyhB may also be a general feature of sRNA regulation, as the same sponge

phenomenon was found to be true for another sRNA, RybB, in the same study (27).

THE IRON-SPARING RESPONSE

Studies over the years have led to the current view that the main role of RyhB is to allow the cell to cope with low-iron conditions. Indeed, plasmidic expression of RyhB in normal conditions leads to an accumulation of free iron in the cell cytoplasm, as seen by electron paramagnetic resonance (28). Likewise, in iron-poor

Table 1   Summary of validated and putative RyhB targets using different set of data[a]

Family	Operon name	Description	Cofactor	Reg.[b]	1[c]	2[c]	3[c]	4[c]	5[c]	6[c]	Exp. Valid.
TCA cycle and aerobic respiration	*acnA*	Aconitate hydrase A	Fe-S	Neg.	+	+		+			(15)
	acnB	Aconitate hydrase B	Fe-S	Neg.	+	+		+	+		(84)
	sdhCDAB	Succinate dehydrogenase	Fe-S/Heme	Neg.	+	C	+	+	CB	DA	(15)
	cydAB	Cytochrome d oxydase	Heme	Neg.	+				A		(85)
	nuoABCCEFGHIJKLMN	NADH dehydrogenase I	Fe-S	Neg.	+	A		+	ACFG		
	fdoGHI	Formate dehydrogenase	Fe-S/ heme/Fe	Neg.	+	G		+			
	fumAC	Fumarase A/Fumarase C	Fe-S	Neg.	+	+	+	+	+	+	(15)
	fumB	Fumarase B	Fe-S	Neg.	+		+	+	+	A	(31)
Anaerobic respiration and metabolism	*frdABCD*	Fumarate reductase	Fe-S	Neg.	+	A		AB	A	A	
	hybOABCDEFG	Hydrogenase II	Fe-S/Ni-Fe/Fe/Ni	Neg.	+	O		OA	O		
	pflA	Pyruvate formate lyase activating enzyme	Fe-S	Neg.	+	+					
	dmsABC	Dimethyl sulfoxide reductase	Fe-S	Neg.	+	A		A	A	A	(31)
	nirBDC-cysG	Nitrite reductase	Fe-S/heme	Neg.	+	B		B	B	B	(33)
	hypABCDE-fhlA	Hydrogenase III	Fe-S/Ni-Fe	Neg.	+	A		B	BE	B	
	narGHJI	Nitrate reductase A	Fe-S/Heme	Neg.	+	G		G	G		
	nrfABCDEFG	Periplasmic nitrite reductase	Fe-S/Heme	Neg.	+	A		A	F		(31)
	napFDAGHBC	Periplasmic nitrate reductase	Fe-S/Heme	Neg.	+		+	+			(31)
Fe/S clusters biogenesis	*iscRSUA*	Fe-S cluster biosynthesis machinery	Fe-S	Neg.	+	S	A	S	S		(22)
	hscAB-fdx-iscX	Fe-S cluster biosynthesis chaperone	Fe-S	Neg.	+			+	A-fdx		
	mrp/apbC	Fe-S cluster formation in *Salmonella*	Fe-S	Neg.	+	+		+	+		
	grxD	Glutaredoxin 4	Fe-S	Neg.	+		+				(27)
	erpA	Fe-S cluster A-type carrier	Fe-S	Neg.	+						(26)
	sufABCDSE	Fe-S cluster biosynthesis machinery	Fe-S	Neg.	+	+		+	B	B	
Fe uptake	*shiA*	Shikimate transporter		Pos.	+	+	+	+	+		(23)
	cirA	Catecholate receptor		Pos.	+		+	+	+		(29)
	bfr	Bacterioferritin	Fe	Neg.	+	+	+	+			(15)
	cysE	Serine acetyltransferase		Neg.	+			+	+		(29)
	fepB	Enterobactin periplasmic binding protein		Neg.	+			+			(31)

Category	Protein	Gene	Cofactor	Regulation	[1]	[2]	[3]	[4]	[5]	[6]	Ref
Heme biosynthesis	Ferrochelatase	hemH	Heme	Neg.							(85)
	Porphobilinogen synthase	hemB	(Zn)	Neg.							(85)
	Coproporphyrinogen III dehydrogenase	hemN	Fe-S	Neg.						+	(15)
Oxidative stress	Iron superoxide dismutase	sodB	Fe	Neg.		+	+	+	+		(36)
	Methionine sulfoxide reductase	msrB	Fe	Neg.		+	+	+			(31)
	Hydroperoxidase	katG	Heme	Neg.			+	+	C		
Metabolism	Putrescine utilization pathway	puuDRCBE		Neg.		C	C				
	Methionine synthase	metH		Neg.		+	+				
	Dihydroxyacetone kinase	dhaKLM	(Mg)	Neg.		+	+				(31)
Motility	Flagella machinery	flgBCDEFGHIJ		Pos.		B	G	A			
	RNA polymerase, sigma 28 factor	fliAZY		Pos.		A	A				
	Flagellar filament structural protein	fliC		Pos.		+	+				
Other	Peptidase B	pepB	(Mn)	Neg.		+	+				
	Metalloprotease	loiP/yggG	(Zn)	Neg.		+	+	+			
	Multicopper oxidase	cueO	(Cu)	Neg.		+	+				
	DNA-binding transcript. regulator	marRAB		Neg.		A	B	B		A	(33)
	Glutamate synthase	gltBDF	Fe-S	Neg.		B	B	B			
Unknown	Serine sensitivity enhancing B	sseB		Neg.		+	+				
	PQQ domain-containing protein	yncE		Pos.		+	+				
	Radical SAM superfamily protein	ygiQ	Fe-S	Neg.		+	+	+			
	Unknown	yeaC		Neg.		+	+				
	Inner membrane protein	ygdQ		Pos.		+	+				
	Putative selenate reductase	ynfEFGGH-dmsD	Fe-S	Neg.		E	F				(31)
	Predicted FAD-linked oxidoreductase	ydiJ-menI-ydiH	Fe-S	Neg.		ydiJ	ydiJ				
	Putative zinc- or iron-chelating protein	ykgJ	Fe-S/Fe	Neg.		+	+				(31)
	Putative pyruvate-flavodoxin oxidoreductase	ydbK/pfo-ompN	Fe-S	Neg.		+	+				
tRNA sponge	Actin family protein	yegD		Neg.			+				(31)
	Glycine, cysteine, leucine tRNA	glyW-cysT-leuZ		Neg.		+	leuZ				(27)

[a] We used available data in the literature to determine the ensemble of RyhB targets. The following set of experiments were used: [1] A transcriptomic approach using RyhB over-expression (30). [2] Fur-CHIP experiments combined with RNA sequencing on WT and ryhB mutant strains (11). [3] An *in vivo* identification of RNAs directly binding RyhB (MAPS) (27). [4] A ribosome profiling approach (Ribo-seq) in cells transiently expressing RyhB (31). [5] A recent global approach on sRNA-mRNA interactants associated with Hfq (32). [6] A sRNA targets prediction program (COPRA RNA) (33).

[b] Regulation (Neg.) and (Pos.) represent genes down-regulated and up-regulated by RyhB, respectively.

[c] + indicates that the target is found in the publication, letters specify which genes of the operon were found in the corresponding study, when applicable.

conditions, wild-type (WT) cells contain 3 times more iron than Δ*ryhB* mutants (29). To date, three main mechanisms have been shown to contribute to this iron-sparing response orchestrated by RyhB: (i) limitation of iron-utilizing proteins, (ii) coordination of Fe-S biogenesis systems, and (iii) promotion of siderophore production (Fig. 2).

RyhB Limits Production of Fe-Utilizing Proteins

To gain a better understanding of the RyhB response, we have here tried to define the ensemble of RyhB targets identified in the literature. An initial transcriptomic approach had identified 56 genes encoded by 20 mRNAs as targets of RyhB (30). This was done using short expression times of the sRNA to avoid including indirect targets as part of the RyhB regulon (e.g., through regulation of a transcription factor). Since then, global approaches either directly on RyhB or on Hfq-binding sRNAs have greatly helped increase the number of potential RyhB targets, although experimental validation of direct interaction (e.g., base-pairing) is mostly missing. We have done a survey on potential RyhB targets using the following set of data (Table 1): [1] the initial transcriptomic approach using short times of RyhB overexpression (30); [2] a recent set of experiments combining Fur-ChIP (chromatin immunoprecipitation) and RNA sequencing on WT and *ryhB* mutant strains, performed in aerobic and anaerobic conditions (11); [3] an *in vivo* identification of RNAs directly binding RyhB using MS2-tagged RyhB affinity purification coupled with RNA sequencing (MAPS) (27); [4] a ribosome profiling approach (Ribo-seq) in cells transiently expressing RyhB (31); [5] a recent global approach on sRNA-mRNA interactants associated with Hfq (RNA interaction by ligation and sequencing [RIL-seq]) (32); and [6] an sRNA targets prediction program that integrates phylogenetic data (COPRA RNA [Comparative Prediction Algorithm for sRNA targets]) (33). In order to have the most accurate picture of the RyhB regulon, we only took into account mRNA targets that were found in at least two approaches and/or that were verified experimentally.

Our survey reveals a total of 56 confirmed or very likely mRNA targets of RyhB, for a total of 143 genes,

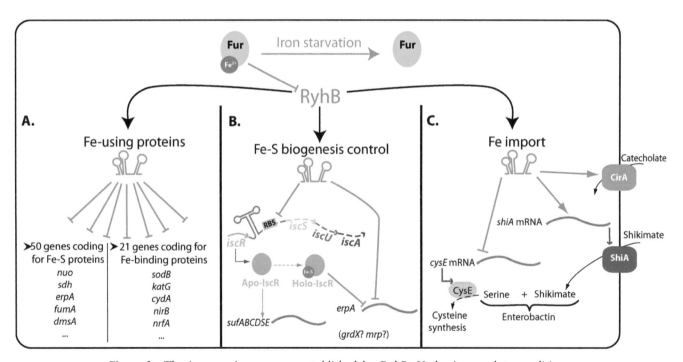

Figure 2 The iron-sparing response established by RyhB. Under iron-replete conditions, Fur-Fe^{2+} represses *ryhB* expression. During iron starvation, Fur repression is abolished and RyhB is rapidly expressed. RyhB mediates an Fe-sparing response through three mechanisms: (A) RyhB represses the expression of mRNAs coding for iron-using proteins; (B) RyhB, together with IscR, orchestrates Fe-S biogenesis systems through regulation of the Isc machinery and *erpA* expression; and (C) RyhB promotes Fe uptake via the upregulation of *shiA* and *cirA* and repression of the *cysE* gene, which leads to serine accumulation used for enterobactin production.

half of which (29 mRNAs) have been validated experimentally. Thus, the RyhB "targetome" may be almost as large as the Fur regulon, strengthening the importance of the sRNA in controlling Fe homeostasis. The majority of RyhB targets are negatively regulated, with only 7 mRNAs probably activated (2 experimentally validated). This may be a consequence of an intrinsic feature of sRNA regulation or may be a bias of the approaches we have considered, which are mostly based on measurements of mRNA levels.

About two-thirds of RyhB targets encode proteins using iron as a cofactor (35 mRNAs encoding a total of 71 iron-containing proteins), Fe-S clusters being the most represented cofactor (28 mRNAs encoding 50 Fe-S cluster-containing proteins). These are all repressed by RyhB, which very much strengthens the idea that RyhB's main role is to repress expression of iron-consuming proteins in order to "preserve" newly imported iron for essential processes (Fig. 2A). As has been noted before, RyhB repression is directed mostly against nonessential Fe-using proteins, redirecting the available iron toward essential processes (30). For instance, mRNAs encoding the Fe-containing ribonucleotide reductase, implicated in DNA synthesis, or IspG/H, two proteins involved in synthesis of the isoprenoid precursor and which are reported to be the only essential Fe-S clusters containing proteins, are not repressed by RyhB (28, 34). However, there are a few repressed targets of RyhB that are essential. For instance, it is the case for *cysE*, involved in cysteine biosynthesis, or *erpA*, an effector of Fe-S biogenesis (29, 35). In both cases, downregulation by RyhB is clearly not as strong as with other targets (such as *sodB* or *fumA*), indicating that repression may be partial to limit levels of these proteins enough to decrease Fe consumption without drastically compromising their activity.

Not surprisingly, most of these targets encode proteins mainly involved in the TCA cycle and aerobic or anaerobic respiration, which are metabolic processes that rely heavily on iron cofactors, in particular Fe-S clusters (Table 1). Other general metabolic functions are also repressed by RyhB. Presumably these actions lead to a major metabolic switch from respiration to fermentation during iron starvation, but this has not been thoroughly characterized thus far. The other major classes of proteins regulated by RyhB belong either to Fe-S clusters and heme biogenesis or to iron import, which are crucial for the redistribution of iron in the cell (see below).

RyhB also targets three mRNAs encoding enzymes that counteract oxidative stress, i.e., the peroxidase SodB, the catalase KatG, and the methionine sulfoxide reductase MsrB, which were all confirmed experimentally (16, 31, 36). Repression of these enzymes by RyhB during iron shortage likely serves to redirect iron usage toward non-iron-utilizing proteins that serve the same functions. This is the case at least for *sodB*: while this mRNA that encodes a superoxide dismutase (SOD) that uses iron is repressed by RyhB during iron starvation, the Mn-using SOD encoded by *sodA*, normally repressed by Fur, replaces it in the same conditions (37).

Finally, two of these global approaches found that RyhB may also target motility genes encoding the flagella machinery and activate their expression (11, 32). While this potential regulation has not been investigated thus far, it is tempting to imagine that it would allow the bacteria to swim toward iron-rich environments.

RyhB Orchestrates Fe-S Biogenesis

Fe-S clusters constitute one of the major pools of Fe in the cells. In *E. coli*, two systems are responsible for the biogenesis of Fe-S clusters: the housekeeping Isc system and the stress-responsive Suf system (38, 39). This second system was found to be essential during iron starvation, although the molecular explanation for this is still partly obscure (40). One tempting hypothesis is that Suf is a "frugal" biogenesis system, either by having a generally slower or diminished activity as compared to Isc or by displaying a preference toward the maturation of essential or more-needed targets. In any case, transition from the Isc to the Suf machinery is a fundamental process of the iron starvation response in which, again, RyhB plays an important role.

The transcriptional regulator IscR is essential to activate Suf expression (38, 41, 42). IscR is itself an Fe-S-containing protein that is encoded by the first gene of the *iscRSUA* operon. In its holo form, IscR represses synthesis of the *isc* operon, acting like a homeostatic regulator of Fe-S biogenesis. Remarkably, *in vivo* maturation of IscR is achieved by the Isc system, the Suf system being inefficient at maturating this protein. Thus, when Isc Fe-S production decreases, IscR is found mostly in its apo form, in which case it will bind and activate the expression of Suf (42). It is noteworthy that *suf* is also under the repression of Fur, thus ensuring that it is not produced when Fe is available (43).

As described above, RyhB base-pairs to the *iscRSUA* mRNA, just upstream of the *iscS* gene, and induces the degradation of the 3′ part of the mRNA, encoding the Isc machinery, while the 5′ part, encoding IscR, remains stable (Fig. 1B) (22). In this way, RyhB likely decreases Isc activity and promotes synthesis and accumulation of apo-IscR, which in turn activates expression of Suf (Fig. 2B). Remarkably, RyhB is not essential in activating the expression of Suf during iron starvation. Indeed, decrease in iron concentration will eventually lead to accumulation of apo-IscR, which will activate Suf expression. Thus, as proposed earlier, the role of RyhB in this case may be to accelerate the switch from Isc to Suf, although precise dynamic measurements of the expression of the systems following stress are still missing to validate this hypothesis (22).

RyhB also organizes Fe-S clusters maturation routes together with IscR. After synthesis by the Isc or Suf systems, Fe-S clusters are carried by A-type proteins (ATCs) that transfer the formed clusters to apoprotein targets. *E. coli* has four ATCs, among which ErpA is essential in aerobic conditions (34, 39). Overexpression of RyhB was recently shown to repress the expression of *erpA* (35, 44). Remarkably, *erpA* is also repressed transcriptionally by the holo, Fe-S cluster-bound, form of IscR. This double regulation seemed somewhat counterintuitive, as it would predict that *erpA* expression would be repressed both in the presence of iron (by holo-IscR) and in the absence of iron (by RyhB). The answer to this apparent paradoxical situation came by looking at expression of *erpA* at different iron concentrations. Indeed, while IscR and RyhB effectively repress expression of *erpA* at high and low iron concentrations, respectively, both repressions are alleviated at intermediate concentrations, allowing for full *erpA* synthesis. In this way, RyhB, together with IscR, allows for expression of the ATC in defined conditions where other ATCs are either absent or nonfunctional (35). Unfortunately, such responses at intermediate Fe concentrations have been only poorly looked at. It would thus be of interest to have a look at less drastic Fe starvation conditions than those commonly used in the laboratory in order to better mimic what is probably happening in nature.

In any case, RyhB's role in Fe-S cluster biogenesis is certainly more important than currently thought. Indeed, in addition to *isc* and *erpA*, our survey indicates that least three other targets that play a role in this pathway are also probably regulated by RyhB (Table 1): *grxD* and *mrp*, which have been shown to participate in biogenesis in other species but whose function is still unclear in *E. coli*; and also the *suf* operon, encoding the Suf system (45). Perhaps most intriguing is the potential regulation of *suf* by RyhB. This would be surprising as Suf is reported to be essential during iron starvation. Conceivably, like *erpA*, regulation of *suf* expression would allow a fine-tuning of Fe-S biogenesis system usage. In line with this, *suf* is also regulated by IscR (42). However, in this case apo-IscR directly activates expression at the *suf* promoter, as opposed to an alle-

viation or repression in the case of *erpA*. Future research on Fe-S cluster biogenesis regulation by RyhB will certainly be key for our comprehension of Fe-S cluster homeostasis.

RyhB Promotes Iron Import

A last role for RyhB in the iron-sparing paradigm is to promote iron entry in the cell (Fig. 2C). As iron is mostly found in the ferric Fe^{3+} form in nature, Fe^{3+}-binding siderophores constitute the major mode of iron import. Enterobactin is the main siderophore of *E. coli*. Its production begins by the conversion of shikimate to chorismate and then 2,3-dihydroxybenzoic acid (DHB) by the action of the enterobactin-producing proteins EntC, EntB, and EntA. DHB is assembled with serine into enterobactin by EntD, EntB, EntE, and EntF. The siderophore is then exported outside the cell by the EntS-TolC complex. Finally, Fe^{3+}-bound siderophores are then reimported inside the cells through a TonB-energized system, and the siderophore molecule is hydrolyzed concomitantly to the reduction of Fe^{3+} that can then be incorporated into proteins.

As discussed above, the only two validated positive targets of RyhB encode proteins involved in entry of the shikimate siderophore precursor (ShiA) and in the energization of siderophore entry (CirA) (Fig. 1C and D and 2C) (23, 24). RyhB has also been shown to play a role in the synthesis of siderophores by activating the transcription of the *entCEBAH* operon (29). However, the mechanism for this induction is not yet elucidated and is most likely indirect.

RyhB has also been shown to favor enterobactin production in another indirect manner. Serine, which is important for the synthesis of enterobactin, is also used as a precursor in the synthesis of cysteine. Interestingly, RyhB was shown to inhibit the translation of CysE, a serine acetyltransferase that uses serine for the biogenesis of cysteine. In this way, RyhB would redirect usage of serine from cysteine production to enterobactin synthesis, thus promoting siderophore production (Fig. 2C).

OTHER BACTERIA, OTHER sRNAs

RyhB Homologs

RyhB homologs have been described in multiple species, mostly enterobacteria. To date, all RyhB homologs that have been identified are regulated by Fur, clearly pointing to a major role of these sRNAs in iron homeostasis regulation. However, homologs display interesting particularities, with RyhB having a role in virulence or biofilm formation in some species, for instance (Table 2).

In uropathogenic *E. coli* (UPEC), a *ryhB* mutant was shown to be affected in murine urinary tract colonization (46) (Table 2). The explanation of this phenotype was a defect in siderophore production. RyhB from *Shigella* is identical in sequence to that of *E. coli*, and given that it shares the same regulator, it regulates a common set of targets (47, 48). In addition to that, *Shigella flexneri* RyhB is involved in adaptation to acid stress through the control of *evgA*, a gene encoding a regulator of extreme acid resistance (48). This trait is likely related to iron availability, as Fe^{3+} is more soluble in acidic environments. In this way, RyhB would promote virulence of *S. flexneri* by helping adaptation of the bacteria to the acidic environments encountered in the stomach during infection.

RyhB is more directly involved in virulence in *Shigella dysenteriae*, where it represses the expression of *virB*, a gene coding for the activator of several virulence-associated traits such as type 3 secretion system pro-

duction and actin-based motility (47, 49). However, for both *virB* and *evgA*, how repression is achieved and whether it is through direct pairing remain elusive.

RyhB is also found in *Vibrio* species, albeit less conserved and larger in size (~200 nt) (50, 51). In addition to controlling iron homeostasis genes, *Vibrio cholerae* RyhB also controls genes involved in motility, chemotaxis, and biofilm formation. However, *ryhB* mutants of *V. cholerae* are not affected in virulence, pointing more toward a role for RyhB in environmental adaptation of this pathogen (50, 51).

Interestingly, some bacterial species encode two copies of RyhB on their chromosomes. This is the case for *Salmonella*, which contains two RyhB homologs called either RfrA/RfrB or RyhB-1/RyhB-2. RyhB-1, which is almost identical to *E. coli* RyhB, is induced preferentially by iron starvation, while RyhB-2, which shares only 53% sequence identity, is induced in stationary phase (52–54). Recently, OxyR, a major transcriptional regulator of the oxidative stress response, has been shown to positively regulate expression of both RyhBs

Table 2 Overview of Fe-sparing response by sRNA in bacteria

Taxonomy	Species	Name[a]	Regulation	Hfq binding	Targets	Reference(s)
Gammaproteobacteria	E. coli	**RyhB**	Fur	Yes	Fe-using proteins, Fe-S cluster formation, Fe acquisition	
	UPEC	**RyhB**	Fur	Yes	Virulence	46
	S. flexneri	**RyhB**	Fur	Yes	Fe-using proteins, acid resistance (*evgA*), virulence	47, 48
	S. dysenteriae	**RyhB**	Fur	Yes	Biofilm formation, virulence (*virB*)	47, 49
	V. cholerae	**RyhB**	Fur	Yes	Fe-using proteins (*sodB*), motility, chemotaxis, biofilm formation	50, 51
	Vibrio parahaemolyticus	**RyhB**	Fur	Yes	Fe acquisition (vibrioferritin)	86
	Salmonella enterica	**RfrA/RfrB**	Fur/OryR	Yes/No	Acid resistance, oxidative stress, antibiotic resistance	52–55
	Y. pestis	**RyhB1/RyhB2**	Fur	Yes/No	Virulence	56
	Klebsiella pneumoniae	**RyhB1/RyhB2**	Fur	Yes	Capsule production, Fe acquisition	57
	P. aeruginosa	PrrF1/PrrF2 PrrH	Fur	Yes	Fe-using proteins, Fe acquisition, quorum sensing (PQS)	58–63
	A. vinelandii	ArrF	Fur	N.D.[b]	Fe-using proteins (*sodB*), oxidative stress (*feSII*)	64, 65
Betaproteobacteria	N. meningitidis	NrrF	Fur	Yes	Fe-using proteins (*sdh*, *petABC*), Fe-S biogenesis (*cyaY*)	66–69
	N. gonorrhoeae	NrrF	Fur	Yes	Fe-using proteins (*sdh*), antibiotic resistance (*mtrF*)	70
Firmicutes	B. subtilis	FsrA	Fur	No (Fbp)	Fe-using proteins (*citB*, *leuCD*, *lutBC*)	71, 72
Cyanobacteria	Synechocystis	IsaR1	Fur	N.D.	Photosynthesis, Fe-S cluster formation (*suf* operon)	73
	Anabaena	α-FurA	High-iron	No	Fe homeostasis via *furA*	74, 75

[a]Bold and regular fonts indicate RyhB homologs and RyhB functional analogs, respectively.
[b]N.D., not determined.

(55). Interestingly, both RyhBs display different sets of targets and play overlapping yet distinct roles in acid resistance and resistance to oxidative stress and antibiotics (52).

Yersinia pestis also encodes two RyhB homologs, RyhB1 and RyhB2, which are both regulated by Fur (56). Expression of both sRNAs increases during lung infection of mice upon intranasal inoculation. While this suggests a role in virulence of the sRNAs, the actual targets of *Yersinia* RyhB1 and RyhB2, as well as their impact on pathogenesis, remain to be evaluated.

Lastly, *Klebsiella* is yet another example of a species containing two RyhB homologs as identified by genome-wide transcription start site profiling (57). In this species, deletion of one *ryhB* allele was shown to affect capsule production and iron acquisition. However, the role of each of the sRNAs in the physiology of this bacterium is not yet elucidated.

RyhB Functional Analogs

In other species devoid of RyhB homologs, sRNAs that display similar functions (i.e., responding to iron starvation and regulating genes involved in iron sparing and/or metabolism) have also been identified. These will be called "functional analogs" of RyhB.

The rrF family of functional analogs of RyhB

Various bacteria encode a class of Fur-regulated sRNAs, termed rrFs (for RNA responsive to Fe), that control iron homeostasis but that are not homologous in sequence to RyhB. As an example, the gammaproteobacterium *Pseudomonas aeruginosa* has two duplicated RNAs involved in iron homeostasis, PrrF1/PrrF2, located in tandem on the chromosome (58). Both sRNAs seem to have redundant functions. PrrFs regulate similar gene targets as RyhB, such as *sodB* and *sdhCDAB*. PrrFs also regulate quorum sensing by preventing the degradation of anthranilate, a precursor of the *Pseudomonas* quinolone signal (PQS). Indeed, the sRNAs repress expression of *antABC*, which encodes the enzymes responsible for degradation of PQS (59). Recently, PrrFs have been shown to be important for virulence, especially during lung acute infection, through control of iron utilization inside the host (60, 61).

Surprisingly, a third sRNA, PrrH, has been identified in *P. aeruginosa* (62). PrrH combines the sequence of both PrrFs and the unique PrrF1-PrrF2 intergenic region. The expression of PrrH would occur by bypassing the PrrF1 Rho-independent terminator. In addition to regulating PrrFs' common targets, PrrH also regulates heme biosynthesis genes (63).

Azotobacter vinelandii codes for the Fur-regulated sRNA ArrF, a PrrF homolog (80% of identity) (64). Like RyhB, ArrF regulates genes coding for iron-containing proteins. ArrF also represses the *feSII* gene coding for a SOD only present in the genome of nitrogen-fixing bacterial species and genes that encode enzymes involved in the synthesis of poly-β-hydroxybutyrate, a carbon storage molecule (64, 65). This suggests that ArrF may have a specific target involved in nitrogen fixation.

Iron homeostasis is regulated by NrrF in the beta-proteobacterium *Neisseria meningitidis* (66). NrrF is Fur regulated, and a recent proteomic analysis revealed that its overexpression changed the levels of various proteins involved in iron uptake, Fe-S cluster biogenesis, and oxidative stress response. More recently, NrrF has been shown to regulate *petABC* mRNA coding for the cytochrome bc_1 (67). Albeit not fully understood yet, NrrF seems to exert its role in an Hfq-dependent or -independent manner depending on its targets (68). Furthermore, in this bacterium, Fur can upregulate *sodB* and *fumB*, which are targets of RyhB in *E. coli* (69). While Hfq is required in these last cases, NrrF is not, suggesting that another yet unidentified sRNA may be responsible for these regulations.

The pathogen *Neisseria gonorrhoeae* carries an *N. meningitidis* NrrF homolog (70). Interestingly, *N. gonorrhoeae* NrrF alters the activity of the antibiotic MtrCDE efflux pump by downregulating the *mtrF* mRNA coding for an accessory protein of the pump.

Other functional analogs

Other species possess Fur-regulated sRNAs that display no sequence similarity with RyhB or PrrF, and no Hfq dependence. For instance, the small RNA FsrA is regulated by Fur in the Gram-positive bacterium *Bacillus subtilis* (71). FsrA, like RyhB, affects the expression of many mRNAs encoding proteins of the TCA cycle that contain Fe-S clusters (72). Uncommonly, FsrA exerts its role in an Hfq-independent manner. However, in some cases, three small proteins also regulated by Fur, FbpA, FbpB, and FbpC, have been hypothesized to replace Hfq in FsrA regulation (71).

Iron cofactors are essential for photosynthetic bacteria. Recently, IsaR1, an sRNA of the cyanobacterium *Synechocystis*, was shown to play a critical role in photosynthesis during iron starvation (73). IsaR1 is functionally similar to *E. coli* RyhB: it is expressed during iron starvation and regulates expression of *sodB* and *acnB*. Moreover, it regulates nearly 15 genes involved in pigment synthesis, electron transfer, and Fe-S cluster biogenesis (73). Interestingly, IsaR1 represses the expression of the *suf* mRNA, encoding the major Fe-S biogenesis system in this bacterium, in a way that is very reminiscent of the Isc regulation by RyhB in *E. coli*. Moreover, regulation of *suf* by IsaR1 leads to the accumulation of an sRNA, named SufZ, whose role has not been elucidated yet. A tempting hypothesis is that SufZ would act like IscR to further control Fe-S biogenesis under iron starvation conditions.

Finally, a last RyhB functional analog operates through an antisense mechanism. This cyanobacterium *Anabaena* sRNA, named α-FurA, is expressed during high-iron conditions, unlike RyhB (74). α-FurA represses the expression of the *furA* mRNA, encoding one of the three putative Fur homologs present in this species (74, 75). In this way, α-FurA may play a role in iron sparing by regulating levels of FurA. In agreement with this, the α-*furA* mutant displays growth defects and disrupted iron homeostasis.

TOWARD A BETTER UNDERSTANDING OF THE IRON-SPARING RESPONSE AND BEYOND

How Does RyhB Redistribute the Cell Iron Pool?

So far, the iron-sparing model for RyhB function has proven true. Most of the targets of RyhB are indeed mRNAs encoding iron-using proteins (see Table 1). Still, there are >200 Fe-using proteins in *E. coli*, while we only found 70 of them as potential RyhB targets in the different global approaches listed above. This likely implies that RyhB does not regulate many nonessential Fe-using proteins. One explanation may be that RyhB preferentially represses "large consumers" of iron, i.e., Fe-using proteins that are present at high levels and/ or protein complexes that use several Fe cofactors. In agreement with this hypothesis, both respiratory complexes, Nuo and Sdh, that use a total of 12 Fe-S cofactors are found as targets of RyhB, even though experimental evidence is lacking as to the regulation of Nuo (30, 76). Likewise, many of the iron-using proteins targeted by RyhB are among the most abundant proteins of *E. coli*, according to a recent global quantification of protein synthesis rates (77). These include SodB (>30,000 copies per cell), GrxD (20,731 copies per cell), ErpA (17,438 copies per cell), and the Isc machinery (up to 14,000 copies per cell). Another striking example is that of the nitrate reductase encoded by the *narGHI* mRNA, which is likely regulated by RyhB (Table 1). Indeed, during anaerobic respiration on nitrate, this complex, which contains five Fe-S clusters and two hemes, represents almost 3% of the total protein content of the cell (78). While it seems reasonable that high Fe consumers may be preferential targets of RyhB, testing this hypothesis would require a thorough investigation on how RyhB redistributes the cellular iron pool that is yet missing.

In addition, the fact that all mRNAs are not regulated with the same apparent efficiency suggests that there is a hierarchy in the targeting of mRNAs by RyhB. Understanding the mechanisms of this hierarchy and its role will certainly broaden our understanding not only of Fe homeostasis control but also of sRNA regulation. The most obvious explanation for a preference for RyhB targeting would be its affinity toward certain mRNAs. Presumably, RyhB-mRNA target affinity itself should mostly derive from the base-pairing free energy. However, things may not be that simple: affinity of the mRNA for the Hfq protein, structural accessibility of the base-pairing site, abundance of the mRNAs, or even spatial distribution (i.e., proximity of the RNAs) may also come into play. Indeed, if only base-pairing was key to sRNA regulation, then target prediction would be much easier and more accurate. For instance, out of the 56 mRNA targets that we have counted, only 17 were predicted by the RNA target prediction program COPRA (Table 1) (33). More-direct *in vivo* approaches, through the quantification of mRNAs bound to RyhB, for example, will be key to a better understanding of target preference. This in turn will allow us to better understand how RyhB reshapes iron utilization in the cell. For instance, what are the key proteins to repress for iron sparing? Are these really the highest iron consumers? What happens first: import of extracellular iron or redistribution of the iron already present in the cell? Such important questions will only be addressed through a better comprehension of RyhB targeting. Understanding how RyhB organizes iron redistribution certainly constitutes one of the major challenges in the near future.

How Is Fe Homeostasis Coupled to Other Stresses via sRNA Regulation?

Iron bioavailability is intimately coupled to the oxidative state of the cell. For instance, Fur was shown to display more targets during anaerobiosis using ChIP-sequencing experiments (11). The reason for this is that Fe^{2+} availability increases under those conditions, driving the formation, or increasing the stability, of holo-Fur (79). The impact of anaerobiosis on RyhB was also evaluated by comparing the expression profiles of *fur* mutants versus the *fur ryhB* double mutant (11). In these conditions, RyhB targets are, as expected, enriched for anaerobic respiration and metabolism processes, while aerobic processes are not regulated. Determining more directly how RyhB affects iron homeostasis in aerobically Fe-deprived cells would certainly help us understand more precisely the role of the sRNA in such conditions, which are actually closer to the environments that may be encountered by bacteria in the gut. The role that RyhB may have in a diametrically opposed condition, namely oxidative stress, also remains unexplored.

It would also be interesting to understand how other stresses that affect Fe availability and/or utilization may connect to regulation of homeostasis via RyhB. For example, acidic pH is known to favor iron solubilization, and oppositely for neutral-basic pH. If and how RyhB may connect to such other stresses to regulate iron homeostasis is still to be studied. In a parallel way, other sRNAs involved in other stress responses may also influence iron homeostasis. An interesting example is that of the two homologous sRNAs, OmrA and OmrB, which are regulated by EnvZ/OmpR, a two-component system that responds to osmolarity stress (80, 81). These sRNAs were found to repress the expression of mRNAs encoding major outer membrane proteins involved in iron uptake (FecA, FepA, and CirA). Another example is that of FnrS, a small RNA expressed under anaerobic conditions due to Fur and ArcA control (82, 83). It was recently shown that this sRNA targets the Fe-S cluster homeostasis regulator IscR, though the physiological relevance of this regulation has not been looked at yet (33). Undoubtedly, understanding these regulatory roles of OmrA and OmrB, FnrS, and possibly other sRNAs will be of critical value toward a better appraisal of iron homeostasis through RNA regulation.

RyhB, Virulence, and Other Functions

It is clear that the main role of RyhB and of its homologs/analogs in other bacteria is primarily that of an iron starvation response. However, since iron homeostasis is centrally connected to multiple other pathways, it is not that surprising that these sRNAs also influence other processes. For instance, RyhB homologs have been found to be implicated in diverse roles such as acid resistance, motility, capsule production, photosynthesis, etc. Whether these effects contribute to an iron-sparing response specific to the ecological niches and lifestyles of the different species is not always clear.

For instance, there are a growing number of bacteria in which RyhB homologs/analogs seem to play a role in virulence (see Table 2). In many cases, it seems reasonable to assume that these sRNAs will influence virulence by allowing the cell to better cope with the iron starvation conditions that are found inside the host.

However, RyhB may have a more direct role in *S. dysenteriae* virulence by regulating *virB* (47, 49). Indeed, in addition to serving as an essential cofactor, iron could also be used as a signal to activate/repress virulence during the course of infection. More studies will be necessary in the future to further understand the role of RyhB and its homologs in other processes connected to iron homeostasis. Nevertheless, RyhB has provided us with lots of exciting stories over the years, and, given this lucky streak, it is certainly likely that the study of RyhB will unveil still more surprises.

Acknowledgments. The authors apologize for all the studies that could not be cited due to length restrictions. We thank F. Barras and B. Py for their critical role in establishing a "brain cell-sparing response" through helpful discussions and critical reading of the manuscript. We also thank E. Massé for exciting discussions. Work by P.M. and S.C. was funded by the Centre National de la Recherche Scientifique and Aix Marseille Université. S.C. is a recipient of a Fondation pour la Recherche Médicale grant (FDT20170436820).

Citation. Chareyre S, Mandin P. 2018. Bacterial iron homeostasis regulation by sRNAs. Microbiol Spectrum 6(2):RWR-0010-2017.

References

1. **Andrews SC, Robinson AK, Rodríguez-Quiñones F.** 2003. Bacterial iron homeostasis. *FEMS Microbiol Rev* 27:215–237.

2. **Frawley ER, Fang FC.** 2014. The ins and outs of bacterial iron metabolism. *Mol Microbiol* 93:609–616.

3. **Py B, Moreau PL, Barras F.** 2011. Fe-S clusters, fragile sentinels of the cell. *Curr Opin Microbiol* 14:218–223.

4. **Miethke M, Marahiel MA.** 2007. Siderophore-based iron acquisition and pathogen control. *Microbiol Mol Biol Rev* 71:413–451.

5. **Fischbach MA, Lin H, Liu DR, Walsh CT.** 2006. How pathogenic bacteria evade mammalian sabotage in the battle for iron. *Nat Chem Biol* 2:132–138.

6. **Imlay JA.** 2006. Iron-sulphur clusters and the problem with oxygen. *Mol Microbiol* 59:1073–1082.

7. **Imlay JA.** 2013. The molecular mechanisms and physiological consequences of oxidative stress: lessons from a model bacterium. *Nat Rev Microbiol* 11:443–454.

8. **Imlay JA.** 2014. The mismetallation of enzymes during oxidative stress. *J Biol Chem* 289:28121–28128.

9. **Braun V.** 2003. Iron uptake by *Escherichia coli*. *Front Biosci* 8:s1409–s1421.

10. **Lee JW, Helmann JD.** 2007. Functional specialization within the Fur family of metalloregulators. *Biometals* 20:485–499.

11. **Beauchene NA, Myers KS, Chung D, Park DM, Weisnicht AM, Keleş S, Kiley PJ.** 2015. Impact of anaerobiosis on expression of the iron-responsive Fur and RyhB regulons. *mBio* 6:e01947–e15.

12. **Seo SW, Kim D, Latif H, O'Brien EJ, Szubin R, Palsson BO.** 2014. Deciphering Fur transcriptional regulatory network

highlights its complex role beyond iron metabolism in *Escherichia coli. Nat Commun* 5:4910.

13. Dubrac S, Touati D. 2000. Fur positive regulation of iron superoxide dismutase in *Escherichia coli*: functional analysis of the *sodB* promoter. *J Bacteriol* 182:3802–3808.

14. Gruer MJ, Guest JR. 1994. Two genetically-distinct and differentially-regulated aconitases (AcnA and AcnB) in *Escherichia coli. Microbiology* 140:2531–2541.

15. Massé E, Gottesman S. 2002. A small RNA regulates the expression of genes involved in iron metabolism in *Escherichia coli. Proc Natl Acad Sci U S A* 99:4620–4625.

16. Massé E, Escorcia FE, Gottesman S. 2003. Coupled degradation of a small regulatory RNA and its mRNA targets in *Escherichia coli. Genes Dev* 17:2374–2383.

17. De Lay N, Schu DJ, Gottesman S. 2013. Bacterial small RNA-based negative regulation: Hfq and its accomplices. *J Biol Chem* 288:7996–8003.

18. Sauer E. 2013. Structure and RNA-binding properties of the bacterial LSm protein Hfq. *RNA Biol* 10:610–618.

19. Gottesman S, Storz G. 2011. Bacterial small RNA regulators: versatile roles and rapidly evolving variations. *Cold Spring Harb Perspect Biol* 3:a003798.

20. Prévost K, Desnoyers G, Jacques JF, Lavoie F, Massé E. 2011. Small RNA-induced mRNA degradation achieved through both translation block and activated cleavage. *Genes Dev* 25:385–396.

21. Morita T, Maki K, Aiba H. 2005. RNase E-based ribonucleoprotein complexes: mechanical basis of mRNA destabilization mediated by bacterial noncoding RNAs. *Genes Dev* 19:2176–2186.

22. Desnoyers G, Morissette A, Prévost K, Massé E. 2009. Small RNA-induced differential degradation of the polycistronic mRNA *iscRSUA. EMBO J* 28:1551–1561.

23. Prévost K, Salvail H, Desnoyers G, Jacques JF, Phaneuf E, Massé E. 2007. The small RNA RyhB activates the translation of *shiA* mRNA encoding a permease of shikimate, a compound involved in siderophore synthesis. *Mol Microbiol* 64:1260–1273.

24. Salvail H, Caron M-P, Bélanger J, Massé E. 2013. Antagonistic functions between the RNA chaperone Hfq and an sRNA regulate sensitivity to the antibiotic colicin. *EMBO J* 32:2764–2778.

25. Battesti A, Majdalani N, Gottesman S. 2011. The RpoS-mediated general stress response in *Escherichia coli. Annu Rev Microbiol* 65:189–213.

26. Mandin P, Gottesman S. 2010. Integrating anaerobic/aerobic sensing and the general stress response through the ArcZ small RNA. *EMBO J* 29:3094–3107.

27. Lalaouna D, Carrier MC, Semsey S, Brouard JS, Wang J, Wade JT, Massé E. 2015. A 3′ external transcribed spacer in a tRNA transcript acts as a sponge for small RNAs to prevent transcriptional noise. *Mol Cell* 58:393–405.

28. Jacques JF, Jang S, Prévost K, Desnoyers G, Desmarais M, Imlay J, Massé E. 2006. RyhB small RNA modulates the free intracellular iron pool and is essential for normal growth during iron limitation in *Escherichia coli. Mol Microbiol* 62:1181–1190.

29. Salvail H, Lanthier-Bourbonnais P, Sobota JM, Caza M, Benjamin JA, Mendieta ME, Lépine F, Dozois CM, Imlay J, Massé E. 2010. A small RNA promotes siderophore production through transcriptional and metabolic remodeling. *Proc Natl Acad Sci U S A* 107:15223–15228.

30. Massé E, Vanderpool CK, Gottesman S. 2005. Effect of RyhB small RNA on global iron use in *Escherichia coli. J Bacteriol* 187:6962–6971.

31. Wang J, Rennie W, Liu C, Carmack CS, Prévost K, Caron MP, Massé E, Ding Y, Wade JT. 2015. Identification of bacterial sRNA regulatory targets using ribosome profiling. *Nucleic Acids Res* 43:10308–10320.

32. Melamed S, Peer A, Faigenbaum-Romm R, Gatt YE, Reiss N, Bar A, Altuvia Y, Argaman L, Margalit H. 2016. Global mapping of small RNA-target interactions in bacteria. *Mol Cell* 63:884–897.

33. Wright PR, Richter AS, Papenfort K, Mann M, Vogel J, Hess WR, Backofen R, Georg J. 2013. Comparative genomics boosts target prediction for bacterial small RNAs. *Proc Natl Acad Sci U S A* 110:E3487–E3496.

34. Loiseau L, Gerez C, Bekker M, Ollagnier-de Choudens S, Py B, Sanakis Y, Teixeira de Mattos J, Fontecave M, Barras F. 2007. ErpA, an iron sulfur (Fe S) protein of the A-type essential for respiratory metabolism in *Escherichia coli. Proc Natl Acad Sci U S A* 104:13626–13631.

35. Mandin P, Chareyre S, Barras F. 2016. A regulatory circuit composed of a transcription factor, IscR, and a regulatory RNA, RyhB, controls Fe-S cluster delivery. *mBio* 7:e00966-16.

36. Bos J, Duverger Y, Thouvenot B, Chiaruttini C, Branlant C, Springer M, Charpentier B, Barras F. 2013. The sRNA RyhB regulates the synthesis of the *Escherichia coli* methionine sulfoxide reductase MsrB but not MsrA. *PLoS One* 8:e63647.

37. Niederhoffer EC, Naranjo CM, Bradley KL, Fee JA. 1990. Control of *Escherichia coli* superoxide dismutase (*sodA* and *sodB*) genes by the ferric uptake regulation (*fur*) locus. *J Bacteriol* 172:1930–1938.

38. Mettert EL, Kiley PJ. 2015. How is Fe-S cluster formation regulated? *Annu Rev Microbiol* 69:505–526.

39. Roche B, Aussel L, Ezraty B, Mandin P, Py B, Barras F. 2013. Iron/sulfur proteins biogenesis in prokaryotes: formation, regulation and diversity. *Biochim Biophys Acta* 1827:455–469.

40. Outten FW, Djaman O, Storz G. 2004. A *suf* operon requirement for Fe-S cluster assembly during iron starvation in *Escherichia coli. Mol Microbiol* 52:861–872.

41. Giel JL, Nesbit AD, Mettert EL, Fleischhacker AS, Wanta BT, Kiley PJ. 2013. Regulation of iron-sulphur cluster homeostasis through transcriptional control of the Isc pathway by [2Fe-2S]-IscR in *Escherichia coli. Mol Microbiol* 87:478–492.

42. Mettert EL, Kiley PJ. 2014. Coordinate regulation of the Suf and Isc Fe-S cluster biogenesis pathways by IscR is essential for viability of *Escherichia coli. J Bacteriol* 196:4315–4323.

43. Yeo WS, Lee JH, Lee KC, Roe JH. 2006. IscR acts as an activator in response to oxidative stress for the *suf*

operon encoding Fe-S assembly proteins. *Mol Microbiol* **61**:206–218.

44. **Giel JL, Rodionov D, Liu M, Blattner FR, Kiley PJ.** 2006. IscR-dependent gene expression links iron-sulphur cluster assembly to the control of O_2-regulated genes in *Escherichia coli*. *Mol Microbiol* **60**:1058–1075.

45. **Py B, Barras F.** 2015. Genetic approaches of the Fe-S cluster biogenesis process in bacteria: historical account, methodological aspects and future challenges. *Biochim Biophys Acta* **1853**:1429–1435.

46. **Porcheron G, Habib R, Houle S, Caza M, Lépine F, Daigle F, Massé E, Dozois CM.** 2014. The small RNA RyhB contributes to siderophore production and virulence of uropathogenic *Escherichia coli*. *Infect Immun* **82**:5056–5068.

47. **Murphy ER, Payne SM.** 2007. RyhB, an iron-responsive small RNA molecule, regulates *Shigella dysenteriae* virulence. *Infect Immun* **75**:3470–3477.

48. **Oglesby AG, Murphy ER, Iyer VR, Payne SM.** 2005. Fur regulates acid resistance in *Shigella flexneri* via RyhB and *ydeP*. *Mol Microbiol* **58**:1354–1367.

49. **Broach WH, Egan N, Wing HJ, Payne SM, Murphy ER.** 2012. VirF-independent regulation of *Shigella virB* transcription is mediated by the small RNA RyhB. *PLoS One* **7**:e38592.

50. **Davis BM, Quinones M, Pratt J, Ding Y, Waldor MK.** 2005. Characterization of the small untranslated RNA RyhB and its regulon in *Vibrio cholerae*. *J Bacteriol* **187**: 4005–4014.

51. **Mey AR, Craig SA, Payne SM.** 2005. Characterization of *Vibrio cholerae* RyhB: the RyhB regulon and role of *ryhB* in biofilm formation. *Infect Immun* **73**:5706–5719.

52. **Kim JN.** 2016. Roles of two RyhB paralogs in the physiology of *Salmonella enterica*. *Microbiol Res* **186–187**: 146–152.

53. **Kim JN, Kwon YM.** 2013. Genetic and phenotypic characterization of the RyhB regulon in *Salmonella* Typhimurium. *Microbiol Res* **168**:41–49.

54. **Padalon-Brauch G, Hershberg R, Elgrably-Weiss M, Baruch K, Rosenshine I, Margalit H, Altuvia S.** 2008. Small RNAs encoded within genetic islands of *Salmonella typhimurium* show host-induced expression and role in virulence. *Nucleic Acids Res* **36**:1913–1927.

55. **Calderón IL, Morales EH, Collao B, Calderón PF, Chahuán CA, Acuña LG, Gil F, Saavedra CP.** 2014. Role of *Salmonella* Typhimurium small RNAs RyhB-1 and RyhB-2 in the oxidative stress response. *Res Microbiol* **165**:30–40.

56. **Deng Z, Meng X, Su S, Liu Z, Ji X, Zhang Y, Zhao X, Wang X, Yang R, Han Y.** 2012. Two sRNA RyhB homologs from *Yersinia pestis* biovar *microtus* expressed *in vivo* have differential Hfq-dependent stability. *Res Microbiol* **163**:413–418.

57. **Huang SH, Wang CK, Peng HL, Wu CC, Chen YT, Hong YM, Lin CT.** 2012. Role of the small RNA RyhB in the Fur regulon in mediating the capsular polysaccharide biosynthesis and iron acquisition systems in *Klebsiella pneumoniae*. *BMC Microbiol* **12**:148.

58. **Wilderman PJ, Sowa NA, FitzGerald DJ, FitzGerald PC, Gottesman S, Ochsner UA, Vasil ML.** 2004. Identifica-

tion of tandem duplicate regulatory small RNAs in *Pseudomonas aeruginosa* involved in iron homeostasis. *Proc Natl Acad Sci U S A* **101**:9792–9797.

59. **Oglesby AG, Farrow JM III, Lee JH, Tomaras AP, Greenberg EP, Pesci EC, Vasil ML.** 2008. The influence of iron on *Pseudomonas aeruginosa* physiology: a regulatory link between iron and quorum sensing. *J Biol Chem* **283**:15558–15567.

60. **Reinhart AA, Powell DA, Nguyen AT, O'Neill M, Djapgne L, Wilks A, Ernst RK, Oglesby-Sherrouse AG.** 2015. The *prrF*-encoded small regulatory RNAs are required for iron homeostasis and virulence of *Pseudomonas aeruginosa*. *Infect Immun* **83**:863–875.

61. **Reinhart AA, Nguyen AT, Brewer LK, Bevere J, Jones JW, Kane MA, Damron FH, Barbier M, Oglesby-Sherrouse AG.** 2017. The *Pseudomonas aeruginosa* PrrF small RNAs regulate iron homeostasis during acute murine lung infection. *Infect Immun* **85**:e00764–16.

62. **Oglesby-Sherrouse AG, Vasil ML.** 2010. Characterization of a heme-regulated non-coding RNA encoded by the *prrF* locus of *Pseudomonas aeruginosa*. *PLoS One* **5**: e9930.

63. **Osborne J, Djapgne L, Tran BQ, Goo YA, Oglesby-Sherrouse AG.** 2014. A method for *in vivo* identification of bacterial small RNA-binding proteins. *Microbiology Open* **3**:950–960.

64. **Jung YS, Kwon YM.** 2008. Small RNA ArrF regulates the expression of *sodB* and *feSII* genes in *Azotobacter vinelandii*. *Curr Microbiol* **57**:593–597.

65. **Pyla R, Kim TJ, Silva JL, Jung YS.** 2010. Proteome analysis of *Azotobacter vinelandii* Δ*arrF* mutant that overproduces poly-β-hydroxybutyrate polymer. *Appl Microbiol Biotechnol* **88**:1343–1354.

66. **Mellin JR, Goswami S, Grogan S, Tjaden B, Genco CA.** 2007. A novel Fur- and iron-regulated small RNA, NrrF, is required for indirect Fur-mediated regulation of the *sdhA* and *sdhC* genes in *Neisseria meningitidis*. *J Bacteriol* **189**:3686–3694.

67. **Pannekoek Y, Huis In 't Veld R, Schipper K, Bovenkerk S, Kramer G, Speijer D, van der Ende A.** 2017. Regulation of *Neisseria meningitidis* cytochrome bc_1 components by NrrF, a Fur-controlled small noncoding RNA. *FEBS Open Bio* **7**:1302–1315.

68. **Mellin JR, McClure R, Lopez D, Green O, Reinhard B, Genco C.** 2010. Role of Hfq in iron-dependent and -independent gene regulation in *Neisseria meningitidis*. *Microbiology* **156**:2316–2326.

69. **Metruccio MM, Fantappiè L, Serruto D, Muzzi A, Roncarati D, Donati C, Scarlato V, Delany I.** 2009. The Hfq-dependent small noncoding RNA NrrF directly mediates Fur-dependent positive regulation of succinate dehydrogenase in *Neisseria meningitidis*. *J Bacteriol* **191**: 1330–1342.

70. **Ducey TF, Jackson L, Orvis J, Dyer DW.** 2009. Transcript analysis of *nrrF*, a Fur repressed sRNA of *Neisseria gonorrhoeae*. *Microb Pathog* **46**:166–170.

71. **Gaballa A, Antelmann H, Aguilar C, Khakh SK, Song KB, Smaldone GT, Helmann JD.** 2008. The *Bacillus subtilis* iron-sparing response is mediated by a Fur-regulated

small RNA and three small, basic proteins. *Proc Natl Acad Sci U S A* **105**:11927–11932.

72. **Smaldone GT, Antelmann H, Gaballa A, Helmann JD.** 2012. The FsrA sRNA and FbpB protein mediate the iron-dependent induction of the *Bacillus subtilis lutABC* iron-sulfur-containing oxidases. *J Bacteriol* **194**:2586–2593.

73. **Georg J, Kostova G, Vuorijoki L, Schön V, Kadowaki T, Huokko T, Baumgartner D, Müller M, Klähn S, Allahverdiyeva Y, Hihara Y, Futschik ME, Aro EM, Hess WR.** 2017. Acclimation of oxygenic photosynthesis to iron starvation is controlled by the sRNA IsaR1. *Curr Biol* **27**:1425–1436.e7.

74. **Hernández JA, Muro-Pastor AM, Flores E, Bes MT, Peleato ML, Fillat MF.** 2006. Identification of a *furA cis* antisense RNA in the cyanobacterium *Anabaena* sp. PCC 7120. *J Mol Biol* **355**:325–334.

75. **Hernández JA, Alonso I, Pellicer S, Luisa Peleato M, Cases R, Strasser RJ, Barja F, Fillat MF.** 2010. Mutants of *Anabaena* sp. PCC 7120 lacking *alr*1690 and α-*furA* antisense RNA show a pleiotropic phenotype and altered photosynthetic machinery. *J Plant Physiol* **167**:430–437.

76. **Gnandt E, Dörner K, Strampraad MF, de Vries S, Friedrich T.** 2016. The multitude of iron-sulfur clusters in respiratory complex I. *Biochim Biophys Acta* **1857**:1068–1072.

77. **Li GW, Burkhardt D, Gross C, Weissman JS.** 2014. Quantifying absolute protein synthesis rates reveals principles underlying allocation of cellular resources. *Cell* **157**:624–635.

78. **Arias-Cartin R, Grimaldi S, Pommier J, Lanciano P, Schaefer C, Arnoux P, Giordano G, Guigliarelli B, Magalon A.** 2011. Cardiolipin-based respiratory complex activation in bacteria. *Proc Natl Acad Sci U S A* **108**:7781–7786.

79. **Beauchene NA, Mettert EL, Moore LJ, Keleş S, Willey ER, Kiley PJ.** 2017. O$_2$ availability impacts iron homeostasis in *Escherichia coli*. *Proc Natl Acad Sci U S A* **114**:12261–12266.

80. **Guillier M, Gottesman S.** 2006. Remodelling of the *Escherichia coli* outer membrane by two small regulatory RNAs. *Mol Microbiol* **59**:231–247.

81. **Guillier M, Gottesman S.** 2008. The 5′ end of two redundant sRNAs is involved in the regulation of multiple targets, including their own regulator. *Nucleic Acids Res* **36**:6781–6794.

82. **Boysen A, Møller-Jensen J, Kallipolitis B, Valentin-Hansen P, Overgaard M.** 2010. Translational regulation of gene expression by an anaerobically induced small non-coding RNA in *Escherichia coli*. *J Biol Chem* **285**:10690–10702.

83. **Durand S, Storz G.** 2010. Reprogramming of anaerobic metabolism by the FnrS small RNA. *Mol Microbiol* **75**:1215–1231.

84. **Benjamin JA, Massé E.** 2014. The iron-sensing aconitase B binds its own mRNA to prevent sRNA-induced mRNA cleavage. *Nucleic Acids Res* **42**:10023–10036.

85. **Li F, Wang Y, Gong K, Wang Q, Liang Q, Qi Q.** 2014. Constitutive expression of RyhB regulates the heme biosynthesis pathway and increases the 5-aminolevulinic acid accumulation in *Escherichia coli*. *FEMS Microbiol Lett* **350**:209–215.

86. **Tanabe T, Funahashi T, Nakao H, Maki J, Yamamoto S.** 2013. The *Vibrio parahaemolyticus* small RNA RyhB promotes production of the siderophore vibrioferrin by stabilizing the polycistronic mRNA. *J Bacteriol* **195**:3692–3703.

Regulating with RNA in Bacteria and Archaea
Edited by Gisela Storz and Kai Papenfort
© 2018 American Society for Microbiology, Washington, DC
doi:10.1128/microbiolspec.RWR-0017-2018

Small RNA-Based Regulation of Bacterial Quorum Sensing and Biofilm Formation

17

Sine Lo Svenningsen[1]

INTRODUCTION

Bacteria use the production, release, and detection of extracellular chemical signaling molecules called autoinducers (AIs) to regulate gene expression at the population level, a widespread phenomenon known as quorum sensing (QS). The first reviews of small RNA (sRNA) control of QS were published in 2006 and 2007 (1, 2). At that time, two QS systems had been described that are entirely dependent on sRNA-based regulation, namely, the QS circuits of the Gram-negative family *Vibrionaceae* and the *agr* QS circuit of the Gram-positive staphylococci. A decade later, these two examples have become paradigms of sRNA-based regulation, and have taught us important regulatory principles relevant not only to QS but to bacterial sRNA-based regulation in general. I begin this review by describing the progress made in understanding sRNA-based execution of the QS response in these two systems. In many other bacteria the core executors of QS are not sRNAs, but sRNAs are nevertheless identified as auxiliary regulators that directly or indirectly affect components of the QS machinery. In particular, sRNA-based regulation of the production of enzymes that catalyze AI synthesis appears as a common trend across numerous unrelated QS circuits, and such examples will be highlighted. Due to space limitations, this review is solely focused on riboregulation carried out by small *trans*-acting RNA molecules that are encoded at loci distinct from those of their target(s), thereby excluding *cis*-acting regulatory RNA elements such as riboswitches and attenuators, as well as antisense RNA transcripts. Notably, QS has only been sparsely connected to *cis*-acting riboregulation in the literature thus far (3, 4).

One of the major benefits of QS is that it enables bacteria to behave collectively as a group and together carry out tasks that would be futile if attempted by an individual bacterium. A key example of such group behavior is biofilm formation, in which bacterial communities embed themselves in a jointly produced extracellular matrix that can provide them with considerable benefits, including protection from biotic, chemical, and mechanical assaults. When bacteria transition from planktonic individuals to members of a sessile community of microorganisms in a biofilm, they undergo profound phenotypic changes. Accordingly, the transcription of genes responsible for biofilm formation and dispersal is highly regulated. The latter part of this review will describe how an impressive number of sRNAs establish an additional, posttranscriptional layer of control over the expression of key biofilm components and regulators, with an emphasis on the well-studied biofilm network of *Escherichia coli*. This overview of sRNA-based regulation of biofilm formation serves to illustrate how sRNAs may be critical for the proper integration of the numerous environmental signals that combine to govern biofilm formation and dispersal.

BACTERIAL QS

The QS response is initiated when an extracellular AI is detected by its cognate sensor. The AI-bound sensor is typically either a membrane-bound protein that triggers a QS signal transduction cascade or a cytoplasmic protein that acts directly as a transcription factor to mediate changes in gene expression. Thus, in its simplest form, the QS phenotype requires only two different

---

[1]Department of Biology, University of Copenhagen, 2200 Copenhagen, Denmark.

proteins: a synthase that produces the AI and a sensor/effector protein that regulates the expression of target genes according to the presence or absence of the minimal threshold concentration of AI required for its detection. Bioluminescence of the marine bacterium *Vibrio fischeri* was the first QS-regulated behavior to be described, and it is controlled by a cytoplasmic synthase/sensor pair: the AI synthase LuxI and the AI-binding transcription factor LuxR (5, 6). For many bacteria in which QS has since been studied in detail, including *V. fischeri*, the complete QS apparatus turns out to be more complex. It is common to find that a single bacterium produces and detects multiple distinct AIs, which may convey different bits of information to the recipient. Additionally, QS pathways are integrated into the global gene regulatory network and may influence and/or be influenced by the metabolic state of the cell and the cellular responses to a multitude of other environmental signals.

A HODGEPODGE OF EXTRACELLULAR SIGNALS

In Gram-negative bacteria, the most widely studied AIs are *N*-acylhomoserine lactones (AHLs) consisting of an invariant homoserine lactone core and a highly variable acyl chain that can differ in terms of length and modifications, depending on the properties of the enzyme(s) that catalyze its synthesis (7). Most known AHL synthases are of the LuxI type mentioned above. Additionally, AIs belonging to the alkylquinolones, α-hydroxyketones, and fatty acid-like diffusible signal factors have been described in Gram-negative bacteria (recently reviewed in references 8 and 9). In Gram-positive bacteria, on the other hand, AIs are typically small oligopeptides that may be posttranslationally modified (reviewed in reference 10). Furthermore, a substantial fraction of sequenced bacterial genomes across phyla harbor the gene *luxS*, which encodes an enzyme that catalyzes the synthesis of 4,5-dihydroxy-2,3-pentanedione (DPD), a highly reactive structure, which gives rise to a collection of interconverting isomers of DPD collectively known as AI-2 (reviewed in reference 11). A number of other AIs have been described that belong to chemical classes other than those described above. For example, the human pathogen *Vibrio cholerae* appears to produce and detect five chemically distinct AIs: a borated form of AI-2 [(2*S*,4*S*)-2-methyl-2,3,3,4-tetrahydroxytetrahydrofuran borate] (12), an α-hydroxyketone called CAI-1 (*S*-3-hydroxytridecan-4-one) (13), a pyrazine called DPO (3,5-dimethylpyrazin-2-ol) (14), and two additional AIs whose chemical composition

has not yet been identified (15, 16). The first information filter in a QS response pathway is the intrinsic ligand specificity of each receptor. This important step in QS information transfer is based on the biochemistry of the interaction between AI ligands and their receptors, which are proteins in all reported cases. Although AIs could in principle serve as ligands for riboregulators, there are no known examples of RNA-based AI detection to date. Thus, the implications of intrinsic ligand-receptor specificity for bacterial QS will not be discussed further here (see reference 9 for a recent review).

Since each AI may differ from the others in terms of its diffusion properties, chemical stability, and the number of species that produce it and thereby contribute to its extracellular concentration in a given environment, different compounds could each relay very different information to the recipient cell about the biotic and even abiotic nature of the surroundings. Further considering the ability of some bacteria to interfere with the QS of other bacteria through the production, uptake, or degradation of their AIs (reviewed in reference 17), it seems a daunting task to detect, sort, and properly integrate such a myriad of information into an appropriate phenotypic response. Nevertheless, QS bacteria appear to be able to make sense of the jumble, given that they have evolved to use QS to regulate vital behaviors such as virulence, DNA exchange, bacteriophage defense mechanisms, biofilm formation, and symbiosis, among others (recently reviewed in references 8, 10, and 18).

RNAIII OF *STAPHYLOCOCCUS AUREUS*

QS in *S. aureus*

S. aureus is a widespread commensal and an opportunistic human pathogen. It can invade a number of bodily tissues to cause a variety of acute as well as chronic infections, ranging in severity from minor skin infections to pneumonia, osteomyelitis, endocarditis, and bacteremia. In addition, some strains of *S. aureus* produce exotoxins that can cause other illnesses such as food poisoning and toxic shock syndrome (19). The versatility of this pathogen is enabled by a large battery of extracellular virulence factors, most of which are regulated by the accessory gene regulator (*agr*) QS system. The *agr* system is required for virulence in several animal models (20–22), and although *agr*-negative isolates have been found in clinical material (23–25), *agr* function seems to be required for *S. aureus* infection of humans too, at least in the early stages of the infection (23, 26). The *agr* QS components are transcribed from an ~3-kb chromosomal region containing two divergent promoters, P2 and P3 (Fig. 1a) (27, 28). P2 controls a transcript known as RNAII, encoding four open reading frames (ORFs) with the genomic organization *agrBDCA* (29). AgrD encodes a 46-amino-acid AI precursor that is processed by the membrane-bound protease AgrB to produce the AI, a 7- to-9-amino-acid-long thiolactone peptide called AIP. When the AIP has reached a minimal extracellular threshold concentration, it triggers the AgrC/AgrA two-component system. AgrC encodes a membrane-bound histidine kinase receptor, which autophosphorylates upon AIP binding, and transfers a phosphoryl group to the response regulator AgrA. Phosphorylated AgrA acts as a transcription factor to positively autoregulate transcription from P2, and also activates transcription of the RNAIII transcript from P3 (30). The *agr* system is highly conserved throughout the staphylococci, but contains a variable region encompassing the last one-third of *agrB*, *agrD*, and the first half of *agrC*, resulting in interspecies and interstrain variations among the AIPs, the enzymes that process it, and the receptors that detect it (31–33). The fascinating consequence of this divergence is that staphylococci can be divided into *agr* subtypes, where the AIPs produced by members of the same group activate *agr* QS, whereas AIPs produced by other groups generally obstruct QS because heterologous AIPs can bind and block the AgrC receptor without activating its autokinase activity (32, 34).

The Posttranscriptional Regulator RNAIII

RNAIII was the first regulatory RNA to be identified as a QS regulator in any bacterium. It is unusually long (514 nucleotides [nt]) (Fig. 1b) and up- or downregulates translation initiation rates and/or mRNA stability of at least nine *S. aureus* transcripts by base-pairing with their 5′ untranslated region (5′ UTR). The confirmed direct targets of RNAIII encode virulence factors (protein A, coagulase, α-toxin, and the adherence proteins SA1000, Sbi, and Map) a cell wall hydrolytic enzyme (LytM), and master transcription factors (Rot and MgrA) (reviewed in reference 35). By contrast to the majority of *trans*-acting sRNAs described in other bacteria, RNAIII does not depend on the RNA chaperone Hfq for stability or for exerting its regulatory functions (36, 37). The only RNA-binding protein (RBP) that has been shown to be important for RNAIII-mediated regulation is the double-strand-specific endoribonuclease III (RNase III), which degrades a subset of RNAIII-targeted mRNAs (38). However, most RBPs in *S. aureus* are uncharacterized, and additional important protein partners for RNAIII could exist (39, 40).

The *agr* regulon includes several hundred genes, the majority of which are believed to be indirectly controlled via RNAIII-dependent regulation of global transcription factors (36, 41–43). RNAIII was the first identified example of a bifunctional bacterial transcript that functions as both an mRNA and an sRNA, since it also harbors a small ORF, *hld*, encoding the δ-hemolysin, a cytolytic peptide in the phenol-soluble modulin (PSM) family (44, 45). *S. aureus* RNAIII is highly structured and contains 14 hairpins (H1 to -14) (Fig. 1b). The long 3′ UTR is characterized by hairpin loops H7, H13, and H14, which contain a conserved C-rich sequence in the loop region (46). Repression of translation initiation of all the confirmed negatively regulated direct targets of RNAIII, namely the *rot*, *lytM*, *Sa1000*, *spa*, *coa*, and *sbi* mRNAs, all depend on base-pairing between one or more of these C-rich regions and the G-rich Shine-Dalgarno sequences of the mRNAs (36, 47–50). The multifaceted RNAIII molecule is also capable of directly activating translation of select mRNA targets. In fact, translational activation of the *hla* transcript encoding α-toxin was the first example of direct positive regulation by a bacterial regulatory RNA (51). RNAIII activates *hla* translation through base-pairing between the 5′ end of RNAIII (H2) and the 5′ UTR of the *hla* mRNA. In the absence of RNAIII, the 5′ UTR of *hla* forms an intramolecular structure that precludes the ribosome binding site. Hybridization with RNAIII prevents the formation of this structure and makes the *hla* transcript accessible for translation (51). This has since proven to be a common mechanism for bacterial sRNA-mediated direct translational activation (52). Two additional target mRNAs, *map* and *mgrA*, are also positively regulated by base-pairing with RNAIII. In the case of *map*, which encodes a surface adhesion protein, the region of pairing appears to be in H4 of RNAIII, because the reduced activity of a *map-lacZ* fusion containing 13 base substitutions in the predicted RNAIII-*map* pairing region was restored by the introduction of an RNAIII variant carrying the complementary mutations in H4 (53). This result supports the notion that the 5′ region of RNAIII is responsible for base-pairing events that lead to target activation, while the 3′ region is primarily responsible for target repression. However, activation of *mgrA*, which occurs by RNAIII-mediated stabilization of the *mgrA* transcript, involves pairing of both the 5′ and the 3′ ends of RNAIII to *mgrA* (43), thus blurring this division.

A handful of dual-function sRNA/mRNA transcripts have been described in bacteria after the discovery of translation of the *hld* coding sequence within RNAIII (54, 55, see also chapter 27). In the case of the dual-

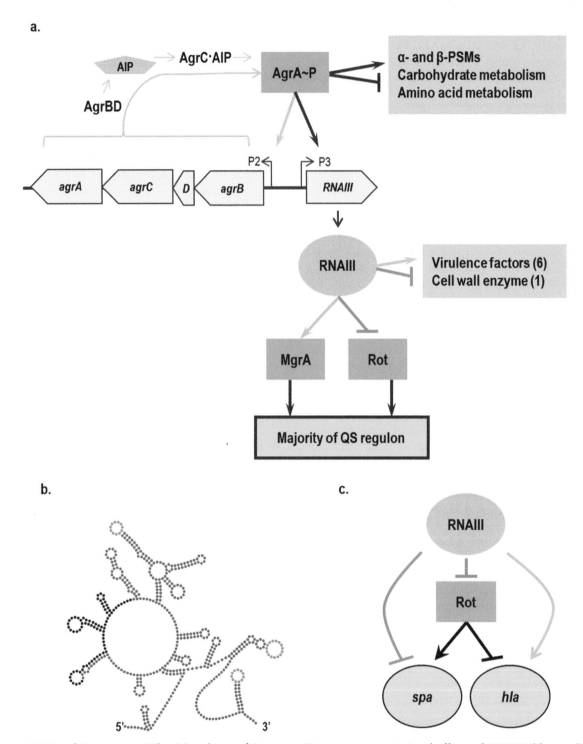

Figure 1 RNAIII of *S. aureus*. (a) The QS pathway of *S. aureus*. Direct posttranscriptional effects of RNAIII (blue oval) on target mRNAs are shown as green arrows (positive regulation) or red blocked arrows (negative regulation). The positive feedback loop of the *agr* system is depicted as light blue arrows from the RNAII operon to AgrA~P. Transcriptional regulation is shown by black arrows. Groups of genes that are *agr* regulated but in an RNAIII-independent manner are included as targets of AgrA~P, although direct transcriptional regulation by AgrA~P has only been demonstrated for RNAII, RNAIII, and the α- and β-PSMs (42, 191). (b) Secondary structure of RNAIII. The *hld* ORF is shown in black. Three C-rich regions involved in base-pairing with target mRNAs are shown in red. Adapted from structure determined by Benito et al. (46) and depicted using Forna (192). (c) The RNA-based double selector switch of *S. aureus*.

function SgrS transcript in *Escherichia coli*, the SgrS sRNA and the encoded small protein, SgrT, play mechanistically different but functionally redundant roles in preventing the accumulation of toxic levels of glucosephosphate. SgrS represses translation of the EIICB^Glc glucose transporter by base-pairing with its mRNA (56), and SgrT appears to interact directly with the EIICB^Glc protein to repress its activity (57–59, see also chapter 14). There is no evidence that the δ-hemolysin encoded by *hld* in the RNAIII transcript similarly complements any part of the regulatory roles of RNAIII in *S. aureus*. In fact, Queck et al. (42) argue that the dual-function transcript most likely originated as an AgrA-activated *hld* gene encoding the PSM δ-hemolysin, which subsequently acquired the regulatory RNAIII function, as opposed to evolution of the *hld* ORF in a preexisting regulatory RNAIII. Notably, they find that phosphorylated AgrA, which was previously thought to only activate the RNAII and RNAIII/*hld* transcripts, strongly and directly activates transcription of two additional PSM operons (encoding PSMα1–4 and PSMβ1–2) (60), suggesting that QS regulation of PSM cytolysins may be a more ancient feature of the agr system than RNAIII-mediated QS control of the remainder of the virulence regulon (42).

RNAIII in the QS Network of *S. aureus*

Intriguingly, two key virulence factors, α-toxin (encoded by *hla* and activated by QS) and protein A (encoded by *spa* and repressed by QS), are regulated both directly by RNAIII binding to the mRNAs and indirectly via RNAIII-mediated repression of Rot, which transcriptionally activates *spa* and represses *hla* (Fig. 1c). Such so-called feedforward motifs are highly overrepresented in gene regulatory networks and have been shown to ensure a tight regulation of gene expression (61, 62). Recently, the Margalit group and collaborators recognized that the combination of these two coherent feedforward loops with opposing effects on their targets (one negative and one positive) effectively serves as a new type of genetic switch. They termed it the double selector switch (DSS) (63). One of the interesting properties of the sRNA-based DSS is that it ensures fine-tuned coordination among targets, so that target 2 (*hla*) will only be activated *after* downregulation of target 1 (*spa*) (63). The general picture of the RNAIII regulon appears to be that RNAIII represses the expression of cell surface proteins (including protein A), which enables *S. aureus* to adhere to host tissue, form biofilm, and evade the host immune system, while RNAIII activates the expression of exotoxins (including α-toxin) and superantigens (reviewed in reference 64). The DSS can therefore be argued to mediate the tran-

sition from a defensive to an offensive lifestyle in response to QS signals (63).

It should be mentioned that many of the studies on the RNAIII regulon were carried out in laboratory strains derived from NCTC8325-4 (65), which produces relatively high amounts of RNAIII (66). Different strains of *S. aureus* have been reported to display wide variance in the levels of RNAIII (23, 66–68). Some of this variation can be explained by differences in the QS dynamics of the four *agr* subtypes (69). Additional variation comes from the effects of the large array of regulatory proteins whose effects impinge on either the *agr* locus itself or on specific downstream targets in response to environmental stress signals and the metabolic status of the cell (reviewed in references 35 and 70). An abundance of additional virulence regulators have also been found among the sRNAs encoded by the core genome and the foreign genetic elements of *S. aureus* (71). One of these, the sRNA *psm-mec*, has been shown to base-pair with the *agr* RNAII transcript and repress translation of AgrA (72). *psm-mec* is encoded by the staphylococcal cassette chromosome of hospital-associated methicillin-resistant *S. aureus* (MRSA), but not community-associated MRSA. Much like RNAIII, *psm-mec* is also an AgrA-activated dual-function transcript that encodes a small cytotoxin, PSMα, in addition to acting as a regulatory RNA. Since AgrA activates *psm-mec* transcription, the *psm-mec*-mediated repression of *agrA* translation closes a negative feedback loop. In fact, several feedback loops between the *agr* locus and targets of RNAIII appear to exist (*agr* → MgrA → SarX ⊣ *agr*) (73, 74) and (*agr* ⊣ Rot → SarA → *agr*) (75). The potential impact of these feedback loops on the dynamic behavior of the QS circuit has not yet been systematically studied.

THE QUORUM REGULATORY RNAs OF THE *VIBRIONACEAE* FAMILY

The quorum regulatory RNAs (Qrr's) were first identified in 2004 (76). The Qrr sequences are found exclusively in members of the *Vibrionaceae* family, and a total of 159 species-assigned Qrr sequences are now collected in the Rfam database of noncoding RNA families (77). Of these, the 107-nt-long Qrr4 of the bioluminescent marine bacterium *Vibrio harveyi* has been most comprehensively characterized (Fig. 2a) (78). The Qrr's act as Hfq-dependent *trans*-acting sRNAs to activate or repress translation of target mRNAs via base-pairing to the 5′ UTR (76, 78–82).

V. harveyi Qrr4 contains four stem-loops, SL1 to SL4. SL1 is involved in base-pairing with a subset of

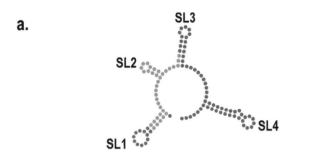

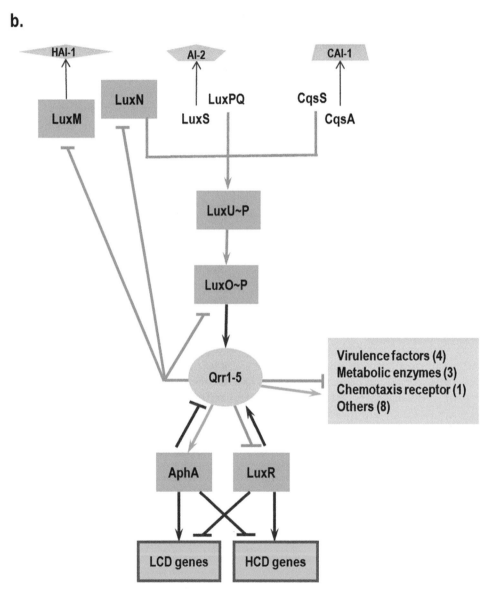

Figure 2 QS regulation by the Qrr sRNAs in *V. harveyi*. (a) Secondary structure of *V. harveyi* Qrr4. The region that is completely conserved among Qrr1 to -5 in *V. harveyi* is shown in red. The 9 nt of stem-loop 1 that are missing in Qrr1 are shown in green. Modified from Shao et al. (81) and depicted using Forna (192). (b) The QS pathway of *V. harveyi*. The symbols and colors are used as in Fig. 1. Direct autorepression by the transcription factors is omitted for clarity but has been demonstrated for all three of the transcription factors shown (LuxO, AphA, and LuxR). LCD genes, genes expressed at low cell density; HCD genes, genes expressed at high cell density.

Qrr targets and is important for Qrr sRNA stability (78, 81, 82). SL2 contains the nucleotides most critical for target specificity but is structurally flexible, meaning that stem-loop formation is not a requirement for Qrr functionality, at least not for regulation of the two mRNA targets tested by Rutherford et al. (78). SL3 plays an accessory role in base-pairing and stability (78), while the fourth stem-loop is the Rho-independent transcription terminator (76).

As depicted in Fig. 2b, the Qrr sRNAs are expressed when the extracellular concentrations of the three AIs of *V. harveyi* (HAI-1, AI-2, and CAI-1) are low. In the ligand-free configuration, three membrane-bound AI receptors (LuxN, LuxQ, and CqsS) act as kinases to phosphorylate the phosphotransfer protein LuxU, which in turn phosphorylates the response regulator LuxO. In its phosphorylated form, LuxO~P acts together with the alternative sigma factor σ^{54} to activate *qrr* sRNA transcription. Conversely, the receptors act as phosphatases in the AI-bound conformation and drain phosphate away from LuxO via LuxU. As a result, *qrr* transcription is turned off at high extracellular concentrations of AIs (83).

The signal transduction pathway consisting of membrane-bound receptors that initiate a phosphorelay, which culminates in transcriptional control of one or more Qrr sRNAs, is conserved in other *Vibrio* species. But there are interesting species-specific variations in the number of synthase/receptor pairs that feed in to the phosphorelay (see, e.g., references 15 and 84); the existence of additional, parallel QS pathways (14, 85); the identity of auxiliary regulators (86, 87); and the identity of the Qrr target mRNAs (88, 89).

A total of 20 target mRNAs have been identified for the Qrr sRNAs in *V. harveyi* (81, 90–93), and more are likely to exist (81, 94). Two of these, *luxR* and *aphA*, encode master transcription factors that control the vast majority of the QS regulon. The Qrr sRNAs modestly activate *aphA* translation by revealing the ribosome binding site of the *aphA* mRNA (79, 91) and strongly repress *luxR* translation by occupying the ribosome binding site and mediating degradation of the *luxR* mRNA (76, 79, 93). Thus, at low cell densities (LCD), when the Qrr sRNAs are abundant, AphA is at its highest level and LuxR is at its lowest level, and vice versa at high cell densities (HCD), when the Qrr sRNAs are absent (95). In this manner, the information relayed by the three AI sensors is integrated into a cell-density-dependent gradient of Qrr sRNAs, which is in turn transformed into opposing gradients of the AphA and LuxR transcriptional regulators. In total, AphA and LuxR regulate >700 genes in *V. harveyi* (91, 94,

96, 97), thereby controlling a diverse set of behaviors such as bioluminescence (98, 99), type III and type VI secretion (91, 94, 95), motility (91), the osmotic stress response (97), and many more.

The Qrr sRNAs Participate Extensively in Regulatory Feedback Loops

Strikingly, the QS signaling network is loaded with negative feedback loops, four of which occur between the Qrr sRNAs and their targets (Fig. 2b). In the first feedback loop, the Qrr sRNAs prevent *luxR* translation, but LuxR activates *qrr* transcription. Importantly, LuxR cannot bypass the need for LuxO~P in *qrr* transcriptional activation, meaning that the *luxR-qrr* feedback is only operative when LuxR and LuxO~P are both present in the cell. This situation occurs shortly after a shift from HCD to LCD conditions, and again when the population has grown to intermediate cell densities and LuxR begins to accumulate (100, 101). Predictably, lack of the *luxR-qrr* feedback results in a delayed regulatory response to LCD conditions in the HCD-to-LCD transition and lowers the AI threshold at which LuxR activates transcription of HCD-appropriate genes in the LCD-to-HCD transition (100, 101). The three additional negative feedback loops together set an upper limit for Qrr accumulation at LCD. First, the Qrr sRNAs activate *aphA* translation, but once accumulated, AphA represses *qrr* transcription (91). Second, LuxO~P activates *qrr* transcription, but once the Qrr sRNAs have accumulated, they repress translation of *luxO* mRNA (90, 102). Third, the Qrr sRNAs destabilize the *luxMN* message encoding the synthase of HAI-1, LuxM, and the HAI-1 receptor, LuxN (92). At LCD, where HAI-1 is below the minimal threshold concentration for detection, Qrr repression of *luxMN* feeds back to lower *qrr* transcription. This is because ligand-free LuxN contributes positively to LuxO~P levels, and by extension to *qrr* transcription levels. As a result of the negative feedback regulation, the Qrr's do not completely shut down *luxR* translation at any cell density. Maximally repressed levels of LuxR, as seen in a triple AI synthase mutant, correspond to ~110 copies of LuxR protein per cell (92, 103).

The Qrr-centered feedback loops must be considered in the context of the whole QS network. This includes additional important feedback loops in the form of direct autorepression by each of the three transcription factors LuxO, AphA, and LuxR, as well as transcriptional repression by AphA on *luxR*, and by LuxR on *aphA*, which will not be described further here (see reference 104 for a recent review). The intricate feedback circuitry of the *V. harveyi* QS signal transduction pathway is responsible for some of its most remarkable features. It expands the dynamic range of the AI input concentrations and enables *V. harveyi* to differentially emphasize the contribution by the species-specific HAI-1 signal and the interspecies AI-2 signal at different cell densities (92). It also compresses the dynamic range of the LuxR output to a mere 6-fold difference between LCD and HCD LuxR levels. This ensures the sensitivity necessary for the network to swiftly adapt in response to abrupt changes in the extracellular AI concentrations, and reduces cell-to-cell heterogeneity with respect to LuxR content (92).

Cross Talk among Qrr Targets

Are there properties of the *V. harveyi* QS network that are uniquely dependent on the central regulators being sRNAs rather than proteins? One striking difference between the two types of regulation is whether the act of regulating a target feeds back to regulate the regulator. Many sRNAs have been reported to function by coupled degradation of the sRNA-mRNA pair, meaning that the outcome of pairing between an sRNA and its mRNA target is degradation of both the sRNA and the mRNA (105). In these cases, a high copy number of a given target mRNA can strongly reduce the level of the sRNA regulator. This reduces the ability of the sRNA to regulate other mRNA targets, resulting in the potential for hierarchical cross talk between target mRNAs (106, 107). The ability of a given target mRNA to affect the level of the sRNA depends on three parameters: the rate of transcription of the sRNA relative to that of the mRNA, the strength of the sRNA-mRNA interaction (106, 108), and the molecular mechanism of regulation (79). Whereas the latter two are only tunable on evolutionary time scales, the first is affected both by changes in the rate of transcription of the sRNA, which in the case of the Qrr's depends on the concentration of LuxO~P, and by the transcription rates of the individual target mRNAs, each of which is typically controlled by transcription factors that respond to other external or intracellular cues. Therefore, a potential exists for highly dynamic temporal changes to the Qrr target hierarchy that could serve to integrate information about environmental factors or intracellular signals with the information about AI levels relayed by LuxO~P. Remarkably, Feng et al. (79) showed that Qrr3 represses three of its mRNA targets by three different mechanisms, showcasing the great versatility of regulatory mechanisms available to a single sRNA. While the *luxMN* mRNA is repressed by coupled degradation (sRNA-mRNA pairing results in degradation of both the mRNA and the sRNA), *luxO* mRNA is repressed

by sequestration of the ribosome binding site, and Qrr3 acts catalytically (one sRNA can be recycled to repress more than one mRNA) to degrade *luxR* mRNA. The different mechanisms for repression give distinct regulatory dynamics for each of the three targets, which is important for the proper functioning of the QS circuit (79). Further, the fact that Qrr3 acts catalytically on *luxR* mRNA means that *luxR* repression minimally affects the concentration of Qrr's available to interact with other targets (79). Thus, the network of potentially cross-talking mRNAs among the >20 Qrr targets is limited to the subset that interact stoichiometrically with the Qrr sRNAs, and does not include *luxR* mRNA.

Sibling Qrr sRNAs

The four *qrr* sRNA genes of *Vibrio cholerae* (named *qrr1–4*) were originally revealed by a bioinformatic search for intergenic sequences flanked by predicted σ^{54}-promoter and Rho-independent terminator sequences (76). They act redundantly, meaning that any one of the Qrr sRNAs is sufficient to fully repress the *V. cholerae* LuxR homolog, HapR (76). Intriguingly, Qrr1–5 of *V. harveyi* do not act redundantly. Instead, they repress *luxR* translation in an additive fashion, in the sense that quadruple mutants harboring only one of the five *qrr* genes each express a distinct residual level of LuxR (93). The redundant phenotype of the *V. cholerae* Qrr sRNAs is due to the indirect control by each Qrr of the transcription of the others via the Qrr-luxO, Qrr-hapR, and presumably Qrr-aphA and Qrr-luxMN feedback loops (102). A natural question to ask is why these same feedback loops don't result in efficient dosage compensation among the *V. harveyi* Qrr sRNAs. The answer may be that some or all of the *qrr* genes of *V. harveyi* are subject to individual regulation by additional species-specific transcription factors (93), which could override the effect of the feedback loops on *qrr* transcription. In fact, the sequences of the regulatory regions upstream of *qrr1–5* are quite distinct outside of the conserved LuxO- and σ^{54}-recognition sites (93), and by contrast to *qrr2–4*, the *qrr1* and *qrr5* promoters are neither repressed by AphA nor activated by LuxR (91, 101).

The existence of sibling families of sRNAs, where two or more homologous sRNAs are encoded in a single genome, is a recurring theme in bacterial sRNA-based regulation (109). Besides the Qrr sRNAs, which are present in four or five copies in many *Vibrio* genomes (110), sRNA families rich in siblings include the widespread GacA-activated Csr/Rsm sRNA family, which functions to titrate the global posttranscriptional regulator CsrA/RsmA and counts up to seven siblings per genome (111); the seven virulence factor regulators LhrC1–7 of *Listeria monocytogenes* (112, 113); and the five CiaR-regulated sRNAs of *Streptococcus pneumoniae* (114). Actually, two of these sRNA families are involved in QS regulation in *V. cholerae*, where the CsrBCD sRNAs repress the LCD state indirectly by limiting *qrr1–4* expression. They do so by repressing CsrA, which in turn appears to stimulate LuxOP-mediated activation of *qrr1–4* transcription by an unknown mechanism (86).

In addition to the large sRNA families, a substantial number of families counting only two sibling sRNA genes per genome have been reported (115–119). The possible advantages of sRNA multiplicity include a reduced risk of inactivation by mutation and a heightened expression potential due to the elevated gene copy number. Furthermore, multicopy sRNA genes have the potential for evolution of nonredundant sibling sRNAs with overlapping but nonidentical target spectra, each of which can be subject to individual transcriptional regulation.

TWO QS PATHWAYS WITH sRNAs AT THE WHEEL

Two types of sRNAs, RNAIII of *S. aureus* and the Qrr sRNAs of *V. harveyi*, have been described in detail here that each form the core of the QS network of their respective bacteria. It seems likely that the approaches developed for the study of either one of these two QS systems could be employed to answer unsolved questions in the other. For example, a proteomics approach employed by the Bassler lab revealed distinct temporal expression profiles of the set of QS targets that are directly regulated by the Qrr sRNAs and the set that is indirectly regulated (94). This pattern enabled them to predict additional likely direct Qrr targets. A similar approach could be useful for identifying additional direct targets of RNAIII in *S. aureus*. Also, the potential effects of the feedback loops in *S. aureus* on the QS dynamics remain to be explored. These could be addressed via the construction of point mutations in promoter sequences or RNA base-pairing regions to specifically eliminate one regulatory link without affecting the others, as has been done in *V. harveyi* (90, 92). Conversely, the widely recognized interstrain and interspecies diversity in the number and strength of interactions between specific nodes in the QS network of the staphylococci could inspire studies to explore and address differences in the QS networks of different *Vibrio* strains and species. For example, the four Qrr sRNAs of *Vibrio anguillarum* strain NB10 appear to reach the peak of their expression at HCD (120), which cannot currently be reconciled with the existing model of Qrr-mediated QS regulation that was mainly developed in *V. harveyi* and *V. cholerae*.

sRNA-BASED CONTROL OF AI SYNTHASES IS FOUND ACROSS DIFFERENT QS SYSTEMS

The extracellular concentration of an AI will only increase proportionally to the population density of AI-producing cells if the growth environment is constant in terms of factors that can influence the stability or diffusion of the AI, and if the rate of AI synthesis is constant. Such a strictly linear relation between population density and AI concentration is presumably difficult to obtain in most natural habitats, and in fact doesn't seem to be the evolutionary "goal" of most QS systems, as many bacteria evidently change the rate of AI synthase production in response to various environmental signals. This section will discuss typical mechanisms of AI synthase regulation and highlight examples from different species where the regulation is sRNA based.

In all canonical LuxI/LuxR-type QS systems where the LuxR protein is a transcriptional activator, the *luxI* gene encoding the AI synthase is a direct target of LuxR, resulting in a positive feedback loop, as first described for the *V. fischeri* LuxI/LuxR QS system depicted in Fig. 3a (121–126). Positive autoregulation of AI synthesis is also observed in some of the two-component QS regulatory systems of Gram-positive bacteria, as described above for the *agr* system of *S. aureus* (reviewed in reference 127). The positive feedback loop has the distinct advantage of synchronizing the QS response across a bacterial population. Synchrony occurs because activation of the QS response in a subset of cells in the population speeds up the extracellular accumulation of the AI, thereby triggering AI detection and activation of the QS response in the remainder of the population (126, 128, 129).

A positive feedback loop can also be mediated by the removal of a repressor of AI synthase expression at HCD. This is the case for *V. harveyi*, where HAI-1 binding to the LuxN receptor leads to termination of *qrr* transcription, which alleviates Qrr sRNA repression of *luxMN* mRNA (Fig. 3b). The result is an elevated rate of production of the LuxM HAI-1 synthase (and its receptor) at high HAI-1 levels (92). Similarly, the Qrr sRNAs repress the synthesis of the CqsS enzyme, which produces CAI-1 in *V. cholerae*, but in this case the repression appears to be indirect, as no pairing between the Qrr sRNAs and the *cqsS* mRNA could be

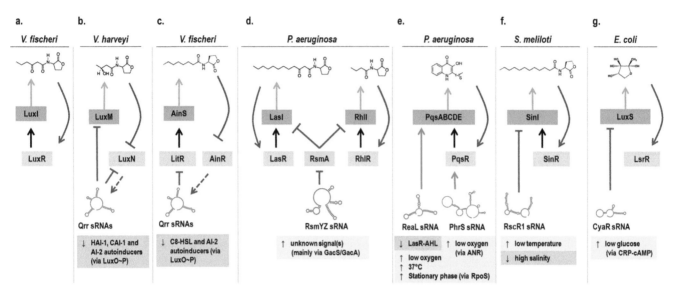

Figure 3 sRNA-based regulation of AI synthase production. AI synthases are shown as red rectangles with a red arrow pointing to the structure of the AI(s) they produce. A curved arrow from the AI to its receptor (brown rectangles) indicates activation of the receptor upon binding to the AI, whereas a curved blocked arrow indicates inactivation of the receptor upon AI binding. Posttranscriptional regulatory interactions are shown by green arrows (increased translation) or red blocked arrows (decreased translation), and transcriptional regulatory interactions are depicted in black. Known signals/conditions that affect sRNA expression are shown below each sRNA. (a) Positive feedback control of 3-oxo-C6-HSL synthesis in *V. fischeri*. (b) Repression of the HAI-1 synthase, LuxM, by the Qrr sRNAs in *V. harveyi*. Blocked arrow from HAI-1 to LuxN indicates that HAI-1 binding blocks the kinase activity of LuxN. The phosphatase activity of LuxN is not affected by AI binding (193). The dashed arrow indicates that LuxN promotes *qrr* transcription indirectly through a phosphorelay when LuxN is acting as a kinase (see Fig. 2b). (c) Positive feedback control of C8-HSL synthesis in *V. fischeri* involves a single Qrr sRNA. Symbols are as in panel b. (d) Positive effect of the RsmYZ sRNAs on the LasI and RhlI AHL synthases of *P. aeruginosa*. Regulatory interactions between the two QS systems are excluded for clarity. (e) Positive effects of the ReaL and PhrS sRNAs on PQS production in *P. aeruginosa*. (f) Repression of *S. meliloti* AHL production by the RscR1 sRNA. SinI produces a range of long-chain AHLs, exemplified here by *N*-(tetrahydro-2-oxo-3-furanyl)-dodecanamide (194). (g) CyaR sRNA represses translation of the AI-2 synthase LuxS in *E. coli*. The depicted isoform of AI-2 is *R*-2-methyl-2,3,3,4-tetrahydroxytetrahydrofuran (R-THMF), which is the AI-2 isoform typically recognized by the Lsr machinery that transports AI-2 into the cytoplasm. AI-2 is phosphorylated before AI-2~P binds and inactivates the LsrR transcriptional repressor. Inactivation of LsrR feeds back to affect extra- and intracellular AI-2 levels because it derepresses production of the Lsr transport system, which leads to increased AI-2 internalization and depletion of extracellular AI-2. That feedback loop is not included here since it does not affect AI-2 synthesis (reviewed in reference 11).

demonstrated (130). Feedback regulation of CqsS synthesis leads to a substantial, 7-fold increase in CAI-1 levels in cultures of a mutant locked in the HCD QS mode compared to a mutant locked in the LCD mode (130). Another variation of positive feedback on AI synthesis in *Vibrio* QS systems is found for the AHL synthase AinS in *V. fischeri*. In addition to the LuxI/LuxR QS system mentioned above, which synthesizes and detects *N*-3-oxohexanoyl-L-homoserine lactone (3-oxo-C6-HSL), *V. fischeri* possesses another AHL-based

QS pathway, which is largely analogous to the one described for *V. harveyi* in Fig. 2b. AinS produces *N*-octanoyl-L-homoserine lactone (C8-HSL), which is recognized by the membrane-bound histidine kinase receptor AinR, resulting in deactivation of the single Qrr sRNA of *V. fischeri*. In the absence of the Qrr sRNA, translation of the QS master transcription factor LitR (LuxR in *V. harveyi*) is derepressed. The autoregulatory loop is completed by LitR activating transcription of the *ainRS* operon (Fig. 3c) (84).

Notably, not all QS systems include positive feedback on the AI synthase(s), and it is not a prerequisite for QS control of gene expression (126, 131). Besides positive autoregulation, AI synthase production is often coordinated with non-QS signaling pathways, and this cross-regulation is frequently mediated by sRNAs. In *Pseudomonas aeruginosa*, translation of the two AHL synthases LasI and RhlI is indirectly activated by sRNAs at HCD (Fig. 3d). Translation of *lasI* and especially *rhlI* is repressed when bound to the RBP RsmA, which in turn is counteracted by the protein-binding sRNAs RsmY and RsmZ, which titrate RsmA away from its mRNA targets (132, 133). Transcription of RsmYZ is upregulated by the GacS/GacA two-component system. A thorough understanding of the regulatory circuits driven by this class of sRNAs is generally hampered by the lack of definitive identification of the molecular signals that activate the GacS sensor, although researchers are closing in on the signal in some organisms. For example, in *P. aeruginosa* PA01, the signal accumulates at HCD and requires GacA for its production, resulting in the positive feedback loop typical of AI production in QS circuits (134). However, the lipophilic signal appears to be unrelated to known AIs, and is likely a Krebs cycle intermediate or derivative thereof (135, 136). The Csr/Rsm sRNA family is found in many gammaproteobacteria, but the number of sibling sRNAs and the metabolites that control their synthesis and degradation appear to differ, presumably reflecting their adaptation to particular niches (137).

Synthesis of a third AI of *P. aeruginosa*, 2-heptyl-3-hydroxy-4-quinolone (PQS), is also subject to RNA-based regulation (Fig. 3e). The sRNA PhrS activates synthesis of PQS by directly activating translation of the *pqsR* regulator (138), which activates the PQS synthesis operon *pqsABCDE* through a positive feedback loop (139). PhrS expression is activated by the oxygen-responsive transcription factor ANR, thereby coupling increased QS signal production to oxygen limitation (138). Recently, it was reported that the sRNA ReaL is likely to base-pair extensively with the *pqsABCDE* mRNA in the intercistronic region between *pqsB* and *pqsC* (140). Consistent with a positive effect of ReaL on *pqsC* translation, PQS synthesis increases upon ReaL overexpression and is reduced in a Δ*reaL* mutant (140). Like PhrS, ReaL is induced by oxygen limitation. In addition, ReaL accumulation was observed during growth at body temperature, and in stationary phase. Interestingly, the LasR-AHL complex was identified as a negative regulator of *reaL* transcription, thereby forming an additional regulatory link between the highly intertwined QS systems of *P. aeruginosa* (140). An additional AI

synthase that is directly targeted by an sRNA is SinI of the legume symbiont *Sinorhizobium meliloti*. The sRNA RcsR1 (rhizobial cold and salinity stress riboregulator 1) base-pairs with the *sinI* transcript and represses its translation (Fig. 3f). RcsR1 was induced at low temperature and strongly reduced at high salinity, thereby apparently providing a link between temperature, salinity stress, and increased QS signal production (141). Finally, in *E. coli*, the *luxS* mRNA encoding the AI-2 synthase is bound and repressed by the sRNA CyaR, which is activated by CRP-cAMP (Fig. 3g) (142). The second messenger cAMP (cyclic AMP) accumulates in the absence of rapidly metabolizable carbon sources, such as glucose, and acts as an inducer of the transcription factor CRP (cAMP receptor protein). CRP-cAMP also activates the Lsr operon encoding the apparatus required for AI-2 internalization and processing (143). Thus, both AI-2 synthesis and AI-2 detection are strongly tied to carbon source availability in *E. coli*. In summary, AI synthesis rates are rarely constant across growth conditions, but are often subject to regulation in response to AI detection (autoregulation) and/or other environmental parameters (salinity, oxygen limitation, temperature, nutrient availability, and growth phase). Coordination of AI synthesis rates with other signaling pathways is frequently obtained by sRNA-based posttranscriptional regulation.

REGULATION OF BIOFILM FORMATION AND DISPERSAL

In the biofilm mode of growth, bacteria are surrounded by a protective, self-produced matrix. The identity of the extracellular polymeric substances (EPS) that make up the matrix is highly variable. Depending on the bacterial species and the growth conditions involved, it may contain proteinaceous material (amyloids, lectins, flagella and other appendages, and enzymes), exopolysaccharides (adhesins, alginate, cellulose, glucans, and more), and in some cases extracellular DNA (eDNA), lipids, and biosurfactants (reviewed in references 144 and 145). Bacteria thus do not simply engage or not engage in biofilm formation, but may also choose between alternative pathways for biofilm formation that result in distinct EPS compositions of the biofilm matrix. Unsurprisingly, extensive regulatory networks have evolved that allow relevant external and intracellular conditions to affect the likelihood of commitment to the costly synthesis and export of matrix components. Some of the most common key parameters include QS signals, levels of the intracellular second messenger cyclic diguanylate (c-di-GMP), nutrient availability, temperature, and carbon metabolism status.

A key role for QS in biofilm regulation was first demonstrated in *P. aeruginosa*, where proper biofilm structure was shown to depend on extracellular *N*-3-oxododecanoyl-L-homoserine lactone (3-oxo-C12-HSL), the AI of the LasI/LasR QS system (146). In *P. aeruginosa*, and in many other bacteria, biofilm formation or maturation is promoted by QS, consistent with the intuitive notion that biofilm formation should be a group behavior. But there are also examples of organisms where QS has a negative effect on biofilm and promotes its dispersal, as in the case of *V. cholerae*. This species-dependent variation has been proposed to be linked to the differing requirements of bacterial pathogens in chronic versus acute biofilm infections (147), where the lifelong *P. aeruginosa* infections of cystic fibrosis patients is an example of the former, and the rapid-onset severe diarrheal disease caused by *V. cholerae* represents the latter. In *V. cholerae*, biofilm formation is under negative regulation by QS and positive control by c-di-GMP (reviewed in reference 148). The key transcriptional activator of biofilm is the VpsT protein, which is activated by binding to c-di-GMP (149). The Qrr sRNAs indirectly activate VpsT transcription at LCD via their activation of *aphA* and repression of *hapR* translation (150, 151). They also upregulate c-di-GMP levels directly by binding to and stabilizing the *vca0939* mRNA, encoding a GGDEF domain-containing protein (89). A new and unrelated QS system has recently been identified in *V. cholerae* that works in parallel with the Qrr-dependent QS systems to repress biofilm formation at HCD. The new AI molecule is 3,5-dimethylpyrazin-2-ol (DPO). DPO binds a cytoplasmic receptor called VqmA, and the VqmA-DPO complex activates the transcription of the Hfq-dependent sRNA VqmR, which represses translation of at least eight target mRNAs, including *vpsT* (14, 152).

The Qrr and VqmR sRNAs are prime examples of sRNAs that function to link information from QS to regulation of biofilm behavior. In the following section, examples of how other sRNAs function to integrate additional inputs besides QS into the networks controlling the biofilm phenotype will be discussed.

Protein-Binding sRNAs That Regulate Biofilm Formation

sRNAs were first directly implicated in a bacterial biofilm phenotype in 1993, when Romeo and colleagues reported that biofilm formation in *E. coli* was deficient in a *csrB* null mutant and stimulated by CsrB overproduction (153). CsrB encodes an Hfq-independent regulatory RNA that, together with its less potent sibling sRNA CsrC, antagonizes the activity of the global RNA-binding regulatory protein CsrA by directly binding and sequestering multiple CsrA dimers (154, 155). The well-conserved CsrA/RsmA family of posttranscriptional regulators repress biofilm formation while promoting motility in several bacterial species, and have been most extensively studied in *E. coli*, *P. aeruginosa*, and *Pseudomonas fluorescens*. The important, multi-tiered negative effect of CsrA on biofilm formation in *E. coli* has been reviewed recently (162 and 195, see also chapter 19), and will not be discussed further here.

Another *Pseudomonas* protein-binding sRNA, CrcZ, has been shown to affect biofilm formation. The CrcZ sRNA was highly upregulated during anoxic biofilm formation in *P. aeruginosa* PA14, suggesting it could be important for biofilm formation under the anaerobic condition, which is believed to occur in the lungs of cystic fibrosis patients. Accordingly, a *crcZ* null mutant produced more biofilm than the corresponding wild type in the anoxic assay, and overexpression of CrcZ repressed biofilm formation (156). By contrast, a *crcZ* deletion mutant of *P. aeruginosa* PA01 showed decreased biofilm formation compared to the wild type in a standard (aerobic) microtiter plate biofilm assay (157), suggesting perhaps that the anoxic condition is required for the CrcZ effect observed with *P. aeruginosa* PA14. Recent work has shed light on the CrcZ mechanism of action. Interestingly, A-rich sRNAs such as CrcZ can simultaneously bind the catabolite repression control protein Crc and Hfq, which allows Crc to affect Hfq-mediated translation repression of target RNAs by a mechanism that is still not completely understood (158). It has not yet been demonstrated whether this mechanism of Hfq-CrcZ-Crc complex formation is responsible for the role of CrcZ in anoxic biofilm formation, but in support of this notion, it was found that Hfq is crucial for anoxic biofilm formation, and CrcZ was identified as the most abundant sRNA on Hfq during anoxic biofilm formation (156).

Control of Biofilm Formation by Base-Pairing sRNAs

E. coli K-12 appears to employ one of two major alternative pathways for biofilm formation, resulting in either a curli-dominated biofilm, which requires the alternative sigma factor σ^S and the CsgD transcriptional activator; or an adhesin-dominated biofilm formed by activation of the *pgaABCD* operon, which catalyzes the synthesis of poly-β-1,6-*N*-acetyl-D-glucosamine (PGA) adhesin (Fig. 4) (159–161). Flagella play a role in the early stages of biofilm formation by mediating surface contact and initial adhesion, but ultimately, motility is the absolute alternative to the sessile biofilm lifestyle.

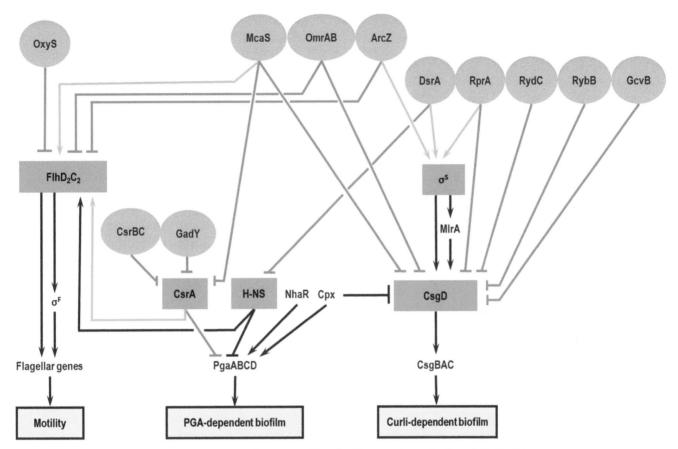

Figure 4 Posttranscriptional control of key biofilm regulators in *E. coli* K-12. Direct posttranscriptional effects of sRNAs (blue ovals) on five protein regulators central to the biofilm/motility switch (colored rectangles) are shown as green arrows (positive regulation) or red blocked arrows (negative regulation). Only direct sRNA-target interactions are shown, although several additional sRNAs indirectly regulate one or more of the five targets (159, 164). Some transcriptional regulation is included to provide context, but many important regulators and interactions are omitted for clarity. These include all proteins involved in c-di-GMP signaling, and interactions concerning σ-factor competition (reviewed in reference 160).

In *E. coli*, $FlhD_2C_2$ is the master activator of the flagellar cascade, which leads to flagella production and hence motility. A multilayered regulatory network ensures that motility and biofilm formation gene expression programs are mutually exclusive. This includes competition between σ^{70}/σ^F (motility) and σ^S (biofilm) for RNA polymerase core enzyme, competition between σ^S (biofilm) and the DNA-binding repressor protein FliZ (motility) for a subset of σ^S-dependent promoters, and, importantly, opposite regulation by c-di-GMP (high for biofilm, low for motility), as expertly reviewed in references 160 and 162. On top of the transcriptional network, extensive regulation at the posttranscriptional level is carried out by numerous *trans*-acting sRNAs that target the key regulators *flhDC*, *rpoS*, and *csgD*,

as well as CsrA and *hns*, which control expression of *pgaABCD* (see Fig. 4). Since most of these sRNAs require Hfq for their action (163), this may explain the severe biofilm defect of an *hfq* mutant.

Four *E. coli* sRNAs—ArcZ, McaS, OmrA, and OmrB—directly regulate both motility and biofilm formation. Two of them, ArcZ and McaS, contribute in opposite ways to regulate the motility and curli-dependent biofilm formation gene expression programs. ArcZ inhibits *flhDC* translation (164) while activating *rpoS* translation (165), and an *arcZ* deletion mutant is defective for curli-dependent biofilm formation (159). In *Salmonella enterica* serovar Typhimurium, ArcZ also promotes biofilm formation through regulation of curli fiber and type 1 fimbriae production, in an unsolved manner that appears independent of RpoS regulation (166). In *E. coli*, very minor effects were detected on type 1 and curli fimbriae in an *arcZ* mutant (163), but an RpoS-independent effect on biofilm formation is nevertheless likely, because the biofilm defect of the *arcZ* mutant was only partly complemented by overexpression of *rpoS* (164). The only known regulator of *arcZ* transcription is the ArcA/ArcB two-component system, which represses *arcZ* in microaerobic and anaerobic conditions. The ArcA/ArcB system is additionally reported to cause repression of *rpoS* at the levels of transcription and σ^S protein stability (167) and cause activation of several class II flagellar genes (168, 169), suggesting that ArcZ posttranscriptionally cements the transcriptional effects of ArcA/B on the biofilm and motility pathways.

McaS activates *flhDC* while repressing *csgD* (170, 171). The upregulation of motility is consistent with the expression profile of McaS, which is activated by CRP-cAMP during growth in poor carbon sources and during the transition into stationary phase (171). *E. coli* motility increases as it is grown on progressively poorer carbon sources, suggesting that the bacterium may employ a foraging strategy when the surroundings become nutrient limited (172, 173). Perplexingly, McaS can also act as a titrator of CsrA, which results in a positive indirect effect of McaS on PGA-dependent biofilm formation and a potential negative indirect effect on *flhDC* (171, 174–176). This ability suggests that McaS could reverse its effects on motility and biofilm formation if growth conditions exist where its function as a CsrA decoy dominates over its role as an Hfq-dependent base-pairing sRNA. Two additional sRNAs, GadY and DsrA, also contribute to the mutually exclusive expression of flagella and biofilm formation, albeit indirectly, via their effects on CsrA activity and *hns*/*rpoS* translation, respectively. These are included in recent reviews (160, 162) and are not discussed here beyond their inclusion in Fig. 4.

Notably, regulatory patterns inconsistent with the idea of a mutually exclusive biofilm/motility switch can also be observed in the *E. coli* sRNA network. For example, OmrA and OmrB repress both flagellar synthesis and curli fiber synthesis, through base-pairing with the 5′ UTRs of *flhDC* and *csgD* mRNAs, respectively (164, 177). This regulatory behavior is consistent with an indiscriminative repression of large, surface-bound structures at high osmolarity (162). Also, RprA activates *rpoS* translation but reduces *csgD* expression directly through binding to the *csgD* mRNA (and indirectly by repression of translation of the *ydaM* mRNA, encoding a c-di-GMP synthase) (178). The net result

of RprA action on biofilm matrix production is negative, suggesting that RprA, which is induced by the Rcs phosphorelay pathway that responds to cell envelope stress (179), has evolved the ability to stimulate the σ^S regulon while specifically excluding curli-dependent biofilm formation. Repression of *csgD* translation during cell envelope stress is further increased by pairing with the sRNA RybB (180, see also chapter 13 for a review of sRNA-mediated stress responses). To date, no less than seven sRNAs (McaS, OmrA, OmrB, RprA, RybB, RydC, and GcvB) have been shown to repress *csgD* posttranscriptionally by base-pairing at partially overlapping sites within the 148 nt of the highly structured *csgD* 5′ UTR. The mechanism of *csgD* repression varies between the sRNAs. OmrA and OmrB, for example, appear to block a ribosome loading site without altering the half-life of the *csgD* mRNA (177), while RprA and RydC directly cover the Shine-Dalgarno region and affect both translation initiation and *csgD* mRNA half-life (178, 181). Clearly, *csgD* mRNA, together with *flhDC* and *rpoS* mRNAs, are three busy hubs for sRNA-mediated integration of environmental signals into biofilm regulation in *E. coli* (160, 162).

Base-pairing sRNAs are also important for biofilm regulation in two of the important pathogens discussed in previous sections: Activation of PQS synthesis by the sRNA PhrS in *P. aeruginosa* (138) results in PQS-stimulated release of eDNA, which contributes to the biofilm matrix (182). And to return to this review's starting point, RNAIII affects biofilm development to varying extents in different isolates of *S. aureus*, where it seems to mainly have an adverse effect on biofilm maturation by promoting dispersal at HCD (reviewed in reference 183). The σ^B-activated sRNA RsaA, on the other hand, was shown to promote biofilm formation in at least four different *S. aureus* strain backgrounds through base-pairing with the *mgrA* mRNA, resulting in repression of translation of the transcription factor MgrA, and therefore derepression of surface proteins and release of eDNA necessary for biofilm maturation (184, 185). Since MgrA also activates the *agr* QS system, and is a direct target of RNAIII, RsaA can be added to a growing list of *S. aureus* sRNAs (35) that may modulate RNAIII execution of the QS response by affecting RNAIII production and/or by pairing with overlapping sets of target mRNAs.

CONCLUDING REMARKS

The study of sRNAs in QS yielded surprising insights into bacterial riboregulation early on, including the concepts of positive regulation of mRNA translation by a base-pairing sRNA, dual-function sRNA/mRNA transcripts, and the redundancy properties of sibling sRNA families. In the past decade, the field has continued to be fertile ground for expanding our understanding of the versatile roles sRNA-based regulation can play in bacterial gene regulation. This review aims at giving an overview of this progress. Space limitations mean that it is by no means complete, and I apologize to those colleagues whose work I could not include here.

The most crucial recent findings span from mechanistic insights into the roles of individual nucleotides for determining the outcome of the sRNA-mRNA pairing event (see, e.g., references 43, 78, 79, and 81) to a network-level understanding of how sRNA-mediated feedback loops, mRNA target cross talk, and feedforward motifs can dictate QS signal transduction dynamics, AI synthesis rates, and global signal integration (63, 79). RNAIII of *S. aureus* and the Qrr sRNAs of *V. harveyi* are the centers of their respective QS networks. Some differences in the regulatory logic that govern these two systems are apparent. Most notably, RNAIII is expressed at HCD, while the Qrr sRNAs are expressed at LCD. The similarities among the two systems are nonetheless striking: RNAIII and the Qrr sRNAs both operate at the same level of the QS response cascades, namely, directly downstream of the response regulator, whose phosphorylation level reflects the extracellular AI concentration, and upstream of the majority of the QS regulon. They also both regulate a limited number of target mRNAs directly, while the remainder of their extensive regulon is reached via translational control of two global transcription factors. One of these is positively regulated (MgrA for RNAIII and AphA for the Qrr sRNAs), and one is negatively regulated (Rot for RNAIII and LuxR for the Qrr sRNAs). Future computational and systems biology approaches may reveal whether this shared setup holds particular advantages for QS execution by sRNAs.

Among the complex behaviors that can be carried out by bacteria, biofilm formation stands out as a remarkable collaborative feat that gives the participating group of microorganisms characteristics resembling a multicellular unit. Posttranscriptional regulation by sRNAs plays a critical role in biofilm development in all the best-studied biofilm model systems. In *E. coli*, the transcripts of a handful of key biofilm/motility regulators are targeted by multiple sRNAs and stand out as pivot points for posttranscriptional biofilm regulation. In other bacteria, such as *P. aeruginosa* and *S. aureus*, sRNA regulation of biofilm appears more scattered. Future research may well change this image as functional roles are assigned to more of the many uncharacterized sRNAs in these species (186, 187). In *E. coli*,

the biofilm regulatory web of posttranscriptional interactions can largely be superimposed on interactions between signal transduction pathways that are established at the transcriptional level. Besides backing up the transcriptional regulatory pattern, however, the sRNA network has an unparalleled capacity for cross talk between these pathways due to the stoichiometric interactions between sRNAs and target mRNAs. For example, a drastic increase in the transcription rate of one mRNA, especially one of the sRNA hubs *csgD*, *flhDC*, and *rpoS*, could result in the rapid depletion of the ensemble of sRNAs that pair with it. This would in turn affect their other mRNA targets, consequently propagating a wave of changes in the cellular RNA landscape, as also pointed out by Mika and Hengge (162, see also chapter 25). Similarly, a sharp increase in the abundance of just one of the dozens of mRNAs targeted by the Qrr sRNAs (or RNAIII for *S. aureus*) could have a dramatic impact on the QS response. The importance of such regulatory "waves" has yet to be explored experimentally in the context of QS and biofilm formation, and represents an intriguing direction for further studies. Taking a step further back, RNAs can potentially affect one another even without shared targets or shared regulators, through competition for the same pool of RBPs. Recent work has demonstrated that the characterization of bacterial RBPs and their RNA ligands is far from complete (188, 189), and new RBPs of importance for sRNA regulation of biofilm formation were recently reported in *P. aeruginosa* (158) and *S.* Typhimurium (190). This clearly demonstrates that we still have much yet to learn about sRNA-based regulation of bacterial collective behaviors.

Acknowledgments. I am grateful to Michael A. Sørensen for critical reading of the manuscript. Current research in the Svenningsen lab is funded by the Danish National Research Foundation (BASP: DNRF120).

Citation. Svenningsen SL. 2018. Small RNA-based regulation of bacterial quorum sensing and biofilm formation. Microbiol Spectrum 6(4):RWR-0017-2018.

References

1. Bejerano-Sagie M, Xavier KB. 2007. The role of small RNAs in quorum sensing. *Curr Opin Microbiol* **10:**189–198.

2. Kay E, Reimmann C, Haas D. 2006. Small RNAs in bacterial cell-cell communication. *Microbe Mag* **1:**63–69.

3. Grosso-Becerra MV, Croda-García G, Merino E, Servín-González L, Mojica-Espinosa R, Soberón-Chávez G. 2014. Regulation of *Pseudomonas aeruginosa* virulence factors by two novel RNA thermometers. *Proc Natl Acad Sci U S A* **111:**15562–15567.

4. Ke X, Miller LC, Ng WL, Bassler BL. 2014. CqsA-CqsS quorum-sensing signal-receptor specificity in *Photobacterium angustum*. *Mol Microbiol* 91:821–833.

5. Engebrecht J, Nealson K, Silverman M. 1983. Bacterial bioluminescence: isolation and genetic analysis of functions from *Vibrio fischeri*. *Cell* 32:773–781.

6. Engebrecht J, Silverman M. 1984. Identification of genes and gene products necessary for bacterial bioluminescence. *Proc Natl Acad Sci U S A* 81:4154–4158.

7. Fuqua C, Greenberg EP. 2002. Listening in on bacteria: acyl-homoserine lactone signalling. *Nat Rev Mol Cell Biol* 3:685–695.

8. Papenfort K, Bassler BL. 2016. Quorum sensing signal-response systems in Gram-negative bacteria. *Nat Rev Microbiol* 14:576–588.

9. Hawver LA, Jung SA, Ng WL. 2016. Specificity and complexity in bacterial quorum-sensing systems. *FEMS Microbiol Rev* 40:738–752.

10. Monnet V, Juillard V, Gardan R. 2016. Peptide conversations in Gram-positive bacteria. *Crit Rev Microbiol* 42:339–351.

11. Pereira CS, Thompson JA, Xavier KB. 2013. AI-2-mediated signalling in bacteria. *FEMS Microbiol Rev* 37:156–181.

12. Chen X, Schauder S, Potier N, Van Dorsselaer A, Pelczer I, Bassler BL, Hughson FM. 2002. Structural identification of a bacterial quorum-sensing signal containing boron. *Nature* 415:545–549.

13. Higgins DA, Pomianek ME, Kraml CM, Taylor RK, Semmelhack MF, Bassler BL. 2007. The major *Vibrio cholerae* autoinducer and its role in virulence factor production. *Nature* 450:883–886.

14. Papenfort K, Silpe JE, Schramma KR, Cong JP, Seyedsayamdost MR, Bassler BL. 2017. A *Vibrio cholerae* autoinducer-receptor pair that controls biofilm formation. *Nat Chem Biol* 13:551–557.

15. Jung SA, Chapman CA, Ng WL. 2015. Quadruple quorum-sensing inputs control *Vibrio cholerae* virulence and maintain system robustness. *PLoS Pathog* 11:e1004837.

16. Shikuma NJ, Fong JC, Odell LS, Perchuk BS, Laub MT, Yildiz FH. 2009. Overexpression of VpsS, a hybrid sensor kinase, enhances biofilm formation in *Vibrio cholerae*. *J Bacteriol* 191:5147–5158.

17. Bassler BL, Losick R. 2006. Bacterially speaking. *Cell* 125:237–246.

18. Marraffini LA. 2017. Sensing danger. *Proc Natl Acad Sci U S A* 114:15–16.

19. Lowy FD. 1998. *Staphylococcus aureus* infections. *N Engl J Med* 339:520–532.

20. Abdelnour A, Arvidson S, Bremell T, Ryden C, Tarkowski A. 1993. The accessory gene regulator (*agr*) controls *Staphylococcus aureus* virulence in a murine arthritis model. *Infect Immun* 61:3879–3885.

21. Gillaspy AF, Hickmon SG, Skinner RA, Thomas JR, Nelson CL, Smeltzer MS. 1995. Role of the accessory gene regulator (*agr*) in pathogenesis of staphylococcal osteomyelitis. *Infect Immun* 63:3373–3380.

22. Cheung AL, Eberhardt KJ, Chung E, Yeaman MR, Sullam PM, Ramos M, Bayer AS. 1994. Diminished virulence of a *sar⁻/agr⁻* mutant of *Staphylococcus aureus* in the rabbit model of endocarditis. *J Clin Invest* 94:1815–1822.

23. Traber KE, Lee E, Benson S, Corrigan R, Cantera M, Shopsin B, Novick RP. 2008. *agr* function in clinical *Staphylococcus aureus* isolates. *Microbiology* 154:2265–2274.

24. Fowler VG Jr, Sakoulas G, McIntyre LM, Meka VG, Arbeit RD, Cabell CH, Stryjewski ME, Eliopoulos GM, Reller LB, Corey GR, Jones T, Lucindo N, Yeaman MR, Bayer AS. 2004. Persistent bacteremia due to methicillin-resistant *Staphylococcus aureus* infection is associated with *agr* dysfunction and low-level in vitro resistance to thrombin-induced platelet microbicidal protein. *J Infect Dis* 190:1140–1149.

25. Sakoulas G, Eliopoulos GM, Moellering RC Jr, Wennersten C, Venkataraman L, Novick RP, Gold HS. 2002. Accessory gene regulator (*agr*) locus in geographically diverse *Staphylococcus aureus* isolates with reduced susceptibility to vancomycin. *Antimicrob Agents Chemother* 46:1492–1502.

26. Wright JS III, Traber KE, Corrigan R, Benson SA, Musser JM, Novick RP. 2005. The *agr* radiation: an early event in the evolution of staphylococci. *J Bacteriol* 187:5585–5594.

27. Peng HL, Novick RP, Kreiswirth B, Kornblum J, Schlievert P. 1988. Cloning, characterization, and sequencing of an accessory gene regulator (*agr*) in *Staphylococcus aureus*. *J Bacteriol* 170:4365–4372.

28. Janzon L, Löfdahl S, Arvidson S. 1989. Identification and nucleotide sequence of the delta-lysin gene, *hld*, adjacent to the accessory gene regulator (*agr*) of *Staphylococcus aureus*. *Mol Gen Genet* 219:480–485.

29. Novick RP, Projan SJ, Kornblum J, Ross HF, Ji G, Kreiswirth B, Vandenesch F, Moghazeh S, Novick RP. 1995. The *agr* P2 operon: an autocatalytic sensory transduction system in *Staphylococcus aureus*. *Mol Gen Genet* 248:446–458.

30. Ji G, Beavis RC, Novick RP. 1995. Cell density control of staphylococcal virulence mediated by an octapeptide pheromone. *Proc Natl Acad Sci U S A* 92:12055–12059.

31. Dufour P, Jarraud S, Vandenesch F, Greenland T, Novick RP, Bes M, Etienne J, Lina G. 2002. High genetic variability of the *agr* locus in *Staphylococcus* species. *J Bacteriol* 184:1180–1186.

32. Ji G, Beavis R, Novick RP. 1997. Bacterial interference caused by autoinducing peptide variants. *Science* 276:2027–2030.

33. Jarraud S, Lyon GJ, Figueiredo AM, Lina G, Vandenesch F, Etienne J, Muir TW, Novick RP. 2000. Exfoliatin-producing strains define a fourth *agr* specificity group in *Staphylococcus aureus*. *J Bacteriol* 182:6517–6522.

34. Canovas J, Baldry M, Bojer MS, Andersen PS, Grzeskowiak PK, Stegger M, Damborg P, Olsen CA, Ingmer H. 2016. Cross-talk between *Staphylococcus aureus* and other staphylococcal species via the *agr* quorum sensing system. *Front Microbiol* 7:1733.

35. Bronesky D, Wu Z, Marzi S, Walter P, Geissmann T, Moreau K, Vandenesch F, Caldelari I, Romby P. 2016. *Staphylococcus aureus* RNAIII and its regulon link quorum sensing, stress responses, metabolic adaptation, and regulation of virulence gene expression. *Annu Rev Microbiol* **70**:299–316.

36. Boisset S, Geissmann T, Huntzinger E, Fechter P, Bendridi N, Possedko M, Chevalier C, Helfer AC, Benito Y, Jacquier A, Gaspin C, Vandenesch F, Romby P. 2007. *Staphylococcus aureus* RNAIII coordinately represses the synthesis of virulence factors and the transcription regulator Rot by an antisense mechanism. *Genes Dev* **21**:1353–1366.

37. Bohn C, Rigoulay C, Bouloc P. 2007. No detectable effect of RNA-binding protein Hfq absence in *Staphylococcus aureus*. *BMC Microbiol* **7**:10.

38. Romilly C, Chevalier C, Marzi S, Masquida B, Geissmann T, Vandenesch F, Westhof E, Romby P. 2012. Loop-loop interactions involved in antisense regulation are processed by the endoribonuclease III in *Staphylococcus aureus*. *RNA Biol* **9**:1461–1472.

39. Zhang X, Zhu Q, Tian T, Zhao C, Zang J, Xue T, Sun B. 2015. Identification of RNAIII-binding proteins in *Staphylococcus aureus* using tethered RNAs and streptavidin aptamers based pull-down assay. *BMC Microbiol* **15**:102.

40. Romilly C, Caldelari I, Parmentier D, Lioliou E, Romby P, Fechter P. 2012. Current knowledge on regulatory RNAs and their machineries in *Staphylococcus aureus*. *RNA Biol* **9**:402–413.

41. Dunman PM, Murphy E, Haney S, Palacios D, Tucker-Kellogg G, Wu S, Brown EL, Zagursky RJ, Shlaes D, Projan SJ. 2001. Transcription profiling-based identification of *Staphylococcus aureus* genes regulated by the *agr* and/or *sarA* loci. *J Bacteriol* **183**:7341–7353.

42. Queck SY, Jameson-Lee M, Villaruz AE, Bach TH, Khan BA, Sturdevant DE, Ricklefs SM, Li M, Otto M. 2008. RNAIII-independent target gene control by the *agr* quorum-sensing system: insight into the evolution of virulence regulation in *Staphylococcus aureus*. *Mol Cell* **32**:150–158.

43. Gupta RK, Luong TT, Lee CY. 2015. RNAIII of the *Staphylococcus aureus agr* system activates global regulator MgrA by stabilizing mRNA. *Proc Natl Acad Sci U S A* **112**:14036–14041.

44. Janzon L, Arvidson S. 1990. The role of the δ-lysin gene (*hld*) in the regulation of virulence genes by the accessory gene regulator (*agr*) in *Staphylococcus aureus*. *EMBO J* **9**:1391–1399.

45. Novick RP, Ross HF, Projan SJ, Kornblum J, Kreiswirth B, Moghazeh S. 1993. Synthesis of staphylococcal virulence factors is controlled by a regulatory RNA molecule. *EMBO J* **12**:3967–3975.

46. Benito Y, Kolb FA, Romby P, Lina G, Etienne J, Vandenesch F. 2000. Probing the structure of RNAIII, the *Staphylococcus aureus agr* regulatory RNA, and identification of the RNA domain involved in repression of protein A expression. *RNA* **6**:668–679.

47. Chunhua M, Yu L, Yaping G, Jie D, Qiang L, Xiaorong T, Guang Y. 2012. The expression of LytM is down-regulated by RNAIII in *Staphylococcus aureus*. *J Basic Microbiol* **52**:636–641.

48. Huntzinger E, Boisset S, Saveanu C, Benito Y, Geissmann T, Namane A, Lina G, Etienne J, Ehresmann B, Ehresmann C, Jacquier A, Vandenesch F, Romby P. 2005. *Staphylococcus aureus* RNAIII and the endoribonuclease III coordinately regulate *spa* gene expression. *EMBO J* **24**:824–835.

49. Chabelskaya S, Bordeau V, Felden B. 2014. Dual RNA regulatory control of a *Staphylococcus aureus* virulence factor. *Nucleic Acids Res* **42**:4847–4858.

50. Chevalier C, Boisset S, Romilly C, Masquida B, Fechter P, Geissmann T, Vandenesch F, Romby P. 2010. *Staphylococcus aureus* RNAIII binds to two distant regions of *coa* mRNA to arrest translation and promote mRNA degradation. *PLoS Pathog* **6**:e1000809.

51. Morfeldt E, Taylor D, von Gabain A, Arvidson S. 1995. Activation of alpha-toxin translation in *Staphylococcus aureus* by the *trans*-encoded antisense RNA, RNAIII. *EMBO J* **14**:4569–4577.

52. Papenfort K, Vanderpool CK. 2015. Target activation by regulatory RNAs in bacteria. *FEMS Microbiol Rev* **39**:362–378.

53. Liu Y, Mu C, Ying X, Li W, Wu N, Dong J, Gao Y, Shao N, Fan M, Yang G. 2011. RNAIII activates *map* expression by forming an RNA-RNA complex in *Staphylococcus aureus*. *FEBS Lett* **585**:899–905.

54. Kumari P, Sampath K. 2015. cncRNAs: bi-functional RNAs with protein coding and non-coding functions. *Semin Cell Dev Biol* **47-48**:40–51.

55. Storz G, Wolf YI, Ramamurthi KS. 2014. Small proteins can no longer be ignored. *Annu Rev Biochem* **83**:753–777.

56. Vanderpool CK, Gottesman S. 2004. Involvement of a novel transcriptional activator and small RNA in post-transcriptional regulation of the glucose phosphoenolpyruvate phosphotransferase system. *Mol Microbiol* **54**:1076–1089.

57. Bobrovskyy M, Vanderpool CK. 2014. The small RNA SgrS: roles in metabolism and pathogenesis of enteric bacteria. *Front Cell Infect Microbiol* **4**:61.

58. Wadler CS, Vanderpool CK. 2007. A dual function for a bacterial small RNA: SgrS performs base pairing-dependent regulation and encodes a functional polypeptide. *Proc Natl Acad Sci U S A* **104**:20454–20459.

59. Kosfeld A, Jahreis K. 2012. Characterization of the interaction between the small regulatory peptide SgrT and the EIICB^{Glc} of the glucose-phosphotransferase system of *E. coli* K-12. *Metabolites* **2**:756–774.

60. Wang R, Braughton KR, Kretschmer D, Bach TH, Queck SY, Li M, Kennedy AD, Dorward DW, Klebanoff SJ, Peschel A, DeLeo FR, Otto M. 2007. Identification of novel cytolytic peptides as key virulence determinants for community-associated MRSA. *Nat Med* **13**:1510–1514.

61. Mangan S, Alon U. 2003. Structure and function of the feed-forward loop network motif. *Proc Natl Acad Sci U S A* **100**:11980–11985.

62. Shen-Orr SS, Milo R, Mangan S, Alon U. 2002. Network motifs in the transcriptional regulation network of *Escherichia coli*. *Nat Genet* **31**:64–68.

63. Nitzan M, Fechter P, Peer A, Altuvia Y, Bronesky D, Vandenesch F, Romby P, Biham O, Margalit H. 2015. A defense-offense multi-layered regulatory switch in a pathogenic bacterium. *Nucleic Acids Res* **43**:1357–1369.

64. Novick R. 2006. *Staphylococcal Pathogenesis and Pathogenicity Factors: Genetics and Regulation*. ASM Press, Washington, DC.

65. Novick R. 1967. Properties of a cryptic high-frequency transducing phage in *Staphylococcus aureus*. *Virology* **33**:155–166.

66. Jelsbak L, Hemmingsen L, Donat S, Ohlsen K, Boye K, Westh H, Ingmer H, Frees D. 2010. Growth phase-dependent regulation of the global virulence regulator Rot in clinical isolates of *Staphylococcus aureus*. *Int J Med Microbiol* **300**:229–236.

67. Cassat J, Dunman PM, Murphy E, Projan SJ, Beenken KE, Palm KJ, Yang SJ, Rice KC, Bayles KW, Smeltzer MS. 2006. Transcriptional profiling of a *Staphylococcus aureus* clinical isolate and its isogenic *agr* and *sarA* mutants reveals global differences in comparison to the laboratory strain RN6390. *Microbiology* **152**:3075–3090.

68. Song J, Lays C, Vandenesch F, Benito Y, Bes M, Chu Y, Lina G, Romby P, Geissmann T, Boisset S. 2012. The expression of small regulatory RNAs in clinical samples reflects the different life styles of *Staphylococcus aureus* in colonization vs. infection. *PLoS One* **7**:e37294.

69. Geisinger E, Chen J, Novick RP. 2012. Allele-dependent differences in quorum-sensing dynamics result in variant expression of virulence genes in *Staphylococcus aureus*. *J Bacteriol* **194**:2854–2864.

70. Priest NK, Rudkin JK, Feil EJ, van den Elsen JM, Cheung A, Peacock SJ, Laabei M, Lucks DA, Recker M, Massey RC. 2012. From genotype to phenotype: can systems biology be used to predict *Staphylococcus aureus* virulence? *Nat Rev Microbiol* **10**:791–797.

71. Felden B, Vandenesch F, Bouloc P, Romby P. 2011. The *Staphylococcus aureus* RNome and its commitment to virulence. *PLoS Pathog* **7**:e1002006.

72. Kaito C, Saito Y, Ikuo M, Omae Y, Mao H, Nagano G, Fujiyuki T, Numata S, Han X, Obata K, Hasegawa S, Yamaguchi H, Inokuchi K, Ito T, Hiramatsu K, Sekimizu K. 2013. Mobile genetic element SCC*mec*-encoded *psm-mec* RNA suppresses translation of *agrA* and attenuates MRSA virulence. *PLoS Pathog* **9**:e1003269.

73. Manna AC, Cheung AL. 2006. Expression of SarX, a negative regulator of *agr* and exoprotein synthesis, is activated by MgrA in *Staphylococcus aureus*. *J Bacteriol* **188**:4288–4299.

74. Ingavale S, van Wamel W, Luong TT, Lee CY, Cheung AL. 2005. Rat/MgrA, a regulator of autolysis, is a regulator of virulence genes in *Staphylococcus aureus*. *Infect Immun* **73**:1423–1431.

75. Reyes D, Andrey DO, Monod A, Kelley WL, Zhang G, Cheung AL. 2011. Coordinated regulation by AgrA, SarA, and SarR to control *agr* expression in *Staphylococcus aureus*. *J Bacteriol* **193**:6020–6031.

76. Lenz DH, Mok KC, Lilley BN, Kulkarni RV, Wingreen NS, Bassler BL. 2004. The small RNA chaperone Hfq and multiple small RNAs control quorum sensing in *Vibrio harveyi* and *Vibrio cholerae*. *Cell* **118**:69–82.

77. Nawrocki EP, Burge SW, Bateman A, Daub J, Eberhardt RY, Eddy SR, Floden EW, Gardner PP, Jones TA, Tate J, Finn RD. 2015. Rfam 12.0: updates to the RNA families database. *Nucleic Acids Res* **43**(Database issue):D130–D137.

78. Rutherford ST, Valastyan JS, Taillefumier T, Wingreen NS, Bassler BL. 2015. Comprehensive analysis reveals how single nucleotides contribute to noncoding RNA function in bacterial quorum sensing. *Proc Natl Acad Sci U S A* **112**:E6038–E6047.

79. Feng L, Rutherford ST, Papenfort K, Bagert JD, van Kessel JC, Tirrell DA, Wingreen NS, Bassler BL. 2015. A Qrr non-coding RNA deploys four different regulatory mechanisms to optimize quorum-sensing dynamics. *Cell* **160**:228–240.

80. Shao Y, Bassler BL. 2014. Quorum regulatory small RNAs repress type VI secretion in *Vibrio cholerae*. *Mol Microbiol* **92**:921–930.

81. Shao Y, Feng L, Rutherford ST, Papenfort K, Bassler BL. 2013. Functional determinants of the quorum-sensing non-coding RNAs and their roles in target regulation. *EMBO J* **32**:2158–2171.

82. Shao Y, Bassler BL. 2012. Quorum-sensing non-coding small RNAs use unique pairing regions to differentially control mRNA targets. *Mol Microbiol* **83**:599–611.

83. Waters CM, Bassler BL. 2005. Quorum sensing: cell-to-cell communication in bacteria. *Annu Rev Cell Dev Biol* **21**:319–346.

84. Lupp C, Ruby EG. 2004. *Vibrio fischeri* LuxS and AinS: comparative study of two signal synthases. *J Bacteriol* **186**:3873–3881.

85. Kuo A, Blough NV, Dunlap PV. 1994. Multiple N-acyl-L-homoserine lactone autoinducers of luminescence in the marine symbiotic bacterium *Vibrio fischeri*. *J Bacteriol* **176**:7558–7565.

86. Lenz DH, Miller MB, Zhu J, Kulkarni RV, Bassler BL. 2005. CsrA and three redundant small RNAs regulate quorum sensing in *Vibrio cholerae*. *Mol Microbiol* **58**:1186–1202.

87. Roh JB, Lee MA, Lee HJ, Kim SM, Cho Y, Kim YJ, Seok YJ, Park SJ, Lee KH. 2006. Transcriptional regulatory cascade for elastase production in *Vibrio vulnificus*: LuxO activates *luxT* expression and LuxT represses *smcR* expression. *J Biol Chem* **281**:34775–34784.

88. Hawver LA, Giulietti JM, Baleja JD, Ng WL. 2016. Quorum sensing coordinates cooperative expression of pyruvate metabolism genes to maintain a sustainable environment for population stability. *mBio* **7**:e01863-16.

89. Hammer BK, Bassler BL. 2007. Regulatory small RNAs circumvent the conventional quorum sensing pathway in pandemic *Vibrio cholerae*. *Proc Natl Acad Sci U S A* **104**:11145–11149.

90. Tu KC, Long T, Svenningsen SL, Wingreen NS, Bassler BL. 2010. Negative feedback loops involving small regula-

tory RNAs precisely control the *Vibrio harveyi* quorum-sensing response. *Mol Cell* **37**:567–579.

91. Rutherford ST, van Kessel JC, Shao Y, Bassler BL. 2011. AphA and LuxR/HapR reciprocally control quorum sensing in vibrios. *Genes Dev* **25**:397–408.

92. Teng SW, Schaffer JN, Tu KC, Mehta P, Lu W, Ong NP, Bassler BL, Wingreen NS. 2011. Active regulation of receptor ratios controls integration of quorum-sensing signals in *Vibrio harveyi*. *Mol Syst Biol* **7**:491.

93. Tu KC, Bassler BL. 2007. Multiple small RNAs act additively to integrate sensory information and control quorum sensing in *Vibrio harveyi*. *Genes Dev* **21**:221–233.

94. Bagert JD, van Kessel JC, Sweredoski MJ, Feng L, Hess S, Bassler BL, Tirrell DA. 2016. Time-resolved proteomic analysis of quorum sensing in *Vibrio harveyi*. *Chem Sci (Camb)* **7**:1797–1806.

95. van Kessel JC, Rutherford ST, Shao Y, Utria AF, Bassler BL. 2013. Individual and combined roles of the master regulators AphA and LuxR in control of the *Vibrio harveyi* quorum-sensing regulon. *J Bacteriol* **195**:436–443.

96. van Kessel JC, Ulrich LE, Zhulin IB, Bassler BL. 2013. Analysis of activator and repressor functions reveals the requirements for transcriptional control by LuxR, the master regulator of quorum sensing in *Vibrio harveyi*. *mBio* **4**:e00378-13.

97. van Kessel JC, Rutherford ST, Cong JP, Quinodoz S, Healy J, Bassler BL. 2015. Quorum sensing regulates the osmotic stress response in *Vibrio harveyi*. *J Bacteriol* **197**:73–80.

98. Martin M, Showalter R, Silverman M. 1989. Identification of a locus controlling expression of luminescence genes in *Vibrio harveyi*. *J Bacteriol* **171**:2406–2414.

99. Showalter RE, Martin MO, Silverman MR. 1990. Cloning and nucleotide sequence of *luxR*, a regulatory gene controlling bioluminescence in *Vibrio harveyi*. *J Bacteriol* **172**:2946–2954.

100. Svenningsen SL, Waters CM, Bassler BL. 2008. A negative feedback loop involving small RNAs accelerates *Vibrio cholerae*'s transition out of quorum-sensing mode. *Genes Dev* **22**:226–238.

101. Tu KC, Waters CM, Svenningsen SL, Bassler BL. 2008. A small-RNA-mediated negative feedback loop controls quorum-sensing dynamics in *Vibrio harveyi*. *Mol Microbiol* **70**:896–907.

102. Svenningsen SL, Tu KC, Bassler BL. 2009. Gene dosage compensation calibrates four regulatory RNAs to control *Vibrio cholerae* quorum sensing. *EMBO J* **28**:429–439.

103. Teng SW, Wang Y, Tu KC, Long T, Mehta P, Wingreen NS, Bassler BL, Ong NP. 2010. Measurement of the copy number of the master quorum-sensing regulator of a bacterial cell. *Biophys J* **98**:2024–2031.

104. Ball AS, Chaparian RR, van Kessel JC. 2017. Quorum sensing gene regulation by LuxR/HapR master regulators in vibrios. *J Bacteriol* **199**:e00105-17.

105. Massé E, Escorcia FE, Gottesman S. 2003. Coupled degradation of a small regulatory RNA and its mRNA targets in *Escherichia coli*. *Genes Dev* **17**:2374–2383.

106. Levine E, Zhang Z, Kuhlman T, Hwa T. 2007. Quantitative characteristics of gene regulation by small RNA. *PLoS Biol* **5**:e229.

107. Massé E, Vanderpool CK, Gottesman S. 2005. Effect of RyhB small RNA on global iron use in *Escherichia coli*. *J Bacteriol* **187**:6962–6971.

108. Mitarai N, Benjamin JA, Krishna S, Semsey S, Csiszovszki Z, Masse E, Sneppen K. 2009. Dynamic features of gene expression control by small regulatory RNAs. *Proc Natl Acad Sci U S A* **106**:10655–10659.

109. Caswell CC, Oglesby-Sherrouse AG, Murphy ER. 2014. Sibling rivalry: related bacterial small RNAs and their redundant and non-redundant roles. *Front Cell Infect Microbiol* **4**:151.

110. Miyashiro T, Wollenberg MS, Cao X, Oehlert D, Ruby EG. 2010. A single *qrr* gene is necessary and sufficient for LuxO-mediated regulation in *Vibrio fischeri*. *Mol Microbiol* **77**:1556–1567.

111. Moll S, Schneider DJ, Stodghill P, Myers CR, Cartinhour SW, Filiatrault MJ. 2010. Construction of an *rsmX* co-variance model and identification of five *rsmX* non-coding RNAs in *Pseudomonas syringae* pv. *tomato* DC3000. *RNA Biol* **7**:508–516.

112. Mollerup MS, Ross JA, Helfer AC, Meistrup K, Romby P, Kallipolitis BH. 2016. Two novel members of the LhrC family of small RNAs in *Listeria monocytogenes* with overlapping regulatory functions but distinctive expression profiles. *RNA Biol* **13**:895–915.

113. Sievers S, Sternkopf Lillebaek EM, Jacobsen K, Lund A, Mollerup MS, Nielsen PK, Kallipolitis BH. 2014. A multicopy sRNA of *Listeria monocytogenes* regulates expression of the virulence adhesin LapB. *Nucleic Acids Res* **42**:9383–9398.

114. Halfmann A, Kovacs M, Hakenbeck R, Bruckner R. 2007. Identification of the genes directly controlled by the response regulator CiaR in *Streptococcus pneumoniae*: five out of 15 promoters drive expression of small non-coding RNAs. *Mol Microbiol* **66**:110–126.

115. Pannekoek Y, Huis In 't Veld RA, Schipper K, Bovenkerk S, Kramer G, Brouwer MC, van de Beek D, Speijer D, van der Ende A. 2017. *Neisseria meningitidis* uses sibling small regulatory RNAs to switch from cataplerotic to anaplerotic metabolism. *mBio* **8**:e02293-16.

116. Torres-Quesada O, Millán V, Nisa-Martínez R, Bardou F, Crespi M, Toro N, Jiménez-Zurdo JI. 2013. Independent activity of the homologous small regulatory RNAs AbcR1 and AbcR2 in the legume symbiont *Sinorhizobium meliloti*. *PLoS One* **8**:e68147.

117. Wilms I, Voss B, Hess WR, Leichert LI, Narberhaus F. 2011. Small RNA-mediated control of the *Agrobacterium tumefaciens* GABA binding protein. *Mol Microbiol* **80**:492–506.

118. Wassarman KM, Repoila F, Rosenow C, Storz G, Gottesman S. 2001. Identification of novel small RNAs using comparative genomics and microarrays. *Genes Dev* **15**:1637–1651.

119. Wilderman PJ, Sowa NA, FitzGerald DJ, FitzGerald PC, Gottesman S, Ochsner UA, Vasil ML. 2004. Identification of tandem duplicate regulatory small RNAs in

Pseudomonas aeruginosa involved in iron homeostasis. *Proc Natl Acad Sci U S A* **101**:9792–9797.

120. Weber B, Lindell K, El Qaidi S, Hjerde E, Willassen NP, Milton DL. 2011. The phosphotransferase VanU represses expression of four *qrr* genes antagonizing VanO-mediated quorum-sensing regulation in *Vibrio anguillarum*. *Microbiology* **157**:3324–3339.

121. Engebrecht J, Silverman M. 1987. Nucleotide sequence of the regulatory locus controlling expression of bacterial genes for bioluminescence. *Nucleic Acids Res* **15**:10455–10467.

122. Devine JH, Shadel GS, Baldwin TO. 1989. Identification of the operator of the *lux* regulon from the *Vibrio fischeri* strain ATCC7744. *Proc Natl Acad Sci U S A* **86**:5688–5692.

123. Winson MK, Camara M, Latifi A, Foglino M, Chhabra SR, Daykin M, Bally M, Chapon V, Salmond GP, Bycroft BW. 1995. Multiple N-acyl-L-homoserine lactone signal molecules regulate production of virulence determinants and secondary metabolites in *Pseudomonas aeruginosa*. *Proc Natl Acad Sci U S A* **92**:9427–9431.

124. Seed PC, Passador L, Iglewski BH. 1995. Activation of the *Pseudomonas aeruginosa lasI* gene by LasR and the *Pseudomonas* autoinducer PAI: an autoinduction regulatory hierarchy. *J Bacteriol* **177**:654–659.

125. Fuqua C, Parsek MR, Greenberg EP. 2001. Regulation of gene expression by cell-to-cell communication: acyl-homoserine lactone quorum sensing. *Annu Rev Genet* **35**:439–468.

126. Scholz RL, Greenberg EP. 2017. Positive autoregulation of an acyl-homoserine lactone quorum-sensing circuit synchronizes the population response. *mBio* **8**:e01079-17.

127. Kleerebezem M, Quadri LE, Kuipers OP, de Vos WM. 1997. Quorum sensing by peptide pheromones and two-component signal-transduction systems in Gram-positive bacteria. *Mol Microbiol* **24**:895–904.

128. Fuqua WC, Winans SC, Greenberg EP. 1994. Quorum sensing in bacteria: the LuxR-LuxI family of cell density-responsive transcriptional regulators. *J Bacteriol* **176**:269–275.

129. Taillefumier T, Wingreen NS. 2015. Optimal census by quorum sensing. *PLoS Comput Biol* **11**:e1004238.

130. Hurley A, Bassler BL. 2017. Asymmetric regulation of quorum-sensing receptors drives autoinducer-specific gene expression programs in *Vibrio cholerae*. *PLoS Genet* **13**:e1006826.

131. Fuqua C, Winans SC, Greenberg EP. 1996. Census and consensus in bacterial ecosystems: the LuxR-LuxI family of quorum-sensing transcriptional regulators. *Annu Rev Microbiol* **50**:727–751.

132. Kay E, Humair B, Denervaud V, Riedel K, Spahr S, Eberl L, Valverde C, Haas D. 2006. Two GacA-dependent small RNAs modulate the quorum-sensing response in *Pseudomonas aeruginosa*. *J Bacteriol* **188**:6026–6033.

133. Pessi G, Williams F, Hindle Z, Heurlier K, Holden MT, Camara M, Haas D, Williams P. 2001. The global post-transcriptional regulator RsmA modulates production of virulence determinants and N-acylhomoserine lactones

in *Pseudomonas aeruginosa*. *J Bacteriol* **183**:6676–6683.

134. Dubuis C, Haas D. 2007. Cross-species GacA-controlled induction of antibiosis in pseudomonads. *Appl Environ Microbiol* **73**:650–654.

135. Lapouge K, Schubert M, Allain FH, Haas D. 2008. Gac/Rsm signal transduction pathway of γ-proteobacteria: from RNA recognition to regulation of social behaviour. *Mol Microbiol* **67**:241–253.

136. Vakulskas CA, Potts AH, Babitzke P, Ahmer BM, Romeo T. 2015. Regulation of bacterial virulence by Csr (Rsm) systems. *Microbiol Mol Biol Rev* **79**:193–224.

137. Pannuri A, Vakulskas CA, Zere T, McGibbon LC, Edwards AN, Georgellis D, Babitzke P, Romeo T. 2016. Circuitry linking the catabolite repression and Csr global regulatory systems of *Escherichia coli*. *J Bacteriol* **198**:3000–3015.

138. Sonnleitner E, Gonzalez N, Sorger-Domenigg T, Heeb S, Richter AS, Backofen R, Williams P, Hüttenhofer A, Haas D, Bläsi U. 2011. The small RNA PhrS stimulates synthesis of the *Pseudomonas aeruginosa* quinolone signal. *Mol Microbiol* **80**:868–885.

139. Gallagher LA, McKnight SL, Kuznetsova MS, Pesci EC, Manoil C. 2002. Functions required for extracellular quinolone signaling by *Pseudomonas aeruginosa*. *J Bacteriol* **184**:6472–6480.

140. Carloni S, Macchi R, Sattin S, Ferrara S, Bertoni G. 2017. The small RNA ReaL: a novel regulatory element embedded in the *Pseudomonas aeruginosa* quorum sensing networks. *Environ Microbiol* **19**:4220–4237.

141. Baumgardt K, Šmídová K, Rahn H, Lochnit G, Robledo M, Evguenieva-Hackenberg E. 2016. The stress-related, rhizobial small RNA RcsR1 destabilizes the autoinducer synthase encoding mRNA *sinI* in *Sinorhizobium meliloti*. *RNA Biol* **13**:486–499.

142. De Lay N, Gottesman S. 2009. The Crp-activated small noncoding regulatory RNA CyaR (RyeE) links nutritional status to group behavior. *J Bacteriol* **191**:461–476.

143. Xavier KB, Bassler BL. 2005. Regulation of uptake and processing of the quorum-sensing autoinducer AI-2 in *Escherichia coli*. *J Bacteriol* **187**:238–248.

144. Flemming HC, Wingender J. 2010. The biofilm matrix. *Nat Rev Microbiol* **8**:623–633.

145. Karatan E, Watnick P. 2009. Signals, regulatory networks, and materials that build and break bacterial biofilms. *Microbiol Mol Biol Rev* **73**:310–347.

146. Davies DG, Parsek MR, Pearson JP, Iglewski BH, Costerton JW, Greenberg EP. 1998. The involvement of cell-to-cell signals in the development of a bacterial biofilm. *Science* **280**:295–298.

147. Nadell CD, Xavier JB, Levin SA, Foster KR. 2008. The evolution of quorum sensing in bacterial biofilms. *PLoS Biol* **6**:e14.

148. Srivastava D, Waters CM. 2012. A tangled web: regulatory connections between quorum sensing and cyclic di-GMP. *J Bacteriol* **194**:4485–4493.

149. Krasteva PV, Fong JC, Shikuma NJ, Beyhan S, Navarro MV, Yildiz FH, Sondermann H. 2010. *Vibrio cholerae*

VpsT regulates matrix production and motility by directly sensing cyclic di-GMP. *Science* **327:**866–868.

150. Srivastava D, Harris RC, Waters CM. 2011. Integration of cyclic di-GMP and quorum sensing in the control of *vpsT* and *aphA* in *Vibrio cholerae*. *J Bacteriol* **193:** 6331–6341.

151. Waters CM, Lu W, Rabinowitz JD, Bassler BL. 2008. Quorum sensing controls biofilm formation in *Vibrio cholerae* through modulation of cyclic di-GMP levels and repression of *vpsT*. *J Bacteriol* **190:**2527–2536.

152. Papenfort K, Förstner KU, Cong JP, Sharma CM, Bassler BL. 2015. Differential RNA-seq of *Vibrio cholerae* identifies the VqmR small RNA as a regulator of biofilm formation. *Proc Natl Acad Sci U S A* **112:**E766–E775.

153. Romeo T, Gong M, Liu MY, Brun-Zinkernagel AM. 1993. Identification and molecular characterization of *csrA*, a pleiotropic gene from *Escherichia coli* that affects glycogen biosynthesis, gluconeogenesis, cell size, and surface properties. *J Bacteriol* **175:**4744–4755.

154. Liu MY, Gui G, Wei B, Preston JF III, Oakford L, Yuksel U, Giedroc DP, Romeo T. 1997. The RNA molecule CsrB binds to the global regulatory protein CsrA and antagonizes its activity in *Escherichia coli*. *J Biol Chem* **272:**17502–17510.

155. Weilbacher T, Suzuki K, Dubey AK, Wang X, Gudapaty S, Morozov I, Baker CS, Georgellis D, Babitzke P, Romeo T. 2003. A novel sRNA component of the carbon storage regulatory system of *Escherichia coli*. *Mol Microbiol* **48:**657–670.

156. Pusic P, Tata M, Wolfinger MT, Sonnleitner E, Häussler S, Bläsi U. 2016. Cross-regulation by CrcZ RNA controls anoxic biofilm formation in *Pseudomonas aeruginosa*. *Sci Rep* **6:**39621.

157. Sonnleitner E, Valentini M, Wenner N, Haichar FZ, Haas D, Lapouge K. 2012. Novel targets of the CbrAB/Crc carbon catabolite control system revealed by transcript abundance in *Pseudomonas aeruginosa*. *PLoS One* **7:** e44637.

158. Sonnleitner E, Wulf A, Campagne S, Pei XY, Wolfinger MT, Forlani G, Prindl K, Abdou L, Resch A, Allain FH, Luisi BF, Urlaub H, Bläsi U. 2017. Interplay between the catabolite repression control protein Crc, Hfq and RNA in Hfq-dependent translational regulation in *Pseudomonas aeruginosa*. *Nucleic Acids Res* **46:**1470–1485.

159. Parker A, Cureoglu S, De Lay N, Majdalani N, Gottesman S. 2017. Alternative pathways for *Escherichia coli* biofilm formation revealed by sRNA overproduction. *Mol Microbiol* **105:**309–325.

160. Mika F, Hengge R. 2014. Small RNAs in the control of RpoS, CsgD, and biofilm architecture of *Escherichia coli*. *RNA Biol* **11:**494–507.

161. Cerca N, Jefferson KK. 2008. Effect of growth conditions on poly-N-acetylglucosamine expression and biofilm formation in *Escherichia coli*. *FEMS Microbiol Lett* **283:**36–41.

162. Mika F, Hengge R. 2013. Small regulatory RNAs in the control of motility and biofilm formation in *E. coli* and *Salmonella*. *Int J Mol Sci* **14:**4560–4579.

163. Bak G, Lee J, Suk S, Kim D, Young Lee J, Kim KS, Choi BS, Lee Y. 2015. Identification of novel sRNAs involved in biofilm formation, motility, and fimbriae formation in *Escherichia coli*. *Sci Rep* **5:**15287.

164. De Lay N, Gottesman S. 2012. A complex network of small non-coding RNAs regulate motility in *Escherichia coli*. *Mol Microbiol* **86:**524–538.

165. Mandin P, Gottesman S. 2010. Integrating anaerobic/aerobic sensing and the general stress response through the ArcZ small RNA. *EMBO J* **29:**3094–3107.

166. Monteiro C, Papenfort K, Hentrich K, Ahmad I, Le Guyon S, Reimann R, Grantcharova N, Romling U. 2012. Hfq and Hfq-dependent small RNAs are major contributors to multicellular development in *Salmonella enterica* serovar Typhimurium. *RNA Biol* **9:**489–502.

167. Mika F, Hengge R. 2005. A two-component phosphotransfer network involving ArcB, ArcA, and RssB coordinates synthesis and proteolysis of σ^S (RpoS) in *E. coli*. *Genes Dev* **19:**2770–2781.

168. Liu X, De Wulf P. 2004. Probing the ArcA-P modulon of *Escherichia coli* by whole genome transcriptional analysis and sequence recognition profiling. *J Biol Chem* **279:** 12588–12597.

169. Kato Y, Sugiura M, Mizuno T, Aiba H. 2007. Effect of the *arcA* mutation on the expression of flagella genes in *Escherichia coli*. *Biosci Biotechnol Biochem* **71:**77–83.

170. Jørgensen MG, Nielsen JS, Boysen A, Franch T, Møller-Jensen J, Valentin-Hansen P. 2012. Small regulatory RNAs control the multi-cellular adhesive lifestyle of *Escherichia coli*. *Mol Microbiol* **84:**36–50.

171. Thomason MK, Fontaine F, De Lay N, Storz G. 2012. A small RNA that regulates motility and biofilm formation in response to changes in nutrient availability in *Escherichia coli*. *Mol Microbiol* **84:**17–35.

172. Liu M, Durfee T, Cabrera JE, Zhao K, Jin DJ, Blattner FR. 2005. Global transcriptional programs reveal a carbon source foraging strategy by *Escherichia coli*. *J Biol Chem* **280:**15921–15927.

173. Zhao K, Liu M, Burgess RR. 2007. Adaptation in bacterial flagellar and motility systems: from regulon members to 'foraging'-like behavior in *E. coli*. *Nucleic Acids Res* **35:**4441–4452.

174. Jorgensen MG, Thomason MK, Havelund J, Valentin-Hansen P, Storz G. 2013. Dual function of the McaS small RNA in controlling biofilm formation. *Genes Dev* **27:**1132–1145.

175. Holmqvist E, Vogel J. 2013. A small RNA serving both the Hfq and CsrA regulons. *Genes Dev* **27:**1073–1078.

176. Wei BL, Brun-Zinkernagel AM, Simecka JW, Prüss BM, Babitzke P, Romeo T. 2001. Positive regulation of motility and *flhDC* expression by the RNA-binding protein CsrA of *Escherichia coli*. *Mol Microbiol* **40:**245–256.

177. Holmqvist E, Reimegard J, Sterk M, Grantcharova N, Romling U, Wagner EG. 2010. Two antisense RNAs target the transcriptional regulator CsgD to inhibit curli synthesis. *EMBO J* **29:**1840–1850.

178. Mika F, Busse S, Possling A, Berkholz J, Tschowri N, Sommerfeldt N, Pruteanu M, Hengge R. 2012. Targeting of *csgD* by the small regulatory RNA RprA links sta-

tionary phase, biofilm formation and cell envelope stress in *Escherichia coli*. *Mol Microbiol* 84:51–65.

179. Majdalani N, Hernandez D, Gottesman S. 2002. Regulation and mode of action of the second small RNA activator of RpoS translation, RprA. *Mol Microbiol* 46:813–826.

180. Serra DO, Mika F, Richter AM, Hengge R. 2016. The green tea polyphenol EGCG inhibits *E. coli* biofilm formation by impairing amyloid curli fibre assembly and downregulating the biofilm regulator CsgD via the σ^E-dependent sRNA RybB. *Mol Microbiol* 101:136–151.

181. Bordeau V, Felden B. 2014. Curli synthesis and biofilm formation in enteric bacteria are controlled by a dynamic small RNA module made up of a pseudoknot assisted by an RNA chaperone. *Nucleic Acids Res* 42:4682–4696.

182. Yang L, Nilsson M, Gjermansen M, Givskov M, Tolker-Nielsen T. 2009. Pyoverdine and PQS mediated subpopulation interactions involved in *Pseudomonas aeruginosa* biofilm formation. *Mol Microbiol* 74:1380–1392.

183. Archer NK, Mazaitis MJ, Costerton JW, Leid JG, Powers ME, Shirtliff ME. 2011. *Staphylococcus aureus* biofilms. *Virulence* 2:445–459.

184. Romilly C, Lays C, Tomasini A, Caldelari I, Benito Y, Hammann P, Geissmann T, Boisset S, Romby P, Vandenesch F. 2014. A non-coding RNA promotes bacterial persistence and decreases virulence by regulating a regulator in *Staphylococcus aureus*. *PLoS Pathog* 10:e1003979.

185. Trotonda MP, Tamber S, Memmi G, Cheung AL. 2008. MgrA represses biofilm formation in *Staphylococcus aureus*. *Infect Immun* 76:5645–5654.

186. Carroll RK, Weiss A, Broach WH, Wiemels RE, Mogen AB, Rice KC, Shaw LN. 2016. Genome-wide annotation, identification, and global transcriptomic analysis of regulatory or small RNA gene expression in *Staphylococcus aureus*. *mBio* 7:e01990-15.

187. Gómez-Lozano M, Marvig RL, Molin S, Long KS. 2012. Genome-wide identification of novel small RNAs in *Pseudomonas aeruginosa*. *Environ Microbiol* 14:2006–2016.

188. Smirnov A, Förstner KU, Holmqvist E, Otto A, Günster R, Becher D, Reinhardt R, Vogel J. 2016. Grad-seq guides the discovery of ProQ as a major small RNA-binding protein. *Proc Natl Acad Sci U S A* 113:11591–11596.

189. Melamed S, Peer A, Faigenbaum-Romm R, Gatt YE, Reiss N, Bar A, Altuvia Y, Argaman L, Margalit H. 2016. Global mapping of small RNA-target interactions in bacteria. *Mol Cell* 63:884–897.

190. Michaux C, Holmqvist E, Vasicek E, Sharan M, Barquist L, Westermann AJ, Gunn JS, Vogel J. 2017. RNA target profiles direct the discovery of virulence functions for the cold-shock proteins CspC and CspE. *Proc Natl Acad Sci U S A* 114:6824–6829.

191. Singh R, Ray P. 2014. Quorum sensing-mediated regulation of staphylococcal virulence and antibiotic resistance. *Future Microbiol* 9:669–681.

192. Kerpedjiev P, Hammer S, Hofacker IL. 2015. Forna (force-directed RNA): simple and effective online RNA secondary structure diagrams. *Bioinformatics* 31:3377–3379.

193. Lorenz N, Shin JY, Jung K. 2017. Activity, abundance, and localization of quorum sensing receptors in *Vibrio harveyi*. *Front Microbiol* 8:634.

194. Marketon MM, Gronquist MR, Eberhard A, González JE. 2002. Characterization of the *Sinorhizobium meliloti sinR/sinI* locus and the production of novel *N*-acyl homoserine lactones. *J Bacteriol* 184:5686–5695.

195. Chambers JR, Sauer K. 2013. Small RNAs and their role in biofilm formation. *Trends Microbiol* 21:39–49.

Regulating with RNA in Bacteria and Archaea
Edited by Gisela Storz and Kai Papenfort
© 2018 American Society for Microbiology, Washington, DC
doi:10.1128/microbiolspec.RWR-0002-2017

Regulatory RNAs in Virulence and Host-Microbe Interactions

18

Alexander J. Westermann[1,2]

INTRODUCTION

Infectious diseases caused by bacterial pathogens still constitute one of the major human health threats (1). Within the host body, pathogenic bacteria face a wide variety of hostile microenvironments and encounter numerous different host cell types as well as commensal bacteria of the resident microbiota. To cope with these environmental changes and respond adequately to any interacting cell, bacterial pathogens have to tightly control their gene expression during infection, in part by means of regulatory RNAs.

Regulatory RNA elements can be involved in transcriptional (e.g., transcriptional riboswitches [2] or *cis*-encoded antisense RNAs [3, 4]) and—more commonly—posttranscriptional control mechanisms (e.g., many *trans*-encoded small RNAs [sRNAs] [5, 6]). As many regulatory RNAs were found to contribute to the adaptation of bacterial physiology to diverse stress conditions *in vitro* (6–8), they also appear suited to cope with the dramatic alterations during infection. Indeed, research of the last decade has revealed that bacterial pathogens possess in the range of fifty to several hundreds of regulatory RNAs, of which many are expressed under infection-relevant conditions (9–15). A number of them were shown to orchestrate diverse virulence mechanisms, including the adaptation to an initial temperature increase and varying nutrient availabilities after entering the host body; adherence to and invasion of target cells; evasion of host defense mechanisms; and transmission or establishment of chronic infections (16–22). These known virulence-related RNAs are summarized in Fig. 1 and Table 1 and will be the topic of the following sections.

CLASSES OF BACTERIAL REGULATORY RNAs WITH VIRULENCE FUNCTIONS

Regulatory RNA Elements within 5′ UTRs

Regulatory RNA elements in extended 5′ untranslated regions (5′ UTRs) of mRNAs may respond to specific triggers by structural rearrangements, thereby affecting expression of their downstream gene. These RNA sensors lend themselves for rapid adjustment of bacterial physiology to the specific stages of an infection (23). Riboswitches, which are widespread particularly in mRNAs encoding metabolic enzymes or transport systems in Gram-positive bacteria, sense small molecules, metabolites, or metal ions, to control expression of their downstream gene at the level of either transcription elongation/termination or translation initiation (2). For example, in the Gram-positive opportunistic enteric pathogen *Clostridium difficile*, the second messenger and circular RNA dinucleotide cyclic di-GMP is recognized by riboswitches that are coupled to flagellar and virulence genes, tying the expression of virulence traits to the levels of this nucleotide-derived alarmone (24, 25). Likewise, in *Mycobacterium tuberculosis*, putative riboswitches were predicted in the 5′ UTRs of *pstS2* and *TB8.4* (26), genes encoding a membrane-associated phosphate transporter that is required for virulence and a T-cell antigen, respectively. Despite their ligands remaining elusive, this suggests that riboswitches may be important regulators of virulence in this major human pathogen.

The temperature increase that accompanies host colonization is sensed by a second group of regulatory 5′ UTR elements, the "RNA thermometers" (27, see also chapter 4). The Gram-negative pathogens *Yersinia*

[1]Institute of Molecular Infection Biology, University of Würzburg; [2]Helmholtz Institute for RNA-Based Infection Research, D-97080 Würzburg, Germany.

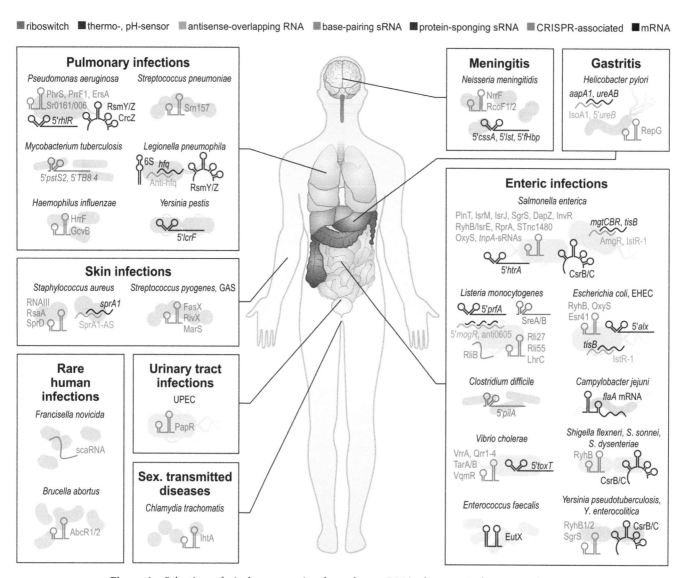

Figure 1 Selection of virulence-associated regulatory RNA elements in human pathogens. Representatives of distinct classes of regulatory RNA are indicated (see color code on top). Illustrated are all RNAs referred to in the main text and in Table 1; absence of RNAs (e.g., CsrB/C) in a given pathogen does not necessarily imply that they are not encoded by that bacterium, but rather that they are not explicitly mentioned in the text. Several bacterial pathogens may colonize multiple niches and cause more than a single disease in their human host, but they are assigned to just one organ and illness here for the sake of simplicity. GAS, group A *Streptococcus*.

pestis, *Pseudomonas aeruginosa*, and *Vibrio cholerae* as well as Gram-positive *Listeria monocytogenes* have each engaged RNA thermometers to link the expression of key virulence transcription factors—LcrF in *Yersinia* (28, 29), RhlR in *Pseudomonas* (30), ToxT in *Vibrio* (31), and PrfA in *Listeria* (32)—to the surrounding temperature. Likewise, the human-specific, meningococcal disease-causing pathogen *Neisseria meningitidis* employs thermosensing to tune the expression of genes involved in

capsule biosynthesis and lipopolysaccharide modification (*cssA*, *lst*, and *fHbp*) during infection (33). In these examples, at low temperature (≤30°C), the respective 5′ UTRs adopt an inactive state by sequestering the Shine-Dalgarno (SD) sequence and/or the start codon in an inaccessible stem-loop structure. Upon shift to mammalian body temperatures (≥37°C), this inhibitory structure "melts," allowing the translation of the messenger. The importance of accurate temperature sensing for bacterial virulence is highlighted by the fact that trapping the *lcrF* thermosensor in a locked conformation renders *Y. pestis* avirulent (29), and thermosensing fosters meningococcal immune evasion in the inflamed host (33).

Besides sensing chemical or temperature changes, 5′ UTRs may also respond to pH fluctuations as a pathogen travels through its host. For example, a pH- and manganese-sensing RNA element was identified in the 5′ UTR of the *Escherichia coli alx* mRNA (34–36). At neutral pH, the *alx* 5′ UTR adopts a translationally inactive structure, which opens up upon a shift to extreme alkaline conditions, allowing translation of a putative redox modulator (36). The Gram-negative enteric pathogen *Salmonella enterica* serovar Typhimurium infects its host via the oral route, and along its passage to the small intestine it faces a wide pH spectrum. Additionally, *Salmonella* can invade intestinal epithelial cells and may be engulfed by macrophages, wherein it finds itself in an acidic vacuolar compartment. Not only has *Salmonella* evolved strategies to cope with low pH, but it even uses it as a trigger to promote its own dissemination. The PhoP/Q system, *Salmonella*'s major regulator of intracellular virulence gene expression, requires low pH for full activation (37). The *mgtCBR* operon, encoding an ATP synthase inhibitor and a magnesium uptake system, is one of the most highly induced PhoP targets inside macrophages (38) and is required for *Salmonella* virulence in mice (39). Interestingly, expression of *mgtCBR* responds to acidic pH even in the absence of PhoP/Q (40). However, unlike *alx* mRNA, which senses pH directly, the 5′ leader of *mgtCBR* rather responds to cytosolic ATP concentrations that are altered in a pH-dependent manner (40).

Salmonella mgt not only responds to altering ATP levels, but its leader sequence also senses low levels of tRNAs charged with proline that prevail in the bacterium's intracellular niche (41). Additionally, *mgt* expression is counteracted by a 1.2-kb-long, *cis*-encoded antisense RNA referred to as AmgR (42). Reciprocally, the *mgt* leader sequence may bind and regulate flagellin mRNA in *trans* (43). The emerging complex regulatory network (44) constitutes a prime example of how bacterial pathogens deploy diverse posttranscriptional control mechanisms to sense defined environmental cues for a rapid adjustment of global gene expression to specific microenvironments inside the host.

cis-Encoded sRNAs

cis-encoded antisense sRNAs were discovered in mobile genetic elements such as plasmids, phages, and transposons decades ago (45). The recent explosion in genome-wide transcriptomics revealed that antisense transcription in bacteria is much more prevalent than previously anticipated (46). This pervasive antisense transcription bears the potential for widespread regulation of gene expression, at the transcriptional (overlapping transcription interference), posttranscriptional (via perfectly complementary RNA-RNA interaction and degradation of the resulting double strand), or translational level (binding and occluding the ribosome binding site [RBS] of mRNAs) (3). While the functional relevance of pervasive antisense transcription is under debate (47–49), several case studies outline the diversity of *cis*-encoded antisense mechanisms and connect them to bacterial pathogenesis. For a more detailed description of cis-antisense RNAs, see chapter 12.

Colonizing the human stomach, the peptic ulcer- and gastric cancer-causative agent *Helicobacter pylori* faces a harsh acidic environment. Expression of the urease, the enzyme hydrolyzing urea into carbon dioxide and ammonia, is critical for pH buffering within this host niche. Yet urease expression is costly (50) and hence must be tightly controlled. A *cis*-encoded sRNA termed 5′ureB-sRNA overlaps with the urease operon, is expressed at neutral pH, and shuts down urease production under this condition by mediating premature transcription termination of *ureAB* mRNA (4, 51). In the Gram-negative, facultative, intracellular lung pathogen *Legionella pneumophila*, an antisense RNA was discovered that overlaps with the gene encoding the global RNA chaperone Hfq (host factor for phage Qβ) (52). Interestingly, binding of this "Anti-hfq" RNA to the 5′ region of *hfq* mRNA is facilitated by Hfq protein in a negative, autoregulatory feedback loop and shuts down virulence gene expression outside *Legionella*'s eukaryotic host.

In the foodborne pathogen *L. monocytogenes*, flagellum biosynthesis is negatively regulated by the transcriptional repressor MogR (53). The repressing effect of MogR, however, manifests itself not only at the transcriptional but also at the posttranscriptional level. *mogR* mRNA comes in two flavors—containing a long or a shortened 5′ UTR—as a result of transcription initiation from two alternative start sites (9). While the *mogR* promoter that gives rise to the short isoform is constitutively expressed, the promoter for the 5′-extended *mogR* version is specifically induced at 37°C. This long 5′ UTR overlaps almost 1.5 kb with three genes on the opposite strand that are required for flagellum synthesis, thereby repressing those genes and reducing *Listeria* motility during host infection (9). Besides *mogR*, other examples of such long antisense RNAs were discovered in *Listeria* (anti0605, anti1846,

Table 1 Virulence-associated regulatory RNA elements in diverse human pathogens

Bacterial organism	Regulatory RNA	RNA class	(Putative) direct target(s)	RBP partner(s)	Virulence-related function(s)	Phenotype(s)	Reference(s)
Staphylococcus aureus	RNAIII	Dual-function sRNA	*hla, spa, SA1000, rot, coa, mgrA*		Quorum sensing-mediated regulation of membrane proteins and toxins, thereby contributing to the transition from colonization to dissemination mode		105, 111, 113, 236, 286–290
	RsaA	Base-pairing sRNA	*mgrA, flr, ssaA*		Virulence suppressor; role in biofilm formation and chronic infections	RsaA attenuates acute systemic *S. aureus* infection and enhances chronic catheter infection in a mouse model	103, 219, 291
	SprD	Base-pairing sRNA	*sbi*		Evasion of the host's immune response	*sprD* deletion attenuates *Staphylococcus* in murine kidneys and rescues survival of infected mice	292
	SprA1-AS	*cis*-encoded antisense RNA	*sprA1*		Repression of the host cytolytic SprA1 protein		64
Streptococcus pyogenes, group A *Streptococcus*	FasX	Base-pairing sRNA	*ska, cpa, prtF1/2*		Activation of the secreted virulence factor streptokinase and repression of pilus formation	*fasX* deletion enhances adherence to human keratinocytes and reduces virulence in a bacteremia model (humanized plasminogen mice)	108–110, 227, 293
	RivX	Base-pairing sRNA (3′-derived)	*scpA, emm, mga, speB*		Affects virulence factor expression	In a *covR* deletion background, RivX is required for virulence in mice	104, 294
	MarS	Base-pairing sRNA	*mga*		Activates expression of the virulence transcription factor Mga	Deletion strain is more susceptible to phagocytosis and shows reduced adherence to keratinocytes (in cell culture); dissemination to liver, kidney, and spleen is reduced in the absence of MarS (in a mouse model)	295
Neisseria meningitidis	NrrF	Base-pairing sRNA	*sdbA, sdbC*		Repression of iron-dependent enzymes under iron-limiting conditions		99, 100
	RcoF1/2 (NmsR$_{A/B}$)	Base-pairing sRNAs	*prpB, prpC, gltA, sucC, sucD, sdbCDAB, fumBC*	Hfq	Contribute to the shift from cataplerotic to anaplerotic metabolism		101, 102
	5′ *cssA*	Thermosensor	*cssA*		Link environmental temperature to capsule biosynthesis and lipopolysaccharide modification; role in immune evasion		33
	5′ *lst*	Thermosensor	*lst*				33
	5′ *fHbp*	Thermosensor	*fHbp*				33

Organism	sRNA	Type	Target	Regulatory protein	Function	References
Pseudomonas aeruginosa	RsmY/Z	Protein-titrating sRNAs	RsmA	RsmA	RsmA sequestration affects biofilm formation and swarming motility	296–302
	CrcZ	Protein-titrating sRNA	Hfq	Hfq	CrcZ has high affinity to Hfq, thereby competing with other members of the Hfq regulon	303
	PhrS	Dual-function sRNA	*pqsR*	Hfq	Couples quorum sensing with oxygen availability	304
	PrrF1	Base-pairing sRNA	*sodB, sdb, PA4880, antR*	Hfq	RyhB homolog; adjusts iron homeostasis under limiting conditions	218, 305, 306
	ErsA	Base-pairing sRNA	*oprD*	Hfq	Increases resistance against carbapenem antibiotics	218
	Sr0161	Base-pairing sRNA	*oprD, exsA*	Hfq	Increases carbapenem resistance; repression of a structural component and an effector of the T3SS	218
	Sr006	Base-pairing sRNA	*pagL*		Reduces the proinflammatory property of PagL and confers polymyxin resistance	218
	5′ *rhlR*	Thermosensor	*rhlR*		Temperature-dependent virulence factor expression	30
Streptococcus pneumoniae	Srn157	Base-pairing sRNA	?		*srn157* deletion causes a fitness defect in the mouse nasopharynx and lung	206
Mycobacterium tuberculosis	5′ *pstS2*	Riboswitch	*pstS2*		Control of a phosphate transporter required for virulence	26
	5′*TB8.4*	Riboswitch	*TB8.4*		Control of a T-cell antigen	26
Legionella pneumophila	6S	Protein-titrating sRNA	RNAP	RNAP	Affects expression of numerous genes, e.g., those of virulence effectors and nutrient acquisition proteins	75
	Anti-hfq	*cis*-encoded antisense RNA	*hfq*	Hfq	Contributes to the switch from replicating, environmental to virulent, transmissive bacteria	52
	RsmY/Z	Protein-titrating sRNAs	RsmA	RsmA	Double deletion (Δ*rsmYZ*) leads to an attenuation in infection of THP-1 macrophages	72
Haemophilus influenzae	HrrF	Base-pairing sRNA	?		Highly induced upon infection of human bronchial epithelial cells	200
	GcvB	Base-pairing sRNA	*oppA*			200
Yersinia pestis	5′*lcrF*	Thermosensor	*lcrF*	Hfq (?)	Ties expression of the virulence transcription factor LcrF to surrounding temperature. *Y. pestis* with stabilized thermometer variants are impaired in the dissemination to Peyer's patches	28, 29

(Continued)

Table 1 (Continued)

Bacterial organism	Regulatory RNA	RNA class	(Putative) direct target(s)	RBP partner(s)	Virulence-related function(s)	Phenotype(s)	Reference(s)
Helicobacter pylori	IsoA1	*cis*-encoded antisense RNA (antitoxin)	*aapA1*		Multiple layers of posttranscriptional control to tighten expression of the AapA1 toxin		60
	5′ *ureB*	*cis*-encoded antisense RNA	*ureAB*		Links urease production to surrounding pH		4, 51
	RepG	Base-pairing sRNA	*tlpB*		Regulates expression of a chemotaxis receptor contributing to stomach colonization of mice		114, 307
Salmonella enterica	PinT	Base-pairing sRNA	*sopE, sopE2, grxA, crp*	Hfq	Induced through the main activator of intracellular *Salmonella* virulence, PhoPQ; represses invasion factors (SopE/E2) at intracellular stages and affects expression of the global transcription factor Crp	Δ*pinT* mutant slightly attenuated during replication inside porcine macrophages, and *pinT* disruption attenuates *Salmonella* in a pig and cattle model	14, 95, 96
	IsrM	Base-pairing sRNA	*sopA, hilE*		Regulation of secreted invasion effectors by direct targeting (SopA) and indirectly through the repression of the global SPI-1 regulator HilE	Deletion interferes with epithelial cell invasion and intramacrophage replication (in cell culture) and confers a growth defect in mouse ileum and spleen	94
	IsrJ	Base-pairing sRNA	?		Activated through the SPI-1 cascade (HilA); affects translocation of the SptP effector protein and invasion into epithelial cells	Δ*isrJ* with invasion defect in HeLa cells	93
	SgrS	Dual-function sRNA	*sopD, ptsG, manXYZ, yigL*	Hfq	sRNA (SgrS) and small protein (SgrT) both act in the response to sugar-phosphate stress; sRNA represses the secreted effector SopD		125, 126, 221
	DapZ	Base-pairing sRNA (3′-derived)	*oppA, dppA*	Hfq	Activated through the SPI-1 cascade (HilD)		121
	RyhB	Base-pairing sRNA	>50 (mostly related to iron utilization)	Hfq	Induced under conditions of iron deficiency (including intracellular stages); role in the response to oxidative stress		14, 95, 245, 308, 309
	IsrE (RyhB2)	Base-pairing sRNA	Most shared with RyhB, but also exclusive targets (*flgJ, fliF, cheY*)	Hfq			14, 95, 244, 309
	InvR	Base-pairing sRNA	*ompD*	Hfq	Activated through the SPI-1 cascade (HilD); might facilitate assembly of the T3SS	*invR* disruption attenuates *Salmonella* in chickens, pigs, and cattle	91, 92, 96
	RprA	Base-pairing sRNA	*rpoS, ricI*	Hfq	Regulates conjugation of the pSLT virulence plasmid		310

Organism	sRNA	Type	Target	Chaperone	Function	Phenotype/comment	References
	tmpA sRNAs	Base-pairing sRNAs (5′-derived)	>70 (direct and/or indirect; including invF)	Hfq (?)	Repression of SPI-1 genes (probably via downregulating the SPI-1 transcriptional activator InvF)	Mutant devoid of these sRNAs exhibits enhanced invasion, and sRNA overexpression reduces invasion within a HeLa model; tmpA-derived sRNAs are required for colonization of mice	119, 120
	5′htrA	Thermosensor	htrA		Role in the response to misfolded proteins		189, 311, 312
	CsrB/C	Protein-titrating sRNAs	CsrA	CsrA	Contribute to the induction of SPI-1 and SPI-2 virulence genes (via CsrA titration off hilD mRNA)		69
	AmgR	cis-encoded antisense RNA	mgtCBR		Activated by the PhoPQ system; times expression of the MgtC virulence protein	Absence of AmgR enhances Salmonella virulence	42
	mgt leader	cis- and trans-acting leader	mgtCBR		Senses cytosolic proline and ATP levels and consequently controls translation elongation into the mgtCBR virulence regulon in cis; binds to and represses fljB mRNA in trans	Mutation of the mgt targeting region within fljB renders Salmonella hypervirulent	41, 43, 313–315
	IstR-1	cis-encoded antisense RNA (antitoxin)	tisB			istR disruption attenuates Salmonella in mice	316
	OxyS	Base-pairing sRNA	>40 mRNAs (in E. coli)	Hfq	Role in the protection against oxidative damage	Gene disruption increases fitness during mouse infection	316
	STnc1480	?	?		Induced inside macrophages in a PhoPQ-dependent manner		14, 317
	Spot-42	Base-pairing sRNA	hilD, mglB	Hfq	Activation of HilD and consequently, induction of effector secretion; transcriptionally repressed by Crp		136, 334
Escherichia coli, EHEC	RyhB	Base-pairing sRNA	>50 (mostly related to iron utilization)	Hfq	See above (Salmonella)		197
	OxyS	Base-pairing sRNA	>40 mRNAs	Hfq	See above (Salmonella)		318
	Esr41 (EcOnc14)	Base-pairing sRNA	fliC, cirA, chuA, bfr, ler (similar to RyhB targetome)	(Hfq)	Specific to EHEC; controls motility, effector secretion and iron transport/storage and confers colicin 1A resistance	Deletion causes fitness advantage under iron-limiting conditions in vitro	149, 214, 319, 335
	IstR-1	cis-encoded antisense RNA (antitoxin)	tisB		TisB/IstR1 contribute to the switch from replicating to persistent bacteria		61, 62, 320

(Continued)

Table 1 (Continued)

Bacterial organism	Regulatory RNA	RNA class	(Putative) direct target(s)	RBP partner(s)	Virulence-related function(s)	Phenotype(s)	Reference(s)
	5' alx	pH sensor	alx		Induces expression of a putative redox modulator under extreme alkaline conditions		34–36
	Spot-42	Base-pairing sRNA	sepL, many metabolic genes	Hfq	Repression of the locus of enterocyte effacement (LEE) which encodes the T3SS of EHEC		336–338
Listeria monocytogenes	5' mogR, anti0605, anti1846, anti0424	Long cis-encoded antisense RNAs (excludon concept)	lmo0675-0677, lmo0605, lmo1846, lmo0424		Ensure mutually exclusive expression of divergently oriented operons that have opposing functions (e.g., virulence vs. motility genes)		9, 54
	Rli27	Base-pairing sRNA	lmo0514		Regulates a cell wall protein at intracellular stages of infection		9, 107
	Rli55	Protein-titrating sRNA (riboswitch-derived)	EutV	EutV	Role in ethanolamine utilization	Listeria strains defective in ethanolamine utilization, or in its regulation by Rli55, are attenuated in mice	9, 77, 321
	RliB	CRISPR-associated	?	PNPase	Upregulated in the intestinal lumen of gnotobiotic mice, in human blood, or under hypoxic conditions	Deletion enhances virulence in mouse livers	9, 133, 321
	LhrC	Base-pairing sRNA (multicopy)	lapB	Hfq	Induced during replication inside HepG2 cells; repression of a virulence-associated adhesin		150, 322
	SreA/B	Riboswitches, base-pairing sRNAs	lmo2419-lmo2418-lmo2417 or lmo0595-lmo0594 (cis), prfA (trans)		Link virulence factor expression to nutrient availability		118
	5' prfA	Thermosensor	prfA		Connects expression of the master regulator of Listeria virulence to ambient temperature		32
Clostridium difficile	5' pilA1	Riboswitch	pilA1		Couples virulence to c-di-GMP levels		24, 25
Campylobacter jejuni	flaA mRNA	Protein-titrating sRNA	CsrA	CsrA	Connects the flagellar and CsrA (and thus, indirectly, virulence) regulons		142, 143
Vibrio cholerae	VrrA	Base-pairing sRNA	ompA, ompT, rbmC, vrp	Hfq	Regulation of protein composition in the outer membrane	ΔvrrA with increased virulence in the murine intestine	79, 80, 323, 324
	Qrr1 to -4	Base-pairing sRNAs	hapR, aphA	Hfq	Control of the expression of HapR and AphA, which are involved in virulence and biofilm formation	Affects virulence in an infant mouse model	82, 325

Organism	sRNA	Type	Target	Protein partner	Function	Phenotype	Reference
	TarA	Base-pairing sRNA	ptsG	Hfq	Activated via the major virulence transcription factor, ToxT; controls glucose transport under virulence conditions	ΔtarA slightly compromised in colonization of an infant mouse model	84, 326
	TarB	Base-pairing sRNA	vspR, tcpF	Hfq	Activated by ToxT; regulates expression of the virulence transcription factor VspR and of the secreted colonization factor TcpF	ΔtarB hypercolonizes the murine small intestine	85, 86, 326
	VqmR	Base-pairing sRNA	At least 8 mRNAs (including vpsT, rtx)	Hfq	Regulates biofilm formation and toxin production upon sensing of the autoinducer 3,5-dimethylpyrazin-2-ol	Likely role during colonization of the human host	87, 88
	5'toxT	Thermosensor	toxT		Links virulence gene expression to ambient temperature		31
Shigella flexneri, S. sonnei, S. dysenteriae	CsrB/C	Protein-titrating sRNAs	CsrA	CsrA	Control carbon consumption during host colonization	Disruption of csrB or csrC causes an increase, and overexpression a decrease, in invasion of Henle cells	71
	RyhB	Base-pairing sRNA	virB	Hfq	Through targeting the transcriptional activator VirB, indirect regulation of T3SS	Overexpression reduces invasion efficiency into epithelial cell cultures	327
Enterococcus faecalis	EutX	Protein-titrating sRNA (riboswitch-derived)	EutV	EutV	Role in ethanolamine utilization		78
Yersinia pseudo-tuberculosis,	RyhB1/2	Base-pairing sRNAs	See above (*Salmonella*)	Hfq	Induced during mouse infection in lymphatic tissues		15
Y. enterocolitica	SgrS	Base-pairing sRNA	See above (*Salmonella*)	Hfq	Highly induced during mouse infection		15
	CsrB/C	Protein-titrating sRNAs	CsrA	CsrA	Strongly repressed during mouse infection, thus releasing CsrA, which is crucial for T3SS expression		15
UPEC	PapR	Base-pairing sRNA	papI	Hfq	Repression of the P-fimbriae phase regulator	Deletion causes a defect in adhesion to bladder epithelial and kidney medullary collecting duct cells	98
Chlamydia trachomatis	IhtA	Base-pairing sRNA	hctA		Contributes to the switch from reticulate to elementary bodies		97
Francisella novicida	scaRNA	CRISPR-associated	FTN_1103	Cas9	Repression of lipoprotein to evade the host's immune response	Deletion mutants are attenuated in spleens and reduce survival rates of infected mice	328
Brucella abortus	AbcR1/2	Base-pairing sRNAs	mRNAs involved in amino acid and polyamine metabolism	Hfq	Likely adapt *Brucella* metabolism to environmental changes occurring during infection	Double mutant (ΔabcR1/2) attenuated in cultured murine macrophages and in a mouse model of chronic infection	329

and anti0424) that overlap with the coding sequences of efflux pump genes on the opposite strand (54). In each case, the overlapping portion of the antisense RNA serves both as the 5′ UTR of the downstream gene(s) and as a repressor for the divergently encoded gene, and thereby ensures mutually exclusive expression of the overlapping operons. These observations imply that such a type of regulation might be more common than previously thought (54, 55), and a concept referred to as "excludon" (56) was introduced to describe genomic loci encoding exceptionally long antisense RNAs (e.g., mRNAs with long 5′ UTRs) that span and repress divergent genes with typically opposing functions.

cis-encoded antisense RNAs also play a central role in toxin-antitoxin (TA) systems (57, see also chapter 11). In type I TA systems, the antitoxin is an antisense RNA that overlaps with and represses the corresponding toxin-encoding mRNA (58). Sublethal concentrations of growth-inhibiting toxins, however, may promote bacterial heterogeneity and thus be beneficial for the pathogen population as a whole. Indeed, TA systems contribute to bacterial persistence and chronic infections (58, 59), and various type I TA systems have been described in bacterial pathogens, including AapA1/IsoA1 in *H. pylori* (60), TisB/IstR-1 in *E. coli* and *S.* Typhimurium (61, 62), and SprA1/Sp rA1-AS in methicillin-resistant *Staphylococcus aureus* (63, 64).

trans-Encoded sRNAs That Titrate Proteins

trans-encoded sRNAs typically control genes expressed from a different genomic locus via one of two mechanisms (6, 7): indirectly, via binding to and titrating a regulatory protein, thereby affecting gene expression of the target set of the respective protein; or directly, by mediating RNA-RNA base-pair interactions with target transcripts, which may or may not depend on the assistance of RNA chaperones. The two mechanisms, however, are not necessarily mutually exclusive (65, 66).

Among the best-characterized systems of protein-targeting sRNAs are the RNA sponges of the regulatory protein CsrA (carbon storage regulator A; also known as RsmA [repressor of secondary metabolites A]) (see chapter 19). Known as CsrB and CsrC, or in pseudomonads as RsmY and RsmZ, these 200-to-400-nucleotide (nt)-long RNAs harbor multiple high-affinity CsrA/RsmA binding sites to efficiently sequester the protein and relieve regulation of CsrA/RsmA target mRNAs, many of which are virulence related (67, 68). In several important enteric pathogens including *S.* Typhimurium, *Shigella flexneri*, and *Yersinia pseudotuberculosis*, CsrB/C homologs are required for the induction of invasion

genes, via a regulatory cascade involving translational derepression of mRNAs encoding the respective master activators of the invasion regulons in these bacteria (69–71). Accordingly, while the deletion of a single RNA of this pair is typically well tolerated, double deletion results in severe virulence defects (72).

Rather than targeting translation, 6S RNA interacts with the housekeeping RNA polymerase holoenzyme (σ^{70}-RNAP complex) and blocks transcription of target genes (73). *ssrS*, the gene encoding 6S RNA, is ubiquitously distributed over the entire bacterial kingdom (74), and 6S RNA plays a main role in the adjustment of global gene expression during the transition from exponential to stationary phase (see chapter 20). More recently, 6S RNA has been implicated in virulence programs of certain bacterial pathogens. In *Legionella*, for example, the 6S RNA seems to predominantly activate—rather than inhibit—gene expression, with various secreted virulence effectors being among its targets, and intracellular replication of a *Legionella* Δ*ssrS* mutant is strongly attenuated (75).

Other, pathogen-specific, examples of protein-sequestering sRNAs exist. For instance, various pathogen species utilize ethanolamine as a carbon source during growth in their host niche (76, 77). In *L. monocytogenes* and *Enterococcus faecalis*, riboswitch-derived sRNAs have been discovered (Rli55 and EutX) that bind to and block the anti-terminator protein EutV, which would otherwise inhibit transcription of the ethanolamine utilization operon (77, 78).

trans-Encoded sRNAs That Act by Base-Pairing

The second mechanism of *trans*-encoded sRNAs, mediating base-pair interactions with target mRNAs, is far more widespread. Base-pairing sRNAs bind sequence-specifically to imperfectly complementary regions typically within the 5′ region of their target mRNAs, which—in Gram-negative species—is frequently facilitated by the RNA chaperone Hfq. This results in the sequestration of the SD sequence and/or the start codon and, hence, translational repression often accompanied by target mRNA degradation. This canonical mechanism is employed by numerous sRNAs in bacterial pathogens to adjust their virulence programs during the infection cycle. For example, *V. cholerae*, the causative agent of cholera, uses VrrA (79, 80), the paralogous sRNAs Qrr1 to -4 (81–83), TarA/B (84–86), and VqmR (87, 88) to fine-tune outer membrane protein expression, synchronize virulence factor production and type VI secretion, and repress biofilm formation during infection. Accordingly, deleting *vrrA* (79), the Qrr sRNA genes (89, 90), *tarA* (84), or *tarB* (85) affects

Vibrio virulence and the outcome of infection in infant mouse models.

S. Typhimurium expresses numerous sRNAs from the *Salmonella*-specific pathogenicity islands. Among them, InvR represses a major outer membrane protein, thereby probably facilitating assembly of the type III secretion system (T3SS) (91, 92). The island-encoded sRNAs IsrJ (93), IsrM (94), and PinT (95) each control the expression of secreted virulence effectors during invasion or intracellular proliferation. Of note, deletion of *isrJ* (93) and *isrM* (94) and disruption of *invR* or *pinT* (96) coincide with virulence defects in cell culture, a systemic mouse model, or gastrointestinal pig and calf models of salmonellosis.

Chlamydia trachomatis, an obligate intracellular bacterium and a leading cause of infectious blindness and chlamydial urogenital infections, encodes IhtA. This sRNA represses an mRNA encoding a histone-like protein and thereby contributes to chlamydial differentiation from the proliferative reticulate to the infectious elementary bodies (97). In uropathogenic *E. coli* (UPEC), the PapR sRNA represses the mRNA encoding the P-fimbriae phase regulator PapI, and *papR* deletion causes a defect in adhesion to both bladder and kidney cell lines (98). *N. meningitidis* produces NrrF sRNA under iron-limiting conditions to repress mRNAs for an iron-dependent enzyme complex (99, 100), and employs two tandem sRNAs, the paralogous RcoF1 and RcoF2 (also known as NmsR$_{A/B}$), for the repression of the mRNA for colonization factor PrpB (101, 102). Interestingly, while the RcoF1/2-mediated regulations depend on Hfq, NrrF-mediated repression of *sdhA* and *sdhC* mRNAs does not require the assistance of this RNA chaperone.

Gram-positive pathogens typically lack Hfq altogether yet nevertheless extensively deploy base-pairing sRNAs to control virulence mechanisms. For instance, RsaA, a staphylococcal sRNA and virulence suppressor, employs two disparate regions to target the mRNA of the central virulence transcription factor MgrA both at the SD region and within its coding sequence (103). Of note, both interactions are required for translational repression of MgrA and contribute to *Staphylococcus* attenuation in mice, favoring chronic infections. Similarly, in the extracellular pathogen of the skin *Streptococcus pyogenes*, an sRNA termed RivX affects virulence in a murine model, despite the direct target(s) remaining elusive (104).

Rather than repressing their target mRNAs, activating sRNAs employ base-pairing interactions to open up inhibitory secondary structures within target 5′ UTRs that otherwise occlude the SD and prevent translation.

This "anti-antisense" mechanism was first discovered for RNAIII, a ~500-nt-long RNA species in *S. aureus* (105) (see below). RNAIII binds to the 5′ UTR of *hla* mRNA, which encodes the α-toxin, and thereby activates its translation. Since this initial observation more than 2 decades ago, similar mechanisms have been described for several additional sRNAs and pathogen species (106). For example, *L. monocytogenes* Rli27 sRNA activates the expression of a cell wall protein-encoding mRNA, *lmo0514*, during its intracellular phase of infection (9, 107). Intriguingly, *lmo0514* mRNA has two isoforms, and only the long mRNA variant that accumulates during infection is subject to Rli27-mediated translational activation. Similarly, in the throat- and skin-infecting pathogen group A *Streptococcus*, binding of FasX sRNA 12 nt upstream of the RBS of the mRNA of the secreted virulence factor streptokinase promotes mRNA stabilization and translation (108). Additionally, FasX negatively regulates pilus expression via base-pairing to the extreme 5′ end of the *cpa* operon, thereby lowering the transcript's stability and shutting down its translation (109). Thus, FasX exerts both positive and negative effects dependent on the context of its target sites, and collectively these mechanisms contribute to *Streptococcus* adherence to human keratinocytes (109) and to full virulence in a bacteremia model of human plasminogen-expressing mice (110).

There are further examples of virulence-related sRNAs that employ multiple modes of regulation. A prime example in this respect is RNAIII from *S. aureus*. Being among the first bacterial sRNAs discovered (111), it is meanwhile known to integrate quorum-sensing signals for the progression from colonization to dissemination mode (112). This is achieved through the concerted action of different portions of this unusually long RNA. In addition to the anti-antisense mechanism that is employed by the 5′ part of RNAIII (see above), its 3′ region interacts with and represses mRNAs of virulence factors and of the transcription factor Rot by binding around their start codon, resulting in translational shutoff and target degradation (113). RepG, a *trans*-encoded sRNA in *H. pylori*, targets phase-variable *tlpB*, which encodes a chemotaxis receptor (114). Interestingly, dependent on the length of a variable homopolymeric G-repeat within the RepG target site, the sRNA either represses or activates chemotaxis receptor synthesis. One of the quorum sensing-associated sRNAs of *V. cholerae*, Qrr3, even employs four distinct regulatory mechanisms to control target gene expression (115). Dependent on the base-pairing strength with its respective target, Qrr3 either activates mRNA translation (via the anti-antisense mechanism) or represses the target through sequestering the mRNA in an inactive form or inducing its degradation either alone or together with Qrr3 itself (i.e., via coupled degradation).

Whereas the above sRNAs are encoded within intergenic regions, i.e., in between two coding genes, and typically are transcribed from their own promoters, UTRs of mRNAs have long been proposed to be additional sRNA reservoirs (116). Indeed, some *cis*-acting riboswitches in the 5′ UTRs of mRNAs might be cleaved off their downstream transcript and fulfill a regulatory function in *trans*. In Gram-positive bacteria, a conserved transcriptional riboswitch senses the coenzyme S-adenosylmethionine (SAM) and controls the expression of dozens of mRNAs in *cis* (117). Interestingly, in *L. monocytogenes*, two SAM riboswitches, termed SreA and SreB, additionally control gene expression in *trans* (118). While SAM binding and the resulting premature termination prevent expression of the riboswitch's downstream gene, the accumulating 5′-UTR termination products (SreA/B) retain regulatory potential, bind to the RBS, and block translation of the mRNA of the virulence regulator PrfA. Reinforcing the relevance of this regulatory mechanism, deletion of *sreA/B* leads to enhanced virulence gene expression (118). Further examples of UTR-derived sRNAs that act in *trans* come from S. Typhimurium. For example, the 5′ UTR of *tnpA*, a transposon-derived mRNA, is processed, giving rise to different RNA species of varying lengths (119). Among the many (direct or indirect) targets of these 5′-UTR-derived sRNAs, there is the mRNA of the transcriptional activator of invasion genes, InvF. Consequently, blocking these sRNAs coincides with an invasion defect in a cell culture model (119) and interferes with the colonization of the mouse gastrointestinal tract (120). In the same bacterium, a large number of 3′-derived sRNAs were discovered (121, 122). For example, DapZ sRNA is encoded within the 3′ UTR of *dapB*, which produces a metabolic enzyme involved in amino acid biosynthesis, and interacts with Hfq to repress the translation of major ABC transporters of short peptides (121). Being transcriptionally activated by HilD (121), the master regulator of *Salmonella* invasion genes, DapZ is highly induced during the entry into epithelial cells and rapidly downregulated thereafter (95), mimicking the expression dynamics of invasion genes.

Some sRNAs contain short open reading frames and encode small proteins. Several of these "dual-output" RNAs have been implicated in bacterial virulence. For instance, in addition to its role in RNA-based regulation (see above), staphylococcal RNAIII can be translated, giving rise to the 26-amino-acid protein hemolysin δ (Hld), with a putative role in biofilm formation (123).

SgrS, a conserved sRNA involved in the response to sugar-phosphate stress (124) and, in *Salmonella*, the control of a secreted effector protein (125), encodes a 43-amino-acid-long protein termed SgrT. Interestingly, reporters used to functionally distinguish SgrS and SgrT demonstrated that both the sRNA and the small protein act within the same pathway to orchestrate the response to overt glucose-phosphate levels (126). For further information on these topics, see chapters 14 and 27.

CRISPR-Associated RNAs

CRISPR (clustered regularly interspaced short palindromic repeats) is a widely distributed bacterial immune system based on CRISPR RNAs and CRISPR-associated (Cas) proteins to combat invading DNA elements (127). Interestingly, a CRISPR-associated RNA termed scaRNA (small CRISPR/Cas-associated RNA) was recently discovered to play a role in endogenous gene expression control in the airborne pathogen *Francisella novicida* (128). In conjunction with Cas9, scaRNA binds and represses a lipoprotein-encoding mRNA via a mechanism distantly resembling that of classic base-pairing sRNAs. As lipoproteins are bacterial surface molecules recognized by host receptors, this regulatory mechanism allows *Francisella* to evade host immune defenses. Interestingly, in *Campylobacter jejuni*, Cas9 targets and degrades endogenous RNA transcripts (129), and *cas9* deletion results in virulence defects (130). Likewise, in *L. pneumophila*, Cas2 has RNase activity, its deletion leads to an attenuation in an amoeba model of infection, and removal of the CRISPR-associated RNA RliB enhances virulence in mice (131–133). Together, these findings indicate that CRISPR-mediated control of bacterial virulence programs could be more prevalent than currently appreciated.

Collectively, these examples illustrate how bacterial pathogens harness the versatility of regulatory RNA to optimize gene expression during infection. However, many RNAs do not act in isolation, but interact with bacterial proteins that increase their stability, facilitate binding to target regions within mRNA transcripts, and/or recruit RNases for target degradation.

PROTEIN PARTNERS OF REGULATORY RNAs AND THEIR CONTRIBUTION TO BACTERIAL VIRULENCE

CsrA (RsmA)

Classically, homodimers of the regulatory protein CsrA (RsmA in pseudomonads) bind within the 5′ UTR of target mRNAs, thereby occluding the RBS and shutting down translation initiation (134–136, see also chapter 19). While in some bacterial organisms *csrA* deletion is lethal, in others it leads to strong fitness defects, including the dysregulaion of virulence programs under infection-relevant conditions (137). The CsrA interactome involves many virulence-related transcripts (70, 137–139). For example, *Salmonella* CsrA binds and translationally represses the mRNA encoding the master regulator of invasion genes, HilD (69), and interacts with mRNAs that encode effectors of both major *Salmonella* pathogenicity islands (136). In enterohemorrhagic *E. coli* (EHEC), CsrA regulates mRNAs encoding translocators, effectors, and virulence transcription factors (140). Mapping its interactome in *L. pneumophila* implicated CsrA in the interface of virulence and metabolism, with many mRNAs encoding global regulators being among the direct CsrA ligands in this pathogen (141). In the foodborne pathogen *C. jejuni*, which lacks any known homolog of CsrB/C, the major flagellin mRNA (*flaA*) steps in as a CsrA sponge (142, 143). In fact, *csrA* and flagellar genes tend to cluster in the genomes of taxonomically diverse bacterial species, suggesting a more general link between CsrA and motility networks (137). Importantly, in *Campylobacter*, while *flaA* mRNA affects the pool of free CsrA, leading to indirect effects on CsrA target mRNAs, CsrA binding to *flaA* mRNA in turn represses flagellin production and affects its subcellular localization (142). This finding blurs the definition of regulator and target within the CsrA network.

Hfq

The global RNA chaperone Hfq is widely conserved within Gram-negative bacteria, though it is missing in many Gram-positive species. Hfq oligomerizes, forming a homohexameric ring with three distinct RNA-binding surfaces (144–146). This property allows Hfq to act as an RNA matchmaker for base-pairing sRNAs and their target mRNAs. Additionally, Hfq promotes sRNA stability and contributes to the recruitment of accessory proteins for target decay. Reminiscent of *csrA* knockouts, bacterial pathogens depleted of Hfq exhibit severe virulence defects, including dysregulated virulence factor expression and secretion, chemotaxis defects, and reduced motility (147). Indeed, in pathogen species in which the global Hfq regulon has been mapped, such as *S.* Typhimurium (121, 136, 148), EHEC (149), UPEC (98), *L. monocytogenes* (150), *P. aeruginosa* (151), or the plant pathogen *Agrobacterium tumefaciens* (152), a high proportion of virulence-related transcripts associate with this protein. Whereas in many cases base-pairing sRNAs may recruit Hfq to these virulence mRNAs, the chaperone was recently shown to also bind and regulate

certain mRNAs in an sRNA-independent manner (153–155); whether or not this noncanonical role might contribute to the virulence phenotype of *hfq* deletion mutants remains unclear.

FinO/ProQ

Besides Hfq, the FinO (fertility inhibition protein) domain-containing proteins have recently emerged as a second family of RNA chaperones widely distributed in beta- and gammaproteobacteria (156), and thus in bacterial classes that contain many medically important human pathogens. Of note, while certain FinO(-like) proteins exhibit a highly specific RNA ligand preference, others are global binders; what molecular features define these divergent substrate specificities is currently unknown. *Salmonella* ProQ (*proP* expression regulator) harbors an N-terminal FinO domain as well as two further putative RNA-binding regions (157), and interacts with hundreds of cellular RNA ligands, including both sRNAs and mRNAs (158, 339). While accumulating evidence suggests a role for FinO/ProQ proteins in the maintenance of genome integrity in diverse bacterial species (159, 160), *proQ* deletion in *Salmonella* and *E. coli* leads to widespread expression changes and *in vitro* phenotypes including defects in biofilm formation (158, 161), as well as the attenuation in a mouse colitis model of salmonellosis (A. J. Westermann, E. Venturini, M. Sellin, K. U. Förstner, W.-D. Hardt, J. Vogel, unpublished data), suggesting a global role for bacterial physiology and pathogenesis.

CspC/E

The latest additions to the list of global bacterial RNA-binding proteins (RBPs) with roles in virulence and host-microbe interaction are the highly conserved members of the cold shock protein family CspC and CspE (162). In *S.* Typhimurium, CspC and CspE constitute a largely overlapping interactome comprising hundreds of RNA ligands (sRNAs and mRNAs) enriched for virulence functions, and Csp binding can protect some of its targets from ribonucleolytic attack (162). Importantly, single deletion of either *cspC* or *cspE* fails to produce any robust phenotype; however, a Δ*cspCE* double-knockout strain is heavily impaired in its resistance to oxidative stress, motility, biofilm formation, and virulence in both cell culture and a systemic mouse model.

Together, the phenotypic effects associated with the removal of each of these global RBPs, despite their different molecular modes of action, are strikingly similar to one another, and several are virulence related (Fig. 2a). For instance, Δ*csrA*, Δ*hfq*, and Δ*cspCE Salmonella* mutants are less motile than wild-type strains, and Δ*csrA*, Δ*hfq*, Δ*proQ*, and Δ*cspCE* show defects in biofilm for-

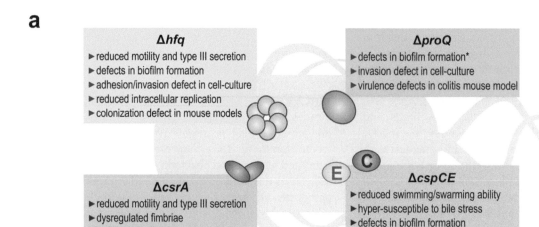

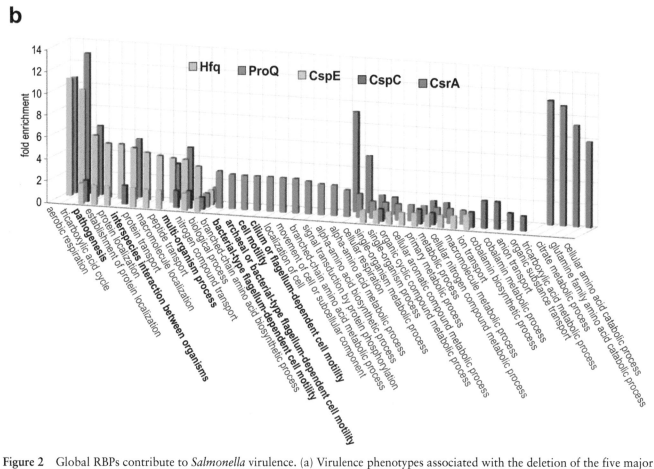

Figure 2   Global RBPs contribute to *Salmonella* virulence. (a) Virulence phenotypes associated with the deletion of the five major RBPs in *S.* Typhimurium: the RNA chaperones Hfq (147, 330) and ProQ (158; Westermann et al., unpublished), the translational regulator CsrA (137), and the cold shock proteins CspC and CspE (162). The asterisk (*) indicates that the biofilm formation phenotype of Δ*proQ* mutants stems from work with *E. coli* (161). (b) The interactome of the same RBPs in *S.* Typhimurium is enriched for virulence-associated mRNAs. Gene ontology enrichment analysis (331, 332) on CLIP-seq data for Hfq and CsrA (136) and RIP-seq data for ProQ (158) and CspC and CspE (162). Fold enrichments of all significantly enriched (*P* < 0.05) pathways are plotted. Virulence-related pathways are in bold.

mation and are attenuated in defined mouse models. While this might partially be explained by the deletion of global posttranscriptional hub proteins resulting in similar pleiotropic effects, transcriptome-wide RBP interactome studies in *Salmonella* have revealed an overlapping set of virulence-associated pathways enriched in the ligand suite of CsrA, Hfq, ProQ, and CspC/E (Fig. 2b).

Additionally, reflecting the situation in eukaryotes (163), there might exist bacterial proteins with secondary RNA-binding functions. As an example, the cytoskeletal protein RodZ was recently shown to act as a membrane-bound RBP, posttranscriptionally controlling the expression of the T3SS in *Shigella sonnei* (164). It will be exciting to extend and functionally characterize the full target set of bacterial RBPs under virulence conditions in the near future.

RNases

Bacterial organisms harbor various RNases with distinct substrate specificities, and many of them contribute to pathogenesis (165). In Gram-negative pathogens, RNase E is an integral part of the degradosome but may also act as a single-strand-specific endonuclease in a degradosome-independent manner, e.g., to initiate sRNA-mediated target decay (166). Lowering RNase E activity in *Y. pseudotuberculosis* skews the secretion of effector proteins (167). Similar effects were observed upon the removal of Hfq (168) or of the phosphorolytic exoribonuclease PNPase (polynucleotide phosphorylase) (169, 170), another constituent of the degradosome, arguing that this phenotype might at least partially depend on Hfq- and RNase E/PNPase-dependent sRNAs (21). Also, in *S.* Typhimurium and *C. jejuni*, PNPase is considered a major regulator of virulence. It affects both major stages of the *Salmonella* infection cycle, invasion and intracellular replication, impacting the outcome of infection in a systemic mouse model (171, 172). Likewise, a *Campylobacter* mutant devoid of PNPase displays severe virulence phenotypes including motility, adhesion, and invasion defects (173).

RNase E, in rare cases, may also mediate the effects of *cis*-antisense RNAs, as exemplified by the RNase E-dependent degradation of the complex consisting of the AmgR antisense RNA and the *mgt* operon in *Salmonella* (42). However, more often, such long, perfectly complementary double-stranded RNA complexes are preferred substrates of RNase III. This endoribonuclease catalyzes the turnover of complexes of *Staphylococcus* RNAIII with its target transcripts (174), and an RNase III-deficient *Salmonella* mutant showed a virulence defect in a wax moth infection model (175). In *S. flexneri*, the exoribonuclease RNase R is required for the

synthesis of a defined set of secreted effector proteins (176); the endoribonuclease YbeY represses the master regulator of effector translocation in *Yersinia enterocolitica* (177) and is critical for *V. cholerae* virulence (178); and *L. monocytogenes* depends on the endoribonuclease RNase HII for full virulence in a mouse model (179). Likewise, the Gram-positive-specific endoribonucleases RNase J1 and J2 (180), as well as RNase Y (181), have been implicated in *Streptococcus* and *Staphylococcus* virulence, respectively. For more information on enzymes involved in bacterial RNA metabolism, see section I.

NEW APPROACHES TO STUDY RNA-MEDIATED REGULATION IN INFECTION BIOLOGY

Notwithstanding the prominent representatives of regulatory RNA described in the previous sections, examples of RNAs involved in virulence and host cell interaction are still relatively sparse. This is due, at least in part, to technical difficulties in the study of bacterial gene expression within the host. Importantly, even once RNA molecules highly expressed under infection-relevant conditions have been identified, their inhibition rarely results in any measurable fitness defect during the infection of the commonly used model systems (Fig. 3). This is in stark contrast to the severe infection phenotypes observed upon the deletion of global RNA chaperones such as Hfq (147) (Fig. 2a), which imply that RNA-based regulation—at least the sum of it—contributes substantially to virulence. Instead, it seems

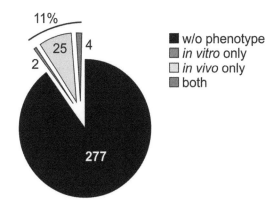

Figure 3 General lack of phenotypes of virulence-associated RNAs. An Rfam search (333) for the query term "virulence" yielded 373 hits (as of February 2017). From these, nonbacterial RNAs and CRISPR RNAs were removed. For the remaining 308 entries, a manual literature search revealed whether or not the respective deletion/disruption mutants exhibit a virulence defect in cell culture (*in vitro* phenotype), live-animal models (*in vivo*), or both.

that the most amenable model systems, such as cell culture or simplistic mouse models, combined with traditional readout measures (e.g., counting viable colonies) provide insufficient sensitivity to resolve the subtle effects arising from the deletion of single RNA genes.

The generic nature of sequencing technology lends this approach to the study of multiorganism interactions, and thus to complex infection systems. Equipped with novel sequencing-based methods (182, 183) superior to traditional approaches in both sensitivity and specificity, we are in a position to identify virulence-related RNA molecules in high-throughput ways and uncover their molecular functions. The sections below provide an overview of these techniques and review some of their key findings from a virulence perspective.

Which Bacterial Transcripts Alter Their Structure in Response to Infection-Relevant Cues?

The identification of RNA thermosensors and riboswitches has been difficult, as these RNA elements are typically only poorly conserved on the primary sequence level and, consequently, challenging to predict computationally based solely on genomic information (184). To map secondary structure changes in eukaryotic transcriptomes, diverse chemical or enzymatic probing approaches followed by genome-wide detection have been established. Tools such as (ic)SHAPE ([*in vivo* click] selective 2′-hydroxyl acylation analyzed by primer extension), PARIS (psoralen analysis of RNA interactions and structures), and PARS (parallel analysis of RNA structures) detect the local structure of RNA molecules (185–188), and ever-improving sensitivity allows their application also to bacterial transcriptomes. For instance, SHAPE was applied to *Salmonella* grown over a temperature gradient in order to characterize the *htrA* thermometer (189), and a PARS-related method was used to map the thermoresponsive transcriptome of *Y. pseudotuberculosis* (190).

SHAPE-based approaches have also been used to map the structure of a thiamine pyrophosphate- and an adenosylcobalamin-sensing riboswitch in *E. coli thiM* and *btuB* mRNAs, respectively (191, 192), as well as of an S-box/SAM-I and a fluoride riboswitch within *yitJ* and *crcB* mRNAs in *Bacillus* species (193, 194). Alternatively, a conditional premature 3′ end is indicative of transcriptional riboswitches and can be detected via Term-seq, a transcriptome-wide 3′-end-mapping protocol for bacterial transcripts (195). Importantly, Term-seq is compatible with RNA samples from complex bacterial consortia, exemplified by its application to the oral microbiota (195). It will be intriguing to transfer this technique to bacterial pathogens during their interaction with host cells and with commensal bacteria of the microbiome.

Which Bacterial Genes Are Differentially Expressed during Infection?

The high sensitivity and resolution of transcriptome sequencing (RNA-seq) resulted in a technological leap allowing the profiling of bacterial gene expression within the host context (196). For example, application of RNA-seq to intracellular *S.* Typhimurium recovered from infected macrophages by selective lysis and differential centrifugation charted the intracellular transcriptome of this bacterium (14). While expression of *Salmonella* mRNAs correlated well with previous microarray data (38), RNA-seq extended those findings by capturing almost 250 (of the annotated 280) sRNAs, which had mostly escaped detection on microarrays. Among them, the two paralogs RyhB and IsrE (also known as RyhB2) (197) were ~100-fold upregulated, suggesting that *Salmonella* faces conditions of iron limitation inside macrophages.

For dual RNA-seq (198), host RNA is no longer depleted prior to sequencing, complementing the information about gene expression of the infecting pathogen with the corresponding host response. The method has been applied to diverse infection systems, consisting of both intra- and extracellular bacterial pathogens in cell culture or animal models (reviewed in reference 199). Its application to human bronchial epithelial cell cultures infected with nontypeable *Haemophilus influenzae* led to the identification of sRNAs strongly expressed during infection, including the iron-responsive HrrF sRNA as well as a homolog of GcvB (200). Transferring dual RNA-seq to *in vivo* models of infection enabled the profiling of gene expression of the extracellular pathogen *Y. pseudotuberculosis* within murine Peyer's patches, including expression information on almost 200 sRNAs (15). Among them, the conserved SgrS was one of the most strongly upregulated transcripts compared to *in vitro* samples (~200-fold induced), but also the *Yersinia* orthologs of RyhB were ~10-fold activated, together implying that *Yersinia* experiences glucose-phosphate stress and iron limitation in its host niche (201).

Dual RNA-seq was also applied to track sRNA expression kinetics of intracellular *S.* Typhimurium during a time course of infection of different immortalized cell lines (95). This identified PinT as one of the most strongly induced *Salmonella* sRNAs inside any host cell type assessed (~100-fold upregulated). Although deleting single sRNAs only rarely causes robust virulence phenotypes, loss (or gain) of function is likely to leave its footprint in the transcriptome of pathogen and host, thereby defining a "molecular" phenotype (202) for a given gene. Comparative dual RNA-seq, i.e., profiling expression differences between wild-type infection and that with a Δ*pinT* strain, was used to characterize the role of the sRNA for *Salmonella* virulence (95). The set of derepressed bacterial genes in Δ*pinT* compared to wild-type samples informed about (direct and indirect) target genes of this sRNA. Additionally, comparative dual RNA-seq pinpointed host processes differentially affected between the two infections. Using interspecies cluster analysis, sRNA-dependent changes in the bacterial transcriptome could be correlated with changes in the host transcriptome, providing valuable information on putative host-*Salmonella* interactions.

Inhibition of Which Bacterial Genes Affects Fitness during Infection?

High expression levels during infection might be indicative of the contribution of a given regulatory RNA to bacterial infection. The traditional approach to test this is by comparing the performance of a defined deletion mutant to that of the isogenic wild-type strain during infection. It may, however, be difficult to find suitable *in vivo* conditions under which a given deletion mutant exhibits a virulence defect. In fact, while in the above-mentioned study of *Yersinia*-infected lymphatic tissues (15) a Δ*sgrS* mutant failed to compete with wild-type *Yersinia* during growth under glucose-phosphate stress conditions *in vitro*, it did not produce a robust fitness defect during mouse infection in any of the tested host organs. Similarly, a Δ*pinT Salmonella* mutant displayed a virulence defect in only one (porcine macrophage-like cells) out of five tested cell culture models (95).

High-throughput approaches to systematically inhibit bacterial genes and evaluate the fitness of the resulting mutants have been established and offer an alternative to tedious manual screening efforts. TIS (transposon insertion sequencing) (203–205) combines random mutagenesis of bacterial genomes by the insertion of transposons with growth under selective pressure, e.g., under infection conditions, followed by massively parallel genome sequencing. For instance, the composition of a *Salmonella* transposon mutant library was quantified before and after its passage through different animal models of salmonellosis—mouse, chicken, pig, and cattle (96). Different sets of genes were found to contribute to the extraintestinal infection of mice and the gastrointestinal infection of chickens, pigs, and cows, reflecting the divergent nature of the diseases caused by *Salmonella* in these host models. Importantly, the transposon insertion density (on average, 1 insertion per 650 bp) was sufficient to capture the contribution of ~40 sRNAs

to infection, such as PinT and InvR, whose disruption each led to an attenuation in the gastrointestinal models (96, 182). Likewise, an average transposon density of 550 to 1,500 bp per insertion enabled the quantification of the relative fitness of individual *Streptococcus pneumoniae* sRNA mutants during colonization of the murine nasopharynx, the lungs, or the bloodstream (206). Most prominently, inhibition of an sRNA termed Srn157 led to reduced fitness in the nasopharynx and the lung, probably due to pleiotropic effects as revealed by proteomics. Interestingly, however, many of the other attenuated sRNA mutants displayed a phenotype in only one of the three host sites, suggesting that pneumococcal sRNAs predominantly contribute to host tissue-specific regulations.

TIS may further be used for high-throughput genetic interaction mapping. To this end, after exposure to a challenge, the composition of a transposon mutant library consisting of single mutants is compared to a second library generated in a defined deletion background (resulting in a pool of double mutants). Proof of concept comes from a study on *M. tuberculosis* that identified genes interacting with the *mce* virulence locus (207). Future studies might be dedicated to the identification of genetic interactions among individual bacterial regulatory RNAs or RBPs during host-pathogen interaction.

To account for population bottlenecks, i.e., the stochastic loss of bacterial mutants during infection (208, 209), the input pool of transposon mutant libraries might be split into several lower complex sublibraries. Alternatively, computational approaches to compensate for bottleneck effects have been introduced that require knowledge of a set of neutral genes to be used as a reference (204). Finally, CRISPRi (CRISPR interference) can be multiplexed, resulting in the parallel knockdown of hundreds of bacterial genes (210, 211), and therefore may be an attractive alternative to TIS for the targeted screening of defined gene subsets (e.g., regulatory RNA genes).

What Are the Molecular Interaction Partners of a Bacterial RNA?

Having identified transcripts affected by the perturbation of an RNA gene, this molecular phenotype may be further investigated by mapping its cellular interaction partners. Several sequencing-based methods have recently been introduced to identify RNA-RNA or RNA-protein interaction on a large scale (212, 213). CLIP-seq (UV cross-linking and immunoprecipitation high-throughput sequencing) (136, 149), CLASH (cross-linking, ligation, and sequencing of hybrids) (214), and RIL-seq (RNA interaction by ligation and sequencing) (215) are based on the pulldown of an RBP followed by the sequencing of its RNA ligands, either alone (CLIP-seq) or themselves bound to interacting RNA transcripts (CLASH or RIL-seq). Conversely, MAPS (MS2 affinity purification coupled with RNA sequencing) (216) and GRIL-seq (global sRNA target identification by ligation and sequencing) (217) pull down RNA directly and sequence the pool of interacting transcripts.

The application of CLIP-seq to *Salmonella* CsrA during growth *in vitro* revealed its binding sites within mRNAs encoding invasion-associated proteins (136). CLASH on the endoribonuclease RNase E charted the sRNA interactome of EHEC and expanded the mRNA target suite of the pathogenicity island-encoded Esr41 sRNA (214), which unexpectedly resembled that of RyhB (197). Likewise, GRIL-seq of *P. aeruginosa* sRNAs PrrF1, ErsA, Sr0161, and Sr006 (217, 218) revealed direct mRNA targets associated with *Pseudomonas* virulence and antibiotic resistance, and MAPS on the *S. aureus* sRNA RsaA supported its role in biofilm formation and chronic infections (219). Together, these pioneering studies, despite being performed *in vitro* under noninfectious conditions, already provide novel insight into RNA-mediated virulence mechanisms in the respective pathogens. Future studies should establish the transfer of RNA-RNA and RNA-protein interactome studies to *in vivo* settings.

Finally, fluorescence *in situ* hybridization (FISH) provides a window into cells to track the fate of specific transcripts (220). As FISH relies on the binding of fluorescently labeled probes to their target RNAs, longer molecules naturally give a stronger signal than short ones, hampering the detection of bacterial sRNAs. The recent development of a super-resolution imaging and modeling platform, however, proved suitable to monitor the target search kinetics of the *Salmonella* sRNA SgrS during growth *in vitro* (221). Ongoing improvement in the method's sensitivity might eventually enable us to follow sRNA-mRNA interactions during infection.

COMMON PRINCIPLES OF VIRULENCE-RELATED REGULATORY RNAs

Virulence-Metabolism Cross Talk

Collectively, the described virulence-related regulatory RNAs—as different as they and their host pathogens are—allow us to deduce some common themes (Fig. 4). Firstly, there is an enrichment of virulence-related RNAs that are associated with metabolic processes (19). Exam-

ples include classic *cis*-acting riboswitches that respond to metabolites or metal ions (2) as well as *trans*-acting riboswitches such as SreA/B (118) and base-pairing (222, 223) or CsrA-titrating sRNAs (137). This overrepresentation relates to the dramatic changes in available nutrient sources that arise when bacterial pathogens invade their hosts, requiring a tight interconnection of virulence and metabolic traits (Fig. 4a). Regulatory RNAs may link virulence and metabolism in different ways. This likely reflects the evolutionary history of the respective RNA element that may have initially been involved exclusively in one process—virulence *or* metabolism—and subsequently acquired novel functions (224). Several examples of RNAs exist whose expression is controlled by metabolic cues and that in turn regulate virulence genes. For instance, the SgrS sRNA is induced under conditions of sugar-phosphate stress and in *S.* Typhimurium represses the horizontally acquired effector protein SopD (125). SAM binding to its cognate riboswitch in *L. monocytogenes* leads to the accumulation of *trans*-acting SreA and SreB, which repress the key virulence regulator PrfA (118). Reciprocally, other RNAs accumulate under defined virulence conditions and control metabolic pathways, such as the TarA sRNA of *V. cholerae*, which is activated by ToxT, the central virulence transcription factor in this organism, and in turn controls glucose uptake (84). An intriguing case of metabolic adaptation upon host sensing was recently described for pathogenic *E. coli*. A *cis* RNA element links translation of the secreted virulence effector NleA to host cell attachment, and upon NleA injection, the corresponding effector chaperone (CesT) antagonizes CsrA, resulting in the widespread posttranscriptional remodeling of metabolic gene expression (225).

Regulation of Lifestyle Transitions

A second prominent feature of virulence-associated RNAs, reminiscent of other stress-related regulatory RNAs (6, 7), relates to their control of transition phases and involvement in feedback regulation (Fig. 4b). In the context of infection, the first major transition occurs upon the colonization of a new host, but also thereafter the infection cycle of pathogens remains associated with ever-changing environments: the bacteria may travel to and occupy different host niches, split into differently behaving subpopulations, interact with distinct host cell types, and accordingly face host immune responses to varying degrees; or they may eventually be released into the surroundings to initiate a new round of infection. Some of the base-pairing sRNAs introduced above, such as *Staphylococcus* RNAIII, *Streptococcus* FasX, *Salmonella* PinT, and *Chlamydia* IhtA,

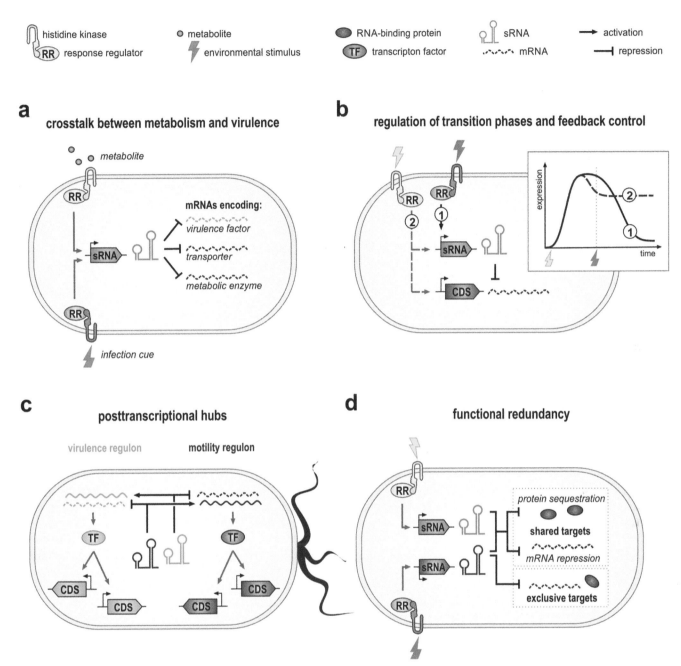

Figure 4 Common principles of sRNA-based virulence control. (a) The tight interconnection between metabolism and virulence is illustrated by the transcription of certain sRNAs in response to metabolic or infection-related cues, which in turn control virulence regulators/effectors, nutrient transporters, and/or metabolic enzymes. (b) Control of major lifestyle transitions. Upon sensing of a specific trigger, a bacterial sRNA may act in concert with transcription factors (not shown) to coordinate the rapid silencing of genes and degradation of mRNAs no longer needed under the new condition (circuit 1). This mechanism differs from feedback control, where sRNA and target are coactivated by the same stimulus (circuit 2). (c) sRNAs target mRNAs for key regulatory proteins, thereby indirectly controlling the expression of genes that are under the transcriptional control of those regulators. Often this RNA-based regulatory network ensures mutually exclusive activation of opposing processes (exemplified here by motility and virulence effector secretion). (d) Functional redundancy between sRNA homologs. sRNA pairs might arise from gene duplication events and thus share sequence homology, allowing the regulation of shared targets. However, functional redundancy might only be partial, when sibling sRNAs are produced under different conditions, interact with distinct cellular protein partners, or base-pair uniquely with specific mRNAs.

which act at the transition of colonization to dissemination mode (RNAIII [226] and FasX [227]) or of extra- to intracellular lifestyle (PinT [95]), or control the switch from elementary to reticulate bodies (IhtA [97]), can be reiterated in this context.

Hubs of Posttranscriptional Control

Regulatory RNAs occasionally affect specific virulence factors, but often even control entire virulence regulons, through positively or negatively affecting key virulence transcription factors. Among the best-characterized "hubs" of posttranscriptional virulence control are the mRNAs of the global activator of *Salmonella* invasion genes, HilD (19, 69, 228–230, 334); of the central biofilm regulator in *Salmonella* and *E. coli*, CsgD (231–234); of the repressor of flagellin and of the key virulence regulator in *Listeria*, MogR and PrfA (235); of the global regulator of virulence genes in *Staphylococcus*, MgrA (103, 236); and of the master regulator of type III secretion in *Yersinia*, LcrF (237). Typically, these mRNAs harbor long UTRs, offering space for various *cis-* and *trans*-acting RNAs to simultaneously act. This regulatory principle—the main regulators of defined virulence traits being themselves controlled by the action of numerous regulatory RNAs—gives rise to a complex network interconnecting the different (often mutually exclusive or hierarchically ordered) processes and ensures that they are only activated once really needed (Fig. 4c). As major virulence programs, e.g., synthesizing flagella or building a secretion apparatus, are energetically costly (238, 239), such RNA-based steering mechanisms provide strong evolutionary benefits.

Functional Redundancy

The sections above touched upon functional redundancy among pairs of sRNAs, such as the CsrA/RsmA-interacting sRNAs, of which most bacteria encode two homologs, but also for the *Vibrio* Qrr sRNAs (240) and *Neisseria* RcoF1/2 (101, 102). To some extent, this redundancy is reminiscent of the repertoire of secreted effector proteins with overlapping functions in the target host cell (241–243), and may help explain why single sRNA (and effector) deletion mutants often lack clear phenotypes (202). While this might argue that functional redundancy has arisen as a safeguard mechanism to maintain crucial regulatory events, it is noteworthy that at least for some of the respective sRNAs exclusive functions have been described (Fig. 4d). The conserved, functionally homologous sRNAs RyhB and IsrE are both induced under conditions of iron deficiency and play a role in the protection of *S.* Typhimurium against oxidative stress by controlling the expression of a set of

shared target genes (244, 245). At the same time, however, IsrE regulates a set of motility genes that are not targeted by its homolog, and vice versa, RyhB exclusively controls the mRNAs of a fimbrial lipoprotein and a metabolic enzyme (244, 246). Similarly, in *Yersinia*, only RyhB1—but not RyhB2—interacts with the Hfq chaperone, likely conferring exclusive functions to this sRNA (247, 248). While Hfq interaction protects RyhB1 from PNPase-catalyzed degradation, RyhB2 is stable even in the absence of this RBP. Interestingly, in both *Salmonella* and *Yersinia*, the variable 5′ regions of the RyhB homologs seem responsible for their selective functions. RyhB arguably belongs to the best-characterized representatives of bacterial sRNAs. Therefore, future studies have to investigate whether complete redundancy among regulatory RNAs actually exists, or—as in the case of RyhB—functional overlap is only partial.

EMERGING TOPICS

Cell-to-Cell Heterogeneity in RNA Expression

Novel concepts are steadily emerging in the field of bacterial infection biology, and a potential involvement of regulatory RNA represents an exciting research avenue for the future. Single-cell approaches, particularly single-cell RNA-seq, have found their way into infection research and highlight the prevalence of heterogeneity in the host responses to bacterial challenge (249, 250). Single-bacterium RNA-seq remains challenging due to insufficient sensitivity and the dependence of current protocols on polyadenylated RNAs (251). However, reporter-based approaches revealed fluctuations in *Salmonella* biofilm formation (252) and in the expression of invasion- (253, 254) and intracellular survival-associated proteins (249), with impact on virulence phenotypes of, and host immune responses induced by, specific bacterial subpopulations. Furthermore, heterogeneous expression of TA systems leads to the development of bacterial persisters (255), thereby contributing to chronic infections. Intriguingly, the *E. coli* TisB/IstR-1 type I TA system controls the transition between a replicating and a persistent state, demonstrating that individual, heterogeneously expressed regulatory RNAs can control phenotype switching (62). Cell-to-cell variability was also observed for the expression of certain *trans*-encoded sRNAs in different bacterial species (256, 257). However, addressing this phenomenon on a genome-wide scale requires the transfer of current eukaryotic single-cell RNA-seq techniques to bacteria (258). Once fully established, genome-wide single-bacteria RNA-seq will be of paramount impact also for the interpretation of pervasive antisense transcription, as it will enable correlation of expression levels of antisense-overlapping genes in the same cell.

Are There Functionally Relevant Extracellular RNAs?

Bacterial species within dense microbial consortia may cooperate, e.g., by trading genes via horizontal gene transfer (HGT). As HGT occurs in a contact-dependent manner, it is common among individual members of the microbiota (259, 260). HGT can sometimes also be observed between commensal and pathogenic species, triggering opportunistic pathogens to cause disease. The question therefore arises of whether bacteria not only exchange genes (i.e., DNA stretches) but also swap their direct products (RNA transcripts). In fact, there is a precedence for secreted enzymes that are shared among cocolonizing bacterial species, enabling the recipient microbes to conduct metabolic tasks they would not be able to perform on their own (261). While proteins are relatively stable, RNA molecules are generally considered more vulnerable to degradation, arguing that their average half-life in an unprotected environment is short. Nevertheless, there are cellular processes that could be envisioned for interspecies RNA transfer. Some Gram-negative bacteria possess T6SSs to pump toxins into competing microbes. It is tempting to speculate that RNA molecules might take the same route; however, the internal diameter of the secretion needle (262) would likely require the RNAs to be translocated in an unfolded state and be refolded upon arrival in the target bacterium.

Alternatively, extracellular membrane vesicles (EVs) may function as shuttles for public goods, including RNP complexes (263–266, 340), within bacterial consortia. However, it is currently unclear whether these RNPs randomly diffuse into vesicles prior to budding or if they are actively sorted into them, and what functions—if any—they serve upon EV fusion with a recipient bacterium. Of note, recent findings suggest that sRNA-containing EVs are employed by bacterial pathogens to manipulate their eukaryotic host (267, 268). It remains to be elucidated whether this occurs passively through the sensing of bacterial RNA molecules by the arsenal of mammalian cytosolic nucleic acid sensors and downstream signaling events, or actively by physical host-pathogen RNA-RNA interaction. It should be considered, however, that to effectively employ a microRNA-like mechanism in the host cell, a bacterial RNA species would have to be highly abundant, such as RNA fragments derived from bacterial housekeeping transcripts (267). Interestingly, there is evidence for microRNA delivery from eukaryotic cells into bacterial members of the intestinal

microbiota (269, 270), suggesting the possibility of RNA-based cross-kingdom communication in both directions. However, irrespective of what route RNA molecules may take, genetic experiments with mutants impaired in secretion systems or vesicle trafficking, and their reconstitution in a heterologous expression model, are essential to convincingly demonstrate the functional importance of extracellular RNA in bacterial infections.

Exploiting Bacterial RNAs and RBPs for Therapeutics

Given the current antibiotic resistance crisis (271, 272), novel therapeutics against bacterial infections are urgently needed. Based on the fact that the vast majority of bacteria populating our bodies are either neutral or health-promoting, these next-generation antibiotics should ideally be highly specific, targeting only pathogenic, while sparing commensal, species. The high degree of pathogen-specific riboregulation (273), as opposed to many protein-centric processes, which tend to be more conserved, provides great potential for RNAs and RBPs in quest of specific drug targets (224). High-throughput screening of a peptide library in *E. coli* identified a cyclic peptide interfering with the Hfq regulon by blocking the chaperone's interaction with RNA ligands (274), and small-molecule inhibitors of *E. coli* RNase E (275) and recently of *Y. pseudotuberculosis* CsrA (276) were discovered. Besides, analogs of riboswitch-binding metabolites have been successfully used to inhibit growth of nonpathogenic *B. subtilis in vitro* (277), or as bactericidals against *S. aureus* in a mouse model of infection (278).

Could *trans*-encoded regulatory RNAs also serve as drug targets? We have seen above that deleting individual sRNAs often fails to produce strong virulence defects. However, a given drug does not necessarily have to kill the respective pathogen directly in order for it to eventually become eradicated from the host. Rather, thanks to bacterial competition within our body, reducing the fitness of a pathogen and/or increasing that of commensal species occupying the same niche may be sufficient to tilt the balance of infection in favor of the host. Moreover, certain sRNAs of pathogenic bacteria are induced upon exposure to antibiotics (279–281) and might represent suitable targets to counteract antibiotic resistance (282, 283, 341). Multiple mRNAs can be simultaneously targeted using a mix of specific antisense nucleic acid derivatives, leading to species-specific eradication within complex bacterial consortia (284), and regulatory RNAs may be knocked down (or trapped in an inactive state) in the same manner. It is conceivable that eventually a cocktail of nucleic acids, inhibiting the pathogen species while simultaneously promoting expansion of competing commensals, may be administered as a highly specific drug combating a given bacterial infection. However, the success of these antivirulence strategies may depend on how effectively individual subpopulations of a pathogen can be targeted (285).

CONCLUSIONS

The last decade has revealed regulatory RNAs with prominent roles in virulence mechanisms in a broad spectrum of bacterial pathogens. Recent technological advances, particularly with respect to RNA-seq-based approaches, suggest, however, that these may just account for the tip of the iceberg and many infection-related regulatory RNAs might have escaped discovery. Further improvement of current methodology, especially the combination of global expression, interaction, and single-cell data with high-throughput functional readout systems, bears great potential to boost this burgeoning field. Many of these techniques are generic in nature, enabling their transfer to complex systems, including a pathogen, its host, and the resident microbiota, to deepen our understanding of—and eventually increase our ability to interfere with—bacterial infections.

Acknowledgments. I thank Sandy R. Pernitzsch (https://www .scigraphix.com/) for producing the figures for this review. Further, I am grateful to Erik Holmqvist, Lars Barquist, Yanjie Chao, the editors, and an anonymous reviewer for critical comments on the manuscript. This review was written while I was a member of the research group of Prof. Jörg Vogel (IMIB Würzburg).

Citation. Westermann AJ. 2018. Regulatory RNAs in virulence and host-microbe interactions. Microbiol Spectrum 6(4):RWR-0002-2017.

References

1. Morens DM, Fauci AS. 2013. Emerging infectious diseases: threats to human health and global stability. *PLoS Pathog* 9:e1003467.

2. Breaker RR. 2012. Riboswitches and the RNA world. *Cold Spring Harb Perspect Biol* 4:a003566.

3. Lasa I, Toledo-Arana A, Gingeras TR. 2012. An effort to make sense of antisense transcription in bacteria. *RNA Biol* 9:1039–1044.

4. Wen Y, Feng J, Sachs G. 2013. *Helicobacter pylori* 5′ *ureB*-sRNA, a *cis*-encoded antisense small RNA, negatively regulates *ureAB* expression by transcription termination. *J Bacteriol* 195:444–452.

5. Storz G, Vogel J, Wassarman KM. 2011. Regulation by small RNAs in bacteria: expanding frontiers. *Mol Cell* 43:880–891.

6. Wagner EG, Romby P. 2015. Small RNAs in bacteria and archaea: who they are, what they do, and how they do it. *Adv Genet* 90:133–208.

7. Gottesman S, Storz G. 2011. Bacterial small RNA regulators: versatile roles and rapidly evolving variations. *Cold Spring Harb Perspect Biol* 3:a003798.

8. Holmqvist E, Wagner EG. 2017. Impact of bacterial sRNAs in stress responses. *Biochem Soc Trans* 45:1203–1212.

9. Toledo-Arana A, Dussurget O, Nikitas G, Sesto N, Guet-Revillet H, Balestrino D, Loh E, Gripenland J, Tiensuu T, Vaitkevicius K, Barthelemy M, Vergassola M, Nahori MA, Soubigou G, Régnault B, Coppée JY, Lecuit M, Johansson J, Cossart P. 2009. The *Listeria* transcriptional landscape from saprophytism to virulence. *Nature* 459:950–956.

10. Sharma CM, Hoffmann S, Darfeuille F, Reignier J, Findeiss S, Sittka A, Chabas S, Reiche K, Hackermuller J, Reinhardt R, Stadler PF, Vogel J. 2010. The primary transcriptome of the major human pathogen *Helicobacter pylori. Nature* 464:250–255.

11. Albrecht M, Sharma CM, Dittrich MT, Muller T, Reinhardt R, Vogel J, Rudel T. 2011. The transcriptional landscape of *Chlamydia pneumoniae. Genome Biol* 12:R98.

12. Kroger C, Colgan A, Srikumar S, Handler K, Sivasankaran SK, Hammarlof DL, Canals R, Grissom JE, Conway T, Hokamp K, Hinton JC. 2013. An infection-relevant transcriptomic compendium for *Salmonella enterica* serovar Typhimurium. *Cell Host Microbe* 14:683–695.

13. Haning K, Cho SH, Contreras LM. 2014. Small RNAs in mycobacteria: an unfolding story. *Front Cell Infect Microbiol* 4:96.

14. Srikumar S, Kroger C, Hebrard M, Colgan A, Owen SV, Sivasankaran SK, Cameron AD, Hokamp K, Hinton JC. 2015. RNA-seq brings new insights to the intramacrophage transcriptome of *Salmonella* Typhimurium. *PLoS Pathog* 11:e1005262.

15. Nuss AM, Beckstette M, Pimenova M, Schmuhl C, Opitz W, Pisano F, Heroven AK, Dersch P. 2017. Tissue dual RNA-seq allows fast discovery of infection-specific functions and riboregulators shaping host-pathogen transcriptomes. *Proc Natl Acad Sci U S A* 114:E791–E800.

16. Papenfort K, Vogel J. 2010. Regulatory RNA in bacterial pathogens. *Cell Host Microbe* 8:116–127.

17. Gripenland J, Netterling S, Loh E, Tiensuu T, Toledo-Arana A, Johansson J. 2010. RNAs: regulators of bacterial virulence. *Nat Rev Microbiol* 8:857–866.

18. Caldelari I, Chao Y, Romby P, Vogel J. 2013. RNA-mediated regulation in pathogenic bacteria. *Cold Spring Harb Perspect Med* 3:a010298.

19. Oliva G, Sahr T, Buchrieser C. 2015. Small RNAs, 5′ UTR elements and RNA-binding proteins in intracellular bacteria: impact on metabolism and virulence. *FEMS Microbiol Rev* 39:331–349.

20. Svensson SL, Sharma CM. 2016. Small RNAs in bacterial virulence and communication. *Microbiol Spectr* 4:VMBF-0028-2015.

21. Heroven AK, Nuss AM, Dersch P. 2017. RNA-based mechanisms of virulence control in *Enterobacteriaceae*. *RNA Biol* **14:**471–487.

22. Quereda JJ, Cossart P. 2017. Regulating bacterial virulence with RNA. *Annu Rev Microbiol* **71:**263–280.

23. Somerville GA, Proctor RA. 2009. At the crossroads of bacterial metabolism and virulence factor synthesis in staphylococci. *Microbiol Mol Biol Rev* **73:**233–248.

24. Sudarsan N, Lee ER, Weinberg Z, Moy RH, Kim JN, Link KH, Breaker RR. 2008. Riboswitches in eubacteria sense the second messenger cyclic di-GMP. *Science* **321:**411–413.

25. Lee ER, Baker JL, Weinberg Z, Sudarsan N, Breaker RR. 2010. An allosteric self-splicing ribozyme triggered by a bacterial second messenger. *Science* **329:**845–848.

26. Miotto P, Forti F, Ambrosi A, Pellin D, Veiga DF, Balazsi G, Gennaro ML, Di Serio C, Ghisotti D, Cirillo DM. 2012. Genome-wide discovery of small RNAs in *Mycobacterium tuberculosis*. *PLoS One* **7:**e51950.

27. Kortmann J, Narberhaus F. 2012. Bacterial RNA thermometers: molecular zippers and switches. *Nat Rev Microbiol* **10:**255–265.

28. Hoe NP, Goguen JD. 1993. Temperature sensing in *Yersinia pestis*: translation of the LcrF activator protein is thermally regulated. *J Bacteriol* **175:**7901–7909.

29. Bohme K, Steinmann R, Kortmann J, Seekircher S, Heroven AK, Berger E, Pisano F, Thiermann T, Wolf-Watz H, Narberhaus F, Dersch P. 2012. Concerted actions of a thermo-labile regulator and a unique intergenic RNA thermosensor control *Yersinia* virulence. *PLoS Pathog* **8:**e1002518.

30. Grosso-Becerra MV, Croda-García G, Merino E, Servín-González L, Mojica-Espinosa R, Soberón-Chávez G. 2014. Regulation of *Pseudomonas aeruginosa* virulence factors by two novel RNA thermometers. *Proc Natl Acad Sci U S A* **111:**15562–15567.

31. Weber GG, Kortmann J, Narberhaus F, Klose KE. 2014. RNA thermometer controls temperature-dependent virulence factor expression in *Vibrio cholerae*. *Proc Natl Acad Sci U S A* **111:**14241–14246.

32. Johansson J, Mandin P, Renzoni A, Chiaruttini C, Springer M, Cossart P. 2002. An RNA thermosensor controls expression of virulence genes in *Listeria monocytogenes*. *Cell* **110:**551–561.

33. Loh E, Kugelberg E, Tracy A, Zhang Q, Gollan B, Ewles H, Chalmers R, Pelicic V, Tang CM. 2013. Temperature triggers immune evasion by *Neisseria meningitidis*. *Nature* **502:**237–240.

34. Bingham RJ, Hall KS, Slonczewski JL. 1990. Alkaline induction of a novel gene locus, *alx*, in *Escherichia coli*. *J Bacteriol* **172:**2184–2186.

35. Stancik LM, Stancik DM, Schmidt B, Barnhart DM, Yoncheva YN, Slonczewski JL. 2002. pH-dependent expression of periplasmic proteins and amino acid catabolism in *Escherichia coli*. *J Bacteriol* **184:**4246–4258.

36. Nechooshtan G, Elgrably-Weiss M, Sheaffer A, Westhof E, Altuvia S. 2009. A pH-responsive riboregulator. *Genes Dev* **23:**2650–2662.

37. Groisman EA. 2001. The pleiotropic two-component regulatory system PhoP-PhoQ. *J Bacteriol* **183:**1835–1842.

38. Hautefort I, Thompson A, Eriksson-Ygberg S, Parker ML, Lucchini S, Danino V, Bongaerts RJ, Ahmad N, Rhen M, Hinton JC. 2008. During infection of epithelial cells *Salmonella enterica* serovar Typhimurium undergoes a time-dependent transcriptional adaptation that results in simultaneous expression of three type 3 secretion systems. *Cell Microbiol* **10:**958–984.

39. Blanc-Potard AB, Groisman EA. 1997. The *Salmonella*-*selC* locus contains a pathogenicity island mediating intramacrophage survival. *EMBO J* **16:**5376–5385.

40. Lee EJ, Groisman EA. 2012. Control of a *Salmonella* virulence locus by an ATP-sensing leader messenger RNA. *Nature* **486:**271–275.

41. Lee EJ, Choi J, Groisman EA. 2014. Control of a *Salmonella* virulence operon by proline-charged tRNA^{Pro}. *Proc Natl Acad Sci U S A* **111:**3140–3145.

42. Lee EJ, Groisman EA. 2010. An antisense RNA that governs the expression kinetics of a multifunctional virulence gene. *Mol Microbiol* **76:**1020–1033.

43. Choi E, Han Y, Cho YJ, Nam D, Lee EJ. 2017. A *trans*-acting leader RNA from a *Salmonella* virulence gene. *Proc Natl Acad Sci U S A* **114:**10232–10237.

44. Lee JW, Lee EJ. 2015. Regulation and function of the *Salmonella* MgtC virulence protein. *J Microbiol* **53:**667–672.

45. Wagner EG, Simons RW. 1994. Antisense RNA control in bacteria, phages, and plasmids. *Annu Rev Microbiol* **48:**713–742.

46. Georg J, Hess WR. 2011. *cis*-antisense RNA, another level of gene regulation in bacteria. *Microbiol Mol Biol Rev* **75:**286–300.

47. Lybecker M, Bilusic I, Raghavan R. 2014. Pervasive transcription: detecting functional RNAs in bacteria. *Transcription* **5:**e944039.

48. Raghavan R, Sloan DB, Ochman H. 2012. Antisense transcription is pervasive but rarely conserved in enteric bacteria. *mBio* **3:**e00156-12.

49. Lloréns-Rico V, Cano J, Kamminga T, Gil R, Latorre A, Chen WH, Bork P, Glass JI, Serrano L, Lluch-Senar M. 2016. Bacterial antisense RNAs are mainly the product of transcriptional noise. *Sci Adv* **2:**e1501363.

50. Belzer C, van Schendel BA, Kuipers EJ, Kusters JG, van Vliet AH. 2007. Iron-responsive repression of urease expression in *Helicobacter hepaticus* is mediated by the transcriptional regulator Fur. *Infect Immun* **75:**745–752.

51. Wen Y, Feng J, Scott DR, Marcus EA, Sachs G. 2011. A *cis*-encoded antisense small RNA regulated by the HP0165-HP0166 two-component system controls expression of *ureB* in *Helicobacter pylori*. *J Bacteriol* **193:**40–51.

52. Oliva G, Sahr T, Rolando M, Knoth M, Buchrieser C. 2017. A unique *cis*-encoded small noncoding RNA is regulating *Legionella pneumophila* Hfq expression in a life cycle-dependent manner. *mBio* **3:**e02182-16.

53. Grundling A, Burrack LS, Bouwer HG, Higgins DE. 2004. *Listeria monocytogenes* regulates flagellar motility gene expression through MogR, a transcriptional repressor required for virulence. *Proc Natl Acad Sci U S A* **101:**12318–12323.

54. Wurtzel O, Sesto N, Mellin JR, Karunker I, Edelheit S, Bécavin C, Archambaud C, Cossart P, Sorek R. 2012. Comparative transcriptomics of pathogenic and non-pathogenic *Listeria* species. *Mol Syst Biol* **8**:583.

55. Schultze T, Hilker R, Mannala GK, Gentil K, Weigel M, Farmani N, Windhorst AC, Goesmann A, Chakraborty T, Hain T. 2015. A detailed view of the intracellular transcriptome of *Listeria monocytogenes* in murine macrophages using RNA-seq. *Front Microbiol* **6**:1199.

56. Sesto N, Wurtzel O, Archambaud C, Sorek R, Cossart P. 2013. The excludon: a new concept in bacterial antisense RNA-mediated gene regulation. *Nat Rev Microbiol* **11**:75–82.

57. Yamaguchi Y, Park JH, Inouye M. 2011. Toxin-antitoxin systems in bacteria and archaea. *Annu Rev Genet* **45**:61–79.

58. Berghoff BA, Wagner EG. 2017. RNA-based regulation in type I toxin-antitoxin systems and its implication for bacterial persistence. *Curr Genet* **63**:1011–1016.

59. Page R, Peti W. 2016. Toxin-antitoxin systems in bacterial growth arrest and persistence. *Nat Chem Biol* **12**:208–214.

60. Arnion H, Korkut DN, Masachis Gelo S, Chabas S, Reignier J, Iost I, Darfeuille F. 2017. Mechanistic insights into type I toxin antitoxin systems in *Helicobacter pylori*: the importance of mRNA folding in controlling toxin expression. *Nucleic Acids Res* **45**:4782–4795.

61. Vogel J, Argaman L, Wagner EG, Altuvia S. 2004. The small RNA IstR inhibits synthesis of an SOS-induced toxic peptide. *Curr Biol* **14**:2271–2276.

62. Berghoff BA, Hoekzema M, Aulbach L, Wagner EG. 2017. Two regulatory RNA elements affect TisB-dependent depolarization and persister formation. *Mol Microbiol* **103**:1020–1033.

63. Beaume M, Hernandez D, Farinelli L, Deluen C, Linder P, Gaspin C, Romby P, Schrenzel J, Francois P. 2010. Cartography of methicillin-resistant *S. aureus* transcripts: detection, orientation and temporal expression during growth phase and stress conditions. *PLoS One* **5**:e10725.

64. Sayed N, Jousselin A, Felden B. 2011. A *cis*-antisense RNA acts in *trans* in *Staphylococcus aureus* to control translation of a human cytolytic peptide. *Nat Struct Mol Biol* **19**:105–112.

65. Jorgensen MG, Thomason MK, Havelund J, Valentin-Hansen P, Storz G. 2013. Dual function of the McaS small RNA in controlling biofilm formation. *Genes Dev* **27**:1132–1145.

66. Holmqvist E, Vogel J. 2013. A small RNA serving both the Hfq and CsrA regulons. *Genes Dev* **27**:1073–1078.

67. Romeo T. 1998. Global regulation by the small RNA-binding protein CsrA and the non-coding RNA molecule CsrB. *Mol Microbiol* **29**:1321–1330.

68. Babitzke P, Romeo T. 2007. CsrB sRNA family: sequestration of RNA-binding regulatory proteins. *Curr Opin Microbiol* **10**:156–163.

69. Martínez LC, Yakhnin H, Camacho MI, Georgellis D, Babitzke P, Puente JL, Bustamante VH. 2011. Integration of a complex regulatory cascade involving the SirA/BarA and Csr global regulatory systems that controls expression of the *Salmonella* SPI-1 and SPI-2 virulence regulons through HilD. *Mol Microbiol* **80**:1637–1656.

70. Heroven AK, Böhme K, Dersch P. 2012. The Csr/Rsm system of *Yersinia* and related pathogens: a post-transcriptional strategy for managing virulence. *RNA Biol* **9**:379–391.

71. Gore AL, Payne SM. 2010. CsrA and Cra influence *Shigella flexneri* pathogenesis. *Infect Immun* **78**:4674–4682.

72. Sahr T, Brüggemann H, Jules M, Lomma M, Albert-Weissenberger C, Cazalet C, Buchrieser C. 2009. Two small ncRNAs jointly govern virulence and transmission in *Legionella pneumophila*. *Mol Microbiol* **72**:741–762.

73. Wassarman KM. 2007. 6S RNA: a small RNA regulator of transcription. *Curr Opin Microbiol* **10**:164–168.

74. Barrick JE, Sudarsan N, Weinberg Z, Ruzzo WL, Breaker RR. 2005. 6S RNA is a widespread regulator of eubacterial RNA polymerase that resembles an open promoter. *RNA* **11**:774–784.

75. Faucher SP, Friedlander G, Livny J, Margalit H, Shuman HA. 2010. *Legionella pneumophila* 6S RNA optimizes intracellular multiplication. *Proc Natl Acad Sci U S A* **107**:7533–7538.

76. Thiennimitr P, Winter SE, Winter MG, Xavier MN, Tolstikov V, Huseby DL, Sterzenbach T, Tsolis RM, Roth JR, Bäumler AJ. 2011. Intestinal inflammation allows Salmonella to use ethanolamine to compete with the microbiota. *Proc Natl Acad Sci U S A* **108**:17480–17485.

77. Mellin JR, Koutero M, Dar D, Nahori MA, Sorek R, Cossart P. 2014. Riboswitches. Sequestration of a two-component response regulator by a riboswitch-regulated noncoding RNA. *Science* **345**:940–943.

78. DebRoy S, Gebbie M, Ramesh A, Goodson JR, Cruz MR, van Hoof A, Winkler WC, Garsin DA, DebRoy S. 2014. Riboswitches. A riboswitch-containing sRNA controls gene expression by sequestration of a response regulator. *Science* **345**:937–940.

79. Song T, Mika F, Lindmark B, Liu Z, Schild S, Bishop A, Zhu J, Camilli A, Johansson J, Vogel J, Wai SN. 2008. A new *Vibrio cholerae* sRNA modulates colonization and affects release of outer membrane vesicles. *Mol Microbiol* **70**:100–111.

80. Song T, Sabharwal D, Wai SN. 2010. VrrA mediates Hfq-dependent regulation of OmpT synthesis in *Vibrio cholerae*. *J Mol Biol* **400**:682–688.

81. Ng WL, Bassler BL. 2009. Bacterial quorum-sensing network architectures. *Annu Rev Genet* **43**:197–222.

82. Bardill JP, Zhao X, Hammer BK. 2011. The *Vibrio cholerae* quorum sensing response is mediated by Hfq-dependent sRNA/mRNA base pairing interactions. *Mol Microbiol* **80**:1381–1394.

83. Shao Y, Bassler BL. 2014. Quorum regulatory small RNAs repress type VI secretion in *Vibrio cholerae*. *Mol Microbiol* **92**:921–930.

84. Richard AL, Withey JH, Beyhan S, Yildiz F, DiRita VJ. 2010. The *Vibrio cholerae* virulence regulatory cascade controls glucose uptake through activation of TarA, a small regulatory RNA. *Mol Microbiol* **78**:1171–1181.

85. Bradley ES, Bodi K, Ismail AM, Camilli A. 2011. A genome-wide approach to discovery of small RNAs involved in regulation of virulence in *Vibrio cholerae*. *PLoS Pathog* **7**:e1002126.

86. Davies BW, Bogard RW, Young TS, Mekalanos JJ. 2012. Coordinated regulation of accessory genetic elements produces cyclic di-nucleotides for *V. cholerae* virulence. *Cell* **149**:358–370.

87. Papenfort K, Forstner KU, Cong JP, Sharma CM, Bassler BL. 2015. Differential RNA-seq of *Vibrio cholerae* identifies the VqmR small RNA as a regulator of biofilm formation. *Proc Natl Acad Sci U S A* **112**:E766–E775.

88. Papenfort K, Silpe JE, Schramma KR, Cong JP, Seyedsayamdost MR, Bassler BL. 2017. A *Vibrio cholerae* autoinducer-receptor pair that controls biofilm formation. *Nat Chem Biol* **13**:551–557.

89. Miller MB, Skorupski K, Lenz DH, Taylor RK, Bassler BL. 2002. Parallel quorum sensing systems converge to regulate virulence in *Vibrio cholerae*. *Cell* **110**:303–314.

90. Zhu J, Miller MB, Vance RE, Dziejman M, Bassler BL, Mekalanos JJ. 2002. Quorum-sensing regulators control virulence gene expression in *Vibrio cholerae*. *Proc Natl Acad Sci U S A* **99**:3129–3134.

91. Pfeiffer V, Sittka A, Tomer R, Tedin K, Brinkmann V, Vogel J. 2007. A small non-coding RNA of the invasion gene island (SPI-1) represses outer membrane protein synthesis from the *Salmonella* core genome. *Mol Microbiol* **66**:1174–1191.

92. Song M, Sukovich DJ, Ciccarelli L, Mayr J, Fernandez-Rodriguez J, Mirsky EA, Tucker AC, Gordon DB, Marlovits TC, Voigt CA. 2017. Control of type III protein secretion using a minimal genetic system. *Nat Commun* **8**:14737.

93. Padalon-Brauch G, Hershberg R, Elgrably-Weiss M, Baruch K, Rosenshine I, Margalit H, Altuvia S. 2008. Small RNAs encoded within genetic islands of *Salmonella typhimurium* show host-induced expression and role in virulence. *Nucleic Acids Res* **36**:1913–1927.

94. Gong H, Vu GP, Bai Y, Chan E, Wu R, Yang E, Liu F, Lu S. 2011. A *Salmonella* small non-coding RNA facilitates bacterial invasion and intracellular replication by modulating the expression of virulence factors. *PLoS Pathog* **7**:e1002120.

95. Westermann AJ, Förstner KU, Amman F, Barquist L, Chao Y, Schulte LN, Müller L, Reinhardt R, Stadler PF, Vogel J. 2016. Dual RNA-seq unveils noncoding RNA functions in host-pathogen interactions. *Nature* **529**:496–501.

96. Chaudhuri RR, Morgan E, Peters SE, Pleasance SJ, Hudson DL, Davies HM, Wang J, van Diemen PM, Buckley AM, Bowen AJ, Pullinger GD, Turner DJ, Langridge GC, Turner AK, Parkhill J, Charles IG, Maskell DJ, Stevens MP. 2013. Comprehensive assignment of roles for *Salmonella typhimurium* genes in intestinal colonization of food-producing animals. *PLoS Genet* **9**:e1003456.

97. Grieshaber NA, Grieshaber SS, Fischer ER, Hackstadt T. 2006. A small RNA inhibits translation of the histone-like protein Hc1 in *Chlamydia trachomatis*. *Mol Microbiol* **59**:541–550.

98. Khandige S, Kronborg T, Uhlin BE, Møller-Jensen J. 2015. sRNA-mediated regulation of P-fimbriae phase variation in uropathogenic *Escherichia coli*. *PLoS Pathog* **11**:e1005109.

99. Mellin JR, Goswami S, Grogan S, Tjaden B, Genco CA. 2007. A novel Fur- and iron-regulated small RNA, NrrF, is required for indirect Fur-mediated regulation of the *sdhA* and *sdhC* genes in *Neisseria meningitidis*. *J Bacteriol* **189**:3686–3694.

100. Mellin JR, McClure R, Lopez D, Green O, Reinhard B, Genco C. 2010. Role of Hfq in iron-dependent and -independent gene regulation in *Neisseria meningitidis*. *Microbiology* **156**:2316–2326.

101. Heidrich N, Bauriedl S, Barquist L, Li L, Schoen C, Vogel J. 2017. The primary transcriptome of *Neisseria meningitidis* and its interaction with the RNA chaperone Hfq. *Nucleic Acids Res* **45**:6147–6167.

102. Pannekoek Y, Huis In 't Veld RA, Schipper K, Bovenkerk S, Kramer G, Brouwer MC, van de Beek D, Speijer D, van der Ende A. 2017. *Neisseria meningitidis* uses sibling small regulatory RNAs to switch from cataplerotic to anaplerotic metabolism. *mBio* **8**:e02293-16.

103. Romilly C, Lays C, Tomasini A, Caldelari I, Benito Y, Hammann P, Geissmann T, Boisset S, Romby P, Vandenesch F. 2014. A non-coding RNA promotes bacterial persistence and decreases virulence by regulating a regulator in *Staphylococcus aureus*. *PLoS Pathog* **10**:e1003979.

104. Roberts SA, Scott JR. 2007. RivR and the small RNA RivX: the missing links between the CovR regulatory cascade and the Mga regulon. *Mol Microbiol* **66**:1506–1522.

105. Morfeldt E, Taylor D, von Gabain A, Arvidson S. 1995. Activation of alpha-toxin translation in *Staphylococcus aureus* by the *trans*-encoded antisense RNA, RNAIII. *EMBO J* **14**:4569–4577.

106. Fröhlich KS, Vogel J. 2009. Activation of gene expression by small RNA. *Curr Opin Microbiol* **12**:674–682.

107. Quereda JJ, Ortega AD, Pucciarelli MG, García-Del Portillo F. 2014. The *Listeria* small RNA Rli27 regulates a cell wall protein inside eukaryotic cells by targeting a long 5′-UTR variant. *PLoS Genet* **10**:e1004765.

108. Ramirez-Peña E, Treviño J, Liu Z, Perez N, Sumby P. 2010. The group A *Streptococcus* small regulatory RNA FasX enhances streptokinase activity by increasing the stability of the *ska* mRNA transcript. *Mol Microbiol* **78**:1332–1347.

109. Liu Z, Treviño J, Ramirez-Peña E, Sumby P. 2012. The small regulatory RNA FasX controls pilus expression and adherence in the human bacterial pathogen group A *Streptococcus*. *Mol Microbiol* **86**:140–154.

110. Danger JL, Cao TN, Cao TH, Sarkar P, Treviño J, Pflughoeft KJ, Sumby P. 2015. The small regulatory RNA FasX enhances group A *Streptococcus* virulence and inhibits pilus expression via serotype-specific targets. *Mol Microbiol* **96**:249–262.

111. Vandenesch F, Kornblum J, Novick RP. 1991. A temporal signal, independent of *agr*, is required for *hla* but not *spa* transcription in *Staphylococcus aureus*. *J Bacteriol* **173**:6313–6320.

112. Novick RP. 2003. Autoinduction and signal transduction in the regulation of staphylococcal virulence. *Mol Microbiol* **48**:1429–1449.

113. Boisset S, Geissmann T, Huntzinger E, Fechter P, Bendridi N, Possedko M, Chevalier C, Helfer AC, Benito Y, Jacquier A, Gaspin C, Vandenesch F, Romby P. 2007. *Staphylococcus aureus* RNAIII coordinately represses the synthesis of virulence factors and the transcription regulator Rot by an antisense mechanism. *Genes Dev* **21**:1353–1366.

114. Pernitzsch SR, Tirier SM, Beier D, Sharma CM. 2014. A variable homopolymeric G-repeat defines small RNA-mediated posttranscriptional regulation of a chemotaxis receptor in *Helicobacter pylori*. *Proc Natl Acad Sci U S A* **111**:E501–E510.

115. Feng L, Rutherford ST, Papenfort K, Bagert JD, van Kessel JC, Tirrell DA, Wingreen NS, Bassler BL. 2015. A Qrr noncoding RNA deploys four different regulatory mechanisms to optimize quorum-sensing dynamics. *Cell* **160**:228–240.

116. Vogel J, Sharma CM. 2005. How to find small noncoding RNAs in bacteria. *Biol Chem* **386**:1219–1238.

117. Winkler WC, Nahvi A, Sudarsan N, Barrick JE, Breaker RR. 2003. An mRNA structure that controls gene expression by binding S-adenosylmethionine. *Nat Struct Biol* **10**:701–707.

118. Loh E, Dussurget O, Gripenland J, Vaitkevicius K, Tiensuu T, Mandin P, Repoila F, Buchrieser C, Cossart P, Johansson J. 2009. A *trans*-acting riboswitch controls expression of the virulence regulator PrfA in *Listeria monocytogenes*. *Cell* **139**:770–779.

119. Ellis MJ, Trussler RS, Charles O, Haniford DB. 2017. A transposon-derived small RNA regulates gene expression in *Salmonella* Typhimurium. *Nucleic Acids Res* **45**:5470–5486.

120. Ellis MJ, Carfrae LA, Macnair CR, Trussler RS, Brown ED, Haniford DB. 2018. Silent but deadly: IS200 promotes pathogenicity in *Salmonella* Typhimurium. *RNA Biol* **15**:176–181.

121. Chao Y, Papenfort K, Reinhardt R, Sharma CM, Vogel J. 2012. An atlas of Hfq-bound transcripts reveals 3′ UTRs as a genomic reservoir of regulatory small RNAs. *EMBO J* **31**:4005–4019.

122. Miyakoshi M, Chao Y, Vogel J. 2015. Regulatory small RNAs from the 3′ regions of bacterial mRNAs. *Curr Opin Microbiol* **24**:132–139.

123. Novick RP, Geisinger E. 2008. Quorum sensing in staphylococci. *Annu Rev Genet* **42**:541–564.

124. Papenfort K, Sun Y, Miyakoshi M, Vanderpool CK, Vogel J. 2013. Small RNA-mediated activation of sugar phosphatase mRNA regulates glucose homeostasis. *Cell* **153**:426–437.

125. Papenfort K, Podkaminski D, Hinton JC, Vogel J. 2012. The ancestral SgrS RNA discriminates horizontally acquired *Salmonella* mRNAs through a single G-U wobble pair. *Proc Natl Acad Sci U S A* **109**:E757–E764.

126. Lloyd CR, Park S, Fei J, Vanderpool CK. 2017. The small protein SgrT controls transport activity of the glucose-specific phosphotransferase system. *J Bacteriol* **199**:e00869-16.

127. Marraffini LA. 2015. CRISPR-Cas immunity in prokaryotes. *Nature* **526**:55–61.

128. Sampson TR, Saroj SD, Llewellyn AC, Tzeng YL, Weiss DS. 2013. A CRISPR/Cas system mediates bacterial innate immune evasion and virulence. *Nature* **497**:254–257.

129. Dugar G, Leenay RT, Eisenbart SK, Bischler T, Aul BU, Beisel CL, Sharma CM. 2018. CRISPR RNA-dependent binding and cleavage of endogenous RNAs by the *Campylobacter jejuni* Cas9. *Mol Cell* **69**:893–905.

130. Louwen R, Horst-Kreft D, de Boer AG, van der Graaf L, de Knegt G, Hamersma M, Heikema AP, Timms AR, Jacobs BC, Wagenaar JA, Endtz HP, van der Oost J, Wells JM, Nieuwenhuis EE, van Vliet AH, Willemsen PT, van Baarlen P, van Belkum A. 2013. A novel link between *Campylobacter jejuni* bacteriophage defence, virulence and Guillain-Barré syndrome. *Eur J Clin Microbiol Infect Dis* **32**:207–226.

131. Gunderson FF, Cianciotto NP. 2013. The CRISPR-associated gene *cas2* of *Legionella pneumophila* is required for intracellular infection of amoebae. *mBio* **4**:e00074-13.

132. Gunderson FF, Mallama CA, Fairbairn SG, Cianciotto NP. 2015. Nuclease activity of *Legionella pneumophila* Cas2 promotes intracellular infection of amoebal host cells. *Infect Immun* **83**:1008–1018.

133. Sesto N, Touchon M, Andrade JM, Kondo J, Rocha EP, Arraiano CM, Archambaud C, Westhof E, Romby P, Cossart P. 2014. A PNPase dependent CRISPR system in *Listeria*. *PLoS Genet* **10**:e1004065.

134. Dubey AK, Baker CS, Romeo T, Babitzke P. 2005. RNA sequence and secondary structure participate in high-affinity CsrA-RNA interaction. *RNA* **11**:1579–1587.

135. Duss O, Michel E, Diarra dit Konté N, Schubert M, Allain FH. 2014. Molecular basis for the wide range of affinity found in Csr/Rsm protein-RNA recognition. *Nucleic Acids Res* **42**:5332–5346.

136. Holmqvist E, Wright PR, Li L, Bischler T, Barquist L, Reinhardt R, Backofen R, Vogel J. 2016. Global RNA recognition patterns of post-transcriptional regulators Hfq and CsrA revealed by UV crosslinking in vivo. *EMBO J* **35**:991–1011.

137. Vakulskas CA, Potts AH, Babitzke P, Ahmer BM, Romeo T. 2015. Regulation of bacterial virulence by Csr (Rsm) systems. *Microbiol Mol Biol Rev* **79**:193–224.

138. Lawhon SD, Frye JG, Suyemoto M, Porwollik S, McClelland M, Altier C. 2003. Global regulation by CsrA in *Salmonella typhimurium*. *Mol Microbiol* **48**:1633–1645.

139. Kusmierek M, Dersch P. 2017. Regulation of host-pathogen interactions via the post-transcriptional Csr/Rsm system. *Curr Opin Microbiol* **41**:58–67.

140. Bhatt S, Edwards AN, Nguyen HT, Merlin D, Romeo T, Kalman D. 2009. The RNA binding protein CsrA is a pleiotropic regulator of the locus of enterocyte effacement pathogenicity island of enteropathogenic *Escherichia coli*. *Infect Immun* **77**:3552–3568.

141. Sahr T, Rusniok C, Impens F, Oliva G, Sismeiro O, Coppée JY, Buchrieser C. 2017. The *Legionella pneumophila* genome evolved to accommodate multiple regulatory mechanisms controlled by the CsrA-system. *PLoS Genet* 13:e1006629.

142. Dugar G, Svensson SL, Bischler T, Wäldchen S, Reinhardt R, Sauer M, Sharma CM. 2016. The CsrA-FliW network controls polar localization of the dual-function flagellin mRNA in *Campylobacter jejuni. Nat Commun* 7:11667.

143. Fields JA, Li J, Gulbronson CJ, Hendrixson DR, Thompson SA. 2016. *Campylobacter jejuni* CsrA regulates metabolic and virulence associated proteins and is necessary for mouse colonization. *PLoS One* 11: e0156932.

144. Vogel J, Luisi BF. 2011. Hfq and its constellation of RNA. *Nat Rev Microbiol* 9:578–589.

145. Peng Y, Curtis JE, Fang X, Woodson SA. 2014. Structural model of an mRNA in complex with the bacterial chaperone Hfq. *Proc Natl Acad Sci U S A* 111:17134–17139.

146. Santiago-Frangos A, Kavita K, Schu DJ, Gottesman S, Woodson SA. 2016. C-terminal domain of the RNA chaperone Hfq drives sRNA competition and release of target RNA. *Proc Natl Acad Sci U S A* 113:E6089–E6096.

147. Chao Y, Vogel J. 2010. The role of Hfq in bacterial pathogens. *Curr Opin Microbiol* 13:24–33.

148. Sittka A, Lucchini S, Papenfort K, Sharma CM, Rolle K, Binnewies TT, Hinton JC, Vogel J. 2008. Deep sequencing analysis of small noncoding RNA and mRNA targets of the global post-transcriptional regulator, Hfq. *PLoS Genet* 4:e1000163.

149. Tree JJ, Granneman S, McAteer SP, Tollervey D, Gally DL. 2014. Identification of bacteriophage-encoded anti-sRNAs in pathogenic *Escherichia coli. Mol Cell* 55: 199–213.

150. Christiansen JK, Nielsen JS, Ebersbach T, Valentin-Hansen P, Sogaard-Andersen L, Kallipolitis BH. 2006. Identification of small Hfq-binding RNAs in *Listeria monocytogenes. RNA* 12:1383–1396.

151. Sonnleitner E, Sorger-Domenigg T, Madej MJ, Findeiss S, Hackermüller J, Hüttenhofer A, Stadler PF, Bläsi U, Moll I. 2008. Detection of small RNAs in *Pseudomonas aeruginosa* by RNomics and structure-based bioinformatic tools. *Microbiology* 154:3175–3187.

152. Möller P, Overlöper A, Förstner KU, Wen TN, Sharma CM, Lai EM, Narberhaus F. 2014. Profound impact of Hfq on nutrient acquisition, metabolism and motility in the plant pathogen *Agrobacterium tumefaciens. PLoS One* 9:e110427.

153. Sonnleitner E, Bläsi U. 2014. Regulation of Hfq by the RNA CrcZ in *Pseudomonas aeruginosa* carbon catabolite repression. *PLoS Genet* 10:e1004440.

154. Ellis MJ, Trussler RS, Haniford DB. 2015. Hfq binds directly to the ribosome-binding site of IS*10* transposase mRNA to inhibit translation. *Mol Microbiol* 96:633–650.

155. Chen J, Gottesman S. 2017. Hfq links translation repression to stress-induced mutagenesis in *E. coli. Genes Dev* 31:1382–1395.

156. Olejniczak M, Storz G. 2017. ProQ/FinO-domain proteins: another ubiquitous family of RNA matchmakers? *Mol Microbiol* 104:905–915.

157. Gonzalez GM, Hardwick SW, Maslen SL, Skehel JM, Holmqvist E, Vogel J, Bateman A, Luisi BF, Broadhurst RW. 2017. Structure of the *Escherichia coli* ProQ RNA-binding protein. *RNA* 23:696–711.

158. Smirnov A, Förstner KU, Holmqvist E, Otto A, Guünster R, Becher D, Reinhardt R, Vogel J. 2016. Grad-seq guides the discovery of ProQ as a major small RNA-binding protein. *Proc Natl Acad Sci U S A* 113:11591–11596.

159. Attaiech L, Boughammoura A, Brochier-Armanet C, Allatif O, Peillard-Fiorente F, Edwards RA, Omar AR, MacMillan AM, Glover M, Charpentier X. 2016. Silencing of natural transformation by an RNA chaperone and a multitarget small RNA. *Proc Natl Acad Sci U S A* 113:8813–8818.

160. Smirnov A, Wang C, Drewry LL, Vogel J. 2017. Molecular mechanism of mRNA repression in *trans* by a ProQ-dependent small RNA. *EMBO J* 36:1029–1045.

161. Sheidy DT, Zielke RA. 2013. Analysis and expansion of the role of the *Escherichia coli* protein ProQ. *PLoS One* 8:e79656.

162. Michaux C, Holmqvist E, Vasicek E, Sharan M, Barquist L, Westermann AJ, Gunn JS, Vogel J. 2017. RNA target profiles direct the discovery of virulence functions for the cold-shock proteins CspC and CspE. *Proc Natl Acad Sci U S A* 114:6824–6829.

163. Beckmann BM, Horos R, Fischer B, Castello A, Eichelbaum K, Alleaume AM, Schwarzl T, Curk T, Foehr S, Huber W, Krijgsveld J, Hentze MW. 2015. The RNA-binding proteomes from yeast to man harbour conserved enigmRBPs. *Nat Commun* 6:10127.

164. Mitobe J, Yanagihara I, Ohnishi K, Yamamoto S, Ohnishi M, Ishihama A, Watanabe H. 2011. RodZ regulates the post-transcriptional processing of the *Shigella sonnei* type III secretion system. *EMBO Rep* 12:911–916.

165. Matos RG, Casinhas J, Bárria C, Dos Santos RF, Silva IJ, Arraiano CM. 2017. The role of ribonucleases and sRNAs in the virulence of foodborne pathogens. *Front Microbiol* 8:910.

166. Mackie GA. 2013. RNase E: at the interface of bacterial RNA processing and decay. *Nat Rev Microbiol* 11: 45–57.

167. Yang J, Jain C, Schesser K. 2008. RNase E regulates the *Yersinia* type 3 secretion system. *J Bacteriol* 190:3774–3778.

168. Schiano CA, Bellows LE, Lathem WW. 2010. The small RNA chaperone Hfq is required for the virulence of *Yersinia pseudotuberculosis. Infect Immun* 78:2034–2044.

169. Rosenzweig JA, Weltman G, Plano GV, Schesser K. 2005. Modulation of *Yersinia* type three secretion system by the S1 domain of polynucleotide phosphorylase. *J Biol Chem* 280:156–163.

170. Rosenzweig JA, Chromy B, Echeverry A, Yang J, Adkins B, Plano GV, McCutchen-Maloney S, Schesser K. 2007. Polynucleotide phosphorylase independently controls

virulence factor expression levels and export in *Yersinia* spp. *FEMS Microbiol Lett* **270**:255–264.

171. Clements MO, Eriksson S, Thompson A, Lucchini S, Hinton JC, Normark S, Rhen M. 2002. Polynucleotide phosphorylase is a global regulator of virulence and persistency in *Salmonella enterica*. *Proc Natl Acad Sci U S A* **99**:8784–8789.

172. Ygberg SE, Clements MO, Rytkonen A, Thompson A, Holden DW, Hinton JC, Rhen M. 2006. Polynucleotide phosphorylase negatively controls *spv* virulence gene expression in *Salmonella enterica*. *Infect Immun* **74**:1243–1254.

173. Haddad N, Tresse O, Rivoal K, Chevret D, Nonglaton Q, Burns CM, Prévost H, Cappelier JM. 2012. Polynucleotide phosphorylase has an impact on cell biology of *Campylobacter jejuni*. *Front Cell Infect Microbiol* **2**:30.

174. Chevalier C, Huntzinger E, Fechter P, Boisset S, Vandenesch F, Romby P, Geissmann T. 2008. *Staphylococcus aureus* endoribonuclease III purification and properties. *Methods Enzymol* **447**:309–327.

175. Viegas SC, Mil-Homens D, Fialho AM, Arraiano CM. 2013. The virulence of *Salmonella enterica* serovar Typhimurium in the insect model *Galleria mellonella* is impaired by mutations in RNase E and RNase III. *Appl Environ Microbiol* **79**:6124–6133.

176. Cheng ZF, Zuo Y, Li Z, Rudd KE, Deutscher MP. 1998. The *vacB* gene required for virulence in *Shigella flexneri* and *Escherichia coli* encodes the exoribonuclease RNase R. *J Biol Chem* **273**:14077–14080.

177. Leskinen K, Varjosalo M, Skurnik M. 2015. Absence of YbeY RNase compromises the growth and enhances the virulence plasmid gene expression of *Yersinia enterocolitica* O:3. *Microbiology* **161**:285–299.

178. Vercruysse M, Köhrer C, Davies BW, Arnold MF, Mekalanos JJ, RajBhandary UL, Walker GC. 2014. The highly conserved bacterial RNase YbeY is essential in *Vibrio cholerae*, playing a critical role in virulence, stress regulation, and RNA processing. *PLoS Pathog* **10**:e1004175.

179. Bigot A, Raynaud C, Dubail I, Dupuis M, Hossain H, Hain T, Chakraborty T, Charbit A. 2009. *lmo1273*, a novel gene involved in *Listeria monocytogenes* virulence. *Microbiology* **155**:891–902.

180. Bugrysheva JV, Scott JR. 2010. The ribonucleases J1 and J2 are essential for growth and have independent roles in mRNA decay in *Streptococcus pyogenes*. *Mol Microbiol* **75**:731–743.

181. Kaito C, Kurokawa K, Matsumoto Y, Terao Y, Kawabata S, Hamada S, Sekimizu K. 2005. Silkworm pathogenic bacteria infection model for identification of novel virulence genes. *Mol Microbiol* **56**:934–944.

182. Barquist L, Vogel J. 2015. Accelerating discovery and functional analysis of small RNAs with new technologies. *Annu Rev Genet* **49**:367–394.

183. Hör J, Gorski SA, Vogel J. 2018. Bacterial RNA biology on a genome scale. *Mol Cell*.

184. Weinberg Z, Wang JX, Bogue J, Yang J, Corbino K, Moy RH, Breaker RR. 2010. Comparative genomics reveals 104 candidate structured RNAs from bacteria, archaea, and their metagenomes. *Genome Biol* **11**:R31.

185. Merino EJ, Wilkinson KA, Coughlan JL, Weeks KM. 2005. RNA structure analysis at single nucleotide resolution by selective 2′-hydroxyl acylation and primer extension (SHAPE). *J Am Chem Soc* **127**:4223–4231.

186. Flynn RA, Zhang QC, Spitale RC, Lee B, Mumbach MR, Chang HY. 2016. Transcriptome-wide interrogation of RNA secondary structure in living cells with icSHAPE. *Nat Protoc* **11**:273–290.

187. Lu Z, Zhang QC, Lee B, Flynn RA, Smith MA, Robinson JT, Davidovich C, Gooding AR, Goodrich KJ, Mattick JS, Mesirov JP, Cech TR, Chang HY. 2016. RNA duplex map in living cells reveals higher-order transcriptome structure. *Cell* **165**:1267–1279.

188. Kertesz M, Wan Y, Mazor E, Rinn JL, Nutter RC, Chang HY, Segal E. 2010. Genome-wide measurement of RNA secondary structure in yeast. *Nature* **467**:103–107.

189. Choi EK, Ulanowicz KA, Nguyen YA, Frandsen JK, Mitton-Fry RM. 2017. SHAPE analysis of the *htrA* RNA thermometer from *Salmonella enterica*. *RNA* **23**:1569–1581.

190. Righetti F, Nuss AM, Twittenhoff C, Beele S, Urban K, Will S, Bernhart SH, Stadler PF, Dersch P, Narberhaus F. 2016. Temperature-responsive in vitro RNA structurome of *Yersinia pseudotuberculosis*. *Proc Natl Acad Sci U S A* **113**:7237–7242.

191. Steen KA, Siegfried NA, Weeks KM. 2011. Selective 2′-hydroxyl acylation analyzed by protection from exoribonuclease (RNase-detected SHAPE) for direct analysis of covalent adducts and of nucleotide flexibility in RNA. *Nat Protoc* **6**:1683–1694.

192. Watters KE, Abbott TR, Lucks JB. 2016. Simultaneous characterization of cellular RNA structure and function with in-cell SHAPE-Seq. *Nucleic Acids Res* **44**:e12.

193. Lu C, Ding F, Chowdhury A, Pradhan V, Tomsic J, Holmes WM, Henkin TM, Ke A. 2010. SAM recognition and conformational switching mechanism in the *Bacillus subtilisyitJ* S box/SAM-I riboswitch. *J Mol Biol* **404**:803–818.

194. Watters KE, Strobel EJ, Yu AM, Lis JT, Lucks JB. 2016. Cotranscriptional folding of a riboswitch at nucleotide resolution. *Nat Struct Mol Biol* **23**:1124–1131.

195. Dar D, Shamir M, Mellin JR, Koutero M, Stern-Ginossar N, Cossart P, Sorek R. 2016. Term-seq reveals abundant ribo-regulation of antibiotics resistance in bacteria. *Science* **352**:aad9822.

196. Colgan AM, Cameron AD, Kröger C. 2017. If it transcribes, we can sequence it: mining the complexities of host-pathogen-environment interactions using RNA-seq. *Curr Opin Microbiol* **36**:37–46.

197. Massé E, Vanderpool CK, Gottesman S. 2005. Effect of RyhB small RNA on global iron use in *Escherichia coli*. *J Bacteriol* **187**:6962–6971.

198. Westermann AJ, Gorski SA, Vogel J. 2012. Dual RNA-seq of pathogen and host. *Nat Rev Microbiol* **10**:618–630.

199. Westermann AJ, Barquist L, Vogel J. 2017. Resolving host-pathogen interactions by dual RNA-seq. *PLoS Pathog* **13**:e1006033.

200. Baddal B, Muzzi A, Censini S, Calogero RA, Torricelli G, Guidotti S, Taddei AR, Covacci A, Pizza M, Rappuoli R, Soriani M, Pezzicoli A. 2015. Dual RNA-seq of non-typeable *Haemophilus influenzae* and host cell transcriptomes reveals novel insights into host-pathogen cross talk. *mBio* 6:e01765-15.

201. Vanderpool CK, Gottesman S. 2004. Involvement of a novel transcriptional activator and small RNA in post-transcriptional regulation of the glucose phosphoenolpyruvate phosphotransferase system. *Mol Microbiol* 54:1076–1089.

202. Barquist L, Westermann AJ, Vogel J. 2016. Molecular phenotyping of infection-associated small non-coding RNAs. *Philos Trans R Soc Lond B Biol Sci* 371:20160081.

203. van Opijnen T, Camilli A. 2013. Transposon insertion sequencing: a new tool for systems-level analysis of microorganisms. *Nat Rev Microbiol* 11:435–442.

204. Chao MC, Abel S, Davis BM, Waldor MK. 2016. The design and analysis of transposon insertion sequencing experiments. *Nat Rev Microbiol* 14:119–128.

205. Barquist L, Boinett CJ, Cain AK. 2013. Approaches to querying bacterial genomes with transposon-insertion sequencing. *RNA Biol* 10:1161–1169.

206. Mann B, van Opijnen T, Wang J, Obert C, Wang YD, Carter R, McGoldrick DJ, Ridout G, Camilli A, Tuomanen EI, Rosch JW. 2012. Control of virulence by small RNAs in *Streptococcus pneumoniae*. PLoS Pathog 8:e1002788.

207. Joshi SM, Pandey AK, Capite N, Fortune SM, Rubin EJ, Sassetti CM. 2006. Characterization of mycobacterial virulence genes through genetic interaction mapping. *Proc Natl Acad Sci U S A* 103:11760–11765.

208. Maier L, Diard M, Sellin ME, Chouffane ES, Trautwein-Weidner K, Periaswamy B, Slack E, Dolowschiak T, Stecher B, Loverdo C, Regoes RR, Hardt WD. 2014. Granulocytes impose a tight bottleneck upon the gut luminal pathogen population during *Salmonella* Typhimurium colitis. *PLoS Pathog* 10:e1004557.

209. Abel S, Abel zur Wiesch P, Chang HH, Davis BM, Lipsitch M, Waldor MK. 2015. Sequence tag-based analysis of microbial population dynamics. *Nat Methods* 12:223–226.

210. Larson MH, Gilbert LA, Wang X, Lim WA, Weissman JS, Qi LS. 2013. CRISPR interference (CRISPRi) for sequence-specific control of gene expression. *Nat Protoc* 8:2180–2196.

211. Hawkins JS, Wong S, Peters JM, Almeida R, Qi LS. 2015. Targeted transcriptional repression in bacteria using CRISPR interference (CRISPRi). *Methods Mol Biol* 1311:349–362.

212. Saliba AE, C Santos S, Vogel J. 2017. New RNA-seq approaches for the study of bacterial pathogens. *Curr Opin Microbiol* 35:78–87.

213. Smirnov A, Schneider C, Hör J, Vogel J. 2017. Discovery of new RNA classes and global RNA-binding proteins. *Curr Opin Microbiol* 39:152–160.

214. Waters SA, McAteer SP, Kudla G, Pang I, Deshpande NP, Amos TG, Leong KW, Wilkins MR, Strugnell R, Gally DL, Tollervey D, Tree JJ. 2017. Small RNA interactome of pathogenic *E. coli* revealed through cross-linking of RNase E. *EMBO J* 36:374–387.

215. Melamed S, Peer A, Faigenbaum-Romm R, Gatt YE, Reiss N, Bar A, Altuvia Y, Argaman L, Margalit H. 2016. Global mapping of small RNA-target interactions in bacteria. *Mol Cell* 63:884–897.

216. Lalaouna D, Carrier MC, Semsey S, Brouard JS, Wang J, Wade JT, Masse E. 2015. A 3′ external transcribed spacer in a tRNA transcript acts as a sponge for small RNAs to prevent transcriptional noise. *Mol Cell* 58:393–405.

217. Han K, Tjaden B, Lory S. 2016. GRIL-seq provides a method for identifying direct targets of bacterial small regulatory RNA by *in vivo* proximity ligation. *Nat Microbiol* 2:16239.

218. Zhang YF, Han K, Chandler CE, Tjaden B, Ernst RK, Lory S. 2017. Probing the sRNA regulatory landscape of *P. aeruginosa*: post-transcriptional control of determinants of pathogenicity and antibiotic susceptibility. *Mol Microbiol* 106:919–937.

219. Tomasini A, Moreau K, Chicher J, Geissmann T, Vandenesch F, Romby P, Marzi S, Caldelari I. 2017. The RNA targetome of *Staphylococcus aureus* non-coding RNA RsaA: impact on cell surface properties and defense mechanisms. *Nucleic Acids Res* 45:6746–6760.

220. Gahlmann A, Moerner WE. 2014. Exploring bacterial cell biology with single-molecule tracking and super-resolution imaging. *Nat Rev Microbiol* 12:9–22.

221. Fei J, Singh D, Zhang Q, Park S, Balasubramanian D, Golding I, Vanderpool CK, Ha T. 2015. RNA biochemistry. Determination of *in vivo* target search kinetics of regulatory noncoding RNA. *Science* 347:1371–1374.

222. Papenfort K, Vogel J. 2014. Small RNA functions in carbon metabolism and virulence of enteric pathogens. *Front Cell Infect Microbiol* 4:91.

223. Bobrovskyy M, Vanderpool CK, Richards GR. 2015. Small RNAs regulate primary and secondary metabolism in gram-negative bacteria. *Microbiol Spectr* 3:MBP-0009-2014.

224. Updegrove TB, Shabalina SA, Storz G. 2015. How do base-pairing small RNAs evolve? *FEMS Microbiol Rev* 39:379–391.

225. Katsowich N, Elbaz N, Pal RR, Mills E, Kobi S, Kahan T, Rosenshine I. 2017. Host cell attachment elicits posttranscriptional regulation in infecting enteropathogenic bacteria. *Science* 355:735–739.

226. Bronesky D, Wu Z, Marzi S, Walter P, Geissmann T, Moreau K, Vandenesch F, Caldelari I, Romby P. 2016. *Staphylococcus aureus* RNAIII and its regulon link quorum sensing, stress responses, metabolic adaptation, and regulation of virulence gene expression. *Annu Rev Microbiol* 70:299–316.

227. Kreikemeyer B, Boyle MD, Buttaro BA, Heinemann M, Podbielski A. 2001. Group A streptococcal growth phase-associated virulence factor regulation by a novel operon (Fas) with homologies to two-component-type regulators requires a small RNA molecule. *Mol Microbiol* 39:392–406.

228. Hung CC, Eade CR, Altier C. 2016. The protein acyl-transferase Pat post-transcriptionally controls HilD to repress *Salmonella* invasion. *Mol Microbiol* **102**:121–136.

229. Gaviria-Cantin T, El Mouali Y, Le Guyon S, Römling U, Balsalobre C. 2017. Gre factors-mediated control of *hilD* transcription is essential for the invasion of epithelial cells by *Salmonella enterica* serovar Typhimurium. *PLoS Pathog* **13**:e1006312.

230. López-Garrido J, Puerta-Fernández E, Casadesús J. 2014. A eukaryotic-like 3′ untranslated region in *Salmonella enterica hilD* mRNA. *Nucleic Acids Res* **42**:5894–5906.

231. Holmqvist E, Reimegård J, Sterk M, Grantcharova N, Römling U, Wagner EG. 2010. Two antisense RNAs target the transcriptional regulator CsgD to inhibit curli synthesis. *EMBO J* **29**:1840–1850.

232. Boehm A, Vogel J. 2012. The *csgD* mRNA as a hub for signal integration via multiple small RNAs. *Mol Microbiol* **84**:1–5.

233. Mika F, Hengge R. 2014. Small RNAs in the control of RpoS, CsgD, and biofilm architecture of *Escherichia coli*. *RNA Biol* **11**:494–507.

234. Ahmad I, Cimdins A, Beske T, Römling U. 2017. Detailed analysis of c-di-GMP mediated regulation of *csgD* expression in *Salmonella typhimurium*. *BMC Microbiol* **17**:27.

235. Lebreton A, Cossart P. 2017. RNA- and protein-mediated control of *Listeria monocytogenes* virulence gene expression. *RNA Biol* **14**:460–470.

236. Gupta RK, Luong TT, Lee CY. 2015. RNAIII of the *Staphylococcus aureus agr* system activates global regulator MgrA by stabilizing mRNA. *Proc Natl Acad Sci U S A* **112**:14036–14041.

237. Schiano CA, Lathem WW. 2012. Post-transcriptional regulation of gene expression in *Yersinia* species. *Front Cell Infect Microbiol* **2**:129.

238. Sturm A, Heinemann M, Arnoldini M, Benecke A, Ackermann M, Benz M, Dormann J, Hardt WD. 2011. The cost of virulence: retarded growth of *Salmonella* Typhimurium cells expressing type III secretion system 1. *PLoS Pathog* **7**:e1002143.

239. Ali SS, Soo J, Rao C, Leung AS, Ngai DH, Ensminger AW, Navarre WW. 2014. Silencing by H-NS potentiated the evolution of *Salmonella*. *PLoS Pathog* **10**:e1004500.

240. Svenningsen SL, Tu KC, Bassler BL. 2009. Gene dosage compensation calibrates four regulatory RNAs to control *Vibrio cholerae* quorum sensing. *EMBO J* **28**:429–439.

241. Dean P, Kenny B. 2009. The effector repertoire of enteropathogenic *E. coli*: ganging up on the host cell. *Curr Opin Microbiol* **12**:101–109.

242. Ham H, Sreelatha A, Orth K. 2011. Manipulation of host membranes by bacterial effectors. *Nat Rev Microbiol* **9**:635–646.

243. LaRock DL, Chaudhary A, Miller SI. 2015. Salmonellae interactions with host processes. *Nat Rev Microbiol* **13**:191–205.

244. Kim JN, Kwon YM. 2013. Identification of target transcripts regulated by small RNA RyhB homologs in *Salmonella*: RyhB-2 regulates motility phenotype. *Microbiol Res* **168**:621–629.

245. Calderón IL, Morales EH, Collao B, Calderón PF, Chahuán CA, Acuña LG, Gil F, Saavedra CP. 2014. Role of *Salmonella* Typhimurium small RNAs RyhB-1 and RyhB-2 in the oxidative stress response. *Res Microbiol* **165**:30–40.

246. Calderón PF, Morales EH, Acuña LG, Fuentes DN, Gil F, Porwollik S, McClelland M, Saavedra CP, Calderón IL. 2014. The small RNA RyhB homologs from *Salmonella typhimurium* participate in the response to S-nitrosoglutathione-induced stress. *Biochem Biophys Res Commun* **450**:641–645.

247. Deng Z, Meng X, Su S, Liu Z, Ji X, Zhang Y, Zhao X, Wang X, Yang R, Han Y. 2012. Two sRNA RyhB homologs from *Yersinia pestis* biovar *microtus* expressed *in vivo* have differential Hfq-dependent stability. *Res Microbiol* **163**:413–418.

248. Deng Z, Liu Z, Bi Y, Wang X, Zhou D, Yang R, Han Y. 2014. Rapid degradation of Hfq-free RyhB in *Yersinia pestis* by PNPase independent of putative ribonucleolytic complexes. *BioMed Res Int* **2014**:798918.

249. Avraham R, Haseley N, Brown D, Penaranda C, Jijon HB, Trombetta JJ, Satija R, Shalek AK, Xavier RJ, Regev A, Hung DT. 2015. Pathogen cell-to-cell variability drives heterogeneity in host immune responses. *Cell* **162**:1309–1321.

250. Saliba AE, Li L, Westermann AJ, Appenzeller S, Stapels DA, Schulte LN, Helaine S, Vogel J. 2016. Single-cell RNA-seq ties macrophage polarization to growth rate of intracellular *Salmonella*. *Nat Microbiol* **2**:16206.

251. Saliba AE, Westermann AJ, Gorski SA, Vogel J. 2014. Single-cell RNA-seq: advances and future challenges. *Nucleic Acids Res* **42**:8845–8860.

252. Grantcharova N, Peters V, Monteiro C, Zakikhany K, Römling U. 2010. Bistable expression of CsgD in biofilm development of *Salmonella enterica* serovar Typhimurium. *J Bacteriol* **192**:456–466.

253. Hautefort I, Proença MJ, Hinton JC. 2003. Single-copy green fluorescent protein gene fusions allow accurate measurement of *Salmonella* gene expression in vitro and during infection of mammalian cells. *Appl Environ Microbiol* **69**:7480–7491.

254. Clark L, Perrett CA, Malt L, Harward C, Humphrey S, Jepson KA, Martinez-Argudo I, Carney LJ, La Ragione RM, Humphrey TJ, Jepson MA. 2011. Differences in *Salmonella enterica* serovar Typhimurium strain invasiveness are associated with heterogeneity in SPI-1 gene expression. *Microbiology* **157**:2072–2083.

255. Helaine S, Cheverton AM, Watson KG, Faure LM, Matthews SA, Holden DW. 2014. Internalization of *Salmonella* by macrophages induces formation of non-replicating persisters. *Science* **343**:204–208.

256. Plener L, Lorenz N, Reiger M, Ramalho T, Gerland U, Jung K. 2015. The phosphorylation flow of the *Vibrio harveyi* quorum-sensing cascade determines levels of phenotypic heterogeneity in the population. *J Bacteriol* **197**:1747–1756.

257. Mars RA, Nicolas P, Ciccolini M, Reilman E, Reder A, Schaffer M, Mäder U, Völker U, van Dijl JM, Denham EL.

2015. Small regulatory RNA-induced growth rate heterogeneity of *Bacillus subtilis*. *PLoS Genet* **11**:e1005046.

258. **Wang J, Chen L, Chen Z, Zhang W.** 2015. RNA-seq based transcriptomic analysis of single bacterial cells. *Integr Biol* **7**:1466–1476.

259. **Aminov RI.** 2011. Horizontal gene exchange in environmental microbiota. *Front Microbiol* **2**:158.

260. **Roberts AP, Kreth J.** 2014. The impact of horizontal gene transfer on the adaptive ability of the human oral microbiome. *Front Cell Infect Microbiol* **4**:124.

261. **Webster NS.** 2014. Cooperation, communication, and co-evolution: grand challenges in microbial symbiosis research. *Front Microbiol* **5**:164.

262. **Silverman JM, Brunet YR, Cascales E, Mougous JD.** 2012. Structure and regulation of the type VI secretion system. *Annu Rev Microbiol* **66**:453–472.

263. **Ghosal A, Upadhyaya BB, Fritz JV, Heintz-Buschart A, Desai MS, Yusuf D, Huang D, Baumuratov A, Wang K, Galas D, Wilmes P.** 2015. The extracellular RNA complement of *Escherichia coli*. *Microbiology Open* **4**:252–266.

264. **Sjöström AE, Sandblad L, Uhlin BE, Wai SN.** 2015. Membrane vesicle-mediated release of bacterial RNA. *Sci Rep* **5**:15329.

265. **Domínguez Rubio AP, Martínez JH, Martínez Casillas DC, Coluccio Leskow F, Piuri M, Pérez OE.** 2017. *Lactobacillus casei* BL23 produces microvesicles carrying proteins that have been associated with its probiotic effect. *Front Microbiol* **8**:1783.

266. **Li M, Lee K, Hsu M, Nau G, Mylonakis E, Ramratnam B.** 2017. *Lactobacillus*-derived extracellular vesicles enhance host immune responses against vancomycin-resistant enterococci. *BMC Microbiol* **17**:66.

267. **Koeppen K, Hampton TH, Jarek M, Scharfe M, Gerber SA, Mielcarz DW, Demers EG, Dolben EL, Hammond JH, Hogan DA, Stanton BA.** 2016. A novel mechanism of host-pathogen interaction through sRNA in bacterial outer membrane vesicles. *PLoS Pathog* **12**:e1005672.

268. **Blenkiron C, Simonov D, Muthukaruppan A, Tsai P, Dauros P, Green S, Hong J, Print CG, Swift S, Phillips AR.** 2016. Uropathogenic *Escherichia coli* releases extracellular vesicles that are associated with RNA. *PLoS One* **11**:e0160440.

269. **Liu S, da Cunha AP, Rezende RM, Cialic R, Wei Z, Bry L, Comstock LE, Gandhi R, Weiner HL.** 2016. The host shapes the gut microbiota via fecal microRNA. *Cell Host Microbe* **19**:32–43.

270. **Duval M, Cossart P, Lebreton A.** 2017. Mammalian microRNAs and long noncoding RNAs in the host-bacterial pathogen crosstalk. *Semin Cell Dev Biol* **65**:11–19.

271. **Ventola CL.** 2015. The antibiotic resistance crisis: part 1: causes and threats. *P&T* **40**:277–283.

272. **Ventola CL.** 2015. The antibiotic resistance crisis: part 2: management strategies and new agents. *P&T* **40**:344–352.

273. **Lindgreen S, Umu SU, Lai AS, Eldai H, Liu W, McGimpsey S, Wheeler NE, Biggs PJ, Thomson NR, Barquist L, Poole AM, Gardner PP.** 2014. Robust identi- fication of noncoding RNA from transcriptomes requires phylogenetically-informed sampling. *PLOS Comput Biol* **10**:e1003907.

274. **El-Mowafi SA, Alumasa JN, Ades SE, Keiler KC.** 2014. Cell-based assay to identify inhibitors of the Hfq-sRNA regulatory pathway. *Antimicrob Agents Chemother* **58**:5500–5509.

275. **Kime L, Vincent HA, Gendoo DM, Jourdan SS, Fishwick CW, Callaghan AJ, McDowall KJ.** 2015. The first small-molecule inhibitors of members of the ribonuclease E family. *Sci Rep* **5**:8028.

276. **Maurer CK, Fruth M, Empting M, Avrutina O, Hossmann J, Nadmid S, Gorges J, Herrmann J, Kazmaier U, Dersch P, Müller R, Hartmann RW.** 2016. Discovery of the first small-molecule CsrA-RNA interaction inhibitors using biophysical screening technologies. *Future Med Chem* **8**:931–947.

277. **Kim JN, Blount KF, Puskarz I, Lim J, Link KH, Breaker RR.** 2009. Design and antimicrobial action of purine analogues that bind Guanine riboswitches. *ACS Chem Biol* **4**:915–927.

278. **Mulhbacher J, Brouillette E, Allard M, Fortier LC, Malouin F, Lafontaine DA.** 2010. Novel riboswitch ligand analogs as selective inhibitors of guanine-related metabolic pathways. *PLoS Pathog* **6**:e1000865.

279. **Yu J, Schneiders T.** 2012. Tigecycline challenge triggers sRNA production in *Salmonella enterica* serovar Typhimurium. *BMC Microbiol* **12**:195.

280. **Howden BP, Beaume M, Harrison PF, Hernandez D, Schrenzel J, Seemann T, Francois P, Stinear TP.** 2013. Analysis of the small RNA transcriptional response in multidrug-resistant *Staphylococcus aureus* after antimicrobial exposure. *Antimicrob Agents Chemother* **57**:3864–3874.

281. **Kim T, Bak G, Lee J, Kim KS.** 2015. Systematic analysis of the role of bacterial Hfq-interacting sRNAs in the response to antibiotics. *J Antimicrob Chemother* **70**:1659–1668.

282. **Lalaouna D, Eyraud A, Chabelskaya S, Felden B, Massé E.** 2014. Regulatory RNAs involved in bacterial antibiotic resistance. *PLoS Pathog* **10**:e1004299.

283. **Dersch P, Khan MA, Mühlen S, Görke B.** 2017. Roles of regulatory RNAs for antibiotic resistance in bacteria and their potential value as novel drug targets. *Front Microbiol* **8**:803.

284. **Mondhe M, Chessher A, Goh S, Good L, Stach JE.** 2014. Species-selective killing of bacteria by antimicrobial peptide-PNAs. *PLoS One* **9**:e89082.

285. **Thanert R, Goldmann O, Beineke A, Medina E.** 2017. Host-inherent variability influences the transcriptional response of *Staphylococcus aureus* during *in vivo* infection. *Nat Commun* **8**:14268.

286. **Novick RP, Ross HF, Projan SJ, Kornblum J, Kreiswirth B, Moghazeh S.** 1993. Synthesis of staphylococcal virulence factors is controlled by a regulatory RNA molecule. *EMBO J* **12**:3967–3975.

287. **Huntzinger E, Boisset S, Saveanu C, Benito Y, Geissmann T, Namane A, Lina G, Etienne J, Ehresmann B, Ehresmann C, Jacquier A, Vandenesch F, Romby P.**

2005. *Staphylococcus aureus* RNAIII and the endoribo-nuclease III coordinately regulate *spa* gene expression. *EMBO J* 24:824–835.

288. Geisinger E, Adhikari RP, Jin R, Ross HF, Novick RP. 2006. Inhibition of *rot* translation by RNAIII, a key feature of *agr* function. *Mol Microbiol* 61:1038–1048.

289. Chevalier C, Boisset S, Romilly C, Masquida B, Fechter P, Geissmann T, Vandenesch F, Romby P. 2010. *Staphylococcus aureus* RNAIII binds to two distant regions of *coa* mRNA to arrest translation and promote mRNA degradation. *PLoS Pathog* 6:e1000809.

290. Chabelskaya S, Bordeau V, Felden B. 2014. Dual RNA regulatory control of a *Staphylococcus aureus* virulence factor. *Nucleic Acids Res* 42:4847–4858.

291. Geissmann T, Chevalier C, Cros MJ, Boisset S, Fechter P, Noirot C, Schrenzel J, Francois P, Vandenesch F, Gaspin C, Romby P. 2009. A search for small noncoding RNAs in *Staphylococcus aureus* reveals a conserved sequence motif for regulation. *Nucleic Acids Res* 37:7239–7257.

292. Chabelskaya S, Gaillot O, Felden B. 2010. A *Staphylococcus aureus* small RNA is required for bacterial virulence and regulates the expression of an immune-evasion molecule. *PLoS Pathog* 6:e1000927.

293. Danger JL, Makthal N, Kumaraswami M, Sumby P. 2015. The FasX small regulatory RNA negatively regulates the expression of two fibronectin-binding proteins in group A *Streptococcus*. *J Bacteriol* 197:3720–3730.

294. Perez N, Trevino J, Liu Z, Ho SC, Babitzke P, Sumby P. 2009. A genome-wide analysis of small regulatory RNAs in the human pathogen group A *Streptococcus*. *PLoS One* 4:e7668.

295. Pappesch R, Warnke P, Mikkat S, Normann J, Wisniewska-Kucper A, Huschka F, Wittmann M, Khani A, Schwengers O, Oehmcke-Hecht S, Hain T, Kreikemeyer B, Patenge N. 2017. The regulatory small RNA MarS supports virulence of *Streptococcus pyogenes*. *Sci Rep* 7:12241.

296. Kay E, Humair B, Dénervaud V, Riedel K, Spahr S, Eberl L, Valverde C, Haas D. 2006. Two GacA-dependent small RNAs modulate the quorum-sensing response in *Pseudomonas aeruginosa*. *J Bacteriol* 188:6026–6033.

297. Mulcahy H, O'Callaghan J, O'Grady EP, Maciá MD, Borrell N, Gómez C, Casey PG, Hill C, Adams C, Gahan CG, Oliver A, O'Gara F. 2008. *Pseudomonas aeruginosa* RsmA plays an important role during murine infection by influencing colonization, virulence, persistence, and pulmonary inflammation. *Infect Immun* 76:632–638.

298. Bordi C, Lamy MC, Ventre I, Termine E, Hachani A, Fillet S, Roche B, Bleves S, Méjean V, Lazdunski A, Filloux A. 2010. Regulatory RNAs and the HptB/RetS signalling pathways fine-tune *Pseudomonas aeruginosa* pathogenesis. *Mol Microbiol* 76:1427–1443.

299. Petrova OE, Sauer K. 2010. The novel two-component regulatory system BfiSR regulates biofilm development by controlling the small RNA *rsmZ* through CafA. *J Bacteriol* 192:5275–5288.

300. O'Callaghan J, Reen FJ, Adams C, O'Gara F. 2011. Low oxygen induces the type III secretion system in *Pseudomonas aeruginosa* via modulation of the small RNAs *rsmZ* and *rsmY*. *Microbiology* 157:3417–3428.

301. Chen R, Weng Y, Zhu F, Jin Y, Liu C, Pan X, Xia B, Cheng Z, Jin S, Wu W. 2016. Polynucleotide phosphorylase regulates multiple virulence factors and the stabilities of small RNAs RsmY/Z in *Pseudomonas aeruginosa*. *Front Microbiol* 7:247.

302. Jean-Pierre F, Tremblay J, Déziel E. 2016. Broth versus surface-grown cells: differential regulation of RsmY/Z small RNAs in *Pseudomonas aeruginosa* by the Gac/HptB system. *Front Microbiol* 7:2168.

303. Sonnleitner E, Prindl K, Bläsi U. 2017. The *Pseudomonas aeruginosa* CrcZ RNA interferes with Hfq-mediated riboregulation. *PLoS One* 12:e0180887.

304. Sonnleitner E, Gonzalez N, Sorger-Domenigg T, Heeb S, Richter AS, Backofen R, Williams P, Hüttenhofer A, Haas D, Bläsi U. 2011. The small RNA PhrS stimulates synthesis of the *Pseudomonas aeruginosa* quinolone signal. *Mol Microbiol* 80:868–885.

305. Wilderman PJ, Sowa NA, FitzGerald DJ, FitzGerald PC, Gottesman S, Ochsner UA, Vasil ML. 2004. Identification of tandem duplicate regulatory small RNAs in *Pseudomonas aeruginosa* involved in iron homeostasis. *Proc Natl Acad Sci U S A* 101:9792–9797.

306. Oglesby AG, Farrow JM III, Lee JH, Tomaras AP, Greenberg EP, Pesci EC, Vasil ML. 2008. The influence of iron on *Pseudomonas aeruginosa* physiology: a regulatory link between iron and quorum sensing. *J Biol Chem* 283:15558–15567.

307. Baldwin DN, Shepherd B, Kraemer P, Hall MK, Sycuro LK, Pinto-Santini DM, Salama NR. 2007. Identification of *Helicobacter pylori* genes that contribute to stomach colonization. *Infect Immun* 75:1005–1016.

308. Kim JN, Kwon YM. 2013. Genetic and phenotypic characterization of the RyhB regulon in *Salmonella* Typhimurium. *Microbiol Res* 168:41–49.

309. Ortega AD, Gonzalo-Asensio J, García-del Portillo F. 2012. Dynamics of *Salmonella* small RNA expression in non-growing bacteria located inside eukaryotic cells. *RNA Biol* 9:469–488.

310. Papenfort K, Espinosa E, Casadesús J, Vogel J. 2015. Small RNA-based feedforward loop with AND-gate logic regulates extrachromosomal DNA transfer in *Salmonella*. *Proc Natl Acad Sci U S A* 112:E4772–E4781.

311. Waldminghaus T, Heidrich N, Brantl S, Narberhaus F. 2007. FourU: a novel type of RNA thermometer in *Salmonella*. *Mol Microbiol* 65:413–424.

312. Klinkert B, Cimdins A, Gaubig LC, Rossmanith J, Aschke-Sonnenborn U, Narberhaus F. 2012. Thermogenetic tools to monitor temperature-dependent gene expression in bacteria. *J Biotechnol* 160:55–63.

313. Park SY, Cromie MJ, Lee EJ, Groisman EA. 2010. A bacterial mRNA leader that employs different mechanisms to sense disparate intracellular signals. *Cell* 142:737–748.

314. Lee EJ, Groisman EA. 2012. Tandem attenuators control expression of the *Salmonella mgtCBR* virulence operon. *Mol Microbiol* 86:212–224.

315. Nam D, Choi E, Shin D, Lee EJ. 2016. tRNA^Pro-mediated downregulation of elongation factor P is required for *mgtCBR* expression during *Salmonella* infection. *Mol Microbiol* 102:221–232.

316. Santiviago CA, Reynolds MM, Porwollik S, Choi SH, Long F, Andrews-Polymenis HL, McClelland M. 2009. Analysis of pools of targeted *Salmonella* deletion mutants identifies novel genes affecting fitness during competitive infection in mice. *PLoS Pathog* 5:e1000477.

317. Colgan AM, Kroger C, Diard M, Hardt WD, Puente JL, Sivasankaran SK, Hokamp K, Hinton JC. 2016. The impact of 18 ancestral and horizontally-acquired regulatory proteins upon the transcriptome and sRNA landscape of *Salmonella enterica* serovar Typhimurium. *PLoS Genet* 12:e1006258.

318. Altuvia S, Weinstein-Fischer D, Zhang A, Postow L, Storz G. 1997. A small, stable RNA induced by oxidative stress: role as a pleiotropic regulator and antimutator. *Cell* 90:43–53.

319. Sudo N, Soma A, Muto A, Iyoda S, Suh M, Kurihara N, Abe H, Tobe T, Ogura Y, Hayashi T, Kurokawa K, Ohnishi M, Sekine Y. 2014. A novel small regulatory RNA enhances cell motility in enterohemorrhagic *Escherichia coli*. *J Gen Appl Microbiol* 60:44–50.

320. Darfeuille F, Unoson C, Vogel J, Wagner EG. 2007. An antisense RNA inhibits translation by competing with standby ribosomes. *Mol Cell* 26:381–392.

321. Mandin P, Repoila F, Vergassola M, Geissmann T, Cossart P. 2007. Identification of new noncoding RNAs in *Listeria monocytogenes* and prediction of mRNA targets. *Nucleic Acids Res* 35:962–974.

322. Sievers S, Sternkopf Lillebaek EM, Jacobsen K, Lund A, Mollerup MS, Nielsen PK, Kallipolitis BH. 2014. A multicopy sRNA of *Listeria monocytogenes* regulates expression of the virulence adhesin LapB. *Nucleic Acids Res* 42:9383–9398.

323. Song T, Sabharwal D, Gurung JM, Cheng AT, Sjöström AE, Yildiz FH, Uhlin BE, Wai SN. 2014. Vibrio cholerae utilizes direct sRNA regulation in expression of a biofilm matrix protein. *PLoS One* 9:e101280.

324. Sabharwal D, Song T, Papenfort K, Wai SN. 2015. The VrrA sRNA controls a stationary phase survival factor Vrp of *Vibrio cholerae*. *RNA Biol* 12:186–196.

325. Lenz DH, Mok KC, Lilley BN, Kulkarni RV, Wingreen NS, Bassler BL. 2004. The small RNA chaperone Hfq and multiple small RNAs control quorum sensing in *Vibrio harveyi* and *Vibrio cholerae*. *Cell* 118:69–82.

326. Bardill JP, Hammer BK. 2012. Non-coding sRNAs regulate virulence in the bacterial pathogen *Vibrio cholerae*. *RNA Biol* 9:392–401.

327. Murphy ER, Payne SM. 2007. RyhB, an iron-responsive small RNA molecule, regulates *Shigella dysenteriae* virulence. *Infect Immun* 75:3470–3477.

328. Sampson TR, Saroj SD, Llewellyn AC, Tzeng YL, Weiss DS. 2013. A CRISPR/Cas system mediates bacterial innate immune evasion and virulence. *Nature* 497:254–257.

329. Caswell CC, Gaines JM, Ciborowski P, Smith D, Borchers CH, Roux CM, Sayood K, Dunman PM, Roop Ii RM. 2012. Identification of two small regulatory RNAs linked to virulence in *Brucella abortus* 2308. *Mol Microbiol* 85:345–360.

330. Ansong C, Yoon H, Porwollik S, Mottaz-Brewer H, Petritis BO, Jaitly N, Adkins JN, McClelland M, Heffron F, Smith RD. 2009. Global systems-level analysis of Hfq and SmpB deletion mutants in *Salmonella*: implications for virulence and global protein translation. *PLoS One* 4:e4809.

331. Ashburner M, Ball CA, Blake JA, Botstein D, Butler H, Cherry JM, Davis AP, Dolinski K, Dwight SS, Eppig JT, Harris MA, Hill DP, Issel-Tarver L, Kasarskis A, Lewis S, Matese JC, Richardson JE, Ringwald M, Rubin GM, Sherlock G, The Gene Ontology Consortium. 2000. Gene ontology: tool for the unification of biology. *Nat Genet* 25:25–29.

332. Gene Ontology Consortium. 2015. Gene Ontology Consortium: going forward. *Nucleic Acids Res* 43(Database issue):D1049–D1056.

333. Nawrocki EP, Burge SW, Bateman A, Daub J, Eberhardt RY, Eddy SR, Floden EW, Gardner PP, Jones TA, Tate J, Finn RD. 2015. Rfam 12.0: updates to the RNA families database. *Nucleic Acids Res* 43(Database issue): D130–D137.

334. El Mouali Y, Gaviria-Cantin T, Sánchez-Romero MA, Gibert M, Westermann AJ, Vogel J, Balsalobre C. 2018. CRP-cAMP mediates silencing of Salmonella virulence at the post-transcriptional level. *PLoS Genet* 14: e1007401.

335. Sudo N, Soma A, Iyoda S, Oshima T, Ohto Y, Saito K, Sekine Y. 2018. Small RNA Esr41 inversely regulates expression of LEE and flagellar genes in enterohaemorrhagic. *Escherichia coli* 164:821–834.

336. Wang D, McAteer SP, Wawszczyk AB, Russell CD, Tahoun A, Elmi A, Cockroft SL, Tollervey D, Granneman S, Tree JJ, Gally DL. 2018. An RNA-dependent mechanism for transient expression of bacterial translocation filaments. *Nucleic Acids Res* 46:3366–3381.

337. Beisel CL, Storz G. 2011. The base-pairing RNA spot 42 participates in a multioutput feedforward loop to help enact catabolite repression in *Escherichia coli*. *Mol Cell* 41:286–297.

338. Møller T, Franch T, Udesen C, Gerdes K, Valentin-Hansen P. 2002. Spot 42 RNA mediates discoordinate expression of the *E. coli* galactose operon. *Genes Dev* 16:1696–1706.

339. Holmqvist E, Li L, Bischler T, Barquist L, Vogel J. 2018. Global maps of ProQ binding in vivo reveal target recognition via RNA structure and stability control at mRNA 3′ ends. *Mol Cell* 70:971–982.

340. Tsatsaronis JA, Franch-Arroyo S, Resch U, Charpentier E. 2018. Extracellular vesicle RNA: a universal mediator of microbial communication? *Trends Microbiol* 26:401–410.

341. Felden B, Cattoir V. 2018. Bacterial adaptation to antibiotics through regulatory RNAs. *Antimicrob Agents Chemother* 62:e02503-17.

Protein Titration and Scaffolding

V

Regulating with RNA in Bacteria and Archaea
Edited by Gisela Storz and Kai Papenfort
© 2018 American Society for Microbiology, Washington, DC
doi:10.1128/microbiolspec.RWR-0009-2017

Global Regulation by CsrA and Its RNA Antagonists

19

Tony Romeo[1] and Paul Babitzke[2,3]

INTRODUCTION

The Csr (carbon storage regulator) or Rsm (repressor of stationary-phase metabolites) system is among the most extensively studied bacterial RNA-based regulatory systems. Its central component, the RNA binding protein CsrA (RsmA), was uncovered by a transposon mutagenesis screen designed to identify regulators of gene expression in the stationary phase of growth, using glycogen biosynthesis and *glgC'-'lacZ* expression as reporters (1). Understanding of RNA binding proteins and their roles in regulation was limited at that time, but included Hfq and ribosomal proteins that mediate negative feedback by binding to their mRNAs (2–4). Soon after its discovery, the regulatory role of CsrA began to emerge, which included repression of other genes similar to *glgC*, which are expressed in stationary phase or under stress conditions (5), and evidence that CsrA activates gene expression that supports robust growth (6). Discoveries that CsrA (RsmA) regulates virulence genes of pathogens associated with plant disease (7) and mammalian cell invasion (8) offered early glimpses of the widespread roles played by CsrA proteins in microbe-host interactions (9). The role of CsrA in biofilm formation (10–14), quorum sensing (15), carbon metabolism (6, 16, 17), motility (18, 19), and stress responses (14, 20–23) is now well documented in *Escherichia coli* and other species. New functions of CsrA are being uncovered at a rapid pace through the use of transcriptomics, proteomics, metabolomics, and other systems approaches (12, 13, 20, 24–35).

Early evidence that CsrA regulates gene expression posttranscriptionally was that it activates *glgC* mRNA decay, which requires the *glgC* translation initiation region but not the promoter region (36), and that CsrA binds to *glgC* mRNA and blocks translation by occluding the Shine-Dalgarno (SD) sequence (37, 38). CsrA activity is regulated by noncoding small RNAs (sRNAs) that compete with mRNAs for CsrA binding. The first of these sRNAs, CsrB, was identified as a component of an RNP complex, isolated by purification of a recombinant CsrA protein (39). The stoichiometry of the CsrA:CsrB RNP complex suggested that CsrA likely bound to CsrB RNA at highly repeated CAGGA(U/A/C)G sequences. These sequences resemble the SD sequence of mRNAs and were located in predicted stem-loops and single-stranded segments of this sRNA. Discovery of CsrC, an sRNA that functions similarly to CsrB, soon followed (40). While the *csrB* and *csrC* loci were both uncovered earlier in a genetic screen designed to identify regulators of *glgC* expression (41), understanding of the function of these regulatory RNAs awaited the discovery of CsrA. Altogether, these findings set the stage for the development of a new paradigm in genetic regulation, in which a sequence-specific RNA binding protein is sequestered and antagonized by a noncoding RNA containing binding sites that mimic its mRNA target sequences.

RNA SEQUENCE AND STRUCTURAL FEATURES OF CsrA BINDING SITES

Early studies with CsrB provided the first suggestion that a GGA motif in the loop of a short hairpin was an important component of a CsrA binding site (39). A consensus sequence for CsrA binding sites was determined using systematic evolution of ligands by exponential enrichment (SELEX). The SELEX-derived consensus sequence for a single high-affinity CsrA bind-

[1]Department of Microbiology and Cell Science, Institute of Food and Agricultural Sciences, University of Florida, Gainesville, FL 32611;
[2]Department of Biochemistry and Molecular Biology; [3]Center for RNA Molecular Biology, The Pennsylvania State University, University Park, PA 16802.

ing site was determined as RUACARGGAUGU, with the GGA motif being 100% conserved. The GGA motif was typically located in a hexaloop (ARGGAU) of an RNA hairpin, with the upstream AC and downstream GU residues always base-paired to one another (Fig. 1A). Mutagenesis of one SELEX-derived RNA target indicated that the GGA motif and the preceding A residue in the loop of the hairpin were critical for *E. coli* CsrA binding (42). This study established that the primary RNA sequence is most important for CsrA binding, and that the presentation of the GGA motif in a loop increases the affinity of CsrA-RNA interaction. More recently, a SELEX method that exploited deep sequencing of RNAs identified a consensus for *Pseudomonas aeruginosa* RsmA and RsmF as CANGGAYG, with the GGA motif typically found in a hexaloop (43). This sequence and structural arrangement is remarkably similar to that determined for *E. coli* CsrA.

CsrA AND CsrA-RNA STRUCTURE

Cross-linking studies with purified CsrA established that *E. coli* CsrA functions as a homodimer (5, 43). Subsequent structural studies demonstrated that each subunit of the dimer contains five β-strands (β_1 to β_5), an α-helix, and a flexible C terminus (44). Strands β_1 and β_5 of one monomer hydrogen bond to β_4 and β_2 of the other monomer, forming a mixed antiparallel β-sheet (44). Similar structures were later determined for RsmA of *Yersinia enterocolitica* (45), RsmE of *Pseudomonas fluorescens* (46), and CsrA of *Geobacillus thermodenitrificans* (47). Alanine scanning mutagenesis of *E. coli* CsrA identified amino acid residues in the β_1

and β_5 strands that are critical for RNA binding (48). The structure of a *P. fluorescens* RsmE-RNA complex confirmed and extended these findings (Fig. 1B). This structure revealed that the RsmE homodimer binds to two RNA molecules simultaneously (46). The RNAs were bound on positively charged surfaces formed by the β_1 and β_5 strands of opposite monomers, with each dimer interacting with all six bases in the hexaloop (ACGGAU). As expected, the GGA motif is specifically recognized by the protein; the Watson-Crick face of both G residues and the Hoogsteen face of the A residue form hydrogen bonds with the protein. In addition, the Watson-Crick face of the A residue preceding the GGA motif makes specific contacts with the protein (46). Interestingly, both of the SELEX studies described above identified an A residue at this position. Even though this upstream A is not as highly conserved as the GGA motif, mutagenesis of the *E. coli* consensus sequence indicated that altering this A residue caused a more severe binding defect than changing the A in the GGA motif itself (42).

Although the SELEX-derived consensus for *E. coli* CsrA and *P. aeruginosa* RsmA/RsmF is virtually identical, this method selected for high-affinity sites. With the exception of the conserved GGA motif, CsrA binding sites within natural RNA targets exhibit considerable sequence variation. To identify the structural basis for recognition of varying RNA sequences, structures of *P. fluorescens* RsmE in complex with a variety of RNA sequences were determined. This study provided an explanation for how the variation of sequence and structural context of the GGA motif modulate the binding affinity (49). Recent structural studies also demonstrated that RsmE binding to multiple sites in its regulatory sRNAs is not random, but rather occurs sequentially in an ordered manner (50).

DIRECT CsrA-MEDIATED REGULATION

Repression of Translation Initiation
Repression of translation initiation was the first detailed regulatory mechanism identified for CsrA (38) (Fig. 2A). This common regulatory strategy has led to CsrA being called a "translational repressor." However, this is an inadequate description of CsrA function, as this protein has been shown to participate in a wide variety of regulatory mechanisms, including mRNA stabilization and destabilization, transcription attenuation, and activation of translation.

The sequence of CsrB provided critical insight into possible modes of CsrA action. The fact that GGA is a

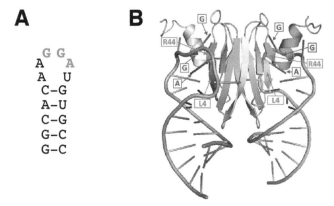

Figure 1 (A) Example of a high-affinity CsrA binding site. The conserved GGA motif is in red. (B) Structure of the CsrA-RNA complex. The GGA motifs are indicated by blue boxes, and the critical L4 and R44 residues are indicated in red. Adapted from reference 46 with permission.

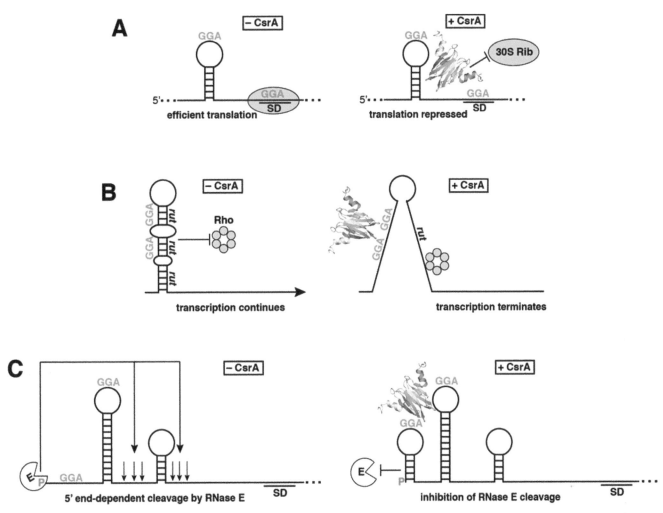

Figure 2 Mechanisms for CsrA-mediated translational repression (A), transcription termination (B), and protection of mRNA from nuclease cleavage. Adapted from reference 9 with permission.

common feature of the SD sequence, an important component of the ribosome binding site, led to the suggestion that CsrA might repress translation of mRNA targets. This prediction was first verified by the demonstration that CsrA represses translation initiation of *glgC*, which encodes a glycogen biosynthetic enzyme (38). CsrA binds to four sites in the 5′ untranslated region (5′ UTR) of *glgCAP* mRNA, one of which overlaps the *glgC* SD sequence, such that bound CsrA prevents 30S ribosomal subunit binding (38, 51). Since CsrA functions as a homodimer with two identical RNA binding surfaces, CsrA is capable of bridging a high-affinity binding site to the lower-affinity site overlapping the *glgC* SD sequence (51). Structural studies confirmed dual-site binding (50). This general mechanism of translational repression has been substantiated

for numerous *E. coli* mRNA targets, including *csrA* itself (5, 9, 11, 14, 20, 22, 23, 52–55). Similar translation repression mechanisms have also been identified in several other bacterial species (31, 56–59). In some cases, CsrA binding sites overlap the start codon (11, 22, 23, 55, 58, 59), or initially translated region (5, 15).

Several translational repression mechanisms differ substantially from the general mechanism just described. RsmA of *P. aeruginosa* represses translation of *psl*, a gene encoding a component of biofilm matrix. In this case, bound RsmA appears to stabilize a structure that sequesters the *psl* SD sequence in an RNA secondary structure (60). In another instance, *E. coli* CsrA indirectly represses translation of *iraD*, a gene encoding an antiadapter protein that inhibits RssB-mediated degradation of RpoS. CsrA represses translation of a leader peptide whose stop codon overlaps with the *iraD* start codon. Thus, translational repression of *iraD* occurs entirely via translational coupling (22). Lastly, unlike other known CsrA-mediated translational repression mechanisms in which CsrA binds to two or more sites, CsrA is capable of repressing *hfq* translation by binding to a single site that overlaps its SD sequence (61). The commonality in all of these examples is that bound CsrA inhibits ribosome binding. Although not always the case, reduced translation often leads to destabilization of the downstream mRNA (9, 53).

Transcription Attenuation

In addition to repressing translation initiation of *pgaA* (11), CsrA participates in a transcription attenuation mechanism in which bound CsrA prevents formation of an RNA secondary structure that would otherwise sequester a Rho binding site (Fig. 2B). Thus, bound CsrA mediates Rho-dependent termination of the nascent *pgaABCD* operon transcript (62). Since CsrA also represses translation of *nhaR*, which encodes an activator of *pgaABCD* operon transcription (14, 63), and the expression of GGDEF domain proteins that synthesize cyclic di-GMP, an allosteric activator of poly-β-1,6-N-acetyl-D-glucosamine (PGA) synthesis (12, 13, 64), CsrA represses the biosynthesis and secretion of the PGA biofilm adhesin by at least four distinct mechanisms.

Activation Mechanisms

The *E. coli moaABCDE* operon is controlled by a MOCO-dependent riboswitch that represses its translation, whereas bound CsrA activates *moaA* translation by altering the RNA structure, thereby increasing the accessibility of the ribosome binding site (65). In another interesting example, RsmA of *P. aeruginosa* activates translation of *phz2*, a phenazine biosynthetic gene cluster. In this case, bound RsmA appears to activate translation by destabilizing an SD-sequestering hairpin (66). Lastly, CsrA activates expression of the *E. coli flhDC* operon, which encodes a DNA binding activator of flagella biosynthesis and chemotaxis. CsrA binds to two sites that are >150 nucleotides (nt) upstream of the *flhD* SD sequence, one of which is positioned at the extreme 5′ end of the transcript. In this case, bound CsrA stabilizes the *flhDC* transcript by blocking 5′ end-dependent cleavage by RNase E (19) (Fig. 2C).

RNA-MEDIATED CsrA ANTAGONISM

Discovery of sRNA Antagonists of CsrA

CsrB of *E. coli* was the first identified sRNA antagonist of CsrA (Table 1) (39). This 369-nt sRNA contains 22

Table 1 CsrA and its antagonists

Organism	CsrA homolog	Antagonist(s)[a]
E. coli	CsrA	CsrB, CsrC
E. coli (EPEC)	CsrA	CesT
P. carotovorum	RsmA	RsmB
S. enterica serovar Typhimurium	CsrA	CsrB, CsrC, *fimA*
P. aeruginosa	RsmA, RsmF (RsmN)	RsmY, RsmZ
P. fluorescens	RsmA, RsmE	RsmX, RsmY, RsmZ
B. subtilis	CsrA	FliW
G. thermodenitrificans	CsrA	FliW
C. jejuni	CsrA	FliW

[a]All of these antagonists are RNAs, with exception of the CesT and FliW proteins.

potential CsrA binding sites and is capable of sequestering ~9 CsrA dimers (39). Soon after the discovery of *E. coli* CsrB, RsmB of *Pectobacterium carotovorum* was identified as an sRNA antagonist of its CsrA homolog, RsmA. This 479-nt sRNA contains 20 GGA motifs (67). A second sRNA, CsrC, was later identified as a redundant *E. coli* CsrA antagonist (40), although recent evidence indicates that these two sRNAs are differentially expressed (68). CsrC contains 13 GGA motifs, many of which were predicted to be in short RNA hairpins, as was previously observed for CsrB (40). Multiple sRNA antagonists have been identified in a variety of species, indicating that this is a common strategy for modulating CsrA/RsmA activity (69–73). However, the sRNAs in *P. aeruginosa* (RsmY and RsmZ) and *P. fluorescens* (RsmX, RsmY, and RsmZ) are much shorter and contain only 5 to 8 GGA motifs (69, 71).

Additional RNAs May Sequester CsrA

CsrB/C RNAs may be dedicated solely to CsrA sequestration; no additional roles for these sRNAs have been identified. Nevertheless, other RNAs appear to bind to and sequester CsrA as a "moonlighting" function, in addition to their other roles (Fig. 3). The 5′ UTR of *fimAICDHF* mRNA of *Salmonella enterica* contains two CAGGAUG sequences that sequester CsrA, along with CsrB/C, as part of a hierarchical control mechanism for fimbriae expression. In this mechanism, abundant expression of the mRNA for type I fimbriae prevents expression of plasmid-encoded fimbriae from *pefACDEF* mRNA, which requires CsrA binding for its expression (74). This sequestration mechanism apparently helps *Salmonella* to avoid the costly expression of plasmid-encoded fimbriae outside of the host. Because CsrA has little effect on *fimA* expression, it appears that high-

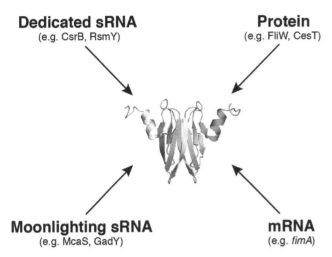

Dedicated sRNA
(e.g. CsrB, RsmY)

Protein
(e.g. FliW, CesT)

Moonlighting sRNA
(e.g. McaS, GadY)

mRNA
(e.g. *fimA*)

Figure 3 Modes of CsrA antagonism. In various species, dedicated sRNAs, moonlighting sRNAs, mRNA, and/or proteins have been found to bind to CsrA and inhibit its activity.

affinity binding of CsrA to *fimA* mRNA acts primarily in hierarchical control.

Two base-pairing sRNAs have been reported to act by sequestering CsrA, McaS, and GadY. These sRNAs contain GGA sequences that allow interaction with CsrA, and upon overexpression, these RNAs are able to activate *pgaA* expression by antagonizing CsrA activity (75, 76). Increased expression of the *pgaABCD* operon promotes both synthesis and secretion of the biofilm adhesin PGA and biofilm formation (77). High-throughput sequencing of RNA isolated by UV cross-linking and immunoprecipitation (CLIP-seq) studies have indicated that CsrA interacts directly with other sRNAs *in vivo* in *Salmonella* and *E. coli* (29, 35). High-affinity binding of CsrA to *E. coli* sRNAs GadY, Spot 42, GcvB, and MicL was confirmed *in vitro*, although biological functions for CsrA binding to the three latter RNAs have not been determined (35).

Ribosome Pausing as a Mechanism Leading to CsrA Sequestration?

Because there appears to be little RNA-free CsrA in the cell (53), any tight-binding RNA expressed at sufficient levels should compete for CsrA binding with lower-affinity transcripts. Recent findings from transcriptomics studies have led to the discovery that CsrA binding occurs at the 3′ UTR of a few transcripts, but occurs predominantly deep within the coding regions of mRNAs (29, 31, 34, 35). Furthermore, CsrA binding within mRNA coding regions is significantly increased in frequency near ribosome pause sites (35). While CsrA was not found to cause ribosome pausing, it is possible that ribosome pausing facilitates CsrA binding

near pause sites. On a transcriptome-wide scale, this might serve as an effective means of restricting CsrA availability when the capacity for translation is limited and pausing is increased. Described below, CsrA activity is inhibited during amino acid limitation by the positive effects of ppGpp and DksA on CsrB/C transcription (20). Conceivably, these two mechanisms might act together, thus increasing the expression of factors that deal with translational stress and other stresses, and decreasing gene expression required for rapid growth, under translational stress.

PROTEIN ANTAGONISTS OF CsrA

Although sRNA antagonism of CsrA/RsmA is a common feature of Csr/Rsm systems, some organisms use a protein to antagonize the activity of CsrA (Fig. 3). CsrA represses translation initiation of *Bacillus subtilis hag*, the gene encoding the flagellar filament protein (56). FliW was identified as the first protein antagonist of CsrA (78). FliW, CsrA, and Hag participate in a protein partner switching mechanism to control Hag synthesis. Following completion of the flagellar hook, secretion of Hag releases FliW from a FliW-Hag complex. Once released, FliW can instead bind to CsrA, thereby relieving CsrA-mediated translational repression of *hag*, so that Hag synthesis is increased precisely when it is needed for synthesis of the flagella. Thus, Hag homeostatically restricts its own translation (78). FliW does not bind to the same residues of CsrA required for *hag* mRNA binding. Some *csrA* mutants abolished CsrA-FliW binding, but others did not, indicating that FliW and RNA interaction is not mutually exclusive (79). Structural analysis of the CsrA-FliW complex from *G. thermodenitrificans* indicates that each CsrA subunit binds to a FliW monomer (47). This structure also revealed that FliW interacts with a C-terminal extension of CsrA, and that FliW allosterically antagonizes CsrA in a noncompetitive manner by excluding RNA from the RNA binding surface of CsrA. An essentially identical FlaA-FliW-CsrA partner switching mechanism is responsible for controlling CsrA-mediated translational repression of *flaA*, which encodes the major flagellin in *Campylobacter jejuni*. Interestingly, *flaA* mRNA is expressed and localized at the cell poles, which depends on the FlaA-FliW-CsrA network (31). Notably, *B. subtilis* and *C. jejuni* lack an sRNA antagonist of CsrA, while *E. coli* lacks FliW and the C-terminal extension of CsrA.

Whereas FliW appears to be the only CsrA antagonist in *B. subtilis* and *C. jejuni*, enteropathogenic *E. coli* (EPEC), whose genome contains the *csrB* and *csrC* genes (80), uses a recently identified protein, CesT, to antagonize CsrA (55). CesT functions as a chaperone for effectors that are injected into host epithelial cells by a type III secretion system. Following effector injection, CesT binds to and antagonizes CsrA. The CesT-CsrA interaction leads to changes in virulence and metabolic gene expression, which is likely required for EPEC's adaptation to life on the epithelial surface (55).

REGULATION OF CsrB/C TRANSCRIPTION

BarA-UvrY TCS and Its Orthologs

The recognition that CsrB and CsrC antagonize CsrA activity in *E. coli* (39, 40) made it crucial to understand how the levels of these RNAs are regulated (Fig. 4). CsrB/C transcription requires the two-component signal transduction system (TCS) BarA-UvrY (40, 81, 82), referred to as GacS-GacA or other names in various species (9, 68). This TCS likely activates transcription of most, though not necessarily all, of the Csr/Rsm sRNAs in the *Gammaproteobacteria*, a notable exception being CsrC of *Yersinia pseudotuberculosis* (83). The membrane-bound sensor kinase BarA is a protein with tripartite architecture, which appears to use a His → Asp → His phosphorelay prior to phosphorylation of its cognate response regulator, UvrY, a FixJ family DNA binding protein (81, 84). In *E. coli*, BarA-UvrY

signaling is activated by short-chain carboxylate compounds such as formate and acetate, which are sensed by BarA (85). In addition, acetyl phosphate can directly phosphorylate and activate UvrY (85, 86). Abundant acetate and other short-chain carboxylates in the mammalian intestinal tract suggests that CsrA activity may be decreased in this environment. Tricarboxylic acid cycle intermediates and citrate appear to serve as signals for orthologous TCS in *P. fluorescens* and *Vibrio fischeri*, respectively (87, 88). The precise signaling mechanisms in all of these cases remain mysterious.

Genomic cross-linking experiments (chromatin immunoprecipitation followed by exonuclease digestion [ChIP-exo]) in *E. coli* and *Salmonella* showed that P-UvrY or P-SirA cross-links to *csrB* and *csrC* DNA at two locations: one far upstream of and another overlapping the promoter (68). The *csrB* upstream site contains an inverted repeat sequence (TGTGAGAGATCTCTTACA) followed by a partial repeat (TGTAGGAGA) in both species; binding to *csrC* occurred at similar sequences. Binding at the promoter apparently represents indirect formaldehyde cross-linking of P-UvrY to DNA, e.g., via RNA polymerase, as only the upstream location is bound by the purified P-UvrY protein *in vitro*. ChIP-exo and ChIP-seq experiments in *E. coli*, *Salmonella*, and *P. aeruginosa* have demonstrated that the Csr/Rsm sRNA genes represent the major or sole binding targets of the response regulator UvrY, SirA, or GacA, respectively, suggesting that the global regulatory effects of these TCSs are mediated largely or entirely via the Csr/Rsm system (26, 68).

Factors that regulate BarA and UvrY expression affect CsrB/C levels. These include two DEAD-box RNA helicases: DeaD (CsdA), which activates translation of *uvrY* by counteracting long-range inhibitory base-pairing interactions between the 5′ UTR and coding regions of this transcript (89); and SrmB, which somehow stimulates the binding of P-UvrY to *csrB* DNA without affecting the levels of this protein (68, 89). Regulation by DeaD helicase may help to support *uvrY* translation under conditions of reduced translation capacity (89). The CsrA protein itself activates *uvrY* transcription and translation indirectly, and is required to switch BarA protein from acting as a UvrY phosphatase to a kinase (90). The autoregulatory circuitry that includes these interactions is discussed below.

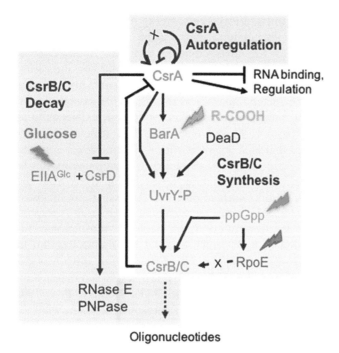

Figure 4 Central regulatory circuitry of the Csr system. Dedicated components of the Csr system are highlighted in red.

Regulation of CsrB/C Transcription by Starvation and Stress Conditions

Several factors in addition to BarA-UvrY regulate CsrB/C sRNA transcription in response to nutrient starvation, extracytoplasmic stress, and other stresses. Furthermore, CsrA mediates reciprocal effects on the genes for a number of the corresponding stress response factors, apparently to fine-tune regulation of stress responses (discussed below). The stringent response system detects amino acid starvation and other stresses and responds by activating the synthesis of the alarmone (p) ppGpp, which in turn activates *csrB/C* transcription (20, 68). Effects of (p)ppGpp are often potentiated by the protein DksA; both of these molecules bind to RNA polymerase and modify its activity (91, 92). Because ppGpp and DksA stimulated expression from a *csrB* promoter reporter *in vitro*, these effects appear to be mediated directly (68). Recent studies revealed that *csrB/C* transcription is also activated by the extracytoplasmic stress response system, which is centered on the sigma factor RpoE or σ^E (23). However, in this case, RpoE indirectly activates transcription of CsrB/C sRNAs from σ^{70} promoters. Physiological implications of this regulation are discussed below. The catabolite repression system also affects CsrB/C expression (21). In contrast to the effects of stringent response and extracytoplasmic response systems, CsrB and CsrC transcription is repressed by the mediator of this carbon starvation stress response, cyclic AMP (cAMP)-cAMP receptor protein (CRP). As discussed below, CsrB/C turnover is also stimulated by active glucose transport, causing complex interplay of the Csr circuitry with cAMP-CRP and other carbon regulatory systems. Finally, we should note that the regulation of Csr/Rsm sRNA transcription can vary greatly in other species. For example, in *P. aeruginosa* three sensor kinases in addition to GacS (BarA) govern the phosphorylation state of GacA (UvrY), none of which is present in *E. coli* (93).

ACTIVATION OF CsrB/C DECAY BY PREFERRED CARBON SOURCES

The intracellular level of an RNA molecule reflects the rates of its synthesis, decay, and dilution by growth. Decay of CsrB/C sRNAs of *E. coli* is tightly regulated, and requires the endonuclease RNase E and the 3′-to-5′ exonuclease polynucleotide phosphorylase (PNPase), housekeeping enzymes widely involved in RNA turnover (94). In contrast to base-pairing sRNAs, turnover of CsrB/C is unaffected by Hfq, but requires a specificity factor referred to as CsrD. The *csrD* gene was uncovered in a transposon screen for mutations that decreased transcription of a *csrB-lacZ* fusion; yet this mutation increased the levels of CsrB (94). This initially puzzling result occurs because the Csr system operates via negative feedback loops. Disruption of *csrD* stabilizes CsrB/C, causing sequestration of CsrA. Because

CsrA activates *csrB/C* transcription (82, 90, 95), this causes *csrB-lacZ* transcription to decrease. Unlike CsrB/C RNA, *lacZ* mRNA stability is not affected by CsrD; thus, in the *csrD* mutant, *csrB-lacZ* expression is decreased. This feedback loop also prevents CsrC RNA from accumulating in the *csrD* mutant, although it is greatly stabilized in this strain, and attenuates the effect of *csrD* disruption on CsrB levels.

The CsrD protein has features suggestive of a signaling protein. It contains two N-terminal membrane-spanning domains, a HAMP-like domain, and degenerate GGDEF and EAL domains (94). All domains are required for its activity, with the exception of the membrane-spanning domains, which are dispensable if the protein is overexpressed. Unlike the classical GGDEF and EAL domain proteins, the domains of CsrD are not involved in synthesis, turnover, or recognition of the signaling molecule cyclic di-GMP, presumably because of the degeneracy of these domains. In addition, CsrD itself is not a nuclease (94, 96). Recent studies demonstrate that its EAL domain mediates a binding interaction that activates CsrD and triggers CsrB/C decay in response to the availability of a preferred carbon substrate, such as glucose (97). The CsrD EAL domain binds only to the unphosphorylated form of EIIA^Glc, which predominates when glucose is being transported via the phosphotransferase system (PTS). In this way, glucose activates CsrB/C decay through its effect on the phosphorylation state of EIIA^Glc. Because EIIA^Glc does not affect CsrD levels, this binding interaction appears to allosterically activate CsrD (Fig. 4). The CsrD-EIIA^Glc decay pathway appears to also operate in *Vibrio cholerae* (97) and perhaps throughout *Enterobacteriaceae*, *Vibrionaceae*, and *Shewanellaceae* families (94). The absence of CsrD in other *Gammaproteobacteria*, which express Csr/Rsm sRNAs, raises questions about its evolution.

While *Gammaproteobacteria* extensively use sRNAs to sequester CsrA, most families of this bacterial class lack a CsrD ortholog (68, 94, 96). Thus, their Csr family sRNAs apparently decay differently than CsrB/C of *E. coli*. Turnover of RsmY sRNA is relatively slow in *P. fluorescens*, 26 min versus ~2 to 4 min for CsrB/C in *E. coli* (94, 98). Furthermore, while *csrA* disruption in *E. coli* has little or no effect on CsrB/C decay in a *csrD* wild-type strain (95, 96), *rsmA* (*csrA*) disruption destabilizes RsmY. Examination of the CsrB decay mechanism helps to explain the distinct turnover patterns in these species, and the reason that RNase E is unable to trigger CsrB decay in the absence of CsrD in *E. coli* (94, 96). RNase E initially cleaves CsrB within an unstructured RNA segment located immediately upstream of the intrinsic terminator (96). CsrA binding to two sites adjacent to this cleavage site blocks RNase E cleavage in the absence of CsrD. In a *csrD* mutant and *in vitro*, CsrA binding to CsrB protects it against turnover and RNase E cleavage, respectively, similar to the effect of RsmA (CsrA) on RsmY sRNA in *P. fluorescens* (96). Therefore, the evolution of CsrD has provided *E. coli* with a means of bypassing CsrA-mediated protection of CsrB/C sRNAs; even in the presence of CsrA their turnover is activated by the availability of glucose or other preferred carbon sources. Therefore, end products of metabolism, such as formate and acetate, activate CsrB/C transcription (85), while preferred carbon activates their turnover (97). Thus, carbon nutritional cues favor enhanced CsrA activity when glycolytic metabolism and other growth-promoting pathways are required and decreased CsrA activity when carbon nutrition is limited. Additional carbon nutrition effects on the behavior of Csr circuitry are discussed below.

FEEDBACK AND AUTOREGULATORY CIRCUITS PERMIT RAPID Csr RESPONSES

Following the discovery that CsrB and CsrC sRNAs act as antagonists of CsrA, CsrA itself was found to indirectly activate CsrB/C transcription in *E. coli* (40, 82, 95). This indirect activation occurs via positive effects of CsrA on *uvrY* expression and the ability to cause BarA to switch from its phosphatase to kinase activity (90). The negative feedback loop that this creates in the Csr system allows CsrB and CsrC to mediate compensatory effects on each other's expression (40, 94) (Fig. 4). A separate negative feedback loop is created in which CsrA represses CsrD, which is needed for turnover of CsrB/C RNAs (13, 94). Negative feedback can cause a number of regulatory outcomes, including acceleration of response times and decreased cell-cell variability (99, 100). In the Csr system, negative feedback has been shown to reduce response times (101). Sequestration of CsrA by CsrB/C allows rapid reduction of CsrA activity without the need for its dilution via growth, while CsrD-mediated decay of CsrB/C rapidly releases CsrA for interaction with other RNAs. Compensatory regulatory interactions between CsrB and CsrC should reduce cell-cell variability in CsrA activity, although this remains to be demonstrated. Following its description in *E. coli*, compensatory sRNA regulation has been demonstrated in other Csr systems (102) and in circuitry involving redundant base-pairing sRNAs (103).

CsrA mediates complex regulation of its own expression, including direct negative and indirect positive effects, which occur simultaneously (Fig. 4) (52). CsrA binds to four sites in its own mRNA leader, preventing ribosome binding to the SD sequence and allowing for rapid inhibition of its translation (52). Transcription of *csrA* involves five promoters, recognized by σ^{70} and/or σ^S RNA polymerase. CsrA indirectly activates transcription from the strong P3 promoter. This promoter responds to σ^S as well as σ^{70} and permits the CsrA protein to accumulate approaching the stationary phase of growth. CsrB/C levels also accumulate at this stage of growth (40, 89, 95). The ability of CsrB/C RNAs to sequester a large number of CsrA dimers allows them to cause a net decrease in CsrA activity upon entry to stationary-phase growth, even as CsrA protein levels are increased. Perhaps the latter increase in CsrA levels positions the Csr system for a robust response when conditions favorable for growth are restored.

INTERACTION OF Csr WITH OTHER GLOBAL REGULATORY SYSTEMS

In the wake of early evidence that CsrA (RsmA) influences the expression of other regulators (18, 104), transcriptomics and other studies have uncovered vast potential for bacterial Csr systems to interact with other regulators and global regulatory systems (see, e.g., references 12, 13, 20, and 24–35). Most of these interactions have not been further investigated, but in a few cases the integration of transcriptional and Csr posttranscriptional regulatory circuitry have been studied. For example, RNA-seq analysis indicated that CsrA binds to mRNAs for stringent response components (RelA, DksA, and SpoT), leading to the discovery of reciprocal circuitry connecting these two regulatory networks (20) (Fig. 5A). Amino acid starvation or other stresses trigger the synthesis of (p)ppGpp by the RelA or SpoT proteins (91, 105). In conjunction with DksA, (p)ppGpp activates transcription of the CsrA antagonists CsrB/C (20, 68). In addition, CsrA represses both the *relA* and *dksA* genes, although the latter effect can be partly masked by DksA autoregulation (20, 105). The positive feedback loop (p)ppGpp → CsrB/C → CsrA ⊣ RelA enhances expression of RelA and perhaps DksA during the stringent response and should allow *relA* expression to be repressed after stress has been eliminated. For the many genes that respond oppositely to CsrA and ppGpp, this circuitry should allow Csr to reinforce the transcriptional effects of (p)ppGpp at the posttranscriptional level (20).

Reciprocal interactions also occur between the Csr and extracytoplasmic function (ECF) systems (23) (Fig. 5B). The *E. coli* ECF system responds to cell envelope stress via the proteolytic inactivation of a cytoplasmic membrane protein, RseA, which acts as an anti-sigma factor

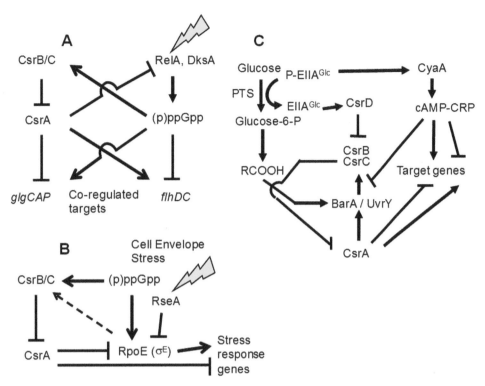

Figure 5 Regulatory interactions of the Csr system with stringent response (A), extracytoplasmic stress (B), and carbon catabolite repression (C) global regulatory systems. Adapted from references 20, 21, and 23 with permission.

of the ECF sigma factor RpoE or σ^E (106). Release of σ^E from its membrane association upon RseA cleavage allows it to access core RNA polymerase and activate transcription of a large set of genes (107, 108). RseA and RpoE mRNAs were identified by copurification with CsrA as well as CLIP-seq analysis (20, 35). Further studies showed that CsrA binds to *rpoE* mRNA at four sites and inhibits *rpoE* translation by directly blocking ribosome access to the translation initiation region of *rpoE* mRNA (23). Reciprocally, Eσ^E activates CsrB/C transcription indirectly (23). The resulting positive feedback loop, RpoE → CsrB/C –| CsrA –| *rpoE*, presumably allows CsrA to reinforce regulation by the ECF system and to assist in restoring basal σ^E levels when membrane damage has been repaired. Full appreciation of the role of Csr in extracytoplasmic stress will require further investigation of CsrA's inhibitory effects on RseA (35). Interestingly, ppGpp activates *rpoE* expression in the stationary phase of growth, independently of envelope stress (109), which should create additional positive feedback between the Csr and ECF systems (Fig. 5B).

Multiple regulatory connections are apparent between CsrA and genes for other carbon metabolism regulators, most of which are not well understood (20, 35). Interactions with the classical catabolite repression system have been investigated, with unexpected results (21) (Fig. 5C). In addition to regulating CsrB/C sRNA decay (described above), the phosphorylation state of EIIA^{Glc} has other vast effects on physiology and metabolism. When glucose is being transported by the PTS system, unphosphorylated EIIA^{Glc} binds to and inhibits transporters of alternative carbon sources, while under carbon limitation P-EIIA^{Glc} activates adenylate cyclase (110). Thus synthesized, cAMP binds to CRP and globally reprograms *E. coli* transcription, the basis of "catabolite repression" (111–113). RNA-seq analysis of transcripts that copurified with CsrA identified the *crp* and *cyaA* transcripts, a finding that was confirmed by *in vitro* binding studies (20, 21). While CsrA exhibited modest, conditional inhibitory effects on *crp* expression, cAMP and CRP exhibited substantial negative effects on transcription of *csrB* and *csrC*. cAMP-Crp repressed *csrC* transcription by directly competing with UvrY-P for binding to the *csrC* promoter region, while it repressed *csrB* transcription indirectly. These findings imply that EIIA^{Glc}-dependent regulatory pathways permit glucose to stimulate both the turnover (97) and synthesis (21) pathways of CsrB/C. In this way, it appears that the presence of glucose has the potential to create a futile cycle that should poise Csr sRNAs for rapid response to changing conditions, while carbon starvation may reduce both arms of this cycle, thereby conserving resources.

CONCLUDING REMARKS

The Csr system is among the best-known RNA-based regulatory systems, with a wide-reaching influence on gene expression and physiology in diverse bacterial species. The structural biology of RNA binding by CsrA is now firmly understood and is contributing to the unraveling of new RNA regulatory mechanisms. Transcriptomics and other systems-based investigations are uncovering the details of its vast influence on bacterial lifestyle decisions, which involve major shifts in physiology. They have also revealed that CsrA binding is not restricted to the 5′ segment of mRNAs, but occurs extensively within mRNA coding regions and is enriched near ribosome pause sites. In addition, CsrA binds not only to sRNAs that appear to serve only to sequester this protein, such as CsrB and CsrC, but also to sRNAs that function in RNA-RNA base-pairing. Understanding the functions of these newly discovered RNA binding activities of CsrA will be crucial for a full appreciation of its regulatory capacity. Regulatory circuitry of the Csr system is well suited for rapid homeostatic responses. Csr interacts in complex ways with other global regulatory systems; in some cases it reinforces transcriptional effects of stress response systems at a posttranscriptional level. While much progress has been made in understanding the regulatory roles of CsrA and its sRNA antagonists, it is clear that much more remains to be learned.

Acknowledgments. Research on the Csr system in the authors' laboratories was supported by the National Institutes of Health: grant number R01GM059969 to P.B. and T.R. and R01AI097116 to T.R.

Citation. Romeo T, Babitzke P. 2018. Global regulation by CsrA and its RNA antagonists. Microbiol Spectrum 6(2):RWR-0009-2017.

References

1. Romeo T, Gong M, Liu MY, Brun-Zinkernagel AM. 1993. Identification and molecular characterization of *csrA*, a pleiotropic gene from *Escherichia coli* that affects glycogen biosynthesis, gluconeogenesis, cell size, and surface properties. *J Bacteriol* 175:4744–4755.

2. Babitzke P, Baker CS, Romeo T. 2009. Regulation of translation initiation by RNA binding proteins. *Annu Rev Microbiol* 63:27–44.

3. Vogel J, Luisi BF. 2011. Hfq and its constellation of RNA. *Nat Rev Microbiol* 9:578–589.

4. Updegrove TB, Zhang A, Storz G. 2016. Hfq: the flexible RNA matchmaker. *Curr Opin Microbiol* 30:133–138.

5. Dubey AK, Baker CS, Suzuki K, Jones AD, Pandit P, Romeo T, Babitzke P. 2003. CsrA regulates translation of the *Escherichia coli* carbon starvation gene, *cstA*, by blocking ribosome access to the *cstA* transcript. *J Bacteriol* 185:4450–4460.

6. Sabnis NA, Yang H, Romeo T. 1995. Pleiotropic regulation of central carbohydrate metabolism in *Escherichia coli* via the gene *csrA*. *J Biol Chem* 270:29096–29104.

7. Chatterjee A, Cui Y, Liu Y, Dumenyo CK, Chatterjee AK. 1995. Inactivation of *rsmA* leads to overproduction of extracellular pectinases, cellulases, and proteases in *Erwinia carotovora* subsp. *Carotovora* in the absence of the starvation/cell density-sensing signal, N-(3-oxohexanoyl)-L-homoserine lactone. *Appl Environ Microbiol* 61:1959–1967.

8. Altier C, Suyemoto M, Lawhon SD. 2000. Regulation of *Salmonella enterica* serovar Typhimurium invasion genes by *csrA*. *Infect Immun* 68:6790–6797.

9. Vakulskas CA, Potts AH, Babitzke P, Ahmer BMM, Romeo T. 2015. Regulation of bacterial virulence by Csr (Rsm) systems. *Microbiol Mol Biol Rev* 79:193–224.

10. Jackson DW, Suzuki K, Oakford L, Simecka JW, Hart ME, Romeo T. 2002. Biofilm formation and dispersal under the influence of the global regulator CsrA of *Escherichia coli*. *J Bacteriol* 184:290–301.

11. Wang X, Dubey AK, Suzuki K, Baker CS, Babitzke P, Romeo T. 2005. CsrA post-transcriptionally represses *pgaABCD*, responsible for synthesis of a biofilm polysaccharide adhesin of *Escherichia coli*. *Mol Microbiol* 56:1648–1663.

12. Jonas K, Edwards AN, Simm R, Romeo T, Römling U, Melefors O. 2008. The RNA binding protein CsrA controls cyclic di-GMP metabolism by directly regulating the expression of GGDEF proteins. *Mol Microbiol* 70:236–257.

13. Jonas K, Edwards AN, Ahmad I, Romeo T, Römling U, Melefors O. 2010. Complex regulatory network encompassing the Csr, c-di-GMP and motility systems of *Salmonella* Typhimurium. *Environ Microbiol* 12:524–540.

14. Pannuri A, Yakhnin H, Vakulskas CA, Edwards AN, Babitzke P, Romeo T. 2012. Translational repression of NhaR, a novel pathway for multi-tier regulation of biofilm circuitry by CsrA. *J Bacteriol* 194:79–89.

15. Yakhnin H, Baker CS, Berezin I, Evangelista MA, Rassin A, Romeo T, Babitzke P. 2011. CsrA represses translation of *sdiA*, which encodes the N-acylhomoserine-L-lactone receptor of *Escherichia coli*, by binding exclusively within the coding region of *sdiA* mRNA. *J Bacteriol* 193:6162–6170.

16. Yang H, Liu MY, Romeo T. 1996. Coordinate genetic regulation of glycogen catabolism and biosynthesis in *Escherichia coli* via the CsrA gene product. *J Bacteriol* 178:1012–1017.

17. Wei B, Shin S, LaPorte D, Wolfe AJ, Romeo T. 2000. Global regulatory mutations in *csrA* and *rpoS* cause

severe central carbon stress in *Escherichia coli* in the presence of acetate. *J Bacteriol* **182**:1632–1640.

18. Wei BL, Brun-Zinkernagel AM, Simecka JW, Prüss BM, Babitzke P, Romeo T. 2001. Positive regulation of motility and *flhDC* expression by the RNA-binding protein CsrA of *Escherichia coli*. *Mol Microbiol* **40**:245–256.

19. Yakhnin AV, Baker CS, Vakulskas CA, Yakhnin H, Berezin I, Romeo T, Babitzke P. 2013. CsrA activates *flhDC* expression by protecting *flhDC* mRNA from Rnase E-mediated cleavage. *Mol Microbiol* **87**:851–866.

20. Edwards AN, Patterson-Fortin LM, Vakulskas CA, Mercante JW, Potrykus K, Vinella D, Camacho MI, Fields JA, Thompson SA, Georgellis D, Cashel M, Babitzke P, Romeo T. 2011. Circuitry linking the Csr and stringent response global regulatory systems. *Mol Microbiol* **80**: 1561–1580.

21. Pannuri A, Vakulskas CA, Zere T, McGibbon LC, Edwards AN, Georgellis D, Babitzke P, Romeo T. 2016. Circuitry linking the catabolite repression and Csr global regulatory systems of *Escherichia coli*. *J Bacteriol* **198**: 3000–3015.

22. Park H, McGibbon LC, Potts AH, Yakhnin H, Romeo T, Babitzke P. 2017. Translational repression of the RpoS antiadapter IraD by CsrA is mediated via translational coupling to a short upstream open reading frame. *MBio* **4**:e01355-17.

23. Yakhnin H, Aichele R, Ades SE, Romeo T, Babitzke P. 2017. Circuitry linking the global Csr and σ^E-dependent cell envelope stress response systems. *J Bacteriol* **199**: e00484-17.

24. Lawhon SD, Frye JG, Suyemoto M, Porwollik S, McClelland M, Altier C. 2003. Global regulation by CsrA in *Salmonella typhimurium*. *Mol Microbiol* **48**:1633–1645.

25. Burrowes E, Baysse C, Adams C, O'Gara F. 2006. Influence of the regulatory protein RsmA on cellular functions in *Pseudomonas aeruginosa* PAO1, as revealed by transcriptome analysis. *Microbiology* **152**:405–418.

26. Brencic A, McFarland KA, McManus HR, Castang S, Mogno I, Dove SL, Lory S. 2009. The GacS/GacA signal transduction system of *Pseudomonas aeruginosa* acts exclusively through its control over the transcription of the RsmY and RsmZ regulatory small RNAs. *Mol Microbiol* **73**:434–445.

27. McKee AE, Rutherford BJ, Chivian DC, Baidoo EK, Juminaga D, Kuo D, Benke PI, Dietrich JA, Ma SM, Arkin AP, Petzold CJ, Adams PD, Keasling JD, Chhabra SR. 2012. Manipulation of the carbon storage regulator system for metabolite remodeling and biofuel production in *Escherichia coli*. *Microb Cell Fact* **11**:79.

28. Tan Y, Liu ZY, Liu Z, Zheng HJ, Li FL. 2015. Comparative transcriptome analysis between *csrA*-disruption *Clostridium acetobutylicum* and its parent strain. *Mol Biosyst* **11**:1434–1442.

29. Holmqvist E, Wright PR, Li L, Bischler T, Barquist L, Reinhardt R, Backofen R, Vogel J. 2016. Global RNA recognition patterns of post-transcriptional regulators Hfq and CsrA revealed by UV crosslinking *in vivo*. *EMBO J* **35**:991–1011.

30. Fields JA, Li J, Gulbronson CJ, Hendrixson DR, Thompson SA. 2016. *Campylobacter jejuni* CsrA regulates metabolic and virulence associated proteins and is necessary for mouse colonization. *PloS One* **11**: e0156932.

31. Dugar G, Svensson SL, Bischler T, Wäldchen S, Reinhardt R, Sauer M, Sharma CM. 2016. The CsrA-FliW network controls polar localization of the dual-function flagellin mRNA in *Campylobacter jejuni*. *Nat Commun* **7**:11667.

32. Morin M, Ropers D, Letisse F, Laguerre S, Portais JC, Cocaign-Bousquet M, Enjalbert B. 2016. The post-transcriptional regulatory system CSR controls the balance of metabolic pools in upper glycolysis of *Escherichia coli*. *Mol Microbiol* **100**:686–700.

33. Sowa SW, Gelderman G, Leistra AN, Buvanendiran A, Lipp S, Pitaktong A, Vakulskas CA, Romeo T, Baldea M, Contreras LM. 2017. Integrative FourD omics approach profiles the target network of the carbon storage regulatory system. *Nucleic Acids Res* **45**:1673–1686.

34. Sahr T, Rusniok C, Impens F, Oliva G, Sismeiro O, Coppée JY, Buchrieser C. 2017. The *Legionella pneumophila* genome evolved to accommodate multiple regulatory mechanisms controlled by the CsrA-system. *PloS Genet* **13**:e1006629.

35. Potts AH, Vakulskas CA, Pannuri A, Yakhnin H, Babitzke P, Romeo T. 2017. Global role of the bacterial post-transcriptional regulator CsrA revealed by integrated transcriptomics. *Nat Commun* **8**:1596.

36. Liu MY, Yang H, Romeo T. 1995. The product of the pleiotropic *Escherichia coli* gene *csrA* modulates glycogen biosynthesis via effects on mRNA stability. *J Bacteriol* **177**:2663–2672.

37. Liu MY, Romeo T. 1997. The global regulator CsrA of *Escherichia coli* is a specific mRNA-binding protein. *J Bacteriol* **179**:4639–4642.

38. Baker CS, Morozov I, Suzuki K, Romeo T, Babitzke P. 2002. CsrA regulates glycogen biosynthesis by preventing translation of *glgC* in *Escherichia coli*. *Mol Microbiol* **44**:1599–1610.

39. Liu MY, Gui G, Wei B, Preston JF III, Oakford L, Yüksel U, Giedroc DP, Romeo T. 1997. The RNA molecule CsrB binds to the global regulatory protein CsrA and antagonizes its activity in *Escherichia coli*. *J Biol Chem* **272**:17502–17510.

40. Weilbacher T, Suzuki K, Dubey AK, Wang X, Gudapaty S, Morozov I, Baker CS, Georgellis D, Babitzke P, Romeo T. 2003. A novel sRNA component of the carbon storage regulatory system of *Escherichia coli*. *Mol Microbiol* **48**: 657–670.

41. Romeo T, Moore J, Smith J. 1991. A simple method for cloning genes involved in glucan biosynthesis: isolation of structural and regulatory genes for glycogen synthesis in *Escherichia coli*. *Gene* **108**:23–29.

42. Dubey AK, Baker CS, Romeo T, Babitzke P. 2005. RNA sequence and secondary structure participate in high-affinity CsrA-RNA interaction. *RNA* **11**:1579–1587.

43. Schulmeyer KH, Diaz MR, Bair TB, Sanders W, Gode CJ, Laederach A, Wolfgang MC, Yahr TL. 2016. Primary and secondary sequence structure requirements

for recognition and discrimination of target RNAs by *Pseudomonas aeruginosa* RsmA and RsmF. *J Bacteriol* **198**:2458–2469.

44. Gutiérrez P, Li Y, Osborne MJ, Pomerantseva E, Liu Q, Gehring K. 2005. Solution structure of the carbon storage regulator protein CsrA from *Escherichia coli*. *J Bacteriol* **187**:3496–3501.

45. Heeb S, Kuehne SA, Bycroft M, Crivii S, Allen MD, Haas D, Cámara M, Williams P. 2006. Functional analysis of the post-transcriptional regulator RsmA reveals a novel RNA-binding site. *J Mol Biol* **355**:1026–1036.

46. Schubert M, Lapouge K, Duss O, Oberstrass FC, Jelesarov I, Haas D, Allain FH-T. 2007. Molecular basis of messenger RNA recognition by the specific bacterial repressing clamp RsmA/CsrA. *Nat Struct Mol Biol* **14**:807–813.

47. Altegoer F, Rensing SA, Bange G. 2016. Structural basis for the CsrA-dependent modulation of translation initiation by an ancient regulatory protein. *Proc Natl Acad Sci U S A* **113**:10168–10173.

48. Mercante J, Suzuki K, Cheng X, Babitzke P, Romeo T. 2006. Comprehensive alanine-scanning mutagenesis of *Escherichia coli* CsrA defines two subdomains of critical functional importance. *J Biol Chem* **281**:31832–31842.

49. Duss O, Michel E, Diarra dit Konté N, Schubert M, Allain FH. 2014. Molecular basis for the wide range of affinity found in Csr/Rsm protein-RNA recognition. *Nucleic Acids Res* **42**:5332–5346.

50. Duss O, Michel E, Yulikov M, Schubert M, Jeschke G, Allain FH. 2014. Structural basis of the non-coding RNA RsmZ acting as a protein sponge. *Nature* **509**:588–592.

51. Mercante J, Edwards AN, Dubey AK, Babitzke P, Romeo T. 2009. Molecular geometry of CsrA (RsmA) binding to RNA and its implications for regulated expression. *J Mol Biol* **392**:511–528.

52. Yakhnin H, Yakhnin AV, Baker CS, Sineva E, Berezin I, Romeo T, Babitzke P. 2011. Complex regulation of the global regulatory gene *csrA*: CsrA-mediated translational repression, transcription from five promoters by Eσ^{70} and Eσ^S, and indirect transcriptional activation by CsrA. *Mol Microbiol* **81**:689–704.

53. Romeo T, Vakulskas CA, Babitzke P. 2013. Post-transcriptional regulation on a global scale: form and function of Csr/Rsm systems. *Environ Microbiol* **15**:313–324.

54. Park H, Yakhnin H, Connolly M, Romeo T, Babitzke P. 2015. CsrA participates in a PNPase autoregulatory mechanism by selectively repressing translation of *pnp* transcripts that have been previously processed by Rnase III and PNPase. *J Bacteriol* **197**:3751–3759.

55. Katsowich N, Elbaz N, Pal RR, Mills E, Kobi S, Kahan T, Rosenshine I. 2017. Host cell attachment elicits post-transcriptional regulation in infecting enteropathogenic bacteria. *Science* **355**:735–739.

56. Yakhnin H, Pandit P, Petty TJ, Baker CS, Romeo T, Babitzke P. 2007. CsrA of *Bacillus subtilis* regulates translation initiation of the gene encoding the flagellin protein (*hag*) by blocking ribosome binding. *Mol Microbiol* **64**:1605–1620.

57. Lapouge K, Sineva E, Lindell M, Starke K, Baker CS, Babitzke P, Haas D. 2007. Mechanism of *hcnA* mRNA recognition in the Gac/Rsm signal transduction pathway of *Pseudomonas fluorescens*. *Mol Microbiol* **66**:341–356.

58. Martínez LC, Yakhnin H, Camacho MI, Georgellis D, Babitzke P, Puente JL, Bustamante VH. 2011. Integration of a complex regulatory cascade involving the SirA/BarA and Csr global regulatory systems that controls expression of the *Salmonella* SPI-1 and SPI-2 virulence regulons through HilD. *Mol Microbiol* **80**:1637–1656.

59. Abbott ZD, Yakhnin H, Babitzke P, Swanson MS. 2015. *csrR*, a paralog and direct target of CsrA, promotes *Legionella pneumophila* resilience in water. *mBio* **6**: e00595.

60. Irie Y, Starkey M, Edwards AN, Wozniak DJ, Romeo T, Parsek MR. 2010. *Pseudomonas aeruginosa* biofilm matrix polysaccharide Psl is regulated transcriptionally by RpoS and post-transcriptionally by RsmA. *Mol Microbiol* **78**:158–172.

61. Baker CS, Eöry LA, Yakhnin H, Mercante J, Romeo T, Babitzke P. 2007. CsrA inhibits translation initiation of *Escherichia coli hfq* by binding to a single site overlapping the Shine-Dalgarno sequence. *J Bacteriol* **189**:5472–5481.

62. Figueroa-Bossi N, Schwartz A, Guillemardet B, D'Heygère F, Bossi L, Boudvillain M. 2014. RNA remodeling by bacterial global regulator CsrA promotes Rho-dependent transcription termination. *Genes Dev* **28**:1239–1251.

63. Goller C, Wang X, Itoh Y, Romeo T. 2006. The cation-responsive protein NhaR of *Escherichia coli* activates *pgaABCD* transcription, required for production of the biofilm adhesin poly-β-1,6-N-acetyl-D-glucosamine. *J Bacteriol* **188**:8022–8032.

64. Steiner S, Lori C, Boehm A, Jenal U. 2013. Allosteric activation of exopolysaccharide synthesis through cyclic di-GMP-stimulated protein-protein interaction. *EMBO J* **32**:354–368.

65. Patterson-Fortin LM, Vakulskas CA, Yakhnin H, Babitzke P, Romeo T. 2013. Dual posttranscriptional regulation via a cofactor-responsive mRNA leader. *J Mol Biol* **425**: 3662–3677.

66. Ren B, Shen H, Lu ZJ, Liu H, Xu Y. 2014. The *phzA2-G2* transcript exhibits direct RsmA-mediated activation in *Pseudomonas aeruginosa* M18. *PloS One* **9**: e89653.

67. Liu Y, Cui Y, Mukherjee A, Chatterjee AK. 1998. Characterization of a novel RNA regulator of *Erwinia carotovora* ssp. *Carotovora* that controls production of extracellular enzymes and secondary metabolites. *Mol Microbiol* **29**:219–234.

68. Zere TR, Vakulskas CA, Leng Y, Pannuri A, Potts AH, Dias R, Tang D, Kolaczkowski B, Georgellis D, Ahmer BM, Romeo T. 2015. Genomic targets and features of BarA-UvrY (-SirA) signal transduction systems. *PloS One* **10**:e0145035.

69. Kay E, Dubuis C, Haas D. 2005. Three small RNAs jointly ensure secondary metabolism and biocontrol in *Pseudomonas fluorescens* CHA0. *Proc Natl Acad Sci U S A* **102**:17136–17141.

70. Lenz DH, Miller MB, Zhu J, Kulkarni RV, Bassler BL. 2005. CsrA and three redundant small RNAs regulate

quorum sensing in *Vibrio cholerae*. *Mol Microbiol* 58: 1186–1202.

71. Kay E, Humair B, Dénervaud V, Riedel K, Spahr S, Eberl L, Valverde C, Haas D. 2006. Two GacA-dependent small RNAs modulate the quorum-sensing response in *Pseudomonas aeruginosa*. *J Bacteriol* 188:6026–6033.

72. Teplitski M, Al-Agely A, Ahmer BM. 2006. Contribution of the SirA regulon to biofilm formation in *Salmonella enterica* serovar Typhimurium. *Microbiology* 152: 3411–3424.

73. Fortune DR, Suyemoto M, Altier C. 2006. Identification of CsrC and characterization of its role in epithelial cell invasion in *Salmonella enterica* serovar Typhimurium. *Infect Immun* 74:331–339.

74. Sterzenbach T, Nguyen KT, Nuccio SP, Winter MG, Vakulskas CA, Clegg S, Romeo T, Bäumler AJ. 2013. A novel CsrA titration mechanism regulates fimbrial gene expression in *Salmonella typhimurium*. *EMBO J* 32: 2872–2883.

75. Jørgensen MG, Thomason MK, Havelund J, Valentin-Hansen P, Storz G. 2013. Dual function of the McaS small RNA in controlling biofilm formation. *Genes Dev* 27:1132–1145.

76. Parker A, Cureoglu S, De Lay N, Majdalani N, Gottesman S. 2017. Alternative pathways for *Escherichia coli* biofilm formation revealed by sRNA overproduction. *Mol Microbiol* 105:309–325.

77. Itoh Y, Rice JD, Goller C, Pannuri A, Taylor J, Meisner J, Beveridge TJ, Preston JF III, Romeo T. 2008. Roles of *pgaABCD* genes in synthesis, modification, and export of the *Escherichia coli* biofilm adhesin poly-β-1,6-N-acetyl-D-glucosamine. *J Bacteriol* 190:3670–3680.

78. Mukherjee S, Yakhnin H, Kysela D, Sokoloski J, Babitzke P, Kearns DB. 2011. CsrA-FliW interaction governs flagellin homeostasis and a checkpoint on flagellar morphogenesis in *Bacillus subtilis*. *Mol Microbiol* 82:447–461.

79. Mukherjee S, Oshiro RT, Yakhnin H, Babitzke P, Kearns DB. 2016. FliW antagonizes CsrA RNA binding by a noncompetitive allosteric mechanism. *Proc Natl Acad Sci U S A* 113:9870–9875.

80. Bhatt S, Edwards AN, Nguyen HT, Merlin D, Romeo T, Kalman D. 2009. The RNA binding protein CsrA is a pleiotropic regulator of the locus of enterocyte effacement pathogenicity island of enteropathogenic *Escherichia coli*. *Infect Immun* 77:3552–3568.

81. Pernestig AK, Melefors O, Georgellis D. 2001. Identification of UvrY as the cognate response regulator for the BarA sensor kinase in *Escherichia coli*. *J Biol Chem* 276: 225–231.

82. Suzuki K, Wang X, Weilbacher T, Pernestig AK, Melefors O, Georgellis D, Babitzke P, Romeo T. 2002. Regulatory circuitry of the CsrA/CsrB and BarA/UvrY systems of *Escherichia coli*. *J Bacteriol* 184:5130–5140.

83. Heroven AK, Sest M, Pisano F, Scheb-Wetzel M, Steinmann R, Böhme K, Klein J, Münch R, Schomburg D, Dersch P. 2012. Crp induces switching of the CsrB and CsrC RNAs in *Yersinia pseudotuberculosis* and links nutritional status to virulence. *Front Cell Infect Microbiol* 2:158.

84. Tomenius H, Pernestig AK, Méndez-Catalá CF, Georgellis D, Normark S, Melefors O. 2005. Genetic and functional characterization of the *Escherichia coli* BarA-UvrY two-component system: point mutations in the HAMP linker of the BarA sensor give a dominant-negative phenotype. *J Bacteriol* 187:7317–7324.

85. Chavez RG, Alvarez AF, Romeo T, Georgellis D. 2010. The physiological stimulus for the BarA sensor kinase. *J Bacteriol* 192:2009–2012.

86. Lawhon SD, Maurer R, Suyemoto M, Altier C. 2002. Intestinal short-chain fatty acids alter *Salmonella typhimurium* invasion gene expression and virulence through BarA/SirA. *Mol Microbiol* 46:1451–1464.

87. Takeuchi K, Kiefer P, Reimmann C, Keel C, Dubuis C, Rolli J, Vorholt JA, Haas D. 2009. Small RNA-dependent expression of secondary metabolism is controlled by Krebs cycle function in *Pseudomonas fluorescens*. *J Biol Chem* 284:34976–34985.

88. Septer AN, Bose JL, Lipzen A, Martin J, Whistler C, Stabb EV. 2015. Bright luminescence of *Vibrio fischeri* aconitase mutants reveals a connection between citrate and the Gac/Csr regulatory system. *Mol Microbiol* 95:283–296.

89. Vakulskas CA, Pannuri A, Cortés-Selva D, Zere TR, Ahmer BM, Babitzke P, Romeo T. 2014. Global effects of the DEAD-box RNA helicase DeaD (CsdA) on gene expression over a broad range of temperatures. *Mol Microbiol* 92:945–958.

90. Camacho MI, Alvarez AF, Chavez RG, Romeo T, Merino E, Georgellis D. 2015. Effects of the global regulator CsrA on the BarA/UvrY two-component signaling system. *J Bacteriol* 197:983–991.

91. Potrykus K, Cashel M. 2008. (p)ppGpp: still magical? *Annu Rev Microbiol* 62:35–51.

92. Ross W, Sanchez-Vazquez P, Chen AY, Lee JH, Burgos HL, Gourse RL. 2016. ppGpp binding to a site at the RNAP-DksA interface accounts for its dramatic effects on transcription initiation during the stringent response. *Mol Cell* 62:811–823.

93. Chambonnier G, Roux L, Redelberger D, Fadel F, Filloux A, Sivaneson M, de Bentzmann S, Bordi C. 2016. The hybrid histidine kinase LadS forms a multicomponent signal transduction system with the GacS/GacA two-component system in *Pseudomonas aeruginosa*. *PloS Genet* 12:e1006032.

94. Suzuki K, Babitzke P, Kushner SR, Romeo T. 2006. Identification of a novel regulatory protein (CsrD) that targets the global regulatory RNAs CsrB and CsrC for degradation by Rnase E. *Genes Dev* 20:2605–2617.

95. Gudapaty S, Suzuki K, Wang X, Babitzke P, Romeo T. 2001. Regulatory interactions of Csr components: the RNA binding protein CsrA activates *csrB* transcription in *Escherichia coli*. *J Bacteriol* 183:6017–6027.

96. Vakulskas CA, Leng Y, Abe H, Amaki T, Okayama A, Babitzke P, Suzuki K, Romeo T. 2016. Antagonistic control of the turnover pathway for the global regulatory sRNA CsrB by the CsrA and CsrD proteins. *Nucleic Acids Res* 44:7896–7910.

97. Leng Y, Vakulskas CA, Zere TR, Pickering BS, Watnick PI, Babitzke P, Romeo T. 2016. Regulation of CsrB/C

sRNA decay by EIIA(^{Glc}) of the phosphoenolpyruvate: carbohydrate phosphotransferase system. *Mol Microbiol* **99**: 627–639.

98. **Valverde C, Lindell M, Wagner EG, Haas D.** 2004. A repeated GGA motif is critical for the activity and stability of the riboregulator RsmY of *Pseudomonas fluorescens*. *J Biol Chem* **279**:25066–25074.

99. **Alon U.** 2007. Network motifs: theory and experimental approaches. *Nat Rev Genet* **8**:450–461.

100. **Beisel CL, Storz G.** 2010. Base pairing small RNAs and their roles in global regulatory networks. *FEMS Microbiol Rev* **34**:866–882.

101. **Adamson DN, Lim HN.** 2013. Rapid and robust signaling in the CsrA cascade via RNA-protein interactions and feedback regulation. *Proc Natl Acad Sci U S A* **110**: 13120–13125.

102. **Heroven AK, Böhme K, Dersch P.** 2012. The Csr/Rsm system of *Yersinia* and related pathogens: a post-transcriptional strategy for managing virulence. *RNA Biol* **9**:379–391.

103. **Svenningsen SL, Tu KC, Bassler BL.** 2009. Gene dosage compensation calibrates four regulatory RNAs to control *Vibrio cholerae* quorum sensing. *EMBO J* **28**:429–439.

104. **Cui Y, Madi L, Mukherjee A, Dumenyo CK, Chatterjee AK.** 1996. The RsmA$^-$ mutants of *Erwinia carotovora* subsp. *Carotovora* strain Ecc71 overexpress *hrpN$_{Ecc}$* and elicit a hypersensitive reaction-like response in tobacco leaves. *Mol Plant Microbe Interact* **9**:565–573.

105. **Chandrangsu P, Lemke JJ, Gourse RL.** 2011. The *dksA* promoter is negatively feedback regulated by DksA and ppGpp. *Mol Microbiol* **80**:1337–1348.

106. **Hayden JD, Ades SE.** 2008. The extracytoplasmic stress factor, σ^E, is required to maintain cell envelope integrity in *Escherichia coli*. *PLoS One* **3**:e1573.

107. **Rhodius VA, Suh WC, Nonaka G, West J, Gross CA.** 2006. Conserved and variable functions of the σ^E stress response in related genomes. *PLoS Biol* **4**:e2.

108. **Shimada T, Tanaka K, Ishihama A.** 2017. The whole set of the constitutive promoters recognized by four minor sigma subunits of *Escherichia coli* RNA polymerase. *PLoS One* **12**:e0179181.

109. **Gopalkrishnan S, Nicoloff H, Ades SE.** 2014. Coordinated regulation of the extracytoplasmic stress factor, sigmaE, with other *Escherichia coli* sigma factors by (p)ppGpp and DksA may be achieved by specific regulation of individual holoenzymes. *Mol Microbiol* **93**:479–493.

110. **Deutscher J, Aké FM, Derkaoui M, Zébré AC, Cao TN, Bouraoui H, Kentache T, Mokhtari A, Milohanic E, Joyet P.** 2014. The bacterial phosphoenolpyruvate: carbohydrate phosphotransferase system: regulation by protein phosphorylation and phosphorylation-dependent protein-protein interactions. *Microbiol Mol Biol Rev* **78**:231–256.

111. **Shimada T, Fujita N, Yamamoto K, Ishihama A.** 2011. Novel roles of cAMP receptor protein (CRP) in regulation of transport and metabolism of carbon sources. *PLoS One* **6**:e20081.

112. **Lee DJ, Busby SJ.** 2012. Repression by cyclic AMP receptor protein at a distance. *mBio* **3**:e00289-12.

113. **You C, Okano H, Hui S, Zhang Z, Kim M, Gunderson CW, Wang YP, Lenz P, Yan D, Hwa T.** 2013. Coordination of bacterial proteome with metabolism by cyclic AMP signaling. *Nature* **500**:301–306.

Regulating with RNA in Bacteria and Archaea
Edited by Gisela Storz and Kai Papenfort
© 2018 American Society for Microbiology, Washington, DC
doi:10.1128/microbiolspec.RWR-0019-2018

6S RNA, a Global Regulator of Transcription

20

Karen M. Wassarman[1]

INTRODUCTION

It is now well established that small RNAs (sRNAs) have diverse and widespread roles in regulating gene expression in all organisms (1–4). Mechanisms of action are varied but can be broadly classified into three categories: (i) sRNAs that act by base-pairing to target RNAs; (ii) sRNAs that act to modulate protein activity through direct RNA-protein interaction; and (iii) sRNAs that have intrinsic function (e.g., catalytic). Two well-studied sRNA families that modulate protein activity include sRNAs that regulate CsrA protein and 6S RNA, which regulates RNA polymerase (RNAP) and is the focus here as well as in several other reviews (5–8). 6S RNA was first identified in *Escherichia coli* (9, 10), which remains the best-understood model of 6S RNA function, although identification of 6S RNAs and their roles in diverse bacterial species has been an active area of research in the past decade. Here, information for *E. coli* will be presented first, followed by discussion of similarities and differences known or postulated for 6S RNAs in diverse species.

E. coli 6S RNA (*Ec*6S RNA) was first discovered as a highly abundant, stable RNA more than 50 years ago (9), but it was not until the discovery that *Ec*6S RNA formed a complex with RNAP more than 30 years later that its biological function and mechanism of action began to be revealed (11). Bacterial RNAP is a multisubunit enzyme, consisting of a core (E: $\alpha\alpha\beta\beta'$) that is transcriptionally competent but requires the addition of a specificity factor (sigma: σ) to form the holoenzyme (Eσ) needed to recognize promoters and initiate transcription (12, 13). All bacteria contain a housekeeping σ, usually referred to as σ^{70} or σ^{A}, which is highly abundant and required at all stages of growth. Different bacterial species have varied numbers of alternative

sigma factors, ranging from zero to several dozen. *Ec*6S RNA interacts specifically and very tightly with the housekeeping holoenzyme form of RNAP (Eσ^{70} in *E. coli*) (11, 14). Early studies demonstrated that *Ec*6S RNA binding to Eσ^{70} resulted in downregulation of σ^{70}-dependent transcription at a tested promoter (*rsd*P2) and that cells lacking *Ec*6S RNA have altered survival phenotypes (11, 15, 16). These findings led to the suggestion that the physiological role of 6S RNA, at least in *E. coli*, is to contribute to regulation of gene expression in response to poor nutrient environments. Research over the past decade has brought considerable insight into 6S RNA function but also raised many questions for future work. This review focuses on several active areas of inquiry including the identification of 6S RNAs in diverse bacterial species; the details of 6S RNA-RNAP interactions; the identification of promoters regulated by 6S RNA; the physiological consequences of 6S RNA-dependent regulation; and finally the discovery and impact of product RNA (pRNA) synthesis, a process in which 6S RNA serves as a template for synthesis of a pRNA. The biogenesis of *Ec*6S RNA has also been an area of interest and active research but will not be discussed here. For information about the biogenesis of 6S RNA, see references 5 and 17–24.

6S RNAs ARE WIDESPREAD THROUGHOUT THE BACTERIAL KINGDOM

Identification of 6S RNA and Candidate 6S RNAs

During the past 15 years, much work on 6S RNA has focused on its interaction with Eσ^{70} and its presence and potential impact in diverse bacteria. Biochemical

---

[1]Department of Bacteriology, University of Wisconsin-Madison, Madison, WI 53562.

and phylogenetic approaches demonstrated that overall secondary structure was critical for 6S RNA-Eσ^{70} interactions (14, 25). Specifically, 6S RNA is primarily double stranded with minor disruptions of bulged nucleotides in addition to a large, central, single-stranded region (Fig. 1). It is this secondary structure, rather than sequence identity, that is required for Ec6S RNA to bind Eσ^{70} and has been the basis for identification of most 6S RNAs and 6S RNA-encoding genes in diverse genomic sequences. Alternative approaches to identify 6S RNAs have included identification of sRNAs that co-immunoprecipitate with RNAP or through identification of 6S RNA-associated pRNA in RNA-sequencing data. Altogether, 6S RNAs have been identified or predicted in a wide range of bacteria (26), making it one of the very highly conserved sRNAs along with transfer-messenger RNA (tmRNA), signal recognition particle RNA, and the Csr/Rsm family of RNAs (27).

Identification of additional 6S RNAs is an ongoing process as more genome sequences become available and as transcriptome sequencing (RNA-seq) approaches to examine RNAs globally are applied to more organisms. However, identification of 6S RNAs globally remains nontrivial, as the parameters defining the structural requirements for interaction with RNAP are not as fixed as structural requirements of some other sRNAs (e.g., tmRNA interaction with ribosomes and alanyl-tRNA synthetases [28]), and recent work suggests that there may be more variation in specific contacts in divergent bacteria than previously predicted. Nevertheless, identification and characterization of diverse 6S RNAs continues to provide important information about 6S RNA and its cellular role generally and in specific species.

Some Species Have Multiple 6S RNAs

Some genomes encode more than one 6S RNA, most notably *Bacillus subtilis*, where the two 6S RNAs (*Bs*6S-1 and *Bs*6S-2 RNAs) (14, 25) have been studied in some detail. *Legionella pneumophila* also has been reported to express two 6S RNAs, *Lp*6S RNA and *Lp*6S-2 RNA, and two candidate 6S RNA-encoding genes were identified in *Hydrogenivirga* (29–32). The presence of two 6S RNAs in divergent bacteria raises interesting questions regarding how common it is to have multiple 6S RNAs, how they are independently regulated, if their activities are redundant or overlapping, and so forth. In *B. subtilis* and *L. pneumophila*, the two 6S RNAs act independently of each other; have different accumulation profiles; and, at least for *B. subtilis*, are known to regulate different genes (14, 25, 33–35; A. T. Cavanagh and K. M. Wassarman, unpublished data).

6S RNA Accumulation Profiles— Insight into Function?

One of the first steps in characterization of many of the newly identified 6S RNAs has been to look at their expression profiles, especially during growth phase or under specialized growth conditions. Ec6S RNA is present at all times of growth in *E. coli*, but is at lower levels in early exponential phase (<1,000 copies per cell) and gradually accumulates during growth until it reaches maximal levels (~10,000 copies per cell) several hours after transition into stationary phase (11). Thus, much work on Ec6S RNA has focused on its role in stationary phase, where it contributes to cell survival, although it should be noted that Ec6S RNA is present in exponential phase, is bound to Eσ^{70}, and can regulate transcription at this time (14, 36, 37). In contrast, many 6S RNAs in divergent species have different expression profiles, which may provide insight into predictions of different physiological roles for these RNAs.

*Bs*6S-1 RNA is an example of a 6S RNA that accumulates in stationary phase with expression profiles similar to Ec6S RNA, and *Lp*6S RNA also accumulates postexponentially (29, 38). In contrast, *Bs*6S-2 RNA levels change only modestly throughout growth (<2- to 3-fold), with maximal levels observed in late exponential phase, although the precise expression profile reported for *Bs*6S-2 RNA has varied between different studies, likely due to differences in strains and growth media and timing of "stationary" phase examined (14, 25, 39). The *Lp*6S-2 RNA expression profile is more complex but is not similar to Ec6S RNA (31). The differences in accumulation of *Bs*6S-1 and *Bs*6S-2 RNAs or *Lp*6S and *Lp*6S-2 RNAs suggest that they likely have different physiological roles, in agreement with observations that gene expression and proteomic profiles for cells lacking *Bs*6S-1 and *Bs*6S-2 RNAs are different (35; Cavanagh and Wassarman, unpublished), and observed mutant phenotypes (i.e., altered timing of sporulation and cell density changes in stationary phase) are associated specifically with the loss of *Bs*6S-1 RNA (33, 35). *Streptomyces coelicolor* 6S RNA also accumulates with a profile similar to Ec6S RNA and influences growth rate (40).

Other 6S RNAs accumulate with different profiles that hint at interesting cellular roles. For example, within identified *Cyanobacteria* 6S RNAs there is an array of different accumulation patterns. Levels of *Prochlorococcus* MED4 6S RNA are cell cycle dependent and change with light, suggesting that 6S RNA may contribute to the high light adaptation of this strain (41). 6S RNA in *Synechocystis* sp. strain PCC 6803 has also been suggested to contribute to light stress as well

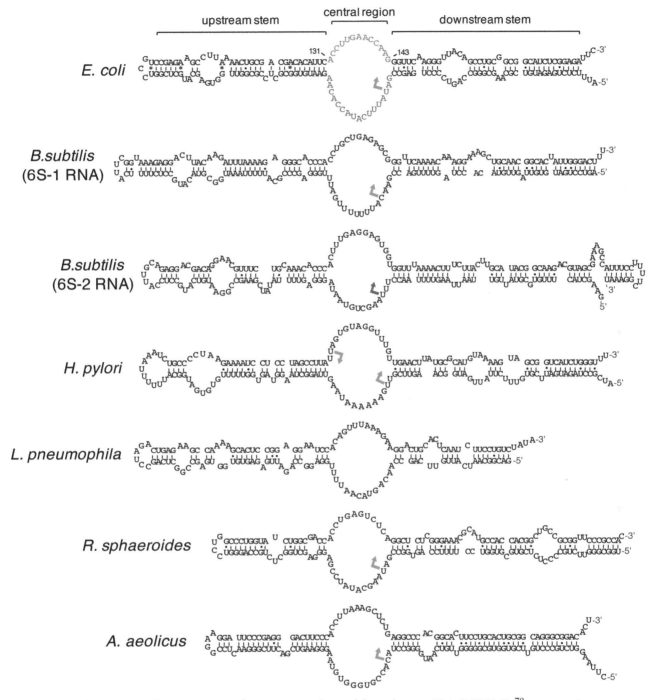

Figure 1 *E. coli* 6S RNA is shown in a secondary structure observed from the cryo-EM 6S RNA-Eσ[70] structure (upstream stem and central region) or predicted from secondary-structure analysis (downstream stem) (14, 61). For reference, the upstream stem, central region, and downstream stem are indicated. The template central region is in green; the site of initiation for pRNA synthesis is indicated by a red arrow; the nontemplate central region is in blue. 6S RNAs from *B. subtilis* (*Bs*6S-1 and *Bs*6S-2), *H. pylori*, *L. pneumophila* (*Lp*6S RNA), *R. sphaeroides*, and *A. aeolicus* are shown in secondary structures expected when in complex with RNAP, with sites of pRNA synthesis initiation indicated by arrows (14, 25, 30, 48, 57, 70, 71). pRNA synthesis has not been examined for *Lp*6S RNA. Note, some 6S RNAs have been demonstrated or predicted to have base pairing within the central region in isolated RNA (*Bs*6S-1 and *Bs*6S-2 RNA, *Hp*6S RNA) (25, 57, 73, 79), but binding studies support that the central region is fully single stranded when bound to RNAP (14, 34, 63, 72). Other alternative structures that contribute to 6S RNA release from RNAP during pRNA synthesis also have been observed (73, 79–81).

as recovery from nitrogen depletion (6, 42). Alternatively, 6S RNA in *Synechococcus* sp. strain PCC 6301 changes during growth, suggesting a response to nutrient status, although in this case 6S RNA is abundant in exponential phase and reduced in stationary phase, perhaps more similar to *Bs*6S-2 RNA (43).

Examples of several alphaproteobacteria that associate with host cells exhibit differential expression dependent on host association. For example, the plant symbiont *Bradyrhizobium japonicum* 6S RNA is higher in root nodules compared to free-living cells (44, 45). *Wolbachia* 6S RNA levels change with host identity (i.e., 6S RNA levels were higher in germ line cells compared to somatic cells) (45, 46). It has been suggested that the *Wolbachia* 6S RNA accumulation increases during fast replication, in contrast to *Ec*6S and *Bs*6S-1 RNAs, which accumulate during slow growth (i.e., stationary phase). *Rhodobacter sphaeroides* and *Caulobacter crescentus* (free-living alphaproteobacteria) 6S RNA accumulation patterns change with cell growth, although the change between exponential- and stationary-phase levels is rather modest (~3-fold) (47, 48; Cavanagh and Wassarman, unpublished). *R. sphaeroides* 6S RNA has been associated with high-salt-stress survival (47).

Several 6S RNAs from pathogenic bacteria or close relatives have been shown to increase under stress, suggesting a potential role in pathogenesis or in host survival. Examples include *Burkholderia cenocepacia* 6S RNA, which increases during oxidative stress (49); *Yersinia pestis* 6S RNA, which has altered levels during lung infection (50); *Rickettsia* 6S RNA, which accumulates many hours postinfection and correlates with intracellular growth kinetics (51); *Coxiella burnetii* 6S RNA, which accumulates in the stress-resistant cellular form (small cell variant; SCV), suggesting a role in stress (52); *Salmonella enterica* serovar Typhimurium 6S RNA, which accumulates at low pH, resulting in altered invasion and stress survival (53); *Clostridium acetobutylicum* 6S RNA, which increases in response to general stress and is reported to promote butanol tolerance (54); and *Borrelia burgdorferi* 6S RNA, which accumulates in ticks with timing suggesting a role in persistence (S. Samuels, L. Hall, and D. Drecktrah, personal communication).

Additional 6S RNAs have been observed to be expressed under at least one condition (e.g., *Bordetella pertussis* [14], *C. crescentus* [55], *Clostridium difficile* [56], *Helicobacter pylori* [57], *Pseudomonas aeruginosa* [58], and *Rhodopseudomonas palustris* [45]), and many others have been predicted from genomic sequences with minimal additional information about expression patterns or function yet available. In fact, 6S RNA candidates have been predicted in the majority of bacterial genomes; see reference 26 for a recent update on the distribution of 6S RNAs throughout bacteria.

Many questions remain about 6S RNAs from diverse bacteria, and future work is anticipated to focus on providing further understanding of their physiological roles in different biological circumstances. In addition, there is the potential that some of these 6S RNAs and candidate 6S RNAs may expand on known mechanistic activities.

How To Define Divergent 6S RNA Candidates as 6S RNAs?

Of particular importance as these divergent 6S RNAs are further studied and as additional 6S RNAs candidates are identified is the question of whether these candidates are all true 6S RNAs and what defines a 6S RNA. We have previously suggested that 6S RNAs should be defined as RNAs that bind their cognate primary holoenzyme form of RNAP in a manner resembling promoter DNA binding, a definition likely to capture a class of RNAs that are mechanistically similar and thereby providing a functionally useful definition (5). It has been suggested to include "directing pRNA synthesis" in the definition (35), but it is not included here, as pRNA synthesis is not required for 6S RNA to regulate transcription (34) and there may be examples where 6S RNAs rely on other mechanisms to cycle off of RNAP but retain similar mechanisms of regulation of transcription. Nevertheless, either definition requires detailed information about RNAP binding, a characteristic that is harder to test, especially in divergent systems where tools are not readily available. In some tested cases, however, there has been quite a large variation in the fraction of 6S RNA that is bound to RNAP (>75% for *Ec*6S RNA, *Bs*6S-1 RNA, and *Bs*6S-2 RNA, compared to <10% for *Lp*6S RNA) (11, 14, 29), which may be due to technical reasons but could represent critical differences in 6S RNA function. As more details about 6S RNA-RNAP interactions in diverse bacteria become available, it will be important to revisit the definition of 6S RNAs, in particular whether there are additional classes of sRNAs that interact with RNAP to function using different mechanisms.

At least one RNA identified based on its structural similarity to 6S RNA (*Ms1* RNA in *Mycobacterium smegmatis*) does not bind to the holoenzyme form of RNAP, although it does interact with core RNAP (59, 60). This example serves as a cautionary tale in relying too heavily on secondary structure as the predictor of 6S RNA candidates, but also raises the intriguing possibility that there are additional classes of sRNAs that function through interaction with RNAP using different mechanisms. Future work must focus on characterization of 6S RNA candidates beyond secondary-structure predictions for definitive 6S RNA identification.

THE 6S RNA-RNAP COMPLEX

*Ec*6S RNA-Eσ^{70} Interactions

The past decade has revealed much about the nature of the 6S RNA-RNAP interaction, including a cryo-electron microscopic (cryo-EM) structure of the *E. coli* 6S RNA-Eσ^{70} complex in the past year (61). As noted above, it has been appreciated for some time that the secondary structure is the key element required for binding, and that there is minimal if any sequence specificity to the interaction (14, 25, 62). It was demonstrated that the central bubble of *Ec*6S RNA resides near the active site, similar to promoter DNA during transcription initiation when the DNA surrounding the start site of transcription is melted to form the "open complex" (63). Consistent with this hypothesis, the *Ec*6S RNA can be used as a template by Eσ^{70} to generate pRNA in a process called pRNA synthesis (63). More about pRNA synthesis below, but the ability to use 6S RNA as a template for pRNA synthesis *in vitro* strongly supported similar binding of 6S RNA and promoter DNA to Eσ^{70} globally. Site-directed cross-link mapping (61) and Fe-BABE [iron(S)-1-(*p*-bromoacetamidobenzyl) EDTA] cleavage mapping (64) further supported a strong correlation between the path of RNA and DNA when bound to Eσ^{70}. The cryo-EM structure provided direct visualization of the 6S RNA-Eσ^{70} complex and its overall similarity to open complexes (61). However, in spite of the overall similarity of architecture, closer inspection also revealed regions where *Ec*6S RNA and promoter DNA binding were quite different.

Template strand central region

One area of great interest is the template strand of the central region in 6S RNA and how similarly it is positioned in RNAP to the open complex DNA template bubble (65), as this is the region where RNA synthesis initiates during both pRNA synthesis and transcription. Perhaps unsurprisingly, the paths of RNA and DNA are quite similar here, as demonstrated by cross-linking to the same residues and directly visualized in the cryo-EM *Ec*6S RNA-Eσ^{70} structure compared to open complex structures, although the most upstream single-stranded region of 6S RNA was not as well resolved in the *Ec*6S RNA-Eσ^{70} structure as the rest of the RNA (61).

Nontemplate central region

The cryo-EM structure also provided information about the path of the nontemplate central region of the 6S RNA. Although most of the nontemplate single-stranded region was similar between *Ec*6S RNA and open complex DNA, one noted difference is that A131 in *Ec*6S RNA, the position equivalent to −11A in promoter DNA, was not flipped and thus did not make interactions with the σ^{70} pocket (66). 6S RNA is a premelted bubble and thus is not expected to require this interaction to initiate and maintain melting, in contrast to open complex formation, where flipping of −11A to interact with σ^{70} is thought to initiate melting and recognition of the −10 element during open complex formation (67). Interestingly, the position equivalent to A131 in *Ec*6S RNA is highly conserved in identified and predicted 6S RNAs, raising questions about whether there is another role for this nucleotide or if there are conditions when it does interact with σ^{70} region 2. However, A131 is not required for *Ec*6S RNA activity, as mutation of this residue has no detectable effect on the kinetics of binding to Eσ^{70} nor on the efficiency of pRNA synthesis (61).

Other contacts between the nontemplate bubble region of *Ec*6S RNA (e.g., U135, G136, and G143) and Eσ^{70} were similar to contacts between DNA and Eσ^{70} in the open complex (61). Intriguingly, however, these contacts likewise are not required for 6S RNA binding to Eσ^{70} nor for efficient pRNA synthesis initiation (61), raising questions about whether they play a role in other aspects of 6S RNA function/activity. Sequences in this region do contribute to timing of pRNA synthesis-mediated release of *Ec*6S RNA from Eσ^{70} through a releasing structure.

Upstream stem interactions with σ^{70} region 4.2

One area where *Ec*6S RNA interactions are quite different than promoter DNA open complexes is the nucleic acid interaction with region 4.2 of σ^{70} (Fig. 2). The fact that this interaction was distinct between DNA and RNA was first revealed by differential binding of Eσ^{70} variants with alanine substitutions in region 4.2 of σ^{70}, which suggested a larger binding surface for 6S RNA (68). The cryo-EM structure provides a high-resolution view of the *Ec*6S RNA interaction with Eσ^{70} in this region and correlates well with the biochemical analysis (61). The cryo-EM structure also demonstrates an unusual structure for the *Ec*6S RNA in this region, an area where secondary-structure mapping and mutagenesis had been largely uninformative (Cavanagh and Wassarman, unpublished), consistent with the fact that

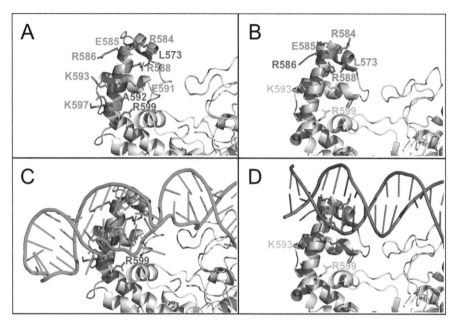

Figure 2 Comparison of 6S RNA and promoter DNA interactions with region 4.2 of σ^{70}. Holoenzyme structures complexed with RNA (PDB ID: 5VT0) (61) (A and C) and promoter DNA (PDB ID: 5VI5) (83) (B and D) are centered on region 4.2 of σ^{70} and shown without (A and B) or with (C and D) the nucleic acids visible. The cryo-EM structure with 6S RNA is *E. coli* holoenzyme, and the crystal structure with DNA in an open complex is *M. smegmatis* holoenzyme, and thus there are small variations in structure due to sequence changes, although region 4.2 of σ^{70} is very highly conserved. Residues within region 4.2 of σ^{70} are labeled (*E. coli* numbering) with color coding based on impact of alanine substitution on binding to RNA (A) or DNA (B) (68): red, strong decrease; blue, moderate decrease; green, increase. A592 (dark gray) is a position where substitution of a positive or negative residue strongly influences 6S RNA binding with little effect on DNA binding (68). Residues labeled in light gray (B and D) are locations where alanine substitution did not influence DNA binding, but are included to assist with comparison of the two structures. Other coloring: 6S RNA, green; promoter DNA, purple; β subunit, cyan; β′ subunit, pink; σ^{70} (outside of region 4.2), light orange.

these techniques are best at predicting canonical structures. The importance of this region of *Ec*6S RNA for binding to Eσ^{70} was also highlighted by random mutagenesis studies (62).

Aspects of 6S RNA structure allow it to adopt an overall architecture that follows the B-form DNA helix of the open complex

Of particular interest from an RNA structure perspective is the region of RNA between the central bubble and the contacts with region 4.2 of σ^{70}, which follows the path and overall architecture of double-stranded, B-form helix DNA rather closely (61). Double-stranded RNA is typically an A-form helix, but the 6S RNA uses a combination of short A-form helices that have some B-form characteristics, in addition to gaps and bulged nucleotides, to adopt this unusual structural mimic of double-stranded DNA.

6S RNA-RNAP Interactions in Diverse Bacteria

Although understanding of the *Ec*6S RNA-Eσ^{70} interaction has increased substantially over the past decade and has provided much insight into how 6S RNAs function mechanistically, detailed biochemical analysis of 6S RNA-RNAP interactions in diverse bacteria remains largely unexplored. In many organisms, 6S RNA-RNAP complexes have been detected by coimmunoprecipitation or *in vitro* analysis (e.g., *B. subtilis* [14, 34, 69], *L. pneumophila* [29], *C. burnetii* [52], *Aquifex aeolicus* [70], and *S. coelicolor* [40]), supporting a general similarity to *Ec*6S RNA. However, detailed understanding of interactions in other organisms is likely to require biochemical or structural analysis to uncover any potential differences from *Ec*6S RNA. Directed *in vitro* studies have been done for *B. subtilis* 6S RNAs, revealing many interesting aspects of these RNAs, much of which is focused on pRNA synthesis initiation and release. Most studies to date have assumed that key features of Ec6S RNA-Eσ^{70} interactions are conserved; however, some observations suggest that this assumption may be premature. Some 6S RNAs, notably from alphaproteobacteria and *A. aeolicus*, have a shortened upstream stem and therefore are lacking the region critical for Ec6S RNA interactions with σ^{70} region 4.2 (25, 68, 71). However, pRNAs have been detected for *A. aeolicus* and *R. sphaeroides* 6S RNAs, strongly supporting an RNAP interaction, and the Aa6S RNA has been demonstrated to bind *B. subtilis* Eσ^{A} *in vitro* (47, 70). The *H. pylori* 6S RNA appears to bind RNAP in either the "forward" or "reverse" orientation based on evidence of pRNAs templated from both strands of the central region (57), in contrast to Ec6S RNA, which binds quite specifically in one orientation and directs pRNA from one strand of the central region (63). Orientation of Ec6S RNA binding to Eσ^{70} has been proposed to be directed through upstream stem contacts with σ^{70} region 4.2; thus, lack of orientation also suggests a change in the requirement or nature of interactions with σ^{70} region 4.2. Additionally, 6S RNA-dependent changes in transcription appear to be different in several tested organisms compared to *E. coli*, perhaps suggesting key or substantial differences in binding that impact regulation of transcription. Likewise, until the potential differences and similarities between divergent 6S RNAs are understood, caution is advised for studies using noncognate 6S RNA-RNAP matches for *in vitro* studies, as species-specific behaviors of RNAP in 6S RNA function have been observed (72).

6S RNA REGULATION OF TRANSCRIPTION

Regulation of Transcription in *E. coli*

The binding of 6S RNA to Eσ^{70}, the major transcriptional machinery of the cell, strongly suggested that 6S RNA would regulate σ^{70}-dependent transcription, as initially confirmed at one tested promoter (*rsd*P2) *in vivo* (11). However, somewhat surprisingly at the time, further experiments revealed that 6S RNA-dependent regulation of transcription is promoter specific (i.e., some promoters are downregulated in the presence of 6S RNA, while others are insensitive to 6S RNA), even during late stationary phase when Ec6S RNA levels are maximal and Eσ^{70} is essentially saturated by Ec6S RNA (11, 15). Work in the last decade on 6S RNA-dependent regulation of transcription has taken two approaches: (i) study of reporter genes with minimal core promoters and mutants to reveal promoter features that

determine 6S RNA sensitivity and (ii) identification of the 6S RNA regulon using global approaches.

Identification of promoter features that determine regulation by Ec6S RNA

The similarity of binding of Ec6S RNA and promoter DNA to Eσ^{70} strongly suggested a direct competition model for regulation of transcription, and *in vitro* binding studies demonstrated that Ec6S RNA blocks binding of promoter DNA (63), as also observed in *B. subtilis* for both Bs6S-1 and Bs6S-2 RNAs (73). However, a set of promoters reported to respond to changes in RNAP concentrations (74) were not uniformly sensitive to Ec6S RNA, nor did 6S RNA sensitivity correlate with affinity for RNAP binding (37). Furthermore, mutagenesis of several studied promoters revealed that strength of the −35 element or the presence of an extended −10 element specifically determined 6S RNA sensitivity, and that changing these parameters alone could interconvert sensitive and insensitive promoters (37). Strength of the core −10 element did not contribute to 6S RNA sensitivity, in conflict with a model of direct competition of 6S RNA and promoter DNA for free Eσ^{70}. Future work will be required to provide a better understanding of the mechanistic details of 6S RNA regulation of transcription in *E. coli* and other bacteria.

Often, *in vitro* transcription assays are used to provide mechanistic insight. However, studies examining Ec6S RNA regulation of transcription *in vitro*, with purified components or in cell lysate, demonstrated a lack of correlation between *in vitro* and *in vivo* observations, which also was observed for Bs6S-1 and Bs6S-2 RNAs. Specifically, *in vitro*, all tested promoters were strongly inhibited in the presence of Ec6S RNA or Bs6S-1 or Bs6S-2 RNAs (14, 73, 75). However, *in vivo*, some promoters were modestly downregulated (~2- to 5-fold) in the presence of Ec6S RNA, while others were insensitive to Ec6S RNA (15, 38, 76); similar results were observed in *B. subtilis* (35; Cavanagh and Wassarman, unpublished). *In vitro* assays in the *E. coli* system also exhibited a strong order-of-addition effect, and both Ec6S RNA and promoter DNA bind to Eσ^{70} very tightly, which led to the hypothesis that standard *in vitro* transcription assays do not represent dynamic exchange that must be occurring *in vivo*. Whether the differences between *in vivo* and *in vitro* observations are due solely to differences in dynamics or whether other factors influence Ec6S RNA regulation *in vivo* that are lacking or inactive in purified and lysate systems remains to be determined.

Identification of the 6S RNA regulon in *E. coli*

Global expression studies revealed that many hundreds of mRNA levels were altered in cells lacking *Ec*6S RNA compared to wild-type cells in exponential phase, early stationary phase, and late stationary phase (37, 76). Global studies identify both those genes altered through direct 6S RNA action as well as those altered through secondary effects. Nevertheless, 6S RNA-dependent changes in late-stationary-phase genes containing mapped promoters correlated fairly well with the promoter features determined through reporter analysis (i.e., weak −35 element or extended −10 element for sensitive promoters) (37). Changes earlier in growth did not correlate well with these identified promoter features (76), which may be due to a decreased response when 6S RNA levels are not maximal, or potentially due to more-secondary effects, as several targets of *Ec*6S RNA regulation are regulators of transcription (e.g., Crp) or generate molecules that regulate transcription (e.g., RelA, a ppGpp synthetase) at these times (36, 37, 76). Additionally, both global studies identified genes that were increased in the presence of 6S RNA, which are likely to be secondary effects and may include 6S RNA influences on alternative σ factor utilization.

Biological role for 6S RNA function

Of note is that both direct and secondary effects of 6S RNA action are relevant for understanding the physiological consequences of regulation, as opposed to mechanistic understanding of 6S RNA function, for which only direct effects are relevant. Phenotypes associated with lack of *Ec*6S RNA are subtle, in part contributing to the long delay between identification of 6S RNA and the first reports of its function in regulating RNAP (9, 11). However, *Ec*6S RNA has been shown to contribute to competitive survival in stationary phase (time scale of days) as well as survival in long-term stationary phase when not in competitive growth (time scale of weeks) (15). Additionally, 6S RNA-dependent changes in transcription of one target gene, *pspF*, have been shown to alter survival at high pH in stationary phase (16), suggesting a connection between cell survival and stress response as a primary role for 6S RNA. Results from the large-scale gene expression studies similarly suggest that 6S RNA is integrated into global pathways including regulation of factors that impact transcription (e.g. Crp, FNR, and ppGpp via RelA) and general translation machinery (36, 37, 76). Detailed understanding of which gene changes are important for various phenotypes awaits future work.

6S RNA Regulation of Transcription in Diverse Bacteria

Global gene expression or proteomic studies to address the role of 6S RNA in transcription have been done in other species (e.g., *B. subtilis* [35; Cavanagh and Wassarman, unpublished], *L. pneumophila* [29], and *Synechocystis* 6803 [42]). Intriguingly, there have been some differences in observations from these studies and the *E. coli* studies (37, 76). Of note, there were many fewer mRNAs with altered levels (e.g., 135 for *L. pneumophila*, fewer in *B. subtilis* and *Synechocystis*) than observed in *E. coli* (>800). Whether this large difference represents a true difference in regulatory impact, a difference in timing of analysis, a difference between maximal 6S RNA levels and maximal binding to RNAP, or a potential difference in direct versus secondary effects remains to be tested. The other key difference was a preponderance of mRNAs that were higher in wild-type cells compared to cells lacking 6S RNA (e.g., 127 out of 135 in *L. pneumophila*), in contrast to *E. coli*, where most were decreased in wild-type compared to mutant cells, consistent with an inhibitory role for 6S RNA. However, in all cases both increased and decreased mRNAs were observed as expected when examining both direct and secondary effects. It remains an open question whether the mechanism of regulation of transcription is common for all of these reported 6S RNAs, although in all cases to date, the changes in gene expression observed suggest a biological role for 6S RNA in response to environmental conditions.

6S RNA—A TEMPLATE FOR pRNA SYNTHESIS

Perhaps one of the most exciting discoveries about *Ec*6S RNA in the past 15 years was that it not only structurally resembled open complex DNA but could be used as a template by RNAP to generate pRNA (63, 75), a process referred to as pRNA synthesis. When bound to 6S RNA, the DNA-dependent RNAP is converted to a specialized RNA-dependent RNAP. Detection of pRNA synthesis *in vitro* provided strong evidence for the similarity of *Ec*6S RNA and open complex DNA interactions with $E\sigma^{70}$. However, it was the determination that pRNA synthesis occurs *in vivo*, along with subsequent work, that demonstrated that pRNA synthesis is one mechanism to regulate *Ec*6S RNA by contributing to the off-rate of 6S RNA from RNAP (34, 63, 77, 78). For bacteria beyond *E. coli* and *B. subtilis*, the specific role of pRNA synthesis has not been investigated directly, but pRNAs have been detected *in vivo* for additional bacterial species (e.g., *H. pylori* [57], *R. sphaeroides* [47], and *A. aeolicus*

[70]), suggesting that pRNA synthesis from 6S RNAs is widespread and potentially ubiquitous.

Mechanism of pRNA Synthesis

Detailed biochemical experiments have examined mechanistic aspects of pRNA synthesis, primarily for *E. coli* and *B. subtilis* 6S RNAs (34, 63, 69, 72, 73, 75, 77, 79–81). Initiation of pRNA synthesis resembles transcription initiation in many ways, including the potential to generate abortive initiation intermediates and release of σ^{70} prior to completion of RNA synthesis. However, unlike transcription, the 6S RNA-RNAP complex is not able to transition into a stable elongation complex, but instead dissociates. In fact, the 6S RNA and pRNA are released as a hybrid (6S RNA-pRNA). This released 6S RNA-pRNA hybrid is unable to re-bind RNAP, thereby providing a release mechanism that cannot be reversed without further action. The fact that both regulation of transcription by 6S RNA and regulation of 6S RNA by pRNA synthesis are mediated by the positioning of 6S RNA within the active site of RNAP provides an elegantly simple mechanism to respond to environmental signals both positively and negatively.

Features of *E. coli* and *B. subtilis* 6S RNAs that contribute to mechanisms of release of 6S RNA-pRNA hybrids from RNAP also have been addressed biochemically (73, 79–81). As pRNA synthesis proceeds, the RNA central bubble is extended, which reveals additional single-stranded RNA sequence that participates in an alternative structure. This structure change facilitates the release of 6S RNA from RNAP as a 6S RNA-pRNA duplex. The details of the alternative structures are not the same for *E. coli* and *B. subtilis* 6S RNAs, but the consequences remain the same: destabilization of the interaction with RNAP, resulting in release and contributing to the length of pRNA generated and the timing of release. *A. aeolicus* 6S RNA forms a release structure similar to *Bs*6S-1 RNA (70), suggesting conservation of this release mechanism, although more information about diverse 6S RNAs is needed before strong conclusions can be made.

Differences in pRNA synthesis have been observed between *E. coli* and *B. subtilis*, most notably in preference for initiating nucleotide identity (72). Although *E. coli* Eσ^{70} has a preference to initiate pRNA synthesis with a purine, as is also observed generally for transcription, this enzyme will initiate readily with any nucleotide in both pRNA synthesis and transcription. In contrast, *B. subtilis* Eσ^{A} demonstrates a much stronger preference for initiating pRNA synthesis with GTP over ATP, UTP, or CTP, although *B. subtilis* contains pro-

moters that direct initiation with any of the four nucleotides, suggesting that this trend does not extend to transcription. *Bs*6S-1 RNA directs pRNA synthesis initiation with GTP, while *Bs*6S-2 RNA directs initiation with ATP, leading to a large difference in efficiency of pRNA synthesis, at least as measured *in vitro* (72). A similar preference for initiation with GTP by *B. subtilis* Eσ^{A} was observed *in vitro* and *in vivo* on *E. coli* 6S RNA, which directs initiation with ATP, compared to a mutant *E. coli* 6S RNA, 6S (iGTP) RNA, that directs initiation with GTP (72). Thus, the difference in preference for initiating nucleotide in pRNA synthesis between *E. coli* and *B. subtilis* originates from differences in RNAP, although the mechanistic details await further study. The impact of the observed difference in pRNA efficiency *in vitro* on *in vivo* function and regulation of *Bs*6S-1 and *Bs*6S-2 RNAs is likely to continue to be an area of active research (34, 82).

Biological Role for pRNA Synthesis

A burst of pRNA synthesis from *Ec*6S RNA and *Bs*6S-1 RNA occurs *in vivo* within minutes of diluting stationary-phase cells into fresh medium (outgrowth) (34, 63, 69, 77), suggesting a role for this process in the transition out of stationary phase to allow restart of growth. The released 6S RNA remains base-paired to the pRNA, thus preventing rebinding to Eσ^{70} or Eσ^{A} due to the altered structure (34, 63, 73, 77, 79). The majority (>90%) of *Ec*6S and *Bs*6S-1 RNAs are degraded during outgrowth, presumably as a consequence of release from RNAP, suggesting that pRNA synthesis may play an important role in determining *Ec*6S RNA and *Bs*6S-1 RNA accumulation profiles. *Ec*6S- and *Bs*6S-1-directed pRNA levels were not detected in late stationary phase in one study (34), supporting a hypothesis that pRNA synthesis initiation is sensitive to nucleotide triphosphate levels, linking pRNA synthesis timing to outgrowth. The influence of nucleotide levels on pRNA synthesis efficiency has been noted *in vitro* (34, 63, 73, 77, 79). However, other studies have detected pRNA from *Bs*6S-1 RNA (and *Bs*6S-2 RNA) in both exponential and earlier stationary phase (82), although timing in stationary phase, strains, and growth conditions varied between different studies. Changes in pRNA length at different stages of growth also have been observed (69), suggesting the potential for additional levels of regulation of pRNA synthesis, such as in pRNA synthesis elongation and release rates in response to nucleotide levels. Certainly further study of these different observations and the potential connections between pRNA synthesis, 6S RNA stability and accumulation profiles, and the impact of pRNA length distribution need to be addressed experimentally in all organisms. More quantitative information about pRNA synthesis efficiency throughout growth is needed, and an examination of whether additional cellular signals or factors regulate pRNA synthesis is required before the contribution of pRNA synthesis in regulating 6S RNA levels throughout growth is fully understood.

The impact of pRNA synthesis on the ability of cells to restart growth upon exit from stationary phase (i.e., outgrowth) was investigated in *E. coli* using mutant *Ec6S* RNAs that retain the ability to bind $E\sigma^{70}$ but lack the ability to be released from $E\sigma^{70}$ through pRNA synthesis (34, 78). In one study, it was found that cells expressing a mutant 6S RNA were delayed in restarting growth, suggesting that pRNA synthesis-mediated release of $E\sigma^{70}$ from *Ec6S* RNA is required for efficient restart of growth after stationary phase (34). Toxic effects of expressing this type of mutant RNA also were observed, although the extent of toxicity was dependent on expression levels (34, 78). However, the nonreleasing 6S RNA mutant regulated transcription similarly to wild-type 6S RNA in stationary phase, both in specificity of promoters sensitive and insensitive to 6S RNA and the extent of regulation (34). Therefore, it has been suggested that pRNA synthesis is one mechanism to regulate 6S RNA accumulation profiles and to facilitate concerted release of $E\sigma^{70}$ during outgrowth but does not otherwise influence mechanisms of 6S RNA action in regulating transcription. Although it is enticing to question whether pRNA itself has a function as an sRNA in addition to the role of its synthesis in regulating 6S RNA levels, currently there is no evidence to support an independent function, and no mutant phenotypes have been revealed in cells expressing mutant *Ec6S* RNAs that direct synthesis of pRNAs with different sequences (Cavanagh and Wassarman, unpublished). In *B. subtilis*, a requirement for pRNA synthesis-mediated release of $E\sigma^{A}$ to promote efficient outgrowth has also been demonstrated (34).

6S RNA—FUTURE QUESTIONS

Many questions about 6S RNA function in *E. coli* and to a lesser extent in *B. subtilis* have been addressed at some level. However, even in these well-studied organisms questions remain, including how specific changes in gene expression contribute to phenotypes associated with loss of 6S RNA function and what specific mechanisms mediate promoter-specific regulation of transcription. Even more questions await investigation in other bacteria and include the following. (i) How many genes are regulated by 6S RNA, and by what mecha-

nism(s)? (ii) How are diverse 6S RNA structures recognized by their cognate RNAPs, and which interactions are conserved or species specific? (iii) What are the details underpinning the relationship between accumulation profile and physiological impact? (iv) What other factors contribute to regulation of 6S RNA activity beyond pRNA synthesis? (v) Are there broader impacts of pRNA synthesis beyond the studied release mechanism?

Acknowledgments. *I thank A. T. Cavanagh for helpful discussions. Work in the Wassarman lab is supported by the National Institutes of Health (GM67955).*

Citation. Wassarman KM. 2018. 6S RNA, a global regulator of transcription. *Microbiol Spectrum* 6(3):RWR-0019-2018.

References

1. **Storz G, Vogel J, Wassarman KM.** 2011. Regulation by small RNAs in bacteria: expanding frontiers. *Mol Cell* **43:**880–891.

2. **Svensson SL, Sharma CM.** 2016. Small RNAs in bacterial virulence and communication, p 169–212. *In* Kudva IT, Cornick NA, Plummer PJ, Zhang Q, Nicholson TL, Bannantine JP, Bellaire BH (ed), *Virulence Mechanisms of Bacterial Pathogens*, 5th ed. ASM Press, Washington, DC.

3. **Holmqvist E, Wagner EG.** 2017. Impact of bacterial sRNAs in stress responses. *Biochem Soc Trans* **45:**1203–1212.

4. **Hess WR, Berghoff BA, Wilde A, Steglich C, Klug G.** 2014. Riboregulators and the role of Hfq in photosynthetic bacteria. *RNA Biol* **11:**413–426.

5. **Cavanagh AT, Wassarman KM.** 2014. 6S RNA, a global regulator of transcription in *Escherichia coli, Bacillus subtilis*, and beyond. *Annu Rev Microbiol* **68:**45–60.

6. **Rediger A, Geissen R, Steuten B, Heilmann B, Wagner R, Axmann IM.** 2012. 6S RNA—an old issue became blue-green. *Microbiology* **158:**2480–2491.

7. **Burenina OY, Elkina DA, Hartmann RK, Oretskaya TS, Kubareva EA.** 2015. Small noncoding 6S RNAs of bacteria. *Biochemistry (Mosc)* **80:**1429–1446.

8. **Steuten B, Hoch PG, Damm K, Schneider S, Köhler K, Wagner R, Hartmann RK.** 2014. Regulation of transcription by 6S RNAs: insights from the *Escherichia coli* and *Bacillus subtilis* model systems. *RNA Biol* **11:**508–521.

9. **Hindley J.** 1967. Fractionation of ^{32}P-labelled ribonucleic acids on polyacrylamide gels and their characterization by fingerprinting. *J Mol Biol* **30:**125–136.

10. **Brownlee GG.** 1971. Sequence of 6S RNA of *E. coli. Nat New Biol* **229:**147–149.

11. **Wassarman KM, Storz G.** 2000. 6S RNA regulates *E. coli* RNA polymerase activity. *Cell* **101:**613–623.

12. **Feklístov A, Sharon BD, Darst SA, Gross CA.** 2014. Bacterial sigma factors: a historical, structural, and genomic perspective. *Annu Rev Microbiol* **68:**357–376.

13. **Browning DF, Busby SJ.** 2016. Local and global regulation of transcription initiation in bacteria. *Nat Rev Microbiol* **14:**638–650.

14. Trotochaud AE, Wassarman KM. 2005. A highly conserved 6S RNA structure is required for regulation of transcription. *Nat Struct Mol Biol* **12**:313–319.

15. Trotochaud AE, Wassarman KM. 2004. 6S RNA function enhances long-term cell survival. *J Bacteriol* **186**:4978–4985.

16. Trotochaud AE, Wassarman KM. 2006. 6S RNA regulation of *pspF* transcription leads to altered cell survival at high pH. *J Bacteriol* **188**:3936–3943.

17. Kim KS, Lee Y. 2004. Regulation of 6S RNA biogenesis by switching utilization of both sigma factors and endoribonucleases. *Nucleic Acids Res* **32**:6057–6068.

18. Chae H, Han K, Kim KS, Park H, Lee J, Lee Y. 2011. Rho-dependent termination of *ssrS* (6S RNA) transcription in *Escherichia coli*: implication for 3′ processing of 6S RNA and expression of downstream *ygfA* (putative 5-formyl-tetrahydrofolate cyclo-ligase). *J Biol Chem* **286**: 114–122.

19. Lee JY, Park H, Bak G, Kim KS, Lee Y. 2013. Regulation of transcription from two *ssrS* promoters in 6S RNA biogenesis. *Mol Cells* **36**:227–234.

20. Hsu LM, Zagorski J, Wang Z, Fournier MJ. 1985. *Escherichia coli* 6S RNA gene is part of a dual-function transcription unit. *J Bacteriol* **161**:1162–1170.

21. Neusser T, Gildehaus N, Wurm R, Wagner R. 2008. Studies on the expression of 6S RNA from *E. coli*: involvement of regulators important for stress and growth adaptation. *Biol Chem* **389**:285–297.

22. Chen H, Dutta T, Deutscher MP. 2016. Growth phase-dependent variation of RNase BN/Z affects small RNAs. *J Biol Chem* **291**:26435–26442.

23. Fadouloglou VE, Lin HT, Tria G, Hernández H, Robinson CV, Svergun DI, Luisi BF. 2015. Maturation of 6S regulatory RNA to a highly elongated structure. *FEBS J* **282**: 4548–4564.

24. Li Z, Pandit S, Deutscher MP. 1998. 3′ exoribonucleolytic trimming is a common feature of the maturation of small, stable RNAs in *Escherichia coli*. *Proc Natl Acad Sci U S A* **95**:2856–2861.

25. Barrick JE, Sudarsan N, Weinberg Z, Ruzzo WL, Breaker RR. 2005. 6S RNA is a widespread regulator of eubacterial RNA polymerase that resembles an open promoter. *RNA* **11**:774–784.

26. Wehner S, Damm K, Hartmann RK, Marz M. 2014. Dissemination of 6S RNA among bacteria. *RNA Biol* **11**: 1467–1478.

27. Gottesman S, Storz G. 2011. Bacterial small RNA regulators: versatile roles and rapidly evolving variations. *Cold Spring Harb Perspect Biol* **3**:a003798.

28. Himeno H, Kurita D, Muto A. 2014. Mechanism of *trans*-translation revealed by *in vitro* studies. *Front Microbiol* **5**:65.

29. Faucher SP, Friedlander G, Livny J, Margalit H, Shuman HA. 2010. *Legionella pneumophila* 6S RNA optimizes intracellular multiplication. *Proc Natl Acad Sci U S A* **107**:7533–7538.

30. Faucher SP, Shuman HA. 2011. Small regulatory RNA and *Legionella pneumophila*. *Front Microbiol* **2**:98.

31. Weissenmayer BA, Prendergast JG, Lohan AJ, Loftus BJ. 2011. Sequencing illustrates the transcriptional response of *Legionella pneumophila* during infection and identifies seventy novel small non-coding RNAs. *PLoS One* **6**: e17570.

32. Lechner M, Nickel AI, Wehner S, Riege K, Wieseke N, Beckmann BM, Hartmann RK, Marz M. 2014. Genome-wide comparison and novel ncRNAs of *Aquificales*. *BMC Genomics* **15**:522.

33. Cavanagh AT, Wassarman KM. 2013. 6S-1 RNA function leads to a delay in sporulation in *Bacillus subtilis*. *J Bacteriol* **195**:2079–2086.

34. Cavanagh AT, Sperger JM, Wassarman KM. 2012. Regulation of 6S RNA by pRNA synthesis is required for efficient recovery from stationary phase in *E. coli* and *B. subtilis*. *Nucleic Acids Res* **40**:2234–2246.

35. Hoch PG, Burenina OY, Weber MH, Elkina DA, Nesterchuk MV, Sergiev PV, Hartmann RK, Kubareva EA. 2015. Phenotypic characterization and complementation analysis of *Bacillus subtilis* 6S RNA single and double deletion mutants. *Biochimie* **117**:87–99.

36. Cavanagh AT, Chandrangsu P, Wassarman KM. 2010. 6S RNA regulation of *relA* alters ppGpp levels in early stationary phase. *Microbiology* **156**:3791–3800.

37. Cavanagh AT, Klocko AD, Liu X, Wassarman KM. 2008. Promoter specificity for 6S RNA regulation of transcription is determined by core promoter sequences and competition for region 4.2 of σ^{70}. *Mol Microbiol* **67**:1242–1256.

38. Suzuma S, Asari S, Bunai K, Yoshino K, Ando Y, Kakeshita H, Fujita M, Nakamura K, Yamane K. 2002. Identification and characterization of novel small RNAs in the *aspS-yrvM* intergenic region of the *Bacillus subtilis* genome. *Microbiology* **148**:2591–2598.

39. Ando Y, Asari S, Suzuma S, Yamane K, Nakamura K. 2002. Expression of a small RNA, BS203 RNA, from the *yocI-yocJ* intergenic region of *Bacillus subtilis* genome. *FEMS Microbiol Lett* **207**:29–33.

40. Mikulík K, Bobek J, Zídková J, Felsberg J. 2014. 6S RNA modulates growth and antibiotic production in *Streptomyces coelicolor*. *Appl Microbiol Biotechnol* **98**:7185–7197.

41. Axmann IM, Holtzendorff J, Voss B, Kensche P, Hess WR. 2007. Two distinct types of 6S RNA in *Prochlorococcus*. *Gene* **406**:69–78.

42. Heilmann B, Hakkila K, Georg J, Tyystjärvi T, Hess WR, Axmann IM, Dienst D. 2017. 6S RNA plays a role in recovery from nitrogen depletion in *Synechocystis* sp. PCC 6803. *BMC Microbiol* **17**:229.

43. Watanabe T, Sugiura M, Sugita M. 1997. A novel small stable RNA, 6Sa RNA, from the cyanobacterium *Synechococcus* sp. strain PCC6301. *FEBS Lett* **416**:302–306.

44. Voss B, Hölscher M, Baumgarth B, Kalbfleisch A, Kaya C, Hess WR, Becker A, Evguenieva-Hackenberg E. 2009. Expression of small RNAs in *Rhizobiales* and protection of a small RNA and its degradation products by Hfq in *Sinorhizobium meliloti*. *Biochem Biophys Res Commun* **390**:331–336.

45. Madhugiri R, Pessi G, Voss B, Hahn J, Sharma CM, Reinhardt R, Vogel J, Hess WR, Fischer HM, Evguenieva-

Hackenberg E. 2012. Small RNAs of the *Bradyrhizobium/Rhodopseudomonas* lineage and their analysis. *RNA Biol* 9:47–58.

46. Darby AC, Armstrong SD, Bah GS, Kaur G, Hughes MA, Kay SM, Koldkjær P, Rainbow L, Radford AD, Blaxter ML, Tanya VN, Trees AJ, Cordaux R, Wastling JM, Makepeace BL. 2012. Analysis of gene expression from the *Wolbachia* genome of a filarial nematode supports both metabolic and defensive roles within the symbiosis. *Genome Res* 22:2467–2477.

47. Elkina D, Weber L, Lechner M, Burenina O, Weisert A, Kubareva E, Hartmann RK, Klug G. 2017. 6S RNA in *Rhodobacter sphaeroides*: 6S RNA and pRNA transcript levels peak in late exponential phase and gene deletion causes a high salt stress phenotype. *RNA Biol* 14:1627–1637.

48. Berghoff BA, Glaeser J, Sharma CM, Vogel J, Klug G. 2009. Photooxidative stress-induced and abundant small RNAs in *Rhodobacter sphaeroides*. *Mol Microbiol* 74:1497–1512.

49. Ghosh S, Dureja C, Khatri I, Subramanian S, Raychaudhuri S, Ghosh S. 2017. Identification of novel small RNAs in *Burkholderia cenocepacia* KC-01 expressed under iron limitation and oxidative stress conditions. *Microbiology* 163:1924–1936.

50. Yan Y, Su S, Meng X, Ji X, Qu Y, Liu Z, Wang X, Cui Y, Deng Z, Zhou D, Jiang W, Yang R, Han Y. 2013. Determination of sRNA expressions by RNA-seq in *Yersinia pestis* grown *in vitro* and during infection. *PLoS One* 8:e74495.

51. Schroeder CL, Narra HP, Sahni A, Rojas M, Khanipov K, Patel J, Shah R, Fofanov Y, Sahni SK. 2016. Identification and characterization of novel small RNAs in *Rickettsia prowazekii*. *Front Microbiol* 7:859.

52. Warrier I, Hicks LD, Battisti JM, Raghavan R, Minnick MF. 2014. Identification of novel small RNAs and characterization of the 6S RNA of *Coxiella burnetii*. *PLoS One* 9:e100147.

53. Ren J, Sang Y, Qin R, Cui Z, Yao YF. 2017. 6S RNA is involved in acid resistance and invasion of epithelial cells in *Salmonella enterica* serovar Typhimurium. *Future Microbiol* 12:1045–1057.

54. Jones AJ, Venkataramanan KP, Papoutsakis T. 2016. Overexpression of two stress-responsive, small, non-coding RNAs, 6S and tmRNA, imparts butanol tolerance in *Clostridium acetobutylicum*. *FEMS Microbiol Lett* 363:fnw063.

55. Landt SG, Abeliuk E, McGrath PT, Lesley JA, McAdams HH, Shapiro L. 2008. Small non-coding RNAs in *Caulobacter crescentus*. *Mol Microbiol* 68:600–614.

56. Soutourina OA, Monot M, Boudry P, Saujet L, Pichon C, Sismeiro O, Semenova E, Severinov K, Le Bouguenec C, Coppée JY, Dupuy B, Martin-Verstraete I. 2013. Genome-wide identification of regulatory RNAs in the human pathogen *Clostridium difficile*. *PLoS Genet* 9:e1003493.

57. Sharma CM, Hoffmann S, Darfeuille F, Reignier J, Findeiss S, Sittka A, Chabas S, Reiche K, Hackermüller J, Reinhardt R, Stadler PF, Vogel J. 2010. The primary transcriptome of the major human pathogen *Helicobacter pylori*. *Nature* 464:250–255.

58. Vogel DW, Hartmann RK, Struck JC, Ulbrich N, Erdmann VA. 1987. The sequence of the 6S RNA gene of *Pseudomonas aeruginosa*. *Nucleic Acids Res* 15:4583–4591.

59. Pánek J, Krásny L, Bobek J, Jezková E, Korelusová J, Vohradský J. 2011. The suboptimal structures find the optimal RNAs: homology search for bacterial non-coding RNAs using suboptimal RNA structures. *Nucleic Acids Res* 39:3418–3426.

60. Hnilicová J, Jirát Matějčková J, Šiková M, Pospíšil J, Halada P, Pánek J, Krásny L. 2014. Ms1, a novel sRNA interacting with the RNA polymerase core in mycobacteria. *Nucleic Acids Res* 42:11763–11776.

61. Chen J, Wassarman KM, Feng S, Leon K, Feklistov A, Winkelman JT, Li Z, Walz T, Campbell EA, Darst SA. 2017. 6S RNA mimics B-form DNA to regulate *Escherichia coli* RNA polymerase. *Mol Cell* 68:388–397.e6.

62. Shephard L, Dobson N, Unrau PJ. 2010. Binding and release of the 6S transcriptional control RNA. *RNA* 16:885–892.

63. Wassarman KM, Saecker RM. 2006. Synthesis-mediated release of a small RNA inhibitor of RNA polymerase. *Science* 314:1601–1603.

64. Steuten B, Setny P, Zacharias M, Wagner R. 2013. Mapping the spatial neighborhood of the regulatory 6S RNA bound to *Escherichia coli* RNA polymerase holoenzyme. *J Mol Biol* 425:3649–3661.

65. Bae B, Feklistov A, Lass-Napiorkowska A, Landick R, Darst SA. 2015. Structure of a bacterial RNA polymerase holoenzyme open promoter complex. *eLife* 4:e08504.

66. Feklistov A, Darst SA. 2011. Structural basis for promoter-10 element recognition by the bacterial RNA polymerase σ subunit. *Cell* 147:1257–1269.

67. Saecker RM, Record MT Jr, Dehaseth PL. 2011. Mechanism of bacterial transcription initiation: RNA polymerase - promoter binding, isomerization to initiation-competent open complexes, and initiation of RNA synthesis. *J Mol Biol* 412:754–771.

68. Klocko AD, Wassarman KM. 2009. 6S RNA binding to Eσ⁷⁰ requires a positively charged surface of σ⁷⁰ region 4.2. *Mol Microbiol* 73:152–164.

69. Beckmann BM, Burenina OY, Hoch PG, Kubareva EA, Sharma CM, Hartmann RK. 2011. In vivo and in vitro analysis of 6S RNA-templated short transcripts in *Bacillus subtilis*. *RNA Biol* 8:839–849.

70. Köhler K, Duchardt-Ferner E, Lechner M, Damm K, Hoch PG, Salas M, Hartmann RK. 2015. Structural and mechanistic characterization of 6S RNA from the hyperthermophilic bacterium *Aquifex aeolicus*. *Biochimie* 117:72–86.

71. Willkomm DK, Minnerup J, Hüttenhofer A, Hartmann RK. 2005. Experimental RNomics in *Aquifex aeolicus*: identification of small non-coding RNAs and the putative 6S RNA homolog. *Nucleic Acids Res* 33:1949–1960.

72. Cabrera-Ostertag IJ, Cavanagh AT, Wassarman KM. 2013. Initiating nucleotide identity determines efficiency of RNA synthesis from 6S RNA templates in *Bacillus subtilis* but not *Escherichia coli*. *Nucleic Acids Res* 41:7501–7511.

73. **Burenina OY, Hoch PG, Damm K, Salas M, Zatsepin TS, Lechner M, Oretskaya TS, Kubareva EA, Hartmann RK.** 2014. Mechanistic comparison of *Bacillus subtilis* 6S-1 and 6S-2 RNAs—commonalities and differences. *RNA* **20**:348–359.

74. **Barker MM, Gaal T, Gourse RL.** 2001. Mechanism of regulation of transcription initiation by ppGpp. II. Models for positive control based on properties of RNAP mutants and competition for RNAP. *J Mol Biol* **305**:689–702.

75. **Gildehaus N, Neusser T, Wurm R, Wagner R.** 2007. Studies on the function of the riboregulator 6S RNA from *E. coli*: RNA polymerase binding, inhibition of *in vitro* transcription and synthesis of RNA-directed *de novo* transcripts. *Nucleic Acids Res* **35**:1885–1896.

76. **Neusser T, Polen T, Geissen R, Wagner R.** 2010. Depletion of the non-coding regulatory 6S RNA in *E. coli* causes a surprising reduction in the expression of the translation machinery. *BMC Genomics* **11**:165.

77. **Wurm R, Neusser T, Wagner R.** 2010. 6S RNA-dependent inhibition of RNA polymerase is released by RNA-dependent synthesis of small *de novo* products. *Biol Chem* **391**:187–196.

78. **Oviedo Ovando M, Shephard L, Unrau PJ.** 2014. In vitro characterization of 6S RNA release-defective mutants uncovers features of pRNA-dependent release from RNA polymerase in *E. coli*. *RNA* **20**:670–680.

79. **Beckmann BM, Hoch PG, Marz M, Willkomm DK, Salas M, Hartmann RK.** 2012. A pRNA-induced structural rearrangement triggers 6S-1 RNA release from RNA polymerase in *Bacillus subtilis*. *EMBO J* **31**:1727–1738.

80. **Steuten B, Wagner R.** 2012. A conformational switch is responsible for the reversal of the 6S RNA-dependent RNA polymerase inhibition in *Escherichia coli*. *Biol Chem* **393**:1513–1522.

81. **Panchapakesan SS, Unrau PJ.** 2012. *E. coli* 6S RNA release from RNA polymerase requires σ^{70} ejection by scrunching and is orchestrated by a conserved RNA hairpin. *RNA* **18**:2251–2259.

82. **Hoch PG, Schlereth J, Lechner M, Hartmann RK.** 2016. *Bacillus subtilis* 6S-2 RNA serves as a template for short transcripts in vivo. *RNA* **22**:614–622.

83. **Hubin EA, Lilic M, Darst SA, Campbell EA.** 2017. Structural insights into the mycobacteria transcription initiation complex from analysis of X-ray crystal structures. *Nat Commun* **8**:16072.

Regulating with RNA in Bacteria and Archaea
Edited by Gisela Storz and Kai Papenfort
© 2018 American Society for Microbiology, Washington, DC
doi:10.1128/microbiolspec.RWR-0023-2018

Bacterial Y RNAs: Gates, Tethers, and tRNA Mimics

21

Soyeong Sim[1] and Sandra L. Wolin[1]

INTRODUCTION

In contrast to most bacterial noncoding RNAs (ncRNAs) (1, 2), Y RNAs were initially characterized in human cells and only later shown to exist in bacteria. The human RNAs were discovered because they are found complexed with the Ro 60-kDa autoantigen (Ro60), a ring-shaped protein that is a clinically important target of autoantibodies in patients with two systemic autoimmune rheumatic diseases, systemic lupus erythematosus and Sjögren's syndrome (3, 4). Y RNAs and their Ro60 protein partner were subsequently shown to be present in all examined animal cells as well as in a subset of bacteria (5–13). The number of distinct Y RNAs varies between species, with most characterized organisms having between two and four (4, 5, 8, 12, 13). Although all experimentally verified Y RNAs are between 69 and 150 nucleotides, homology searches predict that some bacterial Y RNAs may exceed 200 nucleotides (13).

Like many ncRNAs, Y RNAs are modular. All Y RNAs contain a long stem that is formed by base-pairing the 5′ and 3′ ends (Fig. 1). Within this stem is the Ro60 binding site (14–16). Binding by Ro60 stabilizes Y RNAs, as Y RNA levels are dramatically reduced or undetectable in animal cells and bacteria lacking Ro60 (11, 13, 17–20). All characterized Y RNAs also contain a second module, consisting of one or more stem-loops at the other end of the ncRNA (Fig. 1) (6, 7, 9–13, 21–23).

The ways in which Y RNAs function have been studied largely in vertebrate cells and in *Deinococcus radiodurans*, the first sequenced bacterium with a Ro60 ortholog. In these species, most Y RNA roles are intimately linked to that of the Ro60 protein. Ro60 proteins, named for the apparent molecular weight of the human protein (14), range in size from ~43 to 65 kDa. All characterized Ro60 proteins consist of an α-helical HEAT repeat-containing TROVE (Telomerase, Ro, Vault) domain followed by a von Willebrand factor A domain (vWFA) (24, 25). Structural analyses have revealed that the α-helical HEAT repeats form a ring that is closed by the vWFA domain (25, 26). In vertebrate cells, Ro60 is proposed to function in ncRNA quality control, since it is found complexed with misfolded ncRNA precursors in some cell types (18, 27, 28). Structural and biochemical studies revealed that Ro60 binds misfolded ncRNAs that contain both a single-stranded 3′ end and adjacent protein-free helices (25, 29). The 3′ ends of these RNAs insert through the Ro60 central cavity, while nearby helices bind to a basic platform on the outer surface of the ring (29) (Fig. 2A and C). Because the interaction of Ro60 with misfolded ncRNAs is not strongly sequence specific, Ro60 may scavenge ncRNAs that fail to assemble with their correct RNA-binding proteins (29). In contrast, binding of Ro60 to Y RNAs is highly sequence specific (16), with conserved amino acids on the Ro60 outer surface contacting conserved bases in the Y RNA stem (25) (Fig. 2D). Because the binding sites for misfolded ncRNAs and Y RNAs overlap on the Ro60 surface, it was proposed that Y RNAs act as a gate to regulate access of other RNAs to the Ro60 central cavity (25, 29), a hypothesis supported by studies in bacteria (12, 30).

Y RNAs also influence Ro60 function in other ways. In mammalian cells, a bound Y RNA sterically blocks a nuclear accumulation signal on the Ro60 surface, thus retaining Ro60 in the cytoplasm (31). Moreover, in both bacteria and mammalian cells, Y RNAs tether Ro60 to effector proteins to modulate their function (12, 32). In the best-characterized example of this role, a *D. radiodurans* Y RNA tethers the Ro60 ortholog to the ring-shaped 3′-to-5′ exoribonuclease polynucleotide phosphorylase (PNPase), forming a double-ringed RNP machine specialized for structured RNA decay (12). In addition to these Ro60-linked functions, it is reported

[1]RNA Biology Laboratory, Center for Cancer Research, National Cancer Institute, National Institutes of Health, Frederick, MD 21702.

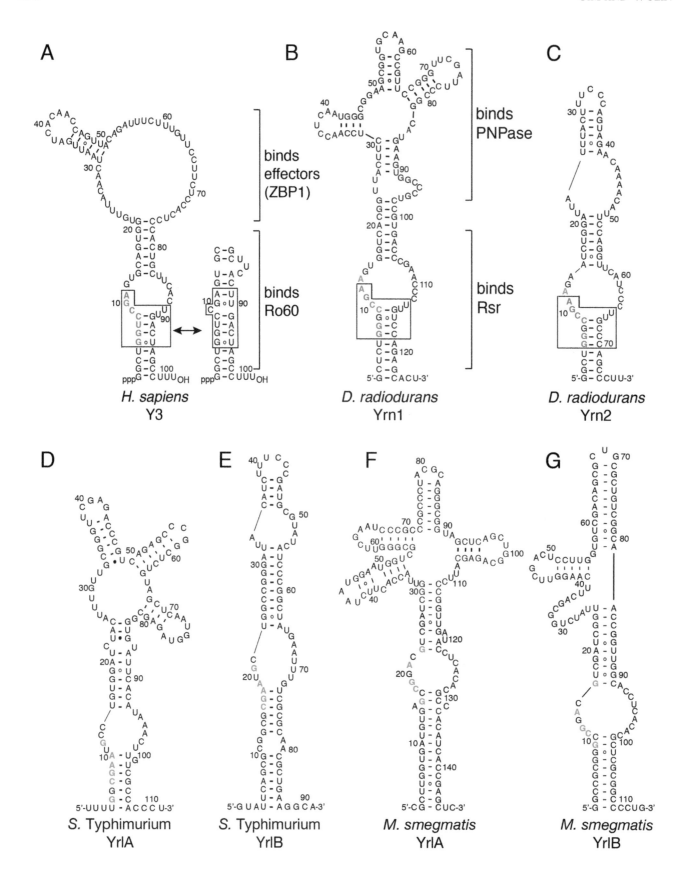

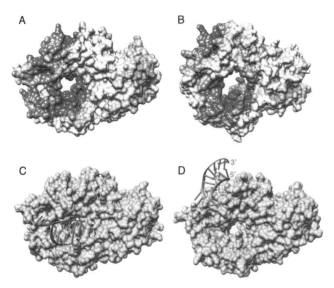

Figure 2 Structures of Ro60 and Rsr proteins. (A) A molecular surface representation of *X. laevis* Ro60 (PDB ID: 1YVR) colored by electrostatic surface potential. (B) A molecular surface representation of *D. radiodurans* Rsr (PDB ID: 2NVO) colored by electrostatic surface potential. For both panels A and B, positive potentials are in blue and negative potentials are in red (−10 kT/e to 10 kT/e). (C) Structure of *X. laevis* Ro60 bound to a misfolded 5S rRNA fragment (PDB ID: 2I91). The helix binds the basic outer surface and the single-stranded 3′ end binds in the hole. (D) Structure of *X. laevis* Ro60 bound to a fragment of Y RNA stem containing the conserved sequences required for Ro60 binding (PDB ID: 1YVP). Positions of the 5′ and 3′ ends are indicated. Biochemical studies support a model in which other portions of the Y RNA contact a basic platform that overlaps with the misfolded RNA-binding site (dashed line) (25, 29).

that vertebrate Y RNAs function independently of Ro60 to initiate DNA replication (33, 34). However, an essential role in vertebrate DNA replication is difficult to reconcile with the findings that although Y RNAs are reduced by ~30-fold in mouse cells and tissues lacking Ro60, these cells have no detectable growth defects and mice lacking Ro60 develop normally (18–20).

The focus of this review is on Y RNAs in bacteria. In part because Ro60 and Y RNA are not present in budding or fission yeast, bacteria have been critical model systems for elucidating functions for Ro60 RNPs. These studies have revealed conserved roles for Ro60 and Y RNAs in assisting cell survival following environmental stress, identified new ways in which ncRNAs function, and demonstrated that some bacterial Y RNAs are tRNA mimics.

DISCOVERY OF Y RNAs IN *D. RADIODURANS*

The first hint that Y RNAs might be present in bacteria came from the finding that the *D. radiodurans* genome encoded a potential Ro60 ortholog (35). *D. radiodurans*, a member of the *Deinococcus-Thermus* phylum, is best known for its extreme resistance to stresses such as ionizing radiation, desiccation, UV irradiation, and oxidative stress (36, 37). Studies of the Ro60 ortholog, named Rsr (Ro sixty-related), revealed that it contributed to *D. radiodurans* survival following UV irradiation (11). This role is conserved, as mouse cells lacking Ro60 are also more sensitive to UV irradiation (18, 19).

Remarkably, both Rsr and four previously unknown ncRNAs (called a, b, c, and d) encoded upstream of Rsr were found to be upregulated after UV irradiation (11). Immunoprecipitations with anti-Rsr antibodies, followed by cDNA sequencing, revealed that Rsr bound these ncRNAs and that the most enriched ncRNA in the immunoprecipitates (RNA c) could be folded to resemble metazoan Y RNAs (Fig. 1B). Specifically, this ncRNA, now called Yrn1 (Y RNA 1), and a second ncRNA, Yrn2 (Y RNA 2), that was discovered subsequently because it cross-linked to Rsr *in vivo* (13) fold to form a long stem that contains sequences known to be important for binding of vertebrate Ro60 proteins to Y RNAs (11, 13) (Fig. 1B and C). Yrn2 is encoded upstream of Yrn1 and is synthesized as a polycistronic precursor with Yrn1 and a tRNA (13). Similar to animal cell Y RNAs, both Yrn1 and Yrn2 RNAs are unstable in *D. radiodurans* lacking Rsr (11, 13). As the remaining three *D. radiodurans* ncRNAs (a, b, and d)

Figure 1 Predicted secondary structures of a human Y RNA and the experimentally identified bacterial Y RNAs. (A) Human Y3 RNA. Modules involved in binding Ro60 and effector proteins are indicated. The portion of the stem containing the Ro60 binding site can form an alternative conformer containing a conserved bulged helix (16). In the structure of *X. laevis* Y3 complexed with Ro60 (25), the bases shown in green (GGUCCGA) are sites of specific interactions with the Ro60 protein. (B, C) *D. radiodurans* Yrn1 and Yrn2. The sequences that can form the conserved helix are boxed, and the conserved "metazoan motif" GGUCCGA is colored in green. An adenine nucleotide that may represent the second A in the "bacterial motif" is colored orange. On Yrn1, regions for Rsr binding and PNPase binding are indicated. (D, E) *S.* Typhimurium YrlA and YrlB. The GNCGAAN$_{0-1}$G motif is in orange. (F, G) *M. smegmatis* YrlA and YrlB. Nucleotides are colored as in panels D and E.

do not appear to form this stem, they have not been designated as Y RNAs. However, these ncRNAs could contribute to the recovery of *D. radiodurans* following UV irradiation.

As more microbial genome sequences were completed, it became clear that Ro60 orthologs and, by inference, Y RNAs are present in numerous bacteria. Approximately 5% of sequenced bacterial genomes contain likely Ro60 orthologs (13, 38). Bacteria containing Ro60 are present in the majority of phyla (13, 38). However, while Ro60 orthologs are easily recognized as 400- to-600-amino-acid proteins containing a TROVE domain adjacent to a vWFA domain (24, 25), bioinformatic identification of Y RNAs lagged behind. For example, our attempts to identify more Y RNAs by searching for primary-sequence or secondary-structure homology revealed likely Yrn1 orthologs in *Deinococcus ficus* and *Deinococcus maricopensis* but not in other bacteria.

IDENTIFICATION OF ADDITIONAL Y RNAs: YrlA RNAs ARE WIDESPREAD

Characterization of Y RNAs in a second bacterium, *Salmonella enterica* serovar Typhimurium, resulted in a breakthrough that allowed identification of Y RNAs in numerous bacteria. Specifically, experiments in which anti-Rsr antibodies were used to immunoprecipitate RNPs from *S.* Typhimurium lysates revealed two discrete ncRNAs in the immunoprecipitates (12). As in *D. radiodurans*, these ncRNAs were encoded within 2 kb of *rsr* (in this case, immediately 3′) and transcribed in the same direction (12). Similar to all characterized Y RNAs, both *S.* Typhimurium ncRNAs could fold into secondary structures in which the 5′ and 3′ ends base-paired to form a long stem (Fig. 1D and E) (12). However, while the putative Ro60/Rsr binding site in metazoan and *D. radiodurans* Y RNAs consists of highly conserved nucleotides on both the 5′ and 3′ strands of the stem, the conservation was less apparent in the *S.* Typhimurium ncRNAs. Because these ncRNAs appeared to differ in some respects from other Y RNAs, we named them YrlA (Y RNA-like A) and YrlB (12).

Remarkably, by performing homology searches using the *S.* Typhimurium RNAs, we identified numerous potential Y RNAs in a wide range of bacteria and also in many bacteriophages (13). Although initial BLASTN searches with YrlB only identified likely YrlB orthologs in other *Salmonella enterica* serovars, the initial search with YrlA identified potential orthologs in the alpha-proteobacterium *Rhizobium etli* and the gammaproteo-bacterium *Pseudomonas fulva* (13). Iterative searches with each newly identified ncRNA yielded putative YrlA RNAs in other Ro60-containing bacteria and also in some mycobacteriophages that encode a Ro60 protein (13, 39). To identify additional YrlA RNAs, we used Infernal (40) to build consensus RNA secondary-structure models. In these experiments, we first collected all bacterial genomes that were annotated in GenBank as containing a TROVE-domain protein and then removed genomes in which this protein lacked a vWFA domain and/or contained other domains. By searching the remaining genomes with the Infernal models, we identified putative YrlA RNAs in >250 bacteria and 22 bacteriophages (13). As more genome sequences have been released, the number of Ro60-containing species with putative YrlA RNAs has continued to increase (41, 42). Moreover, in some bacteria, multiple ncRNAs that resemble YrlA are predicted (13, 41).

As in *D. radiodurans* and *S.* Typhimurium, the vast majority of the putative YrlA orthologs are encoded within 4 kb of the gene encoding Ro60 and are predicted to be transcribed from the same DNA strand (13). Importantly, experiments in which an epitope-tagged Rsr protein was used to immunoprecipitate Rsr-containing RNPs from *Mycobacterium smegmatis* revealed that the predicted YrlA RNA was present in the immunoprecipitates, as was a second ncRNA that was named YrlB (13) (Fig. 1F and G). Additionally, anti-Ro60 antibodies from patients with systemic lupus erythematosus were shown recently to immunoprecipitate RNPs containing the putative YrlA RNA from lysates of *Propionibacterium propionicum* (42). Thus, we consider it likely that most of the predicted YrlA RNAs exist *in vivo*.

Despite the recent success in identifying Y RNAs in numerous bacteria, additional Y RNAs remain to be discovered. Since for some bacteria iterative homology searching failed to identify a likely YrlA (13), these Y RNAs may diverge from current models. Moreover, the finding that biochemical experiments in *D. radiodurans*, *S.* Typhimurium, and *M. smegmatis* all identified a second Y RNA that was not predicted by bioinformatics searches supports the idea that more Y RNAs remain to be found (12, 13). Also, while potential Ro60 orthologs are present in some archaea, no archaeal Y RNAs have yet been characterized. Thus, in addition to the "metazoan-like" Yrn and bacteria-specific Yrl lineages, other Y RNA lineages may exist.

CONSERVED FEATURES OF Y RNAs: THE Ro60 BINDING SITE

Since Y RNAs are defined in part by their sequence-specific binding to Ro60, the nucleotides and amino acids involved in this interaction are expected to be conserved. In the crystal structure of *Xenopus laevis* Ro60 with Y3 RNA, conserved Ro60 amino acids contact the conserved bases GGUCCGA in the 5′ strand of the stem (25). In *D. radiodurans* Yrn1 and Yrn2, the corresponding sequence is GGGCCGA, with only the third nucleotide differing from the metazoan motif (Fig. 1B and C) (11, 13). In YrlA RNAs, the most-conserved sequence in the stem is $GNCGAAN_{0-1}G$, which occurs near the 5′ end (Fig. 1D to G) (13). Although a bacterial Rsr/YrlA structure has not yet been reported, it is plausible that the central CGA of this motif corresponds to the CGA at the end of the metazoan GGUCCGA motif. Consistent with the idea that contacts to this region may be particularly important for Y RNA recognition, two amino acids that contact the CG in the *X. laevis* structure, H187 and D181 (25), are the most conserved of the Ro60 residues that contact Y RNA. Ro60 proteins mutated at D181 have not been reported; however, Ro60 proteins carrying an H187 mutation (H187S) show greatly reduced Y RNA binding *in vitro* and fail to stabilize Y RNAs *in vivo* (25, 30, 31).

Although sequences that can base-pair with the GGUCCGA motif to form a bulged helix are conserved in metazoan and *D. radiodurans* Y RNAs, these 3′ sequences [UUGACC, metazoans; UUG(C/U)CC, *D. radiodurans*] are not well conserved in YrlA and YrlB RNAs. This is consistent with the finding that in the *X. laevis* Ro60/Y3 crystal structure, there are few contacts between Ro60 and these 3′ sequences, suggesting that they are unimportant for Ro60 recognition (25). Consistent with the idea that the ability to form the bulged helix does not contribute strongly to Ro60 recognition, Ro60 binding partially disrupts this helix, widening the RNA major groove to allow contacts with bases on the 5′ strand (16, 25). The ability of the 5′ and 3′ sequences to base-pair could be required for another function, such as the need to maintain the terminal stem required for Exportin-5-mediated nuclear export of the human Y1 RNA (43). However, it remains unclear why some bacterial Y RNAs, such as *D. radiodurans* Yrn1 and Yrn2, have retained both the conserved nucleotides and the ability to form a bulged helix.

THE SECOND Y RNA MODULE INTERACTS WITH OTHER COMPONENTS

In addition to the stem containing the Ro60 binding site, all Y RNAs contain a second module (Fig. 1). A major role of this second domain is to interact with other proteins, thus allowing the Y RNA to scaffold

interactions between Ro60 and additional proteins. A number of proteins interact with one or more mammalian Ro60 RNPs by binding the large internal loops that are a prominent feature of all characterized metazoan Y RNAs (20, 44). These proteins include the multifunctional RNA-binding proteins PUF60, PTBP1, and nucleolin; the zipcode-binding protein ZBP1; and the interferon-inducible protein IFIT5 (32, 45–49). Although in most cases the way in which the tethered protein affects Ro60 RNP function is unknown, binding of ZBP1 to mammalian Y3 RNA adapts the Ro60/Y3 RNP for Crm1-mediated nuclear export (32).

For the best-characterized bacterial Y RNA, *D. radiodurans* Yrn1, structure probing supports a model in which the second module contains three stem-loops (Fig. 1B) (13). This portion of Yrn1 is required for the interaction with PNPase (12). PNPase is a homotrimeric ring topped by S1 and K-homology (KH) single-stranded RNA-binding domains (50). Biochemical and structural data support a model in which single-stranded portions of this second Yrn1 module bind the PNPase S1 and KH domains (12). Interestingly, one stem-loop in this module resembles the T arm found in all tRNAs (13) (discussed below). As this stem-loop is not required for PNPase binding (13), it may contribute to Y RNA structure or function in other ways, such as through stabilizing Y RNA tertiary interactions (51).

YrlA RNAs CONTAIN A DOMAIN THAT MIMICS tRNAs

Remarkably, in YrlA RNAs, the second module contains striking similarities to canonical tRNAs (13). The tRNA mimicry becomes apparent when YrlA RNA is oriented as in Fig. 3A. In this orientation, the middle stem-loop of YrlA RNA corresponds to the tRNA acceptor stem and most conserved nucleotides are located within stem-loops that resemble tRNA D and T arms (Fig. 3A) (13). These include many nucleotides that are invariant in tRNAs and that play critical roles in stabilizing the canonical L-shape. At least seven tertiary interactions that form in tRNA can potentially form in most YrlA RNAs (Fig. 3B) (13). Consistent with the hypothesis that this YrlA domain folds to resemble tRNA, YrlA RNAs from various bacteria contain compensatory changes that maintain the "Levitt base pair" (R15:Y48) that is crucial for stabilizing the tRNA L-shape (13).

In support of the hypothesis that YrlA RNAs are tRNA mimics, these ncRNAs are substrates for some tRNA processing and modification enzymes (13). *S.* Typhimurium YrlA was shown to contain several nucle-

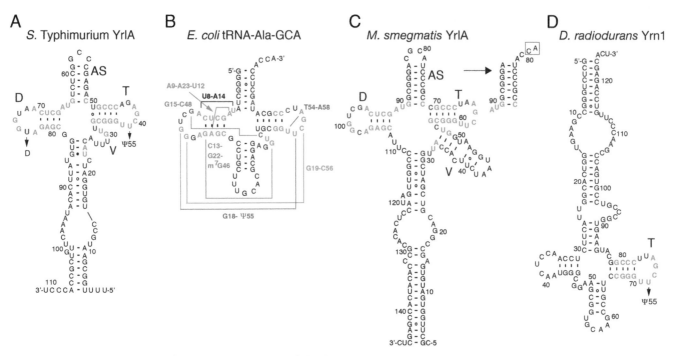

Figure 3 YrlA RNAs contain a module that resembles tRNA. (A) *S.* Typhimurium YrlA presented to resemble a canonical tRNA. Highly conserved nucleotides between YrlA orthologs are colored orange, while conserved purines and pyrimidines are in blue. Bases shown to be modified *in vivo* (11) are indicated. AS, D, T, and V denote the acceptor stem, D arm, T arm, and variable arm, respectively. (B) *E. coli* tRNA-Ala-GCA. Nucleotides that are conserved between YrlA RNAs are in orange. All depicted tertiary interactions can potentially form in YrlA RNAs. (C) The genome-encoded sequence of *M. smegmatis* YrlA drawn to emphasize the resemblance to tRNA. The structure of the acceptor stem after cleavage, end nibbling, and posttranscriptional CA addition *in vivo* (13) is also shown (arrow). Conserved nucleotides are colored as in panel A. (D) *D. radiodurans* Yrn1 presented to resemble tRNA. Nucleotides in the T arm that are conserved between Yrn1 and YrlA RNAs are colored as in panel A.

otide modifications characteristic of tRNAs, such as the pseudouridine that is at position 55 (Ψ55) in the TΨC loop of nearly all tRNAs and dihydrouridine in the D-loop (Fig. 3A) (13). These YrlA RNA modifications are catalyzed by the same enzymes that modify the analogous tRNA sites, as they were not detected in strains containing mutations that disrupt catalysis by TruB, the enzyme that pseudouridylates Ψ55, or by DusA, the enzyme responsible for dihydrouridine formation at the same site in tRNA (Fig. 3A) (13). Moreover, in *M. smegmatis*, the YrlA RNA stem-loop corresponding to the acceptor stem can be cleaved by RNase P, the endonuclease that matures tRNA 5′ ends (13), most likely because this YrlA RNA contains a 7-base-pair acceptor stem, the optimal length for RNase P cleavage (52). Following cleavage, the fragment corresponding to the tRNA 3′ end undergoes posttranscriptional addition of CA, resulting in the CCA tail that is characteristic of all tRNAs (Fig. 3C) (13).

Although the T arm of *D. radiodurans* Yrn1 also contains pseudouridine at the position corresponding to Ψ55 (13), the resemblance to tRNA is less apparent (Fig. 3D). A structure with some similarity to tRNA can be drawn for Yrn1; however, the stem-loop that would correspond to the D arm does not contain dihydrouridine and critical tertiary interactions that stabilize the tRNA L-shape are not predicted to form (Fig. 3D) (13). Thus, the only identified bacterial Y RNAs that are currently predicted to mimic tRNA are members of the YrlA family.

EVOLUTIONARY CONSIDERATIONS

Why do only some bacteria contain Ro60 RNPs? We favor a model in which Rsr and Y RNAs derive from multiple episodes of lateral gene transfer. The patchy distribution of Rsr and Y RNAs, in which they are present in a small fraction of bacteria but a majority of phyla, is consistent with a lateral transfer model, as is the finding that Ro60 and YrlA RNA are present in phages isolated from diverse species, including *Bacillus megaterium*, *Caulobacter crescentus*, *Gordonia malaquae*, *M. smegmatis*, and *Streptomyces griseus* (13, 39). In support of lateral transfer, phylogenetic trees of bacterial Ro60 orthologs are markedly different from standard phylogenetic trees based on bacterial 16S rRNA sequences (Fig. 4) (38).

The striking resemblance of YrlA RNAs to tRNAs makes it likely that these ncRNAs evolved from tRNAs. Since bacterial Y RNAs are encoded adjacent to their Ro60 binding partners (13) and frequently abut tRNA genes, we hypothesize that, following acquisition of a metazoan Ro60 gene, one or more bacteria adopted a tRNA-containing transcript as a binding partner. Moreover, the fact that YrlA RNAs differ from canonical tRNAs in that the T arm is encoded 5′ to the D arm

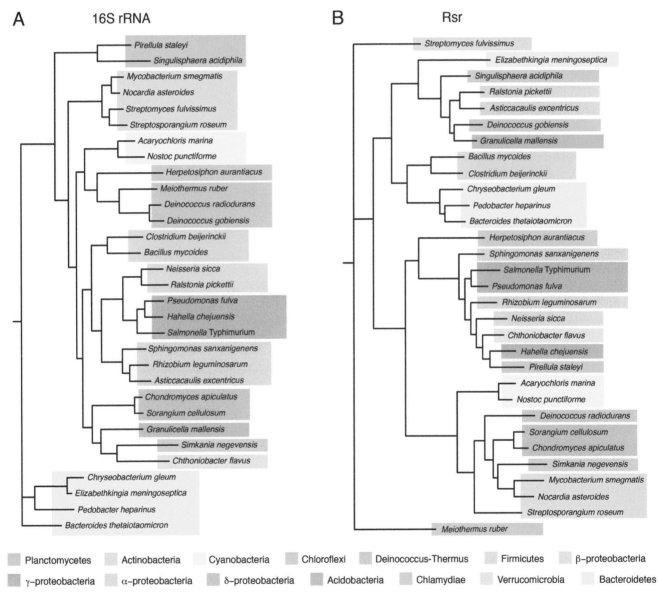

Figure 4 Phylogenetic trees of representative Rsr-containing bacterial species. (A) Phylogenetic tree based on the sequences of 16S rRNAs (70). Each phylum is represented by a distinct color. (B) Phylogenetic tree based on the sequences of Rsr proteins. Sequence alignments were performed using Clustal Omega (71), and trees were drawn with the Phylogeny Interference Package (PHYLIP) using the maximum likelihood method (72).

suggests that the primordial YrlA RNA derived from a dimeric tRNA. In this scenario, the T arm of the first tRNA and the D arm of the second tRNA evolved to form YrlA. This model could provide an explanation for why the acceptor stem of YrlA RNAs varies in sequence and length, while the T and D arms resemble bona fide tRNA (13). If YrlA RNAs derive from tandemly repeated tRNAs, the highly variable gap sequence between the two tRNAs would become the YrlA acceptor stem. In this scenario, YrlA RNAs may have evolved multiple times in bacteria. Alternatively, if YrlA RNAs derive from a single ancestral dimeric tRNA, the variability in the acceptor stem could reflect fewer functional constraints on this stem-loop.

FUNCTIONS OF BACTERIAL Y RNAs AND THEIR Rsr PARTNERS

Rsr and Y RNAs Modulate RNA Metabolism during Environmental Stress

Since bacterial Y RNAs, like their metazoan counterparts, are bound and stabilized by Rsr (11, 13, 30), the functions of Rsr and Y RNAs are entwined. In *D. radiodurans*, a major role of Rsr and Y RNAs is to assist survival following environmental stress. *D. radiodurans* strains lacking Rsr are less resistant to UV irradiation than wild-type strains (11) and are at a competitive disadvantage during growth in stationary phase (53). In addition, both Rsr and Y RNA are upregulated during heat stress (30); growth in stationary phase (53); and recovery from UV radiation, ionizing radiation, and dessication (11, 54). The role of Ro60 RNPs in aiding survival after stress is conserved, as mouse cells lacking Ro60 are sensitive to UV irradiation (18) and nematodes lacking Ro60 have defects in forming dauer larvae, an alternative larval stage that developing worms form upon encountering unfavorable growth conditions (55).

Genetic and biochemical analyses revealed that *D. radiodurans* Rsr and Yrn1 RNA function with exoribonucleases to alter RNA metabolism following environmental stress. In *D. radiodurans*, 23S rRNA maturation is inefficient at the normal growth temperature of 30°C, as ~40% of the 23S rRNA contains 5′ and/or 3′ extensions (30). Maturation of 23S rRNA becomes highly efficient when these cells are shifted to 37°C and requires Rsr and two 3′-to-5′ exoribonucleases, RNase II and RNase PH (30). In this case, the Y RNA-free form of Rsr carries out maturation, and Yrn1 RNA binding to Rsr inhibits 23S rRNA maturation (30).

Additionally, Rsr and the ring-shaped 3′-to-5′ exoribonuclease PNPase function in rRNA degradation

during prolonged growth in stationary phase (53). Rsr and Yrn1 levels increase nearly 30-fold during growth in stationary phase, compared to their levels in logarithmic phase (53). PNPase levels increase 3-fold, as does formation of a complex between Rsr and PNPase (53). Consistent with a role for Rsr as an adaptor that assists rRNA degradation by PNPase, rRNA decay is less complete in strains lacking Rsr or PNPase, and sedimentation of PNPase with partially degraded ribosomal subunits requires Rsr (53).

Although the role of Yrn1 in rRNA decay was not fully explored due to its degradation in *D. radiodurans* lysates, rRNA degradation was also less efficient in strains lacking Yrn1 RNA (53). In support of a role for Yrn1, this RNA exhibits genetic interactions with Rsr and PNPase. Cells lacking PNPase (Δ*pnp*) grow slowly at all temperatures and are sensitive to cold, oxidative stress, and UV irradiation (30). Although cells lacking Yrn1(Δ*yrn1*) grow normally, strains lacking both PNPase and Yrn1 (Δ*pnp* Δ*yrn1*) show enhanced sensitivity to both low temperature and oxidative stress (30). Remarkably, although Δ*rsr* strains also grow similarly to wild-type cells, deletion of Rsr in Δ*pnp* and Δ*pnp* Δ*yrn1* strains (Δ*rsr* Δ*pnp* and Δ*rsr* Δ*pnp* Δ*yrn1*) largely alleviates the sensitivity to cold and oxidative stress, indicating that the growth defects of Δ*pnp* and Δ*pnp* Δ*yrn1* strains are partly due to Rsr (30). One explanation for the genetic interactions is that, in cells lacking PNPase, binding of Y RNA-free Rsr to specific RNAs prevents their degradation by other RNases and inhibits growth (30).

Y RNA Tethers PNPase to Rsr To Form RYPER

To determine how Rsr and Y RNA influence PNPase function, our laboratory purified the Rsr/Y RNA/PNPase complex from *D. radiodurans* (12). Characterization of this complex revealed that it sedimented with a molecular size consistent with one Rsr ring, one Yrn1 RNA, and one PNPase trimer (12). Further analysis revealed that Yrn1 functions as a scaffold, since Rsr and PNPase bind distinct sites on this RNA (12). Specifically, while Rsr binds its high-affinity site on the Yrn1 stem, the S1/KH single-stranded RNA-binding domains of PNPase interact with one or more loops at the other end of Yrn1 (Fig. 5A) (12). This complex was named RYPER (Rsr/Y RNA/PNPase exoribonuclease RNP) (20). Single-particle electron microscopy of RYPER revealed that the Rsr and PNPase rings were bridged by a rod-shaped density that likely represents Y RNA (Fig. 5B) (12). The two rings are configured such that single-stranded RNA can thread through the Rsr ring into the PNPase cavity (Fig. 5B) (12). Biochemical studies revealed that RYPER

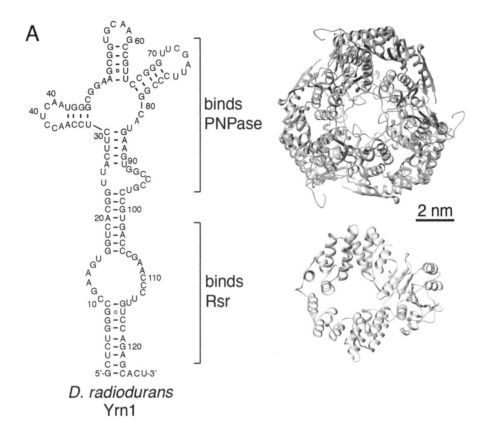

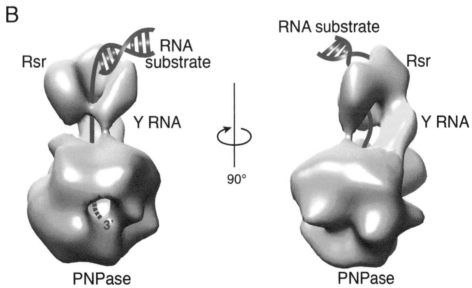

Figure 5 Role of Yrn1 in scaffolding RYPER formation. (A) *D. radiodurans* Yrn1, *D. radiodurans* Rsr (PDB ID: 2NVO) (light blue), and *Streptomyces antibioticus* PNPase (PDB ID: 1E3P) (pink). The Yrn1 modules that bind Rsr and PNPase are indicated. (B) The structure of RYPER predicted by single-particle electron microscopy and three-dimensional reconstruction (12) (EMDB ID: 5389). The density that likely corresponds to Yrn1 is colored in yellow, while densities corresponding to Rsr and PNPase are colored as in panel A. A possible path for degrading a structured RNA substrate, in which the 3′ end threads from Rsr into the PNPase cavity for degradation, is depicted in blue.

degrades structured RNAs more effectively than PNPase but is less active in degrading single-stranded RNAs, possibly because Yrn1-mediated tethering of Rsr to the PNPase S1/KH domains impedes RNA decay substrates from binding these single-stranded RNA-binding domains (12). Additionally, since vertebrate Ro60 binds RNAs containing both a single-stranded tail and helices (29), the replacement of the single-stranded RNA-binding surface of PNPase with the Rsr-binding surface may alter the target specificity of PNPase. As Rsr, PNPase, and YrlA RNA all sediment as part of a complex in *S.* Typhimurium, at least some aspects of RYPER may be conserved (12).

Although RYPER is the only known RNP degradation machine, it resembles the archaeal and eukaryotic RNA degradation machine known as the RNA exosome. In archaea, the exosome consists of a heterohexameric RNase PH domain-containing catalytic ring topped by an RNA-binding ring (56). As PNPase is a trimer in which each monomer includes two RNase PH domains, RYPER also consists of an RNA-binding ring atop an RNase PH domain-containing catalytic ring, although Y RNA tethering replaces the protein-protein interactions that join the two exosome rings (12). In yeast and animal cells, the RNase PH domains of the exosome contain mutations that render them catalytically inactive, and RNA threads through both the RNA-binding ring and the inactive RNase PH domain ring to reach an active exoribonuclease (57). As in these exosomes, where channeling of RNA through the catalytically inactive RNase PH ring contributes to unwinding an RNA duplex (58), threading of RNA through the Rsr ring may assist ATP-independent RNA unwinding (12).

RYPER also exhibits some similarities to the bacterial degradosome, an RNA degradation machine best characterized in *Escherichia coli*. In *E. coli* and many other gammaproteobacteria, the degradosome is a stable complex consisting of PNPase, an RNA helicase, the metabolic enzyme enolase, and the scaffolding endonuclease RNase E (59). In the degradosome, the helicase assists in unwinding structured RNA, while the endonuclease activity of RNase E generates additional ends for PNPase entry (59). Although the activity of RYPER on structured RNAs may be enhanced by endonucleases and/or helicases *in vivo*, it differs from the degradosome in that the increased activity of RYPER on structured RNAs can be observed in the absence of these enzymes (12). Because RYPER has only been well characterized in *D. radiodurans*, an organism in which a PNPase assembly resembling the degradosome has not been described, the extent to which RYPER and the degradosome functionally over-

lap remains unknown. Studies in gammaproteobacteria that contain both RYPER and a degradosome-like complex, such as *S.* Typhimurium (12, 60), are needed to address this question.

Y RNAs Can Regulate Access to Rsr

In addition to acting as tethers, Y RNAs can function as gates to regulate access of other RNAs to the Rsr cavity. One example of this role occurs during the heat stress-induced maturation of 23S rRNA in *D. radiodurans*. As described above, maturation of 23S rRNA is inefficient when cells are grown at 30°C but becomes efficient at 37°C through a process that requires Rsr and two exoribonucleases (30). Consistent with a role for Yrn1 as an inhibitor, 23S rRNA maturation is efficient at all temperatures in strains lacking Yrn1 (30). Maturation is also efficient at all temperatures when Rsr carrying a point mutation that abrogates Y RNA binding is overexpressed in wild-type cells (30). Together, these data indicate that the Y RNA-free form of Ro60 carries out maturation and Y RNAs inhibit this process, presumably by sterically blocking access of the pre-23S rRNA 3′ extensions to the Rsr central cavity.

If a bound Y RNA can prevent entrance of other RNAs to the Ro60 cavity, how is Y RNA binding regulated? For those functions, such as 23S rRNA maturation, that require Y RNA-free Rsr (30), increased synthesis of a single primary transcript encoding both Rsr and Y RNA may result in excess Rsr, compared to Y RNA, since the mRNA encoding Rsr can be translated multiple times. For those functions in which Y RNA acts as a tether, binding of a partner protein such as PNPase to the Y RNA stem-loops may remove this module from the Rsr surface. Although the only reported bacterial Rsr structure lacks Y RNA (26), structural and biochemical experiments revealed that amino acids on the outer edge of the *X. laevis* Ro60 ring contact the Ro60 binding site in the Y RNA stem (25), while the stem-loop-containing second module is predicted to contact a basic platform that overlaps the binding site for misfolded ncRNAs (Fig. 2D) (25, 29). Since our studies of *D. radiodurans* RYPER demonstrated that the stem-loop-containing module interacts with PNPase (12), binding of PNPase may remove this module from the Rsr surface, allowing decay substrates to enter the Rsr cavity.

The Organization of Rsr and Y RNA in Some Genomes Suggests a Role in RNA Repair

In some bacteria, Rsr and Y RNAs are encoded within an "RNA repair" operon. This was first observed in the gammaproteobacterium *S.* Typhimurium, where Rsr, YrlA, and YrlB are encoded within the σ^{54}-regulated *rtcBA* operon, which also encodes the RtcB RNA ligase and the RtcA RNA cyclase (12). In metazoans and archaea, RtcB ligates pre-tRNA halves following intron excision (61–63). The RtcB substrates in bacteria are largely uncharacterized; however, *E. coli* RtcB religates a 16S rRNA 3′ fragment to the rRNA body after cleavage by the MazF toxin (64). Although expression of the *rsr-yrlBA-rtcBA* operon is tightly regulated by the adjacent RtcR transcriptional activator (12, 65), transcription of YrlA and YrlB was detected during infection of human cells (66) and the operon was also reported to be expressed during exposure to the nucleic acid interstrand cross-linker mitomycin C (65).

Comparative genomics revealed that Rsr and Y RNAs are encoded adjacent to RtcB in diverse bacteria, including some proteobacteria, firmicutes, bacteroidetes, planctomycetes, and verrucomicrobia (41, 67). In some bacteria, the operon resembles that of *S.* Typhimurium in that both RtcB and RtcA are encoded downstream of Rsr and Y RNAs; however, in others, *rtcA* is absent (41). Since in a subset of bacteria that lack Rsr, RtcB is encoded adjacent to archease, a protein that enhances RtcB activity (68, 69), it has been speculated that Rsr and Y RNA may similarly augment RtcB function (41). However, while the genomic linkage of Rsr, RtcB, and Y RNAs is relatively frequent in gammaproteobacteria, RtcB is not encoded adjacent to Rsr and Y RNA in the majority of bacteria that encode these components.

CONCLUSIONS AND PERSPECTIVES

Although investigations of bacterial Y RNAs are in their infancy, these studies have resulted in the identification of RYPER, an RNP machine specialized for structured RNA degradation, and in the discovery of new ways in which bacterial ncRNAs function. Future high-resolution structures of RYPER and of Rsr complexed with full-length Y RNA should elucidate both the molecular details that underlie RYPER function and those features of Y RNA structure that allow these RNAs to function as both gates and tethers. Biochemical and genetic studies will be required to elucidate the role of the YrlA RNA tRNA-like domain. Moreover, while all characterized Ro60-containing bacteria contain a second Y RNA (i.e., *D. radiodurans* Yrn2, *S.* Typhimurium YrlB, and *M. smegmatis* YrlB) in which the second module contains a long hairpin closed by a pyrimidine-rich internal loop (Fig. 1), the functions of these other Y RNAs remain unknown. We expect that elucidation of roles for these Y RNAs will continue to reveal new functions for ncRNAs and to provide additional insights into how bacteria adapt to their ever-changing environments.

Acknowledgments. *Work in our laboratory is supported by the Intramural Research Program of the National Institutes of Health, National Cancer Institute, Center for Cancer Research.*

Citation. Sim S, Wolin SL. 2018. Bacterial Y RNAs: gates, tethers, and tRNA mimics. Microbiol Spectrum 6(4):RWR-0023-2018.

References

1. **Gottesman S, Storz G.** 2011. Bacterial small RNA regulators: versatile roles and rapidly evolving variations. *Cold Spring Harb Perspect Biol* **3**:a003798.

2. **Wagner EG, Romby P.** 2015. Small RNAs in bacteria and archaea: who they are, what they do, and how they do it. *Adv Genet* **90**:133–208.

3. **Clark G, Reichlin M, Tomasi TB Jr.** 1969. Characterization of a soluble cytoplasmic antigen reactive with sera from patients with systemic lupus erythmatosus. *J Immunol* **102**:117–122.

4. **Lerner MR, Boyle JA, Hardin JA, Steitz JA.** 1981. Two novel classes of small ribonucleoproteins detected by antibodies associated with lupus erythematosus. *Science* **211**:400–402.

5. **Hendrick JP, Wolin SL, Rinke J, Lerner MR, Steitz JA.** 1981. Ro small cytoplasmic ribonucleoproteins are a subclass of La ribonucleoproteins: further characterization of the Ro and La small ribonucleoproteins from uninfected mammalian cells. *Mol Cell Biol* **1**:1138–1149.

6. **Kato N, Hoshino H, Harada F.** 1982. Nucleotide sequence of 4.5S RNA (C8 or hY5) from HeLa cells. *Biochem Biophys Res Commun* **108**:363–370.

7. **Wolin SL, Steitz JA.** 1983. Genes for two small cytoplasmic Ro RNAs are adjacent and appear to be single-copy in the human genome. *Cell* **32**:735–744.

8. **Reddy R, Tan EM, Henning D, Nohga K, Busch H.** 1983. Detection of a nucleolar 7-2 ribonucleoprotein and a cytoplasmic 8-2 ribonucleoprotein with autoantibodies from patients with scleroderma. *J Biol Chem* **258**:1383–1386.

9. **Farris AD, Gross JK, Hanas JS, Harley JB.** 1996. Genes for murine Y1 and Y3 Ro RNAs have class 3 RNA polymerase III promoter structures and are unlinked on mouse chromosome 6. *Gene* **174**:35–42.

10. **Van Horn DJ, Eisenberg D, O'Brien CA, Wolin SL.** 1995. *Caenorhabditis elegans* embryos contain only one major species of Ro RNP. *RNA* **1**:293–303.

11. **Chen X, Quinn AM, Wolin SL.** 2000. Ro ribonucleoproteins contribute to the resistance of *Deinococcus radiodurans* to ultraviolet irradiation. *Genes Dev* **14**:777–782.

12. **Chen X, Taylor DW, Fowler CC, Galan JE, Wang HW, Wolin SL.** 2013. An RNA degradation machine sculpted by Ro autoantigen and noncoding RNA. *Cell* **153**:166–177.

13. **Chen X, Sim S, Wurtmann EJ, Feke A, Wolin SL.** 2014. Bacterial noncoding Y RNAs are widespread and mimic tRNAs. *RNA* **20**:1715–1724.

14. Wolin SL, Steitz JA. 1984. The Ro small cytoplasmic ribonucleoproteins: identification of the antigenic protein and its binding site on the Ro RNAs. *Proc Natl Acad Sci U S A* **81**:1996–2000.

15. Pruijn GJM, Slobbe RL, van Venrooij WJ. 1991. Analysis of protein-RNA interactions within Ro ribonucleoprotein complexes. *Nucleic Acids Res* **19**:5173–5180.

16. Green CD, Long KS, Shi H, Wolin SL. 1998. Binding of the 60-kDa Ro autoantigen to Y RNAs: evidence for recognition in the major groove of a conserved helix. *RNA* **4**:750–765.

17. Labbé JC, Hekimi S, Rokeach LA. 1999. The levels of the RoRNP-associated Y RNA are dependent upon the presence of ROP-1, the *Caenorhabditis elegans* Ro60 protein. *Genetics* **151**:143–150.

18. Chen X, Smith JD, Shi H, Yang DD, Flavell RA, Wolin SL. 2003. The Ro autoantigen binds misfolded U2 small nuclear RNAs and assists mammalian cell survival after UV irradiation. *Curr Biol* **13**:2206–2211.

19. Xue D, Shi H, Smith JD, Chen X, Noe DA, Cedervall T, Yang DD, Eynon E, Brash DE, Kashgarian M, Flavell RA, Wolin SL. 2003. A lupus-like syndrome develops in mice lacking the Ro 60-kDa protein, a major lupus autoantigen. *Proc Natl Acad Sci U S A* **100**:7503–7508.

20. Wolin SL, Belair C, Boccitto M, Chen X, Sim S, Taylor DW, Wang HW. 2013. Non-coding Y RNAs as tethers and gates: insights from bacteria. *RNA Biol* **10**:1602–1608.

21. O'Brien CA, Harley JB. 1990. A subset of hY RNAs is associated with erythrocyte Ro ribonucleoproteins. *EMBO J* **9**:3683–3689.

22. Perreault J, Perreault JP, Boire G. 2007. Ro-associated Y RNAs in metazoans: evolution and diversification. *Mol Biol Evol* **24**:1678–1689.

23. Mosig A, Guofeng M, Stadler BM, Stadler PF. 2007. Evolution of the vertebrate Y RNA cluster. *Theory Biosci* **126**:9–14.

24. Bateman A, Kickhoefer V. 2003. The TROVE module: a common element in Telomerase, Ro and Vault ribonucleoproteins. *BMC Bioinformatics* **4**:49.

25. Stein AJ, Fuchs G, Fu C, Wolin SL, Reinisch KM. 2005. Structural insights into RNA quality control: the Ro autoantigen binds misfolded RNAs via its central cavity. *Cell* **121**:529–539.

26. Ramesh A, Savva CG, Holzenburg A, Sacchettini JC. 2007. Crystal structure of Rsr, an ortholog of the antigenic Ro protein, links conformational flexibility to RNA binding activity. *J Biol Chem* **282**:14960–14967.

27. O'Brien CA, Wolin SL. 1994. A possible role for the 60-kD Ro autoantigen in a discard pathway for defective 5S rRNA precursors. *Genes Dev* **8**:2891–2903.

28. Shi H, O'Brien CA, Van Horn DJ, Wolin SL. 1996. A misfolded form of 5S rRNA is complexed with the Ro and La autoantigens. *RNA* **2**:769–784.

29. Fuchs G, Stein AJ, Fu C, Reinisch KM, Wolin SL. 2006. Structural and biochemical basis for misfolded RNA recognition by the Ro autoantigen. *Nat Struct Mol Biol* **13**:1002–1009.

30. Chen X, Wurtmann EJ, Van Batavia J, Zybailov B, Washburn MP, Wolin SL. 2007. An ortholog of the Ro autoantigen functions in 23S rRNA maturation in *D. radiodurans*. *Genes Dev* **21**:1328–1339.

31. Sim S, Weinberg DE, Fuchs G, Choi K, Chung J, Wolin SL. 2009. The subcellular distribution of an RNA quality control protein, the Ro autoantigen, is regulated by noncoding Y RNA binding. *Mol Biol Cell* **20**:1555–1564.

32. Sim S, Yao J, Weinberg DE, Niessen S, Yates JR III, Wolin SL. 2012. The zipcode-binding protein ZBP1 influences the subcellular location of the Ro 60-kDa autoantigen and the noncoding Y3 RNA. *RNA* **18**:100–110.

33. Christov CP, Gardiner TJ, Szüts D, Krude T. 2006. Functional requirement of noncoding Y RNAs for human chromosomal DNA replication. *Mol Cell Biol* **26**:6993–7004.

34. Collart C, Christov CP, Smith JC, Krude T. 2011. The midblastula transition defines the onset of Y RNA-dependent DNA replication in *Xenopus laevis*. *Mol Cell Biol* **31**:3857–3870.

35. White O, Eisen JA, Heidelberg JF, Hickey EK, Peterson JD, Dodson RJ, Haft DH, Gwinn ML, Nelson WC, Richardson DL, Moffat KS, Qin H, Jiang L, Pamphile W, Crosby M, Shen M, Vamathevan JJ, Lam P, McDonald L, Utterback T, Zalewski C, Makarova KS, Aravind L, Daly MJ, Minton KW, Fleischmann RD, Ketchum KA, Nelson KE, Salzberg S, Smith HO, Venter JC, Fraser CM. 1999. Genome sequence of the radioresistant bacterium *Deinococcus radiodurans* R1. *Science* **286**:1571–1577.

36. Cox MM, Battista JR. 2005. *Deinococcus radiodurans*—the consummate survivor. *Nat Rev Microbiol* **3**:882–892.

37. Slade D, Radman M. 2011. Oxidative stress resistance in *Deinococcus radiodurans*. *Microbiol Mol Biol Rev* **75**:133–191.

38. Sim S, Wolin SL. 2011. Emerging roles for the Ro 60-kDa autoantigen in noncoding RNA metabolism. *Wiley Interdiscip Rev RNA* **2**:686–699.

39. Pedulla ML, Ford ME, Houtz JM, Karthikeyan T, Wadsworth C, Lewis JA, Jacobs-Sera D, Falbo J, Gross J, Pannunzio NR, Brucker W, Kumar V, Kandasamy J, Keenan L, Bardarov S, Kriakov J, Lawrence JG, Jacobs WR Jr, Hendrix RW, Hatfull GF. 2003. Origins of highly mosaic mycobacteriophage genomes. *Cell* **113**:171–182.

40. Nawrocki EP, Eddy SR. 2013. Infernal 1.1: 100-fold faster RNA homology searches. *Bioinformatics* **29**:2933–2935.

41. Burroughs AM, Aravind L. 2016. RNA damage in biological conflicts and the diversity of responding RNA repair systems. *Nucleic Acids Res* **44**:8525–8555.

42. Greiling TM, Dehner C, Chen X, Hughes K, Iñiguez AJ, Boccitto M, Zegarra Ruiz D, Renfroe SC, Vieira SM, Ruff WE, Sim S, Kriegel C, Glanternik J, Chen X, Girardi M, Degnan P, Costenbader KH, Goodman AL, Wolin SL, Kriegel MA. Commensal orthologs of the human autoantigen Ro60 as triggers of autoimmunity in lupus. *Sci Transl Med* **10**:eaan2306.

43. Gwizdek C, Ossareh-Nazari B, Brownawell AM, Doglio A, Bertrand E, Macara IG, Dargemont C. 2003. Exportin-5 mediates nuclear export of minihelix-containing RNAs. *J Biol Chem* **278**:5505–5508.

44. Teunissen SW, Kruithof MJ, Farris AD, Harley JB, Venrooij WJ, Pruijn GJ. 2000. Conserved features of Y RNAs: a comparison of experimentally derived secondary structures. *Nucleic Acids Res* **28**:610–619.

45. Bouffard P, Barbar E, Brière F, Boire G. 2000. Interaction cloning and characterization of RoBPI, a novel protein binding to human Ro ribonucleoproteins. *RNA* **6**:66–78.

46. Fabini G, Raijmakers R, Hayer S, Fouraux MA, Pruijn GJ, Steiner G. 2001. The heterogeneous nuclear ribonucleoproteins I and K interact with a subset of the Ro ribonucleoprotein-associated Y RNAs *in vitro* and *in vivo*. *J Biol Chem* **276**:20711–20718.

47. Fouraux MA, Bouvet P, Verkaart S, van Venrooij WJ, Pruijn GJ. 2002. Nucleolin associates with a subset of the human Ro ribonucleoprotein complexes. *J Mol Biol* **320**:475–488.

48. Hogg JR, Collins K. 2007. Human Y5 RNA specializes a Ro ribonucleoprotein for 5S ribosomal RNA quality control. *Genes Dev* **21**:3067–3072.

49. Köhn M, Lederer M, Wächter K, Hüttelmaier S. 2010. Near-infrared (NIR) dye-labeled RNAs identify binding of ZBP1 to the noncoding Y3-RNA. *RNA* **16**:1420–1428.

50. Symmons MF, Jones GH, Luisi BF. 2000. A duplicated fold is the structural basis for polynucleotide phosphorylase catalytic activity, processivity, and regulation. *Structure* **8**:1215–1226.

51. Chan CW, Chetnani B, Mondragón A. 2013. Structure and function of the T-loop structural motif in noncoding RNAs. *Wiley Interdiscip Rev RNA* **4**:507–522.

52. Altman S, Kirsebom L, Talbot S. 1993. Recent studies of ribonuclease P. *FASEB J* **7**:7–14.

53. Wurtmann EJ, Wolin SL. 2010. A role for a bacterial ortholog of the Ro autoantigen in starvation-induced rRNA degradation. *Proc Natl Acad Sci U S A* **107**:4022–4027.

54. Tanaka M, Earl AM, Howell HA, Park MJ, Eisen JA, Peterson SN, Battista JR. 2004. Analysis of *Deinococcus radiodurans*'s transcriptional response to ionizing radiation and desiccation reveals novel proteins that contribute to extreme radioresistance. *Genetics* **168**:21–33.

55. Labbé JC, Burgess J, Rokeach LA, Hekimi S. 2000. ROP-1, an RNA quality-control pathway component, affects *Caenorhabditis elegans* dauer formation. *Proc Natl Acad Sci U S A* **97**:13233–13238.

56. Evguenieva-Hackenberg E, Hou L, Glaeser S, Klug G. 2014. Structure and function of the archaeal exosome. *Wiley Interdiscip Rev RNA* **5**:623–635.

57. Zinder JC, Lima CD. 2017. Targeting RNA for processing or destruction by the eukaryotic RNA exosome and its cofactors. *Genes Dev* **31**:88–100.

58. Bonneau F, Basquin J, Ebert J, Lorentzen E, Conti E. 2009. The yeast exosome functions as a macromolecular cage to channel RNA substrates for degradation. *Cell* **139**:547–559.

59. Górna MW, Carpousis AJ, Luisi BF. 2012. From conformational chaos to robust regulation: the structure and function of the multi-enzyme RNA degradosome. *Q Rev Biophys* **45**:105–145.

60. Viegas SC, Pfeiffer V, Sittka A, Silva IJ, Vogel J, Arraiano CM. 2007. Characterization of the role of ribonucleases in *Salmonella* small RNA decay. *Nucleic Acids Res* **35**:7651–7664.

61. Englert M, Sheppard K, Aslanian A, Yates JR III, Söll D. 2011. Archaeal 3′-phosphate RNA splicing ligase characterization identifies the missing component in tRNA maturation. *Proc Natl Acad Sci U S A* **108**:1290–1295.

62. Popow J, Englert M, Weitzer S, Schleiffer A, Mierzwa B, Mechtler K, Trowitzsch S, Will CL, Lührmann R, Söll D, Martinez J. 2011. HSPC117 is the essential subunit of a human tRNA splicing ligase complex. *Science* **331**:760–764.

63. Kosmaczewski SG, Edwards TJ, Han SM, Eckwahl MJ, Meyer BI, Peach S, Hesselberth JR, Wolin SL, Hammarlund M. 2014. The RtcB RNA ligase is an essential component of the metazoan unfolded protein response. *EMBO Rep* **15**:1278–1285.

64. Temmel H, Müller C, Sauert M, Vesper O, Reiss A, Popow J, Martinez J, Moll I. 2017. The RNA ligase RtcB reverses MazF-induced ribosome heterogeneity in *Escherichia coli*. *Nucleic Acids Res* **45**:4708–4721.

65. Hartman CE, Samuels DJ, Karls AC. 2016. Modulating *Salmonella* Typhimurium's response to a changing environment through bacterial enhancer-binding proteins and the RpoN regulon. *Front Mol Biosci* **3**:41.

66. Westermann AJ, Förstner KU, Amman F, Barquist L, Chao Y, Schulte LN, Müller L, Reinhardt R, Stadler PF, Vogel J. 2016. Dual RNA-seq unveils noncoding RNA functions in host-pathogen interactions. *Nature* **529**:496–501.

67. Das U, Shuman S. 2013. 2′-Phosphate cyclase activity of RtcA: a potential rationale for the operon organization of RtcA with an RNA repair ligase RtcB in *Escherichia coli* and other bacterial taxa. *RNA* **19**:1355–1362.

68. Popow J, Jurkin J, Schleiffer A, Martinez J. 2014. Analysis of orthologous groups reveals archease and DDX1 as tRNA splicing factors. *Nature* **511**:104–107.

69. Desai KK, Cheng CL, Bingman CA, Phillips GN Jr, Raines RT. 2014. A tRNA splicing operon: archease endows RtcB with dual GTP/ATP cofactor specificity and accelerates RNA ligation. *Nucleic Acids Res* **42**:3931–3942.

70. Quast C, Pruesse E, Yilmaz P, Gerken J, Schweer T, Yarza P, Peplies J, Glöckner FO. 2013. The SILVA ribosomal RNA gene database project: improved data processing and web-based tools. *Nucleic Acids Res* **41**(Database issue):D590–D596.

71. Sievers F, Wilm A, Dineen D, Gibson TJ, Karplus K, Li W, Lopez R, McWilliam H, Remmert M, Söding J, Thompson JD, Higgins DG. 2011. Fast, scalable generation of high-quality protein multiple sequence alignments using Clustal Omega. *Mol Syst Biol* **7**:539.

72. Felsenstein J. 1989. PHYLIP—Phylogeny Interference Package (Version 3.2). *Cladistics* **5**:164–166.

General Considerations

VI

Regulating with RNA in Bacteria and Archaea
Edited by Gisela Storz and Kai Papenfort
© 2018 American Society for Microbiology, Washington, DC
doi:10.1128/microbiolspec.RWR-0026-2018

Proteins That Chaperone RNA Regulation

22

Sarah A. Woodson[1], Subrata Panja[1], and Andrew Santiago-Frangos[2,3]

INTRODUCTION

Noncoding RNA sequences fold into useful structures that regulate gene expression as ribozymes, metabolite-binding sensors, or antisense RNAs (1–5). These regulatory RNAs are chaperoned by diverse families of RNA-binding proteins, and the loss of RNA chaperone proteins can lead to impaired growth, reduced tolerance to stress, and reduced virulence (6–11). RNA chaperones also facilitate conformational rearrangements during ribosome biogenesis (12) and eukaryotic pre-mRNA splicing (13).

These housekeeping and regulatory functions of RNA chaperones are particularly important at cold temperatures that hyperstabilize RNA structures. For example, the upregulation of cold shock domain RNA-binding proteins during low temperature growth destabilizes RNA structures that would otherwise impair transcription elongation and translation initiation at low temperatures (14). Moreover, overexpression of cold shock proteins and other RNA chaperones buffered deleterious mutations in *Escherichia coli* (15), suggesting that such proteins broadly mitigate the effects of RNA misfolding.

RNA chaperones act by transiently binding and releasing RNA substrates, disrupting the RNA secondary and tertiary structure (unwinding and unfolding), or accelerating base-pairing with a second RNA strand (annealing) (16, 17). RNA helices can be actively unwound by DEAD-box proteins, which couple unfolding of the RNA structure to ATP hydrolysis (18, 19). Bacteria typically encode a handful (0 to 12) of DEAD-box proteins that act mainly in ribosome biogenesis and RNA turnover (12, 20). This review, however, will focus on RNA-binding proteins that "passively" remodel RNA structures without hydrolyzing ATP. In bacteria, this type of passive RNA chaperone includes cold

shock proteins (14), the Sm family protein Hfq (host factor phage Qβ) (21), the FinO/ProQ family of RNA-binding proteins (22), and ribosomal proteins S1 (23–25) and S12 (26). H-NS and StpA, which interact with the bacterial nucleoid, also possess RNA chaperone activity (17). Eukaryotic proteins with ATP-independent RNA chaperone properties include RNA recognition motif (RRM) proteins such as hnRNP A1 (27), viral proteins such as the well-studied retroviral nucleocapsid protein (NCp7) (28, 29), and La and Ro proteins (30, 31).

The diverse biological roles of the above examples highlight the broad importance of proteins that escort, facilitate, or accelerate structural changes in noncoding RNA. This review will discuss how well-studied examples, such as *E. coli* CspA, HIV NCp7, ribosomal protein S1, and *E. coli* Hfq, are beginning to provide a physical picture of how proteins remodel RNA structures during RNA regulation.

RNA FOLDING AND THE NEED FOR RNA CHAPERONES

RNA double helices are stable and long-lived yet able to interchange through the sequential migration of base pairs during strand transfer or branch migration reactions. These features of stability and interchangeability ideally suit RNA for creating metastable structures that can switch gene expression on or off. Owing to the stability of the RNA double helix, RNA regulatory elements may be as small as a single stem-loop or antisense helix or involve more-elaborate tertiary structures (32–34). The potential simplicity of RNA-based regulation offers microbes an expedient means of evolving new regulatory circuits (35, see also chapter 28). It also

---

[1]T.C. Jenkins Department of Biophysics; [2]Program in Cell, Molecular and Developmental Biology and Biophysics, Johns Hopkins University, Baltimore, MD 21218; [3]Department of Microbiology and Immunology, Montana State University, Bozeman, MT 59717.

presents synthetic biologists with an attractive platform for genetic engineering (36, 37, see also chapter 31).

Although stable RNA base pairs create good switches, the small number of natural nucleobases limits the specificity of RNA regulatory interactions and increases the chance of RNA misfolding (16). For example, the free energy of forming a 10-bp stem-loop typically ranges from −10 to −20 kcal/mol at 37°C, yet a mismatched helix of the same length may be only 1 to 2 kcal/mol less stable than the fully matched helix. As an RNA grows longer, the odds that it can form more than one stable secondary structure increase substantially. Tertiary interactions between double helices make folding more specific by favoring folding intermediates in which the double helices are correctly aligned (38, 39). Nevertheless, interactions between nucleotides far apart in the RNA sequence are entropically less favorable than those between nearby nucleotides. As a result, stable local interactions outcompete long-range interactions, frustrating the search for the native conformation and allowing incorrect base-pairing patterns to persist (40, 41).

The potential for misfolding in RNA is exacerbated by the varied time scales for forming secondary and tertiary interactions. In a typical folding pathway (Fig. 1), local stem-loops form in ~10 μs while individual

domains of tertiary structure or intermolecular interactions typically form in 1 to 100 ms, depending on the folding conditions (41–46). This is far less time than the 0.5 to 4 s needed for a bacterial RNA polymerase to synthesize a 50- to 100-nucleotide RNA domain, allowing the 5′ end of a long RNA to form intermediate structures before the 3′ end has been transcribed (47). Because nonnative structures must partially unfold to reach the native structure, refolding can take 1 to 100 s or longer (41). Thus, RNA chaperones are needed to disrupt incorrect secondary or tertiary structure and increase the likelihood that the RNA will achieve its native conformation (16, 48).

Similar challenges complicate the formation of antisense interactions between *trans*-acting regulatory RNAs and their targets. During the association of unstructured oligonucleotides, slow nucleation of the double helix (~10^5 to 10^6 M^{-1}s^{-1}) is followed by more-rapid zippering (≤1 μs/bp) of the remaining base pairs (49, 50). The secondary structure of natural antisense RNAs alters their association kinetics, in some cases necessitating the assistance of a chaperone protein (51). For example, antisense RNAs from ColE1 and R1 plasmids initially base-pair at exposed hairpin loops, forming unstable "kissing complexes" (52, 53) (Fig. 2, middle

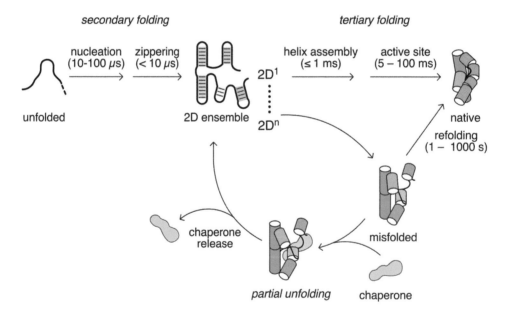

Figure 1 Iterative annealing of RNA by chaperones. Typical kinetic mechanism for forming RNA secondary (2D) structure (left) and native tertiary structure (right). Assembly of the double helices (cylinders) into compact intermediates is followed by further reorganization of tertiary interactions to produce the native RNA. Because the RNA may adopt many secondary structures, some molecules fold directly to the native structure (top path) while others become trapped in nonnative structures. In the classic iterative annealing model, chaperones (gold, bottom) bind and partially unfold misfolded intermediates, then release the unfolded RNA to fold again. Adapted from reference 59 with permission.

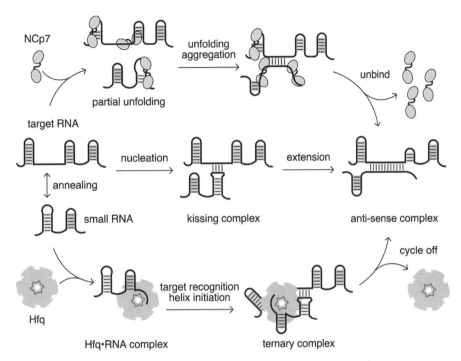

Figure 2 Chaperone-assisted annealing of antisense RNA. Annealing of antisense or *trans*-acting sRNAs with a complementary RNA target typically begins with base-pairing between two hairpin loops (kissing complex) or a loop and a single strand (middle path). This is followed by extension of base-pairing, which often requires refolding of adjacent sequences. HIV nucleocapsid (NCp7) and Rom/Rop promote annealing by disrupting secondary structure in each RNA, lowering the energetic barriers for extending the antisense interactions (top path). NCp7 can also aggregate RNA strands to speed up initiation of base-pairing. Hfq facilitates sRNA-mRNA base-pairing by forming a ternary complex with both RNAs that increases the rate of helix nucleation (bottom path). Hfq can also favor antisense base-pairing by restructuring one or both RNAs.

path). Stable antisense binding depends on the rearrangement of adjacent nucleotides to extend the intermolecular base-pairing (52, 54, 55), which for ColE1 is facilitated by the Rop/Rom protein (56). In an analogous fashion, *trans*-acting small RNAs (sRNAs) depend on chaperone proteins like Hfq to initiate base-pairing, facilitate strand exchange, and destabilize self-structure that can mask complementary regions in the sRNA and the target RNA (9, 21).

THERMODYNAMIC LIMITS TO PASSIVE UNFOLDING OF RNA

To unfold RNA, chaperones must bind single-stranded RNA more strongly than they bind double-stranded RNA (16). This concept was supported by early experiments on hnRNP A1 (27) and StpA (57, 58), which preferentially interact with RNAs that have some unstructured regions. Simple thermodynamic arguments suggest that the more strongly a chaperone binds the

unfolded RNA, the more potent its unfolding activity. If binding is too strong, however, the RNA will not be released, limiting the range of structures on which the chaperone can act (59). In practice, RNA chaperones unfold both native and misfolded substrates, sometimes with little discrimination (60). This suggests that thermodynamic models are inadequate to describe chaperone mechanisms, and that most chaperones take advantage of differences in the dynamics of folded and unfolded RNA to increase the flux through RNA-folding pathways over the short term.

TRANSIENT INTERACTIONS DRIVE ITERATIVE CHAPERONE CYCLES

To facilitate the formation of regulatory RNA structures, molecular chaperones must interact transiently with their substrates, releasing the folded RNA to perform its normal function in the cell. In the classic iterative annealing model for protein or RNA chaperones

(61, 62), the chaperone binds and partially unfolds the substrate before releasing it to fold again (Fig. 1). Repeated rounds of chaperone-induced unfolding and refolding provide the substrate many chances to fold correctly, ultimately resulting in a higher yield of native protein or RNA. On each round of folding, the RNA randomly partitions between folding pathways that lead to the native structure or nonnative structures (41). If the pool of native RNA is depleted through a subsequent biochemical reaction or the addition of other proteins, this cycling will increase the throughput of RNA folding and assembly, even if the chaperone does not discriminate between native and nonnative structures.

The iterative annealing mechanism was validated by experiments on GroEL protein chaperone (61) and on the CYT-19 DEAD-box RNA chaperone (19), both of which use ATP hydrolysis to carry out multiple cycles of substrate unwinding. A recent theoretical analysis of the reaction kinetics of GroEL and CYT-19 ATPases led to the proposal that active protein and RNA chaperones operate far from equilibrium, increasing the amount of native substrate in the short term even if some native protein or RNA is also unfolded by the chaperone (63). In this framework, repeated rounds of unfolding accelerate the rate at which substrates reach the native structure, although the efficiency of the chaperone depends on its ability to discriminate between native and misfolded protein or RNA.

Passive chaperones also cycle on and off their substrates to transiently unfold (and refold) the RNA (59). Analogous cycles of binding and release facilitate annealing between *trans*-acting regulatory RNAs and their targets. The main difference is that the chaperone must simultaneously bind the regulatory or guide RNA and the target RNA, bringing them together in a ternary complex that allows the two RNAs to base-pair (64). After the RNAs base-pair, the chaperone releases the RNA duplex or is recycled when the RNA complex is turned over (65).

Whether the RNA folds upon itself or with another RNA, rapid binding and release of the chaperone is essential for allowing the RNA to search out its most stable structure (17). This has been borne out by experiments on variants of HIV NCp7 (66) and on *E. coli* StpA (58, 67). Similarly, Hfq protein, which promotes annealing between sRNAs and their mRNA targets *in vitro* (58, 68–70), cycles off the sRNA-mRNA duplex (69, 71). Efficient matchmaking requires that the sRNA-Hfq complex can rapidly search among candidate targets until a complementary site is found (72).

RNA UNFOLDING BY CAPTURING SINGLE STRANDS

Recent biophysical studies illuminate how passive RNA chaperones such as CspA and HIV NCp7 weaken RNA secondary structures. CspA is the major cold shock protein in *E. coli* and is strongly induced upon cold shock at 10°C (73). Among other functions, CspA and its relative CspE are proposed to destabilize mRNA secondary structures that may terminate transcription or interfere with mRNA translation at low temperature (74–76). The cold shock domains of CspA and CspE are related to eukaryotic Y-box proteins and form a β-barrel structure that binds RNA along one surface (77, 78). Three aromatic side chains protruding from the surface of the cold shock domain were found to be crucial for RNA chaperone activity (79, 80).

Rennella et al. used time-resolved nuclear magnetic resonance spectroscopy to observe how multiple CspA proteins facilitate the dimerization of two complementary RNA hairpins (81). The slowest unfolding rate equaled the rate of RNA dimerization, suggesting that each hairpin must unzip before the two strands can anneal. CspA increased the hairpin unfolding rates, surprisingly by destabilizing base pairs closest to the loop or at a helix junction. As in previous studies, exposed aromatic residues contributed to destabilization of the RNA. The authors concluded that base pairs in the RNA hairpins are progressively disrupted by stacking interactions with the aromatic side chains, accompanied by hydrogen bonding with surrounding basic residues. This combination of aromatic and basic side chains, found in other RNA-binding proteins (82), seems to offer a flexible yet energetically favorable interaction surface for single-stranded RNA. Importantly, side chains on the exposed surface of CspA are dynamic (83) but are stabilized upon binding to the RNA hairpin (81). Thus, increased motion in the RNA base pairs is accompanied by reduced motion in the protein. As discussed below, many chaperones contain disordered regions, leading to the idea that entropy is transferred from the chaperone to the substrate (84).

The RNA chaperone activities of nucleocapsid proteins are mainly understood from extensive studies of HIV nucleocapsid (NCp7), which can both disrupt RNA structures and promote interactions between RNA strands (29). These chaperone properties contribute to many stages of the retroviral life cycle, including dimerization and packaging of the genomic RNA, tRNA priming, and transactivation (85, 86). Processed from the longer Gag polyprotein, NCp7 contains two small Zn-finger domains that bind and destabilize RNA structures (87–90). Each Zn finger contains a hydrophobic pocket capable of binding an unpaired guanosine nucleobase. A disordered, positively charged N-terminal domain is associated with NCp7's aggregating properties (91).

Like CspA, many copies of NCp7 bind a single RNA, destabilizing its structure (92, 93). Recent force-stretching experiments provided additional insight into how HIV NCp7 destabilizes the transactivation response (TAR) RNA hairpin (94). During reverse transcription of the HIV genome, the TAR RNA hairpin must be destabilized and annealed to a cDNA hairpin (95). In the force-stretching experiments, NCp7 increased the probability of TAR unfolding about 10,000 times. This acceleration was accomplished by moving the position of the transition state for unfolding so that fewer base pairs must open at one time before the entire TAR hairpin unzips (Fig. 2, top path). In agreement with earlier studies (96, 97), the authors concluded that NCp7 acts mainly by preferentially binding guanosines near local defects in the RNA such as GU wobble pairs, bulges, or loops, disrupting their base-pairing interactions. This explains how NCp7 progressively destabilizes large RNA structures at the moderate loading ratios of 1 protein per 7 to 15 nucleotides, at which its chaperone activity is most prominent (98). Moreover, NCp7 interacts weakly with most of its binding sites at moderate protein-to-RNA ratios (66). This ensures that NCp7 dissociates frequently enough that the RNA has a chance to form interactions that are most stable.

MULTIDOMAIN INTERACTIONS BETWEEN CHAPERONES AND RNA

Many RNA chaperones form multiple weak interactions with the RNA that are individually exchangeable yet collectively have a large effect on the RNA's structure. This distributive binding strategy offers several advantages. First, contacts between the chaperone and the RNA can rearrange without the chaperone losing its grip on the RNA substrate. Second, multiple RNA-binding surfaces can be used to flexibly recognize different RNA structures or to select a particular class of substrates. Third, chaperones with annealing activity require multiple surfaces to bring together two RNA strands. Many RNA-binding proteins use multiple domains or multiple copies of the same protein to chaperone RNA refolding or improve substrate selection, although the three-dimensional structure of each protein family is different.

Small chaperone proteins, such as CspA, NCp7, and StpA, restructure the RNA by having many copies of the protein bind at the same time, explaining why these proteins are most effective at moderately high protein-to-RNA ratios (58, 74, 90). Each polypeptide also possesses multiple RNA-binding domains that must work together to unwind or anneal the RNA (67). For example, the chaperone activity of HIV NCp7 requires the combined action of each Zn finger and the basic N-terminal region (87–89, 96, 99). Ribosomal protein S1 also has a well-characterized propensity to unwind RNA helices (23–25). Of the six oligonucleotide-binding (OB) domains in *E. coli* protein S1, at least three are needed for the 30S ribosome to accommodate structured mRNAs (100), suggesting that multiple OB domains must work together to unfold the mRNA. Consistent with this idea, optical tweezer experiments showed that each protein S1 unwinds 10 base pairs in a multistep fashion that requires less energy than would be needed to unfold the entire helix at once (101).

Eukaryotic proteins with chaperone activity also typically contain multiple RNA-binding domains joined by flexible linkers. This "pearl necklace" organization enables more specific recognition of RNA sequence motifs (102) and accommodates conformational changes in the RNA. For example, yeast Prp24 facilitates annealing between U6 and U4 snRNA, an essential step for recycling spliceosomal complexes (103). The four RRM domains of Prp24 wrap around spliceosomal U6 snRNA, making extensive contacts with double- and single-stranded regions of the RNA (104). Reorientation of these domains is presumably needed to allow remodeling of U6 base pairs (105). Yeast La protein binds the 3′ ends of RNA polymerase III transcripts (106, 107) and chaperones misfolded pre-tRNAs (108) and other small nuclear RNAs (109). When La binds the RNA 3′ end, a flexible linker between the La motif and an RRM domain becomes ordered, orienting the two domains and greatly increasing the fidelity of substrate selection (110).

RING-SHAPED CHAPERONES FOR sRNAs

Ring-shaped proteins that bind RNA offer a different solution for the requirement that RNA chaperones contact more than one RNA segment at a time (111). Unlike multidomain RNA-binding proteins, ring-shaped proteins form rigid three-dimensional structures that bind RNA segments on different surfaces. As a result, the RNA is wrapped around or even through the ring, distorting the RNA's structure or aligning two different strands of RNA. The *Bacillus subtilis trp* RNA-binding attenuator protein (TRAP) is a particularly dramatic example, wrapping up to 33 nucleotides of RNA around a large, 11-subunit wheel (112). A smaller example is the Ro antigen, which uses a ring of HEAT repeats to bind and chaperone eukaryotic Y-box sRNAs (31, see also chapter 21).

Hfq is the best-characterized chaperone for sRNAs in bacteria (113). A member of the Sm/Lsm protein family (114), Hfq assembles into a 6-subunit ring that binds sRNAs, protects them from turnover, and aids their annealing with target mRNAs (21) (Fig. 2, bottom path). In addition to supporting posttranscriptional regulation by sRNAs, Hfq directly regulates the translation of its own and other mRNAs (115–119), among other regulatory roles (120).

Hfq uses a mixture of RNA-binding surfaces and disordered domains to recognize a broad family of RNA substrates (121), while ignoring other RNA in the cell. The "proximal" face of Hfq is conserved among Sm/Lsm proteins and interacts with single-stranded U's at the 3′ ends of sRNAs (122–125). These interactions were shown to be important for sRNA stability in the cell (126–129). The opposite, "distal" face of E. coli Hfq binds a single-stranded AAN triplet motif present in mRNA targets of Hfq and sRNA regulation (130–132), as well as a minor class of sRNA (129). AAN recognition was found to be important for sRNA upregulation of rpoS translation in E. coli (131, 133, 134) and downregulation of many targets by Spot 42 (135). Thus, one Hfq ring can bring together an sRNA and mRNA, promoting their association (68, 114, 130, 136). In addition to these sequence-specific binding surfaces, the lateral edge or rim of the Hfq hexamer contains patches of basic residues (usually arginines) that interact with complementary regions of each RNA (137, 138). The importance of these multilateral interactions for Hfq-dependent sRNA regulation has been supported by additional mutational analyses and in-cell cross-linking and coimmunoprecipitation (see, for example, references 117, 129, 130, and 139–142).

A variety of biophysical experiments showed that Hfq accelerates base-pairing between complementary RNA strands (69, 70, 143, 144), and this annealing activity requires the basic patch on the rim of the hexamer (138, 145). Like other RNA chaperone proteins, Hfq binds substrate RNA within a few seconds but then cycles off the sRNA-mRNA duplex once base-pairing is complete (69, 71). Single-molecule fluorescence resonance energy transfer (FRET) studies using short RNA substrates showed that stable annealing is preceded by transient RNA binding to Hfq, and that Hfq dissociates from the RNA duplex soon after base-pairing occurs (146). Although Hfq is reported to bind ATP (147, 148), Hfq lacks a RecA-like domain, and ATP hydrolysis is not needed for its annealing activity (69, 143, 144).

It is not known precisely how the arginine side chains on the rim of the Hfq hexamer facilitate RNA base-pairing, but several features of the annealing reaction have been established. First, interactions between the rim and UA motifs in the body of the sRNA (127) predispose the seed region for base-pairing with a complementary strand (137), perhaps by increasing the flexibility of the bound sRNA (149). Second, the arginines directly stabilize a helix initiation complex, either by overcoming electrostatic repulsion of the two RNA strands or by hydrogen bonding (150). Third, by simultaneously interacting with the RNA via its proximal, distal, and lateral rim surfaces, Hfq can refold the RNA into a structure that is more amenable to pairing with its complement. For example, the distal face of Hfq is recruited to an $(AAN)_4$ motif in an upstream domain of the rpoS mRNA 5′ untranslated region (131, 151), while the rim interacts with a U-rich loop downstream of the sRNA target site (141). SHAPE (selective 2′-hydroxyl acylation analyzed by primer extension) modification and SAXS (small-angle X-ray scattering) experiments showed that this multisite recognition folds the rpoS 5′ untranslated region into a distorted structure that is poised to base-pair with sRNAs that regulate rpoS expression (152). Other mRNA targets of sRNA and Hfq regulation similarly contain multiple Hfq recognition sites (117, 153–155), suggesting that distortion of the mRNA conformation is a common strategy for enhancing the efficiency of sRNA regulation.

RNA CHAPERONES DRIVE sRNA COMPETITION AND TARGET SELECTION

Small, noncoding RNAs must identify their proper targets among a myriad of spurious candidates in the cell. The target recognition kinetics depends on colocalization, a search among potential targets, and accurate recognition of the cognate target sequence. CRISPR (clustered regularly interspaced short palindromic repeat) guide RNAs and eukaryotic microRNAs are stably packaged with Cas and RISC protein, respectively, which rapidly scan DNA or RNA substrates for potential target sites while increasing the fidelity of target-site recognition (156). By contrast, overexpression experiments show that bacterial sRNAs must compete for a limited pool of Hfq hexamers in the cell (157, 158). To compete, sRNAs rapidly cycle on and off the protein.

Initially, it was not clear how Hfq-bound RNAs exchange, given that sRNA and mRNA containing U-rich or AAN recognition motifs have dissociation constants of 1 to 30 nM (159). Fender et al. showed that the sRNA dissociation kinetics depends on the pool of free RNA (160), and they proposed that sRNAs actively displace each other from the proximal face of the Hfq hexamer. RNA competition for the proximal face of Hfq was also observed in single-molecule FRET experiments (146). Regardless of the mechanism of sRNA exchange, mRNAs are typically targeted within 1 to 2 min of sRNA induction. Rapid cycling is needed to match sRNAs with their complementary targets within this time frame (157) and to quickly retool regulatory circuits as growth conditions change (65).

INTRINSICALLY DISORDERED DOMAINS IN RNA CHAPERONES

In addition to structured RNA-binding domains, RNA chaperones commonly contain intrinsically disordered peptide regions that are essential for their chaperone activity (84). The frequent presence of these regions in RNA chaperones has led to the "entropy transfer" hypothesis, in which folding of the chaperone is coupled to increased disorder in the RNA substrate (84, 161, 162). These disordered regions likely play different roles, depending on their charge. Basic N-terminal polypeptides frequently increase the RNA affinity of the chaperone and the ability to remodel RNA structures. For example, the basic N-terminal domain of HIV NCp7 contributes substantially to acceleration of TAR hairpin annealing with cDNA during minus-strand transfer (163). The Mss116p and CYT-19 DEAD-box helicases possess an unstructured, arginine-rich C-terminal extension that recruits the helicase to the RNA substrate (164, 165) and may even help loosen the RNA structure (166). Disordered regions may simply allow a chaperone to bind a variety of RNA structures (167), as proposed for La protein (168). Polyamines or oligo-Lys also accelerate refolding of misfolded RNA, however, suggesting that the distributed positive charge associated with basic peptides lowers the energetic barrier for refolding (169, 170).

In contrast to the examples above, many DNA- and RNA-binding proteins contain acidic peptides that mimic nucleic acid and inhibit nucleic acid binding to the core of the protein (171–174). Recent experiments on Hfq illustrate the importance of these acidic peptides for RNA chaperone activity (175, 176). The last 30 residues of E. coli Hfq are disordered in solution and protrude from the edge of the hexameric ring formed by the stable Sm domain (177, 178). Acidic residues at the tip of the C-terminal domain (CTD), which are moderately conserved in bacterial Hfq sequences, interact with the arginine patches on the rim of the hexamer (175), displacing double-stranded RNA and inhibiting the binding of nonspecific RNA and DNA (176). In an analogous fashion, acidic residues at the C terminus of NCp7 were also found to increase the rate of dissociation from nucleic acid substrates (179).

Displacement of RNA by the CTD has several consequences for Hfq's role in sRNA regulation. First, the CTD prevents separation of sRNA-mRNA strands (the reverse of annealing) while recycling Hfq to bind another RNA. Second, it creates a hierarchy of sRNA regulation because the CTDs do not displace AAN RNA from the distal face of Hfq, enabling RNAs containing AAN motifs to outcompete other sRNAs for access to Hfq in the cell (176). Because the length of the CTD and number of acidic residues vary among bacterial species, these features may be used to fine-tune the permissiveness and turnover rate of Hfq in different organisms (180).

CONCLUSION

Proteins that bind and chaperone noncoding RNA sequences are ubiquitous in biology and are necessary for many forms of gene control, stress response, viral replication, and normal homeostasis of RNA metabolism. This review has outlined some of the common physical mechanisms by which such proteins help refold regulatory RNA structures or facilitate antisense interactions between regulatory RNAs and their targets. Further understanding of RNA chaperone mechanisms will require not only additional high-resolution structures but advanced physical methods for tracking the dynamical motions of the RNA-protein complexes in real time.

Acknowledgments. The authors acknowledge the support of their research from the National Institutes of Health (R01 GM120425) and thank G. Storz and S. Gottesman for helpful discussion and comments on the manuscript.

Citation. Woodson SA, Panja S, Santiago-Frangos A. 2018. Proteins that chaperone RNA regulation. Microbiol Spectrum 6(4):RWR-0026-2018.

References

1. **Serganov A, Patel DJ.** 2007. Ribozymes, riboswitches and beyond: regulation of gene expression without proteins. *Nat Rev Genet* 8:776–790.
2. **Grundy FJ, Henkin TM.** 2006. From ribosome to riboswitch: control of gene expression in bacteria by RNA structural rearrangements. *Crit Rev Biochem Mol Biol* 41:329–338.
3. **Winkler WC, Breaker RR.** 2005. Regulation of bacterial gene expression by riboswitches. *Annu Rev Microbiol* 59:487–517.

4. Beisel CL, Storz G. 2010. Base pairing small RNAs and their roles in global regulatory networks. *FEMS Microbiol Rev* 34:866–882.

5. Wagner EG, Romby P. 2015. Small RNAs in bacteria and archaea: who they are, what they do, and how they do it. *Adv Genet* 90:133–208.

6. Chao Y, Vogel J. 2010. The role of Hfq in bacterial pathogens. *Curr Opin Microbiol* 13:24–33.

7. Gottesman S, McCullen CA, Guillier M, Vanderpool CK, Majdalani N, Benhammou J, Thompson KM, FitzGerald PC, Sowa NA, FitzGerald DJ. 2006. Small RNA regulators and the bacterial response to stress. *Cold Spring Harb Symp Quant Biol* 71:1–11.

8. Romeo T, Vakulskas CA, Babitzke P. 2013. Post-transcriptional regulation on a global scale: form and function of Csr/Rsm systems. *Environ Microbiol* 15:313–324.

9. Sobrero P, Valverde C. 2012. The bacterial protein Hfq: much more than a mere RNA-binding factor. *Crit Rev Microbiol* 38:276–299.

10. Lucchetti-Miganeh C, Burrowes E, Baysse C, Ermel G. 2008. The post-transcriptional regulator CsrA plays a central role in the adaptation of bacterial pathogens to different stages of infection in animal hosts. *Microbiology* 154:16–29.

11. Romby P, Vandenesch F, Wagner EG. 2006. The role of RNAs in the regulation of virulence-gene expression. *Curr Opin Microbiol* 9:229–236.

12. Redder P, Hausmann S, Khemici V, Yasrebi H, Linder P. 2015. Bacterial versatility requires DEAD-box RNA helicases. *FEMS Microbiol Rev* 39:392–412.

13. Staley JP, Woolford JL Jr. 2009. Assembly of ribosomes and spliceosomes: complex ribonucleoprotein machines. *Curr Opin Cell Biol* 21:109–118.

14. Phadtare S, Severinov K. 2010. RNA remodeling and gene regulation by cold shock proteins. *RNA Biol* 7:788–795.

15. Rudan M, Schneider D, Warnecke T, Krisko A. 2015. RNA chaperones buffer deleterious mutations in *E. coli*. *eLife* 4:e04745.

16. Herschlag D. 1995. RNA chaperones and the RNA folding problem. *J Biol Chem* 270:20871–20874.

17. Doetsch M, Schroeder R, Fürtig B. 2011. Transient RNA-protein interactions in RNA folding. *FEBS J* 278:1634–1642.

18. Jankowsky E, Gross CH, Shuman S, Pyle AM. 2001. Active disruption of an RNA-protein interaction by a DExH/D RNA helicase. *Science* 291:121–125.

19. Bhaskaran H, Russell R. 2007. Kinetic redistribution of native and misfolded RNAs by a DEAD-box chaperone. *Nature* 449:1014–1018.

20. Iost I, Bizebard T, Dreyfus M. 2013. Functions of DEAD-box proteins in bacteria: current knowledge and pending questions. *Biochim Biophys Acta* 1829:866–877.

21. Updegrove TB, Zhang A, Storz G. 2016. Hfq: the flexible RNA matchmaker. *Curr Opin Microbiol* 30:133–138.

22. Olejniczak M, Storz G. 2017. ProQ/FinO-domain proteins: another ubiquitous family of RNA matchmakers? *Mol Microbiol* 104:905–915.

23. Bear DG, Ng R, Van Derveer D, Johnson NP, Thomas G, Schleich T, Noller HF. 1976. Alteration of polynucleotide secondary structure by ribosomal protein S1. *Proc Natl Acad Sci U S A* 73:1824–1828.

24. Hajnsdorf E, Boni IV. 2012. Multiple activities of RNA-binding proteins S1 and Hfq. *Biochimie* 94:1544–1553.

25. Kolb A, Hermoso JM, Thomas JO, Szer W. 1977. Nucleic acid helix-unwinding properties of ribosomal protein S1 and the role of S1 in mRNA binding to ribosomes. *Proc Natl Acad Sci U S A* 74:2379–2383.

26. Coetzee T, Herschlag D, Belfort M. 1994. *Escherichia coli* proteins, including ribosomal protein S12, facilitate in vitro splicing of phage T4 introns by acting as RNA chaperones. *Genes Dev* 8:1575–1588.

27. Herschlag D, Khosla M, Tsuchihashi Z, Karpel RL. 1994. An RNA chaperone activity of non-specific RNA binding proteins in hammerhead ribozyme catalysis. *EMBO J* 13:2913–2924.

28. Levin JG, Guo J, Rouzina I, Musier-Forsyth K. 2005. Nucleic acid chaperone activity of HIV-1 nucleocapsid protein: critical role in reverse transcription and molecular mechanism. *Prog Nucleic Acid Res Mol Biol* 80:217–286.

29. Rein A, Henderson LE, Levin JG. 1998. Nucleic-acid-chaperone activity of retroviral nucleocapsid proteins: significance for viral replication. *Trends Biochem Sci* 23:297–301.

30. Bayfield MA, Yang R, Maraia RJ. 2010. Conserved and divergent features of the structure and function of La and La-related proteins (LARPs). *Biochim Biophys Acta* 1799:365–378.

31. Sim S, Wolin SL. 2011. Emerging roles for the Ro 60-kDa autoantigen in noncoding RNA metabolism. *Wiley Interdiscip Rev RNA* 2:686–699.

32. Fürtig B, Nozinovic S, Reining A, Schwalbe H. 2015. Multiple conformational states of riboswitches fine-tune gene regulation. *Curr Opin Struct Biol* 30:112–124.

33. Lau MW, Ferré-D'Amaré AR. 2016. Many activities, one structure: functional plasticity of ribozyme folds. *Molecules* 21:E1570.

34. Krajewski SS, Narberhaus F. 2014. Temperature-driven differential gene expression by RNA thermosensors. *Biochim Biophys Acta* 1839:978–988.

35. Peer A, Margalit H. 2014. Evolutionary patterns of *Escherichia coli* small RNAs and their regulatory interactions. *RNA* 20:994–1003.

36. Kang Z, Zhang C, Zhang J, Jin P, Zhang J, Du G, Chen J. 2014. Small RNA regulators in bacteria: powerful tools for metabolic engineering and synthetic biology. *Appl Microbiol Biotechnol* 98:3413–3424.

37. Trausch JJ, Batey RT. 2015. Design of modular "plug-and-play" expression platforms derived from natural riboswitches for engineering novel genetically encodable RNA regulatory devices. *Methods Enzymol* 550:41–71.

38. Behrouzi R, Roh JH, Kilburn D, Briber RM, Woodson SA. 2012. Cooperative tertiary interaction network guides RNA folding. *Cell* 149:348–357.

39. Chauhan S, Woodson SA. 2008. Tertiary interactions determine the accuracy of RNA folding. *J Am Chem Soc* 130:1296–1303.

40. Thirumalai D, Hyeon C. 2005. RNA and protein folding: common themes and variations. *Biochemistry* **44**: 4957–4970.

41. Thirumalai D, Woodson SA. 1996. Kinetics of folding of protein and RNA. *Acc Chem Res* **29**:433–439.

42. Crothers DM. 2001. RNA conformational dynamics, p 61–70. *In* Söll D, Nishimura S, Moore P (ed), *RNA*. Elsevier, Oxford, United Kingdom.

43. Draper DE. 1996. Strategies for RNA folding. *Trends Biochem Sci* **21**:145–149.

44. Zarrinkar PP, Williamson JR. 1994. Kinetic intermediates in RNA folding. *Science* **265**:918–924.

45. Pan T, Sosnick TR. 1997. Intermediates and kinetic traps in the folding of a large ribozyme revealed by circular dichroism and UV absorbance spectroscopies and catalytic activity. *Nat Struct Biol* **4**:931–938.

46. Sclavi B, Sullivan M, Chance MR, Brenowitz M, Woodson SA. 1998. RNA folding at millisecond intervals by synchrotron hydroxyl radical footprinting. *Science* **279**: 1940–1943.

47. Lai D, Proctor JR, Meyer IM. 2013. On the importance of cotranscriptional RNA structure formation. *RNA* **19**: 1461–1473.

48. Schroeder R, Barta A, Semrad K. 2004. Strategies for RNA folding and assembly. *Nat Rev Mol Cell Biol* **5**: 908–919.

49. Craig ME, Crothers DM, Doty P. 1971. Relaxation kinetics of dimer formation by self complementary oligonucleotides. *J Mol Biol* **62**:383–401.

50. Pörschke D. 1974. A direct measurement of the unzippering rate of a nucleic acid double helix. *Biophys Chem* **2**:97–101.

51. Nordström K, Wagner EG. 1994. Kinetic aspects of control of plasmid replication by antisense RNA. *Trends Biochem Sci* **19**:294–300.

52. Tomizawa J. 1984. Control of ColE1 plasmid replication: the process of binding of RNA I to the primer transcript. *Cell* **38**:861–870.

53. Persson C, Wagner EG, Nordström K. 1990. Control of replication of plasmid R1: formation of an initial transient complex is rate-limiting for antisense RNA-target RNA pairing. *EMBO J* **9**:3777–3785.

54. Tamm J, Polisky B. 1985. Characterization of the ColE1 primer-RNA1 complex: analysis of a domain of ColE1 RNA1 necessary for its interaction with primer RNA. *Proc Natl Acad Sci U S A* **82**:2257–2261.

55. Kolb FA, Engdahl HM, Slagter-Jäger JG, Ehresmann B, Ehresmann C, Westhof E, Wagner EG, Romby P. 2000. Progression of a loop-loop complex to a four-way junction is crucial for the activity of a regulatory antisense RNA. *EMBO J* **19**:5905–5915.

56. Tomizawa J. 1990. Control of ColE1 plasmid replication. Interaction of Rom protein with an unstable complex formed by RNA I and RNA II. *J Mol Biol* **212**:695–708.

57. Grossberger R, Mayer O, Waldsich C, Semrad K, Urschitz S, Schroeder R. 2005. Influence of RNA structural stability on the RNA chaperone activity of the *Escherichia coli* protein StpA. *Nucleic Acids Res* **33**:2280–2289.

58. Mayer O, Rajkowitsch L, Lorenz C, Konrat R, Schroeder R. 2007. RNA chaperone activity and RNA-binding properties of the *E. coli* protein StpA. *Nucleic Acids Res* **35**:1257–1269.

59. Woodson SA. 2010. Taming free energy landscapes with RNA chaperones. *RNA Biol* **7**:677–686.

60. Tijerina P, Bhaskaran H, Russell R. 2006. Nonspecific binding to structured RNA and preferential unwinding of an exposed helix by the CYT-19 protein, a DEAD-box RNA chaperone. *Proc Natl Acad Sci U S A* **103**: 16698–16703.

61. Todd MJ, Lorimer GH, Thirumalai D. 1996. Chaperonin-facilitated protein folding: optimization of rate and yield by an iterative annealing mechanism. *Proc Natl Acad Sci U S A* **93**:4030–4035.

62. Hyeon C, Thirumalai D. 2013. Generalized iterative annealing model for the action of RNA chaperones. *J Chem Phys* **139**:121924.

63. Chakrabarti S, Hyeon C, Ye X, Lorimer GH, Thirumalai D. 2017. Molecular chaperones maximize the native state yield on biological times by driving substrates out of equilibrium. *Proc Natl Acad Sci U S A* **114**:E10919–E10927.

64. Storz G, Opdyke JA, Zhang A. 2004. Controlling mRNA stability and translation with small, noncoding RNAs. *Curr Opin Microbiol* **7**:140–144.

65. Wagner EG. 2013. Cycling of RNAs on Hfq. *RNA Biol* **10**:619–626.

66. Cruceanu M, Gorelick RJ, Musier-Forsyth K, Rouzina I, Williams MC. 2006. Rapid kinetics of protein-nucleic acid interaction is a major component of HIV-1 nucleocapsid protein's nucleic acid chaperone function. *J Mol Biol* **363**:867–877.

67. Cusick ME, Belfort M. 1998. Domain structure and RNA annealing activity of the *Escherichia coli* regulatory protein StpA. *Mol Microbiol* **28**:847–857.

68. Moll I, Leitsch D, Steinhauser T, Bläsi U. 2003. RNA chaperone activity of the Sm-like Hfq protein. *EMBO Rep* **4**:284–289.

69. Lease RA, Woodson SA. 2004. Cycling of the Sm-like protein Hfq on the DsrA small regulatory RNA. *J Mol Biol* **344**:1211–1223.

70. Rajkowitsch L, Schroeder R. 2007. Dissecting RNA chaperone activity. *RNA* **13**:2053–2060.

71. Hopkins JF, Panja S, Woodson SA. 2011. Rapid binding and release of Hfq from ternary complexes during RNA annealing. *Nucleic Acids Res* **39**:5193–5202.

72. Adamson DN, Lim HN. 2011. Essential requirements for robust signaling in Hfq dependent small RNA networks. *PLoS Comput Biol* **7**:e1002138.

73. Goldstein J, Pollitt NS, Inouye M. 1990. Major cold shock protein of *Escherichia coli*. *Proc Natl Acad Sci U S A* **87**:283–287.

74. Jiang W, Hou Y, Inouye M. 1997. CspA, the major cold-shock protein of *Escherichia coli*, is an RNA chaperone. *J Biol Chem* **272**:196–202.

75. Phadtare S, Inouye M, Severinov K. 2002. The nucleic acid melting activity of *Escherichia coli* CspE is critical

for transcription antitermination and cold acclimation of cells. *J Biol Chem* **277**:7239–7245.

76. Phadtare S, Tadigotla V, Shin WH, Sengupta A, Severinov K. 2006. Analysis of *Escherichia coli* global gene expression profiles in response to overexpression and deletion of CspC and CspE. *J Bacteriol* **188**:2521–2527.

77. Newkirk K, Feng W, Jiang W, Tejero R, Emerson SD, Inouye M, Montelione GT. 1994. Solution NMR structure of the major cold shock protein (CspA) from *Escherichia coli*: identification of a binding epitope for DNA. *Proc Natl Acad Sci U S A* **91**:5114–5118.

78. Schindelin H, Jiang W, Inouye M, Heinemann U. 1994. Crystal structure of CspA, the major cold shock protein of *Escherichia coli*. *Proc Natl Acad Sci U S A* **91**:5119–5123.

79. Phadtare S, Tyagi S, Inouye M, Severinov K. 2002. Three amino acids in *Escherichia coli* CspE surface-exposed aromatic patch are critical for nucleic acid melting activity leading to transcription antitermination and cold acclimation of cells. *J Biol Chem* **277**:46706–46711.

80. Phadtare S, Inouye M, Severinov K. 2004. The mechanism of nucleic acid melting by a CspA family protein. *J Mol Biol* **337**:147–155.

81. Rennella E, Sára T, Juen M, Wunderlich C, Imbert L, Solyom Z, Favier A, Ayala I, Weinhäupl K, Schanda P, Konrat R, Kreutz C, Brutscher B. 2017. RNA binding and chaperone activity of the *E. coli* cold-shock protein CspA. *Nucleic Acids Res* **45**:4255–4268.

82. Hall KB. 2017. RNA and proteins: mutual respect. *F1000 Res* **6**:345.

83. Feng W, Tejero R, Zimmerman DE, Inouye M, Montelione GT. 1998. Solution NMR structure and backbone dynamics of the major cold-shock protein (CspA) from *Escherichia coli*: evidence for conformational dynamics in the single-stranded RNA-binding site. *Biochemistry* **37**:10881–10896.

84. Tompa P, Kovacs D. 2010. Intrinsically disordered chaperones in plants and animals. *Biochem Cell Biol* **88**:167–174.

85. Darlix JL, de Rocquigny H, Mély Y. 2016. The multiple roles of the nucleocapsid in retroviral RNA conversion into proviral DNA by reverse transcriptase. *Biochem Soc Trans* **44**:1427–1440.

86. Rein A, Datta SA, Jones CP, Musier-Forsyth K. 2011. Diverse interactions of retroviral Gag proteins with RNAs. *Trends Biochem Sci* **36**:373–380.

87. Guo J, Wu T, Kane BF, Johnson DG, Henderson LE, Gorelick RJ, Levin JG. 2002. Subtle alterations of the native zinc finger structures have dramatic effects on the nucleic acid chaperone activity of human immunodeficiency virus type 1 nucleocapsid protein. *J Virol* **76**:4370–4378.

88. Heath MJ, Derebail SS, Gorelick RJ, DeStefano JJ. 2003. Differing roles of the N- and C-terminal zinc fingers in human immunodeficiency virus nucleocapsid protein-enhanced nucleic acid annealing. *J Biol Chem* **278**:30755–30763.

89. Williams MC, Gorelick RJ, Musier-Forsyth K. 2002. Specific zinc-finger architecture required for HIV-1

nucleocapsid protein's nucleic acid chaperone function. *Proc Natl Acad Sci U S A* **99**:8614–8619.

90. Williams MC, Rouzina I, Wenner JR, Gorelick RJ, Musier-Forsyth K, Bloomfield VA. 2001. Mechanism for nucleic acid chaperone activity of HIV-1 nucleocapsid protein revealed by single molecule stretching. *Proc Natl Acad Sci U S A* **98**:6121–6126.

91. Le Cam E, Coulaud D, Delain E, Petitjean P, Roques BP, Gérard D, Stoylova E, Vuilleumier C, Stoylov SP, Mély Y. 1998. Properties and growth mechanism of the ordered aggregation of a model RNA by the HIV-1 nucleocapsid protein: an electron microscopy investigation. *Biopolymers* **45**:217–229.

92. Azoulay J, Clamme JP, Darlix JL, Roques BP, Mély Y. 2003. Destabilization of the HIV-1 complementary sequence of TAR by the nucleocapsid protein through activation of conformational fluctuations. *J Mol Biol* **326**:691–700.

93. Heilman-Miller SL, Wu T, Levin JG. 2004. Alteration of nucleic acid structure and stability modulates the efficiency of minus-strand transfer mediated by the HIV-1 nucleocapsid protein. *J Biol Chem* **279**:44154–44165.

94. McCauley MJ, Rouzina I, Manthei KA, Gorelick RJ, Musier-Forsyth K, Williams MC. 2015. Targeted binding of nucleocapsid protein transforms the folding landscape of HIV-1 TAR RNA. *Proc Natl Acad Sci U S A* **112**:13555–13560.

95. You JC, McHenry CS. 1994. Human immunodeficiency virus nucleocapsid protein accelerates strand transfer of the terminally redundant sequences involved in reverse transcription. *J Biol Chem* **269**:31491–31495.

96. Godet J, Ramalanjaona N, Sharma KK, Richert L, de Rocquigny H, Darlix JL, Duportail G, Mély Y. 2011. Specific implications of the HIV-1 nucleocapsid zinc fingers in the annealing of the primer binding site complementary sequences during the obligatory plus strand transfer. *Nucleic Acids Res* **39**:6633–6645.

97. Grohman JK, Gorelick RJ, Lickwar CR, Lieb JD, Bower BD, Znosko BM, Weeks KM. 2013. A guanosine-centric mechanism for RNA chaperone function. *Science* **340**:190–195.

98. Darlix JL, Godet J, Ivanyi-Nagy R, Fossé P, Mauffret O, Mély Y. 2011. Flexible nature and specific functions of the HIV-1 nucleocapsid protein. *J Mol Biol* **410**:565–581.

99. Wu H, Mitra M, Naufer MN, McCauley MJ, Gorelick RJ, Rouzina I, Musier-Forsyth K, Williams MC. 2014. Differential contribution of basic residues to HIV-1 nucleocapsid protein's nucleic acid chaperone function and retroviral replication. *Nucleic Acids Res* **42**:2525–2537.

100. Duval M, Korepanov A, Fuchsbauer O, Fechter P, Haller A, Fabbretti A, Choulier L, Micura R, Klaholz BP, Romby P, Springer M, Marzi S. 2013. *Escherichia coli* ribosomal protein S1 unfolds structured mRNAs onto the ribosome for active translation initiation. *PLoS Biol* **11**:e1001731.

101. Qu X, Lancaster L, Noller HF, Bustamante C, Tinoco I Jr. 2012. Ribosomal protein S1 unwinds double-stranded RNA in multiple steps. *Proc Natl Acad Sci U S A* **109**:14458–14463.

102. Cléry A, Blatter M, Allain FH. 2008. RNA recognition motifs: boring? Not quite. *Curr Opin Struct Biol* **18**:290–298.

103. Raghunathan PL, Guthrie C. 1998. A spliceosomal recycling factor that reanneals U4 and U6 small nuclear ribonucleoprotein particles. *Science* **279**:857–860.

104. Montemayor EJ, Curran EC, Liao HH, Andrews KL, Treba CN, Butcher SE, Brow DA. 2014. Core structure of the U6 small nuclear ribonucleoprotein at 1.7-Å resolution. *Nat Struct Mol Biol* **21**:544–551.

105. Didychuk AL, Montemayor EJ, Brow DA, Butcher SE. 2016. Structural requirements for protein-catalyzed annealing of U4 and U6 RNAs during di-snRNP assembly. *Nucleic Acids Res* **44**:1398–1410.

106. Belair C, Sim S, Wolin SL. 2017. Noncoding RNA surveillance: the ends justify the means. *Chem Rev* **118**:4422–4447.

107. Maraia RJ, Lamichhane TN. 2011. 3′ processing of eukaryotic precursor tRNAs. *Wiley Interdiscip Rev RNA* **2**:362–375.

108. Pannone BK, Xue D, Wolin SL. 1998. A role for the yeast La protein in U6 snRNP assembly: evidence that the La protein is a molecular chaperone for RNA polymerase III transcripts. *EMBO J* **17**:7442–7453.

109. Blewett NH, Maraia RJ. 2018. La involvement in tRNA and other RNA processing events including differences among yeast and other eukaryotes. *Biochim Biophys Acta* **1861**:361–372.

110. Kotik-Kogan O, Valentine ER, Sanfelice D, Conte MR, Curry S. 2008. Structural analysis reveals conformational plasticity in the recognition of RNA 3′ ends by the human La protein. *Structure* **16**:852–862.

111. Wilusz CJ, Wilusz J. 2005. Eukaryotic Lsm proteins: lessons from bacteria. *Nat Struct Mol Biol* **12**:1031–1036.

112. Babitzke P. 2004. Regulation of transcription attenuation and translation initiation by allosteric control of an RNA-binding protein: the *Bacillus subtilis* TRAP protein. *Curr Opin Microbiol* **7**:132–139.

113. Waters LS, Storz G. 2009. Regulatory RNAs in bacteria. *Cell* **136**:615–628.

114. Zhang A, Wassarman KM, Ortega J, Steven AC, Storz G. 2002. The Sm-like Hfq protein increases OxyS RNA interaction with target mRNAs. *Mol Cell* **9**:11–22.

115. Vecerek B, Moll I, Afonyushkin T, Kaberdin V, Bläsi U. 2003. Interaction of the RNA chaperone Hfq with mRNAs: direct and indirect roles of Hfq in iron metabolism of *Escherichia coli*. *Mol Microbiol* **50**:897–909.

116. Vecerek B, Moll I, Bläsi U. 2005. Translational autocontrol of the *Escherichia coli hfq* RNA chaperone gene. *RNA* **11**:976–984.

117. Desnoyers G, Massé E. 2012. Noncanonical repression of translation initiation through small RNA recruitment of the RNA chaperone Hfq. *Genes Dev* **26**:726–739.

118. Chen J, Gottesman S. 2017. Hfq links translation repression to stress-induced mutagenesis in *E. coli*. *Genes Dev* **31**:1382–1395.

119. Ellis MJ, Trussler RS, Haniford DB. 2015. Hfq binds directly to the ribosome-binding site of IS*10* transposase mRNA to inhibit translation. *Mol Microbiol* **96**:633–650.

120. Kavita K, de Mets F, Gottesman S. 2018. New aspects of RNA-based regulation by Hfq and its partner sRNAs. *Curr Opin Microbiol* **42**:53–61.

121. Weichenrieder O. 2014. RNA binding by Hfq and ring-forming (L)Sm proteins: a trade-off between optimal sequence readout and RNA backbone conformation. *RNA Biol* **11**:537–549.

122. Schumacher MA, Pearson RF, Møller T, Valentin-Hansen P, Brennan RG. 2002. Structures of the pleiotropic translational regulator Hfq and an Hfq-RNA complex: a bacterial Sm-like protein. *EMBO J* **21**:3546–3556.

123. Brescia CC, Mikulecky PJ, Feig AL, Sledjeski DD. 2003. Identification of the Hfq-binding site on DsrA RNA: Hfq binds without altering DsrA secondary structure. *RNA* **9**:33–43.

124. Otaka H, Ishikawa H, Morita T, Aiba H. 2011. PolyU tail of rho-independent terminator of bacterial small RNAs is essential for Hfq action. *Proc Natl Acad Sci U S A* **108**:13059–13064.

125. Sauer E, Weichenrieder O. 2011. Structural basis for RNA 3′-end recognition by Hfq. *Proc Natl Acad Sci U S A* **108**:13065–13070.

126. Sledjeski DD, Whitman C, Zhang A. 2001. Hfq is necessary for regulation by the untranslated RNA DsrA. *J Bacteriol* **183**:1997–2005.

127. Ishikawa H, Otaka H, Maki K, Morita T, Aiba H. 2012. The functional Hfq-binding module of bacterial sRNAs consists of a double or single hairpin preceded by a U-rich sequence and followed by a 3′ poly(U) tail. *RNA* **18**:1062–1074.

128. Fei J, Singh D, Zhang Q, Park S, Balasubramanian D, Golding I, Vanderpool CK, Ha T. 2015. RNA biochemistry. Determination of *in vivo* target search kinetics of regulatory noncoding RNA. *Science* **347**:1371–1374.

129. Zhang A, Schu DJ, Tjaden BC, Storz G, Gottesman S. 2013. Mutations in interaction surfaces differentially impact *E. coli* Hfq association with small RNAs and their mRNA targets. *J Mol Biol* **425**:3678–3697.

130. Mikulecky PJ, Kaw MK, Brescia CC, Takach JC, Sledjeski DD, Feig AL. 2004. *Escherichia coli* Hfq has distinct interaction surfaces for DsrA, *rpoS* and poly(A) RNAs. *Nat Struct Mol Biol* **11**:1206–1214.

131. Soper TJ, Woodson SA. 2008. The *rpoS* mRNA leader recruits Hfq to facilitate annealing with DsrA sRNA. *RNA* **14**:1907–1917.

132. Link TM, Valentin-Hansen P, Brennan RG. 2009. Structure of *Escherichia coli* Hfq bound to polyriboadenylate RNA. *Proc Natl Acad Sci U S A* **106**:19292–19297.

133. Soper T, Mandin P, Majdalani N, Gottesman S, Woodson SA. 2010. Positive regulation by small RNAs and the role of Hfq. *Proc Natl Acad Sci U S A* **107**:9602–9607.

134. Updegrove T, Wilf N, Sun X, Wartell RM. 2008. Effect of Hfq on RprA-*rpoS* mRNA pairing: Hfq-RNA binding and the influence of the 5′ *rpoS* mRNA leader region. *Biochemistry* **47**:11184–11195.

135. Beisel CL, Updegrove TB, Janson BJ, Storz G. 2012. Multiple factors dictate target selection by Hfq-binding small RNAs. *EMBO J* **31**:1961–1974.

136. Møller T, Franch T, Højrup P, Keene DR, Bächinger HP, Brennan RG, Valentin-Hansen P. 2002. Hfq: a bacterial Sm-like protein that mediates RNA-RNA interaction. *Mol Cell* **9**:23–30.

137. Sauer E, Schmidt S, Weichenrieder O. 2012. Small RNA binding to the lateral surface of Hfq hexamers and structural rearrangements upon mRNA target recognition. *Proc Natl Acad Sci U S A* **109**:9396–9401.

138. Panja S, Schu DJ, Woodson SA. 2013. Conserved arginines on the rim of Hfq catalyze base pair formation and exchange. *Nucleic Acids Res* **41**:7536–7546.

139. Papenfort K, Said N, Welsink T, Lucchini S, Hinton JCD, Vogel J. 2009. Specific and pleiotropic patterns of mRNA regulation by ArcZ, a conserved, Hfq-dependent small RNA. *Mol Microbiol* **74**:139–158.

140. Tree JJ, Granneman S, McAteer SP, Tollervey D, Gally DL. 2014. Identification of bacteriophage-encoded anti-sRNAs in pathogenic *Escherichia coli*. *Mol Cell* **55**:199–213.

141. Peng Y, Soper TJ, Woodson SA. 2014. Positional effects of AAN motifs in *rpoS* regulation by sRNAs and Hfq. *J Mol Biol* **426**:275–285.

142. Schu DJ, Zhang A, Gottesman S, Storz G. 2015. Alternative Hfq-sRNA interaction modes dictate alternative mRNA recognition. *EMBO J* **34**:2557–2573.

143. Hopkins JF, Panja S, McNeil SA, Woodson SA. 2009. Effect of salt and RNA structure on annealing and strand displacement by Hfq. *Nucleic Acids Res* **37**:6205–6213.

144. Arluison V, Hohng S, Roy R, Pellegrini O, Régnier P, Ha T. 2007. Spectroscopic observation of RNA chaperone activities of Hfq in post-transcriptional regulation by a small non-coding RNA. *Nucleic Acids Res* **35**:999–1006.

145. Zheng A, Panja S, Woodson SA. 2016. Arginine patch predicts the RNA annealing activity of Hfq from Gram negative and Gram positive bacteria. *J Mol Biol* **428**:2259–2264.

146. Hwang W, Arluison V, Hohng S. 2011. Dynamic competition of DsrA and *rpoS* fragments for the proximal binding site of Hfq as a means for efficient annealing. *Nucleic Acids Res* **39**:5131–5139.

147. Sukhodolets MV, Garges S. 2003. Interaction of *Escherichia coli* RNA polymerase with the ribosomal protein S1 and the Sm-like ATPase Hfq. *Biochemistry* **42**:8022–8034.

148. Wang W, Wang L, Zou Y, Zhang J, Gong Q, Wu J, Shi Y. 2011. Cooperation of *Escherichia coli* Hfq hexamers in DsrA binding. *Genes Dev* **25**:2106–2117.

149. Ribeiro EA Jr, Beich-Frandsen M, Konarev PV, Shang W, Vecerek B, Kontaxis G, Hämmerle H, Peterlik H, Svergun DI, Bläsi U, Djinović-Carugo K. 2012. Structural flexibility of RNA as molecular basis for Hfq chaperone function. *Nucleic Acids Res* **40**:8072–8084.

150. Panja S, Paul R, Greenberg MM, Woodson SA. 2015. Light-triggered RNA annealing by an RNA chaperone. *Angew Chem Int Ed Engl* **54**:7281–7284.

151. Updegrove TB, Wartell RM. 2011. The influence of *Escherichia coli* Hfq mutations on RNA binding and sRNA•mRNA duplex formation in *rpoS* riboregulation. *Biochim Biophys Acta* **1809**:532–540.

152. Peng Y, Curtis JE, Fang X, Woodson SA. 2014. Structural model of an mRNA in complex with the bacterial chaperone Hfq. *Proc Natl Acad Sci U S A* **111**:17134–17139.

153. Lease RA, Belfort M. 2000. Riboregulation by DsrA RNA: *trans*-actions for global economy. *Mol Microbiol* **38**:667–672.

154. Bordeau V, Felden B. 2014. Curli synthesis and biofilm formation in enteric bacteria are controlled by a dynamic small RNA module made up of a pseudoknot assisted by an RNA chaperone. *Nucleic Acids Res* **42**:4682–4696.

155. Salim NN, Feig AL. 2010. An upstream Hfq binding site in the *fhlA* mRNA leader region facilitates the OxyS-*fhlA* interaction. *PLoS One* **5**:e13028.

156. Gorski SA, Vogel J, Doudna JA. 2017. RNA-based recognition and targeting: sowing the seeds of specificity. *Nat Rev Mol Cell Biol* **18**:215–228.

157. Hussein R, Lim HN. 2011. Disruption of small RNA signaling caused by competition for Hfq. *Proc Natl Acad Sci U S A* **108**:1110–1115.

158. Moon K, Gottesman S. 2011. Competition among Hfq-binding small RNAs in *Escherichia coli*. *Mol Microbiol* **82**:1545–1562.

159. Olejniczak M. 2011. Despite similar binding to the Hfq protein regulatory RNAs widely differ in their competition performance. *Biochemistry* **50**:4427–4440.

160. Fender A, Elf J, Hampel K, Zimmermann B, Wagner EG. 2010. RNAs actively cycle on the Sm-like protein Hfq. *Genes Dev* **24**:2621–2626.

161. Boehr DD, Nussinov R, Wright PE. 2009. The role of dynamic conformational ensembles in biomolecular recognition. *Nat Chem Biol* **5**:789–796.

162. Uversky VN. 2015. The multifaceted roles of intrinsic disorder in protein complexes. *FEBS Lett* **589**(19 Pt A):2498–2506.

163. Vo MN, Barany G, Rouzina I, Musier-Forsyth K. 2009. HIV-1 nucleocapsid protein switches the pathway of transactivation response element RNA/DNA annealing from loop-loop "kissing" to "zipper". *J Mol Biol* **386**:789–801.

164. Grohman JK, Del Campo M, Bhaskaran H, Tijerina P, Lambowitz AM, Russell R. 2007. Probing the mechanisms of DEAD-box proteins as general RNA chaperones: the C-terminal domain of CYT-19 mediates general recognition of RNA. *Biochemistry* **46**:3013–3022.

165. Mohr G, Del Campo M, Mohr S, Yang Q, Jia H, Jankowsky E, Lambowitz AM. 2008. Function of the C-terminal domain of the DEAD-box protein Mss116p analyzed *in vivo* and *in vitro*. *J Mol Biol* **375**:1344–1364.

166. Busa VF, Rector MJ, Russell R. 2017. The DEAD-box protein CYT-19 uses arginine residues in its C-tail to tether RNA substrates. *Biochemistry* **56**:3571–3578.

167. Russell R, Jarmoskaite I, Lambowitz AM. 2013. Toward a molecular understanding of RNA remodeling by DEAD-box proteins. *RNA Biol* **10**:44–55.

168. Kucera NJ, Hodsdon ME, Wolin SL. 2011. An intrinsically disordered C terminus allows the La protein to assist the biogenesis of diverse noncoding RNA precursors. *Proc Natl Acad Sci U S A* **108**:1308–1313.

169. Koculi E, Lee NK, Thirumalai D, Woodson SA. 2004. Folding of the *Tetrahymena* ribozyme by polyamines: importance of counterion valence and size. *J Mol Biol* **341**:27–36.

170. Koculi E, Thirumalai D, Woodson SA. 2006. Counterion charge density determines the position and plasticity of RNA folding transition states. *J Mol Biol* **359**:446–454.

171. Hauk G, McKnight JN, Nodelman IM, Bowman GD. 2010. The chromodomains of the Chd1 chromatin remodeler regulate DNA access to the ATPase motor. *Mol Cell* **39**:711–723.

172. Kozlov AG, Cox MM, Lohman TM. 2010. Regulation of single-stranded DNA binding by the C termini of *Escherichia coli* single-stranded DNA-binding (SSB) protein. *J Biol Chem* **285**:17246–17252.

173. Tretter EM, Berger JM. 2012. Mechanisms for defining supercoiling set point of DNA gyrase orthologs: I. A nonconserved acidic C-terminal tail modulates *Escherichia coli* gyrase activity. *J Biol Chem* **287**:18636–18644.

174. Wang C, Uversky VN, Kurgan L. 2016. Disordered nucleiome: abundance of intrinsic disorder in the DNA- and RNA-binding proteins in 1121 species from Eukaryota, Bacteria and Archaea. *Proteomics* **16**:1486–1498.

175. Santiago-Frangos A, Jeliazkov JR, Gray JJ, Woodson SA. 2017. Acidic C-terminal domains autoregulate the RNA chaperone Hfq. *eLife* **6**:e27049.

176. Santiago-Frangos A, Kavita K, Schu DJ, Gottesman S, Woodson SA. 2016. C-terminal domain of the RNA chaperone Hfq drives sRNA competition and release of target RNA. *Proc Natl Acad Sci U S A* **113**:E6089–E6096.

177. Beich-Frandsen M, Vecerek B, Konarev PV, Sjöblom B, Kloiber K, Hämmerle H, Rajkowitsch L, Miles AJ, Kontaxis G, Wallace BA, Svergun DI, Konrat R, Bläsi U, Djinovic-Carugo K. 2011. Structural insights into the dynamics and function of the C-terminus of the *E. coli* RNA chaperone Hfq. *Nucleic Acids Res* **39**:4900–4915.

178. Vincent HA, Henderson CA, Stone CM, Cary PD, Gowers DM, Sobott F, Taylor JE, Callaghan AJ. 2012. The low-resolution solution structure of *Vibrio cholerae* Hfq in complex with Qrr1 sRNA. *Nucleic Acids Res* **40**:8698–8710.

179. Qualley DF, Stewart-Maynard KM, Wang F, Mitra M, Gorelick RJ, Rouzina I, Williams MC, Musier-Forsyth K. 2010. C-terminal domain modulates the nucleic acid chaperone activity of human T-cell leukemia virus type 1 nucleocapsid protein via an electrostatic mechanism. *J Biol Chem* **285**:295–307.

180. Santiago-Frangos A, Woodson SA. 2018. Hfq chaperone brings speed dating to bacterial sRNA. *Wiley Interdiscip Rev RNA* **9**:e1475.

Regulating with RNA in Bacteria and Archaea
Edited by Gisela Storz and Kai Papenfort
© 2018 American Society for Microbiology, Washington, DC
doi:10.1128/microbiolspec.RWR-0015-2017

Epitranscriptomics: RNA Modifications in Bacteria and Archaea

23

Katharina Höfer[1] and Andres Jäschke[1]

INTRODUCTION

Today, RNA is no longer seen as a mere intermediary, transferring information from genes to proteins. It has become more and more obvious that RNA plays additional biological roles in living organisms (1, 2). Such roles are highly diverse, including catalytic and gene-regulatory functions. While the catalytic roles in present-day biology appear limited, regulatory RNAs are abundant in all kingdoms of life and control central biological processes from cell cycle progression, differentiation, adaptation, and stress response to pathological processes, such as carcinogenesis or inflammation (3, 4).

This increasingly complex functionality of RNA is contrasted by its simple chemical composition. RNA is generally built from only four different nucleotides (adenine, guanine, cytosine, uracil), but a set of >160 chemical modifications decorate RNA molecules to support their diverse coding, structural, catalytic, and regulatory functions. The location, abundance, and distribution of various types of RNA modifications are dependent on organism as well as on environmental conditions. Many modified nucleosides are conserved throughout bacteria, archaea, and eukaryotes, while some are unique to each branch of life (2). Most known modifications occur at internal positions, while there is limited diversity at the termini (5, 6).

The main goal of this review is to provide an overview of bacterial and archaeal mRNA modifications. For the sake of completeness, we only briefly summarize rRNA and tRNA modifications that have been extensively reviewed elsewhere (7, 8). Thus, this review aims to bring together the latest developments in the identification and characterization of internal as well as 5′-terminal mRNA modifications in prokaryotes and archaea.

The difference in profile and type of RNA modifications is an important feature that distinguishes the three kingdoms of life. Our general understanding of the molecular functions of RNA modifications is largely based on investigations on tRNAs and rRNAs (9, 10). Compared to other RNA species, tRNAs possess the highest number of RNA modifications, on average, 14 modifications per tRNA. A comparison of modified tRNA species from *Archaea*, *Bacteria*, and *Eukarya* reveals a core set of 18 modifications that occur in tRNA in all three domains of life (11). Currently, ~60 rRNA and tRNA modifications are known in bacteria (5, 12). These include base isomerization, base alteration, ribose 2′-hydroxyl group methylation, and complex modifications involving the sequential addition of modifications or large chemical groups (5). While most modifications seem constitutive, certain tRNA modifications are dynamic and depend on growth rate/phase or defined stress conditions (10).

Intensive research into tRNA and rRNA modifications has extended their scope and contributed heavily to the development of new methods to identify as well as to characterize RNA modifications.

The detection and identification of RNA modifications has a long history and reaches back to the early days of molecular biology. Traditionally, RNA modifications were identified by thin-layer chromatography (TLC), anion-exchange chromatography, or UV spectrophotometry (Fig. 1A) (13–15). Early discoveries of modified nucleotides relied on the complete digestion of a given RNA molecule into mononucleotides or dinucleoside monophosphates by use of different types

[1]Institute of Pharmacy and Molecular Biotechnology, Im Neuenheimer Feld 364, Heidelberg University, 69120 Heidelberg, Germany.

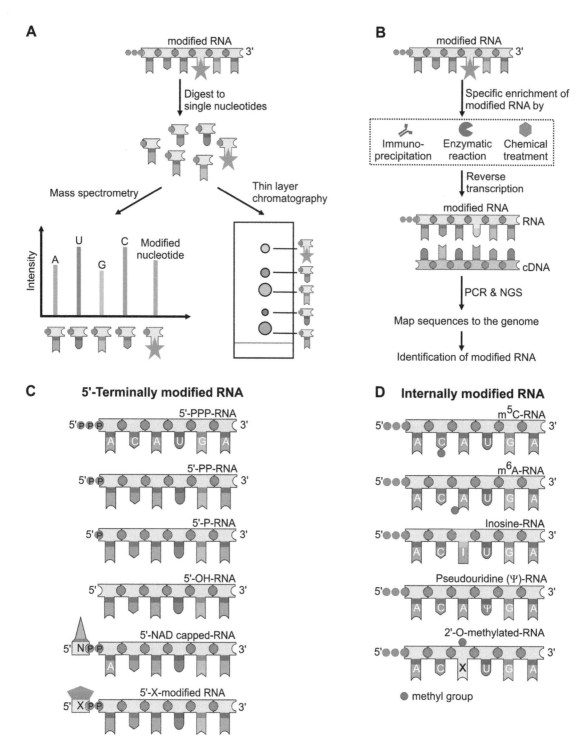

Figure 1 Identification of RNA modifications. (A) Identification of RNA modifications by a combination of digest to single nucleotides and chromatography. Total RNA is digested by nucleases (nuclease P1) and single building blocks separated by TLC or MS coupled with LC. (B) Identification of RNA modifications and their associated transcripts. Modified RNA is specifically enriched by a protocol that is based on antibody treatment (immunoprecipitation), an enzymatic reaction, or a chemical treatment. Afterwards, enriched, modified RNA is converted into cDNA to produce a library for NGS. The reads are mapped to the genome to identify the sequence of the transcripts that bear a specific RNA modification. Known or possible 5′-terminal (C) or internal (D) mRNA modifications in bacteria and archaea.

of nucleases. The resulting hydrolysates were analyzed by TLC. A two-dimensional TLC on cellulose plates is the most frequently applied version, which has been used to identify at least 100 chemically distinct nucleotides and dinucleotides (16). The use of pre- or post-radioactive labeling procedures greatly facilitated the identification of even tiny amounts of modified nucleotides. More recently, liquid chromatography coupled with mass spectrometry (LC-MS) has been used to discriminate modified nucleosides based on both retention time and mass-to-charge ratios (13, 17–19). However, one major drawback of both techniques is the relatively large amount (often micrograms) of pure and homogeneous sample that is required for reliable detection. Such amounts are feasible for highly abundant RNAs like tRNA, rRNA, or transfer-messenger RNA but hard to obtain for less abundant RNAs like mRNA or small regulatory RNA (sRNA) (16). Another important drawback of the digestion of RNA to single nucleotides is the unavoidable loss of the sequence information of those RNAs that actually carry an RNA modification. To overcome these bottlenecks, the rapid development of novel high-throughput or next-generation transcriptome-sequencing techniques (RNA-Seq) (20) helped researchers to identify the location of RNA modifications and to reveal distinct distribution patterns of these modifications throughout the transcriptome (Fig. 1B).

RNA-Seq usually addresses the complexity of prokaryotic and eukaryotic transcriptomes and can be defined as the massively parallel analysis of cDNA molecules by high-throughput sequencing. In comparison to the previously established microarray technology, RNA-Seq offers great advantages including lower cost; transcript profiling at single-nucleotide resolution; as well as a high dynamic range, sensitivity, and discriminatory power (21). In classical RNA-Seq experiments, RNA is converted to a cDNA library and PCR amplified. Following next-generation sequencing (NGS), the obtained short sequence reads are mapped onto a reference genome (20, 22).

Depending on the scientific question, different RNA-Seq protocols were established that use either poly(A)-tailing of RNAs combined with oligo-d(T)-priming of cDNA, or cDNA synthesis from a ligated RNA adapter (23). In recent years, RNA-Seq generally helped to clarify the nature of transcriptomes, to determine gene expression changes, and to identify coregulated genes. With this technique, extensive information on transcriptional start sites (TSSs) (21, 24), untranslated regions (UTRs) of mRNA genes, sRNA genes, and unknown open reading frames was provided (22). Aside from expression profiling, RNA-Seq has been applied to study RNA-protein interaction and the sum of actively translated mRNAs by so-called ribosome profiling (25).

In addition to transcriptome-wide analysis, RNA-Seq protocols were developed that allow the identification of biologically relevant internal RNA modifications. RNA-Seq has been combined with immunoprecipitation techniques (26–29), chemical derivatization (30, 31), or enzymatic treatments (32, 33) (Fig. 1B), facilitating the identification of several RNA modifications in prokaryotes, archaea, and eukaryotic organisms (34). The multitude of established RNA modifications were first designated as "RNA epigenetics" in 2010 (35). The dynamic nature of internal RNA modifications and newly discovered regulatory functions of some of these RNA modifications gave birth to a new field, now often referred to as "epitranscriptomics" (36–38).

In this review, several mRNA modifications that were identified at the 5′ terminus of RNA or at internal positions of mRNAs will be discussed in detail (Fig. 1C and D). Much attention has been drawn by eukaryotic mRNA modifications. However, transcriptome-wide studies to characterize internal or 5′-terminal mRNA modifications in bacteria and archaea have surfaced only in the past 5 years. This review describes on the one hand the well-known 5′-triphosphorylated (5′-PPP), 5′-monophosphorylated (5′-P), and 5′-hydroxylated (5′-OH) termini. On the other hand, the newly identified 5′-diphosphorylated RNA (5′-PP) and NAD-capped RNA (5′-NAD-RNA) are highlighted (Fig. 1C). In contrast to 5′-terminal RNA modifications, a significantly larger number of internal RNA modifications are known. Therefore, this review focuses on five mRNA modifications (N^6-methyladenosine [m6A], 5-methylcytosine [m5C], inosine [I], pseudouridine [Ψ], and 2′-O-methylated nucleotides [Nm]) that have been identified in bacterial or archaeal mRNAs or are very likely to be present in these organisms.

HIGH-THROUGHPUT SEQUENCING APPROACHES TO IDENTIFY INTERNAL RNA MODIFICATIONS

The combination of RNA-Seq with immunoprecipitation or chemical treatments facilitated the identification of several internal RNA modifications: m6A, m5C, I, Ψ, and Nm (Table 1; Fig. 1) (5). The existence of these universal modifications has been proven already in large numbers of archaeal, bacterial, and eukaryal tRNAs (reviewed in references 8, 10, 11, 39, and 40). However, the presence of these modifications in mRNA is poorly understood. In this section, the cur-

Table 1   Internal RNA modifications

Modification	Structure	Detection	Enzymes	Molecular roles[a]
Adenosine to Inosine editing	Inosine	ICE-Seq	Adenosine deaminases that act on RNA (ADAR)	Amino acid alterations, alternative splicing, nuclear retention, nonsense-mediated decay, RNAi, 3'-UTRs variation, translation, miRNA processing repression
5-methylcytosine (m^5C)	m^5C	Bisulfite sequencing	NSUN2, Dnmt2	tRNA function and stability, translational fidelity of ribosome
N6-methyladenosine (m^6A)	m^6A	m6A-Seq, MeRIP-Seq	Writers/methylases: METTL3, METTL14, WTAP Erasers/demethylases: FTO, ALKBH5 Readers/effectors: YTHDF1, YTHDF2, YTHDF3	mRNA stability, RNA-protein interactions, miRNA processing, splicing translation initiation
Pseudouridine Ψ	Ψ	Ψ-Seq Pseudo-Seq PSI-Seq	Site-specific and snoRNA-guided pseudouridine synthases	tRNA decoding, rRNA folding, codon recoding, mRNA stability
2'-O-methylated (Nm) nucleotides	Nm	2OMe-Seq RiboMeth-Seq Nm-Seq RibOxi-Seq	2'-O-methyltransferase Synthesis by small (nucleolar) box C/D guide ribonucleoproteins (s(no)RNPs)	High frequencies in rRNAs and small RNAs, eukaryotic 5'-cap and mRNAs estimated to be present in bacteria

[a]Abbreviations: RNAi, RNA interference; miRNA, microRNA.

rent scientific knowledge about m6A, m5C, inosine, Ψ, and Nm, which are present in all kingdoms of life, is summarized.

I

Adenosine-to-inosine (A-to-I) editing is one of the most prevalent types of RNA editing, especially in higher eukaryotes. Inosine is well studied in tRNAs in all domains of life and is mainly present at position 34, which is the first nucleotide of the anticodon (wobble position); position 37 (following the anticodon); and position 57 (at the TΨC-loop) of the tRNA. Interestingly, at position 34, inosine is the final modified base, while at positions 37 and 57, inosine is further modified by methylation. A-to-I editing is generally performed by two groups of RNA adenosine deaminases that are active on either tRNA (adenosine deaminases acting on tRNAs [ADATs]) or mRNA (adenosine deaminases acting on mRNAs [ADARs]) (34, 41, 42). Inosine at position 57 has only been identified in archaea. In contrast, A-to-I editing at position 34 has been confirmed in bacteria and eukarya.

The conversion of adenosine to inosine in mRNA plays numerous roles in modulating gene expression, including recoding codons, altering alternative splicing, and regulating microRNA biogenesis and function (43). Inosine base-pairs with cytidine, rather than thymidine, in reverse transcription and is thus read as guanosine in cDNA. Based on this phenomenon, high-throughput RNA-Seq was performed and enabled transcriptome-wide identification of A-to-I edited sites. However, the simple conversion of RNA into cDNA and genome mapping of A to C results in background noise caused by single-nucleotide polymorphisms, somatic mutations, pseudogenes, and sequencing errors. Therefore, a new biochemical technique called inosine chemical erasing (ICE)-Seq was established to selectively identify A-to-I edited sites (Table 1; Fig. 2D). In this approach, acrylonitrile is used to selectively react with inosines in RNA, forming N^1-cyanoethylinosine (ce^{1}I) (44). Because ce^{1}I inhibits first-strand cDNA extension, inosine-containing RNA is eliminated, while unmodified RNA gives rise to full-length cDNA that is analyzed by high-throughput sequencing. Using ICE-Seq, Sakurai et al. have successfully identified >19,000 novel editing sites in the human adult brain transcriptome (45). ICE-Seq is suitable for identifying A-to-I editing in any organism. Its application to RNA isolated from bacteria or archaea may contribute to the identification of editing sites in mRNAs isolated from these organisms.

Recently, Bar-Yaacov et al. identified 15 novel A-to-G RNA-editing events in prokaryotic mRNAs (46). Instead of using ICE-Seq, they sequenced in parallel RNA and DNA isolated from *Escherichia coli* and identified editing events as base difference between DNA and RNA sequence. The editing itself, performed by the tRNA-specific adenosine deaminase TadA, recodes the protein sequence, which potentially affects protein function and cell physiology (46).

m5C

Another recently characterized RNA modification is m5C. The methylation of cytosine in DNA (m5dC) is a widespread epigenetic marker and has been extensively studied in genomic eukaryotic DNA (47). This modification is usually detected in DNA by bisulfite treatment (48). Bisulfite causes the deamination of unmethylated cytosine to uracil, while leaving methylated cytosine intact. However, the conversion of m5dC is performed under harsh reaction conditions (high pH) that are

detrimental to the stability of phosphodiester bonds in RNA. Thus, a modified bisulfite treatment protocol was developed that allowed m5C-site detection in tRNA and rRNA (49) (Table 1; Fig. 2A). In 2012, the first m5C methylome was obtained in a transcriptome-wide manner by combining the modified bisulfite treatment and high-throughput sequencing (50). m5C has been identified in noncoding RNAs including tRNA and rRNA. Especially in tRNAs, m5C stabilizes the secondary structure and changes the anticodon stem-loop conformation (39). Moreover, >8,000 potential m5C sites in human mRNA were identified. However, due to possible incomplete conversion of regular cytosine in double-stranded RNA regions and to the presence of other cytosine modifications resistant to bisulfite treatment, it has been suggested that these sites include potential false positives from stochastic nonconversion events (51, 52).

The first studies to identify m5C in bacteria and archaea were reported in 2013. RNA bisulfite sequencing was used on total RNA isolated from both Gram-positive (*Bacillus subtilis*) and Gram-negative (*E. coli*) bacteria, an archaeon (*Sulfolobus solfataricus*), and a eukaryote (*Saccharomyces cerevisiae*) (28), followed by massively parallel sequencing to map m5C. In this study, the known m5C residues, identified in rRNAs in previous studies, were confirmed for all organisms. However, m5C seemed to be absent from any of the *E. coli* and *B. subtilis* mRNAs. Intriguingly, m5C has been mapped in *S. solfataricus* mRNAs. In addition, a first consensus motif (AUCGANGU) was identified that directs methylation in this archaeon. These results were the first evidence for mRNA modifications in archaea, suggesting that this mode of posttranscriptional regulation extends beyond the eukaryotic domain (28).

Methylation of the cytosine is performed by specific m5C methyltransferases (m5C-MTases), which transfer methyl groups from *S*-adenosylmethionine (SAM) to form m5C (53). The first characterized m5C-MTase family member is RsmB, which mainly modifies bacterial 16S rRNA (54). Thus far, >30 homologous proteins have been identified by sequence analysis. Two RNA methyltransferases, NSun2 and DNMT2, have been found to catalyze m5C methylation in higher eukaryotes (55, 56). Recently a cysteine-to-alanine mutation (C271A) (57, 58) in human NSUN2 has been applied to identify m5C sites (33). The mutation inhibits the release of the enzyme from the protein-RNA complex, resulting in a stable covalent bond between NSun2 and its RNA targets (33). Combining the methyltransferase activity with cross-linking by UV light successfully identifies the targets of NSun2 throughout

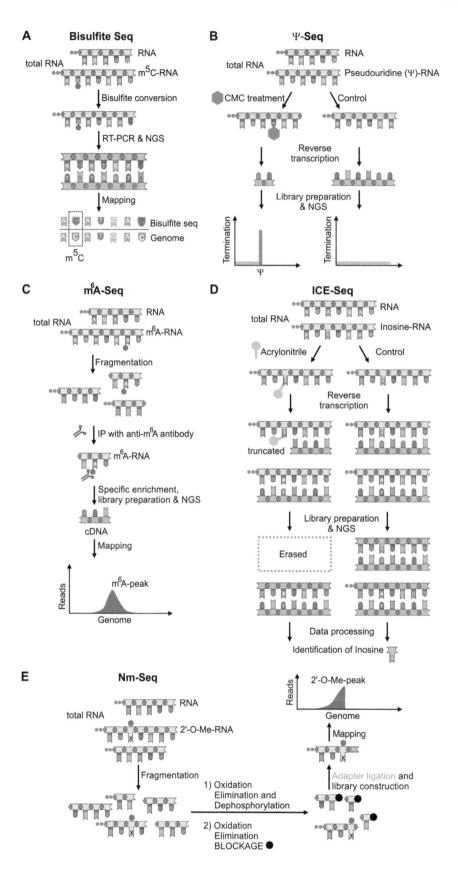

the transcriptome. In addition to the described protocols, m5C sites could be analyzed and verified by m5C RNA immunoprecipitation (m5C-RIP) approaches (28). Using a monoclonal antibody that specifically binds 5-methylcytosine, m5C sites of *S. solfataricus* mRNA, identified by bisulfite sequencing, could be validated. Such immunoprecipitation procedures allow the detection of low-abundant methylated RNAs without the requirement of extremely deep sequencing and can confirm the results obtained by high-throughput sequencing.

m6A

The most abundant internal modification in eukaryotic mRNA is methylation at the N^6 position of adenosine. In fact, 0.2 to 0.6% of all adenosines in mammalian mRNA are methylated at position 6 (59). This modification was identified using a combination of radioactive labeling and chromatography >40 years ago (60). However, a transcriptome-wide profile of m6A in mammalian cells was not generated until the development of new methods in 2012, which are based on m6A-specific immunoprecipitation and high-throughput sequencing called m6A-Seq (26) and MeRIP-Seq (27) (Table 1; Fig. 2C). In both methods, purified mRNA is sheared into 100-to-150-nucleotide RNA fragments. m6A-containing RNAs are immunoprecipitated by an m6A-specific antibody. The enriched m6A-containing RNA fragments are then subjected to library construction and high-throughput sequencing. Both methods have identified ~10,000 m6A peaks in the mammalian

transcriptome and showed for the first time that m6A modifications are enriched in 3′ UTRs, peaking sharply near the stop codon. However, identifying the precise location of m6A within a transcript remained a challenge (34). Recently, UV-induced RNA-antibody cross-linking strategies have been adapted into the m6A-Seq and MeRIP-Seq protocols, allowing identification of base-resolution m6A methylomes. This technique, called photo-cross-linking-assisted m6A sequencing (PA-m6A-Seq) (61), is based on the incorporation of 4-thiouridine (^{4S}U) into RNA during cell growth. After m6A immunoprecipitation, the recovered m6A-containing RNA can be cross-linked to the anti-m6A antibody under UV light (366 nm) specific for ^{4S}U. Moreover, m6A-CLIP (cross-linking and immunoprecipitation) and miCLIP methods were developed that are based on photo-cross-linking as well (62, 63). Briefly, RNA fragments are immunoprecipitated and cross-linked to the antibody by UV light (254 nm). The protein-RNA cross-linking sites lead to patterned mutational or truncation profiles during reverse transcription, thereby revealing the precise position of m6A. These approaches have enabled important discoveries in the field of epitranscriptomics. Over the last years the biosynthesis and the removal of m6A by writers (METTL3 or METTL14) or erasers (FTO [64] and ALKBH5 [65]) have been studied in eukaryotic organisms in detail.

In contrast, studies describing the potential presence of m6A modifications in bacterial or archaeal RNA were lacking. m6A has been well described in the

Figure 2 Methods to detect internal RNA modifications. (A) Identification of m5C-modified mRNA by bisulfite sequencing. Selective conversion of cytosine to uracil by bisulfite ions, whereas m5C remains as cytosine. After reverse transcription, a cDNA library is prepared, analyzed by high-throughput sequencing, and mapped to the genome. Cytosine is mapped as thymine, whereas m5C residues read as cytosine. (B) Ψ-Seq: identification of pseudouridine-modified RNA. mRNA is treated with CMC, which selectively reacts with Ψ and causes a stop during reverse transcription. cDNA libraries are amplified and sequenced. The reads from a CMC-treated and nontreated control are compared to map pseudouridine-modified RNA. (C) Identification of methylated adenosine by m6A-Seq. Total RNA is fragmented to 100-nucleotide-long RNAs. m6A-specific antibodies are used to immunoprecipitate RNA. RNA is reverse-transcribed to cDNA and analyzed by high-throughput sequencing. These reads produce a peak whose summit reflects an underlying m6A residue. (D) Identification of adenosine-to-inosine editing: ICE-Seq. ICE is based on cyanoethylation of inosine by acrylonitrile combined with reverse transcription and high-throughput sequencing. Inosine pairs with cytosine (control), whereas the chemical modification of inosine results in a stop during reverse transcription. Erased G signals originate from inosines in the sequence map of cDNAs and are finally used to identify A-to-I editing sites. (E) Mapping of 2′-O-methylated nucleotides (Nm) by Nm-Seq. Fragmented RNA is subjected to iterative oxidation-elimination-dephosphorylation cycles that remove 2′-hydroxylated nucleotides in the 3′-to-5′ direction. Internal 2′-O-methylation sites stay intact, and fragments ending with 2′-hydroxyl are finally blocked by an incomplete oxidation-elimination cycle. 2′-O-methylated RNA fragments are ligated to an adapter. After library construction and high-throughput sequencing, reads are mapped to the genome. At 2′-OMe sites, an asymmetric coverage profile is observed whose uniform 3′ end corresponds to the methylation position.

rRNA in bacteria (66). In *E. coli*, A1618 and A2030 of 23S rRNA are methylated by methyltransferases RlmF and RlmJ, respectively (67, 68). The deletion and overexpression of RlmF results in a loss of cell fitness and growth (67), while an RlmJ mutant shows mild phenotypes under various growth conditions. In order to identify m6A modifications in bacterial mRNA, Deng et al. performed a highly sensitive LC-MS-based measurement of m6A/A ratios on purified mRNAs isolated from seven model bacteria (*E. coli*, *Pseudomonas aeruginosa*, *Pseudomonas syringae*, *Staphylococcus aureus*, *B. subtilis*, *Anabaena* sp. PCC 7120, and *Synechocystis* sp. PCC 6803) (63). The highest m6A/A ratios were detected in the Gram-negative species *E. coli*, *P. aeruginosa*, and *P. syringae* (>0.2%). In *E. coli*, 265 m6A peaks were identified, corresponding to 213 mRNAs and 15 sRNAs. In *P. aeruginosa*, 105 m6A peaks were determined that were mapped to 68 mRNAs and 2 sRNAs. Interestingly, the majority (70%) of m6A sites occur within open reading frames and have a GCCAG consensus sequence. This contrasts with eukaryotes, where m6A sites are mapped around stop codons and in 3′ UTRs with an RRACU motif (26, 27, 69). Moreover, the majority of identified m6A-modified genes in *E. coli* and *P. aeruginosa* are involved in energy metabolism or belong to the class of sRNAs. It has been suggested that m6A plays a potential functional role in these processes (63). In contrast to eukaryotic organisms, the bacterial or archaeal RNA methyltransferases are still unknown. In *E. coli*, no METTL3 or METTL14 homologs have been identified yet (63). Moreover, unlike in eukaryotes, where m6A is a dynamic modification (64), m6A/A ratios seem to be stable under a variety of growth conditions. This suggest that prokaryotes and eukaryotes regulate m6A modification by different mechanisms that have to be explored (70).

Ψ, an Abundant Internal mRNA Modification

The most abundant RNA modification is Ψ, which is often referred to as "the fifth ribonucleoside" (71, 72). Ψ is generated via isomerization of uridine, catalyzed by two distinct mechanisms: the RNA-dependent mechanism with the box H/ACA RNPs and the RNA-independent mechanism with Ψ synthases (34). Historically, Ψ was first identified in rRNA and tRNA in all kingdoms of life (reviewed in references 8 and 72). Further occurrences of Ψ are known in snRNAs of various eukaryotes (5). Ψ's functional relevance is well documented in rRNAs, where pseudouridylation is required for ribosome biogenesis and translational fidelity. Moreover, Ψ-modified snRNAs have been shown to con-

tribute to pre-mRNA splicing (73). Many Ψ residues present in rRNAs and snRNAs are highly conserved across species. However, it has been shown that pseudouridylation can be induced in response to cellular stress and differentiation in yeast (74). These findings suggest that inducible pseudouridylation could be a lot more widespread and may provide a dynamic regulatory mechanism for RNA function in all kingdoms of life (2, 75).

To identify and to map Ψ-modified positions across the entire transcriptome with single-nucleotide resolution, different high-throughput sequencing methods (Pseudo-Seq, Ψ-Seq, PSI-Seq, and CeU-Seq) have been published (34, 75) (Table 1; Fig. 2B). These methods are based on selective chemical labeling of Ψ by the carbodiimide *N*-cyclohexyl-*N′*-β-(4-methylmorpholino) ethylcarbodiimide *p*-toluenesulfonate, known as CMC. The modified CMC-Ψ adduct results in a termination of cDNA synthesis. Next, cDNA strands are processed into cDNA libraries and analyzed by high-throughput sequencing. Finally, reads are mapped to the genome to detect Ψ sites across the entire transcriptome at single-base resolution.

Using these methods, Ψ was also found to be present in eukaryotic mRNA (76), although the biological function of such mRNA pseudouridylations remains enigmatic. Nevertheless, Ψ is abundant in mammalian mRNA, with a Ψ/U ratio of about 0.2 to 0.6% in human cells and mouse tissues (77). However, proof for the existence of Ψ modifications in bacterial or archaeal mRNA is lacking (70).

Nm

Next to the already described Ψ, the methylation of the 2′ hydroxyl group of the ribose (Nm, for unspecified 2′-O-methyl nucleoside) moiety is the most abundant RNA modification in rRNA (78). 2′-O-methylation changes the biophysical and biochemical properties of the nucleotides that carry this modification. For instance, 2′-O-methylation alters the hydration sphere around the oxygen and impacts sugar edge interactions and sugar pucker conformation. Moreover, 2′-O-methylation increases the stability of RNA against alkaline hydrolysis. 2′-O-methylation has been extensively studied for a number of years in order to identify functional and mechanistic links between this modification and specific biological pathways. For instance, 2′-O-methylation in RNA modulates the biogenesis and activity of the ribosome (79). However, the detailed roles of 2′-O-methylation in rRNA are not yet well understood (80).

The identification of 2′-O-methylation sites traditionally relied on the property of Nm to pause reverse transcription in the presence of limiting amounts of dNTPs (81). This mapping method was recently adapted to a high-throughput sequencing format to identify eukaryotic 2′-O-methylated rRNAs (2OMe-seq) (82).

In addition, a new method, called RiboMeth-Seq, was developed that allows the detection of relatively small changes (>10%) in Nm profiles (83). RiboMeth-Seq is based on the resistance of 2′-O-methylated ribose to alkaline hydrolysis, leading to underrepresentation of these positions among read starts or ends to provide a negative readout of the methylation landscape. However, other methods to detect Nm positions, present in relatively rare RNA molecules (e.g., mRNA), Nm-Seq and RibOxi-Seq (84, 85), were established based on the different chemical properties of nucleosides with 2′-OH and 2′-OMe (Table 1; Fig. 2E). The Nm-Seq protocol applies classical periodate oxidation of ribose 2′,3′ vicinal diols to yield a dialdehyde intermediate that undergoes spontaneous β-elimination under mildly basic conditions (85). After dephosphorylation of the resulting 3′ monophosphate, another oxidation-elimination-dephosphorylation (OED) is carried out. The progressive shortening process comes to a halt when a 2′-O-methylation site is reached, as it lacks vicinal diols. After several iterative cycles, the library is enriched in fragments ending with Nm, which are preferentially ligated to the 3′ adapter and analyzed by high-throughput sequencing.

The development of a variety of different sequencing techniques facilitated the identification of Nm modifications in eukaryotes. Ribose-methylated bases have been identified in rRNA, tRNA, snRNA, microRNA, and mRNA (59). In addition, Nm has been found to be associated with the eukaryotic m7G caps (86) and is involved in host pathogen responses (87, 88).

However, as of yet, 2′-O-ribose methylation has not been observed in archaeal and bacterial mRNA. Only modifications in tRNA (88) or rRNA have been described. In archaea, 2′-O-methylation of nascent rRNA molecules is triggered by RNP complexes containing small C/D box s(no)RNAs, which identify targets for methylation of the ribose. It has been hypothesized that 2′-O-methylation contributes to the folding, structural stabilization, assembly, and function of the 23S rRNA within the large subunit of the ribosome in archaea (89).

In bacteria, 2′-O-methylation modifications are relatively rare and introduced by dedicated site-specific or region-specific methyltransferases. A 2′-O-methylation of guanosine at position 18 (Gm18) of *E. coli* tRNA^{Tyr} has attracted great interest. This methylation was found to suppress innate immune activation in human innate immune cells (88). However, little is known about Nm sites in archaeal and bacterial mRNA. The application of Nm-Seq or RibOxi-Seq to new species will contribute to a better understanding of the biological relevance of this RNA modification.

5′-TERMINAL MODIFICATIONS OF RNA

In eukaryotes as well as in prokaryotes, the 5′ end of newly synthesized RNAs bears a triphosphate derived from the first transcribed nucleotide. Depending on the cellular processing, the 5′ termini can be converted into a diphosphate, monophosphate, or hydroxyl function. Recent evidence indicates that the 5′ status is an important determinant for molecular recognition by RNA-processing enzymes (90–92), for instance, the 5′-monophosphate-assisted RNA cleavage by RNases (93).

Efforts toward detecting modifications at the 5′ terminus of RNA started with the identification of the eukaryotic cap. In the mid-1970s, biochemical analyses revealed that certain viral and eukaryotic mRNAs have a cap—a methylated guanosine residue linked to the 5′ end of mRNA through an inverted 5′-to-5′ triphosphate bridge (94, 95). Functions associated with the 5′ cap are pre-mRNA splicing, 3′-poly(A) addition, overall stability, nuclear exit to the cytoplasm, protein synthesis, and mRNA turnover induced by decapping (96). The methylated guanosine and the neighboring nucleotides can carry further modifications (e.g., O- and N-methylations). Decapping and the conversion to 5′-monophosphorylated RNA inhibit translation initiation and trigger RNA degradation by 5′-to-3′ exonucleases (e.g., Xrn1) (97–100). In contrast to the eukaryotic kingdom, it was believed that most primary transcripts in bacteria and archaea possess a 5′-end triphosphate moiety that is derived from the first transcribed nucleotide (101, 102). In addition, certain primary bacterial transcripts were identified that possess a 5′ hydroxyl group and are generated by primer-dependent transcription initiation in *E. coli* and *Vibrio cholerae* (103, 104).

Primary Transcripts: 5′-PPP-RNA, 5′-P-RNA, and 5′-OH-RNA

The annotation of TSSs is essential for analyzing promoters, 5′ UTRs, and operon architecture and for discovering novel transcripts. Differential RNA-Seq (dRNA-Seq) approaches, selectively analyzing primary transcripts with 5′-PPP termini, have been developed in recent years. These TSS mapping techniques generally combine enzymatic treatments and adapter ligation with RNA-Seq techniques (21, 24) to identify primary transcripts (5′-PPP-RNA) (Fig. 3A). Briefly, total RNA

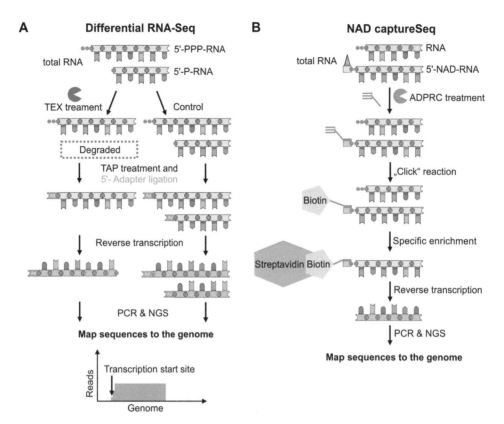

Figure 3 Identification of 5′-terminal RNA modifications. (A) Differential RNA-Seq to identify primary transcripts. Total RNA is treated with a 5′-P-dependent exonuclease that degrades specifically 5′-P-RNA. 5′-PPP-RNA is converted into 5′-P-RNA enzymatically by TAP or similar enzymes. The 5′ end is ligated to an adapter sequence and the RNA reverse-transcribed into cDNA, which is analyzed by high-throughput sequencing. The reads are mapped to the genome to identify TSSs. (B) Schematic representation of the NAD captureSeq protocol that allows the identification of NAD-capped RNAs. Total RNA is treated with ADPRC from *A. californica*, which specifically catalyzes the transglycosylation reaction of NAD with 4-pentyn-1-ol. The product of the reaction is biotinylated by CuAAC (click reaction). The NAD-capped RNA is captured as well as enriched on streptavidin beads and ligated to an adapter. After on-bead reverse transcription and a second adapter ligation, the obtained cDNA is amplified by PCR and submitted to high-throughput sequencing.

is treated with 5′-P-dependent terminator exonuclease (TEX), which specifically degrades 5′-P-RNAs. 5′-PPP-RNAs are not degraded and are therefore enriched in the sample. Afterwards, the RNA is treated with tobacco acid pyrophosphatase (TAP), converting 5′-PPP-RNA into 5′-P-RNA. The 5′-P-RNA is finally ligated to an adapter and analyzed by high-throughput sequencing. The sequencing of dRNA-Seq libraries yields characteristic enrichment patterns of the cDNA coverages at the 5′ end of primary transcripts, which can be used to annotate TSS. dRNA-Seq has been applied to >30 organisms, including diverse bacteria and some archaea (21). In addition, Ettwiller et al. developed another technique, called Cappable-seq, to identify TSS at single-base resolution (102). They perform an enzymatic labeling of 5′-PPP termini with biotin, which enables the subsequent enrichment and high-throughput sequencing of primary transcripts.

In addition to 5′-PPP termini, primary transcripts can carry 5′-OH or 5′-P termini that are formed by primer-dependent transcription initiation. RNA polymerases can use 2-to-~4-nucleotide RNAs, so-called "nanoRNAs" (105), to prime transcription initiation *in vivo* (106). For the identification of these transcripts carrying 5′ monophosphate or a 5′ hydroxyl group, several NGS strategies were developed. 5′-Hydroxyl RNA sequencing was established, which enables the specific capture of 5′-OH termini (107). The *E. coli* RtcB RNA ligase attaches an oligonucleotide linker to 5′-hydroxylated RNAs that are finally analyzed by

NGS. Moreover, 5′-P-RNAs are sequenced by 5′-RNA-Seq methods, which are based on the ligation of an adapter to 5′-monophosphorylated RNAs (104, 108).

5′-NAD-Capped RNA

The absence of 5′-capped RNA was considered as one of the hallmarks of prokaryotic gene expression. Until recently, the only known structure in prokaryotic RNA that resembles a cap was the covalent RNA adenylate intermediate involved in RNA ligation pathways (5′-AppRNA), in which an adenosine is linked to the RNA fragment via a 5′,5′ pyrophosphate bridge (109). Biological roles of these AppRNAs, besides ligation, have not been reported.

In 2009, the Liu group provided evidence that RNA can contain nucleotide-linked cofactors like the ubiquitous redox cofactor NAD (110) or coenzyme A (CoA) (Table 2) (111). Using an MS-based technique that requires complete hydrolysis of RNAs to the single-nucleotide level, they identified NAD in total RNA isolates from *E. coli* and *Streptomyces venezuelae* (110). Another publication in 2009 reported that CoA is linked to RNA species, and experimental evidence suggested attachment at the 5′ end of the RNA (111). However, both studies combined hydrolysis of the RNA with MS to detect the nucleotide-linked cofactors. Therefore, the sequences that bear the NAD modification initially remained unknown. Approaches that specifically enrich for NAD-modified RNA from total RNA were lacking.

The techniques that had been successfully applied to identify and analyze internal RNA modifications, such as cross-linking or immunoprecipitation, could not be directly transferred to identify NAD-modified RNA.

Antibodies offer too weak binding affinities for NAD, and cross-linking approaches generally result in a non-specific enrichment of RNA. Thus, to isolate NAD-RNA from biological samples, a new capture technique was required, which had to be specific for the NAD moiety without damaging the RNA molecules. Inspired by the identification of cofactor-modified RNA in 2009 (110), Cahová et al. developed a NAD-specific chemo-enzymatic capture approach, called NAD captureSeq (Fig. 3B), which enabled specific enrichment of NAD-modified RNA from total RNA and the analysis of the enriched RNA by high-throughput sequencing (32, 112).

The NAD captureSeq protocol combines enzymatic transglycosylation with "click" chemistry (113, 114) and can be split into three parts: (i) isolation of NAD-modified RNA, (ii) cDNA preparation, and (iii) validation of the NGS hits. To enrich for 5′-NAD-modified RNA, total RNA is treated with ADP-ribosyl cyclase (ADPRC) from *Aplysia californica*. This enzyme, previously described to replace nicotinamide with other N-heterocyclic nucleophiles (115), was surprisingly found to be capable of catalyzing the transglycosylation reaction of NAD with alkynyl alcohols. This enzyme can be applied to introduce selectively a "clickable" pentynyl handle in place of the nicotinamide. The product of the reaction is biotinylated by copper-catalyzed azide-alkyne cycloaddition (CuAAC). To ensure specific isolation of NAD-modified RNA from total RNA, the biotinylated, formerly NAD-modified RNA is captured and enriched on streptavidin beads. In the next step, enriched RNAs that were 5′-NAD-modified before are converted into cDNA. Finally, a combination of adapter ligation and on-bead reverse transcription can be applied

Table 2 Cofactor-capped RNAs

Modification	Structure	Detection	Enzymes	Molecular roles
NAD cap		NAD captureSeq	Incorporation into RNA by RNA polymerase	Stabilization of the RNA, first RNA cap in bacteria
			Decapping by the Nudix hydrolase NudC	Identification of sRNA and 5′-mRNA termini
CoA modification		Mass-spectrometry	Incorporation into RNA by RNA polymerase	RNA sequences remain unknown

to obtain a cDNA library that is submitted to high-throughput sequencing and mapped to the genome (112).

Based on the identification of NAD in total RNA isolated from the well-studied bacterium *E. coli* by Chen et al. in 2009 (110), the NAD captureSeq protocol was applied to *E. coli* total RNA for the first time. To identify enriched NAD-RNA in the NAD capture Seq sample, two negative controls that lack either ADPRC (minus ADPRC) or the pentynol were applied. In both cases a "clickable" pentynol cannot be attached to the NAD moiety, which makes biotinylation by CuAAC reaction impossible. The comparison of the NGS reads from the fully treated sample with those from appropriate controls revealed strong enrichment of a specific set of sRNAs, e.g., GadY, GcvB, ChiX, McaS, RNAI, and CopA. Those enriched candidates were reported to act by different pathways (116–126). Interestingly two sRNAs were enriched (RNAI and CopA) that are encoded on plasmids and control the plasmid replication, a mechanism that has been studied extensively since the 1980s (118, 120). Intriguingly, other sRNAs were not found to be enriched. Another set of enriched sequences represented 5′ fragments of different mRNAs, e.g., *gatY*, *pgk*, *hdeD*, and leader peptides *ilvL* and *hisL*, encoding either enzymes involved in cellular metabolism or leader peptides with known regulatory functions (32). These 5′ fragments were homogeneous in size and likely represent unknown sRNAs that exist in the cell in a NAD-modified state (32). Interestingly, tRNAs and rRNAs were not enriched in the NAD captureSeq approach.

The results of the NAD capture experiments were confirmed by different biochemical analyses, including enzymatic digests and MS. Especially for RNAI, the NAD modification was validated by LC-MS and quantified by a biochemical, ligation-based assay to 15% (32).

To identify characteristic features of the enriched RNAs, the NAD-modified RNAs were analyzed for common structural and sequence properties (32). The only feature that most enriched genes share is the TSS (+1 position) (32). The promoters predicted to generate the enriched NAD-modified RNAs possess an adenosine at the start site, which is in agreement with published TSS mapping data (24, 127). However, using bioinformatic predictions and database scans, no common sequence or structural features in promoters, transcripts, or transcription factor binding sites were identified for the NAD-modified RNAs (32).

Inspired by the identification of NAD-modified RNA in *E. coli*, the NAD captureSeq protocol was applied to total RNA isolated from yeast (128) and human cells (129). Intriguingly, a subset of eukaryotic

RNAs (mRNA, snoRNAs, and small Cajal body RNAs [scaRNAs]) has been reported to possess a NAD-cap, which was confirmed by a biochemical shift assay. However, the presence of eukaryotic NAD-capped RNA has not yet been validated by MS.

The identification of NAD-RNA in archaea or bacterial species other than *E. coli* is still lacking and needs to be investigated.

Biosynthesis of NAD-Capped RNA

RNA is a central molecule in all kingdoms of life, as it connects the storage of genetic information in the form of DNA to its translation into proteins. DNA is thereby transcribed into RNA via DNA-dependent RNA polymerases. In eukaryotes, RNA modifications including the synthesis of the mRNA cap are generally generated post- or cotranscriptionally (Fig. 4A) (130, 131). In contrast to the eukaryotic m7G-cap synthesis, the incorporation of NAD seems to occur during transcription initiation (128, 132, 133). One observation in support of this assumption is the nucleotide sequence of those RNAs that carry a NAD modification. All NAD-RNAs that were identified by the NAD captureSeq approach possess as first nucleotide an adenosine. NAD contains an adenosine substructure that is recognized by the polymerase. During transcription initiation, NAD competes with ATP for the incorporation into the RNA chain at the +1 position. Systematic investigation revealed that the promoter nucleoside immediately upstream of +1 is a key determinant of coenzyme incorporation (132). Crystal structures of RNA polymerase in the presence of NAD and biochemical analysis confirmed that the promoter influences the NAD-capping efficiency (132, 134).

Interestingly, a recent report from Julius et al. provides further insight into the mechanism of cofactor capping (135). The authors show *in vitro* that adenine-containing cofactors (NAD, NADH, FAD, and CoA) can all be incorporated at the +1 position by the *E. coli* RNA polymerase. Moreover, they expanded the repertoire of potential capping molecules to include uridine-containing precursors of oligosaccharide and cell wall biosynthesis (UDP-glucose and UDP-*N*-acetylglucosamine) into their capping studies. However, the existence of NADH-RNA, FAD-RNA, UDP-glucose-RNA, and UDP-*N*-acetylglucosamine-RNA *in vivo* remains to be shown.

Removal of 5′-Terminal RNA Modifications

RNA capping is considered a hallmark of eukaryotic gene expression, in which a 5′,5′-triphosphate-linked 7-methylguanosine protects mRNA from degradation

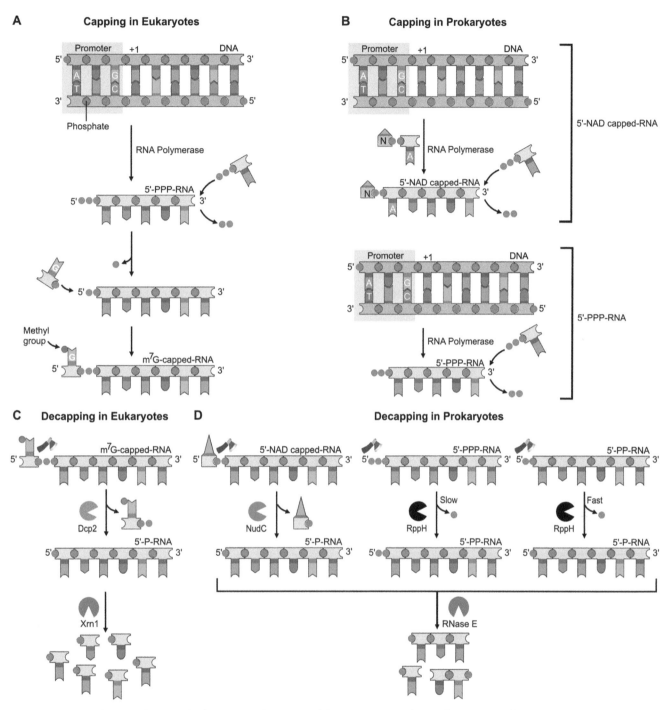

Figure 4 Biosynthesis and removal of 5′-terminal RNA modification. (A) Synthesis of m7G-capped RNA in eukaryotes. 5′-PPP-RNA is synthesized by the RNA polymerase. The 5′-γ-phosphate of the nascent pre-mRNA is hydrolyzed by an RNA triphosphatase to 5′-PP-RNA. In the next step, a guanine monophosphate nucleoside is transferred to the 5′-diphosphate mRNA end by RNA guanylyltransferase to generate G-PPP-RNA. Finally, the guanine N7 position (blue) is methylated by SAM to form m7G-PPP-RNA by SAM. (B) Schematic representation of the synthesis of primary transcripts and cofactor-capped RNA in *E. coli*. Bacterial RNA polymerase is able to initiate transcription with a nucleotide triphosphate or an adenosine-containing cofactor, such as NAD, to generate 5′-PPP-RNA or NAD/cofactor-capped RNA. (C) Decapping of m7G-capped RNA in eukaryotic organisms. A decapping enzyme complex including Dcp2 removes the cap and converts the RNA in 5′-P-RNA that is targeted for degradation by 5′-dependent exonucleases like Xrn1. (D) 5′-End processing in *E. coli*. NAD-RNA is decapped by NudC into 5′-P-RNA. RppH processes primary transcripts and 5′-PP-RNA into 5′-P-RNA, which triggers RNase E-mediated degradation.

and modulates maturation, localization, and translation (96). The eukaryotic m7G-cap is removed by various decapping enzymes, such as Dcp2/Dcp1, thereby initiating different RNA decay pathways by exonucleases (136) (Fig. 4C). Dcp2 belongs to the class of Nudix enzymes. The acronym "Nudix" was coined in 1996 and is derived from the chemical nature of the substrates of the catalyzed reaction, namely, the hydrolysis of nucleoside diphosphates (NDP) linked to another compound X, generating nucleoside monophosphates (NMPs) and the phosphorylated compound X (137). Nudix hydrolases occur in all kingdoms of life and are usually small proteins (16 to 21 kDa). They are known to hydrolyze nucleoside triphosphates, nucleotide sugars, dinucleoside polyphosphates, dinucleotide coenzymes, and capped RNAs (138–140).

In *E. coli*, 13 different Nudix hydrolases are known to date. The first Nudix enzyme that has been described to be active on 5′-terminal RNA modifications is the RNA pyrophosphohydrolase RppH (Fig. 4D) (140, 141). *E. coli* RppH was discovered to convert triphosphorylated RNA to monophosphorylated RNA *in vitro* (141). It was assumed that the conversion of a 5′-terminal triphosphate to a monophosphate occurs in a single step by pyrophosphate removal (142). However, a recent study from Luciano et al. shows that 5′-PPP-RNA is first transformed (likely by another, not yet identified hydrolase) into diphosphorylated RNA, which accumulates to a high cellular concentration. The 5′-PP-RNA is the preferred substrate of RppH and converted into 5′-monophosphorylated RNA (142). These results indicate that a previously unrecognized event modulates 5′-end-dependent mRNA degradation in *E. coli* and suggest an important role for 5′-PP-RNAs in other cellular processes and organisms other than *E. coli*.

The conversion to 5′-P-RNA triggers their degradation by RNase E, which is an endoribonuclease (101). RNase E was recently identified to sense specifically the 5′-modification status of RNAs (triphosphate versus monophosphate), which determines their stability and processing (92, 143, 144). It has been shown that, similarly to triphosphorylated RNA, 5′-NAD-modified RNA is much less susceptible to cleavage by RNase E than 5′-P-RNA. Thus, NAD represents the first prokaryotic RNA cap (70, 145) that protects the RNA against processing by Nudix RNA pyrophosphatase RppH. Another Nudix enzyme, the NADH-pyrophosphohydrolase NudC (146), was identified to specifically convert NAD-RNA into 5′-P-RNA, thereby triggering RNase E-mediated decay (32) (Fig. 4D). Crystallographic and biochemical mutation analysis identified the conserved

Nudix motif as the catalytic center of NudC, which needs to be homodimeric, as the catalytic pocket is composed of amino acids from both protomers. NudC is single-strand specific and has a purine preference for the 5′-terminal nucleotide (147, 148). The enzyme strongly prefers NAD-RNA over NAD and binds to a diverse set of cellular RNAs in an unspecific manner (147). The decapping activity of NudC could provide the bacterium *E. coli* with an additional mechanism to selectively initiate degradation for a subset of cellular RNAs orthogonal to RppH processing of 5′-PPP and PP-RNAs.

Characterization and Quantification of 5′-Terminal RNA Modifications

The discovery of differently modified 5′ termini opens up new questions regarding their biological role(s) (37, 70, 145, 149). To examine the modified RNA itself and the enzymes associated with these 5′-terminal RNA modifications, several *in vitro* and *in vivo* assays were established in the past (32, 141, 142, 150, 151).

The relative quantification of 5′-PP-RNAs *in vivo* is performed by a phosphorylation assay by capping outcome (PACO) (142) (Fig. 5A). This method is based on the substrate specificity of the RNA guanylyltransferase that is usually part of the eukaryotic capping machinery. The guanylyltransferase transfers a GMP selectively to RNAs with a 5′-PP end (152). After successful capping of 5′-PP-RNA to 5′-G-PPP-RNA, residual 5′-P-RNA and 5′-PPP-RNA is dephosphorylated to 5′-OH-RNA by alkaline phosphatase treatment. Afterwards, the 5′-G-PPP-RNA is enzymatically decapped by pyrophosphohydrolases, such as "cap-clip," thereby generating a 5′-P end. Thus, only those RNAs that previously carried a 5′-PP end now have a 5′-P terminus and can therefore be ligated to an adapter. After adapter ligation, a gel shift of the original 5′-PP-RNAs can be detected by Northern blot (142). By allowing simultaneous detection of both the ligation product and the unligated transcript, Northern hybridization makes it possible to quantify their relative abundance and thus to calculate the percentage of the transcript that is diphosphorylated *in vivo* and *in vitro*. In a similar fashion, 5′-P-RNAs are detected and relatively quantified *in vivo* by a PABLO assay (phosphorylation assay by ligation of oligonucleotides) (150).

To study cofactor-capped RNA *in vitro* and *in vivo*, acryloylaminophenyl boronic acid (APB) gel electrophoresis was recently applied to separate NAD- or FAD-modified RNAs from unmodified RNAs (PPP-/PP-/P-RNA) (151) (Fig. 5B). APB is copolymerized with polyacrylamide and forms a relatively stable complex

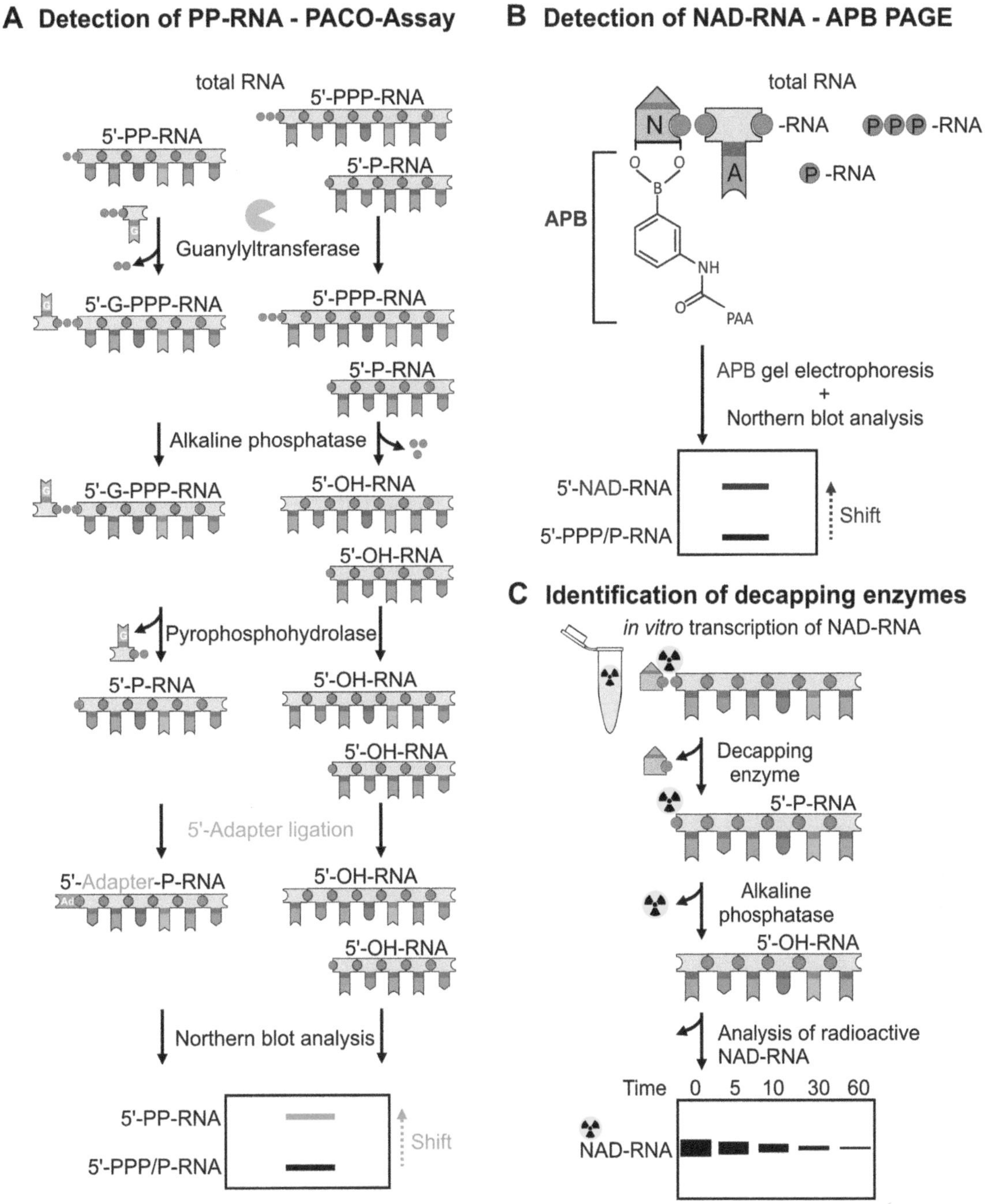

A Detection of PP-RNA - PACO-Assay

B Detection of NAD-RNA - APB PAGE

C Identification of decapping enzymes

Figure 5 Methods to relatively quantify and characterize 5′-terminally modified RNAs and their decapping enzymes *in vitro* and *in vivo*. (A) Schematic representation of the identification and relative quantification of 5′-PP-RNA (PACO assay) *in vivo*. The 5′ end of 5′-PP-RNA is capped with a guanosine. Subsequent treatment with alkaline phosphatase removes the exposed 5′-terminal phosphates of 5′-PPP-RNA and 5′-P-RNA but not the protected phosphates of the guanylylated G-PPP-RNA, which is then converted to 5′-P-RNA by treatment with a pyrophosphohydrolase. After adapter ligation and Northern blot analysis, the amount of 5′-PP-RNA can be quantified by a shift. (B) APB PAGE to quantify 5′-NAD-capped RNA *in vivo*. APB interacts with *cis*-diols of the ribose, which results in retardation of NAD-capped RNA during gel electrophoresis. (C) Identification of decapping enzymes by specific radioactive labeling of RNA *in vitro*. NAD-capped RNA is transcribed by the bacteriophage T7 RNA polymerase in the presence of radioactively labeled NAD. This technology enables specific radioactive labeling of each RNA with a single radioactive mark specifically located at the 5′ termini. After a decapping reaction, e.g., by NudC, 5′-P-RNA is generated. The accessible radioactive phosphate is removed by alkaline phosphatase treatment. The RNA is analyzed by PAGE.

with 1,2-*cis*-diols (153), which occur naturally at the 3′ end of RNA and in the nicotinamide riboside of NAD-modified RNA at the 5′ end. The transient formation of diesters between the immobilized boronic acid and the diols causes lower mobility of the modified RNAs, which is visualized by a shift on the gel (151). APB affinity gels can be used to study cofactor-modified RNAs with small amounts of material, and to rapidly screen for their relative abundance in total RNA while avoiding complex sample treatments (151).

Another method to study and to identify enzymes that are processing 5′-modified RNA is the specific radioactive labeling of 5′ termini (32, 154). For this purpose, RNA can be prepared by *in vitro* transcription using the bacteriophage T7 RNA polymerase radioactive nucleotides. To study the processing of 5′-modifications, compounds that contain radioactively labeled phosphorus (^{32}P or ^{33}P) are best suited. Today, various nucleotides like α-AMP (^{32}P-A), γ-ATP (^{32}PPP-A), or NAD ($5′$-N-P^{32}P-A) are commercially available that are specifically incorporated at the 5′ end of the RNA by the T7 RNA polymerase (155). These radioactive initiator nucleotides enable specific tagging of each RNA with a single radioactive label specifically located at the 5′ terminus (32, 154, 156, 157). Thus, a cleavage or processing of the 5′ end of the RNA by an enzyme can be directly followed by a decrease of the radioactive signal. In the context of the identification of decapping enzymes, a specific radioactive labeling of NAD-capped RNA was used (Fig. 5C). To study enzymes that might convert NAD-RNA into 5′-^{32}P-RNA, alkaline phosphatase was added to the reaction mixture to remove the [^{32}P]phosphate, which is only accessible after enzymatic processing of the NAD-RNA. The decapping activity of NudC and different NudC mutants was examined by such classical biochemical assays *in vitro* (32, 147, 154).

INFLUENCE OF RNA MODIFICATIONS ON THE TRANSLATION OF PROTEINS

RNA is a central molecule in all kingdoms of life, as it connects the storage of genetic information in the form of DNA to its translation into proteins. In eukaryotes, the 5′ terminus of mRNAs is generally modified by a methylated guanosine (m7G) cap, which is crucial for the initiation of protein synthesis (158–160). The protein synthesis itself can be divided into four parts—initiation, elongation, termination, and ribosome recycling—and is regulated by the interplay of the ribosome, tRNAs, mRNA, ribosome, RNA binding proteins, as well as noncoding RNAs. Studies that

examine the influence of co- and posttranscriptional (m)RNA modification on protein synthesis have evolved in recent years. Since the identification of m6A, the modifications m5C, Ψ, and 2′-O-methyl have been found within the open reading frames of mRNAs, where their presence suggests an influence on the fate of the mRNA, ranging from maturation to its translation and degradation (161, 162).

The influence of the bacterial NAD-cap on translation is not known yet. In eukaryotic cells, NAD-capped mRNA was not translated (129).

Recently, tRNA modifications were identified to be linked to translation efficiency and decoding fidelity (163, 164) and to be involved in fine-tuning of stress-related genes by driving codon-biased translation (163). To investigate the influence of RNA modifications on translation, *in vitro* translation assays are generally performed (161, 165). In studies published by Hoernes et al., m5C, m6A, or 2-O-methylated nucleotides were introduced at each of the three positions within a codon of a bacterial mRNA and analyzed concerning their influence on translation. The data indicate astonishingly versatile effects on protein synthesis depending not only on the type of the RNA modification but also on the codon position (165). Incorporation of single modifications in specific codon contexts reduced protein production in *E. coli* by 20 to 70% or resulted in truncated proteins through site-specific ribosome stalling at modified codons. To affect gene function through regulated rewiring of the genetic code is one exciting potential for mRNA modifications. So far, few studies describing an influence of RNA modification on translation are published. However, the development of several approaches to identify the precise location of the RNA modification sets the foundation to study the effect of epitranscriptomics on translation *in vitro* and *in vivo* (52).

OUTLOOK

The recent advances in the field of RNA modifications clearly show that the epitranscriptome and its modifying enzymes form a complex constellation that holds widely diverse functions. Posttranscriptional RNA modifications allow additional control of gene expression, serving as powerful mechanisms that eventually affect protein translation. The boost in the development of techniques to detect RNA modifications has revealed complex networks of modified RNAs and proteins that hold widely diverse and partly unknown functions.

Despite these major achievements, the epitranscriptome has mostly been studied in eukaryotic organisms.

In contrast, detailed studies about RNA modifications and their writers, readers, and erasers in archaea or bacteria have received little attention. Because there are substantial differences between prokaryotic and eukaryotic mRNA metabolism, the functional consequences of mRNA modifications are likely to differ as well. m6A has been reported in mRNA from *E. coli* and *P. aeruginosa*, while m5C has been mapped in *S. solfataricus* mRNAs, and I was detected in 15 mRNAs from *E. coli*. 2-O-methylated nucleotides and Ψ have so far been identified in bacteria only in rRNAs and tRNAs. To elucidate the function of internal as well as 5′-terminal RNA modifications, knowledge about the exact location (detection) as well as the amount present (absolute and relative quantification) is required. Right now, there are hurdles to overcome in the fast-developing field of epitranscriptomics. First, there is a need for orthogonal methods to detect the RNA modifications. Moreover, we need robust and sensitive methods that need less input and are able to detect even less abundant RNA modifications. However, it is obvious that the development of such tools remains challenging especially in the context of >160 known RNA modifications. But is there a technique to identify different epitranscriptomic marks within the same transcript? Recently, two single-molecule methods—SMRT sequencing and the Oxford Nanopore Technology (ONT)—have demonstrated specific and base-resolution detection of m6A in synthetic RNA molecules (34, 166, 167). These approaches could potentially be used to detect multiple RNA modifications simultaneously in a single transcript (167). However, the rapid development of techniques to determine RNA modifications has resulted in a boost of data that are mapped to the genome and have to be independently validated. Still, the error rate of Illumina, SMRT, and ONT sequencing might lead to misinterpretation of the data. Moreover, chemical treatments of RNA as well as the application of antibodies are prone to creation and inclusion of experimental artifacts. To overcome these bottlenecks, careful validation of the RNA-Seq data and direct proof of the identified RNA modification are needed. The development of MS technology (13, 168, 169), different quantification tools like SCARLET (170), ligation-based assays like PABLO or PACO (142, 150), and gel-retardation assays (151) set the foundation to determine modification stoichiometry (9).

Finally, the main question that remains to be answered is the analysis of the biological function(s) of an RNA modification. The majority of functional studies have been performed in eukaryotic organisms. Studies of readers, writers, and erasers in bacteria and archaea

are generally the exception. Thus, research on classical model organisms like *E. coli*, *Salmonella enterica* serovar Typhimurium, or *B. subtilis* should be expedited to identify the relevance of RNA modifications in these kingdoms of life as well.

Citation. Höfer K, Jäschke A. 2018. Epitranscriptomics: RNA modifications in bacteria and archaea. Microbiol Spectrum 6(3):RWR-0015-2017.

References

1. Higgs PG, Lehman N. 2015. The RNA World: molecular cooperation at the origins of life. *Nat Rev Genet* **16**: 7–17.

2. Yi C, Pan T. 2011. Cellular dynamics of RNA modification. *Acc Chem Res* **44**:1380–1388.

3. Clark MB, Choudhary A, Smith MA, Taft RJ, Mattick JS. 2013. The dark matter rises: the expanding world of regulatory RNAs. *Essays Biochem* **54**:1–16.

4. Morris KV, Mattick JS. 2014. The rise of regulatory RNA. *Nat Rev Genet* **15**:423–437.

5. Machnicka MA, Milanowska K, Osman Oglou O, Purta E, Kurkowska M, Olchowik A, Januszewski W, Kalinowski S, Dunin-Horkawicz S, Rother KM, Helm M, Bujnicki JM, Grosjean H. 2013. MODOMICS: a database of RNA modification pathways—2013 update. *Nucleic Acids Res* **41**(Database issue):D262–D267.

6. Helm M, Alfonzo JD. 2014. Posttranscriptional RNA modifications: playing metabolic games in a cell's chemical Legoland. *Chem Biol* **21**:174–185.

7. Shepherd J, Ibba M. 2015. Bacterial transfer RNAs. *FEMS Microbiol Rev* **39**:280–300.

8. Lorenz C, Lünse CE, Mörl M. 2017. tRNA modifications: impact on structure and thermal adaptation. *Biomolecules* **7**:e35.

9. Schaefer M, Kapoor U, Jantsch MF. 2017. Understanding RNA modifications: the promises and technological bottlenecks of the 'epitranscriptome'. *Open Biol* **7**:170077.

10. Björk GR, Hagervall TG. 2014. Transfer RNA modification: presence, synthesis, and function. *Ecosal Plus* **6**:ESP-0007-2013.

11. Jackman JE, Alfonzo JD. 2013. Transfer RNA modifications: nature's combinatorial chemistry playground. *Wiley Interdiscip Rev RNA* **4**:35–48.

12. Cantara WA, Crain PF, Rozenski J, McCloskey JA, Harris KA, Zhang X, Vendeix FA, Fabris D, Agris PF. 2011. The RNA Modification Database, RNAMDB: 2011 update. *Nucleic Acids Res* **39**(Database issue): D195–D201.

13. Kellner S, Burhenne J, Helm M. 2010. Detection of RNA modifications. *RNA Biol* **7**:237–247.

14. Dubin DT, Taylor RH. 1975. The methylation state of poly A-containing messenger RNA from cultured hamster cells. *Nucleic Acids Res* **2**:1653–1668.

15. Schibler U, Perry RP. 1977. The 5′-termini of heterogeneous nuclear RNA: a comparison among molecules of different sizes and ages. *Nucleic Acids Res* **4**:4133–4149.

16. Grosjean H, Keith G, Droogmans L. 2004. Detection and quantification of modified nucleotides in RNA using thin-layer chromatography. *Methods Mol Biol* **265:**357–391.

17. Pomerantz SC, McCloskey JA. 1990. Analysis of RNA hydrolyzates by liquid chromatography-mass spectrometry. *Methods Enzymol* **193:**796–824.

18. Apffel A, Chakel JA, Fischer S, Lichtenwalter K, Hancock WS. 1997. Analysis of oligonucleotides by HPLC-electrospray ionization mass spectrometry. *Anal Chem* **69:**1320–1325.

19. Suzuki T, Ikeuchi Y, Noma A, Suzuki T, Sakaguchi Y. 2007. Mass spectrometric identification and characterization of RNA-modifying enzymes. *Methods Enzymol* **425:**211–229.

20. Wang Z, Gerstein M, Snyder M. 2009. RNA-Seq: a revolutionary tool for transcriptomics. *Nat Rev Genet* **10:**57–63.

21. Sharma CM, Vogel J. 2014. Differential RNA-seq: the approach behind and the biological insight gained. *Curr Opin Microbiol* **19:**97–105.

22. Croucher NJ, Thomson NR. 2010. Studying bacterial transcriptomes using RNA-seq. *Curr Opin Microbiol* **13:**619–624.

23. Head SR, Komori HK, LaMere SA, Whisenant T, Van Nieuwerburgh F, Salomon DR, Ordoukhanian P. 2014. Library construction for next-generation sequencing: overviews and challenges. *Biotechniques* **56:**61–64, 66, 68, passim.

24. Thomason MK, Bischler T, Eisenbart SK, Förstner KU, Zhang A, Herbig A, Nieselt K, Sharma CM, Storz G. 2015. Global transcriptional start site mapping using differential RNA sequencing reveals novel antisense RNAs in *Escherichia coli*. *J Bacteriol* **197:**18–28.

25. Jackson R, Standart N. 2015. The awesome power of ribosome profiling. *RNA* **21:**652–654.

26. Dominissini D, Moshitch-Moshkovitz S, Schwartz S, Salmon-Divon M, Ungar L, Osenberg S, Cesarkas K, Jacob-Hirsch J, Amariglio N, Kupiec M, Sorek R, Rechavi G. 2012. Topology of the human and mouse m6A RNA methylomes revealed by m6A-seq. *Nature* **485:**201–206.

27. Meyer KD, Saletore Y, Zumbo P, Elemento O, Mason CE, Jaffrey SR. 2012. Comprehensive analysis of mRNA methylation reveals enrichment in 3′ UTRs and near stop codons. *Cell* **149:**1635–1646.

28. Edelheit S, Schwartz S, Mumbach MR, Wurtzel O, Sorek R. 2013. Transcriptome-wide mapping of 5-methylcytidine RNA modifications in bacteria, archaea, and yeast reveals m5C within archaeal mRNAs. *PLoS Genet* **9:**e1003602.

29. Dominissini D, Nachtergaele S, Moshitch-Moshkovitz S, Peer E, Kol N, Ben-Haim MS, Dai Q, Di Segni A, Salmon-Divon M, Clark WC, Zheng G, Pan T, Solomon O, Eyal E, Hershkovitz V, Han D, Doré LC, Amariglio N, Rechavi G, He C. 2016. The dynamic N^1-methyladenosine methylome in eukaryotic messenger RNA. *Nature* **530:**441–446.

30. Suzuki T, Ueda H, Okada S, Sakurai M. 2015. Transcriptome-wide identification of adenosine-to-inosine editing using the ICE-seq method. *Nat Protoc* **10:**715–732.

31. Schaefer M. 2015. RNA 5-methylcytosine analysis by bisulfite sequencing. *Methods Enzymol* **560:**297–329.

32. Cahová H, Winz ML, Höfer K, Nübel G, Jäschke A. 2015. NAD captureSeq indicates NAD as a bacterial cap for a subset of regulatory RNAs. *Nature* **519:**374–377.

33. Hussain S, Sajini AA, Blanco S, Dietmann S, Lombard P, Sugimoto Y, Paramor M, Gleeson JG, Odom DT, Ule J, Frye M. 2013. NSun2-mediated cytosine-5 methylation of vault noncoding RNA determines its processing into regulatory small RNAs. *Cell Rep* **4:**255–261.

34. Li X, Xiong X, Yi C. 2016. Epitranscriptome sequencing technologies: decoding RNA modifications. *Nat Methods* **14:**23–31.

35. He C. 2010. Grand challenge commentary: RNA epigenetics? *Nat Chem Biol* **6:**863–865.

36. Schwartz S. 2016. Cracking the epitranscriptome. *RNA* **22:**169–174.

37. Jäschke A, Höfer K, Nübel G, Frindert J. 2016. Cap-like structures in bacterial RNA and epitranscriptomic modification. *Curr Opin Microbiol* **30:**44–49.

38. Saletore Y, Meyer K, Korlach J, Vilfan ID, Jaffrey S, Mason CE. 2012. The birth of the Epitranscriptome: deciphering the function of RNA modifications. *Genome Biol* **13:**175.

39. Motorin Y, Helm M. 2010. tRNA stabilization by modified nucleotides. *Biochemistry* **49:**4934–4944.

40. Phillips G, de Crécy-Lagard V. 2011. Biosynthesis and function of tRNA modifications in Archaea. *Curr Opin Microbiol* **14:**335–341.

41. Nishikura K. 2010. Functions and regulation of RNA editing by ADAR deaminases. *Annu Rev Biochem* **79:**321–349.

42. Bass BL. 2002. RNA editing by adenosine deaminases that act on RNA. *Annu Rev Biochem* **71:**817–846.

43. Wulff BE, Sakurai M, Nishikura K. 2011. Elucidating the inosinome: global approaches to adenosine-to-inosine RNA editing. *Nat Rev Genet* **12:**81–85.

44. Sakurai M, Yano T, Kawabata H, Ueda H, Suzuki T. 2010. Inosine cyanoethylation identifies A-to-I RNA editing sites in the human transcriptome. *Nat Chem Biol* **6:**733–740.

45. Sakurai M, Ueda H, Yano T, Okada S, Terajima H, Mitsuyama T, Toyoda A, Fujiyama A, Kawabata H, Suzuki T. 2014. A biochemical landscape of A-to-I RNA editing in the human brain transcriptome. *Genome Res* **24:**522–534.

46. Bar-Yaacov D, Mordret E, Towers R, Biniashvili T, Soyris C, Schwartz S, Dahan O, Pilpel Y. 2017. RNA editing in bacteria recodes multiple proteins and regulates an evolutionarily conserved toxin-antitoxin system. *Genome Res* **27:**1696–1703.

47. Suzuki MM, Bird A. 2008. DNA methylation landscapes: provocative insights from epigenomics. *Nat Rev Genet* **9:**465–476.

48. Frommer M, McDonald LE, Millar DS, Collis CM, Watt F, Grigg GW, Molloy PL, Paul CL. 1992. A

genomic sequencing protocol that yields a positive display of 5-methylcytosine residues in individual DNA strands. *Proc Natl Acad Sci U S A* **89**:1827–1831.

49. Schaefer M, Pollex T, Hanna K, Lyko F. 2009. RNA cytosine methylation analysis by bisulfite sequencing. *Nucleic Acids Res* **37**:e12.

50. Squires JE, Patel HR, Nousch M, Sibbritt T, Humphreys DT, Parker BJ, Suter CM, Preiss T. 2012. Widespread occurrence of 5-methylcytosine in human coding and non-coding RNA. *Nucleic Acids Res* **40**:5023–5033.

51. Hussain S, Aleksic J, Blanco S, Dietmann S, Frye M. 2013. Characterizing 5-methylcytosine in the mammalian epitranscriptome. *Genome Biol* **14**:215.

52. Gilbert WV, Bell TA, Schaening C. 2016. Messenger RNA modifications: form, distribution, and function. *Science* **352**:1408–1412.

53. Bujnicki JM, Feder M, Ayres CL, Redman KL. 2004. Sequence-structure-function studies of tRNA:m5C methyltransferase Trm4p and its relationship to DNA:m5C and RNA:m5U methyltransferases. *Nucleic Acids Res* **32**:2453–2463.

54. Gu XR, Gustafsson C, Ku J, Yu M, Santi DV. 1999. Identification of the 16S rRNA m5C967 methyltransferase from *Escherichia coli*. *Biochemistry* **38**:4053–4057.

55. Brzezicha B, Schmidt M, Makalowska I, Jarmolowski A, Pienkowska J, Szweykowska-Kulinska Z. 2006. Identification of human tRNA:m5C methyltransferase catalysing intron-dependent m5C formation in the first position of the anticodon of the pre-tRNA Leu (CAA). *Nucleic Acids Res* **34**:6034–6043.

56. Goll MG, Kirpekar F, Maggert KA, Yoder JA, Hsieh CL, Zhang X, Golic KG, Jacobsen SE, Bestor TH. 2006. Methylation of tRNAAsp by the DNA methyltransferase homolog Dnmt2. *Science* **311**:395–398.

57. King MY, Redman KL. 2002. RNA methyltransferases utilize two cysteine residues in the formation of 5-methylcytosine. *Biochemistry* **41**:11218–11225.

58. Redman KL. 2006. Assembly of protein-RNA complexes using natural RNA and mutant forms of an RNA cytosine methyltransferase. *Biomacromolecules* **7**:3321–3326.

59. Roundtree IA, He C. 2016. RNA epigenetics—chemical messages for posttranscriptional gene regulation. *Curr Opin Chem Biol* **30**:46–51.

60. Desrosiers R, Friderici K, Rottman F. 1974. Identification of methylated nucleosides in messenger RNA from Novikoff hepatoma cells. *Proc Natl Acad Sci U S A* **71**:3971–3975.

61. Chen K, Lu Z, Wang X, Fu Y, Luo GZ, Liu N, Han D, Dominissini D, Dai Q, Pan T, He C. 2015. High-resolution N^6-methyladenosine (m6A) map using photo-crosslinking-assisted m6A sequencing. *Angew Chem Int Ed Engl* **54**:1587–1590.

62. Linder B, Grozhik AV, Olarerin-George AO, Meydan C, Mason CE, Jaffrey SR. 2015. Single-nucleotide-resolution mapping of m6A and m6Am throughout the transcriptome. *Nat Methods* **12**:767–772.

63. Deng X, Chen K, Luo GZ, Weng X, Ji Q, Zhou T, He C. 2015. Widespread occurrence of N^6-methyladenosine in bacterial mRNA. *Nucleic Acids Res* **43**:6557–6567.

64. Jia G, Fu Y, Zhao X, Dai Q, Zheng G, Yang Y, Yi C, Lindahl T, Pan T, Yang YG, He C. 2011. N^6-Methyladenosine in nuclear RNA is a major substrate of the obesity-associated FTO. *Nat Chem Biol* **7**:885–887.

65. Zheng G, Dahl JA, Niu Y, Fedorcsak P, Huang CM, Li CJ, Vågbø CB, Shi Y, Wang WL, Song SH, Lu Z, Bosmans RP, Dai Q, Hao YJ, Yang X, Zhao WM, Tong WM, Wang XJ, Bogdan F, Furu K, Fu Y, Jia G, Zhao X, Liu J, Krokan HE, Klungland A, Yang YG, He C. 2013. ALKBH5 is a mammalian RNA demethylase that impacts RNA metabolism and mouse fertility. *Mol Cell* **49**:18–29.

66. Tanaka T, Weisblum B. 1975. Systematic difference in the methylation of ribosomal ribonucleic acid from gram-positive and gram-negative bacteria. *J Bacteriol* **123**:771–774.

67. Sergiev PV, Serebryakova MV, Bogdanov AA, Dontsova OA. 2008. The *ybiN* gene of *Escherichia coli* encodes adenine-N6 methyltransferase specific for modification of A1618 of 23 S ribosomal RNA, a methylated residue located close to the ribosomal exit tunnel. *J Mol Biol* **375**:291–300.

68. Golovina AY, Dzama MM, Osterman IA, Sergiev PV, Serebryakova MV, Bogdanov AA, Dontsova OA. 2012. The last rRNA methyltransferase of *E. coli* revealed: the *yhiR* gene encodes adenine-N6 methyltransferase specific for modification of A2030 of 23S ribosomal RNA. *RNA* **18**:1725–1734.

69. Schwartz S, Agarwala SD, Mumbach MR, Jovanovic M, Mertins P, Shishkin A, Tabach Y, Mikkelsen TS, Satija R, Ruvkun G, Carr SA, Lander ES, Fink GR, Regev A. 2013. High-resolution mapping reveals a conserved, widespread, dynamic mRNA methylation program in yeast meiosis. *Cell* **155**:1409–1421.

70. Marbaniang CN, Vogel J. 2016. Emerging roles of RNA modifications in bacteria. *Curr Opin Microbiol* **30**:50–57.

71. Karijolich J, Yi C, Yu YT. 2015. Transcriptome-wide dynamics of RNA pseudouridylation. *Nat Rev Mol Cell Biol* **16**:581–585.

72. Spenkuch F, Motorin Y, Helm M. 2014. Pseudouridine: still mysterious, but never a fake (uridine)! *RNA Biol* **11**:1540–1554.

73. Karijolich J, Yu YT. 2010. Spliceosomal snRNA modifications and their function. *RNA Biol* **7**:192–204.

74. Wu G, Xiao M, Yang C, Yu YT. 2011. U2 snRNA is inducibly pseudouridylated at novel sites by Pus7p and snR81 RNP. *EMBO J* **30**:79–89.

75. Zaringhalam M, Papavasiliou FN. 2016. Pseudouridylation meets next-generation sequencing. *Methods* **107**:63–72.

76. Carlile TM, Rojas-Duran MF, Zinshteyn B, Shin H, Bartoli KM, Gilbert WV. 2014. Pseudouridine profiling reveals regulated mRNA pseudouridylation in yeast and human cells. *Nature* **515**:143–146.

77. Li X, Zhu P, Ma S, Song J, Bai J, Sun F, Yi C. 2015. Chemical pulldown reveals dynamic pseudouridylation of the mammalian transcriptome. *Nat Chem Biol* **11**:592–597.

78. Yang J, Sharma S, Kötter P, Entian KD. 2015. Identification of a new ribose methylation in the 18S rRNA of *S. cerevisiae*. *Nucleic Acids Res* 43:2342–2352.

79. Tollervey D, Lehtonen H, Jansen R, Kern H, Hurt EC. 1993. Temperature-sensitive mutations demonstrate roles for yeast fibrillarin in pre-rRNA processing, pre-rRNA methylation, and ribosome assembly. *Cell* 72:443–457.

80. Aschenbrenner J, Marx A. 2016. Direct and site-specific quantification of RNA 2′-O-methylation by PCR with an engineered DNA polymerase. *Nucleic Acids Res* 44: 3495–3502.

81. Maden BE, Corbett ME, Heeney PA, Pugh K, Ajuh PM. 1995. Classical and novel approaches to the detection and localization of the numerous modified nucleotides in eukaryotic ribosomal RNA. *Biochimie* 77:22–29.

82. Incarnato D, Anselmi F, Morandi E, Neri F, Maldotti M, Rapelli S, Parlato C, Basile G, Oliviero S. 2017. High-throughput single-base resolution mapping of RNA 2′-O-methylated residues. *Nucleic Acids Res* 45:1433–1441.

83. Marchand V, Blanloeil-Oillo F, Helm M, Motorin Y. 2016. Illumina-based RiboMethSeq approach for mapping of 2′-O-Me residues in RNA. *Nucleic Acids Res* 44:e135.

84. Zhu Y, Pirnie SP, Carmichael GG. 2017. High-throughput and site-specific identification of 2′-O-methylation sites using ribose oxidation sequencing (RibOxi-seq). *RNA* 23:1303–1314.

85. Dai Q, Moshitch-Moshkovitz S, Han D, Kol N, Amariglio N, Rechavi G, Dominissini D, He C. 2017. Nm-seq maps 2′-O-methylation sites in human mRNA with base precision. *Nat Methods* 14:695–698.

86. Lee J, Harris AN, Holley CL, Mahadevan J, Pyles KD, Lavagnino Z, Scherrer DE, Fujiwara H, Sidhu R, Zhang J, Huang SC, Piston DW, Remedi MS, Urano F, Ory DS, Schaffer JE. 2016. Rpl13a small nucleolar RNAs regulate systemic glucose metabolism. *J Clin Invest* 126: 4616–4625.

87. Daffis S, Szretter KJ, Schriewer J, Li J, Youn S, Errett J, Lin TY, Schneller S, Zust R, Dong H, Thiel V, Sen GC, Fensterl V, Klimstra WB, Pierson TC, Buller RM, Gale M Jr, Shi PY, Diamond MS. 2010. 2′-O methylation of the viral mRNA cap evades host restriction by IFIT family members. *Nature* 468:452–456.

88. Rimbach K, Kaiser S, Helm M, Dalpke AH, Eigenbrod T. 2015. 2′-O-Methylation within bacterial RNA acts as suppressor of TLR7/TLR8 activation in human innate immune cells. *J Innate Immun* 7:482–493.

89. Dennis PP, Tripp V, Lui L, Lowe T, Randau L. 2015. C/D box sRNA-guided 2′-O-methylation patterns of archaeal rRNA molecules. *BMC Genomics* 16:632.

90. Hornung V, Ellegast J, Kim S, Brzózka K, Jung A, Kato H, Poeck H, Akira S, Conzelmann KK, Schlee M, Endres S, Hartmann G. 2006. 5′-Triphosphate RNA is the ligand for RIG-I. *Science* 314:994–997.

91. Goubau D, Schlee M, Deddouche S, Pruijssers AJ, Zillinger T, Goldeck M, Schuberth C, Van der Veen AG, Fujimura T, Rehwinkel J, Iskarpatyoti JA, Barchet W, Ludwig J, Dermody TS, Hartmann G, Reis E Sousa

C. 2014. Antiviral immunity via RIG-I-mediated recognition of RNA bearing 5′-diphosphates. *Nature* 514: 372–375.

92. Mackie GA. 1998. Ribonuclease E is a 5′-end-dependent endonuclease. *Nature* 395:720–723.

93. Richards J, Belasco JG. 2016. Distinct requirements for 5′-monophosphate-assisted RNA cleavage by *Escherichia coli* RNase E and RNase G. *J Biol Chem* 291:20825.

94. Wei CM, Gershowitz A, Moss B. 1975. Methylated nucleotides block 5′ terminus of HeLa cell messenger RNA. *Cell* 4:379–386.

95. Furuichi Y, Muthukrishnan S, Shatkin AJ. 1975. 5′-Terminal $m^7G(5')ppp(5')G^mp$ *in vivo*: identification in reovirus genome RNA. *Proc Natl Acad Sci U S A* 72: 742–745.

96. Topisirovic I, Svitkin YV, Sonenberg N, Shatkin AJ. 2011. Cap and cap-binding proteins in the control of gene expression. *Wiley Interdiscip Rev RNA* 2:277–298.

97. Nagarajan VK, Jones CI, Newbury SF, Green PJ. 2013. XRN 5′→3′ exoribonucleases: structure, mechanisms and functions. *Biochim Biophys Acta* 1829:590–603.

98. Valkov E, Jonas S, Weichenrieder O. 2017. *Mille viae* in eukaryotic mRNA decapping. *Curr Opin Struct Biol* 47: 40–51.

99. Grudzien-Nogalska E, Kiledjian M. 2017. New insights into decapping enzymes and selective mRNA decay. *Wiley Interdiscip Rev RNA* 8:e1379.

100. Li Y, Kiledjian M. 2010. Regulation of mRNA decapping. *Wiley Interdiscip Rev RNA* 1:253–265.

101. Schoenberg DR. 2007. The end defines the means in bacterial mRNA decay. *Nat Chem Biol* 3:535–536.

102. Ettwiller L, Buswell J, Yigit E, Schildkraut I. 2016. A novel enrichment strategy reveals unprecedented number of novel transcription start sites at single base resolution in a model prokaryote and the gut microbiome. *BMC Genomics* 17:199.

103. Vvedenskaya IO, Sharp JS, Goldman SR, Kanabar PN, Livny J, Dove SL, Nickels BE. 2012. Growth phase-dependent control of transcription start site selection and gene expression by nanoRNAs. *Genes Dev* 26: 1498–1507.

104. Druzhinin SY, Tran NT, Skalenko KS, Goldman SR, Knoblauch JG, Dove SL, Nickels BE. 2015. A conserved pattern of primer-dependent transcription initiation in *Escherichia coli* and *Vibrio cholerae* revealed by 5′ RNA-seq. *PLoS Genet* 11:e1005348.

105. Nickels BE, Dove SL. 2011. NanoRNAs: a class of small RNAs that can prime transcription initiation in bacteria. *J Mol Biol* 412:772–781.

106. Goldman SR, Sharp JS, Vvedenskaya IO, Livny J, Dove SL, Nickels BE. 2011. NanoRNAs prime transcription initiation *in vivo*. *Mol Cell* 42:817–825.

107. Peach SE, York K, Hesselberth JR. 2015. Global analysis of RNA cleavage by 5′-hydroxyl RNA sequencing. *Nucleic Acids Res* 43:e108.

108. Vvedenskaya IO, Goldman SR, Nickels BE. 2015. Preparation of cDNA libraries for high-throughput RNA

sequencing analysis of RNA 5′ ends. *Methods Mol Biol* **1276**:211–228.

109. Smith P, Wang LK, Nair PA, Shuman S. 2012. The adenylyltransferase domain of bacterial Pnkp defines a unique RNA ligase family. *Proc Natl Acad Sci U S A* **109**:2296–2301.

110. Chen YG, Kowtoniuk WE, Agarwal I, Shen Y, Liu DR. 2009. LC/MS analysis of cellular RNA reveals NAD-linked RNA. *Nat Chem Biol* **5**:879–881.

111. Kowtoniuk WE, Shen Y, Heemstra JM, Agarwal I, Liu DR. 2009. A chemical screen for biological small molecule-RNA conjugates reveals CoA-linked RNA. *Proc Natl Acad Sci U S A* **106**:7768–7773.

112. Winz ML, Cahová H, Nübel G, Frindert J, Höfer K, Jäschke A. 2017. Capture and sequencing of NAD-capped RNA sequences with NAD captureSeq. *Nat Protoc* **12**:122–149.

113. Rostovtsev VV, Green LG, Fokin VV, Sharpless KB. 2002. A stepwise Huisgen cycloaddition process: copper (I)-catalyzed regioselective "ligation" of azides and terminal alkynes. *Angew Chem Int Ed Engl* **41**:2596–2599.

114. Tornøe CW, Christensen C, Meldal M. 2002. Peptidotriazoles on solid phase: [1,2,3]-triazoles by regiospecific copper(I)-catalyzed 1,3-dipolar cycloadditions of terminal alkynes to azides. *J Org Chem* **67**:3057–3064.

115. Preugschat F, Tomberlin GH, Porter DJ. 2008. The base exchange reaction of NAD⁺ glycohydrolase: identification of novel heterocyclic alternative substrates. *Arch Biochem Biophys* **479**:114–120.

116. Opdyke JA, Fozo EM, Hemm MR, Storz G. 2011. RNase III participates in GadY-dependent cleavage of the gadX-gadW mRNA. *J Mol Biol* **406**:29–43.

117. Opdyke JA, Kang JG, Storz G. 2004. GadY, a small-RNA regulator of acid response genes in *Escherichia coli*. *J Bacteriol* **186**:6698–6705.

118. Lacatena RM, Cesareni G. 1981. Base pairing of RNA I with its complementary sequence in the primer precursor inhibits ColE1 replication. *Nature* **294**:623–626.

119. Masukata H, Tomizawa J. 1986. Control of primer formation for ColE1 plasmid replication: conformational change of the primer transcript. *Cell* **44**:125–136.

120. Gerhart E, Wagner H, Nordström K. 1986. Structural analysis of an RNA molecule involved in replication control of plasmid R1. *Nucleic Acids Res* **14**:2523–2538.

121. Blomberg P, Wagner EG, Nordström K. 1990. Control of replication of plasmid R1: the duplex between the antisense RNA, CopA, and its target, CopT, is processed specifically *in vivo* and *in vitro* by RNase III. *EMBO J* **9**:2331–2340.

122. Jin Y, Watt RM, Danchin A, Huang JD. 2009. Small noncoding RNA GcvB is a novel regulator of acid resistance in *Escherichia coli*. *BMC Genomics* **10**:165.

123. Miyakoshi M, Chao Y, Vogel J. 2015. Cross talk between ABC transporter mRNAs via a target mRNA-derived sponge of the GcvB small RNA. *EMBO J* **34**:1478–1492.

124. Thomason MK, Fontaine F, De Lay N, Storz G. 2012. A small RNA that regulates motility and biofilm formation in response to changes in nutrient availability in *Escherichia coli*. *Mol Microbiol* **84**:17–35.

125. van Nues RW, Castro-Roa D, Yuzenkova Y, Zenkin N. 2016. Ribonucleoprotein particles of bacterial small non-coding RNA IsrA (IS61 or McaS) and its interaction with RNA polymerase core may link transcription to mRNA fate. *Nucleic Acids Res* **44**:2577–2592.

126. Schu DJ, Zhang A, Gottesman S, Storz G. 2015. Alternative Hfq-sRNA interaction modes dictate alternative mRNA recognition. *EMBO J* **34**:2557–2573.

127. Mendoza-Vargas A, Olvera L, Olvera M, Grande R, Vega-Alvarado L, Taboada B, Jimenez-Jacinto V, Salgado H, Juárez K, Contreras-Moreira B, Huerta AM, Collado-Vides J, Morett E. 2009. Genome-wide identification of transcription start sites, promoters and transcription factor binding sites in *E. coli*. *PLoS One* **4**:e7526.

128. Walters RW, Matheny T, Mizoue LS, Rao BS, Muhlrad D, Parker R. 2017. Identification of NAD⁺ capped mRNAs in *Saccharomyces cerevisiae*. *Proc Natl Acad Sci U S A* **114**:480–485.

129. Jiao X, Doamekpor SK, Bird JG, Nickels BE, Tong L, Hart RP, Kiledjian M. 2017. 5′ end nicotinamide adenine dinucleotide cap in human cells promotes RNA decay through DXO-mediated deNADding. *Cell* **168**:1015–1027.e10.

130. McCracken S, Fong N, Rosonina E, Yankulov K, Brothers G, Siderovski D, Hessel A, Foster S, Shuman S, Bentley DL. 1997. 5′-Capping enzymes are targeted to pre-mRNA by binding to the phosphorylated carboxy-terminal domain of RNA polymerase II. *Genes Dev* **11**:3306–3318.

131. Cho EJ, Takagi T, Moore CR, Buratowski S. 1997. mRNA capping enzyme is recruited to the transcription complex by phosphorylation of the RNA polymerase II carboxy-terminal domain. *Genes Dev* **11**:3319–3326.

132. Bird JG, Zhang Y, Tian Y, Panova N, Barvík I, Greene L, Liu M, Buckley B, Krásný L, Lee JK, Kaplan CD, Ebright RH, Nickels BE. 2016. The mechanism of RNA 5′ capping with NAD⁺, NADH and desphospho-CoA. *Nature* **535**:444–447.

133. Malygin AG, Shemyakin MF. 1979. Adenosine, NAD and FAD can initiate template-dependent RNA synthesis catalyzed by *Escherichia coli* RNA polymerase. *FEBS Lett* **102**:51–54.

134. Höfer K, Jäschke A. 2016. Molecular biology: a surprise beginning for RNA. *Nature* **535**:359–360.

135. Julius C, Yuzenkova Y. 2017. Bacterial RNA polymerase caps RNA with various cofactors and cell wall precursors. *Nucleic Acids Res* **45**:8282–8290.

136. Arribas-Layton M, Wu D, Lykke-Andersen J, Song H. 2013. Structural and functional control of the eukaryotic mRNA decapping machinery. *Biochim Biophys Acta* **1829**:580–589.

137. Bessman MJ, Frick DN, O'Handley SF. 1996. The MutT proteins or "Nudix" hydrolases, a family of versatile, widely distributed, "housecleaning" enzymes. *J Biol Chem* **271**:25059–25062.

138. Song MG, Bail S, Kiledjian M. 2013. Multiple Nudix family proteins possess mRNA decapping activity. *RNA* **19**:390–399.

139. McLennan AG. 2006. The Nudix hydrolase superfamily. *Cell Mol Life Sci* **63**:123–143.

140. McLennan AG. 2013. Substrate ambiguity among the nudix hydrolases: biologically significant, evolutionary remnant, or both? *Cell Mol Life Sci* **70**:373–385.

141. Deana A, Celesnik H, Belasco JG. 2008. The bacterial enzyme RppH triggers messenger RNA degradation by 5′ pyrophosphate removal. *Nature* **451**:355–358.

142. Luciano DJ, Vasilyev N, Richards J, Serganov A, Belasco JG. 2017. A novel RNA phosphorylation state enables 5′ end-dependent degradation in *Escherichia coli*. *Mol Cell* **67**:44–54.e6.

143. Bandyra KJ, Bouvier M, Carpousis AJ, Luisi BF. 2013. The social fabric of the RNA degradosome. *Biochim Biophys Acta* **1829**:514–522.

144. Mackie GA. 2013. RNase E: at the interface of bacterial RNA processing and decay. *Nat Rev Microbiol* **11**:45–57.

145. Luciano DJ, Belasco JG. 2015. NAD in RNA: unconventional headgear. *Trends Biochem Sci* **40**:245–247.

146. Frick DN, Bessman MJ. 1995. Cloning, purification, and properties of a novel NADH pyrophosphatase. Evidence for a nucleotide pyrophosphatase catalytic domain in MutT-like enzymes. *J Biol Chem* **270**:1529–1534.

147. Höfer K, Li S, Abele F, Frindert J, Schlotthauer J, Grawenhoff J, Du J, Patel DJ, Jäschke A. 2016. Structure and function of the bacterial decapping enzyme NudC. *Nat Chem Biol* **12**:730–734.

148. Zhang D, Liu Y, Wang Q, Guan Z, Wang J, Liu J, Zou T, Yin P. 2016. Structural basis of prokaryotic NAD-RNA decapping by NudC. *Cell Res* **26**:1062–1066.

149. Güell M, Yus E, Lluch-Senar M, Serrano L. 2011. Bacterial transcriptomics: what is beyond the RNA horizome? *Nat Rev Microbiol* **9**:658–669.

150. Celesnik H, Deana A, Belasco JG. 2008. PABLO analysis of RNA: 5′-phosphorylation state and 5′-end mapping. *Methods Enzymol* **447**:83–98.

151. Nübel G, Sorgenfrei FA, Jäschke A. 2017. Boronate affinity electrophoresis for the purification and analysis of cofactor-modified RNAs. *Methods* **117**:14–20.

152. Shuman S. 2001. Structure, mechanism, and evolution of the mRNA capping apparatus. *Prog Nucleic Acid Res Mol Biol* **66**:1–40.

153. Igloi GL, Kössel H. 1985. Affinity electrophoresis for monitoring terminal phosphorylation and the presence of queuosine in RNA. Application of polyacrylamide containing a covalently bound boronic acid. *Nucleic Acids Res* **13**:6881–6898.

154. Höfer K, Abele F, Schlotthauer J, Jäschke A. 2016. Synthesis of 5′-NAD-capped RNA. *Bioconjug Chem* **27**:874–877.

155. Seelig B, Jäschke A. 1999. Ternary conjugates of guanosine monophosphate as initiator nucleotides for the enzymatic synthesis of 5′-modified RNAs. *Bioconjug Chem* **10**:371–378.

156. Huang F. 2003. Efficient incorporation of CoA, NAD and FAD into RNA by *in vitro* transcription. *Nucleic Acids Res* **31**:e8.

157. Huang F, Bugg CW, Yarus M. 2000. RNA-catalyzed CoA, NAD, and FAD synthesis from phosphopantetheine, NMN, and FMN. *Biochemistry* **39**:15548–15555.

158. Shatkin AJ, Manley JL. 2000. The ends of the affair: capping and polyadenylation. *Nat Struct Biol* **7**:838–842.

159. Shatkin AJ. 1976. Capping of eucaryotic mRNAs. *Cell* **9**:645–653.

160. Mitchell SF, Walker SE, Algire MA, Park EH, Hinnebusch AG, Lorsch JR. 2010. The 5′-7-methylguanosine cap on eukaryotic mRNAs serves both to stimulate canonical translation initiation and to block an alternative pathway. *Mol Cell* **39**:950–962.

161. Hoernes TP, Hüttenhofer A, Erlacher MD. 2016. mRNA modifications: dynamic regulators of gene expression? *RNA Biol* **13**:760–765.

162. Hoernes TP, Erlacher MD. 2017. Translating the epitranscriptome. *Wiley Interdiscip Rev RNA* **8**:e1375.

163. Endres L, Dedon PC, Begley TJ. 2015. Codon-biased translation can be regulated by wobble-base tRNA modification systems during cellular stress responses. *RNA Biol* **12**:603–614.

164. Manickam N, Joshi K, Bhatt MJ, Farabaugh PJ. 2016. Effects of tRNA modification on translational accuracy depend on intrinsic codon-anticodon strength. *Nucleic Acids Res* **44**:1871–1881.

165. Hoernes TP, Clementi N, Faserl K, Glasner H, Breuker K, Lindner H, Hüttenhofer A, Erlacher MD. 2016. Nucleotide modifications within bacterial messenger RNAs regulate their translation and are able to rewire the genetic code. *Nucleic Acids Res* **44**:852–862.

166. Novoa EM, Mason CE, Mattick JS. 2017. Charting the unknown epitranscriptome. *Nat Rev Mol Cell Biol* **18**:339–340.

167. Garalde DR, Snell EA, Jachimowicz D, Sipos B, Lloyd JH, Bruce M, Pantic N, Admassu T, James P, Warland A, Jordan M, Ciccone J, Serra S, Keenan J, Martin S, McNeill L, Wallace EJ, Jayasinghe L, Wright C, Blasco J, Young S, Brocklebank D, Juul S, Clarke J, Heron AJ, Turner DJ. 2018. Highly parallel direct RNA sequencing on an array of nanopores. *Nat Methods* **15**:201–206.

168. Kellner S, Ochel A, Thüring K, Spenkuch F, Neumann J, Sharma S, Entian KD, Schneider D, Helm M. 2014. Absolute and relative quantification of RNA modifications via biosynthetic isotopomers. *Nucleic Acids Res* **42**:e142.

169. Boccaletto P, Machnicka MA, Purta E, Piatkowski P, Baginski B, Wirecki TK, de Crécy-Lagard V, Ross R, Limbach PA, Kotter A, Helm M, Bujnicki JM. 2018. MODOMICS: a database of RNA modification pathways. 2017 update. *Nucleic Acids Res* **46**(D1):D303–D307.

170. Liu N, Parisien M, Dai Q, Zheng G, He C, Pan T. 2013. Probing N^6-methyladenosine RNA modification status at single nucleotide resolution in mRNA and long noncoding RNA. *RNA* **19**:1848–1856.

Regulating with RNA in Bacteria and Archaea
Edited by Gisela Storz and Kai Papenfort
© 2018 American Society for Microbiology, Washington, DC
doi:10.1128/microbiolspec.RWR-0024-2018

RNA Localization in Bacteria

24

Jingyi Fei[1] and Cynthia M. Sharma[2]

INTRODUCTION

Spatial and temporal localization of macromolecules, including RNAs, reflects the compartmentalization of living cells and plays important roles in gene expression and regulation. In eukaryotic cells, physical separation between the transcription and translation machineries in the nucleus and cytoplasm, respectively, naturally results in the synthesis, processing, and translation of mRNA to be spatially disconnected. Both mRNA localization and localized translation can be important regulatory mechanisms underlying embryonic patterning, asymmetric cell division, epithelial polarity, cell migration, and neuronal morphogenesis (1, 2). RNAs can be transported in the eukaryotic cell in several ways, such as (i) vectorial movement of mRNA by direct coupling to motor proteins, (ii) transport of mRNA by hitchhiking on another cargo, (iii) random transport of mRNA-motor complexes and local enrichment of mRNAs at target sites, or (iv) diffusion and motor-driven cytoplasmic flows with subsequent localized anchorage of the mRNA (3). Moreover, localized translation induction by phosphorylation and activation of translation initiation factors and their regulators in response to localized signals have been reported to impact gene regulation in eukaryotes (4).

In contrast, due to a lack of canonical membrane-bound organelles and a nuclear compartment, prokaryotic cells were long assumed to lack complex subcellular localization of macromolecules, and spatial localization has not been considered to play a significant role in expression and posttranscriptional regulation of bacterial mRNAs. This is also reflected by the classical picture of cotranscriptional translation of bacterial mRNAs, where transcription and protein synthesis are not spatially or temporally separated. Moreover, due to the much smaller size of bacterial cells compared to their eukaryotic counterparts, it has been more challenging to determine the subcellular organization of bacteria, to observe the subcellular distribu-

tion of biomolecules in bacterial cells, and to relate such organization and distribution to biological functions.

With the development of numerous labeling and imaging techniques as well as advanced microscopy approaches that can break the diffraction limit, it is now clear that the bacterial cells are also compartmentalized (5–7). Emerging evidence for differential localization of bacterial mRNAs indicates that the spatial organization in the cell can also impact gene expression and posttranscriptional regulation in prokaryotes (8–10). Commonly, localization patterns of mRNAs in bacteria include the nucleoid region, the cytoplasm, the cell poles, and the inner membrane. Frequently observed organizations of bacterial biomolecules include uniform distribution, distinct foci, and a putative helical structure, often in the vicinity of the cell envelope. Along with the visualization of transcript localization, it has also been shown that many enzymes and complexes involved in RNA metabolism, such as RNA polymerase (RNAP), ribosomes, and the degradosome, show distinct subcellular distributions, providing further support for the role of spatial organization in genetic information flow. Certain observations and conclusions in the study of bacterial RNA localization are still controversial, and the mechanisms underlying observed examples of subcellular localized transcripts remain to be further explored. However, it has nonetheless become clear that spatial control of RNA and related cellular machineries is likely important for gene expression and regulation in prokaryotes, just as it has been a well-established concept in higher organisms. These preliminary observations of distinct RNA localization patterns have brought attention to new phenomena and questions in bacterial posttranscriptional control, such as transcription-coupled versus transcription-uncoupled translation, translation-dependent and translation-independent mRNA localization, as well as localized degradation, stabilization, or

[1]Department of Biochemistry and Molecular Biology, Institute for Biophysical Dynamics, The University of Chicago, Chicago, IL 60637; [2]Chair of Molecular Infection Biology II, Institute of Molecular Infection Biology (IMIB), University of Würzburg, 97080 Würzburg, Germany.

regulation by small regulatory RNAs (sRNAs) and RNA-metabolizing complexes.

Posttranscriptional regulation by regulatory RNAs, RNA-binding proteins (RBPs), and RNases is a central layer of gene expression control in all kingdoms of life. Bacterial sRNAs (typically 50 to 300 nucleotides in length) can control specific genes and/or coordinate expression of distinct regulons with clear physiological outcomes (11). While most sRNAs act as antisense RNAs by short and imperfect base-pairing, several can also directly bind to proteins and control their activity. sRNA-mediated regulation requires numerous and dynamic interplay with various cellular machineries, including RNAP, ribosomes, and degradosomes, and perturbs these machineries in the pathways of mRNA metabolism to broadly affect gene expression. The RNA chaperone Hfq serves as a key player in the sRNA regulatory pathways, where it functions in two main aspects: stabilization of sRNAs from degradation and promotion of the annealing between sRNAs and their target mRNAs (12). Base-pairing of sRNAs to their target mRNAs with the help of Hfq often leads to changes in translation and/or mRNA stability (positive or negative). Translation inhibition is often associated with RNase-mediated codegradation of the sRNA-mRNA pair. While posttranscriptional regulation in bacteria has mainly been studied at the population level in batch cultures, little is known about sRNA-mediated regulation at the single-cell level and even less about the extent and impact of subcellular localization of RNAs on regulatory processes in these organisms. Due to their important biological function, the subcellular localization of sRNAs and their interactions with target mRNAs, Hfq, and RNase E have become an intriguing research topic.

In this review, we describe recent advances in methods that allow for the investigation of RNA localization in bacterial systems, as well as findings regarding mRNA and sRNA localizations in these organisms. We discuss the models and mechanisms revealed by these examples of spatial control of RNA. In addition, we introduce new labeling and imaging methods recently developed in eukaryotic cells, most of which have not yet been applied to bacteria but have the potential to reveal new insights about prokaryotic transcript localization. Finally, we conclude by laying out open questions and future challenges in the field.

APPROACHES TO STUDY RNA LOCALIZATION

Biochemically, cell fractionation methods have been routinely used to study protein localization to the outer membrane, periplasm, inner membrane, and cytoplasm (13, 14). Similar approaches have also been used to investigate RNA localization. Particularly, when cell fractionation methods are combined with high-throughput RNA sequencing, the relative distribution of transcripts between the membrane and the cytoplasm can be estimated at the whole-transcriptome level (15). Compared to fractionation-based approaches, visualization of RNAs by light microscopy techniques provides the most direct information on subcellular localization of individual RNA species. It is worth mentioning that three-dimensional cryo-electron tomography provides another remarkable imaging category with enhanced spatial resolution compared to conventional light microscopy, and has been applied to imaging subcellular organization of bacteria (reviewed in reference 16). However, compared to light microscopy, as in general no specific tags are introduced to the specific biomolecules of interest, cryo-electron tomography usually cannot provide selectivity of the specific biomolecules of interest during imaging.

Since many studies on bacterial RNA localization are fluorescence imaging based, here, we discuss labeling and imaging methods used in bacterial systems (see also several recent reviews on RNA imaging methods [2, 17, 18]).

FISH

Single-molecule fluorescence *in situ* hybridization (smFISH) is one of the most widely used RNA-labeling strategies in both eukaryotic and bacterial systems (19–21). By fixing and permeabilizing the cells, DNA oligonucleotides covalently linked with fluorophores can access the interior and hybridize to the RNAs of interest, thereby labeling them (Fig. 1A). To enhance the signal-to-noise and target specificity, a few to tens of labeled oligos are used for each RNA, tiling along the nucleotide sequence. Enhanced fluorescent signals from multiple oligos on the same RNA appear as a single spot under the diffraction-limited fluorescence microscope, providing single-molecule sensitivity. Despite the limitation of FISH to "dead" cells, many important details can be gleaned from this approach, including the expression levels and localization of RNAs, which can be further used to derive the kinetic mechanisms of transcription or degradation (20, 22–24). FISH has been applied to both mRNAs and sRNAs in bacteria (for references, see the sections on mRNA and sRNA localization below). However, due to the relatively short length of sRNAs, the application of FISH to sRNAs may be case dependent, and be more applicable to sRNAs existing in high copy number.

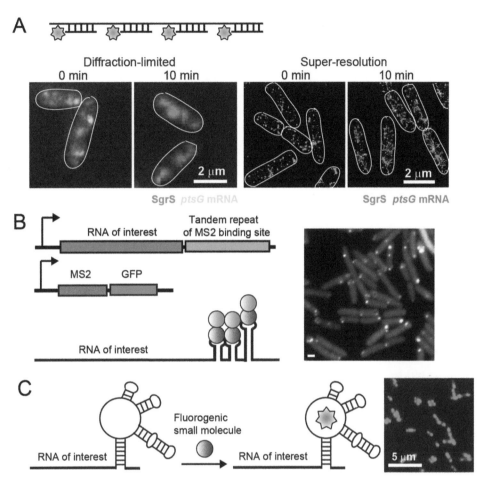

Figure 1 Methods to visualize bacterial RNAs. (A) smFISH and its application to imaging SgrS sRNA and its target mRNA *ptsG*. Images from diffraction-limited and super-resolution microscopes are shown for comparison. Adapted from reference 105. (B) Illustration of the FP reporter approaches with a FP-RBP and a corresponding RNA motif, using the MS2 system as an example, and its application to track mRNAs at the single-molecule level in live *E. coli* cells. Image adapted from reference 67. The scale bar in the image represents 1 μm. (C) The Spinach aptamer and its application to image mRNAs in live *E. coli* cells, in which a Spinach aptamer is fused to the RNA and with fluorescence detection upon addition of the organic ligand 3,5-difluoro-4-hydroxybenzylidene imidazolinone (DFHBI). Image adapted from reference 58 (licensed under a Creative Commons Attribution 4.0 International License [http://creativecommons.org/licenses/by/4.0/]).

FISH does not require genetic manipulation of the RNAs of interest, and normally does not perturb their function or certain features, such as lifetime. However, for fixed-cell imaging, fixation and permeabilization conditions can potentially affect the native localization of biomolecules and/or the accessibility of the labeling reagents (25). While the accessibility issue is less of a concern for short FISH probes (usually ~20 nucleotides) compared to sizable antibodies in immunostaining protocol, it is still recommended that imaging results are validated by using multiple fixation and permeabilization methods. In addition, negative con-trols (such as a knockout strain of the RNA of interest) should always be used to examine the level of nonspecific binding of the FISH probes.

Fluorescent Protein Reporters

Live-cell imaging of RNAs allows for a direct observation of transcript motion, as well as the kinetics of RNA-associated activities. While such approaches are more challenging and require genetic manipulation, several methodologies have been developed for live-cell RNA imaging. One category of approaches relies on orthogonal protein-RNA interactions, consisting of an

RNA-binding motif engineered into the transcript of interest, together with the cognate RBP fused to a fluorescent protein (FP) (2). The FP-fused RBP recognizes and binds to the RNA motif, thereby labeling the RNA. Repetitive RNA motifs of the same kind are often inserted into the RNA of interest to recruit multiple FP-fused RBPs, thereby enhancing the signal, even to single-transcript sensitivity (Fig. 1B). Commonly used RNA-RBP pairs include the MS2 and PP7 phage coat proteins with their respective RNA motifs, as well as the λ phage N-protein–*boxB* hairpin pair (26–28). There are several derivatives of this approach. To increase the signal brightness and photostability, FPs can be replaced by other visualizable tags on the RBPs, such as the SNAP-tag or an *Escherichia coli* dihydrofolate reductase tag (eDHFR-tag) (29–32). In these approaches, the tags can be labeled by the addition of organic dyes introduced into the cell. Such methods rely on the ability to introduce the dyes into cells by either their natural membrane permeability or manual delivery via transfection or microinjection. Since FP-fused RBPs are constitutively expressed and fluorescent in the cell, even if not bound to the target RNAs, they can result in significant background signals. To lower the background, another derivative of the FP live-cell approach employs complementation of the fluorescent reporter upon RNA binding. In this method, FPs are expressed as two independent halves, fused to either two RBPs or halves of the same RBP, respectively. When the two FP halves are brought into close proximity through binding to the same RNA molecule, a complete FP is reconstituted. For example, the RNA-binding eukaryotic translation initiation factor eIF4A can be expressed as two independent halves, each fused to a half of a split enhanced green fluorescent protein (EGFP) (33, 34). Upon binding of an eIF4A target inserted into a transcript of interest, fluorescent EGFP is reconstituted. It should be noted that the split eIF4A reporter system can only be applied to bacterial cells, as they natively lack eIF4A. Similarly, split EGFP has been fused to two different Pumilio homology domains (PUM-HD) of human PUMILIO1 (35), and the FP Venus has been split into two domains, fused to either the PP7 or MS2 coat proteins, and has been used to detect transcripts expressing adjacent PP7 and MS2 binding motifs (36).

The FP reporter systems use indirect labeling methods, in which significant modifications have to be introduced to the RNAs of interest. Therefore, it is important to know the potential pitfalls of these methods. First, FPs have a propensity to oligomerize (37, 38), and wild-type MS2 coat proteins, but not the V75EA81G double mutant, also have such propensity (39). Therefore, a careful choice of the FP-fused RBP is necessary to avoid the formation of artificial, RNA-independent foci. An independent labeling method, such as FISH, is recommended to be used to verify whether the fluorescence protein foci indeed also contain the RNA of interest. Second, the FP reporter systems have potential risks of changing the mRNA processing, lifetime, and localization, due to protection by the bound proteins or due to the tandem array of the inserted RNA motif itself (40–44). Therefore, while this approach is very useful in revealing transcriptional activity, results have to be carefully interpreted when applied to studies of RNA degradation and localization. Recently, a reengineered MS2 system has been developed that has minimal effect on mRNA half-life (45). In addition, an mRNA degradation assay using rifampin can provide a good test of potential effects of the labeling method on mRNA turnover. Hereby, the signal of the fluorescent reporter on the mRNA should decay with the same kinetics as the native mRNA upon rifampin treatment, e.g., to rule out accumulation of a reconstituted or aggregated reporter with the mRNA part already being degraded. A comparatively long-lived fluorescent signal is therefore unlikely to reflect the correct localization of the mRNA. A strong overexpression, e.g., from an artificial promoter, might also impact on the transcript's properties, such as stability, function, or localization. Therefore, preference should be given to expression from a native promoter.

Fluorescent RNA Aptamers

RNA-protein interaction-based methods often raise the concern of changes of molecular and functional properties of the tagged RNA, due to the binding of multiple bulky FPs to the inserted repetitive and often structured RNA motifs. This approach may be especially problematic for short or structured transcripts such as sRNAs. The development of RNA aptamer-based imaging methods provides another possibility of live-cell RNA imaging. This approach utilizes RNA aptamer sequences added to the transcript of interest and fluorogenic small molecules that can freely diffuse into the cell and become fluorescent upon binding to the RNA aptamer (for recent reviews, see references 17 and 46–48). While not quite suitable for live-cell imaging, earlier developments of malachite green, thiazole orange, and dimethyl indole red aptamers, etc., provided proof of concept and illustrated the feasibility of such approaches (49–51). The Spinach system (Fig. 1C) is an example of the first generation of RNA aptamers for live-cell imaging (52). Since then, new aptamer systems have been developed, including Spinach 2 (53), Mango (54), Broccoli (55, 56), and Corn (57). Repetitive Spinach aptamers have been shown to increase the brightness by a maximum of 17-fold (with 64 repeats) compared to a single Spinach monomer at a cost of a significant lengthening of the aptamer sequence (58). Alternatively, the fluorogenic small molecules have been designed into a covalently linked fluorophore-quencher pair, and the aptamers are selected to bind either the quencher or the fluorophore. In this design, the fluorophore is quenched by the linked quencher in the absence of the aptamer. Once either the quencher or the fluorophore binds to the aptamer, they are more physically separated and the fluorophore is dequenched (59, 60).

Despite their relatively broad applications in the area of metabolite sensing through engineering to various riboswitches (for reviews, see references 17 and 46–48), the use of RNA aptamers in single-molecule RNA imaging is still considered to be limited. This limited application of the aptamer-based method for single-molecule RNA imaging might be related to, e.g., the limited brightness of the tag and the requirement for correct folding of the aptamer *in vivo*. Further engineering of smaller and more stable aptamers, brighter fluorophores, and tandem arrays of the aptamer could collectively help to achieve single-RNA sensitivity. Similar to the FP reporter systems, the fusion of an aptamer sequence to a transcript of interest might also affect its properties, such as stability or function. Therefore, certain functional validation of the tagged RNA is required.

Super-Resolution Microscopy

With the various labeling methods described above, RNA can be directly visualized under the fluorescence microscope. However, due to the diffraction limit of visible light, conventional optical microscopy has a limited resolution of ~200 to 300 nm in the lateral direction and 500 to 700 nm in the axial direction.

While FISH can provide single-molecule sensitivity on RNA imaging, because of the small size of a bacterium (a few hundred nanometers to a few microns), only individual RNAs with very low abundance can be resolved in bacteria. Several super-resolution techniques that can break the diffraction limit have been developed. Among these techniques, single-molecule localization microscopies are most frequently applied in bacterial systems, in which photoactivatable FPs and photoswitchable dyes are used to label biomolecules, and can push the spatial resolution into the range of 10 to 20 nm (61–63). Therefore, the utilization of such super-resolution microscopy techniques can provide finer details on the RNA localization.

LOCALIZATION OF mRNAs

Besides distinct subcellular localizations of proteins and the nucleoid, diverse subcellular localizations of mRNAs have also been reported in various bacteria. Commonly observed patterns of distribution of bacterial mRNAs include uniform expression throughout the cytoplasm, localization into distinct foci close to the nucleoid, formation of helical patterns along the cell axis, enrichment at the inner membrane, or concentration at the cell poles or the septum during cell division (Fig. 2). For example, using the split eIF4A approach in live cells, Broude and coworkers observed that *lacZ* mRNA was evenly distributed along the *E. coli* cell (Fig. 2A) (33). In contrast, another study observed limited dispersion of *lacZ* mRNA in *E. coli* as well as of several mRNAs in *Caulobacter crescentus* from their site of transcription using FISH in fixed cells (Fig. 2B) (20, 64). The *cat* mRNA, encoding a cytoplasmic chloramphenicol acetyltransferase, demonstrates a helix-like pattern in the cytoplasm in *E. coli* (Fig. 2C) (65). The *lacY* mRNA, encoding a membrane-bound lactose permease, localizes at or near the inner membrane (Fig. 2D) (65). In the Gram-positive bacterium *Bacillus subtilis*, MS2-GFP tagging revealed that *comE* mRNA, encoding the late competence operon, localizes to the cell poles (Fig. 2E) and the nascent septum that separates the daughter cells (Fig. 2F) (66). The sometimes disparate results of observed subcellular localizations of the same transcript might be due to the different RNA visualization technologies and expression systems used.

mRNA Accumulation Near the Site of Transcription

The observed nucleoid-associated transcript foci led to the proposal that bacterial mRNAs may remain localized close to their genomic site of transcription (64). Using a combination of MS2 tagging and validation by FISH, one study reported that the mRNAs of *groESL*, *creS*, *divJ*, *ompA*, and *fljK* of *C. crescentus* and the *lacZ* transcript in *E. coli* show limited dispersion from their site of transcription and are visible as distinct foci in the vicinity of the genomic DNA locus from where they are transcribed (64). Similarly, visualization of a transcript containing repeated MS2-binding motif sequences revealed that most transcripts moved randomly in a restricted spot near the midpoint or quarter-point of the cells, which corresponds to the localization of the F-plasmid, which was used for expression of the RNAs, whereas a minority diffused throughout the cell or moved as chains (67). In this study, the observed distinct spots were suggested to be mRNAs that are still

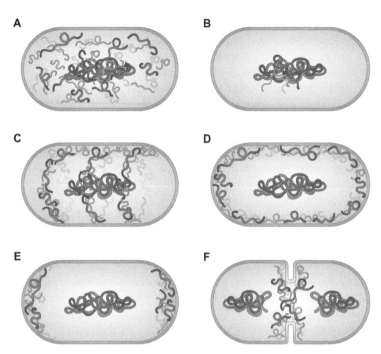

Figure 2 Diverse patterns of subcellular mRNA localization in bacteria. Schematic drawings of diverse mRNA localization patterns commonly reported in different bacteria. RNA molecules are shown in green, and the nucleoid in gray. (A) Distribution throughout the cytoplasm. (B) Localization at the site of transcription in the nucleoid. (C) Helical localization. (D) Enrichment at the inner membrane. (E) Localization at the cell poles and (F) septum.

tethered to DNA by RNAP, while freely diffusing transcripts are completed transcripts.

mRNA Accumulation at the Sites of Protein Requirement

The observations that mRNAs accumulate at various other subcellular places in addition to close association with their genomic site of transcription (Fig. 2) indicate that other cellular processes can affect the localization of the mRNA. For example, using a split aptamer approach for RNA labeling in live *E. coli* cells combined with independent validation by FISH, distinct spots of the endogenous *ptsC* mRNA, encoding an integral transmembrane transporter, could be detected, which do not colocalize with bulk DNA (68). Similarly, FISH analysis revealed distinct, uneven patterns of localization and specific foci at one or both poles for the dinitrogenase reductase-encoding *nifH* transcripts of *Klebsiella oxytoca* and *Azotobacter vinelandii* (69). One intriguing hypothesis is that mRNAs can be targeted to the subcellular domains where their encoded protein products are required (e.g., the cytoplasm, poles, or inner membrane) (65, 70, 71). For example, while the *cat* mRNA, encoding the cytoplasmic chloramphenicol acetyltransferase, was observed in a helix-

like pattern in the cytoplasm of *E. coli*, the mRNA of *lacY*, encoding the membrane-bound lactose permease, was preferentially detected near the cytoplasmic membrane (65). In line with these imaging observations, upon fractionation of *E. coli*, the authors found that the *cat* and *lacY* mRNAs are enriched in cytosolic and membrane samples, respectively (65). Moreover, the same study reported the polycistronic *bglGFB* mRNA, which encodes both membrane-bound (BglF sugar permease) and soluble (BglG transcription factor and BglB phospho-β-glucosidase) components, was enriched at the cell membrane, indicating that envelope-targeting signals may be dominant. Furthermore, inhibition of translation with kasugamycin or chloramphenicol revealed that this membrane localization occurs in a translation-independent manner (65). This observation was further supported by introducing mutations that abolish *bglF* translation, which did not affect localization, indicating that *cis*-acting signals in the RNA itself dictate membrane localization (65). The *bglF* membrane-targeting signal, which is present in the sequence encoding the first two transmembrane helices of BglF, was found to be dominant over other operon components, because *bglG* mRNA expressed alone localizes to cell poles and *bglB* mRNA expressed alone is cytoplasmic (65). It has also been reported that *cis*-encoded RNA elements in the early coding region of mRNAs encoding Yop effector proteins (e.g., YopE and YopN) are required for the secretion of these effector proteins by type III secretion systems in *Yersinia* (72, 73). However, it is unknown if these transcripts are localized to the membrane or in proximity to the secretion apparatus. Similarly, N-terminal domains are required for secretion of flagellar proteins in diverse bacteria, and both mRNA and peptide signals are recognized by the secretion apparatus and contribute to secretion efficiency (74). For example, secretion of heterologous proteins was facilitated by fusing the 173-bp 5′ untranslated region (5′ UTR) of the *fliC* gene (encoding flagellin) as well as a *fliC* transcriptional terminator (75).

While most studies have mainly looked at the localization of only a single or small set of mRNAs, recently Zhuang and coworkers investigated the spatial organization of mRNAs in *E. coli* on the transcriptome scale (70). They designed complex FISH probe sets to visualize defined sets of mRNAs, categorized by the subcellular localizations of their encoded proteins, allowing examination of ~27% of all *E. coli* mRNAs. This revealed that mRNAs encoding inner membrane proteins tended to be preferentially located at the inner membrane, whereas transcripts encoding cytoplasmic, periplasmic, and outer membrane proteins were generally detected more uniformly throughout the cytoplasm. Moreover, labeling of specific subgroups of mRNAs as a function of genomic localization of their genes revealed that spatial genome organization does not play a major role in the shaping of the cellular distribution of the transcriptome. Because inhibition of translation initiation using kasugamycin treatment disrupted envelope localization of membrane protein-encoding transcripts, Moffitt et al. concluded that membrane localization of these transcripts is translation dependent (70). Most inner membrane proteins are secreted via the signal recognition particle (SRP) pathway, which cotranslationally inserts proteins into the membrane, while outer membrane proteins are posttranslationally secreted via the SecB pathway (76). Examination of localization by FISH of fluorescent reporter fusions to SRP or SecB signal peptides confirmed that mRNAs with SRP signal sequences were found enriched at membranes, suggesting that these transcripts are recruited to the envelope during cotranslational secretion of the ribosome-bound nascent signal peptide (70). Furthermore, introduction of stop codon mutations into the *bglF* mRNA abolished its localization at the membrane, indicating that translation of its SRP signal sequence is required for membrane localization (70). Whether the

protein signal sequence or solely the act of translation of its mRNA is required for directing the transcripts to the membrane is so far unknown. Another study also reported, based on cell fractionation, RNA-sequencing, and quantitative PCR, that mRNAs of inner membrane proteins are enriched at the membrane and depletion of the SRP receptor FtsY reduced the amounts of all mRNAs at the membrane (15). However, mRNAs encoding inner membrane proteins were also found in the soluble ribosome-free fraction, which might represent a stage prior to their translation and targeting to the membrane. Mathematical modeling has suggested that translation initiation rates, the availability of secretory apparatuses, and the composition of the coding region define mRNA abundance and residence time near the membrane (77). In addition, modeling suggested that formation of membrane protein clusters might be facilitated by bursts of proteins translated from a single mRNA anchored to the membrane, and therefore spatiotemporal dynamics of mRNAs might strongly influence the organization of membrane protein complexes.

Recently, the mRNA encoding the major flagellin FlaA of the foodborne pathogen *Campylobacter jejuni* was also observed, using FISH, to localize to the cell poles in a translation-dependent manner (71). *C. jejuni* has two polar flagella, and polar *flaA* mRNA localization was primarily detected in short cells, which likely correspond to cells that have just divided and are building a new flagellum at the nascent cell pole. While polar localization was abolished for a translation-incompetent *flaA* mRNA with stop codon mutations before the N-terminal peptide, translation of the first 100 codons partially restored polar localization, suggesting that an N-terminal amino acid signal, or its translation, is sufficient for mRNA localization. Similarly, translation of the N-terminal peptide of type III secretion effector protein mRNAs has been shown to be required for secretion (72, 78).

mRNA Localization by Specific *trans*-Acting Factors

Emerging examples have also indicated that there are specific *trans*-acting factors that posttranscriptionally regulate mRNA localization. For example, overexpression of the cold shock protein CspE increased the fraction of membrane protein encoding mRNAs in the ribosome-free fraction and their amount on the membrane and positively affected their translation, indicating potential regulation of subcellular RNA localization (15). Moreover, the 98-amino-acid protein ComN, a posttranscriptional regulator of competence gene expression, localizes to the division site and cell poles via direct interaction with DivIVA, a key protein involved in cell pole differentiation in *B. subtilis* (66). ComN-DivIVA interaction promotes accumulation of *comE* mRNA to septal and polar sites, indicating that localized regulators can also impact mRNA localization in bacteria. Furthermore, localized mRNA translation of ComE proteins might be required for efficient competence development. During a global analysis of the direct RNA regulon of the translational regulator CsrA by RNA immunoprecipitation sequencing (RIP-seq), *flaA* mRNA was revealed as the major translationally repressed target of CsrA in *C. jejuni* and the abovementioned polar *flaA* mRNA localization was detected to be posttranscriptionally regulated by the CsrA-FliW regulatory system (71). Deletion of the CsrA protein antagonist FliW releases CsrA and in turn allows translational repression of *flaA* mRNA, abrogating its polar localization. This observation suggests both that the *flaA* mRNA localization is translation dependent and that spatial control of bacterial transcripts can be regulated posttranscriptionally. It remains to be seen how many other flagellar mRNAs localize to the flagella apparatus and whether polar localization of the *flaA* mRNA or membrane localization of inner membrane protein mRNAs has any effects on the efficiency of secretion and/or assembly of larger complexes, or is just a by-product of cotranslational secretion.

LOCALIZED mRNA TRANSLATION AND DEGRADATION

While transcription and translation occur at distinct places in the nucleus and cytoplasm in eukaryotes, translation can already initiate cotranscriptionally in the cytoplasm of bacteria and translation can continue after transcription completion and release of the mRNA. Furthermore, the abovementioned examples show that mRNAs can migrate outside the nucleoid and translation can take place at distinct locations in the cell, such as at the membrane during cotranslational secretion of inner membrane proteins. Moreover, it has been reported that the RNAP transcription machinery and ribosomes occupy partially different subcellular regions within different bacterial cells (79–81). For example, in *B. subtilis*, RNAP has been primarily detected inside and ribosomes outside the nucleoid, respectively (82), indicating that transcription and translation are spatially separated. Cryo-electron tomography indicated that 70S ribosomes of *Spiroplasma melliferum* are distributed throughout the cytoplasm, with 15% in close proximity to the membrane (83). Similar electron cryotomography analysis of the cellular ultrastructure of

logarithmically growing cultures of *C. jejuni* revealed ribosome exclusion zones at cell poles (80). Also consistent with extranucleoid distribution of ribosomes, 5S rRNA was detected as an array of fluorescent particles distributed along the cell or the cell poles in *E. coli* (33), indicating specific localization of ribosomes outside the nucleoid in *E. coli*, which is in agreement with enrichment of ribosomal proteins at the cell periphery and cell poles in both *E. coli* (84) and *B. subtilis* (82, 85). While several ribosomal proteins (L1, S2, and L7/L12) are enriched at either of the cell poles and the translation factor EF-Tu colocalizes with the bacterial cytoskeleton protein MreB, it remains unclear how many of these localized translation factors are incorporated into actively translated ribosomes (reviewed in reference 86). It should be noted that the cellular localization can differ for unbound subunits versus actively translating ribosomes or transcribing RNAP. For example, it has been reported that mRNA-free ribosome subunits are not fully excluded from the nucleoid, thereby allowing for translation initiation on nascent mRNAs throughout the nucleoid and cotranscriptional translation (81). Moreover, a large fraction of RNAPs, presumably the transcribing population, has been reported to be primarily located at the periphery of the nucleoid and thus is close to the pool of ribosomes excluded from the nucleoid (87). In *C. crescentus* and another alphaproteobacterium, *Sinorhizobium meliloti*, ribosomes are detected throughout the cell, including in the nucleoid region (64, 90), which is different from the ribosome/nucleoid segregation observed in *E. coli* and *B. subtilis* (82, 84). Hereby, it was suggested that mRNA-bound ribosomal subunits show limited mobility in *C. crescentus* due to the observed limited dispersion of their mRNA targets from their site of transcription (64, 90, 91).

The specific organization of ribosomes has also been reported to change during different growth conditions in some bacteria (88). The distinct subcellular localization of ribosomes, mRNAs, and RNAP indicates that transcription and translation are not necessarily coupled in bacteria and localized translation by specific ribosomes at subcellular locations might also play a role in bacterial gene expression. For example, inner membrane-bound ribosomes in *E. coli* that are actively engaged in translation (89) might play a role in specific translation of inner membrane proteins.

Similar to the specific and dynamic localization of ribosomes and transcripts in bacterial cells, distinct subcellular distribution of RNA-degrading proteins and complexes has also been reported in prokaryotes (92–94). The *E. coli* degradosome initiates most of the RNA decay in bacteria and contains RNase E, the RNA helicase RhlB, polynucleotide phosphorylase (PNPase), and enolase (95). The *B. subtilis* degradosome consists of PNPase; RNases J1, J2, and Y; as well as the DEAD-box RNA helicase CshA, enolase, and phosphofructokinase (96). Components of the *E. coli* degradosome have been reported to associate with the membrane or to assemble into helical filaments (97–100). Using super-resolution imaging of 24 FP fusions to RNA degradation and processing enzymes in *E. coli*, Moffitt et al. detected that only the four proteins (RNase E, PNPase, RhlB, and the polyadenylation enzyme PAPI) were enriched at the membrane, whereas the other tested fusions were mainly uniformly distributed throughout the cytoplasm (70). The membrane localization of the degradosome is mediated by membrane anchoring of segment A of RNase E (98). In line with a colocalization of inner membrane protein-coding mRNAs and degradosome components at the cell envelope, it has been observed that these transcripts had shorter half-lives than the mRNAs of cytoplasmic, periplasmic, or outer membrane proteins (70). Moreover, artificial targeting of mRNAs to the membrane via fusion to SRP signal sequences destabilizes these mRNAs. Thus, cocolocalization of certain mRNAs with RNA degradation components can lead to their specific degradation. Interestingly, the degradosome localization can be impacted by growth conditions: a redistribution of RNase E/enolase from membrane-associated patterns under aerobic to diffuse patterns under anaerobic conditions results in stabilization of DicF sRNA and filamentation of the bacteria (101). For more information on the *E. coli* degradosome, see chapter 1.

In *C. crescentus*, RNase E shows a patchy localization pattern and the clustered localization of this enzyme is determined by the location of DNA, independent of its mRNA substrates (64, 90). Hereby, the localization of RNase E clusters was found to correlate with two subcellular chromosomal positions that encode the highly expressed rRNA genes, indicating that RNase E-mediated rRNA processing occurs at the site of rRNA synthesis (90). Although mediated by apparently different mechanisms compared to *E. coli*, the association of RNase E with DNA in *C. crescentus* also indicates that there is spatially organized mRNA decay in this organism. Such a subcellular organization of the RNA degradation machinery likely also applies to other prokaryotes. For example, the membrane-bound RNase Y of Gram-positive bacteria (e.g., *B. subtilis* and *Staphylococcus aureus*) appears to be assembled at similar cellular locations as other RNA degradation enzymes (93, 102). Recently, it has been reported that RNase Y is recruited to lipid

rafts via flotillin in *S. aureus* and that flotillin increases RNase Y function (103), indicating localized RNA degradation in bacterial membrane microdomains. How many and which mRNAs are mainly targeted by such localized degradation remains to be seen.

LOCALIZATION OF sRNAs

Localization of bacterial sRNAs has been investigated predominantly with FISH. The housekeeping transcript transfer-messenger RNA (tmRNA, also known as SsrA), involved in stalled ribosome rescue, was one of the first sRNAs investigated regarding its subcellular location (104). tmRNA contains both tRNA-like and mRNA properties. It forms an RNP complex (tmRNP) with the small protein B (SmpB) to function in *trans*-translation. During *trans*-translation, the tmRNP binds to stalled ribosomes and adds a proteolysis signaling tag to the stalled peptide, thereby recycling the ribosome and facilitating the degradation of the aberrant protein product. By FISH labeling of tmRNA and immunostaining of several relevant proteins, including SmpB and RNase R, in *C. crescentus*, Russell et al. demonstrated a helix-like pattern of tmRNA, SmpB, and RNase R in the swarmer cells (G1 phase), whereas in S phase after initiation of DNA replication, tmRNA molecules are largely degraded by RNase R, with the remaining transcripts becoming homogeneously distributed in the cytoplasm (104). Both tmRNA and SmpB show a high degree of colocalization, consistent with the formation of a tmRNP, whereas the helical structures formed by the tmRNP and RNase R are mostly distinct from each other. The helical organization of tmRNA relies on SmpB; however, the underlying molecular basis of this distinct organization remains unclear. In addition, such helical organization is not disrupted upon translation inhibition, suggesting that it is not related to *trans*-translation activity. These results indicate that the most likely biological relevance of such a helical organization is to protect tmRNA from RNase R-mediated degradation during G1, as they are localized "out of phase," whereas in the S phase, proteolysis of SmpB would release tmRNA from its helical location away from RNase R, allowing it to be degraded in order to regulate the abundance and function of tmRNA in a cell-cycle-dependent manner.

Compared to mRNAs, studies of several sRNAs suggest that the distribution of these transcripts is less compartmentalized. With smFISH and super-resolution imaging, Fei et al. showed that SgrS, an sRNA involved in the glucose-phosphate stress response, is roughly homogeneously distributed in the cytoplasm when its copy number is high, whereas when its expression is lower, it appears to be specifically absent from the central nucleoid region (105). A more comprehensive study of the localization of several sRNAs, including GlmZ, OxyS, RyhB, and SgrS, found equal preference for localization within the cytoplasm and nucleoid region and no preferential localization at the membrane or cell poles, as often observed for mRNAs by quantifying the signal overlap between RNA FISH and 4′,6-diamidino-2-phenylindole (DAPI) staining of DNA (106). Interestingly, the ability of a particular transcript to freely diffuse into the nucleoid region was correlated with the length and the translation activity of the RNA, because shortening and reducing translation of *gfp* mRNA led to the same localization as sRNAs and reduced nucleoid exclusion. Therefore, in general, longer RNAs, including coding RNAs, have a lower propensity toward nucleoid localization compared to shorter, often noncoding transcripts. So far, the available studies on a handful of examples of sRNAs suggest an unbiased cellular localization (105, 106). It should be noted that investigation of sRNA localizations is predominantly focused on noncoding regulatory RNAs. Even though SgrS is a dual-functional sRNA, which also encodes the small protein SgrT (107) under the conditions of previous investigations, SgrT was not actively translated. Therefore, SgrS is still considered a regulatory RNA only under the investigated conditions, and demonstrates the same localization behavior as other tested regulatory sRNAs (108).

The unbiased localization of regulatory sRNAs investigated so far seems to be consistent with the fact that typically each sRNA species can regulate multiple mRNA targets (11, 109), which themselves might each adopt different cellular localizations. Therefore, the availability of sRNAs throughout the cell will ensure that all targets can be regulated. On the other hand, the localization of the target mRNA might kinetically affect regulation by the local availability of its sRNA regulator or other protein factors. Computational simulations have shown that a biased localization of an sRNA (such as membrane versus cytoplasmic localization) can lead to a distinct regulation of different target mRNAs, providing an opportunity for regulatory hierarchy among different targets (110). For example, a membrane-localized sRNA would regulate a membrane-localized target mRNA more efficiently compared to a cytoplasmic counterpart. However, the observation that sRNAs tested so far all exhibit an unbiased localization suggests that mRNA targets, regardless of their localization, would be equally sampled by the same sRNA species. Therefore, any regulatory hierarchy would be attributed to the other factors, such as localization of Hfq and RNase E, and the strength of base-pairing interactions, etc. As an example, the SgrS-mediated degradation of the mRNA encoding a fusion of *ptsG* to *crp* has been shown to be significantly reduced upon elimination of the transmembrane encoding domains of PtsG (111). This observation is reminiscent of the observed faster endogenous turnover of membrane-localized mRNAs compared to mRNAs with other cellular localizations (70) and of the membrane localization of RNase E (see also chapter 1).

Hfq is an important chaperone for sRNAs in bacteria and also facilitates sRNA-mediated gene regulation via base-pairing (12, 112). Therefore, understanding the localization of Hfq can provide insight into spatiotemporal themes of sRNA-based regulation. However, results from various studies using different labeling and imaging approaches to study Hfq localization are conflicting, and therefore its localization in the cell remains controversial. Hfq has been observed to adopt a diffuse cytoplasmic localization outside of the nucleoid region by immunofluorescence staining (79), as well as preferential membrane localization by electron microscopy (113), which was recently recapitulated in an *in vitro* system using artificial vesicles (113, 114). A helical organization of Hfq along the longitudinal direction of the cell has also been observed by immunofluorescence staining (99, 113, 115). In contrast with all the fixed-cell experiments, single-molecule tracking of FP-tagged Hfq in a live cell reveals that Hfq is essentially freely diffusing inside the cell, with the diffusion rate slowed down when Hfq binds to the newly transcribed RNA attached to the nucleoid (116). Future experiments are needed to resolve this discrepancy in observation of Hfq localization. Despite the absence of Hfq significantly affecting the stability and abundance of sRNAs (105), it has minimal effect on the localization of tested sRNAs (106). As mentioned above, localization of the *ptsG* mRNA to the inner membrane has been reported to strongly contribute to its efficient repression by the sRNA SgrS, together with Hfq, during phosphosugar stress (111). So far, it remains unclear if Hfq can actively localize SgrS to the membrane to facilitate sRNA-*ptsG* mRNA interactions or if, in contrast, SgrS-*ptsG* mRNA complexes might instead lead to Hfq localization. Moreover, membrane localization of Hfq might also be due to interactions with the degradosome.

EMERGING STRATEGIES AND APPROACHES

Recently, new developments in chemical biology and molecular engineering have expanded the toolkit for imaging RNA localization and measuring their activities *in vivo*. These tools have been developed mostly for RNA imaging in eukaryotic systems, but some of them might have the great potential to be applied to bacterial systems as well.

First, various new developments in FISH methods have allowed for signal amplification and high-throughput imaging. For example, in the *in situ* hybridization chain reaction (HCR) (117, 118), two fluorescently labeled

oligonucleotides that alone exist as metastable hairpin structures are designed (Fig. 3A). A third initiator probe is then introduced to bind to the target RNA of interest. This initiator probe has an overhang region that can open one of the two fluorescently labeled hairpins by branch migration, making the first hairpin-containing oligo capable of hybridizing to the second hairpin oligo. The two fluorescent oligos bind to each other in an alternating fashion in the hybridization

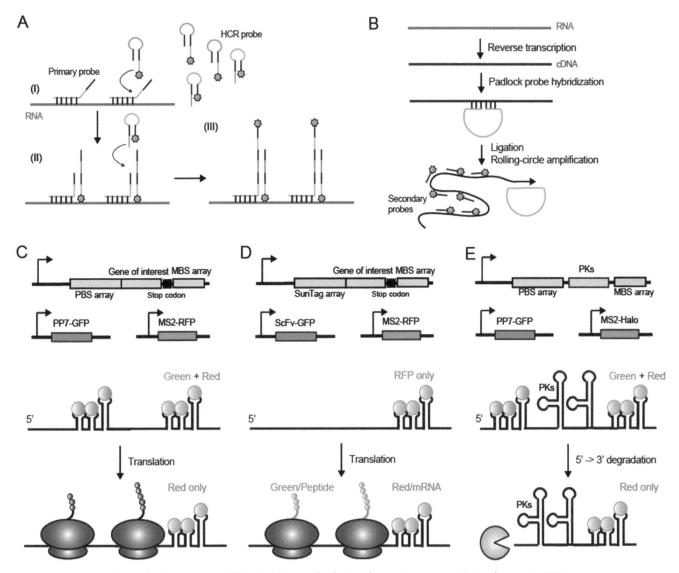

Figure 3 Emerging mRNA imaging methods in eukaryotic systems. (A) In the *in situ* HCR, binding of the primary probe initiates the alternating binding of two HCR probes, thereby amplifying the signal. (B) In the *in situ* PCR, a cDNA is first generated from the RNA of interest. Padlock probes are hybridized to the cDNA and ligated to be circular DNAs. Fluorophore-labeled secondary probes are then hybridized to the products generated from rolling circle amplification of these circular DNA templates. (C) Schematic representation of the TRICK reporter construct. (D) Schematic representation of the SunTag construct. (E) Schematic representation of the TREAT reporter construct.

chain reaction, thereby linking multiple fluorescently labeled probes to the RNA of interest and amplifying the signal. *In situ* PCR has also been used for FISH signal amplification (119–122). In this method, padlock probes (long oligonucleotides whose ends are complementary to the target RNA, where hybridization results in circularization of the probe) are hybridized to cDNAs generated from endogenous RNAs of interest *in situ*, which generates nicked single-stranded DNAs that are further ligated to be circular DNAs (Fig. 3B). Rolling circle amplification from these circular DNA templates by DNA polymerase can generate products carrying repetitive sequences for the hybridization of the fluorophore-labeled secondary probes. Such FISH strategies with signal amplification can be adapted to bacterial cells, and could be beneficial for imaging sRNAs, to which it is difficult to attach multiple conventional FISH probes due to their short length. To increase the throughput of RNA imaging, various multiplexed FISH imaging methods have also been developed in eukaryotic cells. Multiplexed imaging can be achieved by fluorescent barcoding, in which different combinations of fluorophore-labeled oligos hybridize to separate RNA species to generate pseudocolors (123), or repetitive hybridization and imaging cycles (124–127). However, such multiplexed imaging methods rely on the ability to spatially resolve individual transcripts. With diffraction-limited microscopy, only RNAs with very low abundance can be resolved individually in bacterial cells due to their small size, making the application of the multiplexed imaging method practically difficult in bacterial systems.

Recently, CRISPR (clustered regularly interspaced short palindromic repeat)-Cas systems have been engineered to label both DNA (128–130) and RNA (131, 132) in live cells. An early example of RNA labeling with CRISPR-Cas systems originated from the finding that Cas9, an RNA-guided DNase, can also target single-stranded RNA by providing the PAM (protospacer adjacent motif) as part of an oligonucleotide (PAMmer) (133). Thereby, a catalytically inactive Cas9 (dCas9), fused to an FP and in complex with the PAMmer and single guide RNA (sgRNA), can target and label the RNA of interest in a programmable fashion (131). Recently, Cas13a, an RNA-guided RNA-targeting CRISPR-Cas effector, has been applied for RNA tracking in mammalian cells (132). Compared to the MS2/PP7 systems mentioned above, one advantage of the CRISPR-based imaging is that no additional tagging sequence must be fused to the RNA of interest, as the sgRNA can be flexibly programmed to target any RNA sequence of interest. Nevertheless, similar to other

FP reporter systems, CRISPR-based labeling might affect mRNA processing, degradation, or localization for a specific RNA of interest. Moreover, to distinguish target RNA-bound fusion proteins from those that are unbound, which could contribute to high level of background fluorescence, additional modifications are necessary. One such strategy could be to utilize multiple sgRNAs to tile along the RNA sequence of interest to enhance the signal on the RNA compared to background level, in a similar fashion as is employed in FISH labeling. Another strategy would be to reduce the unbound fraction of the fluorescent Cas proteins. For example, in a recent approach, a zinc-finger binding domain and the transcription repressor KRAB have been fused to Cas13a-FP, and the zinc-finger binding site was inserted into the promoter controlling the expression of the fusion protein (132, 134). In this way, a negative feedback loop is created, and the unbound Cas13a-FP therefore autorepresses its expression to control the background fluorescence.

Single-molecule RNA tracking has also been combined with additional reporters to correlate their translation or degradation activities in eukaryotic cells. A biosensor called TRICK (translating RNA imaging by coat protein knockoff) has been developed that reports the first round of translation (135). In this imaging system, cassettes of the PP7 binding site (PBS) and the MS2 binding sites (MBS) are integrated into the coding region and the 3′ UTR, respectively (Fig. 3C). Newly synthesized mRNAs carry both GFP-fused PP7 proteins and red fluorescent protein (RFP)-fused MS2 proteins, generating colocalized green and red signals under the microscope. During the first round of translation after export of the mRNA to the cytoplasm, ribosomes displace the GFP-PP7 fusion from transcripts, leaving only RFP signals. The SunTag system allows for real-time tracking of multiple cycles of translation on individual mRNAs (136–140). In this technique, a tandem array of a sequence coding for a short peptide is inserted into the open reading frame of interest. Once translated, the resulting polypeptide, containing multiple such short peptide epitopes, is recognized and bound by a specific single-chain variable fragment (scFv) antibody fused with GFP coexpressed in the cell (Fig. 3D). In parallel, the mRNA of interest is also labeled by either the MS2 or PP7 system with a different FP (e.g., RFP) at the 3′ UTR. Therefore, actively translating mRNAs generate two fluorescence signals, whereas untranslated mRNAs only generate an RFP signal. In addition to reporters of translation, a biosensor called TREAT (3′-RNA end accumulation during turnover) has been developed to track mRNA turnover in eukaryotes (126, 141). In this approach, the 3′ UTR is engineered to include two viral pseudoknots (PKs) flanked by PP7 and MS2 binding sites (Fig. 3E). Since the PK structures block 5′-to-3′ degradation of the transcript by the 5′-3′ exoribonuclease Xrn1 (142), TREAT allows one to distinguish between intact and partially degraded mRNAs. The feasibility of applying such functional reporters to bacterial systems remains to be tested. For example, FP reporters in these systems are generally tagged with a nuclear localization signal sequence. This sequence can effectively reduce the amount of unbound FPs in the cytoplasm and thereby reduce the imaging background. Such a segregation of unbound fluorescent reporter from the mRNA-bound signal cannot be applied in bacteria because they lack a separation of nucleus versus cytoplasm. In addition, given that there are fundamental differences in mRNA metabolism between prokaryotes and eukaryotes, such as transcription-coupled translation in bacteria versus spatially separated transcription and translation in eukaryotes, and the much shorter lifetimes of mRNA in bacteria compared to the ones of eukaryotic mRNAs, these functional reporters may need to be significantly reengineered for their applications in bacterial cells.

In addition to imaging of mRNAs, a new method, FASTmiR, has been developed for the imaging of eukaryotic microRNAs (miRNAs) in live cells (143). FASTmiR was designed based on the Spinach system, in which a sensory domain that can base-pair with the miRNA of interest is fused to a modified Spinach aptamer. Binding of a miRNA to the FASTmiR helps the folding of the Spinach aptamer and forming the DHFBI binding pocket, thereby generating a fluorescence signal. The concept of FASTmiR may be further developed into a platform for imaging bacterial sRNAs. However, given that sRNAs are in general longer than miRNAs, and most contain secondary structures, the structural design of FASTmiR might need to be significantly modified to allow efficient sRNA detection in bacteria.

Finally, in addition to optical approaches that enhance the spatial resolution, expansion microscopy utilizes swellable polymer networks to physically expand the specimen. Such physical magnification leads to an enhanced effective spatial resolution (126, 141). In addition to immunofluorescence, expansion microscopy has recently been combined with FISH imaging of RNA (144), and iterative expansion microscopy further enhances resolution from ~70 nm to ~25 nm (145). Expansion microscopy has already been applied to bacteria (146), and should allow for studies of biomolecular location with finer resolution.

OPEN QUESTIONS AND FUTURE CHALLENGES

Despite the emerging evidence of subcellular RNA localization in bacteria, we are still only at the beginning of understanding the underlying mechanisms and potential functions of this process. One of the major questions is what are the deterministic factors or the driving forces for mRNA localization in bacteria. So far, several cis- and trans-acting factors have been suggested to affect transcript localization. For example, many mRNAs encoding inner membrane proteins are preferentially localized at the membrane, and this localization can be mediated in a translation-dependent manner during synthesis and secretion of N-terminal signal peptides. However, for certain mRNAs encoding inner membrane proteins or secreted effectors, it has been reported that signals within the mRNA itself are instead required for localization/secretion or that the mRNAs have multiple localization signals, including RNA sequence-based signals and those encoded in the N-terminal peptide amino acid sequence (10, 72, 74, 147). It remains to be seen how many bacterial mRNAs have a eukaryotic-like RNA zip code, and what the sequence and/or structural features of these elements are. Likewise, the features embedded in N-terminal signal sequences of the encoded proteins that might mediate localization of their cognate mRNA are unknown. While eukaryotic mRNAs can carry their zip codes at either the 3′ UTR, 5′ UTR, or coding region (1), the few examples of RNA elements directing localization of bacterial mRNAs have been reported to be located at 5′ ends of mRNAs or within open reading frames (10). It has been suggested that an enrichment for uracils in mRNAs encoding integral membrane proteins might be a physiologically relevant signature of this group of mRNAs (148). Furthermore, it will also be interesting to see which (protein) factors bind to bacterial RNA zip codes and how they transport RNAs in the cell. In eukaryotes, many mRNAs are actively transported by RNA-motor complexes along polarized cytoskeleton structures and localize at local anchor signals such as the dynein-1 motor (3). Similar mechanisms involving binding to cytoskeleton proteins, or even cell division factors, might apply in bacteria. In addition, localized protection from RNases, as observed in eukaryotes, might mediate subcellular localization of bacterial transcripts.

It also remains unclear what the functions and physiological roles of localized RNAs and/or localized translation are in the cell, and how perturbation of such localization affects phenotypes related to the encoded protein. The organization and coexpression of functionally related genes in operons in bacteria might be in line with the observation that RNAs stay close to the site of transcription and are also translated close to the nucleoid so that protein complexes can be assembled. On the other hand, localized translation, e.g., at the membrane, might increase the efficiency of assembly of larger protein complexes at the future site of action, especially for those that might be hydrophobic. Such a localized translation at the membrane especially would make sense for secreted factors such as type III secretion system effectors or flagellins that might be translationally repressed until the secretion machinery is completed, and their translation might only be activated upon completion of the secretion machinery. Nevertheless, it is also possible that the membrane localization of mRNAs is just a by-product of secretion of the nascent N-terminal peptide of the translated protein. Moreover, it is still unclear whether the localization of certain mRNAs is regulated during changing growth or stress conditions or whether this is connected to cell division. In eukaryotes, alternative splicing and polyadenylation can control incorporation of localization signals into mature transcripts, and after nuclear export, diverse RBPs, adaptors, and cytoskeleton motors are recruited to localizing mRNAs (1, 2). It has been observed that posttranscriptional regulators, such as the competence regulator ComN in *B. subtilis* or the CsrA-FliW regulatory network of *C. jejuni*, can impact RNA localization in bacteria (66, 71). Although this can be an indirect effect via regulation of translation, these first examples show that posttranscriptional regulatory networks can impact RNA localization.

Compared to mRNA localization, even less is known about sRNA localization in bacteria. Several sRNAs also encode small proteins (for a review, see reference 149). It still remains to be studied whether these small-protein-encoding sRNAs might show a more defined cellular localization compared to solely noncoding RNAs. And if yes, it remains to be seen whether the final cellular locations of the encoded small-protein products affect the localization of their encoding dual-function sRNAs. Localization of sRNAs and their cognate mRNAs to certain sites in the cell might impact the efficiency and hierarchy of target gene regulation and in turn affect the functionality/outcome of genetic circuits. Therefore, considerations about subcellular localization might also be relevant for the design of synthetic gene regulatory circuits, since differential localization of the encoded protein and/or mRNA (membrane versus cytoplasmic) might impact efficiency of regulation if not all components are expressed in close vicinity.

The ongoing technological developments listed above, together with additional new approaches that emerge, will help in the further study of RNA localization in bacteria. So far, it is more challenging to apply existing mRNA imaging methods to sRNAs. For example, smFISH has a significantly compromised signal-to-noise ratio for sRNA imaging because of the much shorter length of sRNAs compared to mRNAs and resulting lower number of probes that can be applied. Similarly, live-cell imaging of sRNAs remains challenging because tagging with FPs or RNA aptamers could impact their regulatory properties and/or localization. Beyond simple detection of transcripts, ideally one would also simultaneously image RNA and translation/protein localization *in vivo* to decipher whether localization and localized translation impact translation, protein abundance, and physiology. Considering the sometimes controversial observations that have been reported so far regarding bacterial RNA localization, it is also recommended that findings are validated using different approaches, and that the experimental designs maintain transcript expression and characteristics as close to native as possible. Certainly, the continued technological advancements for studying RNA localization in these small organisms will reveal the previously unappreciated extent of bacterial compartmentalization and its contribution to physiology.

Acknowledgments. We thank Dr. Sarah Svensson, Eric McLean, and Dr. Seongjin Park for critical comments and Dr. Sandy Pernitzsch (www.scigraphix.com) for help with preparing Fig. 2. Work in the lab of C.M.S. is supported by the Deutsche Forschungsgemeinschaft (DFG) (Sh580/4-1, GRK2157, SPP2002: Sh580/7-1 and Sh580/8-1), the BMBF (Infect-ERA [ERA-Net], 2nd call, CampyRNA), and a HIRI (Helmholtz Institute of RNA-Based Infection Research, Würzburg, Germany) seed grant (Project-No 6) through funds from the Bavarian Ministry of Economic Affairs and Media, Energy, and Technology (grant allocation numbers 0703/68674/5/2017 and 0703/89374/3/2017). J.F. acknowledges the support from the Searle Scholars Program and the NIH Director's New Innovator Award (1DP2GM128185-01).

Citation. Fei J, Sharma CM. 2018. RNA localization in bacteria. Microbiol Spectrum 6(5):RWR-0024-2018.

References

1. Chin A, Lécuyer E. 2017. RNA localization: making its way to the center stage. *Biochim Biophys Acta* **1861** (11 Pt B):2956–2970.

2. Buxbaum AR, Haimovich G, Singer RH. 2015. In the right place at the right time: visualizing and understanding mRNA localization. *Nat Rev Mol Cell Biol* **16**: 95–109.

3. Mofatteh M, Bullock SL. 2017. SnapShot: subcellular mRNA localization. *Cell* **169**:178–178.e1.

4. Jung H, Gkogkas CG, Sonenberg N, Holt CE. 2014. Remote control of gene function by local translation. *Cell* **157**:26–40.

5. Govindarajan S, Nevo-Dinur K, Amster-Choder O. 2012. Compartmentalization and spatiotemporal organization of macromolecules in bacteria. *FEMS Microbiol Rev* **36**: 1005–1022.

6. Nevo-Dinur K, Govindarajan S, Amster-Choder O. 2012. Subcellular localization of RNA and proteins in prokaryotes. *Trends Genet* **28**:314–322.

7. Campos M, Jacobs-Wagner C. 2013. Cellular organization of the transfer of genetic information. *Curr Opin Microbiol* **16**:171–176.

8. Keiler KC. 2011. RNA localization in bacteria. *Curr Opin Microbiol* **14**:155–159.

9. Buskila AA, Kannaiah S, Amster-Choder O. 2014. RNA localization in bacteria. *RNA Biol* **11**:1051–1060.

10. Kannaiah S, Amster-Choder O. 2014. Protein targeting via mRNA in bacteria. *Biochim Biophys Acta* **1843**: 1457–1465.

11. Storz G, Vogel J, Wassarman KM. 2011. Regulation by small RNAs in bacteria: expanding frontiers. *Mol Cell* **43**:880–891.

12. Vogel J, Luisi BF. 2011. Hfq and its constellation of RNA. *Nat Rev Microbiol* **9**:578–589.

13. McLean R, Inglis GD, Mosimann SC, Uwiera RR, Abbott DW. 2017. Determining the localization of carbohydrate active enzymes within gram-negative bacteria. *Methods Mol Biol* **1588**:199–208.

14. Fontaine F, Fuchs RT, Storz G. 2011. Membrane localization of small proteins in *Escherichia coli*. *J Biol Chem* **286**:32464–32474.

15. Benhalevy D, Biran I, Bochkareva ES, Sorek R, Bibi E. 2017. Evidence for a cytoplasmic pool of ribosome-free mRNAs encoding inner membrane proteins in *Escherichia coli*. *PLoS One* **12**:e0183862.

16. Milne JLS, Subramaniam S. 2009. Cryo-electron tomography of bacteria: progress, challenges and future prospects. *Nat Rev Microbiol* **7**:666–675.

17. Chakraborty K, Veetil AT, Jaffrey SR, Krishnan Y. 2016. Nucleic acid-based nanodevices in biological imaging. *Annu Rev Biochem* **85**:349–373.

18. van Gijtenbeek LA, Kok J. 2017. Illuminating messengers: an update and outlook on RNA visualization in bacteria. *Front Microbiol* **8**:1161.

19. Femino AM, Fay FS, Fogarty K, Singer RH. 1998. Visualization of single RNA transcripts in situ. *Science* **280**: 585–590.

20. So LH, Ghosh A, Zong C, Sepúlveda LA, Segev R, Golding I. 2011. General properties of transcriptional time series in *Escherichia coli*. *Nat Genet* **43**:554–560.

21. Raj A, van den Bogaard P, Rifkin SA, van Oudenaarden A, Tyagi S. 2008. Imaging individual mRNA molecules using multiple singly labeled probes. *Nat Methods* **5**:877–879.

22. Raj A, Peskin CS, Tranchina D, Vargas DY, Tyagi S. 2006. Stochastic mRNA synthesis in mammalian cells. *PLoS Biol* **4**:e309.

23. Neuert G, Munsky B, Tan RZ, Teytelman L, Khammash M, van Oudenaarden A. 2013. Systematic identification of signal-activated stochastic gene regulation. *Science* **339**:584–587.

24. Jones DL, Brewster RC, Phillips R. 2014. Promoter architecture dictates cell-to-cell variability in gene expression. *Science* **346**:1533–1536.

25. Schnell U, Dijk F, Sjollema KA, Giepmans BN. 2012. Immunolabeling artifacts and the need for live-cell imaging. *Nat Methods* **9**:152–158.

26. Bertrand E, Chartrand P, Schaefer M, Shenoy SM, Singer RH, Long RM. 1998. Localization of ASH1 mRNA particles in living yeast. *Mol Cell* **2**:437–445.

27. Lange S, Katayama Y, Schmid M, Burkacky O, Bräuchle C, Lamb DC, Jansen RP. 2008. Simultaneous transport of different localized mRNA species revealed by live-cell imaging. *Traffic* **9**:1256–1267.

28. Hocine S, Raymond P, Zenklusen D, Chao JA, Singer RH. 2013. Single-molecule analysis of gene expression using two-color RNA labeling in live yeast. *Nat Methods* **10**:119–121.

29. Miller LW, Cai Y, Sheetz MP, Cornish VW. 2005. *In vivo* protein labeling with trimethoprim conjugates: a flexible chemical tag. *Nat Methods* **2**:255–257.

30. Keppler A, Gendreizig S, Gronemeyer T, Pick H, Vogel H, Johnsson K. 2003. A general method for the covalent labeling of fusion proteins with small molecules *in vivo*. *Nat Biotechnol* **21**:86–89.

31. Sun X, Zhang A, Baker B, Sun L, Howard A, Buswell J, Maurel D, Masharina A, Johnsson K, Noren CJ, Xu MQ, Corrêa IR Jr. 2011. Development of SNAP-tag fluorogenic probes for wash-free fluorescence imaging. *Chembiochem* **12**:2217–2226.

32. Carrocci TJ, Hoskins AA. 2014. Imaging of RNAs in live cells with spectrally diverse small molecule fluorophores. *Analyst (Lond)* **139**:44–47.

33. Valencia-Burton M, McCullough RM, Cantor CR, Broude NE. 2007. RNA visualization in live bacterial cells using fluorescent protein complementation. *Nat Methods* **4**:421–427.

34. Valencia-Burton M, Shah A, Sutin J, Borogovac A, McCullough RM, Cantor CR, Meller A, Broude NE. 2009. Spatiotemporal patterns and transcription kinetics of induced RNA in single bacterial cells. *Proc Natl Acad Sci U S A* **106**:16399–16404.

35. Ozawa T, Natori Y, Sato M, Umezawa Y. 2007. Imaging dynamics of endogenous mitochondrial RNA in single living cells. *Nat Methods* **4**:413–419.

36. Wu B, Chen J, Singer RH. 2014. Background free imaging of single mRNAs in live cells using split fluorescent proteins. *Sci Rep* **4**:3615.

37. Wang S, Moffitt JR, Dempsey GT, Xie XS, Zhuang X. 2014. Characterization and development of photoactivatable fluorescent proteins for single-molecule-based superresolution imaging. *Proc Natl Acad Sci U S A* **111**: 8452–8457.

38. Landgraf D, Okumus B, Chien P, Baker TA, Paulsson J. 2012. Segregation of molecules at cell division reveals native protein localization. *Nat Methods* **9**:480–482.

39. LeCuyer KA, Behlen LS, Uhlenbeck OC. 1995. Mutants of the bacteriophage MS2 coat protein that alter its cooperative binding to RNA. *Biochemistry* **34**:10600–10606.

40. Golding I, Paulsson J, Zawilski SM, Cox EC. 2005. Real-time kinetics of gene activity in individual bacteria. *Cell* **123**:1025–1036.

41. Haimovich G, Zabezhinsky D, Haas B, Slobodin B, Purushothaman P, Fan L, Levin JZ, Nusbaum C, Gerst JE. 2016. Use of the MS2 aptamer and coat protein for RNA localization in yeast: a response to "MS2 coat proteins bound to yeast mRNAs block 5′ to 3′ degradation and trap mRNA decay products: implications for the localization of mRNAs by MS2-MCP system." *RNA* **22**:660–666.

42. Garcia JF, Parker R. 2015. MS2 coat proteins bound to yeast mRNAs block 5′ to 3′ degradation and trap mRNA decay products: implications for the localization of mRNAs by MS2-MCP system. *RNA* **21**:1393–1395.

43. Heinrich S, Sidler CL, Azzalin CM, Weis K. 2017. Stem-loop RNA labeling can affect nuclear and cytoplasmic mRNA processing. *RNA* **23**:134–141.

44. Garcia JF, Parker R. 2016. Ubiquitous accumulation of 3′ mRNA decay fragments in *Saccharomyces cerevisiae* mRNAs with chromosomally integrated MS2 arrays. *RNA* **22**:657–659.

45. Tutucci E, Vera M, Biswas J, Garcia J, Parker R, Singer RH. 2018. An improved MS2 system for accurate reporting of the mRNA life cycle. *Nat Methods* **15**:81–89.

46. You M, Jaffrey SR. 2015. Structure and mechanism of RNA mimics of green fluorescent protein. *Annu Rev Biophys* **44**:187–206.

47. Ouellet J. 2016. RNA fluorescence with light-up aptamers. *Front Chem* **4**:29.

48. Dolgosheina EV, Unrau PJ. 2016. Fluorophore-binding RNA aptamers and their applications. *Wiley Interdiscip Rev RNA* **7**:843–851.

49. Babendure JR, Adams SR, Tsien RY. 2003. Aptamers switch on fluorescence of triphenylmethane dyes. *J Am Chem Soc* **125**:14716–14717.

50. Constantin TP, Silva GL, Robertson KL, Hamilton TP, Fague K, Waggoner AS, Armitage BA. 2008. Synthesis of new fluorogenic cyanine dyes and incorporation into RNA fluoromodules. *Org Lett* **10**:1561–1564.

51. Sando S, Narita A, Hayami M, Aoyama Y. 2008. Transcription monitoring using fused RNA with a dye-binding light-up aptamer as a tag: a blue fluorescent RNA. *Chem Commun (Camb)* (33):3858–3860.

52. Paige JS, Wu KY, Jaffrey SR. 2011. RNA mimics of green fluorescent protein. *Science* **333**:642–646.

53. Strack RL, Disney MD, Jaffrey SR. 2013. A superfolding Spinach2 reveals the dynamic nature of trinucleotide repeat-containing RNA. *Nat Methods* **10**:1219–1224.

54. Dolgosheina EV, Jeng SC, Panchapakesan SS, Cojocaru R, Chen PS, Wilson PD, Hawkins N, Wiggins PA, Unrau PJ. 2014. RNA Mango aptamer-fluorophore: a bright, high-affinity complex for RNA labeling and tracking. *ACS Chem Biol* **9**:2412–2420.

55. Filonov GS, Moon JD, Svensen N, Jaffrey SR. 2014. Broccoli: rapid selection of an RNA mimic of green fluorescent protein by fluorescence-based selection and directed evolution. *J Am Chem Soc* **136**:16299–16308.

56. Filonov GS, Jaffrey SR. 2016. RNA imaging with dimeric Broccoli in live bacterial and mammalian cells. *Curr Protoc Chem Biol* **8**:1–28.

57. Warner KD, Sjekloća L, Song W, Filonov GS, Jaffrey SR, Ferré-D'Amaré AR. 2017. A homodimer interface without base pairs in an RNA mimic of red fluorescent protein. *Nat Chem Biol* **13**:1195–1201.

58. Zhang J, Fei J, Leslie BJ, Han KY, Kuhlman TE, Ha T. 2015. Tandem Spinach array for mRNA imaging in living bacterial cells. *Sci Rep* **5**:17295.

59. Sunbul M, Jäschke A. 2013. Contact-mediated quenching for RNA imaging in bacteria with a fluorophore-binding aptamer. *Angew Chem Int Ed Engl* **52**:13401–13404.

60. Arora A, Sunbul M, Jäschke A. 2015. Dual-colour imaging of RNAs using quencher- and fluorophore-binding aptamers. *Nucleic Acids Res* **43**:e144.

61. Betzig E, Patterson GH, Sougrat R, Lindwasser OW, Olenych S, Bonifacino JS, Davidson MW, Lippincott-Schwartz J, Hess HF. 2006. Imaging intracellular fluorescent proteins at nanometer resolution. *Science* **313**:1642–1645.

62. Rust MJ, Bates M, Zhuang X. 2006. Sub-diffraction-limit imaging by stochastic optical reconstruction microscopy (STORM). *Nat Methods* **3**:793–795.

63. Huang B, Wang W, Bates M, Zhuang X. 2008. Three-dimensional super-resolution imaging by stochastic optical reconstruction microscopy. *Science* **319**:810–813.

64. Montero Llopis P, Jackson AF, Sliusarenko O, Surovtsev I, Heinritz J, Emonet T, Jacobs-Wagner C. 2010. Spatial organization of the flow of genetic information in bacteria. *Nature* **466**:77–81.

65. Nevo-Dinur K, Nussbaum-Shochat A, Ben-Yehuda S, Amster-Choder O. 2011. Translation-independent localization of mRNA in *E. coli*. *Science* **331**:1081–1084.

66. dos Santos VT, Bisson-Filho AW, Gueiros-Filho FJ. 2012. DivIVA-mediated polar localization of ComN, a posttranscriptional regulator of *Bacillus subtilis*. *J Bacteriol* **194**:3661–3669.

67. Golding I, Cox EC. 2004. RNA dynamics in live *Escherichia coli* cells. *Proc Natl Acad Sci U S A* **101**:11310–11315.

68. Toran P, Smolina I, Driscoll H, Ding F, Sun Y, Cantor CR, Broude NE. 2014. Labeling native bacterial RNA in live cells. *Cell Res* **24**:894–897.

69. Pilhofer M, Pavlekovic M, Lee NM, Ludwig W, Schleifer KH. 2009. Fluorescence *in situ* hybridization for intracellular localization of *nifH* mRNA. *Syst Appl Microbiol* **32**:186–192.

70. Moffitt JR, Pandey S, Boettiger AN, Wang S, Zhuang X. 2016. Spatial organization shapes the turnover of a bacterial transcriptome. *Elife* **5**:313065.

71. Dugar G, Svensson SL, Bischler T, Wäldchen S, Reinhardt R, Sauer M, Sharma CM. 2016. The CsrA-FliW network controls polar localization of the dual-function flagellin mRNA in *Campylobacter jejuni*. *Nat Commun* **7**:11667.

72. Sorg JA, Miller NC, Schneewind O. 2005. Substrate recognition of type III secretion machines—testing the RNA signal hypothesis. *Cell Microbiol* **7**:1217–1225.

73. Anderson DM, Schneewind O. 1997. A mRNA signal for the type III secretion of Yop proteins by *Yersinia enterocolitica*. *Science* 278:1140–1143.

74. Singer HM, Erhardt M, Hughes KT. 2014. Comparative analysis of the secretion capability of early and late flagellar type III secretion substrates. *Mol Microbiol* 93:505–520.

75. Majander K, Anton L, Antikainen J, Lång H, Brummer M, Korhonen TK, Westerlund-Wikström B. 2005. Extracellular secretion of polypeptides using a modified *Escherichia coli* flagellar secretion apparatus. *Nat Biotechnol* 23:475–481.

76. Driessen AJ, Nouwen N. 2008. Protein translocation across the bacterial cytoplasmic membrane. *Annu Rev Biochem* 77:643–667.

77. Korkmazhan E, Teimouri H, Peterman N, Levine E. 2017. Dynamics of translation can determine the spatial organization of membrane-bound proteins and their mRNA. *Proc Natl Acad Sci U S A* 114:13424–13429.

78. Anderson DM, Schneewind O. 1999. *Yersinia enterocolitica* type III secretion: an mRNA signal that couples translation and secretion of YopQ. *Mol Microbiol* 31:1139–1148.

79. Azam TA, Hiraga S, Ishihama A. 2000. Two types of localization of the DNA-binding proteins within the *Escherichia coli* nucleoid. *Genes Cells* 5:613–626.

80. Müller A, Beeby M, McDowall AW, Chow J, Jensen GJ, Clemons WM Jr. 2014. Ultrastructure and complex polar architecture of the human pathogen *Campylobacter jejuni*. *MicrobiologyOpen* 3:702–710.

81. Sanamrad A, Persson F, Lundius EG, Fange D, Gynnå AH, Elf J. 2014. Single-particle tracking reveals that free ribosomal subunits are not excluded from the *Escherichia coli* nucleoid. *Proc Natl Acad Sci U S A* 111:11413–11418.

82. Lewis PJ, Thaker SD, Errington J. 2000. Compartmentalization of transcription and translation in *Bacillus subtilis*. *EMBO J* 19:710–718.

83. Ortiz JO, Förster F, Kürner J, Linaroudis AA, Baumeister W. 2006. Mapping 70S ribosomes in intact cells by cryoelectron tomography and pattern recognition. *J Struct Biol* 156:334–341.

84. Bakshi S, Choi H, Weisshaar JC. 2015. The spatial biology of transcription and translation in rapidly growing *Escherichia coli*. *Front Microbiol* 6:636.

85. Mascarenhas J, Weber MH, Graumann PL. 2001. Specific polar localization of ribosomes in *Bacillus subtilis* depends on active transcription. *EMBO Rep* 2:685–689.

86. Keiler KC. 2011. Localization of the bacterial RNA infrastructure. *Adv Exp Med Biol* 722:231–238.

87. Stracy M, Lesterlin C, Garza de Leon F, Uphoff S, Zawadzki P, Kapanidis AN. 2015. Live-cell super-resolution microscopy reveals the organization of RNA polymerase in the bacterial nucleoid. *Proc Natl Acad Sci U S A* 112:E4390–E4399.

88. Chai Q, Singh B, Peisker K, Metzendorf N, Ge X, Dasgupta S, Sanyal S. 2014. Organization of ribosomes and nucleoids in *Escherichia coli* cells during growth and in quiescence. *J Biol Chem* 289:11342–11352.

89. Herskovits AA, Bibi E. 2000. Association of *Escherichia coli* ribosomes with the inner membrane requires the signal recognition particle receptor but is independent of the signal recognition particle. *Proc Natl Acad Sci U S A* 97:4621–4626.

90. Bayas CA, Wang J, Lee MK, Schrader JM, Shapiro L, Moerner WE. 2018. Spatial organization and dynamics of RNase E and ribosomes in *Caulobacter crescentus*. *Proc Natl Acad Sci U S A* 115:E3712–E3721.

91. Montero Llopis P, Sliusarenko O, Heinritz J, Jacobs-Wagner C. 2012. In vivo biochemistry in bacterial cells using FRAP: insight into the translation cycle. *Biophys J* 103:1848–1859.

92. Evguenieva-Hackenberg E, Roppelt V, Lassek C, Klug G. 2011. Subcellular localization of RNA degrading proteins and protein complexes in prokaryotes. *RNA Biol* 8:49–54.

93. Redder P. 2016. How does sub-cellular localization affect the fate of bacterial mRNA? *Curr Genet* 62:687–690.

94. Mackie GA. 2013. RNase E: at the interface of bacterial RNA processing and decay. *Nat Rev Microbiol* 11:45–57.

95. Carpousis AJ. 2007. The RNA degradosome of *Escherichia coli*: an mRNA-degrading machine assembled on RNase E. *Annu Rev Microbiol* 61:71–87.

96. Lehnik-Habrink M, Lewis RJ, Mäder U, Stülke J. 2012. RNA degradation in *Bacillus subtilis*: an interplay of essential endo- and exoribonucleases. *Mol Microbiol* 84:1005–1017.

97. Taghbalout A, Rothfield L. 2007. RNaseE and the other constituents of the RNA degradosome are components of the bacterial cytoskeleton. *Proc Natl Acad Sci U S A* 104:1667–1672.

98. Khemici V, Poljak L, Luisi BF, Carpousis AJ. 2008. The RNase E of *Escherichia coli* is a membrane-binding protein. *Mol Microbiol* 70:799–813.

99. Taghbalout A, Yang Q, Arluison V. 2014. The *Escherichia coli* RNA processing and degradation machinery is compartmentalized within an organized cellular network. *Biochem J* 458:11–22.

100. Strahl H, Turlan C, Khalid S, Bond PJ, Kebalo JM, Peyron P, Poljak L, Bouvier M, Hamoen L, Luisi BF, Carpousis AJ. 2015. Membrane recognition and dynamics of the RNA degradosome. *PLoS Genet* 11:e1004961.

101. Murashko ON, Lin-Chao S. 2017. *Escherichia coli* responds to environmental changes using enolasic degradosomes and stabilized DicF sRNA to alter cellular morphology. *Proc Natl Acad Sci U S A* 114:E8025–E8034.

102. Cascante-Estepa N, Gunka K, Stülke J. 2016. Localization of components of the RNA-degrading machine in *Bacillus subtilis*. *Front Microbiol* 7:1492.

103. Koch G, Wermser C, Acosta IC, Kricks L, Stengel ST, Yepes A, Lopez D. 2017. Attenuating *Staphylococcus aureus* virulence by targeting flotillin protein scaffold activity. *Cell Chem Biol* 24:845–857.e6.

104. Russell JH, Keiler KC. 2009. Subcellular localization of a bacterial regulatory RNA. *Proc Natl Acad Sci U S A* 106:16405–16409.

105. Fei J, Singh D, Zhang Q, Park S, Balasubramanian D, Golding I, Vanderpool CK, Ha T. 2015. RNA biochemistry. Determination of *in vivo* target search kinetics of regulatory noncoding RNA. *Science* 347:1371–1374.

106. Sheng H, Stauffer WT, Hussein R, Lin C, Lim HN. 2017. Nucleoid and cytoplasmic localization of small RNAs in *Escherichia coli*. *Nucleic Acids Res* 45:2919–2934.

107. Wadler CS, Vanderpool CK. 2007. A dual function for a bacterial small RNA: SgrS performs base pairing-dependent regulation and encodes a functional polypeptide. *Proc Natl Acad Sci U S A* 104:20454–20459.

108. Wadler CS, Vanderpool CK. 2009. Characterization of homologs of the small RNA SgrS reveals diversity in function. *Nucleic Acids Res* 37:5477–5485.

109. Wagner EG, Romby P. 2015. Small RNAs in bacteria and archaea: who they are, what they do, and how they do it. *Adv Genet* 90:133–208.

110. Teimouri H, Korkmazhan E, Stavans J, Levine E. 2017. Sub-cellular mRNA localization modulates the regulation of gene expression by small RNAs in bacteria. *Phys Biol* 14:056001.

111. Kawamoto H, Morita T, Shimizu A, Inada T, Aiba H. 2005. Implication of membrane localization of target mRNA in the action of a small RNA: mechanism of post-transcriptional regulation of glucose transporter in *Escherichia coli*. *Genes Dev* 19:328–338.

112. Updegrove TB, Zhang A, Storz G. 2016. Hfq: the flexible RNA matchmaker. *Curr Opin Microbiol* 30:133–138.

113. Diestra E, Cayrol B, Arluison V, Risco C. 2009. Cellular electron microscopy imaging reveals the localization of the Hfq protein close to the bacterial membrane. *PLoS One* 4:e8301.

114. Malabirade A, Jiang K, Kubiak K, Diaz-Mendoza A, Liu F, van Kan JA, Berret JF, Arluison V, van der Maarel JR. 2017. Compaction and condensation of DNA mediated by the C-terminal domain of Hfq. *Nucleic Acids Res* 45:7299–7308.

115. Malabirade A, Morgado-Brajones J, Trépout S, Wien F, Marquez I, Seguin J, Marco S, Velez M, Arluison V. 2017. Membrane association of the bacterial riboregulator Hfq and functional perspectives. *Sci Rep* 7:10724.

116. Persson F, Lindén M, Unoson C, Elf J. 2013. Extracting intracellular diffusive states and transition rates from single-molecule tracking data. *Nat Methods* 10:265–269.

117. Choi HM, Beck VA, Pierce NA. 2014. Next-generation *in situ* hybridization chain reaction: higher gain, lower cost, greater durability. *ACS Nano* 8:4284–4294.

118. Shah S, Lubeck E, Zhou W, Cai L. 2017. Editorial note to: In situ transcription profiling of single cells reveals spatial organization of cells in the mouse hippocampus. *Neuron* 94:745–746.

119. Qian X, Lloyd RV. 2003. Recent developments in signal amplification methods for in situ hybridization. *Diagn Mol Pathol* 12:1–13.

120. Bagasra O. 2007. Protocols for the *in situ* PCR-amplification and detection of mRNA and DNA sequences. *Nat Protoc* 2:2782–2795.

121. Larsson C, Koch J, Nygren A, Janssen G, Raap AK, Landegren U, Nilsson M. 2004. *In situ* genotyping individual DNA molecules by target-primed rolling-circle amplification of padlock probes. *Nat Methods* 1:227–232.

122. Larsson C, Grundberg I, Söderberg O, Nilsson M. 2010. *In situ* detection and genotyping of individual mRNA molecules. *Nat Methods* 7:395–397.

123. Lubeck E, Cai L. 2012. Single-cell systems biology by super-resolution imaging and combinatorial labeling. *Nat Methods* 9:743–748.

124. Lubeck E, Coskun AF, Zhiyentayev T, Ahmad M, Cai L. 2014. Single-cell *in situ* RNA profiling by sequential hybridization. *Nat Methods* 11:360–361.

125. Jungmann R, Avendaño MS, Woehrstein JB, Dai M, Shih WM, Yin P. 2014. Multiplexed 3D cellular super-resolution imaging with DNA-PAINT and Exchange-PAINT. *Nat Methods* 11:313–318.

126. Chen KH, Boettiger AN, Moffitt JR, Wang S, Zhuang X. 2015. RNA imaging. Spatially resolved, highly multiplexed RNA profiling in single cells. *Science* 348: aaa6090.

127. Moffitt JR, Hao J, Wang G, Chen KH, Babcock HP, Zhuang X. 2016. High-throughput single-cell gene-expression profiling with multiplexed error-robust fluorescence in situ hybridization. *Proc Natl Acad Sci U S A* 113:11046–11051.

128. Takei Y, Shah S, Harvey S, Qi LS, Cai L. 2017. Multiplexed dynamic imaging of genomic loci by combined CRISPR imaging and DNA sequential FISH. *Biophys J* 112:1773–1776.

129. Guan J, Liu H, Shi X, Feng S, Huang B. 2017. Tracking multiple genomic elements using correlative CRISPR imaging and sequential DNA FISH. *Biophys J* 112: 1077–1084.

130. Chen B, Gilbert LA, Cimini BA, Schnitzbauer J, Zhang W, Li GW, Park J, Blackburn EH, Weissman JS, Qi LS, Huang B. 2013. Dynamic imaging of genomic loci in living human cells by an optimized CRISPR/Cas system. *Cell* 155:1479–1491.

131. Nelles DA, Fang MY, O'Connell MR, Xu JL, Markmiller SJ, Doudna JA, Yeo GW. 2016. Programmable RNA tracking in live cells with CRISPR/Cas9. *Cell* 165:488–496.

132. Abudayyeh OO, Gootenberg JS, Essletzbichler P, Han S, Joung J, Belanto JJ, Verdine V, Cox DB, Kellner MJ, Regev A, Lander ES, Voytas DF, Ting AY, Zhang F. 2017. RNA targeting with CRISPR-Cas13. *Nature* 550: 280–284.

133. O'Connell MR, Oakes BL, Sternberg SH, East-Seletsky A, Kaplan M, Doudna JA. 2014. Programmable RNA recognition and cleavage by CRISPR/Cas9. *Nature* 516: 263–266.

134. Gross GG, Junge JA, Mora RJ, Kwon HB, Olson CA, Takahashi TT, Liman ER, Ellis-Davies GC, McGee AW, Sabatini BL, Roberts RW, Arnold DB. 2013. Recombinant probes for visualizing endogenous synaptic proteins in living neurons. *Neuron* 78:971–985.

135. Halstead JM, Lionnet T, Wilbertz JH, Wippich F, Ephrussi A, Singer RH, Chao JA. 2015. Translation. An

RNA biosensor for imaging the first round of translation from single cells to living animals. *Science* 347: 1367–1671.

136. Tanenbaum ME, Gilbert LA, Qi LS, Weissman JS, Vale RD. 2014. A protein-tagging system for signal amplification in gene expression and fluorescence imaging. *Cell* 159:635–646.

137. Yan X, Hoek TA, Vale RD, Tanenbaum ME. 2016. Dynamics of translation of single mRNA molecules *in vivo*. *Cell* 165:976–989.

138. Wu B, Eliscovich C, Yoon YJ, Singer RH. 2016. Translation dynamics of single mRNAs in live cells and neurons. *Science* 352:1430–1435.

139. Wang C, Han B, Zhou R, Zhuang X. 2016. Real-time imaging of translation on single mRNA transcripts in live cells. *Cell* 165:990–1001.

140. Morisaki T, Lyon K, DeLuca KF, DeLuca JG, English BP, Zhang Z, Lavis LD, Grimm JB, Viswanathan S, Looger LL, Lionnet T, Stasevich TJ. 2016. Real-time quantification of single RNA translation dynamics in living cells. *Science* 352:1425–1429.

141. Horvathova I, Voigt F, Kotrys AV, Zhan Y, Artus-Revel CG, Eglinger J, Stadler MB, Giorgetti L, Chao JA. 2017. The dynamics of mRNA turnover revealed by single-molecule imaging in single cells. *Mol Cell* 68: 615–625.e9.

142. Kieft JS, Rabe JL, Chapman EG. 2015. New hypotheses derived from the structure of a flaviviral Xrn1-resistant RNA: conservation, folding, and host adaptation. *RNA Biol* 12:1169–1177.

143. Huang K, Doyle F, Wurz ZE, Tenenbaum SA, Hammond RK, Caplan JL, Meyers BC. 2017. FASTmiR: an RNA-based sensor for *in vitro* quantification and live-cell localization of small RNAs. *Nucleic Acids Res* 45:e130.

144. Chen F, Wassie AT, Cote AJ, Sinha A, Alon S, Asano S, Daugharthy ER, Chang JB, Marblestone A, Church GM, Raj A, Boyden ES. 2016. Nanoscale imaging of RNA with expansion microscopy. *Nat Methods* 13:679–684.

145. Chang JB, Chen F, Yoon YG, Jung EE, Babcock H, Kang JS, Asano S, Suk HJ, Pak N, Tillberg PW, Wassie AT, Cai D, Boyden ES. 2017. Iterative expansion microscopy. *Nat Methods* 14:593–599.

146. Zhang YS, Chang JB, Alvarez MM, Trujillo-de Santiago G, Aleman J, Batzaya B, Krishnadoss V, Ramanujam AA, Kazemzadeh-Narbat M, Chen F, Tillberg PW, Dokmeci MR, Boyden ES, Khademhosseini A. 2016. Hybrid microscopy: enabling inexpensive high-performance imaging through combined physical and optical magnifications. *Sci Rep* 6:22691.

147. Ramamurthi KS, Schneewind O. 2005. A synonymous mutation in *Yersinia enterocolitica yopE* affects the function of the YopE type III secretion signal. *J Bacteriol* 187:707–715.

148. Prilusky J, Bibi E. 2009. Studying membrane proteins through the eyes of the genetic code revealed a strong uracil bias in their coding mRNAs. *Proc Natl Acad Sci U S A* 106:6662–6666.

149. Storz G, Wolf YI, Ramamurthi KS. 2014. Small proteins can no longer be ignored. *Annu Rev Biochem* 83:753–777.

Regulating with RNA in Bacteria and Archaea
Edited by Gisela Storz and Kai Papenfort
© 2018 American Society for Microbiology, Washington, DC
doi:10.1128/microbiolspec.RWR-0021-2018

Sponges and Predators in the Small RNA World

25

Nara Figueroa-Bossi[1] and Lionello Bossi[1]

LESSONS FROM EUKARYOTIC miRNAs: FROM SPONGES TO THE ceRNA HYPOTHESIS

MicroRNAs (miRNAs) are 20- to-24-nucleotide (nt)-long RNAs that guide Argonaute proteins to silence mRNA expression in animal and plant cells (1–3). Similarly to bacterial *trans*-encoded small RNAs (sRNAs), miRNAs act by establishing imperfect base-pair interactions with seed sequences that can be as short as 6 to 8 nt. Seeking ways to selectively control miRNA activity *in vivo*, a decade ago Ebert and coworkers engineered transcripts containing multiple tandemly arranged target sites for one or more miRNAs and had these constructs expressed at high levels in transfected mammalian cells (4). They found the exogenous RNAs to have the ability to sequester ("soak up") the miRNAs, relieving the regulation of their natural targets. The authors termed the artificial transcripts "microRNA sponges." At about the same time, a study on the mechanism responsible for inhibiting the activity of a miRNA (miR399) in plant cells identified an endogenous noncoding RNA, named IPS1, that could base-pair with miR399 and compete for its binding to the primary target (5). This indicated that a natural RNA could have sponge-like activity and that target site amplification was not required for this effect. Following these initial findings, several examples of miRNA target mimicry have been described involving different types of coding and noncoding RNAs (6, 7), including some of viral origin (8, 9). Particularly noteworthy is the case of the circular antisense RNA named CDR1as, highly expressed in human and mouse brain, which harbors as many as 74 potential target sites for the miR-7 miRNA and thus closely fulfills the original definition of a sponge (10). Recent evidence showed CDR1as to be a highly efficient miR-7 sponge *in vivo*: in cells lacking CDR1as, deregulation of miR-7 networks leads to profound defects in brain development and function (11).

The discovery of sponges opened new perspectives into the complexity of the miRNA targetome. Seitz argued that among the multitude of mRNA species that are potentially controlled by any given miRNA, only a tiny fraction produce clear phenotypic changes in response to miRNA regulation (which typically involves less than a 2-fold variation in mRNA levels) (12). He proposed that only the members of this minority are authentic miRNA targets; the remainder are "pseudo-targets" whose miRNA binding sites have evolved to limit miRNA availability, so as to render regulation of the true targets more robust. Further elaboration on these ideas led to the so-called competitive endogenous RNA (ceRNA) hypothesis, which extends the sponge concept to the whole-transcriptome level (13). The ceRNA hypothesis views miRNA binding sites, also called miRNA response elements, as the letters of a language through which RNAs communicate with each other—that is, affect each other's expression—through competition for shared miRNAs. The model conveys two innovative concepts: (i) that mRNAs are not merely a repository of protein-coding information but can act directly to regulate other mRNAs; and (ii) that portions of the transcriptome generally considered nonfunctional, such as pseudogene transcripts, play an active role by engaging in cross talk with their protein-coding counterparts (13).

Some experimental evidence, together with computational simulations, have challenged the central predicate of the ceRNA hypothesis by showing that effective cross talk would require a large excess of miRNA binding sites in the competing RNAs, possible only under artificial and/or unphysiological conditions (14–16). It is noteworthy that in these studies, as in the original

[1]Institute for Integrative Biology of the Cell (I2BC), CEA, CNRS, University of Paris-Sud, University of Paris-Saclay, Gif-sur-Yvette, France.

formulation of the ceRNA hypothesis, miRNA partitioning among competitors is solely dictated by the equilibrium dissociation constant for each binding site and by the number of sites. The situation would drastically change if the miRNA were degraded upon binding to one of the competitors (8, 9). Introducing a channel of "stoichiometric decay" (17) might allow most current inconsistencies to be reconciled. We will see below that such a decay channel is a predominant feature of mechanisms regulating the activity of prokaryotic sRNAs.

TARGET MIMICS AND SPONGE-LIKE INHIBITORS OF sRNAs IN BACTERIA

Bacterial *trans*-encoded regulatory sRNAs differ from miRNAs in many respects, but a particularly relevant difference in the context of this review is that most, if not all, sRNAs are susceptible to cleavage upon pairing with an mRNA target (18–20). This brings a new variable into the sponging landscape: a competitive inhibitor of a bacterial sRNA might not need to stably "soak up" the sRNA; rather, it would suffice if it were efficient at capturing and promoting the destruction of the sRNA (provided that it is made in excess to the sRNA) (21). It is not a sponge, therefore, in the strict sense of the word, but rather something more like a "predator." Since sRNA cleavage is not an obligate outcome of pairing (22), it seems possible that an RNA capable of base-pairing with the sRNA could function as a sponge and/or a predator depending on structural features of the sRNA-RNA duplex. Below we discuss representative examples of both types of mechanisms. We limit our coverage to RNA-RNA interactions. Classical and more recent examples of RNAs that regulate gene expression through sponging of regulatory proteins, namely, CsrA and Hfq, were reviewed recently (23, see also chapter 19).

Predatory Mimicry in the Regulation of Chitosugar Uptake

In *Salmonella enterica* and *Escherichia coli*, an ~80-nt sRNA named ChiX (also named MicM) represses the synthesis of outer membrane chitoporin ChiP (also named ybfM) by base-pairing with a 12-nt sequence within the ribosome binding site of *chiP* mRNA and blocking translation of this mRNA (22, 24). Repression is relieved in the presence of the chitin-derived sugars chitobiose and chitotriose, consistent with the fact that ChiP is needed for the chitosugars to cross the outer membrane. The mechanism responsible for the relief of *chiP* repression was elucidated in 2009. Two parallel studies showed that growth in the presence of chitosugars induces the transcription of an RNA recognized

as target by ChiX, which upon base-pairing with ChiX promotes cleavage and rapid degradation of the sRNA (24) (Fig. 1). This decoy target originates from the *chbBCARFG* operon, which encodes the components of the chitosugars' inner membrane transport system (*chbBCA*), the operon's transcriptional activator (ChbR), and two catabolic enzymes (ChbFG) (25). Thus, the ChiX predation mechanism effectively couples the outer membrane entry of chitosugars with their active transport across the inner membrane. The decoy sequence is encoded within the *chbB-chbC* intercistronic spacer; however, the actual form responsible for capturing ChiX is not known. ChiX cleavage correlates with the appearance, in wild-type but not in RNase E mutant cells, of an ~400-nt RNA from the *chbBCA* portion of the transcript (24). This suggests that the decoy sequence acts as part of a much longer RNA that undergoes RNase E cleavage upon pairing with ChiX, generating the ~400-nt intermediate. Nonetheless, the

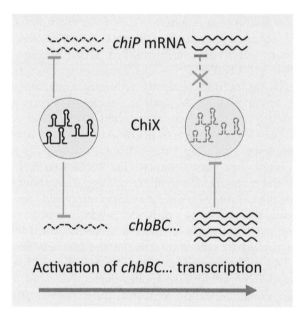

Figure 1 Regulation of chitosugar uptake in *Salmonella* and *E. coli*. The *chiP* gene and the *chbBCARFG* operon encode proteins involved in the uptake and utilization of chitin-derived sugars. When no chitosugars are available, ChiP synthesis is prevented by constitutively made ChiX sRNA, which represses translation of *chiP* mRNA (made at a relatively high basal level), while the *chbBCARF* operon is repressed transcriptionally by the NagC repressor (not shown). ChiX further lowers the uninduced levels of the *chb* mRNA by pairing with a sequence in the *chbB-chbC* intercistronic region. In the presence of chitosugars, transcriptional activation of *chbBCARF* operon produces a large accumulation of the polycistronic mRNA. Now in excess over ChiX, this mRNA titrates out ChiX through base-pairing and promotes its degradation. ChiX depletion results in the derepression of the *chiP* mRNA.

chbBC spacer sequence maintains its ability to capture ChiX and induce its cleavage even when removed from its natural context and expressed ectopically. By comparison, the *chiP* leader is much less effective at causing such cleavage (24). This is consistent with evidence indicating that ChiX action on *chiP* mRNA is, at least partially, catalytic (i.e., the sRNA is recycled a number of times before being degraded) (22). Thus, the ChiX-*chiP* and ChiX-*chbBC* RNA hybrids must differ in some features responsible for the differential fate of ChiX. Our initial proposal that destabilization of ChiX in the ChiX-*chbBC* hybrid results from the partial melting of a CG-rich stem-loop structure at the 3′ end of ChiX could not be experimentally confirmed. More-recent evidence suggests that binding affinity is the determining factor. The ChiX-*chbBC* RNA hybrid (19 bp) is predicted to be significantly more stable than the ChiX-*chiP* hybrid (12 bp). The 19-bp ChiX-*chbBC* RNA duplex is interrupted by two mismatches at adjacent positions. We found that eliminating these two mismatches by mutation—thus making the *chbBC* RNA fully complementary to ChiX over 21 consecutive nucleotides—stimulated ChiX cleavage even further (our unpublished observations). This raises the question as to why the wild-type sequence is not a perfect 21-bp match. The answer might be found in the observation that the pairing between ChiX and the *chbBC* spacer mRNA, while inactivating ChiX under inducing conditions (*chbBC* mRNA in excess), actually represses *chbC* gene expression when the operon is uninduced and transcribed at its basal level (ChiX in excess; Fig. 1) (26). A continuous 21-bp interaction is expected to make this repression even tighter. This might interfere with the inducibility of the entire network, as some low-level production of the transport system may be required when bacteria first encounter chitosugars to allow some of these molecules to leak into the cytoplasm and prime the induction cascade.

sRNA Sponging by tRNA Spacer Sequences

The sRNAs RybB and RyhB control two major homeostatic networks in *E. coli* and *S. enterica*, the σ^E-dependent envelope stress response and iron homeostasis, respectively. The *rybB* gene is transcribed upon activation of alternative sigma factor σ^E triggered by folding defects in outer membrane proteins (OMPs) (27, 28). RybB downregulates some of the major OMPs, and in doing so, it suppresses the σ^E-activating signal and thus its own transcription. The *ryhB* gene is repressed by the Fur repressor and becomes derepressed when the intracellular iron is depleted. RyhB silences the expression of nonessential iron-binding proteins while upre-

gulating iron uptake systems (29, 30). Thus, RyhB activity contributes to replenishing the intracellular iron pool, which, in turn, restores Fur-mediated repression of the *ryhB* gene. Recently, the RybB and RyhB regulons were chosen in the implementation of a new method for sRNA target identification based on high-throughput sequencing of transcripts copurified with the sRNA of interest (31). The method retrieved many of the previously known targets of RybB and RyhB, but in addition, it uncovered a peculiar new target that, surprisingly, copurified with either of the two sRNAs. The shared target is an ~50-nt RNA originating from the 3′ external transcribed spacer (3′ ETS) of the tricistronic *glyW-cysT-leuZ* tRNA precursor. The 3′ETS^{leuZ} RNA, cleaved off the tRNA precursor by RNase E during tRNA maturation, contains sequences complementary to the pairing domain of both RybB and RyhB and can engage in a base-pair interaction with each of the two sRNAs (Fig. 2). The interaction does not promote the degradation of the sRNAs, and it actually stabilizes 3′ETS^{leuZ} RNA, suggesting that the RybB-3′ETS^{leuZ} and RyhB-3′ETS^{leuZ} hybrids are relatively long-lived (31). Thus, unlike what is observed in ChiX regulation (above), 3′ETS^{leuZ} sequesters the sRNAs as opposed to destroying them; that is, it acts as a true sponge. The sponging activity serves to absorb RybB and RyhB molecules made adventitiously due to noise in promoter activity (RybB) or incomplete Fur repression during normal growth. In doing so, 3′ETS^{leuZ} sets a threshold level of expression that each of the two sRNAs must attain to begin acting on the respective targets (Fig. 2). Above the threshold, 3′ETS^{leuZ} does not hamper further accumulation of either RybB or RyhB, presumably because the absorbing capacity of the sponge is saturated (31). One might then predict that full induction of either of the two sRNAs will free the 3′ETS^{leuZ}-bound fraction of the other sRNA, causing it to increase to a higher basal level. This suggests that the ability of 3′ETS^{leuZ} to target both sRNAs is designed to link iron homeostasis to the σ^E-dependent envelope stress response. Indeed, the use of a tRNA processing product as a sponge for RybB and RyhB might have evolved to allow the activities of the two sRNAs to be modulated as a function of the physiological state of the cell. The *glyW-cysT-leuZ* operon is predicted to be susceptible to the stringent control and the growth rate-dependent regulation. This implies that the sponging activity of 3′ETS^{leuZ} will be maximal under fast growth conditions but should drop abruptly if ppGpp levels increase or, more generally, growth slows down. Allowing the free fraction of RybB and RyhB to increase under these conditions could be important for "preadapting" the

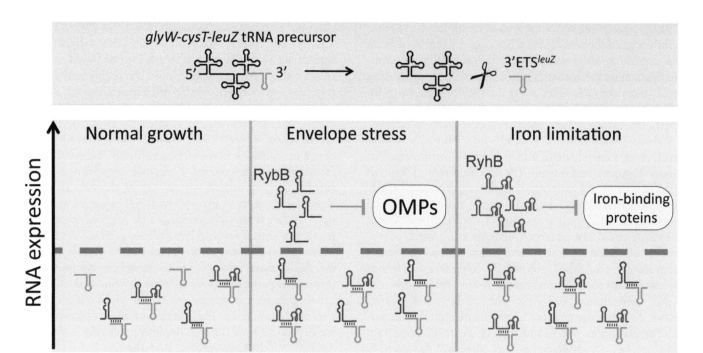

Figure 2 sRNA sponging by a tRNA spacer sequence. The sRNAs RybB (blue) and RyhB (purple) are made in response to envelope stress or iron limitation, respectively. An ~50-nt RNA, named 3′ETS^{leuZ} (red), released by RNase E cleavage of the *glyW-cysT-leuZ* tRNA precursor (top) can form stable base-pair interactions with both RybB and RyhB. This allows 3′ ETS^{leuZ} to capture and sequester RybB and RyhB molecules that are made adventitiously (in the absence of any stress) due to transcriptional noise (left). Under inducing conditions (envelope stress or iron limitation), accumulation of either RybB or RyhB saturates the sponging capacity of 3′ETS^{leuZ}. This sets the threshold concentration (dotted line) that either of the two sRNAs must attain to begin performing its regulatory task: downregulation of OMPs for RybB (middle) or of nonessential iron-binding proteins for RyhB (right).

cells to an incoming stress. On one hand, (p)ppGpp is known to stimulate σ^E activity (directly by acting on σ^E promoter and indirectly by favoring the ability of σ^E to compete with σ^{70} for binding to core RNA polymerase) (32) and is thought to be the primary signal responsible for induction of the σ^E response in stationary phase (33). On the other hand, iron limitation was reported to induce SpoT-dependent ppGpp accumulation (34). One could easily see how ppGpp-mediated relief of RybB and RyhB sponging would help the cells to more rapidly integrate the activities of these two sRNAs in the respective regulatory networks.

The SroC RNA: Target-Derived, but Not a Target Mimic

GcvB is a conserved, ~200-nt sRNA that downregulates >40 different mRNAs, most of them encoding proteins involved in amino acid uptake (35, 36). Regulation is maximally exerted during exponential growth in nutrient-rich media. The precise role of this control

is unclear, but it is probably aimed at balancing metabolic fluxes connected with the glycine cleavage system, whose main regulator, GcvA, activates the *gcvB* transcript (37). The glycine cleavage pathway is the major source of one-carbon units used, among others, in the synthesis of purines, thus directly linking amino acid and nucleotide metabolisms (38).

Among the GcvB-regulated mRNAs is the polycistronic transcript of the *gltIJKL* operon, which encodes the glutamate/aspartate ABC transporter. GcvB represses expression of the *gltIJKL* operon by pairing with a sequence within the leader region of the first cistron, *gltI* (35). As early as in 2003, Vogel and coworkers discovered that the *gltIJKL* operon also encoded an sRNA in the intercistronic region between *gltI* and *gltJ* (39). This sRNA, named SroC, is made from transcripts that don't extend all the way to the end of the operon but terminate at a leaky Rho-independent terminator ~150 nt downstream from the stop codon of the *gltI* gene. This ~150-nt tail is clipped off by RNase E, and the released RNA fragment, SroC, associates with Hfq and is stably maintained in the cell, suggesting that it might have a function of its own (40). Indeed, this study revealed that SroC can bind to and inactivate GcvB, and thus drive a feedforward regulatory loop resulting in the derepression of the entire GcvB regulon (Fig. 3). SroC base-pairs simultaneously with two separate segments in the GcvB sequence, distant from each other, neither of which corresponds to the domain used by GcvB to pair with most of its targets. The interaction exposes GcvB to cleavage by RNase E, whereas SroC is not cleaved and can be recycled (40). The use of a specific binding domain implies that SroC does not need to compete with most GcvB targets for binding to GcvB. Combined with recycling, lack of competition is expected to make SroC a particularly effective GcvB predator even at low concentrations. Interestingly, GcvB-dependent regulation and its reversal by SroC appear to be entirely recapitulated in cells growing in LB medium. GcvB accumulates maximally during the exponential phase when SroC levels are low; then it declines rapidly when cells enter stationary phase (35). The decline coincides with a vast increase in the intracellular concentration of SroC (39). One might envision that the activity of the Rho-independent terminator that generates the SroC precursor in the first place is increased in stationary phase. The recent finding that transcription termination at the Rho-independent terminators of sRNA genes is enhanced under stress conditions and in stationary phase lends support to this possibility (41). Yet the physiological role of SroC in the GcvB network remains elusive. The link to glutamate, a molecule at the crossroads of key metabolic pathways, suggests that the action of SroC aims at rewiring the GcvB regulon in a manner more adapted to the metabolism of stationary phase and/or of stress conditions.

Some of the transcriptomic changes that result from the overexpression of the *sroC* gene in *Salmonella* persist in a Δ*gcvB* background and thus are likely to

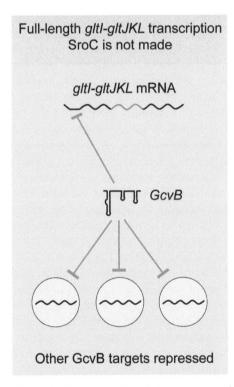
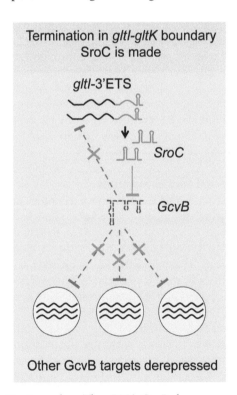

Figure 3 Target-mediated derepression of the GcvB regulon. The sRNA GcvB downregulates several mRNAs encoding amino acid and small peptide transporters. Among these is the *gltIJKL* mRNA (left). Presence of a leaky Rho-independent transcription terminator in the spacer between *gltI* and *gltJ* causes a fraction of transcripts initiating at the *gltI* promoter to terminate prematurely in the spacer region (right). RNase E cleavage of the prematurely terminated transcripts generates SroC, an ~150-nt RNA, which captures GcvB through a base-pairing interaction and destabilizes it. As a result, all of the GcvB targets become derepressed. Since the SroC precursor RNA itself is one of these targets, SroC activity drives a feedforward regulatory loop.

reflect regulation of genes outside the GcvB network. In particular, a handful of sRNAs were found downregulated in cells overproducing SroC. These effects were ascribed to SroC overaccumulation draining the Hfq pool (i.e., SroC acting as an Hfq sponge) and the decline in Hfq availability causing the destabilization of other Hfq-binding sRNAs (40). It turns out that for one of the downregulated sRNAs, Hfq depletion might not be the sole cause of destabilization. This is the case for MgrR, an sRNA that depends on the PhoP-PhoQ two-component system for expression (42). A recent study presented evidence that SroC can base-pair with MgrR and that this interaction stimulates MgrR turnover (43). Most importantly, by promoting MgrR decay, SroC alleviates the repression of MgrR's primary target, the lipopolysaccharide-modifying enzyme EptB. The EptB-directed modification decreases the bacterial susceptibility to the antimicrobial peptide polymyxin B in both *E. coli* and *Salmonella* (42, 43). The PhoP-PhoQ two-component system is also known to contribute functions enhancing polymyxin B resistance (44). Hence the PhoP-PhoQ requirement for MgrR expression is counterintuitive, since in repressing EptB, MgrR is expected to decrease, not increase, polymyxin B resistance. The discovery that SroC can inactivate MgrR helps solve this conundrum. One may speculate that SroC production or action is somehow potentiated during polymyxin B exposure so as to prevent the EptB repression. This hypothesis remains to be tested.

Prophage-Encoded GcvB Sponges

Two 60-nt sRNAs with anti-GcvB activity, encoded in the genome of distinct prophages, have been identified in enteropathogenic *E. coli* O157:H7 (45). The two sRNAs, named AgvB1 and AgvB2, bind Hfq *in vivo* and *in vitro* and can base-pair with the main seed sequence of GcvB, the so-called R1 region. When overexpressed, AgvB1 alleviates the repression of at least one GcvB target (*ddpA*, the only one tested), presumably by competing with the *dppA* mRNA for pairing with GcvB. These findings, together with the observation that neither the phage-encoded sRNAs nor GcvB are destabilized upon formation of the hybrid, indicate that, unlike SroC, AgvB1 and AgvB2 sequester rather than inactivate GcvB (45). Deleting the *agvB1* and *agvB2* genes from the *E. coli* O157:H7 genome did not significantly affect the growth of the strain in laboratory media, but it had a fitness cost when bacteria were grown in mucus from the bovine terminal rectum, suggesting that the relief of GcvB repression confers a growth advantage in the host.

Glimpses of Pervasive Sponging from Global Snapshots

Two new methodologies—CLASH (cross-linking, ligation, and sequencing of hybrids [46]) and RIL-seq (RNA interaction by ligation and sequencing [47])—have made it possible to profile RNA-RNA interactions at transcriptome-wide levels (48). Both methods exploit high-throughput sequencing of RNA UV-cross-linked to a pertinent protein, affinity purified, and subjected to proximity ligation. The ligation step captures RNA molecules in the act of base-pairing at the time of RNA extraction, allowing their identification as chimeric reads in transcriptome sequencing output. RIL-seq and CLASH were recently implemented in *E. coli* with RNA cross-linked to Hfq (47) and RNase E (49), respectively. The results of both studies retrieve a global picture of the bacterial interactome of unprecedented depth. While confirming most of the known interactions, these studies unveil a plethora of putative new interactions, greatly expanding the networks of virtually all known sRNAs and identifying new ones. Included in the sRNA networks are not only mRNAs but also tRNAs, tRNA precursors, and sRNAs, suggesting that sRNA sponging by tRNA-associated sequences or other sRNAs is a pervasive phenomenon. Particularly striking is the expansion of the ChiX network, found to include not less than 20 unrelated mRNAs (47, 49), at least as many tRNAs (47, 49), and a handful of sRNAs (49). In the absence of a coherent biological framework to accommodate most of the new interactions, such unsuspected complexity is somewhat disconcerting. The CLASH data suggest that ChiX uses a different seed sequence to interact with most of the newly identified targets (49). Unlike the sequence used to base-pair with *chiP* and *chbBC*, and previously identified *dpiBA* mRNA (50), this second pairing region lies in a portion of the ChiX sequence that is not conserved between *E. coli* and *Salmonella*, raising doubt as to the existence of the same interactions in *Salmonella*. This would be surprising given that the vast majority of known sRNA networks are conserved between the two bacterial species. Thus, it seems important that the novel interactions be directly verified experimentally and characterized in quantitative terms. One such verification was performed in the RIL-seq study to confirm the discovery of a new sponge system (47). An sRNA reportedly processed from the 3′ end of the *pspG* mRNA (encoding phage shock protein G) and named PspH was shown to interact with, and destabilize, the sRNA Spot 42 (Spf). The latter is a key regulator of carbohydrate metabolism. Activated by growth on glucose, Spot 42 tightens catabolite repression by downregulating genes involved in uptake and catabolism of nonfavored carbon sources (51). The ability of PspH to destabilize Spot 42 (causing the derepression of Spot 42 targets) was assessed with PspH overexpressed from a plasmid, and it is currently unclear what might trigger this regulation under physiological conditions, in particular whether PspH accumulates during the phage shock response. We notice, however, that the sRNA encoded at the *pspH* locus in *Salmonella* (named STnc880) is transcribed from its own promoter and likely regulated independently of *pspG* (52). STnc880 is considerably smaller (79 nt) than what one can infer to be the PspH size from reference 47; however, it is highly conserved in the shared portion of the sequence and 100% identical to PspH in the region involved in the Spot 42 interaction.

BROADENING THE SPONGE CONCEPT: A SYSTEMS BIOLOGY PERSPECTIVE

A widespread routine for assessing the regulatory activity of an sRNA involves overexpressing the sRNA gene cloned in a multicopy plasmid in cells carrying a reporter gene fusion to a presumptive target and assessing a change in the reporter expression. We suspect that if one were to apply the exact same procedure but switching the roles, that is, overexpressing the target mRNA in the presence of cognate sRNA, one might observe a decrease in the sRNA's regulatory activity on other targets. This is because the overproduced mRNA could sequester or promote cleavage of the cognate sRNA. In other words, many functional mRNAs that are targets of sRNA regulation might have the potential for acting as sRNA sponges. This capacity has interesting implications in the systems biology of sRNA networks. Indeed, one can imagine that a surge in concentration of a transcript at the interface of two networks, resulting from a regulatory change occurring in one network, could affect the linked network via a sponging activity (Fig. 4). For example, one predicts that all newly identified targets of ChiX (47, 49) will be derepressed upon activation of chitobiose operon transcription in *E. coli*. This type of mechanism was invoked to explain the silencing of the RprA sRNA network when *E. coli* cells enter stationary phase or during initial biofilm formation (53). Both conditions coincide with a strong transcriptional activation of the *csgD* gene, which encodes a positive regulator of curli and cellulose biosynthesis in *E. coli* and *Salmonella*. The *csgD* mRNA, a high-affinity target of RprA, was proposed to sequester RprA via the extensive base-pairing interaction and, in doing so, relieve the regulation of the other targets by this sRNA (53). Since *csgD*

is also the target of additional sRNAs (e.g., OmrA and OmrB) (54), its sponging activity could conceivably extend to other networks.

Other candidates for mRNA sponge activity are found in the regulation of quorum sensing in *Vibrio* species. Quorum sensing involves production and extracellular release of effector molecules, called autoinducers, which upon binding to specific receptors in the cell membrane signal the density state of the bacterial population, triggering a transcriptional response (55). At low cell densities (no autoinducers bound), phosphorelay-mediated activation of the transcription factor LuxO leads to the synthesis of five homologous sRNAs, the Qrr sRNAs, which repress several targets including the mRNA for master regulator LuxR. At high cell densities, autoinducer binding reverses the phosphate flow, inactivating LuxO and abolishing Qrr sRNA production. Three relevant Qrr-downregulated mRNAs are those encoding LuxO, LuxR, and the autoinducer synthase LuxM (55). Work by Bassler's group recently showed that Qrr sRNAs form different base-pair interactions with each of the three mRNAs and that the nature of the interaction determines the fate of the sRNA-mRNA hybrid. Specifically, a representative Qrr sRNA, Qrr3, induces the degradation of *luxR* mRNA and is then recycled (catalytic degradation), forms a stable hybrid with *luxO* mRNA (sequestration), and induces the degradation of *luxM* mRNA while being degraded at the same time (coupled degradation) (56). Catalytic degradation produced the most robust regulatory response. Its use in *luxR* repression could be rationalized considering that high- to-low-cell-density transitions can be abrupt in nature (e.g., bacterial excretion from a host), requiring a rapid shutoff of the entire *luxR* regulon. Sequestration of *luxO* mRNA by Qrr3 was proposed to serve as a negative feedback mechanism buffering fluctuations in LuxO protein levels. Since Qrr3 is itself sequestered in the process, an alternative (or additional) interpretation is that *luxO* mRNA acts as a sponge to absorb surges in Qrr sRNA levels that result from transcriptional noise. Finally, the use of coupled degradation to regulate *luxM* allows this mRNA to turn into a regulator and contribute to the elimination of Qrr sRNAs at the onset of the high-cell-density program (56).

Overall, the above examples offer clues to better appreciate the evolutionary significance of sRNA-mediated regulation: sRNAs might have evolved not only to allow rapid responses, as usually thought, but also to achieve a continuous physiological balance between distinct cellular processes, with the target mRNA nodes acting as the connectors of communicating vessels (Fig. 4).

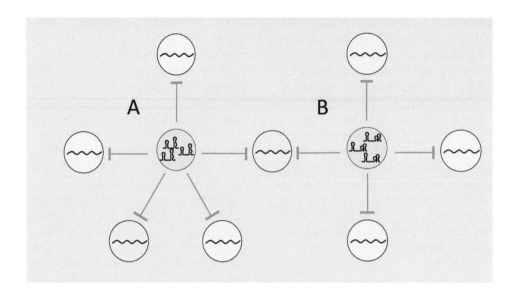

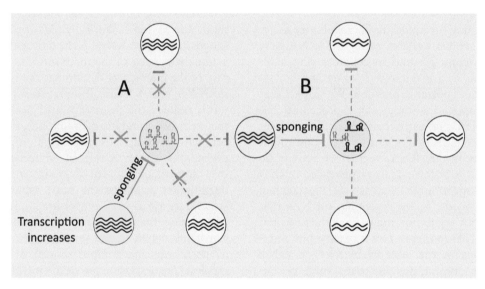

Figure 4   A sponging relay model. Depicted are two hypothetical sRNA networks (A and B) linked by an mRNA node (cyan-filled circle). (Top) The two sRNAs downregulate their respective targets. (Bottom) A transcriptional regulatory event leads to a large increase in the concentration of one of the mRNAs in network A (yellow-filled circle). The accumulated mRNA sequesters and destabilizes its cognate sRNA, resulting in the derepression of the entire network A, including the nodal mRNA. In turn, the latter acts as a sponge for cognate sRNA in network B, thus relieving, or attenuating (depicted here), the repression of the B network.

OUTLOOK

Since its discovery, sRNA-mediated regulation continues to attract the interest of a wide scientific audience and constitutes a fertile and dynamic field of research. Recent advances in the development of high-throughput methods have allowed a major leap forward in the characterization of the bacterial sRNA interactome. Still, in some ways, the recent breakthroughs epitomize

the so-called knowledge paradox: as the body of knowledge increases, so does the area we become aware of not knowing. The scores of unsuspected sRNA interactions whose biological rationale remains unexplained highlight the gaps that still exist in our understanding of the inner workings of the bacterial cells. Filling these gaps will require validating the newly discovered interactions individually as well as applying the high-throughput screen to sRNA knockout or overproducing strains and to mutants of transcriptional regulators whose networks feed into sRNA networks (57, 58). This work, which could include proteomic and metabolomic analyses, seems an obligatory step to obtain a more complete picture of how sRNA activities are integrated in the physiological control of gene expression in bacterial cells.

Acknowledgments. We apologize to our colleagues whose work has not been cited due to unintentional oversight. This work was supported by the Centre National de la Recherche Scientifique (CNRS) and by the Agence Nationale de la Recherche (ANR-3-BSV3-0005).

Citation. Figueroa-Bossi N, Bossi L. 2018. Sponges and predators in the small RNA world. Microbiol Spectrum 6(4): RWR-0021-2018.

References

1. Ambros V. 2004. The functions of animal microRNAs. *Nature* 431:350–355.

2. Bartel DP. 2009. MicroRNAs: target recognition and regulatory functions. *Cell* 136:215–233.

3. Jones-Rhoades MW, Bartel DP, Bartel B. 2006. Micro RNAS and their regulatory roles in plants. *Annu Rev Plant Biol* 57:19–53.

4. Ebert MS, Neilson JR, Sharp PA. 2007. MicroRNA sponges: competitive inhibitors of small RNAs in mammalian cells. *Nat Methods* 4:721–726.

5. Franco-Zorrilla JM, Valli A, Todesco M, Mateos I, Puga MI, Rubio-Somoza I, Leyva A, Weigel D, García JA, Paz-Ares J. 2007. Target mimicry provides a new mechanism for regulation of microRNA activity. *Nat Genet* 39: 1033–1037.

6. Cesana M, Cacchiarelli D, Legnini I, Santini T, Sthandier O, Chinappi M, Tramontano A, Bozzoni I. 2011. A long noncoding RNA controls muscle differentiation by functioning as a competing endogenous RNA. *Cell* 147:358–369.

7. Hansen TB, Jensen TI, Clausen BH, Bramsen JB, Finsen B, Damgaard CK, Kjems J. 2013. Natural RNA circles function as efficient microRNA sponges. *Nature* 495:384–388.

8. Marcinowski L, Tanguy M, Krmpotic A, Rädle B, Lisnić VJ, Tuddenham L, Chane-Woon-Ming B, Ruzsics Z, Erhard F, Benkartek C, Babic M, Zimmer R, Trgovcich J, Koszinowski UH, Jonjic S, Pfeffer S, Dölken L. 2012. Degradation of cellular mir-27 by a novel, highly abundant viral transcript is important for efficient virus replication *in vivo*. *PLoS Pathog* 8:e1002510.

9. Cazalla D, Yario T, Steitz JA. 2010. Down-regulation of a host microRNA by a *Herpesvirus saimiri* noncoding RNA. *Science* 328:1563–1566.

10. Memczak S, Jens M, Elefsinioti A, Torti F, Krueger J, Rybak A, Maier L, Mackowiak SD, Gregersen LH, Munschauer M, Loewer A, Ziebold U, Landthaler M, Kocks C, le Noble F, Rajewsky N. 2013. Circular RNAs are a large class of animal RNAs with regulatory potency. *Nature* 495:333–338.

11. Piwecka M, Glažar P, Hernandez-Miranda LR, Memczak S, Wolf SA, Rybak-Wolf A, Filipchyk A, Klironomos F, Cerda Jara CA, Fenske P, Trimbuch T, Zywitza V, Plass M, Schreyer L, Ayoub S, Kocks C, Kühn R, Rosenmund C, Birchmeier C, Rajewsky N. 2017. Loss of a mammalian circular RNA locus causes miRNA deregulation and affects brain function. *Science* 357:eaam8526.

12. Seitz H. 2009. Redefining microRNA targets. *Curr Biol* 19:870–873.

13. Salmena L, Poliseno L, Tay Y, Kats L, Pandolfi PP. 2011. A ceRNA hypothesis: the Rosetta Stone of a hidden RNA language? *Cell* 146:353–358.

14. Denzler R, Agarwal V, Stefano J, Bartel DP, Stoffel M. 2014. Assessing the ceRNA hypothesis with quantitative measurements of miRNA and target abundance. *Mol Cell* 54:766–776.

15. Denzler R, McGeary SE, Title AC, Agarwal V, Bartel DP, Stoffel M. 2016. Impact of microRNA levels, target-site complementarity, and cooperativity on competing endogenous RNA-regulated gene expression. *Mol Cell* 64:565–579.

16. Jens M, Rajewsky N. 2015. Competition between target sites of regulators shapes post-transcriptional gene regulation. *Nat Rev Genet* 16:113–126.

17. Figliuzzi M, Marinari E, De Martino A. 2013. MicroRNAs as a selective channel of communication between competing RNAs: a steady-state theory. *Biophys J* 104:1203–1213.

18. Gottesman S, Storz G. 2011. Bacterial small RNA regulators: versatile roles and rapidly evolving variations. *Cold Spring Harb Perspect Biol* 3:a003798.

19. Storz G, Vogel J, Wassarman KM. 2011. Regulation by small RNAs in bacteria: expanding frontiers. *Mol Cell* 43:880–891.

20. Wagner EG, Romby P. 2015. Small RNAs in bacteria and archaea: who they are, what they do, and how they do it. *Adv Genet* 90:133–208.

21. Levine E, Zhang Z, Kuhlman T, Hwa T. 2007. Quantitative characteristics of gene regulation by small RNA. *PLoS Biol* 5:e229.

22. Overgaard M, Johansen J, Møller-Jensen J, Valentin-Hansen P. 2009. Switching off small RNA regulation with trap-mRNA. *Mol Microbiol* 73:790–800.

23. Bossi L, Figueroa-Bossi N. 2016. Competing endogenous RNAs: a target-centric view of small RNA regulation in bacteria. *Nat Rev Microbiol* 14:775–784.

24. Figueroa-Bossi N, Valentini M, Malleret L, Fiorini F, Bossi L. 2009. Caught at its own game: regulatory small RNA inactivated by an inducible transcript mimicking its target. *Genes Dev* 23:2004–2015.

25. Plumbridge J, Pellegrini O. 2004. Expression of the chitobiose operon of *Escherichia coli* is regulated by three transcription factors: NagC, ChbR and CAP. *Mol Microbiol* 52:437–449.

26. Plumbridge J, Bossi L, Oberto J, Wade JT, Figueroa-Bossi N. 2014. Interplay of transcriptional and small RNA-dependent control mechanisms regulates chitosugar uptake in *Escherichia coli* and *Salmonella*. *Mol Microbiol* 92:648–658.

27. Johansen J, Rasmussen AA, Overgaard M, Valentin-Hansen P. 2006. Conserved small non-coding RNAs that belong to the σ^E regulon: role in down-regulation of outer membrane proteins. *J Mol Biol* 364:1–8.

28. Papenfort K, Pfeiffer V, Mika F, Lucchini S, Hinton JC, Vogel J. 2006. σ^E-dependent small RNAs of *Salmonella* respond to membrane stress by accelerating global *omp* mRNA decay. *Mol Microbiol* 62:1674–1688.

29. Massé E, Vanderpool CK, Gottesman S. 2005. Effect of RyhB small RNA on global iron use in *Escherichia coli*. *J Bacteriol* 187:6962–6971.

30. Jacques JF, Jang S, Prévost K, Desnoyers G, Desmarais M, Imlay J, Massé E. 2006. RyhB small RNA modulates the free intracellular iron pool and is essential for normal growth during iron limitation in *Escherichia coli*. *Mol Microbiol* 62:1181–1190.

31. Lalaouna D, Carrier MC, Semsey S, Brouard JS, Wang J, Wade JT, Massé E. 2015. A 3′ external transcribed spacer in a tRNA transcript acts as a sponge for small RNAs to prevent transcriptional noise. *Mol Cell* 58: 393–405.

32. Costanzo A, Nicoloff H, Barchinger SE, Banta AB, Gourse RL, Ades SE. 2008. ppGpp and DksA likely regulate the activity of the extracytoplasmic stress factor σ^E in *Escherichia coli* by both direct and indirect mechanisms. *Mol Microbiol* 67:619–632.

33. Gopalkrishnan S, Nicoloff H, Ades SE. 2014. Co-ordinated regulation of the extracytoplasmic stress factor, σ^E, with other *Escherichia coli* sigma factors by (p)ppGpp and DksA may be achieved by specific regulation of individual holoenzymes. *Mol Microbiol* 93:479–493.

34. Vinella D, Albrecht C, Cashel M, D'Ari R. 2005. Iron limitation induces SpoT-dependent accumulation of ppGpp in *Escherichia coli*. *Mol Microbiol* 56:958–970.

35. Sharma CM, Darfeuille F, Plantinga TH, Vogel J. 2007. A small RNA regulates multiple ABC transporter mRNAs by targeting C/A-rich elements inside and upstream of ribosome-binding sites. *Genes Dev* 21:2804–2817.

36. Sharma CM, Papenfort K, Pernitzsch SR, Mollenkopf HJ, Hinton JC, Vogel J. 2011. Pervasive post-transcriptional control of genes involved in amino acid metabolism by the Hfq-dependent GcvB small RNA. *Mol Microbiol* 81: 1144–1165.

37. Urbanowski ML, Stauffer LT, Stauffer GV. 2000. The *gcvB* gene encodes a small untranslated RNA involved in expression of the dipeptide and oligopeptide transport systems in *Escherichia coli*. *Mol Microbiol* 37:856–868.

38. Stauffer GV. 1996. Biosynthesis of serine, glycine and one carbon units, p 506–513. *In* Neidhardt FC, Curtiss R III, Ingraham JL, Lin EC, Low KB, Magasanik B, Reznikoff WS, Riley M, Schaechter M, Umbarger HE (ed), Escherichia coli *and* Salmonella: *Cellular and Molecular Biology*, 2nd ed. ASM Press, Washington, DC.

39. Vogel J, Bartels V, Tang TH, Churakov G, Slagter-Jäger JG, Hüttenhofer A, Wagner EG. 2003. RNomics in *Escherichia coli* detects new sRNA species and indicates parallel transcriptional output in bacteria. *Nucleic Acids Res* 31:6435–6443.

40. Miyakoshi M, Chao Y, Vogel J. 2015. Cross talk between ABC transporter mRNAs via a target mRNA-derived sponge of the GcvB small RNA. *EMBO J* 34:1478–1492.

41. Morita T, Ueda M, Kubo K, Aiba H. 2015. Insights into transcription termination of Hfq-binding sRNAs of *Escherichia coli* and characterization of readthrough products. *RNA* 21:1490–1501.

42. Moon K, Gottesman S. 2009. A PhoQ/P-regulated small RNA regulates sensitivity of *Escherichia coli* to antimicrobial peptides. *Mol Microbiol* 74:1314–1330.

43. Acuña LG, Barros MJ, Peñaloza D, Rodas PI, Paredes-Sabja D, Fuentes JA, Gil F, Calderón IL. 2016. A feedforward loop between SroC and MgrR small RNAs modulates the expression of *eptB* and the susceptibility to polymyxin B in *Salmonella* Typhimurium. *Microbiology* 162:1996–2004.

44. Moon K, Six DA, Lee HJ, Raetz CR, Gottesman S. 2013. Complex transcriptional and post-transcriptional regulation of an enzyme for lipopolysaccharide modification. *Mol Microbiol* 89:52–64.

45. Tree JJ, Granneman S, McAteer SP, Tollervey D, Gally DL. 2014. Identification of bacteriophage-encoded anti-sRNAs in pathogenic *Escherichia coli*. *Mol Cell* 55:199–213.

46. Kudla G, Granneman S, Hahn D, Beggs JD, Tollervey D. 2011. Cross-linking, ligation, and sequencing of hybrids reveals RNA-RNA interactions in yeast. *Proc Natl Acad Sci U S A* 108:10010–10015.

47. Melamed S, Peer A, Faigenbaum-Romm R, Gatt YE, Reiss N, Bar A, Altuvia Y, Argaman L, Margalit H. 2016. Global mapping of small RNA-target interactions in bacteria. *Mol Cell* 63:884–897.

48. Hör J, Vogel J. 2017. Global snapshots of bacterial RNA networks. *EMBO J* 36:245–247.

49. Waters SA, McAteer SP, Kudla G, Pang I, Deshpande NP, Amos TG, Leong KW, Wilkins MR, Strugnell R, Gally DL, Tollervey D, Tree JJ. 2017. Small RNA interactome of pathogenic *E. coli* revealed through crosslinking of RNase E. *EMBO J* 36:374–387.

50. Mandin P, Gottesman S. 2009. A genetic approach for finding small RNAs regulators of genes of interest identifies RybC as regulating the DpiA/DpiB two-component system. *Mol Microbiol* 72:551–565.

51. Beisel CL, Storz G. 2011. The base-pairing RNA Spot 42 participates in a multioutput feedforward loop to help enact catabolite repression in *Escherichia coli*. *Mol Cell* 41:286–297.

52. Kröger C, Colgan A, Srikumar S, Händler K, Sivasankaran SK, Hammarlöf DL, Canals R, Grissom JE, Conway T, Hokamp K, Hinton JC. 2013. An infection-relevant transcriptomic compendium for *Salmonella enterica* serovar Typhimurium. *Cell Host Microbe* 14:683–695.

53. Mika F, Busse S, Possling A, Berkholz J, Tschowri N, Sommerfeldt N, Pruteanu M, Hengge R. 2012. Targeting of *csgD* by the small regulatory RNA RprA links stationary phase, biofilm formation and cell envelope stress in *Escherichia coli*. *Mol Microbiol* 84:51–65.

54. Holmqvist E, Reimegård J, Sterk M, Grantcharova N, Römling U, Wagner EG. 2010. Two antisense RNAs

target the transcriptional regulator CsgD to inhibit curli synthesis. *EMBO J* **29:**1840–1850.

55. **Rutherford ST, Bassler BL.** 2012. Bacterial quorum sensing: its role in virulence and possibilities for its control. *Cold Spring Harb Perspect Med* **2:**a012427.

56. **Feng L, Rutherford ST, Papenfort K, Bagert JD, van Kessel JC, Tirrell DA, Wingreen NS, Bassler BL.** 2015. A Qrr noncoding RNA deploys four different regulatory mecha-

nisms to optimize quorum-sensing dynamics. *Cell* **160:**228–240.

57. **Beisel CL, Storz G.** 2010. Base pairing small RNAs and their roles in global regulatory networks. *FEMS Microbiol Rev* **34:**866–882.

58. **Nitzan M, Rehani R, Margalit H.** 2017. Integration of bacterial small RNAs in regulatory networks. *Annu Rev Biophys* **46:**131–148.

Regulating with RNA in Bacteria and Archaea
Edited by Gisela Storz and Kai Papenfort
© 2018 American Society for Microbiology, Washington, DC
doi:10.1128/microbiolspec.RWR-0014-2017

Bacterial Small RNAs in Mixed Regulatory Networks

26

Anaïs Brosse[1] and Maude Guillier[1]

INTRODUCTION

Regulatory RNAs have emerged as important regulators of gene expression in all kingdoms of life, and many advances toward the understanding of their biology have been achieved in bacteria. Bacterial regulatory RNAs are often also referred to as small RNAs (sRNAs), as most of them range in size from 50 to 400 nucleotides. While it was recognized early on that these sRNAs can ensure extremely diverse biological functions, such as *trans*-translation (transfer-messenger RNA), ribonucleolytic activity (RNA moiety of the RNase P), or even involvement in protein secretion (4.5S RNA), the shared efforts of multiple groups in the last 2 decades have led to the identification of a plethora of sRNAs in virtually all bacteria. Many of these act as posttranscriptional regulators of gene expression and generally function by imperfectly base-pairing to target mRNA(s), leading to changes in their translation and/or stability.

As more and more bacterial sRNAs are investigated in detail, it appears that regulatory events mediated by sRNAs are intimately intertwined with the previously characterized transcriptional bacterial network. First, the ability of sRNAs to act as regulators is mediated by the fact that their synthesis or activity somehow responds to the environment. In the very vast majority of cases, this is achieved by transcriptional control of sRNA synthesis, mediated by transcriptional regulators (TRs). The second characteristic that makes sRNAs important actors of regulatory networks is that they target multiple genes involved in all kinds of cellular functions, including genes for TRs, possibly creating a feedback circuit (i.e., the sRNA regulates expression of its own TR).

Thus, the expression of many bacterial genes is in fact controlled by mixed regulatory networks combining transcriptional and posttranscriptional regulatory steps mediated by proteins and sRNAs, respectively. Different cases have been reported where transcriptional and posttranscriptional controls can occur independently, or rather "in cascade" (Fig. 1A). In this review, we will discuss more specifically bacterial circuits in which an sRNA is regulated by a TR and/or regulates synthesis of a TR. Even though sRNAs of many kinds participate in such networks, we will focus on the class of imperfectly base-pairing sRNAs, as they have been the most studied so far.

TRANSCRIPTIONAL CONTROL OF sRNA SYNTHESIS BY TRs

In the vast majority of cases, changes in sRNA levels in response to the environment are achieved by transcriptional control. Interestingly, the TRs involved are "classic" regulators, i.e., regulators that are not dedicated to regulation of sRNA genes but had been previously characterized and generally possess much larger regulons. These TRs can be response regulators from two-component systems (TCSs), alternative sigma factors, or neither of these, hereafter referred to as transcription factors, such as Fur, Lrp, or FNR, for instance. Known examples of connections between sRNAs and TRs in *Escherichia coli* and *Salmonella* are listed in Table 1 and in Fig. 1B, and in other species in Table 2. As a result, most, if not all, major TRs include at least one sRNA in their regulon, and in cases where it was addressed in detail, the sRNAs often appear among the most highly regulated targets. For instance, the gene for the RyhB sRNA, whose transcription is under direct control by the Fur repressor (1), is one of the most strongly deregulated genes in a Δ*fur* mutant (2). RyhB thus accumulates under iron starvation and participates in iron homeostasis

[1]CNRS UMR8261, Associated with University Paris Diderot, Sorbonne Paris Cité, Institut de Biologie Physico-Chimique, 75005 Paris, France.

A.

B.

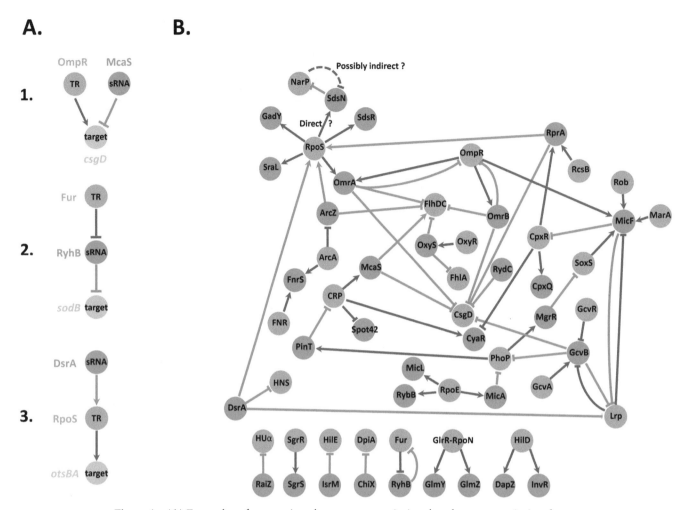

Figure 1 (A) Examples of connections between transcriptional and posttranscriptional control. Transcriptional regulators/transcriptional regulations and sRNAs/posttranscriptional regulations are in blue and red, respectively. Green nodes indicate nonregulatory target genes. Examples of TR, sRNA, or target found in the depicted circuits are given on the side, and positive or negative regulatory interactions were chosen here based on these examples. Note, however, that all regulations can be positive or negative. (B) Known regulatory interactions between imperfectly base-pairing sRNAs and TRs in *E. coli* and *Salmonella*. For clarity, target genes that do not encode regulators are not shown. Unless otherwise indicated, only direct or likely direct interactions are shown. See Table 1 for details.

by repressing targets encoding dispensable iron-storage or iron-containing proteins and activating synthesis of proteins important for iron scavenging (3, see also chapter 16). Similarly, a detailed analysis of promoters from the RpoE (σ^E) regulon showed that the MicA, RybB, and MicL sRNA promoters were among the most active and the most regulated by RpoE (4, 5). The RpoE sigma factor is activated upon membrane stress via a proteolytic cascade responding to an accumulation of unfolded outer membrane proteins or unassembled lipopolysaccharides in the periplasm. Among its sRNA targets, both MicA and RybB repress synthesis of multiple outer membrane

proteins, while MicL represses that of Lpp, a very abundant lipoprotein involved in the stabilization and integrity of the outer membrane. These controls participate in relieving the membrane stress responsible for RpoE signaling, and accordingly, these three sRNAs were shown to lower RpoE activation (5–8).

While many sRNAs are transcribed from freestanding genes as primary transcripts, they can also be synthesized through processing of longer transcripts, such as mRNAs or longer forms of sRNAs. These processed sRNAs are, logically, transcriptionally controlled by the same regulator as their precursor RNA: for instance,

Table 1 Examples of imperfectly base-pairing small RNAs regulated by and/or regulating TRs in *E. coli* and *Salmonella*[a]

sRNA	Regulatory TR[b]	Reference(s)	Regulated TR[c]	Reference(s)
ArcZ (RyhA/SraH)	ArcB-ArcA (aerobic condition) [−]	16	RpoS (σ^S) [+], (ArcB) [−], FlhD$_2$C$_2$ [−]	16, 23
ChiX (MicM/ RybC/SroB)			DpiB-DpiA [−]	74
CpxQ	CpxA-CpxR (cell envelope stress) [+]	9		
CyaR (RyeE)	CRP (low glucose) [+]	75–77		
	CpxA-CpxR (cell envelope stress) [−]			
DapZ	HilD (in *Salmonella*) (host cell infection) [+]	78		
DsrA			H-NS [−], RpoS (σ^S) [+], Lrp [−]	14, 29, 30, 34
FnrS (RydD)	FNR (anaerobic condition) [+]	79		
	(ArcA) [+]			
GadY (IS183) (not in *Salmonella*)	RpoS (σ^S) (stationary phase) [+]	80	GadX [+], GadW[d] [+]	80
GcvB	GcvA (high glycine levels) [+]	32, 81	PhoQ-PhoP [−], Lrp [−], CsgD (direct?) [−]	32, 34, 41
	Lrp (leucine levels) [−]			
GlmY (Tke1/SroF)	GlrK-GlrR with RpoN (σ^{54}) [+]	82, 83		
GlmZ (RyiA/SraJ)	GlrK-GlrR with RpoN (σ^{54}) (not in *E. coli*) [+]	83		
InvR (not in *E. coli*)	HilD (host cell infection) [+]	84		
McaS (not in *Salmonella*)	CRP (low glucose) [+]	20	CsgD [−], FlhD$_2$C$_2$ [+]	20, 21, 23
MgrR	PhoQ-PhoP (low magnesium) [+]	85	SoxS [−]	34
MicA (SraD)	RpoE (σ^E) (cell envelope stress) [+]	6, 86–88	PhoQ-PhoP [−]	40
MicF	Lrp (direct?) (leucine levels) [−]	32, 89 (review)	CpxA-CpxR (direct?) [−], Lrp [−]	31, 76, 90
	Rob, SoxS, MarA (oxidative stress) [+]			
	EnvZ-OmpR (high osmolarity or acid pH) [+]			
MicL	RpoE (σ^E) (cell envelope stress) [+]	5		
OmrA (RygA/SraE)	EnvZ-OmpR (high osmolarity or acid pH) [+]	9, 44, 91	FlhD$_2$C$_2$ [−], EnvZ-OmpR [−], CsgD [−]	18, 23, 44, 93
	RpoS (σ^S) (stationary phase) [+]			
OmrB (RygB)	EnvZ-OmpR (high osmolarity or acid pH) [+]	44	FlhD$_2$C$_2$ [−], EnvZ-OmpR [−], CsgD [−]	18, 23, 44, 93
OxyS	OxyR (oxidative stress) [+]	94	RpoS (σ^S) (indirect), FhlA [−], FlhD$_2$C$_2$ [−]	17, 23, 94, 95
RaiZ			HUα [−]	96
RprA	CpxA-CpxR (cell envelope stress) [+]	15, 76	RpoS (σ^S) [+], CsgD [−]	15, 19
	RcsCDB (cell envelope stress) [+]			
RybB	RpoE (σ^E) (cell envelope stress) [+]	6, 87, 97		
RydC			CsgD (direct?) [−]	22
RyhB (SraI)	Fur (iron starvation) [−]	1	Fur (direct and indirect) [−]	98, 99
SdsN	RpoS (σ^S) (stationary phase) [+]	39	NarQ-NarP [−]	39
SdsR (RyeB)	RpoS (σ^S) (stationary phase) [+]	100		
SgrS (RyaA)	SgrR (phosphosugar stress) [+]	101		
Spot 42 (Spf)	CRP (low glucose) [−]	102		
SraL (RyjA)	RpoS (σ^S) (stationary phase) [+]	103		
IsrM (not in *E. coli*)			HilE [−]	104
PinT (not in *E. coli*)	PhoP-PhoQ (low magnesium) [+]	105	CRP (likely direct) [−]	105

[a]TCSs are in blue, sigma factors in red, and other TRs in black.
[b]Only direct or presumably direct regulators are listed. Known stimuli for the TRs are indicated by brackets, and [−] or [+] signs specify negative or positive controls of the sRNA, respectively.
[c]Unless otherwise indicated, direct interaction between the sRNA and the TR target was shown. [−] or [+] signs indicate repression or activation of targets, respectively.
[d]Note that GadY is perfectly complementary to the *gadXW* mRNA, as it is encoded on the complementary strand.

the synthesis of the CpxQ sRNA, processed from the *cpxP* mRNA, responds, like *cpxP*, to the CpxA-CpxR TCS (9). However, whether the processing of precursors into mature sRNAs is also a regulated step is still poorly characterized in general.

Integrating sRNAs into a TR regulon can provide several advantages (see "Properties of sRNAs in Mixed Networks," below). An important one, although not strictly restricted to sRNAs, is that such a cascade of regulators may switch a repressor protein into an indi-

Table 2 Examples of connections between imperfectly base-pairing sRNAs and TRs in species other than *E. coli* or *Salmonella*[a]

RNA name	Organism	Regulatory TR[b]	Reference(s)	Regulated TR[c]	Reference(s)
Gram-negative bacteria					
Qrr 1-4 (*Vibrio cholerae*)	*Vibrio cholerae*	LuxO with RpoN (σ^N) (cell density) [+]	146 and references therein	LuxR (in *V. harveyi*)/HapR (*V. cholerae*), LuxO, AphA [−]	146 and references therein
Qrr 1-5 (*Vibrio harveyi*)	*Vibrio harveyi*	AphA (cell density) for Qrr2/3/4 [+]; LuxR (cell density) [+]			
VrrA	*Vibrio cholerae*	RpoE (σ^E) (cell envelope stress) [+]	106		
TarA	*Vibrio cholerae*	ToxT (host cell infection) [+]	107		
TarB	*Vibrio cholerae*	ToxT (host cell infection) [+]	108	VspR (likely direct) [−]	109
VqmR	*Vibrio cholerae*	VqmA (cell density) [+]	110	VspT [−]	110
TfoR	*Vibrio cholerae*	TfoS (chitin) [+]. TfoS is a transmembrane regulator, itself controlled (possibly indirectly) by the orphan kinase ChiS.	111, 112	TfoX [+]	113
PhrS	*Pseudomonas aeruginosa*	ANR (FNR homolog) (anaerobic condition) [+]	114	PqsR [+]	114
PrrF1/PrrF2 (RyhB analogs)	*Pseudomonas aeruginosa*	Fur (iron starvation) [−]	115		
CrcZ	*Pseudomonas aeruginosa*	CbrA-CbrB (nitrogen availability) [+]	116		
NrsZ	*Pseudomonas aeruginosa*	NtrB-NtrC with RpoN (σ^N) (nitrogen availability) [+]	117		
Sr0161	*Pseudomonas aeruginosa*			ExsA [−]	118
LprA	*Legionella pneumophila*	RpoS (σ^S) (stationary phase); OxyR (oxidative stress) [+]	119		
PcrZ	*Rhodobacter sphaeroides*	PrrB-PrrA (nitrite respiration) [+]	120		
Gram-positive bacteria					
RNAIII	*Staphylococcus aureus*	AgrC-AgrA (cell density) [+]	121, 122	Rot [−], MgrA [+]	123, 124
ArtR	*Staphylococcus aureus*	AgrC-AgrA (cell density) [−]	125	SarT [−]	125
RoxS (RsaE)	*Bacillus subtilis*, *Staphylococcus aureus*	ResE-ResD (SrrB-SrrA in *S. aureus*) (anaerobic condition, nitric oxide) [+]; Rex (in *B. subtilis*, oxidative stress, NAD/NADH ratio) [−]	126, 127		

sRNA	Organism	TR (stimuli) [sign]		TR target [sign]	
RsaA	Staphylococcus aureus	SigB (σ^B) (stationary phase) [+]	128	MgrA [−]	129
RsaF	Staphylococcus aureus	SigB (σ^B) (stationary phase) [+]	128		
SbrA/SbrB/SbrC	Staphylococcus aureus	SigB (σ^B) (stationary phase) [+]	130		
Psm-mec	Staphylococcus aureus			AgrC-AgrA [−]	131
SprC (Srn_3610)	Staphylococcus aureus	SarA (microaerobic environment, other signals?) [−]	132		
Srn_9340	Staphylococcus aureus	SarA (microaerobic environment, other signals?) [−]	132		
FsrA	Bacillus subtilis	Fur (iron starvation) [+]	133		
SR1	Bacillus subtilis	CcpN [−], CcpA (glucose, other sugars) [−]	134	AhrC [−]	135
RnaC/S1022 (IGR yrhK-yrhJ)	Bacillus subtilis	SigD (σ^D) (sporulation) [+]	136	AbrB [−]	137
CsfG	Bacillus subtilis	SigF (σ^F), SigG (σ^G) (sporulation) [+]	138		
LhrC1-5	Listeria monocytogenes	LisK-LisR (Diverse signals, among which cell envelope stress, antimicrobials) [+]	139		
Rli22	Listeria monocytogenes	LisK-LisR (Diverse signals, among which cell envelope stress, antimicrobials) [+]	140		
Rli33-1	Listeria monocytogenes	SigB (σ^B) (stationary phase) [+]	140		
SreA/SreB	Listeria monocytogenes			PrfA [−]	141
MarS	Streptococcus pyogenes			Mga [+]	142
Srn206	Streptococcus pneumoniae			ComD-ComE (direct?) [−]	143
csRNA1/2/3/4/5	Streptococcus pneumoniae	CiaH-CiaR (TCS involved in diverse functions, among which competence, virulence or antimicrobial resistance) [+]	144		
VR-RNA	Clostridium perfringens	VirR-VirS (host cell infection) [+]	145		

[a] TCSs are in blue, sigma factors in red, and other TRs in black.

[b] Only direct or presumably direct regulators are listed. Known stimuli for the TRs are listed. Known stimuli for the TRs are indicated by brackets, and [−] or [+] signs specify negative or positive controls of the sRNA, respectively.

[c] Unless otherwise indicated, direct interaction between the sRNA and the TR target was shown. [−] or [+] signs indicate repression or activation of targets, respectively.

rect activator, or conversely an activator protein into an indirect repressor. This is observed, for instance, with the Fur repressor, which indirectly activates expression of genes involved in iron storage, such as genes of the *sdh* operon, *sodB*, and many others, via RyhB sRNA (see above). Conversely, the RpoE-dependent sRNAs allow this alternative sigma factor to repress expression of several membrane proteins. In these two examples, the sRNAs explained experimental observations that Fur and RpoE activated or decreased expression of some genes, respectively (2, 10, 11). These two examples are also a very nice illustration of a clear physiological link between sRNA expression pattern and their targets. However, in some other examples, the link between the conditions in which an sRNA is produced and the function of its targets is less obvious.

POSTTRANSCRIPTIONAL CONTROL OF TR SYNTHESIS BY sRNAs

Control of Transcription Factors and Sigma Factors

sRNAs regulate expression of extremely diverse genes and thereby participate in the control of important cellular functions, including, for instance, virulence (see references 12 and 13 for reviews), quorum sensing, or iron homeostasis (respectively detailed in chapters 18, 17, and 16). As mentioned above, sRNAs also commonly regulate synthesis of all kinds of TRs. The first example came with the recognition that translation of *rpoS*, encoding the alternative sigma factor involved in many stress responses, is directly activated by the DsrA sRNA (14). In this case, pairing of the sRNA with the 5′ untranslated region (UTR) of *rpoS* mRNA prevents formation of an inhibitory structure sequestering the Shine-Dalgarno sequence. Remarkably, two additional sRNAs, RprA and ArcZ, were later found to act on *rpoS* mRNA via the same mechanism (15, 16), while the OxyS sRNA was shown to repress *rpoS* expression, most likely indirectly (17).

Control of TRs by sRNAs is not restricted to sigma factors. In particular, several concomitant studies pointed to an important role of sRNAs in controlling the switch between planktonic/sessile and biofilm/adhesive lifestyle. This is largely due to the remarkable number of sRNAs that modulate the synthesis of the master regulators CsgD and FlhDC, which control curli and flagella synthesis, respectively. *csgD* translation is repressed by at least six sRNAs: OmrA, OmrB, McaS, and RprA, which directly interact with different regions of the *csgD* mRNA (18–21); and GcvB and RydC, for which base-

pairing with *csgD* mRNA is also suspected (21, 22). *flhDC* mRNA is also a "hot spot" for sRNA binding, as five sRNAs modulate *flhDC* translation through direct base-pairing: four negatively (OmrA, OmrB, ArcZ, and OxyS) (23) and one positively (McaS) (20). In addition, two sRNAs, GadY and SdsR, repress expression of *flhDC*, most likely via an indirect mechanism (23). All these sRNAs are themselves controlled by different TRs and thus are produced under different conditions. Furthermore, CsgD and FlhDC levels are in general inversely regulated, as they control pathways that should not be concomitantly expressed (24). Thus, while it seems logical that the McaS sRNA represses *csgD* while activating *flhDC*, it is rather surprising that other sRNAs such as OmrA/B repress both *csgD* and *flhDC*. Finally, some of these sRNAs can also impact biofilm formation and motility independently of their pairing to *csgD* or *flhDC* mRNAs: RprA represses expression of *ydaM*, encoding a diguanylate cyclase synthesizing c-di-GMP and acting as a cofactor for *csgD* transcription activation (19); while McaS activates the synthesis of PgaA, a porin required for the export of a polysaccharide important for the biofilm structure (25). And even more sRNAs are involved in the control of biofilm or motility, possibly by acting in different pathways; however, details of the regulation remain to be addressed in some cases (26–28).

Yet other master TRs are controlled by sRNAs. This is notably true for HNS (histone-like nucleoid structuring protein, which has a global role in repressing gene expression) and Lrp (leucine responsive protein, which regulates genes involved in amino acid biosynthesis and metabolism, among others), both pleiotropic modulators of transcription of an impressive number of genes and whose mRNAs were recognized as sRNA targets early on. More precisely, DsrA is a direct inhibitor of *hns* translation (29, 30), while GcvB and MicF repress that of *lrp* (31–34). Many TR targets of sRNAs were first identified serendipitously by the detailed study of chosen sRNAs, typically by microarray analysis following the pulse induction of the sRNA or by bioinformatics and pairing prediction. More recently, the study of RNA-RNA interactions on a global scale also suggested that more TRs could be targeted by sRNAs (35, 36). However, because mRNAs encoding TRs are often not very abundant, they may have been missed as sRNA targets in some of these approaches. This prompted several studies based on the reverse approach, i.e., aiming at finding sRNA regulators of a given TR, for instance, by using gene fusions and screening for regulators using a dedicated library of sRNAs. When applied to *hns* or *lrp*, this approach recovered the previously identified regulators DsrA (for *hns*) and MicF and GcvB (for *lrp*), but also revealed that DsrA can base-pair to the *lrp* mRNA to repress its translation (34). In addition, while this study showed that regulation by sRNAs also applies to other TRs (e.g., SoxS), this may not prove to be a general rule, as the synthesis of the major TRs CRP (cyclic AMP receptor protein) and FNR (fumarate and nitrate reductase) was not significantly modulated by any of the sRNAs tested. At this stage, however, these results do not rule out possible control of CRP or FNR synthesis by sRNAs that were not present in the tested library.

Note that in some cases, sRNAs control synthesis of their own TR, thus creating a feedback circuit whose properties are discussed below. Finally, control of sigma factors or of transcription factors by sRNAs can also be indirect. The above-mentioned RpoE-dependent sRNAs were, for instance, found to limit RpoE activity indirectly by repressing *de novo* synthesis of outer membrane proteins that play an important role in activating RpoE (5).

Intuitively, controlling the levels of a TR could be a way to considerably expand the number of targets of an sRNA. Consistent with this, control of Lrp by MicF, GcvB, or DsrA results in downregulation of transcription of Lrp-positive targets, or conversely in upregulation of transcription of Lrp-negative targets, at least under conditions of sRNA overproduction (31, 34). Other examples of sRNA indirectly affecting gene expression through control of a transcription factor are given by the phenotypes observed upon overexpression of the sRNAs that regulate *csgD* and/or *flhDC*: several of the *csgD* repressors decrease curli formation (18, 20, 21), while regulators of *flhDC* affect motility (23). However, even under such conditions where the sRNAs are overproduced, sRNA control of TRs does not systematically appear to affect the downstream targets, and whether regulation of those targets is affected when deleting the endogenous copy of the sRNA remains poorly addressed. This is thus rather different from what was observed in the situation where TRs regulate an sRNA but is in line with the fact that regulation exerted by sRNAs is often more modest than regulation exerted by proteins. Consequently, changes in TR levels upon sRNA control may not be sufficient to affect all TR targets, or an effect on the TR targets may be restricted to specific conditions only. This question may be even more complex for sRNAs that control synthesis of TCS proteins, i.e., regulators that require a posttranslational modification step (phosphorylation) to perform their biological function. This is discussed in more detail in the next section.

Regulating the Synthesis of Response Regulators with sRNAs

Control of TRs by sRNAs also applies to TCS response regulators. An early example was provided by the Qrr sRNAs that lie at the core of the quorum-sensing regulatory cascade in vibrios and regulate several TRs involved in this process, including the LuxO response regulator (see below and chapter 17 for more details). As transcription of the Qrr sRNAs is itself directly activated by LuxO (under its phosphorylated form, LuxO-P) (37), this also results in a feedback circuit that was shown to allow indirect autoregulation of the Qrr sRNAs, indicating that the Qrr sRNAs decrease LuxO-P levels by repressing *luxO* expression (38).

In enterobacteria, expression of the *narP* gene, encoding the regulator of the NarQ-NarP TCS, was found to be repressed by SdsN, one of the RpoS-dependent sRNAs. In addition to *narP*, *nfsA* and *hmpA* were also identified as direct negative targets of SdsN (39). While NarQ-NarP is a TCS involved in the nitrate/nitrite control of anaerobic respiration, *nfsA* and *hmpA* also encode proteins associated with the metabolism of and response to nitrogen compounds. As in the above example, a feedback circuit exists between SdsN and *narP*, as SdsN levels appeared to be repressed by NarP, although this might not be due to direct transcriptional regulation (39). Finally, whether control of *narP* by SdsN affected the downstream genes was not further investigated.

In some examples, controlling response regulator synthesis with sRNAs led to unsuspected consequences on the TCS regulon. This was the case, for instance, for the PhoP regulator, involved in the response to low magnesium and presence of antimicrobial peptides, and the OmpR regulator, involved in the response to high osmolarity and acidic pH. Synthesis of both regulators was shown to be repressed by multiple sRNAs. More precisely, the sRNAs MicA and GcvB, expressed under envelope stress conditions (via RpoE) or glycine accumulation (via GcvA/GcvR), respectively, both pair with the translation initiation region of *phoP*, the first cistron of *phoPQ* mRNA. This results in repression of its translation by competition with binding of the 30S ribosomal subunit (40, 41). Surprisingly, however, despite the fact that these two sRNAs similarly decrease the levels of *phoP* mRNA, they have completely different effects on expression of PhoP targets: while MicA overproduction represses expression of several positive targets of PhoP, such as *ompT*, *mgrR*, *mgtS* (previously *yneM*) (42), or *mgtA*, this is not the case with GcvB, which either has no effect on expression of the same targets or even increases expression of some of them.

While the molecular explanation for this puzzling observation is still missing, the use of different mutants of GcvB suggested that its inability to control PhoP-regulated genes is related to the other targets of this sRNA. Indeed, overproduction of a GcvB variant that still controls *phoP* but not most of the other GcvB targets resulted in regulation of PhoP-dependent genes (41). Thus, possibly depending on the expression of one or several GcvB targets that could themselves modify PhoP levels and/or activity, regulation of *phoP* by GcvB may or may not affect expression of the downstream genes, providing an additional layer of regulation to the PhoP regulon.

In the case of OmpR, its synthesis is repressed by two apparently redundant sRNAs, OmrA and OmrB, again creating a negative feedback circuit as OmrA/B transcription is directly activated by OmpR. These two sRNAs pair with the *ompR* translation initiation region on the *ompR-envZ* mRNA via their almost identical 5′ ends and induce a decrease in OmpR protein levels, presumably by competition with ribosome binding. In complete agreement with a previous report indicating that the levels of the phosphorylated form of OmpR, OmpR-P, were insensitive to a large range of changes in *ompR* or *envZ* expression (43), the control of *ompR* by OmrA/B does not affect the amount of OmpR-P (44). In line with this, overexpression of either OmrA or OmrB sRNA only modestly affects the transcription of OmpR-P-regulated targets such as *ompC* or *ompF*. Interestingly, however, we found that transcription of some other OmpR targets was, in contrast, sensitive to changes in OmpR levels and thus to OmrA/B overproduction (Fig. 2). These targets, which include the OmrA/B sRNAs themselves, are thus regulated as well by the nonphosphorylated form of OmpR, according to a still uncharacterized molecular mechanism (44).

A remaining question of these different studies is whether the sRNAs also modulate the levels of the cognate sensor kinases. In the case of *luxOU*, *phoPQ*, and *ompR-envZ* mRNAs, the organization of genes for the regulator and the kinase in bicistronic operons suggests that sRNA-mediated control of the regulator may lead to control of the kinase as well through translational coupling and/or degradation of the bicistronic target mRNA; this was not directly addressed, however. The situation is different in the case of the NarQ-NarP TCS, since *narQ* and *narP* genes are located at distant loci. Thus, at least in some cases, sRNA control of genes for response regulators might change the kinase/regulator ratio, which could affect the cross talk with other TCSs. sRNA control was also reported for sensor kinases: the ArcZ sRNA and the *arcB* mRNA,

made from convergent genes, were found to reciprocally control their own levels. This again results in an indirect feedback circuit, as *arcB* encodes the kinase of the ArcB-ArcA TCS, which controls *arcZ* transcription (16).

Together, these results highlight the fact that it is extremely difficult to predict how the control of a TR by sRNAs will affect the TR regulon, in particular when the TR is a TCS whose activity depends not only on the protein levels but also on its phosphorylation status. In some cases, it is even clear that controlling the synthesis of TCS will not affect the levels of the phosphorylated form of the response regulator. Such controls are thus likely to have more-limited effects on the TCS regulon than directly controlling the activity of a TCS, as was shown for the EutW-EutV system in *Firmicutes*. In this exquisite mechanism, a riboswitch-controlled transcript made under specific conditions interacts with the phosphorylated form of EutV, an unusual response regulator that interacts with RNA and promotes antitermination, thereby allowing expression of genes for ethanolamine utilization when both ethanolamine and the cofactor, adenosylcobalamin (AdoCbl), are present (45, 46, see also chapter 8).

PROPERTIES OF sRNAs IN MIXED NETWORKS

Regulating Gene Expression with sRNAs

To understand why it could be useful to include sRNAs in regulatory networks, it may be worth considering the major differences between sRNA- and protein-mediated controls. First, sRNAs are mainly posttranscriptional regulators, whereas the regulatory proteins act mostly at the transcriptional level. This is an important difference that has several consequences. In particular, posttranscriptional control allows for a faster response, since it targets mRNA and not the DNA (47). This is especially important for the repression of genes whose mRNAs have long half-lives and could thus be subject to several rounds of translation even after their synthesis has been repressed by transcriptional control (48, 49). Furthermore, sRNAs may be synthesized rapidly given their short size; this may contribute to the high speed of response, which is probably one of the major reasons why sRNAs are so widespread in adaptive response of bacteria to diverse stresses. Importantly, however, posttranscriptional regulators may be proteins as well, e.g., the CsrA global regulator or the ribosomal proteins that control their own synthesis (see chapters 19 and 7, respectively). Even though one can argue that, in theory, sRNA synthesis could be faster than synthesis

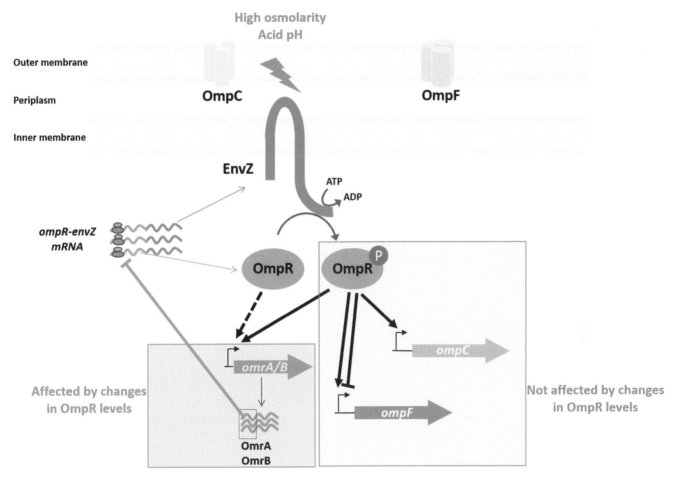

Figure 2 Feedback circuit between the EnvZ-OmpR TCS and OmrA/B sRNAs. *E. coli* OmrA/B sRNAs are transcribed from two adjacent genes and repress expression of multiple mRNAs, including the *ompR-envZ* mRNA, encoding the EnvZ-OmpR TCS. This control decreases the levels of OmpR, but without affecting that of OmpR-P, its phosphorylated form. Because *omrA/B* transcription is directly activated by OmpR-P but also responds to the nonphosphorylated OmpR, this feedback circuit allows OmrA/B to indirectly modulate their own synthesis, while expression of other OmpR targets such as *ompC* or *ompF* porin genes that are regulated only by OmpR-P remains unchanged. See reference 44 for details.

of proteins that are translated from longer transcripts, this argument is clearly not valid in many cases. Indeed, proteins acting at the posttranscriptional level are often already present when the input signal for control appears: for instance, free ribosomal proteins will bind to their own mRNA and repress its expression when they cannot be assembled within ribosomes as not enough free rRNA is available (50). Regarding CsrA, it is mostly its activity that is controlled via the CsrB and CsrC sRNAs (see chapter 19). In those cases, regulation by proteins can thus be as fast as regulation by sRNAs, if not faster.

Posttranscriptional regulation could also be more efficient at limiting fluctuations in the levels of target proteins than transcriptional control; indeed, in the case of

transcriptional control, leaks in transcription will be amplified by translation and give rise to bursts of protein synthesis (51). Posttranscriptional control may thus be beneficial, for instance, for genes whose products might be deleterious under specific conditions. Posttranscriptional control was also reported to discoordinate expression among genes of a same operon, as is the case for regulation of genes from the *galETKM* and *iscRSUA* operons by the sRNAs Spot 42 and RyhB, respectively (52, 53).

Another key feature of sRNAs is that they are commonly degraded together with their targets (54). This stoichiometric action contrasts with the catalytically acting TRs and explains several important properties of sRNA control. First, sRNAs may allow a faster recovery time when the signal inducing the synthesis of the regulator has been turned off compared to TRs that may be highly stable. One exception to this is if the TR is a phosphorylated response regulator, i.e., a short-lived species, which might also allow for fast recovery (47).

Another consequence of their stoichiometric action is that sRNAs mediate a linear threshold response in which repression depends on the relative transcription rates of sRNAs (α_s) and their targets (α_m) (55). If $\alpha_s > \alpha_m$, then expression of the target is expected to be strongly silenced; in contrast, when $\alpha_m > \alpha_s$, expression of the target is expected to linearly increase with the difference between α_m and α_s. Near the threshold (i.e., when α_m and α_s are close), sRNA regulation is ultrasensitive and small changes in sRNA or target mRNA levels will have strong effects on the control. Note that α_s and α_m can change with environmental conditions, making this threshold dynamically tunable. In the "strongly repressed" state, i.e., when α_s is considerably larger than α_m, the sRNAs suppress fluctuations in gene expression and filter transient signals. This is likely to have key implications for the control of TRs: since TR mRNAs are commonly of low abundance, they can easily fall into this window of strong repression by sRNAs. In this case, preventing fluctuations in their expression and filtering transient signals may contribute to limiting induction of TR to extremely specific conditions. This might be crucial for TRs controlling group behavior and/or costly processes such as curli or flagella synthesis, and it is thus not surprising to find multiple sRNAs regulating the different quorum-sensing regulators in vibrios or the CsgD or FlhDC TRs lying at the top of the biofilm or motility regulatory cascade, respectively. In addition, TRs are commonly autoregulated, sometimes positively, which may allow them to sustain two stable states of expression and lead to a bimodal distribution, with the concentration of the TR being low in some cells and high in others (56). The low expression state is, however, typically unstable, as bursts in expression will switch the system to the high expression state through the positive autoregulation. By preventing fluctuations in expression, repression by sRNAs is thus expected in such cases to both stabilize this two-state behavior and increase the probability of it remaining in the low expression state (51). In this regard, CsgD is itself a positively autoregulated transcription factor, and it would be especially interesting to determine whether the observed bistability of *csgD* expression (57) is somehow related to its control by multiple sRNAs.

Another crucial property of sRNAs that likely participates in their frequent usage in bacterial regulation is the capacity of these molecules to rapidly evolve. Even if there are some exceptions, imperfectly base-pairing sRNAs are often conserved only in closely related bacteria, consistent with high evolvability. Furthermore, while the sRNA sites involved in pairing to targets are evolutionarily conserved, this is not the case for the mRNA sites involved in pairing to sRNAs. This suggests that sRNAs can not only rapidly appear but also rapidly acquire or lose targets, thereby allowing establishment of regulation specific to a restricted group of organisms and environmental niches (58, 59, see also chapter 28).

At last, another possible advantage of sRNA-mediated control is the cross talk they establish between their different targets, leading to possible connections between otherwise unrelated pathways. In other words, the ability of an sRNA to regulate a given target depends on the expression of its other targets. This was nicely illustrated by the role of the ChiX sRNA in the induction of ChiP chitoporin synthesis in response to chitin degradation products. ChiX is a constitutively expressed sRNA that tightly represses *chiPQ* expression; under inducing conditions, however, the *chbBCARFG* operon is expressed, producing another target of ChiX. Interestingly, ChiX is degraded upon pairing to this latter target, and the resulting decrease in ChiX levels allows derepression of *chiP* and synthesis of the chitoporin when chitooligosaccharides are present (60). Thus, a target of an sRNA can turn out to be a regulator of this sRNA as well, and more examples of so-called "sponge" RNAs affecting the levels of sRNAs and their ability to regulate their other targets have been described (61, 62, see also chapter 25). More generally, the competition between the different targets of a single sRNA leads to a hierarchy in their control that depends on many factors, among which are the abundance of the different RNAs and their affinity for the sRNA, as well as their ability to promote degradation of the sRNA or not (63).

Note that there is also a competition between different sRNAs for RNA-binding protein(s) required for their action, such as Hfq, for instance (64–66). Regulation by sRNAs will thus also depend on expression of several other sRNAs, which is another source for unsuspected connections.

sRNAs in Mixed Regulatory Circuits

As described initially for TRs (56), recurrent regulatory patterns, usually referred to as network motifs, are observed in regulatory networks involving sRNAs. For a detailed description of these motifs and of their possible properties, we refer the reader to previous and excellent reviews (67, 68), and we will here focus more specifically on two examples of motifs that combine sRNAs and TRs.

Mixed networks, i.e., networks that use both sRNAs and TRs, often include feedback circuit (FC) motifs in which an sRNA regulates and is regulated by a TR. These FCs can be positive (i.e., the two regulatory events are either repressing or activating) or negative (i.e., one event is activating while the other is repressing). This is an interesting observation, as it contrasts with purely transcriptional networks in which negative FCs are extremely rare. Instead, TRs can negatively autoregulate themselves, which was reported to increase response time and decrease cell-to-cell variations (56).

Among known examples of mixed negative FCs are the Qrr sRNA-containing FCs. Qrr's are homologous sRNAs that are central to the quorum-sensing regulatory circuit in vibrios (37). Quorum sensing designates a cell-cell communication process allowing bacteria to sense the population density and to regulate their gene expression accordingly. In this circuit, the Qrr sRNAs regulate expression of multiple targets. In particular, they activate the synthesis of the low-cell-density master regulator AphA while repressing that of the high-cell-density master regulator LuxR/HapR and of the LuxO response regulator, itself at the top of the quorum-sensing regulatory cascade and whose phosphorylation depends on the cell density. This results in mixed FCs between the Qrr's and these different quorum-sensing TRs, LuxO, AphA, and LuxR/HapR, which, together with the negative autoregulation of the different TRs, minimize fluctuations and coordinate the cell response within the population (see chapter 17 for more detail). More specifically, the LuxO-Qrr FC participates in fine-tuning the threshold at which transition from low cell density to high cell density occurs (38). In this case, control of the TR by the sRNAs and the direct autocontrol of the TR appear to have similar roles, leaving open the question of whether the sRNAs provide specific properties to these particular regulatory circuits. The situation is different in the LuxR/HapR-Qrr FC, where the negative autoregulation of LuxR/HapR controls the transition from low to high cell density, while the sRNA-mediated feedback accelerates the transition from high to low cell density (69, 70). Interestingly, the Qrr's use different mechanisms to regulate this set of targets: luxO and luxR expression is repressed by sequestration and catalytic degradation, respectively, while activation of aphA is accompanied by the degradation of the Qrr's. These different mechanisms ensure specific variations in the level and the dynamics of expression of the target genes and are important for the observed quorum-sensing dynamics (71).

As mentioned above, sRNAs also participate in positive FCs, such as the ones existing between MicF and Lrp, and between GcvB and Lrp. Positive FCs exist in transcriptional networks and were shown to provide two stable steady states: either both regulators being "on" or "off" at the same time (for double-positive FC) or one regulator "on" and the other "off" (for double-negative FC). Consistent with this property of double-negative FCs, levels of MicF and Lrp were found to be inversely correlated (31), but again, whether properties of these circuits differ between purely transcriptional and mixed versions remains unclear.

Such a difference was suggested by sRNAs involved in other mixed circuit motifs: the feedforward motifs (FFMs). In such motifs, a regulator A controls expression of gene X both directly and indirectly by acting on a regulator B. Depending on the positive or negative direction of each regulatory interaction, eight different FFMs have been defined: four coherent (i.e., where direct and indirect regulation of X by A are both positive or negative) and four incoherent (i.e., direct regulation of X by A is positive, while indirect regulation is negative, or conversely). Their properties have been investigated for purely transcriptional networks and differ for the different FFMs. Some coherent FFMs were shown to induce a delay in response to induction or deinduction and to filter transient signals, while some incoherent FFMs function as pulse generators and accelerators of response (56). Interestingly, integration of posttranscriptional regulators such as sRNAs in certain types of coherent FFMs was proposed to provide a tighter repression and a faster recovery following deactivation compared to purely transcriptional FFMs (47). In addition, a faster kinetics of recovery was experimentally observed in the mixed CRP-Spot 42 FFM, while this was not predicted for such a purely transcriptional FFM, thus indicating that sRNAs could provide specific properties to regulatory motifs (72). More recently, a mixed FFM with an sRNA on top of the two regulatory arms was reported: in this case, the RprA sRNA controls the synthesis of RicI, a protein involved in inhibition of plasmid transfer, both directly and via the RpoS sigma factor (73). Because all regulations are positive and given that this FFM functions with AND logic (i.e., both RprA and RpoS are required for RicI activation), this ensures that synthesis of RicI is delayed following RprA induction and can be rapidly turned off. In addition, this provides some specialization within the RpoS regulon, as a target such as RicI, even though RpoS dependent, will not respond to the stresses that activate the RpoS response but have no effect on RprA. Again, however, whether the sRNA brings specific properties to this circuit in comparison with a TR remains unclear.

CONCLUSIONS

Many aspects of sRNA biology likely contributed to the widespread use of this class of regulators in bacterial regulatory circuits, among which are their rapid evolvability, their action at the posttranscriptional level, their fast dynamics of regulation, and their stoichiometric mode of action. Note, however, that some of these properties do not apply to all imperfectly base-pairing sRNAs. More generally, features of regulation will be different for other classes of regulatory RNAs, such as the ones acting by interaction with a protein, for instance, that also control TR synthesis and/or activity. Thus, properties of mixed regulatory networks will have to be precisely determined for each considered case, both in steady-state and in non-steady-state conditions, as sRNAs may provide specific dynamics to the circuit.

As more examples of TR-sRNA regulatory interactions will be analyzed in diverse bacteria, it will also be highly interesting to determine whether certain classes of sRNAs and/or of TRs are more represented in those mixed regulatory circuits. In parallel with a better understanding of the different properties of the various classes of sRNAs, this is likely to lead to still unsuspected facets of bacterial physiology.

Acknowledgments. We thank C. Chiaruttini, C. Condon, S. Durand, J. Jagodnik, and M. Springer for comments on the manuscript. Work in our lab is supported by the CNRS, the ANR (Grant ANR-14-CE10-0004-01 to M.G.), and the "Initiative d'Excellence" program from the French State (Grant "DYNAMO," ANR-11-LABX-0011).

Citation. Brosse A, Guillier M. 2018. Bacterial small RNAs in mixed regulatory networks. *Microbiol Spectrum* 6(3):RWR-0014-2017.

References

1. **Massé E, Gottesman S.** 2002. A small RNA regulates the expression of genes involved in iron metabolism in *Escherichia coli. Proc Natl Acad Sci U S A* 99:4620–4625.

2. **Seo SW, Kim D, Latif H, O'Brien EJ, Szubin R, Palsson BO.** 2014. Deciphering Fur transcriptional regulatory network highlights its complex role beyond iron metabolism in *Escherichia coli. Nat Commun* 5:4910.

3. **Salvail H, Massé E.** 2012. Regulating iron storage and metabolism with RNA: an overview of posttranscriptional controls of intracellular iron homeostasis. *Wiley Interdiscip Rev RNA* 3:26–36.

4. **Mutalik VK, Nonaka G, Ades SE, Rhodius VA, Gross CA.** 2009. Promoter strength properties of the complete σ^E regulon of *Escherichia coli* and *Salmonella enterica. J Bacteriol* 191:7279–7287.

5. **Guo MS, Updegrove TB, Gogol EB, Shabalina SA, Gross CA, Storz G.** 2014. MicL, a new σ^E-dependent sRNA, combats envelope stress by repressing synthesis of Lpp, the major outer membrane lipoprotein. *Genes Dev* 28:1620–1634.

6. **Papenfort K, Pfeiffer V, Mika F, Luchhini S, Hinton JCD, Vogel J.** 2006. σ^E-Dependent small RNAs of *Salmonella* respond to membrane stress by accelerating global *omp* mRNA decay. *Mol Microbiol* 62:1674–1688.

7. **Gogol EB, Rhodius VA, Papenfort K, Vogel J, Gross CA.** 2011. Small RNAs endow a transcriptional activator with essential repressor functions for single-tier control of a global stress regulon. *Proc Natl Acad Sci U S A* 108:12875–12880.

8. **Bossi L, Maloriol D, Figueroa-Bossi N.** 2008. Porin biogenesis activates the σE response in *Salmonella hfq* mutants. *Biochimie* 90:1539–1544.

9. **Chao Y, Vogel J.** 2016. A 3′ UTR-derived small RNA provides the regulatory noncoding arm of the inner membrane stress response. *Mol Cell* 61:352–363.

10. **Rhodius VA, Suh WC, Nonaka G, West J, Gross CA.** 2006. Conserved and variable functions of the σ^E stress response in related genomes. *PLoS Biol* 4:e2.

11. **Bury-Moné S, Nomane Y, Reymond N, Barbet R, Jacquet E, Imbeaud S, Jacq A, Bouloc P.** 2009. Global analysis of extracytoplasmic stress signaling in *Escherichia coli. PLoS Genet* 5:e1000651.

12. **Quereda JJ, Cossart P.** 2017. Regulating bacterial virulence with RNA. *Annu Rev Microbiol* 71:263–280.

13. **Caldelari I, Chao Y, Romby P, Vogel J.** 2013. RNA-mediated regulation in pathogenic bacteria. *Cold Spring Harb Perspect Med* 3:a010298.

14. **Majdalani N, Cunning C, Sledjeski D, Elliott T, Gottesman S.** 1998. DsrA RNA regulates translation of RpoS message by an anti-antisense mechanism, independent of its action as an antisilencer of transcription. *Proc Natl Acad Sci U S A* 95:12462–12467.

15. **Majdalani N, Hernandez D, Gottesman S.** 2002. Regulation and mode of action of the second small RNA activator of RpoS translation, RprA. *Mol Microbiol* 46:813–826.

16. **Mandin P, Gottesman S.** 2010. Integrating anaerobic/aerobic sensing and the general stress response through the ArcZ small RNA. *EMBO J* 29:3094–3107.

17. **Zhang A, Altuvia S, Tiwari A, Argaman L, Hengge-Aronis R, Storz G.** 1998. The OxyS regulatory RNA represses *rpoS* translation and binds the Hfq (HF-1) protein. *EMBO J* 17:6061–6068.

18. **Holmqvist E, Reimegård J, Sterk M, Grantcharova N, Römling U, Wagner EG.** 2010. Two antisense RNAs target the transcriptional regulator CsgD to inhibit curli synthesis. *EMBO J* 29:1840–1850.

19. **Mika F, Busse S, Possling A, Berkholz J, Tschowri N, Sommerfeldt N, Pruteanu M, Hengge R.** 2012. Targeting of *csgD* by the small regulatory RNA RprA links

stationary phase, biofilm formation and cell envelope stress in *Escherichia coli*. *Mol Microbiol* 84:51–65.

20. Thomason MK, Fontaine F, De Lay N, Storz G. 2012. A small RNA that regulates motility and biofilm formation in response to changes in nutrient availability in *Escherichia coli*. *Mol Microbiol* 84:17–35.

21. Jørgensen MG, Nielsen JS, Boysen A, Franch T, Møller-Jensen J, Valentin-Hansen P. 2012. Small regulatory RNAs control the multi-cellular adhesive lifestyle of *Escherichia coli*. *Mol Microbiol* 84:36–50.

22. Bordeau V, Felden B. 2014. Curli synthesis and biofilm formation in enteric bacteria are controlled by a dynamic small RNA module made up of a pseudoknot assisted by an RNA chaperone. *Nucleic Acids Res* 42:4682–4696.

23. De Lay N, Gottesman S. 2012. A complex network of small non-coding RNAs regulate motility in *Escherichia coli*. *Mol Microbiol* 86:524–538.

24. Mika F, Hengge R. 2013. Small regulatory RNAs in the control of motility and biofilm formation in *E. coli* and *Salmonella*. *Int J Mol Sci* 14:4560–4579.

25. Jørgensen MG, Thomason MK, Havelund J, Valentin-Hansen P, Storz G. 2013. Dual function of the McaS small RNA in controlling biofilm formation. *Genes Dev* 27:1132–1145.

26. Bak G, Lee J, Suk S, Kim D, Young Lee J, Kim KS, Choi BS, Lee Y. 2015. Identification of novel sRNAs involved in biofilm formation, motility, and fimbriae formation in *Escherichia coli*. *Sci Rep* 5:15287.

27. Parker A, Cureoglu S, De Lay N, Majdalani N, Gottesman S. 2017. Alternative pathways for *Escherichia coli* biofilm formation revealed by sRNA overproduction. *Mol Microbiol* 105:309–325.

28. Parker A, Gottesman S. 2016. Small RNA regulation of TolC, the outer membrane component of bacterial multidrug transporters. *J Bacteriol* 198:1101–1113.

29. Lease RA, Cusick M, Belfort M. 1998. Riboregulation in *Escherichia coli*: DsrA RNA acts by RNA:RNA interactions at multiple loci. *Proc Natl Acad Sci U S A* 95:12456–12461.

30. Lalaouna D, Morissette A, Carrier MC, Massé E. 2015. DsrA regulatory RNA represses both *hns* and *rbsD* mRNAs through distinct mechanisms in *Escherichia coli*. *Mol Microbiol* 98:357–369.

31. Holmqvist E, Unoson C, Reimegård J, Wagner EG. 2012. A mixed double negative feedback loop between the sRNA MicF and the global regulator Lrp. *Mol Microbiol* 84:414–427.

32. Modi SR, Camacho DM, Kohanski MA, Walker GC, Collins JJ. 2011. Functional characterization of bacterial sRNAs using a network biology approach. *Proc Natl Acad Sci U S A* 108:15522–15527.

33. Sharma CM, Papenfort K, Pernitzsch SR, Mollenkopf HJ, Hinton JC, Vogel J. 2011. Pervasive post-transcriptional control of genes involved in amino acid metabolism by the Hfq-dependent GcvB small RNA. *Mol Microbiol* 81:1144–1165.

34. Lee HJ, Gottesman S. 2016. sRNA roles in regulating transcriptional regulators: Lrp and SoxS regulation by sRNAs. *Nucleic Acids Res* 44:6907–6923.

35. Melamed S, Peer A, Faigenbaum-Romm R, Gatt YE, Reiss N, Bar A, Altuvia Y, Argaman L, Margalit H. 2016. Global mapping of small RNA-target interactions in bacteria. *Mol Cell* 63:884–897.

36. Waters SA, McAteer SP, Kudla G, Pang I, Deshpande NP, Amos TG, Leong KW, Wilkins MR, Strugnell R, Gally DL, Tollervey D, Tree JJ. 2017. Small RNA interactome of pathogenic *E. coli* revealed through crosslinking of RNase E. *EMBO J* 36:374–387.

37. Lenz DH, Mok KC, Lilley BN, Kulkarni RV, Wingreen NS, Bassler BL. 2004. The small RNA chaperone Hfq and multiple small RNAs control quorum sensing in *Vibrio harveyi* and *Vibrio cholerae*. *Cell* 118:69–82.

38. Tu KC, Long T, Svenningsen SL, Wingreen NS, Bassler BL. 2010. Negative feedback loops involving small regulatory RNAs precisely control the *Vibrio harveyi* quorum-sensing response. *Mol Cell* 37:567–579.

39. Hao Y, Updegrove TB, Livingston NN, Storz G. 2016. Protection against deleterious nitrogen compounds: role of σ^S-dependent small RNAs encoded adjacent to *sdiA*. *Nucleic Acids Res* 44:6935–6948.

40. Coornaert A, Lu A, Mandin P, Springer M, Gottesman S, Guillier M. 2010. MicA sRNA links the PhoP regulon to cell envelope stress. *Mol Microbiol* 76:467–479.

41. Coornaert A, Chiaruttini C, Springer M, Guillier M. 2013. Post-transcriptional control of the *Escherichia coli* PhoQ-PhoP two-component system by multiple sRNAs involves a novel pairing region of GcvB. *PLoS Genet* 9:e1003156.

42. Wang H, Yin X, Wu Orr M, Dambach M, Curtis R, Storz G. 2017. Increasing intracellular magnesium levels with the 31-amino acid MgtS protein. *Proc Natl Acad Sci U S A* 114:5689–5694.

43. Batchelor E, Goulian M. 2003. Robustness and the cycle of phosphorylation and dephosphorylation in a two-component regulatory system. *Proc Natl Acad Sci U S A* 100:691–696.

44. Brosse A, Korobeinikova A, Gottesman S, Guillier M. 2016. Unexpected properties of sRNA promoters allow feedback control via regulation of a two-component system. *Nucleic Acids Res* 44:9650–9666.

45. DebRoy S, Gebbie M, Ramesh A, Goodson JR, Cruz MR, van Hoof A, Winkler WC, Garsin DA. 2014. Riboswitches. A riboswitch-containing sRNA controls gene expression by sequestration of a response regulator. *Science* 345:937–940.

46. Mellin JR, Koutero M, Dar D, Nahori MA, Sorek R, Cossart P. 2014. Riboswitches. Sequestration of a two-component response regulator by a riboswitch-regulated noncoding RNA. *Science* 345:940–943.

47. Shimoni Y, Friedlander G, Hetzroni G, Niv G, Altuvia S, Biham O, Margalit H. 2007. Regulation of gene expression by small non-coding RNAs: a quantitative view. *Mol Syst Biol* 3:138.

48. Hershey JW, Sonenberg N, Mathews MB. 2012. Principles of translational control: an overview. *Cold Spring Harb Perspect Biol* 4:a011528.

49. Springer M. 1996. Translational control of gene expression in *E. coli* and bacteriophage, p 85–126. *In* Lin EC,

Lynch AS (ed), *Regulation of Gene Expression in* Escherichia coli. R G Landes Co, Austin, TX.

50. **Nomura M, Gourse R, Baughman G.** 1984. Regulation of the synthesis of ribosomes and ribosomal components. *Ann Rev Biochem* **53:**73–117.

51. **Levine E, Hwa T.** 2008. Small RNAs establish gene expression thresholds. *Curr Opin Microbiol* **11:**574–579.

52. **Møller T, Franch T, Udesen C, Gerdes K, Valentin-Hansen P.** 2002. Spot 42 RNA mediates discoordinate expression of the *E. coli* galactose operon. *Genes Dev* **16:**1696–1706.

53. **Desnoyers G, Morissette A, Prévost K, Massé E.** 2009. Small RNA-induced differential degradation of the polycistronic mRNA *iscRSUA*. *EMBO J* **28:**1551–1561.

54. **Massé E, Escorcia FE, Gottesman S.** 2003. Coupled degradation of a small regulatory RNA and its mRNA targets in *Escherichia coli*. *Genes Dev* **17:**2374–2383.

55. **Levine E, Zhang Z, Kuhlman T, Hwa T.** 2007. Quantitative characteristics of gene regulation by small RNA. *PLoS Biol* **5:**e229.

56. **Alon U.** 2007. Network motifs: theory and experimental approaches. *Nat Rev Genet* **8:**450–461.

57. **Grantcharova N, Peters V, Monteiro C, Zakikhany K, Römling U.** 2010. Bistable expression of CsgD in biofilm development of *Salmonella enterica* serovar Typhimurium. *J Bacteriol* **192:**456–466.

58. **Gottesman S, Storz G.** 2011. Bacterial small RNA regulators: versatile roles and rapidly evolving variations. *Cold Spring Harb Perspect Biol* **3:**a003798.

59. **Peer A, Margalit H.** 2014. Evolutionary patterns of *Escherichia coli* small RNAs and their regulatory interactions. *RNA* **20:**994–1003.

60. **Figueroa-Bossi N, Valentini M, Malleret L, Fiorini F, Bossi L.** 2009. Caught at its own game: regulatory small RNA inactivated by an inducible transcript mimicking its target. *Genes Dev* **23:**2004–2015.

61. **Lalaouna D, Carrier MC, Semsey S, Brouard JS, Wang J, Wade JT, Massé E.** 2015. A 3′ external transcribed spacer in a tRNA transcript acts as a sponge for small RNAs to prevent transcriptional noise. *Mol Cell* **58:**393–405.

62. **Miyakoshi M, Chao Y, Vogel J.** 2015. Cross talk between ABC transporter mRNAs via a target mRNA-derived sponge of the GcvB small RNA. *EMBO J* **34:**1478–1492.

63. **Bossi L, Figueroa-Bossi N.** 2016. Competing endogenous RNAs: a target-centric view of small RNA regulation in bacteria. *Nat Rev Microbiol* **14:**775–784.

64. **Fender A, Elf J, Hampel K, Zimmermann B, Wagner EG.** 2010. RNAs actively cycle on the Sm-like protein Hfq. *Genes Dev* **24:**2621–2626.

65. **Moon K, Gottesman S.** 2011. Competition among Hfq-binding small RNAs in *Escherichia coli*. *Mol Microbiol* **82:**1545–1562.

66. **Hussein R, Lim HN.** 2011. Disruption of small RNA signaling caused by competition for Hfq. *Proc Natl Acad Sci U S A* **108:**1110–1115.

67. **Beisel CL, Storz G.** 2010. Base pairing small RNAs and their roles in global regulatory networks. *FEMS Microbiol Rev* **34:**866–882.

68. **Nitzan M, Rehani R, Margalit H.** 2017. Integration of bacterial small RNAs in regulatory networks. *Annu Rev Biophys* **46:**131–148.

69. **Tu KC, Waters CM, Svenningsen SL, Bassler BL.** 2008. A small-RNA-mediated negative feedback loop controls quorum-sensing dynamics in *Vibrio harveyi*. *Mol Microbiol* **70:**896–907.

70. **Svenningsen SL, Waters CM, Bassler BL.** 2008. A negative feedback loop involving small RNAs accelerates *Vibrio cholerae*'s transition out of quorum-sensing mode. *Genes Dev* **22:**226–238.

71. **Feng L, Rutherford ST, Papenfort K, Bagert JD, van Kessel JC, Tirrell DA, Wingreen NS, Bassler BL.** 2015. A Qrr noncoding RNA deploys four different regulatory mechanisms to optimize quorum-sensing dynamics. *Cell* **160:**228–240.

72. **Beisel CL, Storz G.** 2011. The base-pairing RNA Spot 42 participates in a multioutput feedforward loop to help enact catabolite repression in *Escherichia coli*. *Mol Cell* **41:**286–297.

73. **Papenfort K, Espinosa E, Casadesús J, Vogel J.** 2015. Small RNA-based feedforward loop with AND-gate logic regulates extrachromosomal DNA transfer in *Salmonella*. *Proc Natl Acad Sci U S A* **112:**E4772–4781.

74. **Mandin P, Gottesman S.** 2009. A genetic approach for finding small RNAs regulators of genes of interest identifies RybC as regulating the DpiA/DpiB two-component system. *Mol Microbiol* **72:**551–565.

75. **De Lay N, Gottesman S.** 2009. The Crp-activated small noncoding regulatory RNA CyaR (RyeE) links nutritional status to group behavior. *J Bacteriol* **191:**461–476.

76. **Vogt SL, Evans AD, Guest RL, Raivio TL.** 2014. The Cpx envelope stress response regulates and is regulated by small noncoding RNAs. *J Bacteriol* **196:**4229–4238.

77. **Johansen J, Eriksen M, Kallipolitis B, Valentin-Hansen P.** 2008. Down-regulation of outer membrane proteins by noncoding RNAs: unraveling the cAMP-CRP- and σ^E-dependent CyaR-*ompX* regulatory case. *J Mol Biol* **383:**1–9.

78. **Chao Y, Papenfort K, Reinhardt R, Sharma CM, Vogel J.** 2012. An atlas of Hfq-bound transcripts reveals 3′ UTRs as a genomic reservoir of regulatory small RNAs. *EMBO J* **31:**4005–4019.

79. **Durand S, Storz G.** 2010. Reprogramming of anaerobic metabolism by the FnrS small RNA. *Mol Microbiol* **75:**1215–1231.

80. **Opdyke JA, Kang JG, Storz G.** 2004. GadY, a small-RNA regulator of acid response genes in *Escherichia coli*. *J Bacteriol* **186:**6698–6705.

81. **Urbanowski ML, Stauffer LT, Stauffer GV.** 2000. The *gcvB* gene encodes a small untranslated RNA involved in expression of the dipeptide and oligopeptide transport systems in *Escherichia coli*. *Mol Microbiol* **37:**856–868.

82. **Reichenbach B, Gopel Y, Gorke B.** 2009. Dual control by perfectly overlapping σ54- and σ70-promoters adjusts small RNA GlmY expression to different environmental signals. *Mol Microbiol* **74:**1054–1070.

83. Gopel Y, Luttmann D, Heroven AK, Reichenbach B, Dersch P, Gorke B. 2011. Common and divergent features in transcriptional control of the homologous small RNAs GlmY and GlmZ in *Enterobacteriaceae*. *Nucleic Acids Res* 39:1294–1309.

84. Pfeiffer V, Sittka A, Tomer R, Tedin K, Brinkmann V, Vogel J. 2007. A small non-coding RNA of the invasion gene island (SPI-1) represses outer membrane protein synthesis from the *Salmonella* core genome. *Mol Microbiol* 66:1174–1191.

85. Moon K, Gottesman S. 2009. A PhoQ/P-regulated small RNA regulates sensitivity of *Escherichia coli* to antimicrobial peptides. *Mol Microbiol* 74:1314–1330.

86. Udekwu KI, Wagner EG. 2007. Sigma E controls biogenesis of the antisense RNA MicA. *Nucleic Acids Res* 35:1279–1288.

87. Johansen J, Rasmussen AA, Overgaard M, Valentin-Hansen P. 2006. Conserved small non-coding RNAs that belong to the σ^E regulon: role in down-regulation of outer membrane proteins. *J Mol Biol* 364:1–8.

88. Figueroa-Bossi N, Lemire S, Maloriol D, Balbontin R, Casadesús J, Bossi L. 2006. Loss of Hfq activates the σ^E-dependent envelope stress response in *Salmonella enterica*. *Mol Microbiol* 62:838–852.

89. Delihas N, Forst S. 2001. *MicF*: an antisense RNA gene involved in response of *Escherichia coli* to global stress factors. *J Mol Biol* 313:1–12.

90. Corcoran CP, Podkaminski D, Papenfort K, Urban JH, Hinton JC, Vogel J. 2012. Superfolder GFP reporters validate diverse new mRNA targets of the classic porin regulator, MicF RNA. *Mol Microbiol* 84:428–445.

91. Peano C, Wolf J, Demol J, Rossi E, Petiti L, De Bellis G, Geiselmann J, Egli T, Lacour S, Landini P. 2015. Characterization of the *Escherichia coli* σ^S core regulon by Chromatin Immunoprecipitation-sequencing (ChIP-seq) analysis. *Sci Rep* 5:10469.

92. Lévi-Meyrueis C, Monteil V, Sismeiro O, Dillies MA, Monot M, Jagla B, Coppée JY, Dupuy B, Norel F. 2014. Expanding the RpoS/σ^S-network by RNA sequencing and identification of σ^S-controlled small RNAs in *Salmonella*. *PLoS One* 9:e96918.

93. Guillier M, Gottesman S. 2008. The 5′ end of two redundant sRNAs is involved in the regulation of multiple targets, including their own regulator. *Nucleic Acids Res* 36:6781–6794.

94. Altuvia S, Weinstein-Fischer D, Zhang A, Postow L, Storz G. 1997. A small stable RNA induced by oxidative stress: role as a pleiotropic regulator and antimutator. *Cell* 90:43–53.

95. Altuvia S, Zhang A, Argaman L, Tiwari A, Storz G. 1998. The *Escherichia coli* OxyS regulatory RNA represses *fhlA* translation by blocking ribosome binding. *EMBO J* 17:6069–6075.

96. Smirnov A, Wang C, Drewry LL, Vogel J. 2017. Molecular mechanism of mRNA repression in *trans* by a ProQ-dependent small RNA. *EMBO J* 36:1029–1045.

97. Thompson KM, Rhodius VA, Gottesman S. 2007. σ^E regulates and is regulated by a small RNA in *Escherichia coli*. *J Bacteriol* 189:4243–4256.

98. Massé E, Vanderpool CK, Gottesman S. 2005. Effect of RyhB small RNA on global iron use in *Escherichia coli*. *J Bacteriol* 187:6962–6971.

99. Vecerek B, Moll I, Bläsi U. 2007. Control of Fur synthesis by the non-coding RNA RyhB and iron-responsive decoding. *EMBO J* 26:965–975.

100. Fröhlich KS, Papenfort K, Berger AA, Vogel J. 2012. A conserved RpoS-dependent small RNA controls the synthesis of major porin OmpD. *Nucleic Acids Res* 40:3623–3640.

101. Vanderpool CK, Gottesman S. 2007. The novel transcription factor SgrR coordinates the response to glucose-phosphate stress. *J Bacteriol* 189:2238–2248.

102. Polayes DA, Rice PW, Garner MM, Dahlberg JE. 1988. Cyclic AMP-cyclic AMP receptor protein as a repressor of transcription of the *spf* gene of *Escherichia coli*. *J Bacteriol* 170:3110–3114.

103. Silva IJ, Ortega AD, Viegas SC, García-Del Portillo F, Arraiano CM. 2013. An RpoS-dependent sRNA regulates the expression of a chaperone involved in protein folding. *RNA* 19:1253–1265.

104. Gong H, Vu GP, Bai Y, Chan E, Wu R, Yang E, Liu F, Lu S. 2011. A *Salmonella* small non-coding RNA facilitates bacterial invasion and intracellular replication by modulating the expression of virulence factors. *PLoS Pathog* 7:e1002120.

105. Westermann AJ, Förstner KU, Amman F, Barquist L, Chao Y, Schulte LN, Müller L, Reinhardt R, Stadler PF, Vogel J. 2016. Dual RNA-seq unveils noncoding RNA functions in host-pathogen interactions. *Nature* 529:496–501.

106. Song T, Mika F, Lindmark B, Liu Z, Schild S, Bishop A, Zhu J, Camilli A, Johansson J, Vogel J, Wai SN. 2008. A new *Vibrio cholerae* sRNA modulates colonization and affects release of outer membrane vesicles. *Mol Microbiol* 70:100–111.

107. Richard AL, Withey JH, Beyhan S, Yildiz F, DiRita VJ. 2010. The *Vibrio cholerae* virulence regulatory cascade controls glucose uptake through activation of TarA, a small regulatory RNA. *Mol Microbiol* 78:1171–1181.

108. Bradley ES, Bodi K, Ismail AM, Camilli A. 2011. A genome-wide approach to discovery of small RNAs involved in regulation of virulence in *Vibrio cholerae*. *PLoS Pathog* 7:e1002126.

109. Davies BW, Bogard RW, Young TS, Mekalanos JJ. 2012. Coordinated regulation of accessory genetic elements produces cyclic di-nucleotides for *V. cholerae* virulence. *Cell* 149:358–370.

110. Papenfort K, Forstner KU, Cong JP, Sharma CM, Bassler BL. 2015. Differential RNA-seq of *Vibrio cholerae* identifies the VqmR small RNA as a regulator of biofilm formation. *Proc Natl Acad Sci U S A* 112:E766–E775.

111. Yamamoto S, Mitobe J, Ishikawa T, Wai SN, Ohnishi M, Watanabe H, Izumiya H. 2014. Regulation of natural competence by the orphan two-component system sensor kinase ChiS involves a non-canonical transmembrane regulator in *Vibrio cholerae*. *Mol Microbiol* 91:326–347.

112. **Dalia AB, Lazinski DW, Camilli A.** 2014. Identification of a membrane-bound transcriptional regulator that links chitin and natural competence in *Vibrio cholerae*. *mBio* 5:e01028-13.

113. **Yamamoto S, Izumiya H, Mitobe J, Morita M, Arakawa E, Ohnishi M, Watanabe H.** 2011. Identification of a chitin-induced small RNA that regulates translation of the *tfoX* gene, encoding a positive regulator of natural competence in *Vibrio cholerae*. *J Bacteriol* 193:1953–1965.

114. **Sonnleitner E, Gonzalez N, Sorger-Domenigg T, Heeb S, Richter AS, Backofen R, Williams P, Hüttenhofer A, Haas D, Bläsi U.** 2011. The small RNA PhrS stimulates synthesis of the *Pseudomonas aeruginosa* quinolone signal. *Mol Microbiol* 80:868–885.

115. **Wilderman PJ, Sowa NA, FitzGerald DJ, FitzGerald PC, Gottesman S, Ochsner UA, Vasil ML.** 2004. Identification of tandem duplicate regulatory small RNAs in *Pseudomonas aeruginosa* involved in iron homeostasis. *Proc Natl Acad Sci U S A* 101:9792–9797.

116. **Sonnleitner E, Haas D.** 2011. Small RNAs as regulators of primary and secondary metabolism in *Pseudomonas* species. *Appl Microbiol Biotechnol* 91:63–79.

117. **Wenner N, Maes A, Cotado-Sampayo M, Lapouge K.** 2014. NrsZ: a novel, processed, nitrogen-dependent, small non-coding RNA that regulates *Pseudomonas aeruginosa* PAO1 virulence. *Environ Microbiol* 16:1053–1068.

118. **Zhang YF, Han K, Chandler CE, Tjaden B, Ernst RK, Lory S.** 2017. Probing the sRNA regulatory landscape of *P. aeruginosa*: post-transcriptional control of determinants of pathogenicity and antibiotic susceptibility. *Mol Microbiol* 106:919–937.

119. **Faucher SP, Friedlander G, Livny J, Margalit H, Shuman HA.** 2010. *Legionella pneumophila* 6S RNA optimizes intracellular multiplication. *Proc Natl Acad Sci U S A* 107:7533–7538.

120. **Mank NN, Berghoff BA, Hermanns YN, Klug G.** 2012. Regulation of bacterial photosynthesis genes by the small noncoding RNA PcrZ. *Proc Natl Acad Sci U S A* 109:16306–16311.

121. **Janzon L, Löfdahl S, Arvidson S.** 1989. Identification and nucleotide sequence of the delta-lysin gene, *hld*, adjacent to the accessory gene regulator (*agr*) of *Staphylococcus aureus*. *Mol Gen Genet* 219:480–485.

122. **Novick RP, Ross HF, Projan SJ, Kornblum J, Kreiswirth B, Moghazeh S.** 1993. Synthesis of staphylococcal virulence factors is controlled by a regulatory RNA molecule. *EMBO J* 12:3967–3975.

123. **Boisset S, Geissmann T, Huntzinger E, Fechter P, Bendridi N, Possedko M, Chevalier C, Helfer AC, Benito Y, Jacquier A, Gaspin C, Vandenesch F, Romby P.** 2007. *Staphylococcus aureus* RNAIII coordinately represses the synthesis of virulence factors and the transcription regulator Rot by an antisense mechanism. *Genes Dev* 21:1353–1366.

124. **Gupta RK, Luong TT, Lee CY.** 2015. RNAIII of the *Staphylococcus aureus agr* system activates global regulator MgrA by stabilizing mRNA. *Proc Natl Acad Sci U S A* 112:14036–14041.

125. **Xue T, Zhang X, Sun H, Sun B.** 2014. ArtR, a novel sRNA of *Staphylococcus aureus*, regulates α-toxin expression by targeting the 5′ UTR of *sarT* mRNA. *Med Microbiol Immunol* 203:1–12.

126. **Durand S, Braun F, Lioliou E, Romilly C, Helfer AC, Kuhn L, Quittot N, Nicolas P, Romby P, Condon C.** 2015. A nitric oxide regulated small RNA controls expression of genes involved in redox homeostasis in *Bacillus subtilis*. *PLoS Genet* 11:e1004957.

127. **Durand S, Braun F, Helfer AC, Romby P, Condon C.** 2017. sRNA-mediated activation of gene expression by inhibition of 5′-3′ exonucleolytic mRNA degradation. *eLife* 6:e23602.

128. **Geissmann T, Chevalier C, Cros MJ, Boisset S, Fechter P, Noirot C, Schrenzel J, François P, Vandenesch F, Gaspin C, Romby P.** 2009. A search for small noncoding RNAs in *Staphylococcus aureus* reveals a conserved sequence motif for regulation. *Nucleic Acids Res* 37:7239–7257.

129. **Romilly C, Lays C, Tomasini A, Caldelari I, Benito Y, Hammann P, Geissmann T, Boisset S, Romby P, Vandenesch F.** 2014. A non-coding RNA promotes bacterial persistence and decreases virulence by regulating a regulator in *Staphylococcus aureus*. *PLoS Pathog* 10:e1003979.

130. **Nielsen JS, Christiansen MH, Bonde M, Gottschalk S, Frees D, Thomsen LE, Kallipolitis BH.** 2011. Searching for small σB-regulated genes in *Staphylococcus aureus*. *Arch Microbiol* 193:23–34.

131. **Kaito C, Saito Y, Ikuo M, Omae Y, Mao H, Nagano G, Fujiyuki T, Numata S, Han X, Obata K, Hasegawa S, Yamaguchi H, Inokuchi K, Ito T, Hiramatsu K, Sekimizu K.** 2013. Mobile genetic element SCC*mec*-encoded *psm-mec* RNA suppresses translation of *agrA* and attenuates MRSA virulence. *PLoS Pathog* 9:e1003269.

132. **Mauro T, Rouillon A, Felden B.** 2016. Insights into the regulation of small RNA expression: SarA represses the expression of two sRNAs in *Staphylococcus aureus*. *Nucleic Acids Res* 44:10186–10200.

133. **Gaballa A, Antelmann H, Aguilar C, Khakh SK, Song KB, Smaldone GT, Helmann JD.** 2008. The *Bacillus subtilis* iron-sparing response is mediated by a Fur-regulated small RNA and three small, basic proteins. *Proc Natl Acad Sci U S A* 105:11927–11932.

134. **Licht A, Preis S, Brantl S.** 2005. Implication of CcpN in the regulation of a novel untranslated RNA (SR1) in *Bacillus subtilis*. *Mol Microbiol* 58:189–206.

135. **Heidrich N, Moll I, Brantl S.** 2007. *In vitro* analysis of the interaction between the small RNA SR1 and its primary target *ahrC* mRNA. *Nucleic Acids Res* 35:4331–4346.

136. **Schmalisch M, Maiques E, Nikolov L, Camp AH, Chevreux B, Muffler A, Rodriguez S, Perkins J, Losick R.** 2010. Small genes under sporulation control in the *Bacillus subtilis* genome. *J Bacteriol* 192:5402–5412.

137. **Mars RA, Nicolas P, Ciccolini M, Reilman E, Reder A, Schaffer M, Mader U, Völker U, van Dijl JM, Denham EL.** 2015. Small regulatory RNA-induced growth rate heterogeneity of *Bacillus subtilis*. *PLoS Genet* 11:e1005046.

138. Marchais A, Duperrier S, Durand S, Gautheret D, Stragier P. 2011. CsfG, a sporulation-specific, small non-coding RNA highly conserved in endospore formers. *RNA Biol* 8:358–364.

139. Sievers S, Sternkopf Lillebaek EM, Jacobsen K, Lund A, Mollerup MS, Nielsen PK, Kallipolitis BH. 2014. A multicopy sRNA of *Listeria monocytogenes* regulates expression of the virulence adhesin LapB. *Nucleic Acids Res* 42:9383–9398.

140. Mollerup MS, Ross JA, Helfer AC, Meistrup K, Romby P, Kallipolitis BH. 2016. Two novel members of the LhrC family of small RNAs in *Listeria monocytogenes* with overlapping regulatory functions but distinctive expression profiles. *RNA Biol* 13:895–915.

141. Loh E, Dussurget O, Gripenland J, Vaitkevicius K, Tiensuu T, Mandin P, Repoila F, Buchrieser C, Cossart P, Johansson J. 2009. A *trans*-acting riboswitch controls expression of the virulence regulator PrfA in *Listeria monocytogenes*. *Cell* 139:770–779.

142. Pappesch R, Warnke P, Mikkat S, Normann J, Wisniewska-Kucper A, Huschka F, Wittmann M, Khani A, Schwengers O, Oehmcke-Hecht S, Hain T, Kreikemeyer B, Patenge N. 2017. The regulatory small RNA MarS supports virulence of *Streptococcus pyogenes*. *Sci Rep* 7:12241.

143. Acebo P, Martin-Galiano AJ, Navarro S, Zaballos A, Amblar M. 2012. Identification of 88 regulatory small RNAs in the TIGR4 strain of the human pathogen *Streptococcus pneumoniae*. *RNA* 18:530–546.

144. Halfmann A, Kovács M, Hakenbeck R, Brückner R. 2007. Identification of the genes directly controlled by the response regulator CiaR in *Streptococcus pneumoniae*: five out of 15 promoters drive expression of small non-coding RNAs. *Mol Microbiol* 66:110–126.

145. Shimizu T, Yaguchi H, Ohtani K, Banu S, Hayashi H. 2002. Clostridial VirR/VirS regulon involves a regulatory RNA molecule for expression of toxins. *Mol Microbiol* 43:257–265.

Regulating with RNA in Bacteria and Archaea
Edited by Gisela Storz and Kai Papenfort
© 2018 American Society for Microbiology, Washington, DC
doi:10.1128/microbiolspec.RWR-0032-2018

Dual-Function RNAs

27

Medha Raina,[1] Alisa King,[2] Colleen Bianco,[2] and Carin K. Vanderpool[2]

INTRODUCTION

Bacteria have evolved elaborate responses to sense, protect against, and help recovery from stressful fluctuations in environmental conditions. In the past decade, small regulatory RNAs (sRNAs) have emerged as important players in the posttranscriptional regulation of various stress responses. Advances in deep sequencing have led to the identification of hundreds of these sRNAs, which range from 50 to 350 nucleotides (nt) in length, thereby greatly increasing the numbers of known sRNAs (1). Usually, these sRNA regulators are thought to be noncoding and are generally presumed to act by modulating the stability and translation of mRNAs through short base-pairing interactions or by binding to and modulating the activities of RNA-binding proteins.

The majority of the base-pairing sRNAs prevent translation of their mRNA targets, at times leading to degradation of the sRNA-mRNA complex (2). For this category of sRNAs, base-pairing interactions occlude the ribosome binding site (RBS) of the mRNA and thus inhibit translation initiation (3, 4). There is also a small subset of sRNAs that stabilize transcripts that are otherwise prone to degradation by cellular RNases (5, 6). In addition, sRNAs that base-pair within the 5′ untranslated region (5′ UTR) of the mRNA can open up an inhibitory secondary structure that normally prevents translation of the mRNA (7). Irrespective of the mechanism of regulation, many sRNAs require the Sm-like RNA chaperone protein Hfq. Without Hfq, sRNAs tend to be unstable and unable to regulate their respective mRNA targets (8). The sRNAs also generally require Hfq to promote sRNA-mRNA binding and to remodel RNA secondary structures. Thus far, Hfq-dependent sRNAs and their mRNA targets have been characterized through genome-wide and bioinformatic searches and coimmunoprecipitation experiments with Hfq (9–11).

The majority of sRNAs do not contain an open reading frame (ORF) and are thus thought to be noncoding and function exclusively through sRNA-mRNA or sRNA-protein interactions. However, a few sRNAs contain short ORFs, and in a small number of cases, translation of these ORFs has been shown. Of these, there are fewer still that have an experimentally demonstrated function (12, 13). These sRNAs encoding small proteins form a special class of sRNAs refered to as "dual-function sRNAs." The small proteins tended to be overlooked due to challenges related to their annotation and biochemical detection. However, recent work suggests that the prevalence of dual-function sRNAs may be greater than currently appreciated. Computational analyses of the genomes of 14 phylogenetically diverse bacteria predicted many sRNAs encoding proteins between 10 and 50 amino acids (aa). These analyses took into account sequence features and comparative genomics to quantify the sRNA ORFs under natural selection to maintain protein-coding function (14). A second major problem with respect to studies of small proteins is barriers to finding functions of this class of proteins, particularly given that tags commonly used to characterize larger proteins can have adverse effects on proteins, which may be smaller than the tag itself.

The characterized dual-function sRNAs that act by sRNA-mRNA interactions and encode small proteins will be the focus of this review. Other classes of multifunctional RNAs will also be considered briefly. One class of dual-function RNAs comprises known mRNAs that contain sRNAs derived from the 3′ UTRs of these mRNAs either by processing or by transcription from an independent promoter located within the 3′ end of the ORF (15). Another variety of dual-function sRNAs carries out regulation of mRNA targets via base-pairing but can also bind proteins and regulate the activities

[1]Division of Molecular and Cellular Biology, Eunice Kennedy Shriver National Institute of Child Health and Human Development, Bethesda, MD 20892; [2]Department of Microbiology, University of Illinois, Urbana, IL 61801.

of these proteins (16). There is also a growing list of examples of tRNA fragments derived from processing of pre-tRNA transcripts by RNases that have been shown to base-pair with sRNAs or other RNA targets to carry out a variety of regulatory effects (17).

BASE-PAIRING sRNAs THAT ENCODE CHARACTERIZED SMALL PROTEINS

Despite the challenges in identifying and characterizing the function of small proteins encoded by known sRNAs, to date, five sRNAs have been defined as dual-function RNAs with characterized base-pairing and protein-coding functions. These include SgrS from enteric species (18, 19), RNAIII and Psm-mec from *Staphylococcus aureus* (20), Pel RNA from *Streptococcus pyogenes* (21), and SR1 from *Bacillus subtilis* (22). All of these and other potential dual-function RNAs are involved in myriad physiological responses including quorum sensing, virulence regulation, and metabolic regulation. Below we provide an overview of the

individual dual-function sRNAs, discussing in detail what is known about their base-pairing function and the function of the encoded protein, as well as the interplay between the two different functions, thereby providing insight into how a dual-function sRNA coordinates the two different activities to carry out its role in the cell.

Enterobacterial SgrS

SgrS has been well characterized in *Escherichia coli* and *Salmonella* and is currently the only known dual-function sRNA in Gram-negative bacteria (Fig. 1). Orthologs of SgrS are also found in numerous gammaproteobacteria such as *Shigella* spp., *Klebsiella pneumoniae*, *Erwinia amylovora*, *Citrobacter koseri*, and *Yersinia pestis* (23). The 227-nt SgrS sRNA was first discovered in a computational screen for sRNAs in *E. coli* (24) and later found to play an important role in mediating the cellular response to glucose-phosphate stress (19). Bacteria use specific phosphoenolpyruvate phosphotransferase systems (PTS) to take up sugars, such as glucose and mannose,

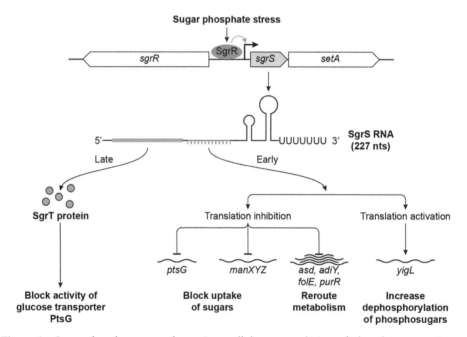

Figure 1 Sugar-phosphate stress due to intracellular accumulation of phosphosugars triggers expression of the transcription factor SgrR. SgrR, in turn, induces transcription of the 227-nt sRNA SgrS, which also encodes a small, 43-aa protein, SgrT (blue). The other features of this sRNA include the base-pairing region (red) and the Hfq-binding region [poly(U) tail]. To relieve the sugar-phosphate stress, SgrS represses translation of mRNAs coding for sugar transporters (PtsG and ManXYZ) and other mRNAs involved in various metabolic pathways (Asd, AdiY, FolE, and PurR) to help restore metabolic homeostasis during stress conditions. SgrS also activates translation of a phosphatase (YigL) that dephosphorylates the phosphosugars for export out of the cell. SgrT, meanwhile, is expressed from SgrS later and inhibits the activity of the glucose transporter PtsG; thereby, both the sRNA and the encoded small protein act together in the same pathway to combat sugar-phosphate stress.

and then phosphorylate them (see chapter 14). Glucose-phosphate stress occurs when sugar transport and metabolism become uncoupled. This can occur in certain mutant strains (25, 26) or in wild-type strains that take up the nonmetabolizable sugar analogs α-methylglucoside (αMG) and 2-deoxyglucose (2DG) (3, 19). Accumulation of sugar-phosphates or their analogs and depletion of other glycolytic intermediates results in growth inhibition (27). This stress condition is sensed by the transcription factor SgrR and allows for *sgrS* transcription. Both SgrS and SgrR are required for growth under glucose-phosphate stress (28).

SgrS regulation of three different mRNAs reduces the amount of sugar-phosphate accumulation in the cell by reducing synthesis of sugar transporters and enhancing dephosphorylation of the sugar-phosphates. The sugar analogs αMG and 2DG are taken into the cell and then phosphorylated by PTS proteins PtsG (EIICB^{Glc}) and ManY (EIIC^{Man}), respectively. Upon accumulation of phosphorylated αMG and 2DG in the cytoplasm, SgrS is expressed and represses translation of the two mRNA targets, *ptsG* (19) and *manXYZ* (29). SgrS base-pairs with *ptsG* mRNA through a short region of complementarity that includes the RBS of *ptsG* mRNA. This base-pairing interaction prevents translation of the encoded PtsG protein and allows for subsequent RNase E-dependent degradation of *ptsG* mRNA (19, 30, 31). SgrS again uses direct base-pairing to repress the *manXYZ* polycistronic mRNA. Here, SgrS is able to bind *manXYZ* mRNA at two distinct sites, one within the *manX* ORF and the other in the untranslated region between *manX* and *manY*. Similar to the *ptsG* mRNA, SgrS binding impedes *manXYZ* mRNA translation and facilitates mRNA degradation by RNase E (29, 32). SgrS activates a third mRNA target, *yigL*, which encodes a haloacid dehalogenase-like phosphatase that dephosphorylates the phosphorylated sugars prior to their efflux. The *yigL* gene is in an operon with an upstream gene, *pldB*, and the dicistronic *pldB-yigL* mRNA is processed by RNase E, which results in the '*pldB-yigL* mRNA that is susceptible to further degradation. SgrS stabilizes '*pldB-yigL* mRNA by base-pairing to and occluding a specific RNase E cleavage site upstream of the *yigL* coding region to prevent further mRNA degradation (6).

Very recently, SgrS was shown to regulate four more mRNAs: *asd*, *adiY*, *folE*, and *purR* (33). In typical sRNA-mediated repression, SgrS base-pairs within the 5′ UTR of *adiY* and *folE*, which encode arginine decarboxylase gene activator and GTP cyclohydrolase I, respectively, to occlude the RBS and prevent translation initiation of these mRNAs. Similar to *manXYZ*, SgrS

binds *asd* mRNA (encoding aspartate semialdehyde dehydrogenase) at two distinct sites, one that directly occludes the RBS and the other in the *asd* ORF. SgrS binding to both sites is required for full translational repression. Lastly, SgrS binds within the *purR* (encoding a repressor of purine biosynthesis) coding sequence and recruits the RNA chaperone Hfq, where it can interfere with ribosome binding and directly repress *purR* translation. Interestingly, Hfq alone can repress *purR* translation, while SgrS alone has a very modest effect. These enzymes and transcription factors belong to a diverse set of metabolic pathways, and while their specific role in glucose-phosphate stress has not been defined, their repression by SgrS may help restore metabolic homeostasis during stress conditions.

Along with functioning as a base-pairing RNA, the SgrS sRNA encodes a 43-aa protein, SgrT, which also functions in the glucose-phosphate stress response (18). The *E. coli sgrT* coding sequence is located at the 5′ end of SgrS (nucleotides 22 through 153), with the middle region responsible for base-pairing with mRNAs (approximately nucleotides 167 through 187) and the 3′ end responsible for binding Hfq (18). Ectopic overexpression of *sgrS* alleles that possess only base-pairing activity or only produced SgrT was sufficient to rescue cells from glucose-phosphate stress, indicating that the sRNA and small protein have redundant functions. SgrT by itself was found to have no effect on *ptsG* mRNA stability or translation, indicating that it must affect recovery from glucose-phosphate stress using a different mechanism (4, 18). In the years following the discovery of SgrT, the mechanism by which it allowed recovery from glucose-phosphate stress was unknown. However, it has now been shown that SgrT acts to specifically inhibit the transport activity of the major glucose permease PtsG (34). The primary transporter of glucose is the glucose-PTS complex, comprising two subunits: the transmembrane EIICB^{Glc} protein, which is a glucose permease and is encoded by *ptsG*, and the cytoplasmic EIIA^{Glc} protein, which functions as an intermediate phosphotransfer protein and is encoded by *crr* (35). Super-resolution microscopy, which showed colocalization of SgrT and PtsG in a PtsG-dependent manner, indicated that SgrT binds the EIIC^{Glc} domain of PtsG to inhibit further glucose transport (34). This model was also supported by the observation of strong SgrT-mediated growth inhibition when cells were grown in glucose media, which requires the IIC domain of PtsG for glucose transport. The inhibition of the EIIC^{Glc} domain of PtsG by SgrT possibly leads to accumulation of phosphorylated EIIA^{Glc}, which can no longer bind and block other transport proteins (relieving inducer exclusion). Thus, together, SgrS and SgrT block further glucose-6-phosphate accumulation, diminish intracellular glucose-6-phosphate levels, and promote the utilization of alternative carbon sources. It is interesting to note that though the base-pairing activity of SgrS represses the synthesis of both PtsG and ManXYZ, SgrT had no effect on ManXYZ transport activity (34). Overall, these new results support the model in which the base-pairing activity of SgrS acts to inhibit new PtsG synthesis while SgrT inhibits the activity of preexisting PtsG.

Little is known about how the base-pairing function of SgrS and translation of *sgrT* are coordinated. The 5′ end of SgrS contains the ORF for SgrT, and the nucleotides used for the base-pairing function of SgrS are 15 nt downstream of the SgrT stop codon. Thus, a ribosome located at the *sgrT* stop codon would block the important base-pairing nucleotides for SgrS. This block, as well as the fact that sRNA binding to mRNA targets leads to coupled SgrS-mRNA degradation, led to the hypothesis that the translation and base-pairing functions of SgrS are mutually exclusive. Thus, a given SgrS molecule could serve as a substrate for *sgrT* translation or as an sRNA with base-pairing ability, but not both simultaneously. In *Salmonella*, mutations that impair *sgrT* translation have no effect on the ability of SgrS to regulate mRNA targets via base-pairing (4). However, mutations that impaired base-pairing interactions of SgrS lead to increased SgrT production. This is consistent with the model that SgrS molecules engaged in base-pairing to mRNA are unavailable for translation, and thus removing base-pairing ability would increase the pool of SgrS molecules available for translation of *sgrT*. Lastly, it was shown that SgrT and SgrS base-pairing functions act at different times during glucose-phosphate stress. SgrS RNA is produced rapidly in response to initial stress and the base-pairing function acts immediately, whereas SgrT is not detected until 30 min following the onset of stress and is required to inhibit activity of preexisting PtsG (Fig. 1).

S. aureus RNAIII

RNAIII, the first dual-function sRNA identified, is an important virulence regulator in the human pathogen *S. aureus* (36). Specifically, RNAIII is the effector of the staphylococcal accessory gene regulator (*agr*) quorum-sensing system, which has been assigned a central role in the pathogenesis of *S. aureus* (Fig. 2) (see also chapter 18). The *agr* system decreases the expression of several cell surface proteins and increases the expression of many secreted virulence factors in the transition from late exponential growth to stationary phase *in vitro* (37, 38). The *agr* locus comprises two operons, P1 and

P2, encoding transcripts RNAII and RNAIII, respectively. The RNAII transcript produces four proteins—AgrB, AgrD, AgrC, and AgrA—that make up the *agr* sensing mechanism. AgrD and AgrB constitute the cell density-sensing cassette involved in producing the autoinducing peptide (AIP). AgrD encodes pro-AIP, and the transmembrane protein AgrB processes pro-AIP to AIP, transporting it to the external cellular space. In late exponential phase, upon reaching critical concentration, AIP then activates the two-component sensory transduction system comprising AgrC and AgrA. AIP binds to AgrC, a kinase embedded within the membrane, which in turn phosphorylates the DNA-binding regulator AgrA, which is then responsible for the increased transcription of RNAII and RNAIII (39). RNAIII produced at the end of exponential phase is responsible for repressing the synthesis of early virulence genes and surface proteins (which are needed during early infection) and increasing the production of secreted factors (which are necessary for late infection) (36). Nearly all *S. aureus* clinical isolates from acute infections produce RNAIII, highlighting its importance as a central virulence regulator.

RNAIII is 514 nt long and has a complex secondary structure composed of 14 stem-loop (SL) motifs referred to, from 5′ to 3′, as SL1 through SL14 (40). To date, RNAIII is known to be directly involved in base-pairing-dependent regulation of 12 mRNA targets, including 2 transcription factors. Generally, stem-loops in the 5′ UTR of RNAIII are involved in target activation while stem-loops in the 3′ region of RNAIII are involved in target repression; however, there are exceptions. The 3′ end of RNAIII is more conserved among different *S. aureus* isolates compared to other regions of the sRNA (41). This 3′ end contains several CU-rich domains present in apical loops and unpaired regions that were found to base-pair with the RBS of several target mRNAs. This base-pairing blocks ribosome binding, prevents translation initiation, and in some cases facilitates the subsequent degradation of the mRNA by RNase III. Using this mechanism, RNAIII represses synthesis of multiple targets, all of which are involved in the early stages of infection. These targets include surface protein A (*spa*), coagulase (*coa*), fibrinogen-binding protein (SA1000), homologs of staphylococcal secretory antigen SsaA (*sa2353* and *sa2093*), immunoglobulin-binding protein (*sbi*), lipoteichoic acid synthase (*ltaS*), and the major cell wall autolysin (*lytM*). In the case of *spa* repression, annealing of RNAIII to *spa* mRNA can inhibit the formation of the translation initiation complex, but recruitment of the double-strand-specific endoribonuclease III (RNase III) by RNAIII is

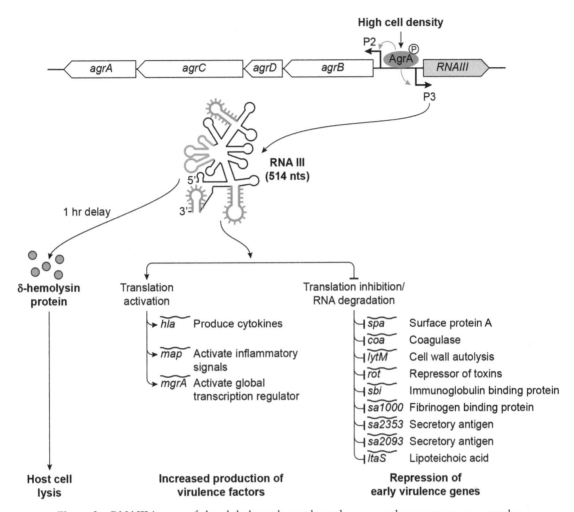

Figure 2 RNAIII is part of the global regulatory locus known as the accessory gene regulator (*agr*) locus, which encodes the components of an autoregulatory quorum-sensing system. The *agr* locus consists of two divergent transcripts, RNAII and RNAIII, which initiate from promoters P2 and P3, respectively. Increases in cell density lead to phosphorylation and activation of the DNA-binding response regulator AgrA. Phosphorylated AgrA in turn activates transcription from the P2 and P3 promoters, P3 activation leading to expression of RNAIII, the major effector molecule of the *agr* response. The secondary structure of the 514-nt RNAIII consists of 14 stem-loop structures with multiple base-pairing regions (red). RNAIII encodes a 26-aa δ-hemolysin protein (blue, *hld*) but also acts as a posttranscriptional regulator of several mRNAs, most of which impact virulence. The RNA activates expression of Map, α-hemolysin, and MgrA proteins by either promoting a more open secondary structure surrounding the RBS by base-pairing in the case of *map* and *hla* mRNAs or by stabilizing the RNA in the case of *mgrA*. RNAIII is also involved in translation inhibition and RNA degradation of various mRNAs involved in the early stages of infection.

essential to degrade the mRNA and completely repress *spa* translation (42). RNAIII also represses another important target, *rot* mRNA, which encodes the transcription factor Rot (repressor of toxins) (43). Through its regulation of Rot, RNAIII is able to indirectly control the transcription of many secondary targets.

RNAIII activates three mRNA targets: α-hemolysin (*hla*), extracellular adherence Map protein (*map*), and

the global transcription regulator MgrA (*mgrA*). In the absence of RNAIII, translation of *hla* is prevented by a stable intrinsic RNA hairpin that sequesters the RBS. Base-pairing between the 5′ end of RNAIII (SL2 and SL3) and the *hla* 5′ UTR releases the hairpin structure to allow ribosome binding and *hla* translation (44). A similar mechanism is predicted for activation of *map* translation after interaction with SL4 of RNAIII (45).

In contrast to *hla* and *map*, which interact with the 5′ end of RNAIII, *mgrA* mRNA interacts with both the 5′ end (SL2) and 3′ end (SL13) of RNAIII (46). MgrA is a global transcriptional regulator that controls >350 genes involved in virulence, antibiotic resistance, and biofilm formation (47–51). MgrA is transcribed from two promoters, both of which result in transcripts with unusually long 5′ UTRs. RNAIII base-pairs within the 5′ UTR of only the longer transcript to stabilize the mRNA against degradation by an unknown RNase and allows for increased production of MgrA. MgrA is the second master global regulator through which RNAIII can indirectly control the expression of multiple secondary targets, which significantly increases the influence of RNAIII (46).

RNAIII also contains an ORF that encodes the 26-aa cytotoxic peptide δ-hemolysin (*hld*), which targets host cell membranes and causes cell lysis (52). In contrast to other hemolytic peptides, δ-hemolysin is not active against bacteria. The ORF encoding δ-hemolysin is located toward the 5′ end of RNAIII and encompasses SL3, SL4, and SL5. Because δ-hemolysin synthesis is controlled by the levels of RNAIII, δ-hemolysin is produced in the late exponential phase. Intriguingly, the translation of *hld* is delayed by 1 h after RNAIII transcription (53). The 5′ and 3′ ends of RNAIII are in close proximity, suggesting that this delay may be due to intramolecular interactions of the 3′ end of RNAIII with the RBS of *hld*, resulting in the RBS of *hld* being occluded in a secondary structure (40, 53). Deletion of the 3′ end of RNAIII eliminates the delay between RNAIII production and the appearance of δ-hemolysin, again suggesting the presence of a translation-inhibitory structure between the 3′ end of RNAIII and the *hld* RBS (53). The factors mediating the conformational change in RNAIII secondary structure to allow δ-hemolysin production are not currently known.

S. aureus Psm-mec

The *psm-mec* gene is encoded on staphylococcal cassette chromosome (SCC*mec*), the mobile genetic element that confers antibiotic resistance to methicillin-resistant *S. aureus* (MRSA). The importance of the *psm-mec* RNA first came to light when researchers investigated why community-acquired (CA) MRSA, which infects otherwise healthy people outside of the hospital, was more virulent and produced larger amounts of exotoxins than hospital-associated (HA) MRSA, which infects immunocompromised patients in hospitals. They determined that the SCC*mec* of CA-MRSA did not contain the *psm-mec* locus that exists in the HA-MRSA SCC*mec* (20). They also found that the *psm-mec* transcript

reduced the expression of phenol-soluble modulin α (PSMα), a cytolytic toxin of *S. aureus*, resulting in decreased extracellular toxin production (20). However, other effects of the *psm-mec* locus in *S. aureus* on biofilm formation, cell spreading, and the expression of PSMα virulence factors are highly strain dependent.

The *psm-mec* sRNA is about 143 to 157 nt in length, with the *agrA* gene being the only known target encoding a positive transcription factor for the *agr* quorum-sensing system (Fig. 3) (54, 55). *psm-mec* RNA represses translation of the *agrA* transcript by directly binding within the *agrA* coding region (~200 nt downstream of the start codon), which results in decreased extracellular toxin production, specifically PSMα. The exact mechanism by which *psm-mec* RNA base-pairing causes *agrA* translation inhibition is still unknown. Base-pairing with Psm-mec decreases the stability of *agrA* mRNA in an RNase III-dependent manner; however, *psm-mec* RNA base-pairing can also decrease *agrA* translation independently of this decrease in *agrA* mRNA stability and RNase III (55). The transcription factor AgrA activates the transcription of multiple genes, including *rnaIII* and *psmα*, so *psm-mec* RNA-mediated regulation of *agrA* translation may indirectly control the transcription of many secondary targets. Lastly, the regulatory effects of the *psm-mec* RNA are highly strain dependent for reasons that are not yet well understood (55, 56). For example, deletion of *psm-mec* resulted in an increase of AgrA compared to the parental strain in only 14 out of 18 isolates (55).

psm-mec RNA also encodes a 22-aa PSM named PSM-mec (57), with the PSM-mec ORF making up most of the transcript. This protein is expressed with an N-terminal formyl-methionine and secreted without a signal peptide (58). This small protein is a staphylococcal cytolytic toxin that plays an important role in *S. aureus* infection and immune evasion. The fact that PSM-mec is encoded on the SCC*mec* mobile genetic element makes it, to date, the only known PSM that is not encoded on the staphylococcal core genome (57). Regulation of PSM-mec production is not well understood, but it has been observed that PSM-mec, like all other PSMs, is positively regulated by the Agr quorum-sensing system (54, 56). In contrast to all other known PSMs, which completely lack cysteine, PSM-mec contains one cysteine residue. This cysteine residue may be important for PSM-mec secondary structure, as it is necessary for proinflammatory and cytolytic activities of PSM-mec (57). PSM-mec expression varies widely among MRSA isolates, which may be why the phenotypes associated with the protein, such as adhesion to surfaces and biofilm formation, also differ among isolates (56, 57).

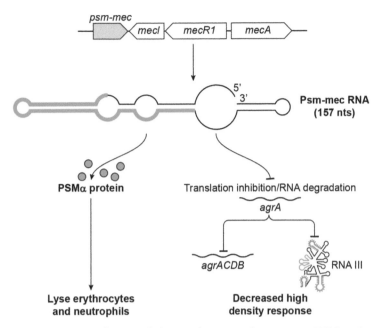

Figure 3 Psm-mec is located on staphylococcal cassette chromosome (SCC*mec*), next to the *mecI*/*mecR1*/*mecA* genes, which confer methicillin resistance and its regulation. The 143- to 157-nt sRNA also encodes PSM-mec, a 22-aa cytolytic toxin with the ORF making up most of the transcript (blue). The protein plays a role in *S. aureus* infection and immune evasion, while the sRNA represses the translation of *agrA* mRNA by inhibiting translation and affecting the stability of the mRNA.

B. subtilis SR1

SR1, the first dual-function sRNA discovered in *B. subtilis*, was identified via a bioinformatics approach that searched for sRNAs in intergenic regions of the *B. subtilis* chromosome. The 205-nt sRNA is expressed under gluconeogenic and repressed under glycolytic conditions, mainly by CcpN but also to a minor extent by CcpA; both regulators are involved in carbon catabolite repression (Fig. 4) (22, 59–61).

SR1 transcription is induced by L-arginine and its degradation product L-ornithine through an unknown mechanism and is involved in regulation of arginine catabolism by base-pairing with its target, the *ahrC* mRNA (62). AhrC is the transcriptional activator of the *rocABC* and *rocDEF* arginine catabolic operons. SR1 base-pairs within the coding region of *ahrC* mRNA potentially using seven complementary regions that make up the 3′ half of SR1 and the central and 3′ end of *ahrC* mRNA (62, 63). This base-pairing interaction induces structural changes ~20 nt downstream of the *ahrC* start codon, which inhibits translation initiation by preventing binding of the ribosomal 30S subunit. Decreased AhrC consequently leads to decreased expression of the *rocABC* and *rocDEF* mRNA. Consistent with the regulation, deletion of SR1 causes increased levels of the arginine catabolism proteins RocA, RocD,

and RocF (13, 62). The regulation of *ahrC* by SR1 was found to not require Hfq, although the Hfq is able to bind both SR1 and *ahrC* mRNA. Instead, Hfq is required for proper *ahrC* translation possibly by opening secondary structures that otherwise inhibit binding of the 30S initiation complex. This model is supported by the identification of an Hfq-binding site (5′ AAAUA) immediately upstream of the *ahrC* RBS and the observation that the *ahrC* mRNA is not translated in an *hfq* knockout strain.

In addition to its base-pairing activity, SR1 encodes a 39-aa protein called SR1P. Both functions of SR1 are highly conserved, as SR1 and SR1P homologs with high structural identity have been identified in 23 species of the *Bacillales* order (13, 64). During the search for additional targets of SR1, *gapA* mRNA, encoding one of the two glyceraldehyde-3P-dehydrogenases of *B. subtilis*, was found to be increased in the presence of SR1. Moreover, it was observed that *gapA* mRNA is rapidly degraded in the absence of SR1, indicating that *gapA* could be a potential target of SR1. However, it was also observed that SR1 did not affect transcription or translation of the *gapA* operon. Instead, it was shown that there is a direct interaction between SR1P and GapA and that this interaction inhibited the degradation of *gapA* operon mRNAs by an unknown mecha-

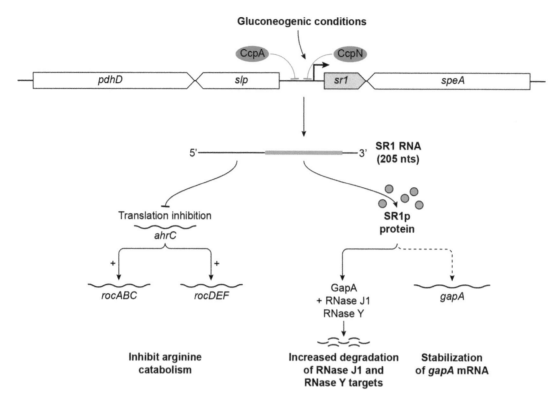

Figure 4 The SR1 gene is encoded between *pdhD* and *speA*. Its transcription is repressed by CcpA and CcpN under glycolytic conditions. The 205-nt sRNA expressed under gluconeogenic conditions and in the presence of L-arginine also encodes a small, 39-aa protein, SR1P (blue). The ORF and the base-pairing region overlap on this sRNA. In the presence of arginine, SR1 represses translation of the *ahrC* mRNA, the transcriptional activator of two arginine catabolic operons, *rocABC* and *rocDEF*. The small protein SR1P plays a role in gluconeogenic conditions by binding to GapA and stabilizing the *gapA* operon mRNA from degradation by an unknown mechanism. It also binds RNase J1 and enhances its activity. Thus, the activities of the small protein and base-pairing RNA affect different pathways.

nism (65). In 2016, the function of this interaction was determined. In *B. subtilis*, GapA plays a role in RNA degradation by binding RNase J1 and the main endonuclease RNase Y (66). SR1P binds to GapA and promotes GapA binding of RNase J1, which enhances RNase J1 activity (66). Since SR1P is only expressed under gluconeogenic conditions where the metabolic activity of GapA is not needed, SR1P modulation of the GapA ability to recruit RNases links RNA degradation to the metabolic state of the cell. Given that the SR1 RNA affects arginine catabolism and SR1P affects RNA stability, SR1 is an example of a dual-function sRNA, where the regulatory RNA and the small protein have functions in very different pathways.

S. pyogenes Pel RNA

Pel/SagA (pleiotropic effect locus/streptolysin-associated gene A) RNA was one of the first sRNAs studied in *S.*

pyogenes. Pel is a 469-nt sRNA expressed from the first gene of the *sagABCDEFGHI* operon, involved in the processing and export of streptolysin S in a growth phase-dependent manner (Fig. 5) (21, 67). Deletion of *pel* sRNA resulted in increased transcription of various virulence factor genes, such as those coding for streptococcal inhibitor of complement (*sic*), NAD-glycohydrolase (*nga*), and M-protein (*emm*), and is associated with delayed maturation of the cysteine protease SpeB (21). However, the mechanism as to how Pel transcriptionally or posttranscriptionally regulates these targets remains elusive. Due to the observed regulation at the transcriptional level, Pel RNA conceivably could base-pair with the mRNA of a global transcriptional regulator; however, this has yet to be reported. Moreover, the effects of the Pel RNA seem to be strain specific, as *pel* deletion in four M1T1 *S. pyogenes* isolates had no effect on the transcription of the previously identified targets (68).

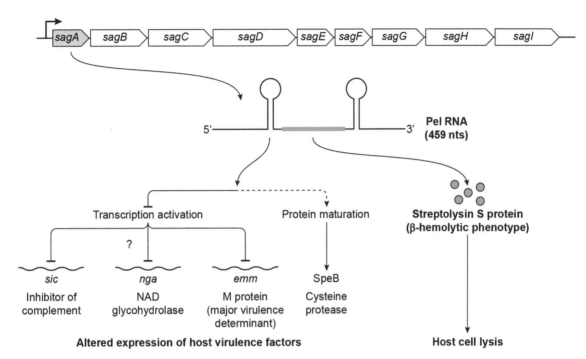

Figure 5 Pel/SagA sRNA is expressed from the pleiotropic effect locus of *S. pyogenes*, comprising the *sagABCEDEFGHI* operon. This 459-nt sRNA also encodes a 53-aa protein called streptolysin S (purple). Pel sRNA activates transcription of various mRNAs coding for different virulence factors, like Sic, Nga, and M protein, by an unknown mechanism. The sRNA also modulates maturation of cysteine protease SpeB.

Pel has much similarity to RNAIII, not only with respect to its length but also because base-pairing and mRNA functions are very similar. Like RNAIII, Pel encodes a hemolytic peptide, the 53-aa streptolysin S, a virulence factor responsible for the β-hemolytic phenotype of *S. pyogenes* (69). The protein is synthesized as a prepropeptide with a Gly-Gly proteolytic cleavage site that has been predicted to release a 30-aa propeptide from the 23-aa leader sequence. The remaining genes in the operon are required for efficient posttranslational modification and export of streptolysin S peptide (70).

DUAL-FUNCTION sRNAs WITH UNCHARACTERIZED PEPTIDE FUNCTIONS

Other sRNAs with established base-pairing functions could also encode small proteins. However, translation of the protein has only been observed for one: PhrS from *Pseudomonas aeruginosa*. As an sRNA, PhrS, whose expression requires the oxygen-responsive regulator ANR, activates translation of the quorum-sensing and virulence regulator PqsR (71). PhrS binds in the 5′ UTR of an ORF upstream of the *pqsR* gene, whose translation is coupled to that of *pqsR* (71). The physiological relevance of the 37-aa protein encoded by PhrS

is currently unknown. Interestingly, the ORF in *phrS* is more conserved than the sRNA sequence (71). Other potential dual-function sRNAs include VR-RNA from *Clostridium perfringens*, which controls a variety of different mRNAs resulting in activation of virulence and contains a 72-aa ORF denoted *hyp7*, the translation of which has not been reported (72). The sRNA RivX in *S. pyogenes* controls transcriptional activation of the streptococcal virulence regulator Mga and also encodes a 32-aa protein (73). The *Streptomyces coelicolor* sRNA scr5239 inhibits translation of the *dagA* (encodes agarase) and *metE* (encodes methionine synthase) mRNAs and contains a 33-codon ORF (74, 75). Another potential dual-function sRNA in Gram-negative bacteria is the sRNA RSs0019 in *Rhodobacter sphaeroides*. RSs0019 is produced in response to photooxidative stress; however, no target mRNAs have been identified and translation of the 50-codon ORF has not been observed (76–78).

OTHER TYPES OF MULTIFUNCTIONAL sRNAs/mRNAs

Although the sRNAs discussed thus far in this review function dually as both sRNA and mRNA, there are a few other sRNAs with multiple mechanisms of action. One example is the McaS sRNA of *E. coli*, which represses synthesis of CsgD, the master regulator of curli biogenesis, and activates synthesis of FlhD, the master regulator of flagellar biogenesis as a typical base-pairing sRNA (16, 79). However, McaS also binds the global RNA-binding protein regulator CsrA, which negatively regulates *pga* translation. Through McaS sequestration of CsrA, *pga* expression increases, leading to heightened production of poly-β-1,6-*N*-acetyl-D-glucosamine (PGA) and increased biofilm formation (see also chapter 19). As such, McaS is a unique sRNA with two different functions: base-pairing and protein titration. Other sRNAs, such as GadY, Spot 42, GcvB, and MicL, have been shown to bind CsrA by UV cross-linking and immunoprecipitation (CLIP-seq) analysis. These sRNAs have primary base-pairing functions, and hence binding to CsrA could indicate that, like McaS, these sRNAs might also be involved in the titration of CsrA (80).

Another example of a multifunctional sRNA is the Gifsy-1 prophage-encoded IsrK sRNA of *Salmonella*. Encoded on a *Salmonella* pathogenicity island, the IsrK RNA exists in two isoforms: the transcriptionally inactive *isrK-orf45-anrP* long mRNA transcript and a shorter IsrK transcript (81, see also chapter 29). Normally, the secondary structure in the longer isoform silences translation. However, the short IsrK isoform binds to the translationally inactive long isoform, allowing *anrP* translation and therefore production of the antirepressor AnrP. As a result, the lysogenic phage repressor is inactivated and expression of the antiterminator protein AntQ increased. This interference with transcription termination ultimately leads to bacterial growth arrest followed by cell death. Thus, IsrK exists as two isoforms with separate functions.

Initially, it was assumed that sRNAs are encoded by independent genes; however, transcriptome sequencing approaches have revealed that there are large numbers of sRNAs that originate from either the 5′ or 3′ end of various mRNAs. The first evidence of 5′- and 3′-derived sRNAs came from cloning-based searches for sRNA (82, 83), but the first hints that these mRNA fragments could have functions came from studies of the *S*-adenosylmethionine (SAM) riboswitch of *Listeria monocytogenes* (84). This riboswitch, present in the 5′ UTR of genes involved in methionine and cysteine metabolism, regulates expression of the downstream genes by binding to SAM. Binding of SAM to the riboswitch results in premature termination of the transcript, leading to repression of the downstream genes. However, this truncated transcript (SreA) can moonlight as an sRNA and regulate the expression of the *prfA* transcript in *trans* by base-pairing with the 5′ UTR and affecting translation and stability (84). Other characterized 5′-UTR sRNAs include the 5′ UTR of *irvA* mRNA in *Streptococcus mutans*, which modulates the synthesis of a critical surface-exposed lectin (*gbpC*) that serves as an adhesin for biofilm development. The *irvA* 5′ UTR base-pairs with *gbpC* mRNA and prevents its degradation by an RNase J2-mediated pathway (85). Another example is the sRNA derived from the 5′ UTR of the IS200 transposase mRNA (*tnpA*), which base-pairs with and represses the mRNA encoding a transcriptional regulator (*invF*) required for the expression of several genes encoding type III secretion system SPI1 effector proteins (86).

Currently, there are two classifications of 3′ UTR-derived sRNAs: type I, in which the sRNA is transcribed from a promoter embedded inside the ORF; and type II, in which the sRNA is derived from processing of the mRNA. For both types, the sRNAs share the terminator with the parental mRNA (15). *E. coli* MicL is an example of type I 3′-UTR-derived sRNA, which is transcribed from a promoter located within the coding sequence of the *cutC* gene and inhibits the synthesis of lipoprotein Lpp, thereby reducing envelope stress (87). Another example of this type of sRNA is DapZ of *Salmonella enterica*, transcribed from the promoter in the 3′ UTR of *dapB* mRNA, which represses the translation of mRNAs of the *dpp* and *opp* operons, encoding major ABC transporters (10). Two examples of 3′-UTR-derived sRNAs generated by cleavage are the 3′ UTR of *S. coelicolor sodF*, which represses the expression of *sodN* mRNA, thereby shutting off the synthesis of nickel-containing superoxide dismutase during nickel starvation (88); and the 3′ UTR of *S. enterica cpxP*, which produces CpxQ sRNA, which represses multiple mRNAs encoding extracytoplasmic proteins to potentially reduce synthesis of problematic proteins to combat envelope stress (89).

It is also worth noting that, like IsrK, a number of sRNAs, particularly those derived from UTRs, have been found to be processed into multiple sizes. It is possible that the different versions act on different sets of mRNA targets or by different mechanisms (87, 90). Hints of such nuanced regulation are coming from global analyses of sRNA-mRNA pairs in which different sets of targets are observed for different isoforms of sRNAs (80, 91). In general, the deep-sequencing studies highlight how much remains to be learned about different cellular networks and how even the smallest fragments of RNA can have important functions in the cell. Further identification of more multifunctional RNAs and studies into uncovering their roles in the cell will help us to further understand the complex regulatory networks controlling cellular physiology (92).

PERSPECTIVES

As we have discussed, there currently are only five dual-function RNAs in bacteria with reported functions. While most sRNAs identified are thought to be noncoding and this is likely in the majority of cases, there are enough examples of sRNAs for which ORFs have been noted that it is prudent to be cautious about calling an sRNA noncoding.

A significant challenge is documenting whether ORFs found in sRNAs are in fact translated. The challenge of identifying and detecting small proteins stems from the difficulties in their annotation and biochemical detection due to their small size (93). The short amino acid sequences make the computational identification of coding ORFs difficult due to a lack of sufficient sequence for reliable domain and homology determination, a problem compounded by the possibility of an ORF starting with a non-AUG codon and the chances of start and stop codons existing in sRNA sequences by chance (14). Detection of these small proteins is limited by difficulties inherent in using standard proteomic techniques to isolate and identify proteins <10 kDa in size. For example, mass spectrometry-based methods to detect small proteins still suffer from false negatives due to proteolytic cleavage of the small proteins into products too small to be detected by mass spectrometry. Despite these challenges, we anticipate that the number of confirmed dual-function sRNAs will grow given the emergence of new techniques such as improved peptidomics and ribosome profiling. When more dual-function RNAs have been found and characterized, a better understanding of the properties of these RNAs likely will facilitate the development of additional tools for identifying and characterizing a putative dual-function RNA.

With the increasing discoveries of dual-function sRNAs, many questions remain. One important issue is the interplay between riboregulation and translation of the sRNA. Further understanding of this interplay undoubtedly will provide vital information regarding expression and function of the small protein. Some of the unique features that have come to light through studies of the few examples of dual-function sRNAs include instances where the riboregulation function can influence the translation of the small protein. For example, the translation of *hld* is delayed by 1 h after RNAIII transcription potentially because of an inhibitory structure between the 3′ end of RNAIII and the *hld* RBS (53). As another example, it was noted that SRP1 is not

translated under the same conditions that most strongly induce *sr1* transcription (62). Further studies regarding the mechanisms by which riboregulation and translation influence each other and identification of other factors involved will shed light on the interplay between the sRNA and protein function. The spatial location of the base-pairing versus the coding region of the sRNA likely is an important factor. It is also possible that other regulatory sRNAs might bind these dual-function sRNAs to influence the translation of the encoded small proteins. Moreover, little is known about the role of RNA chaperones in the interplay between the two different roles of these dual-function sRNAs. Binding of RNA chaperones to these dual-function sRNAs might influence one role versus the other by either sequestering the RNA for its role as a riboregulator or remodeling secondary structures that affect the translation of the ORF. Moreover, there might be an inherent competition between the RNA chaperones and ribosomes for binding to these sRNAs that might influence the roles of these dual-function sRNAs.

While the evolution of sRNAs is beginning to be studied (see also chapter 28), not much is known about the evolution of these multifunctional sRNAs. For example, did the RNAs first evolve as sRNAs or as mRNAs? SgrS was first thought to be a noncoding regulatory sRNA until the later discovery that it also encodes SgrT. In contrast, RNAIII and Pel/SagA mRNA were first found to be protein coding until the realization that the RNAs also have a regulatory function. Looking at the conservation across different species might give insight into the original function of these dual sRNAs.

Considering synthetic biology applications, dual-function RNAs are good candidates for exploitation. For example, a dual-function RNA that can inhibit both the synthesis and activity of a protein could allow for tightly regulated control of a process, and a dual-function RNA that has both base-pairing and protein-titrating activity can be used to control two different classes of genes. All in all, dual-function RNAs present an opportunity to tightly modulate gene expression using two distinct functions under specific conditions.

Clearly, studies of dual-functional sRNAs have revealed how little we understand bacterial physiology and modes of regulation. Previously thought to be junk, even the smallest RNA fragments might have important functions in the cell. Many questions remain unanswered but likely will be answered with increasing discoveries of new dual-function sRNAs and studies of the their roles and mechanism of regulation. Further breakthroughs in methodology regarding protein study and computational approaches for identification of these sRNAs and small proteins also will aid in expanding this interesting field.

Acknowledgments. M.R. is supported by the Intramural Research Program of the Eunice Kennedy Shriver National Institute of Child Health and Human Development. C.B. and A.K. are supported by NIH R01 GM092830.

Citation. Raina M, King A, Bianco C, Vanderpool CK. 2018. Dualfunction RNAs. Microbiol Spectrum 6(5):RWR-0032-2018.

References

1. Storz G, Vogel J, Wassarman KM. 2011. Regulation by small RNAs in bacteria: expanding frontiers. *Mol Cell* **43**:880–891.

2. Waters LS, Storz G. 2009. Regulatory RNAs in bacteria. *Cell* **136**:615–628.

3. Bobrovskyy M, Vanderpool CK. 2014. The small RNA SgrS: roles in metabolism and pathogenesis of enteric bacteria. *Front Cell Infect Microbiol* **4**:61.

4. Balasubramanian D, Vanderpool CK. 2013. Deciphering the interplay between two independent functions of the small RNA regulator SgrS in *Salmonella. J Bacteriol* **195**:4620–4630.

5. Fröhlich KS, Papenfort K, Fekete A, Vogel J. 2013. A small RNA activates CFA synthase by isoform-specific mRNA stabilization. *EMBO J* **32**:2963–2979.

6. Papenfort K, Sun Y, Miyakoshi M, Vanderpool CK, Vogel J. 2013. Small RNA-mediated activation of sugar phosphatase mRNA regulates glucose homeostasis. *Cell* **153**:426–437.

7. McCullen CA, Benhammou JN, Majdalani N, Gottesman S. 2010. Mechanism of positive regulation by DsrA and RprA small noncoding RNAs: pairing increases translation and protects *rpoS* mRNA from degradation. *J Bacteriol* **192**:5559–5571.

8. Schumacher MA, Pearson RF, Møller T, Valentin-Hansen P, Brennan RG. 2002. Structures of the pleiotropic translational regulator Hfq and an Hfq-RNA complex: a bacterial Sm-like protein. *EMBO J* **21**:3546–3556.

9. Zhang A, Wassarman KM, Rosenow C, Tjaden BC, Storz G, Gottesman S. 2003. Global analysis of small RNA and mRNA targets of Hfq. *Mol Microbiol* **50**:1111–1124.

10. Chao Y, Papenfort K, Reinhardt R, Sharma CM, Vogel J. 2012. An atlas of Hfq-bound transcripts reveals 3′ UTRs as a genomic reservoir of regulatory small RNAs. *EMBO J* **31**:4005–4019.

11. Vogel J, Luisi BF. 2011. Hfq and its constellation of RNA. *Nat Rev Microbiol* **9**:578–589.

12. Vanderpool CK, Balasubramanian D, Lloyd CR. 2011. Dual-function RNA regulators in bacteria. *Biochimie* **93**:1943–1949.

13. Gimpel M, Brantl S. 2017. Dual-function small regulatory RNAs in bacteria. *Mol Microbiol* **103**:387–397.

14. Friedman RC, Kalkhof S, Doppelt-Azeroual O, Mueller SA, Chovancová M, von Bergen M, Schwikowski B. 2017. Common and phylogenetically widespread coding for peptides by bacterial small RNAs. *BMC Genomics* **18**:553.

15. Miyakoshi M, Chao Y, Vogel J. 2015. Regulatory small RNAs from the 3′ regions of bacterial mRNAs. *Curr Opin Microbiol* **24**:132–139.

16. Jørgensen MG, Thomason MK, Havelund J, Valentin-Hansen P, Storz G. 2013. Dual function of the McaS small RNA in controlling biofilm formation. *Genes Dev* **27**:1132–1145.

17. Lalaouna D, Carrier MC, Massé E. 2015. Every little piece counts: the many faces of tRNA transcripts. *Transcription* **6**:74–77.

18. Wadler CS, Vanderpool CK. 2007. A dual function for a bacterial small RNA: SgrS performs base pairing-dependent regulation and encodes a functional polypeptide. *Proc Natl Acad Sci U S A* **104**:20454–20459.

19. Vanderpool CK, Gottesman S. 2004. Involvement of a novel transcriptional activator and small RNA in post-transcriptional regulation of the glucose phosphoenolpyruvate phosphotransferase system. *Mol Microbiol* **54**:1076–1089.

20. Kaito C, Saito Y, Nagano G, Ikuo M, Omae Y, Hanada Y, Han X, Kuwahara-Arai K, Hishinuma T, Baba T, Ito T, Hiramatsu K, Sekimizu K. 2011. Transcription and translation products of the cytolysin gene *psm-mec* on the mobile genetic element SCC*mec* regulate *Staphylococcus aureus* virulence. *PLoS Pathog* **7**:e1001267.

21. Mangold M, Siller M, Roppenser B, Vlaminckx BJ, Penfound TA, Klein R, Novak R, Novick RP, Charpentier E. 2004. Synthesis of group A streptococcal virulence factors is controlled by a regulatory RNA molecule. *Mol Microbiol* **53**:1515–1527.

22. Licht A, Preis S, Brantl S. 2005. Implication of CcpN in the regulation of a novel untranslated RNA (SR1) in *Bacillus subtilis. Mol Microbiol* **58**:189–206.

23. Horler RS, Vanderpool CK. 2009. Homologs of the small RNA SgrS are broadly distributed in enteric bacteria but have diverged in size and sequence. *Nucleic Acids Res* **37**:5465–5476.

24. Wassarman KM, Repoila F, Rosenow C, Storz G, Gottesman S. 2001. Identification of novel small RNAs using comparative genomics and microarrays. *Genes Dev* **15**:1637–1651.

25. Kimata K, Tanaka Y, Inada T, Aiba H. 2001. Expression of the glucose transporter gene, *ptsG*, is regulated at the mRNA degradation step in response to glycolytic flux in *Escherichia coli. EMBO J* **20**:3587–3595.

26. Morita T, El-Kazzaz W, Tanaka Y, Inada T, Aiba H. 2003. Accumulation of glucose 6-phosphate or fructose 6-phosphate is responsible for destabilization of glucose transporter mRNA in *Escherichia coli. J Biol Chem* **278**:15608–15614.

27. Richards GR, Patel MV, Lloyd CR, Vanderpool CK. 2013. Depletion of glycolytic intermediates plays a key role in glucose-phosphate stress in *Escherichia coli. J Bacteriol* **195**:4816–4825.

28. Vanderpool CK, Gottesman S. 2007. The novel transcription factor SgrR coordinates the response to glucose-phosphate stress. *J Bacteriol* **189**:2238–2248.

29. Rice JB, Vanderpool CK. 2011. The small RNA SgrS controls sugar-phosphate accumulation by regulating multiple PTS genes. *Nucleic Acids Res* **39:**3806–3819.

30. Kawamoto H, Koide Y, Morita T, Aiba H. 2006. Base-pairing requirement for RNA silencing by a bacterial small RNA and acceleration of duplex formation by Hfq. *Mol Microbiol* **61:**1013–1022.

31. Maki K, Uno K, Morita T, Aiba H. 2008. RNA, but not protein partners, is directly responsible for translational silencing by a bacterial Hfq-binding small RNA. *Proc Natl Acad Sci U S A* **105:**10332–10337.

32. Rice JB, Balasubramanian D, Vanderpool CK. 2012. Small RNA binding-site multiplicity involved in translational regulation of a polycistronic mRNA. *Proc Natl Acad Sci U S A* **109:**E2691–E2698.

33. Bobrovskyy M, Vanderpool CK. 2016. Diverse mechanisms of post-transcriptional repression by the small RNA regulator of glucose-phosphate stress. *Mol Microbiol* **99:**254–273.

34. Lloyd CR, Park S, Fei J, Vanderpool CK. 2017. The small protein SgrT controls transport activity of the glucose-specific phosphotransferase system. *J Bacteriol* **199:**e0086e9-16.

35. Jahreis K, Pimentel-Schmitt EF, Brückner R, Titgemeyer F. 2008. Ins and outs of glucose transport systems in eubacteria. *FEMS Microbiol Rev* **32:**891–907.

36. Novick RP, Ross HF, Projan SJ, Kornblum J, Kreiswirth B, Moghazeh S. 1993. Synthesis of staphylococcal virulence factors is controlled by a regulatory RNA molecule. *EMBO J* **12:**3967–3975.

37. Vuong C, Götz F, Otto M. 2000. Construction and characterization of an *agr* deletion mutant of *Staphylococcus epidermidis. Infect Immun* **68:**1048–1053.

38. Novick RP. 2003. Autoinduction and signal transduction in the regulation of staphylococcal virulence. *Mol Microbiol* **48:**1429–1449.

39. Novick RP, Geisinger E. 2008. Quorum sensing in staphylococci. *Annu Rev Genet* **42:**541–564.

40. Benito Y, Kolb FA, Romby P, Lina G, Etienne J, Vandenesch F. 2000. Probing the structure of RNAIII, the *Staphylococcus aureus agr* regulatory RNA, and identification of the RNA domain involved in repression of protein A expression. *RNA* **6:**668–679.

41. Fechter P, Caldelari I, Lioliou E, Romby P. 2014. Novel aspects of RNA regulation in *Staphylococcus aureus. FEBS Lett* **588:**2523–2529.

42. Huntzinger E, Boisset S, Saveanu C, Benito Y, Geissmann T, Namane A, Lina G, Etienne J, Ehresmann B, Ehresmann C, Jacquier A, Vandenesch F, Romby P. 2005. *Staphylococcus aureus* RNAIII and the endoribonuclease III coordinately regulate *spa* gene expression. *EMBO J* **24:**824–835.

43. Boisset S, Geissmann T, Huntzinger E, Fechter P, Bendridi N, Possedko M, Chevalier C, Helfer AC, Benito Y, Jacquier A, Gaspin C, Vandenesch F, Romby P. 2007. *Staphylococcus aureus* RNAIII coordinately represses the synthesis of virulence factors and the transcription regulator Rot by an antisense mechanism. *Genes Dev* **21:**1353–1366.

44. Morfeldt E, Taylor D, von Gabain A, Arvidson S. 1995. Activation of alpha-toxin translation in *Staphylococcus aureus* by the *trans*-encoded antisense RNA, RNAIII. *EMBO J* **14:**4569–4577.

45. Liu Y, Mu C, Ying X, Li W, Wu N, Dong J, Gao Y, Shao N, Fan M, Yang G. 2011. RNAIII activates map expression by forming an RNA-RNA complex in *Staphylococcus aureus. FEBS Lett* **585:**899–905.

46. Gupta RK, Luong TT, Lee CY. 2015. Correction for Gupta et al., RNAIII of the *Staphylococcus aureus agr* system activates global regulator MgrA by stabilizing mRNA. *Proc Natl Acad Sci U S A* **112:**E7306.

47. Ingavale S, van Wamel W, Luong TT, Lee CY, Cheung AL. 2005. Rat/MgrA, a regulator of autolysis, is a regulator of virulence genes in *Staphylococcus aureus. Infect Immun* **73:**1423–1431.

48. Trotonda MP, Tamber S, Memmi G, Cheung AL. 2008. MgrA represses biofilm formation in *Staphylococcus aureus. Infect Immun* **76:**5645–5654.

49. Crosby HA, Schlievert PM, Merriman JA, King JM, Salgado-Pabón W, Horswill AR. 2016. The *Staphylococcus aureus* global regulator MgrA modulates clumping and virulence by controlling surface protein expression. *PLoS Pathog* **12:**e1005604.

50. Luong TT, Newell SW, Lee CY. 2003. Mgr, a novel global regulator in *Staphylococcus aureus. J Bacteriol* **185:**3703–3710.

51. Luong TT, Dunman PM, Murphy E, Projan SJ, Lee CY. 2006. Transcription profiling of the *mgrA* regulon in *Staphylococcus aureus. J Bacteriol* **188:**1899–1910.

52. Verdon J, Girardin N, Lacombe C, Berjeaud JM, Héchard Y. 2009. δ-Hemolysin, an update on a membrane-interacting peptide. *Peptides* **30:**817–823.

53. Balaban N, Novick RP. 1995. Translation of RNAIII, the *Staphylococcus aureus agr* regulatory RNA molecule, can be activated by a 3′-end deletion. *FEMS Microbiol Lett* **133:**155–161.

54. Qin L, McCausland JW, Cheung GY, Otto M. 2016. PSM-mec—a virulence determinant that connects transcriptional regulation, virulence, and antibiotic resistance in staphylococci. *Front Microbiol* **7:**1293.

55. Kaito C, Saito Y, Ikuo M, Omae Y, Mao H, Nagano G, Fujiyuki T, Numata S, Han X, Obata K, Hasegawa S, Yamaguchi H, Inokuchi K, Ito T, Hiramatsu K, Sekimizu K. 2013. Mobile genetic element SCC*mec*-encoded *psm-mec* RNA suppresses translation of *agrA* and attenuates MRSA virulence. *PLoS Pathog* **9:**e1003269.

56. Chatterjee SS, Chen L, Joo HS, Cheung GY, Kreiswirth BN, Otto M. 2011. Distribution and regulation of the mobile genetic element-encoded phenol-soluble modulin PSM-mec in methicillin-resistant *Staphylococcus aureus. PLoS One* **6:**e28781.

57. Queck SY, Khan BA, Wang R, Bach TH, Kretschmer D, Chen L, Kreiswirth BN, Peschel A, Deleo FR, Otto M. 2009. Mobile genetic element-encoded cytolysin connects virulence to methicillin resistance in MRSA. *PLoS Pathog* **5:**e1000533.

58. Wang R, Braughton KR, Kretschmer D, Bach TH, Queck SY, Li M, Kennedy AD, Dorward DW, Klebanoff SJ,

Peschel A, DeLeo FR, Otto M. 2007. Identification of novel cytolytic peptides as key virulence determinants for community-associated MRSA. *Nat Med* **13**:1510–1514.

59. Licht A, Golbik R, Brantl S. 2008. Identification of ligands affecting the activity of the transcriptional repressor CcpN from *Bacillus subtilis*. *J Mol Biol* **380**:17–30.

60. Licht A, Brantl S. 2006. Transcriptional repressor CcpN from *Bacillus subtilis* compensates asymmetric contact distribution by cooperative binding. *J Mol Biol* **364**:434–448.

61. Licht A, Brantl S. 2009. The transcriptional repressor CcpN from *Bacillus subtilis* uses different repression mechanisms at different promoters. *J Biol Chem* **284**:30032–30038.

62. Heidrich N, Chinali A, Gerth U, Brantl S. 2006. The small untranslated RNA SR1 from the *Bacillus subtilis* genome is involved in the regulation of arginine catabolism. *Mol Microbiol* **62**:520–536.

63. Heidrich N, Moll I, Brantl S. 2007. *In vitro* analysis of the interaction between the small RNA SR1 and its primary target *ahrC* mRNA. *Nucleic Acids Res* **35**:4331–4346.

64. Gimpel M, Preis H, Barth E, Gramzow L, Brantl S. 2012. SR1—a small RNA with two remarkably conserved functions. *Nucleic Acids Res* **40**:11659–11672.

65. Gimpel M, Heidrich N, Mäder U, Krügel H, Brantl S. 2010. A dual-function sRNA from *B. subtilis*: SR1 acts as a peptide encoding mRNA on the *gapA* operon. *Mol Microbiol* **76**:990–1009.

66. Gimpel M, Brantl S. 2016. Dual-function sRNA encoded peptide SR1P modulates moonlighting activity of *B. subtilis* GapA. *RNA Biol* **13**:916–926.

67. Betschel SD, Borgia SM, Barg NL, Low DE, De Azavedo JC. 1998. Reduced virulence of group A streptococcal Tn*916* mutants that do not produce streptolysin S. *Infect Immun* **66**:1671–1679.

68. Perez N, Treviño J, Liu Z, Ho SC, Babitzke P, Sumby P. 2009. A genome-wide analysis of small regulatory RNAs in the human pathogen group A *Streptococcus*. *PLoS One* **4**:e7668.

69. Nizet V, Beall B, Bast DJ, Datta V, Kilburn L, Low DE, De Azavedo JC. 2000. Genetic locus for streptolysin S production by group A streptococcus. *Infect Immun* **68**:4245–4254.

70. Datta V, Myskowski SM, Kwinn LA, Chiem DN, Varki N, Kansal RG, Kotb M, Nizet V. 2005. Mutational analysis of the group A streptococcal operon encoding streptolysin S and its virulence role in invasive infection. *Mol Microbiol* **56**:681–695.

71. Sonnleitner E, Gonzalez N, Sorger-Domenigg T, Heeb S, Richter AS, Backofen R, Williams P, Hüttenhofer A, Haas D, Bläsi U. 2011. The small RNA PhrS stimulates synthesis of the *Pseudomonas aeruginosa* quinolone signal. *Mol Microbiol* **80**:868–885.

72. Shimizu T, Yaguchi H, Ohtani K, Banu S, Hayashi H. 2002. Clostridial VirR/VirS regulon involves a regulatory RNA molecule for expression of toxins. *Mol Microbiol* **43**:257–265.

73. Roberts SA, Scott JR. 2007. RivR and the small RNA RivX: the missing links between the CovR regulatory cascade and the Mga regulon. *Mol Microbiol* **66**:1506–1522.

74. Vockenhuber MP, Heueis N, Suess B. 2015. Identification of *metE* as a second target of the sRNA scr5239 in *Streptomyces coelicolor*. *PLoS One* **10**:e0120147.

75. Vockenhuber MP, Suess B. 2012. *Streptomyces coelicolor* sRNA scr5239 inhibits agarase expression by direct base pairing to the *dagA* coding region. *Microbiology* **158**:424–435.

76. Berghoff BA, Glaeser J, Sharma CM, Vogel J, Klug G. 2009. Photooxidative stress-induced and abundant small RNAs in *Rhodobacter sphaeroides*. *Mol Microbiol* **74**:1497–1512.

77. Hess WR, Berghoff BA, Wilde A, Steglich C, Klug G. 2014. Riboregulators and the role of Hfq in photosynthetic bacteria. *RNA Biol* **11**:413–426.

78. Müller KM, Berghoff BA, Eisenhardt BD, Remes B, Klug G. 2016. Characteristics of Pos19—a small coding RNA in the oxidative stress response of *Rhodobacter sphaeroides*. *PLoS One* **11**:e0163425.

79. Thomason MK, Fontaine F, De Lay N, Storz G. 2012. A small RNA that regulates motility and biofilm formation in response to changes in nutrient availability in *Escherichia coli*. *Mol Microbiol* **84**:17–35.

80. Potts AH, Vakulskas CA, Pannuri A, Yakhnin H, Babitzke P, Romeo T. 2017. Global role of the bacterial post-transcriptional regulator CsrA revealed by integrated transcriptomics. *Nat Commun* **8**:1596.

81. Hershko-Shalev T, Odenheimer-Bergman A, Elgrably-Weiss M, Ben-Zvi T, Govindarajan S, Seri H, Papenfort K, Vogel J, Altuvia S. 2016. Gifsy-1 prophage IsrK with dual function as small and messenger RNA modulates vital bacterial machineries. *PLoS Genet* **12**:e1005975.

82. Kawano M, Reynolds AA, Miranda-Rios J, Storz G. 2005. Detection of 5′- and 3′-UTR-derived small RNAs and *cis*-encoded antisense RNAs in *Escherichia coli*. *Nucleic Acids Res* **33**:1040–1050.

83. Vogel J, Bartels V, Tang TH, Churakov G, Slagter-Jäger JG, Hüttenhofer A, Wagner EG. 2003. RNomics in *Escherichia coli* detects new sRNA species and indicates parallel transcriptional output in bacteria. *Nucleic Acids Res* **31**:6435–6443.

84. Loh E, Dussurget O, Gripenland J, Vaitkevicius K, Tiensuu T, Mandin P, Repoila F, Buchrieser C, Cossart P, Johansson J. 2009. A *trans*-acting riboswitch controls expression of the virulence regulator PrfA in *Listeria monocytogenes*. *Cell* **139**:770–779.

85. Liu N, Niu G, Xie Z, Chen Z, Itzek A, Kreth J, Gillaspy A, Zeng L, Burne R, Qi F, Merritt J. 2015. The *Streptococcus mutans irvA* gene encodes a *trans*-acting riboregulatory mRNA. *Mol Cell* **57**:179–190.

86. Ellis MJ, Trussler RS, Charles O, Haniford DB. 2017. A transposon-derived small RNA regulates gene expression in *Salmonella* Typhimurium. *Nucleic Acids Res* **45**:5470–5486.

87. Guo MS, Updegrove TB, Gogol EB, Shabalina SA, Gross CA, Storz G. 2014. MicL, a new σ^E-dependent sRNA,

combats envelope stress by repressing synthesis of Lpp, the major outer membrane lipoprotein. *Genes Dev* **28**: 1620–1634.

88. Kim HM, Shin JH, Cho YB, Roe JH. 2014. Inverse regulation of Fe- and Ni-containing SOD genes by a Fur family regulator Nur through small RNA processed from 3′ UTR of the *sodF* mRNA. *Nucleic Acids Res* **42**:2003–2014.

89. Chao Y, Vogel J. 2016. A 3′ UTR-derived small RNA provides the regulatory noncoding arm of the inner membrane stress response. *Mol Cell* **61**:352–363.

90. Hao Y, Updegrove TB, Livingston NN, Storz G. 2016. Protection against deleterious nitrogen compounds: role of σ^S-dependent small RNAs encoded adjacent to *sdiA*. *Nucleic Acids Res* **44**:6935–6948.

91. Melamed S, Peer A, Faigenbaum-Romm R, Gatt YE, Reiss N, Bar A, Altuvia Y, Argaman L, Margalit H. 2016. Global mapping of small RNA-target interactions in bacteria. *Mol Cell* **63**:884–897.

92. Papenfort K, Espinosa E, Casadesús J, Vogel J. 2015. Small RNA-based feedforward loop with AND-gate logic regulates extrachromosomal DNA transfer in *Salmonella*. *Proc Natl Acad Sci U S A* **112**:E4772–E4781.

93. Storz G, Wolf YI, Ramamurthi KS. 2014. Small proteins can no longer be ignored. *Annu Rev Biochem* **83**:753–777.

Regulating with RNA in Bacteria and Archaea
Edited by Gisela Storz and Kai Papenfort
© 2018 American Society for Microbiology, Washington, DC
doi:10.1128/microbiolspec.RWR-0004-2017

Origin, Evolution, and Loss of Bacterial Small RNAs

28

H. Auguste Dutcher[1] and Rahul Raghavan[1]

INTRODUCTION

As our understanding of the transcriptional landscape of bacteria continues to expand, it has become clear that noncoding small RNAs (sRNAs) play a pivotal regulatory role (1–3). Typically 50 to 400 nucleotides in length, sRNAs posttranscriptionally regulate gene expression, usually by base-pairing with one or more mRNA targets (4). sRNAs likely provide certain advantages over protein regulators because they act quickly, are relatively metabolically inexpensive, and provide an additional way to respond to environmental signals (1). Beyond these basic characteristics, however, the roles of bacterial sRNAs are extremely diverse: they are capable of upregulating or downregulating translation, stabilizing mRNAs or targeting them for degradation, sharing targets, and/or targeting multiple mRNAs. Variability in their sequence, structure, and how and when they are transcribed allows them to meet a wide range of nuanced regulatory needs based on the diverse environments to which bacteria must adapt.

Despite their regulatory importance, we lack a full understanding of how sRNAs originate and evolve. This topic was recently reviewed with particular attention to the feature requirements that shape sRNA evolution (5). These requirements typically include an environmentally regulated promoter; a Rho-independent terminator; double-stranded regions that allow for stable secondary structure; and an unstructured seed region, where the sRNA base-pairs with its target (5). These features do impose limitations on sequence evolution, but the marked divergence observed among sRNAs also suggests that these constraints are not so limiting as to prevent rapid and significant change (2).

Indeed, the fast pace at which sRNA sequences change is one of several factors that make systematic studies of bacterial sRNA evolution particularly challenging. High degrees of both intra- and interspecies polymorphism yield low sequence similarity, especially as compared to protein-coding genes, and sRNAs on the whole are known to be poorly conserved across broad evolutionary distances (6). Comparing the contents of Rfam (7), a database that houses sequence and structural information for known sRNA families, to Pfam (8), an analogous database for protein families, only 60% of RNA families were found to be conserved between species of the same taxonomic family, as opposed to 90% for proteins (9). In evolutionary analyses, where a clear picture of relationships between gene sequences through time is critical in order to draw meaningful conclusions, this presents the unique challenge of distinguishing between an sRNA's absence in a given lineage and the possibility that the sequence or structure of the gene has simply diverged too drastically to accurately trace its ancestry.

Not only do sRNA genes evolve rapidly; they also are arguably more difficult to identify than protein-coding genes. Whereas open reading frames (ORFs) serve as signposts for coding sequences, sRNAs have no such singular reliable identifier, and in fact, many of the features they do require are found readily throughout bacterial genomes. In *Escherichia coli*, for example, sigma-70 promoter-like sequences are widely distributed and highly redundant (10); more broadly, secondary structure alone is not enough to reliably identify sRNAs (11), and algorithms that predict noncoding RNAs often have high false-positive rates (12). Moreover, without codons as guides, homologous sRNAs are notoriously difficult to align (13), presenting obstacles to accurate quantification of their sequence evolution.

sRNAs are known to play key roles in regulatory networks (2, 3), and so are crucial to our understanding of the evolution of metabolic pathways, environ-

[1]Department of Biology and Center for Life in Extreme Environments, Portland State University, Portland, OR 97201.

mental adaptation, virulence, speciation, and countless other processes in bacteria. Furthermore, their evolutionary trajectories are inextricably linked with that of their targets, and as such, our understanding of the evolution of protein-coding genes is not complete without inquiries into that of these key regulators. Given the existence of their regulatory analogs in eukaryotes and archaea, insights into bacterial sRNA evolution could also lead to a better understanding of gene regulation in all domains of life. Finally, since posttranscriptional controls are now understood to be more than a peripheral consideration when it comes to regulating gene expression—rather, they are a key part of it—successful identification of sRNAs in bacterial genomes has become all the more critical. A better understanding of sRNA evolution could lead to improved methodologies for identifying novel sRNAs and facilitate enhanced annotation of bacterial genomes to reflect these key genetic components.

With these goals in mind, this review explores the evolutionary considerations associated with sRNA origin, functional divergence, and loss from bacterial genomes. While many of the dynamics discussed here are broadly applicable to noncoding RNAs in general, this article will focus on *trans*-acting sRNAs, which are most notably characterized by their requirement for only partial mRNA complementarity and their being encoded in genetic loci often distant from that of their targets. Other works, including articles that accompany this one, provide ample examples of sRNAs and their full range of diverse characteristics and functions. Consequently, we will focus less on detailing the features of known sRNAs and instead turn our attention to broad evolutionary patterns and the resulting questions that emerge. Tracing a "life cycle" of sRNA evolution, we will begin by documenting their origins, both *de novo* and from preexisting genetic elements; detail what is known about their functional evolution; and end by discussing the dynamics associated with sRNA loss from bacterial genomes.

sRNA ORIGINS

Despite the challenges associated with identifying sRNAs and tracking them over evolutionary time, several mechanisms for their emergence have been investigated. It is possible that sRNA genes have, on the whole, more diverse origins as compared to protein-coding genes, for which duplication events and horizontal gene transfer (HGT) are considered to be the dominant forces (14, 15), though *de novo* origination of coding sequences is becoming increasingly well studied (16, 17). For

microRNAs (miRNAs), which are the functional analogs of sRNAs in eukaryotes, it was recently shown that the requirements for formation are somewhat less stringent than protein-coding genes in part due to their small size: protein-coding genes are significantly longer than miRNA genes and require ORFs in order to be functional, whereas the hairpins that form potential miRNAs are pervasive in mammalian transcriptomes (18). Similarly, provided that certain structural requirements are met, a bacterial sRNA requires only a short seed sequence in order to acquire a viable mRNA target (19, 20). Particularly since the seed sequence need only be partially complementary to the target mRNA, the existence (or emergence by chance) of such a target is likely (21). Thus, it is highly likely that functional sRNAs may emerge more readily and evolve more quickly than their protein-coding counterparts.

Documented bacterial sRNA origins include duplication events, HGT, and *de novo* emergence (Fig. 1), mirroring known eukaryotic gene origins (22), but altogether less well investigated to date. The mechanisms briefly discussed below, though by no means entirely discrete, raise new questions regarding the selection pressures that result in sRNA emergence. Additionally, gene origination source has been shown to correlate with trends in network integration and functional gain in eukaryotes. For instance, in *Saccharomyces*, genes that originated *de novo* gained function more quickly, were more likely to reflect responses to environmental changes, and formed new interactions faster than genes that arose via duplication events (23). Hence, the characterization of sRNAs by their evolutionary origins may help illuminate patterns in their highly varied interactions and expression levels within bacterial genomes.

Emergence of New sRNAs through Duplication

Since duplication events are a major driver behind protein origination and evolution, it comes as no surprise that homologous sRNAs, with one or more copies thought to have arisen via gene duplication, are relatively common in bacterial genomes. There are several examples of bacterial sRNAs that have persisted as closely related but discrete copies (24), including PrrF1 and PrrF2 in *Pseudomonas*, which regulate iron metabolism via the ferric uptake regulator (Fur) repressor (25); the functionally analogous RyhB1 and RyhB2 in *Salmonella* (26); a suite of Qrr sRNAs in *Vibrio* that control quorum sensing (27); and AbcR1 and AbcR2 in *Rhizobiales*, involved in ABC transporter regulation (28–30). Duplicated sRNAs exhibit varying degrees of functional overlap. Some, such as OmrA and OmrB

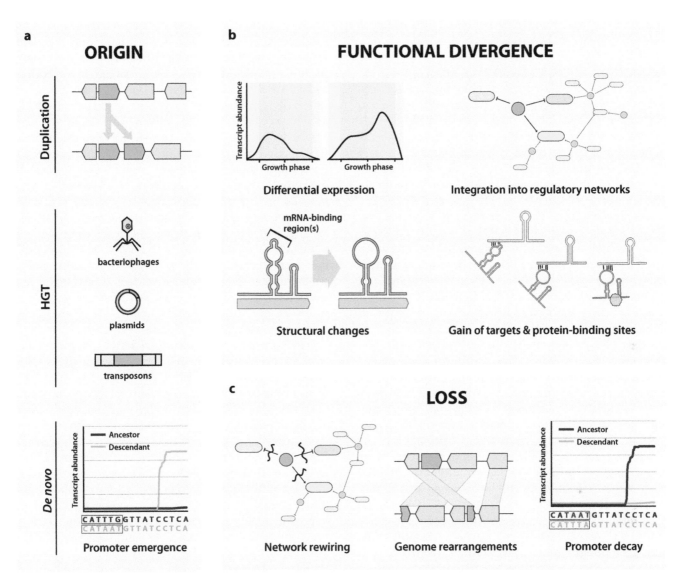

Figure 1 sRNA origin, functional divergence, and loss. (a) sRNA sources include duplication events, HGT, and *de novo* origination via promoter emergence. (b) Sequence and structural changes are often accompanied by differential sRNA gene expression and accumulation of mRNA targets and/or protein-binding regions, causing the sRNA to become fully integrated into regulatory networks. (c) sRNA loss occurs through mutations that erode promoter sequences, genome rearrangements that split sRNA-containing IGRs, and selective pressures that prompt shifts in network interactions.

in *E. coli*, show apparent target redundancy (31); others appear to have significantly diverged since their emergence—either in sequence, target, regulation, or a combination of these characteristics (5). The evolutionary considerations regarding sRNA functional divergence will be discussed later in this review.

For those sRNA duplicates that share mRNA targets, an interesting question emerges regarding the utility of such redundancy (24, 32), and thus the selection

to maintain it. It has been suggested that multicopy sRNAs enable more efficient or nuanced regulatory control (27, 32), or ensure proper regulation in the case of particularly critical functions (24). Redundancy also enables differential regulation of the sRNAs themselves (2), perhaps allowing their regulatory cascades to be activated under an additional set of environmental conditions (24). This scenario is easy to envision, as a small sequence change in the course of an sRNA's rapid evolution might allow it to be regulated by a different transcription factor (5).

Acquisition of New sRNAs via HGT

Another common source of sRNAs in bacterial genomes is HGT via plasmids, transposable elements, and bacteriophages. Some of the best-studied plasmid-derived sRNAs are associated with virulence, and include the antitoxins of many toxin-antitoxin systems (33) and the island-encoded sRNAs in *Salmonella enterica* serovar Typhimurium and *Staphylococcus aureus* (26, 34–36). For example, the HGT-derived sRNA IsrJ in *Salmonella* enables the use of a type III secretion system to invade intestinal epithelial cells (26), and IsrM likewise regulates proteins required for intracellular invasion and replication (35). Prophage-origin IsrK is also encoded in a *Salmonella* pathogenicity island and is notable as an example of dual-function RNAs (37), which are the subject of an accompanying review. The short IsrK isoform activates the longer isoform, an inactive mRNA, ultimately causing expression of anti-terminator protein AntQ. The dual-function phenomenon warrants further study in the context of sRNA evolution; for example, did mRNAs and sRNAs with shared genetic loci coemerge? Given that "overprinting" is relatively common in viral genomes (38), might there be a relationship between viral origination of an sRNA gene and the likelihood that it is associated with an overlapping mRNA? Other evolutionary implications of dual functionality will be discussed later in this review.

In addition to the now well-documented phenomenon of sRNAs encoded on pathogenicity islands, there may also be structural considerations related to the likelihood of sRNA emergence from bacteriophages or transposable elements. Two sRNA genes in *Coxiella burnetii*, the causative agent of Q fever, were found to be encoded by families of a transposon class called miniature inverted-repeat transposable elements (MITEs) (M. Minnick, personal communication). Transposons, including MITES in particular, are a known source of miRNAs in eukaryotes (39). Given that these elements are known to readily form stable stem-loop structures (40), it is possible that MITEs along with other types of transposons are particularly suited for sRNA origination.

De Novo Origination of sRNAs

As more bacterial sRNAs are identified and characterized and their origins elucidated, *de novo* origination from apparently nonfunctional transcripts has also been observed. Genome rearrangements, such as those that occur via homologous recombination, can form new intergenic regions (IGRs), from which new sRNAs can emerge (41). A survey of novel IGRs in *E. coli* and *Salmonella* Typhimurium identified one such rearrangement in the latter species that resulted in the origination of the SesR2 sRNA (41). The formation of this IGR is believed to have been mediated by HGT, as evidenced by plasmid-derived flanking genes and a nearby prophage gene. The sRNA encoded in that IGR likely arose via a point mutation that created a functional sigma-70 promoter region (Fig. 1), which explains the significant transcription observed in *Salmonella* but not in the corresponding regions in *E. coli*. This example highlights the nondiscrete nature of sRNA origination sources, in that the emergence of SesR2 was ultimately facilitated by by both HGT and *de novo* promoter emergence. For the purposes of this discussion, we consider it to have originated *de novo* given that the horizontally transferred IGR from which it was born was presumed to be nonfunctional prior to the point mutation that created a promoter region.

The same study that identified SesR2's origins found that an *E. coli*-specific IGR gave rise to the sRNA EcsR2 (41), and this sRNA was later determined to have originated from a degraded bacteriophage gene (21). EcsR2 is also believed to have originated *de novo* via promoter-like sequence emergence, and exhibits low expression compared to older sRNAs in the same species. In reconstructing the likely evolution of EcsR2, it appears that its mRNA-binding region has become less structured and its terminator more structured over time, suggesting that secondary structure changes are a key component of bacterial sRNA evolution, as has been observed in eukaryotes (42) and suggested in bacteria (43). Though EcsR2 is one of only a handful of bacterial sRNAs for which a source has been determined, the phenomena of noncoding genes emerging *de novo* and originating from selfish genetic elements have been observed in eukaryotes (22, 42, 44, 45). Further comparative genomics and transcriptomics studies in bacteria are required to identify all modes of sRNA origination.

GAIN OF FUNCTION AND FUNCTIONAL DIVERGENCE

Although there are few comprehensive surveys of related sRNAs, with most focusing on well-studied enteric bacteria (21, 32, 46), there is evidence that newly born sRNAs evolve quickly (21). This, coupled with the observation that constraints on sRNA evolution appear to be loose enough to accommodate rapid structure and sequence change (2), points to significant functional divergence in relatively short spans of evolutionary time. This corroborates the most likely evolutionary scenario of bacterial sRNAs: emergence of the initial sRNA-mRNA interaction induces selection to maintain the sRNA, followed by the accumulation of additional mRNA-binding sites (21) (Fig. 1). Of course, this initial selection to maintain the sRNA is predicated on the beneficial nature of the interaction. The same short-seed-sequence and partial-complementarity requirements that promote the emergence of functional sRNAs also create many opportunities for deleterious effects. In a recent study, Kacharia et al. showed that recently emerged sRNAs in *E. coli* and *Salmonella* were expressed at significantly lower levels than older sRNAs, indicating that low expression of incipient sRNAs might help to mitigate the risk of any fitness costs while maintaining the sRNA long enough for beneficial interactions to arise (21). This pattern of low expression coupled with rapid evolution has been observed in young miRNAs as well (18, 47, 48), suggesting a possible mechanism by which newly emerged noncoding RNAs can gain function and be successfully integrated into regulatory networks.

A high rate of evolution among sRNA genes would help to explain the now well-documented observation that even closely related sRNAs often exhibit distinct characteristics in expression patterns, targets, and transcriptional regulation (5, 24). For example, the sRNA GcvB is expressed at high levels during exponential phase and low levels during stationary phase in *Salmonella*; in *Vibrio*, this expression pattern is reversed (49). Given that the abundance of mRNA targets as well as the consequence of mRNA regulation will be different based on the organism's stage of growth, it is highly likely that differential expression patterns indicate diverged sRNA function. This is supported by the observation that differences in gene expression affect bacterial fitness and adaptation during exposure to stress (50).

Homologous sRNAs are likewise known to be involved in different regulatory circuits; for example, RyhB regulates the expression of different genes in *Vibrio* as compared to its homolog in *E. coli* (51). Perhaps the starkest examples of divergence come from examination of intraorganismal homologs, which show varying levels of functional redundancy (24). One such example is GlmY and GlmZ in *Enterobacteriaceae*, which act hierarchically to regulate the enzyme GlmS as part of the cell wall synthesis pathway (52). Despite sharing 66% sequence identity in *E. coli* along with extremely similar secondary structures, only GlmZ has a high affinity for Hfq, allowing it to directly activate

glmS. However, due to their structural similarity, GlmY is able to act as a decoy for the endonuclease-recruiting adapter protein RapZ, reducing the chances of GlmZ degradation and thereby indirectly regulating *glmS* (53).

In addition to being an interesting example of the regulatory complexity that can result from sRNA gene duplication and subsequent divergence, GlmY/GlmZ also underscores the importance of examining sRNA functional divergence in the context of the other regulatory elements, such as proteins, with which they interact. Understanding the nuances of protein binding may aid in our understanding of highly specific RNA degradation mechanisms (53), which in turn could shed light on how such closely related sRNAs come to occupy such distinct regulatory niches. Interestingly, GlmY/GlmZ is also an example of differential transcriptional control, in that this pair of sRNA genes is transcribed via different combinations of sigma-54 and sigma-70 promoter sequences in closely related species: sigma-54 promoter for *glmY* and *glmZ* in *Yersinia pseudotuberculosis*, overlapping sigma-54 and sigma-70 promoters for *glmY* and *glmZ* in *Salmonella* Typhimurium and *glmY* in *E. coli*, and sigma-70 promoters for *glmZ* in *E. coli* (52). This observation identifies the evolution of transcription factor or sigma factor binding site as yet another means by which the regulatory roles of sRNAs can diverge.

Conversely, examining sRNAs across large phylogenetic distances has yielded some likely examples of parallel evolution, in which discrete sRNAs have independently evolved nearly identical functions. PrrF1 and PrrF2 in *Pseudomonas aeruginosa* and RyhB in *E. coli* provide one such example. PrrF sRNAs and RyhB are both regulated by the Fur repressor and thus transcribed when iron concentrations are low, and all act by inhibiting ribosome binding on an overlapping set of mRNA targets (5, 25, 54). Similarly, MicA in *E. coli* and VrrA in *Vibrio cholerae* both regulate OmpA (outer membrane protein A), although Hfq appears to be required for MicA function but not for VrrA (5, 55, 56). The lack of sequence similarity between functional analogs in each of these cases points to likely convergent evolution; however, we cannot rule out the possibility that these sRNAs shared a common ancestor but have diverged so significantly that this relationship is obscured.

Despite ample anecdotal evidence for diversification of sRNA expression and function, the mechanisms by which these changes take place have yet to be systematically investigated. Because an sRNA requires only short, partial complementarity to its mRNA target in order to be functional, it seems likely that both target gain and target loss—perhaps especially the former—could occur fairly readily via point mutations. *E. coli* sRNAs, for example, appear to have gradually accumulated mRNA-binding sites since their emergence (46). Aside from binding sites, other mutations within sRNA genes could increase sRNA stability, indirectly allowing for acquisition of new targets by prolonged sRNA survival into a different stage of growth where alternative mRNAs are available (Fig. 1). A study examining the relationships between sRNA structure, function, and stability found that mutations in structural elements could result in increased sRNA abundance and induce regulation of nonspecific targets (43). The same study found that in sRNAs with short seed sequences, mutations in this region could be used to fine-tune an sRNA's effect on its target, whereas sRNAs with longer seed sequences were less susceptible to these mutation-induced changes. This finding begs a more thorough investigation into the relationship between seed sequence characteristics and evolutionary patterns among sRNAs. Additionally, given that mRNA-binding sequences are the most conserved regions of sRNAs (19, 21) (Fig. 2), this characteristic can be exploited to identify potential seed sequences in novel sRNAs and aid in mRNA target prediction (21).

Just as they promote target gain, sequence changes can also catalyze target loss. The loss of mRNA targets in turn could cause complete loss of sRNA function; however, it is difficult to clearly draw the line between functional loss and functional divergence unless a given sRNA is unequivocally deemed nonfunctional. For example, sequence changes were the probable cause for the loss of original function observed in two Pxr paralogs, which typically regulate fruiting body development in *Cystobacter* species (57), but whether the Pxr sRNAs in question have functionally diverged or become nonfunctional entirely is unclear. This question is further complicated by the fact that sRNA pseudogenes may be more likely than most protein-coding pseudogenes to reacquire function via the mechanisms previously described. Not constrained by the requirement of an ORF, even "decommissioned" sRNA genes may not permanently remain so.

Since the initial discovery of bacterial sRNAs, increasing numbers of what we may consider noncanonical regulatory mechanisms have been elucidated, which highlights the varied means by which sRNAs can gain (or regain) function. For example, the *E. coli* sRNAs SgrS and DicF were recently found to bind the mannose transporter *manX* mRNA target outside regions that would allow for direct translational repression, with

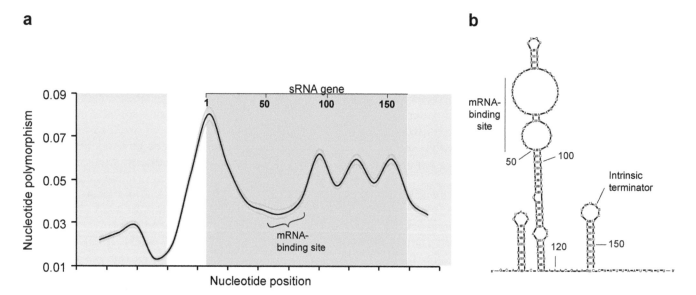

Figure 2 Sequence conservation and structure of an sRNA gene. (a) Sequence conservation within an sRNA gene (orange) and flanking protein-coding genes (blue). The black line represents nucleotide diversity index, π, calculated using a sliding-window analysis; the flanking green lines indicate the 95% confidence interval. Lowest nucleotide polymorphism within sRNA genes is observed in mRNA-binding regions. (b) Predicted structure of an sRNA, showing a single-stranded mRNA-binding site and a terminator hairpin. Adapted from reference 21 with permission.

Hfq instead blocking the ribosome binding site (RBS) (58). Evolutionary analysis suggests that the SgrS-*manX* mRNA interaction—facilitated by a relatively less well-conserved region of SgrS and a coding region of the *manX* mRNA—evolved much later than the sRNA's association with the sugar transporter *ptsG* mRNA, whose RBS binds with the most highly conserved region of SgrS (46). This illustrates the capacity of sRNAs to gain targets over evolutionary time. In another example, Qrr3, one of the quorum-sensing sRNAs in *Vibrio* species, has four known targets and can bind mRNAs in two regions, which in turn allows the sRNA to act via four discrete mechanisms (59). mRNA targets that bind the first stem-loop of Qrr3 induce degradation of the sRNA, resulting in either mRNA activation or repression; targets that bind the second stem-loop sequester Qrr3, with downstream effects dependent in part on binding strength, which in turn is affected by even small sequence changes. While there has not been an analogous analysis of Qrr evolution, it seems likely that only one stem-loop structure was critical for an ancestral Qrr, with additional functions emerging later via sequence mutations. The evolutionary implications of sRNA mechanism diversity and divergence warrant further investigation.

Clearly, the evolutionary trajectories of sRNAs must be considered along with those of the other genetic elements with which they interact, including sRNA-binding proteins, target mRNAs, and "sponge" sRNAs (2). The evolutionary relationships between sRNAs and their mRNA targets were the subject of a survey of the relatively well-studied sRNAs in enteric bacteria (46), which linked the accumulation of new sRNAs in *Enterobacteriales* with the evolution of a longer C-terminal region in Hfq, a protein known to bind and stabilize many sRNAs in this lineage. This is significant given that the C-terminal region of Hfq, which differs between Gram-negative and Gram-positive bacteria, likely modulates access to RBSs (60). While Hfq is the best known, it is not the only sRNA-binding protein that may be critical to sRNA regulatory networks (4). For example, RapZ protein in *E. coli* acts as an adapter, recruiting an endoribonuclease and thereby inducing sRNA degradation (53). Expression of ProQ, a protein that regulates osmolarity via interactions with the ProP transporter, significantly affects sRNA abundance in *Salmonella* and has been shown to aid in translation inhibition both by increasing sRNA stability and by blocking ribosome binding on target mRNAs (61, 62). Moreover, potential coevolutionary considerations are not limited to sRNA-binding proteins: a recent comparison of *Listeria* genomes identified 12 sRNAs that appear to have coevolved with protein-coding genes linked to virulence (63). Further study is needed to better understand these coevolutionary dynamics, but their import is becoming increasingly clear.

MODES AND CONSEQUENCES OF sRNA LOSS

From the few studies that have examined sRNA evolution more broadly across species, strains, or other closely related taxa, it appears that sRNA loss in a given lineage is a fairly common, but understudied, phenomenon. It is worth noting here that given the aforementioned relative ease with which sRNAs should be able to emerge in bacteria, and the high likelihood that their initial interactions will be deleterious, sRNAs are very likely born and lost at a rapid rate. The "death rate" of miRNAs in eukaryotes has been quantified by comparing an estimated birth rate with the number of persisting miRNAs (18); for bacterial sRNAs, no such estimate is feasible, as the rate of sRNA emergence has yet to be quantified. The discussion of loss here will thus be limited to those sRNAs determined through phylogenetic analysis to be ancestral to a given lineage but lost in one or more clades, and the considerations that come to bear on selective pressures to retain or lose sRNA genes.

One such consideration stems from the aforementioned phenomenon of dual-function sRNAs. Logically, the encoding of sRNAs and mRNAs at intersecting loci should affect the retention and evolution of these genetic elements in the genome. For example, TnpA, a transposable element-derived sRNA in *Salmonella*, shares its promoter with an mRNA, which may have resulted in increased selective pressure to maintain the transposon within which it is encoded (64). It stands to reason that even if its target mRNA is lost, the sRNA may persist long enough to acquire a new target. Additionally, when it comes to sequence diversity, we would expect the evolution of dual-encoded sRNAs to be more restricted as compared to regions that encode a single RNA, but this question has yet to be systematically investigated. These dynamics regarding conservation and retention would also apply to sRNAs that are the product of mRNA processing of 3′ untranslated regions and the so-called intraRNAs whose promoters are found within protein-coding genes (65–67).

Despite the varied selective pressures that may act to maintain them, sRNA loss has been documented in groups of highly related, well-studied organisms such as the *E. coli*-*Shigella* clade. Pathogenic strains of *E. coli*, categorized as *Shigella* and enteroinvasive *E. coli*, are believed to have evolved multiple times from non-pathogenic ancestors via the acquisition of a large virulence plasmid, pINV (68–70). The transition to an intracellular, pathogenic lifestyle was accompanied by loss of protein-coding genes from the chromosomes of *Shigella* and enteroinvasive *E. coli* strains (71, 72). For example, the independent loss of *cadA* from multiple *Shigella* species is believed to have increased the virulence of these pathogens (72). This suggests that gene loss could be an adaptive process in which beneficial pseudogenizations or deletions are the product of positive selection rather than a neutral process associated with relaxed selection and genetic drift typically observed among organisms with a newly acquired intracellular lifestyle (73). The mechanism by which gene loss contributes to bacterial fitness is thought to be through the "rewiring" of regulatory networks, perhaps resetting metabolic regulation in order to develop a system more suited to the new environment (66) (Fig. 1). Even in cases when existing genes would hypothetically provide a fitness advantage in the new environment, existing regulatory network structure might limit the ability of those genes to confer the advantage (74). Given that sRNAs play a key role in these regulatory networks, it follows that sRNA gene loss could likewise be part of this rewiring, and thus an adaptive process in certain scenarios. There is possible evidence of this phenomenon in the *E. coli-Shigella* clade, where approximately one-third of sRNAs were found to have variable distribution across strains, with sRNA loss observed to be most frequent in *Shigella* and *E. coli* strains with a restricted host range (32).

Due to the paucity of studies on the topic, it is difficult to draw many broad conclusions about sRNA loss within particular bacterial lineages, though other discrete examples of absence could perhaps be explained using an adaptive framework. For example, the IGR encoding EcsR1 in *E. coli* is absent in *Salmonella* due to a genome rearrangement event that fragmented the sRNA gene and caused it to be effectively lost from the bacterium (41). EcsR1 was shown to inhibit biofilm formation, a function that would likely not be advantageous to a pathogen like *Salmonella*, and it is therefore plausible to consider that perhaps this sRNA was lost due to positive selection. Although ectopic expression of EscsR1 in *Salmonella* did not cause a significant difference in biofilm formation, it was correlated with increased expression of genes associated with pathogenicity, which could ultimately be disruptive to the highly specialized virulence-related regulatory networks *Salmonella* has evolved since its split from *E. coli* (41). McaS, another sRNA that regulates biofilm formation in

E. coli, is also missing from *Salmonella* and a few other enteric bacteria, but the evolutionary significance of its absence is not known (75, 76).

CONCLUDING REMARKS

There is clearly much left to be investigated regarding the origination, evolution, and loss of bacterial sRNAs; however, some patterns emerge from what has been studied to date. sRNAs arise from diverse origins and appear to evolve quickly, acquiring targets and integrating into regulatory circuits via mutations that form regulatory motifs, seed sequences, and protein-binding regions. They share broad structural characteristics, but are varied in their expression patterns, distribution across bacterial lineages, and network interactions. In particular, sRNAs of different ages or at different stages of network integration are likely regulated differently and/or expressed at different levels. On the whole, sRNAs likely evolve faster than protein-coding genes, but a majority are still under purifying selection (21), indicative of their critical role in bacterial fitness. Similarly, at least a few cases of sRNA loss or pseudogenization are likely adaptive, as opposed to being a consequence of genome reduction or neutral processes.

Thorough, systematic investigation of these dynamics is necessary if we are to fully understand sRNA evolution. This includes continued improvement of our ability to identify and characterize novel sRNAs—particularly in groups other than well-studied enteric bacteria—and enhanced capacities to detect homology across species and genera. Furthermore, it is essential that we examine the evolution of sRNAs along with that of their targets. The observation that seed sequences change at lower rates than other sRNA gene regions can be put to work in batch identification of such sequences, helping both to trace the evolution of sRNA-mRNA interactions and to discern sRNA functions and network interactions. Drawing guidance from the study of eukaryotic miRNAs, it would be useful to more broadly characterize sRNA emergence and loss dynamics in a large number of bacterial genomes, enabling us to quantify the rates at which they emerge, evolve, and persist. Studying sRNAs in the context of these evolutionary forces will help us better understand not only sRNA emergence and change but also how these elements fit in the larger context of bacterial posttranscriptional regulation and adaptation to environmental stress.

Acknowledgments. Research in the Raghavan lab is supported by National Institutes of Health grants AI123464 and AI126385.

Citation. Dutcher HA, Raghavan R. 2018. Origin, evolution, and loss of bacterial small RNAs. Microbiol Spectrum 6(2): RWR-0004-2017.

References

1. Beisel CL, Storz G. 2010. Base pairing small RNAs and their roles in global regulatory networks. *FEMS Microbiol Rev* **34:**866–882.

2. Gottesman S, Storz G. 2011. Bacterial small RNA regulators: versatile roles and rapidly evolving variations. *Cold Spring Harb Perspect Biol* **3:**a003798.

3. Nitzan M, Rehani R, Margalit H. 2017. Integration of bacterial small RNAs in regulatory networks. *Annu Rev Biophys* **46:**131–148.

4. Storz G, Vogel J, Wassarman KM. 2011. Regulation by small RNAs in bacteria: expanding frontiers. *Mol Cell* **43:**880–891.

5. Updegrove TB, Shabalina SA, Storz G. 2015. How do base-pairing small RNAs evolve? *FEMS Microbiol Rev* **39:**379–391.

6. Hoeppner MP, Gardner PP, Poole AM. 2012. Comparative analysis of RNA families reveals distinct repertoires for each domain of life. *PLoS Comput Biol* **8:** e1002752.

7. Nawrocki EP, Burge SW, Bateman A, Daub J, Eberhardt RY, Eddy SR, Floden EW, Gardner PP, Jones TA, Tate J, Finn RD. 2015. Rfam 12.0: updates to the RNA families database. *Nucleic Acids Res* **43**(Database issue):D130–D137.

8. Finn RD, Coggill P, Eberhardt RY, Eddy SR, Mistry J, Mitchell AL, Potter SC, Punta M, Qureshi M, Sangrador-Vegas A, Salazar GA, Tate J, Bateman A. 2016. The Pfam protein families database: towards a more sustainable future. *Nucleic Acids Res* **44**(D1):D279–D285.

9. Lindgreen S, Umu SU, Lai AS, Eldai H, Liu W, McGimpsey S, Wheeler NE, Biggs PJ, Thomson NR, Barquist L, Poole AM, Gardner PP. 2014. Robust identification of noncoding RNA from transcriptomes requires phylogenetically-informed sampling. *PLoS Comput Biol* **10:**e1003907.

10. Huerta AM, Francino MP, Morett E, Collado-Vides J. 2006. Selection for unequal densities of σ^{70} promoter-like signals in different regions of large bacterial genomes. *PLoS Genet* **2:**e185.

11. Rivas E, Eddy SR. 2000. Secondary structure alone is generally not statistically significant for the detection of noncoding RNAs. *Bioinformatics* **16:**583–605.

12. Zhang Y, Huang H, Zhang D, Qiu J, Yang J, Wang K, Zhu L, Fan J, Yang J. 2017. A review on recent computational methods for predicting noncoding RNAs. *BioMed Res Int* **2017:**9139504.

13. Gardner PP, Wilm A, Washietl S. 2005. A benchmark of multiple sequence alignment programs upon structural RNAs. *Nucleic Acids Res* **33:**2433–2439.

14. Zhang J. 2003. Evolution by gene duplication: an update. *Trends Ecol Evol* **18:**292–298.

15. Ochman H, Lawrence JG, Groisman EA. 2000. Lateral gene transfer and the nature of bacterial innovation. *Nature* **405:**299–304.

16. Schlötterer C. 2015. Genes from scratch—the evolutionary fate of de novo genes. *Trends Genet* **31:**215–219.

17. Tautz D, Domazet-Lošo T. 2011. The evolutionary origin of orphan genes. *Nat Rev Genet* **12:**692–702.

18. Lyu Y, Shen Y, Li H, Chen Y, Guo L, Zhao Y, Hungate E, Shi S, Wu CI, Tang T. 2014. New microRNAs in *Drosophila*—birth, death and cycles of adaptive evolution. *PLoS Genet* **10:**e1004096.

19. Peer A, Margalit H. 2011. Accessibility and evolutionary conservation mark bacterial small-RNA target-binding regions. *J Bacteriol* **193:**1690–1701.

20. Richter AS, Backofen R. 2012. Accessibility and conservation: general features of bacterial small RNA-mRNA interactions? *RNA Biol* **9:**954–965.

21. Kacharia FR, Millar JA, Raghavan R. 2017. Emergence of new sRNAs in enteric bacteria is associated with low expression and rapid evolution. *J Mol Evol* **84:**204–213.

22. Kaessmann H. 2010. Origins, evolution, and phenotypic impact of new genes. *Genome Res* **20:**1313–1326.

23. Capra JA, Pollard KS, Singh M. 2010. Novel genes exhibit distinct patterns of function acquisition and network integration. *Genome Biol* **11:**R127.

24. Caswell CC, Oglesby-Sherrouse AG, Murphy ER. 2014. Sibling rivalry: related bacterial small RNAs and their redundant and non-redundant roles. *Front Cell Infect Microbiol* **4:**151.

25. Wilderman PJ, Sowa NA, FitzGerald DJ, FitzGerald PC, Gottesman S, Ochsner UA, Vasil ML. 2004. Identification of tandem duplicate regulatory small RNAs in *Pseudomonas aeruginosa* involved in iron homeostasis. *Proc Natl Acad Sci U S A* **101:**9792–9797.

26. Padalon-Brauch G, Hershberg R, Elgrably-Weiss M, Baruch K, Rosenshine I, Margalit H, Altuvia S. 2008. Small RNAs encoded within genetic islands of *Salmonella typhimurium* show host-induced expression and role in virulence. *Nucleic Acids Res* **36:**1913–1927.

27. Lenz DH, Mok KC, Lilley BN, Kulkarni RV, Wingreen NS, Bassler BL, Way I. 2004. The small RNA chaperone Hfq and multiple small RNAs control quorum sensing in *Vibrio harveyi* and *Vibrio cholerae*. *Cell* **118:**69–82.

28. del Val C, Rivas E, Torres-Quesada O, Toro N, Jiménez-Zurdo JI. 2007. Identification of differentially expressed small non-coding RNAs in the legume endosymbiont *Sinorhizobium meliloti* by comparative genomics. *Mol Microbiol* **66:**1080–1091.

29. Wilms I, Voss B, Hess WR, Leichert LI, Narberhaus F. 2011. Small RNA-mediated control of the *Agrobacterium tumefaciens* GABA binding protein. *Mol Microbiol* **80:** 492–506.

30. Torres-Quesada O, Millán V, Nisa-Martínez R, Bardou F, Crespi M, Toro N, Jiménez-Zurdo JI. 2013. Independent activity of the homologous small regulatory RNAs AbcR1 and AbcR2 in the legume symbiont *Sinorhizobium meliloti*. *PLoS One* **8:**e68147.

31. Papenfort K, Vogel J. 2009. Multiple target regulation by small noncoding RNAs rewires gene expression at the post-transcriptional level. *Res Microbiol* **160:**278–287.

32. Skippington E, Ragan MA. 2012. Evolutionary dynamics of small RNAs in 27 *Escherichia coli* and *Shigella* genomes. *Genome Biol Evol* 4:330–345.

33. Brantl S, Jahn N. 2015. sRNAs in bacterial type I and type III toxin-antitoxin systems. *FEMS Microbiol Rev* 39:413–427.

34. Chabelskaya S, Gaillot O, Felden B. 2010. A *Staphylococcus aureus* small RNA is required for bacterial virulence and regulates the expression of an immune-evasion molecule. *PLoS Pathog* 6:e1000927.

35. Gong H, Vu GP, Bai Y, Chan E, Wu R, Yang E, Liu F, Lu S. 2011. A *Salmonella* small non-coding RNA facilitates bacterial invasion and intracellular replication by modulating the expression of virulence factors. *PLoS Pathog* 7:e1002120.

36. Lee YH, Kim S, Helmann JD, Kim BH, Park YK. 2013. RaoN, a small RNA encoded within *Salmonella* pathogenicity island-11, confers resistance to macrophage-induced stress. *Microbiology* 159:1366–1378.

37. Hershko-Shalev T, Odenheimer-Bergman A, Elgrably-Weiss M, Ben-Zvi T, Govindarajan S, Seri H, Papenfort K, Vogel J, Altuvia S. 2016. Gifsy-1 prophage IsrK with dual function as small and messenger RNA modulates vital bacterial machineries. *PLoS Genet* 12:e1005975.

38. Sabath N, Wagner A, Karlin D. 2012. Evolution of viral proteins originated de novo by overprinting. *Mol Biol Evol* 29:3767–3780.

39. Piriyapongsa J, Jordan IK. 2007. A family of human microRNA genes from miniature inverted-repeat transposable elements. *PLoS One* 2:e203.

40. Chen Y, Zhou F, Li G, Xu Y. 2008. A recently active miniature inverted-repeat transposable element, *Chunjie*, inserted into an operon without disturbing the operon structure in *Geobacter uraniireducens* Rf4. *Genetics* 179:2291–2297.

41. Raghavan R, Kacharia FR, Millar JA, Sislak CD, Ochman H. 2015. Genome rearrangements can make and break small RNA genes. *Genome Biol Evol* 7:557–566.

42. Heinen TJ, Staubach F, Häming D, Tautz D. 2009. Emergence of a new gene from an intergenic region. *Curr Biol* 19:1527–1531.

43. Peterman N, Lavi-Itzkovitz A, Levine E. 2014. Large-scale mapping of sequence-function relations in small regulatory RNAs reveals plasticity and modularity. *Nucleic Acids Res* 42:12177–12188.

44. Piriyapongsa J, Mariño-Ramírez L, Jordan IK. 2007. Origin and evolution of human microRNAs from transposable elements. *Genetics* 176:1323–1337.

45. Smalheiser NR, Torvik VI. 2005. Mammalian microRNAs derived from genomic repeats. *Trends Genet* 21:322–326.

46. Peer A, Margalit H. 2014. Evolutionary patterns of *Escherichia coli* small RNAs and their regulatory interactions. *RNA* 20:994–1003.

47. Chen K, Rajewsky N. 2007. The evolution of gene regulation by transcription factors and microRNAs. *Nat Rev Genet* 8:93–103.

48. Jovelin R, Cutter AD. 2014. Microevolution of nematode miRNAs reveals diverse modes of selection. *Genome Biol Evol* 6:3049–3063.

49. Papenfort K, Förstner KU, Cong JP, Sharma CM, Bassler BL. 2015. Differential RNA-seq of *Vibrio cholerae* identifies the VqmR small RNA as a regulator of biofilm formation. *Proc Natl Acad Sci U S A* 112:E766–E775.

50. Otoupal PB, Erickson KE, Escalas-Bordoy A, Chatterjee A. 2017. CRISPR perturbation of gene expression alters bacterial fitness under stress and reveals underlying epistatic sonstraints. *ACS Synth Biol* 6:94–107.

51. Massé E, Vanderpool CK, Gottesman S. 2005. Effect of RyhB small RNA on global iron use in *Escherichia coli*. *J Bacteriol* 187:6962–6971.

52. Göpel Y, Lüttmann D, Heroven AK, Reichenbach B, Dersch P, Görke B. 2011. Common and divergent features in transcriptional control of the homologous small RNAs GlmY and GlmZ in *Enterobacteriaceae*. *Nucleic Acids Res* 39:1294–1309.

53. Göpel Y, Papenfort K, Reichenbach B, Vogel J, Görke B. 2013. Targeted decay of a regulatory small RNA by an adaptor protein for RNase E and counteraction by an anti-adaptor RNA. *Genes Dev* 27:552–564.

54. Massé E, Gottesman S. 2002. A small RNA regulates the expression of genes involved in iron metabolism in *Escherichia coli*. *Proc Natl Acad Sci U S A* 99:4620–4625.

55. Udekwu KI, Darfeuille F, Vogel J, Reimegård J, Holmqvist E, Wagner EG. 2005. Hfq-dependent regulation of OmpA synthesis is mediated by an antisense RNA. *Genes Dev* 19:2355–2366.

56. Song T, Mika F, Lindmark B, Liu Z, Schild S, Bishop A, Zhu J, Camilli A, Johansson J, Vogel J, Wai SN. 2008. A new *Vibrio cholerae* sRNA modulates colonization and affects release of outer membrane vesicles. *Mol Microbiol* 70:100–111.

57. Chen IK, Velicer GJ, Yu YN. 2017. Divergence of functional effects among bacterial sRNA paralogs. *BMC Evol Biol* 17:199.

58. Azam MS, Vanderpool CK. 2017. Translational regulation by bacterial small RNAs via an unusual Hfq-dependent mechanism. *Nucleic Acids Res*.

59. Feng L, Rutherford ST, Papenfort K, Bagert JD, van Kessel JC, Tirrell DA, Wingreen NS, Bassler BL. 2015. A Qrr noncoding RNA deploys four different regulatory mechanisms to optimize quorum-sensing dynamics. *Cell* 160:228–240.

60. Robinson KE, Orans J, Kovach AR, Link TM, Brennan RG. 2014. Mapping Hfq-RNA interaction surfaces using tryptophan fluorescence quenching. *Nucleic Acids Res* 42:2736–2749.

61. Smirnov A, Förstner KU, Holmqvist E, Otto A, Günster R, Becher D, Reinhardt R, Vogel J. 2016. Grad-seq guides the discovery of ProQ as a major small RNA-binding protein. *Proc Natl Acad Sci U S A* 113:11591–11596.

62. Smirnov A, Wang C, Drewry LL, Vogel J. 2017. Molecular mechanism of mRNA repression in *trans* by a ProQ-dependent small RNA. *EMBO J* 36:1029–1045.

63. Cerutti F, Mallet L, Painset A, Hoede C, Moisan A, Bécavin C, Duval M, Dussurget O, Cossart P, Gaspin C, Chiapello H. 2017. Unraveling the evolution and co-

evolution of small regulatory RNAs and coding genes in *Listeria*. *BMC Genomics* **18**:882.

64. Ellis MJ, Trussler RS, Charles O, Haniford DB. 2017. A transposon-derived small RNA regulates gene expression in *Salmonella* Typhimurium. *Nucleic Acids Res* **45**: 5470–5486.

65. Lybecker M, Bilusic I, Raghavan R. 2014. Pervasive transcription: detecting functional RNAs in bacteria. *Transcription* **5**:e944039.

66. Bilusic I, Popitsch N, Rescheneder P, Schroeder R, Lybecker M. 2014. Revisiting the coding potential of the *E. coli* genome through Hfq co-immunoprecipitation. *RNA Biol* **11**:641–654.

67. Chao Y, Papenfort K, Reinhardt R, Sharma CM, Vogel J. 2012. An atlas of Hfq-bound transcripts reveals 3′ UTRs as a genomic reservoir of regulatory small RNAs. *EMBO J* **31**:4005–4019.

68. Rolland K, Lambert-Zechovsky N, Picard B, Denamur E. 1998. *Shigella* and enteroinvasive *Escherichia coli* strains are derived from distinct ancestral strains of *E. coli*. *Microbiology* **144**:2667–2672.

69. van den Beld MJ, Reubsaet FA. 2012. Differentiation between *Shigella*, enteroinvasive *Escherichia coli* (EIEC) and noninvasive *Escherichia coli*. *Eur J Clin Microbiol Infect Dis* **31**:899–904.

70. Lan R, Alles MC, Donohoe K, Martinez MB, Reeves PR, Martinez MB. 2004. Molecular evolutionary relationships of enteroinvasive *Escherichia coli* and *Shigella* spp. *Infect Immun* **72**:5080–5088.

71. Parsot C. 2005. *Shigella* spp. and enteroinvasive *Escherichia coli* pathogenicity factors. *FEMS Microbiol Lett* **252**:11–18.

72. Day WA Jr, Fernández RE, Maurelli AT. 2001. Pathoadaptive mutations that enhance virulence: genetic organization of the *cadA* regions of *Shigella* spp. *Infect Immun* **69**:7471–7480.

73. Koskiniemi S, Sun S, Berg OG, Andersson DI. 2012. Selection-driven gene loss in bacteria. *PLoS Genet* **8**: e1002787.

74. Hottes AK, Freddolino PL, Khare A, Donnell ZN, Liu JC, Tavazoie S. 2013. Bacterial adaptation through loss of function. *PLoS Genet* **9**:e1003617.

75. Jørgensen MG, Thomason MK, Havelund J, Valentin-Hansen P, Storz G. 2013. Dual function of the McaS small RNA in controlling biofilm formation. *Genes Dev* **27**:1132–1145.

76. Thomason MK, Fontaine F, De Lay N, Storz G. 2012. A small RNA that regulates motility and biofilm formation in response to changes in nutrient availability in *Escherichia coli*. *Mol Microbiol* **84**:17–35.

Emerging Topics

VII

Regulating with RNA in Bacteria and Archaea
Edited by Gisela Storz and Kai Papenfort
© 2018 American Society for Microbiology, Washington, DC
doi:10.1128/microbiolspec.RWR-0027-2018

Cross-Regulation between Bacteria and Phages at a Posttranscriptional Level

29

Shoshy Altuvia[1], Gisela Storz[2], and Kai Papenfort[3]

INTRODUCTION

The impact that the study of phages, both in their lytic form and as prophages integrated into bacterial chromosomes, has had on molecular biology and microbiology is hard to overstate. The ease of phage manipulation helped establish several of the central dogmas in molecular biology. For example, characterization of various phage DNA polymerases contributed to the understanding of replication (1, 2), and models of transcription regulation were greatly influenced by studies of *c*I, the phage λ repressor (3, 4). Phages also have continually provided important tools such as transduction, the phage-assisted movement of DNA from one bacterium to another, which has been an essential tool since the early years of molecular biology (5, 6). As another example, the development of chain termination DNA-sequencing approaches benefited from single-stranded DNA cloning vectors derived from phage M13 (7).

Aside from their benefits as models and tools, the study of phages is important given their enormous impact on bacterial genome evolution, both as prophages integrated into the genomes and through mechanisms related to their ability to transduce genes. For instance, ~10% of the genome of *Streptococcus pyogenes*, including several pathogenicity factors, is of prophage origin, whereas ~16% of the genetic information of *Escherichia coli* O157:H7 strain Sakai traces back to 18 prophages (8). Since some prophage sequences are similar, recombination between the integrated sequences can lead to chromosomal inversions and deletions. Phage sequences also can serve as precursors of new genes (9). Additionally, an estimated 10^{25} phage infections occur worldwide every second (10). There are many mechanisms by which gene transfer takes place during these infections. These include (i) specialized transduction, whereby DNA located adjacent to an integrated prophage is transferred after imprecise excision of a prophage; (ii) gene transfer agents, prophage-like elements that package random bacterial DNA but cannot package enough to enable the transmission of their own genes; and (iii) phage-inducible chromosomal islands, which hijack helper phages to assist in their high-efficiency transfer to neighboring bacteria (11, 12). In one medically important example, the *Staphylococcus aureus* pathogenicity islands, which produce superantigens, utilize phages for effective transduction (13).

The transferred DNA can dramatically modify the recipient organism by encoding a wide range of genes, including virulence factors, toxins, secretion systems, and regulators. For bacteria to survive, the expression of the prophages or other foreign genes must be integrated into existing regulatory circuits (14–16). The diversity of the gene products encoded by phages, together with the rapidity with which these genes are integrated and transferred, results in great evolutionary pressure on the phages, prophages, and bacteria, leading to rapid changes in both the gene products and the regulatory circuits.

All of these concepts—the value of studying phage-prophage-bacterial interactions, the impact of rapid evolution, and the interwoven regulatory circuits—are

[1]Department of Microbiology and Molecular Genetics, IMRIC, The Hebrew University-Hadassah Medical School, Jerusalem, Israel; [2]Division of Molecular and Cellular Biology, Eunice Kennedy Shriver National Institute of Child Health and Human Development, Bethesda, MD 20892; [3]Faculty of Biology, Department of Microbiology, Ludwig-Maximilians-University of Munich, 82152 Martinsried, Germany.

applicable to RNA-based regulation, as we will discuss in this review.

THE CONTRIBUTION OF PHAGE BIOLOGY TO THE DEVELOPMENT OF TOOLS AND NEW CONCEPTS IN RNA-BASED REGULATION

The study of phages and the cross talk between phages and bacteria has led to a number of critical RNA-based tools for molecular biology. The most prominent tools come from the CRISPR (clustered regularly interspaced short palindromic repeat) phage defense systems, whose exploitation for genome engineering is changing molecular biology forever (17). A second class of important tools takes advantage of the high-affinity RNA-binding proteins encoded by phages. Probably the most widely utilized protein is the MS2 coat protein of the RNA phage R17, which binds a 19-nucleotide (nt) RNA hairpin, denoted MBS, with nanomolar affinity (18). MS2 and other phage RNA-binding proteins are employed in techniques that rely on these proteins for detecting and isolating correspondingly tagged RNAs in complex with their associated molecules (19).

The study of phages has also led to the identification of key factors of small RNA (sRNA)-based regulation. For example, the OOP RNA, encoded by DNA phage λ, was one of the first characterized sRNAs (20, 21). Studies of this antisense sRNA showed that it base-pairs with the cII-O mRNA, leading to degradation in a process involving RNase III (an endoribonuclease that recognizes double-stranded RNA, see also chapter 2) and possibly another RNases, ultimately resulting in decreased levels of the cII activator (20, 22). As another example, Hfq, critical to the function of base-pairing sRNAs in many bacteria, was first identified as a host factor required for RNA phage Qβ replication (23). In the Qβ context, Hfq has been proposed to alter the structure of the phage RNA, possibly allowing the 3′ end to be brought into the proximity of the replicase (24).

sRNAs REGULATING PROPHAGE-ENCODED VIRULENCE FACTORS

In pathogens, phage-mediated gene transfer has resulted in the acquisition of virulence genes. For example, the *Vibrio cholerae* toxin originates from the filamentous phage CTXΦ (25), and the emergence of new epidemic strains of *Salmonella enterica* has involved phages carrying virulence factors (26).

The role of regulatory sRNAs in the interplay of core genomic elements and the horizontally acquired

virulence genes has been particularly well studied in *S. enterica*, a model for enteric infections. This bacterium utilizes specialized protein secretion systems encoded within *S. enterica* pathogenicity islands 1 and 2 (SPI-1 and SPI-2) to deliver effector proteins that manipulate mammalian cell signaling cascades (27). Several effectors, including SopE, SspH1, SseI, and SopE2, are encoded by phages or phage remnants (8, 26).

Two core *S. enterica* genome-encoded sRNAs that impact both core-encoded and horizontally acquired virulence genes are SgrS and RprA. SgrS induction is triggered by the accumulation of nonmetabolizable, phosphorylated sugars in the cell. The SgrS role in responding to phosphosugar stress, repressing the synthesis of sugar transporters and increasing the level of a phosphatase, is conserved across several enteric bacteria (28). However, SgrS has broadened its regulatory repertoire in *S. enterica* to also repress expression of the horizontally acquired SopD effector protein (29). Phosphosugar induction of SgrS, with the concomitant repression of SopD, might help *Salmonella* adjust effector protein production to changes in carbon source availability during the infection process. The core-encoded sRNA RprA is induced by both the Rcs and Cpx two-component systems in response to cell envelope stress and activates the expression of the core-encoded stationary sigma factor σ^S. Like SgrS, RprA has broadened its selection of targets to include two prophage-derived transcripts of *S. enterica* (SL2594 and SL2705) and several mRNAs encoded by the virulence plasmid pSLT (30). Interestingly, RprA activation of *ricI*, one of the pSLT-encoded targets that inhibits plasmid transfer, involves regulation of both core-encoded (indirectly through the stationary-phase sigma factor σ^S, which activates *ricI* transcription) and horizontally acquired (directly through activation of *ricI* translation) genes.

In contrast to SgrS and RprA, two other *S. enterica* sRNAs that control the synthesis of prophage-encoded virulence factors are themselves encoded on horizontally acquired genes specific to *S. enterica*. The first of these, PinT, is strongly induced by the PhoPQ two-component system, a key regulator of *S. enterica* virulence, when the bacteria are internalized in the mammalian cells (31). By base-pairing with the corresponding mRNAs, PinT blocks further synthesis of the prophage-encoded SopE and SopE2 effectors, which are expressed early in infection to facilitate bacterial invasion. PinT also base-pairs with the mRNA that encodes Crp, the transcription factor that controls central carbon metabolism and activates transcription of genes encoding SPI-2 proteins. By regulating the temporal expression of both SPI-1 effectors and SPI-2 virulence genes, PinT facilitates *Salmonella*'s transition from an invasive state to a state capable of intracellular replication (Fig. 1A). Another *S. enterica* island-encoded sRNA, IsrM, is also expressed during infection and inhibits the production of other horizontally acquired genes, such as the SopA effector protein and HilE, a global regulator of SPI-1 transcription (32).

In two examples from other bacteria, both the sRNAs and their target genes were acquired together within a single horizontally transferred module (33). The AfaR sRNA of extraintestinal pathogenic *E. coli* is expressed from the intergenic region between the *afaABCD* and *afaE* transcription units, adjacent to a prophage locus encoding a family of afimbrial adhesins (34). AfaR base-pairs close to the *afaD* translational start site, promoting cleavage by RNase E (an endoribonuclease that recognizes single-stranded RNA and is part of the degradosome, see also chapter 1), thus reducing the pro-duction of AfaD VIII invasin protein while leaving *afaABC* unaffected (35). In another example from *V. cholerae*, the expression of the TarB sRNA from the horizontally acquired *Vibrio* pathogenicity island (VPI) is activated by the master virulence regulator ToxT (36). Upon its induction, TarB inhibits the expression of the VPI-encoded *tcpF* mRNA, which codes for an essential colonization factor of *V. cholerae* (37). It has been speculated that, similar to PinT, TarB helps to coordinate the timing of steps in the infection process by repressing TcpF expression prior to penetration of the mucous barrier of the small intestine.

sRNAs REGULATING PROPHAGE GENES ENCODING TOXINS

Prophages also encode proteins that are toxic to bacteria when synthesized at high levels. sRNA-based mech-

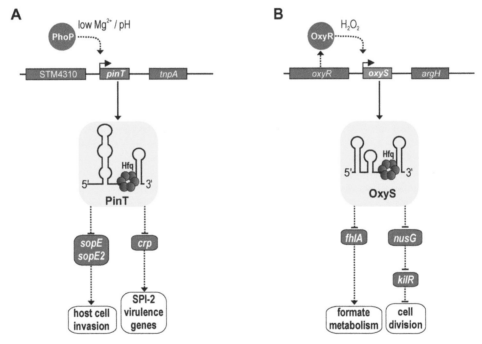

Figure 1 Repression of both prophage- and bacterial-encoded mRNAs by sRNAs encoded by horizontally acquired elements and the bacterial core genome. (A) Following host-cell invasion, the prophage-encoded sRNA PinT (purple) is activated by the core genome-encoded transcription factor PhoP (blue). PinT is an Hfq-binding sRNA that regulates multiple target genes through direct base-pairing. These include the mRNAs of the two horizontally acquired effector proteins, SopE and SopE2, as well as the core genome-encoded *crp* mRNA. The Crp protein acts as an activator of SPI-2 (intracellular) virulence genes of *S. enterica*. (B) The core genome-encoded (blue) OxyS sRNA is activated by the OxyR transcription factor under conditions of oxidative stress. OxyS associates with Hfq to regulate at least two targets: the mRNA encoding the FhlA transcription regulator of formate metabolism and the transcript encoding NusG, an important transcription termination factor. OxyS repression of NusG, which normally blocks expression of the prophage-encoded (purple) KilR protein together with the Rho termination factor, results in increased production of KilR, which transiently inhibits cell division.

anisms have evolved to modulate expression of some of these prophage toxins. One example of indirect induction of a toxin involves *E. coli* OxyS, a conserved sRNA characterized early on (38, see also chapter 13). In fact, it was studies on OxyS that revealed that Hfq functions to facilitate base-pairing of *trans*-encoded sRNAs with their targets (39). Transcription of OxyS is induced by the OxyR transcription factor in response to hydrogen peroxide, and the sRNA was found to repress mutagenesis by an unknown mechanism (38). Only a limited number of OxyS targets, such as *fhlA*, were known (40), and none could explain the toxic phenotype of OxyS overexpression. To identify additional targets, a computational search for mRNAs encoded by essential genes that could base-pair with a predominantly single-stranded section of OxyS, a region encompassing point mutations known to exacerbate or suppress the toxic phenotype, was carried out (41). This approach led to the identification of *nusG*, encoding a vital transcription termination factor, as a target of OxyS. Mutational and probing experiments confirmed that OxyS base-pairs with the transcript and blocks NusG synthesis.

NusG inhibits the production of toxic gene products encoded on horizontally acquired genomic elements, including the *rac* prophage, which carries the *kilR* gene (42–44). The KilR protein blocks cell division by interfering with the function of FtsZ, which forms the tubulin-like ring required for division (45, 46). Consistent with the model that OxyS repression of NusG results in cell killing by KilR, the effects of OxyS overexpression on cell viability and cell elongation were decreased in a strain lacking the KilR toxin. Given that the OxyS-mediated antimutator phenotype was similarly lost in the *kilR* mutant strain, it seemed plausible that induction of OxyS results in a transient reduction of NusG production, which consequently increases KilR expression from the *rac* prophage and triggers temporary growth arrest (Fig. 1B). The growth arrest, like cell cycle checkpoints in response to DNA damage in eukaryotic cells, allows for DNA repair before normal growth is resumed.

An example of an sRNA that indirectly induces expression of a toxic protein is IsrK, encoded by Gifsy-1 prophage of *Salmonella* (47). The *isrK* promoter directs the synthesis of two distinct RNA species: a short IsrK sRNA and a long mRNA, which encodes an open reading frame of unknown function (*orf45*) and an antirepressor (*anrP*) but is translationally inactive. IsrK sRNA base-pairs with the translationally inactive *orf45-anrP* mRNA to increase translation of the antirepressor protein, AnrP. AnrP in turn activates the transcription of the prophage-encoded antiterminator AntQ. Increased

levels of AntQ protein globally impact bacterial transcription termination, leading to growth arrest and ultimately cell death.

A more direct way of controlling toxin production and cell growth is exemplified by *cis*-encoded antisense RNAs that base-pair directly with the toxin mRNAs transcribed from the opposite DNA strand. While a significant number of these classic so-called type I toxin-antitoxin systems (see chapter 11) are known to be encoded by plasmids and the core bacterial genome, they are also found in phage and prophage sequences (48). We speculate that large numbers of *cis*-encoded antitoxin sRNAs remain to be characterized. For example, a distinct antisense transcript detected in *E. coli* is encoded opposite the *ymfL* gene of the lambdoid prophage e14 (49) and may silence expression of the protein. While the function of YmfL is currently unknown, overexpression of this region of e14 leads to cell filamentation (44). In a slightly different variation, the RalA sRNA is encoded in *trans*, downstream of the toxic *ralR* gene of the *rac* prophage, but shares 16 nt of complementarity and can block synthesis of the RalR endonuclease in an Hfq-dependent manner (50). Many other examples of *trans*-encoded sRNAs controlling toxin production are anticipated.

PROPHAGE sRNAs REGULATING TRANSCRIPTS ENCODED ON THE CORE GENOME

Given their sophisticated regulatory networks and the concise genomes, it is not surprising that phages and also prophages encode sRNAs (33, 51). These sRNAs are being found not only to regulate phage and prophage genes but also to modulate the expression of genes transcribed from the core genome (Table 1). One of the earliest sRNAs to be discovered in *E. coli* is DicF, an sRNA processed from a polycistronic transcript of the defective lambdoid prophage Qin/Kim (52). Expression of the transcript, which also encodes five small proteins (YdfA, YdfB, YdfC, DicB, and YdfD), is under the control of a *cI*-like repressor (53). The DicF sRNA accumulates as two isoforms in the cell (Fig. 2A). RNase E-mediated processing of the polycistronic transcript generates the 5′ end of both, while alternative Rho-independent transcription termination and RNase III-mediated processing produce the 53-nt and 72-nt variants, respectively (54).

Of the proteins encoded on the polycistronic RNA, only the DicB protein has been reported to have a biological function. The protein interacts with MinC and ZipA and thereby inhibits FtsZ polymerization and consequently cell division (55, 56). Overproduction of

Table 1 Examples of posttranscriptional cross-regulation between bacteria and phages

Species	Phage	sRNA	Bacterial target	Bacterial/phage process affected	Reference(s)
Prophage sRNAs regulating transcripts encoded on the core genome					
E. coli	Qin	DicF	ftsZ	Cell division	57, 58
			xylR, pykA, manX	Metabolism	57, 61
E. coli	Degraded prophage	EcsR2	ansB	Fumarate production	64, 65
EHEC	SpLE1	Esr41	fliC	Motility	66, 67
			cirA, bfr, chuA	Iron metabolism	68
Prophage sRNAs regulating sRNAs encoded on the core genome					
EHEC	Sp5	AgvB	GcvB	Niche colonization in cattle	66
	Sp10, SP11	AsxR	FnrS	Iron release from heme	66
Phage sRNAs regulating transcripts encoded on the core genome					
E. coli	PA-2	IpeX	ompC	Porin regulation	73
E. coli	Φ24B	24B_1	d_ant	Phage production	79
P. aeruginosa	PAK_P3	sRNA2	TΨC-tRNA loop	Translation	82
Phage proteins impacting host posttranscriptional regulation					
E. coli	T4	SrD	RNase E	Bacterial RNA decay	85
E. coli	T7	0.7	RNase E	Phage RNA stabilization	86
E. coli	T7	0.7	RNase III	Phage RNA maturation	87
P. aeruginosa	ΦKZ	Dip	RNase E	Phage RNA stabilization	88

the DicF sRNA similarly inhibits cell division, in this case by pairing with the *ftsZ* mRNA to repress translation initiation (57, 58). Consistent with this base-pairing role, DicF associates with the Hfq chaperone *in vivo* (59) and *in vitro* (60).

Two recent studies revealed possible metabolic functions for DicF. High levels of DicF inhibit the expression of metabolic genes encoding a transcription factor involved in D-xylose degradation (*xylR*), pyruvate kinase A (*pykA*), and a mannose transporter (*manX*) (57, 61). While repression of *xylR* requires the DicF 5′ end, repression of *ftsZ* and *manX* involves the very 3′ end of DicF (57, 61), which is unusual for Hfq-mediated base-pairing interactions (62) and might suggest an alternative mechanism of gene regulation. Indeed, rather than sequestering the *manX* ribosome binding site (RBS) directly, base-pairing of DicF with the *manX* coding sequence recruits Hfq to the RBS to inhibit translation initiation (61). Whether DicF affects metabolic fluxes as a consequence is currently unknown; however, another link between DicF and metabolism comes from the report that DicF is stabilized by enolase under anaerobic conditions (63). Enolase is central to the glycolytic pathway and is also part of the degradosome complex, which is key for bulk RNA turnover in *E. coli* and related organisms (see chapter 1).

Degraded prophage genes are the sources of the *E. coli* EcsR2 and *Salmonella* SesR2 sRNAs (64). Whereas very little is known about the biological function of SesR2, posttranscriptional gene regulation by EcsR2 has been studied in *E. coli*. The sRNA is expressed at low levels from the *yagU-ykgJ* intergenic region, and phylogenetic analysis suggests that the sRNA originated from a vestigial phage gene (65). Two independent experimental approaches revealed that the mRNA for a periplasmic L-asparaginase (*ansB*) is a direct interaction partner of EcsR2 and suggested that base-pairing with *ansB* requires an unstructured and conserved sequence element in the center of the sRNA. Because of its recent appearance in the *E. coli* genome, EcsR2 is considered a "young" sRNA and could therefore serve as a model to study sRNA evolution in bacteria (see chapter 28).

As discussed above, horizontal gene transfer, especially through infecting phages and lysogens, has a major impact on the evolution of pathogenic microbes. The sRNAs encoded on these virulence-related prophages were previously often overlooked; however, they are now being recognized as a source of posttranscriptional regulators. One such example is Esr41/EcOnc14 (66) from the Sakai prophage-like element (SpLE1) of enterohemorrhagic *E. coli* (EHEC) (Fig. 2B). Esr41 is ~70 nt long and was initially found to stimulate flagellin (*fliC*) expression, and consequently motility, when expressed from a multicopy plasmid (67). It is currently not known if this phenotype requires base-pairing of Esr41 with the *fliC* mRNA or rather is an indirect effect involving additional factors.

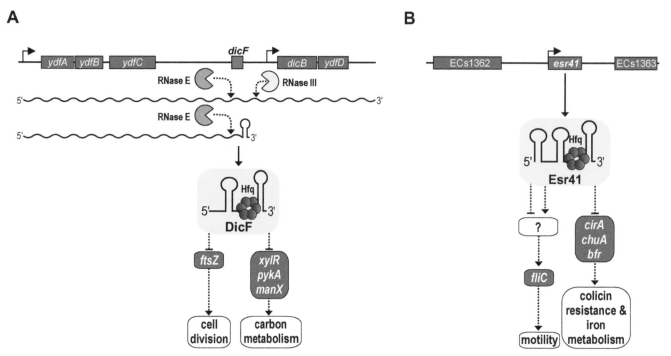

Figure 2 Prophage-encoded sRNAs that regulate the expression of host genes. (A) The prophage-encoded (purple) sRNA DicF is processed from a polycistronic transcript by RNase E, and, for the second DicF isoform, by RNase III. DicF associates with Hfq to repress synthesis of the core genome-encoded (blue) FtsZ protein, required for cell division, as well as XylR, PykA, and ManX, all involved in carbon metabolism. (B) Esr41 is a prophage-encoded (purple) sRNA that binds Hfq to inhibit translation of the core genome-encoded (blue) *cirA*, *chuA*, and *bfr* mRNAs. The gene products of the mRNAs are involved in iron metabolism, and repression of *cirA* results in colicin resistance. Esr41 also leads to increased motility by upregulation of FliC; however, the molecular mechanism underlying this process has not yet been determined.

A recent study applying CLASH (cross-linking, ligation, and sequencing of hybrids) to RNase E revealed multiple direct interaction partners of Esr41 in EHEC (68). Three mRNAs encoding an iron-siderophore complex uptake receptor (*cirA*), bacterioferritin (*bfr*), and outer membrane heme receptor (*chuA*) were confirmed to form duplexes with Esr41. In all three cases, base-pairing with Esr41 involved the RBS of the target mRNA, suggesting that Esr41 inhibits translation initiation (68). These findings were confirmed at the phenotypic level by showing that Esr41 overexpression renders *E. coli* resistant to pore-forming colicin 1A, which enters the cell through the CirA receptor. Additionally, EHEC cells deficient for *esr41* gained a fitness advantage in iron-limited medium, which might be attributable to derepression of iron transporters in the mutant (68). It is interesting to note that some of the Esr41 targets overlap with validated interaction partners of the core-encoded RyhB sRNA (see chapter 16), and that RyhB paralogs have been discovered in the

horizontally acquired elements of other enteric pathogens such as *S. enterica* (69).

PROPHAGE sRNAs REGULATING THE ACTIVITY OF CORE GENOME-ENCODED sRNAs

As discussed in detail in chapter 25, it has become clear that the activities of sRNAs themselves can be regulated by other RNAs, often referred to as "sponge RNAs," that base-pair with and block the activities of the target sRNA. Prophage examples of this type of sponge RNAs have now also been found through studies of Hfq-binding sRNAs in EHEC (66). In this study, transcripts that bound to Hfq were identified upon UV-induced cross-linking of the RNAs to affinity-tagged Hfq, followed by isolation of Hfq and deep sequencing. A surprising finding from this study was that very short sRNAs of 51 to 60 nt encoded by lambdoid prophages were among the most frequently recovered sRNAs. The sRNAs were found to be encoded in similar locations in eight different prophages, and the four most abundant transcripts carry variable 5′ regions of 14 to 18 nt together with highly conserved 3′ regions of 42 nt, which encompass a Rho-independent terminator.

Two of the abundant sRNAs, AsxR and AgvB, were characterized in more detail and found to act as "anti-sRNAs" or sponges against the core genome-encoded FnrS and GcvB sRNAs, respectively (Table 1). Transcriptomic studies examining the consequences of short-term overexpression of AsxR showed that AsxR increases the levels of the *chuAS* mRNA (encoding a heme outer membrane receptor and a heme oxygenase, respectively). These effects were at the posttranscriptional level, but the lack of homology between AsxR and the *chuAS* mRNA did not support a mechanism involving direct base-pairing. Instead, complementarity was observed between the 5′ end of AsxR and the FnrS sRNA, whose expression is highest under anaerobic conditions (70). FnrS does base-pair with and repress *chuAS* translation. Thus, by titrating the negative regulator FnrS and promoting its decay, AsxR indirectly promotes expression of *chuAS*, and potentially other FnrS target genes. AgvB, similarly, was found to alleviate GcvB sRNA-mediated repression of the *dpp* mRNA (encoding a dipeptide transporter). GcvB is highly conserved among the enterobacteria and controls a large regulon of genes coding for amino acid and peptide transporters (71). The 5′ end of AgvB is partially complementary to the conserved R1 seed region by which GcvB recognizes most of its targets, and base-pairing of AgvB with this region antagonizes its function (66). Interestingly, the core-encoded sRNA SroC uses a similar mechanism to counteract GcvB function (Fig. 3) (72). However, different from AgvB, SroC base-pairs with two distinct sequence elements to achieve GcvB degradation. It is worth noting that *E. coli* O157 strain Sakai encodes two copies of *agvB*, which might act additively to curtail GcvB function.

As for other examples discussed thus far, expression of these prophage sponge sRNAs was proposed to impact EHEC virulence. Thus, repression of FnrS, which conceivably is upregulated in the microaerobic environments of the gastrointestinal tract, by AsxR (expressed from the same prophage as Shiga toxin-2), would lead to increased synthesis of heme utilization proteins of benefit in the low iron environment of the infected mammalian cell.

PHAGE sRNAs REGULATING EXPRESSION OF TRANSCRIPTS ENCODED BY THE CORE GENOME

Similar to prophage sRNAs, sRNAs of infecting phages can control core genome expression (Table 1). One example is the IpeX sRNA, transcribed downstream of the *lc* porin gene of phage PA-2. An identical sequence is also present on the genome of the cryptic phage *qsr*

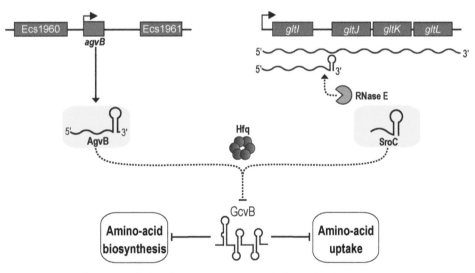

Figure 3 Prophage-encoded and core genome-encoded sRNAs that act as sponges to block the activities of core-encoded sRNAs. The prophage-encoded (purple) AgvB sRNA, as well as the core genome-encoded sRNA (blue) SroC use Hfq to base-pair with the GcvB sRNA to inhibit the function of the GcvB global regulator of amino acid uptake and metabolism. SroC is generated from RNase E-mediated endonucleolytic processing of a polycistronic transcript, while AgvB is transcribed from a freestanding gene.

(DLP12). Infection of *E. coli* with PA-2 results in the reduction of OmpC production (73), which was attributed to IpeX activity since IpeX overexpression from a plasmid also inhibits OmpC and OmpF production in *E. coli* (74). However, the absence of convincing sequence complementarity between IpeX and *ompC* suggests that regulation might require additional factors. Similar to DicF, IpeX production was reported to require processing from a larger transcript, a feature that has now been reported for several other sRNAs from diverse microbes (75–78).

An unusual phage-derived sRNA is 24B_1, encoded on the genome of the Shiga toxin-converting phage Φ24B. Different from the bacterial sRNAs described above, 24B_1 is processed from an ~80-nt precursor and accumulates as a transcript of only ~20 nt in the cell (79). As such, 24B_1 has been suggested to resemble the ubiquitous eukaryotic microRNAs (80), a class of posttranscriptional regulators proposed to also exist in bacteria (81). Deletion of 24B_1 from the Φ24B genome has multiple effects on the physiology of the phage, including more efficient prophage induction, increased phage production, and differential bacterial cell adsorption (79). The molecular mechanisms underlying these phenotypes are yet to be discovered, and it will be interesting to explore if these microRNA-sized transcripts also work through base-pairing with mRNA targets or employ alternative regulatory mechanisms.

Unconventional types of gene regulation might also be employed by sRNA1 and sRNA2, which are expressed from the genome of the PAK_P3 phage infecting *Pseudomonas aeruginosa* (82). Both sRNAs accumulate as ~100-nt transcripts and are differentially regulated during the infection process, with expression peaking during late stages of infection. Although complementarity between sRNA2 and bacterial tRNAs has been noted, it again is not clear if sRNA1 and sRNA2 function by base-pairing and whether the true targets are of bacterial and/or phage origin.

Phage- and prophage-encoded sRNA-sized transcripts have now been detected in the transcriptomes of other microbes, including relevant pathogens such as *Mycobacterium* and *Listeria* species (83, 84). These sRNAs await functional characterization, which will be key to understanding their biological functions during phage replication.

PHAGE PROTEINS IMPACTING HOST POSTTRANSCRIPTIONAL REGULATION

To promote their own proliferation, phages have evolved a multitude of mechanisms to exploit the core bacterial

machineries, increasing the expression of phage genes while limiting the expression of bacterial genes. The mechanisms include using and modifying bacterial RNA polymerases and, as we will discuss next, the bacterial machinery for degrading RNA (Table 1). For example, infection of *E. coli* by T4 phage results in rapid degradation of bacterial mRNAs. Consequently, bacterial gene expression ceases while the associated generation of ribonucleotides and free ribosomes facilitates transcription and translation of T4 genes (85). A 29-kDa phage protein denoted Srd (due to its similarity with RpoD) was suggested to be responsible for the differential degradation. The association of Srd and RNase E *in vivo*, the involvement of Srd in the turnover of the unrelated bacterial mRNAs *lpp* and *ompA*, and the importance of Srd to phage proliferation led to the suggestion that Srd stimulates RNase E activity, thus leading to rapid bacterial RNA degradation (85).

Unlike T4, the phage T7 achieves differential RNA stability by inhibiting RNase E. The protein kinase domain of gene *0.7* of T7 phage phosphorylates RNase E and the associated RNA helicase RhlB, which results in the stabilization of mRNAs that are synthesized by T7 RNA polymerase but not those synthesized by *E. coli* RNA polymerase, by mechanisms that to our knowledge are not yet understood (86). Phosphorylation by the phage T7 gp0.7 kinase conversely has been reported to stimulate the activity of RNase III (87).

A more recent study showed that the activity of the *P. aeruginosa* RNA degradosome is inhibited by Gp37/Dip (degradosome interacting protein) encoded by the unusually large phage ΦKZ. Structural studies revealed that acidic patches on the convex outer surface of Dip contact two RNA-binding sites on RNase E, thus preventing RNAs from being bound and degraded by the RNA degradosome (88). The three different phage proteins mentioned, Srd, gp0.7, and Dip, modulate RNase E by very different mechanisms. As the RNA degradation machinery is central to bacterial and phage growth and is broadly conserved, we propose that there are other phage proteins engaged in modulating RNase activity by additional mechanisms that remain to be identified.

VAST POTENTIAL TO IDENTIFY NEW RNA-BINDING PROTEINS, REGULATORY RNAS, AND UNIQUE FUNCTIONS

In general, one aspect of phage and prophage biology that is particularly exciting, also for investigators studying regulatory RNAs, is the vast numbers of unknown genes encoded by these elements. There are estimates of

10^{31} phage particles worldwide (89). Although some phage genes are similar and functionally conserved, the number of potential genes encoded by this many phages is incomprehensible. These worldwide estimates likely even underestimate the number of phages with either single-stranded RNA or double-stranded RNA genomes, or those that dominate unique environments, as these classes may not be identified by standard phage isolation and genomic sequencing approaches (90, 91), or are underannotated because of their divergence from more well-characterized phages (92). One of the few characterized single-stranded RNA phages is the enterobacterial phage Qβ, whose study led to the discovery of Hfq (23). There is no doubt many of the as-yet uncharacterized genes and activities required for the proliferation of these viruses have RNA-related functions. It is worth noting that all of the examples mentioned here come from only three phage taxa (*Caudovirales*, *Inoviridae*, and *Leviviridae*), while virtually nothing is known about RNA-mediated processes for cystoviruses, plasmaviruses, tectiviruses, and microviruses, as well as the viruses that infect archaeal cells.

Likely there are a number of uncharacterized protein families encoded by phages or prophages or by bacterial genomes to modulate phage functions that carry out activities similar to those that are already known. For instance, for organisms that do not have recognizable RNA chaperone proteins such as Hfq, an analogous activity may be found among other proteins that are required for phage replication. Similarly, there is likely to be plethora of unidentified phage- and prophage-encoded sRNAs with standard base-pairing functions. Given the size of the phage metagenome, surprisingly few phage regulatory RNAs, such as the λ OOP and PA-2 IpeX sRNAs (20), have been characterized. The expectation is that there are many such sRNA regulators, especially since sRNA regulators might provide a selective advantage given that phage genomes are small and genes are densely packed. Structured RNAs encoded by phages and prophages, such as the *cis*-acting BOXA and BOXB RNA structures of phage λ and the PUT RNAs of HK022 phage, have long been known to impact transcription elongation (93, see also chapter 8). Other structured regions of phage and prophage transcripts undoubtedly will be found to have roles in transcription antitermination or additional phage functions, similar to the structured internal ribosome entry site elements required for translation of eukaryotic RNA viruses (94). Some of these elements may bind proteins or tRNAs or molecules like T-box RNAs or riboswitches (see chapters 5 and 6). It is also possible that the structured sequences are sources for sRNA species cleaved from the longer transcripts.

In addition to uncovering different permutations of known mechanisms, the study of phage and prophage RNAs and their associated proteins could lead to the discovery of unexpected new mechanisms. There still are no known homologs for many phage-encoded proteins. Proteins that are conserved across multiple phage species (core genes) are probably required for phage propagation and thus likely impact conserved bacterial processes, while those that are only present in a more limited number of species (accessory genes) may have predominantly regulatory roles. The characterization of proteins in both categories may uncover new RNA-based mechanisms. It is also worth noting that some of the largest noncoding RNAs of uncharacterized function, such as the GOLLD and ROOL RNAs, are encoded by phages and prophages (95, 96, see also chapter 30). Since their size and predicted structural complexity rival those of ribosomes, there are expectations that these large RNAs may have novel ribozyme activities.

INSIGHTS INTO THE EVOLUTION OF RNA-BASED REGULATION FROM PHAGES AND PROPHAGES

Given the evolutionary constraints imposed on the interactions between bacteria, phages, and prophages, especially during phage-mediated horizontal gene transfer, the phage-bacteria interplay is an attractive system for studying the evolution of regulatory RNAs, a relatively unexplored topic (97, see also chapter 28). Interestingly, early studies noted that the genes encoding the *E. coli* CyaR and *Salmonella* SdsR/RyeB correspond to phage attachment or integration sites (98, 99) due to features that are not yet understood but are possibly shared with tRNA genes given that they also frequently overlap attachment sites (100). As already mentioned, the *E. coli* EcsR2 and *Salmonella* SesR2 sRNAs appear to have evolved from degraded prophage genes (65). Further comparisons among genomes should lead to additional examples and better understanding of the origin and evolution of regulatory RNAs.

The bacteria-phage/prophage systems are also useful for examining the evolution of base-pairing between sRNAs and new mRNA targets. For example, if the main role of OxyS sRNA-mediated repression of *nusG* translation is to promote induction of the prophage KilR, which brings about cell stasis and allows DNA damage repair (41), did this base-pairing only evolve in bacteria that carry the *rac* prophage? OxyS is fairly broadly conserved; did OxyS pairing with mRNAs encoding other proteins affecting cell stasis evolve in organisms without the *rac* prophage? Similar questions are relevant for horizontally acquired targets such as *sopD*, which is repressed by SgrS. Interestingly, while SgrS effectively represses *sopD*, the sRNA does not repress the homolog *sopD2*, although the sequence with potential to base-pair with SgrS only differs by one nucleotide (29). Regions of base-pairing between sRNAs and their targets generally are relatively short, and frequently single nucleotide differences can make or break an interaction and therefore gene regulation, but the rules for productive versus nonproductive base-pairing still are not well understood. Beyond base-pairing-based regulation, comparisons of structured RNA elements across phage species should give insights into the evolution of these features. Given the strong selective forces, only those nucleotides or secondary or tertiary structures that are essential for regulation will be maintained.

Finally, phage- and prophage-associated regulatory RNAs could be useful systems for exploring the evolution of the protein requirements for RNA function. While many bacterial sRNAs require Hfq for function, ProQ and other proteins with a FinO domain have recently been shown to facilitate sRNA-target mRNA pairing in some bacteria (101). Genes encoding potential FinO domain proteins have been discovered in genomes of many phages, and ligands of *Salmonella* ProQ include several phage-associated sRNAs (102). Future experiments will show if phage- or prophage-specific functions of ProQ exist and how these are integrated into the intrinsic gene regulatory networks orchestrated by Hfq and ProQ. A related issue is how many phage base-pairing sRNAs even require RNA chaperones for function. Conceivably, some phage-associated sRNAs, which evolved as *cis*-acting regulators, subsequently adapted to control expression of bacterial genes in *trans* without a requirement for a chaperone protein. Indeed, the Gifsy-1 IsrK sRNA does not require Hfq for function (47).

TAPPING INTO THE PHAGE AND PROPHAGE GOLD MINE OF RNA REGULATION

What are the best approaches to tap into the potential provided by phage and prophage genes? As the transcriptomes of more and more organisms are being determined, it will be critical to annotate the transcripts originating from phage or prophage sequences. This also is true for studies in which RNAs that associate with particular proteins or base-pair with specific RNAs are determined by deep sequencing. Computational searches to identify phage and prophage genes that encode RNAs with predicted secondary structures similar to known regulatory RNAs or proteins with known RNA binding

motifs, as well as genes that are syntenic with other genes encoding RNAs or proteins with known functions in RNA metabolism, could also be productive. Likely this will require iterative searches with each newly identified homolog, as was carried out to find phage-encoded transcripts with Y-RNA-like structures and homologs of Ro60 RNA-binding proteins (103, see also chapter 21). Regardless of the method for identification, detailed functional characterization of phage/prophage regulatory RNAs and RNA-binding proteins will be required to uncover their roles in phage and host physiology.

The expanding and unforeseen biological functions and molecular mechanisms uncovered by the studies of phage/prophage regulatory RNAs are expected to lead to new tools in biotechnology as well as advances in synthetic biology and phage therapy. For example, the large burst sizes and sometimes promiscuous replication mechanisms have made phages useful tools for mutational analysis, and they are now being exploited for directed evolution experiments (104). Either alone or assisted by computational predictions, phages might be used to evolve RNAs with dedicated biological functions or to construct synthetic gene regulatory circuits (105, 106). Finally, phages might be used as a potential treatment to change the human microbiome. In light of the rapidly worsening problem of multidrug-resistant bacterial pathogens, phage-directed antimicrobial therapies are currently experiencing a renaissance (107). Clinical applications of phage therapeutics require a thorough understanding of phage-controlled gene regulatory mechanisms, including RNA-based regulation (108). Thus, the continued characterization of the intricate and sophisticated RNA-based regulatory systems controlling phages and their cross talk with bacteria promises to be a fruitful direction for research for many years to come.

Acknowledgments. We thank K. Fröhlich, S. Gottesman, and S. Krishnamurthy for comments on the manuscript. Work in the lab of S.A. is supported by GIF (G-1311-416.13/ 2015); the Israel Science Foundation founded by the Israel Academy of Sciences and Humanities (711/13); the Israel Centers of Research Excellence (ICORE), Chromatin and RNA (1796/12); and DIP (AM 441/1-1 SO 568/1-1); work in the lab of G.S. is supported by the Intramural Program of the Eunice Kennedy Shriver National Institute of Child Health and Human Development; and work in the lab of K.P. is supported by DFG (Exc114-2, GRK2062, SPP2002, PA2820/1, and TRR174), the Human Frontiers Science Program (CDA00024/2016-C), GIF (G-2411-416.13/2016), and the European Research Council (StG-758212).

Citation. Altuvia S, Storz G, Papenfort K. 2018. Cross-regulation between bacteria and phages at a posttranscriptional level. Microbiol Spectrum 6(4):RWR-0027-2018.

References

1. Rittie L, Perbal B. 2008. Enzymes used in molecular biology: a useful guide. *J Cell Commun Signal* 2:25–45.

2. Benkovic SJ, Spiering MM. 2017. Understanding DNA replication by the bacteriophage T4 replisome. *J Biol Chem* 292:18434–18442.

3. Dodd IB, Shearwin KE, Egan JB. 2005. Revisited gene regulation in bacteriophage λ. *Curr Opin Genet Dev* 15:145–152.

4. Herskowitz I. 1973. Control of gene expression in bacteriophage lambda. *Annu Rev Genet* 7:289–324.

5. Zinder ND, Lederberg J. 1952. Genetic exchange in *Salmonella*. *J Bacteriol* 64:679–699.

6. Salmond GP, Fineran PC. 2015. A century of the phage: past, present and future. *Nat Rev Microbiol* 13:777–786.

7. Yanisch-Perron C, Vieira J, Messing J. 1985. Improved M13 phage cloning vectors and host strains: nucleotide sequences of the M13mp18 and pUC19 vectors. *Gene* 33:103–119.

8. Brüssow H, Canchaya C, Hardt WD. 2004. Phages and the evolution of bacterial pathogens: from genomic rearrangements to lysogenic conversion. *Microbiol Mol Biol Rev* 68:560–602.

9. Daubin V, Ochman H. 2004. Start-up entities in the origin of new genes. *Curr Opin Genet Dev* 14:616–619.

10. Pedulla ML, Ford ME, Houtz JM, Karthikeyan T, Wadsworth C, Lewis JA, Jacobs-Sera D, Falbo J, Gross J, Pannunzio NR, Brucker W, Kumar V, Kandasamy J, Keenan L, Bardarov S, Kriakov J, Lawrence JG, Jacobs WR Jr, Hendrix RW, Hatfull GF. 2003. Origins of highly mosaic mycobacteriophage genomes. *Cell* 113:171–182.

11. Soucy SM, Huang J, Gogarten JP. 2015. Horizontal gene transfer: building the web of life. *Nat Rev Genet* 16:472–482.

12. García-Aljaro C, Ballesté E, Muniesa M. 2017. Beyond the canonical strategies of horizontal gene transfer in prokaryotes. *Curr Opin Microbiol* 38:95–105.

13. Novick RP, Christie GE, Penadés JR. 2010. The phage-related chromosomal islands of Gram-positive bacteria. *Nat Rev Microbiol* 8:541–551.

14. Dorman CJ. 2014. H-NS-like nucleoid-associated proteins, mobile genetic elements and horizontal gene transfer in bacteria. *Plasmid* 75:1–11.

15. Feiner R, Argov T, Rabinovich L, Sigal N, Borovok I, Herskovits AA. 2015. A new perspective on lysogeny: prophages as active regulatory switches of bacteria. *Nat Rev Microbiol* 13:641–650.

16. Canchaya C, Fournous G, Brüssow H. 2004. The impact of prophages on bacterial chromosomes. *Mol Microbiol* 53:9–18.

17. Wright AV, Nuñez JK, Doudna JA. 2016. Biology and applications of CRISPR systems: harnessing nature's toolbox for genome engineering. *Cell* 164:29–44.

18. Bardwell VJ, Wickens M. 1990. Purification of RNA and RNA-protein complexes by an R17 coat protein affinity method. *Nucleic Acids Res* 18:6587–6594.

19. Chen Y, Varani G. 2013. Engineering RNA-binding proteins for biology. *FEBS J* 280:3734–3754.

20. Krinke L, Wulff DL. 1987. OOP RNA, produced from multicopy plasmids, inhibits λ *c*II gene expression through an RNase III-dependent mechanism. *Genes Dev* 1:1005–1013.

21. Hayes S, Szybalski W. 1973. Control of short leftward transcripts from the immunity and *ori* regions in induced coliphage lambda. *Mol Gen Genet* 126:275–290.

22. Krinke L, Wulff DL. 1990. RNase III-dependent hydrolysis of λcII-O gene mRNA mediated by λ OOP antisense RNA. *Genes Dev* 4:2223–2233.

23. Franze de Fernandez MT, Eoyang L, August JT. 1968. Factor fraction required for the synthesis of bacteriophage Qβ-RNA. *Nature* 219:588–590.

24. Barrera I, Schuppli D, Sogo JM, Weber H. 1993. Different mechanisms of recognition of bacteriophage Qβ plus and minus strand RNAs by Qβ replicase. *J Mol Biol* 232:512–521.

25. Waldor MK, Mekalanos JJ. 1996. Lysogenic conversion by a filamentous phage encoding cholera toxin. *Science* 272:1910–1914.

26. Ehrbar K, Hardt WD. 2005. Bacteriophage-encoded type III effectors in *Salmonella enterica* subspecies 1 serovar Typhimurium. *Infect Genet Evol* 5:1–9.

27. LaRock DL, Chaudhary A, Miller SI. 2015. Salmonellae interactions with host processes. *Nat Rev Microbiol* 13:191–205.

28. Bobrovskyy M, Vanderpool CK. 2013. Regulation of bacterial metabolism by small RNAs using diverse mechanisms. *Annu Rev Genet* 47:209–232.

29. Papenfort K, Podkaminski D, Hinton JC, Vogel J. 2012. The ancestral SgrS RNA discriminates horizontally acquired *Salmonella* mRNAs through a single G-U wobble pair. *Proc Natl Acad Sci U S A* 109:E757–E764.

30. Papenfort K, Espinosa E, Casadesús J, Vogel J. 2015. Small RNA-based feedforward loop with AND-gate logic regulates extrachromosomal DNA transfer in *Salmonella*. *Proc Natl Acad Sci U S A* 112:E4772–E4781.

31. Westermann AJ, Förstner KU, Amman F, Barquist L, Chao Y, Schulte LN, Müller L, Reinhardt R, Stadler PF, Vogel J. 2016. Dual RNA-seq unveils noncoding RNA functions in host-pathogen interactions. *Nature* 529:496–501.

32. Gong H, Vu GP, Bai Y, Chan E, Wu R, Yang E, Liu F, Lu S. 2011. A *Salmonella* small non-coding RNA facilitates bacterial invasion and intracellular replication by modulating the expression of virulence factors. *PLoS Pathog* 7:e1002120.

33. Fröhlich KS, Papenfort K. 2016. Interplay of regulatory RNAs and mobile genetic elements in enteric pathogens. *Mol Microbiol* 101:701–713.

34. Pichon C, du Merle L, Caliot ME, Trieu-Cuot P, Le Bouguénec C. 2012. An *in silico* model for identification of small RNAs in whole bacterial genomes: characterization of antisense RNAs in pathogenic *Escherichia coli* and *Streptococcus agalactiae* strains. *Nucleic Acids Res* 40:2846–2861.

35. Pichon C, du Merle L, Lequeutre I, Le Bouguénec C. 2013. The AfaR small RNA controls expression of the

AfaD-VIII invasin in pathogenic *Escherichia coli* strains. *Nucleic Acids Res* **41**:5469–5482.

36. Bradley ES, Bodi K, Ismail AM, Camilli A. 2011. A genome-wide approach to discovery of small RNAs involved in regulation of virulence in *Vibrio cholerae*. *PLoS Pathog* **7**:e1002126.

37. Kirn TJ, Bose N, Taylor RK. 2003. Secretion of a soluble colonization factor by the TCP type 4 pilus biogenesis pathway in *Vibrio cholerae*. *Mol Microbiol* **49**:81–92.

38. Altuvia S, Weinstein-Fischer D, Zhang A, Postow L, Storz G. 1997. A small, stable RNA induced by oxidative stress: role as a pleiotropic regulator and antimutator. *Cell* **90**:43–53.

39. Zhang A, Wassarman KM, Ortega J, Steven AC, Storz G. 2002. The Sm-like Hfq protein increases OxyS RNA interaction with target mRNAs. *Mol Cell* **9**:11–22.

40. Altuvia S, Zhang A, Argaman L, Tiwari A, Storz G. 1998. The *Escherichia coli* OxyS regulatory RNA represses *fhlA* translation by blocking ribosome binding. *EMBO J* **17**:6069–6075.

41. Barshishat S, Elgrably-Weiss M, Edelstein J, Georg J, Govindarajan S, Haviv M, Wright PR, Hess WR, Altuvia S. 2017. OxyS small RNA induces cell cycle arrest to allow DNA damage repair. *EMBO J* **37**:413–426.

42. Cardinale CJ, Washburn RS, Tadigotla VR, Brown LM, Gottesman ME, Nudler E. 2008. Termination factor Rho and its cofactors NusA and NusG silence foreign DNA in *E. coli*. *Science* **320**:935–938.

43. Conter A, Bouché JP, Dassain M. 1996. Identification of a new inhibitor of essential division gene *ftsZ* as the *kil* gene of defective prophage Rac. *J Bacteriol* **178**:5100–5104.

44. Burke C, Liu M, Britton W, Triccas JA, Thomas T, Smith AL, Allen S, Salomon R, Harry E. 2013. Harnessing single cell sorting to identify cell division genes and regulators in bacteria. *PLoS One* **8**:e60964.

45. Hernández-Rocamora VM, Alfonso C, Margolin W, Zorrilla S, Rivas G. 2015. Evidence that bacteriophage λ Kil peptide inhibits bacterial cell division by disrupting FtsZ protofilaments and sequestering protein subunits. *J Biol Chem* **290**:20325–20335.

46. Haeusser DP, Hoashi M, Weaver A, Brown N, Pan J, Sawitzke JA, Thomason LC, Court DL, Margolin W. 2014. The Kil peptide of bacteriophage λ blocks *Escherichia coli* cytokinesis via ZipA-dependent inhibition of FtsZ assembly. *PLoS Genet* **10**:e1004217.

47. Hershko-Shalev T, Odenheimer-Bergman A, Elgrably-Weiss M, Ben-Zvi T, Govindarajan S, Seri H, Papenfort K, Vogel J, Altuvia S. 2017. Gifsy-1 prophage IsrK with dual function as small and messenger RNA modulates vital bacterial machineries. *PLoS Genet* **12**:e1005975.

48. Fozo EM, Makarova KS, Shabalina SA, Yutin N, Koonin EV, Storz G. 2010. Abundance of type I toxin-antitoxin systems in bacteria: searches for new candidates and discovery of novel families. *Nucleic Acids Res* **38**:3743–3759.

49. Thomason MK, Bischler T, Eisenbart SK, Förstner KU, Zhang A, Herbig A, Nieselt K, Sharma CM, Storz G. 2015. Global transcriptional start site mapping using differential RNA sequencing reveals novel antisense RNAs in *Escherichia coli*. *J Bacteriol* **197**:18–28.

50. Guo Y, Quiroga C, Chen Q, McAnulty MJ, Benedik MJ, Wood TK, Wang X. 2014. RalR (a DNase) and RalA (a small RNA) form a type I toxin-antitoxin system in *Escherichia coli*. *Nucleic Acids Res* **42**:6448–6462.

51. Shinhara A, Matsui M, Hiraoka K, Nomura W, Hirano R, Nakahigashi K, Tomita M, Mori H, Kanai A. 2011. Deep sequencing reveals as-yet-undiscovered small RNAs in *Escherichia coli*. *BMC Genomics* **12**:428.

52. Bouché F, Bouché JP. 1989. Genetic evidence that DicF, a second division inhibitor encoded by the *Escherichia coli dicB* operon, is probably RNA. *Mol Microbiol* **3**:991–994.

53. Bejar S, Bouché F, Bouché JP. 1988. Cell division inhibition gene *dicB* is regulated by a locus similar to lambdoid bacteriophage immunity loci. *Mol Gen Genet* **212**:11–19.

54. Faubladier M, Cam K, Bouché JP. 1990. *Escherichia coli* cell division inhibitor DicF-RNA of the *dicB* operon. Evidence for its generation *in vivo* by transcription termination and by RNase III and RNase E-dependent processing. *J Mol Biol* **212**:461–471.

55. Johnson JE, Lackner LL, Hale CA, de Boer PA. 2004. ZipA is required for targeting of DMinC/DicB, but not DMinC/MinD, complexes to septal ring assemblies in *Escherichia coli*. *J Bacteriol* **186**:2418–2429.

56. Zhou H, Lutkenhaus J. 2005. MinC mutants deficient in MinD- and DicB-mediated cell division inhibition due to loss of interaction with MinD, DicB, or a septal component. *J Bacteriol* **187**:2846–2857.

57. Balasubramanian D, Ragunathan PT, Fei J, Vanderpool CK. 2016. A prophage-encoded small RNA controls metabolism and cell division in *Escherichia coli*. *mSystems* **1**:e00021-15.

58. Tétart F, Bouché JP. 1992. Regulation of the expression of the cell-cycle gene *ftsZ* by DicF antisense RNA. Division does not require a fixed number of FtsZ molecules. *Mol Microbiol* **6**:615–620.

59. Zhang A, Wassarman KM, Rosenow C, Tjaden BC, Storz G, Gottesman S. 2003. Global analysis of small RNA and mRNA targets of Hfq. *Mol Microbiol* **50**:1111–1124.

60. Olejniczak M. 2011. Despite similar binding to the Hfq protein regulatory RNAs widely differ in their competition performance. *Biochemistry* **50**:4427–4440.

61. Azam MS, Vanderpool CK. 2017. Translational regulation by bacterial small RNAs via an unusual Hfq-dependent mechanism. *Nucleic Acids Res* **46**:2585–2599.

62. Vogel J, Luisi BF. 2011. Hfq and its constellation of RNA. *Nat Rev Microbiol* **9**:578–589.

63. Murashko ON, Lin-Chao S. 2017. *Escherichia coli* responds to environmental changes using enolasic degradosomes and stabilized DicF sRNA to alter cellular morphology. *Proc Natl Acad Sci U S A* **114**:E8025–E8034.

64. Raghavan R, Kacharia FR, Millar JA, Sislak CD, Ochman H. 2015. Genome rearrangements can make and break small RNA genes. *Genome Biol Evol* **7**:557–566.

65. Kacharia FR, Millar JA, Raghavan R. 2017. Emergence of new sRNAs in enteric bacteria is associated with low expression and rapid evolution. *J Mol Evol* **84**:204–213.

66. Tree JJ, Granneman S, McAteer SP, Tollervey D, Gally DL. 2014. Identification of bacteriophage-encoded anti-sRNAs in pathogenic *Escherichia coli*. *Mol Cell* **55**:199–213.

67. Sudo N, Soma A, Muto A, Iyoda S, Suh M, Kurihara N, Abe H, Tobe T, Ogura Y, Hayashi T, Kurokawa K, Ohnishi M, Sekine Y. 2014. A novel small regulatory RNA enhances cell motility in enterohemorrhagic *Escherichia coli*. *J Gen Appl Microbiol* **60**:44–50.

68. Waters SA, McAteer SP, Kudla G, Pang I, Deshpande NP, Amos TG, Leong KW, Wilkins MR, Strugnell R, Gally DL, Tollervey D, Tree JJ. 2017. Small RNA interactome of pathogenic *E. coli* revealed through crosslinking of RNase E. *EMBO J* **36**:374–387.

69. Padalon-Brauch G, Hershberg R, Elgrably-Weiss M, Baruch K, Rosenshine I, Margalit H, Altuvia S. 2008. Small RNAs encoded within genetic islands of *Salmonella typhimurium* show host-induced expression and role in virulence. *Nucleic Acids Res* **36**:1913–1927.

70. Durand S, Storz G. 2010. Reprogramming of anaerobic metabolism by the FnrS small RNA. *Mol Microbiol* **75**:1215–1231.

71. Sharma CM, Papenfort K, Pernitzsch SR, Mollenkopf HJ, Hinton JC, Vogel J. 2011. Pervasive post-transcriptional control of genes involved in amino acid metabolism by the Hfq-dependent GcvB small RNA. *Mol Microbiol* **81**:1144–1165.

72. Miyakoshi M, Chao Y, Vogel J. 2015. Cross talk between ABC transporter mRNAs via a target mRNA-derived sponge of the GcvB small RNA. *EMBO J* **34**:1478–1492.

73. Schnaitman C, Smith D, de Salsas MF. 1975. Temperate bacteriophage which causes the production of a new major outer membrane protein by *Escherichia coli*. *J Virol* **15**:1121–1130.

74. Castillo-Keller M, Vuong P, Misra R. 2006. Novel mechanism of *Escherichia coli* porin regulation. *J Bacteriol* **188**:576–586.

75. Chao Y, Papenfort K, Reinhardt R, Sharma CM, Vogel J. 2012. An atlas of Hfq-bound transcripts reveals 3′ UTRs as a genomic reservoir of regulatory small RNAs. *EMBO J* **31**:4005–4019.

76. Chao Y, Vogel J. 2016. A 3′ UTR-derived small RNA provides the regulatory noncoding arm of the inner membrane stress response. *Mol Cell* **61**:352–363.

77. Papenfort K, Förstner KU, Cong JP, Sharma CM, Bassler BL. 2015. Differential RNA-seq of *Vibrio cholerae* identifies the VqmR small RNA as a regulator of biofilm formation. *Proc Natl Acad Sci U S A* **112**:E766–E775.

78. Davis BM, Waldor MK. 2007. RNase E-dependent processing stabilizes MicX, a *Vibrio cholerae* sRNA. *Mol Microbiol* **65**:373–385.

79. Nejman-Faleńczyk B, Bloch S, Licznerska K, Dydecka A, Felczykowska A, Topka G, Węgrzyn A, Węgrzyn G. 2015. A small, microRNA-size, ribonucleic acid regulating gene expression and development of Shiga toxin-converting bacteriophage Φ24$_B$. *Sci Rep* **5**:10080.

80. Bartel DP. 2004. MicroRNAs: genomics, biogenesis, mechanism, and function. *Cell* **116**:281–297.

81. Bloch S, Węgrzyn A, Węgrzyn G, Nejman-Faleńczyk B. 2017. Small and smaller—sRNAs and microRNAs in the regulation of toxin gene expression in prokaryotic cells: a mini-review. *Toxins (Basel)* **9**:E181.

82. Chevallereau A, Blasdel BG, De Smet J, Monot M, Zimmermann M, Kogadeeva M, Sauer U, Jorth P, Whiteley M, Debarbieux L, Lavigne R. 2016. Next-generation "-omics" approaches reveal a massive alteration of host RNA metabolism during bacteriophage infection of *Pseudomonas aeruginosa*. *PLoS Genet* **12**:e1006134.

83. Mraheil MA, Billion A, Mohamed W, Mukherjee K, Kuenne C, Pischimarov J, Krawitz C, Retey J, Hartsch T, Chakraborty T, Hain T. 2011. The intracellular sRNA transcriptome of *Listeria monocytogenes* during growth in macrophages. *Nucleic Acids Res* **39**:4235–4248.

84. Dedrick RM, Marinelli LJ, Newton GL, Pogliano K, Pogliano J, Hatfull GF. 2013. Functional requirements for bacteriophage growth: gene essentiality and expression in mycobacteriophage Giles. *Mol Microbiol* **88**:577–589.

85. Qi D, Alawneh AM, Yonesaki T, Otsuka Y. 2015. Rapid degradation of host mRNAs by stimulation of RNase E activity by Srd of bacteriophage T4. *Genetics* **201**:977–987.

86. Marchand I, Nicholson AW, Dreyfus M. 2001. Bacteriophage T7 protein kinase phosphorylates RNase E and stabilizes mRNAs synthesized by T7 RNA polymerase. *Mol Microbiol* **42**:767–776.

87. Mayer JE, Schweiger M. 1983. RNase III is positively regulated by T7 protein kinase. *J Biol Chem* **258**:5340–5343.

88. Van den Bossche A, Hardwick SW, Ceyssens PJ, Hendrix H, Voet M, Dendooven T, Bandyra KJ, De Maeyer M, Aertsen A, Noben JP, Luisi BF, Lavigne R. 2016. Structural elucidation of a novel mechanism for the bacteriophage-based inhibition of the RNA degradosome. *eLife* **5**:e16413.

89. Hatfull GF. 2015. Dark matter of the biosphere: the amazing world of bacteriophage diversity. *J Virol* **89**:8107–8110.

90. Decker CJ, Parker R. 2014. Analysis of double-stranded RNA from microbial communities identifies double-stranded RNA virus-like elements. *Cell Rep* **7**:898–906.

91. Kauffman KM, Hussain FA, Yang J, Arevalo P, Brown JM, Chang WK, VanInsberghe D, Elsherbini J, Sharma RS, Cutler MB, Kelly L, Polz MF. 2018. A major lineage of non-tailed dsDNA viruses as unrecognized killers of marine bacteria. *Nature* **554**:118–122.

92. Krishnamurthy SR, Janowski AB, Zhao G, Barouch D, Wang D. 2016. Hyperexpansion of RNA bacteriophage diversity. *PLoS Biol* **14**:e1002409.

93. Weisberg RA, Gottesman ME. 1999. Processive antitermination. *J Bacteriol* **181**:359–367.

94. Martinez-Salas E, Francisco-Velilla R, Fernandez-Chamorro J, Embarek AM. 2017. Insights into structural

and mechanistic features of viral IRES elements. *Front Microbiol* **8**:2629.

95. Weinberg Z, Lünse CE, Corbino KA, Ames TD, Nelson JW, Roth A, Perkins KR, Sherlock ME, Breaker RR. 2017. Detection of 224 candidate structured RNAs by comparative analysis of specific subsets of intergenic regions. *Nucleic Acids Res* **45**:10811–10823.

96. Weinberg Z, Perreault J, Meyer MM, Breaker RR. 2009. Exceptional structured noncoding RNAs revealed by bacterial metagenome analysis. *Nature* **462**:656–659.

97. Updegrove TB, Shabalina SA, Storz G. 2015. How do base-pairing small RNAs evolve? *FEMS Microbiol Rev* **39**:379–391.

98. Wassarman KM, Repoila F, Rosenow C, Storz G, Gottesman S. 2001. Identification of novel small RNAs using comparative genomics and microarrays. *Genes Dev* **15**:1637–1651.

99. Balbontín R, Figueroa-Bossi N, Casadesús J, Bossi L. 2008. Insertion hot spot for horizontally acquired DNA within a bidirectional small-RNA locus in *Salmonella enterica*. *J Bacteriol* **190**:4075–4058.

100. Reiter WD, Palm P, Yeats S. 1989. Transfer RNA genes frequently serve as integration sites for prokaryotic genetic elements. *Nucleic Acids Res* **17**:1907–1914.

101. Olejniczak M, Storz G. 2017. ProQ/FinO-domain proteins: another ubiquitous family of RNA matchmakers? *Mol Microbiol* **104**:905–915.

102. Smirnov A, Förstner KU, Holmqvist E, Otto A, Günster R, Becher D, Reinhardt R, Vogel J. 2016. Grad-seq guides the discovery of ProQ as a major small RNA-binding protein. *Proc Natl Acad Sci U S A* **113**:11591–11596.

103. Chen X, Sim S, Wurtmann EJ, Feke A, Wolin SL. 2014. Bacterial noncoding Y RNAs are widespread and mimic tRNAs. *RNA* **20**:1715–1724.

104. Esvelt KM, Carlson JC, Liu DR. 2011. A system for the continuous directed evolution of biomolecules. *Nature* **472**:499–503.

105. Rodrigo G, Landrain TE, Jaramillo A. 2012. De novo automated design of small RNA circuits for engineering synthetic riboregulation in living cells. *Proc Natl Acad Sci U S A* **109**:15271–15276.

106. Brödel AK, Isalan M, Jaramillo A. 2017. Engineering of biomolecules by bacteriophage directed evolution. *Curr Opin Biotechnol* **51**:32–38.

107. Lu TK, Koeris MS. 2011. The next generation of bacteriophage therapy. *Curr Opin Microbiol* **14**:524–531.

108. Citorik RJ, Mimee M, Lu TK. 2014. Sequence-specific antimicrobials using efficiently delivered RNA-guided nucleases. *Nat Biotechnol* **32**:1141–1145.

Regulating with RNA in Bacteria and Archaea
Edited by Gisela Storz and Kai Papenfort
© 2018 American Society for Microbiology, Washington, DC
doi:10.1128/microbiolspec.RWR-0005-2017

Large Noncoding RNAs in Bacteria

30

Kimberly A. Harris[1,2] and Ronald R. Breaker[1,2,3]

DIVERSITY OF LARGE ncRNA FUNCTIONS

Although bacteria harbor far fewer long noncoding RNAs (ncRNAs) than eukaryotes, the known classes of large, structured ncRNAs in bacteria perform essential roles in the core processes of information transfer, metabolism, and physiological adaptation (1). For example, many classes are central to genetic information processing: rRNAs act as ribozymes (2) to translate mRNAs, RNase P ribozymes process precursor tRNAs (3, 4), transfer-messenger RNAs (tmRNAs) rescue stalled ribosomes (5, 6), and riboswitches bind ions and metabolites to regulate gene expression (7–10). Furthermore, most of the large, structured ncRNA classes whose functions are known operate as ribozymes that perform essential chemical reactions such as peptide bond formation (2), RNA splicing (11, 12), and RNA cleavage (3). Two of these ribozyme classes, namely group I and group II introns, are sometimes components of selfish genetic elements that both splice mRNAs and mobilize to various regions in DNA genomes (13, 14). Of course, many self-splicing ribozymes also carry protein-coding regions, located either in their exon flanks or inserted into noncritical portions of their ribozyme structure. However, these coding regions are usually incidental to the main functions performed by the ncRNA's structure. Collectively, these ncRNAs have an enormous influence on both genetic and cellular processes, which suggests the intriguing possibility that newly found large ncRNA classes may also serve fundamental roles in biology.

Given what we currently know about bacterial ncRNAs, these additional classes can be expected to possess functions ranging from catalytic activity to gene regulation, but also may have biological and biochemical functions that have yet to be observed for RNA. It is not practical at this time to put boundaries around the possible set of functions for these ncRNAs. However, it seems highly unlikely that large ncRNAs that have extensive and nonrepetitive conserved sequences, and that have complexly folded structures, will prove to perform only simple biochemical tasks like base-pairing to another RNA or serving as a passive binding site for a protein factor.

Notably, where are the ribozymes that promote chemical transformations typical of metabolic enzymes? To date, ribosomes are the only natural ribozymes known to perform chemistry other than phosphoester transfer or hydrolysis. If a newly found ribozyme class performs a critical task but is a legacy biocatalyst from the RNA world (15) that has persisted in deeply branching bacterial lineages, the RNA might have a role that has been replaced by proteins in eukaryotic organisms. Or it is also possible that the ncRNA serves a purpose for which an RNA molecule is well suited, and, therefore, the ncRNA has emerged more recently in evolution.

To date, seven classes of large, structured ncRNAs (consistently of >200 nucleotides) with established biochemical functions are known to exist in bacteria (Fig. 1). These RNA classes are defined by their distinct biochemical functions and/or distinct consensus sequence and structural models. The signal recognition particle (SRP) RNA is excluded here because many representatives are <200 nucleotides long (16). Three additional putative classes of large ncRNAs, called OLE (17), GOLLD (18), and HEARO (18), have been investigated to some degree, but their biochemical roles are not well understood. Another 18 possible large ncRNA classes were recently discovered by using bioinformatics and await experimental validation (19). If all of these candidates indeed represent additional classes of large ncRNAs, establishing their functions would increase the number of large ncRNA activities in bacteria by 4 fold.

Based on the considerable size and structural sophistication of these ncRNA candidates, we believe that opportunities exist to discover entirely new biological and

[1]Howard Hughes Medical Institute; [2]Department of Molecular, Cellular and Developmental Biology; [3]Department of Molecular Biophysics and Biochemistry, Yale University, New Haven, CT 06520.

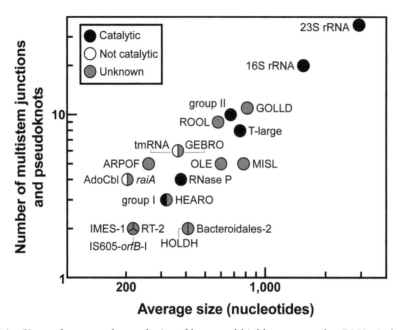

Figure 1 Size and structural complexity of large and highly structured ncRNAs in bacteria. Structural complexity is represented by the number of multistem junctions and pseudoknots present in the predicted secondary-structure models, as described previously (18). Overlapping points representing different ncRNAs are depicted with split circles. Narrowly distributed ncRNAs and ncRNAs with <2 multistem junctions and pseudoknots were omitted. For example, noncoding RNAs such as large sRNAs and clustered regularly interspaced short palindromic repeat (CRISPR) RNAs are commonly >200 nucleotides long but have repetitive and simple hairpin secondary structures that are bound by proteins. Although 23S rRNA forms the active site for the peptidyltransferase reaction catalyzed by ribosomes, 16S rRNA functions in complex with the catalytic RNA component and is classified accordingly.

biochemical roles for RNA in modern cells. Importantly, all structured bacterial ncRNAs of >350 nucleotides with known functions are catalytic RNAs or function in catalytic complexes (Fig. 1). Moreover, by using the number of multistem junctions and pseudoknots as a metric for structural complexity, we also note that ncRNAs with the most sophisticated structures tend to be ribozymes. These observations support the hypothesis that some of the additional mysterious large ncRNAs with sizes and structural complexities similar to those of known ribozymes might possess hidden catalytic abilities.

DISCOVERY OF NOVEL ncRNAs

Computational search strategies have been very productive for revealing hundreds of structured ncRNA classes in bacteria. For example, comparative sequence analysis algorithms have been used to identify multiple novel classes of riboswitches (19–23) and ribozymes (24, 25), as well as exceptionally large and structurally complex ncRNAs (17, 18; A. Roth, Z. Weinberg, K.

Vanderschuren, M. H. Murdock, E. Poiata, and R. R. Breaker, unpublished data). Each of the ncRNA classes discussed below was uncovered computationally through such phylogenetic analyses. This general method identifies nucleotide positions that exhibit strong sequence conservation and secondary-structure features that are supported by nucleotide covariation indicative of Watson-Crick base pairing.

Bacterial genomic and metagenomic DNA sequences are particularly amenable for searches directed toward the discovery of structured RNA molecules with functions other than coding for proteins. First, the abundance of bacterial genomic sequence data provides a deep data set for conducting searches for novel ncRNA classes by using comparative sequence analysis. Second, structured RNAs such as ribozymes and riboswitches have never been observed to reside entirely within the coding regions of mRNAs. Noncoding nucleotides comprise only a small fraction of the total nucleotides within most bacterial genomes. Thus, undiscovered classes of structured RNAs are enriched in these noncoding regions, which reduces the demands placed on computer algorithms that use comparative sequence analysis as a search mechanism. Third, the expansive evolutionary separations between many diverse bacterial species allow researchers to gain confidence in novel RNA classes that remain exceptionally well conserved.

These features of bacterial genomes permit the use of computational search strategies to uncover the most common ncRNA classes. However, such RNAs that are exceedingly rare cannot easily be identified by existing comparative sequence analysis algorithms. These searches fail when they do not have sufficiently distinct representatives of a ncRNA class for comparison. Fortunately, sequencing technologies, sequence databases, and computational resources continue to grow and improve. Therefore, the future of ncRNA discovery via computational searching is promising, and offers the opportunity to discover ever-rarer ncRNAs that exist only in certain biological niches.

LARGE ncRNAs WITH UNKNOWN FUNCTIONS

Among the most common functions for the highly structured ncRNA classes, which excludes less structured classes such as bacterial small RNAs (sRNAs) (26), are RNA self-cleavage (24, 25, 27) and riboswitch-mediated ligand binding and gene control (28). Most representatives of these natural self-cleaving ribozyme and riboswitch classes are <200 nucleotides long. Thus, the discovery of a novel structured ncRNA class whose representatives are consistently >200 nucleotides long should encourage researchers to consider possible biochemical functions other than RNA self-cleavage or riboswitch regulation. Below are brief descriptions of large bacterial RNAs whose functions have yet to be established.

OLE RNA

The OLE (ornate, large, extremophilic) RNA class was first described in 2006 based on the discovery of 15 representatives from bacterial genomes (17). Currently, there are 657 unique representatives known that reside exclusively in the genomes of extremophilic species and environmental metagenomes (K. A. Harris, Z. Zhou, M. L. Peters, S. G. Wilkins, and R. R. Breaker, submitted for publication). Strikingly, this RNA is found in a wide range of species in *Firmicutes*, wherein about half of its ~600 nucleotides are conserved, with covariation supporting a complex secondary structure comprising several multistem junctions (Fig. 2) (17, 29, 30). The intricate network of bulges and loops and the positioning of conserved nucleotides suggest that this RNA

forms a complex tertiary structure that is critical for its function.

Because OLE RNAs are so widespread in anaerobic extremophiles, it is tempting to speculate that they may have a role in protecting these species from the extreme environments in which they thrive. Transcriptome analysis in *Bacillus halodurans* revealed that *ole*, the gene for OLE RNA, is one of the most highly expressed. Excluding rRNA and tRNA transcripts, the OLE RNA is the 16th-most-abundant transcript under normal growth conditions. OLE RNA abundance further increases when cells are exposed to short-chain alcohols, including ethanol, which is produced during anaerobic growth (30). Again excluding RNAs responsible for translation, OLE RNA becomes the 5th most common in cells grown in the presence of 5% ethanol. OLE RNA is surpassed in abundance only by SRP RNA, tmRNA, and two mRNAs. Under these conditions, OLE RNA transcripts are processed and remain relatively stable, with a half-life of ~3 h (30).

The *ole* gene is commonly embedded in a large operon that contains genes involved in isoprenoid biosynthesis, DNA repair, coenzyme metabolism, and transcription regulation (17). Directly downstream of *ole* is a gene of unknown function that encodes the OLE-associated protein (OAP). The position of the tandem-arranged *ole* and *oap* genes immediately downstream of the *ispA* gene and immediately upstream of the *dxs* gene is highly conserved in almost all of the bacterial genomes carrying OLE RNA (K. A. Harris, Z. Zhou, M. L. Peters, S. G. Wilkins, and R. R. Breaker, submitted for publication). This suggests that the role of OLE RNA might be related to cell membrane biochemistry, given that the IspA protein (geranyltranstransferase) and the Dxs protein (1-deoxy-D-xylulose-5-phosphate synthase) are key enzymes in the isoprenoid biosynthesis pathway (31).

OAP is a 21-kDa, predicted four-helix transmembrane protein that specifically binds OLE RNA *in vitro* (29). The complex has an apparent 2:1 OAP/OLE RNA stoichiometry, suggesting that the protein might function as a dimer. Because OAP is predicted to be a transmembrane protein and binds OLE RNA, the ability of OLE RNA to localize to cell membranes was examined by using fluorescence *in situ* hybridization (FISH) microscopy. Indeed, OLE RNA localizes to the cell membrane, but only in the presence of OAP (29). This finding again suggests that the function of OLE RNA might be related to the biochemistry of membranes, or perhaps cell walls.

Knockouts of both *ole* and *oap* show that they are not essential in *B. halodurans* (30) when grown under

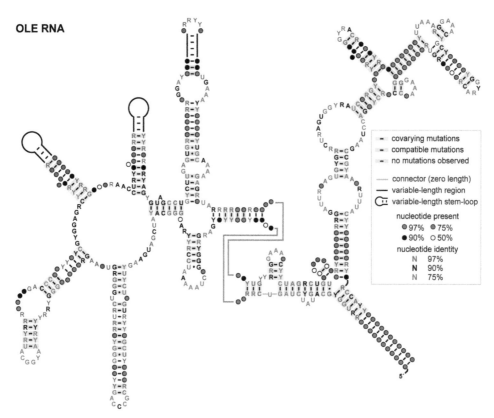

Figure 2 Consensus sequence and secondary-structure model for OLE RNAs. This model is based on the alignment of 657 unique representatives from genomic sequences from RefSeq version 63 and metagenomic sequences as described in reference 25. R and Y represent purine and pyrimidine nucleotides, respectively.

normal conditions. However, growth of *B. halodurans* strains lacking the *ole* and/or *oap* genes is reduced 5 fold compared to wild-type cells in the presence of 5% ethanol. Alcohols can cause a range of stresses to bacterial cells, such as increased membrane permeability that allows ions and small molecules to enter the cytoplasm (32). Therefore, cells have ethanol-induced responses to upregulate specific pathways and make changes to protein and lipid composition of the cell membrane (33–35). Because OLE RNA and OAP localize to cell membranes (29), the RNP may have a role in a response mechanism to this stress, such as stabilizing or producing membrane components to fortify against leakage. Furthermore, *B. halodurans* cells lacking *ole* and/or *oap* genes are less tolerant of growth in cold temperatures (30) and other growth conditions (K. A. Harris and R. R. Breaker, unpublished data). Still, it is not known how OLE RNAs help cells adapt to these stresses.

OLE RNA is currently the most prevalent structured ncRNA class of >500 nucleotides whose function is unknown. This distinction alone makes it a particularly attractive target for further analysis. Moreover, OLE

RNAs are among the most complex and well-conserved ncRNAs known to exist in bacteria (Fig. 1). Nearly all RNAs that are similar in size and structural complexity to OLE RNAs whose functions are already known (e.g., RNase P and group I and II self-splicing RNAs) function as ribozymes with biologically important activities. Therefore, establishing the biochemical function of OLE RNAs will likely reveal the action of a new ribozyme, or meaningful knowledge about a fundamental aspect of the cells that carry this molecule.

GOLLD RNA

With an average of >800 nucleotides, GOLLD (giant, ornate, lake- and *Lactobacillales*-derived) RNA is the third-largest bacterial ncRNA discovered to date, behind only 16S and 23S rRNA (18). A common arrangement originally reported for GOLLD RNAs includes numerous RNA substructures that are indicative of the formation of an exceedingly complex tertiary structure (Fig. 3). A total of 391 representatives have been identified in *Lactobacillales* and *Actinomycetales* orders and among environmental DNA sequences. The motif con-

GOLLD RNA

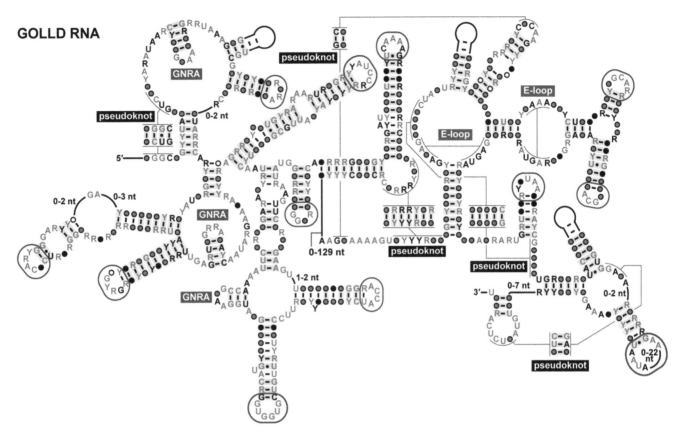

Figure 3 Consensus sequence and secondary-structure model for GOLLD RNAs. This model is based on the alignment of sequences identified in reference 18. Notable predicted substructures include 2 E-loops, 3 GNRA tetraloops, and 5 pseudoknots. Of the 20 hairpin loops, 5 form pseudoknots or represent GNRA tetraloops. A total of 12 of the remaining 15 hairpin loops carry highly conserved nucleotides, suggesting that they might be involved in forming RNA tertiary contacts that are important for the function of GOLLD RNA. Other annotations are as described for Fig. 2.

sists of distinct 5′ and 3′ domains. The 3′ half is highly conserved, contains most of the long-range interactions, and is present in all GOLLD RNAs. The 5′ half appears to diverge into variant structures where some substructures are absent or substituted (18). This type of domain variation is not uncommon for complex, structured RNAs (36).

GOLLD RNAs are commonly encoded by bacteriophages and are frequently located adjacent to tRNA genes (18). However, genes for GOLLD RNAs are sometimes present in bacterial genomes and unaffiliated with bacteriophages. This suggests that GOLLD RNAs have a biochemical function that is beneficial to both bacterial cells and the viruses that infect them or, alternatively, are a type of selfish genetic element.

Approximately 15% of the examples carry tRNAs embedded within a variable region of the motif. The significance of these overlapping arrangements is unclear.

For example, the biochemical function of GOLLD RNAs might somehow relate to the processing or activity of tRNAs. Alternatively, GOLLD RNAs might have a function that is completely independent from tRNAs, and they are only occasionally coexpressed as an efficient way to produce large amounts of specific ncRNAs that are separated by posttranscriptional processing. As observed in the initial report (17), *Lactobacillus* bacteriophages and prophages often have large noncoding regions surrounding their tRNA genes. Bacteriophages also exhibit high rates of host-parasite recombination and are capable of both horizontal and vertical genetic transfer (37, 38). Consequently, there is a possibility that GOLLD RNAs are not relevant to the bacteriophage life cycle.

Experiments performed in *Lactobacillus brevis*, which harbors the *golld* gene in a prophage, revealed that when the prophage is induced, GOLLD RNA expression levels correlated with bacteriophage particle production. Mapping of the 5′ and 3′ termini of these RNAs demonstrated that the entire predicted structure of GOLLD RNA is produced (18). Given the wide distribution of *golld* among a number of bacteriophages, it is plausible that GOLLD RNA has a useful function for phage reproduction. However, initial analysis of a bacteriophage carrying a *golld* knockout resulted in mutant phages that reproduced without evidence of a replication defect (39). Therefore, the biochemical function of GOLLD RNA does not seem to be essential for the replication of phages, at least in certain bacterial hosts.

HEARO RNA

The proposed HEARO (HNH endonuclease-associated RNA and ORF) RNAs are highly structured molecules of ~350 nucleotides surrounding an embedded open reading frame (ORF) (18). The motif does not have many highly conserved nucleotides but exhibits plentiful evidence for covariation to support the formation of many base-paired regions. HEARO representatives are located in species from 10 different bacterial phyla, predominantly *Firmicutes*, *Proteobacteria*, *Cyanobacteria*, and *Actinobacteria*. One example in the archaean *Methanosarcina mazei* has previously been reported (18). However, in some bacterial species, dozens of *hearo* genes are present. This pattern of *hearo* distribution is strongly indicative of a function as a selfish genetic element. Such mobile elements can operate as RNA or DNA (40). It is not yet clear that HEARO functions as a structured ncRNA or whether its function is manifested as a structured, single-stranded DNA (ssDNA) element.

In most instances, the motif contains an embedded ORF that encodes a putative HNH endonuclease. The presence of this protein-coding region indicates that the motif is at least occasionally transcribed. Indeed, expression of the HEARO RNA was detected in one bacterium (18). HNH endonucleases are a family of homing endonucleases, which are commonly embedded within group I and group II introns and are involved in the transfer of these elements (41). Close relatives of the HEARO ORF include the ORF associated with IS*605* selfish genetic elements. These are known to exploit small, structured DNA motifs as part of their replicative cycle (42). Each HEARO representative is much larger and more complex than these insertion element ssDNA motifs, and so it is unclear if they are related. Experiments to seek self-splicing activity of HEARO RNA transcripts have yielded no positive results (18). Further research is needed to determine if the RNA is processed, interacts with any proteins, and

ultimately is functional as an RNA polymer beyond coding for an endonuclease.

T-Large Ribozymes Are Permuted Group II Introns

In rare instances, computational approaches to discover structured ncRNAs reveal distant variants of known ncRNA classes (see, e.g., references 43–46). A large RNA called "T-large" was found (A. Roth, Z. Weinberg, K. Vanderschuren, M. H. Murdock, E. Poiata, and R. R. Breaker, unpublished data) that resembles group II self-splicing introns but is circularly permuted. This RNA does promote phosphoester transfer reactions like those required for normal RNA splicing and even exploits the same splice sites. However, these splice sites are present in the precursor RNA in reverse order, such that the ribozyme yields a circularized rather than linear exon product and yields a branched T-like product rather than a lariat intron. T-large representatives are fairly widespread in *Proteobacteria*, *Cyanobacteria*, and *Deinococcus-Thermus* phyla and are common in environmental DNA samples. Unfortunately, the biological role of the unusual RNA splicing reaction products produced by T-large ribozymes has yet to be established.

OTHER NEWLY DISCOVERED LARGE ncRNAs

With the exception of T-large, the ncRNA classes noted above have unknown biochemical functions. All the RNAs have unclear biological roles, and we are even uncertain whether the complex structure of the HEARO motif is relevant as an RNA molecule or as an ssDNA. Although we have not studied each of these ncRNA candidates continuously for the last decade since their discoveries (17, 18), the fact that their biochemical and biological functions remain mysterious is both intriguing and cautionary. Perhaps new functional features of RNA will be revealed and new biological processes exposed upon establishing the roles these RNAs serve. However, it is inherently challenging to study the biochemical functions of an RNA whose connection to a specific biological process remains obscure.

Intriguingly, there is reason to believe that additional large bacterial ncRNA classes will continue to be revealed as new genomes are sequenced and new bioinformatics methods are developed. Recently, at least 16 candidate large ncRNA classes have been reported as part of a collection of 224 structured RNA motifs (19), which add to the list of candidates reported previously (18). The most noteworthy classes are discussed below, whereas the remaining motifs are listed in Table 1. Full descriptions of all these ncRNAs can be found in the previous publications describing their initial identification (17–19).

IMES-1 RNA

Several IMES (identified in marine environmental sequences) ncRNA classes that are abundantly expressed were uncovered by comparative sequence analysis (18). With an average length of ~220 nucleotides, the IMES-1 class (also reported elsewhere as groups 3, 4, 11, and 19 putative small RNA [psRNA]) (47) is the longest of the four IMES RNA classes identified. Its secondary structure contains one pseudoknot and one multistem junction and features a large number of highly conserved nucleotides.

Metatranscriptome data from cells isolated from the Pacific Ocean revealed the presence of IMES-1 RNA with a 5-fold-higher abundance than that measured for 5S rRNA (18, 47, 48). It is unclear why these RNAs are so highly expressed, but RNAs with this high level of expression are extremely unusual. Unfortunately, the >400 examples of IMES-1 have been located only in environmental metagenomes, making it challenging to study this ncRNA and establish its biological importance.

ROOL RNA

The ROOL (rumen-originating, ornate, large) RNAs were originally identified (19) among metagenomic DNA sequences isolated from cow rumen. A total of 397 distinct examples have been identified, predominantly in the *Lactobacillales* and *Clostridiales* orders. The predicted secondary structure of the ROOL motif (19) makes it one of the most complex ncRNAs (Fig. 1). However, it has fewer highly conserved nucleotides than most other structured ncRNAs of similar size.

ROOL RNAs share some contextual similarities with GOLLD RNAs, such as frequent proximity to tRNA genes, association with bacteriophages and prophages, and occurrence in species of *Lactobacillales*. ROOL RNAs, however, do not have any recognized sequence or structural similarities to GOLLD RNAs, other than the fact that both RNA classes exhibit intricately folded structures with several pseudoknots. GOLLD and ROOL RNAs are the most complexly structured bacterial large ncRNAs of unknown function (Fig. 1). It is possible that they have similar biological functions, but certainly their distinct structures merit separate classifications.

raiA Motif RNA

Discovered in *Firmicutes* and *Actinobacteria*, the *raiA* RNA class has 1,347 representatives that exhibit a mod-

erately complex secondary structure containing two pseudoknots and two multistem junctions (19). This RNA motif is named for its frequent occurrence in the 5′ untranslated region of *raiA* genes. This gene encodes the RaiA protein, which binds ribosomes to halt translation during cell stress (49). The *raiA* motif RNA potentially functions as a *cis*-regulatory ncRNA. However, these RNAs are occasionally more than 600 base pairs upstream of the neighboring protein-coding region and are sometimes closer to the 3′ end of the gene located immediately upstream, commonly *comFC*. Bacterial *cis*-regulatory RNAs rarely reside in the 3′ untranslated regions of the genes they control. Therefore, it seems unlikely that *raiA* motif RNAs control expression of ComFC proteins, which are involved in genetic competence for DNA uptake (50).

The *raiA* motif RNA is also found upstream of genes encoding periplasmic binding proteins that transport a variety of substrates. Again, it seems possible that *raiA* motif RNAs function as riboswitches or another type of *cis*-regulatory domain, but ligand candidates that trigger changes in gene expression are not immediately apparent.

Additional Candidates

A major challenge when initially evaluating candidate structured ncRNAs is to build confidence in the hypothesis that they even represent RNAs. For example, there are three candidates that are striking with regard to their size and structural complexity, but it is possible that they actually function as ssDNAs. The ARRPOF (area required for replication in a plasmid of *Fusobacterium*) motif and the GEBRO (GC–enriched, between replication origins) motif (19) are two of the most complexly structured ncRNA candidates of unknown function (Fig. 1). They each form multiple pseudoknots and carry a large number of conserved nucleotides. These complex nucleic acid structures presumably aid in plasmid replication or perform regulatory roles associated with plasmids. Similarly, the PAGEV (plasmid-associated *Gammaproteobacteria*, especially *Vibrionales*) motif also frequently is present in plasmids (19). It is not yet certain what polynucleotide form these structures use to carry out their biological functions.

A fourth ncRNA candidate that might actually function as a ssDNA, called IS605-*orfB*-I, appears to form one multistem junction and one pseudoknot (19). Representatives of this class reside 3′ of genes encoding a transposase of the IS605 OrfB family. The IS605 family of transposases use ssDNA as a transposition intermediate (42, 51). Thus, it is possible that the IS605-*orfB*-I motif is functional as ssDNA but employs a secondary structure that is much larger and more complex than analogous elements in other IS605 representatives.

An additional 10 large ncRNA motif candidates are summarized in Table 1. These ncRNAs may be new large ncRNA classes, but it is not certain that they form structures sophisticated enough to perform challenging biochemical functions. The Bacteroidales-2, HOLDH (human oral, large, distant to HINT), MISL (mostly independently structured, large), and RT-2 (reverse transcriptase 2) motifs possess some distinct secondary-structure features. However, we do not currently have any additional clues to speculate on their functions beyond what has been stated previously (19). Clostridiales-3, EGFOA-assoc-1, ilvB-OMG, lysM-Actino, RT-7, and throat-1 are potential new large ncRNAs, but they do not appear to form complex structures (18).

EXPERIMENTAL VALIDATION OF NOVEL ncRNA FUNCTIONS

Due to the tremendous outputs of bioinformatics search pipelines, there is a growing number of interesting large ncRNA candidates to study. However, the challenge of assigning functions to these ncRNA classes remains

a difficult barrier to surpass. In past decades, novel ncRNAs were sometimes discovered by researchers who were studying a particular biological or biochemical process. As a result, they had strong clues regarding the possible function of the RNA, as was the case for RNase P (3), group I (11) and group II (12) introns, and ribosomes (2). The discovery of a novel ncRNA class by bioinformatics sometimes provides fewer clues regarding its function, and so additional experimental tactics are needed to define its biological and biochemical activities.

Below we briefly describe some of these possible experimental approaches, which have been useful for those elucidating the functions of other biomolecules, such as proteins or eukaryotic long ncRNAs (lncRNAs). This is hardly an exhaustive list, and surely each ncRNA class will need tailored experiments for in-depth analysis. As the field moves forward, new approaches will undoubtedly be developed to characterize these challenging ncRNAs.

Bacterial Genetics

Bacteria tend to cluster genes into operons that code for proteins involved in a single biochemical or physiological process, such as a metabolic pathway or a stress response. Therefore, it might be possible to infer the

Table 1   Large ncRNAs in bacteria with unpublished functions[a]

RNA	Avg size (nucleotides)	No.[b]	Nucleotide conservation (%)[c]	Taxa
ARRPOF	260	78	35	Fusobacteria
Bacteroidales-2	419	355	56	*Bacteroidales*
Clostridiales-3	252	559	40	*Clostridiales*
EGFOA-assoc-1	251	23	64	Environmental
GEBRO	349	66	49	*Streptococcus* (*Firmicutes*)
GOLLD	829	391	34	*Firmicutes* and *Actinobacteria*
HEARO	350	3,283	29	*Firmicutes*, *Proteobacteria*, *Cyanobacteria*, and *Actinobacteria*
HOLDH	401	22	52	Environmental
ilvB-OMG	209	39	70	OMG group (*Gammaproteobacteria*)
IMES-1	217	491	59	Marine environmental
IS605-orfB-I	213	444	37	*Enterococcus* (*Firmicutes*)
lysM-Actino	211	359	54	*Actinomycetales*
MISL	782	55	33	*Verrucomicrobia*
OLE	596	657	40	*Firmicutes*
PAGEV	223	123	33	*Gammaproteobacteria*
raiA	211	1,347	43	*Actinobacteria* and *Firmicutes*
ROOL	581	397	17	*Firmicutes*, fusobacteria, and *Tenericutes*
RT-2	214	482	21	*Coriobacteriales* (*Actinobacteria*) and *Clostridiales* (*Firmicutes*)
RT-7	201	202	34	*Bacteroidales*
T-large	765	291	33	*Proteobacteria*, *Cyanobacteria*, and *Deinococcus-Thermus*
throat-1	294	63	71	Throat and tongue metagenomes

[a]The initial reports of these RNA motifs were published elsewhere (references 17–19).
[b]Number of examples, with data derived from the microbial data set of RefSeq version 63.
[c]Nucleotide conservation is computed as a percentage of the average size in nucleotides divided by the total number of nucleotides conserved in 75% or more of the representatives for each motif.

function of a ncRNA that frequently clusters with genes for a given biological pathway. The information derived from the genomic location of a ncRNA has been particularly useful for determining the functions of *cis*-acting ncRNA regulatory elements, especially if the gene association of the ncRNA gene is highly conserved. For example, the location of a ncRNA sequence upstream of a metabolite synthase gene is a strong indication that the RNA functions as a riboswitch that responds to the metabolite made by the enzyme encoded immediately downstream.

If the large ncRNA of interest resides in a genetically tractable, culturable organism, genetics-based methods can be very powerful. Expression levels of the ncRNA can be determined by transcriptomics analysis under specific growth conditions (52). Gene deletions and overexpression constructs can be used to determine if the ncRNA is essential, or if there are particular phenotypes associated with the ncRNA (30, 53). Additionally, plasmids can be transformed into a knockout strain to express mutated or truncated versions of the ncRNA. If deleterious ncRNA mutations are identified, genetic screens may be used to identify gene mutations that rescue the phenotypes of defective cells or binding partners (54). Transcriptome sequencing (RNA-seq) of the knockout strains or cells grown under stress conditions may provide insight into genes upregulated to compensate for the loss of the ncRNA.

Biochemistry and Chemical Biology

Experiments that probe direct binding interactions of the ncRNA with small molecules, metabolites, or ions can be useful in determining if ncRNAs have a ligand or cofactor (55, 56). Modified nucleotides are critical for the structure and function of some of the known ncRNAs, such as rRNAs and tRNAs. Unfortunately, it is not yet known if any of the large ncRNAs described herein (Table 1) carry such modification. If a candidate ncRNA is naturally modified, this might hinder the biochemical analysis of RNAs made by *in vitro* transcription using only the four standard nucleotides. There are a myriad of RNA pulldown and copurification methods paired with mass spectrometry or sequencing to identify proteins, DNAs, or RNAs that interact with the ncRNA of interest (57, 58). Techniques such as gradient profiling by sequencing (Grad-seq), which captures RNAs based on their biochemical profiles and protein interactions, can provide valuable clues to implicate the ncRNA in particular biochemical pathways if the RNA binds proteins of known functions (59). Cells also can be treated with antibiotics or chemical inhibitors that target specific biosynthetic pathways to

provide additional phenotypic insights. For example, genetic knockout cells lacking the ncRNA might exhibit unusual growth characteristics when exposed to unusual nutrient sources, or otherwise sublethal doses of antibiotics or other toxic agents (53).

Structural Biology and Biophysics

High-resolution structure models, such as those generated by X-ray crystallography analyses, are important for ascertaining the molecular details of ncRNAs. Unfortunately, with large ncRNAs that likely interact with proteins, it has not been practical to make attempts to crystallize a potentially floppy, long ncRNA that might lack critical binding partners. Perhaps cryo-electron microscopy methods employed with ncRNAs gently removed from their cellular environments could yield useful data regarding the fine structures of ncRNAs and their biochemical partners. Otherwise, careful structural analysis might need to await the outcomes of other experiments seeking to establish fundamental details of the functions of candidate large ncRNAs.

Unlike eukaryotic lncRNAs that can be visualized in cells by FISH microscopy, bacterial cells are substantially smaller, making it difficult to precisely determine the cellular localization of RNAs in bacteria. However, with high-resolution microscopes and mathematical modeling, subcellular localization of RNAs can be determined with techniques such as FISH or live-cell MS2-GFP (bacteriophage MS2 coat protein fused to green fluorescent protein) (60). Such methods can be particularly informative if the ncRNA is naturally abundant and localizes to a prominent feature in a bacterial cell, such as the cell membrane (29).

PROSPECTS FOR LARGE ncRNA DISCOVERIES AND IMPACTS

Most of the large ncRNAs discussed herein have functions that remain elusive. Because of their striking complexity, sequence conservation, and differences from known ncRNA classes, their biochemical and biological roles are likely to be novel. The key challenge in this area will be to decipher the precise functions of these ncRNAs. New large ncRNA classes appear to present more difficulties for those seeking to establish functions than most previously discovered ribozymes and ncRNAs, which were typically discovered more serendipitously. The first ribozyme to be experimentally validated, a self-splicing group I intron, was uncovered in an mRNA that was known to be spliced (11). The RNase P RNA was known to be a component of a pre-tRNA-processing RNP complex before ribozyme activity was demonstrated (3). By contrast, large bacterial ncRNAs such as OLE RNA, GOLLD RNA, and a variety of others do not yet exhibit an obvious link to a known biochemical process.

Indeed, the biological functions of some ncRNA classes, such as tmRNAs (6) and 6S RNAs (61), took decades from their initial discovery date to elucidate. It seems likely that this lag time between discovery and functional validation might also occur for many of these newly found ncRNA classes, which are all less widespread and perhaps have less prominent functions than tmRNAs and 6S RNAs. For instance, OLE RNAs were first reported 12 years ago (17), and GOLLD RNAs (18) 9 years ago. Some progress in elucidating the structural and functional characteristics has so far only been reported for OLE RNAs (29, 30).

The accelerating rate of genome sequencing and the ever-expanding amounts of metagenomic data ensure that more novel ncRNAs will continue to be found. Despite the challenging tasks of experimentally validating their functions, each new ncRNA offers the possibility of identifying processes that have never previously been observed in biology. These and future discoveries will continue to alter the landscape of known RNA functions. If the known large ncRNAs, many of which have revolutionized our understanding of biology, are any indication of the impact of these classes of unknown function, then it will be well worth the effort.

Acknowledgments. We thank Adam Roth, Danielle Widner, and other members of the Breaker Laboratory for helpful discussions. ncRNA research in the Breaker Laboratory is supported by the National Institutes of Health grant F32GM116426 to K.A.H. and P01GM022778 to R.R.B. R.R.B is also an investigator with the Howard Hughes Medical Institute.

Citation. Harris KA, Breaker RR. 2018. Large noncoding RNAs in bacteria. Microbiol Spectrum 6(4):RWR-0005-2017.

References

1. Cech TR, Steitz JA. 2014. The noncoding RNA revolution—trashing old rules to forge new ones. *Cell* **157**:77–94.

2. Nissen P, Hansen J, Ban N, Moore PB, Steitz TA. 2000. The structural basis of ribosome activity in peptide bond synthesis. *Science* **289**:920–9370.

3. Guerrier-Takada C, Gardiner K, Marsh T, Pace N, Altman S. 1983. The RNA moiety of ribonuclease P is the catalytic subunit of the enzyme. *Cell* **35**:849–857.

4. Ellis JC, Brown JW. 2009. The RNase P family. *RNA Biol* **6**:362–369.

5. Keiler KC, Waller PR, Sauer RT. 1996. Role of a peptide tagging system in degradation of proteins synthesized from damaged messenger RNA. *Science* **271**:990–993.

6. Janssen BD, Hayes CS. 2012. The tmRNA ribosome-rescue system. *Adv Protein Chem Struct Biol* **86**:151–191.

7. Mandal M, Breaker RR. 2004. Gene regulation by riboswitches. *Nat Rev Mol Cell Biol* **5**:451–463.

8. Roth A, Breaker RR. 2009. The structural and functional diversity of metabolite-binding riboswitches. *Annu Rev Biochem* **78**:305–334.

9. Serganov A, Nudler E. 2013. A decade of riboswitches. *Cell* **152**:17–24.

10. Lotz TS, Suess B. 2018. Small molecule binding riboswitches. *Microbiol Spectr* **6**:RWR-0025-2018.

11. Kruger K, Grabowski PJ, Zaug AJ, Sands J, Gottschling DE, Cech TR. 1982. Self-splicing RNA: autoexcision and autocyclization of the ribosomal RNA intervening sequence of *Tetrahymena*. *Cell* **31**:147–157.

12. Peebles CL, Perlman PS, Mecklenburg KL, Petrillo ML, Tabor JH, Jarrell KA, Cheng HL. 1986. A self-splicing RNA excises an intron lariat. *Cell* **44**:213–223.

13. Hausner G, Hafez M, Edgell DR. 2014. Bacterial group I introns: mobile RNA catalysts. *Mob DNA* **5**:8.

14. Toro N, Jiménez-Zurdo JI, García-Rodríguez FM. 2007. Bacterial group II introns: not just splicing. *FEMS Microbiol Rev* **31**:342–358.

15. Benner SA, Ellington AD, Tauer A. 1989. Modern metabolism as a palimpsest of the RNA world. *Proc Natl Acad Sci U S A* **86**:7054–7058.

16. Rosenblad MA, Larsen N, Samuelsson T, Zwieb C. 2009. Kinship in the SRP RNA family. *RNA Biol* **6**:508–516.

17. Puerta-Fernandez E, Barrick JE, Roth A, Breaker RR. 2006. Identification of a large noncoding RNA in extremophilic eubacteria. *Proc Natl Acad Sci U S A* **103**:19490–19495.

18. Weinberg Z, Perreault J, Meyer MM, Breaker RR. 2009. Exceptional structured noncoding RNAs revealed by bacterial metagenome analysis. *Nature* **462**:656–659.

19. Weinberg Z, Lünse CE, Corbino KA, Ames TD, Nelson JW, Roth A, Perkins KR, Sherlock ME, Breaker RR. 2017. Detection of 224 candidate structured RNAs by comparative analysis of specific subsets of intergenic regions. *Nucleic Acids Res* **45**:10811–10823.

20. Barrick JE, Corbino KA, Winkler WC, Nahvi A, Mandal M, Collins J, Lee M, Roth A, Sudarsan N, Jona I, Wickiser JK, Breaker RR. 2004. New RNA motifs suggest an expanded scope for riboswitches in bacterial genetic control. *Proc Natl Acad Sci U S A* **101**:6421–6426.

21. Corbino KA, Barrick JE, Lim J, Welz R, Tucker BJ, Puskarz I, Mandal M, Rudnick ND, Breaker RR. 2005. Evidence for a second class of *S*-adenosylmethionine riboswitches and other regulatory RNA motifs in alphaproteobacteria. *Genome Biol* **6**:R70.

22. Weinberg Z, Barrick JE, Yao Z, Roth A, Kim JN, Gore J, Wang JX, Lee ER, Block KF, Sudarsan N, Neph S, Tompa M, Ruzzo WL, Breaker RR. 2007. Identification of 22 candidate structured RNAs in bacteria using the CMfinder comparative genomics pipeline. *Nucleic Acids Res* **35**:4809–4819.

23. Weinberg Z, Wang JX, Bogue J, Yang J, Corbino K, Moy RH, Breaker RR. 2010. Comparative genomics reveals

104 candidate structured RNAs from bacteria, archaea, and their metagenomes. *Genome Biol* 11:R31.

24. Roth A, Weinberg Z, Chen AG, Kim PB, Ames TD, Breaker RR. 2014. A widespread self-cleaving ribozyme class is revealed by bioinformatics. *Nat Chem Biol* 10:56–60.

25. Weinberg Z, Kim PB, Chen TH, Li S, Harris KA, Lünse CE, Breaker RR. 2015. New classes of self-cleaving ribozymes revealed by comparative genomics analysis. *Nat Chem Biol* 11:606–610.

26. Nitzan M, Rehani R, Margalit H. 2017. Integration of bacterial small RNAs in regulatory networks. *Annu Rev Biophys* 46:131–148.

27. Jimenez RM, Polanco JA, Lupták A. 2015. Chemistry and biology of self-cleaving ribozymes. *Trends Biochem Sci* 40:648–661.

28. McCown PJ, Corbino KA, Stav S, Sherlock ME, Breaker RR. 2017. Riboswitch diversity and distribution. *RNA* 23:995–1011.

29. Block KF, Puerta-Fernandez E, Wallace JG, Breaker RR. 2011. Association of OLE RNA with bacterial membranes via an RNA-protein interaction. *Mol Microbiol* 79:21–34.

30. Wallace JG, Zhou Z, Breaker RR. 2012. OLE RNA protects extremophilic bacteria from alcohol toxicity. *Nucleic Acids Res* 40:6898–6907.

31. Julsing MK, Rijpkema M, Woerdenbag HJ, Quax WJ, Kayser O. 2007. Functional analysis of genes involved in the biosynthesis of isoprene in *Bacillus subtilis*. *Appl Microbiol Biotechnol* 75:1377–1384.

32. Ingram LO. 1990. Ethanol tolerance in bacteria. *Crit Rev Biotechnol* 9:305–319.

33. Huffer S, Clark ME, Ning JC, Blanch HW, Clark DS. 2011. Role of alcohols in growth, lipid composition, and membrane fluidity of yeasts, bacteria, and archaea. *Appl Environ Microbiol* 77:6400–6408.

34. Yang S, Giannone RJ, Dice L, Yang ZK, Engle NL, Tschaplinski TJ, Hettich RL, Brown SD. 2012. *Clostridium thermocellum* ATCC27405 transcriptomic, metabolomic and proteomic profiles after ethanol stress. *BMC Genomics* 13:336.

35. Williams TI, Combs JC, Lynn BC, Strobel HJ. 2007. Proteomic profile changes in membranes of ethanol-tolerant *Clostridium thermocellum*. *Appl Microbiol Biotechnol* 74:422–432.

36. Michel F, Westhof E. 1990. Modelling of the three-dimensional architecture of group I catalytic introns based on comparative sequence analysis. *J Mol Biol* 216:585–610.

37. Mazodier P, Davies J. 1991. Gene transfer between distantly related bacteria. *Annu Rev Genet* 25:147–171.

38. Chibani-Chennoufi S, Bruttin A, Dillmann ML, Brüssow H. 2004. Phage-host interaction: an ecological perspective. *J Bacteriol* 186:3677–3686.

39. Chen AG. 2014. Functional investigation of ribozymes and ribozyme candidates in viruses, bacteria and eukaryotes. Ph.D. thesis. Yale University, New Haven, CT.

40. Curcio MJ, Derbyshire KM. 2003. The outs and ins of transposition: from mu to kangaroo. *Nat Rev Mol Cell Biol* 4:865–877.

41. Stoddard BL. 2005. Homing endonuclease structure and function. *Q Rev Biophys* 38:49–95.

42. He S, Corneloup A, Guynet C, Lavatine L, Caumont-Sarcos A, Siguier P, Marty B, Dyda F, Chandler M, Ton Hoang B. 2015. The IS*200*/IS*605* family and "peel and paste" single-strand transposition mechanism. *Microbiol Spectr* 3:MDNA3-0039-2014.

43. Webb CH, Riccitelli NJ, Ruminski DJ, Lupták A. 2009. Widespread occurrence of self-cleaving ribozymes. *Science* 326:953.

44. Perreault J, Weinberg Z, Roth A, Popescu O, Chartrand P, Ferbeyre G, Breaker RR. 2011. Identification of hammerhead ribozymes in all domains of life reveals novel structural variations. *PLoS Comput Biol* 7:e1002031.

45. McCown PJ, Liang JJ, Weinberg Z, Breaker RR. 2014. Structural, functional, and taxonomic diversity of three preQ$_1$ riboswitch classes. *Chem Biol* 21:880–889.

46. Weinberg Z, Nelson JW, Lünse CE, Sherlock ME, Breaker RR. 2017. Bioinformatic analysis of riboswitch structures uncovers variant classes with altered ligand specificity. *Proc Natl Acad Sci U S A* 114:E2077–E2085.

47. Shi Y, Tyson GW, DeLong EF. 2009. Metatranscriptomics reveals unique microbial small RNAs in the ocean's water column. *Nature* 459:266–269.

48. Frias-Lopez J, Shi Y, Tyson GW, Coleman ML, Schuster SC, Chisholm SW, DeLong EF. 2008. Microbial community gene expression in ocean surface waters. *Proc Natl Acad Sci U S A* 105:3805–3810.

49. Agafonov DE, Kolb VA, Spirin AS. 2001. Ribosome-associated protein that inhibits translation at the aminoacyl-tRNA binding stage. *EMBO Rep* 2:399–402.

50. Marchler-Bauer A, Derbyshire MK, Gonzales NR, Lu S, Chitsaz F, Geer LY, Geer RC, He J, Gwadz M, Hurwitz DI, Lanczycki CJ, Lu F, Marchler GH, Song JS, Thanki N, Wang Z, Yamashita RA, Zhang D, Zheng C, Bryant SH. 2015. CDD: NCBI's conserved domain database. *Nucleic Acids Res* 43(Database issue):D222–D226.

51. Barabas O, Ronning DR, Guynet C, Hickman AB, Ton-Hoang B, Chandler M, Dyda F. 2008. Mechanism of IS*200*/IS*605* family DNA transposases: activation and transposon-directed target site selection. *Cell* 132:208–220.

52. Creecy JP, Conway T. 2015. Quantitative bacterial transcriptomics with RNA-seq. *Curr Opin Microbiol* 23:133–140.

53. Bochner BR. 2009. Global phenotypic characterization of bacteria. *FEMS Microbiol Rev* 33:191–205.

54. Shuman HA, Silhavy TJ. 2003. The art and design of genetic screens: *Escherichia coli*. *Nat Rev Genet* 4:419–431.

55. Regulski EE, Breaker RR. 2008. In-line probing analysis of riboswitches. *Methods Mol Biol* 419:53–67.

56. Rice GM, Busan S, Karabiber F, Favorov OV, Weeks KM. 2014. SHAPE analysis of small RNAs and riboswitches. *Methods Enzymol* 549:165–187.

57. **McHugh CA, Russell P, Guttman M.** 2014. Methods for comprehensive experimental identification of RNA-protein interactions. *Genome Biol* **15:**203.

58. **Simon MD.** 2016. Insight into lncRNA biology using hybridization capture analyses. *Biochim Biophys Acta* **1859:** 121–127.

59. **Smirnov A, Förstner KU, Holmqvist E, Otto A, Günster R, Becher D, Reinhardt R, Vogel J.** 2016. Grad-seq guides the discovery of ProQ as a major small RNA-binding protein. *Proc Natl Acad Sci U S A* **113:**11591–11596.

60. **Montero Llopis P, Jackson AF, Sliusarenko O, Surovtsev I, Heinritz J, Emonet T, Jacobs-Wagner C.** 2010. Spatial organization of the flow of genetic information in bacteria. *Nature* **466:**77–81.

61. **Wassarman KM, Storz G.** 2000. 6S RNA regulates *E. coli* RNA polymerase activity. *Cell* **101:**613–623.

Regulating with RNA in Bacteria and Archaea
Edited by Gisela Storz and Kai Papenfort
© 2018 American Society for Microbiology, Washington, DC
doi:10.1128/microbiolspec.RWR-0007-2017

Synthetic Biology of Small RNAs and Riboswitches

31

Jordan K. Villa,[1,*] Yichi Su,[2,*] Lydia M. Contreras,[1,3]
and Ming C. Hammond[2,4]

INTRODUCTION: COMPARISON OF REGULATORY MECHANISMS OF sRNAs AND RIBOSWITCHES

RNAs have been known to perform a vast amount of regulatory functions in bacteria and archaea. Small RNAs (sRNAs) and riboswitches are two extensively studied classes of regulatory RNAs. sRNAs are *trans*-acting RNA elements between ~50 and 500 nucleotides (nt) in length that are either independently transcribed or processed from a nontarget mRNA, and contain imperfect complementarity to the target mRNA to perform posttranscriptional regulatory functions. On the contrary, riboswitches are *cis*-regulatory structured RNA elements in the untranslated regions of mRNAs, capable of regulating downstream gene expression through small-molecule ligand-induced conformational switching. These regulatory RNAs have revealed the precise and sophisticated nature of natural gene regulatory networks and have inspired efforts to mimic these mechanisms and functions by engineering RNA tools for an increasing number of synthetic biology applications.

There have been several recent reviews on the synthetic biology of sRNAs and riboswitches that have focused on one regulatory RNA class and mainly have been addressed to the synthetic biology field (1–6). Our approach in this review is to provide comparisons between these two classes of regulatory RNAs in bacteria, in terms of their regulatory features, current challenges to their paradigms, and synthetic biology and biotechnology applications. We hope this review sparks productive dialogue and creative exchange of ideas between synthetic biologists and researchers working on the two natural classes.

Initial Discoveries and Regulatory Models of sRNAs and Riboswitches

Several important discoveries and technical advances paved the way for understanding the importance of RNA elements in cellular gene regulation (Fig. 1). Regulatory sRNAs were first described as chromosomally encoded noncoding RNAs called antisense RNAs (asRNAs) capable of regulating expression of mRNA targets in *Escherichia coli* (7). Initially, sRNAs were discovered from [^{32}P]orthophosphate-labeled total RNA analysis via polyacrylamide gel electrophoresis (4.5S, 6S, Spot 42, transfer-messenger RNA, and RNase P RNA), found as cloned genomic fragments that modulated expression levels of a particular target (MicF, DicF, and DsrA), or identified by conditions that suggested function (CsrB and OxyS) (7–9). In contrast to sRNAs, riboswitches are embedded within the untranslated regions (UTRs) of mRNAs and thereby cannot be separately isolated as independent transcripts or genes. Instead, two contemporaneous advances set the stage for riboswitch discovery. First, *in vitro* evolution from random sequence libraries showed that short RNA sequences were capable of binding small molecules selectively (10, 11), and these "aptamers" could even allosterically regulate ribozyme activity (12, 13). Second, there was genetic evidence for regulatory sequences within the 5′ UTRs of biosynthetic pathway genes that were responsible for feedback inhibition by the metabolite product (14–16). Discovery that these 5′ UTRs contained complex RNA structures capable of direct binding to adenosylcobalamin (B$_{12}$), thiamine pyrophosphate (TPP), and flavin mononucleotide (FMN) led to the coining of the term "riboswitches"

[1]Institute of Cellular and Molecular Biology, The University of Texas at Austin, Austin, TX 78712; [2]Department of Chemistry, University of California, Berkeley, Berkeley, CA 94720; [3]Department of Chemical Engineering, The University of Texas at Austin, Austin, TX 78712; [4]Department of Molecular & Cell Biology, University of California, Berkeley, Berkeley, CA 94720. *These authors contributed equally.

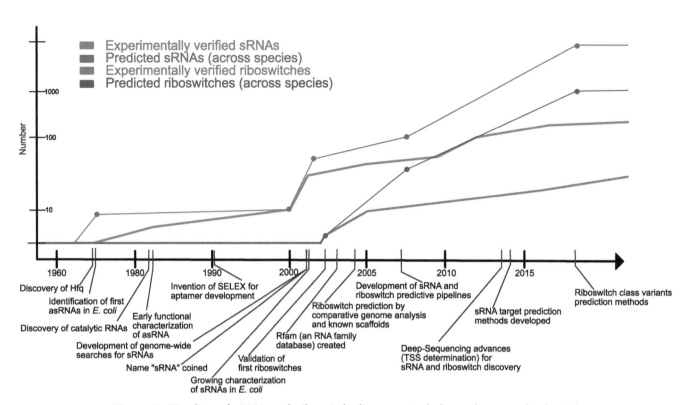

Figure 1 Timeline of sRNA and riboswitch discovery, including relevant technological advances that aided identification and verification of regulatory RNAs. The development of high-throughput, deep-sequencing techniques in particular has led to an explosion of sRNA and riboswitch discovery. However, although identification of sRNAs and riboswitches has rapidly expanded, verification of function still lags behind.

for these metabolite-sensing regulatory RNA elements (14, 17–20).

For most discovered sRNAs, interaction with its mRNA target involves repression of translation by antisense binding to mRNAs in a way that blocks the ribosome binding site (RBS) or altering mRNA stability by revealing an RNAse E degradation site (Fig. 2A) (4, 8, 9, 21). Likewise, the most widespread mechanisms of gene regulation for riboswitches are transcription termination or translation inhibition by RBS occlusion upon ligand binding, turning off gene expression (Fig. 2B). Both Rho-independent and Rho-dependent transcription termination (22) mechanisms have been described for riboswitches, although the former is more readily identified by the intrinsic terminator stem (23). Regulation by sRNAs can also involve transcription repression, as in the case of the role of 6S RNA on the RNA polymerase and the sigma factor σ^{70} (9, 24). However, there are examples in each class that turn on gene expression instead. sRNAs can activate translation by releasing a hairpin structure that blocks the RBS (as demonstrated by DsrA), enhance mRNA stability by blocking RNAse E degradation sites, or suppress Rho-

dependent transcription termination (Fig. 2A) (8, 9, 21, 25, 26). Ligand binding to a riboswitch leading to gene activation by transcription antitermination or translation activation has been demonstrated, with the first examples in the adenine riboswitch class (27) (Fig. 2B). While riboswitches that respond to metabolic end products usually turn off genes, those that respond to signaling molecules (e.g., cyclic dinucleotides) are more likely to turn on genes (28, 29).

Studies of the initial exemplars revealed key aspects of the target binding mechanism for sRNAs and riboswitches that still comprise the canonical models for these two regulatory RNA classes. For sRNAs, two key mechanistic binding aspects were unveiled from initial *E. coli* studies of the sRNAs MicF and SgrS (30–32): (i) imperfect complementarity to the target region and (ii) short target region (~6 nt of complementarity) required for proper binding and regulation (31). These small "seed" regions of required complementarity to targets (typically between 6 and 12 nt in length) are often present in unstructured regions at the 5′ end of the sRNA and are frequently the most conserved regions of sRNAs (30). Recent biophysical models have

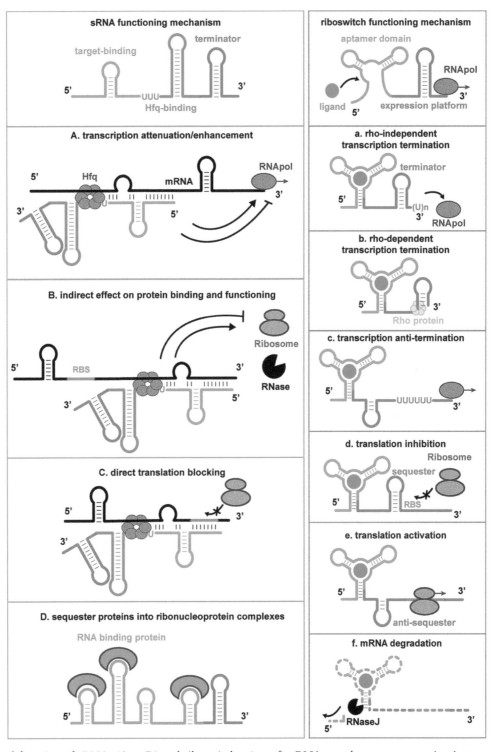

Figure 2 General function of sRNAs (A to D) and riboswitches (a to f). sRNAs regulate gene expression in *trans* through several functions enacted by antisense interactions, including transcription attenuation/enhancement through interactions with the RNA polymerase (A), inhibition of protein or ribosome binding either indirectly (B) or directly (C), and sequestration of protein factors (such as CsrA) (D). Riboswitches regulate gene expression in *cis* through a ligand-induced conformational change in the expression platform. The resulting gene expression consequences include Rho-dependent/independent transcription termination (a, b), transcription antitermination (c), translation activation (d), translation inhibition (e), and mRNA degradation (f).

demonstrated that accessibility of these small regions to the target mRNA is critical for sRNA-mRNA antisense interaction (33). For riboswitches, their canonical model contains (i) a ligand-binding aptamer domain and (ii) an expression platform for downstream gene regulation. Aptamer domains are evolutionarily conserved in secondary structure and specific nucleotides involved in ligand binding or folding. However, expression platforms typically exhibit little sequence conservation due to variability in regulatory logic between genes/operons and regulatory mechanisms between organisms, including control of transcription, translation, splicing, and mRNA stability.

Discovery of sRNAs and Riboswitches in the Genomics Era

Starting in 2001, deep-sequencing techniques rapidly expanded the collection of identified sRNAs and riboswitches in bacteria and archaea (9, 34–37) (Fig. 1). The development of RNA sequencing (RNA-seq), which provides a strand-specific readout of the transcriptome, has been one of the most useful tools in finding novel sRNAs in a range of bacteria (9, 36–41), and adaptions of this technique, including differential RNA sequencing (dRNA-seq) (38, 42), have aided the specific genomic search for sRNAs. Despite the fact that some sRNAs maintain the same function across species, the high sequence divergence of sRNAs has challenged sRNA prediction and discovery via phylogenetic conservation (9). However, machine learning approaches have made advances toward predicting new members in broad classes of sRNAs in a diverse set of organisms (35, 43).

De novo riboswitch discovery has also been made possible by the sequencing of many microbial genomes. Early genome-scale computational riboswitch prediction works relied on comparative genomic analysis. An initial pairwise, BLAST-based analysis of intergenic regions of similar genes among 91 prokaryotic genomes resulted in discovery of six orphan riboswitch families that were eventually validated experimentally (44). Later, advances in RNA motif prediction algorithms led to a powerful pipeline that is structure oriented and applicable to unaligned or even poorly conserved sequences. RNA motif prediction was integrated with RNA homology search to further refine structural alignments (45, 46). This pipeline was then expanded to be independent of protein-coding genes for clustering intergenic regions (47). To date, many of the predicted orphan riboswitches have had their ligands identified, making the total number of validated riboswitch classes reach around 40. However, the ligands of some rarer variants remain elusive, and this prediction pipeline

may fail to discover extremely rare riboswitch classes (48). Some alternative strategies, such as using RNA-seq and transcription start site (TSS) profiling to discover potential regulatory elements in the 5′ UTR of mRNA, might offer a solution to identify species-specific riboswitches (49, 50).

CHALLENGES TO THE PARADIGMS FOR sRNAs AND RIBOSWITCHES

While there are general principles for designing sRNAs or riboswitches for synthetic biology applications (Table 1), as more sRNAs and riboswitches are discovered in diverse organisms, an increasing number remain functionally uncharacterized, while some of those that are studied have proven to be exceptions to the general rules. In this section, we focus on specific challenges to the paradigms that have arisen from fundamental studies on sRNAs and riboswitches that should be instructive for those interested in engineering regulatory RNAs.

Challenges to Binding Target Identification

Identification of target genes for sRNAs and target small-molecule ligands for riboswitches remains an enormous challenge, often lagging far behind discovery of the sRNA gene or prediction of the conserved riboswitch structure in a UTR (Fig. 1). For sRNAs, the difficulty is in part due to the sRNA acting in *trans* to target multiple mRNAs. Moreover, sRNAs are known to have multiple targets that can be differentially regulated (or not) based on specific cellular conditions. Furthermore, combinations of different regulatory mechanisms are also possible, as one sRNA can regulate multiple mRNA targets with different mechanisms. For example, DsrA activates translation of the alternative sigma factor *rpoS* upon binding by relieving a hairpin that blocks the RBS, but represses translation of its other target, *hns*, by blocking the RBS (51, 52). Additionally, DsrA prevents Rho-dependent transcription termination of *rpoS* by sharing the same binding site with the Rho terminator (25). Several new sequencing-based methods have been developed to aid in the rapid, high-throughput identification of sRNA targets (34, 53–55) to complement computational tools (56, 57) and validation of hypothesized sRNA-mRNA interactions by additional fluorescence (58), genetics (deletion and overexpression), and biochemical assays (59). Determination of the target network aids in engineering sRNAs to regulate multiple targets under specific conditions without off-target effects.

For riboswitches, the difficulty in ligand identification is in part due to incomplete knowledge of gene

Table 1 General considerations for synthetic design: comparison of general factors to be considered in synthetic applications of small regulatory RNAs and riboswitches

Factors	sRNAs	Riboswitches
Protein factors	Several, but not all, require protein chaperone Hfq[a]; Hfq scaffolds can be added to synthetic sRNAs	No chaperone required, but interacts with transcriptional or translational machinery[b]
Length	50–500 nt with ~12 nt "seed" region for binding mRNA target	~30–120 nt for aptamer domain; variable for expression platform
Sequence complementarity to target; structural considerations	Minimal mismatches in sRNA seed binding region	Accurate secondary structure model is critical; tertiary structure is most helpful[c]
Target/ligand recognition	Specificity for desired mRNA target(s) to minimize off-target effects	Specificity for desired ligand to minimize off-target binding to other cellular metabolites
Cellular stability	Designs that avoid recognition by intracellular RNases	Designs that fold stably in cells
Kinetics of binding	Short- or long-term interaction with target RNA depending on application[d]	Ligand on-rate vs. thermodynamic binding stability depending on regulatory mechanism[e]
Design methods to alter targets/ligands	Modify the sRNA binding region and location on mRNA target, by modification of sRNA scaffolds (through redesign/transport of sRNA seed regions) or high-throughput screening of synthetic sRNAs	Structure-guided mutagenesis, computational modeling, or high-throughput screening/selection

[a]The "requirement" for Hfq chaperone is typically species and RNA dependent, and both presence and absence of Hfq should be tested in application.
[b]Riboswitch-based fluorescent biosensors do not function via gene regulation.
[c]While computational modeling of three-dimensional RNA structure is advancing rapidly, *de novo* prediction of small-molecule ligands has not yet been achieved.
[d]Additional design aspects include utilizing different promoter strengths to express the sRNA of interest at varying levels, and inducible promoters to alter the timing of expression.
[e]Kinetics of ligand binding is important for transcriptional control, whereas thermodynamic stability is important for translational control.

function outside of major metabolic pathways. Furthermore, gene annotations for most bacterial genomes are based on homology and may not be accurate or supported by experimental evidence. In fact, the riboswitches for cyclic di-AMP, fluoride, and guanidine were discovered as conserved regulatory structured RNA elements several years before their ligands were known to be physiologically relevant metabolites (44). The identification of the riboswitch ligand provided insights into signaling pathways for cyclic di-AMP and cyclic AMP-GMP (60–63) and the function of transporters for fluoride, prequeuosine₁ (PreQ₁), and guanidine (64–67).

Like gene annotations, the target ligands for most riboswitches annotated in bacterial genomes are based on homology to a known riboswitch. However, there is a growing realization that hidden diversity of riboswitch-ligand pairs within structural classes is more prevalent than previously assumed. This is exemplified by riboswitches for adenine (68), 2'-deoxyguanosine (2'-dG) (69), and cyclic AMP-GMP (61, 62). These variant classes appear to have diverged from the parent riboswitch classes to evolve an altered ligand specificity. Two complementary strategies have been demonstrated for revealing such hidden diversity. A "bottom-up" strategy biochemically examines the structure-activity relationship to identify key nucleotide positions, which guides the refinement of the consensus sequence and covariance

model to search for rare variants among the parent riboswitch class (62). A more generalizable "top-down" strategy utilizes the X-ray crystal structures of riboswitches to identify key positions and their associated gene annotations to predict novel riboswitch functions. The resulting refined bioinformatics search led to discovery of riboswitch subclasses that await biochemical validation (70). These two strategies have led to the discovery of variants among glycine, FMN, and the two cyclic di-GMP riboswitch classes. These many examples of natural riboswitch reprogramming showcase the potential for engineering riboswitch scaffolds to specifically target new metabolite ligands.

Noncoding RNAs That Still Code and Riboswitches That Are Not Switches

Although sRNAs are traditionally described as being noncoding RNAs, in multiple systems, it is now evident that sRNAs can encode small peptides as dual-function sRNAs (71, see also chapter 27). The first bacterial dual-function sRNA described is RNAIII from *Staphylococcus aureus*, which regulates several virulence factors. The 5′ region of RNAIII contains an open reading frame (ORF), *hld*, that encodes a secreted 26-amino-acid peptide, δ-hemolysin, that causes lysis of host cell membranes (71). Regulation of the *hld* ORF occurs by the partial overlap by one of RNAIII's mRNA targets, *map* (a surface adhesion protein), which could serve to regulate expression of δ-hemolysin (8). Other dual-function sRNAs appear to have virulence-related functions (72); however, not all dual-function sRNAs appear to regulate virulence factors (71). In fact, some dual-function sRNAs are capable of regulating one system through two distinct mechanisms, as demonstrated by SgrS. This sRNA negatively regulates the glucose transporter *ptsG* mRNA under glucose-phosphate stress; contains an ORF, *sgrT*, translated in *E. coli* under glucose-phosphate stress; and interferes with PtsG transport activity (73, 74). The ability of the cell to regulate translation of the ORF within an sRNA is still to be determined for some dual-function sRNAs; however, some dual-function sRNAs appear to be translationally dependent on σ^S (75). The presence of dual-function sRNAs indicates the need to verify ORFs contained in an sRNA as part of the discovery and validation process as the small peptide encoded could aid sRNA function or be separately regulated by the mRNA targets. Additionally, dual-function sRNAs suggest an opportunity to engineer multifunction sRNAs to obtain both the regulation of an sRNA and the enzymatic activity of a small peptide in a promising new research frontier.

Similar to the inaccuracy of naming sRNAs as noncoding, the term "riboswitch" propagates the classic induced fit model that riboswitch conformation consists of two main states, "on" and "off," and ligand binding switches the RNA from one conformation to the other. However, this view has been challenged by biophysical studies tracking single-molecule trajectories for RNA folding (76–78) and using advanced nuclear magnetic resonance spectroscopy techniques to detect highly transient states (79). In the conformational selection model, a riboswitch can fold into and dynamically sample different conformational states without the ligand. Presence of ligand drives the equilibrium toward a favored conformational state (80). For example, a recent single-molecule fluorescence resonance energy transfer study of the full *S*-adenosyl-L-methionine (SAM)-I riboswitch revealed that there are four discrete conformational states that are populated even in the absence of SAM (81). Note that such dynamics is revealed *in vitro* under thermal equilibrium conditions for full-length riboswitches. It is expected that riboswitch functional dynamics *in vivo* would be even more complicated due to cotranscriptional folding. For example, cotranscriptional folding of a fluoride riboswitch appeared highly dynamic, such that addition of a newly transcribed single nucleotide might greatly alter the folding landscape of the nascent riboswitch (82). Undoubtedly, rational riboswitch engineering would greatly benefit from a better understanding of such riboswitch dynamics.

Not All sRNAs Depend on Hfq Chaperoning, but All RNA Regulators Depend on Intracellular Protein Interactions

Many sRNAs (but not all) require an RNA chaperone protein, Hfq, for proper function (including sRNAs OxyS and DsrA). Hfq acts as a "meeting platform" to mediate a number of steps in sRNA-mRNA interaction, including exposing the seed region for sRNA-mRNA hybridization, neutralizing negative charges of the RNAs, protecting RNAs from degradation, and assisting annealing of the RNA strands (83). Crystal structures of Hfq suggest that each of the four faces of the homohexameric toroid protein has a different binding preference, and Hfq-dependent sRNAs have been classified into two groups depending on the binding preference to Hfq (84). Moreover, different sides of Hfq have also been attributed to aiding release of the sRNA-mRNA pairs to allow rapid Hfq cycling to other sRNA-mRNA pairs (85) and to interact with bacterial membranes (86). However, it is now known that Hfq is not present in all bacteria, even in those that contain sRNA-based regulation. While some bacteria contain Hfq homologs, these are often not required for proper sRNA function (although Hfq can aid function) and are functionally distinct between organisms (87). Hfq is largely not required in Gram-positive bacteria, with only a few exceptions (87, 88). Recent efforts have identified FinO-domain proteins, like ProQ, as another form of specific sRNA chaperone (89). However, the FinO-domain proteins do not appear to be alternatives to Hfq, as almost all bacterial families that lack Hfq homologs also lack FinO-domain proteins (89). Although there appear to be some general trends of Hfq-based interactions with sRNAs (such as binding U- or A-rich sequences of RNA) (83), the understanding of what makes an sRNA Hfq dependent is still lacking, and there is even less understanding about the importance of FinO-domain RNA chaperones.

In contrast to sRNAs, a hallmark of riboswitches is their ability to directly bind a metabolite ligand without assistance from protein factors. However, for gene regulation to occur, the riboswitch must interact with or recruit protein/RNP factors required for transcription termination, translation initiation, or mRNA degradation, e.g., RNA polymerase, Rho, ribosome, or RNase. Thus, understanding riboswitch-protein interactions is important. For example, riboswitches fold while being transcribed by RNA polymerase. Thus, the kinetics of cotranscriptional RNA folding and the transcription speed of RNA polymerase need to match well for riboswitch function, and in fact, natural riboswitches have been shown to harbor transcriptional "pause" sequences (90). Furthermore, binding of ribosomes to the RBS also contributes to the overall thermodynamic energy equilibrium, along with the intramolecular RNA-RNA and intermolecular ligand-RNA interactions (91). Similar considerations of sRNA-protein interactions (e.g., interactions with the cellular machinery) are important for fully understanding sRNA function *in vivo* (e.g., transcription, degradation, target binding, and ribosome binding). Ignoring these riboswitch-protein and sRNA-protein interactions would result in misunderstanding of the native biological functions of these regulators and in failure of engineering them.

Complex Network Regulation Involves Cross-Talk, Competition, and Coordination

RNA networks involve multiple levels of regulation, and recent studies have demonstrated that levels of individual mRNA targets, or shared resources (like Hfq), can affect the regulatory ability of an sRNA across entire regulatory networks. These competing endogenous RNAs (ceRNAs) (which can be another sRNA or mRNA) interfere with sRNA target binding, through either competition/sequestration of sRNA binding sites via mimicry or alternative binding, or competition for Hfq, such that the level of sRNA-based regulation is dependent on the level of the ceRNA (92, 93). In this manner, communication occurs between the numerous regulatory pathways that sRNAs can mediate in an organism. Based on this hypothesis of cross-talk between sRNAs and their RNA targets and protein mediators, it is important to consider the effect across a network when engineering changes to a system. This competition and cross-talk can result in unexpected results when modulating a desired system. For example, when engineering acid tolerance in *E. coli*, researchers overexpressed three sRNAs that all translationally activate target *rpoS* (DsrA, RprA, and ArcZ) but found the benefit of overexpression to be supra-additive instead of the expected linear trend (94).

In contrast to the competition of sRNA-based networks, natural riboswitches arranged cooperatively in tandem have been discovered, which reveals the great potential of riboswitches to form complex biocomputational logic circuits. In nature, some riboswitches contain tandem aptamer domains for the same ligand, such as a glycine riboswitch in both *Vibrio cholerae* and *Bacillus subtilis* (95), or tandem complete riboswitches from the same class, such as a TPP riboswitch in *Bacillus anthracis* (96) and a triple-tandem cyclic di-GMP riboswitch in *Bacillus thuringiensis* (97). The multivalent organization yields a more digital response to the cognate ligand due to a steeper ligand-dependent binding curve. Tandem arrangements with riboswitches for

distinct ligands result in a logic gate, which is exemplified by the *metE* UTR in *Bacillus clausii* (98). This tandem arrangement contains a SAM riboswitch and an adenosylcobalamin (AdoCbl) riboswitch and functions as a two-input Boolean NOR logic gate. Only when both SAM and AdoCbl concentrations are low in the cells will expression of the downstream gene *metE*, which encodes an AdoCbl-independent SAM synthase, be turned on. These natural examples are inspiring for constructing artificial RNA-based logic gates.

Furthermore, several natural riboswitch-ribozyme mechanisms have been discovered to date. Ligand binding to the *glmS* riboswitch catalyzes site-specific cleavage that triggers mRNA degradation. The *glmS* sequence acts as both the riboswitch and the ribozyme by placing the ligand's nucleophilic amine group in position to attack a specific phosphodiester bond (99–101). In contrast, a c-di-GMP-II riboswitch in *Clostridium difficile* is found adjacent to a group I self-splicing intron that acts as its expression platform. The ligand-bound riboswitch induces rearrangement of the group I intron conformation that allows GTP to attack and activates the splicing event, resulting in a translatable mRNA (29). The eukaryotic TPP riboswitches found in plants and filamentous fungi also regulate gene expression by splicing mechanisms, albeit through the spliceosome rather than a self-splicing element (102). These natural examples reveal quite different mechanisms from those of artificial aptazymes that previously have been engineered by fusing an aptamer to a ribozyme to effect ligand-induced self-cleavage (12, 13).

Finally, riboswitches and sRNAs can work together. One example is the ethanolamine (EA) utilization (*eut*) locus of *Enterococcus faecalis*, in which the EutX sRNA contains an AdoCbl riboswitch that, in the absence of AdoCbl (a cofactor required for EA catabolism), sequesters a binding site for the response regulator EutV. EutV prevents transcription termination upon binding EutX; however, EA is required for EutV to be functionally activated (103). In this manner, the presence of both AdoCbl and EA is required to (i) release the EutV binding site and (ii) activate EutV to prevent transcriptional termination (103). In another example, two SAM riboswitches, SreA and SreB in *Listeria monocytogenes*, regulate gene expression not only in *cis* (via binding of SAM ligand and termination of transcription) but also independently in *trans* as sRNAs by binding to the 5′ UTR of the mRNA target, *prfA*, to downregulate expression of the virulence regulator (104). Such dual-acting regulatory RNA elements should find applications in constructing more complex regulation networks in synthetic biology.

COMPARISON OF SYNTHETIC BIOLOGY APPLICATIONS FOR sRNAs AND RIBOSWITCHES

Even as regulatory RNA functions in native systems continue to be elucidated, a plethora of applications of noncoding RNA have already been demonstrated, although it is not possible to provide a comprehensive list here. Earlier sRNA applications focused on metabolic engineering, and earlier riboswitch applications focused on orthogonal gene regulation. However, recent approaches have expanded the use of noncoding RNAs for novel *in vivo* applications such as characterization of cellular RNA-RNA interactions, cellular metabolism, and signaling and construction of synthetic logic gates (Fig. 3).

sRNAs are primarily known to interact through antisense interactions with a target mRNA, so a native or synthetic sRNA can be designed to bind and regulate desired mRNA targets. Although native sRNAs are frequently useful (especially in an overexpressed or deleted manner), these efforts are primarily used to alter an sRNA's native pathway in its native organism. To modulate other targets, or use an sRNA in another organism, synthetic sRNAs are frequently designed. While some sRNAs (e.g., DsrA and MicF) are uniquely portable in their ability to be utilized in other organisms (105, 106), many native sRNAs cannot easily be transported between different organisms. Portability of sRNAs can be enhanced by an alternative approach that incorporates specific seed regions of sRNAs into an unrelated sRNA and maintains the original targeting despite the new RNA context (107, 108). As many sRNAs are associated with cellular factors (e.g., Hfq), design of sRNAs should account for the effect (aid or hindrance) of these

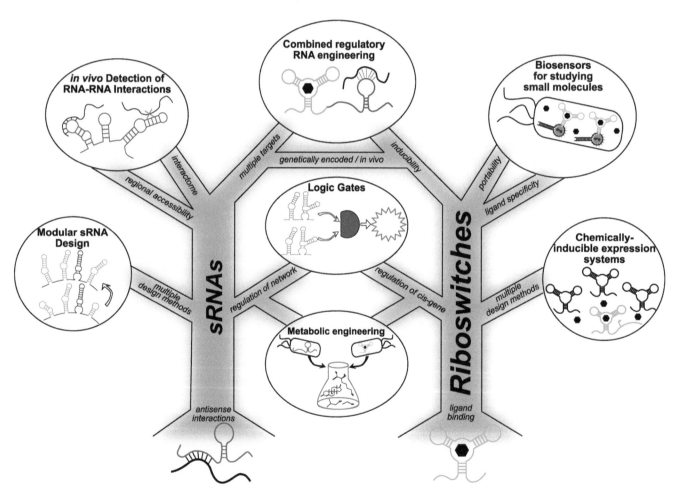

Figure 3 Examples of applications of sRNAs and riboswitches. Applications of these regulatory RNAs are rooted in their unique functional characteristics (antisense interactions for sRNA and ligand binding for riboswitches). Recent applications of these systems have begun to interweave these mechanisms to provide more complex engineering strategies.

factors in the system (92, 106). Additionally, interaction of the sRNA with other mRNAs or sRNAs in a competitive manner should also be considered and decoy sequences should be avoided (93). Often the accessibility (ability of RNA to interact with another RNA) of the sRNA and RNA target should be considered in design to ensure that the sRNA and target will be able to functionally interact (2, 3, 33). High-throughput screening of a large set of synthetic sRNAs has also proved useful at designing sRNAs for target regulation (109–111). Factors to consider are summarized in Table 1, but for detailed discussion on synthetic sRNA engineering strategies, please see references 2, 3, 106, 112, and 113.

Thus far, synthetic sRNAs have been derived *de novo* from artificial libraries (109, 110, 113), by modularizing and redesigning natural sRNA scaffolds (105, 106, 111, 114), and by rationally engineering the affinity with its targets (115). Two specific examples are the rational engineering of altered affinity of CsrB to CsrA (and thus the downstream gene expression of CsrA targets) (115) and the modification of the DsrA scaffold to switch its targeting regions (for *rpoS* and *hns* in *E. coli*) to a heterologous target from *n*-butanol synthesis in *Clostridium acetobutylicum* (105). However, it is important to note that while these methods may have the advantage of a direct target choice, synthetic sRNA systems might not be as robust as native regulatory systems.

By comparison, riboswitches are known to confer small molecule-dependent control of gene expression, so a natural or synthetic riboswitch can be placed downstream of a native promoter to regulate the target transgene in *cis*. For metabolic engineering or metabolite reporter/biosensor applications, typically an endogenous metabolite is the target of interest, so synthetic riboswitch designs in these cases have focused on expression platform engineering to convert from a turn-off to a turn-on system (and vice versa), to implement a different mechanism of gene regulation or biosensor output, or to adapt a riboswitch to function between different organisms. For orthogonal gene regulation, the goal is usually to utilize a chemical-inducible riboswitch in place of a chemical-inducible promoter to control gene expression, since the latter requires introduction of an orthogonal transcription factor. Thus, the synthetic designs have focused on riboswitch aptamer engineering to reprogram the ligand-binding pocket to bind nonnative, cell-permeable small molecules. Alternatively, the *in vitro*-selected aptamer for theophylline (116) has demonstrated high portability for orthogonal gene regulation in many different bacteria, although in almost every case, optimization of the expression platform through

screening has been necessary (117, 118). Thus far, synthetic riboswitches have been derived via rational or structure-based design (119, 120), medium-throughput selection or screening strategies (117, 121), and computational modeling (91, 122). Factors to consider are summarized in Table 1, but for detailed discussion on riboswitch engineering strategies, please see references 1 and 5.

Applications to Metabolic Engineering

sRNAs provide network-level manipulation of complex phenotypes

A unique characteristic of sRNAs is their ability to regulate multiple mRNA targets in response to changes in environmental conditions. This provides an excellent building point for metabolic/network engineering by being able to control entire genetic circuits with one sRNA. Additionally, sRNAs have been determined to be useful regulators in less conventional organisms (such as many industrially relevant strains) where genetic manipulation and understanding of regulatory mechanisms are less available. Moreover, examples in this section involve engineering further robustness for stress tolerance of a specific environmental threat (e.g., acid, ethanol, butanol, or overproduction of a compound) that arises in biotechnology applications.

For example, *C. acetobutylicum* and *Zymomonas mobilis*, two model organisms for acetone-butanol-ethanol fermentation and ethanol and/or farnesene production, respectively, do not have many genetic tools available for metabolic engineering efforts or traditional transcriptional modifications. However, deletion and overexpression of sRNAs (both native and heterologous) have proven to be useful tools to improve product yields in these organisms (123–125). Likewise, manipulation of well-characterized sRNAs (DsrA, ArcZ, and RprA) has led to improved acid stress resistance in *E. coli* (94). For many of these systems, discovery of important sRNAs for the desired phenotype is often in conjunction with development of new strains. Typically, analysis of transcriptome changes under desired stress or production conditions reveals native sRNAs that could be manipulated (by deletion or overexpression) to enhance stress resistance or production of desired product. Recent examples of this method of sRNA-based metabolic engineering by overexpression of native sRNAs include increasing butanol tolerance in *C. acetobutylicum* (126), ethanol and butanol tolerance in cyanobacterium *Synechocystis* sp. PCC 6803 (127) and *Z. mobilis* (125), and steroid intermediate production in *Mycobacterium neoaurum* (128). Deletion of sRNAs that interfere with the desired pathway or produce undesired phenotypes or growth rates has also become a useful engineering method to ensure high production of final products (such as subtilisin in *Bacillus licheniformis* [129]).

Although many sRNA-based systems utilize native sRNAs, synthetic sRNAs can be designed to interact with and regulate a specific target; alternatively, a conserved sRNA can be adapted for use in another organism. For example, the *E. coli* sRNA system IS*10* RNA-IN/OUT regulator was adapted for use in *Synechococcus* sp. strain PCC 7002 to achieve 70% knockdown of the desired mRNA target (130). Another example utilized an inducible *E. coli* sRNA and Hfq construct to knock down UDP-glucose pyrophosphorylase (UGPase), which is involved in cellulose synthesis to permit fine-tuning of cellulose production (131). The addition of Hfq with the heterologous sRNA is capable of improving efficiency of targeted gene knockouts. This has been demonstrated by the use of an *E. coli* MicC scaffold in combination with expression of *E. coli* Hfq in *C. acetobutylicum* PJC4BK to increase butanol production (123). This system in *C. acetobutylicum* was found to be more efficient than other asRNA knockdown systems lacking Hfq (123). In a similar way, synthetic sRNAs based on known scaffolds have been designed to improve tyrosine production in *E. coli* (114).

Riboswitches control metabolic flux to favor target metabolite production

Riboswitches also have been applied to control metabolic flux to maximize the yield of target products. For example, an industrially relevant lysine-fermenting bacterium, *Corynebacterium glutamicum*, was engineered for improved lysine yield. In this organism, oxaloacetate can be converted either into the desired product, lysine, by a multistep biosynthesis pathway or into citrate by citrate synthase in the tricarboxylic acid cycle. To redirect the citrate synthase-mediated metabolic flux to favor lysine production, a natural lysine-OFF riboswitch from *E. coli* was inserted into the genome to regulate expression of the citrate synthase gene. The net effect of this engineering is to titrate expression of citrate synthase in response to lysine levels, which led to an increase of lysine yield by 63% (132). A further 21% improvement in yield was achieved by engineering the *E. coli* lysine-OFF riboswitch into a synthetic lysine-ON riboswitch and inserting it into the *C. glutamicum* genome to regulate expression of a natural lysine transporter (133). This engineering released the feedback inhibition that lowers lysine yield.

Riboswitches as sensors for screening enzymes or strains to favor metabolite production

Another application of riboswitches in metabolic engineering is to act as sensors to screen for enzymes or strains that give higher yield of a target product. For instance, to improve lysine fermentation efficiency, a natural *E. coli* lysine riboswitch controlling a tetracycline resistance gene was used to select for a chimeric aspartate kinase that was not feedback inhibited by the end product, lysine (134). For strain engineering, a similar riboswitch-regulated selection cassette was used to select for pathway-optimized *E. coli* strains from a library of 10^9 to 10^{10} variants with different expression levels of phosphoenolpyruvate carboxylase, an enzyme involved in oxaloacetate biosynthesis and thus lysine production (135). Another example of strain engineering used an engineered *E. coli* strain as a cell-based reporter to screen a *B. subtilis* production strain library for riboflavin production. The *E. coli* reporter strain carries an engineered riboswitch-ribozyme based on the endogenous *E. coli* FMN riboswitch to turn on green fluorescent protein (GFP) expression in response to FMN. By coencapsulating *E. coli* reporter cells with individual *B. subtilis* cells in nanoliter reactors, the entire production strain library could be sorted using fluorescence to correlate to levels of released riboflavin. The winner *Bacillus* strain displayed up to 150% improvement in riboflavin production compared to the industrially optimized parent strain (136).

Applications of Riboswitches to Chemically Inducible Expression Systems

Protein-based chemically inducible expression systems, such as the isopropyl-β-D-thiogalactopyranoside-inducible *lac* operon or the arabinose-inducible *araC* promoter, have proven to be important tools for conditional gene expression in bacteria. However, such protein-based systems are cumbersome to implement in new species due to the requirement of multiple protein factors. Furthermore, protein-based systems also respond to environmental changes and/or cellular stress, which make them not entirely bio-orthogonal. By contrast, riboswitches do not require extra biomolecular factors and can be encoded within a few hundred nucleotides. To engineer riboswitch-based inducible expression systems, one strategy is to reprogram the ligand specificity of the aptamer domain of natural riboswitches. For example, screening structure-based mutants of purine riboswitches with a small library of purine analogs resulted in orthogonal riboswitches that respond to pyrimido[4,5-*d*]pyrimidine-2,4-diamine (PPDA) and 2-aminopyrimido[4,5-*d*]pyrimidin-4(3*H*)-one (PPAO), respectively (137, 138). The PPDA riboswitch called M6″ was used to control *cheZ* expression and caused dose-dependent change in cell motility in *E. coli*. A similar strategy applied to PreQ$_1$ riboswitches identified a C17U riboswitch mutant that selectively responds to DPQ$_0$, a diamino analog of PreQ$_0$. The DPQ$_0$ riboswitch was used for chemically inducible regulation of a gene involved in cell morphology in *B. subtilis* (139).

An alternative strategy involves engineering *in vitro*-selected RNA aptamers into functional riboswitches in live cells. One of the most successful examples is the theophylline aptamer, due to the stability, cell permeability, and bio-orthogonality of theophylline and the specificity of the aptamer (117, 118, 140). In particular, synthetic theophylline riboswitches with engineered expression platforms to control ribosome binding have enabled research studies in a number of bacterial species that otherwise lacked inducible expression systems, including *Streptococcus pyogenes*, *Francisella tularensis*, *Francisella novicida*, *Mycobacterium tuberculosis*, *Streptomyces coelicolor*, cyanobacteria, and *Bdellovibrio bacteriovorus* (141–148). However, more orthogonal riboswitches are still needed as they can have broad applications in the study and engineering of non-model organisms in particular.

Applications to Studying Biological Systems

Utilizing sRNA-based antisense interactions for *in vivo* molecular studies

Recent techniques have utilized mimicking the natural mechanism of sRNA-mRNA binding for fundamental molecular characterization studies in cells. In these systems, antisense RNA probes that bind a specific region within a target RNA of interest are used to understand RNA-RNA interactions *in vivo*. Using a similar logic as split-protein complementation (149), researchers developed a Split-Broccoli system, adapted from the dye-binding aptamer Broccoli and a three-way junction RNA. RNA-RNA interactions can be detected *in vivo* by coexpression of the predicted RNA-interacting partners as fusions with the two parts (top and bottom) of the Split-Broccoli system (150). Validation of this system with an RNA Toehold switch demonstrated that both binding of the RNAs (visualized by the Broccoli fusion) and translational regulation (by an mCherry fusion with the Toehold portion of RNA) could be observed from the single construct with relatively low background fluorescence (150).

Another method has been demonstrated to characterize the interactions of an entire RNA molecule. The *in vivo* RNA Structural Sensing System (iRS^3) utilizes a fluorescent reporter-based system to determine tertiary interactions of an RNA molecule *in vivo* based on *in vivo* hybridization profiles (151). To determine RNA-RNA interactions, a set of antisense probes are designed that cover the RNA sequence space. These probes are then attached to a fluorescence reporter such that upon binding of the probe the RBS of a fluorescent reporter is released and thereby can be measured by an increase in fluorescence. Through a set of probes, the entire accessibility of an RNA molecule can be determined and linked to its structure and predicted RNA-RNA interactions (151). Currently, this method has been demonstrated to contribute to further understanding the *Tetrahymena* group I intron *in vivo* (151) and the accessibility of CsrB regions to CsrA (33). Utilization of this method has also provided a biophysical model to predict sRNA-mRNA binding interactions by considering the availability of RNA interactions using the suboptimal structures (33).

Using riboswitches as *in vivo* reporters and biosensors for target small molecules

Riboswitches can function as reporters or be engineered as fluorescent biosensors for sensing target small molecules *in vivo*, which make them ideal tools to study biological pathways in living organisms. Riboswitch-based reporters can be constructed by replacing the downstream gene of a well-characterized riboswitch with a reporter gene such as LacZ for colorimetric measurement, GFP for fluorescence, or luciferase for luminescence. For instance, the coding sequence of the *E. coli btuB* operon was replaced by a GFP reporter gene and the resulting B$_{12}$ reporter was applied to study the biological roles of several *E. coli* membrane transporters in maintaining B$_{12}$ homeostasis (152). This reporter also was used to screen >100 BtuC$_2$D$_2$F transporter mutants to identify key residues in the substrate-binding pocket (153). In another case, c-di-GMP riboswitch-based reporters have enabled screening and validation of predicted c-di-GMP-metabolizing enzymes from *C. difficile* (154) and membrane-bound diguanylate cyclases from *B. thuringiensis* and *Xanthomonas oryzae* (97).

Riboswitch-based fluorescent biosensors provide the ability both to screen and to study the temporal dynamics of biological pathways in single cells. Fluorescent biosensors can be constructed by fusing the aptamer domains of natural riboswitches to a dye-binding aptamer, such as Spinach (155), so that ligand binding results in a conformational change that directly activates the fluorescence of the RNA-dye complex. Thus, these biosensors do not require translation of a protein and do not require oxygen for fluorophore maturation, both of which delay signal turn-on in fluorescent reporter systems. For example, biosensors based on the S-adenosyl-L-homocysteine and cyclic di-GMP riboswitches have been used for near real-time imaging of *in vivo* enzyme activity related to autoinducer biosynthesis and cyclic di-GMP signaling, respectively (156–158). Since the fluorescence activation mechanism is not dependent on the gene expression machinery of the host (other than requiring a host promoter to drive biosensor expression), these biosensors should be more portable between different bacterial species. For instance, a biosensor based on the cyclic di-AMP riboswitch was shown to function in both *E. coli* and *Listeria monocytogenes* (159). More recently, it has been shown that riboswitch-based fluorescent biosensors can be reprogrammed to sense new ligands (160). In the past 5 years, these biosensors have been applied to high-throughput screening of enzyme activity (63), imaging enzyme dynamics (161), and anaerobic imaging (157). With the increasing number of discovered natural riboswitches and the power of *in vitro* selection, riboswitch-based biosensors are promising to study enzyme activity, metabolism, and signaling in bacteria and potentially other organisms.

Synthetic Integration of Regulatory RNAs: Applications to Biocomputation

In an sRNA's native context, sRNAs can be activated using AND-gate decision logic, as demonstrated by the requirement of the *Salmonella* sRNA RprA and σ^S for the mRNA target, *ricI*, activation (162, 163). However, sRNAs as such have not been utilized by themselves in logic circuits; instead, the principles governing sRNA function (particularly antisense interactions) have been utilized to make complex riboregulators. In particular, Toehold switches have been utilized to develop complex synthetic biological circuits (163). Toehold switches combine the ligand binding and conformational change of a riboswitch with the requirement of antisense base pairing of an sRNA, by the release of the Toehold hairpin repression via the antisense binding of the "trigger" RNA (164). These riboregulators can be utilized as sensors of particular gene expression or to regulate translation of endogenous genes (164). A similar approach of using antisense interactions to regulate circuits involves designing synthetic RNAs to disrupt transcription terminators placed upstream of reporter genes (165). These small transcription activating RNAs (STARs) are capable of creating diverse RNA-only logic gates (165).

Natural riboswitches function as basic computational units for a complex biocomputation network in the living organism. The input of ligand concentration is converted by the riboswitch into gene expression (ON or OFF) as an output signal. By combining these basic units, higher-level architectures such as Boolean logic gates can be constructed. Some natural examples of such riboswitch-based logic circuits have been presented above ("see Complex Network Regulation Involves Cross-Talk, Competition, and Coordination"). Inspired by natural tandems, theophylline riboswitches engineered to regulate transcription antitermination were placed in tandem to decrease leaky expression. The resulting expression was less leaky, but maximum expression level was compromised (166). In another study, previously engineered theophylline and tetracycline riboswitches were arranged in tandem to construct an AND logic gate that should activate gene expression only when both ligands are present. Interestingly, it was found that the sequence order of the two riboswitches affected the performance of the logic gate (167). AND and NAND Boolean logic gates were also realized via a cell-based selection using a theophylline aptamer followed by a TPP riboswitch (168). For the AND logic gate, ligand-dependent conformational change of the theophylline riboswitch rearranged the TPP riboswitch to allow for TPP binding, which reorganizes the downstream expression platform to initiate translation. Therefore, an 18-fold turn-on of gene expression requires high concentrations of both theophylline and thiamine. These exciting examples suggest that riboswitches can be powerful building blocks for constructing complex logic networks. However, it should also be noted that the current way to design a riboswitch logic gate is not yet fully rational. Semirational cell-based screen or selection methods still play an important role in developing such logic circuits. It is expected that growing understanding of natural systems and experience gained from engineering studies will help provide general rules for designing such complex RNA-based computational networks.

CONCLUSIONS AND FUTURE HORIZONS OF RNA ENGINEERING

In this review we have discussed the history of regulatory RNA discovery and innovations, giving insight into current discoveries and recent applications. Regulatory RNAs are expected to expand as useful tools for synthetic biology applications (Fig. 3), even beyond the applications demonstrated so far. For example, we envision novel synthetic systems being developed for medicine and nanotechnology. While not discussed here in detail, the inspiration of sRNA-based regulation in bacteria has already been utilized in many fields, including clustered regularly interspaced short palindromic repeat (CRISPR) systems (reviewed in references 169 and 170). Regulatory RNAs have also been described as key points for antibiotic targeting in bacterial infections, as several elements of virulence are controlled by sRNAs (reviewed in references 171 and 172) and compounds targeting riboswitches have demonstrated antibiotic activity (173). Recently, insights from RNA folding have led to an entire new field of RNA nanotechnology for biological applications (174). While many of the applications are in eukaryotes, the basis of RNA-based interactions and folding remains the same as in bacteria and can serve as inspiration for a wide variety of applications. These studies have produced examples of RNA structures such as lattices and tubular structures (175), triangles from a tetra-uracil motif (176), and catalytic triangles and squares based on the *Tetahymena* group I intron (177). Construction of these elaborate RNA origami structures has been largely aided by computational methods that model the structures formed by RNA interactions (122), and it is anticipated that the next steps in this field would incorporate functional regulatory elements like sRNAs and riboswitches. Additional applications include biocontainment (178) and use of logic gates for diagnostics. An example of the latter has already been reported using Toehold switches in the context of paper-based diagnostics to detect viral RNAs (179). It is therefore expected that elucidation of the mechanistic diversity of sRNAs and riboswitches would lead to improved applications in these areas spanning biotechnology and medicine.

Acknowledgments. This work was supported by the Welch Foundation (F-1756) and the Air Force Office of Scientific Research Young Investigator program (FA9550-16-1-0174) for L.M.C. and J.K.V., and the National Institutes of Health (R01 GM124589) and Office of Naval Research (N000141712638) for Y.S. and M.C.H. Additionally, J.K.V. is supported by the University of Texas at Austin Provost's Graduate Excellence Fellowship and the National Science Foundation Graduate Research Fellowship, and Y.S. is supported by the UC Cancer Research Coordinating Committee Predoctoral Fellowship.

Citation. Villa JK, Su Y, Contreras LM, Hammond MC. 2018. Synthetic biology of small RNAs and riboswitches. Microbiol Spectrum 6(3):RWR-0007-2017.

References

1. **Hallberg ZF, Su Y, Kitto RZ, Hammond MC.** 2017. Engineering and in vivo applications of riboswitches. *Annu Rev Biochem* **86:**515–539.

2. **Vazquez-Anderson J, Contreras LM.** 2013. Regulatory RNAs: charming gene management styles for synthetic biology applications. *RNA Biol* **10:**1778–1797.

3. **Cho SH, Haning K, Contreras LM.** 2015. Strain engineering via regulatory noncoding RNAs: not a one-blueprint-fits-all. *Curr Opin Chem Eng* **10:**25–34.

4. **Saberi F, Kamali M, Najafi A, Yazdanparast A, Moghaddam MM.** 2016. Natural antisense RNAs as mRNA regulatory elements in bacteria: a review on function and applications. *Cell Mol Biol Lett* **21:**6.

5. **Etzel M, Mörl M.** 2017. Synthetic riboswitches: from plug and pray toward plug and play. *Biochemistry* **56:** 1181–1198.

6. **Kushwaha M, Rostain W, Prakash S, Duncan JN, Jaramillo A.** 2016. Using RNA as molecular code for programming cellular function. *ACS Synth Biol* **5:**795–809.

7. **Wassarman KM, Zhang A, Storz G.** 1999. Small RNAs in *Escherichia coli. Trends Microbiol* **7:**37–45.

8. **Papenfort K, Vanderpool CK.** 2015. Target activation by regulatory RNAs in bacteria. *FEMS Microbiol Rev* **39:**362–378.

9. **Wagner EG, Romby P.** 2015. Small RNAs in bacteria and archaea: who they are, what they do, and how they do it. *Adv Genet* **90:**133–208.

10. **Ellington AD, Szostak JW.** 1990. *In vitro* selection of RNA molecules that bind specific ligands. *Nature* **346:** 818–822.

11. **Tuerk C, Gold L.** 1990. Systematic evolution of ligands by exponential enrichment: RNA ligands to bacteriophage T4 DNA polymerase. *Science* **249:**505–510.

12. **Soukup GA, Breaker RR.** 1999. Engineering precision RNA molecular switches. *Proc Natl Acad Sci U S A* **96:** 3584–3589.

13. **Araki M, Okuno Y, Hara Y, Sugiura Y.** 1998. Allosteric regulation of a ribozyme activity through ligand-induced conformational change. *Nucleic Acids Res* **26:**3379–3384.

14. **Gelfand MS, Mironov AA, Jomantas J, Kozlov YI, Perumov DA.** 1999. A conserved RNA structure element involved in the regulation of bacterial riboflavin synthesis genes. *Trends Genet* **15:**439–442.

15. **Nou X, Kadner RJ.** 1998. Coupled changes in translation and transcription during cobalamin-dependent regulation of *btuB* expression in *Escherichia coli. J Bacteriol* **180:** 6719–6728.

16. **Miranda-Ríos J, Navarro M, Soberón M.** 2001. A conserved RNA structure (*thi* box) is involved in regulation of thiamin biosynthetic gene expression in bacteria. *Proc Natl Acad Sci U S A* **98:**9736–9741.

17. **Mironov AS, Gusarov I, Rafikov R, Lopez LE, Shatalin K, Kreneva RA, Perumov DA, Nudler E.** 2002. Sensing small molecules by nascent RNA: a mechanism to control transcription in bacteria. *Cell* **111:**747–756.

18. **Nahvi A, Sudarsan N, Ebert MS, Zou X, Brown KL, Breaker RR.** 2002. Genetic control by a metabolite binding mRNA. *Chem Biol* **9:**1043–1049.

19. **Winkler WC, Cohen-Chalamish S, Breaker RR.** 2002. An mRNA structure that controls gene expression by binding FMN. *Proc Natl Acad Sci U S A* **99:**15908–15913.

20. **Winkler W, Nahvi A, Breaker RR.** 2002. Thiamine derivatives bind messenger RNAs directly to regulate bacterial gene expression. *Nature* **419:**952–956.

21. Bandyra KJ, Said N, Pfeiffer V, Górna MW, Vogel J, Luisi BF. 2012. The seed region of a small RNA drives the controlled destruction of the target mRNA by the endoribonuclease RNase E. *Mol Cell* 47:943–953.

22. Hollands K, Proshkin S, Sklyarova S, Epshtein V, Mironov A, Nudler E, Groisman EA. 2012. Riboswitch control of Rho-dependent transcription termination. *Proc Natl Acad Sci U S A* 109:5376–5381.

23. Serganov A, Nudler E. 2013. A decade of riboswitches. *Cell* 152:17–24.

24. Cavanagh AT, Wassarman KM. 2014. 6S RNA, a global regulator of transcription in *Escherichia coli*, *Bacillus subtilis*, and beyond. *Annu Rev Microbiol* 68:45–60.

25. Sedlyarova N, Shamovsky I, Bharati BK, Epshtein V, Chen J, Gottesman S, Schroeder R, Nudler E. 2016. sRNA-mediated control of transcription termination in *E. coli*. *Cell* 167:111–121.e13.

26. Sedlyarova N, Rescheneder P, Magán A, Popitsch N, Rziha N, Bilusic I, Epshtein V, Zimmermann B, Lybecker M, Sedlyarov V, Schroeder R, Nudler E. 2017. Natural RNA polymerase aptamers regulate transcription in *E. coli*. *Mol Cell* 67:30–43.e6.

27. Lemay JF, Desnoyers G, Blouin S, Heppell B, Bastet L, St-Pierre P, Massé E, Lafontaine DA. 2011. Comparative study between transcriptionally- and translationally-acting adenine riboswitches reveals key differences in riboswitch regulatory mechanisms. *PLoS Genet* 7:e1001278.

28. Sudarsan N, Lee ER, Weinberg Z, Moy RH, Kim JN, Link KH, Breaker RR. 2008. Riboswitches in eubacteria sense the second messenger cyclic di-GMP. *Science* 321:411–413.

29. Lee ER, Baker JL, Weinberg Z, Sudarsan N, Breaker RR. 2010. An allosteric self-splicing ribozyme triggered by a bacterial second messenger. *Science* 329:845–848.

30. Gorski SA, Vogel J, Doudna JA. 2017. RNA-based recognition and targeting: sowing the seeds of specificity. *Nat Rev Mol Cell Biol* 18:215–228.

31. Kawamoto H, Koide Y, Morita T, Aiba H. 2006. Base-pairing requirement for RNA silencing by a bacterial small RNA and acceleration of duplex formation by Hfq. *Mol Microbiol* 61:1013–1022.

32. Mizuno T, Chou MY, Inouye M. 1984. A unique mechanism regulating gene expression: translational inhibition by a complementary RNA transcript (micRNA). *Proc Natl Acad Sci U S A* 81:1966–1970.

33. Vazquez-Anderson J, Mihailovic MK, Baldridge KC, Reyes KG, Haning K, Cho SH, Amador P, Powell WB, Contreras LM. 2017. Optimization of a novel biophysical model using large scale *in vivo* antisense hybridization data displays improved prediction capabilities of structurally accessible RNA regions. *Nucleic Acids Res* 45:5523–5538.

34. Wassarman KM, Repoila F, Rosenow C, Storz G, Gottesman S. 2001. Identification of novel small RNAs using comparative genomics and microarrays. *Genes Dev* 15:1637–1651.

35. Carter RJ, Dubchak I, Holbrook SR. 2001. A computational approach to identify genes for functional RNAs in genomic sequences. *Nucleic Acids Res* 29:3928–3938.

36. Livny J, Waldor MK. 2007. Identification of small RNAs in diverse bacterial species. *Curr Opin Microbiol* 10:96–101.

37. Babski J, Maier LK, Heyer R, Jaschinski K, Prasse D, Jäger D, Randau L, Schmitz RA, Marchfelder A, Soppa J. 2014. Small regulatory RNAs in Archaea. *RNA Biol* 11:484–493.

38. Sharma CM, Vogel J. 2009. Experimental approaches for the discovery and characterization of regulatory small RNA. *Curr Opin Microbiol* 12:536–546.

39. Tsai CH, Liao R, Chou B, Palumbo M, Contreras LM. 2015. Genome-wide analyses in bacteria show small-RNA enrichment for long and conserved intergenic regions. *J Bacteriol* 197:40–50.

40. Gelderman G, Contreras LM. 2013. Discovery of post-transcriptional regulatory RNAs using next generation sequencing technologies. *Methods Mol Biol* 985:269–295.

41. DeJesus MA, Gerrick ER, Xu W, Park SW, Long JE, Boutte CC, Rubin EJ, Schnappinger D, Ehrt S, Fortune SM, Sassetti CM, Ioerger TR. 2017. Comprehensive essentiality analysis of the *Mycobacterium tuberculosis* genome via saturating transposon mutagenesis. *mBio* 8: e02133-16.

42. Sharma CM, Hoffmann S, Darfeuille F, Reignier J, Findeiss S, Sittka A, Chabas S, Reiche K, Hackermüller J, Reinhardt R, Stadler PF, Vogel J. 2010. The primary transcriptome of the major human pathogen *Helicobacter pylori*. *Nature* 464:250–255.

43. Fakhry CT, Kulkarni P, Chen P, Kulkarni R, Zarringhalam K. 2017. Prediction of bacterial small RNAs in the RsmA (CsrA) and ToxT pathways: a machine learning approach. *BMC Genomics* 18:645.

44. Barrick JE, Corbino KA, Winkler WC, Nahvi A, Mandal M, Collins J, Lee M, Roth A, Sudarsan N, Jona I, Wickiser JK, Breaker RR. 2004. New RNA motifs suggest an expanded scope for riboswitches in bacterial genetic control. *Proc Natl Acad Sci U S A* 101:6421–6426.

45. Weinberg Z, Barrick JE, Yao Z, Roth A, Kim JN, Gore J, Wang JX, Lee ER, Block KF, Sudarsan N, Neph S, Tompa M, Ruzzo WL, Breaker RR. 2007. Identification of 22 candidate structured RNAs in bacteria using the CMfinder comparative genomics pipeline. *Nucleic Acids Res* 35:4809–4819.

46. Yao Z, Barrick J, Weinberg Z, Neph S, Breaker R, Tompa M, Ruzzo WL. 2007. A computational pipeline for high-throughput discovery of *cis*-regulatory noncoding RNA in prokaryotes. *PLoS Comput Biol* 3:e126.

47. Weinberg Z, Wang JX, Bogue J, Yang J, Corbino K, Moy RH, Breaker RR. 2010. Comparative genomics reveals 104 candidate structured RNAs from bacteria, archaea, and their metagenomes. *Genome Biol* 11: R31.

48. McCown PJ, Corbino KA, Stav S, Sherlock ME, Breaker RR. 2017. Riboswitch diversity and distribution. *RNA* 23:995–1011.

49. Rosinski-Chupin I, Sauvage E, Sismeiro O, Villain A, Da Cunha V, Caliot ME, Dillies MA, Trieu-Cuot P,

Bouloc P, Lartigue MF, Glaser P. 2015. Single nucleotide resolution RNA-seq uncovers new regulatory mechanisms in the opportunistic pathogen *Streptococcus agalactiae*. *BMC Genomics* 16:419.

50. Rosinski-Chupin I, Soutourina O, Martin-Verstraete I. 2014. Riboswitch discovery by combining RNA-seq and genome-wide identification of transcriptional start sites. *Methods Enzymol* 549:3–27.

51. Sledjeski D, Gottesman S. 1995. A small RNA acts as an antisilencer of the H-NS-silenced *rcsA* gene of *Escherichia coli*. *Proc Natl Acad Sci U S A* 92:2003–2007.

52. Lalaouna D, Massé E. 2016. The spectrum of activity of the small RNA DsrA: not so narrow after all. *Curr Genet* 62:261–264.

53. Melamed S, Peer A, Faigenbaum-Romm R, Gatt YE, Reiss N, Bar A, Altuvia Y, Argaman L, Margalit H. 2016. Global mapping of small RNA-target interactions in bacteria. *Mol Cell* 63:884–897.

54. Liu T, Zhang K, Xu S, Wang Z, Fu H, Tian B, Zheng X, Li W. 2017. Detecting RNA-RNA interactions in *E. coli* using a modified CLASH method. *BMC Genomics* 18:343.

55. Wang J, Rennie W, Liu C, Carmack CS, Prévost K, Caron MP, Massé E, Ding Y, Wade JT. 2015. Identification of bacterial sRNA regulatory targets using ribosome profiling. *Nucleic Acids Res* 43:10308–10320.

56. Bourqui R, Dutour I, Dubois J, Benchimol W, Thébault P. 2017. rNAV 2.0: a visualization tool for bacterial sRNA-mediated regulatory networks mining. *BMC Bioinformatics* 18:188.

57. Wang J, Liu T, Zhao B, Lu Q, Wang Z, Cao Y, Li W. 2016. sRNATarBase 3.0: an updated database for sRNA-target interactions in bacteria. *Nucleic Acids Res* 44(D1): D248–D253.

58. Ivain L, Bordeau V, Eyraud A, Hallier M, Dreano S, Tattevin P, Felden B, Chabelskaya S. 2017. An *in vivo* reporter assay for sRNA-directed gene control in Gram-positive bacteria: identifying a novel sRNA target in *Staphylococcus aureus*. *Nucleic Acids Res* 45:4994–5007.

59. Jagodnik J, Brosse A, Le Lam TN, Chiaruttini C, Guillier M. 2017. Mechanistic study of base-pairing small regulatory RNAs in bacteria. *Methods* 117:67–76.

60. Nelson JW, Sudarsan N, Furukawa K, Weinberg Z, Wang JX, Breaker RR. 2013. Riboswitches in eubacteria sense the second messenger c-di-AMP. *Nat Chem Biol* 9:834–839.

61. Nelson JW, Sudarsan N, Phillips GE, Stav S, Lünse CE, McCown PJ, Breaker RR. 2015. Control of bacterial exoelectrogenesis by c-AMP-GMP. *Proc Natl Acad Sci U S A* 112:5389–5394.

62. Kellenberger CA, Wilson SC, Hickey SF, Gonzalez TL, Su Y, Hallberg ZF, Brewer TF, Iavarone AT, Carlson HK, Hsieh YF, Hammond MC. 2015. GEMM-I riboswitches from *Geobacter* sense the bacterial second messenger cyclic AMP-GMP. *Proc Natl Acad Sci U S A* 112:5383–5388.

63. Hallberg ZF, Wang XC, Wright TA, Nan B, Ad O, Yeo J, Hammond MC. 2016. Hybrid promiscuous (Hypr) GGDEF enzymes produce cyclic AMP-GMP (3′, 3′-cGAMP). *Proc Natl Acad Sci U S A* 113:1790–1795.

64. Meyer MM, Roth A, Chervin SM, Garcia GA, Breaker RR. 2008. Confirmation of a second natural preQ$_1$ aptamer class in Streptococcaceae bacteria. *RNA* 14: 685–695.

65. Baker JL, Sudarsan N, Weinberg Z, Roth A, Stockbridge RB, Breaker RR. 2012. Widespread genetic switches and toxicity resistance proteins for fluoride. *Science* 335: 233–235.

66. Nelson JW, Atilho RM, Sherlock ME, Stockbridge RB, Breaker RR. 2017. Metabolism of free guanidine in bacteria is regulated by a widespread riboswitch class. *Mol Cell* 65:220–230.

67. Sherlock ME, Breaker RR. 2017. Biochemical validation of a third guanidine riboswitch class in bacteria. *Biochemistry* 56:359–363.

68. Mandal M, Breaker RR. 2004. Adenine riboswitches and gene activation by disruption of a transcription terminator. *Nat Struct Mol Biol* 11:29–35.

69. Kim JN, Roth A, Breaker RR. 2007. Guanine riboswitch variants from *Mesoplasma florum* selectively recognize 2′-deoxyguanosine. *Proc Natl Acad Sci U S A* 104: 16092–16097.

70. Weinberg Z, Nelson JW, Lünse CE, Sherlock ME, Breaker RR. 2017. Bioinformatic analysis of riboswitch structures uncovers variant classes with altered ligand specificity. *Proc Natl Acad Sci U S A* 114:E2077–E2085.

71. Kumari P, Sampath K. 2015. cncRNAs: bi-functional RNAs with protein coding and non-coding functions. *Semin Cell Dev Biol* 47–48:40–51.

72. Bronsard J, Pascreau G, Sassi M, Mauro T, Augagneur Y, Felden B. 2017. sRNA and *cis*-antisense sRNA identification in *Staphylococcus aureus* highlights an unusual sRNA gene cluster with one encoding a secreted peptide. *Sci Rep* 7:4565.

73. Wadler CS, Vanderpool CK. 2007. A dual function for a bacterial small RNA: SgrS performs base pairing-dependent regulation and encodes a functional polypeptide. *Proc Natl Acad Sci U S A* 104:20454–20459.

74. Lloyd CR, Park S, Fei J, Vanderpool CK. 2017. The small protein SgrT controls transport activity of the glucose-specific phosphotransferase system. *J Bacteriol* 199: 1–14.

75. Lago M, Monteil V, Douche T, Guglielmini J, Criscuolo A, Maufrais C, Matondo M, Norel F. 2017. Proteome remodelling by the stress sigma factor RpoS/σ^S in *Salmonella*: identification of small proteins and evidence for post-transcriptional regulation. *Sci Rep* 7:2127.

76. Savinov A, Perez CF, Block SM. 2014. Single-molecule studies of riboswitch folding. *Biochim Biophys Acta* 1839:1030–1045.

77. Haller A, Rieder U, Aigner M, Blanchard SC, Micura R. 2011. Conformational capture of the SAM-II riboswitch. *Nat Chem Biol* 7:393–400.

78. Heppell B, Blouin S, Dussault AM, Mulhbacher J, Ennifar E, Penedo JC, Lafontaine DA. 2011. Molecular insights into the ligand-controlled organization of the SAM-I riboswitch. *Nat Chem Biol* 7:384–392.

79. Zhao B, Guffy SL, Williams B, Zhang Q. 2017. An excited state underlies gene regulation of a transcriptional riboswitch. *Nat Chem Biol* **13**:968–974.

80. Hammond MC. 2011. RNA folding: a tale of two riboswitches. *Nat Chem Biol* **7**:342–343.

81. Manz C, Kobitski AY, Samanta A, Keller BG, Jäschke A, Nienhaus GU. 2017. Single-molecule FRET reveals the energy landscape of the full-length SAM-I riboswitch. *Nat Chem Biol* **13**:1172–1178.

82. Watters KE, Strobel EJ, Yu AM, Lis JT, Lucks JB. 2016. Cotranscriptional folding of a riboswitch at nucleotide resolution. *Nat Struct Mol Biol* **23**:1124–1131.

83. Updegrove TB, Zhang A, Storz G. 2016. Hfq: the flexible RNA matchmaker. *Curr Opin Microbiol* **30**:133–138.

84. Schu DJ, Zhang A, Gottesman S, Storz G. 2015. Alternative Hfq-sRNA interaction modes dictate alternative mRNA recognition. *EMBO J* **34**:2557–2573.

85. Santiago-Frangos A, Kavita K, Schu DJ, Gottesman S, Woodson SA. 2016. C-terminal domain of the RNA chaperone Hfq drives sRNA competition and release of target RNA. *Proc Natl Acad Sci U S A* **113**:E6089–E6096.

86. Malabirade A, Morgado-Brajones J, Trépout S, Wien F, Marquez I, Seguin J, Marco S, Velez M, Arluison V. 2017. Membrane association of the bacterial riboregulator Hfq and functional perspectives. *Sci Rep* **7**:10724.

87. Bouloc P, Repoila F. 2016. Fresh layers of RNA-mediated regulation in Gram-positive bacteria. *Curr Opin Microbiol* **30**:30–35.

88. Nielsen JS, Lei LK, Ebersbach T, Olsen AS, Klitgaard JK, Valentin-Hansen P, Kallipolitis BH. 2010. Defining a role for Hfq in Gram-positive bacteria: evidence for Hfq-dependent antisense regulation in *Listeria monocytogenes*. *Nucleic Acids Res* **38**:907–919.

89. Olejniczak M, Storz G. 2017. ProQ/FinO-domain proteins: another ubiquitous family of RNA matchmakers? *Mol Microbiol* **104**:905–915.

90. Wickiser JK, Winkler WC, Breaker RR, Crothers DM. 2005. The speed of RNA transcription and metabolite binding kinetics operate an FMN riboswitch. *Mol Cell* **18**:49–60.

91. Espah Borujeni A, Mishler DM, Wang J, Huso W, Salis HM. 2016. Automated physics-based design of synthetic riboswitches from diverse RNA aptamers. *Nucleic Acids Res* **44**:1–13.

92. Adamson DN, Lim HN. 2011. Essential requirements for robust signaling in Hfq dependent small RNA networks. *PLoS Comput Biol* **7**:e1002138.

93. Bossi L, Figueroa-Bossi N. 2016. Competing endogenous RNAs: a target-centric view of small RNA regulation in bacteria. *Nat Rev Microbiol* **14**:775–784.

94. Gaida SM, Al-Hinai MA, Indurthi DC, Nicolaou SA, Papoutsakis ET. 2013. Synthetic tolerance: three noncoding small RNAs, DsrA, ArcZ and RprA, acting supra-additively against acid stress. *Nucleic Acids Res* **41**:8726–8737.

95. Mandal M, Lee M, Barrick JE, Weinberg Z, Emilsson GM, Ruzzo WL, Breaker RR. 2004. A glycine-dependent riboswitch that uses cooperative binding to control gene expression. *Science* **306**:275–279.

96. Welz R, Breaker RR. 2007. Ligand binding and gene control characteristics of tandem riboswitches in *Bacillus anthracis*. *RNA* **13**:573–582.

97. Zhou H, Zheng C, Su J, Chen B, Fu Y, Xie Y, Tang Q, Chou SH, He J. 2016. Characterization of a natural triple-tandem c-di-GMP riboswitch and application of the riboswitch-based dual-fluorescence reporter. *Sci Rep* **6**:20871.

98. Sudarsan N, Hammond MC, Block KF, Welz R, Barrick JE, Roth A, Breaker RR. 2006. Tandem riboswitch architectures exhibit complex gene control functions. *Science* **314**:300–304.

99. Winkler WC, Nahvi A, Roth A, Collins JA, Breaker RR. 2004. Control of gene expression by a natural metabolite-responsive ribozyme. *Nature* **428**:281–286.

100. Klein DJ, Ferré-D'Amaré AR. 2006. Structural basis of *glmS* ribozyme activation by glucosamine-6-phosphate. *Science* **313**:1752–1756.

101. Cochrane JC, Lipchock SV, Smith KD, Strobel SA. 2009. Structural and chemical basis for glucosamine 6-phosphate binding and activation of the *glmS* ribozyme. *Biochemistry* **48**:3239–3246.

102. Cheah MT, Wachter A, Sudarsan N, Breaker RR. 2007. Control of alternative RNA splicing and gene expression by eukaryotic riboswitches. *Nature* **447**:497–500.

103. DebRoy S, Gebbie M, Ramesh A, Goodson JR, Cruz MR, van Hoof A, Winkler WC, Garsin DA. 2014. Riboswitches. A riboswitch-containing sRNA controls gene expression by sequestration of a response regulator. *Science* **345**:937–940.

104. Loh E, Dussurget O, Gripenland J, Vaitkevicius K, Tiensuu T, Mandin P, Repoila F, Buchrieser C, Cossart P, Johansson J. 2009. A *trans*-acting riboswitch controls expression of the virulence regulator PrfA in *Listeria monocytogenes*. *Cell* **139**:770–779.

105. Lahiry A, Stimple SD, Wood DW, Lease RA. 2017. Retargeting a dual-acting sRNA for multiple mRNA transcript regulation. *ACS Synth Biol* **6**:648–658.

106. Hoynes-O'Connor A, Moon TS. 2016. Development of design rules for reliable antisense RNA behavior in *E. coli*. *ACS Synth Biol* **5**:1441–1454.

107. Papenfort K, Bouvier M, Mika F, Sharma CM, Vogel J. 2010. Evidence for an autonomous 5′ target recognition domain in an Hfq-associated small RNA. *Proc Natl Acad Sci U S A* **107**:20435–20440.

108. Fröhlich KS, Papenfort K, Fekete A, Vogel J. 2013. A small RNA activates CFA synthase by isoform-specific mRNA stabilization. *EMBO J* **32**:2963–2979.

109. Noro E, Mori M, Makino G, Takai Y, Ohnuma S, Sato A, Tomita M, Nakahigashi K, Kanai A. 2017. Systematic characterization of artificial small RNA-mediated inhibition of *Escherichia coli* growth. *RNA Biol* **14**:206–218.

110. Sharma V, Yamamura A, Yokobayashi Y. 2012. Engineering artificial small RNAs for conditional gene silencing in *Escherichia coli*. *ACS Synth Biol* **1**:6–13.

111. Wasmuth EV, Lima CD. 2017. The Rrp6 C-terminal domain binds RNA and activates the nuclear RNA exosome. *Nucleic Acids Res* **45**:846–860.

112. Lee YJ, Moon TS. 2018. Design rules of synthetic non-coding RNAs in bacteria. *Methods* S1046-2023(17) 30338-9.

113. Man S, Cheng R, Miao C, Gong Q, Gu Y, Lu X, Han F, Yu W. 2011. Artificial *trans*-encoded small non-coding RNAs specifically silence the selected gene expression in bacteria. *Nucleic Acids Res* **39**:e50.

114. Na D, Yoo SM, Chung H, Park H, Park JH, Lee SY. 2013. Metabolic engineering of *Escherichia coli* using synthetic small regulatory RNAs. *Nat Biotechnol* **31**: 170–174.

115. Leistra AN, Amador P, Buvanendiran A, Moon-Walker A, Contreras LM. 2017. Rational modular RNA engineering based on *in vivo* profiling of structural accessibility. *ACS Synth Biol* **6**:2228–2240.

116. Jenison RD, Gill SC, Pardi A, Polisky B. 1994. High-resolution molecular discrimination by RNA. *Science* **263**:1425–1429.

117. Desai SK, Gallivan JP. 2004. Genetic screens and selections for small molecules based on a synthetic riboswitch that activates protein translation. *J Am Chem Soc* **126**:13247–13254.

118. Suess B, Fink B, Berens C, Stentz R, Hillen W. 2004. A theophylline responsive riboswitch based on helix slipping controls gene expression *in vivo*. *Nucleic Acids Res* **32**:1610–1614.

119. Goler JA, Carothers JM, Keasling JD. 2014. Dual-selection for evolution of *in vivo* functional aptazymes as riboswitch parts. *Methods Mol Biol* **1111**:221–235.

120. Dixon N, Duncan JN, Geerlings T, Dunstan MS, McCarthy JE, Leys D, Micklefield J. 2010. Reengineering orthogonally selective riboswitches. *Proc Natl Acad Sci U S A* **107**:2830–2835.

121. Nomura Y, Yokobayashi Y. 2007. Reengineering a natural riboswitch by dual genetic selection. *J Am Chem Soc* **129**:13814–13815.

122. Sparvath SL, Geary CW, Andersen ES. 2017. Computer-aided design of RNA origami structures. *Methods Mol Biol* **1500**:51–80.

123. Cho C, Lee SY. 2017. Efficient gene knockdown in *Clostridium acetobutylicum* by synthetic small regulatory RNAs. *Biotechnol Bioeng* **114**:374–383.

124. Venkataramanan KP, Jones SW, McCormick KP, Kunjeti SG, Ralston MT, Meyers BC, Papoutsakis ET. 2013. The *Clostridium* small RNome that responds to stress: the paradigm and importance of toxic metabolite stress in *C. acetobutylicum*. *BMC Genomics* **14**:849.

125. Cho SH, Lei R, Henninger TD, Contreras LM. 2014. Discovery of ethanol-responsive small RNAs in *Zymomonas mobilis*. *Appl Environ Microbiol* **80**:4189–4198.

126. Jones AJ, Venkataramanan KP, Papoutsakis T. 2016. Overexpression of two stress-responsive, small, non-coding RNAs, 6S and tmRNA, imparts butanol tolerance in *Clostridium acetobutylicum*. *FEMS Microbiol Lett* **363**:1–6.

127. Pei G, Sun T, Chen S, Chen L, Zhang W. 2017. Systematic and functional identification of small non-coding RNAs associated with exogenous biofuel stress in cyanobacterium *Synechocystis* sp. PCC 6803. *Biotechnol Biofuels* **10**:57.

128. Liu M, Zhu ZT, Tao XY, Wang FQ, Wei DZ. 2016. RNA-seq analysis uncovers non-coding small RNA system of *Mycobacterium neoaurum* in the metabolism of sterols to accumulate steroid intermediates. *Microb Cell Fact* **15**:64.

129. Hertel R, Meyerjürgens S, Voigt B, Liesegang H, Volland S. 2017. Small RNA mediated repression of subtilisin production in *Bacillus licheniformis*. *Sci Rep* **7**:5699.

130. Zess EK, Begemann MB, Pfleger BF. 2016. Construction of new synthetic biology tools for the control of gene expression in the cyanobacterium *Synechococcus* sp. strain PCC 7002. *Biotechnol Bioeng* **113**:424–432.

131. Florea M, Hagemann H, Santosa G, Abbott J, Micklem CN, Spencer-Milnes X, de Arroyo Garcia L, Paschou D, Lazenbatt C, Kong D, Chughtai H, Jensen K, Freemont PS, Kitney R, Reeve B, Ellis T. 2016. Engineering control of bacterial cellulose production using a genetic toolkit and a new cellulose-producing strain. *Proc Natl Acad Sci U S A* **113**:E3431–E3440.

132. Zhou LB, Zeng AP. 2015. Exploring lysine riboswitch for metabolic flux control and improvement of L-lysine synthesis in *Corynebacterium glutamicum*. *ACS Synth Biol* **4**:729–734.

133. Zhou LB, Zeng AP. 2015. Engineering a lysine-ON riboswitch for metabolic control of lysine production in *Corynebacterium glutamicum*. *ACS Synth Biol* **4**:1335–1340.

134. Wang J, Gao D, Yu X, Li W, Qi Q. 2015. Evolution of a chimeric aspartate kinase for L-lysine production using a synthetic RNA device. *Appl Microbiol Biotechnol* **99**:8527–8536.

135. Yang J, Seo SW, Jang S, Shin SI, Lim CH, Roh TY, Jung GY. 2013. Synthetic RNA devices to expedite the evolution of metabolite-producing microbes. *Nat Commun* **4**:1413.

136. Meyer A, Pellaux R, Potot S, Becker K, Hohmann HP, Panke S, Held M. 2015. Optimization of a whole-cell biocatalyst by employing genetically encoded product sensors inside nanolitre reactors. *Nat Chem* **7**:673–678.

137. Vincent HA, Robinson CJ, Wu MC, Dixon N, Micklefield J. 2014. Generation of orthogonally selective bacterial riboswitches by targeted mutagenesis and *in vivo* screening. *Methods Mol Biol* **1111**:107–129.

138. Robinson CJ, Vincent HA, Wu MC, Lowe PT, Dunstan MS, Leys D, Micklefield J. 2014. Modular riboswitch toolsets for synthetic genetic control in diverse bacterial species. *J Am Chem Soc* **136**:10615–10624.

139. Wu MC, Lowe PT, Robinson CJ, Vincent HA, Dixon N, Leigh J, Micklefield J. 2015. Rational re-engineering of a transcriptional silencing PreQ$_1$ riboswitch. *J Am Chem Soc* **137**:9015–9021.

140. Topp S, Reynoso CM, Seeliger JC, Goldlust IS, Desai SK, Murat D, Shen A, Puri AW, Komeili A, Bertozzi CR, Scott JR, Gallivan JP. 2010. Synthetic riboswitches that

induce gene expression in diverse bacterial species. *Appl Environ Microbiol* **76**:7881–7884.

141. Bugrysheva JV, Froehlich BJ, Freiberg JA, Scott JR. 2011. The histone-like protein Hlp is essential for growth of *Streptococcus pyogenes*: comparison of genetic approaches to study essential genes. *Appl Environ Microbiol* **77**:4422–4428.

142. Reynoso CM, Miller MA, Bina JE, Gallivan JP, Weiss DS. 2012. Riboswitches for intracellular study of genes involved in *Francisella* pathogenesis. *mBio* **3**:e00253-12.

143. Seeliger JC, Topp S, Sogi KM, Previti ML, Gallivan JP, Bertozzi CR. 2012. A riboswitch-based inducible gene expression system for mycobacteria. *PLoS One* **7**:e29266.

144. Rudolph MM, Vockenhuber MP, Suess B. 2015. Conditional control of gene expression by synthetic riboswitches in *Streptomyces coelicolor*. *Methods Enzymol* **550**:283–299.

145. Rudolph MM, Vockenhuber MP, Suess B. 2013. Synthetic riboswitches for the conditional control of gene expression in *Streptomyces coelicolor*. *Microbiology* **159**:1416–1422.

146. Nakahira Y, Ogawa A, Asano H, Oyama T, Tozawa Y. 2013. Theophylline-dependent riboswitch as a novel genetic tool for strict regulation of protein expression in cyanobacterium *Synechococcus elongatus* PCC 7942. *Plant Cell Physiol* **54**:1724–1735.

147. Dwidar M, Yokobayashi Y. 2017. Controlling *Bdellovibrio bacteriovorus* gene expression and predation using synthetic riboswitches. *ACS Synth Biol* **6**:2035–2041.

148. Ma AT, Schmidt CM, Golden JW. 2014. Regulation of gene expression in diverse cyanobacterial species by using theophylline-responsive riboswitches. *Appl Environ Microbiol* **80**:6704–6713.

149. Gelderman G, Sivakumar A, Lipp S, Contreras L. 2015. Adaptation of tri-molecular fluorescence complementation allows assaying of regulatory Csr RNA-protein interactions in bacteria. *Biotechnol Bioeng* **112**:365–375.

150. Alam KK, Tawiah KD, Lichte MF, Porciani D, Burke DH. 2017. A fluorescent split aptamer for visualizing RNA-RNA assembly *in vivo*. *ACS Synth Biol* **6**:1710–1721.

151. Sowa SW, Vazquez-Anderson J, Clark CA, De La Peña R, Dunn K, Fung EK, Khoury MJ, Contreras LM. 2015. Exploiting post-transcriptional regulation to probe RNA structures *in vivo* via fluorescence. *Nucleic Acids Res* **43**:e13.

152. Fowler CC, Brown ED, Li Y. 2010. Using a riboswitch sensor to examine coenzyme B_{12} metabolism and transport in *E. coli*. *Chem Biol* **17**:756–765.

153. Fowler CC, Sugiman-Marangos S, Junop MS, Brown ED, Li Y. 2013. Exploring intermolecular interactions of a substrate binding protein using a riboswitch-based sensor. *Chem Biol* **20**:1502–1512.

154. Gao X, Dong X, Subramanian S, Matthews PM, Cooper CA, Kearns DB, Dann CE III. 2014. Engineering of *Bacillus subtilis* strains to allow rapid characterization of heterologous diguanylate cyclases and phosphodiesterases. *Appl Environ Microbiol* **80**:6167–6174.

155. Paige JS, Wu KY, Jaffrey SR. 2011. RNA mimics of green fluorescent protein. *Science* **333**:642–646.

156. Kellenberger CA, Wilson SC, Sales-Lee J, Hammond MC. 2013. RNA-based fluorescent biosensors for live cell imaging of second messengers cyclic di-GMP and cyclic AMP-GMP. *J Am Chem Soc* **135**:4906–4909.

157. Wang XC, Wilson SC, Hammond MC. 2016. Next-generation RNA-based fluorescent biosensors enable anaerobic detection of cyclic di-GMP. *Nucleic Acids Res* **44**:e139.

158. Su Y, Hickey SF, Keyser SG, Hammond MC. 2016. *In vitro* and *in vivo* enzyme activity screening via RNA-based fluorescent biosensors for S-adenosyl-L-homocysteine (SAH). *J Am Chem Soc* **138**:7040–7047.

159. Kellenberger CA, Chen C, Whiteley AT, Portnoy DA, Hammond MC. 2015. RNA-based fluorescent biosensors for live cell imaging of second messenger cyclic di-AMP. *J Am Chem Soc* **137**:6432–6435.

160. Bose D, Su Y, Marcus A, Raulet DH, Hammond MC. 2016. An RNA-based fluorescent biosensor for high-throughput analysis of the cGAS-cGAMP-STING pathway. *Cell Chem Biol* **23**:1539–1549.

161. Yeo J, Dippel AB, Wang XC, Hammond MC. 2018. *In vivo* biochemistry: single-cell dynamics of cyclic di-GMP in *Escherichia coli* in response to zinc overload. *Biochemistry* **57**:108–116.

162. Papenfort K, Espinosa E, Casadesús J, Vogel J. 2015. Small RNA-based feedforward loop with AND-gate logic regulates extrachromosomal DNA transfer in *Salmonella*. *Proc Natl Acad Sci U S A* **112**:E4772–E4781.

163. Green AA, Kim J, Ma D, Silver PA, Collins JJ, Yin P. 2017. Complex cellular logic computation using ribocomputing devices. *Nature* **548**:117–121.

164. Green AA, Silver PA, Collins JJ, Yin P. 2014. Toehold switches: de-novo-designed regulators of gene expression. *Cell* **159**:925–939.

165. Chappell J, Takahashi MK, Lucks JB. 2015. Creating small transcription activating RNAs. *Nat Chem Biol* **11**:214–220.

166. Wachsmuth M, Domin G, Lorenz R, Serfling R, Findeiß S, Stadler PF, Mörl M. 2015. Design criteria for synthetic riboswitches acting on transcription. *RNA Biol* **12**:221–231.

167. Domin G, Findeiß S, Wachsmuth M, Will S, Stadler PF, Mörl M. 2017. Applicability of a computational design approach for synthetic riboswitches. *Nucleic Acids Res* **45**:4108–4119.

168. Sharma V, Nomura Y, Yokobayashi Y. 2008. Engineering complex riboswitch regulation by dual genetic selection. *J Am Chem Soc* **130**:16310–16315.

169. Jakočiūnas T, Jensen MK, Keasling JD. 2016. CRISPR/Cas9 advances engineering of microbial cell factories. *Metab Eng* **34**:44–59.

170. Haeussler M, Concordet JP. 2016. Genome editing with CRISPR-Cas9: can it get any better? *J Genet Genomics* **43**:239–250.

171. Dersch P, Khan MA, Mühlen S, Görke B. 2017. Roles of regulatory RNAs for antibiotic resistance in bacteria and their potential value as novel drug targets. *Front Microbiol* **8**:803.

172. Jakobsen TH, Warming AN, Vejborg RM, Moscoso JA, Stegger M, Lorenzen F, Rybtke M, Andersen JB, Petersen R, Andersen PS, Nielsen TE, Tolker-Nielsen T, Filloux A, Ingmer H, Givskov M. 2017. A broad range quorum sensing inhibitor working through sRNA inhibition. *Sci Rep* 7:9857.

173. Howe JA, Wang H, Fischmann TO, Balibar CJ, Xiao L, Galgoci AM, Malinverni JC, Mayhood T, Villafania A, Nahvi A, Murgolo N, Barbieri CM, Mann PA, Carr D, Xia E, Zuck P, Riley D, Painter RE, Walker SS, Sherborne B, de Jesus R, Pan W, Plotkin MA, Wu J, Rindgen D, Cummings J, Garlisi CG, Zhang R, Sheth PR, Gill CJ, Tang H, Roemer T. 2015. Selective small-molecule inhibition of an RNA structural element. *Nature* 526:672–677.

174. Jasinski D, Haque F, Binzel DW, Guo P. 2017. Advancement of the emerging field of RNA nanotechnology. *ACS Nano* 11:1142–1164.

175. Stewart JM, Subramanian HK, Franco E. 2017. Self-assembly of multi-stranded RNA motifs into lattices and tubular structures. *Nucleic Acids Res* 45:5449–5457.

176. Bui MN, Brittany Johnson M, Viard M, Satterwhite E, Martins AN, Li Z, Marriott I, Afonin KA, Khisamutdinov EF. 2017. Versatile RNA tetra-U helix linking motif as a toolkit for nucleic acid nanotechnology. *Nanomedicine (Lond)* 13:1137–1146.

177. Oi H, Fujita D, Suzuki Y, Sugiyama H, Endo M, Matsumura S, Ikawa Y. 2017. Programmable formation of catalytic RNA triangles and squares by assembling modular RNA enzymes. *J Biochem* 161: 451–462.

178. Gallagher RR, Patel JR, Interiano AL, Rovner AJ, Isaacs FJ. 2015. Multilayered genetic safeguards limit growth of microorganisms to defined environments. *Nucleic Acids Res* 43:1945–1954.

179. Pardee K, Green AA, Ferrante T, Cameron DE, DaleyKeyser A, Yin P, Collins JJ. 2014. Paper-based synthetic gene networks. *Cell* 159:940–954.

Resources

VIII

Regulating with RNA in Bacteria and Archaea
Edited by Gisela Storz and Kai Papenfort
© 2018 American Society for Microbiology, Washington, DC
doi:10.1128/microbiolspec.RWR-0033-2018

Functional Transcriptomics for Bacterial Gene Detectives

32

Blanca M. Perez-Sepulveda[1] and Jay C. D. Hinton[1]

INTRODUCTION

Transcriptional profiling is a valuable part of the functional genomics toolbox. Since the developments in nanotechnology and imaging that led to the invention of next-generation sequencing (1), study of the bacterial transcriptome at the level of the individual nucleotide has proved fruitful. Scientists are now generating increasing amounts of transcriptomic data that need to be managed, analyzed, and stored in an appropriate manner (2). The current need for systematic and accessible approaches for the analysis of gene expression has focused bioinformatic efforts into developing tools for processing transcriptomic data.

Since the amount of transcriptomic data being generated is increasing exponentially, and already exceeds the interpretive capacity of the human brain, we need to improve the ways in which we interact with complex information (3) and choose the best methods for data display and interpretation. "Big data" visualization constitutes a significant challenge, as inappropriate approaches can lead to conclusion bias and other errors (2, 4, 5).

In recent years, the molecular microbiological community has focused on approaches to simplify data visualization and analysis. The creation of online resources by some research labs has provided access to large data sets. A good example is Listeriomics (https://listeriomics.pasteur.fr), a user-friendly online platform that includes curated genomic, transcriptomic, and proteomic data sets generated from *Listeria* species (6). Other valuable online resources include PneumoBrowse (https://veeninglab.com/pneumobrowse) for the analysis of the *Streptococcus pneumoniae* D39V genome (7) and AcinetoCom (http://bioinf.gen.tcd.ie/acinetocom) for investigation of the primary transcriptome of *Acinetobacter baumannii* ATCC 17978 (8). Researchers can interrogate and explore the data in these online resources in creative ways. However, other online data repositories are less intuitive, and it can be difficult to access the transcriptomic data needed to make appropriate biological conclusions. The lack of ongoing technical support or suitable funding has led to the demise of some online platforms, leading to valuable resources becoming inaccessible to the wider scientific community.

Here, we will discuss the challenges of visualizing transcriptomic expression data and explain the role that "gene detectives" can play in the interrogation of online resources to develop hypotheses and answer biological questions. We hope to inspire the reader to use online resources to explore transcriptomic data sets.

ENVIRONMENTAL REGULATION OF GENE EXPRESSION

Bacterial pathogens face intense competition in the microbial world. Enterobacterial species that inhabit the gastrointestinal tract have evolved the ability to sense the environment and to use physicochemical information to control gene expression. Environmental modulation of gene expression enables the bacterium to express "the right gene in the right place at the right time," a prerequisite for fitness (9). John Mekalanos commented in 1992 that to understand how bacterial virulence genes are regulated we must first understand when and where the genes are expressed (10).

Transcriptional gene fusions were an invaluable tool in the early days of molecular microbiology, but only allowed the examination of individual genes (11). Soon after the first bacterial genomes were published, it became possible to monitor the expression of all genes in a single experiment. In subsequent years, these

---

[1]Institute of Integrative Biology, University of Liverpool, Liverpool, United Kingdom.

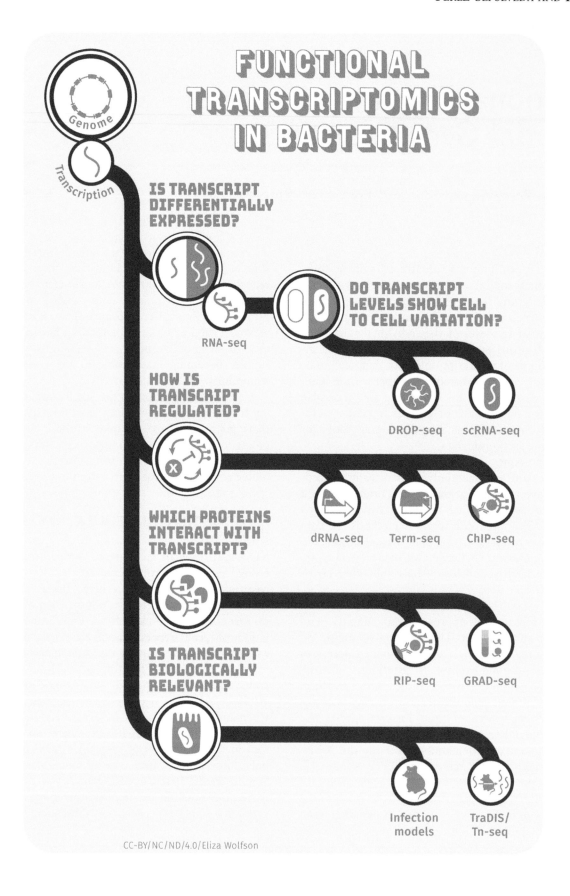

transcriptome-based approaches (Fig. 1) have evolved alongside proteomic, metabolomic, and comparative genomic techniques.

TECHNOLOGICAL DEVELOPMENTS THAT HAVE ENABLED FUNCTIONAL TRANSCRIPTOMICS

The transcriptomic revolution began with the advent of microarrays in the late 1990s, which involved the hybridization of target bacterial mRNA transcripts to specific DNA probes (12). DNA microarray technology was developed in the late 1990s by Pat Brown and Joe DeRisi at Stanford University, making it possible to profile the expression of all genes in a single experiment (12, 13). Microarrays became a popular high-throughput tool for studying functional transcriptomics in bacteria. A PubMed (https://www.ncbi.nlm.nih.gov/pubmed) search of the term "bacteria microarray" shows that the number of publications using microarray technology peaked in 2012 with 918 hits, followed by a slow decline in subsequent years.

Microarrays have been largely superseded by high-throughput RNA sequencing (RNA-seq), which was developed in 2008 (14). RNA-seq allows the visualization of every bacterial transcript at the nucleotide level and can provide accurate differential gene expression data from whole-genome transcriptomic analyses. To generate RNA-seq sequence reads, bacteria are grown under defined conditions, and RNA is extracted and then reverse-transcribed to cDNA before high-throughput sequencing (15, 16). Levels of gene expression are assessed by quantifying the RNA-seq reads that map to annotated coding or noncoding genes (tRNA, rRNA, transfer messenger RNA [tmRNA], small RNA [sRNA], etc.). Importantly, RNA-seq allows the identification of elements such as promoters and gene boundaries and the discovery of regulatory RNA species, such as sRNAs and antisense RNAs (17). RNA-seq is now an established technique that can be performed in many sequencing centers worldwide, making it relatively easy to generate transcriptomic data.

RNA-seq has stimulated functional transcriptomic research by enabling a plethora of innovative technologies that have broadened the number of biological questions that can be addressed (see chapter 23 [18], Fig. 1). These technologies have been the subject of two excellent reviews (15, 19).

THE CHALLENGE OF ACCESSING AND UNDERSTANDING TRANSCRIPTOMIC DATA

Powerful software tools have been developed for the analysis of RNA-seq-based transcriptomic data (20). These approaches are excellent for the generation of lists of differentially expressed genes that meet certain statistical criteria. However, a key barrier to understanding gene expression profiles at a global level has been the huge tab-delimited spreadsheets that can languish untouched in the Supplementary Data section of journals.

The Gene Expression Omnibus (GEO; www.ncbi.nlm.nih.gov/geo) and the EBI ArrayExpress (www.ebi.ac.uk/arrayexpress) are curated public repositories of transcriptomic data from all organisms, including results from microarray and RNA-seq experiments. Although the idea of sharing produced expression data in a public repository has been a positive step forward, GEO is not intuitive to use and data mining can be a challenge. Bioinformatic expertise is too often required to examine and interpret complex data sets or even to successfully download and interpret a transcriptomic experiment from a public repository.

Individual RNA-seq experiments are generally planned with a particular hypothesis in mind, but the entirety of the data collected can subsequently be mined or reinterpreted by the wider community to answer new research questions. It is time to rethink the way in which the worldwide microbiological community manages and accesses our data sets.

Figure 1 Bacterial functional transcriptomics is facilitated by RNA-seq technology. The development of RNA-seq has expanded the range of transcriptome-based techniques that address a variety of biological questions. DROP-seq, RNA-seq of single cells compartmentalized in a droplet; scRNA, single-cell RNA-seq; dRNA-seq, differential RNA-seq; Term-seq, global mapping of 3′ ends of transcripts; ChIP-seq, chromatin immunoprecipitation followed by sequencing; RIP-seq, native RNA immunoprecipitation followed by RNA-seq; GRAD-seq, gradient profiling by RNA-seq; TraDIS, transposon-directed insertion site sequencing; Tn-seq, transposon sequencing. See reference 20 for more details of these techniques. Image by Eliza Wolfson (https://lizawolfson.co.uk) is used under the terms of a creative commons CC-BY-NC-ND license (https://creativecommons.org/licenses/by-nc-nd/4.0/legalcode).

IMPROVING THE SHARING AND VISUALIZATION OF BACTERIAL TRANSCRIPTOMIC DATA

To facilitate collaborative analysis, online resources are needed to put global expression data into the hands of biologists. Many RNA-seq-based transcriptomic data sets have now been generated for a range of microbial model systems including yeasts, *Synechocystis*, *Bacillus*, *Escherichia coli*, *Pseudomonas aeruginosa*, *Listeria*, and *Salmonella enterica* (21). Depending on the RNA-seq-based experimental approach, different features can be visualized. These include strand-specific information for the correct annotation of coding sequences, and the identification of transcriptional start sites (TSSs) and 3′ ends (22, 23). To extract accurate differential gene expression values, it is important to normalize RNA-seq data, and several algorithms are available for this purpose (20). Ideally, statistically significant changes at the transcriptional level are generated from biological replicates, allowing mechanistic deductions to be made.

However, the number of online resources or platforms that are available to analyze bacterial expression data does not reflect the amount of published experiments. Of the gene expression data sets that can currently be browsed interactively, many are presented in a way that is not informative, perhaps reflecting the fact that creating, curating, and maintaining online platforms for expression data analysis is a significant effort. Good examples are provided by other research areas that have extremely large data sets, such as the human online resource for transcriptomic analysis, the Genotype-Tissue Expression (GTEx) project (24), which allows intuitive exploration of a variety of features.

The intelligent use of online resources for transcriptomic data analysis can allow gene expression profiles to be used to identify important environmental responses that could otherwise be overlooked. For example, transcriptomic information for 1,292 *E. coli* microarray experiments can be visualized in the Gen ExpDB (*E. coli* Gene Expression Database; https://genexpdb.okstate.edu), developed by Joe Grissom and Tyrrell Conway.

TOWARD A GLOBAL UNDERSTANDING OF *SALMONELLA* TRANSCRIPTIONAL REGULATION

The analysis and reconstruction of genome-scale transcriptional regulatory networks represents the next frontier in microbial bioinformatics (25), and requires information about the expression of every transcript and the regulatory inputs for every promoter. This challenging aim involves the combination of "-omics" data, including transcriptomic information and a range of chromatin immunoprecipitation (ChIP)-based strategies. Using methods detailed in Fig. 1, binding sites for key regulatory proteins can be identified, and an appreciation of the role of every RNA-binding protein may also be involved. A range of ChIP studies have already been conducted, focused on proteins like H-NS, as well as regulatory factors that control *Salmonella* pathogenicity island 1 and 2 (SPI1 and SPI2) (26–28).

An overarching analytical approach is needed to integrate different types of transcriptional regulatory data together into a single transcriptional network (25). This necessitates an intense bioinformatic effort, which is already being addressed by a few laboratories across the world (20). Specifically, network inference-based bioinformatic approaches have been used to study transcriptional networks and shed light upon *Salmonella* gene function (29–31), and the SalmoNet network, which integrates metabolic, transcriptional, and protein-protein interaction data, is now available (32). The end result of this long-term, systems-level approach will be a genome-scale regulatory network that will be a significant challenge to visualize effectively (3).

It will be some time before we have a comprehensive transcriptional network that reveals the entire regulatory complexity of *Salmonella*. However, an alternative strategy can already be used to gain insights to the transcriptional world of *Salmonella*. This intuitive approach relies on the ability of the human brain to find patterns within complex data sets (33), coupled with the prior knowledge of molecular microbiologists.

DETECTIVE WORK: INFERRING sRNA FUNCTION FROM GENE EXPRESSION DATA

The development of RNA-seq technology has led to the identification of sRNAs at a global level in many bacterial species (34). The characteristic expression profiles of individual sRNAs (35) allow an understanding of the environmental stresses that modulate sRNA expression to provide functional insights. For *S. enterica* serovar Typhimurium, transcriptional signatures can be visualized in the SalComMac compendium (http://tinyurl.com/SalComMac), which shows the expression of all coding genes and sRNAs in 21 environmental conditions (35, 36). To investigate the impact of infection-relevant environmental stress upon the *Salmonella* transcriptional network, we devised a suite of *in vitro* conditions that reflect particular aspects of the infection process. The conditions include exposing the bacteria to oxidative stress, osmotic shock, acid shock, anaerobic shock, and a low-iron environment (36), plus growth of *S.* Typhimurium within mammalian macrophages (35).

The SalComMac data show that the level of expression of the majority of coding genes and sRNAs is environmentally regulated. Genes that encode key *Salmonella* virulence systems have characteristic transcriptional signatures that we have reported previously (35, 36). For example, coding genes that encode both the structural subunits and the associated effector proteins of the SPI1 type III secretion system show an "SPI1-like" pattern of expression, focused on early stationary phase in rich media. In parallel, we reported a distinct "SPI2-like" transcriptional signature in defined acidic, low-phosphate media (35). More recently, we identified 13 *S.* Typhimurium sRNAs with either an SPI1-like or SPI2-like expression profile (37).

The SalComMac resource reveals changes in the expression of individual sRNAs by particular environmental stressors, and can be used to visualize the global sRNA expression landscape. This approach can generate hypotheses about gene function for experimental investigation. To demonstrate how transcriptomic data can be used to investigate sRNA function, we have explored the expression profiles of 14 well-characterized sRNAs using two data repositories that are currently available for *S.* Typhimurium.

USING ONLINE RESOURCES TO DETERMINE *S.* TYPHIMURIUM sRNA FUNCTION

The availability of comprehensive RNA-seq data sets, coupled with the work of many outstanding geneticists and RNA biologists, has made *S. enterica* a model organism for the study of RNA-mediated regulation. The pathogen *S.* Typhimurium is responsible for hundreds of thousands of deaths each year, largely caused by systemic infections (38). RNA-seq-based transcriptomics involving *in vitro* infection-relevant growth conditions have identified characteristic transcriptional signatures (35). To determine whether *in vitro* transcriptional patterns are relevant to intracellular infection, bacterial RNA has been isolated from murine macrophages infected with *S.* Typhimurium and used for RNA-seq and functional transcriptomic analysis (36). These data allow the transcriptional differences observed under *in vitro* conditions to be related to the infection of mammalian cells.

One way to address gene function is to integrate transcriptomic data with global transposon mutagenesis of bacterial pathogens during infection, a concept

that has been discussed previously (39). These types of experiments can answer the crucial question: Are these transcripts biologically relevant? (Fig. 1).

As well as investigating the effect of environmental stress upon *Salmonella* gene expression, we also studied the impact of 18 regulatory proteins upon expression of all coding genes and sRNAs. These data are available in a separate online resource, SalComRegulon (http://tinyurl.com/SalComRegulon), that presents transcriptomic data for a selection of mutants that lack transcriptional activators, repressors, sigma factors, and the Hfq RNA chaperone, reflecting both direct and indirect regulatory effects (37).

Following "traditional" data analysis, heat maps of unfiltered gene lists can be a valuable tool for subsequent transcriptomic data exploration. Figures 2, 3, and 4 show the impact of 21 environmental conditions and key regulatory factors upon the expression of three groups of sRNAs. In each of these three figures, panel A shows the "absolute" level of expression of each gene, in TPM (transcripts per million) units, which have been defined elsewhere (36). A TPM value of <10 indicates that a particular gene is not expressed, and the color scale (yellow through orange to red) indicates the level of expression. Panel B shows the "relative" level of expression between a stress environment and a control culture. The use of relative values allows the effect of environmental stress upon gene expression to be visualized as distinct colored patterns. Panel C also uses a "relative" approach that highlights differences in gene expression between wild-type and mutant bacteria cultivated in particular growth conditions.

Figure 2 shows sRNAs that respond to low iron levels or oxidative stress. OxyS was one of the first sRNAs found to be environmentally regulated (40), being induced by almost 300-fold upon exposure to hydrogen peroxide (peroxide) (Fig. 2B). RyhB is a key iron-regulated sRNA, described in chapter 16 (41). *S.* Typhimurium carries two RyhB paralogs called RyhB-1 and RyhB-2 (42). The RyhB-1 sRNA shares extensive sequence similarity to the *E. coli* RyhB ortholog, and has been reported to be highly induced by iron starvation, upregulated by peroxide, and activated by OxyR (43, 44). Consistent with this, the SalComMac data show that RyhB-1 is induced by 165-fold in the low-iron environment and upregulated by 700-fold in response to peroxide. In contrast, the RyhB-2 paralog, which is more distantly related to *E. coli* RyhB, is only upregulated by 30-fold in the low-iron environment and is peroxide-induced at a lower level than RyhB-1. These findings are consistent with the literature (see chapter 16 [41]). Figure 2B shows that the levels of expression of

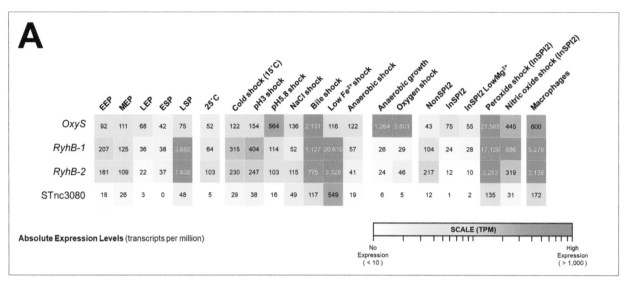

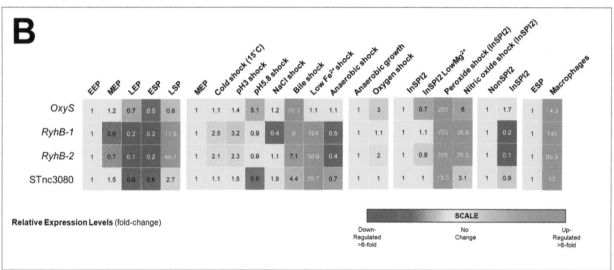

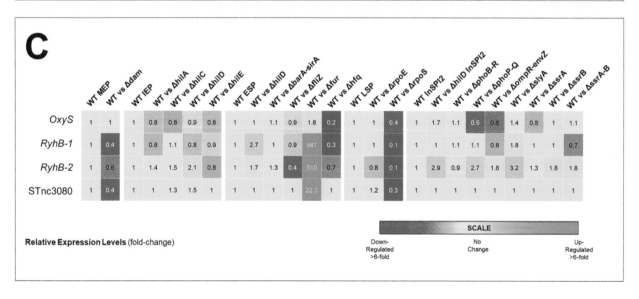

RyhB-1 and RyhB-2 also increase in response to treatment with nitric oxide and bile and during the infection of macrophages (Fig. 2B). These new findings could be worthy of study in the future.

STnc3080 is an *S.* Typhimurium sRNA that has previously been shown to be upregulated in a low-iron environment (36), and here we observe that STnc3080 is induced by peroxide (Fig. 2B). The impact of various regulatory factors upon the expression of OxyS, RyhB-1, RyhB-2, and STnc3080 is apparent (Fig. 2C). The most significant regulatory input appears to be the action of Fur to repress expression of the three iron-responsive sRNAs. The *E. coli* RyhB has previously been shown to be Fur repressed (45), and this is the first time that a role for Fur in the expression of STnc3080 has been suggested. Expression of both RyhB-1 and RyhB-2 is reduced in the absence of RpoS, suggesting that the σ^{38} sigma factor could activate expression of the two paralogs.

Figure 3 focuses on the expression of four sRNAs involved in the envelope stress response, namely RybB, RyeF, MicA, and RprA. Figure 3A shows that three of the four sRNAs are expressed in all environmental conditions, at relatively high levels. In contrast, RyeF is only expressed in 4 of the 21 conditions at medium to high levels. Figure 3B shows that three of the sRNAs are induced in a low-magnesium environment, a condition known to induce bacterial envelope stress (46). It is likely that the upregulation of the same three sRNAs during macrophage infection also reflects an envelope stress response. Two of the envelope stress-associated sRNAs were osmoinducible, with MicA and RprA showing between 15- and 21-fold upregulation following a 10-min exposure to 0.3 M NaCl. The RyeF (MicL) and MicA sRNAs were significantly upregulated following cold shock at 15°C, which, to our knowledge, has not been reported previously.

In terms of transcriptional inputs, the SalComRegulon data suggest that RybB, RyeF, and MicA are activated by

RpoE and are repressed by Fur (Fig. 3C). Activation of MicA and RyeF (MicL) by RpoE is consistent with the literature (46, 47). The data summarized in Fig. 3C raise the possibility of cross talk between PhoPQ/OmpR-EnvZ and MicA, and it is noteworthy that the *phoP* gene is a confirmed target of MicA (48).

Figure 4 presents the expression profiles of six sRNAs that respond to oxygen or osmolarity. MicA is included in both Fig. 3 and 4 for comparative purposes. Two of the sRNAs, MicA and MntS (RybA), are expressed at medium to high levels in all environmental conditions. In contrast, FnrS is only upregulated under conditions of oxygen limitation. MicA and SraL also show a low level of upregulation under anaerobic conditions. Four of the sRNAs are osmoregulated, showing between 12- and 37-fold induction following a 10-min exposure to 0.3 M NaCl (Fig. 4B). As well as being osmoregulated, MntS is induced by cold shock and by pH 3 shock, which could have implications for manganese homeostasis (49).

The SraL sRNA is conserved in many enteric bacteria and is upregulated when *S.* Typhimurium is grown under conditions that induce expression of the SPI2 pathogenicity island (50). As well as visualizing this finding, Fig. 4B shows that SraL is both osmoregulated and induced under anaerobic conditions. Because SraL has not been studied extensively in *Salmonella*, SalCom Regulon adds a useful perspective (Fig. 4C). The data suggest that RpoS, PhoPQ, and OmpR-EnvZ activate SraL, ideas that need to be tested experimentally. In contrast, MntS shows a distinct pattern of expression and does not appear to be RpoE dependent. The Hfq chaperone does not bind to the MntS transcript (51), and so is unlikely to play a stabilizing role.

Figure 4 also shows the expression profiles of two uncharacterized *S.* Typhimurium sRNAs that are highly osmoregulated (STnc1330 and STnc4260). The Sal ComRegulon data suggest that STnc1330 has an RpoS-

Figure 2 Environmental and genetic regulation of four sRNAs that are iron responsive and/or induced by oxidative stress. Gene expression data are presented for the sRNAs OxyS, RyhB-1, RyhB-2, and STnc3080 (these data can be visualized online at https://tinyurl.com/ya7s466m and https://tinyurl.com/yb5wz7dt). Data are shown as differential expression profiles involving six discrete heat-map blocks, each block being normalized to the condition on the left-hand side. The heat maps show differential expression, a strategy that lacks accuracy when expression levels are extremely low. Absolute (A) and relative (B) expression levels of *S.* Typhimurium grown under 21 different conditions (SalComMac). (C) Relative expression levels of the wild-type (WT) and mutant *S.* Typhimurium 4/74 grown under different conditions (SalComRegulon). Before experimental validation is considered, it should be ensured that the levels of absolute expression of particular sRNAs are above the expression threshold of 10 TPM units (35–37). EEP, early exponential phase; MEP, mid-exponential phase; LEP, late exponential phase; ESP, early stationary phase; LSP, late stationary phase; InSPI2, SPI2-inducing minimal media.

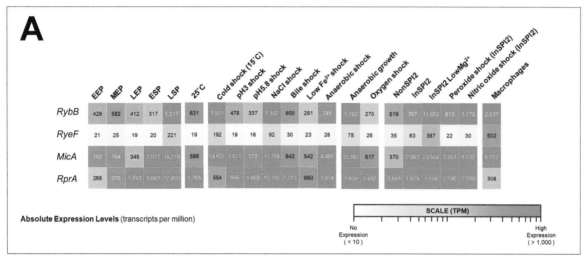

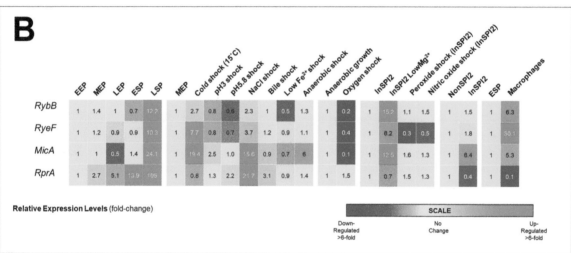

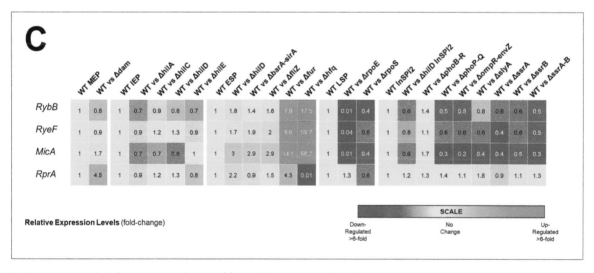

Figure 3   Environmental and genetic regulation of four sRNAs involved in the envelope stress response. Gene expression data are shown for the sRNAs RybB, RyeF, MicA, and RprA (these data can be visualized online at https://tinyurl.com/y9mskb6j and https://tinyurl.com/ybnr6jja). Panels A, B, and C are as described in the legend to Fig. 2.

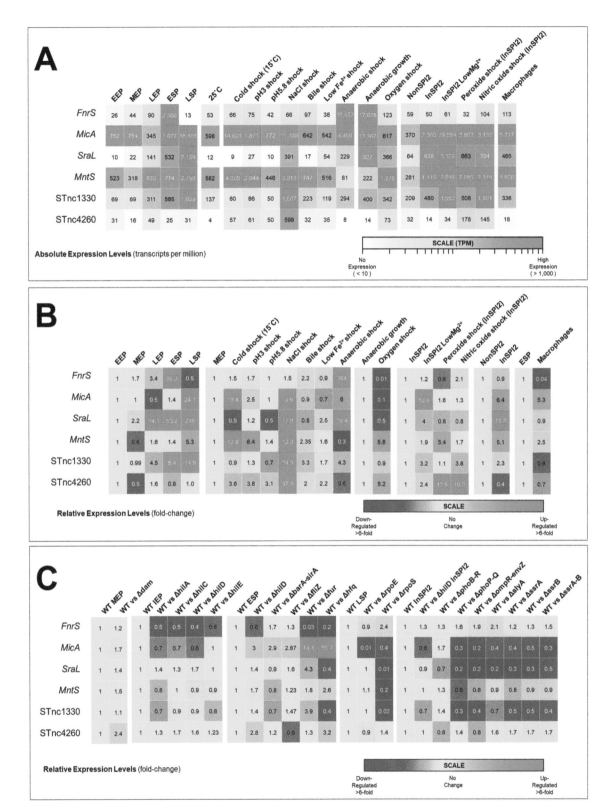

Figure 4 Environmental and genetic regulation of six sRNAs that respond to oxygen or osmolarity. Gene expression data are shown for the sRNAs FnrS, MicA, SraL, MntS (RybA), STnc1330, and STnc4260 (these data can be visualized online at https://tinyurl.com/yat8qrql and https://tinyurl.com/y8f533gy). Panels A, B, and C are as described in the legend to Fig. 2.

activated, PhoPQ-activated pattern of expression (Fig. 4C). Taken together, the panels of Fig. 4 show that sRNAs that respond to osmolarity and/or anaerobic growth are controlled by a variety of transcriptional inputs.

As well as facilitating data visualization as heat maps, SalComMac and SalComRegulon can also show RNA-seq sequence reads in a genomic context. This feature can be accessed via the "view in JBrowse" option on the SalComMac and SalComRegulon Web pages, which is located between the Absolute and Relative gene expression panels.

JBrowse is an open-source, JavaScript-based genome browser that is fast and zoomable, ideal for the analysis of multiple sets of mapped transcriptomic sequence reads (52, 53). The JBrowse feature of SalComMac and SalComRegulon allows transcripts of interest to be seen in the context of neighboring genes and TSSs. For example, the transcript of STnc1330, an *S.* Typhimurium

141-nucleotide sRNA (54), can be seen in the context of neighboring TSSs. Figure 5 shows the appearance of the STnc1330 transcript during growth in different environmental conditions; STnc1330 is expressed at low levels during growth at mid-exponential phase, and is highly abundant following osmotic or anaerobic shock (Fig. 5A). STnc1330 is also expressed in SPI2-inducing minimal media and at increased levels under magnesium limitation (Fig. 5B). The impact of regulatory factors is apparent in Fig. 5C and D, which show that PhoPQ and RpoS are required for expression.

THE USE OF EXPRESSION DATA TO INFER REGULATORY INTERACTIONS REQUIRES CAUTION!

Genome-scale expression data always have caveats. It should be noted that monitoring the levels of individual

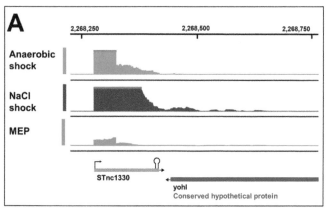
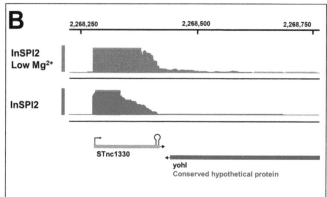
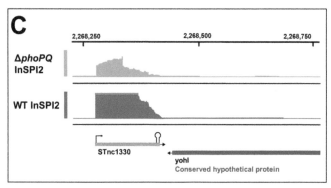
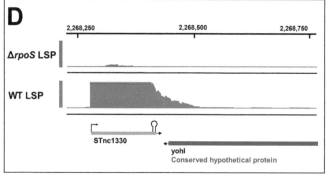

Figure 5 Visualization of the STnc1330 sRNA transcript. RNA-seq reads are mapped to the *S.* Typhimurium 4/74 genome (plus strand), showing STnc1330 expression under different conditions (35, 37). (A) MEP, anaerobic shock, and NaCl shock (https://tinyurl.com/STnc1330-NaCl); (B) InSPI2 and InSPI2 low Mg2+ (https://tinyurl.com/STnc1330-LowMg); (C) WT InSPI2 versus Δ*phoPQ* (https://tinyurl.com/STnc1330-PhoPQ); (D) WT LSP versus Δ*rpoS* (https://tinyurl.com/STnc1330-RpoS). Height of colored tracks represents the normalized sequencing reads at that locus (scale, 0 to 100). All arrows indicate the direction of transcription; TSSs are indicated by bent arrows and predicted Rho (ρ)-independent terminators are denoted by stem-loop structures.

mRNAs and sRNAs by conventional transcriptomics shows the average expression profiles in a large population of bacteria. Furthermore, the expression level of an individual transcript represents a combination of the positive impact of transcriptional initiation and elongation versus the negative effect of RNA decay. Consequently, the stability of individual transcripts can influence global gene expression levels.

Some of the transcriptomic data shown in SalCom Mac and SalComRegulon are based on RNA harvested between 10 and 30 min after an environmental perturbation (36). As environmental shocks were studied at a single time point, and individual stressors are just used at one concentration, these experimental conditions can differ from those used in other publications. Some coding genes and sRNAs may not show the "expected" patterns of expression. Despite these caveats, the expression of most well-known genes in SalComMac and SalComRegulon is consistent with the literature (36).

CONCLUSIONS AND FUTURE DIRECTIONS

We have come a long way since the first bacterial transcriptomic experiment was reported (12). Nowadays, many laboratories are generating informative RNA-seq data, but key interpretive challenges remain to be solved. Making the microbiology community's valuable gene expression data available for intuitive interrogation by current and future generations of microbiologists must be a priority. Currently, a variety of online resources are available for a small selection of model organisms. We need to improve and optimize these resources to ensure that the availability of crucial data sets will be maintained in the future, and not depend exclusively on particular scientists or laboratories.

Transcriptomic experiments can yield surprising biological insights. Recently, we showed that in African *S.* Typhimurium ST313, one nucleotide change was responsible for the upregulation of the *pgtE* transcript and the PgtE protein. This single regulatory difference accounted for the ability of the African *Salmonella* to resist killing by human serum (55), a discovery that was made by the comparison of transcriptomes of two pathovariants of *S.* Typhimurium.

For the future, a new era of comparative transcriptomics will necessitate moving beyond the analysis of individual bacterial strains. To understand key host-specificity phenotypes, the comparative transcriptomic analysis of closely related bacterial pathogens that have different lifestyles will be necessary. The continued development of tools to visualize and interrogate gene expression data will further our understanding of bacterial functional transcriptomics and, hopefully, inspire a new generation of gene detectives.

Acknowledgments. We are grateful to many people for contributing to the visualization of Salmonella *transcriptomic data over recent years, particularly members of the Hinton lab (including Aoife Colgan, Rocío Canals Carsten Kröger and Nicolas Wenner), Tyrrell Conway, and Karsten Hokamp. This work was supported by funding from a Wellcome Trust Senior Investigator award (to J.C.D.H.) (Grant 106914/Z/ 15/Z).*

Citation. Perez-Sepulveda B, Hinton JCD. 2018. Functional transcriptomics for bacterial gene detectives. Microbiol Spectrum 6(5):RWR-0033-2018.

References

1. **MacLean D, Jones JD, Studholme DJ.** 2009. Application of 'next-generation' sequencing technologies to microbial genetics. *Nat Rev Microbiol* **7:**287–296.

2. **Stephens ZD, Lee SY, Faghri F, Campbell RH, Zhai C, Efron MJ, Iyer R, Schatz MC, Sinha S, Robinson GE.** 2015. Big data: astronomical or genomical? *PLoS Biol* **13:**e1002195.

3. **Wang R, Perez-Riverol Y, Hermjakob H, Vizcaíno JA.** 2015. Open source libraries and frameworks for biological data visualisation: a guide for developers. *Proteomics* **15:** 1356–1374.

4. **Toker L, Feng M, Pavlidis P.** 2016. Whose sample is it anyway? Widespread misannotation of samples in transcriptomics studies. *F1000 Res* **5:**2103.

5. **Heiss JA, Just AC.** 2018. Identifying mislabeled and contaminated DNA methylation microarray data: an extended quality control toolset with examples from GEO. *Clin Epigenetics* **10:**73.

6. **Bécavin C, Koutero M, Tchitchek N, Cerutti F, Lechat P, Maillet N, Hoede C, Chiapello H, Gaspin C, Cossart P.** 2017. Listeriomics: an interactive web platform for systems biology of *Listeria*. *mSystems* **2:**e00186-e16.

7. **Slager J, Aprianto R, Veening JW.** 2018. Deep genome annotation of the opportunistic human pathogen *Streptococcus pneumoniae* D39. *Nucleic Acids Res.*

8. **Kröger C, MacKenzie KD, Alshabib EY, Kirzinger MW, Suchan DM, Chao TC, Akulova V, Miranda-CasoLuengo AA, Monzon VA, Conway T, Sivasankaran SK, Hinton JC, Hokamp K, Cameron AD.** The primary transcriptome, small RNAs, and regulation of antimicrobial resistance in *Acinetobacter baumannii* ATCC 17978. *Nucleic Acids Res.*

9. **Ilyas B, Tsai CN, Coombes BK.** 2017. Evolution of *Salmonella*-host cell interactions through a dynamic bacterial genome. *Front Cell Infect Microbiol* **7:**428.

10. **Mekalanos JJ.** 1992. Environmental signals controlling expression of virulence determinants in bacteria. *J Bacteriol* **174:**1–7.

11. **Silhavy TJ.** 2000. Gene fusions. *J Bacteriol* **182:**5935–5938.

12. **DeRisi JL, Iyer VR, Brown PO.** 1997. Exploring the metabolic and genetic control of gene expression on a genomic scale. *Science* **278:**680–686.

13. Schena M, Shalon D, Davis RW, Brown PO. 1995. Quantitative monitoring of gene expression patterns with a complementary DNA microarray. *Science* 270:467–470.

14. Nagalakshmi U, Wang Z, Waern K, Shou C, Raha D, Gerstein M, Snyder M. 2008. The transcriptional landscape of the yeast genome defined by RNA sequencing. *Science* 320:1344–1349.

15. Colgan AM, Cameron AD, Kröger C, Colgan AM, Srikumar S, Händler K, Sivasankaran SK, Hammarlöf DL, Canals R, Grissom JE, Conway T, Hokamp K, Hinton JC. 2017. If it transcribes, we can sequence it: mining the complexities of host-pathogen-environment interactions using RNA-seq. *Curr Opin Microbiol* 36:37–46.

16. Wang Z, Gerstein M, Snyder M. 2009. RNA-Seq: a revolutionary tool for transcriptomics. *Nat Rev Genet* 10:57–63.

17. Mäder U, Nicolas P, Richard H, Bessières P, Aymerich S. 2011. Comprehensive identification and quantification of microbial transcriptomes by genome-wide unbiased methods. *Curr Opin Biotechnol* 22:32–41.

18. Höfer K, Jäschke A. 2018. Epitranscriptomics: RNA modifications in bacteria and archaea. *Microbiol Spectr* 6: RWR-0015-2017.

19. Saliba AE, C Santos S, Vogel J. 2017. New RNA-seq approaches for the study of bacterial pathogens. *Curr Opin Microbiol* 35:78–87.

20. Conesa A, Madrigal P, Tarazona S, Gomez-Cabrero D, Cervera A, McPherson A, Szcześniak MW, Gaffney DJ, Elo LL, Zhang X, Mortazavi A. 2016. A survey of best practices for RNA-seq data analysis. *Genome Biol* 17:1–19.

21. Aikawa C, Maruyama F, Nakagawa I. 2010. The dawning era of comprehensive transcriptome analysis in cellular microbiology. *Front Microbiol* 1:118.

22. Creecy JP, Conway T. 2015. Quantitative bacterial transcriptomics with RNA-seq. *Curr Opin Microbiol* 23:133–140.

23. Levin JZ, Yassour M, Adiconis X, Nusbaum C, Thompson DA, Friedman N, Gnirke A, Regev A. 2010. Comprehensive comparative analysis of strand-specific RNA sequencing methods. *Nat Methods* 7:709–715.

24. The GTEx Consortium. 2013. The Genotype-Tissue Expression (GTEx) project. *Nat Genet* 45:580–585.

25. Faria JP, Overbeek R, Xia F, Rocha M, Rocha I, Henry CS. 2014. Genome-scale bacterial transcriptional regulatory networks: reconstruction and integrated analysis with metabolic models. *Brief Bioinform* 15:592–611.

26. Lucchini S, Rowley G, Goldberg MD, Hurd D, Harrison M, Hinton JC. 2006. H-NS mediates the silencing of laterally acquired genes in bacteria. *PLoS Pathog* 2:e81.

27. Smith C, Stringer AM, Mao C, Palumbo MJ, Wade JT. 2016. Mapping the regulatory network for *Salmonella enterica* serovar Typhimurium invasion. *mBio* 7:e01024-16.

28. Tomljenovic-Berube AM, Mulder DT, Whiteside MD, Brinkman FS, Coombes BK. 2010. Identification of the regulatory logic controlling *Salmonella* pathoadaptation by the SsrA-SsrB two-component system. *PLoS Genet* 6: e1000875.

29. Cloots L, Marchal K. 2011. Network-based functional modeling of genomics, transcriptomics and metabolism in bacteria. *Curr Opin Microbiol* 14:599–607.

30. McDermott JE, Yoon H, Nakayasu ES, Metz TO, Hyduke DR, Kidwai AS, Palsson BO, Adkins JN, Heffron F. 2011. Technologies and approaches to elucidate and model the virulence program of *Salmonella*. *Front Microbiol* 2:121.

31. Yoon H, McDermott JE, Porwollik S, McClelland M, Heffron F. 2009. Coordinated regulation of virulence during systemic infection of *Salmonella enterica* serovar Typhimurium. *PLoS Pathog* 5:e1000306.

32. Métris A, Sudhakar P, Fazekas D, Demeter A, Ari E, Olbei M, Branchu P, Kingsley RA, Baranyi J, Korcsmáros T. 2017. SalmoNet, an integrated network of ten *Salmonella enterica* strains reveals common and distinct pathways to host adaptation. *NPJ Syst Biol Appl* 3:31.

33. Laczny CC, Sternal T, Plugaru V, Gawron P, Atashpendar A, Margossian HH, Coronado S, der Maaten L, Vlassis N, Wilmes P. 2015. VizBin—an application for reference-independent visualization and human-augmented binning of metagenomic data. *Microbiome* 3:1–7.

34. Barquist L, Vogel J. 2015. Accelerating discovery and functional analysis of small RNAs with new technologies. *Annu Rev Genet* 49:367–394.

35. Srikumar S, Kröger C, Hébrard M, Colgan A, Owen SV, Sivasankaran SK, Cameron AD, Hokamp K, Hinton JC. 2015. RNA-seq brings new insights to the intramacrophage transcriptome of *Salmonella* Typhimurium. *PLoS Pathog* 11:e1005262.

36. Kröger C, Colgan A, Srikumar S, Händler K, Sivasankaran SK, Hammarlöf DL, Canals R, Grissom JE, Conway T, Hokamp K, Hinton JC. 2013. An infection-relevant transcriptomic compendium for *Salmonella enterica* serovar Typhimurium. *Cell Host Microbe* 14:683–695.

37. Colgan AM, Kröger C, Diard M, Hardt WD, Puente JL, Sivasankaran SK, Hokamp K, Hinton JC. 2016. The impact of 18 ancestral and horizontally-acquired regulatory proteins upon the transcriptome and sRNA landscape of *Salmonella enterica* serovar Typhimurium. *PLoS Genet* 12:e1006258.

38. Havelaar AH, Kirk MD, Torgerson PR, Gibb HJ, Hald T, Lake RJ, Praet N, Bellinger DC, de Silva NR, Gargouri N, Speybroeck N, Cawthorne A, Mathers C, Stein C, Angulo FJ, Devleesschauwer B, World Health Organization Food borne Disease Burden Epidemiology Reference Group. 2015. World Health Organization global estimates and regional comparisons of the burden of foodborne disease in 2010. *PLoS Med* 12:e1001923.

39. Hammarlöf DL, Canals R, Hinton JC. 2013. The FUN of identifying gene function in bacterial pathogens; insights from *Salmonella* functional genomics. *Curr Opin Microbiol* 16:643–651.

40. Altuvia S, Weinstein-Fischer D, Zhang A, Postow L, Storz G. 1997. A small, stable RNA induced by oxidative stress: role as a pleiotropic regulator and antimutator. *Cell* 90:43–53.

41. Chareyre S, Mandin P. 2018. Bacterial iron homeostasis regulation by sRNAs. *Microbiol Spectr* 6:RWR-0010-2017.

42. Kim JN. 2016. Roles of two RyhB paralogs in the physiology of *Salmonella enterica*. *Microbiol Res* 186-187:146–152.

43. Calderón IL, Morales EH, Collao B, Calderón PF, Chahuán CA, Acuña LG, Gil F, Saavedra CP. 2014. Role of *Salmonella* Typhimurium small RNAs RyhB-1 and RyhB-2 in the oxidative stress response. *Res Microbiol* 165:30–40.

44. Padalon-Brauch G, Hershberg R, Elgrably-Weiss M, Baruch K, Rosenshine I, Margalit H, Altuvia S. 2008. Small RNAs encoded within genetic islands of *Salmonella* Typhimurium show host-induced expression and role in virulence. *Nucleic Acids Res* 36:1913–1927.

45. Massé E, Gottesman S. 2002. A small RNA regulates the expression of genes involved in iron metabolism in *Escherichia coli*. *Proc Natl Acad Sci U S A* 99:4620–4625.

46. Coornaert A, Lu A, Mandin P, Springer M, Gottesman S, Guillier M. 2010. MicA sRNA links the PhoP regulon to cell envelope stress. *Mol Microbiol* 76:467–479.

47. Guo MS, Updegrove TB, Gogol EB, Shabalina SA, Gross CA, Storz G. 2014. MicL, a new σ^E-dependent sRNA, combats envelope stress by repressing synthesis of Lpp, the major outer membrane lipoprotein. *Genes Dev* 28:1620–1634.

48. Gogol EB, Rhodius VA, Papenfort K, Vogel J, Gross CA. 2011. Small RNAs endow a transcriptional activator with essential repressor functions for single-tier control of a global stress regulon. *Proc Natl Acad Sci U S A* 108:12875–12880.

49. Waters LS, Sandoval M, Storz G. 2011. The *Escherichia coli* MntR miniregulon includes genes encoding a small protein and an efflux pump required for manganese homeostasis. *J Bacteriol* 193:5887–5897.

50. Silva IJ, Ortega ÁD, Viegas SC, García-Del Portillo F, Arraiano CM. 2013. An RpoS-dependent sRNA regulates the expression of a chaperone involved in protein folding. *RNA* 19:1253–1265.

51. Chao Y, Papenfort K, Reinhardt R, Sharma CM, Vogel J. 2012. An atlas of Hfq-bound transcripts reveals 3′ UTRs as a genomic reservoir of regulatory small RNAs. *EMBO J* 31:4005–4019.

52. Skinner ME, Uzilov AV, Stein LD, Mungall CJ, Holmes IH. 2009. JBrowse: a next-generation genome browser. *Genome Res* 19:1630–1638.

53. Westesson O, Skinner M, Holmes I. 2013. Visualizing next-generation sequencing data with JBrowse. *Brief Bioinform* 14:172–177.

54. Kröger C, Dillon SC, Cameron AD, Papenfort K, Sivasankaran SK, Hokamp K, Chao Y, Sittka A, Hébrard M, Händler K, Colgan A, Leekitcharoenphon P, Langridge GC, Lohan AJ, Loftus B, Lucchini S, Ussery DW, Dorman CJ, Thomson NR, Vogel J, Hinton JC. 2012. The transcriptional landscape and small RNAs of *Salmonella enterica* serovar Typhimurium. *Proc Natl Acad Sci U S A* 109:E1277–E1286.

55. Hammarlöf DL, Kröger C, Owen SV, Canals R, Lacharme-Lora L, Wenner N, Schager AE, Wells TJ, Henderson IR, Wigley P, Hokamp K, Feasey NA, Gordon MA, Hinton JC. 2018. Role of a single noncoding nucleotide in the evolution of an epidemic African clade of *Salmonella*. *Proc Natl Acad Sci U S A* 115:E2614–E2623.

Regulating with RNA in Bacteria and Archaea
Edited by Gisela Storz and Kai Papenfort
© 2018 American Society for Microbiology, Washington, DC
doi:10.1128/microbiolspec.RWR-0001-2017

Structure and Interaction Prediction in Prokaryotic RNA Biology

33

Patrick R. Wright[1,*], Martin Mann[1,*], and Rolf Backofen[1,2,*]

INTRODUCTION

For over a decade, prokaryotic and eukaryotic RNA biology exploration has unveiled the multifaceted and central contribution of RNA-based control in all domains of life. RNA interactions are at the core of many regulative processes and have hence been heavily studied by wet-lab and biocomputational researchers alike. Within this review, we focus on biocomputational methods and outline the technical details of standard algorithms for RNA secondary structure and RNA-RNA interaction prediction. Furthermore, we highlight their application in the context of prokaryotic RNA biology.

Similar to DNA, RNA molecules can undergo stable base pairing when stretches of complementary nucleotides are present and form a so-called duplex. The generalized model assumes that adenine (A) can form base pairs with uracil (U) while guanine (G) can pair with cytosine (C) or U. *In vivo*, further interactions are possible (1–3), which are, however, not considered for the methods presented in this review. Base pairs in RNA or DNA can be established due to complementary nucleotides forming hydrogen bonds (4). Two types of base pairing are conceivable for RNA molecules. Firstly, base pairing can occur within a single RNA molecule. These intramolecular interactions give rise to RNA secondary structures that are often important for an RNA molecule's function or regulation and are thus central to cellular physiology.

The second type of base pairing is referred to as intermolecular interaction or RNA-RNA interaction. These interactions occur between RNAs that are present as individual molecules and play a key role in processes that employ RNA molecules as regulators of other RNA molecules. Examples are prokaryotic *trans*-acting small RNAs (sRNAs) or eukaryotic microRNAs

(miRNAs). Both sRNAs and miRNAs are posttranscriptional regulators that often bind target RNAs and thereby modulate the target's function in a positive or negative manner (5–7). Laboratory-based identification and verification of RNA-RNA interactions is a cumbersome task. In accordance with this, and taking the pivotal role of RNA-RNA interactions in the regulatory network of cellular systems into account, *in silico* prediction of such interactions has been intensely studied and several predictive approaches are available.

INTRAMOLECULAR RNA STRUCTURE PREDICTION

RNA molecules are usually chain-like polymers of nucleotides that differ in their base composition and length. They are therefore typically represented by their base sequence in the $5' \rightarrow 3'$ direction, where $5'$ denotes the $5'$ phosphate group and $3'$ denotes the $3'$ hydroxyl group of the first and last nucleotide, respectively. Thus, an RNA molecule of length n is encoded by its sequence $\mathcal{R} \in \Sigma^n$, where $\Sigma = \{A, C, G, U\}$ encodes the possible bases. A structure $\mathcal{P}$ for a given RNA $\mathcal{R}$ can be encoded by its set of base pairs $\mathcal{P} = \{(i, j) | 1 \leq i < j \leq n\}$.

In the basic RNA secondary structure model, each base can form only a single base pairing within the molecule. A valid secondary structure $\mathcal{P}$ fulfills the following criteria: (i) unique base pairing ($\forall (i, j) \neq (k, l) \in \mathcal{P} : i \notin \{k, l\} \wedge j \notin \{k, l\}$), (ii) sequence complementarity ($\forall (i, j) \in \mathcal{P} : \{\mathcal{R}_i, \mathcal{R}_j\} \in \{\{A, U\}, \{C, G\}, \{G, U\}\}$), and (iii) minimal base-pair span s_l ($\forall (i, j) \in \mathcal{P} : i + s_l < j$). Therefore, an empty secondary structure $\mathcal{P}^{oc} = \emptyset$ corresponds to the open chain without intramolecular base pairings. The minimal base-pair span s_l, also called loop size, incorporates steric bending constraints into the struc-

[1]Bioinformatics Group; [2]Center for Biological Signaling Studies (BIOSS), University of Freiburg, Freiburg, Germany. *These authors contributed equally to the manuscript.

ture model and is usually set to $s_l = 3$. If the base pairs within a structure are nested, i.e., it holds $\nexists_{(i,j)} \neq_{(k,l) \in \mathcal{P}} : i < k < j < l$, then we call $\mathcal{P}$ a *nested structure*. Otherwise, it is called a *crossing or pseudoknot structure*. Nested models enable a unique decomposition of $\mathcal{P}$ into secondary structure elements, which facilitates efficient RNA structure and interaction prediction methods. Therefore, we will focus on nested models only in the following.

A nested secondary structure $\mathcal{P}$ can be uniquely decomposed into structural elements. These elements are called *loops*. Each loop is defined by an enclosing base pair $(i, j) \in \mathcal{P}$. The loop type is determined by the directly enclosed base pairs $(k, l) \in \mathcal{P}$, with $i < k < l < j$. A base pair (k, l) is directly enclosed by (i, j), if there is no other (k, l)-enclosing base pair $(\breve{i}, \breve{j}) \in \mathcal{P}$ that is enclosed by (i, j) too. We distinguish the loop types that are depicted in Fig. 1.

hairpin loop:	no enclosed base pair
stacking:	adjacent enclosed base pair, i.e., $i + 1 = k$ and $j - 1 = l$
bulge loop:	only one side adjacent, i.e., $(i + 1 = k \wedge j - 1 > l)$ or $(i + 1 < k \wedge j - 1 = l)$
interior loop:	no stacked enclosed base pair, i.e., $(i + 1 < k \wedge j - 1 < l)$
multiloop:	more than one directly enclosed base pair

Individual Structure Prediction

One of the first and most fundamental algorithms for the prediction of nested RNA secondary structures was introduced by Ruth Nussinov and coworkers (8). It applies dynamic programming techniques to efficiently

identify a stable structure $\mathcal{P}$ for an RNA molecule with sequence $\mathcal{R}$ by maximizing the number of base pairs $|\mathcal{P}|$. To this end, the Nussinov algorithm recursively computes the maximum number of base pairs for each subsequence $\mathcal{R}_i..\mathcal{R}_j$ and stores this information in the dynamic programming matrix $N_{i,j}$. It can be filled by employing equation 1 (depicted in Fig. 2), which is a variant of the original recursions formulated in reference 8 to enable the relation to other approaches in the following. The maximum number of base pairs is found in the matrix entry $N_{1,n}$. The corresponding structure can be identified via traceback.

$$N_{i,j} = max \begin{cases} 0 & : \text{if } (i + s_l) \geq j \\ N_{i+1,j-1} + 1 & : \text{if } \mathcal{R}_i, \mathcal{R}_j \text{ can form} \\ & \quad \text{base pair} \quad (1) \\ max_{i \leq k < j} & : \text{decomposition} \\ \{N_{i,k} + N_{k+1,j}\} & \end{cases}$$

While efficient and in theory applicable to the folding of RNAs, this simple optimization scheme shows poor prediction accuracy for several reasons. Firstly, it does not account for differences in base-pairing strengths. In general, base-pair complementarity is at the core of every RNA interaction prediction, but a distinction needs to be made between stronger and weaker base pairs. While a G-C pair incorporates three hydrogen bonds, G-U and A-U base pairs only form two hydrogen bonds and are less stable compared to a G-C pair. Furthermore, the stability influence of loop sizes, base-pair stackings, loop context, multiloop formations, etc., is not considered. The stacking of base pairs, for instance, is central to RNA helix stability (9). Nevertheless, the algorithmic idea was transferred into more sophisticated optimization schemes that are discussed in the following.

Current RNA secondary structure prediction algorithms are usually energy minimization methods. Typically, they use the aforementioned loop decomposition of a structure (see Fig. 1) in combination with loop-specific energy contributions. This enables an incorporation of empirically determined loop type- and context-specific contributions (10, 11). Thus, this so-called *nearest-neighbor model* (12, 13) considers the directly neighboring bases and base pairs for each interaction. The overall energy $E(\mathcal{P})$ of a nested structure $\mathcal{P}$ can therefore be computed by the summation over energies for each loop enclosed by $(i, j) \in \mathcal{P}$ (equation 2).

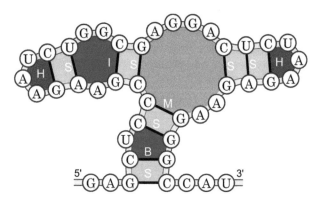

Figure 1 Loop decomposition of a nested RNA structure into hairpin loops (H; dark gray), multiloops (M; light gray), stackings (S; light blue), bulges (B; dark blue), and interior loops (I; dark blue). Initials of the loop types are placed in white next to the enclosing base pair (black).

$$E(\mathcal{P}) = \sum_{(i,j) \in \mathcal{P}} \begin{cases} E^{loop}(i, j, i, j) & : \text{if hairpin loop} \\ E^{loop}(i, j, k, l) & : \text{if stacking, bulge,} \\ & \quad \text{or interior loop} \quad (2) \\ E^{loop}_{multi}(i, j, h, u) & : \text{if multiloop} \end{cases}$$

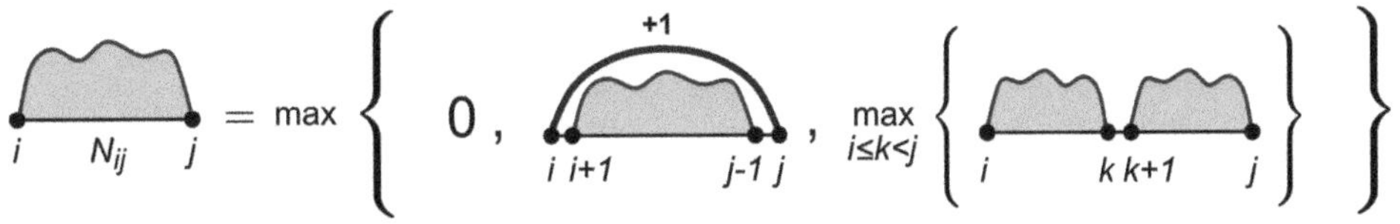

Figure 2 Graphical depiction of the Nussinov-like recursion from equation 1.

where E^{loop} provides the loop's energy contribution for stackings, bulges, and interior and hairpin loops and E^{loop}_{multi} gives the energy for a multiloop element. To reduce complexity, multiloop energies are typically estimated by $E^{loop}_{multi}(i,j,h,u) = E^m_c + hE^m_b + uE^m_u$. This uses empirically identified constants E^m_c, E^m_b, and E^m_u, with E^m_c penalizing the multiloop closure by (i,j), E^m_b accounting for the $h \geq 2$ directly enclosed base pairs (i.e., branching helices), and E^m_u weighting the u directly enclosed unpaired bases. For instance, the multiloop in Fig. 1 results in $h = 2$ and $u = 6$. To respect the outer context of nonenclosed loops, so-called *dangling end contributions* have to be added as well, which is neglected in equation 2 and in the following presentations for simplicity. Throughout the last decades, several parameter sets for the nearest-neighbor model have been derived from experimental data (10, 14, 15).

Given such an energy decomposition, a dynamic programming scheme to compute the *minimal free energy (mfe) structure* $\mathcal{P}^{mfe}$ for an RNA $\mathcal{R}$ was introduced by Michael Zuker and Patrick Stiegler (16). It uses three matrices to store results for distinct subproblems: $V_{i,j}$ provides the mfe for all possible structures that can be formed by the subsequence $R_i..R_j$ under the assumption that $\mathcal{R}_i$ and $\mathcal{R}_j$ form a base pair; $W^M_{i,j}$ handles multiloop decompositions, where $R_i..R_j$ is enclosed in the multiloop; and W_i encodes the mfe for the prefix $\mathcal{R}_1..\mathcal{R}_i$. Given the following recursions (equations 3 to 6), the mfe of the whole RNA can be found in W_n. Note, the multiloop decomposition in the $W^M_{i,j}$ recursion is not unique, which makes it unsuitable for suboptimal structure prediction, as addressed in reference 17. When restricting the maximally allowed interior loop size in equation 5, this algorithm runs in $O(n^3)$ time and $O(n^2)$ space. Note, since this transfer of a base pair-maximizing recursion (Nussinov algorithm) to energy minimization using the nearest-neighbor model (Zuker algorithm) is generic, we will restrict where appropriate the algorithm presentations to a Nussinov-like form.

$$W_0 = 0 \qquad (3)$$

$$W_j = min \begin{cases} W_{j-1} & : j \text{ is unpaired} \\ \min_{1 \leq k < j} \{W_{k-1} + V_{k,j}\} & : \text{base pair } (k,j) \end{cases} \quad (4)$$

$$V_{i,j} = min \begin{cases} \infty & : \mathcal{R}_i, \mathcal{R}_j \text{ cannot} \\ & \quad \text{pair or} \\ & \quad (i+s_l) \geq j \\ E^{loop}(i,j,i,j) & : (i,j) \text{ closes} \\ & \quad \text{hairpin} \\ \min_{i<k<l<j} & : (i,j),(k,l) \\ \{E^{loop}(i,j,k,l) + V_{k,l}\} & \quad \text{stacking,} \\ & \quad \text{bulge,} \\ \min_{i<k<j} & : (i,j) \text{ closes} \\ \{E^m_c + W^M_{i+1,k} + W^M_{k+1,j-1}\} & \quad \text{multiloop} \end{cases} \quad (5)$$

$$W^M_{i,j} = min \begin{cases} \infty & : (i+s_l) \geq j \\ E^m_u + W^M_{i+1,j} & : i \text{ unpaired} \\ E^m_u + W^M_{i,j-1} & : j \text{ unpaired} \\ E^m_b + V_{i,j} & : (i,j) \text{ directly} \\ & \quad \text{enclosed} \\ \min_{i<k<j} & : \text{decompose} \\ \{W^M_{i,k-1} + V_{k,j} + E^m_b\} & \quad \text{and } (k,j) \\ & \quad \text{directly} \\ & \quad \text{enclosed} \end{cases} \quad (6)$$

The advanced versions of the Zuker algorithm are implemented in the standard folding programs UNAFold (18) (formerly Mfold [19]) and RNAfold (20, 21). Both implementations are being successfully used by the research community and show good prediction accuracy for RNAs such as Spot 42 (22) and FnrS (23), to name just two.

The time complexity of the Nussinov and Zuker algorithms is $O(n^3)$, at least when restricting the interior loop size in Zuker. This is, however, still high for long RNA sequences as, for instance, large mRNAs, long noncoding RNAs, and viral RNAs. For that reason, attempts have been made to reduce the overall time complexity on average to $O(n^2)$. When revisiting the recursion in equation 1, the split in the final decomposition case is the one causing the high complexity. Many of these splits will not lead to the optimal solution. This observation sparked the idea of sparsification techniques, first introduced by Ydo Wexler and colleagues (24), which is discussed in the supplementary material at http://www.asmscience.org/files/Backofen_Supplemental.pdf.

Finally, more and more tools now allow for the inclusion of structure probing data (25, 26), which can guide a more sophisticated structure prediction process based on experimentally established constraints (27, 28). Examples are RNAstructure (29), RNAsc (30), and RNAfold (31, 32).

Comparative Structure Prediction

To increase the prediction quality, it is often useful to take not only one but a set of evolutionarily related molecules into account. That is, one wants to compute a common structure for a set of sequences, which requires an alignment that takes both sequence and structure features into account. Paul R. Gardner and Robert Giegerich classified such approaches according to the applied plan (33): Plan A, "first align then fold"; Plan B, "simultaneously align and fold"; and Plan C, "first fold then align."

The class of methods that is referred to as Plan A is basically an extension of the individual structure prediction to alignments. That is, first all sequences are aligned based on their sequence similarity. This multiple sequence alignment can be successively folded into a consensus structure that is compatible with all sequences. Common approaches are RNAalifold (34), Pfold (35), and PETfold (36). Such approaches are as efficient as individual structure prediction and work well for data sets with high sequence similarity. RNA alifold, for instance, has been used to determine the conservation of the CyaR sRNA (37). If sequence identity within the data drops below 60%, Plan A approaches have been shown to fail (38). Here, the other two plans are more promising, since they utilize the observation that an RNA's structure is often more strongly conserved than its sequence on an evolutionary scale (38).

The first practical implementations of Plan C were RNAforester (39) and MARNA (multiple alignment of RNAs) (40). They generate a multiple structure alignment for a set of given input structures. The latter are either known or stem from individual structure prediction. Thus, Plan C approaches depend on the accuracy of the input structures. Since there is only a limited number of known RNA structures, the overall alignment quality is often flawed by the individual structure predictions used instead.

Thus, the current state-of-the-art approaches are applying Plan B, which was first introduced by David Sankoff (41). Here, sequences are simultaneously folded

and aligned, leading to an algorithm with $O(n^6)$ time and $O(n^4)$ space complexity. The key idea of Sankoff is to simultaneously find two equivalent structures $\mathcal{P}^1$ and $\mathcal{P}^2$ for the given sequences $\mathcal{R}^1$ and $\mathcal{R}^2$, respectively, in combination with a compatible sequence alignment of $\mathcal{R}^1$ and $\mathcal{R}^2$. That is, we have to optimize the combination: $E(\mathcal{P}^1) + E(\mathcal{P}^2) + S$, where $S_{i,j,i',j'}$ provides the alignment score for the respective subsequences. The two structures are *equivalent* if they are of the same size ($|\mathcal{P}^1|=|\mathcal{P}^2|$) and show the same nesting. The sequence alignment is *compatible* with both structures if equivalent base pairs $(i, j) \in \mathcal{P}^1$ and $(i', j') \in \mathcal{P}^2$ are aligned to each other.

In the following, we present a reduced Nussinov-like version of the algorithm using a base-pair maximization scheme. Here, the Nussinov matrix N (equation 1) for one sequence is extended to a four-dimensional matrix F (equation 7) that encodes the optimal Sankoff-like score for two aligned subsequences $\mathcal{R}^1_{i..j}$ and $\mathcal{R}^2_{i'..j'}$.

$$F_{i,j,i',j'} = max \begin{cases} S_{i,j,i',j'} & : \text{alignment of} \\ & \quad \text{no structure} \\ F_{i+1,j-1,i'+1,j'-1} & : \text{if } (\mathcal{R}^1_i, \mathcal{R}^1_j) \text{ and} \\ + 2 + S_{i,i',i',i'} + S_{j,j,j',j'} & \quad (\mathcal{R}^2_{i'}, \mathcal{R}^2_{j'}) \text{ are} \\ & \quad \text{complementary,} \\ & \quad \text{alignment of} \\ & \quad \text{base pairs} \\ & \quad (i, j), (i', j') \\ \underset{k,k'}{max} & : \text{decomposition} \\ \{F_{i,k,i',k'} + F_{k+1,k'+1,j,j'}\} \end{cases}$$

(7)

where the sequence alignment contributions are computed by

$$S_{i,j,i',j'} = max \begin{cases} S_{i+1,j,i'+1,j'} & : \text{align } i \text{ and } i' \text{ if } \mathcal{R}^1_i, \\ & \quad \mathcal{R}^2_{i'} \text{ match} \\ S_{i+1,j,i'+1,j'}-s_m & : \text{align } i \text{ and } i' \text{ if } \mathcal{R}^1_i, \\ & \quad \mathcal{R}^2_{i'} \text{ mismatch} \\ S_{i+1,j,i',j'}-s_g & : \text{align } \mathcal{R}^1_i \text{ with gap} \\ S_{i,j,i'+1,j'}-s_g & : \text{align } \mathcal{R}^2_{i'} \text{ with gap} \end{cases}$$

(8)

using the penalties s_m and s_g (both ≥ 0) for mismatch and gap alignments, respectively. The introduction of base pairs into both structures at once ensures their equivalence, and the direct inclusion of respective alignment scores ensures the compatibility of the alignment with the structures. As for the Nussinov-like recursion in equation 1, the Sankoff-like recursion from equation 7 can be extended to the nearest-neighbor energy model just like the Zuker algorithm.

To reduce the computational complexity, several simplifications have been introduced. One class of variants uses sequence-based heuristics to restrict the search space. Programs of this class are, for instance, Foldalign (42, 43), Dynalign (44), and Stemloc (45). Another class of approaches, e.g., implemented in PMcomp (46), LocARNA (47), and FoldalignM (48), does not restrict the alignment search space but uses a simplified energy model based on base-pair probabilities to reduce the considered structural search space and such computational complexity and runtime. Here, instead of directly considering loop energies, as done by Sankoff, energy terms are indirectly encoded within base-pair probabilities that are efficiently precomputed using the algorithm by John S. McCaskill (49). Both classes of simplifications are combined by RAF (RNA Alignment and Folding) (50), which fuses heuristics based on subsequence alignment quality with the simplified energy model of PMcomp, an approach first introduced in Stemloc.

RNA-RNA INTERACTION PREDICTION

Key Components in RNA-RNA Interactions

Recalling the commonly known double-helical structure of DNA, one may naively assume that base-pair complementarity is the sole component needed to form a stable interaction between two RNAs. However, further factors influence RNA-RNA interactions. One of these factors is the previously discussed intramolecular structure (see "Intramolecular RNA Structure Prediction" above) that can be formed by each individual copy of the interacting RNAs before they *meet*. Given the hypothetical RNA sequence, where 5′ denotes the 5′ phosphate group and 3′ denotes the 3′ hydroxyl group, 5′-GGGGGGGGGGGCCCCCCCCCC-3′: if no intramolecular base pairs are assumed, one may be tempted to see a perfect duplex forming between two individual RNAs of this type (Fig. 3A). However, each individual RNA of this type is also capable of forming a hairpin structure incorporating a strong G-C stem enclosing a small loop. Hence, a more realistic assumption is that only the unpaired hairpin loop regions will interact and the duplex is not as long as naively expected (Fig. 3B). This theoretical scenario is also reflected by *cis*-antisense RNAs *in vivo* (51). Neglecting or considering the component of intramolecular structures splits the prediction approaches that will be discussed in this review into two major groups. They will be reviewed in the two sections that follow.

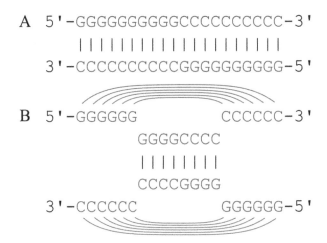

Figure 3 Potential interactions for two identical RNA molecules (blue and green) as predicted by the RNAhybrid web server (A) (52) and IntaRNA web server (B) (53) (intramolecular base pairs subsequently added, which form a kissing hairpin interaction). Inter- and intramolecular base pairs are indicated by vertical pipe symbols and arches, respectively.

RNA-RNA Interaction Prediction Not Considering Intramolecular Base Pairing

As previously mentioned, approaches to predict RNA-RNA interactions can be split into two major groups. The first group, which will be discussed here, neglects the impact of intramolecular base pairing within the interaction partners. The algorithmic solutions are either purely sequence- or structure-based approaches. While the sequence-based models solely search for stretches of extensive base-pair complementarity, a physical energy model is employed by structure-based approaches.

To find base-pair complementarity, the Basic Local Alignment Search Tool (BLAST) algorithm (54) is appropriate. Yet, next to the canonical Watson-Crick base pairs G-C and A-U, also the non-Watson-Crick base pair G-U can form within RNA-RNA interactions and must thus be considered. GUUGle (55) is an approach that incorporates the G-U wobble base pairs, and was mainly developed as filtering scheme to reduce the search space for more complex algorithms. The advantage of these sequence-based approaches is that they immediately inherit a method to calculate p-values from local alignment approaches (56).

TargetRNA (57), which was developed to predict the targets of bacterial sRNAs, approaches the problem from different angles. Two scoring schemes for an interaction of RNAs are proposed. Firstly, TargetRNA allows for a purely sequence-based solution using a variant of the Smith-Waterman alignment algorithm (58). Therein, it searches for base-pair complementarity rather than sequence similarity. Furthermore, loops within the interaction are penalized, while G-C and A-U base pairs are favored over G-U base pairs. The second solution in TargetRNA uses an energy model. Here, the free energy of an RNA duplex is considered when scoring an interaction. A lower energy represents a more stable interaction. This second type of scoring resembles the energy model used for individual RNA structure prediction (see "Intramolecular RNA Structure Prediction" above). RNAhybrid (59), which was developed prior to Target RNA, also uses minimal free energy scoring primarily in order to predict eukaryotic miRNA targets. Yet it has also been frequently applied in the prediction of prokaryotic sRNA target interactions (37, 60, 61). The scoring in these approaches strongly depends on the energies of stacked back-to-back base pairs, interior loops, and bulges. The stacking energies, which were originally derived for intramolecular structures, are available from experimental testing (see "Intramolecular RNA Structure Prediction" above). Energies for small interior loops and bulges are also available from experimental data.

Both RNAhybrid and TargetRNA restrict the length for long interior loops (i.e., considered values for p, q in equations 9 and 10), as these structures increase the computational complexity. The rationale behind this is that long interior loops do not represent structures that are favorable, and thus may be disregarded in the interaction prediction due to their limited *real-world* relevance. RNAplex (62), on the other hand, deviates from RNAhybrid's and TargetRNA's type of treatment for long interior loops by using an affine function for scoring long interior loops and bulges within the interaction, while its energy model is similar to those applied in the previously mentioned approaches. Furthermore, RNAplex can avoid disproportionately long duplex predictions, which often occur in RNAhybrid, by imposing a penalty for every nucleotide in the interaction. Thereby, RNAplex provides a more realistic estimation of the potential *in vivo* duplex, which is especially helpful when predicting the targets of prokaryotic sRNAs. The constraint on duplex lengths can be regarded as a step toward consideration of intramolecular structures without specifically addressing them. A prefiltering algorithm for RNA-RNA interaction prediction on the genomic scale employing a simplified Turner energy model is RIsearch (63).

The above-mentioned energy-based approaches predict RNA-RNA interactions by minimizing the free energy of the resulting duplex. Specifically, this can be solved in polynomial time using dynamic programming. In principle, one can consider all possible interaction sites $i..k$ on the first sequence $\mathcal{R}^1$ together with all possible interaction sites $j..l$ on the second RNA $\mathcal{R}^2$,

and store the minimal duplex energy in a matrix, $D_{j..l}^{i..k}$. Note that we number the first sequence in $5' \rightarrow 3'$ and the second sequence in reverse orientation $(3' \rightarrow 5')$ since we consider only sense-antisense interactions. To guarantee that these interaction sites are actually covered by a duplex, we have to enforce that i, j, k, l are occupied by intermolecular base pairs. By the noncrossing condition for the intermolecular base pairs, this is only possible if i is paired to k and j to l. Then we can simply calculate all possible interactions by the following recursion:

$$
D_{j..l}^{i..k} = min \begin{cases} +\infty & \text{if } \mathcal{R}_i^1, \mathcal{R}_j^2 \text{ can } not \\ & \text{pair} \\ E^{\text{Init}}(i,j) & \text{if } i = k \text{ and } j = l \\ \underset{p,q}{min} & \text{if } i < k \text{ and } j < l \\ \{E^{loop}(i,j,p,q) + D_{q..l}^{p..k}\} \end{cases}
$$

(9)

Here, $E^{\text{Init}}(i,j)$ is the free energy for the first intermolecular base pair in a duplex. Following reference 64, this comprises the dangling end contributions for the initial base pair (i,j) and the intermolecular initiation free energy (usually 4.10 kcal/mol). $E^{loop}(i,j,p,q)$ is the energy contribution for the loop enclosed by the base pairs (i,j) and (p,q), where $i < p$ and $j < q$. This can either be a stacking $(i = p - 1$ and $j = q - 1)$, bulge $(i = p - 1$ or $j = q - 1$, but not both), or interior loop $(i < p -1$ and $j < q - 1)$ (see "Intramolecular RNA Structure Prediction" above). When restricting the maximal size of interior loops (typically not more than 30 bases), this gives rise to an algorithm with complexity $O(n^4)$ time and space. The best duplex interaction of $\mathcal{R}^1$ and $\mathcal{R}^2$ can be found by starting a traceback from the minimal $D_{j..l}^{i..k}$ entry, where (i,j) and (k,l) define the duplex's left- and rightmost intermolecular base pairs. The recursion is depicted in Fig. 4A.

When neglecting the intramolecular base pairing, information about the interacting sites as used in $D_{j..l}^{i..k}$ is actually not required to determine the best duplex. Instead, one can use a matrix, C_j^i, that provides the minimal duplex energies for the suffixes $\mathcal{R}_i^1 \mathcal{R}_n^1$ and $\mathcal{R}_j^2 \mathcal{R}_{n'}^2$ (see Fig. 4B), where n and n' are the respective sequence lengths. The simplified recursion is then given by

$$
C_j^i = min \begin{cases} +\infty & \text{if } \mathcal{R}_i^1, \mathcal{R}_j^2 \text{ can } not \\ & \text{pair} \\ E^{\text{Init}}(i,j) & \\ \underset{p,q}{min}\{E^{loop}(i,j,p,q) + C_q^p\} \end{cases}
$$

(10)

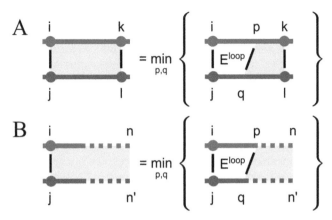

Figure 4 Recursion depiction of interaction prediction via $D_{j,l}^{i,k}$ (A) and C_j^i (B).

The actual interaction sites can be subsequently generated via traceback starting at the minimal C_j^i entry, where (i, j) marks the leftmost intermolecular base pair. Assuming $n' \in O(n)$, this simplified recursion has an $O(n^2)$ time and space complexity and according variants are used in RNAhybrid, RNAduplex (21), and TargetRNA.

Evidently, the approaches described in the current section can be split into purely sequence-based methods and methods that incorporate an energy model for RNA-RNA interaction prediction. While the methods that solely rely on sequence complementarity are a useful initial approximation for potential interactions of RNA molecules, their disregard of the energetic properties of RNA duplexes represents a major pitfall. Hence, the minimum free energy-based algorithms have several advantages. Firstly, they have the same runtime complexity as the sequence-based methods and their use of experimentally derived energy values for specific structure elements allows a more realistic approximation of RNA duplexes. Furthermore, the incorporation of a thermodynamic context allows the consideration of temperature, which is a key factor when defining the structural states of molecules such as RNA. The temperature is a dynamic parameter and can thus be adjusted to accommodate for the investigated system's temperature properties. The latter may be especially helpful for predictions on systems that are not assumed to be at 37°C (human body temperature), like the native environment of organisms such as the thermophile archaeons belonging to the *Sulfolobus* genus (65).

The inclusion of thermodynamic parameters into the prediction of RNA-RNA interactions represents an advance in this field of research when compared to the purely sequence-based methods. Yet by not directly addressing the influence of intramolecular structures within the interaction partners, a major *in vivo* factor of RNA-RNA interactions is neglected and can cause duplex predictions that may never occur. Thus, a common artifact of the methods that disregard intramolecular structures can be disproportionately long duplex predictions (62, 66), as these are generally favored by the energy model. To counter such effects, recent versions of RNAplex (67) (and its web server RNA predator [68]), TargetRNA2 (69), and RIsearch2 (70) incorporate the accessibility of the interacting sites for intermolecular base-pairing prediction. This concept is more closely discussed in the following section.

RNA-RNA Interaction Prediction Accounting for Intramolecular Base Pairing

Intramolecular base pairing plays a key role in the *in vivo* interplay of distinct RNA molecules. Hence, *in silico* predictions taking intramolecular base pairs into account currently belong to the most sophisticated and successful approaches in this field of biocomputational research. The algorithms can be split into concatenation-based, accessibility-based, and general joint structure approaches, which is also the order in which the algorithms will be presented.

Concatenation-based approaches

In general, the concatenation-based approaches make extensive use of the predefined algorithms for single RNA secondary structure prediction (see "Intramolecular RNA Structure Prediction" above) by concatenating the putatively interacting RNAs, usually interspaced with a so-called linker sequence (20, 64). Tools that allow RNA-RNA interaction prediction in this manner are, e.g., Mfold/UNAfold (71), PairFold (72), RNAcofold (73), and one of NUPACK's (Nucleic Acid Package) utilities (74). The approaches record the position of the linker and fold the concatenated sequences, thus returning the joint minimum free energy structure for the two sequences and the linker. The main difference between the concatenation-based approaches and general RNA folding is that they need special handling of loops containing the linker sequence, as the linker is an artificially introduced entity. Hence, the energy contributions added by the structures including the linker need to be adjusted. Figure 5A shows a "hairpin context." Instead of treating the structure formed by the linker sequence as a bulge or hairpin, respectively, one rather treats them as structure ends. As a consequence, the high energy penalties for the embedding of unpaired regions can be appropriately reduced. Technically, this is solved by an extension of the Zuker recursions (see

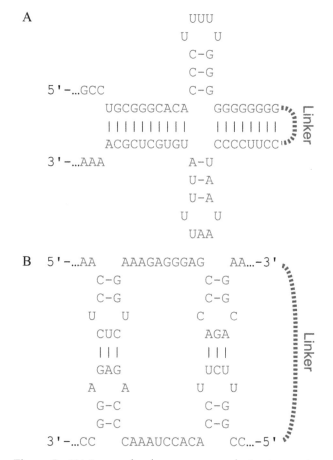

Figure 5 (A) Intramolecular structure-enclosing interaction predictable by concatenation-based approaches. (B) Double kissing hairpin interaction that *cannot* be predicted since it forms a pseudoknot when linked. The red dotted line denotes the linker, and blue and green denote the first and second sequences, respectively. Base pairs are indicated by pipe or dash symbols.

"Intramolecular RNA Structure Prediction" above). Here, for every matrix (see equations 3 to 6), one has to treat the case that the linker position is covered by the current loop in addition to the normal case (see reference 72 for details). Still, the computational complexity of Zuker's algorithm is retained.

Conveniently, the nature of the concatenation-based approaches also allows for the calculation of the partition function and base-pair probabilities for the joint structures under application of the principles described by McCaskill (49). Furthermore, interactions that form a multiloop (Fig. 5A) can also be considered. A downside, however, is the inability of these approaches to detect interactions that represent pseudoknots in the concatenated model, namely kissing hairpin interactions (75, 76), shown in Fig. 3B and 5B. Here, comple-

mentary nucleotides within the hairpin loops, which are not entangled in intramolecular base pairs, form an intermolecular duplex between sRNA and mRNA, thereby tying the two RNA molecules together. This common kind of RNA-RNA interaction represents an unpredictable pseudoknot structure in the nested concatenation-based approach, and thus is a central limitation of these approaches. An experimentally verified example of an interaction involving a pseudoknot in the concatenation context is the interplay of the *Escherichia coli* sRNA RyhB and its target mRNA encoded by the *sodB* gene. Here, the second loop of RyhB interacts with the translation initiation region of its target mRNA and thereby represses its translation into the superoxide dismutase protein (77).

Accessibility-based approaches

Accessibility-based approaches like IntaRNA (66, 80) and RNAup (78) can predict kissing hairpin interactions while still considering the contribution of intramolecular structures. These approaches specifically evaluate the structuredness of putative RNA-RNA interaction sites within the interaction partners and penalize intermolecular duplexes that require the breakup of intramolecular base pairs. Hence, interactions between commonly unstructured or accessible regions, like hairpin loops, are favored.

For this, both RNAup and IntaRNA incorporate the hybridization energy (E^{hybrid}) for the interacting RNAs $\mathcal{R}^1$ and $\mathcal{R}^2$ and the unfolding energies ($ED^{\mathcal{R}^1}$, $ED^{\mathcal{R}^2}$) required to make the interacting regions in both RNAs accessible. The general strategy of these ensemble-based algorithms is given in Fig. 6. E^{hybrid} is calculated by employing the energy model from RNAhybrid and the energy parameters from Mathews et al. (10) (see $D_{j..l}^{i..k}$ in equation 9), while the unfolding energy is derived under application of a partition function approach. The partition function $Z_{\mathfrak{P}}$ for all possible structures $\mathfrak{P}$ of a given RNA is defined as:

$$Z_{\mathfrak{P}} = \sum_{\mathcal{P} \in \mathfrak{P}} e^{\frac{-E(\mathcal{P})}{RT}} \qquad (11)$$

where $E(\mathcal{P})$ is the free energy of a specific structure $\mathcal{P} \in \mathfrak{P}$ that the given RNA can form (see equation 2), R is the gas constant, and T is the temperature of the system. McCaskill introduced an efficient algorithm for the partition function computation (49) that adapts the structure prediction approach of Zuker (see "Intramolecular RNA Structure Prediction" above) with equal complexity. Given the loop decomposition of the energy function, i.e., $E(\mathcal{P}) = \sum_{(i,j) \in \mathcal{P}} E(\text{loop}\ (i,j))$ (see equation 2),

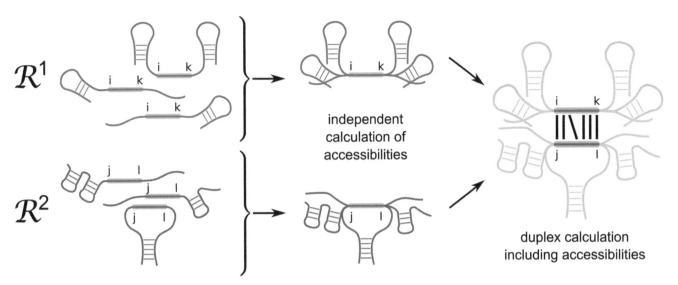

Figure 6 The ensemble-based approach for interaction prediction. Instead of considering only a single individual structure for the RNAs $\mathcal{R}^1$ and $\mathcal{R}^2$, RNAup and IntaRNA introduce a sequence-specific accessibility term, which represents all structures with an accessible (i.e., not covered by intramolecular base pairs) interaction site $i..k$ and $j..l$. These are incorporated into a modified duplex calculation.

the Boltzmann weight $e^{\frac{-E(\mathcal{P})}{RT}}$ in equation 11 can be replaced by a loop-based product:

$$e^{\frac{-E(\mathcal{P})}{RT}} = e^{-\frac{1}{RT}\sum\limits_{(i,j)\in\mathcal{P}} E(loop(i,j))} = \prod\limits_{(i,j)\in\mathcal{P}} e^{\frac{-E(loop(i,j))}{RT}} \quad (12)$$

Based on that, one can replace energy summations within the Zuker recursions with Boltzmann weight multiplication (following equation 12) and change the minimization strategy to a weight summation of structural alternatives (equation 11), which results in the computation of $Z_\mathfrak{P}$. Once $Z_\mathfrak{P}$ is identified, the free energy of the ensemble (E^{ens}) of all possible structures $\mathfrak{P}$ of the given RNA can be calculated by $E^{\text{ens}}(\mathfrak{P}) = -RT \, ln(Z_\mathfrak{P})$.

Finally, the $ED^\mathcal{R}$ value, to make a stretch of bases $i..k$ within an RNA $\mathcal{R}$ unpaired, can be computed by subtracting the ensemble energy E^{ens} of all possible structures $\mathfrak{P}$ from the ensemble energy of all structures with $i..k$ unpaired $\mathfrak{P}_{i..k}^{\text{unpaired}}$, which can also be calculated using an extension of McCaskill's algorithm (78, 79). Note from the following that ED values can be derived either from the ensemble energies or via the unpaired probabilities Pr^{unpaired} $(i..k)$ of subregions.

$$\begin{aligned} ED^\mathcal{R}(i,k) &= E^{\text{ens}}(\mathfrak{P}_{i..k}^{\text{unpaired}}) - E^{\text{ens}}(\mathfrak{P}) \\ &= -RT \, ln(Z_{\mathfrak{P}_{i..k}^{\text{unpaired}}}/Z_\mathfrak{P}) \\ &= -RT \, ln(Pr^{\text{unpaired}}(i..k)) \end{aligned} \quad (13)$$

The ED^R term from equation 13 is positive by definition (since $\mathfrak{P}_{i..k}^{\text{unpaired}} \subseteq \mathfrak{P}$) and thus represents a penalty in the RNA-RNA interaction context, as smaller energies are considered to be favorable. Given an RNA-RNA interaction between the closing base pairs (i, j) and (k, l), where i and k denote the outermost bases of a stretch of RNA $\mathcal{R}^1$ that is written in the $5' \rightarrow 3'$ direction, and j and l denote the outermost bases of a stretch of an RNA $\mathcal{R}^2$ that is written in the $3' \rightarrow 5'$ direction, the extended hybridization energy computed by RNAup (78) is given as follows:

$$E^{\text{RNAup}}(i,j,k,l) = D_{j..l}^{i..k} + ED^{\mathcal{R}^1}(i,k) + ED^{\mathcal{R}^2}(j,l) \quad (14)$$

where $D_{j..l}^{i..k}$ is the duplex energy as calculated by equation 9. To get the interaction details for the E^{RNAup} entry that minimizes equation 14, one has to traceback the according hybridization entry $D_{j..l}^{i..k}$, which directly defines the duplex's boundaries (i, j) and (k, l). This gives rise to an $O(n^4)$ time and space complexity, which can be reduced to $O(n^2w^2)$ when using a maximal interaction length w. Nevertheless, this complexity is too high for genomic screens. For that reason, IntaRNA (66, 80) was introduced to replace the exhaustive recursions from the presented approach with a heuristic. Therein, for each possible leftmost interaction base pair only one respective best interaction (and its right endpoint) is stored and considered. Thus, the recursion is similar to the hybrid-only recursion in equation 10 but additionally considers *ED* values. This reduces the complexity to $O(n^2)$ but retains the predictive power of the accessibility-based model.

IntaRNA also enforces a seed region as a necessary constraint for two RNAs to be able to interact. This means that a stretch of perfectly complementary base pairs, the seed, needs to be present within the potentially interacting RNAs in order to make a prediction. This constraint is biologically warranted for both sRNAs (81) and miRNAs (82). The seed length is usually assumed to be between 6 and 8 nucleotides. To speed up genome-wide target prediction, tools like RIsearch2 (70) or RIblast (83) apply suffix array-based screens to identify seed regions that are subsequently extended in both directions using an accessibility-based prediction approach.

IntaRNA (version ≥ 2.0) (80) can emulate most hybrid-only and accessibility-based approaches that have been previously discussed. While IntaRNA and similar approaches can predict interactions between single hairpin loops (Fig. 3B), they cannot be used to predict interaction patterns forming multiple kissing hairpins (Fig. 5B). An *in vivo*-verified example of double kissing hairpins is the interaction of the enterobacterial sRNA OxyS pairing with its target mRNA encoded by the *fhlA* gene (84). Furthermore, interactions that would represent a multiloop structure within the interacting region (Fig. 5A) cannot be predicted. Such interactions are conceivable for RNAs such as the glucose-activated enterobacterial sRNA Spot 42 (22), which could potentially interact with targets using its highly accessible unstructured regions I and III simultaneously. These aspects highlight that both concatenation-based approaches and accessibility-based approaches have intrinsic limitations that restrict the extent of interactions that can be predicted. General joint structure prediction algorithms, which will be discussed next, attempt to tackle these limitations.

General joint structure approaches

Alkan and coworkers have shown that unrestricted prediction of RNA-RNA interactions is an NP-hard problem (85). Nevertheless, they were able to identify topological constraints that enable efficient prediction schemes that are also satisfied by the following approaches.

Dmitri D. Pervouchine introduced and applied the first intermolecular RNA interaction search (IRIS) method (86) that can predict *general duplex structures* incorporating the structural context of both interacting RNAs. The supplementary material (http://www.asmscience.

org/files/Backofen_Supplemental.pdf) provides further details on the underlying recursions. IRIS was applied to computationally reconstruct certain known interactions of prokaryotic RNAs such as OxyS with the *fhlA* mRNA and several further examples (86), which form a double kissing hairpin interaction as shown in Fig. 5B. While the joint structure prediction enables a wide spectrum of predictable interaction patterns, its extreme runtime complexity of $O(n^6)$ renders it inapplicable to genome-wide screens or similar problems.

The previously mentioned approaches predict a single optimal interaction. However, as in the case of the folding of a single RNA sequence, the *mathematically* optimal structure is not necessarily the biologically functional one. For that reason, the partition function version of RNA-RNA interaction was independently introduced (87, 88), leading to an $O(n^6)$ time and $O(n^4)$ space algorithm. This partition function version of RNA-RNA interaction prediction not only allows the prediction of suboptimal interactions and their probabilities but also allows the computation of probabilities for intermolecular interactions and melting curves, which can be used to assess the stability of the interaction.

Due to the high complexity of these methods, several approaches for reducing the requirements of these algorithms have been introduced. The idea of sparsification (see supplementary material http://www.asmscience. org/files/Backofen_Supplemental.pdf) was applied to this problem (89). An approach was introduced that extends the accessibility-based approaches to multiple binding sites (90, 91).

As already discussed for intramolecular RNA structure prediction (see "Intramolecular RNA Structure Prediction" above), one can increase the prediction quality when considering sets of evolutionarily related sequences instead of solely considering individual examples. The following section discusses such comparative approaches for RNA-RNA interaction prediction.

Comparative RNA-RNA Interaction Prediction

Currently, one of the standard applications for RNA-RNA interaction prediction algorithms is whole-genome target prediction. This is usually the first step toward characterizing the function of an RNA regulator that exerts its function by directly base-pairing with its target RNA (124). Unfortunately, the pool of potential targets can be huge. The bacterium *E. coli*, for instance, has >4,000 protein-coding genes, which must all be considered as putative targets. This number is significantly higher in eukaryotes. Due to the fact that an RNA-RNA duplex can be predicted between most RNA molecules, the magnitude of potential interaction partners often leads to many false-positive predictions and thus represents the central limitation of RNA target prediction algorithms. Specifically, this means that predictions for real targets may be lost on the genomic scale due to the noise created by the high abundance of false positives. Simply put, the lists of putative targets are very long, and oftentimes the real targets are not in the top ranks.

An explanation for false-positive predictions is that the fundamental principles used by most RNA-RNA interaction algorithms generally neglect external factors such as proteins or other RNAs. Therefore, the system that most algorithms imply is an *in vitro* system in which only the two potentially interacting RNAs are present. Clearly this is far from a realistic *in vivo* setting, in which RNAs are usually densely covered, for example, by factors like proteins, RNAs, or small ligands. Hence, regions of the interacting RNAs may appear accessible within the thermodynamic model even though they are blocked due to additional factor binding. This means that sites considered accessible *in silico* may be inaccessible *in vivo* or vice versa (see Fig. 7).

To at least partially resolve this issue, data from more than one source or organism can be used, which can greatly aid in the reduction of false-positive predictions. In fact, it has been stressed that RNA target predictions should be carried out in a comparative manner if possible (59). A selection of homologous input sequences for comparative predictions can be obtained by using tools such as GLASSgo (92), RNAlien (93), or GotohScan (94). There are two major approaches for comparative RNA-RNA interaction prediction. The first one, as implemented in PETcofold (95, 96) and ripalign (97), uses the same ideas as applied for comparative structure prediction (see above). That is, instead of predicting the interaction of two single sequences, one predicts the interaction for two alignments. This assumes not only a conserved interaction site but also a conserved interaction structure, which is a strong signal. Furthermore, TargetRNA2 also optionally incorporates a phylogenetic target prediction based on an assessment of conservation within the regions of the sRNA input (69). However, as shown in reference 98, interaction sites are not necessarily conserved. This implies that the potential of two RNAs to interact might be conserved without a strictly conserved interaction site, which is not in agreement with most alignment-based assumptions.

For that reason, the second major approach to comparative RNA-RNA interaction prediction combines individual RNA-RNA interaction predictions without enforcing a strict consensus in order to obtain a more reliable result, given that the regulation is also conserved throughout the considered systems. In principle, this is like asking several people a question and making a joint conclusion or decision without closely investigating how each individual reached his or her answer. Such a joint conclusion is most likely better when compared to the conclusion derived from the answer of a single person. In the original RNAhybrid publication (59), Rehmsmeier et al. present a scheme that uses distinct RNAhybrid predictions for homologous miRNAs from different organisms on orthologous targets of said organisms in order to achieve superior predictions.

Duplex energies cannot be directly used to make a combined prediction, because they are strongly influenced by the GC content and dinucleotide frequency of the organism they are made for. For instance, organisms with higher GC content will generally produce duplex predictions with lower energies. Hence, the duplex energies predicted by RNAhybrid need to be transformed to p-values, which are then comparable. In the following, we will introduce how p-values can be derived and how p-values from different organisms can be combined to enable comparative RNA-RNA interaction prediction.

A p-value represents a statistical measure for the quality of a given prediction and, if correctly estimated, also enables comparability. Following the conclusions from extreme value theory (99), it is appropriate to regard the results of RNA-RNA interaction predictions as extreme value distributed. The density function f of the generalized extreme value (GEV) distribution is introduced and discussed in the supplementary material.

A p-value is the probability that a certain event x or something more extreme ($\geq x$) is observed for a specific background model. Given the density function f of the events, a p-value can be computed by the integral $\int_x^\infty f(x)\mathrm{d}x$. The cumulative distribution F for the GEV distribution (see supplementary material) provides the

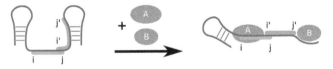

Figure 7 (Left) *In silico* accessibility scenario without consideration of RNA-binding factors (A, B). Here, region *i..j* is accessible, while *ĭ..ĭ* is blocked by intramolecular base pairs. (Right) Putative *in vivo* situation with bound factors A and B. The *in silico* accessible site *i..j* is blocked by A, while *ĭ..ĭ* becomes accessible due to structural reconfiguration upon binding of A and B.

integral $F(x) = \int_{-\infty}^{x} f(x)dx$ for events $\leq x$, such that we can compute the p-value by $1 - F(x)$.

Since the significant p-values depend on the right tail of the distribution, its correct estimation via the parameters μ, σ, and ε is central to their quality. An appropriate volume of background predictions for estimating the parameters of the GEV can be obtained by dinucleotide shuffling sequences that are actually present in the real search space (e.g., putative target sequences that are present in the investigated genome) and predicting RNA-RNA interactions for these shuffled sequences. It is important to retain the dinucleotide frequency because the duplex prediction depends on base-pair stacking and mononucleotide shuffling would thus no longer yield random sequences that still appropriately represent the properties of the nonrandom system. If target sequences of differing lengths are used, the energy scores need to be normalized with $E_n = -E/\ln(mn)$ to prevent inappropriately high abundance of better predictions for longer sequences.

The negative normalized energy is denoted as E_n, while the unnormalized energy is denoted as E, with m and n being the lengths of the target and the binding RNA, respectively (59). The parameters for the GEV can then be derived by fitting a GEV to the empiric background's energy scores after duplex prediction and if necessary length normalization. A p-value for a given energy score within the search space can then be inferred from the GEV's cumulative distribution function. For whole-genome target predictions, it has been shown that the GEV's parameters can also be estimated by using all the predictions on real putative target sequences and fitting the GEV to these (100). These predictions are clearly not all completely random due to functionally correct predictions presumably being present. Yet the majority of predicted duplexes can be assumed as not functionally relevant *in vivo*. While the p-value quality might be inferior compared to a shuffled background model, this strategy leads to strongly reduced runtimes, which is important when performing predictions on a genomic scale.

The individual p-values (p_1) for an orthologous putative target can then be combined to a joint p-value (p_{joint}) by selecting the biggest individual p-value and raising it to the power of the amount of participating organisms (k) (equation 15).s

$$p_{joint} = (max\{p_1,, p_k\})^k \qquad (15)$$

This, however, assumes complete statistical autonomy of the individual p-values, which is not the case for the given biological scenario. Here, the investigated species are assumed to be descended from a common ancestor, which intuitively implies a certain degree of mutual dependence. Consequently, smaller evolutionary distance between organisms leads to higher statistical dependence of individual results. Hence, the effective number of organisms or sequences (k_{eff}) needs to be assessed. k_{eff} lies between 1 and the number of participating organisms (k).

k_{eff} can be estimated by shuffling the homologous sRNA or miRNA sequences while retaining the original dinucleotide frequencies followed by duplex predictions for homologous targets. This supplies a background set of optimal duplexes for each participating homologous target, from which extreme value distribution parameters can be assessed. With these parameters, the duplex energies can be transformed into p-values. These p-values are then joined following equation 15 using several consecutive k' values instead of one single k. k' can lie between 1 and k. Finally, the k' yielding the most uniform distribution of joint p-values is set as k_{eff}. A small k_{eff} implies high dependence between the putative target sequences. Specifically, $k_{eff} = 1$ would mean that no additional information could be gained by incorporation of predictions for homologous sequences. The method for joining the p-values (equation 15) can be regarded as very conservative due to the fact that it always selects the highest/worst individual p-value for p-value combination. In other words, this means that a putative target needs to return a good prediction for each organism participating in the comparative analysis to be considered as a true target. This most likely leads to a lowered number of false positives but may also cause many false negatives depending on the set of organisms that is used.

The comparative prediction algorithm for sRNA targets CopraRNA (100, 101), which was developed to make whole-genome target predictions, also uses the concept of transforming energy scores to p-values. Copra RNA, unlike other comparative methods that enforce a consensus interaction site within the homologous putative targets (95), is very unrestrictive. It solely enforces an interaction (as predicted by IntaRNA) to be present anywhere in the putative target sequence without demanding a consistent duplex pattern throughout the homologs. Also, homologs of a putative target need to be present in at least half of the participating organisms to ensure a sensible degree of conservation. Missing single-organism p-values are sampled from a multivariate normal distribution, which is based on clusters of homologous genes that contain a homolog from every organism participating in the comparative analysis.

The first step in CopraRNA is to compute individual whole-genome target predictions with the homologous sRNAs for each organism participating in the analysis. The energy scores for each putative target are computed by IntaRNA. Following the logic that most duplexes predicted by IntaRNA are likely to be nonfunctional (i.e., they represent a random background), the whole-genome prediction can be used to estimate extreme value distribution parameters for each of the homologous sRNAs. Based on these parameters, the energy scores can be transformed to p-values. The next step, like in RNAhybrid, is the combination of p-values for homologous putative targets. CopraRNA employs Hartung's method for the combination of dependent p-values (102) to calculate a joint p-value for a cluster of homologous genes with size K. For this, the initial p-values need to be transformed to probits (t_i), which is done based on the inverse of the cumulative distribution function. The combination of the probits t_{joint} is computed following equation 16.

$$t_{\text{joint}} = \frac{\sum_{i=1}^{K} \lambda_i t_i}{\sqrt{(1-\rho)\sum_{i=1}^{K} \lambda_i^2 + \rho\left(\sum_{i=1}^{K} \lambda_i\right)^2}} \quad (16)$$

The result of equation 16, t_{joint}, can then be transformed back to a p-value. Hartung's method includes both a correction for the dependence in the data (ρ) and a weighting (λ_i) for each individual single p-value. The rationale for the dependence correction is the same as described previously for the comparative approach in RNAhybrid, and ρ is assessed in a similar manner as k_{eff}. ρ can adopt values between 0 and 1, and higher values for ρ indicate higher dependence in the data. The optimal ρ is the ρ that yields the most uniform distribution of joint CopraRNA p-values. CopraRNA uses the organisms' 16S rDNA to construct a phylogenetic tree. Organisms that are very similar need to have a lower individual weight λ_i compared to an organism that is evolutionarily very distant. The weights are calculated by a recursive scheme that computes the relative weights of an organism in all subtrees and then multiplies these. The weights are subsequently subjected to a root function to reduce overly strong effects of outlier organisms.

CopraRNA was originally benchmarked on a set of 101 experimentally verified enterobacterial sRNA-target interactions and significantly outperformed its competitor algorithms while also rivaling experimental target predictions derived from microarray experiments. An independent comprehensive benchmark has since confirmed this finding (103). Furthermore, CopraRNA has

been successfully applied in nonenterobacterial systems (104–108), and it has been shown that it can benefit from the incorporation of Hfq binding data (109). As a concept, CopraRNA may also be promising in a eukaryotic setting but has not yet been implemented to accommodate for such a context. Due to CopraRNA using IntaRNA as a background RNA-RNA interaction model, it partly inherits IntaRNA's limitations. Yet the general concept of phylogeny-guided p-value combination is detached from IntaRNA and allows the application of other interaction prediction algorithms. For the future, this means that CopraRNA can benefit from advances in single-organism target prediction.

OUTLOOK

The RNA structure and RNA-RNA interaction prediction approaches that have been discussed here are typically used via either according web servers or command-line interfaces of local installations. To enhance reproducibility and to accommodate for large-scale application, pipeline and workflow systems like Galaxy (110, 111) and bioconda (112) have been developed. Recently, the "RNA workbench" extension of Galaxy was published (113), which features many of the approaches outlined here. This enables their high-throughput application in sophisticated (partially predefined) workflows for nonexpert scientists (114).

All algorithms discussed here are tailored for linear RNA molecules. However, circular noncoding RNAs also exist and have been reported in, e.g., eukaryotic cells (115, 116) and archaea (117). To tackle this new class of noncoding RNAs, some of the sketched approaches have been appropriately adapted (118, 119).

Overall, a lot of prokaryotic RNA research has been intensely focusing on the one-by-one functional classification of newly identified RNA-based regulators, and many such projects are still ongoing. One of the central pillars in most of these studies is RNA structure and RNA-RNA interaction elucidation aided by the computational tools that have been mentioned in this review. However, more recently, transcriptome-wide RNA interactomics data have been produced (25, 26, 120–123), which can be expected to strongly shift interest toward large-scale projects. The opportunities and information within the newly acquired data might be able to answer long-standing questions in the predictive community and allow for the development of more-sophisticated data-driven algorithms.

Citation. Wright PR, Mann M, Backofen R. 2018. Structure and interaction prediction in prokaryotic RNA biology. Microbiol Spectrum 6(2):RWR-0001-2017.

References

1. Crick FH. 1966. Codon–anticodon pairing: the wobble hypothesis. *J Mol Biol* **19**:548–555.

2. Gerber AP, Keller W. 1999. An adenosine deaminase that generates inosine at the wobble position of tRNAs. *Science* **286**:1146–1149.

3. Murphy FV IV, Ramakrishnan V. 2004. Structure of a purine-purine wobble base pair in the decoding center of the ribosome. *Nat Struct Mol Biol* **11**:1251–1252.

4. Watson JD, Crick FH. 1953. Molecular structure of nucleic acids; a structure for deoxyribose nucleic acid. *Nature* **171**:737–738.

5. Wagner EG, Romby P. 2015. Small RNAs in bacteria and archaea: who they are, what they do, and how they do it. *Adv Genet* **90**:133–208.

6. Ameres SL, Zamore PD. 2013. Diversifying microRNA sequence and function. *Nat Rev Mol Cell Biol* **14**:475–488.

7. Waters LS, Storz G. 2009. Regulatory RNAs in bacteria. *Cell* **136**:615–628.

8. Nussinov R, Pieczenik G, Griggs JR, Kleitman DJ. 1978. Algorithms for loop matchings. *SIAM J Appl Math* **35**:68–82.

9. Devoe H, Tinoco I Jr. 1962. The stability of helical polynucleotides: base contributions. *J Mol Biol* **4**:500–517.

10. Mathews DH, Sabina J, Zuker M, Turner DH. 1999. Expanded sequence dependence of thermodynamic parameters improves prediction of RNA secondary structure. *J Mol Biol* **288**:911–940.

11. Turner DH, Mathews DH. 2010. NNDB: the nearest neighbor parameter database for predicting stability of nucleic acid secondary structure. *Nucleic Acids Res* **38** (Database issue):D280–D282.

12. Tinoco I Jr, Borer PN, Dengler B, Levin MD, Uhlenbeck OC, Crothers DM, Bralla J. 1973. Improved estimation of secondary structure in ribonucleic acids. *Nat New Biol* **246**:40–41.

13. Borer PN, Dengler B, Tinoco I Jr, Uhlenbeck OC. 1974. Stability of ribonucleic acid double-stranded helices. *J Mol Biol* **86**:843–853.

14. Andronescu M, Condon A, Hoos HH, Mathews DH, Murphy KP. 2007. Efficient parameter estimation for RNA secondary structure prediction. *Bioinformatics* **23**:i19–i28.

15. Turner DH, Sugimoto N, Jaeger JA, Longfellow CE, Freier SM, Kierzek R. 1987. Improved parameters for prediction of RNA structure. *Cold Spring Harb Symp Quant Biol* **52**:123–133.

16. Zuker M, Stiegler P. 1981. Optimal computer folding of large RNA sequences using thermodynamics and auxiliary information. *Nucleic Acids Res* **9**:133–148.

17. Wuchty S, Fontana W, Hofacker IL, Schuster P. 1999. Complete suboptimal folding of RNA and the stability of secondary structures. *Biopolymers* **49**:145–165.

18. Markham NR, Zuker M. 2008. UNAFold: software for nucleic acid folding and hybridization. *Methods Mol Biol* **453**:3–31.

19. Zuker M. 1994. Prediction of RNA secondary structure by energy minimization. *Methods Mol Biol* **25**:267–294.

20. Hofacker IL, Fontana W, Stadler PF, Bonhoeffer S, Tacker M, Schuster P. 1994. Fast folding and comparison of RNA secondary structures. *Monatsh Chem* **125**:167–188.

21. Lorenz R, Bernhart SH, Höner Zu Siederdissen C, Tafer H, Flamm C, Stadler PF, Hofacker IL. 2011. ViennaRNA Package 2.0. *Algorithms Mol Biol* **6**:26.

22. Møller T, Franch T, Udesen C, Gerdes K, Valentin-Hansen P. 2002. Spot 42 RNA mediates discoordinate expression of the *E. coli* galactose operon. *Genes Dev* **16**:1696–1706.

23. Durand S, Storz G. 2010. Reprogramming of anaerobic metabolism by the FnrS small RNA. *Mol Microbiol* **75**:1215–1231.

24. Wexler Y, Zilberstein C, Ziv-Ukelson M. 2007. A study of accessible motifs and RNA folding complexity. *J Comput Biol* **14**:856–872.

25. Cordero P, Lucks JB, Das R. 2012. An RNA Mapping DataBase for curating RNA structure mapping experiments. *Bioinformatics* **28**:3006–3008.

26. Norris M, Kwok CK, Cheema J, Hartley M, Morris RJ, Aviran S, Ding Y. 2017. FoldAtlas: a repository for genome-wide RNA structure probing data. *Bioinformatics* **33**:306–308.

27. Low JT, Weeks KM. 2010. SHAPE-directed RNA secondary structure prediction. *Methods* **52**:150–158.

28. Washietl S, Hofacker IL, Stadler PF, Kellis M. 2012. RNA folding with soft constraints: reconciliation of probing data and thermodynamic secondary structure prediction. *Nucleic Acids Res* **40**:4261–4272.

29. Deigan KE, Li TW, Mathews DH, Weeks KM. 2009. Accurate SHAPE-directed RNA structure determination. *Proc Natl Acad Sci U S A* **106**:97–102.

30. Zarringhalam K, Meyer MM, Dotu I, Chuang JH, Clote P. 2012. Integrating chemical footprinting data into RNA secondary structure prediction. *PLoS One* **7**:e45160.

31. Lorenz R, Hofacker IL, Stadler PF. 2016. RNA folding with hard and soft constraints. *Algorithms Mol Biol* **11**:8.

32. Lorenz R, Luntzer D, Hofacker IL, Stadler PF, Wolfinger MT. 2016. SHAPE directed RNA folding. *Bioinformatics* **32**:145–147.

33. Gardner PP, Giegerich R. 2004. A comprehensive comparison of comparative RNA structure prediction approaches. *BMC Bioinformatics* **5**:140.

34. Hofacker IL, Fekete M, Stadler PF. 2002. Secondary structure prediction for aligned RNA sequences. *J Mol Biol* **319**:1059–1066.

35. Knudsen B, Hein J. 2003. Pfold: RNA secondary structure prediction using stochastic context-free grammars. *Nucleic Acids Res* **31**:3423–3428.

36. Seemann SE, Gorodkin J, Backofen R. 2008. Unifying evolutionary and thermodynamic information for RNA folding of multiple alignments. *Nucleic Acids Res* **36**:6355–6362.

37. Papenfort K, Pfeiffer V, Lucchini S, Sonawane A, Hinton JCD, Vogel J. 2008. Systematic deletion of *Salmonella* small RNA genes identifies CyaR, a conserved CRP-dependent riboregulator of OmpX synthesis. *Mol Microbiol* **68**:890–906.

38. Gardner PP, Wilm A, Washietl S. 2005. A benchmark of multiple sequence alignment programs upon structural RNAs. *Nucleic Acids Res* **33**:2433–2439.

39. Höchsmann M, Töller T, Giegerich R, Kurtz S. 2003. Local similarity in RNA secondary structures. *Proc IEEE Comput Soc Bioinform Conf* **2**:159–168.

40. Siebert S, Backofen R. 2005. MARNA: multiple alignment and consensus structure prediction of RNAs based on sequence structure comparisons. *Bioinformatics* **21**:3352–3359.

41. Sankoff D. 1985. Simultaneous solution of the RNA folding, alignment and protosequence problems. *SIAM J Appl Math* **45**:810–825.

42. Gorodkin J, Heyer LJ, Stormo GD. 1997. Finding the most significant common sequence and structure motifs in a set of RNA sequences. *Nucleic Acids Res* **25**:3724–3732.

43. Havgaard JH, Lyngsø RB, Stormo GD, Gorodkin J. 2005. Pairwise local structural alignment of RNA sequences with sequence similarity less than 40%. *Bioinformatics* **21**:1815–1824.

44. Mathews DH, Turner DH. 2002. Dynalign: an algorithm for finding the secondary structure common to two RNA sequences. *J Mol Biol* **317**:191–203.

45. Bradley RK, Pachter L, Holmes I. 2008. Specific alignment of structured RNA: stochastic grammars and sequence annealing. *Bioinformatics* **24**:2677–2683.

46. Hofacker IL, Bernhart SH, Stadler PF. 2004. Alignment of RNA base pairing probability matrices. *Bioinformatics* **20**:2222–2227.

47. Will S, Reiche K, Hofacker IL, Stadler PF, Backofen R. 2007. Inferring noncoding RNA families and classes by means of genome-scale structure-based clustering. *PLoS Comput Biol* **3**:e65.

48. Torarinsson E, Havgaard JH, Gorodkin J. 2007. Multiple structural alignment and clustering of RNA sequences. *Bioinformatics* **23**:926–932.

49. McCaskill JS. 1990. The equilibrium partition function and base pair binding probabilities for RNA secondary structure. *Biopolymers* **29**:1105–1119.

50. Do CB, Foo CS, Batzoglou S. 2008. A max-margin model for efficient simultaneous alignment and folding of RNA sequences. *Bioinformatics* **24**:i68–i76.

51. Georg J, Hess WR. 2011. *cis*-Antisense RNA, another level of gene regulation in bacteria. *Microbiol Mol Biol Rev* **75**:286–300.

52. Kruger J, Rehmsmeier M. 2006. RNAhybrid: microRNA target prediction easy, fast and flexible. *Nucleic Acids Res* **34**(Web Server issue):W451–W454.

53. Wright PR, Georg J, Mann M, Sorescu DA, Richter AS, Lott S, Kleinkauf R, Hess WR, Backofen R. 2014. CopraRNA and IntaRNA: predicting small RNA targets, networks and interaction domains. *Nucleic Acids Res* **42**(Web Server issue):W119–W123.

54. Altschul SF, Gish W, Miller W, Myers EW, Lipman DJ. 1990. Basic local alignment search tool. *J Mol Biol* **215**:403–410.

55. Gerlach W, Giegerich R. 2006. GUUGle: a utility for fast exact matching under RNA complementary rules including G-U base pairing. *Bioinformatics* **22**:762–764.

56. Karlin S, Altschul SF. 1990. Methods for assessing the statistical significance of molecular sequence features by using general scoring schemes. *Proc Natl Acad Sci U S A* **87**:2264–2268.

57. Tjaden B, Goodwin SS, Opdyke JA, Guillier M, Fu DX, Gottesman S, Storz G. 2006. Target prediction for small, noncoding RNAs in bacteria. *Nucleic Acids Res* **34**:2791–2802.

58. Smith TF, Waterman MS. 1981. Identification of common molecular subsequences. *J Mol Biol* **147**:195–197.

59. Rehmsmeier M, Steffen P, Höchsmann M, Giegerich R. 2004. Fast and effective prediction of microRNA/target duplexes. *RNA* **10**:1507–1517.

60. Gong H, Vu GP, Bai Y, Chan E, Wu R, Yang E, Liu F, Lu S. 2011. A *Salmonella* small non-coding RNA facilitates bacterial invasion and intracellular replication by modulating the expression of virulence factors. *PLoS Pathog* **7**:e1002120.

61. Papenfort K, Sun Y, Miyakoshi M, Vanderpool CK, Vogel J. 2013. Small RNA-mediated activation of sugar phosphatase mRNA regulates glucose homeostasis. *Cell* **153**:426–437.

62. Tafer H, Hofacker IL. 2008. RNAplex: a fast tool for RNA-RNA interaction search. *Bioinformatics* **24**:2657–2663.

63. Wenzel A, Akbasli E, Gorodkin J. 2012. RIsearch: fast RNA-RNA interaction search using a simplified nearest-neighbor energy model. *Bioinformatics* **28**:2738–2746.

64. Mathews DH, Burkard ME, Freier SM, Wyatt JR, Turner DH. 1999. Predicting oligonucleotide affinity to nucleic acid targets. *RNA* **5**:1458–1469.

65. Brock TD, Brock KM, Belly RT, Weiss RL. 1972. *Sulfolobus*: a new genus of sulfur-oxidizing bacteria living at low pH and high temperature. *Arch Mikrobiol* **84**:54–68.

66. Busch A, Richter AS, Backofen R. 2008. IntaRNA: efficient prediction of bacterial sRNA targets incorporating target site accessibility and seed regions. *Bioinformatics* **24**:2849–2856.

67. Tafer H, Amman F, Eggenhofer F, Stadler PF, Hofacker IL. 2011. Fast accessibility-based prediction of RNA-RNA interactions. *Bioinformatics* **27**:1934–1940.

68. Eggenhofer F, Tafer H, Stadler PF, Hofacker IL. 2011. RNApredator: fast accessibility-based prediction of sRNA targets. *Nucleic Acids Res* **39**(Web Server issue):W149–W154.

69. Kery MB, Feldman M, Livny J, Tjaden B. 2014. TargetRNA2: identifying targets of small regulatory RNAs in bacteria. *Nucleic Acids Res* **42**(Web Server issue):W124–129.

70. Alkan F, Wenzel A, Palasca O, Kerpedjiev P, Rudebeck AF, Stadler PF, Hofacker IL, Gorodkin J. 2017. RIsearch2: suffix array-based large-scale prediction of

RNA-RNA interactions and siRNA off-targets. *Nucleic Acids Res* 45:e60.

71. Zuker M. 2003. Mfold web server for nucleic acid folding and hybridization prediction. *Nucleic Acids Res* 31:3406–3415.

72. Andronescu M, Zhang ZC, Condon A. 2005. Secondary structure prediction of interacting RNA molecules. *J Mol Biol* 345:987–1001.

73. Bernhart SH, Tafer H, Mückstein U, Flamm C, Stadler PF, Hofacker IL. 2006. Partition function and base pairing probabilities of RNA heterodimers. *Algorithms Mol Biol* 1:3.

74. Dirks RM, Bois JS, Schaeffer JM, Winfree E, Pierce NA. 2007. Thermodynamic analysis of interacting nucleic acid strands. *SIAM Rev* 49:65–88.

75. Chang KY, Tinoco I Jr. 1997. The structure of an RNA "kissing" hairpin complex of the HIV TAR hairpin loop and its complement. *J Mol Biol* 269:52–66.

76. Salim N, Lamichhane R, Zhao R, Banerjee T, Philip J, Rueda D, Feig AL. 2012. Thermodynamic and kinetic analysis of an RNA kissing interaction and its resolution into an extended duplex. *Biophys J* 102:1097–1107.

77. Vecerek B, Moll I, Afonyushkin T, Kaberdin V, Bläsi U. 2003. Interaction of the RNA chaperone Hfq with mRNAs: direct and indirect roles of Hfq in iron metabolism of *Escherichia coli*. *Mol Microbiol* 50:897–909.

78. Mückstein U, Tafer H, Hackermüller J, Bernhart SH, Stadler PF, Hofacker IL. 2006. Thermodynamics of RNA-RNA binding. *Bioinformatics* 22:1177–1182.

79. Bernhart SH, Mückstein U, Hofacker IL. 2011. RNA accessibility in cubic time. *Algorithms Mol Biol* 6:3.

80. Mann M, Wright PR, Backofen R. 2017. IntaRNA 2.0: enhanced and customizable prediction of RNA-RNA interactions. *Nucleic Acids Res* 45(W1):W435–W439.

81. Balbontín R, Fiorini F, Figueroa-Bossi N, Casadesús J, Bossi L. 2010. Recognition of heptameric seed sequence underlies multi-target regulation by RybB small RNA in *Salmonella enterica*. *Mol Microbiol* 78:380–394.

82. Brennecke J, Stark A, Russell RB, Cohen SM. 2005. Principles of microRNA-target recognition. *PLoS Biol* 3:e85.

83. Fukunaga T, Hamada M. 2017. RIblast: an ultrafast RNA-RNA interaction prediction system based on a seed-and-extension approach. *Bioinformatics* 33:2666–2674.

84. Argaman L, Altuvia S. 2000. *fhlA* repression by OxyS RNA: kissing complex formation at two sites results in a stable antisense-target RNA complex. *J Mol Biol* 300:1101–1112.

85. Alkan C, Karakoç E, Nadeau JH, Sahinalp SC, Zhang K. 2006. RNA-RNA interaction prediction and antisense RNA target search. *J Comput Biol* 13:267–282.

86. Pervouchine DD. 2004. IRIS: intermolecular RNA interaction search. *Genome Inform* 15:92–101.

87. Chitsaz H, Salari R, Sahinalp SC, Backofen R. 2009. A partition function algorithm for interacting nucleic acid strands. *Bioinformatics* 25:i365–i373.

88. Huang FW, Qin J, Reidys CM, Stadler PF. 2009. Partition function and base pairing probabilities for RNA-RNA interaction prediction. *Bioinformatics* 25:2646–2654.

89. Salari R, Möhl M, Will S, Sahinalp SC, Backofen R. 2010. Time and space efficient RNA-RNA interaction prediction via sparse folding, p 473–490. *In* Berger B (ed), *Research in Computational Molecular Biology. RECOMB 2010. Lecture Notes in Computer Science*, vol 6044. Springer, Berlin, Germany.

90. Chitsaz H, Backofen R, Sahinalp SC. 2009. biRNA: fast RNA-RNA binding sites prediction, p 25–36. *In* Salzberg S, Warnow T (ed), *Algorithms in Bioinformatics. Lecture Notes in Computer Science*, vol 5724. Springer, Berlin, Germany.

91. Salari R, Backofen R, Sahinalp SC. 2010. Fast prediction of RNA-RNA interaction. *Algorithms Mol Biol* 5:5.

92. Lott S, Schäfer R, Mann M, Hess W, Voß B, Georg J. 2018. GLASSgo-automated and reliable detection of sRNA homologs from a single input sequence. *Front Genet* 9:124.

93. Eggenhofer F, Hofacker IL, Höner Zu Siederdissen C. 2016. RNAlien—unsupervised RNA family model construction. *Nucleic Acids Res* 44:8433–8441.

94. Hertel J, de Jong D, Marz M, Rose D, Tafer H, Tanzer A, Schierwater B, Stadler PF. 2009. Non-coding RNA annotation of the genome of *Trichoplax adhaerens*. *Nucleic Acids Res* 37:1602–1615.

95. Seemann SE, Richter AS, Gesell T, Backofen R, Gorodkin J. 2011. PETcofold: predicting conserved interactions and structures of two multiple alignments of RNA sequences. *Bioinformatics* 27:211–219.

96. Seemann SE, Menzel P, Backofen R, Gorodkin J. 2011. The PETfold and PETcofold web servers for intra- and intermolecular structures of multiple RNA sequences. *Nucleic Acids Res* 39(Web Server issue):W107–W111.

97. Li AX, Marz M, Qin J, Reidys CM. 2011. RNA-RNA interaction prediction based on multiple sequence alignments. *Bioinformatics* 27:456–463.

98. Richter AS, Backofen R. 2012. Accessibility and conservation: general features of bacterial small RNA-mRNA interactions? *RNA Biol* 9:954–965.

99. Gumbel EJ. 1958. *Statistics of Extremes*. Columbia University Press, New York, NY.

100. Wright PR, Richter AS, Papenfort K, Mann M, Vogel J, Hess WR, Backofen R, Georg J. 2013. Comparative genomics boosts target prediction for bacterial small RNAs. *Proc Natl Acad Sci U S A* 110:E3487–E3496.

101. Wright PR. 2016. *Predicting small RNA targets in prokaryotes—a challenge beyond the barriers of thermodynamic models. Ph.D. thesis*. Albert-Ludwigs-University, Freiburg, Germany.

102. Hartung J. 1999. A note on combining dependent tests of significance. *Biom J* 41:849–855.

103. Pain A, Ott A, Amine H, Rochat T, Bouloc P, Gautheret D. 2015. An assessment of bacterial small RNA target prediction programs. *RNA Biol* 12:509–513.

104. Georg J, Dienst D, Schürgers N, Wallner T, Kopp D, Stazic D, Kuchmina E, Klähn S, Lokstein H, Hess WR, Wilde A. 2014. The small regulatory RNA SyR1/PsrR1

controls photosynthetic functions in cyanobacteria. *Plant Cell* 26:3661–3679.

105. Overlöper A, Kraus A, Gurski R, Wright PR, Georg J, Hess WR, Narberhaus F. 2014. Two separate modules of the conserved regulatory RNA AbcR1 address multiple target mRNAs in and outside of the translation initiation region. *RNA Biol* 11:624–640.

106. Robledo M, Frage B, Wright PR, Becker A. 2015. A stress-induced small RNA modulates alpha-rhizobial cell cycle progression. *PLoS Genet* 11:e1005153.

107. Klähn S, Schaal C, Georg J, Baumgartner D, Knippen G, Hagemann M, Muro-Pastor AM, Hess WR. 2015. The sRNA NsiR4 is involved in nitrogen assimilation control in cyanobacteria by targeting glutamine synthetase inactivating factor IF7. *Proc Natl Acad Sci U S A* 112: E6243–E6252.

108. Durand S, Braun F, Lioliou E, Romilly C, Helfer AC, Kuhn L, Quittot N, Nicolas P, Romby P, Condon C. 2015. A nitric oxide regulated small RNA controls expression of genes involved in redox homeostasis in *Bacillus subtilis*. *PLoS Genet* 11:e1004957.

109. Holmqvist E, Wright PR, Li L, Bischler T, Barquist L, Reinhardt R, Backofen R, Vogel J. 2016. Global RNA recognition patterns of post-transcriptional regulators Hfq and CsrA revealed by UV crosslinking in vivo. *EMBO J* 35:991–1011.

110. Afgan E, Goecks J, Baker D, Coraor N, Nekrutenko A, Taylor J. 2011. Galaxy: a gateway to tools in e-science, p 145–177. *In* Yang X, Wang L, Jie W (ed), *Guide to e-Science: Next Generation Scientific Research and Discovery*. Springer London, London, United Kingdom.

111. Afgan E, Baker D, van den Beek M, Blankenberg D, Bouvier D, Čech M, Chilton J, Clements D, Coraor N, Eberhard C, Grüning B, Guerler A, Hillman-Jackson J, Von Kuster G, Rasche E, Soranzo N, Turaga N, Taylor J, Nekrutenko A, Goecks J. 2016. The Galaxy platform for accessible, reproducible and collaborative biomedical analyses: 2016 update. *Nucleic Acids Res* 44(W1):W3–W10.

112. Grüning B, Dale R, Sjödin A, Rowe J, Chapman BA, Tomkins-Tinch CH, Valieris R, The Bioconda Team, Köster J. 2017. Bioconda: a sustainable and comprehensive software distribution for the life sciences. *bioRxiv*.

113. Grüning BA, Fallmann J, Yusuf D, Will S, Erxleben A, Eggenhofer F, Houwaart T, Batut B, Videm P, Bagnacani A, Wolfien M, Lott SC, Hoogstrate Y, Hess WR, Wolkenhauer O, Hoffmann S, Akalin A, Ohler U, Stadler PF, Backofen R. 2017. The RNA workbench: best practices for RNA and high-throughput sequencing bioinformatics in Galaxy. *Nucleic Acids Res* 35(W1):W560–W566.

114. Grüning BA, Rasche E, Rebolledo-Jaramillo B, Eberhard C, Houwaart T, Chilton J, Coraor N, Backofen R, Taylor J, Nekrutenko A. 2017. Jupyter and Galaxy: easing entry barriers into complex data analyses for biomedical researchers. *PLoS Comput Biol* 13:e1005425.

115. Chen L, Huang C, Wang X, Shan G. 2015. Circular RNAs in eukaryotic cells. *Curr Genomics* 16:312–318.

116. Koch L. 2017. RNA: translated circular RNAs. *Nat Rev Genet* 18:272–273.

117. Danan M, Schwartz S, Edelheit S, Sorek R. 2012. Transcriptome-wide discovery of circular RNAs in Archaea. *Nucleic Acids Res* 40:3131–3142.

118. Hofacker IL, Stadler PF. 2006. Memory efficient folding algorithms for circular RNA secondary structures. *Bioinformatics* 22:1172–1176.

119. Hofacker IL, Reidys CM, Stadler PF. 2012. Symmetric circular matchings and RNA folding. *Discrete Math* 312: 100–112.

120. Melamed S, Peer A, Faigenbaum-Romm R, Gatt YE, Reiss N, Bar A, Altuvia Y, Argaman L, Margalit H. 2016. Global mapping of small RNA-target interactions in bacteria. *Mol Cell* 63:884–897.

121. Waters SA, McAteer SP, Kudla G, Pang I, Deshpande NP, Amos TG, Leong KW, Wilkins MR, Strugnell R, Gally DL, Tollervey D, Tree JJ. 2016. Small RNA interactome of pathogenic *E. coli* revealed through crosslinking of RNase E. *EMBO J* 36:374–387.

122. Sharma E, Sterne-Weiler T, O'Hanlon D, Blencowe BJ. 2016. Global mapping of human RNA-RNA interactions. *Mol Cell* 62:618–626.

123. Lu Z, Zhang QC, Lee B, Flynn RA, Smith MA, Robinson JT, Davidovich C, Gooding AR, Goodrich KJ, Mattick JS, Mesirov JP, Cech TR, Chang HY. 2016. RNA duplex map in living cells reveals higher-order transcriptome structure. *Cell* 165:1267–1279.

124. Wright PR, Georg J. 2018. Workflow for a computational analysis of an sRNA candidate in bacteria. *Methods Mol Biol* 1737:3–30.

Index